## Multiplication Notations for SI Units

| MULTIPLICATION FACTOR | PREFIX | SYMBOL |
|---|---|---|
| $10^9$ | giga | G |
| $10^6$ | mega | M |
| $10^3$ | kilo | k |
| $10^{-2}$ | centi | c |
| $10^{-3}$ | milli | m |
| $10^{-6}$ | micro | $\mu$ |

Note that conversion factors are approximations. The given multiplication factors are those commonly used in engineering mechanics. For further details, see ASTM Metric Practice, E380-76.

**TABLE 2**

Approximate Coefficients of Static Friction

| MATERIALS IN CONTACT | | $\mu_s$ |
|---|---|---|
| Aluminum | Metal (dry) | 0.6–1.05 |
| | Metal (greasy) | 0.1 |
| Asbestos (brake lining) | Cast iron (dry) | 0.4 |
| | Cast iron (wet) | 0.2 |
| Asphalt (dry) | Rubber tires | 0.7 |
| Cast iron | Cast iron | 1.10 |
| | Leather | 0.2 |
| | Rubber | 0.6–0.9 |
| | Wood | 0.3 |
| Leather | Metal | 0.4 |
| | Wood | 0.6 |
| Nylon | Metal | 0.3 |
| Rubber | Metal | 0.6–0.9 |
| | Slippery roads | 0.1 |
| | Wood | 0.4 |
| Steel | Clay | 0.6 |
| | Graphite | 0.1 |
| | Ice | 0.1 |
| | Metal (dry) | 0.7 |
| | Metal (greasy) | 0.1 |
| | Teflon | 0.04 |
| Teflon | Metal | 0.04 |
| Wood | Metal (dry) | 0.4 |
| | Metal (wet) | 0.2 |

# Engineering Mechanics
# Statics and Dynamics

2nd Edition

# Engineering Mechanics Statics and Dynamics

## Bela I. Sandor

*University of Wisconsin-Madison*

with

## Karen J. Richter

PRENTICE-HALL, INC., ENGLEWOOD CLIFFS, N.J. 07632

*Library of Congress Cataloging-in-Publication Data*

SANDOR, BELA, IMRE.
  "Combined volume."
  Includes index.
  1. Mechanics, Applied.   I. Richter, Karen J.
II. Title.
TA350.S29 1987      620.1      86-22678
ISBN  0-13-279092-0

Editorial/production supervision: Virginia Huebner
Interior & Cover design: Janet Schmid
Cover Photograph: © Cincinnati Milacron
Manufacturing buyer: Rhett Conklin
Page layout: Diane Korohmas

Printed in the United States of America

10  9  8  7  6  5  4  3  2  1

ISBN 0-13-279092-0      01

Prentice-Hall International (UK) Limited, *London*
Prentice-Hall of Australia Pty. Limited, *Sydney*
Prentice-Hall Canada Inc., *Toronto*
Prentice-Hall Hispanoamericana, S.A., *Mexico*
Prentice-Hall of India Private Limited, *New Delhi*
Prentice-Hall of Japan, Inc., *Tokyo*
Prentice-Hall of Southeast Asia Pte. Ltd., *Singapore*
Editora Prentice-Hall do Brasil, Ltda., *Rio de Janeiro*

# Contents

PREFACE                                                                    xi

## STATICS—VOLUME 1

INTRODUCTION TO ENGINEERING MECHANICS                     3        CHAPTER 1

    Overview                                              3
1-1  Engineering Mechanics                                  4
1-2  Basic Concepts of Mechanics                            5
1-3  Newton's Laws                                          5
1-4  Scalars and Vectors                                    7
1-5  Units                                                  8
1-6  Modeling, Approximations, Required Accuracies         12
1-7  Recurring Examples for Progressive Analysis           13

FORCES ON PARTICLES                                          17        CHAPTER 2

    Overview                                             17
2-1  Two-Dimensional Force Systems. Resultants of Concurrent Forces   18
2-2  Vector Components of a Force                          19
2-3  Unit Vectors                                         27
2-4  Three-Dimensional Force Systems. Resultants and
    Components of Forces                                 34
2-5  Method for Determining Force Vector from Scalar Information   36
2-6  Scalar Product of Two Vectors. Angles between Vectors.
    Projection of a Vector                               44
2-7  Equilibrium of a Particle. Free-Body Diagrams         50
    Summary                                              63

v

## CHAPTER 3    FORCES ON RIGID BODIES    69

|       | Overview | 69 |
|-------|----------|----|
| 3-1   | External and Internal Forces | 71 |
| 3-2   | Equivalent Forces | 72 |
| 3-3   | Moment of a Force | 72 |
| 3-4   | Vector Product of Two Vectors. Moment of a Force about a Point | 77 |
| 3-5   | Triple Products of Three Vectors | 89 |
| 3-6   | Moment of a Force about a Line | 89 |
| 3-7   | Moment of a Couple | 96 |
| 3-8   | Equivalent Couples | 98 |
| 3-9   | Components and Addition of Couples | 99 |
| 3-10  | Transformation of a Force into a Force–Couple System | 100 |
| 3-11  | Simplification of Many Forces to a Force and a Couple | 108 |
| 3-12  | Common Simple Force Systems | 109 |
|       | Summary | 118 |

## CHAPTER 4    EQUILIBRIUM OF RIGID BODIES    123

|       | Overview | 123 |
|-------|----------|-----|
| 4-1   | Complete Equilibrium Conditions | 124 |
| 4-2   | Free-Body Diagrams | 125 |
| 4-3   | Calculation of Unknown Forces and Moments in Two Dimensions | 129 |
| 4-4   | Two-Force Members | 131 |
| 4-5   | Calculation of Unknown Forces and Moments in Three Dimensions | 148 |
| 4-6   | Interrelated Free-Body Diagrams. Newton's Third Law | 156 |
|       | Summary | 166 |

## CHAPTER 5    FORCES AND MOMENTS IN BEAMS    171

|       | Overview | 171 |
|-------|----------|-----|
| 5-1   | Classification of Supports | 172 |
| 5-2   | Internal Forces and Moments | 173 |
| 5-3   | Sign Conventions | 176 |
| 5-4   | Shear and Bending Moment Diagrams | 187 |
|       | Summary | 202 |

## CHAPTER 6    ANALYSIS OF SIMPLE STRUCTURES AND MACHINES    207

|       | Overview | 207 |
|-------|----------|-----|
| 6-1   | Trusses | 208 |
| 6-2   | Method of Joints | 210 |
| 6-3   | Special Cases of Joints | 213 |
| 6-4   | Formal Procedure in Joint Analysis | 221 |
| 6-5   | Graphical Analysis of Trusses. Maxwell Diagram | 222 |
| 6-6   | Method of Sections | 226 |
| 6-7   | Space Trusses | 227 |
| 6-8   | Frames and Machines | 237 |
| 6-9   | Cables with Concentrated Loads | 254 |
|       | Summary | 261 |

## DISTRIBUTED FORCES — CHAPTER 7 — 266

| | | |
|---|---|---|
| | Overview | 266 |
| 7-1 | Center of Gravity | 267 |
| 7-2 | Centroids of Lines, Areas, and Volumes | 269 |
| 7-3 | Composite Objects | 272 |
| 7-4 | Theorems of Pappus | 282 |
| 7-5 | Distributed Loads on Structures. External Reactions | 289 |
| 7-6 | Distributed Loads on Beams. Internal Reactions | 292 |
| 7-7 | Distributed Loads on Cables | 305 |
| 7-8 | Fluid Statics | 313 |
| 7-9 | Buoyancy | 316 |
| 7-10 | Elements of Stress Analysis | 323 |
| | Summary | 329 |

## FRICTION — CHAPTER 8 — 336

| | | |
|---|---|---|
| | Overview | 336 |
| 8-1 | Dry Friction | 337 |
| 8-2 | Angles of Friction | 339 |
| 8-3 | Classification of Problems of Dry Friction | 340 |
| 8-4 | Wedges and Screws | 351 |
| 8-5 | Disk Friction | 360 |
| 8-6 | Axle Friction | 361 |
| 8-7 | Rolling Resistance | 362 |
| 8-8 | Belt Friction | 369 |
| 8-9 | Fluid Friction | 371 |
| | Summary | 375 |

## MOMENTS OF INERTIA — CHAPTER 9 — 382

| | | |
|---|---|---|
| | Overview | 382 |
| 9-1 | Moment of Inertia of a Mass | 384 |
| 9-2 | Moment of Inertia of an Area | 385 |
| 9-3 | Polar Moment of Inertia of an Area | 386 |
| 9-4 | Parallel-Axis Theorem | 392 |
| 9-5 | Moments of Inertia of Complex Areas and Bodies | 393 |
| 9-6 | Products of Inertia | 400 |
| 9-7 | Principal Axes and Moments of Inertia of an Area | 402 |
| 9-8 | Transformation of Moments of Inertia of a Mass | 405 |
| | Summary | 414 |

## WORK AND POTENTIAL ENERGY — CHAPTER 10 — 418

| | | |
|---|---|---|
| | Overview | 418 |
| 10-1 | Work of a Force | 419 |
| 10-2 | Work of a Couple | 422 |
| 10-3 | Virtual Work | 429 |
| 10-4 | Mechanical Efficiency of Real Systems | 430 |
| 10-5 | Potential Energy | 435 |
| 10-6 | Potential Energy at Equilibrium | 437 |
| 10-7 | Stability of Equilibrium | 438 |
| | Summary | 445 |

# DYNAMICS—VOLUME 2

**CHAPTER 11**      KINEMATICS OF PARTICLES: SCALAR ANALYSIS    451

| | | |
|---|---|---|
| | Overview | 451 |
| 11-1 | Rectilinear Position, Velocity, and Acceleration | 452 |
| 11-2 | Measured and Derived Quantities as Functions of Time | 454 |
| 11-3 | Useful Expressions Based on Acceleration | 457 |
| 11-4 | Relative Motion of Two Particles along the Same Line | 470 |
| | Summary | 479 |

**CHAPTER 12**      KINEMATICS OF PARTICLES: VECTOR ANALYSIS    483

| | | |
|---|---|---|
| | Overview | 483 |
| 12-1 | Curvilinear Motion. Rectangular Cartesian Coordinates | 484 |
| 12-2 | Physical Aspects of Velocity and Acceleration in Curvilinear Motion | 485 |
| 12-3 | Tangential and Normal Components | 486 |
| 12-4 | Motion of a Particle in Polar Coordinates | 498 |
| 12-5 | Cylindrical Coordinates | 501 |
| 12-6 | Spherical Coordinates | 501 |
| 12-7 | Relative Motion of Two Particles. Translation of Coordinates | 514 |
| | Summary | 520 |

**CHAPTER 13**      KINETICS OF PARTICLES: NEWTON'S SECOND LAW    528

| | | |
|---|---|---|
| | Overview | 528 |
| 13-1 | Newton's Second Law of Motion | 529 |
| 13-2 | Systems of Units | 530 |
| 13-3 | Equations of Motion | 531 |
| 13-4 | Problem Solving in Particle Kinetics | 533 |
| | Summary | 548 |

**CHAPTER 14**      KINETICS OF PARTICLES: WORK AND ENERGY METHODS    551

| | | |
|---|---|---|
| | Overview | 551 |
| 14-1 | Work of a Force | 552 |
| 14-2 | Potential Energy and Work | 554 |
| 14-3 | Kinetic Energy. Principle of Work and Energy | 557 |
| 14-4 | Power and Efficiency | 558 |
| 14-5 | Opportunities and Limitations of Using Energy Methods | 559 |
| 14-6 | Conservative Systems. Potential Functions | 573 |
| 14-7 | Conservation of Energy | 575 |
| | Summary | 585 |

**CHAPTER 15**      KINETICS OF PARTICLES: MOMENTUM METHODS    590

| | | |
|---|---|---|
| | Overview | 590 |
| 15-1 | Linear Momentum. Time Rate of Change of Linear Momentum | 591 |
| 15-2 | Angular Momentum | 592 |
| 15-3 | Time Rate of Change of Angular Momentum | 593 |
| 15-4 | Newton's Law of Universal Gravitation | 600 |

15-5   Kepler's Laws of Planetary Motion                                601
15-6   Impulse and Momentum                                             609
15-7   Conservation of Total Momentum of Particles                      612
15-8   Collision of Deformable Bodies                                   620
15-9   Direct Central Impact                                            620
15-10  Oblique Central Impact                                           624
       Summary                                                          632

KINETICS OF PARTICLES: COMPILATION OF METHODS.                          **CHAPTER 16**
STRATEGIES FOR SOLVING PROBLEMS                                  637

       Overview                                                         637
16-1   Overview of Kinetics Analysis of Particles                       638


KINETICS OF SYSTEMS OF PARTICLES                                 656     **CHAPTER 17**

       Overview                                                         656
17-1   Newton's Second Law Applied to a System of Particles             657
17-2   Motion of the Center of Mass                                     658
17-3   Gravitational Potential Energy of a System of Particles          664
17-4   Kinetic Energy of a System of Particles                          664
17-5   Work and Energy. Conservation of Energy                          667
17-6   Moments of Forces on a System of Particles                       672
17-7   Linear and Angular Momenta of a System of Particles              672
17-8   Angular Momentum about the Center of Mass                        673
17-9   Conservation of Momentum of a System of Particles                676
17-10  Impulse and Momentum of a System of Particles                    677
17-11  Definitions of Fluid Flow                                        685
17-12  Steady Flow of Mass                                              685
17-13  Systems with Variable Mass                                       689
       Summary                                                          699


KINEMATICS OF RIGID BODIES                                       705     **CHAPTER 18**

       Overview                                                         705
18-1   Translation                                                      708
18-2   Rotation about a Fixed Axis                                      709
18-3   Kinematics Equations for Rotating Rigid Bodies                   711
18-4   General Plane Motion. Analysis of Velocities                     719
18-5   Instantaneous Center of Rotation                                 722
18-6   General Plane Motion. Analysis of Accelerations                  731
18-7   General Motion                                                   741
18-8   Time Derivative of a Vector Using a Rotating Frame               748
18-9   Analysis of Velocities and Accelerations Using a Rotating Frame  749
       Summary                                                          758


KINETICS OF RIGID BODIES IN PLANE MOTION                         765     **CHAPTER 19**

       Overview                                                         765
19-1   Equation of Translational Motion                                 766
19-2   Equation of Rotational Motion                                    767
19-3   Application of Equations of Motion                               769
19-4   Rotation about a Fixed Axis                                      780
19-5   General Plane Motion                                             790
       Summary                                                          800

**CHAPTER 20**

ENERGY AND MOMENTUM METHODS
FOR RIGID BODIES IN PLANE MOTION                                          805

    Overview                                                              805
20-1  Work of a Force on a Rigid Body                                        806
20-2  Kinetic Energy of a Rigid Body                                         808
20-3  Principle of Work and Energy                                           810
20-4  Conservation of Energy                                                 811
20-5  Power                                                                  812
20-6  Impulse and Momentum of a Rigid Body                                   825
20-7  Impulse and Momentum of a System of Rigid Bodies                       827
20-8  Conservation of Momentum                                              828
    Summary                                                           841

**CHAPTER 21**

KINETICS OF RIGID BODIES IN THREE DIMENSIONS                             846

    Overview                                                              846
21-1  Angular Momentum in Three Dimensions                                   847
21-2  Impulse and Momentum of a Rigid Body                                   850
21-3  Kinetic Energy of a Rigid Body in Three-Dimensional Motion            851
21-4  Equations of Motion in Three Dimensions                                861
21-5  Euler's Equations of Motion                                            862
21-6  Solution of Basic Problems in Three-Dimensional Motion                863
21-7  Gyroscopic Motion                                                      870
21-8  Special Cases of Gyroscopic Motion                                     873
    Summary                                                           883

**CHAPTER 22**

VIBRATIONS                                                                888

    Overview                                                              888
22-1  Simple Harmonic Motion                                                 889
22-2  Free Vibrations of Particles Without Damping                           893
22-3  Free Vibrations of Rigid Bodies                                        893
22-4  Energy Methods in Vibration Analysis                                   901
22-5  Forced Vibrations Without Damping                                      905
22-6  Damping of Vibrations                                                  911
22-7  Damped Free Vibrations                                                 911
22-8  Damped Forced Vibrations                                               913
22-9  Electrical Analogues of Mechanical Vibrations                          915
    Summary                                                           919

APPENDIX A: VECTOR DERIVATIVES                                            922

APPENDIX B: COMMENTS ON THE USE OF UNITS                                  924

APPENDIX C: INERTIA FORCE AND DYNAMIC
EQUILIBRIUM                                                               925

APPENDIX D: CORIOLIS EFFECTS                                              927

ANSWERS TO EVEN-NUMBERED PROBLEMS                                         A1

LIST OF SYMBOLS                                                           L1

INDEX                                                                     I1

# Preface

Statics and dynamics are fundamental engineering subjects; their concepts and methods are useful in subsequent courses and in engineering practice. It is essential for the student to develop a working knowledge in these areas. The material in this text facilitates the learning of statics and dynamics and the development of a clear, concise method for solving problems in the following ways:

**Perspectives.**   Concise statements at the beginning of each chapter describe the need for learning the material in that chapter. A few sections that may be omitted in statics and dynamics courses are included to serve ambitious students and to provide for special assignments by the instructor.

**Photographs and modeling.**   The text is interspersed with photographs to enhance the realistic aspects of the material and motivate the students by arousing their curiosity toward modern practical problems. Line drawings accompany the photographs to illustrate the analytical modeling of actual structures and devices, which is important in engineering work.

**Examples.**   Groups of worked-out examples illustrate the applications of the equations and the methods developed in the preceding sections of the text. These realistic examples show the relevance of the subject to engineering practice and demonstrate a straightforward procedure for analyzing a problem. Over one quarter of the 227 examples are new in the second edition.

**Judgment of the results.**   Many example problems are concluded by a brief section showing judgment of the results. The numerical answer to a problem may or may not be reasonable, so thought and experience are required to assess the answer for acceptability. Whenever possible, the judgment should be based on calculations, preferably using a different method of solution. Even apparently simplistic judgments (for example, knowing that the answer makes sense for the given problem) are important in engineering practice.

**Common errors and tips.**  As a new feature of the second edition, the text is interspersed with boxes offset in color and titled "Common Error" or "Tip". Many years of teaching experience have shown us the most frequent mistakes made by students, so we pass on these words of caution and hints to aid the beginning student and to serve as a reminder to those reviewing the material.

**Homework problems.**  Problem sets of a wide range and of increasing difficulty follow each group of examples and require the use of methods from the preceding sections. The review problems at the end of the chapter are especially challenging and encompass all of the methods learned previously. A few of the problems require plotting the results as a function of a given parameter. Although this is time consuming, it is important for two reasons. First, a graphic display of the results enhances the learning process. The patterns of the data may quickly reinforce one's knowledge or provoke thought. Second, engineers often work and communicate in graphic ways. The practice of plotting results and sketching reasonable diagrams is useful preparation for the student. The total number of homework problems is 1764, of which over fifteen percent are new or revised from the first edition. Answers are given to all even-numbered problems at the back of the book.

**Summary sections.**  Each chapter is concluded with a graphically illustrated summary of the essential methods and equations. These summaries are very useful at all stages in the process of learning and in the preparation for examinations.

## ORGANIZATION OF CHAPTERS

Many topics in elementary mechanics can be studied in various orders, each with some advantages and disadvantages. The compromises in this text were made in an effort to ease the student's overall progress through the course. For example, bending moments in beams are first discussed in Ch. 5, but bending moments in beams under distributed loadings are presented in Ch. 7, because the latter topic is rather difficult for many students. Therefore, they benefit from the early presentation of the basic concept (which is simple) and from the considerable additional practice with free-body diagrams in Ch. 6 (trusses and frames) before studying distributed loadings. However, such an order of topics is by no means sacred. Those instructors who strongly wish to rearrange some of the presentations can do so without great difficulty for themselves or harm to the students.

In the second edition an even greater emphasis has been placed on free-body diagrams than in the first edition. In Ch. 2, the section on particle equilibrium has been expanded to include problem-solving techniques, stressing the use of free-body diagrams. Chapter 4 on the equilibrium of rigid bodies has been rearranged in order to cover two-dimensional and three-dimensional problems separately, affording an opportunity for broader and more detailed coverage of each. A new table illustrates the common three-dimensional supports and their use in free-body diagrams.

The important topic of interrelated free-body diagrams for both two-dimensional and three-dimensional problems (Sec. 4-6) is retained from the first edition. In Ch. 7 a new section has been added to cover the determination of external support reactions on structures with distributive loading. This is useful for dealing with the difficult topic of internal reactions due to distributive loads. Section 13-4 on problem-solving techniques has been added in particle kinetics, stressing the importance of free-body diagrams in dynamics.

The organization of chapters in the dynamics portion of this text follows the standard one of presenting the kinematics and kinetics of particles and then the kinematics and kinetics of rigid bodies. This second edition departs from the standard by adding a new chapter following the chapters on particle kinetics, work-energy methods, and momentum methods (Chs. 13–15). Chapter 16 compares these different methods in a comprehensive manner and presents example problems worked with alternative methods. This extraordinary chapter will help students sharpen their problem-solving skills by developing a systematic way of choosing the most efficient method to arrive at a solution. The challenging problems in this chapter are truly review problems and a test of the student's cumulative knowledge.

**Appendices.**   The four Appendices at the end of the text should be read as they become appropriate. The following comments about these are worth noting:

*Appendix A. Vector derivatives.*   This Appendix is devoted to the derivation of an important equation (Eq. A-7). The derivation involves the important definition of the angular velocity vector. Although the details are not necessary to remember, Eq. A-7 and its special cases are useful in the discussion of several topics.

*Appendix B. Comments on the use of units.*   The two correct systems of units to be used in engineering are discussed in the main body of the text. In addition, the reader should be aware that other units, even incorrect ones may be encountered in practice. Although it is necessary to mention these, they are removed from the mainstream of topics to reduce the chance of confusion about them.

*Appendix C. Inertia force and dynamic equilibrium.*   All students of dynamics should be aware of this topic. However, there are practical reasons (which are given in App. C) for removing this topic from the mainstream of elementary engineering mechanics.

*Appendix D. Coriolis effects.*   The understanding of Coriolis acceleration is rather difficult for many students, although its calculation is quite simple. A detailed discussion of Coriolis effects is presented in this Appendix in order to streamline the main body of the text where several accelerations must be considered simultaneously.

**Mathematics.**   Vector methods are introduced in the text and become increasingly important in dynamics analysis. A working knowledge of geometry, trigonometry, algebra, and elementary calculus is required. The example problems in Sec. 7-6 of this edition illustrate alternative methods of drawing shear and bending moment diagrams using calculus directly.

**Units.** The SI and U.S. customary units are used approximately to the same extent throughout. There are enough examples and homework problems to allow a total concentration on either system of units if desired. Engineers in the United States will need to know both systems of units in the foreseeable future.

BELA I. SANDOR
KAREN J. RICHTER

*Madison, Wisconsin*

# Acknowledgements

We gratefully acknowledge the valuable contributions of the following professors who reviewed all or portions of this text for the first or the second edition:

George G. Adams—Northeastern University
David Barnett—Stanford University
Leonard Berkowitz—Cal Poly Pomona
Mark Berrio—California State Polytechnic University, San Luis Obispo
William Clausen—Ohio State University
Robert D. Culp—University of Colorado
Nicholas P. Daddario—GMI Engineering & Management Institute
Stephen F. Felszeghy—California State University at Los Angeles
E. T. Kornhauser—Brown University
William M. Lee—U.S. Naval Academy—Annapolis
Josue Njock Libii—Purdue University at Fort Wayne
G. W. May—University of New Mexico
David McGill—Georgia Institute of Technology
G. Allen Pugh—Purdue University at Fort Wayne
Robert Schaller—U.S. Air Force Academy
R. A. Smith—Cambridge University

In addition, we are indebted to professors Robert D. Cook and A. L. Schlack (University of Wisconsin—Madison) for many important suggestions, to Dr. Brian Bartholomeusz for extensive checking of the problem solutions, and to graduate student David T. Lohr (University of Wisconsin-Madison) for major work on the problem solutions and final refinements of the text. We also appreciate the fine work of our student assistants and the helpful comments of the many instructors who have used the book in class.

Special thanks are due the project team at Prentice-Hall, in particular Melissa Halverstadt, Doug Humphrey, Virginia Huebner, Niels Nielsen, and Janet Schmid.

B. I. S.
K. J. R.

VOLUME 1

# Statics

# Introduction to Engineering Mechanics

Students of several engineering specialties begin their engineering studies in the sophomore year by taking a course in statics, one of the fundamental subjects in engineering mechanics. Statics courses are based on the mechanics concepts of physics and provide the foundation for many other courses and engineering practice. Statics is the prerequisite for two common courses, dynamics and mechanics of materials (also called strength of materials). These in turn open the door to many other courses, as illustrated in the following flow diagram.

Statics

Dynamics           Mechanics of materials

| Courses in | Courses in |
|---|---|
| Advanced dynamics | Advanced mechanics of materials |
| Mechanics of fluids | Machine design |
| Machine design | Structural mechanics |
| Vibrations | Experimental stress analysis |
| etc. | etc. |

Both dynamics and mechanics of materials are more difficult than statics for the average student. Therefore, a firm foundation in statics is important to assure satisfactory progress in subsequent courses.

3

## STUDY GOALS

The remaining sections of this introductory chapter present definitions, basic concepts, systems of measurement, and some practical suggestions that are important for all fundamental or advanced courses in engineering mechanics. An overview of these sections is given here.

**SECTION 1-1** defines the major branches of engineering mechanics. Most engineers know the fundamentals of all of these subjects, but at the advanced level people tend to specialize in the dynamics, the materials, or the fluids area.

**SECTIONS 1-2 and 1-3** present the terminology and basic definitions of Newtonian mechanics, which is the foundation of engineering mechanics. The student should try to remember these sections because a concise review of the basic concepts is desirable from time to time throughout statics and dynamics.

**SECTION 1-4** defines the two kinds of quantities used in mechanics: scalars and vectors. Various aspects and details of manipulations of vectors are presented throughout statics and dynamics. At all levels, these should be studied with intensity to facilitate the solution of complex problems.

**SECTION 1-5** describes the SI system of units for the specific measure of the quantities used in engineering mechanics. The student should read this section to understand the scheme of basic and derived units and to note the statements concerning proper practices. For solving problems, the student will find the table inside the front cover useful for checking the formulas of units and conversion factors between the SI and the U.S. customary units.

**SECTION 1-6** introduces the method of analytical modeling, which is used throughout statics and dynamics to facilitate the solving of problems. In many cases the modeling will appear very simple. However, sufficiently accurate and yet practical analytical models may require careful thought and should always have high priority in setting up solutions and checking the results.

**SECTION 1-7** introduces recurring practical examples which show that the fundamental concepts of engineering mechanics and the necessary simplifications in modeling are indispensable in working on small or large projects. These examples will give the student a perspective view of engineering analysis in general.

## 1-1

## ENGINEERING MECHANICS

Mechanics is a branch of physics and has the longest history among the physical sciences. Archimedes (287–212 B.C.) is considered the first systematic investigator of this subject, mainly for his work on the principles of levers and of buoyancy. The most prominent of numerous other scientists in this area are Galileo (1564–1642), who made the first fundamental analyses and experiments in dynamics, and Newton (1642–1727), who formulated the laws of motion and of gravitation.

Engineering mechanics mainly concerns the practical applications of the fundamental principles of mechanics. It is the foundation of many engineering sciences and is used daily by many people. The vocabulary of engineering mechanics is part of the language that most engineers must learn. The large area of engineering mechanics is divided into three parts:

1. *Mechanics of rigid bodies.* Here it is assumed that all solids are perfectly rigid and do not deform even under large forces.
   a. *Statics* deals with rigid bodies at rest or under conditions that are static for certain practical purposes.
   b. *Dynamics* deals with rigid bodies in motion.
2. *Mechanics of deformable bodies,* or *mechanics of materials,* or *strength of materials.* This area deals with real solids which are never perfectly rigid. Even small forces cause deformations in strong materials.
3. *Mechanics of fluids.* This field concerns the behavior of liquids and gases.

The student should appreciate that often, the distinctions between the various areas are not sharp. For example, mechanics of materials may be considered in the realms of statics or dynamics.

## BASIC CONCEPTS OF MECHANICS    1-2

Engineering mechanics is based on Newtonian mechanics, in which relativistic effects are negligible. In Newtonian mechanics the basic concepts are space, time, and mass. These are absolute concepts because they are independent of each other. They are also quite abstract. It is best to accept them in engineering work on the basis of personal experiences of the physical world, and intuition if necessary. Other concepts, such as force, are dependent on the basic concepts. *Space* is a geometric region in which events involving bodies occur. The region of interest may be one-, two-, or three-dimensional. The bodies may be *particles* of negligible dimensions or bodies of finite dimensions. *Reference frames* are used to describe the relative positions or coordinates of objects with respect to each other and in the space under consideration. Frequently, the frame of reference chosen is stationary with respect to the earth. Sometimes it is necessary to use an astronomical frame of reference in which the earth itself is moving. *Time* is an absolute quantity in Newtonian mechanics, and in this system it requires no special definition. *Mass* is the quantitative measure of *inertia,* which is the property of all matter. Inertia is the resistance to change in motion of matter. *Force* is the action of a body on another. When an external force acts on a body, the body tends to move in the direction of that force.

## NEWTON'S LAWS    1-3

Sir Isaac Newton formulated the laws that are the basis for analysis in statics and dynamics. These laws may be stated as follows:

**FIRST LAW**
> A particle remains at rest (if originally at rest) or moves in a straight line with constant velocity if the net force on it is zero.

**SECOND LAW**
> A particle with a net force acting on it has an acceleration proportional to the magnitude of the force and in the direction of that force.

This law is formulated as

$$\mathbf{F} = m\mathbf{a}$$

1-1

where $\mathbf{F}$, $m$, and $\mathbf{a}$ are force, mass, and acceleration, respectively. The boldface letters indicate vectors, which are described in Sec. 1-4. Equation 1-1 is first used extensively in Ch. 13 of the associated Dynamics volume.

### THIRD LAW

The forces of action and reaction between interacting bodies are equal in magnitude, opposite in direction, and have the same line of action.

This principle is valid at all times and is essential in understanding the nature of forces. For example, consider a rope-pulling contest between two persons, as shown in Fig. 1-1. Assume that person $A$ is pulling the rope to the left with a force $F$. Intuitively, person $B$ must be pulling the rope to the right with a force of equal magnitude $F$ if there is a stalemate. Next, assume that person $B$ slips and pulls the rope with a different force $F_1$ at a particular instant (the rope remains taut). According to Newton's Third law, person $A$ pulls the rope in the opposite direction with the same force $F_1$ at that particular instant. This is true regardless of who or what caused person $B$ to slip.

## Law of Gravitation

Newton expressed the mutual gravitational attraction between two particles by the equation

$$F = G \frac{Mm}{r^2}$$

1-2

where

$F$ = magnitude of force of attraction
$M$ and $m$ = masses of the particles
$G$ = universal constant of gravitation
$r$ = distance between the centers of the particles

The mutual forces $F$ determined from this equation must satisfy all of the conditions of the Third Law (collinear forces, equal in magnitude, opposite in direction).

The force of attraction between the earth and any body near its surface is called the *weight W* of that body. This force causes the acceleration of falling bodies. The *gravitational acceleration* is denoted by $g$, and it can be determined by experiments with freely falling bodies in a vacuum. Alternatively, it can be obtained from Eqs. 1-1 and 1-2, using the experimental value of the constant $G$. The average value of $g$ is 9.81 m/s$^2$ or 32.2 ft/s$^2$. It varies slightly depending on location on the surface of the earth because the earth is not a perfect, homogeneous sphere.

Equation 1-1 can be written using the quantities $W$ and $g$,

$$W = mg$$

1-3

This formula is frequently used in practice and is worth remembering.

**FIGURE 1-1**

Rope-pulling contest illustrates Newton's Third Law of action and reaction

Two kinds of quantities are distinguished in mechanics. *Scalar* quantities have only magnitudes. Examples are density, energy, mass, speed, temperature, time, and volume. *Vector* quantities have magnitude and direction. Unlike scalars, vectors add according to a special rule called the *parallelogram law*, as shown in Sec. 2-1. Examples of vectors are acceleration, displacement, force, moment, momentum, and velocity. Vectors are commonly represented by arrows in drawings. In graphical solutions when working with vectors, the lengths of the arrows should be proportional to the magnitudes (and drawn exactly within tolerable error). In many cases, however, an engineer or student would tend to work with freehand sketches of vectors with the magnitudes given numerically alongside the arrows. The arrowhead indicates the direction of the vector by defining its sense along the *line of action* of the vector in space (Fig. 1-2). In this text a boldface letter designates a vector, and the italicized letter stands for its magnitude.* Occasionally, vectors, as directed line segments between two points in space, will be represented as pairs of letters with an arrow above them, as in Fig. 1-3.

**FIGURE 1-2**

Vector **F** of magnitude *F*

The following classification of vectors is useful. A *free vector* is not uniquely associated with any given point or line in space. For example, consider the body of a car that moves in a straight path. At any given instant the velocity vector for point *A* is the same as for point *B*, or any other point in the car (Fig. 1-4). Thus, a single vector drawn anywhere in the body of the car is adequate to describe its motion, a fact which is convenient in analysis. Note that a free vector is not entirely free; only parallel lines of action are acceptable for choice.

**FIGURE 1-3**

Vector $\overrightarrow{AB}$ between points *A* and *B*

A *sliding vector* may be moved along its line of action but not to another line even if it is parallel to the original line of action. For example, consider the laboratory test of an automobile whose wheels are on rigid supporting platens. It makes no difference whether a sideways force **F** is pushing or **F′** of equal magnitude is pulling the platen as in Fig. 1-5. This is called the *principle of transmissibility*. The same force applied to the spindle ($P = F$ in magnitude) would make a significant difference in the vehicle's response because of the relatively flexible tire between the platen and the spindle. This does not rule out using **P** instead of **F** with proper corrections, but it has to be recognized that these vectors are not the same in effect.

**FIGURE 1-4**

A free vector may be placed anywhere in a parallel position in a rigid body

**FIGURE 1-5**

One wheel of an automobile in a laboratory setup for simulating forces from the road

---

\* Common notations for vectors in longhand writing are $\vec{P}$, $\underline{P}$, and $\mathbb{P}$. Common notations for the magnitude of vectors in longhand writing are $|\vec{P}|$ and $|\underline{P}|$.

A *fixed vector* has the most restrictions, with both its line of action and its point of application specified. Such a vector must be used when the deformation of a nonrigid body has to be involved. For example, consider the identical, collinear forces **V** (acting on the bottom of the tire) and **V′** (acting on the spindle) in Fig. 1-5. These vectors could be considered as sliding vectors if the tire were rigid. As it is, the point of application of a force makes a large difference in the vehicle's response to that force.

More detailed considerations and manipulations of vectors are presented in Ch. 2.

## 1-5 UNITS

Several systems have evolved for the specific measure of the basic and derived quantities. Engineers have traditionally preferred to use a gravitational system with length, time, and force as fundamental quantities, and mass as a derived quantity. In physics mass is taken as fundamental and force is derived; there is no practical difference between the systems if they are used properly.

The system of measurement that is favored now is the International System of Units (SI). This is a modern version of the MKS (meter, kilogram, second) system. Its details are established and controlled by an international treaty organization. Many agencies and technical societies publish these in full or in part. One example is the Standard for Metric Practice E380, which is published by the American Society for Testing and Materials. This standard presents the SI system and a small number of non-SI units that are strongly needed and have been widely accepted.

The SI system of units is intended for worldwide standardization, and it is the basic system in this text. A reasonably detailed description of the SI system is presented here. However, inevitably, engineers will encounter other systems of units for a long time, especially the customary English system, which may persist in the United States. This system is described only briefly because of its familiarity to Americans. Since nearly half of the problems herein are in U.S. customary units, a conversion table for these is given inside the front cover.

### SI Units

SI units are divided into three classes: base, supplementary, and derived units. The three base units are arbitrarily defined, using the *meter* (m) for length, the *kilogram* (kg) for mass, and the *second* (s) for time. For example, the kilogram is the mass of a platinum standard kept in France (one kilogram is approximately equal to the mass of 0.001 $m^3$ of water). The most important derived unit is that of force, called the *newton* (N). It is expressed from Eq. 1-1 using the unit magnitudes of mass and acceleration:

$$\text{Force} = (\text{mass})(\text{acceleration})$$

$$1 \text{ N} = (1 \text{ kg})(1 \text{ m/s}^2) = 1 \text{ kg·m/s}^2 \qquad \boxed{1\text{-}4}$$

The most relevant SI units for statics and dynamics are presented here.

| QUANTITY | | UNIT | SYMBOL | FORMULA |
|---|---|---|---|---|
| Base units | Length | meter or metre | m | |
| | Mass | kilogram | kg | |
| | Time | second | s | |
| Supplementary unit | Plane angle | radian | rad | |
| Derived units | Energy, work | joule | J | N·m |
| | Force | newton | N | kg·m/s$^2$ |
| | Frequency (of a periodic phenomenon) | hertz | Hz | 1/s |
| | Power | watt | W | J/s |
| | Pressure, stress | pascal | Pa | N/m$^2$ |

Other commonly used and acceptable derived units of SI are as follows:

| QUANTITY | UNIT | SYMBOL |
|---|---|---|
| Acceleration | meter per second squared | m/s$^2$ |
| Angular acceleration | radian per second squared | rad/s$^2$ |
| Angular velocity | radian per second | rad/s |
| Area | square meter | m$^2$ |
| Density, mass | kilogram per cubic meter | kg/m$^3$ |
| Moment of force | newton meter | N·m |
| Velocity | meter per second | m/s |
| Volume | cubic meter | m$^3$ |

The following non-SI units may be used with the SI system according to the ASTM standard E380:

| QUANTITY | UNIT | SYMBOL | DEFINITION |
|---|---|---|---|
| Time | minute | min | min = 60 s |
| | hour | h | 1 h = 60 min |
| | day | d | 1 d = 24 h |
| Plane angle | degree$^a$ | ° | $1° = \dfrac{\pi}{180}$ rad |
| Volume | liter or litre | $\ell$ | $1\,\ell = 10^{-3}$ m$^3$ |
| Mass | metric ton | t | 1 t = 10$^3$ kg |

$^a$ Decimal submultiples are permissible, but minutes and seconds are discouraged.

Correct usage of the SI system is important to avoid a gradual degradation that has occurred in the older measurement systems. The following aspects of the proper practices should be noted.

*Prefixes* are used to denote orders of magnitude, and spaces are used in long strings of digits as shown for some common numbers in engineering mechanics.

| MULTIPLICATION FACTOR | PREFIX | SYMBOL |
|---|---|---|
| $10^9 = 1\ 000\ 000\ 000$ | giga | G |
| $10^6 = 1\ 000\ 000$ | mega | M |
| $10^3 = 1000$ | kilo | k |
| $10^{-2} = 0.01$ | centi | c |
| $10^{-3} = 0.001$ | milli | m |
| $10^{-6} = 0.000\ 001$ | micro | $\mu$ |

It is best to choose prefixes so that the numerical value is between 0.1 and 1000. Prefixes representing powers of 1000 are preferred. In calculations the powers-of-10 notation should be used instead of the prefixes.

The meanings of the words *mass* and *weight* should be clarified, especially since SI has a principal difference from the gravitational system of metric units with respect to these quantities. *Mass* is a scalar quantity and is a property of matter. *Weight* is a vector quantity and is equal to the *force of gravity*, which is mass times the local acceleration of gravity. The appropriate units in correct technical practice are kilograms for mass and newtons for weight. Most students will need plenty of practice with numerical problems before fully understanding and accepting these units and the various derived ones.

## U.S. Customary Units

The base units of the English system still widely used in American industry are the *foot* (ft) for length, the *pound* (lb) for force, and the *second* (s) for time. It is convenient to define these in terms of SI units. The foot is defined as equal to 0.3048 m. The pound is the *weight* of a platinum standard (at the National Bureau of Standards, Washington, D.C.) whose mass is 0.453 592 43 kg. The second for time is the same in both systems of units.

The derived unit of mass in the U.S. customary system is the *slug*. This is the mass that is accelerated at 1 ft/s² when a force of 1 lb is applied to it. Mass can be expressed from Eq. 1-1 using the unit magnitudes of force and acceleration,

$$1 \text{ slug} = \frac{1 \text{ lb}}{1 \text{ ft/s}^2} = 1 \text{ lb·s}^2/\text{ft} \qquad \boxed{1\text{-}5}$$

## Comments on the SI and U.S. Customary Systems

Conversions between these major systems of units are readily made, but their differences in points of view should be emphasized.

**1.** The SI system is an *absolute* system of units because the three base units are chosen to be independent of the locations where the quantities are measured. This is true for any location in the known universe.

**2.** The U.S. customary system is *not* an absolute system but a *gravitational* system of units because the weight of a body depends on the gravitational attraction of the earth, which varies somewhat with location even on the surface of the earth. The variation of weight of a given mass with location is negligible in many engineering problems, but the possibility of even large variations should be kept in mind. If there is any doubt concerning the accuracy of weight or mass using U.S. customary units, the value of the acceleration of gravity, $g$, should be verified for the given location.

## Conversions of Units

Engineers occasionally have to convert numerical quantities from one system of units to another. The conversions are straightforward using the table given inside the front cover. Some of the most common conversions are presented in detail here to illustrate the procedures.

**Length.** By definition,

$$1 \text{ ft} = 0.3048 \text{ m} \qquad \boxed{1\text{-}6}$$

The U.S. customary unit of *mile* (mi) is defined in terms of feet and converted into SI units as follows:

$$1 \text{ mi} = 5280 \text{ ft} = 5280 \, (0.3048 \text{ m}) = 1609 \text{ m}$$

or

$$1 \text{ mi} = 1.609 \text{ km} \qquad \boxed{1\text{-}7}$$

Similarly for the unit of *inch* (in.),

$$1 \text{ in.} = \frac{1}{12} \text{ ft} = \frac{1}{12} \, (0.3048 \text{ m}) = 0.0254 \text{ m}$$

or

$$1 \text{ in.} = 25.4 \text{ mm} \qquad \boxed{1\text{-}8}$$

**Force.** The U.S. customary unit of pound is defined as the weight of a mass of approximately 0.4536 kg at a location where $g = 9.807 \text{ m/s}^2$. Thus, from Eq. 1-3,

$$W = mg$$

$$1 \text{ lb} = (0.4536 \text{ kg})(9.807 \text{ m/s}^2) = 4.448 \text{ kg·m/s}^2$$

From Eq. 1-4,

$$1 \text{ lb} = 4.448 \text{ N} \qquad \boxed{1\text{-}9}$$

A force of 1000 lb is called the *kilopound* (kip), and 2000 lb equals the *ton* in the U.S. customary system. It should be noted that the *metric ton* in the SI system represents a mass of 1000 kg. Thus, ton can be an ambiguous unit if not specified precisely.

**Mass.** The U.S. customary unit of mass, the slug, can be converted into kilograms as follows. From Eqs. 1-5, 1-6, and 1-9,

$$1 \text{ slug} = 1 \text{ lb·s}^2/\text{ft} = \frac{4.448 \text{ N}}{0.3048 \text{ m/s}^2} = 14.59 \text{ N·s}^2/\text{m}$$

From Eq. 1-4,

$$1 \text{ slug} = 14.59 \text{ kg} \qquad \boxed{1\text{-}10}$$

**Moment of a force.**   The moment of a force has units of force times length in both systems of units. For example *pound·foot* (lb·ft) is used in the U.S. customary system, and *newton·meter* (N·m) in the SI system. The conversion is written using Eqs. 1-6 and 1-9,

$$1 \text{ lb·ft} = (0.3048 \text{ m})(4.448 \text{ N}) = 1.356 \text{ N·m} \qquad \boxed{1\text{-}11}$$

To illustrate the reverse procedure in a more complex conversion, assume that 500 N·m should be converted into pound·inches. Again using Eqs. 1-6 and 1-9, and the fact that 1 ft = 12 in.,

$$500 \text{ N·m} = 500 \text{ N·m} \left( \frac{1 \text{ lb}}{4.448 \text{ N}} \right) \left( \frac{1 \text{ ft}}{0.3048 \text{ m}} \right) \left( \frac{12 \text{ in.}}{1 \text{ ft}} \right)$$

Note that each quantity in parentheses has a magnitude of unity; the purpose of these quantities is to change the units of the original quantity to the desired final set of units without changing the value of the original quantity. After the computations and the cancelation of newtons, meters, and feet, the result is

$$500 \text{ N·m} = 4426 \text{ lb·in}$$

## Need for Displaying Units

The student should always remember that in most cases numbers have meaning only if the correct units are displayed with the numbers (exceptions are ratios, which have no units). In performing calculations there is often a temptation to work only with the numbers and write the "expected" units with the final result. This method of economizing in the computation may be reasonable in some cases, but in general it increases the chances of error. The ideal method is to show the correct units with each number throughout the solution. This helps in detecting errors both in the concepts and details of the solution.

## 1-6   MODELING, APPROXIMATIONS, REQUIRED ACCURACIES

Modeling for analysis means that a real system of bodies, no matter how complex, is assumed to have only a few parts, that these are of simple shape, and that only a few major forces are acting on them. This is always necessary to do in analyzing any system because it is not possible to include everything that has an effect on the system. For example, the gravitational effects of the other planets at the surface of the earth are real but negligible in comparison with the effect of the earth.

It is always best to start with the simplest possible model that reasonably represents the system under consideration. Sometimes this may involve difficult choices regarding the inclusion of certain items. Even then it is best to start with a simple model and gradually improve it as experience is gained concerning the given problem.

For examples of analytical models of complex systems, consider skyscrapers and suspension systems of cars. Figure 1-6 shows the simplest reasonable models for skyscrapers. Of course, other models of similar or greater complexity may be drawn. The model for the suspension system

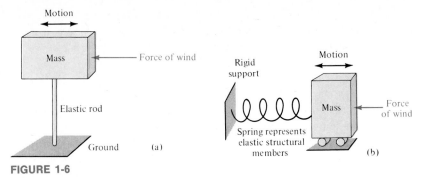

**FIGURE 1-6**

Simple models of skyscrapers

in Fig. 1-7 includes a shock absorber because its effect is quite significant on the spring and mass system.

In making analytical models the important assumptions must be explicitly stated, preferably in writing. This allows rapid reevaluation and changes for improvements when the results from the original model appear (or are proved by experiment) to be wrong. Work in modeling complex systems is quite challenging and interesting mainly because of the judgments and evaluations that are necessary.

Since assumptions are always involved in solving engineering problems, the numerical accuracy of the answers must be in some doubt. In most cases, especially in course work, it is reasonable to round the answers within a few percent. Accuracies within 1% are easily obtained from electronic calculators but may be meaningless in practice. Rounding the answers reasonably requires understanding the practical aspects of engineering work. The student should constantly strive to develop a physical sense of the problem beyond obtaining a facility in manipulating symbols.

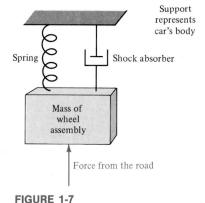

**FIGURE 1-7**

Simple model of a suspension system of an automobile

## RECURRING EXAMPLES FOR PROGRESSIVE ANALYSIS

Many students in introductory courses such as Statics find it difficult to visualize how the basic concepts, models, and methods form an integral part of solving more complex and interesting problems in engineering work. To assist the student in developing a coherent view of engineering analysis, several examples in this text reappear from time to time for the analysis of different problems. It is hoped that these examples will gradually become familiar to the student so that the various analyses might make sense as necessary parts in the task of completing a project. Two recurring examples are introduced in this section to establish the context for progressive analysis of all such projects.

### Skyscrapers with Tuned Mass Dampers

Very tall buildings can vibrate in a wind as a reed does. The amplitude (maximum displacement of the building during oscillations) depends on the geometry and materials of construction of the building, and on the wind. Modern steel-and-glass buildings are relatively susceptible to large

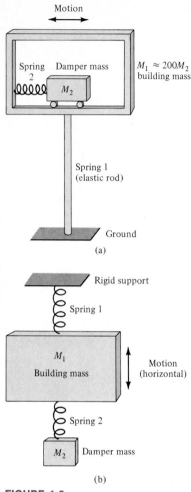

Motion

Spring 2

Damper mass

$M_2$

$M_1 \approx 200M_2$ building mass

Spring 1 (elastic rod)

Ground

(a)

Rigid support

Spring 1

$M_1$ Building mass

Motion (horizontal)

Spring 2

$M_2$ Damper mass

(b)

**FIGURE 1-8**

Models of a skyscraper with a tuned mass damper

displacements in wind, with amplitudes exceeding 1 m possible. These displacements would not be dangerous from a structural point of view, but are undesirable with respect to the comfort of the occupants.

There are several ways to keep the oscillations of a building within tolerable limits, although the motions cannot be completely eliminated. One of the viable solutions to the problem uses a *tuned mass damper* system. Such dampers have been used extensively to reduce the vibrations of bridges, tall antennas, power line cables, automotive and aircraft piston engines, and others. They are new in the area of skyscraper design.

The essence of a tuned mass damper in a building is illustrated in Fig. 1-8. A large mass $M_2$ is located near the top of the building of mass $M_1$. The damper mass $M_2$ can move only in a horizontal plane and has a spring and other constraining devices attached to it. The damper system's mass, frictional resistance to motion, and spring characteristics are designed to make the damper mass $M_2$ move opposite to the direction of displacement of the main mass $M_1$ at every instant. This reactive oscillating motion is called out-of-phase vibration, which can be adjusted or "tuned" by choosing the system's components to minimize the displacements of the building. An application of this device is in the John Hancock Tower in Boston. This building has two damper masses $A$ and $B$ as illustrated in Fig. 1-9. Each mass has a weight of $6 \times 10^5$ lb and a maximum displacement of $\pm 81$ in. relative to the building.

The design of a tuned mass damper is mainly in the realm of vibrations, an advanced area of dynamics, but the design also involves statics and elementary dynamics. There are numerous aspects of such a system that can be analyzed progressively as the student's knowledge of engineering mechanics gradually increases. A few relatively simple problem areas of a tuned mass damper are outlined as follows.

**1.** The damper mass $M_2$ in Fig. 1-8 is supported by linear bearings that allow it to move horizontally. Concentrated vertical forces are applied through the bearings to the mass even when it is stationary. Elementary statics is useful in analyzing these forces.

**2.** The constraining devices of the damper mass must apply moments, and these can also be analyzed early in a statics course.

**3.** There is friction involved in the motion of the damper mass. The conditions of interest are among the most common in statics and dynamics.

**4.** The energy methods of elementary engineering mechanics can be readily exemplified with the tuned mass damper.

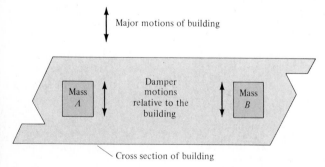

Major motions of building

Damper motions relative to the building

Mass $A$

Mass $B$

Cross section of building

**FIGURE 1-9**

Tuned-mass-damper locations on 58th floor in John Hancock Tower in Boston

**5.** The complete system of a tuned mass damper includes over-travel bumpers for safety. The elastic impacts of the damper mass with these bumpers are among the common problems in elementary dynamics.

**6.** The oscillating motion of the tuned mass damper (ignoring motion of the building) can be easily characterized using the basic concepts of vibrations which are included in most dynamics texts (although seldom covered in elementary dynamics courses).

## Road Simulation in the Laboratory

The complete road testing of cars and trucks is very expensive and time consuming, and the results are often difficult to evaluate. The modern method is to use road simulation under the controlled conditions of the laboratory. Simulators are used to study the endurance life of components, subassemblies, and the vehicle as a whole. Other testing is aimed at studying shock and vibration characteristics, safety, and ride comfort.

Two kinds of simulators are common. Figure 1-10 shows a firmly held vehicle frame, with three electronically controlled *hydraulic actuators*

**FIGURE 1-10**

Automobile in a laboratory test. The vehicle's frame is firmly fixed while three independent forces may be applied to the wheel spindle *A*. (Courtesy MTS Systems Corporation, Minneapolis, Minnesota.)

(hydraulic rams) directly coupled to a *wheel spindle* (axle) *A*. These actuators apply forces to the spindle on command. Three independently controlled actuators can simulate a wide range of vertical, fore-and-aft, and sideways forces on the wheel. Figure 1-11 shows an arrangement for independently controlled vertical force input through each tire. In this case the vehicle can bounce, twist, and vibrate on top of the hydraulic actuators.

A road simulator for automotive vehicles offers numerous examples for progressive analysis involving elementary engineering mechanics, as outlined in the following.

**1.** The minimal requirement of each hydraulic actuator that applies the vertical force to a wheel is to hold the vehicle stationary. These requirements can be established using elementary statics.

**2.** The force applied by each hydraulic actuator can be measured quite accurately and easily. Thus, each vertical actuator in Fig. 1-11 can be considered as an electronic scale for measuring weight, which allows for the experimental determination of the center of gravity of the vehicle. This topic is covered in statics.

**3.** The linkages and brackets between the hydraulic actuators and the wheel spindle in Fig. 1-10 provide simple examples for vector manipulations, moments of forces, equilibrium, and two-force members, all topics of statics.

**4.** When a vehicle is assumed to be a rigid body, the road simulation provides examples for dynamics analysis.

**5.** The basic equipment and techniques of road simulation for automotive vehicles are readily extended to exemplify other kinds of important engineering analysis, such as the impact testing of packages, locomotives, or electronic equipment. Thus, the student will find that the principles and methods of elementary engineering mechanics have wide-ranging practical application, regardless of how complex the overall project may be.

Motion

Hydraulic actuator

**FIGURE 1-11**

Automobile in a laboratory test. Hydraulic actuators apply realistic vertical forces to each wheel while the vehicle's frame is not constrained.

# Forces on Particles

Basic Concepts in Modeling Forces on Bodies

Forces on bodies are described by their *magnitudes*, *directions*, and *points of application.* The body being studied is called a *particle* if it has negligible dimensions compared to other dimensions in the same problem, or if its dimensions are not important in the solution. Particles are useful models in many practical problems. Since a particle has negligible dimensions, all forces on it are assumed to have the same point of application.

**SECTIONS 2-1 AND 2-2** present the methods for adding and subtracting vectors or resolving vectors into components. Force vectors are used to exemplify the methods, but these operations are also essential for other vector quantities of statics and dynamics.

**SECTION 2-3** introduces unit vectors, which provide a concise scheme of notation for all vector quantities. Unit vectors are especially useful in complex vector operations.

**SECTIONS 2-4 AND 2-5** extend the methods of Secs. 2-1 and 2-2 to three-dimensional force systems. Some advantages of using unit vectors are demonstrated here. Other advantages will become obvious in several areas of statics and especially in dynamics.

**SECTION 2-6** defines the scalar product of two vectors and illustrates several useful applications of scalar products in mechanics.

**SECTION 2-7** presents the definition of equilibrium of a particle and the method of modeling using free-body diagrams. The concept of equilibrium is extremely important in engineering mechanics because it is used in determining unknown forces. Free-body diagrams are indispensable in solving most problems, so they should be mastered in this section. Students are strongly advised to practice the use of free-body diagrams as much as possible.

## TWO-DIMENSIONAL FORCE SYSTEMS. RESULTANTS OF CONCURRENT FORCES

Consider a body that is modeled as a particle at point $O$ in Fig. 2-1a. Assume that two separate forces **F** and **P** are acting on the particle as shown.* The two forces **F** and **P** may be replaced by a single resultant force **R** that has the same effect on the particle. This resultant force lies along the diagonal through point $O$ of the parallelogram constructed using the original two vectors as in Fig. 2-1a. The procedure is expressed in a *vector equation* as **F** + **P** = **R**. This method is called the *parallelogram law* for the addition of two vectors, which can be proved only experimentally. In the case of forces, the proof of this law requires force measuring devices; in other cases such as for displacement or velocity vectors, the proof may be more intuitively obvious to some people. The parallelogram method is valid for *any* two vectors that are *concurrent* (acting at the same point). The resultant vector is always the diagonal of the parallelogram, as illustrated in Fig. 2-2a.

The parallelogram law has led to an alternative method of vector addition, the *triangle rule*. In this method one of the vectors is redrawn parallel to its original position, with its tail at the arrowhead of the other

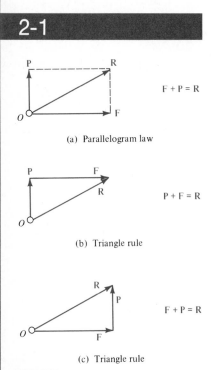

**(a) Parallelogram law**

F + P = R

**(b) Triangle rule**

P + F = R

**(c) Triangle rule**

F + P = R

**FIGURE 2-1**

Addition of perpendicular, concurrent vectors **P** and **F**

(a) Parallelogram law     F + P = R

(b) Triangle rule     P + F = R

F + P = R

(c) Triangle rule

**FIGURE 2-2**

Addition of arbitrary, concurrent vectors **P** and **F**

---

* Vectors are distinguished here by boldface type (**P**). Common notations for vectors in longhand writing are $\vec{P}$, $\underline{P}$, and $\mathbb{P}$.

vector. The vector that completes the triangle is the resultant of the original two vectors. The various possibilities are shown in Fig. 2-1b and c and in Fig. 2-2b and c. The results are identical from both routes in each figure.

The vector addition of two vectors is readily extended to that of any number of vectors, as shown for three, **A**, **B**, and **C**, in Fig. 2-3a. **P** is the resultant of vectors **A** and **B**, and **R** is the resultant of **P** and **C**. The addition can be started with any pair of these. The final resultant **R** should be the same for a given set of vectors. Thus, the addition of vectors is *commutative*. Formally, with respect to Fig. 2-3a,

$$\mathbf{A} + \mathbf{B} = \mathbf{B} + \mathbf{A} = \mathbf{P} \qquad \boxed{2\text{-}1a}$$

$$\mathbf{A} + \mathbf{C} = \mathbf{C} + \mathbf{A} \qquad \boxed{2\text{-}1b}$$

$$\mathbf{B} + \mathbf{C} = \mathbf{C} + \mathbf{B} \qquad \boxed{2\text{-}1c}$$

$$\mathbf{A} + \mathbf{B} + \mathbf{C} = (\mathbf{A} + \mathbf{B}) + \mathbf{C} = \mathbf{A} + (\mathbf{B} + \mathbf{C}) \qquad \boxed{2\text{-}1d}$$
$$= (\mathbf{B} + \mathbf{C}) + \mathbf{A} = \mathbf{B} + (\mathbf{C} + \mathbf{A})$$
$$= \mathbf{R}$$

The last of these, which could be further rearranged, shows that vector addition is also *associative*.

The triangle rule leads to the convenient *polygon rule* for graphically adding more than two concurrent, *coplanar* vectors (they must be in the same plane). This involves rearranging the vectors in a head-to-tail sequence as in Fig. 2-3b, and drawing the resultant arrow from the tail of the sequence to close the polygon.

The subtraction of a vector from another vector can be visualized as the summation of a positive and a negative vector. For appreciating this method, note that vectors **A** and −**A** have the same magnitude but opposite directions. These facts are used in the illustration of vector addition and subtraction in Fig. 2-4. Part a of this figure shows the addition of vectors **P** and **F**, **P** + **F** = **R**. Part b shows the subtraction of vector **F** from vector **R**, as **R** + (−**F**) = **P**.

**FIGURE 2-3**

Addition of concurrent, coplanar vectors **A**, **B**, and **C**

**FIGURE 2-4**

Addition and subtraction of two vectors

## VECTOR COMPONENTS OF A FORCE

2-2

Frequently, there is a need to replace a given force by other forces that have the same effect on the body but are more conveniently oriented for analysis. This *resolving of a force into vector components* is done as a reversal of the vector addition procedure. For any force there is an infinite number of possible sets of components, but in practice only a few specific sets may be of interest. Two frequently encountered cases of specifying components are worth noting:

1. *The lines of action of the components are given.* For an illustration of this, assume that the components of a given force **F** are required along lines *m* and *n* in Fig. 2-5. There is only one set of two components that satisfies the parallelogram law for **F** and the two lines. These components are $\mathbf{F}_m$ and $\mathbf{F}_n$. Their resultant in vector addition must be **F**. Note

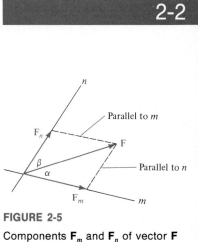

**FIGURE 2-5**

Components $\mathbf{F}_m$ and $\mathbf{F}_n$ of vector **F** that the magnitude of the components can be found graphically or by using the law of sines if **F** and the angles $\alpha$ and $\beta$ are given.

**2. One component is completely specified.** For example, assume that force **F** in Fig. 2-6 is given and that one of its two components must be force $\mathbf{F}_1$. The other component $\mathbf{F}_2$ is determined by applying the parallelogram law or the triangle rule. Using the triangle rule, $\mathbf{F}_2'$ is a vector from the tip of $\mathbf{F}_1$ to the tip of **F**, but both components $\mathbf{F}_1$ and $\mathbf{F}_2$ are applied at point $O$.

**FIGURE 2-6**

Determination of component $\mathbf{F}_2$ if force **F** and one of its components $\mathbf{F}_1$ are given

There are many other possibilities in specifying conditions for the components of a vector, but *any set of components must satisfy the parallelogram law.* In most engineering problems the components of interest are along orthogonal axes, in which case the components are called *rectangular components.*

The following trigonometric relationships are often helpful in determining components of forces.

**LAW OF SINES**

Referring to Fig. 2-5,

$$\frac{F_n}{\sin \alpha} = \frac{F_m}{\sin \beta} = \frac{F}{\sin \left[ 180^\circ - (\alpha + \beta) \right]}$$

**LAW OF COSINES**

Referring to Fig. 2-5,

$$F^2 = F_n^2 + F_m^2 - 2F_n F_m \cos \left[ 180^\circ - (\alpha + \beta) \right]$$

Determine the resultant **R** of forces **F** and **P** applied concurrently to the bracket in Fig. 2-7a. Use a graphical and a trigonometric method of solution.

# EXAMPLE 2-1

## SOLUTION

**GRAPHICAL METHOD:** A parallelogram is constructed with sides **F** and **P** drawn to scale in Fig. 2-7b. The magnitude and direction of the resultant **R** are measured from the drawing. The results can be stated in scalar or vector forms as follows.

*Answer in scalar form:* $R = 1500$ N, $\alpha = 43°$

*Answer in vector form:* $R = 1500$ N $\quad$ 43°

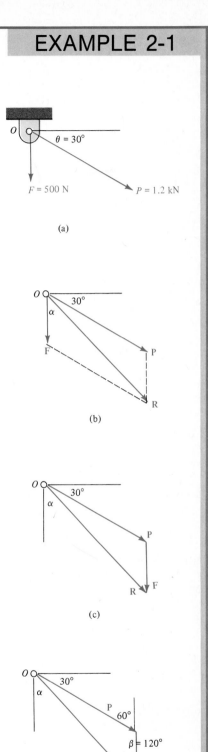

(a)

The same results can be obtained using the triangle rule, as shown in Fig. 2-7c. Note the repositioning of force vector **F** to a parallel position at the tip of force vector **P**.

**TRIGONOMETRIC METHOD:** The law of cosines can be applied to the triangle of the three forces in Fig. 2-7d (from the triangle rule).

$$R^2 = F^2 + P^2 - 2FP \cos \beta$$
$$= (500 \text{ N})^2 + (1200 \text{ N})^2 - 2(500 \text{ N})(1200 \text{ N})(\cos 120°)$$

$$R = 1513 \text{ N}$$

(b)

The direction of **R** is obtained by applying the law of sines.

$$\frac{\sin \beta}{R} = \frac{\sin \gamma}{P} = \frac{\sin \alpha}{P}$$

$$\sin \alpha = \frac{P}{R} \sin \beta = \frac{1200 \text{ N}}{1513 \text{ N}} \sin 120°$$

$$\alpha = \text{arc sin } (0.6869) = 43.4°$$

*Answer in vector form:* **R** = 1513 N $\quad$ 43°

(c)

## JUDGMENT OF THE RESULTS

The magnitude and direction of the resultant **R** obtained trigonometrically agree with those from the graphical construction. In practice, the trigonometric solution is often preferable since it is more accurate. The values of the answer are judged reasonable at a glance (or from a freehand sketch) using Fig. 2-7a.

**FIGURE 2-7**

The resultant **R** of vectors **F** and **P**

(d)

EXAMPLE 2-1    21

# EXAMPLE 2-2

In a mechanism of a road simulator (for a description of this, see Sec. 1-7) the maximum intended force to the wheel spindle is **P**, as modeled in Fig. 2-8a. What is the required force capacity **F** of the hydraulic actuator that applies the force **P** through pin $B$ of the rigid rocker arm $AB$? What is the component **A** of the force **F** toward pin $A$? Note that **F** is applied to cause a specified force **P** at the same point, so **P** is a known component of the unknown force **F**.

## SOLUTION

**GRAPHICAL METHOD:**  From Fig. 2-8b,

$$\mathbf{F} = 5500 \text{ lb}$$

$$\mathbf{A} = 1900 \text{ lb}$$

**TRIGONOMETRIC METHOD:**  Using the law of sines,

$$\frac{\sin 20°}{A} = \frac{\sin 65°}{P}$$

$$A = \frac{\sin 20°}{\sin 65°} (5000 \text{ lb}) = 1887 \text{ lb}$$

Using the law of sines again yields

$$\frac{F}{\sin 95°} = \frac{5000}{\sin 65°}$$

$$F = 5500 \text{ lb}$$

Thus, the forces **F** and **A** agree reasonably with those obtained from the graphical solution.

(a)

(b)

**FIGURE 2-8**

Determination of a required force **F** to produce a given component **P**

EXAMPLE 2-3

Consider a model of eight radial spokes with equal *tensions* (pulling forces) at the hub of a stationary bicycle wheel. The magnitudes of the tensions are $T_1 = T_2 = T_3 = \ldots = T_8$ in Fig. 2-9a. Determine graphically the net force on the hub by adding one force at a time to $\mathbf{T}_1$. Check the result by using the polygon rule for the eight forces at once.

## SOLUTION

The resultants $\mathbf{R}_1$, $\mathbf{R}_2$, ... for the successive pairs of vectors are drawn in Fig. 2-9b. Note that the last addition is of $\mathbf{T}_8$ and $\mathbf{R}_6$, which are equal in magnitude but opposite in direction, so the net resultant is zero. The same is shown by the polygon construction in Fig. 2-9c. The lesson from the latter is that *whenever the polygon of the given forces is closed* (*the last arrowhead touches the tail of the first vector*), *the net force is zero.*

## JUDGMENT OF THE RESULTS

It is reasonable that the resultant force on the hub is zero in the absence of external forces on the wheel, since the spokes pull on the stationary hub evenly in all directions.

(a)

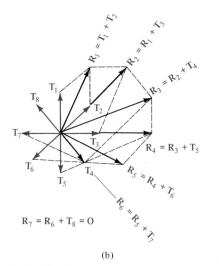

$R_7 = R_6 + T_8 = 0$

(b)

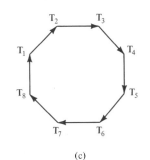

(c)

**FIGURE 2-9**

Summation of eight concurrent forces

EXAMPLE 2-3    23

In the following problems all forces are in the plane of the paper.

**2–1** Determine graphically the resultant of the two forces if $F_1 = F_2 = 3$ kN and $\theta_1 = \theta_2 = 20°$. Use the parallelogram law and the triangle rule.

**Fig. P2-1**

**2–2** In Fig. P2-1, $F_1 = 500$ lb, $F_2 = 800$ lb, and $\theta_1 = \theta_2 = 30°$. Determine graphically the resultant of $\mathbf{F}_1$ and $\mathbf{F}_2$.

**2–3** Determine by trigonometry the resultant of the two forces if $F_1 = 800$ N, $F_2 = 1.4$ kN, $\theta_1 = 30°$, and $\theta_2 = 45°$.

**Fig. P2-3**

**2–4** In Fig. P2-3, $F_1 = 300$ lb at $\theta_1 = 25°$. Determine graphically the magnitude of force $\mathbf{F}_2$ at $\theta_2 = 15°$ to make the resultant of $\mathbf{F}_1$ and $\mathbf{F}_2$ vertical. What is the magnitude of the resultant in this case?

**2–5** A boat is towed in a narrow channel by applying forces $\mathbf{F}$ and $\mathbf{P}$ through ropes attached to the boat. Determine by trigonometry the ratio $P/F$ for which the resultant force is along line $l$. Write an expression for the magnitude of the resultant.

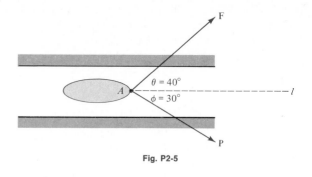

**Fig. P2-5**

**2–6** Block $C$ is pulled from the ground by cranes $A$ and $B$. The tension in the cable at $A_1$ is 20 kN, at $B_1$ it is 15 kN. (a) Determine the magnitude and direction of the resultant force $\mathbf{R}$ applied by the cranes at point $C$ and (b) calculate the angle $\theta$ to make the resultant force $\mathbf{R}$ vertical.

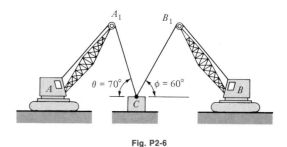

**Fig. P2-6**

**2–7** Two ropes are attached to a post at point $A$. For the given tensions in the ropes, determine the range of the angle $\theta$ between the ropes to limit the maximum resultant force at $A$ to 200 lb.

**Fig. P2-7**

**2–8**  Determine the resultant of the three coplanar forces using a graphical method. $F_1 = 500$ N, $F_2 = 800$ N, $F_3 = 900$ N, $\theta = 30°$, and $\phi = 50°$.

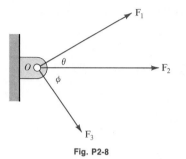

Fig. P2-8

**2–9**  Determine graphically the additional force $\mathbf{F}_4$ that could be added to the bracket in Fig. P2-8 in order to make the resultant of the four coplanar forces zero ($\mathbf{R} = 0$). $F_1 = 200$ lb, $F_2 = 250$ lb, $F_3 = 400$ lb, $\theta = 20°$, and $\phi = 70°$.

**2–10**  Determine by trigonometry the resultant of the three coplanar forces acting on the welded bracket. $F_1 = 4$ kN, $F_2 = 5$ kN, $F_3 = 7.5$ kN, $\theta = 45°$, and $\phi = 60°$.

Fig. P2-10

**2–11**  Assume that force $\mathbf{F}_2$ is reversed in direction in Fig. P2-10 while $\mathbf{F}_1$ and $\mathbf{F}_3$ act as shown. Determine by trigonometry the resultant of the three coplanar forces. $F_1 = 1000$ lb, $F_2 = 1500$ lb, $F_3 = 800$ lb, $\theta = 90°$, and $\phi = 20°$.

**2–12**  Determine by trigonometry the resultant of the four coplanar forces acting on the welded bracket.

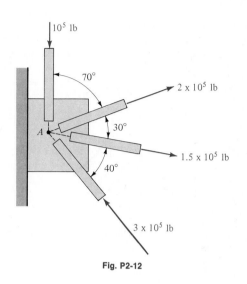

Fig. P2-12

**2–13**  Draw the polygon for the four forces and their resultant in Fig. P2-12. Assume that the lowest bar suddenly fails so that the force of $3 \times 10^5$ lb becomes zero. Show the new resultant of the remaining three forces on the same diagram.

**2–14**  Determine by trigonometry the resultant of the three coplanar forces.

Fig. P2-14

**2–15**  Determine graphically the additional force $\mathbf{P}$ that would make the resultant force on the bracket in Fig. P2-14 zero ($\mathbf{R} = 0$).

**2–16** One of the constraining forces acting on the block of a tuned mass damper (see Sec. 1-7) is inclined with the horizontal as shown. Determine the horizontal and vertical components of this force.

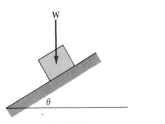

Fig. P2-16

**2–17** (a) Determine the horizontal and vertical components of the force acting on the test platform. (b) Determine the component perpendicular to the arm *AB*.

Fig. P2-17

**2–18** The mass of the block resting on the inclined plane is 20 kg. Determine the components of the weight along the plane and perpendicular to the plane for $\theta = 10°$.

Fig. P2-18

**2–19** Plot the magnitude of the components of the force along bar *AB* and along bar *BC* for $\theta = 0$, 30°, and 60°. Enter all data on the same diagram and sketch curves through the appropriate points for each bar.

Fig. P2-19

**2–20** Resolve the horizontal force into components along lines *AB* and *AC*.

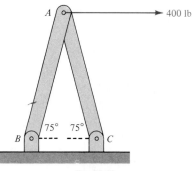

Fig. P2-20

There are many situations where the directions of forces remain constant but their magnitudes change. For example, a force of magnitude $0.9P$ increases to a magnitude of $2.2P$ while acting in the same direction. This change can be illustrated as a *uniaxial* vector addition (meaning that there is only one line of action) as in Fig. 2-10a. The change of magnitude in the uniaxial vector addition is equivalent to a scalar addition. Thus, the operation $0.9P + 1.3P = 2.2P$ completely describes the change of the vector in the example. The same change can be expressed as a multiplication of the magnitude of the vector by a scalar, $2.444(0.9P) = 2.2P$.

A decrease in the magnitude of a vector can be considered as a uniaxial vector subtraction (Fig. 2-10b). This can be described by a scalar subtraction or a division of the magnitude of the vector by a scalar. For example, the above vector of magnitude $2.2P$ decreases to $1.7P$. In scalar subtraction this is stated as $2.2P - 0.5P = 1.7P$. Using the scalar division procedure, $2.2P \div 1.294 = 1.7P$.

Note that the direction of a vector is reversed when it is multiplied by a negative scalar. For example, force $-0.5\mathbf{P}$ in Fig. 2-10b is in the opposite direction to force $2.2\mathbf{P}$ because their signs are different. This can be written as $2.2\mathbf{P}(-0.227) = -0.5\mathbf{P}$.

The multiplication or division of a vector by a scalar is important in the application of *unit vectors*. By definition, *a unit vector $\mathbf{n}$ is a quantity that has a direction; its magnitude is unity*. It is very useful that any vector $\mathbf{F}$ may be expressed as the unit vector $\mathbf{n}$, taken in the same direction as $\mathbf{F}$, multiplied by the magnitude $F$. This is illustrated in Fig. 2-11, where $\mathbf{n}$ is simply superimposed on $\mathbf{F}$ so that $\mathbf{F} = F\mathbf{n}$. If a unit vector $\mathbf{n}'$ is in the opposite direction but along the same line of action as $\mathbf{F}$, $\mathbf{F}$ can be written as $(-1)F\mathbf{n}'$. The magnitude of this vector is the non-negative quantity $F$; the scalar, $-1$, reverses the direction from that of $\mathbf{n}'$.

The concept of the unit vector can be extended to the components of vectors. The unit vectors along rectangular, Cartesian axes are so frequently used that they have widely accepted names, $\mathbf{i}$, $\mathbf{j}$, and $\mathbf{k}$, as shown in Fig. 2-12. Each one of these has a magnitude of unity. These unit vectors, called *unit coordinate vectors*, provide a concise, written scheme of notation for vectors in a rectangular coordinate system. The scheme is illustrated for a force with two-dimensional components in Fig. 2-13. The *vector component* $\mathbf{F}_x$ can be written as $F_x$ times the unit vector $\mathbf{i}$, and similarly for the $y$ component,

$$\mathbf{F}_x = F_x\mathbf{i} \quad \mathbf{F}_y = F_y\mathbf{j} \qquad \boxed{2\text{-}2}$$

The original force $\mathbf{F}$ is completely stated by

$$\mathbf{F} = F_x\mathbf{i} + F_y\mathbf{j} \qquad \boxed{2\text{-}3}$$

The quantities $F_x$ and $F_y$ are the *scalar components* of the vector $\mathbf{F}$. They are calculated from the magnitude of $\mathbf{F}$ and the angle $\theta$ in Fig. 2-13,

$$F_x = F\cos\theta \quad F_y = F\sin\theta \qquad \boxed{2\text{-}4}$$

The angle $\theta$ can be readily calculated if the components are known,

$$\tan\theta = \frac{F_y}{F_x} \qquad \boxed{2\text{-}5}$$

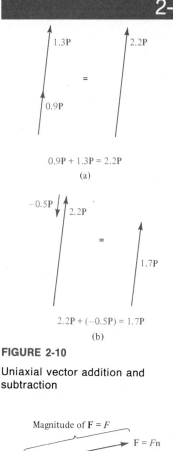

$$0.9P + 1.3P = 2.2P$$
(a)

$$2.2P + (-0.5P) = 1.7P$$
(b)

**FIGURE 2-10**

Uniaxial vector addition and subtraction

Magnitude of $\mathbf{F} = F$

$\mathbf{F} = F\mathbf{n}$

$\mathbf{n}$

Magnitude of $\mathbf{n} = 1$

**FIGURE 2-11**

Representation of a vector **F** by its own magnitude $F$ multiplied by the unit vector **n** of the same direction as **F**

**FIGURE 2-12**

Unit vectors in Cartesian coordinates

**FIGURE 2-13**

Unit vector notation for the components of vector **F**

The magnitude $F$ of **F** can be determined from

$$F = \sqrt{F_x^2 + F_y^2} \qquad \boxed{2\text{-}6}$$

which is derived from Fig. 2-13 using the Pythagorean theorem.

The unit vector notation is convenient in vector summation. Since the addition or subtraction of uniaxial vectors can be performed as scalar addition or subtraction (Fig. 2-10), all components denoted by the same unit vector can be summed as scalars. Thus, if **A**, **B**, and **C** are coplanar vectors and **R** is their resultant, $\mathbf{A} + \mathbf{B} + \mathbf{C} = \mathbf{R}$ can be written as $A_x\mathbf{i} + A_y\mathbf{j} + B_x\mathbf{i} + B_y\mathbf{j} + C_x\mathbf{i} + C_y\mathbf{j} = (A_x + B_x + C_x)\mathbf{i} + (A_y + B_y + C_y)\mathbf{j} = R_x\mathbf{i} + R_y\mathbf{j}$. Using summation notation for any number of vectors,

*Scalar components:* $\quad R_x = \sum (x \text{ components}) \qquad \boxed{2\text{-}7\text{a}}$

$$R_y = \sum (y \text{ components})$$

*Vector components:* $\quad \mathbf{R}_x = \sum \mathbf{F}_x = \sum F_x\,\mathbf{i} \qquad \boxed{2\text{-}7\text{b}}$

$$\mathbf{R}_y = \sum \mathbf{F}_y = \sum F_y\,\mathbf{j}$$

Note that in practice the intermediate step such as $A_x\mathbf{i} + B_x\mathbf{i} + \dots$ is frequently skipped and $(A_x + B_x + \dots)\mathbf{i}$ is written directly.

**Sign conventions.** It is necessary to discuss the sign conventions for unit vectors and scalar components of vectors in a rectangular coordinate system.

**1.** *The unit coordinate vectors* **i**, **j**, *and* **k** are directed along the corresponding positive coordinate axes, $x$, $y$, and $z$.

**2.** *Scalar components* are given signs according to the directions that the corresponding vector components have in the coordinate system.

The application of signs is illustrated with respect to Fig. 2-13 as follows.

| QUANTITY | INDICATES |
|---|---|
| $F_x\mathbf{i}$ | Vector component $\mathbf{F}_x$ of magnitude $F_x$ is directed in positive $x$ direction |
| $-P_x\mathbf{i}$ (not shown) | Vector component $\mathbf{P}_x$ of magnitude $P_x$ is directed in negative $x$ direction |
| $F_x$ | Scalar component of vector $\mathbf{F}_x$ which is directed in positive $x$ direction |
| $-P_x$ (not shown) | Scalar component of vector $\mathbf{P}_x$ which is directed in negative $x$ direction |
| $F_y\mathbf{j}$ | Vector component $\mathbf{F}_y$ of magnitude $F_y$ is directed in positive $y$ direction |
| $-P_y\mathbf{j}$ (not shown) | Vector component $\mathbf{P}_y$ of magnitude $P_y$ is directed in negative $y$ direction |

## Unit Vectors in Parallel Coordinate Systems

The **ijk** unit vectors of an $xyz$ coordinate system (such as in Fig. 2-12) are also valid for any other $x'y'z'$ coordinate system if the respective axes are parallel to one another ($x$ is parallel to $x'$, etc.), because the unit vectors indicate directions in space. This is especially useful in situations where the forces are not concurrent.

EXAMPLE 2-4

Write an expression for the resultant **R** of the two forces in Fig. 2-14a using force components with unit vectors.

## SOLUTION

For the rectangular coordinates $x$ and $y$, use the unit coordinate vectors **i** and **j**. Write the $x$ and $y$ components of the forces as functions of the given angles using Fig. 2-14b and c.

$$\mathbf{F}_{1x} = F_1 \cos \theta \, \mathbf{i} \quad \mathbf{F}_{2x} = -F_2 \sin \phi \, \mathbf{i}$$

$$\mathbf{F}_{1y} = F_1 \sin \theta \, \mathbf{j} \quad \mathbf{F}_{2y} = F_2 \cos \phi \, \mathbf{i}$$

The resultant force has two components, which are determined using Eq. 2-7b,

$$\mathbf{R}_x = \sum \mathbf{F}_x = F_1 \cos \theta \, \mathbf{i} - F_2 \sin \phi \, \mathbf{i}$$

$$\mathbf{R}_y = \sum \mathbf{F}_y = F_1 \sin \theta \, \mathbf{j} + F_2 \cos \phi \, \mathbf{j}$$

The resultant is the vector sum of the components,

$$\mathbf{R} = R_x\mathbf{i} + R_y\mathbf{j} = (F_1 \cos \theta - F_2 \sin \phi)\mathbf{i} + (F_1 \sin \phi + F_2 \cos \phi)\mathbf{j}$$

(a)

(b)

(c)

**FIGURE 2-14**

Summation of two forces using components and unit vectors

EXAMPLE 2-4    29

# EXAMPLE 2-5

Consider an anchoring block of a tall antenna tower that holds six guy wires for steadying the tower (Fig. 2-15). The initial tension in each wire is 23 kN when there is no wind. Assume a simple model of this junction with the forces concurrent. Determine the resultant force of the six tensions in terms of its magnitude $R$ and angle $\theta$ with respect to the horizontal.

## SOLUTION

First choose horizontal and vertical $x$ and $y$ axes with unit coordinate vectors **i** and **j**. Then determine the $x$ and $y$ components of the tension forces using Eq. 2-4 and the given angles. The resultant force is $\mathbf{R} = R_x\mathbf{i} + R_y\mathbf{j}$, where the components $R_x$ and $R_y$ are determined using Eq. 2-7b.

$$\mathbf{R}_x = \sum \mathbf{F}_x = (23 \text{ kN})(\cos 15° + \cos 30° + \cos 40° + \cos 50°)$$
$$+ \cos 57.5° + \cos 65°)\mathbf{i}$$

$$\mathbf{R}_x = (23 \text{ kN})(4.20)\mathbf{i} = 96.6 \text{ kN } \mathbf{i}$$

$$\mathbf{R}_y = \sum \mathbf{F}_y = (23 \text{ kN})(\sin 15° + \sin 30° + \sin 40° + \sin 50°$$
$$+ \sin 57.5° + \sin 65°)\mathbf{j}$$

$$\mathbf{R}_y = (23 \text{ kN})(3.92)\mathbf{j} = 90.1 \text{ kN } \mathbf{j}$$

Using Fig. 2-13, we obtain

$$R = \sqrt{R_x^2 + R_y^2} = 132.1 \text{ kN} \qquad \tan \theta = \frac{R_y}{R_x} = \frac{90.1 \text{ kN}}{96.6 \text{ kN}} = 0.933$$

$$\theta = 43.0°$$

## JUDGMENT OF THE RESULTS

The magnitude and direction of the resultant force appear reasonable by inspection of Fig. 2-15.

**FIGURE 2-15**

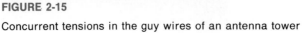

Concurrent tensions in the guy wires of an antenna tower

EXAMPLE 2-6

Reconsider the wheel hub in Ex. 2-3 and determine the net force **R** on it using the addition of force components with unit vectors.

## SOLUTION

Use the $x$ and $y$ axes as shown in Fig. 2-16 with the unit coordinate vectors **i** and **j**. Denote the magnitude of each tension by $T$, and go clockwise starting from the top.

$$\mathbf{R} = (0\mathbf{i} + T\mathbf{j}) + \left(\frac{\sqrt{2}}{2}T\mathbf{i} + \frac{\sqrt{2}}{2}T\mathbf{j}\right) + (T\mathbf{i} + 0\mathbf{j})$$

$$+ \left(\frac{\sqrt{2}}{2}T\mathbf{i} - \frac{\sqrt{2}}{2}T\mathbf{j}\right) + (0\mathbf{i} - T\mathbf{j}) + \left(-\frac{\sqrt{2}}{2}T\mathbf{i} - \frac{\sqrt{2}}{2}T\mathbf{j}\right)$$

$$+ (-T\mathbf{i} + 0\mathbf{j}) + \left(-\frac{\sqrt{2}}{2}T\mathbf{i} + \frac{\sqrt{2}}{2}T\mathbf{j}\right)$$

$$\mathbf{R} = \left(\frac{\sqrt{2}}{2} + 1 + \frac{\sqrt{2}}{2} - \frac{\sqrt{2}}{2} - 1 - \frac{\sqrt{2}}{2}\right)T\mathbf{i}$$

$$+ \left(1 + \frac{\sqrt{2}}{2} - \frac{\sqrt{2}}{2} - 1 - \frac{\sqrt{2}}{2} + \frac{\sqrt{2}}{2}\right)T\mathbf{j} = 0\mathbf{i} + 0\mathbf{j} = \mathbf{0}$$

## JUDGMENT OF THE RESULTS

It agrees with the previous result in Ex. 2-3. Since **R** is a vector, it is correct to give the answer as the *zero vector* **0**.

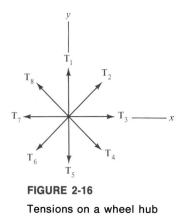

**FIGURE 2-16**

Tensions on a wheel hub

EXAMPLE 2-6    31

# PROBLEMS

In the following problems all forces are in the $xy$ plane.

**2-21** Determine the resultant **R** of the two forces using unit vectors. $F_1 = F_2 = 3$ kN and $\theta_1 = \theta_2 = 20°$.

**2-22** In Fig. P2-21, $F_1 = 500$ lb, $F_2 = 800$ lb, and $\theta_1 = \theta_2 = 30°$. Express the resultant **R** in terms of unit vectors.

**2-23** Express the resultant **R** of the two forces in terms of the magnitude **R** and the angle $\phi$ of **R** with the $x$ axis. $F_1 = 800$ N, $F_2 = 1.4$ kN, $\theta_1 = 30°$, and $\theta_2 = 45°$.

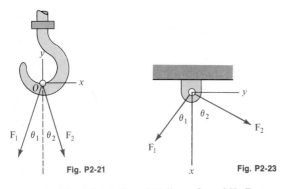

Fig. P2-21    Fig. P2-23

**2-24** In Fig. P2-23, $F_1 = 300$ lb at $\theta_1 = 25°$. Determine the magnitude of force $F_2$ at $\theta_2 = 15°$ to make the resultant of $F_1$ and $F_2$ vertical. Use unit vectors.

**2-25** A boat is towed in a narrow channel by applying forces $F = 4$ kN and $P = 6$ kN through ropes attached to the boat. Express the resultant force using unit vectors. What is the angle that the resultant force makes with the $x$ axis?

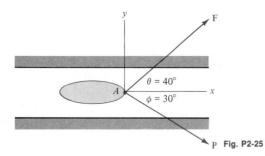

Fig. P2-25

**2-26** Block $C$ is pulled from the ground by cranes $A$ and $B$. The tension in the cable at $A_1$ is 20 kN; at $B_1$ it is 15 kN. (a) Determine the magnitude and direction of the resultant force **R** applied by the cranes at point $C$ and (b) calculate the angle $\theta$ to make the resultant force **R** vertical.

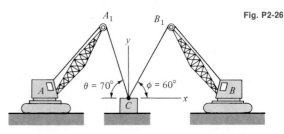

Fig. P2-26

**2-27** Two ropes are attached to a post at point $A$. For the given tensions in the ropes, plot the $x$ component of the resultant force **R** as a function of $\theta$ for $0 < \theta < 90°$ (plot $R_x$ vs. $\theta$). In all cases the 200-lb force is in the $x$-direction.

Fig. P2-27

**2-28** Determine the resultant of the three coplanar forces using unit vectors. $F_1 = 500$ N, $F_2 = 800$ N, $F_3 = 900$ N, $\theta = 30°$, and $\phi = 50°$.

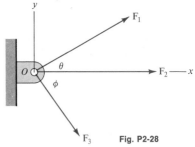

Fig. P2-28

**2-29** Determine by rectangular components the smallest additional force $\mathbf{F}_4$ to make the resultant component $R_y = 0$ in Fig. P2-28. $F_1 = 200$ lb, $F_2 = 250$ lb, $F_3 = 400$ lb, $\theta = 20°$, and $\phi = 70°$.

**2-30** Determine the resultant of the three coplanar forces in terms of its magnitude $R$ and the angle $\alpha$ of **R** with the $x$ axis. $F_1 = 4$ kN, $F_2 = 5$ kN, $F_3 = 7.5$ kN, $\theta = 45°$, and $\phi = 60°$. Use rectangular components.

Fig. P2-30

**2–31** Determine by rectangular components the smallest additional force $\mathbf{F}_4$ to make the resultant component $R_y = 0$ in Fig. P2-30. $F_1 = 1000$ lb, $F_2 = 1500$ lb, $F_3 = 800$ lb, $\theta = 40°$, and $\phi = 50°$.

**2–32** Determine the resultant $\mathbf{R}$ of the four coplanar forces using unit vectors in the $xy$ coordinate system shown.

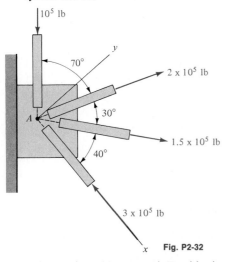

Fig. P2-32

**2–33** Solve Prob. 2–32 as stated. Double the magnitude of each force and solve for the new resultant $\mathbf{R}'$. Write the ratio $R/R'$.

**2–34** Determine the resultant force $\mathbf{R}$ using unit vectors in the $xy$ coordinate system shown.

Fig. P2-34

**2–35** Determine the smallest additional force $\mathbf{P}$ that would make the resultant component $R_x = 0$ in Fig. P2-34.

**2–36** Express the resultant $\mathbf{R}$ of the three coplanar forces in terms of unit vectors of the $xy$ coordinate system.

Fig. P2-36

**2–37** The anchoring block for four guy wires of a radio antenna tower is modeled as a simple bracket with coplanar, concurrent forces. Determine the resultant force $\mathbf{R}$ if $\mathbf{F}_1 = (-15\mathbf{i} + 5\mathbf{j})$ kN, $\mathbf{F}_2 = (-12\mathbf{i} + 7\mathbf{j})$ kN, $\mathbf{F}_3 = (-10\mathbf{i} + 9\mathbf{j})$ kN, and $\mathbf{F}_4 = (-8\mathbf{i} + 10\mathbf{j})$ kN.

Fig. P2-37

**2–38** Three coplanar forces acting at the top of the pole are parallel to the $xz$ plane. Determine the resultant force $\mathbf{R}$.

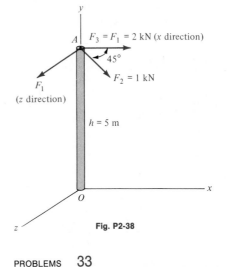

Fig. P2-38

# THREE-DIMENSIONAL FORCE SYSTEMS. RESULTANTS AND COMPONENTS OF FORCES

Unit vectors are most useful when applied to three-dimensional force systems. The analysis of a force in two dimensions can be readily extended to three dimensions.

Consider force **F** in Fig. 2-17 and a box for which **F** is a diagonal line. Each edge of the box is along or parallel to one of the rectangular coordinate axes, $x$, $y$, and $z$. The three vector components of **F** are equal in magnitude to the respective dimensions of the box. From Fig. 2-17, $F_x = OH$, $F_y = OC$, and $F_z = OE$. A detailed explanation of these components can be made using two-dimensional components in succession. First, the vector components $\mathbf{F}_y$ and $\mathbf{F}_G$ are drawn in the plane $OGAC$. $\mathbf{F}_y$ is one of the major components sought, so it is completely determined. Next, $\mathbf{F}_G$ is seen as the diagonal of the rectangle $OEGH$ in the $xz$ plane. $\mathbf{F}_x$ and $\mathbf{F}_z$ are the vector components of $\mathbf{F}_G$ in this plane. The force **F** can be written in terms of the scalar components and the unit coordinate vectors,

$$\mathbf{F} = F_x\mathbf{i} + F_y\mathbf{j} + F_z\mathbf{k} \qquad \boxed{2\text{-}8}$$

The scalar components are also given by

$$F_x = F \cos \theta_x \quad F_y = F \cos \theta_y \quad F_z = F \cos \theta_z \qquad \boxed{2\text{-}9}$$

The angles $\theta_x$, $\theta_y$, and $\theta_z$ are defined in Fig. 2-18. $\theta_x$ is between **F** and the $x$ axis, $\theta_y$ is between **F** and the $y$ axis, and $\theta_z$ is between **F** and the $z$ axis. The cosines of these angles are the *direction cosines* of the force **F**. The angles are always between 0 and 180°. They are measured from the positive coordinate axes as in Fig. 2-18. The direction cosines are positive or negative depending on the size of the angles.*

Another relation between the magnitude of **F** and its rectangular scalar components is obtained from the triangles $OAC$ and $OGH$ in Fig. 2-17. According to the Pythagorean theorem,

$$F^2 = F_y^2 + F_G^2$$

$$F_G^2 = F_x^2 + F_z^2$$

From these equations, the magnitude of the vector **F** is

$$F = \sqrt{F_x^2 + F_y^2 + F_z^2} \qquad \boxed{2\text{-}10}$$

The substitution of Eq. 2-9 into Eq. 2-8 gives

$$\mathbf{F} = F(\cos \theta_x \, \mathbf{i} + \cos \theta_y \, \mathbf{j} + \cos \theta_z \, \mathbf{k}) = F\mathbf{n} \qquad \boxed{2\text{-}11}$$

where $\cos \theta_x$, $\cos \theta_y$, and $\cos \theta_z$ are the scalar components of the vector **n** whose magnitude is equal to 1. The unit vector **n** is collinear with **F** as

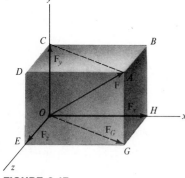

**FIGURE 2-17**

Three-dimensional components of vector **F**

**FIGURE 2-18**

Angles used to define the direction of vector **F**

* Note that a notational inconsistency may occur in the use of the symbols $F_x$, $F_y$, and $F_z$. It is common in engineering practice to use these symbols for both the *magnitude* of a vector component and the corresponding *scalar component*. Confusion can be avoided by recognizing the context in which the symbols are used and remembering that a magnitude is always a non-negative quantity denoting the length of the vector component but that a scalar component is a signed quantity, depending on the angle between the vector component and the corresponding coordinate axis.

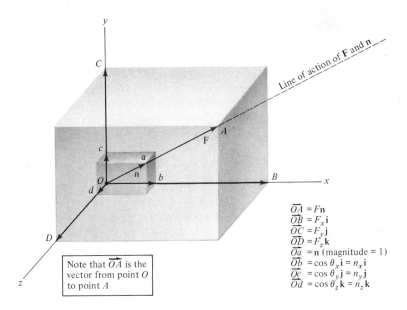

$$\overrightarrow{OA} = F\mathbf{n}$$
$$\overrightarrow{OB} = F_x\mathbf{i}$$
$$\overrightarrow{OC} = F_y\mathbf{j}$$
$$\overrightarrow{OD} = F_z\mathbf{k}$$
$$\overrightarrow{Oa} = \mathbf{n} \text{ (magnitude = 1)}$$
$$\overrightarrow{Ob} = \cos\theta_x\mathbf{i} = n_x\mathbf{i}$$
$$\overrightarrow{Oc} = \cos\theta_y\mathbf{j} = n_y\mathbf{j}$$
$$\overrightarrow{Od} = \cos\theta_z\mathbf{k} = n_z\mathbf{k}$$

Note that $\overrightarrow{OA}$ is the vector from point $O$ to point $A$

**FIGURE 2-19**

Rectangular components of force **F** and unit vector **n**

in Fig. 2-19. The components of the unit vector are equal to the corresponding direction cosines of **F**,

$$n_x = \cos\theta_x \quad n_y = \cos\theta_y \quad n_z = \cos\theta_z \qquad \boxed{2\text{-}12}$$

These expressions can be combined as in Eq. 2-10. Noting that the magnitude of **n** is 1, the result is

$$\boxed{n_x^2 + n_y^2 + n_z^2 = 1} \qquad \boxed{2\text{-}13}$$

It is seen in Fig. 2-19 that the respective sides and diagonals of the two boxes are proportional to one another. These proportions can be expressed formally as

$$\frac{n_x}{F_x} = \frac{n_y}{F_y} = \frac{n_z}{F_z} = \frac{1}{F} \qquad \boxed{2\text{-}14}$$

In practice it is not necessary to draw either of the boxes to exactly define a vector **F** as in Fig. 2-19. The analysis can be performed if the coordinates of two points on the line of action are known. This is demonstrated for a two-dimensional vector for simplicity, and can be extended for a three-dimensional vector.

Consider the problem of defining a force **F** on the line $AB$ in Fig. 2-20. Assume that the coordinates of points $A$ and $B$ are known and that

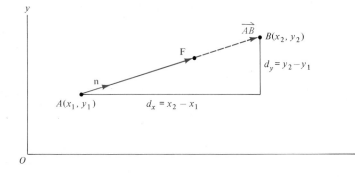

**FIGURE 2-20**

Direction of force **F** defined by points $A$ and $B$ on its line of action

$\overrightarrow{AB}$ is a vector from $A$ to $B$. The differences $d_x$ and $d_y$ between the coordinates are the scalar components of $\overrightarrow{AB}$, so

$$\overrightarrow{AB} = d_x\mathbf{i} + d_y\mathbf{j} = d\mathbf{n} \qquad \boxed{\text{2-15}}$$

where the distance $d$ is the magnitude of the vector $\overrightarrow{AB}$, $d = \sqrt{d_x^2 + d_y^2}$. The real force $\mathbf{F}$ is given in terms of its scalar components,

$$\mathbf{F} = F_x\mathbf{i} + F_y\mathbf{j} = F\mathbf{n} \qquad \boxed{\text{2-16}}$$

where $\mathbf{n} = \dfrac{d_x\mathbf{i} + d_y\mathbf{j}}{d}$.

The respective terms in Eq. 2-15 and 2-16 are proportional, and the results are also valid for the third dimension $z$ (on the basis of Fig. 2-19). Therefore,

$$\frac{F_x}{d_x} = \frac{F_y}{d_y} = \frac{F_z}{d_z} = \frac{F}{d} \qquad \boxed{\text{2-17}}$$

Finally, from Eqs. 2-14 and 2-17,

$$\frac{n_x}{d_x} = \frac{n_y}{d_y} = \frac{n_z}{d_z} = \frac{1}{d} \qquad \boxed{\text{2-18}}$$

### Addition of Forces

The unit vector notation is the most convenient for the addition of concurrent forces. The method of addition that was already used for two-dimensional force systems is equally valid and even more advantageous for three-dimensional forces. In general, each rectangular component of the resultant force equals the sum of the corresponding individual components. As in Eq. 2-7,

$$\text{Scalar components:} \quad R_x = \sum F_x \qquad \boxed{\text{2-19a}}$$
$$R_y = \sum F_y$$
$$R_z = \sum F_z$$

$$\text{Vector components:} \quad \mathbf{R}_x = \sum \mathbf{F}_x = \sum F_x\mathbf{i} \qquad \boxed{\text{2-19b}}$$
$$\mathbf{R}_y = \sum \mathbf{F}_y = \sum F_y\mathbf{j}$$
$$\mathbf{R}_z = \sum \mathbf{F}_z = \sum F_z\mathbf{k}$$

## 2-5 METHOD FOR DETERMINING FORCE VECTOR FROM SCALAR INFORMATION

In some cases a force may be given in terms of its magnitude $F$, its sense of direction, and the coordinates of two points on its line of action. It is often advantageous in the solution of problems to express such a force as a vector quantity. The transformation of the scalar quantities into the appropriate vector can be made using the following procedure based on Fig. 2-20 as a simple model (the third dimension $z$ is added for completeness).

**Given scalar information:**

1. $F$, magnitude of force
2. $(x_1, y_1, z_1)$, coordinates of point $A$ on line of action of force $\mathbf{F}$
3. $(x_2, y_2, z_2)$, coordinates of point $B$ on line of action of force $\mathbf{F}$
4. Direction of force $\mathbf{F}$ from point $A$ to $B$

**Unknown to be determined:** Vector expression

$$\mathbf{F} = F_x\mathbf{i} + F_y\mathbf{j} + F_z\mathbf{k}$$

**Method of solution:**

1. From Eq. 2-11,

$$\mathbf{F} = F\mathbf{n} \qquad \boxed{\text{2-20}}$$

   where $F$ is known.
2. Define a vector $\overrightarrow{AB}$ between the given points $A$ and $B$ as in Eq. 2-15,

$$\overrightarrow{AB} = d_x\mathbf{i} + d_y\mathbf{j} + d_z\mathbf{k} = d\mathbf{n} \qquad \boxed{\text{2-21}}$$

   where $d = \sqrt{d_x^2 + d_y^2 + d_z^2}$ = magnitude of $\overrightarrow{AB}$.
3. Express the unit vector

$$\mathbf{n} = \frac{\text{vector } \overrightarrow{AB}}{\text{magnitude of } \overrightarrow{AB}} = \frac{d_x\mathbf{i} + d_y\mathbf{j} + d_z\mathbf{k}}{d} \qquad \boxed{\text{2-22}}$$

4. Substitute Eq. 2-22 into Eq. 2-20,

$$\mathbf{F} = F\frac{(d_x\mathbf{i} + d_y\mathbf{j} + d_z\mathbf{k})}{d} \qquad \boxed{\text{2-23}}$$

5. With $F_x = Fd_x/d$, etc.,

$$\mathbf{F} = \frac{Fd_x}{d}\mathbf{i} + \frac{Fd_y}{d}\mathbf{j} + \frac{Fd_z}{d}\mathbf{k} \qquad \boxed{\text{2-24}}$$

The desired vector $\mathbf{F}$ can be expressed either by Eq. 2-23 or 2-24.

## Notations for Units of a Vector

There are several acceptable ways to write the units of a vector when the vector is given in terms of its rectangular components and the corresponding unit vectors. It is only necessary to use the same units for all components. For example,

$$\begin{aligned} \mathbf{F} &= 50\text{ N }\mathbf{i} - 80\text{ N }\mathbf{j} + 130\text{ N }\mathbf{k} \\ &= (50\mathbf{i} - 80\mathbf{j} + 130\mathbf{k})\text{ N} \\ &= 50\mathbf{i} - 80\mathbf{j} + 130\mathbf{k}\text{ N} \end{aligned}$$

The last expression is the simplest in writing but the least precise of the three. It is used because its meaning is generally understood. The student will see and practice these different notations throughout statics and dynamics.

# EXAMPLE 2-7

Figure 2-21a and b are the side and top views of a tall antenna tower, showing only the three guy wires $AA'$, $BB'$, and $CC'$. The origin $O$ of the coordinate system is at the bottom of the tower. Anchoring points $A$, $B$, and $C$ are in the $xy$ plane (the ground), and the $z$ axis is vertical along the tower. The distances on the ground from the tower to the anchoring points are $OA = OB = OC = 100$ m. Determine the angles $\theta_x$, $\theta_y$, and $\theta_z$ of wires $AA'$ and $BB'$ with respect to the coordinate axes. Use unit vector notation, and assume that $A'$, $B'$, and $C'$ are practically at the same point.

## SOLUTION

The angles of the wires with respect to the axes are determined using Eqs. 2-12 and 2-22. Placing a unit vector from $A$ toward $A'$,

for wire $AA'$

$$n_x = \cos\theta_x = \frac{d_x}{d} = \frac{x_{A'} - x_A}{\sqrt{d_y^2 + d_z^2}}$$

$$= \frac{(0-0)\text{ m}}{\sqrt{100^2 + 100^2}\text{ m}} = 0, \qquad \theta_x = 90°$$

$$n_y = \cos\theta_y = \frac{d_y}{d} = \frac{y_{A'} - y_A}{d}$$

$$= \frac{100\text{ m} - 0}{141.4\text{ m}} = 0.7071, \qquad \theta_y = 45°$$

$$n_z = \cos\theta_z = \frac{d_z}{d} = \frac{z_{A'} - z_A}{d}$$

$$= \frac{0 - 100\text{ m}}{141.4\text{ m}} = -0.7071, \qquad \theta_z = 135°$$

Placing a unit vector from $B$ toward $B'$,

for wire $BB'$

$$n_x = \cos\theta_x = \frac{d_x}{d} = \frac{x_{B'} - x_B}{d}$$

$$= \frac{0 - 100\text{ m}\cos 30°}{141.4\text{ m}} = -0.6125, \qquad \theta_x = 127.8°$$

$$n_y = \cos\theta_y = \frac{d_y}{d} = \frac{y_{B'} - y_B}{d}$$

$$= \frac{100\text{ m} - 0}{141.4\text{ m}} = 0.7071, \qquad \theta_y = 45°$$

$$n_z = \cos\theta_z = \frac{d_z}{d} = \frac{z_{B'} - z_B}{d}$$

$$= \frac{0 - (-100\text{ m}\sin 30°)}{141.4\text{ m}} = 0.3536, \qquad \theta_z = 69.3°$$

## JUDGMENT AND INTERPRETATION OF THE RESULTS

The values for $\theta_y$ in both cases are correct by inspection. For wire $BB'$, the complementary angles could be stated as the final results for $\theta_x$ and $\theta_z$ ($\theta_x = 52.2°$, $\theta_z = 110.7°$). For interpreting the angle between two nonintersecting lines such as $BB'$ and the $x$ axis, imagine a line $x'$ that is parallel to $x$ but intersects line $BB'$.

(a) Side view of tower

(b) Top view of tower

**FIGURE 2-21**

Three guy wires of an antenna tower

EXAMPLE 2-8

Two cranes are being used to salvage a wrecked vehicle by applying forces as shown in Fig. 2-22. Crane $A$ exerts a force of 1000 lb on the vehicle, and crane $B$ exerts a 650-lb force. Determine the force vectors acting on the vehicle at point $O$.

## SOLUTION

The scalar $y$ component of force $\mathbf{F}_{OA}$ equals (1000 lb)(cos 30°) by inspection, and the magnitude of the vector component in the $xz$ plane is (1000 lb)(sin 30°). The direction cosines for the 1000-lb force are then

$$n_x = \cos \theta_x = (\sin 30°)(\cos 120°) = -0.25$$

$$n_y = \cos \theta_y = \cos 30° = 0.866$$

$$n_z = \cos \theta_z = (\sin 30°)(\cos 30°) = 0.433$$

The force vector, $\mathbf{F}_{OA}$, is determined from Eq. 2-11,

$$\begin{aligned} \mathbf{F}_{OA} &= (1000 \text{ lb})(\cos \theta_x \mathbf{i} + \cos \theta_y \mathbf{j} + \cos \theta_z \mathbf{k}) \\ &= (1000 \text{ lb})(-0.25\mathbf{i} + 0.866\mathbf{j} + 0.433\mathbf{k}) \\ &= (-250\mathbf{i} + 866\mathbf{j} + 433\mathbf{k}) \text{ lb} \end{aligned}$$

The force vector, $\mathbf{F}_{OB}$, is determined using Eq. 2-23,

$$\begin{aligned} \mathbf{F}_{OB} &= (650 \text{ lb}) \frac{d_x \mathbf{i} + d_y \mathbf{j} + d_z \mathbf{k}}{d} \\ &= (650 \text{ lb}) \frac{12\mathbf{i} + 3\mathbf{j} - 4\mathbf{k}}{12^2 + 3^2 + 4^2} \\ &= (650 \text{ lb})(0.923\mathbf{i} + 0.231\mathbf{j} - 0.308\mathbf{k}) \\ &= (600\mathbf{i} + 150\mathbf{j} - 200\mathbf{k}) \text{ lb} \end{aligned}$$

## JUDGMENT OF THE RESULTS

Using Eq. 2-13, a quick check is made on the direction cosines for $\mathbf{F}_{OA}$,

$$0.25^2 + 0.866^2 + 0.433^2 = 1$$

as they should.

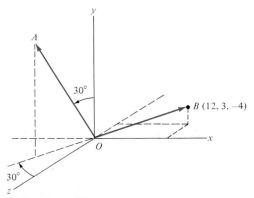

**FIGURE 2-22**

Schematic of forces applied by two cranes being used to upright an overturned vehicle

EXAMPLE 2-8    39

EXAMPLE 2-9

Consider the mast of a sailboat with three guy wires as shown in Fig. 2-23a. Each initial tension in the wires at $B$ and $C$ is planned to be 200 lb. Determine whether the same tension in the third wire would make the horizontal resultant $R_h$ at $A$ equal to zero.

## SOLUTION

From Fig. 2-23b,

$$\mathbf{R} = \mathbf{T}_{AB} + \mathbf{T}_{AC} + \mathbf{T}_{AD}$$

The three tensions of known magnitude (200 lb) must be written as vectors. Using Eqs. 2-20 and 2-23,

$$\mathbf{T}_{AB} = (\text{tension } AB)(\text{unit vector } A \text{ to } B)$$

$$= 200 \text{ lb } \mathbf{n}_{AB} = 200 \text{ lb } \frac{(d_x\mathbf{i} + d_y\mathbf{j} + d_z\mathbf{k})}{d}$$

With the data in Fig. 2-23a,

$$\mathbf{T}_{AB} = \frac{200 \text{ lb}}{\sqrt{5^2 + 10^2 + 4^2}}(-5\mathbf{i} - 10\mathbf{j} + 4\mathbf{k})\frac{\text{ft}}{\text{ft}}$$

$$= -84.2 \text{ lb } \mathbf{i} - 168.4 \text{ lb } \mathbf{j} + 67.4 \text{ lb } \mathbf{k}$$

Similarly for the other two tensions,

$$\mathbf{T}_{AC} = \frac{200 \text{ lb}}{11.87 \text{ ft}}(5\mathbf{i} - 10\mathbf{j} + 4\mathbf{k}) \text{ ft} = 84.2 \text{ lb } \mathbf{i} - 168.4 \text{ lb } \mathbf{j} + 67.4 \text{ lb } \mathbf{k}$$

$$\mathbf{T}_{AD} = \frac{200 \text{ lb}}{11.66 \text{ ft}}(0\mathbf{i} - 10\mathbf{j} - 6\mathbf{k}) \text{ ft} = -171.5 \text{ lb } \mathbf{j} - 102.9 \text{ lb } \mathbf{k}$$

The resultant of the three tensions is written according to Eq. 2-19b,

$$\begin{aligned}\mathbf{R} &= \sum F_x\mathbf{i} + \sum F_y\mathbf{j} + \sum F_z\mathbf{k} \\ &= (-84.2 + 84.2 + 0) \text{ lb } \mathbf{i} + (-168.4 - 168.4 - 171.5) \text{ lb } \mathbf{j} \\ &\quad + (67.4 + 67.4 - 102.9) \text{ lb } \mathbf{k} \\ &= 0 \text{ lb } \mathbf{i} - 508 \text{ lb } \mathbf{j} + 31.9 \text{ lb } \mathbf{k}\end{aligned}$$

The horizontal resultant of interest is obtained from

$$R_h = (R_x^2 + R_z^2)^{1/2} = (0^2 + 31.9^2)^{1/2} \text{ lb} = 31.9 \text{ lb} \neq 0$$

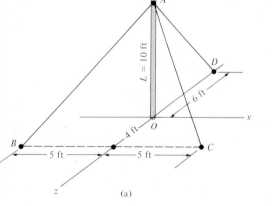

**FIGURE 2-23**

Model for mast of a sailboat with steadying guy wires

**2–39** The wire under tension has $xyz$ coordinates ($a$, $b$, $c$) at point $A$. Determine the unit vector at point $O$ in the direction of $A$.

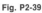

**Fig. P2-39**

**2–40** The magnitude of the tension in Prob. 2–39 is 6 kN. Express the force **T** in terms of rectangular vector components if $a = 0.5$ m, $b = 0.3$ m, and $c = 0.15$ m.

**2–41** The chain under tension has $xyz$ coordinates ($-a$, $-b$, $c$) at point $A$. Determine the unit vector at point $O$ in the direction of $A$.

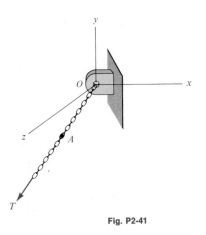

**Fig. P2-41**

**2–42** The magnitude of the tension in Prob. 2–41 is 350 lb. Express the force **T** in terms of rectangular vector components if $a = 2$ ft, $b = 5$ ft, and $c = 7$ ft.

**2–43** The wire under tension is in the $xy$ plane. Determine its direction cosines with respect to the three coordinate axes.

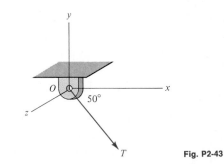

**Fig. P2-43**

**2–44** The rope under tension makes an angle of 25° with the vertical axis as shown. Determine its direction cosines with respect to the $x$, $y$, and $z$ axes.

**Fig. P2-44**

**2–45** Determine the angles $\theta_x$, $\theta_y$, $\theta_z$ of wire $CC'$ with respect to the coordinate axes if point $C$ is in the $xz$ plane. Use unit vector notation. (This is part of Ex. 2-7 and Fig. 2-21.)

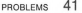

**Fig. P2-45**

**2–46** Determine the angles $\theta_x$ and $\theta_z$ of wire $AB$. (This is part of Ex. 2-9 and Fig. 2-23.)

**Fig. P2-46**

**2–47** Determine the three direction cosines of wire $AD$. (This is part of Ex. 2-9 and Fig. 2-23.)

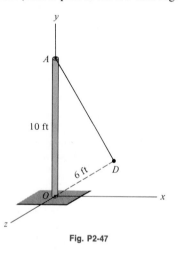

**Fig. P2-47**

**2–48** A truck pulls an airplane by applying an axial force of magnitude $P = 6$ kN through the drawbar $AB$. Determine the force vector acting at $A$ toward $B$ if the coordinates of the points are $A$: (10, 0.2, 0) m and $B$: (14, 0.25, 0) m.

**Fig. P2-48**

**2–49** The hood of a car is supported by bar $AB$ in which the axial force is 10 lb. Determine the force vector acting at $B$ toward $A$ if the coordinates of these points are $A$: (15, 0, 0) in. and $B$: (8, 22, −3) in.

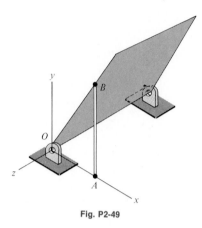

**Fig. P2-49**

**2–50** Determine the angles $\theta_x$, $\theta_y$, $\theta_z$ of wires $AB$ and $AC$ using unit vector notation and any reasonable shortcuts. The coordinates of the three points are $A$: (2, 0, 0) m, $B$: (0, 1.5, 1) m, and $C$: (0, 1.5, −1) m.

**Fig. P2-50**

**2–51** In Fig. P2-50, the coordinates of $A$, $B$, and $C$ are $A$: (7, 0, 0) ft, $B$: (0, 4, 3) ft, and $C$: (0, 5, −2) ft. The tension in each wire ($AB$ and $AC$) is 250 lb. Determine the vector components of the resultant force $\mathbf{R}$ that the two wires apply at point $A$.

**2–52** The coordinates of points $A$, $B$, and $C$ are $A$: $(0, 5, 0)$ m, $B$: $(-2, 0, 1.5)$ m, and $C$: $(2, 0, 1.5)$ m. The tension in each wire ($AB$ and $AC$) is 2 kN. Determine the horizontal and vertical vector components of the resultant force that the two wires apply at point $A$.

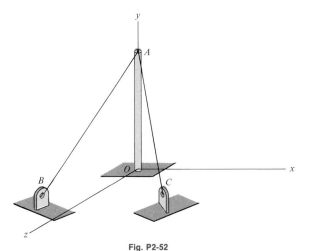

**Fig. P2-52**

**2–53** In Fig. P2-52, the coordinates are $A$: $(0, 20, 0)$ ft, $B$: $(-8, 0, 6)$ ft, and $C$: $(7, 2, 5)$ ft. The tension is 400 lb in wire $AB$, and 500 lb in wire $AC$. Determine the vector components of the resultant force $R$ that the two wires apply at point $A$.

**2–54** Three forces are concurrent at the top of an electric utility pole. Determine the horizontal vector components of the resultant force $R$ if $F_1 = -0.8F\mathbf{i} - 0.1F\mathbf{j}$, $F_2 = -0.7F\mathbf{i} - 0.1F\mathbf{j} + 0.25F\mathbf{k}$, $F_3 = 1.2F\mathbf{i} - 0.05F\mathbf{j} + 0.2F\mathbf{k}$.

**Fig. P2-54**

**2–55** Reconsider Prob. 2–54. Determine the smallest force $\mathbf{F}_4$ that should be added at point $O$ in Fig. P2-54 to make the component $R_x$ of the resultant zero.

**2–56** Three concurrent forces are applied at the top of a foundation block: $\mathbf{F}_1 = 10^4(3\mathbf{i} - 4\mathbf{j} - \mathbf{k})$ lb, $\mathbf{F}_2 = 10^4(-2\mathbf{i} - 3\mathbf{j} + 2\mathbf{k})$ lb, $\mathbf{F}_3 = 10^4(-3\mathbf{i} - 2.5\mathbf{j} - \mathbf{k})$ lb. Determine the resultant force $\mathbf{R}$ of the three forces.

**Fig. P2-56**

**2–57** Determine the scalar components of the force $\mathbf{F}_4$ that should be added to $\mathbf{F}_1$, $\mathbf{F}_2$, and $\mathbf{F}_3$ in Prob. 2–56 to make all components of the resultant force $\mathbf{R}$ zero.

## 2-6

# SCALAR PRODUCT OF TWO VECTORS.
# ANGLES BETWEEN VECTORS.
# PROJECTION OF A VECTOR

**FIGURE 2-24**

The scalar product of two vectors:
$\mathbf{A} \cdot \mathbf{B} = AB \cos \phi$

The *scalar product* of two concurrent vectors **A** and **B** is, *by definition*, the product of the magnitudes of **A** and **B** and the cosine of the angle between them. This product of **A** and **B** is denoted by **A** · **B** and is also called the *dot product*. The formal expression for the vectors in Fig. 2-24 is

$$\mathbf{A} \cdot \mathbf{B} = AB \cos \phi \qquad \boxed{2\text{-}25}$$

The scalar product of two vectors is *commutative*,

$$\mathbf{A} \cdot \mathbf{B} = \mathbf{B} \cdot \mathbf{A} \qquad \boxed{2\text{-}26}$$

since $AB \cos \phi = BA \cos \phi$.

The scalar product is also *distributive*,

$$\mathbf{A} \cdot (\mathbf{B} + \mathbf{C}) = \mathbf{A} \cdot \mathbf{B} + \mathbf{A} \cdot \mathbf{C} \qquad \boxed{2\text{-}27}$$

The proof of this is left as an exercise for the student (Prob. 2-58).

There is no basis to discuss the *associative* property of scalar products since **A** · **B** is a scalar, so (**A** · **B**) · **C** is a scalar dotted with a vector, which is undefined.

The various possible scalar products of unit vectors are simple and worth remembering. Since $\cos 0 = 1$ and $\cos 90° = 0$, the basic set of products is

$$\mathbf{i} \cdot \mathbf{i} = \mathbf{j} \cdot \mathbf{j} = \mathbf{k} \cdot \mathbf{k} = 1 \qquad \boxed{2\text{-}28}$$

$$\mathbf{i} \cdot \mathbf{j} = \mathbf{j} \cdot \mathbf{k} = \mathbf{k} \cdot \mathbf{i} = 0$$

The other possibilities, such as **j** · **i**, give the same results because of the commutative property.

Following are several useful applications of scalar products in mechanics.

## Magnitude of a Vector

Equation 2-10 can also be proved using the scalar product. In general,

$$\mathbf{A} \cdot \mathbf{B} = (A_x\mathbf{i} + A_y\mathbf{j} + A_z\mathbf{k}) \cdot (B_x\mathbf{i} + B_y\mathbf{j} + B_z\mathbf{k})$$

This can be reduced using Eqs. 2-27 and 2-28,

$$\mathbf{A} \cdot \mathbf{B} = A_xB_x + A_yB_y + A_zB_z$$

For the special case when $A = B$,

$$\mathbf{A} \cdot \mathbf{A} = A_x^2 + A_y^2 + A_z^2 = A^2$$

which is equivalent to Eq. 2-10.

## Angle between Vectors

The angle between two vectors can be easily calculated when the rectangular components of both are known. From

$$\mathbf{A} \cdot \mathbf{B} = AB \cos \phi = A_x B_x + A_y B_y + A_z B_z \qquad \boxed{2\text{-}29}$$

$$\phi = \text{arc} \cos \frac{A_x B_x + A_y B_y + A_z B_z}{AB}$$

## Projection of a Vector

It is frequently necessary to investigate the effect of a vector in a direction different from its line of action. This "side effect" can be determined by drawing a line from the tip of the vector and perpendicular to the line of interest. This is shown for a two-dimensional vector **F** in Fig. 2-25. The *projection* **F**′ has a length

$$F' = F \cos \phi \qquad \boxed{2\text{-}30}$$

In two dimensions the projection is easily obtained using the graphical or trigonometric method. Note that in practice the term "projection" may mean a vector or a scalar quantity.

In three-dimensional situations it may be very difficult to determine the angle $\phi$ between two lines. The scalar product is convenient to use in such problems. The idea is to place a unit vector on the line of interest as in Fig. 2-26. $\mathbf{F} \cdot \mathbf{n} = F(1) \cos \phi = F'$, as in Eq. 2-25. The advantage is that the angle $\phi$ is not involved in the calculation if the rectangular coordinates of **F** and **n** are known. With $\mathbf{n} = \cos \theta_x \mathbf{i} + \cos \theta_y \mathbf{j} + \cos \theta_z \mathbf{k}$,

$$F' = \mathbf{F} \cdot \mathbf{n} \qquad \boxed{2\text{-}31}$$

$$F' = F_x \cos \theta_x + F_y \cos \theta_y + F_z \cos \theta_z$$

where $\cos \theta_x, \ldots$ are the direction cosines of the unit vector along the line of interest (they are not related to the angle $\phi$ between **F** and **n**).

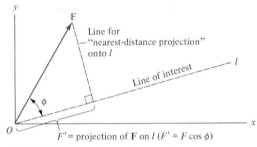

**FIGURE 2-25**

Projection of vector **F** on line *l*

**FIGURE 2-26**

Scalar product of **F** and **n** for determining angle $\phi$

# EXAMPLE 2-10

The laboratory test of a structure is modeled in Fig. 2-27. Forces $\mathbf{F}_1$ and $\mathbf{F}_2$ are applied by hydraulic actuators (see Sec. 1-7) at point $O$. Determine the force vectors and the angle between them if $F_1 = 1.4$ kN and $F_2 = 5.6$ kN. Point $A$ is in the $yz$ plane, and $B$ is in the $xy$ plane.

## SOLUTION

First, the forces must be expressed in vector notation. This is done using unit vectors $\mathbf{n}_{AO}$ and $\mathbf{n}_{BO}$ as follows.

$$\mathbf{F}_1 = F_1 \, \mathbf{n}_{AO} = F_1 \frac{\text{vector } \overrightarrow{AO}}{\text{magnitude of } \overrightarrow{AO}}$$

$$= (1.4 \text{ kN}) \frac{(0\mathbf{i} - 0.6\mathbf{j} - 0.8\mathbf{k}) \text{ m}}{\sqrt{0.6^2 + 0.8^2} \text{ m}}$$

$$= (-0.84\mathbf{j} - 1.12\mathbf{k}) \text{ kN}$$

$$\mathbf{F}_2 = F_2 \, \mathbf{n}_{BO} = F_2 \frac{\text{vector } \overrightarrow{BO}}{\text{magnitude of } \overrightarrow{BO}}$$

$$= (5.6 \text{ kN}) \frac{(0.9\mathbf{i} - 0.7\mathbf{j} + 0\mathbf{k}) \text{ m}}{\sqrt{0.9^2 + 0.7^2} \text{ m}}$$

$$= (4.42\mathbf{i} - 3.44\mathbf{j}) \text{ kN}$$

The angle $\theta$ between $\mathbf{F}_1$ and $\mathbf{F}_2$ is calculated using Eq. 2-29,

$$\theta = \text{arc cos} \left( \frac{F_{1x}F_{2x} + F_{1y}F_{2y} + F_{1z}F_{2z}}{F_1 F_2} \right)$$

$$= \text{arc cos} \left[ \frac{(0)(4.42) + (-0.84)(-3.44) + (-1.12)(0)}{(1.4)(5.6)} \right]$$

$$= 68.4°$$

## JUDGMENT OF THE RESULT

The angle appears reasonable by inspection since it should be a little less than 90° (which would be the angle if the distances $AA'$ and $BB'$ were both zero).

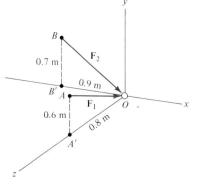

**FIGURE 2-27**

Determination of two force vectors and the angle between them from scalar information

EXAMPLE 2-11

Part of a mechanism consists of slider $S$ on rod $OA$, as shown in Fig. 2-28. The coordinates of point $A$ are $(10, 7, -4)$ in. Force $\mathbf{F} = 3\mathbf{i} - 0.5\mathbf{j} - 2\mathbf{k}$ lb acts on the slider. Determine the magnitude of the projection $\mathbf{F}'$ of force $\mathbf{F}$ on line $OA$ (this is the effective force for moving the slider along line $OA$).

## SOLUTION

First, a unit vector $\mathbf{n}_{OA}$ along line $OA$ is determined using Eq. 2-22,

$$\mathbf{n}_{OA} = \frac{10\mathbf{i} + 7\mathbf{j} - 4\mathbf{k}}{\sqrt{10^2 + 7^2 + 4^2}} = 0.78\mathbf{i} + 0.55\mathbf{j} - 0.31\mathbf{k}$$

The projection $F'$ is calculated using Eq. 2-31.

$$F' = \mathbf{F} \cdot \mathbf{n}_{AO}$$
$$F' = (3)(0.78) + (-0.5)(0.55) + (-2)(-0.31) \text{ lb}$$
$$= 2.69 \text{ lb}$$

## JUDGMENT OF THE RESULT

Because the projection $F'$ is the nearest-distance projection, $F$ must be greater than $F'$. Indeed, $F = \sqrt{3^2 + 0.5^2 + 2^2}$ lb $= 3.64$ lb $> 2.69$ lb.

**FIGURE 2-28**

Calculation of projection of vector
**F** on line $OA$

EXAMPLE 2-11    47

**2-58**  Prove Eq. 2-27,

$$\mathbf{A} \cdot (\mathbf{B} + \mathbf{C}) = \mathbf{A} \cdot \mathbf{B} + \mathbf{A} \cdot \mathbf{C}$$

**2-59**  Determine the angle between lines $AC$ and $BC$ using vector methods.

**2-62**  Calculate the projection of force $\mathbf{F}$ in Prob. 2-61 on line $AB$, which is parallel to the $x$ axis. Use the dot product of two vectors.

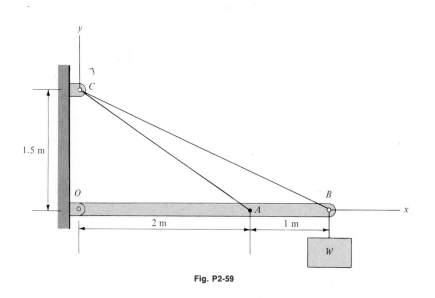

**Fig. P2-59**

**2-60**  The tension in wire $AC$ in Fig. P2-59 is 2 kN. Calculate the projection of tension $\mathbf{T}_{CA}$ on line $CB$ using the scalar product of two vectors.

**2-61**  Use vector methods to determine the angle $\phi$ between force $\mathbf{F} = (200\mathbf{i} + 170\mathbf{j})$ lb and arm $OA$ of the testing platform.

**2-63**  Determine the angle between wires $AB$ and $AC$.

**Fig. P2-61**

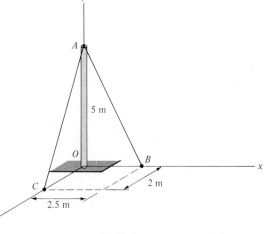

**Fig. P2-63**

**2–64**  The tension in wire $AB$ in Fig. P2-63 is 1.5 kN. Calculate the projection of tension $\mathbf{T}_{AB}$ on line $AC$.

**2–65**  The coordinates of $A$, $B$, and $C$ are $(0, 0, 2)$ m, $(-1, 1.5, 0)$ m, and $(1.2, 1.5, 0.2)$ m, respectively. Calculate the angle between lines $AB$ and $AC$.

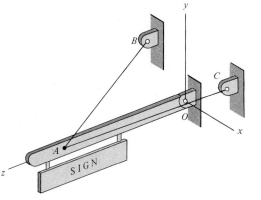

**Fig. P2-65**

**2–66**  In Prob. 2–65, the tension in each wire is 900 N. Determine the projections of the tensions in $AB$ and $AC$ on line $BC$.

Problems 2–67 to 2–71 are based on Fig. P2-67. Use the scalar product of two vectors to determine the required quantity in each problem.

**2–67**  Determine the angle between lines $OA$ and $OE$.

**2–68**  Determine the angle between lines $OA$ and $BG$.

**2–69**  A force $\mathbf{F}$ is acting from $O$ to $A$. Determine the projection of this force on line $AB$.

**2–70**  A force $\mathbf{P}$ is acting from $D$ to $G$. Determine the projection of this force on line $EF$.

**2–71**  A 100-lb force is acting from $O$ to $A$. Determine the projection of this force on the $x$ axis in two ways:
- (a) directly, going from $OA$ to $OG$
- (b) first going from $OA$ to $OF$, then from $OF$ to $OG$ (i.e., obtain the projection of another projection)

In both cases, the dimensions in Fig. P2-67 are $a = 2b = 1.5c$.

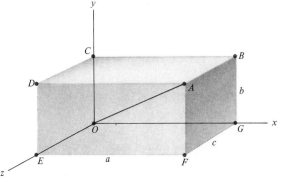

**Fig. P2-67**

# EQUILIBRIUM OF A PARTICLE.
# FREE-BODY DIAGRAMS

In previous discussions the resultant of the force system was occasionally zero. Such a situation is defined as *equilibrium*. According to strict definition:

> A particle is in equilibrium when the resultant of all forces acting on it is zero

The concept of equilibrium is enormously useful in engineering because conditions of equilibrium are very common. These conditions are discussed in the following paragraphs for uniaxial, biaxial, and triaxial force systems acting on a particle.

## Uniaxial Equilibrium

The forces acting on a particle that is in uniaxial equilibrium can be reduced to two forces. These forces must be equal in magnitude and opposite in direction, either compressing the particle or pulling it apart, as in Fig. 2-29. An alternative statement of this condition is often useful. When it is known that *only two forces* are acting on a particle that is in equilibrium, they must be opposite in direction. This is the only way that the resultant of two forces can be zero.

## Biaxial Equilibrium

A particle that is free to move in two dimensions may have any number of forces on it and still be in equilibrium if the resultant of the forces is zero. The condition can be stated or checked graphically using the polygon rule of vector addition. The algebraic method of summing all rectangular scalar components of the forces provides the necessary and sufficient conditions for biaxial equilibrium of a particle:

$$\sum F_x = 0 \quad \sum F_y = 0 \qquad \boxed{2\text{-}32}$$

The *orthogonal xy* axes of reference may be chosen arbitrarily without changing the mathematical statement of equilibrium. The two axes must be perpendicular, or the equations will not be independent.

## Triaxial Equilibrium

The conditions of equilibrium of a particle in three dimensions are extensions of the simpler cases. The major practical difference is that the polygon rule is not convenient to use for a three-dimensional problem. The algebraic method of summing rectangular scalar components is valid and convenient to use with the following formal statement of the necessary and sufficient conditions of a equilibrium:

$$\sum F_x = 0 \quad \sum F_y = 0 \quad \sum F_z = 0 \qquad \boxed{2\text{-}33}$$

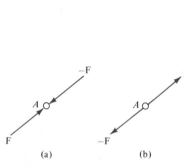

**FIGURE 2-29**

Two forces in equilibrium must be collinear, equal in magnitude, and opposite in direction

## Determination of Unknown Forces

Knowledge of the existence of equilibrium is very useful in determining unknown forces. The polygon rule is easy to use for two-dimensional systems of forces. For example, all of the known forces $F_1$ to $F_4$ that act on particle $A$ in Fig. 2-30a are drawn as shown in Fig. 2-30b. The resultant of the four original forces is then $R$. Equilibrium is possible only if there is a fifth force $P$ on the particle. $P$ must be equal in magnitude and opposite in direction to $R$ to close the polygon and obtain a zero resultant force on the particle. A force such as $P$, which acts on a particle to hold it in equilibrium when any number of other forces is applied, is called a *reaction* to those forces.

(a)  (b)

**FIGURE 2-30**

Graphical determination of an unknown force $P$ for equilibrium of a particle

## Free-Body Diagrams

Simple sketches of the physical situations relevant to a problem are often drawn in the process of solution. More important, however, an isolated diagram is made of the body, part of the body, or system of bodies involved. On this diagram the effects of other bodies on the body of interest are simply shown by drawing the forces that they cause. Thus, one draws a *free-body diagram*, which is a simple model of the body under consideration. Since this model is imagined to be isolated from all other bodies, the effects of the other bodies are represented by showing appropriate force vectors on the free-body diagram. The correct direction of all known forces and the assumed direction of all unknown forces must be indicated. It is important to show all significant forces in order to allow a proper determination of unknown forces.* For example, the free-body diagram for particle $A$ of Fig. 2-30 is shown in Fig. 2-31.

**FIGURE 2-31**

Free-body diagram of point $A$ with unknown reaction $P$

## Problem Solving with Particle Equilibrium

The problems of particle statics involve determining the unknown forces acting on a particle in order to maintain or ensure equilibrium. The first step is to determine which body or group of bodies to isolate as a particle. Even very large bodies may be treated as particles as long as the forces acting on them can be considered concurrent and acting at the center of mass. Once a particle is isolated, a complete free-body drawing of it is drawn.

Solving a problem may require drawing many free-body diagrams for different parts or groups of bodies. For each separate free-body diagram either one, two, or three independent scalar equilibrium equations may be obtained, depending on whether the free body is in uniaxial,

---

* Judgment is required in determining the significant forces in each problem. For example, the weight of a member may be neglected if it is much smaller than other forces acting on that member. The magnetic field of the earth is ignored in most cases but may have to be considered for a delicate electromagnetic device.

biaxial, or triaxial equilibrium. The most convenient *orthogonal* axes should be chosen for which the equilibrium equations are written. Starting with a particle on which at least one force is known, the equations can be solved in an orderly fashion until all unknown forces are determined.

The number of unknown forces that can be determined in any problem is limited to the number of *independent* equilibrium equations available. It is important to remember that each force, as a vector, is actually two unknown quantities: the magnitude and the direction. Often, the direction of a force is predetermined, such as the tension in a cable of known orientation. When the direction is unknown, the sense of the vector components can be assumed along the coordinate axes and drawn on the free-body diagram. The solutions of the equilibrium equations give the correct magnitudes of each component. Positive answers indicate that the assumptions regarding the directions were correct. Negative answers indicate that the unknown vector components were assumed in a direction opposite to the correct direction.

For example, the components of the unknown force **P** in the free-body diagram of Fig. 2-31 can be determined algebraically if the rectangular components of the other forces are known. The particle $A$ is in biaxial equilibrium, so Eqs. 2-32 apply for the $xy$ axes shown in Fig. 2-31. The directions of the two unknown components, $\mathbf{P}_x$ and $\mathbf{P}_y$, are assumed as shown in Fig. 2-32a. When the force summations are made, a negative answer for $P_x$ indicates that $\mathbf{P}_x$ was assumed in the wrong direction. The correct directions of the vector components are shown in Fig. 2-32b and agree with the direction of the reaction **P** in Fig. 2-30.

(a)
Assumed
directions

(b)
Correct
directions

**FIGURE 2-32**

**TIP**

For clarity in the statements of answers to equilibrium problems (as in the back of this book), the following guidelines are recommended:

(1) If a problem has a defined coordinate system, give answers as scalar components in that system.
(2) If no coordinate system is defined, give the answer as the magnitude of a vector component with an accompanying arrow designating the correct direction.

EXAMPLE 2-12

Determine the magnitude and direction of the force **P** that must oppose the tensions in the guy wires shown in the photograph in order to maintain the block in equilibrium.

## SOLUTION

First draw the free-body diagram as shown in Fig. 2-33. Here the weight of the block is included in force **P** as a force that opposes the tensions. Recall from Ex. 2-5 that the resultant of the forces due to the wires was **R** = 96.6 kN **i** + 90.1 kN **j**. Since there is equilibrium, $\sum \mathbf{F} = 0$. Thus, **P** + **R** = 0, from which **P** = −**R**, and

$$\mathbf{P} = -96.6 \text{ kN } \mathbf{i} - 90.1 \text{ kN } \mathbf{j}$$

**FIGURE 2-33**

Free-body diagram of an anchoring block of six guy wires

EXAMPLE 2-12   53

# EXAMPLE 2-13

A body of weight $W$ is held by a rope going over two pulleys as shown in Fig. 2-34a. Determine the tension $T$ in the rope and the force $\mathbf{F}_B$ on the axle of pulley $B$ for **(a)** $\theta = 0$, and **(b)** $\theta = 60°$. Assume that the tension in a rope is the same on the two sides of each pulley. This can be proved using the concept of moment, which is presented in the next chapter.

## SOLUTION

Assume that each pulley is a particle with three concurrent forces acting on it. In each case, the solution should start at pulley $A$ because a known force, $\mathbf{W}$, is acting there.

**(a)** $\theta = 0$. The free-body diagram of pulley $A$ is drawn in Fig. 2-34b. Since the rope tensions $\mathbf{T}$ are concurrent, parallel, and equal in magnitude, they are shown as a single force $2\mathbf{T}$, which must be equal in magnitude and opposite in direction to $\mathbf{W}$. Thus, $T = W/2$.

A similar procedure is applied at pulley $B$ as shown in Fig. 2-34c. Here the combined tensions $2\mathbf{T}$ in the ropes are acting downward, and $T = W/2$ as at pulley $A$, by Newton's law of action and reaction. The other required force is $F_B = W$ by inspection (this is the magnitude of the force vector $\mathbf{F}_B = -\mathbf{W}$).

**(b)** $\theta = 60°$. The free-body diagram of pulley $A$ in Fig. 2-34b is still applicable since the center of pulley $B$ is fixed and the ropes at $A$ remain parallel. Thus, $T = W/2$ at pulley $A$.

To determine $F_B$, consider Fig. 2-34d. The tensions $\mathbf{T}_1$ and $\mathbf{T}_2$ have equal magnitudes, $T$, so their resultant $\mathbf{R}$ bisects the 60° angle between them. $\mathbf{F}_B$ must be equal and opposite to $\mathbf{R}$, making a 30° angle with the vertical as shown. For equilibrium,

$$\sum F_x = 0$$

$$-F_B \sin 30° + T_2 \cos 30° = 0$$

Substituting $T_2 = T = W/2$,

$$F_B = \frac{W \cos 30°}{2 \sin 30°} = W \cos 30° = 0.866W$$

## JUDGMENT OF THE LAST RESULT

The biaxial forces in Fig. 2-34 can be reduced to uniaxial ones by summing forces along the line of action of $\mathbf{F}_B$. Then $F_B = T_1 \cos 30° + T_2 \cos 30° = 2T \cos 30° = W \cos 30°$, in agreement with the previous result.

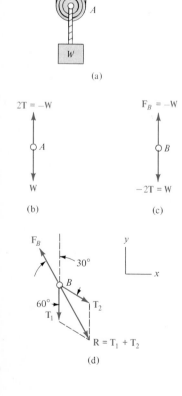

(a)

2T = —W

$A$

W

(b)

$F_B$ = —W

$B$

— 2T = W

(c)

$F_B$

30°

$B$

60°

$T_1$

$T_2$

R = $T_1$ + $T_2$

(d)

**FIGURE 2-34**

Determination of unknown forces
**T** and $\mathbf{F}_B$ in a pulley system

EXAMPLE 2-14

Suppose that you wish to hold a 200 kg cart on a 10° incline by applying a force **F** to the cart parallel to the incline. The cart is modeled in Fig. 2-35a. Determine the required magnitude of **F** along line *l* to hold the cart in equilibrium assuming that there is no friction between the cart and the incline. An important and reasonable assumption in such a problem is that all forces between the wheels and the inclined plane are *perpendicular* to the incline. These forces are commonly called *normal forces*.

## SOLUTION

First draw the free-body diagram in Fig. 2-35b. Since the *mass* of the cart is given, the *weight* of the cart must be determined from Eq. 1-3. Using the acceleration due to gravity in SI units of 9.81 m/s², $W = (200 \text{ kg})(9.81 \text{ m/s}^2) = 1962$ N. One graphical and two analytic methods of solution are presented for comparison:

(a) Assuming that the cart is a particle, a graphical solution is rapid for the three forces. The force **F** has a magnitude of 340 N in the force triangle of Fig. 2-35c.

(b) The analytic solution for the *x* (horizontal) and *y* (vertical) components of the forces is obtained from the two scalar equilibrium equations that can be written,

$$\sum F_x = 0$$

$$P \sin 10° - F \cos 10° = 0$$

$$\sum F_y = 0$$

$$P \cos 10° - W + F \sin 10° = 0$$

From these equations,

$$F = P \tan 10°$$

$$P(\cos 10° + \sin 10° \tan 10°) = W$$

$$P = \frac{W}{1.015} = 1933 \text{ N}$$

$$F = 341 \text{ N}$$

(c) There is a simpler analytic solution if the coordinate axes are chosen more advantageously. It is generally best to orient the coordinates parallel and perpendicular to the unknown force as *x'* and *y'* in Fig. 2-35d. With these,

$$\sum F_{x'} = 0$$

$$W \sin 10° - F = 0$$

$$F = 341 \text{ N}$$

## JUDGMENT OF THE RESULT

The magnitude of the force **F** must approach zero as the angle of the incline becomes zero, so a small *F* is reasonable. Furthremore, the three different solutions lead to the same answer.

**FIGURE 2-35**

Determination of unknown force **F** that holds the cart in equilibrium

EXAMPLE 2-14   55

# EXAMPLE 2-15

A helicopter is to lift a cargo of 900-kg mass. The cables are attached as in Fig. 2-36a, with point $A$ on the $y$ axis and points $B$, $C$, and $D$ in the $xz$ plane. Determine the tension in each cable.

(a)

(b)

$W = -8.83\mathbf{j}\,\text{kN}$

(c)

**FIGURE 2-36**

Determination of unknown tensions $T_B$, $T_C$, and $T_D$

## SOLUTION

First consider the free-body diagram of point $A$ as in Fig. 2-36b. Since point $A$ is in triaxial equilibrium, three scalar equilibrium equations are available to solve for three unknowns. To eliminate the fourth unknown, force $\mathbf{F}$, consider the free-body diagram of the cargo, Fig. 2-36c. The cargo is in uniaxial equilibrium, so $\mathbf{F} = (900 \text{ kg})(9.81 \text{ m/s}^2)\mathbf{j} = 8.83\mathbf{j}$ kN.

The three tensions are unknown in magnitude but known in direction. Write each tension force in terms of its vector components using Eqs. 2-20 and 2-22.

$$\mathbf{T}_B = T_B\,\mathbf{n}_B \qquad \mathbf{T}_C = T_C\,\mathbf{n}_C \qquad \mathbf{T}_D = T_D\,\mathbf{n}_D$$

$$\mathbf{n}_B = \frac{\text{vector } \overrightarrow{AB}}{\text{magnitude of } \overrightarrow{AB}} = \frac{-\mathbf{i} - \mathbf{j} + 0\mathbf{k}}{\sqrt{1^2 + 1^2 + 0^2}}$$

$$\mathbf{T}_B = T_B(-0.707\mathbf{i} - 0.707\mathbf{j})$$

$$\mathbf{n}_C = \frac{\text{vector } \overrightarrow{AC}}{\text{magnitude of } \overrightarrow{AC}} = \frac{1.2\mathbf{i} - \mathbf{j} + 0.6\mathbf{k}}{\sqrt{1.2^2 + 1^2 + 0.6^2}}$$

$$\mathbf{T}_C = T_C(0.717\mathbf{i} - 0.598\mathbf{j} + 0.359\mathbf{k})$$

$$\mathbf{n}_D = \frac{\text{vector } \overrightarrow{AD}}{\text{magnitude of } \overrightarrow{AD}} = \frac{1.4\mathbf{i} - \mathbf{j} - 0.5\mathbf{k}}{\sqrt{1.4^2 + 1^2 + 0.5^2}}$$

$$\mathbf{T}_D = T_D(0.781\mathbf{i} - 0.558\mathbf{j} - 0.279\mathbf{k})$$

**COMMON ERROR**

> It is a common error to write force $\mathbf{T}_D$, say, as $T_D(1.4\mathbf{i} - \mathbf{j} - 0.5\mathbf{k})$, forgetting to divide through by the *magnitude* of vector $\overrightarrow{AD}$. It is important to make sure that the unit vector, $\mathbf{n}$, is always equal to unity in magnitude.

For equilibrium, $\mathbf{T}_B + \mathbf{T}_C + \mathbf{T}_D + \mathbf{F} = 0$. Summing scalar components according to Eq. 2-33,

$$\sum F_x = 0$$
$$-0.707T_B + 0.717T_C + 0.781T_D = 0$$
$$\sum F_y = 0$$
$$-0.707T_B + 0.598T_C + 0.558T_D + 8.83 \text{ kN} = 0$$
$$\sum F_z = 0$$
$$0T_B + 0.359T_C - 0.279T_D = 0$$

From the simultaneous solution of these three equations,

$$T_B = 7.08 \text{ kN} \qquad T_C = 2.91 \text{ kN} \qquad T_D = 3.74 \text{ kN}$$

## JUDGMENT OF THE RESULTS

The tension values are acceptable since the sum of the vertical scalar components of $\mathbf{T}_B$, $\mathbf{T}_C$, and $\mathbf{T}_D$ is 8.83 kN.

EXAMPLE 2-15    57

# PROBLEMS

**UNIAXIAL EQUILIBRIUM**
In Probs. 2–72 to 2–81, assume that all the ropes are essentially vertical, that the pulleys are frictionless, and that the system is in equilibrium.

**2–72** Determine the tension in the continuous parts $a$, $b$, $c$, and $d$ of the rope if $W = 2$ kN.

W    **Fig. P2-72**

**2–73** Determine the magnitude of the force **P** acting on pulley $C$ and the force of bar $BD$ on pulley $B$ if $W = 600$ lb in Fig. P2-72.

**2–74** The rope segment $a$ is attached to the center shaft of pulley $A$, where the force **W** is also acting. The rope is continuous from part $a$ through $b$ to $c$. Calculate the tension $T$ if $W = 225$ lb.

W    **Fig. P2-74**

**2–75** Determine the force of bar $BC$ on pulley $B$ in Prob. 2–74.

**2–76** The rope segment $a$ is attached to the center shaft of pulley $A$, where the force **W** is also acting. The rope is continuous from part $a$ through $d$. Calculate the tension $T$ if $W = 5$ kN.

W    **Fig. P2-76**

**2–77** Determine the forces of bars $BC$ and $DE$ on pulleys $B$ and $D$, respectively, in Prob. 2–76.

**2–78** The rope segment $a$ is attached to the center shaft of pulley $A$, where the force **W** is also acting. The rope is continuous from part $a$ through $d$. Calculate the force on the axle of pulley $C$ due to bar $BC$ and the force on the axle of pulley $B$ due to bar $BD$. $W = 4$ kN.

W    **Fig. P2-78**

**2–79** Determine the load $W$ that a tension $T = 100$ lb can support in Prob. 2–78.

**2–80** The rope segment $e$ is attached to the center shaft of pulley $D$, where bar $AD$ is also connected. Determine the forces on the axles of pulleys $C$ and $D$ and the force on $B$ due to $BE$ if a tension $T = 400$ N is required to support a load $W$.

**Fig. P2-80**

**2–81** Calculate the tension in the rope segment $e$ in Prob. 2–80 if $W = 500$ lb.

## BIAXIAL EQUILIBRIUM

**2–82** A wire is attached at points $A$ and $B$ and supports a block of mass $m = 50$ kg. Determine the tensions in segments $AC$ and $BC$ if $\theta = \phi = 20°$.

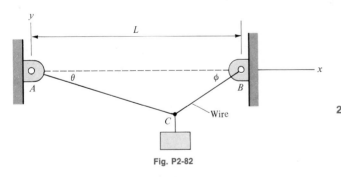

**Fig. P2-82**

**2–83** In Fig. P2-82, $\theta = \phi$. Plot the tension $T$ in the wire $AB$ as a function of $\theta = 5°$, $10°$, and $15°$ if the load is $W$. Draw a smooth curve through the three points and extrapolate the curve to $\theta = 3°$.

**2–84** In Fig. P2-82, $\theta = 10°$ and $\phi = 15°$. Calculate the tensions in wires $AC$ and $BC$ if the weight of the block is 100 lb.

**2–85** Determine the tension $T$ and the reactions on the axle of pulley $B$ if pulley $A$ supports a load of mass $m = 50$ kg and $\theta = 30°$. The weights of the pulleys are negligible in comparison with the load and the tension force.

**Fig. P2-85**

**2–86** The rope is continuous from $a$ through $d$ as it passes over three pulleys. Segments $b$, $c$, and $d$ are parallel to one another and make an angle $\theta = 5°$. Determine the supporting reactions at point $C$ if $\mathbf{T}_1 = 300\mathbf{i}$ N.

**Fig. P2-86**

**2–87** A 100-lb block is supported by force $\mathbf{F}$ on an inclined plane. The block is essentially frictionless, which means that only a normal force can act between the block and the incline. Plot the magnitude of the force $\mathbf{F}$ necessary to maintain equilibrium of the block as a function of the angle, using $\theta = 60°$, $90°$, and $120°$.

**Fig. P2-87**

**2–88**  A block of 5-kg mass is to be held in equilibrium on a frictionless incline by force **F**. Draw the free-body diagram and determine the magnitude of the force using two different coordinate systems.

**Fig. P2-88**

**2–89**  The cable junction structure of an electric trolley bus system is essentially in the horizontal plane. The buses pass below this network and obtain electric power through arms that rub against the wires ($w$). The tension in supporting cable $C$ is $T_C = 1.2$ kN. Determine the tension in cable $A$ and in cable $B$ using the angles in the diagram.

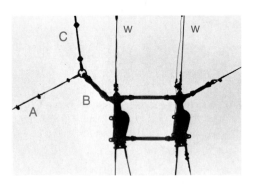

Network of horizontal cables of electric trolley bus system in Vancouver, B.C. (Photo is from the ground, in the vertical direction)

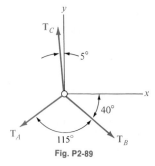

**Fig. P2-89**

**2–90**  Four bars are welded to a bracket. Calculate the unknown reaction at $O$ necessary to hold the bracket in equilibrium if $A = 1000$ lb, $B = 2000$ lb, $C = 500$ lb, and $D = 6000$ lb.

**Fig. P2-90**

**2–91**  The following forces on a welded bracket have fixed directions as shown: $A = 2$ kN, $B = 3$ kN, and $C = 5$ kN. Forces **F** and **P** have unknown magnitudes and directions, but their lines of action are fixed as shown. Determine **F** and **P** in terms of vector components for equilibrium of the bracket.

**Fig. P2-91**

## TRIAXIAL EQUILIBRIUM

**2–92**  A tripod is designed to support a camera of weight $W$. The camera is at height $h$ above the ground when the legs form an equilateral triangle $ABC$ of side $a$. Determine the magnitude of the forces in the legs assuming that they share the load equally and that the forces are axial (that is, longitudinal) in the legs.

**Fig. P2-92**

**2-93** The rectangular crate is lifted by four ropes of equal length $l$. Determine the net force applied by the crane hook to the junction $A$ of the four ropes if the tension in each rope is 2.5 kN. What is the mass of the crate?

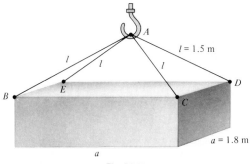

**Fig. P2-93**

**2-94** Four identical wires are suspended from a horizontal ceiling at points $A$, $B$, $C$, and $D$. The wires are concurrent at $E$ and support a load $W = 200$ lb. Calculate the tension in the wires if $a = 2$ ft, $b = 4$ ft, and $l = 3$ ft. Assume that the wires share the load symmetrically.

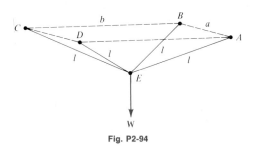

**Fig. P2-94**

**2-95** Solve Prob. 2-94 using $a = 4$ ft, 2 ft, and 0. There are four wires in each case. Plot the tension $T$ in wire $BE$ as a function of $a$, and explain the observed general tendency.

**2-96** A proposed light fixture for a theater has a mass of 80 kg. It is to be suspended by 16 wires of length $l = 10$ m. The wires are evenly spaced at the supports, which are on a horizontal circle of 2.5-m radius on the ceiling. Determine the tension in the wires if they share the load uniformly.

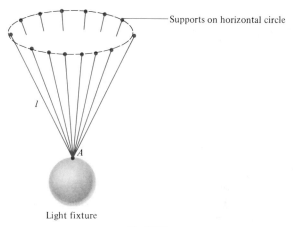

Light fixture

**Fig. P2-96**

**2-97** The two wires are attached to a single bolt at $A$. The coordinates of the three points are $A:(2,0,0)$ m, $B:(0, 1.5, 1)$ m, and $C:(0, 1.5, -1)$ m. The tension in $AB$ is 800 N; in $AC$ it is 850 N. The mass of the sign is 80 kg and the sign is directly attached to the bolt at $A$. Determine the force that the boom must exert on the bolt at $A$ in order to maintain equilibrium.

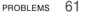

**Fig. P2-97**

**2–98** The three wires are attached to a small collar $A$ on the pole. The tension in each wire is 300 lb, and the coordinates are $A$: (0, 15, 0) ft, $B$: (−1, 0, 5) ft, $C$: (4, 0, 4) ft, and $D$: (0, 0, −3) ft. Determine the reaction of the pole acting on the collar.

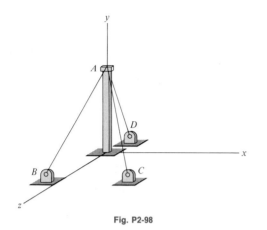

**Fig. P2-98**

**2–99** Determine the tension in each of the wires in Prob. 2–98 if it is known that the pole exerts only an upward force of 900 lb on the collar.

**2–100** The tensions in chains $AB$ and $AC$ and the force **F** support the load at point $A$. The coordinates of the three points are $A$: (10, 0, 0) ft, $B$: (0, 7, −4) ft, and $C$: (0, 7, 4) ft, and **F** is in the $xy$ plane. Determine $F$ if the load weighs 500 lb and $\theta$ must be zero.

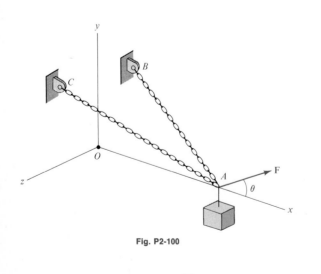

**Fig. P2-100**

**2–101** Solve Prob. 2–100 using $\theta = 0, 45°$, and $90°$. Plot $F$ as a function of $\theta$ and estimate the angle at which $F$ is a minimum.

**2–102** Determine the tension in the chains and the mass of the load held by a 3-kN force **F** in Fig. P2–100. The coordinates of the points are $A$: (3, 0, 0) m, $B$: (0, 2, −1) m, and $C$: (0, 2, 1) m. $\theta = 30°$.

**2–103** A large balloon for atmospheric research is tethered by three cables. The coordinates of these cables are $A$: (0, 60, 0) ft, $B$: (−40, 3, 30) ft, $C$: (38, 0, 33) ft, and $D$: (0, −4, −55) ft. Determine the tensions in the cables if the net upward lift of the balloon is 500 lb.

**Fig. P2-103**

**PARALLELOGRAM LAW:**

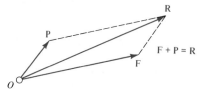

$$F + P = R$$

**TRIANGLE RULE:**

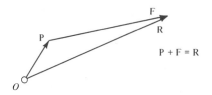

$$P + F = R$$

**ADDITION OF VECTORS F AND P:**  $\mathbf{F} + \mathbf{P} = \mathbf{R}$

**SUBTRACTION OF VECTOR F FROM R:**  $\mathbf{R} + (-\mathbf{F}) = \mathbf{P}$

**POLYGON RULE:**

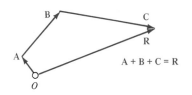

$$A + B + C = R$$

Section 2-2

**VECTOR COMPONENTS F$_m$ AND F$_n$ OF FORCE F:**

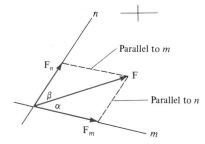

**LAW OF SINES**

$$\frac{F_n}{\sin \alpha} = \frac{F_m}{\sin \beta} = \frac{F}{\sin \left[180° - (\alpha + \beta)\right]}$$

**LAW OF COSINES**

$$F^2 = F_n^2 + F_m^2 - 2F_nF_m \cos \left[180° - (\alpha + \beta)\right]$$

**UNIT VECTORS n AND i, j, k:**

**1.** Each has a magnitude of 1.
**2.** Vector **F** of magnitude $F$ can be written as

$$\mathbf{F} = F\mathbf{n} \qquad \text{where } \mathbf{F} \text{ and } \mathbf{n} \text{ have the same direction}$$

In terms of $xyz$ components,

$$\mathbf{F} = F_x\mathbf{i} + F_y\mathbf{j} + F_z\mathbf{k}$$

**VECTOR SUMMATION TO OBTAIN RESULTANT R = $R_x$i + $R_y$j + $R_z$k:**

*Scalar Components:* $\quad R_x = \sum F_x \quad R_y = \sum F_y$

$$R_z = \sum F_z$$

*Vector Components:* $\quad \mathbf{R}_x = \sum F_x\mathbf{i} \quad \mathbf{R}_y = \sum F_y\mathbf{j}$

$$\mathbf{R}_z = \sum F_z\mathbf{k}$$

**SIGN CONVENTIONS:**

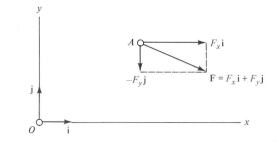

**MAGNITUDE $F$ OF VECTOR F = $F_x$i + $F_y$j + $F_z$k:**

$$F = \sqrt{F_x^2 + F_y^2 + F_z^2}$$

**SCALAR COMPONENTS OF VECTOR F IN TERMS OF ANGLES:**

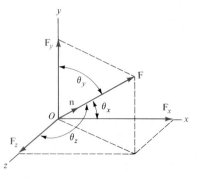

$$F_x = F \cos \theta_x \quad F_y = F \cos \theta_y \quad F_z = F \cos \theta_z$$

**DIRECTION COSINES OF COLLINEAR VECTORS F AND n:**

$$n_x = \cos\theta_x \quad n_y = \cos\theta_y \quad n_z = \cos\theta_z$$

**USEFUL EXPRESSIONS:**

$$\mathbf{n} = n_x\mathbf{i} + n_y\mathbf{j} + n_z\mathbf{k}$$

$$n_x^2 + n_y^2 + n_z^2 = 1$$

Section 2-5

**FORCE VECTOR F FROM SCALAR INFORMATION:**
Necessary information:

1. $F$, magnitude of force
2. Direction of force from point $A$ to $B$
3. Coordinates of points $A$ and $B$ on the line of action of force

The required force vector $\mathbf{F}$ is expressed as

$$\mathbf{F} = F(n_x\mathbf{i} + n_y\mathbf{j} + n_z\mathbf{k}) = F\frac{d_x\mathbf{i} + d_y\mathbf{j} + d_z\mathbf{k}}{d}$$

where $n_x, \ldots$ are direction cosines of line $A$ to $B$
$d_x, \ldots$ are coordinate differences from $A$ to $B$
$d =$ distance $A$ to $B = \sqrt{d_x^2 + d_y^2 + d_z^2}$

Section 2-6

**SCALAR PRODUCT (OR DOT PRODUCT) OF VECTORS A AND B**

$$\mathbf{A} \cdot \mathbf{B} = AB\cos\phi$$

**USEFUL PROPERTIES:**

$$\mathbf{A} \cdot \mathbf{B} = \mathbf{B} \cdot \mathbf{A}$$

$$\mathbf{A} \cdot (\mathbf{B} + \mathbf{C}) = \mathbf{A} \cdot \mathbf{B} + \mathbf{A} \cdot \mathbf{C}$$

$$\mathbf{i} \cdot \mathbf{i} = \mathbf{j} \cdot \mathbf{j} = \mathbf{k} \cdot \mathbf{k} = 1$$

$$\mathbf{i} \cdot \mathbf{j} = \mathbf{j} \cdot \mathbf{k} = \mathbf{k} \cdot \mathbf{i} = 0$$

**ANGLE $\phi$ BETWEEN VECTORS A AND B:**

$$\phi = \text{arc cos} \frac{A_x B_x + A_y B_y + A_z B_z}{AB}$$

**PROJECTION F' OF VECTOR F ON LINE /:**

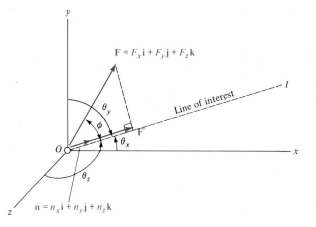

$$
\begin{aligned}
F' &= \mathbf{F} \cdot \mathbf{n} \\
&= F_x n_x + F_y n_y + F_z n_z \\
&= F_x \cos \theta_x + F_y \cos \theta_y + F_z \cos \theta_z
\end{aligned}
$$

Section 2-7

**SCALAR EXPRESSIONS FOR EQUILIBRIUM OF A PARTICLE:**

$$\sum F_x = 0 \quad \sum F_y = 0 \quad \sum F_z = 0$$

where the mutually orthogonal $xyz$ axes may be chosen for convenience.

**FREE-BODY DIAGRAMS:** Make a simple sketch of a body (or part of a body, or system of bodies) to be analyzed. Imagine this body to be isolated from all other bodies. Show *all* significant forces on the free body, including unknown forces. Use the equilibrium equations to solve for unknown forces.

**2–104**  Calculate the horizontal and vertical components of the force acting on the bracket of a road simulator (see Sec. 1-7). Also calculate the components normal and tangent to line *AB*.

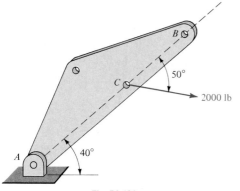

Fig. P2-104

**2–105**  Plot the magnitude of the components of the horizontal force along bars *AB* and *AC* for the following pairs of angles:

| $\theta$ | $\phi$ |
| --- | --- |
| 0 | 40° |
| 10° | 50° |
| 20° | 60° |

Enter all data on the same diagram and sketch curves through the appropriate points for each bar.

Fig. P2-105

**2–106**  In a laboratory test of the strut *l* and wheel *w* of an aircraft a force $\mathbf{F} = (500\mathbf{i} + 700\mathbf{j} - 200\mathbf{k})$ lb is applied at point *O*. The orientation of the strut is given by the vector $\mathbf{l} = \mathbf{i} + 8\mathbf{j} - 2\mathbf{k}$. Calculate the angle between $\mathbf{F}$ and $\mathbf{l}$ and determine the projection of force $\mathbf{F}$ along $\mathbf{l}$.

Fig. P2-106

**2–107**  A bridge is supported at point *O* by two cables. The forces in the cables are $\mathbf{F}_1 = (20\mathbf{i} + 28\mathbf{j} + 2\mathbf{k})$ kips and $\mathbf{F}_2 = (-18\mathbf{i} + 36\mathbf{j} + 1\mathbf{k})$ kips. (a) Calculate the angle between the two cables, and (b) determine the net horizontal and vertical forces applied by the cables to the bridge.

Fig. P2-107

**2–108**  The forces in the supporting guy wires of a tower are $\mathbf{F}_A = (-20\mathbf{i} - 35\mathbf{j} + 18\mathbf{k})$ kN, $\mathbf{F}_B = (20\mathbf{i} - 32\mathbf{j} + 19\mathbf{k})$ kN, and $\mathbf{F}_C = (-36\mathbf{j} - 40\mathbf{k})$ kN. (a) Determine the angle $\theta_{AB}$ (between wires *A* and *B*), and (b) calculate the resultant force $\mathbf{R}$ of the three forces acting at point *D*.

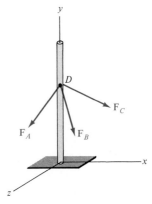

Fig. P2-108

**2–109** A block of weight $W$ is held in equilibrium by a rope going over three pulleys. Assume that the tension is the same on the two sides of each pulley and that the pulleys are particles with concurrent forces acting on them. (a) Draw the free-body diagram of each pulley, and (b) determine the tensions $T$, $T_a$, $T_b$, and $T_c$. Neglect the weights of the pulleys.

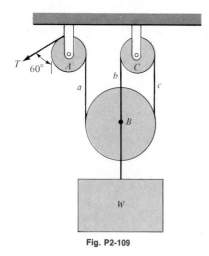

Fig. P2-109

**2–110** A 10-lb block is to be held in equilibrium on a smooth plane by force $\mathbf{F} = (-12\mathbf{i} + 3\mathbf{j})$ lb and force $\mathbf{P}$ which acts along line $l$. Draw the free-body diagram and determine the force $\mathbf{P}$.

Fig. P2-110

**2–111** The tower exerts a force of 225 kN vertically upward at point $D$. Determine the tension in the three cables $AD$, $BD$, and $CD$.

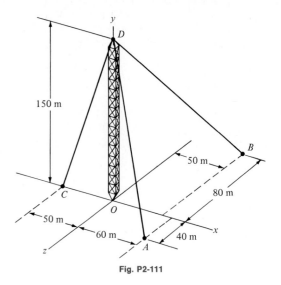

Fig. P2-111

# Forces on Rigid Bodies

## Basic Concepts in Modeling Rigid Bodies

There are several important concepts and principles of engineering mechanics that must be kept in mind when dealing with sizable members. The most commonly used concept is the assumption of rigidity. All solid materials deform when forces are applied to them, but for some purposes the deformations are negligible. A *rigid body* is a useful model in the solution of many different problems in mechanics. This is true at least in the early part of the analysis, which eventually may have to involve the deformability of the material. The various other principles discussed in this chapter will be applied to rigid-body models.

Another important concept in modeling rigid bodies is that the applied forces are generally not concurrent at the center of mass of the body. (If they were, the body would be modeled as a particle.) Because of this fact, the external forces tend to *rotate* as well as translate the rigid body. This tendency toward rotation is what distinguishes rigid-body analysis from particle analysis.

For an example of modeling rigid bodies, consider the movable *wave board* of a complex ocean-wave-simulating facility, shown in Fig. 3-1a and b. Such a board is the end wall of a long channel of water, and its controlled motion about hinges *A* can generate a wide variety of waves. The two-piece wave board shown has two hinges at the bottom (*A*) and two hinges (*B*) between the two boards. Motion of the boards is caused by

**FIGURE 3-1**

Two-piece wave board for the generation of water waves for research purposes. (Courtesy MTS Systems Corporation, Minneapolis, Minnesota.)

the two *hydraulic actuators* (hydraulic rams) shown in Fig. 3-1b. Hydraulic actuator *C* applies a nearly horizontal force to the lower board, causing it to rotate about the hinges at *A*. Hydraulic actuator *D* applies a force to move the upper board relative to the lower board. Depending on the displacements of these actuators, the wave board may be considered either as a single rigid body pivoted at the fixed axis *A*, or as two rigid bodies that are connected at axis *B* and supported at axis *A*. Such wave boards will be useful examples in numerous fundamental mechanics analyses.

**SECTION 3-1** presents the important distinction between external and internal forces that may act on a rigid body. Understanding these forces is essential in drawing free-body diagrams of rigid bodies.

**SECTION 3-2** defines equivalent forces for the purpose of altering the statements of some problems to advantage.

**SECTIONS 3-3 AND 3-4** introduce the concept of moment of a force. A moment is a "turning effect" of a force. It is one of the most frequently used concepts in engineering mechanics and should be thoroughly studied. The vector definition and evaluation of moment are particularly useful in solving complex problems.

**SECTIONS 3-5 AND 3-6** present the method of evaluating the moment of a force about a line. In the most general case this is the most complex problem involving moments. Vector methods are especially useful for solving such problems.

**SECTIONS 3-7 TO 3-9** present couples (a special case of moments) and the basic operations concerning them. These are advantageous in certain computations.

**SECTION 3-10** shows the procedures for transforming a force into a force–couple system, or vice versa. These possibilities facilitate solving problems.

**SECTIONS 3-11 AND 3-12** extend the force–couple transformations for the simplified analysis of any complex system of forces and moments. These transformations are advantageous both in the theoretical and experimental investigation of mechanical systems.

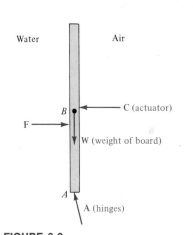

**FIGURE 3-2**

Free-body diagram of a rigid wave board in a simple configuration

## EXTERNAL AND INTERNAL FORCES

The two-piece wave board shown in Fig. 3-1 can be used in two distinct configurations. When actuator $D$ is "locked" in any particular position, the upper board does not move relative to the lower one, so the two boards may be analyzed as a single rigid body. There are a number of *external forces* acting on this member at any time, as shown in the free-body diagram of Fig. 3-2. Here **W** is the weight of the complete board and attached mechanisms that move as a unit about the hinges on line $A–A$. This force **W** is always vertical. Force **C** is from the horizontal actuator and force **F** is the net resultant of the forces caused by the water. Reaction **A** is exerted on the wave board by the two hinges on the bottom.

Figure 3-3 shows a more general configuration of the single wave board. The magnitudes and directions of the forces (except **W**) depend on the shape, position, and movement of the board. The points of application of forces **C** and **A** are fixed by the geometries of the actuator and lower hinge connections. **F** can move up or down depending on the level and movement of the water. **W** can move left or right depending on the relative position $\theta$ of the two boards; it always acts at the *center of gravity* of the whole body shown in the free-body diagram. This concept is discussed in detail in Ch. 7.

The forces acting between the upper and lower parts of the wave board are *internal forces* when the whole board is considered as a rigid body. These internal forces completely cancel each other and must be ignored when drawing the whole free-body diagram as in Fig. 3-2. However, the free-body diagram may be drawn for any body of interest (for

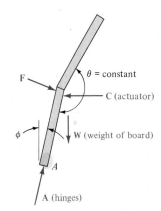

**FIGURE 3-3**

Free-body diagram of a rigid wave board in a general configuration example, the upper piece alone of the wave board). In that case the forces from the hinges at line $B$–$B$ and from actuator $D$ are external forces and must be shown as in Fig. 3-4. Here $\mathbf{F}_1$ and $\mathbf{W}_1$ are the parts of $\mathbf{F}$ and $\mathbf{W}$ acting only on the upper board, while $\mathbf{C}$ and $\mathbf{A}$ are not shown because they are not directly acting on this member. The details of analyzing inter-related free-body diagrams are presented in Ch. 4.

**FIGURE 3-4**

Upper part of a two-piece
wave board

## 3-2 EQUIVALENT FORCES

*Sliding vectors* and the *principle of transmissibility* have been introduced in Sec. 1-4. These concepts are especially important in dealing with external forces on rigid bodies, but the assumption of rigidity must be carefully checked. For example, it does not make any difference in theory whether the actuator force is pushing ($\mathbf{D}$) or pulling ($\mathbf{D}'$) on the wave board in Fig. 3-5. However, it may make a significant difference in practice, depending on the deformations of the tubular structure through which the force is applied to the board (Fig. 3-1b).

Two forces $\mathbf{D}$ and $\mathbf{D}'$ acting at different points are *equivalent* if they have the same magnitude, direction, line of action, and the same effect on a given rigid body in static equilibrium.

**FIGURE 3-5**

Equivalent forces **D** and **D**′ acting
on a rigid body ($D = D'$)

## 3-3 MOMENT OF A FORCE

The *moment $M_A$* of a force $\mathbf{F}$ (also called *torque*) about a point $A$ is defined as a scalar quantity

$$M_A = Fd \qquad \boxed{3\text{-}1}$$

where $d$ is the nearest distance from $A$ to the line of action of $\mathbf{F}$ as in Fig. 3-6. The terms *lever arm* or *moment arm* are often used for the distance $d$.

The concept of the moment describes the turning effect of the force with respect to the point of interest. The sense of the rotational tendency gives the sense of the moment, clockwise or counterclockwise. In Fig. 3-6 the sense of the moment of force $\mathbf{F}$ about point $A$ is clockwise.

The *scalar definition of moment* given in Eq. 3-1 is valid for rigid bodies of any shape and in three dimensions, as well, but care must be taken that $d$ is the nearest distance to the line of action. The unit of moment is force times length.

The statement in Sec. 3-2 concerning the equivalence of two forces can be expanded to include a formal statement of equivalence with respect to turning effects. Forces $\mathbf{F}$ and $\mathbf{F}'$ are equivalent if

$$\mathbf{F} = \mathbf{F}' \quad \text{and} \quad M_A = M'_A \qquad \boxed{3\text{-}2}$$

These are the necessary and sufficient conditions for equivalence as illustrated in Fig. 3-7.

**FIGURE 3-6**

Force **F** has a moment of magnitude
$M_A = Fd$ with respect to point $A$.

**FIGURE 3-7**

Equivalent forces **F** and **F**′ have the
same moment about point $A$

# EXAMPLE 3-1

Figure 3-8a shows a test setup for applying vertical forces to the ball joint of a car. Determine the moments of force **F** about points *A*, *B*, *C*, and *D*. The locations of these points are given in Fig. 3-8b using the *xy* coordinate system for reference.

## SOLUTION

Assume that the magnitude *F* is given in pound units.

$$M = Fd \qquad \text{where } d \text{ is a horizontal distance in each case since } \mathbf{F} \text{ is vertical}$$

$$M_A = F\,(14)\ \text{lb·in.}$$

$$M_B = F\,(19)\ \text{lb·in.}$$

$$M_C = F\,(21)\ \text{lb·in.}$$

$$M_D = F\,(40)\ \text{lb·in.}$$

In each case the moment is counterclockwise.

## JUDGMENT OF THE RESULTS

Point *D* is farthest from the applied force, hence the moment about *D* is largest as expected; $M_D > M_C > M_B > M_A$. It is emphasized that the vertical distances to a vertical force were not needed (law of transmissibility).

**FIGURE 3-8**

Test setup of a ball joint of a car. (Photograph courtesy MTS Systems Corporation, Minneapolis, Minnesota.)

EXAMPLE 3-1    73

# EXAMPLE 3-2

A relatively small, one-piece wave board is modeled in Fig. 3-9a. The maximum magnitude of the force **C** from the hydraulic actuator is always the same 150 kN, regardless of the angular positions of the board. The distance $AB$ is 0.8 m. Determine the moment of **C** about point $A$ for **(a)** $\theta = \phi = 0$, and **(b)** $\theta = 4°$, $\phi = 2°$.

## SOLUTION

**(a)** For $\theta = \phi = 0$, $d = 0.8$ m. Using Eq. 3-1,

$$M_A = Fd = (150 \text{ kN})(0.8 \text{ m}) = 120 \text{ kN·m} \quad \text{(counterclockwise)}$$

**(b)** From Fig. 3-9b,

$$d = 0.8 \cos(\phi + \theta) = 0.8 \cos 6° = 0.796 \text{ m}$$

$$M_A = Fd = (150 \text{ kN})(0.796 \text{ m}) = 119.3 \text{ kN·m} \quad \text{(counterclockwise)}$$

## JUDGMENT OF THE RESULTS

As expected, the moment of the force applied at a given point depends on its angle of application. The difference here is small because the change in angle is small.

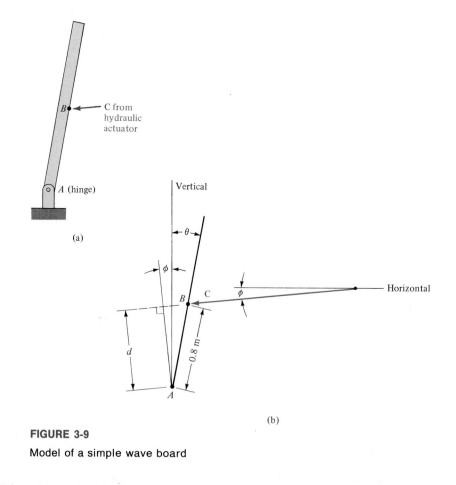

(a)

(b)

**FIGURE 3-9**

Model of a simple wave board

**3-1** Determine the moment of the force about point $A$.

**Fig. P3-1**

**3-2** Determine the moment of $\mathbf{F}$ about point $O$ if $L = 5$ ft, $F = 300$ lb, and $\theta = 60°$.

**Fig. P3-2**

**3-3** Plot $M_O$ for the pole in Fig. P3-2 using $F$ and $L$ as constants and $\theta = 0$, $30°$, $60°$, and $90°$.

**3-4** Determine the moments $M_A$ and $M_B$ of the force $\mathbf{F}$.

**Fig. P3-4**

**3-5** Plot the moments $M_A$ and $M_B$ of the force $\mathbf{F}$ in Fig. P3-4. Let $F = 500$ lb, $L = 4$ ft, and $\theta = 20°$, $30°$, and $40°$. Use a single diagram for showing both moments.

**3-6** Determine the moments $M_A$ and $M_C$ of the force $\mathbf{F}$ if $F = 3$ kN, $L = 4$ m, and $\theta = 20°$.

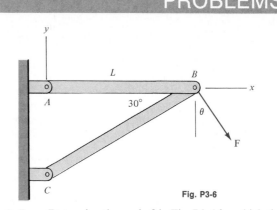

**Fig. P3-6**

**3-7** Determine the angle $\theta$ in Fig. P3-6 for which the moments of the force $\mathbf{F}$ are $M_A = M_C$.

**3-8** Calculate the moments $M_A$ and $M_B$ of the force if $F = 2000$ lb, $L = 5$ ft, $\theta = 70°$, $\phi = 50°$, and $\alpha = 60°$.

**Fig. P3-8**

**3-9** Determine the angle $\alpha$ in Fig. P3-8 for which the moments of the force $F = 10$ kN are $M_A = M_B$. $L = 1.5$ m, $\theta = 70°$, and $\phi = 45°$.

**3-10** Calculate the moments $M_1$ and $M_2$ of forces $\mathbf{F}_1$ and $\mathbf{F}_2$ about $A$. Also determine the resultant $\mathbf{R}$ of $\mathbf{F}_1$ and $\mathbf{F}_2$, and the moment of $\mathbf{R}$ about $A$. $F_1 = 2$ kN, $F_2 = 3$ kN, $L = 1$ m, $\theta = 60°$, and $\phi = 40°$.

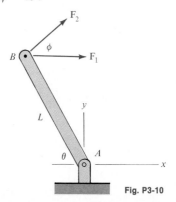

**Fig. P3-10**

**3–11** Assume that $F_1 = 500$ lb is always horizontal and to the right in Fig. P3-10. Determine the angle $\phi$ and the smallest value of $F_2$ to apply a moment $M_{2A}$ which is equal in magnitude but opposite in direction to $M_{1A}$. $L = 4$ ft and $\theta = 50°$.

**3–12** The street lamp $A$ has a mass of 10 kg. The force $\mathbf{F}$ of the wind has magnitude $F = 300$ N. Determine the moments of $\mathbf{F}$ and $\mathbf{W}$ about $O$. Also calculate the resultant $\mathbf{R}$ of $\mathbf{F}$ and $\mathbf{W}$, and the moment of $\mathbf{R}$ about $O$. $d = 2$ m and $h = 5$ m.

**Fig. P3-12**

**3–13** Assume that the force $\mathbf{F}$ of the wind on lamp $A$ in Fig. P3-12 is directly into the paper. $W = 50$ lb and $F = 40$ lb. Determine the moments of $\mathbf{F}$ and $\mathbf{W}$ about $O$. $d = 7$ ft and $h = 15$ ft.

**3–14** The force on the hand crank acts parallel to the $x$ axis. Determine the moments about points $O$, $A$, and $B$ if $F = 16$ lb.

**Fig. P3-14**

**3–15** Determine the moments of $\mathbf{F}$ and $\mathbf{P}$ about $A$ and $B$ individually (four moments). Calculate the resultant $\mathbf{R}$ of $\mathbf{F}$ and $\mathbf{P}$, and the moments of $\mathbf{R}$ about $A$ and $B$. $F = 800$ N, $P = 500$ N, $L = 3$ m, $\theta = 70°$, and $\phi = 30°$.

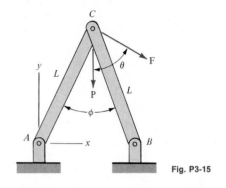

**Fig. P3-15**

**3–16** The ladder is modeled as a rigid bar pinned at end $A$. Write an expression for the sum of the moments $M_A$ of the three forces. Use the distances $AB$, $AC$, and $AD$.

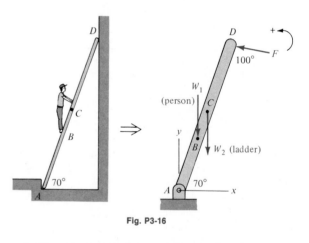

**Fig. P3-16**

**3–17** The test platform and the test object have a combined mass of 300 kg. The total weight, $\mathbf{W}$, can be assumed to act at a horizontal distance of 1 m from $B$. Determine the sum of the moments of $\mathbf{F}$ and $\mathbf{W}$ about $A$ if $F = 6$ kN and $d = 0.8$ m.

**Fig. P3-17**

The scalar method of determining the moment of a force is the best in solving simple problems where the nearest distance $d$ to the line of action is easily found. The solutions of more complex problems are simplified by using additional mathematical tools.

The *vector product* of two concurrent vectors **A** and **B** is, by definition, the vector **V** that has the following properties:

**1.** **V** is perpendicular to the plane of vectors **A** and **B**.
**2.** The sense of **V** obeys the right-hand rule:

**RIGHT-HAND RULE**

> The fingers of the right hand are pointed in the direction of vector **A** in such a way that they can be curled toward the vector **B**. The extended thumb points in the direction of vector $V = A \times B$.

The right-hand rule is illustrated in Fig. 3-10.

**3.** The magnitude of **V** is the product of the magnitudes of **A**, **B**, and $\sin \theta$, where $\theta$ is the angle between **A** and **B**:

$$V = AB \sin \theta \qquad \boxed{3\text{-}3}$$

The commonly used mathematical notation for the vector product is

$$V = A \times B \qquad \boxed{3\text{-}4}$$

which is also called the *cross product of* **A** and **B**.

According to the second property given above, vector products are *not commutative*, so

$$A \times B \neq B \times A \qquad \boxed{3\text{-}5}$$

but

$$A \times B = -(B \times A) \qquad \boxed{3\text{-}6}$$

Vector products are *distributive*, so for three vectors **A**, **B**, and **C**,

$$A \times (B + C) = A \times B + A \times C \qquad \boxed{3\text{-}7}$$

This is intuitively acceptable to most people. A rigorous proof is left as an exercise for the reader.

The distributive property is the basis for quantitative determination of the vector product in terms of vector components. For the general three-dimensional case,

$$V = A \times B = (A_x i + A_y j + A_z k) \times (B_x i + B_y j + B_z k) \qquad \boxed{3\text{-}8}$$
$$= (A_x i) \times (B_x i) + (A_x i) \times (B_y j) + \ldots + (A_z k) \times (B_z k)$$

**FIGURE 3-10**

Illustration of right-hand rule for vector product of vectors **A** and **B**

**FIGURE 3-11**

Right-handed coordinate system

Equation 3-8 can be simplified because all the possible vector products of the unit vectors are, according to definition,

$$\mathbf{i} \times \mathbf{i} = \mathbf{j} \times \mathbf{j} = \mathbf{k} \times \mathbf{k} = 0 \qquad \text{since } \sin 0 = 0 \qquad \boxed{3\text{-}9}$$

$$\mathbf{i} \times \mathbf{j} = \mathbf{k} \qquad\qquad\qquad \text{since } \sin 90° = 1$$

$$\mathbf{i} \times \mathbf{k} = -\mathbf{j}$$

$$\mathbf{j} \times \mathbf{i} = -\mathbf{k}$$

$$\mathbf{j} \times \mathbf{k} = \mathbf{i}$$

$$\mathbf{k} \times \mathbf{i} = \mathbf{j}$$

$$\mathbf{k} \times \mathbf{j} = -\mathbf{i}$$

The positive or negative signs of these are obtained from the right-hand rule, assuming a right-handed coordinate system as in Fig. 3-11.* With these, Eq. 3-8 becomes

$$\mathbf{V} = (A_yB_z - A_zB_y)\mathbf{i} + (A_zB_x - A_xB_z)\mathbf{j} + (A_xB_y - A_yB_x)\mathbf{k} \qquad \boxed{3\text{-}10}$$

The terms in parentheses are the rectangular components of the vector **V**,

$$V_x = A_yB_z - A_zB_y \qquad\qquad \boxed{3\text{-}11}$$

$$V_y = A_zB_x - A_xB_z$$

$$V_z = A_xB_y - A_yB_x$$

Equation 3-11 can also be obtained with the method used for calculating a determinant. This method should be favored in routine solutions because it is the easiest to remember.

$$\mathbf{V} = \begin{vmatrix} \mathbf{i} & \mathbf{j} & \mathbf{k} \\ A_x & A_y & A_z \\ B_x & B_y & B_z \end{vmatrix} \qquad \boxed{3\text{-}12}$$

Equation 3-12 can be evaluated as follows.[†] Write the sum of the products along the solid diagonals, and subtract from this the sum of the products along the dashed diagonals. Each term must be the product of two different scalar components and a unit vector. In practice it is not necessary to write these as done here to illustrate the method. It is possible to write directly from Eq. 3-12,

$$\mathbf{V} = A_yB_z\mathbf{i} + A_zB_x\mathbf{j} + A_xB_y\mathbf{k} - A_yB_x\mathbf{k} - A_xB_z\mathbf{j} - A_zB_y\mathbf{i} \qquad \boxed{3\text{-}13}$$

---

* In this text right-handed coordinate systems are used exclusively.

† Note that this particular method is easy to remember after a little practice, but it is not valid for higher-order determinants.

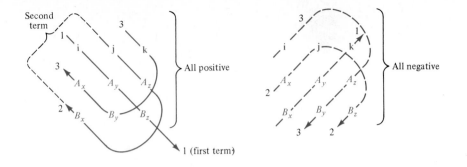

Second term — All positive — 1 (first term) — All negative

The vector product has great practical utility in determining moments of forces. The moment of a force $\mathbf{F}$ about a point $O$ is the vector product of the *position vector* $\mathbf{r}$ (from $O$ to *any* point $A$ on the line of action of $\mathbf{F}$) and $\mathbf{F}$,

$$\boxed{\mathbf{M}_O = \mathbf{r} \times \mathbf{F}} \qquad \boxed{3\text{-}14}$$

This is the *vector definition of moment*. The moment vector is denoted by a double-headed arrow in drawings in this text.

For a simple example of validating the vector definition of moment with the intuitively more acceptable scalar definition, consider Fig. 3-12, which shows a two-dimensional arrangement. The force $\mathbf{F}$ is applied at point $A$ with coordinates $a$ and $b$. From the scalar definition,

$$M_O = Fa \qquad \text{counterclockwise}$$

From the vector product,

$$\mathbf{M}_O = \mathbf{r} \times \mathbf{F} = (a\mathbf{i} + b\mathbf{j}) \times F\mathbf{j} = Fa\mathbf{k}$$

The unit vector $\mathbf{k}$ is correct for $\mathbf{M}_O$ since it means "counterclockwise" in this diagram according to the right-hand rule.

**TIP**

> The direction of the moment vector is easily determined by curling the fingers of the right hand in the direction that the force would tend to rotate the body. The extended thumb points in the direction of the moment vector. In Fig. 3-12, curling the fingers counterclockwise results in a moment vector direction *out* of the paper, that is, in the positive $z$ direction.

**FIGURE 3-12**

To determine the moment $\mathbf{M}_O$ of force $\mathbf{F}$ use a position vector from point $O$ to any point on the line of action $\mathbf{F}$

Note that the definition leading to Eq. 3-14 states that $\mathbf{r}$ is a position vector from point $O$ to *any* point $A$ on the line of action of $\mathbf{F}$. To check this, the moment $\mathbf{M}_O$ is recalculated using $\mathbf{r}'$ in Fig. 3-12,

**FIGURE 3-13**

Rectangular components of a moment $\mathbf{M}_O = \mathbf{r} \times \mathbf{F}$

$$\mathbf{M}_O = \mathbf{r}' \times \mathbf{F} = \left( a\mathbf{i} + \frac{b}{3}\mathbf{j} \right) \times F\mathbf{j} = Fa\mathbf{k}$$

It is clear that for any $r_y$ component, $r_y\mathbf{j} \times F\mathbf{j} = 0$, and $M_O$ is always $Fa\mathbf{k}$. The vector method is valid for the general case of three-dimensional problems as well. Generally, *the more complex the arrangement, the more advantageous the vector method becomes.*

### Rectangular Components of a Moment

In the general case where both $\mathbf{r}$ and $\mathbf{F}$ have three-dimensional components, the moment of $\mathbf{F}$ about a point also has three components. From Fig. 3-13,

$$\begin{aligned}
\mathbf{M}_O = \mathbf{r} \times \mathbf{F} &= (r_x\mathbf{i} + r_y\mathbf{j} + r_z\mathbf{k}) \times (F_x\mathbf{i} + F_y\mathbf{j} + F_z\mathbf{k}) \quad \boxed{3\text{-}15} \\
&= (r_yF_z - r_zF_y)\mathbf{i} + (r_zF_x - r_xF_z)\mathbf{j} + (r_xF_y - r_yF_x)\mathbf{k} \\
&= M_x\mathbf{i} + M_y\mathbf{j} + M_z\mathbf{k}
\end{aligned}$$

The moment scalar components $M_x$, $M_y$, and $M_z$ show the turning effect of the force $\mathbf{F}$ about the corresponding coordinate axes. Note that a single force $\mathbf{F}$ causes only one moment about a given point. However, since the moment is a vector, it can have three components with respect to the chosen coordinate axes.

### Varignon's Theorem

The distributive property of vector products is sometimes advantageous in calculating the resultant moment of several concurrent forces as in Fig. 3-14. On the basis of Eq. 3-7, the distributive property of the vector product,

$$\boxed{\mathbf{M}_O = \mathbf{r} \times (\mathbf{F}_1 + \mathbf{F}_2 + \ldots) = \mathbf{r} \times \mathbf{F}_1 + \mathbf{r} \times \mathbf{F}_2 + \ldots} \quad \boxed{3\text{-}16}$$

This procedure is related to the preceding one, where the three rectangular components of a force were used to determine the moment (Eq. 3-15). The statement of Eq. 3-16 is useful for dealing with any number of concurrent forces or for resolving a given force into components that are not parallel to the chosen coordinate axes.

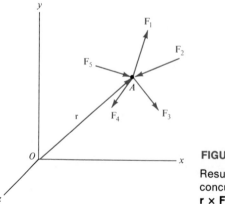

**FIGURE 3-14**

Resultant moment of several concurrent forces: $\mathbf{M}_O = \mathbf{r} \times \mathbf{F}_1 + \mathbf{r} \times \mathbf{F}_2 + \ldots$

EXAMPLE 3-3

Particle $A$ and point $O$ in Fig. 3-15 are in a vertical plane. Use the scalar and vector methods to determine the moment of the particle's weight $\mathbf{W}$ about point $O$ for the position given.

## SOLUTION

**SCALAR METHOD:**   Using Eq. 3-1,

$$M_O = Fd$$
$$d = r \sin 50°$$
$$M_O = Wr \sin 50° \qquad \text{(clockwise)}$$

**VECTOR METHOD:**   Using Eq. 3-14,

$$\mathbf{M}_O = \mathbf{r} \times \mathbf{F}$$

with $\mathbf{F} = -W\mathbf{j}$ and $\mathbf{r} = r \sin 50° \,\mathbf{i} - r \cos 50° \,\mathbf{j}$,

$$\mathbf{M}_O = (r \sin 50° \,\mathbf{i} - r \cos 50° \,\mathbf{j}) \times (-W\mathbf{j})$$
$$= -Wr \sin 50° \,(\mathbf{i} \times \mathbf{j})$$
$$= -Wr \sin 50° \,\mathbf{k} \qquad \text{(clockwise by right-hand rule)}$$

## JUDGMENT OF THE RESULTS

Both methods yield the same result. For this problem the scalar method is the easiest since the calculation of $d$ is simple.

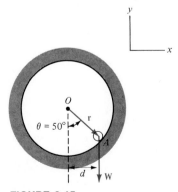

**FIGURE 3-15**
Moment of a particle's weight

EXAMPLE 3-3    81

# EXAMPLE 3-4

The torque wrench in Fig. 3-16a has an arm of constant length $L$ but a variable socket length $d$ because of interchangeable tool sizes. Determine how the moment applied at point $O$ depends on the length $d$ for a constant force **F** from the hand. Use the coordinate system shown in Fig. 3-16b.

## SOLUTION

Using $\mathbf{M}_O = \mathbf{r} \times \mathbf{F}$ with $\mathbf{r} = L\mathbf{i} + d\mathbf{j}$ and $\mathbf{F} = F\mathbf{k}$,

$$\mathbf{M}_O = (L\mathbf{i} + d\mathbf{j}) \times F\mathbf{k} = Fd\mathbf{i} - FL\mathbf{j}$$

## JUDGMENT OF THE RESULT

According to a visual analysis the wrench should turn clockwise, so the $-\mathbf{j}$ component of the moment is justified. Looking at the wrench from the positive $x$ direction, point $A$ has a tendency to rotate counterclockwise. Thus, the $\mathbf{i}$ component is correct using the right-hand rule.

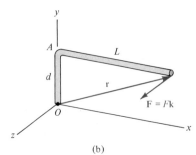

(b)

**FIGURE 3-16**

Moment applied using a torque wrench. (Photograph courtesy MTS Systems Corporation, Minneapolis, Minnesota.)

EXAMPLE 3-5

Consider two sets of three guy wires on a tall antenna tower as shown in Fig. 3-17. The initial tension is 50,000 lb in each wire. The ground distance $OA$ to each anchor $A$ is 1000 ft. The wires are attached to the tower at elevations $B = 150$ ft and $C = 1200$ ft. Assume that a wire breaks at $B$ or $C$ (but not both at once). In each case the remaining two wires cause a net moment about point $O$. Determine which failure causes the larger moment about $O$.

## SOLUTION

There is no resultant moment about $O$ when all six cables are intact. Since all tensions are equal and the cable distribution is symmetric, there is no resultant force in the $x$ or $z$ directions and the resultant force in the $y$ direction goes through the point $O$; hence, $M_O = F_y d = 0$ $(d = 0)$.

**CASE 1:** For ease of calculation, assume that cable $BA_3$ breaks. Tension forces $T_{A_1}$ and $T_{A_2}$ are then symmetric about the $z$ axis, so there is no resultant force in the $x$ direction (Fig. 3-17b). Using Varignon's theorem,

$$\mathbf{M}_O = \mathbf{r}_{OB} \times \mathbf{T}_{A_1} + \mathbf{r}_{OB} \times \mathbf{T}_{A_2} = \mathbf{r}_{OB} \times (\mathbf{T}_{A_1} + \mathbf{T}_{A_2})$$
$$= \mathbf{r}_{OB} \times T(\mathbf{n}_{BA_1} + \mathbf{n}_{BA_2})$$
$$= (150 \text{ ft } \mathbf{j}) \times (50{,}000 \text{ lb})(-0.856\mathbf{i} - 0.148\mathbf{j} + 0.494\mathbf{k}$$
$$+ 0.856\mathbf{i} - 0.148\mathbf{j} + 0.494\mathbf{k})$$
$$= (150 \text{ ft } \mathbf{j}) \times (50{,}000 \text{ lb})(0\mathbf{i} - 0.296\mathbf{j} + 0.988\mathbf{k})$$
$$= 7.41 \times 10^6 \text{ lb·ft } \mathbf{i}$$

**CASE 2:** Assume that cable $CA_3$ breaks.

$$\mathbf{M}_O = \mathbf{r}_{OC} \times (\mathbf{T}_{A_1} + \mathbf{T}_{A_2})$$
$$= \mathbf{r}_{OC} \times T(\mathbf{n}_{CA_1} + \mathbf{n}_{CA_2})$$
$$= (1200 \text{ ft } \mathbf{j}) \times (50{,}000 \text{ lb})(-0.554\mathbf{i} - 0.768\mathbf{j}$$
$$+ 0.32\mathbf{k} + 0.554\mathbf{i} - 0.768\mathbf{j} + 0.32\mathbf{k})$$
$$= (1200 \text{ ft } \mathbf{j}) \times (50{,}000 \text{ lb})(0\mathbf{i} - 1.536\mathbf{j} + 0.64\mathbf{k})$$
$$= 38.4 \times 10^6 \text{ lb·ft } \mathbf{i}$$

## JUDGMENT OF THE RESULTS

If the cable breaks at $C$, there is a larger moment caused about $O$. The cause for this larger moment is the much larger moment arm to point $C$. The moment is not as large as might be expected because there is a smaller resultant force component in the $z$ direction which is the only component of the force causing a moment about $O$. There is no $x$ component and the $y$ component goes through $O$. Note that the magnitudes of the two moments are the same regardless of whether the cables at $A_1$, $A_2$, or $A_3$ break—only the directions of the moments change.

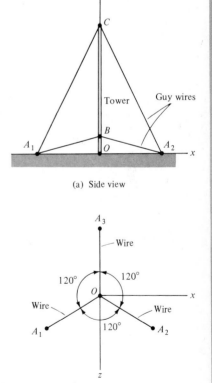

(a) Side view

(b) Top view

**FIGURE 3-17**

Model of antenna tower

EXAMPLE 3-5   83

**EXAMPLE 3-6**

The designers of a new goal post (American football) wish to prevent damage to the post by overly excited spectators. It is estimated that at most about 10 people can hang onto a rope thrown over the post. The rope can be pulled in many different directions, but the maximum magnitude of the force is assumed to be 2500 lb regardless of the direction. What is the moment of this force about the base $O$ of the post for the situation sketched in Fig. 3-18? Use the vector method.

## SOLUTION

$$\mathbf{M}_O = \mathbf{r}_{OD} \times \mathbf{F}_D = (8\mathbf{i} + 12\mathbf{j} - 2\mathbf{k}) \times (1768\mathbf{i} - 1768\mathbf{j})$$

$$= \begin{vmatrix} \mathbf{i} & \mathbf{j} & \mathbf{k} \\ 8 & 12 & -2 \\ 1768 & -1768 & 0 \end{vmatrix}$$

$$= (-3536\mathbf{i} - 3536\mathbf{j} - 35{,}360\mathbf{k}) \text{ lb·ft}$$

The magnitude of the moment is

$$M = 35{,}700 \text{ lb·ft}$$

**TIP**

To determine the position vector $\mathbf{r}_{OD}$, consider a path *from* point $O$ *to* point $D$ along lines parallel to the coordinate axes. For convenience, move 12 ft in the positive $y$ direction, 2 ft in the negative $z$ direction, and 8 ft in the positive $x$ direction, resulting in $\mathbf{r}_{OD} = (8\mathbf{i} + 12\mathbf{j} - 2\mathbf{k})$ ft, as above.

## JUDGMENT OF THE RESULT

The relative magnitudes of the moment's components are intuitively correct. The goal post would tend to bend around the $z$ axis much more than it would tend to bend about the $x$ or $y$ axes. A check with the right-hand rule shows that the sign of each component is also correct. It is left as an exercise for the reader to check the result using scalars (perpendicular distances and force components).

$OA = 12$ ft
$AB = 2$ ft
$CB = BD = 8$ ft

**FIGURE 3-18**

Model of a goal post

Use the vector product in each problem.

**3–18** Prove Eq. 3-7,

$$\mathbf{A} \times (\mathbf{B} + \mathbf{C}) = \mathbf{A} \times \mathbf{B} + \mathbf{A} \times \mathbf{C}$$

**3–19** Show that the magnitude of the vector product of vectors **A** and **B** equals the area of the parallelogram with sides $A$ and $B$. Let $A = 3$ and $B = 2$ arbitrary units and (a) $\theta = 90°$, (b) $\theta = 30°$, and (c) $\theta = 150°$.

**Fig. P3-19**

**3–20** Determine the moment of force **F** about $O$ using vector methods.

**Fig. P3-20**

**3–21** Calculate the moment of force **F** about $O$ using vector methods.

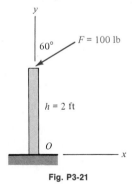

**Fig. P3-21**

**3–22** Determine the moment $\mathbf{M}_A$ of force **F** at $\theta = 20°$ if $F = 10$ lb, $a = 4$ in., and $b = 3$ in.

**Fig. P3-22**

**3–23** Calculate the moment $\mathbf{M}_A$ of force **F** in Fig. P3-22 if $\theta = 120°$, $F = 200$ N, $a = 0.5$ m, and $b = 0.3$ m.

**3–24** Calculate the moment $\mathbf{M}_O$ using $a = b = 3c = 1$ ft, and $\mathbf{F} = (120\mathbf{i} + 70\mathbf{j})$ lb.

**Fig. P3-24**

**3–25** Calculate the moment $\mathbf{M}_A$ of the 10-kN force.

**Fig. P3-25**

**3–26** Determine the moments $\mathbf{M}_A$ and $\mathbf{M}_B$ of the force $\mathbf{F}$.

F = 5 kN

20°

C

L

$y$

L = 2 m

$\theta = 30°$

A $\qquad x$ B

**Fig. P3-26**

**3–27** The boom of the simple crane is supported by a single wire $AB$ in which the tension is 2 kN. Calculate the moment $\mathbf{M}_O$ caused by the resultant force of the tension and the weight of the block at $A$ if the mass of the block is 101.9 kg.

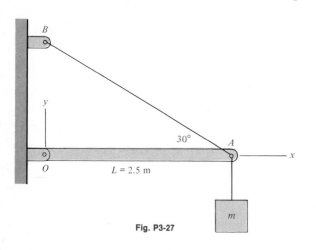

B

$y$

30°

A

O $\qquad x$

L = 2.5 m

$m$

**Fig. P3-27**

**3–28** Use Varignon's theorem to determine the moment sum of $\mathbf{F}$ and $\mathbf{P}$ about $A$ and about $B$. $F = 800$ N, $P = 500$ N (vertical), $L = 3$ m, $\theta = 70°$, and $\phi = 30°$.

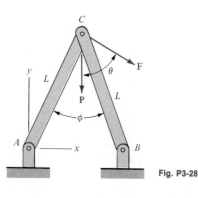

C

$y$ L

$\theta$

F

P

L

$\phi$

A $\qquad x$ B

**Fig. P3-28**

**3–29** The street lamp $A$ has a mass of 10 kg. The force of the wind is $\mathbf{F} = 300$ N $\mathbf{i}$. Calculate the moment $\mathbf{M}_O$ of the resultant of the force of the wind and the weight of the lamp. $d = 2$ m and $h = 5$ m.

$y$

F

A

W

$d$

$h$

O $\qquad x$

**Fig. P3-29**

**3–30** Assume that the force of wind on lamp $A$ in Fig. P3-29 is $\mathbf{F} = -40$ lb $\mathbf{k}$. $\mathbf{W} = -50$ lb $\mathbf{j}$. Determine the moment $\mathbf{M}_O$ of the resultant of the two forces. $d = 7$ ft and $h = 15$ ft.

**3–31** A crate is lifted by a fork lift. Calculate the moments $\mathbf{M}_A$ for the two extreme positions $\mathbf{r}_1 = (3\mathbf{i} + 2\mathbf{j})$ ft and $\mathbf{r}_2 = (2\mathbf{i} + 7\mathbf{j})$ ft.

$y$

$\mathbf{r}_2$

W

$\mathbf{r}_1$

W = −500 j lb

A

Front wheel of fork lift

**Fig. P3-31**

**3–32** In a road simulator for cars a force $\mathbf{F} = (-300\mathbf{i} + 1000\mathbf{j} - 250\mathbf{k})$ lb is applied to a wheel at point $A$, which has coordinates $(-1, -1.3, 0.4)$ ft. Determine the moment $\mathbf{M}_O$ of the force.

**Fig. P3-32**

**3–33** The wheel of a car in a test is supported at point $O$ in Fig. P3-32. Calculate $M_x$, $M_y$, and $M_z$, the components of $\mathbf{M}_O$, using $\mathbf{F} = (4\mathbf{j} + 6\mathbf{k})$ kN acting at point $A$, which has coordinates $(-0.3, -0.4, 0)$ m.

**3–34** In a laboratory test for the landing gear of a small aircraft a force $\mathbf{F} = (300\mathbf{i} + 200\mathbf{j} - 300\mathbf{k})$ lb is applied at point $A$, which has coordinates $(1, -4, 0)$ ft. Calculate the moment $\mathbf{M}_O$ of the force.

**Fig. P3-34**

**3–35** A force $\mathbf{F} = (1.5\mathbf{i} + 20\mathbf{j} - 0.8\mathbf{k})$ kN is applied at point $A$ of coordinates $(0.3, -1, 0.25)$ m in Fig. P3-34. Determine the scalar components, $M_x$, $M_y$, and $M_z$, of the moment $\mathbf{M}_O$ of the force $\mathbf{F}$.

**3–36** A jet aircraft is flying horizontally when the left engine fails. Determine the moment $\mathbf{M}_C$ of the thrust $\mathbf{T}$ of the right engine. $\mathbf{T}$ is applied at point $A$, which has coordinates $(18, -4, -2)$ ft.

**Fig. P3-36**

**3–37** A mast is supported by several guy wires (2 are shown). (a) Calculate the moment $\mathbf{M}_O$ of a 5-kN tension in wire $AB$. The coordinates of $A$ and $B$ are $(10, 0, 5)$ m, and $(0, 12, 0)$ m, respectively. (b) Determine the perpendicular distance between point $O$ and the wire $AB$.

**Fig. P3-37**

**3–38** Bolt $A$ is located at $(0.3, 1.2, -0.1)$ m, and point $B$ of the wrench is at $(0.7, 1.3, 0.2)$ m. Determine the moments at points $A$ and $O$ caused by the force $\mathbf{F} = (80\mathbf{i} - 70\mathbf{k})$ N.

**Fig. P3-38**

**3–39** A force $\mathbf{F} = (100\mathbf{i} - 40\mathbf{j} + 20\mathbf{k})$ N is acting at the middle of handle $BC$ (0.1 m from $B$) of the crank. Determine the moments $\mathbf{M}_O$, $\mathbf{M}_A$, and $\mathbf{M}_B$ of the force when the crank is in the $yz$ plane.

**Fig. P3-39**

**3–40** The following forces are acting on the highway sign: $\mathbf{W} = -140\mathbf{j}$ lb and $\mathbf{F} = -200\mathbf{k}$ lb, both acting at point $A$ of coordinates (5, 18, 0) ft. Determine the scalar components $M_x$, $M_y$, and $M_z$, of $\mathbf{M}_O$.

**Fig. P3-40**

**3–41** The coordinate system is drawn with $y$ in the vertical direction and with the wing spars (supporting brackets) on the $z$ axis. Assume that the engine weight $\mathbf{W} = -6000\mathbf{j}$ lb and thrust $\mathbf{T} = (400\mathbf{i} + 700\mathbf{j} + 15,000\mathbf{k})$ lb are acting at point $A$ of coordinates $(12, -5, -3)$ ft. Determine the scalar components of the resultant moment, $M_x$, $M_y$, and $M_z$, caused by the two forces.

**Fig. P3-41**

**3–42** The test fixture shown in the photograph is attached to the right front wheel axle $A$ of a car. $A$ has coordinates (5, $-0.3$, 3) ft with respect to point $O$ at the center of the car. Three forces are applied to the axle $A$: $\mathbf{F}_1 = 1500$ lb $\mathbf{j}$, $\mathbf{F}_2 = -700$ lb $\mathbf{i}$, and $\mathbf{F}_3 = -1000$ lb $\mathbf{k}$, as illustrated in Fig. P3-42b. Determine the scalar components, $M_x$, $M_y$, and $M_z$, of the resultant moment of these simultaneous forces. $x$ is the longitudinal centerline of the car.

(a)
Test fixture attached to a car axle.
(Courtesy MTS Systems Corporation, Minneapolis, Minnesota)

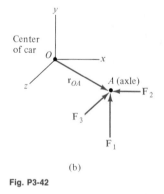

(b)

**Fig. P3-42**

**3–43** Assume that in Prob. 3–42 each force can vary randomly from zero to the given value. Determine the largest possible scalar component of the resultant moment about the $x$ axis of the car as the three forces vary randomly.

Various products of three vectors may be considered. The *vector triple products* are mathematically valid because $\mathbf{A} \times \mathbf{B}$ is a vector, so $(\mathbf{A} \times \mathbf{B}) \times \mathbf{C}$ can be calculated. However, the associative property applies to this product only for unit vectors, and the result is trivial,

$$(\mathbf{i} \times \mathbf{j}) \times \mathbf{k} = \mathbf{i} \times (\mathbf{j} \times \mathbf{k}) = 0 \qquad \boxed{3\text{-}17}$$

since

$$\mathbf{i} \times \mathbf{j} = \mathbf{k} \quad \text{and} \quad \mathbf{j} \times \mathbf{k} = \mathbf{i}$$

In other cases,

$$(\mathbf{A} \times \mathbf{B}) \times \mathbf{C} \neq \mathbf{A} \times (\mathbf{B} \times \mathbf{C}) \qquad \boxed{3\text{-}18}$$

The vector triple product is used in dynamics analysis.

A useful product of three vectors in the area of statics is the *mixed triple product*, which is the scalar product $\mathbf{A}$ with the vector product of $\mathbf{B}$ and $\mathbf{C}$,

$$\mathbf{A} \cdot (\mathbf{B} \times \mathbf{C}) \qquad \boxed{3\text{-}19}$$

The result of this triple product is a scalar.

The mixed triple product is calculated by first performing the vector product part of it,

$$\mathbf{A} \cdot (\mathbf{B} \times \mathbf{C}) = \mathbf{A} \cdot \mathbf{D} \qquad \boxed{3\text{-}20}$$

where $\mathbf{D}$ is $\mathbf{B} \times \mathbf{C}$. In terms of rectangular components,

$$\mathbf{A} \cdot (\mathbf{B} \times \mathbf{C}) = A_x D_x + A_y D_y + A_z D_z$$

On the basis of Eq. 3-10, the complete details are

$$\mathbf{A} \cdot (\mathbf{B} \times \mathbf{C}) = A_x(B_y C_z - B_z C_y) + A_y(B_z C_x - B_x C_z) \qquad \boxed{3\text{-}21}$$
$$+ A_z(B_x C_y - B_y C_x)$$

Equation 3-21 is the same form as the expansion of the determinant,

$$\mathbf{A} \cdot (\mathbf{B} \times \mathbf{C}) = \begin{vmatrix} A_x & A_y & A_z \\ B_x & B_y & B_z \\ C_x & C_y & C_z \end{vmatrix} \qquad \boxed{3\text{-}22}$$

which is the easier form to remember.

## MOMENT OF A FORCE ABOUT A LINE

3-6

Commonly in practice, a body is constrained to rotate about a given axis regardless of the directions of forces acting on it. If the resultant of the forces acts in a plane that is perpendicular to the axis, the concept of moment about a point can be used directly. The restriction on using this simple approach is explained with the aid of Fig. 3-19. A force $\mathbf{F}$ tends to cause rotation of a rigid body about the axis $l$. The axis may be fixed

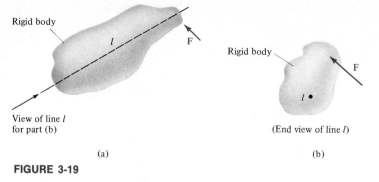

Rigid body

*l*

F

View of line *l*
for part (b)

(a)

Rigid body

F

*l* •

(End view of line *l*)

(b)

**FIGURE 3-19**

Moment of a force about a line

or moving. Imagine that the situation in Fig. 3-19a is viewed from one end of the line *l* so that the line appears as a point (Fig. 3-19b). In this end view the force may or may not appear in its full length. The concept of moment about a point is valid and easy to use for any line *l* if **F** is seen in its full length in Fig. 3-19b. The situation is more complicated if **F** is not in the plane of the paper in Fig. 3-19b. This is discussed in detail as follows.

By definition, the moment of an arbitrarily oriented force about a line is the projection, on the line, of the force's moment about any point on the same line. Consider Fig. 3-20 for a more formal description of this statement. **F** is an arbitrary force whose line of action does not intersect the line of interest *l*. $\mathbf{M}_O$ is the moment of **F** about an arbitrary point *O* on the line *l*. $\mathbf{M}_O$ is perpendicular to the parallelogram *OABC* according to the vector product of **r** and **F**. $\mathbf{M}_l$ is the projection of $\mathbf{M}_O$ on the line *l*. This projection is the moment of **F** about *l*. There may be other components of $\mathbf{M}_O$, but these components do not have a moment about the line *l*, only about point *O*. For example, if $\mathbf{M}_O$ is resolved into rectangular components $\mathbf{M}_l$ and $\mathbf{M}'$, the latter would tend to rotate *l* itself, not have a moment *about l*. $M_l$ alone is what tends to cause rotation about *l*.

The mixed triple product is a convenient mathematical tool for calculating the moment of a force about a line. Using **n**, the unit vector along the line *l* in Fig. 3-20, the magnitude of the projection of $\mathbf{M}_O$ on the line *l* is

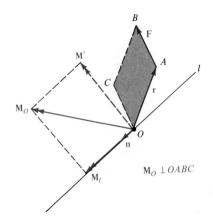

$B$

F

$M'$

$A$

$C$

*l*

r

$M_O$

$O$

n

$M_O \perp OABC$

$M_l$

**FIGURE 3-20**

Moment $\mathbf{M}_O$ of force **F** about line *l*

$$\boxed{M_l = \mathbf{n} \cdot \mathbf{M}_O = \mathbf{n} \cdot (\mathbf{r} \times \mathbf{F})} \qquad \boxed{\text{3-23}}$$

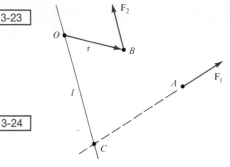

This can be written as a determinant as in Eq. 3-22,

$$M_l = \begin{vmatrix} n_x & n_y & n_z \\ r_x & r_y & r_z \\ F_x & F_y & F_z \end{vmatrix} \qquad \boxed{\text{3-24}}$$

**FIGURE 3-21**

A force has no moment about an intersecting line or one parallel to its line of action

where  $n_x, n_y, n_z$ = rectangular components of $\mathbf{n}$ (also called *direction cosines of* $\mathbf{n}$)

$r_x, r_y, r_z$ = components of position vector from $O$ to point $A$ where $\mathbf{F}$ is applied

$F_x, F_y, F_z$ = rectangular components of force $\mathbf{F}$

## Special Cases

**1.** The moment about a line is zero when the line of action of the force intersects the line of interest $l$ as force $\mathbf{F}_1$ does in Fig. 3-21. This can be appreciated by inspection since the moment arm is zero in this case.

**2.** The moment about a line is zero when the line of action of the force is parallel to the line of interest $l$ as is the case for $\mathbf{F}_2$ in Fig. 3-21. In this case the moment $\mathbf{M}_O$ is perpendicular to the plane of $l$, $r$, and $\mathbf{F}_2$, and the projection of $\mathbf{M}_O$ on $l$ is a point of zero length.

The various possibilities of moments about lines are illustrated for a door in Fig. 3-22.

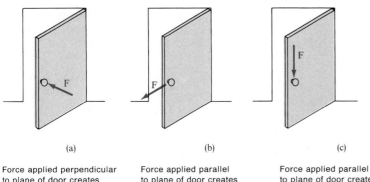

|     (a)     |     (b)     |     (c)     |

Force applied perpendicular to plane of door creates large moment about line through hinges because of large moment arm; the door turns easily

Force applied parallel to plane of door creates zero moment about line through hinges because the force has no moment arm; the door does not turn

Force applied parallel to plane of door creates zero moment about line through hinges because the force is parallel to that line; the door does not turn

**FIGURE 3-22**

Illustration of moments about lines

# EXAMPLE 3-7

The wheel assembly tester shown in Fig. 3-23a allows changing the position of the wheel in many ways with respect to the large driving drum as in part b of the figure. The position of the wheel affects the force $\mathbf{F}$ applied to the supporting structure (or the vehicle in real life) because of dimensional changes and tire sliding mixed with rolling. Moments about two lines are of interest: $x$ and $y$ are the transverse and longitudinal center lines of the vehicle (Fig. 3-23c). Determine $M_x$ and $M_y$ for $\mathbf{F} = (-300\mathbf{j} + 1000\mathbf{k})$ lb applied at $A$: $(-3, 6, -2)$ ft.

**FIGURE 3-23**

Test equipment for automotive wheel assemblies. (Photographs courtesy MTS Systems Corporation, Minneapolis, Minnesota.)

## SOLUTION

Using Eq. 3-14,

$$\mathbf{M}_O = \mathbf{r}_{OA} \times \mathbf{F}$$

$$\mathbf{r}_{OA} = (-3\mathbf{i} + 6\mathbf{j} - 2\mathbf{k}) \text{ ft}$$

$$\mathbf{F} = (-300\mathbf{j} + 1000\mathbf{k}) \text{ lb}$$

$$\mathbf{M}_O = \mathbf{r}_{OA} \times \mathbf{F} = \begin{vmatrix} \mathbf{i} & \mathbf{j} & \mathbf{k} \\ -3 \text{ ft} & 6 \text{ ft} & -2 \text{ ft} \\ 0 & -300 \text{ lb} & 1000 \text{ lb} \end{vmatrix}$$

$$= (5400\mathbf{i} + 3000\mathbf{j} + 900\mathbf{k}) \text{ lb·ft}$$

From Eq. 3-23,

$$M_x = \mathbf{i} \cdot \mathbf{M}_O = 5400 \text{ lb·ft}$$

$$M_y = \mathbf{j} \cdot \mathbf{M}_O = 3000 \text{ lb·ft}$$

(a)

(b)

(c)

EXAMPLE 3-8

In bending and torsion tests of pipe systems a hydraulic actuator applies the desired force **F** through an end plate as shown in Fig. 3-24a. **F** = 500**i** kN, and it is applied at $A$: (4, 1, 2) m with respect to the chosen coordinate system. Determine the moment of **F** about a line $l$ defined by two points, $B$: (0, 2, 0) m, $C$: (1, 4, 0.5) m. The situation is sketched in Fig. 3-24b.

SOLUTION

From Eq. 3-23,

$$M_{BC} = \mathbf{n}_{BC} \cdot (\mathbf{r} \times \mathbf{F})$$

From Eq. 2-22,

$$\mathbf{n}_{BC} = \frac{(d_x\mathbf{i} + d_y\mathbf{j} + d_z\mathbf{k})}{d_{BC}}$$

$$d_{BC} = [(x_C - x_B)\mathbf{i} + (y_C - y_B)\mathbf{j} + (z_C - z_B)\mathbf{k}] \text{ m}$$
$$= [(1 - 0)\mathbf{i} + (4 - 2)\mathbf{j} + (0.5 - 0)\mathbf{k}] \text{ m}$$

$$\mathbf{n}_{BC} = \frac{(\mathbf{i} + 2\mathbf{j} + 0.5\mathbf{k})}{[(1)^2 + (2)^2 + (0.5)^2]^{1/2}}$$

$$= 0.436\mathbf{i} + 0.873\mathbf{j} + 0.218\mathbf{k}$$

(a)

The criterion for choosing vector **r** is that it join any point on the line about which the moment is taken and any point on the line of action of the force. Choosing points $B$ and $A$ on these lines, respectively,

$$\mathbf{r}_{BA} = (x_A - x_B)\mathbf{i} + (y_A - y_B)\mathbf{j} + (z_A - z_B)\mathbf{k}$$
$$= (4 \text{ m} - 0)\mathbf{i} + (1 \text{ m} - 2 \text{ m})\mathbf{j} + (2 \text{ m} - 0)\mathbf{k}$$
$$= (4\mathbf{i} - \mathbf{j} + 2\mathbf{k}) \text{ m}$$

$$\mathbf{F} = 500\mathbf{i} \text{ kN}$$

$$M_{BC} = \mathbf{n}_{BC} \cdot (\mathbf{r}_{BA} \times \mathbf{F}) = \begin{vmatrix} 0.436 & 0.873 & 0.218 \\ 4 \text{ m} & -1 \text{ m} & 2 \text{ m} \\ 500 \text{ kN} & 0 & 0 \end{vmatrix}$$

$$= (873 + 109) \text{ kN·m} = 982 \text{ kN·m}$$

(b)

**FIGURE 3-24**

Test setup of a large pipe. (Photograph courtesy MTS Systems Corporation, Minneapolis, Minnesota.)

Note that if the unit vector going from $C$ to $B$ ($\mathbf{n}_{CB} = -\mathbf{n}_{BC}$) had been chosen, an answer of $-982$ kN·m would have been obtained. However, both answers mean the same, remembering that by the right-hand rule the thumb is pointed along the *positive* sense of the axis about which the moment is taken. In the first case, pointing the thumb along the line going from $B$ to $C$, the curl of the fingers gives the sense of rotation about this line (since the result was positive). In the second case, with the thumb pointing from $C$ to $B$, the fingers curl in the opposite direction, but the negative sign of the answer means that this is opposite to the direction of rotation, hence both results yield the same sense of rotation about the line $BC$.

EXAMPLE 3-8    93

**3-44**  Calculate the vector triple product $\mathbf{V} = (\mathbf{A} \times \mathbf{B}) \times \mathbf{C}$ of $\mathbf{A} = 2a\mathbf{i} + 4b\mathbf{j} - 3c\mathbf{k}$, $\mathbf{B} = -b\mathbf{j} + 5c\mathbf{k}$, and $\mathbf{C} = 6a\mathbf{i} - 8b\mathbf{j}$.

**3-45**  Calculate the vector triple product $\mathbf{V} = (\mathbf{A} \times \mathbf{B}) \times \mathbf{C}$ of $\mathbf{A} = 10\mathbf{i} - 5\mathbf{j} - 12\mathbf{k}$, $\mathbf{B} = -\mathbf{i} + 10\mathbf{j}$, and $\mathbf{C} = 7\mathbf{k}$. Show that $(\mathbf{A} \times \mathbf{B}) \times \mathbf{C} \neq \mathbf{A} \times (\mathbf{B} \times \mathbf{C})$.

**3-46**  Calculate $\mathbf{A} \cdot (\mathbf{B} \times \mathbf{C})$ if $\mathbf{A} = \mathbf{i} + \mathbf{j} + \mathbf{k}$, $\mathbf{B} = 3\mathbf{i} + 5\mathbf{k}$, and $\mathbf{C} = -4\mathbf{i} - 3\mathbf{j}$.

**3-47**  Calculate $\mathbf{A} \cdot (\mathbf{B} \times \mathbf{C})$ if $\mathbf{A} = -5\mathbf{i} - 4\mathbf{j} - 10\mathbf{k}$, $\mathbf{B} = -12\mathbf{i} + 10\mathbf{j} + 6\mathbf{k}$, and $\mathbf{C} = 20\mathbf{i} - 15\mathbf{j} + 30\mathbf{k}$.

**3-48**  The rectangular plate is hinged at the $x$ axis and lies in the $xz$ plane. Determine the moments $M_x$, $M_y$, and $M_z$ of the force $\mathbf{F} = (-50\mathbf{i} + 180\mathbf{j} + 30\mathbf{k})$ lb, using the mixed triple product.

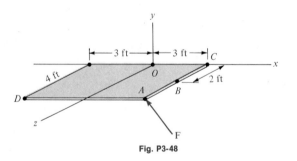

**Fig. P3-48**

**3-49**  What is the smallest force $\mathbf{F}$ that can be applied at point $B$ in Fig. P3-48 to have the same moment $M_x$ that the force $\mathbf{F}$ has in Prob. 3–48?

**3-50**  Determine the moment of the force $\mathbf{F}$ in Prob. 3–48 about the diagonal of the plate, $CD$.

**3-51**  The tension in guy wire $AC$ is 200 lb. Determine the moment of this tension at $A$ about the line $BD$.

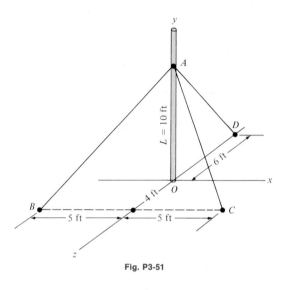

**Fig. P3-51**

**3-52**  The coordinates of point $A$ of the adjustable lamp are $(0.5, 0.7, 0.2)$ m. Determine the moments $M_x$, $M_y$, and $M_z$ of the weight $\mathbf{W}$ of the lamp about the $x$, $y$, and $z$ axes, respectively, if the lamp has a mass of 1.5 kg. Use a vector method of solution and check the results using scalar methods.

**Fig. P3-52**

**3–53** Force $\mathbf{F} = (-20\mathbf{i} - 30\mathbf{j} + 200\mathbf{k})$ N is applied to the rigid linkage $OACB$, which is hinged at $O$ and $B$. Determine the moment $M_x$ of $\mathbf{F}$ if $\theta$ in the $yz$ plane is 30°.

Fig. P3-53

Fig. P3-55

**3–54** Determine the moment $M_l$ of a force $\mathbf{F} = (100\mathbf{i} + 300\mathbf{j} - 600\mathbf{k})$ N in Fig. P3-53. Line $OA$ makes an angle $\theta = 20°$ with the $y$ axis. The line of interest $l$ is through points $O$ and $D$; the latter has coordinates $(1, 0, 0.5)$ m.

**3–55** The force acting on the crank arm is $\mathbf{F} = (120\mathbf{i} - 30\mathbf{j} - 20\mathbf{k})$ N. Calculate the moments of the force about the $x$, $y$, and $z$ axes, assuming that the crank is in the $yz$ plane.

**3–56** Define a line $l$ through points $O$ and $D$ in Fig. P3-55; $D$ has coordinates $(1, 1, 0)$ m. Determine the moment $M_l$ of $\mathbf{F}$ if the crank is in the $yz$ plane.

**3–57** Force $\mathbf{F} = (400\mathbf{i} + 1000\mathbf{j} - 500\mathbf{k})$ lb is acting at point $A$ of coordinates $(1.2, -1, 0)$ ft on the wheel. Determine the moment $M_l$ of the force about line $l$ which passes through $O$ and point $B$ of coordinates $(-0.5, 0, 0.2)$ ft.

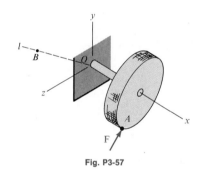

Fig. P3-57

## MOMENT OF A COUPLE

A special case is defined when a moment is caused not by a single force but by two forces that are equal in magnitude, parallel in lines of action, and opposite in direction. Such a pair of forces is called a *couple*. An example is the moment applied to the test specimen by the elevated-temperature extensometer (sensitive device for the measurement of deformations) shown in Fig. 3-25a. On the basis of intuition, the weight of the device would cause it to rotate clockwise, and this causes tension at section $A$ on the specimen and compression at $B$ (Fig. 3-25b). These forces are equal in magnitude (this is proved in Ch. 4), parallel, and opposite in direction. Thus, they have no net force on the specimen as a whole, but slightly tend to rotate it in the clockwise direction (at high temperatures even this small moment may be important).

The magnitude of the moment of a couple is given by

$$M = Fd \qquad \boxed{3\text{-}25}$$

where $d$ is the distance between the lines of action of the forces of magnitude $F$. To prove this, sum the moments about any of the three points in Fig. 3-25b:

For $A$,   $M_A = Fd$   (only the lower force has a moment)

For $B$,   $M_B = Fd$   (only the upper force has a moment)

For $C$,   $M_C = 2\left(F\dfrac{d}{2}\right)$

$\qquad\qquad = Fd$   (both forces have moments but the moment arm is $d/2$)

The distinction between moments caused by a single force or a couple is illustrated as follows. The single force $\mathbf{F}$ causes a moment of magnitude $Fd/2$ on the steering wheel of diameter $d$ in Fig. 3-26. It also acts as a net force $\mathbf{F}$ to the left on the steering column assembly. Exactly the same turning effect about the steering column is obtained by applying a pair of forces of magnitude $F/2$ at a distance $d$ as shown. In this case the net force on the steering column is zero, and the turning effect is not dependent on where the forces are applied on the wheel.

**FIGURE 3-25**

Extensometer attached to a metal specimen. (Photograph courtesy MTS Systems Corporation, Minneapolis, Minnesota.)

(a)

(b)

(a)

(b)

**FIGURE 3-26**

Moment of a single force **F** and

moment of a couple, **F**/2 and $-\dfrac{\mathbf{F}}{2}$.

(Photograph courtesy MTS Systems Corporation, Minneapolis, Minnesota.)

For a simple vector analysis of a couple, consider Fig. 3-27. Taking moments about the arbitrary point $O$, with all points and vectors in the $xy$ plane,

$$\mathbf{M}_O = \mathbf{r}_A \times \mathbf{F} + \mathbf{r}_B \times (-\mathbf{F}) = (\mathbf{r}_A - \mathbf{r}_B) \times \mathbf{F} \qquad \boxed{\text{3-26}}$$
$$= \mathbf{d} \times \mathbf{F} \qquad (\text{since } \mathbf{r}_A = \mathbf{r}_B + \mathbf{d})$$

For the given arrangement,

$$\mathbf{M}_O = \begin{vmatrix} \mathbf{i} & \mathbf{j} & \mathbf{k} \\ 0 & d & 0 \\ F & 0 & 0 \end{vmatrix} = -Fd\mathbf{k}$$

The sign is correct according to the right-hand rule. Note that the cross product should involve a force lying at the head of the position vector.

The same result is obtained if three-dimensional components are used for the forces and position vectors. An important aspect of this analysis is that the position of the chosen reference point $O$ does not affect the moment of the couple, which is thus a *free vector* and can be applied anywhere in the given plane.

It should be noted that a single vector **M** representing a couple (sometimes called a *couple vector*) is perpendicular to the plane of the two forces as in Fig. 3-28a. In general, the vector **M** may have three vector components, $\mathbf{M}_x$, $\mathbf{M}_y$, and $\mathbf{M}_z$. The turning effect of **M** or any of its components is established using the right-hand rule. For example, $\mathbf{M}_z$ in Fig. 3-28a means a counterclockwise rotation when viewed from the positive $z$ direction. Care must be taken in each problem so that the moment vector is not confused with any other vector quantity. Sometimes it is helpful to use a curled arrow in the plane of the two forces instead of the couple vector **M**, as in Fig. 3-28b.

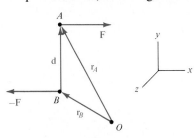

**FIGURE 3-27**

Vector analysis of a couple

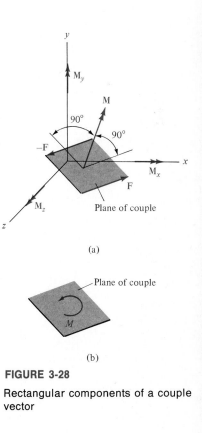

(a)

(b)

**FIGURE 3-28**

Rectangular components of a couple vector

## EQUIVALENT COUPLES

For considerations of space requirements and strengths of members, it may be necessary to decide where and how to apply a given couple to a member. Evaluation of a few possible alternatives shows the principles involved and the most reasonable directions in design.

Figure 3-29 shows three of the infinite number of ways of applying

**FIGURE 3-29**

Equivalent couples

**FIGURE 3-30**

Antiyaw (antirotation) mechanism of a tuned mass damper (see Sec. 1-7). (Courtesy MTS Systems Corporation, Minneapolis, Minnesota.)

(a) Couple $C$ can be moved anywhere in plane $ABCD$

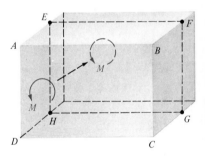

(b) Couple $C$ can be moved from plane $ABCD$ to plane $EFGH$; it cannot be moved to planes $AFGD$, $EFCD$, etc.

**FIGURE 3-31**

Equivalence of couples in the same plane or in parallel planes

the same moment of a couple to a block. These are *equivalent* in theory since they have the same turning effect, and the net force is also the same, zero in each case. Of course, they are different to the extent that those in Fig. 3-29a and c may require considerable space for a mechanism, while that in part b of the figure requires much stronger linkages to apply the forces. A compromise is often made as shown in Fig. 3-30, where bars $A$ and $B$ apply the force couple to the large block $C$.

The generalization about the equivalence of couples is based on the previous vector analysis associated with Fig. 3-27, and is summarized with respect to Fig. 3-31:

**1.** A couple can be moved anywhere in its own plane without changing the moment acting on a given rigid body (Fig. 3-31a).

**2.** A couple can be moved to any other plane that is parallel to its original plane in the same rigid body (Fig. 3-31b). Conversely: a couple cannot be moved to change the position of its plane without changing its effect on the rigid body.

## COMPONENTS AND ADDITION OF COUPLES

The vector representing a couple can be resolved into components as can be done for any other vector. For example, the antiyaw mechanism shown in Fig. 3-30 applies a moment of variable direction to the block as illustrated in Fig. 3-32. The moment **M** is always perpendicular to the plane of the two linkage bars represented by $BC$ (this is based on the concept of two-force members which is proved in Ch. 4). The angle $\theta$ varies as the block moves left and right. The block can move only horizontally, so the

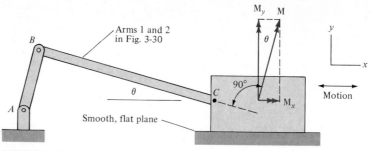

**FIGURE 3-32**

Rectangular components of a moment acting on a tuned mass damper (see Sec. 1-7)

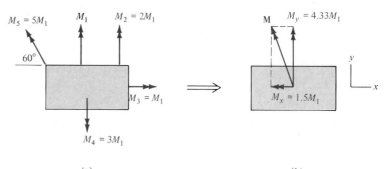

(a)                                                        (b)

**FIGURE 3-33**

Addition of couples by vector addition. Moment vectors $\mathbf{M}_1$, $\mathbf{M}_2$, . . . are all in the $xy$ plane

effective antiyaw moment is $\mathbf{M}_y$ in Fig. 3-32. Of course, $\mathbf{M}_x$ is present, tending to cause rotation about the $x$ axis. This moment is felt by the horizontal plane on which the block glides ($M_x$ is small and does not tip the block).

The addition of couples obeys the rule of vector addition. The moment vectors of couples in parallel planes can be moved to have a single line of action, so such couples can be algebraically added. Couples that are in different, nonparallel planes can be resolved into rectangular components to facilitate their addition. The simple procedure is illustrated in Fig. 3-33. In general, the resultant $\mathbf{M}$ of couples $\mathbf{M}_1$, $\mathbf{M}_2$, . . . is obtained by vector addition,

$$\boxed{\mathbf{M} = \mathbf{M}_1 + \mathbf{M}_2 + \ldots} \qquad \boxed{3\text{-}27}$$

Any number of couples can be reduced to a single resultant $\mathbf{M}$ that may have up to three rectangular vector components, $\mathbf{M}_x$, $\mathbf{M}_y$, and $\mathbf{M}_z$.

## 3-10  TRANSFORMATION OF A FORCE INTO A FORCE–COUPLE SYSTEM

Consider the use of a torque wrench for bolts arranged on a circle as in Fig. 3-16. The general situation is depicted in Fig. 3-34a. The force $\mathbf{F}$ may have distinct effects at two points of interest, at a bolt $A$, and at center

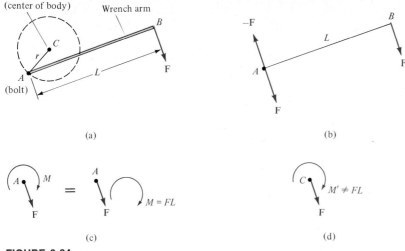

(a)

(b)

(c)

(d)

**FIGURE 3-34**

Transformation of a force into a force–couple system

$C$ of the whole rigid body. These effects are most clear if the force is transformed into an equivalent system consisting of a couple and a force acting at the point of interest.

To deal with point $A$ first, add two collinear but opposite forces of magnitude $F$ at $A$. This does not change the situation in any way, regardless of the orientation of the line of action of these forces, but it is advantageous to put the forces parallel to the original force at $B$ as in Fig. 3-34b. Here the upward force at $A$ and the downward force at $B$ make a couple of magnitude $FL$. Thus, the single force at $B$ can be replaced by an identical force acting at $A$ and a couple $\mathbf{M}$ which can be moved anywhere in the plane of $\mathbf{F}$ and $L$ (Fig. 3-34c).

A similar procedure can be applied in going from point $A$ to $C$ resulting in an identical force $\mathbf{F}$ but a different moment $\mathbf{M}'$ (Fig. 3-34d). Several of these interesting aspects of moments can be demonstrated using common T-shaped and L-shaped tire tools on the wheel-lug bolts of a jacked-up car. In such experiments the small motions of the wheel should be noted as the tool is used to loosen or tighten the bolts.

The manipulations of moments shown in Fig. 3-34 can be stated in vector notation as follows.

$$\mathbf{M}_A = \mathbf{r}_{AB} \times \mathbf{F} \quad \text{and} \quad \mathbf{M}_C = \mathbf{r}_{CB} \times \mathbf{F} = \mathbf{M}_A + \mathbf{r}_{CA} \times \mathbf{F} \quad \boxed{3\text{-}28}$$

The force–couple systems described here are such that the force and moment vectors are mutually perpendicular in each case. For example, the moment vector $\mathbf{M}$ is into the paper in Fig. 3-34c, and $\mathbf{M}'$ is into the paper in part d of the figure.

The procedures described above can be applied in the reverse order to replace a force–couple system by a single force that has the same total effect on a rigid body. The idea is to move the force $\mathbf{F}$ in its plane perpendicular to the moment vector $\mathbf{M}$ until it causes the same moment as $\mathbf{M}$ would cause. For example, starting with Fig. 3-34c, $\mathbf{F}$ is moved to the right a distance $L$, parallel to itself. At this stage $\mathbf{F}$ alone must be shown as in Fig. 3-34a.

## EXAMPLE 3-9

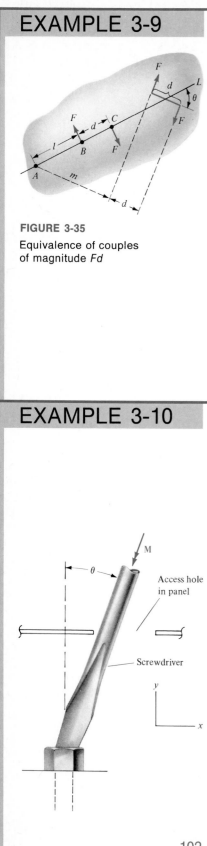

**FIGURE 3-35**

Equivalence of couples
of magnitude *Fd*

Prove that a couple of magnitude $Fd$ acting on a rigid body can be moved anywhere in the same plane along line $L$, even to make the angle $\theta$ with $L$ as in Fig. 3-35, and its moment about point $A$ remains the same if $F$ and $d$ are constant.

### SOLUTION

With the forces perpendicular to $L$ and taking clockwise as positive, the net moment about $A$ is $M_A = F(l + d) - Fl = Fd$, independent of $l$. For the position with the angle $\theta$, $M_A = F(m + d) - Fm = Fd$, independent of $m$ or $\theta$.

Note that in this relatively simple problem the scalar method of calculating moments is the easiest and allows visualizing that the results are correct. Using vector products is slightly more complicated in this case but provides the same results. This is left as an exercise for the reader.

## EXAMPLE 3-10

In a machine the access hole for tightening a bolt is misaligned because of a mistake in fabrication (Fig. 3-36). The moment on the handle of the screwdriver is applied as shown. Determine the effective moment that turns the bolt as a function of $\theta$. At what angle is the effective moment 90% of the applied moment?

### SOLUTION

The effective moment is the $M_y$ component of **M**, and

$$M_y = M \cos \theta$$

For the effective moment, $M_y$, to be 90% of $M$,

$$M_y = 0.9M$$

Then

$$M \cos \theta = 0.9M \qquad \text{when } \theta = 25.8°$$

**FIGURE 3-36**

Effective moment $M_y$ for tightening
a bolt

EXAMPLE 3-11

The boom of a crane is modeled in Fig. 3-37a. It is supported at points $A$ and $B$, and an 8-kN force is applied at point $C$ in the direction shown. **(a)** Determine the force and couple at point $B$ equivalent to the 8-kN force, and **(b)** determine the smallest forces acting at points $A$ and $B$ that would create an equivalent couple to that found in part (a).

## SOLUTION

**(a)** First determine the vector components of the 8-kN force,

$$\mathbf{F} = (8 \text{ kN})(\sin 30° \, \mathbf{i} - \cos 30° \, \mathbf{j})$$
$$= (4\mathbf{i} - 6.93\mathbf{j}) \text{ kN}$$

The moment of the 8-kN force about point $B$ is

$$\mathbf{M}_B = \mathbf{r}_{BC} \times \mathbf{F}$$
$$= (3\mathbf{i} + 2\mathbf{j}) \text{ m} \times (4\mathbf{i} - 6.93\mathbf{j}) \text{ kN}$$
$$= -28.8\mathbf{k} \text{ kN·m}$$

The equivalent force and couple at point $B$ are shown in Fig. 3-37b,

$$\mathbf{F} = (4\mathbf{i} - 6.93\mathbf{j}) \text{ kN}$$

$$\mathbf{M} = -28.8\mathbf{k} \text{ kN·m}$$

**(b)** The equivalent couple formed by forces applied at points $A$ and $B$ must have a magnitude of 28.8 kN·m and create a clockwise moment. Since $M = Fd$, the smallest forces would be those applied perpendicular to line $AB$, which is the largest perpendicular distance. The magnitude of the forces is

$$F = \frac{M}{d} = \frac{28.8 \text{ kN·m}}{2\text{m}}$$

$$= 14.4 \text{ kN}$$

In order for the moment of the couple to be clockwise,

$$\mathbf{F}_A = 14.4\mathbf{i} \text{ kN}$$

$$\mathbf{F}_B = -14.4\mathbf{i} \text{ kN}$$

These forces are shown in Fig. 3-37c.

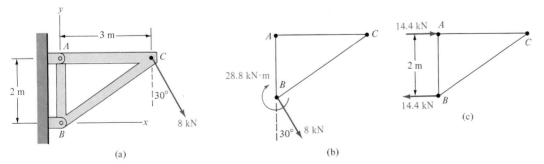

(a)     (b)     (c)

**FIGURE 3-37**

Model of the boom of a crane

EXAMPLE 3-11     103

# PROBLEMS

**3-58** Use vector products to prove that the results obtained in Ex. 3-9 are correct.

**3-59** Two couples are acting on the rectangular block. Determine the resultant moment **M** of these couples.

**Fig. P3-59**

**3-60** Determine the net moment **M** caused by the two couples on the rectangular block. All forces are parallel to the $x$ axis.

**Fig. P3-60**

**3-61** Determine the net moment **M** caused by the two couples on the rectangular block. All forces are parallel to the $y$ axis.

**Fig. P3-61**

**3-62** Determine the net moment **M** caused by the two couples on the rectangular block. All forces are parallel to the $z$ axis.

**Fig. P3-62**

**3-63** A moment $\mathbf{M} = M_x\mathbf{i} + M_y\mathbf{j} + M_z\mathbf{k}$ is acting on the rectangular block. Determine the directions and magnitudes of horizontal and vertical forces acting at points $A$, $B$, $C$ to have the same effect on the block as **M** does. Is your set of forces the only set possible?

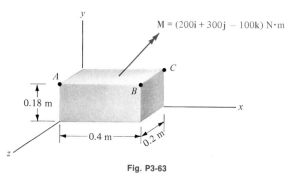

**Fig. P3-63**

**3-64** In an assembly operation five lug nuts on a car's wheel are tightened simultaneously. The same torque $M$ is applied to each nut. Calculate the net moment $\mathbf{M}_O$ acting on the wheel if $x$ is perpendicular to the plane of the nuts.

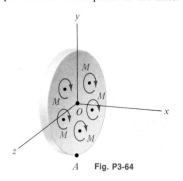

**Fig. P3-64**

**3–65** In Prob. 3–64, $M = 100$ lb·ft. Calculate the magnitude of the force $\mathbf{F} = F\mathbf{k}$ applied at point $A$ that would cause the same moment as $\mathbf{M}_O$. The coordinates of $A$ are $(0, -14$ in., $0)$.

**3–66** In an assembly operation the following moments are applied simultaneously: $\mathbf{M}_A = -200$ N·m $\mathbf{k}$, $\mathbf{M}_B = -250$ N·m $\mathbf{i}$, $\mathbf{M}_C = 150$ N·m $\mathbf{k}$. Determine the net moment $\mathbf{M}_O$ acting on the block. Calculate the magnitude of the smallest force $\mathbf{F}$ that could cause the same moment when applied at a distance of 0.5 m from $O$.

**Fig. P3-66**

**3–67** The shaft $S$ is subjected to a torsion test. The rotary hydraulic actuator $A$ applies a moment of total magnitude $M = 2000$ lb·in. through the bolts $A$ to $H$. Determine the magnitude of the force $\mathbf{P}$ acting on each bolt assuming that the bolts share the total load equally.

(a)

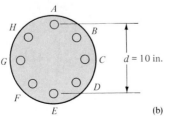

(b)

**Fig. P3-67**

Torsion test setup of a shaft. (Photograph courtesy MTS Systems Corporation, Minneapolis, Minnesota.)

**3–68** The turning effect of the 10-kN force on the bar is to be duplicated by a couple of magnitude $Fd$. What is $F$ if the distance $d$ between the forces must be 0.2 m?

**Fig. P3-68**

**3–69** Assume that $F = 5$ kN and $d = 0.7$ m. The turning effect of $\mathbf{F}$ about point $A$ is to be duplicated by a couple of magnitude $Pl$. Calculate $P$ if $l$ must be 0.1 m.

**Fig. P3-69**

**3–70** Assume that $F = 100$ lb and $AC = 4$ ft. The turning effect of this force about point $A$ is to be represented by a couple of magnitude $Pd$. If $P$ must be 200 lb, determine the distance $d$ of the couple.

**Fig. P3-70**

**3–71** (a) Represent the turning effect of force **P** at supports $A$ and $B$ by a couple. (b) Calculate the magnitude of the horizontal forces acting at $A$ and $B$ that form an equivalent couple.

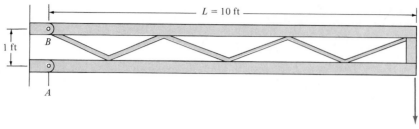

Fig. P3-71

**3–72** Part of a tire testing machine is shown. A rotary hydraulic actuator is considered to apply a moment at $A$ to replace the linear actuator $BC$, which can apply a force of 500 lb. $BC$ and the wheel's spindle are in the same plane. What is the required capacity (moment) of the rotary actuator? $AB = 1.33$ ft and $\theta = 93°$.

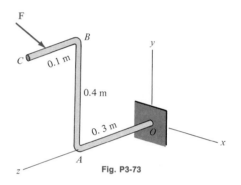

Fig. P3-73

**3–74** Two linear hydraulic actuators of a tire testing machine are visible in the photograph. The maximum force from the upper actuator is $\mathbf{F}_1 = (-3\mathbf{i} + 0.5\mathbf{j} - \mathbf{k})$ kN acting at $A$ of coordinates $(-0.05, 0.8, -0.3)$ m. Determine the equivalent force-couple system at point $O$ of coordinates $(0, 0, 0)$ m.

Motion of simulated road

Fig. P3-72

**3–73** Force $\mathbf{F} = (200\mathbf{i} - 30\mathbf{j} - 20\mathbf{k})$ N. Determine the equivalent force–couple system at point $O$.

Fig. P3-74

**3–75** At one stage in the assembly of a bicycle the following couples and force are applied at the points indicated:

$$\mathbf{M}_A = -30\mathbf{k} \text{ lb·ft at } A: (0, 1, 0.15) \text{ ft}$$

$$\mathbf{M}_B = -35\mathbf{k} \text{ lb·ft at } B: (3, 1, 0.15) \text{ ft}$$

$$\mathbf{M}_C = (-10\mathbf{i} - 25\mathbf{j}) \text{ lb·ft at } C: (1, 3, 0) \text{ ft}$$

$$\mathbf{F}_D = (-15\mathbf{i} - 40\mathbf{j}) \text{ lb at } D: (2.5, 3, 0) \text{ ft}$$

Determine the equivalent force–couple system at $O$ and the resultant moments $M_x$, $M_y$, and $M_z$.

Fig. P3-76

**3–77** Three coplanar forces acting at the top of the pole are parallel to the $xz$ plane. Calculate the equivalent force–couple system at point $O$ of the resultant of the three forces.

Fig. P3-75

Fig. P3-77

**3–76** Determine the equivalent force–couple system at point $A$ of the resultant of the two forces if $F_1 = 500$ N, $F_2 = 300$ N, $L = 0.6$ m, $\theta = 70°$, and $\phi = 40°$.

# SIMPLIFICATION OF MANY FORCES TO A FORCE AND A COUPLE

There have been some instances in the preceding discussions where more or less complex systems of forces were simplified without altering the net effects on rigid bodies. These procedures can be formally stated at this stage to cover all possible systems of any complexity. The following statements are for a three-dimensional system of forces such as the one in Fig. 3-38a.

In an attempt to simplify this complex system, it is necessary to establish a point of reference. The center of mass, point $O$, may be convenient. Some forces may have a moment about $O$, and each of these may be replaced by an identical force at $O$ and the appropriate couple (for example, $\mathbf{M}_5 = \mathbf{r}_5 \times \mathbf{F}_5$). With all the forces made concurrent at $O$, they have a resultant $\mathbf{R}$ in Fig. 3-38b. Also, all the individual moments of the forces in their original positions ($\mathbf{M}_i = \mathbf{r}_i \times \mathbf{F}_i$) can be reduced to a single resultant moment, $\mathbf{M}_R$. Thus, the equivalent force–couple system for Fig. 3-38a is shown schematically in part b of the figure, and the corresponding equations are

$$\mathbf{R} = \sum_{i=1}^{7} \mathbf{F}_i \quad \text{and} \quad \mathbf{M}_R = \sum_{i=1}^{7} (\mathbf{r}_i \times \mathbf{F}_i) \qquad \boxed{3\text{-}29}$$

Note that $\mathbf{M}_R$ is calculated with respect to the arbitrarily chosen point $O$. The result would be different for another reference point $O'$, but it could be calculated equally easily.

In the most general case of a system of forces a resultant force $\mathbf{R}$ and resultant moment $\mathbf{M}_R$ can be determined, but these will not be mutually perpendicular. This fact limits the chances for further simplification of the equivalent system.

## Equivalent Force Systems

Equation 3-29 is the basis for checking the equivalence of two force systems. For all cases involving rigid bodies, *two systems of forces are equivalent if they can be reduced to the same resultant force and resultant moment with respect to any arbitrary point*. In concise notation, a system of forces $\mathbf{F}_1, \mathbf{F}_2, \ldots, \mathbf{F}_i \ldots$ is equivalent to another system consisting of $\mathbf{P}_1, \mathbf{P}_2, \ldots, \mathbf{P}_j \ldots$ if

$$\boxed{\sum_{i=1}^{m} \mathbf{F}_i = \sum_{j=1}^{n} \mathbf{P}_j \quad \text{and} \quad \sum_{i=1}^{m} (\mathbf{r}_i \times \mathbf{F}_i) = \sum_{j=1}^{n} (\mathbf{r}_j \times \mathbf{P}_j)} \qquad \boxed{3\text{-}30}$$

for any point of reference. Here $i$ and $j$ are not unit vectors but subscripts to distinguish individual vectors.

(a)

$$= \qquad \mathbf{M}_R = \Sigma\, \mathbf{M}_i \qquad \mathbf{R} = \Sigma\, \mathbf{F}_i$$

$O$

(b)

**FIGURE 3-38**

Simplification of many forces of a tuned mass damper to a force and a couple. $F_1$ and $F_2$ from hydraulic actuators; $F_3 = $ weight; $F_4 \neq F_5$, track reactions; $F_6$ and $F_7$, antiyaw couple

It is of considerable advantage in analyzing complex problems of rigid bodies that any force system can be reduced to an equivalent system of a resultant force **R** and a resultant moment $\mathbf{M}_R$. There are special cases of these resultant systems that are quite common. They are classified and listed here to allow a convenient overview and comparison of them.

  **1. *Resultant force is zero.*** A nontrivial case is when **R** is zero but there is a resultant moment $\mathbf{M}_R$. Sometimes this $\mathbf{M}_R$ is called the *resultant couple* of the system.

  **2. *Concurrent forces.*** When all the forces act at the same point $O$, they have a resultant **R** but the moment $\mathbf{M}_R$ is zero with respect to $O$.

  **3. *Coplanar forces.*** When all the forces act in the same plane, they can be reduced to a resultant force **R** in the same plane and a resultant moment $\mathbf{M}_R$ about a chosen reference point $O$. $\mathbf{M}_R$ is perpendicular to the plane of the forces as in Fig. 3-39a. The force–couple system could be replaced by the force **R** alone by moving **R** from point $O$ parallel to itself in the $xy$ plane through a distance $d$ until its moment about $O$ equals $\mathbf{M}_R$. In this situation $\mathbf{M}_R$ should not be shown since **R** (solid line in Fig. 3-39b) represents the total effect of $\mathbf{F}_1$, $\mathbf{F}_2$, and $\mathbf{F}_3$ with respect to point $O$.

(a)

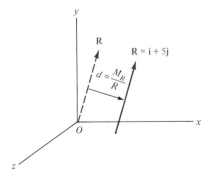

(b)

**FIGURE 3-39**

Reduction of a coplanar force system

**4. Parallel forces.** All such forces have parallel lines of action but their directions may or may not be the same as in Fig. 3-40a. In this example their resultant **R** is also parallel to the $y$ axis (Fig. 3-40b). The resultant moment $\mathbf{M}_R$ about point $O$ is perpendicular to **R**, as shown in Fig. 3-40c.

The magnitude of **R** is obtained by algebraic summation. Its location can be determined in several ways without introducing any new concepts. Essentially, it amounts to writing the same moment in two different ways, with one of them involving an unknown coordinate, and then equating the two expressions. This is illustrated with finding coordinate $x$ of **R** in Fig. 3-40b. This coordinate is the moment arm for **R** inducing rotation about the $z$ axis. Equating the moments about the $z$ axis in Fig. 3-40a and b,

$$F_1 a_x + F_2 b_x - F_3 c_x = R_x x$$

$$3(1) + 2(4) - 1.5(1) = 3.5x$$

$$x = \frac{9.5}{3.5} = 2.71$$

(a)

(b)

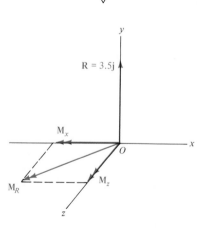

(c)

**FIGURE 3-40**

Reduction of a general parallel force system

EXAMPLE 3-12

Assume that the points of application of the major forces acting on the tall crane modeled in Fig. 3-41 are coplanar. Force $\mathbf{F}$ on the right is normally vertical, but it may make an angle up to $5°$ with the vertical. The other major loads are the counterweight $\mathbf{P}$ at the top left and the weight $\mathbf{W}$ of the vertical column whose centerline is the $y$ axis. Determine the equivalent force–couple system acting at the base $O$ of the column for a given height $L$ of the crane, with **(a)** $\mathbf{F}$ making a $5°$ angle with the vertical and acting in the $xy$ plane as shown, and **(b)** $\mathbf{F}$ making a $5°$ angle with the vertical, acting in the $z$ direction in the $yz$ plane.

## SOLUTION

To determine the equivalent force–couple system at point $O$, use Eq. 3-29.

**(a)** When $\mathbf{F}$ acts in the $xy$ plane,

$$\mathbf{F} = F(\sin 5° \,\mathbf{i} - \cos 5° \,\mathbf{j}) = 0.087F\mathbf{i} - 0.996F\mathbf{j}$$

$$\mathbf{R} = \sum_{i=1}^{3} \mathbf{F}_i = 0.087F\mathbf{i} - 0.996F\mathbf{j} - P\mathbf{j} - W\mathbf{j}$$

$$\mathbf{M}_R = \sum \mathbf{M}_O = \sum_{i=1}^{3} (\mathbf{r}_i \times \mathbf{F}_i)$$

$$= (0.9L\mathbf{i} + L\mathbf{j}) \times (0.087F\mathbf{i} - 0.996F\mathbf{j})$$
$$+ (-0.45L\mathbf{i} + L\mathbf{j}) \times (-P\mathbf{j}) + (0.5L\mathbf{j}) \times (-W\mathbf{j})$$
$$= -0.983FL\mathbf{k} + 0.45PL\mathbf{k}$$

Thus, the force–couple system with respect to point $O$ is

$$\mathbf{R} = 0.087F\mathbf{i} - (0.996F + P + W)\mathbf{j}$$

$$\mathbf{M}_R = (0.45P - 0.983F)L\mathbf{k}$$

**(b)** When $\mathbf{F}$ acts in the $yz$ plane,

$$\mathbf{F} = F(-\cos 5° \,\mathbf{j} + \sin 5° \,\mathbf{k})$$
$$= -0.996F\mathbf{j} + 0.087F\mathbf{k}$$

$$\mathbf{R} = \sum \mathbf{F}_i = -0.996F\mathbf{j} + 0.087F\mathbf{k} - P\mathbf{j} - W\mathbf{j}$$

$$\mathbf{M}_R = \sum \mathbf{M}_O = \sum (\mathbf{r}_i \times \mathbf{F}_i)$$
$$= (0.9L\mathbf{i} + L\mathbf{j}) \times (-0.996F\mathbf{j} + 0.087F\mathbf{k})$$
$$+ (-0.45L\mathbf{i} + L\mathbf{j}) \times (-P\mathbf{j})$$
$$= -0.896LF\mathbf{k} - 0.0783LF\mathbf{j} + 0.087LF\mathbf{i} + 0.45LP\mathbf{k}$$
$$= 0.087LF\mathbf{i} - 0.0783LF\mathbf{j} + (0.45P - 0.896F)L\mathbf{k}$$

Note that the weight was not included in the calculation. Since $\mathbf{W}$ intersects point $O$, $\mathbf{W}$ creates no moment about point $O$, as shown in part **(a)**. The force–couple system with respect to point $O$ is

$$\mathbf{R} = -(P + W + 0.996F)\mathbf{j} + 0.087F\mathbf{k}$$

$$\mathbf{M}_R = 0.087LF\mathbf{i} - 0.0783LF\mathbf{j} + (0.45P - 0.896F)L\mathbf{k}$$

## JUDGMENT OF THE RESULTS

Since all forces intersect or are parallel to the $x$ and $y$ axes in part **(a)**, the equivalent couple at $O$ would not have an $x$ or $y$ component. In part **(b)**, however, the small component of $\mathbf{F}$ in the $z$ direction would create couple components about both the $x$ and $y$ axes.

**FIGURE 3-41**

Model of a crane

EXAMPLE 3-12    111

# EXAMPLE 3-13

A rail test setup such as in Fig. 3-42a can be imagined as a bar in Fig. 3-42b with coplanar forces. Determine the equivalent force–couple system at $O$, where there is a weldment. Also determine the location of the single resultant force that would have the same effect as the force–couple system.

## SOLUTION

To determine the equivalent force–couple system at point $O$, use Eq. 3-29.

$$\mathbf{R} = \sum_{i=1}^{4} \mathbf{F}_i = (-50 \text{ kN} + 10 \text{ kN} + 20 \text{ kN} - 5 \text{ kN})\mathbf{j}$$

$$= -25 \text{ kN } \mathbf{j}$$

The scalar method is convenient for calculating the resultant moment since all forces are coplanar and parallel. Taking counterclockwise moments as positive, the moments about $O$ are

$$M_R = \sum_{i=1}^{4} F_i d_i$$

$$= (50 \text{ kN})(1.5 \text{ m}) - (10 \text{ kN})(0.8 \text{ m})$$
$$- (20 \text{ kN})(2.3 \text{ m}) + (5 \text{ kN})(2.5 \text{ m})$$
$$= 33.5 \text{ kN·m} \qquad \text{counterclockwise}$$

$$\mathbf{M}_R = 33.5 \text{ kN·m } \mathbf{k}$$

These results are shown in Fig. 3-42c.

To obtain the location of the resultant force to cause the same resultant moment, assume that $\mathbf{R}$ acts at a distance $l\mathbf{i}$ with respect to point $O$. Equating the moment $\mathbf{M}_R$ (from the individual forces) with the moment of the resultant force $\mathbf{R}$, the unknown distance $l$ is determined.

$$\mathbf{M}_R = l\mathbf{i} \times \mathbf{R}$$

$$33.5 \text{ kN·m } \mathbf{k} = l\mathbf{i} \times (-25 \text{ kN } \mathbf{j}) = -25l \text{ kN } \mathbf{k}$$

$$l = -1.34 \text{ m}$$

The negative sign of $l$ indicates that $\mathbf{R}$ acts to the left of point $O$, as expected and as shown in Fig. 3-42d.

**TIP**

Note that the force–couple system in this example can be reduced to a single equivalent force because the resultant force vector and the resultant moment vector are perpendicular. Also recall Ex. 3-12. The force–couple system in part (a) could be further reduced, but no further simplification can be made in part (b). When the force $\mathbf{F}$ is in the $yz$ plane, the resultant force and resultant moment at point $O$ are not perpendicular.

(a)

(b)

(c)                                    (d)

**FIGURE 3-42**

Test setup of railroad tracks. (Photograph courtesy MTS Systems
Corporation, Minneapolis, Minnesota.)

EXAMPLE 3-13    113

EXAMPLE 3-14

Reduce the system of forces acting on the bracket in Fig. 3-43a to an equivalent force–couple system at point $O$.

## SOLUTION

The resultant force and moment are determined using Eq. 3-29.

$$\mathbf{R} = \sum \mathbf{F}_i = 50 \text{ lb } \mathbf{i} - 100 \text{ lb } \mathbf{j} - 25 \text{ lb } \mathbf{j} + 25 \text{ lb } \mathbf{j}$$
$$= (50\mathbf{i} - 100\mathbf{j}) \text{ lb}$$

Note that the forces at $C$ and $D$ form a couple and do not contribute to the resultant force. Since the couple moment is a free vector, the only dimension important in its contribution to the resultant moment equation is the distance between the two forces forming the couple. The direction of the couple moment is determined by inspection.

$$\mathbf{M}_R = \sum \mathbf{M}_O = \sum (\mathbf{r}_i \times \mathbf{F}_i)$$
$$= 4 \text{ ft } \mathbf{i} \times (-100 \text{ lb } \mathbf{j}) + (25 \text{ lb})(6 \text{ ft})\mathbf{i}$$
$$= (150\mathbf{i} - 400\mathbf{k}) \text{ lb·ft}$$

The equivalent force–couple system is shown in Fig. 3-43b.

(a)

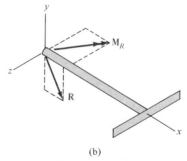

(b)

**FIGURE 3-43**

**3–78** Calculate the equivalent force–couple system of forces **F** and **P** at point $O$ on the column.

**Fig. P3-78**

**3–79** Consider Fig. P3-78 again. Let the equivalent force and couple at point $O$ be $\mathbf{F}_O$ and $\mathbf{M}_O$, respectively. Prove that if $\mathbf{F}_O = 0$, $\mathbf{M}_O \neq 0$, and if $\mathbf{M}_O = 0$, $\mathbf{F}_O \neq 0$.

**3–80** Determine the equivalent force–couple system at point $O$ if all forces are in the $xy$ plane on the beam.

**Fig. P3-80**

**3–81** Calculate the equivalent force–couple system of $\mathbf{F}_A = -400\ \text{N}\ \mathbf{j}$ and $\mathbf{F}_B = -300\ \text{N}\ \mathbf{k}$ at point $O$; $a = 0.3$ m and $b = 0.4$ m.

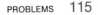

**Fig. P3-81**

**3–82** Two forces are acting on the bracket of a road simulator (see Sec. 1-7). Determine the equivalent force–couple system at point $O$. $OA = AB = 0.5$ m. The bracket lies in the $xy$ plane.

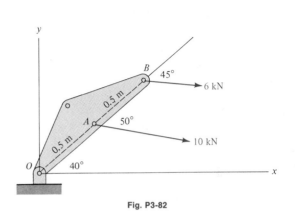

**Fig. P3-82**

**3–83** The large crane stands on rails laid in the $z$ direction. The weight of the structure is assumed to act at point $A$ of coordinates $(40, 45, -9)$ m. Force $\mathbf{F} = (80\mathbf{i} - 400\mathbf{j} + 50\mathbf{k})$ kN is acting at point $B$ of coordinates $(35, 50, 6)$ m. Determine the equivalent force–couple system at $O$ and the resultant moments $M_x$, $M_y$, and $M_z$ caused by the two forces if the mass of the structure is $1.5 \times 10^5$ kg.

**Fig. P3-83**

**3–84**  Assume that $W = 30$ lb, $F = 100$ lb, and $AB = BC = 2$ ft. (a) Determine the equivalent force–couple system acting at point $A$, and (b) determine the distance $d$ along bar $AC$ where the resultant force should act in order to create the same moment about point $A$ as the couple found in part (a).

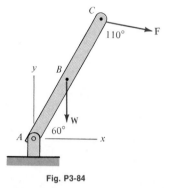

**Fig. P3-84**

**3–85**  A mechanical hoe is modeled with forces $W_1 = -1.2$ kN $\mathbf{j}$ at $A$ of coordinates $(0.5, 1.5, 0)$ m, $W_2 = -1.6$ kN $\mathbf{j}$ at $B$ of coordinates $(1.4, 1.2, 0)$ m, and $\mathbf{F} = (800\mathbf{i} - 900\mathbf{j})$ N at $C$ of coordinates $(1.7, -0.6, 0)$ m. Determine the location where the equivalent resultant of the three forces intersects the $x$ axis.

**Fig. P3-85**

**3–86**  The vertical forces acting on an airplane wing are approximated by four forces given in pounds. The forces intersect the $x$ axis at the indicated distances which are given in feet. (a) Calculate the equivalent force–couple system of these forces at $O$, and (b) determine the location $d$ measured from point $O$ along the $x$ axis where the resultant force $\mathbf{R}$ of the four forces appears to be acting.

**Fig. P3-86**

**3–87**  Consider a variation of Prob. 3–86 using a swept wing. In Fig. P3-87, the forces are perpendicular to the plane of the paper while the distances are along line $l$ in the $xz$ plane. Determine the equivalent force–couple system at point $O$.

**Fig. P3-87**

**3–88**  Three forces are acting on the frame of a simple crane. (a) Resolve this coplanar system of forces into an equivalent force–couple system at point $O$, and (b) replace this coplanar system of forces by an equivalent force system consisting of a vertical force at $O$ and horizontal forces at $O$ and $B$.

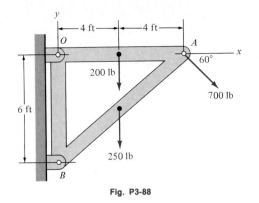

**Fig. P3-88**

$\mathbf{F}_A = 40\mathbf{j}$ kN at $A$, $(2, -0.7, 1.4)$ m

$\mathbf{F}_B = 0$  (temporarily)

$\mathbf{F}_C = 120\mathbf{j}$ kN at $C$, $(-0.5, -0.7, 1.3)$ m

$\mathbf{F}_D = 80\mathbf{j}$ kN at $D$, $(-0.5, -0.7, -1.3)$ m

$\mathbf{F}_E = -40\mathbf{j}$ kN (weight of tractor) at $E$, $(2, 0, 0)$ m

$\mathbf{F}_G = -160\mathbf{j}$ kN (load in trailer) at $G$, $(0, 2, 1)$ m

(b) Determine the $x$ and $z$ coordinates of the resultant of the six forces.

**3–89** In a laboratory test of a motorcycle three forces are applied simultaneously: $\mathbf{F}_1 = 500$ lb $\mathbf{j}$, $\mathbf{F}_2 = 450$ lb $\mathbf{j}$, $\mathbf{F}_3 = -200$ lb $\mathbf{j}$. The coordinates are $A$: $(-2.5, -1.7, 0)$ ft, $B$: $(2.3, -1.7, 0)$ ft, and $C$: $(0.7, 1, 0)$ ft. Calculate the $x$ intercept of the resultant force $\mathbf{R}$ in order for $\mathbf{R}$ to have the same net effect on the motorcycle.

(a)

**Fig. P3-91**
(a) Laboratory test of a tractor-trailer truck. (Courtesy MTS Systems Corporation, Minneapolis, Minnesota);

**Fig. P3-89**

**3–90** In Prob. 3–89, an additional force $\mathbf{F}_4 = 50$ lb $\mathbf{i} + 40$ lb $\mathbf{k}$ is applied at point $A$. Determine the equivalent force–couple system of the four forces at point $O$.

**3–91** Consider the tractor-trailer truck in a laboratory test. The origin $O$ of the $xyz$ frame is at the pivot connection of the tractor and the trailer. (a) Determine the equivalent force–couple system at point $O$ for the given forces (this is important for making the pivot connection strong enough).

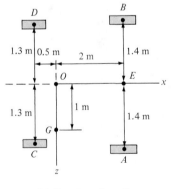

(b) Top view schematic

**3–92** In Prob. 3–91, an additional force $\mathbf{P} = -10\mathbf{k}$ kN is acting at point $A$. Determine the components $M_x$, $M_y$, and $M_z$ of the equivalent moment of all of the forces (these are important in dealing with the "ride characteristics" of the cab).

# SUMMARY

Section 3-1

**EXTERNAL AND INTERNAL FORCES:** *External forces* are those forces exerted on the rigid body by another body. When the line of action of an external force does not pass through the center of mass of the rigid body, the force tends to cause the body to rotate as well as translate. *Internal forces* are those forces that act between parts of the rigid body and cancel out when considering the rigid body as a whole.

Section 3-2

**EQUIVALENT FORCES:** Two forces acting at different points on a given rigid body are equivalent if they have the same magnitude, line of action, and direction. Note certain practical limitations concerning the equivalence of forces.

Section 3-3

**MOMENT OF A FORCE, SCALAR DEFINITION:**

$$M_A = Fd$$

Section 3-4

**VECTOR PRODUCT V OF TWO VECTORS A AND B:**

$$\mathbf{V} = \mathbf{A} \times \mathbf{B}$$

$$V = AB \sin \theta$$

**USEFUL PROPERTIES:**

$$\mathbf{A} \times \mathbf{B} \neq \mathbf{B} \times \mathbf{A}$$

$$\mathbf{A} \times \mathbf{B} = -(\mathbf{B} \times \mathbf{A})$$

$$\mathbf{A} \times (\mathbf{B} + \mathbf{C}) = \mathbf{A} \times \mathbf{B} + \mathbf{A} \times \mathbf{C}$$

$$\mathbf{i} \times \mathbf{i} = \mathbf{j} \times \mathbf{j} = \mathbf{k} \times \mathbf{k} = 0$$

$$\mathbf{i} \times \mathbf{j} = \mathbf{k} \qquad \mathbf{j} \times \mathbf{i} = -\mathbf{k} \qquad \mathbf{k} \times \mathbf{i} = \mathbf{j}$$

$$\mathbf{i} \times \mathbf{k} = -\mathbf{j} \qquad \mathbf{j} \times \mathbf{k} = \mathbf{i} \qquad \mathbf{k} \times \mathbf{j} = -\mathbf{i}$$

$$\mathbf{V} = \mathbf{A} \times \mathbf{B} = \begin{vmatrix} \mathbf{i} & \mathbf{j} & \mathbf{k} \\ A_x & A_y & A_z \\ B_x & B_y & B_z \end{vmatrix}$$

**MOMENT OF A FORCE, VECTOR DEFINITION:**

$$\mathbf{M}_O = \mathbf{r}_{OA} \times \mathbf{F} \qquad (A \text{ is any point on the line of action of } \mathbf{F})$$

**RECTANGULAR COMPONENTS OF A MOMENT:**

$$\mathbf{M}_O = M_x\mathbf{i} + M_y\mathbf{j} + M_z\mathbf{k}$$

Section 3-5

**VECTOR TRIPLE PRODUCT:**   $\mathbf{V} = (\mathbf{A} \times \mathbf{B}) \times \mathbf{C}$ (used in dynamics)

**MIXED TRIPLE PRODUCT:**

$$\mathbf{A} \cdot (\mathbf{B} \times \mathbf{C}) = \begin{vmatrix} A_x & A_y & A_z \\ B_x & B_y & B_z \\ C_x & C_y & C_z \end{vmatrix}$$

Section 3-6

**MOMENTS OF A FORCE ABOUT A LINE:**

$$M_l = \mathbf{n} \cdot \mathbf{M}_O = \mathbf{n} \cdot (\mathbf{r} \times \mathbf{F}) = \begin{vmatrix} n_x & n_y & n_z \\ r_x & r_y & r_z \\ F_x & F_y & F_z \end{vmatrix}$$

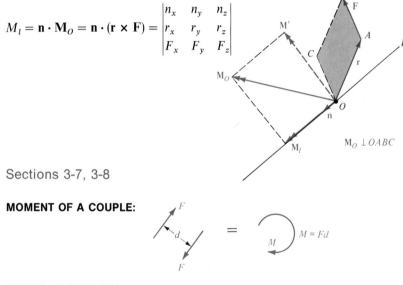

Sections 3-7, 3-8

**MOMENT OF A COUPLE:**

$M = Fd$

**USEFUL PROPERTIES:**

1. A couple can be moved anywhere in its own plane in the same rigid body.
2. A couple can be moved to any other parallel plane in the same rigid body.

Section 3-9

**COMPONENTS AND ADDITION OF COUPLES:**

1. Couples in the same plane or in parallel planes in a rigid body can be summed algebraically.
2. In general, couples are resolved into rectangular components for vector addition.
3. Any number of couples $\mathbf{M}_1, \mathbf{M}_2, \ldots$ can be reduced to a single resultant $\mathbf{M}$ with $xyz$ components,

$$\mathbf{M}_1 + \mathbf{M}_2 + \cdots = \mathbf{M} = M_x\mathbf{i} + M_y\mathbf{j} + M_z\mathbf{k}$$

### FORCE–COUPLE TRANSFORMATIONS:

1. A force $\mathbf{F}$ acting at $B$ on a rigid body can be replaced by the same force $\mathbf{F}$ acting at $A$ and a moment $\mathbf{M}_A = \mathbf{r} \times \mathbf{F}$ about $A$.
2. A force $\mathbf{F}$ and moment $\mathbf{M}_A$ acting at $A$ can be replaced by a force $\mathbf{F}$ acting at $B$ for the same total effect on the rigid body.

Section 3-11

### REDUCTION OF COMPLEX FORCE SYSTEMS:
Any system of forces acting on a rigid body can be reduced to a resultant force $\mathbf{R}$ and a resultant moment $\mathbf{M}_R$,

$$\mathbf{R} = \sum_{i=1}^{n} \mathbf{F}_i \quad \mathbf{M}_R = \sum_{i=1}^{n} \mathbf{M}_i$$

### EQUIVALENT FORCE SYSTEMS:
A system of forces $\mathbf{F}_1, \mathbf{F}_2, \ldots, \mathbf{F}_i$ is equivalent to another system of forces $\mathbf{P}_1, \mathbf{P}_2, \ldots, \mathbf{P}_j$ if

$$\sum_{i=1}^{m} \mathbf{F}_i = \sum_{j=1}^{n} \mathbf{P}_j \qquad \text{(equivalence of forces)}$$

$$\sum_{i=1}^{m} (\mathbf{r}_i \times \mathbf{F}_i) = \sum_{j=1}^{n} (\mathbf{r}_j \times \mathbf{P}_j) \quad \text{(equivalence of moments)}$$

Section 3-12

### COMMON FORCE SYSTEMS:

1. Resultant force is zero, but there is a resultant moment,

$$\mathbf{R} = 0 \quad \mathbf{M}_R \neq 0$$

2. Concurrent forces have a resultant force, but there is no resultant moment,

$$\mathbf{R} \neq 0 \quad \mathbf{M}_R = 0$$

3. Coplanar forces may have a resultant force and a resultant moment about a chosen point,

$$\mathbf{R} \neq 0 \quad \mathbf{M}_R \neq 0$$

$\mathbf{M}_R$ is perpendicular to the plane of the forces.

4. Parallel forces may have a resultant force and a resultant moment about a chosen point

$$\mathbf{R} \neq 0 \quad \mathbf{M}_R \neq 0$$

$\mathbf{M}_R$ is perpendicular to $\mathbf{R}$.

**3-93** A garage door is modeled with three forces acting on it: $F = 70$ lb, $P = 200$ lb, and $W = 180$ lb. Determine the sum of the moments of **F**, **P**, and **W** about $O$ if $OA = 2$ ft, $OB = 4$ ft, and $OC = 8$ ft.

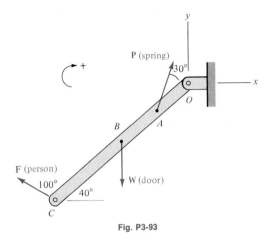

Fig. P3-93

**3-94** Determine the moments of the force **F** about points $A$ and $B$ of the rigid arm $AC$ which lies in the $xy$ plane. Use (a) a scalar method and (b) a vector method.

Fig. P3-94

**3-95** The coordinates of points $A$ and $B$ of the wrench are $A$: (0, 4, 0) in. and $B$: (10, 4, $-3$) in. The force applied at $B$ is $\mathbf{F} = (20\mathbf{i} - 4\mathbf{j} + 35\,\mathbf{k})$ lb. Determine the moments of the force **F** about (a) point $A$, and (b) point $O$.

Fig. P3-95

**3-96** The coordinates of points $A$ and $B$ are $A$: (40, 37, $-6$) ft and $B$: (55, $-2$, 7) ft. The tension in the cable $AB$ is 1700 lb. Determine the magnitude of the moment of this tension about point $O$.

Fig. P3-96

**3-97** The shake table for testing electronic equipment has a hydraulic actuator $A$ attached to it. Pivot $B$ of the actuator and the lower pivots of the four supporting arms are attached to the same base plate. For a simple modeling of an arm, consider the arrangement in Fig. P3-97(b) where $l$ and $r$ are constant. Determine $\mathbf{M}_O$ of the force **F** for $\theta = 40°$, $60°$, and $80°$. $F$ is constant.

Fig. P3-97
Shake table for shock and vibration testing of electronic equipment. (Photograph courtesy MTS Systems Corporation, Minneapolis, Minnesota.)

**3–98** For any angle $\theta$ in Prob. 3–97, prove that only the component of **F** perpendicular to arm $OC$ causes a moment $\mathbf{M}_O$.

**3–99** Determine the moments of the force $\mathbf{F} = (-100\mathbf{i} - 20\mathbf{j} + 10\mathbf{k})$ N about joints $A$, $B$, and $C$. Pipes $BC$ and $CD$ are parallel to the $y$ and $z$ axes, respectively. Will pipe $AB$ tend to tighten or loosen where it is screwed into the wall at $A$? The pipe has a right-handed thread.

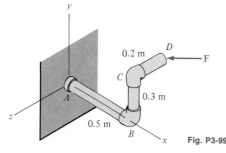

Fig. P3-99

**3–100** In Prob. 3–99 determine the moment of force **F** about line $AC$.

**3–101** The steering wheel is mounted at point $A$ on a column which is fixed at point $O$. The coordinates of $A$ are $(0.7, 0.3, 0)$ m. A force $\mathbf{F} = (60\mathbf{i} - 40\mathbf{j} - 30\mathbf{k})$ N is applied to the wheel at point $B$ of coordinates $(0.8, 0.6, -0.2)$ m. Calculate the moment of the force about the steering column defined by the line $OA$.

Fig. P3-101

**3–102** The supporting pads $A$ and $B$ of the crane are located on the $z$ axis when the coordinates of point $C$ are $(-5, 8, 1)$ m. Determine the moments of the tension $\mathbf{T} = (-4\mathbf{i} - 12\mathbf{j} + 3\mathbf{k})$ kN about the $x$, $y$, and $z$ axes.

Fig. P3-102

**3–103** Three tensions in the wires of an antenna tower are $\mathbf{T}_1 = (-50\mathbf{i} - 60\mathbf{j} + 50\mathbf{k})$ kN, $\mathbf{T}_2 = (50\mathbf{i} - 55\mathbf{j} + 53\mathbf{k})$ kN, and $\mathbf{T}_3 = (-70\mathbf{j} - 65\mathbf{k})$ kN. Determine the equivalent force–couple system at point $O$ if $OA = 30$ m.

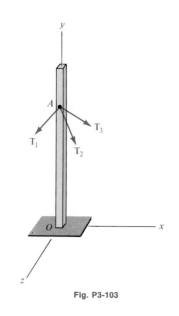

Fig. P3-103

**3–104** An acrobat applies the system of forces shown to the rigid vertical pole. (a) Resolve this system into an equivalent force–couple system at point $O$, and (b) calculate the vertical distance $d$ from $O$ where the resultant force appears to be acting. All forces are in the $xy$ plane.

Fig. P3-104

# Equilibrium of Rigid Bodies

# 4

## Advantages of Using Equilibrium Conditions

The concept of equilibrium is enormously useful in analyzing engineering problems. The reason is that equilibrium conditions are the basis for determining unknown forces and moments. The unknowns may be external or internal to a rigid body.

**SECTION 4-1** presents the most useful set of equations in statics. The equilibrium equations express the necessary and sufficient conditions for equilibrium of any rigid body. They are essential in mechanical design.

**SECTION 4-2** describes the method of drawing proper free-body diagrams that are essential in the practical application of the equilibrium equations.

**SECTION 4-3** introduces the application of equilibrium equations for determining unknown forces and moments for problems in two dimensions. This method is extremely useful in engineering mechanics, and it should be practiced in as many different problems as possible. Such preparation is important for solving new problems.

**SECTION 4-4** describes two-force members, which are the simplest of common structural members. The recognition of some components as two-force members in any mechanical system simplifies the analysis of that system.

**SECTION 4-5** extends the application of equilibrium equations for determining unknown forces and moments to three-dimensional problems.

**SECTION 4-6** presents the method of analyzing connected rigid bodies. This method is based on Newton's Third Law (action and reaction). It is particularly useful for determining unknown forces that are internal to a rigid body or a system of bodies.

123

## COMPLETE EQUILIBRIUM CONDITIONS

The equilibrium equations that describe the state of balance for forces are valid for particles and for rigid bodies. In general, static equilibrium means that no motion of the body occurs. For analyzing such a condition, it is convenient to state that a rigid body not in equilibrium *could* have six distinct motions relative to an arbitrary $xyz$ reference frame. There could be linear displacements along the $x$, $y$, or $z$ axes (this is called *translation*), and there could be angular displacements about the $x$, $y$, or $z$ axes (*rotation*). Any two or more (up to six) of these motions may occur simultaneously. Static equilibrium means that none of these six motions is occurring.

For example, the helicopter in Fig. 4-1 may translate with respect to each of the $xyz$ coordinates, and it may rotate with respect to each of these axes, as well. Complete static equilibrium of the helicopter's fuselage means that it hovers in place without any rotation (of course, rotors and internal machinery rotate). Such a situation is possible if the external forces have neither a resultant force nor a resultant moment on the body.

The formal statement of equilibrium is that a *rigid body is in equilibrium when the equivalent force–couple system of the external forces acting on it is zero*. The equations that express the necessary and sufficient conditions for equilibrium of any rigid body are

$$\sum \mathbf{F} = 0 \quad \text{and} \quad \sum \mathbf{M}_O = \sum (\mathbf{r} \times \mathbf{F}) = 0 \qquad \boxed{4\text{-}1}$$

where $O$ is an arbitrary point of reference.

In practice it is most convenient to write Eq. 4-1 in terms of rectangular scalar components,

$$\sum F_x = 0 \qquad \sum M_x = 0 \qquad \boxed{4\text{-}2}$$
$$\sum F_y = 0 \qquad \sum M_y = 0$$
$$\sum F_z = 0 \qquad \sum M_z = 0$$

These scalar equilibrium equations are easy to use, but occasionally it is necessary to be cautious and judiciously check the results as shown in Ex. 4-1.

### Useful Notes Concerning the Equilibrium Equations

Problems concerning equilibrium of rigid bodies may involve one, two, or three dimensions. In each case it is necessary to start the solution of the problem by drawing complete free-body diagrams. The equilibrium equations are written for each rigid body as follows.

**1. *One-dimensional problem.*** Only one force summation is necessary using Eq. 4-1 or 4-2,

$$\sum \mathbf{F} = 0 \quad \text{or} \quad \sum F = 0 \qquad \boxed{4\text{-}3}$$

It makes no difference what subscript ($x$, $y$, or $z$) is used for the force $F$.

**FIGURE 4-1**

A helicopter may have six distinct motions in the *xyz* frame. (Courtesy MTS Systems Corporation, Minneapolis, Minnesota.)

**2. Two-dimensional problem.** At most, three independent summations involving forces and moments can be used from Eq. 4-2. The possibilities for a force system in the $xy$ plane are as follows:

**a.** Two summations for forces and one summation for moments, such as

$$\sum F_x = 0 \qquad \sum F_y = 0 \qquad \sum M_A = 0 \qquad \boxed{\text{4-4}}$$

Note that summing moments about any point in the $xy$ plane is equivalent to summing moments about the $z$ axis.

**b.** One summation for forces and two summations for moments, such as

$$\sum F_x = 0 \qquad \sum M_A = 0 \qquad \sum M_B = 0 \qquad \boxed{\text{4-5}}$$

where the line joining points $A$ and $B$ must not be parallel to the $y$ axis.

**c.** Three summations for moments, such as

$$\sum M_A = 0 \qquad \sum M_B = 0 \qquad \sum M_C = 0 \qquad \boxed{\text{4-6}}$$

where the points $A$, $B$, and $C$ must not form a straight line.

**3. Three-dimensional problem.** At most all six summations in Eq. 4-2 can be used.

The details of these methods are exemplified in subsequent sections.

## FREE-BODY DIAGRAMS 4-2

The equilibrium equations are necessary for a determination of unknown forces and moments acting on a rigid body in equilibrium. The first step in this process is the drawing of a proper *free-body diagram* as was done in Ch. 2. Such a diagram is for a mechanical system (a body or group of bodies, rigid or nonrigid) that is imagined to be isolated from all other bodies. All significant effects caused by the other bodies *on* the free body must be shown in the free-body diagram. Internal forces and moments are ignored because they occur in pairs and cancel one another. *Unknown forces and moments* must be shown at the appropriate places on the diagram. The directions of the unknowns may be assumed arbitrarily if the directions are not known.

### Common Forces and Constraints

The constraining forces or moments acting on a body are often called *reactions* or *supports*. In the majority of the problems these are unknown and must be determined analytically by applying the equilibrium equations to the appropriate free-body diagram. Drawing free-body diagrams always involves making simplifying assumptions. These are facilitated by knowledge of the common characteristics of mechanical systems. The following list and the schematic drawings in Fig. 4-2 are representative of frequently encountered two-dimensional problems. Common forces and supports for three-dimensional problems are illustrated in Fig. 4-3.

| Description of members | Sketch of members | Representation for use in free-body diagrams |
|---|---|---|
| (a) Flexible tension members | Wires / $W$ | $T$ $T$ / $W$ |
| (b) Two-force members | Fixed / Wire | $F$ / $F$ |
| (c) Rolling members (no friction) | | $F$ $F$ |
| (d) Frictionless sliders | Block / Fixed / Slider | $F$ |
| (e) Pin connections and knife edges (no friction) | No friction at pin / Fixed | $F_x$ $F_y$ $F_x$ $F_y$ |
| (f) Sharp edges or sliders on rough surfaces | Block / Fixed | $F_x$ $F_y$ $F_x$ $F_y$ |
| (g) Pin connections with friction | Fixed / Bar | $F_x$ $M_z$ $F_y$ |
| (h) Fixed or built-in members | Fixed wall / $O$ | $M_z$ $F_x$ $O$ $F_y$ |

**FIGURE 4-2**
Common forces, supports, and their representations in free-body diagrams for two-dimensional problems

| Types of forces | Unknowns |
|---|---|
| *Cables, chains, belts, or ropes* are flexible, so they can have only axial tension in them.<br>    The *weight* of the body is often included in the diagram. Its direction is always toward the center of the earth. The weight can be neglected only if it is much smaller than the other forces acting on the body. | Forces with known lines of action; only one force component is possible |
| *Two-force members* are rigid members with all external forces applied at two separate connections. For such a member the net force at each connection has a line of action that goes through both connections. | |
| *Rolling contacts* allow motion parallel to the surfaces in contact, so there is only one force between the bodies and it is perpendicular to these surfaces. The force is called a *normal force*. | |
| Some sliding members have very low friction, so these tend to have only normal forces as rolling members do. | |
| *Pin connections* and *knife edges* allow no translation of a body, but rotation is possible. In a freely hinged connection there is no moment acting between the members. | Forces with unknown lines of action; two force components are possible |
| *Rough surfaces* apply both a normal force and a tangential force to the member at the point of contact. | |
| Friction in a pin connection prevents free rotation of the member about the axis of the pin. | Two force components and a moment are possible |
| *Fixed or built-in members* are restrained from translation and rotation. | |

**FIGURE 4.2** Continued

| Description of members | Sketch of members | Representation for use in free-body diagrams |
|---|---|---|
| (a) Ball-and-socket joints (no friction) | | |
| (b) Sharp points on rough surfaces | | |
| (c) Thrust bearings (no friction) | | |
| (d) Sleeve bearings and hinges (no friction) | | |
| (e) Ball or roller bearings (no friction) | Cross section | |
| (f) Fixed or built-in members | Fixed wall | |

**FIGURE 4-3**

Common forces, supports, and their representations in free-body diagrams for three-dimensional problems

### Notes on three-dimensional forces and supports:

**1.** Each case should be analyzed carefully. Sometimes a particular force or moment is possible in a device, but it must be neglected for most practical purposes. For example, a very short sleeve bearing cannot support significant moments. A roller bearing may be designed to carry much larger loads perpendicular to the shaft than along the shaft.

**2.** Combinations of supports are common. For example, a hinge may also act as a thrust bearing to carry loads along the shaft and perpendicular to the shaft.

The number of unknown quantities that can be determined in a problem is the same as the number of independent equilibrium equations that can be written using the available information. For a two-dimensional problem there are three independent equilibrium equations, so three unknowns can be determined. The student should realize that any number of equilibrium equations can be written for a given problem, but only a few of them are independent, and it is only fruitful to work with these. However, there are many different sets of independent equations, and each can be used to obtain a solution. This is illustrated for the beam in Fig. 4-4a.

The beam is called *simply supported*. The roller at $B$ can apply only a normal-force reaction $B_y$ as shown in the free-body diagram in Fig. 4-4b. The pin at $A$ can apply both a horizontal force $A_x$ and a vertical force $A_y$. In this problem the applied force is known and the reactions at the supports, $A_x$, $A_y$, and $B_y$, are the unknowns to be determined. Three independent equilibrium equations are obtained from Eq. 4-4:

$$\sum F_x = 0 \qquad \sum F_y = 0 \qquad \boxed{4\text{-}7}$$
$$A_x + 9 \text{ kN} = 0 \qquad A_y + B_y - 12 \text{ kN} = 0$$

$$\sum M_A = 0$$
$$(3 \text{ m})B_y - (1 \text{ m})(12 \text{ kN}) = 0$$

From these, $A_x = -9$ kN, $B_y = 4$ kN, and $A_y = 8$ kN. The negative sign for $A_x$ indicates that the correct direction of $\mathbf{A}_x$ is opposite that shown on the free-body diagram. To support the applied load, $\mathbf{A}_x$ must act in the negative $x$ direction. Note that in the moment equation the small width dimension of the beam is neglected.

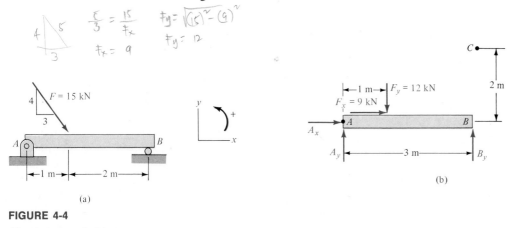

(a)

**FIGURE 4-4**

Simply supported beam

An equally valid set of equations is obtained from Eq. 4-5:

$$\sum F_x = 0 \qquad \sum M_A = 0 \qquad \boxed{4\text{-}8}$$
$$A_x + 9\ \text{kN} = 0 \qquad (3\ \text{m})B_y - (1\ \text{m})(12\ \text{kN}) = 0$$
$$\sum M_B = 0$$
$$(3\ \text{m})A_y - (2\ \text{m})(12\ \text{kN}) = 0$$

from which the same results are obtained. It is important to note that the force summation in the $y$ direction could not be used as an independent equation with the two moment equations. Addition of the two moment equations results in the force equation for the $y$ direction, and no new information is obtained.

An equally valid but slightly less convenient set of equations is obtained from Eqs. 4-6:

$$\sum M_A = 0 \qquad \sum M_B = 0 \qquad \sum M_C = 0 \qquad \boxed{4\text{-}9}$$

Equation 4-8 is more convenient than Eq. 4-9 because the force summation in the $x$ direction involves only one unknown. The moment summation about point $C$ involves two unknowns. Generally, equations with one unknown in each are the best to use. An important part of the problem solving for two-dimensional problems is choosing the most convenient set of equations. Taking moments about a point where an unknown force acts is usually advantageous. Note that the moment center does not have to be on the rigid body.

The two supports shown in Fig. 4-4a are the minimum number of supports with which this beam can be in equilibrium. The beam would not remain stationary if either the support at $A$ or at $B$ were removed. Members in such situations, with the minimum number of constraints for equilibrium, are called *statically determinate*.

Any extra supports beyond the minimum are *redundant* in theory; they could be removed without changing the state of equilibrium. It may not be reasonable to remove redundant supports in practice since they may be necessary for extra stiffness or safety of the structure. The solution of problems is more difficult when there are redundant supports. For example, consider the addition of a spring support to the beam at midspan as in Fig. 4-5. Four independent equations are needed to solve for the four reactions, $A_x$, $A_y$, $B_y$, and $D_y$. However, since this is a two-dimensional problem, only three independent equilibrium equations are available. Such members with any redundant supports are called *statically indeterminate*. They can be solved only with the methods of statics and mechanics of materials together.

Another configuration that is common is the built-in member. The simplest built-in member is the cantilever beam (Fig. 4-6a), which is statically determinate. The free-body diagram for this beam is drawn on the basis that the external force **F** tends to cause a translation and a rotation at the support $A$. These effects of the force **F** are shown by the equivalent force–couple system at point $A$ in Fig. 4-6b. The wall at $A$ resists the applied force and couple with an equal and opposite force and couple if there is equilibrium. The complete free-body diagram is shown in Fig. 4-6c. Here the reactions $\mathbf{A}_x$, $\mathbf{A}_y$, and $\mathbf{M}_A$ are considered as an unknown, external force–couple system to the beam. Their directions may be drawn

**FIGURE 4-5**

Statically indeterminate beam

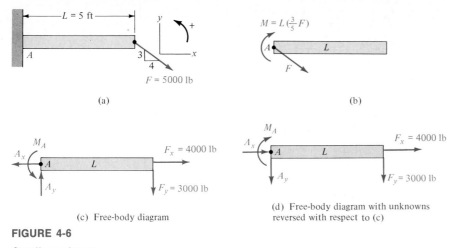

(a)

(b)

(c) Free-body diagram

(d) Free-body diagram with unknowns reversed with respect to (c)

FIGURE 4-6

Cantilever beam

correctly in the free-body diagram, knowing that they oppose the applied force–couple system shown in Fig. 4-6b. The equations of equilibrium are written for Fig. 4-6c,

$$\sum F_x = 0 \quad \Big| \quad \sum F_y = 0 \quad \Big| \quad \sum M_A = 0 \qquad \boxed{4\text{-}10}$$
$$-A_x + F_x = 0 \quad \Big| \quad A_y - F_y = 0 \quad \Big| \quad M_A - L(\tfrac{3}{5}F) = 0$$
$$A_x = 4000 \text{ lb} \quad \Big| \quad A_y = 3000 \text{ lb} \quad \Big| \quad M_A = 15,000 \text{ lb·ft}$$

Note that changing the directions of $\mathbf{A}_x$, $\mathbf{A}_y$, and $\mathbf{M}_A$ while they are considered as unknowns is permissible. Applying the given sign convention to Fig. 4-6d, the equilibrium equations become

$$A_x + F_x = 0 \quad \Big| \quad -A_y - F = 0 \quad \Big| \quad -M_A - L(\tfrac{3}{5}F) = 0 \qquad \boxed{4\text{-}11}$$
$$A_x = -4000 \text{ lb} \quad \Big| \quad A_y = -3000 \text{ lb} \quad \Big| \quad M_A = -15,000$$

The negative signs indicate that the directions of the unknown force and couple were assumed incorrectly. This is such a simple matter that it is not reasonable to spend time in figuring out the correct directions before the calculations are made.

## TWO-FORCE MEMBERS

4-4

In many engineering problems particular conditions of constraint must be created. An interesting example is the tire-couple road simulation of any vehicle (see Fig. 1-11 in Sec. 1-7). In typical tests only vertical forces are intentionally applied to each wheel, but the vehicle could suddenly or gradually move horizontally during a test if not constrained properly. There are stories of vehicles that moved off their actuator platens and crashed in the laboratory. Such horizontal motions can be prevented but it must be done with a minimum of interference with the vertical motions. The best solution is to attach horizontal wires or pinned rods to the vehicle at several points. Wire $A$ and frame $B$ (and others in the background) in Fig. 4-7 show such a system.

(a)

(b)

**FIGURE 4-7**

Laboratory test setup of a bus. Horizontal wires such as *A* apply horizontal constraint while allowing vertical motion of the bus. (Photograph courtesy MTS Systems Corporation, Minneapolis, Minnesota.)

The constraining wires and pinned rods are the simplest structural components because they have only axial forces acting on them. The wires can have only tension, since they collapse under compression. The rods, being rigid members, can have either tensile or compressive forces acting on them. Both members are called *two-force members* because their free-body diagrams can have only two forces as shown in Fig. 4-8. These forces must be collinear if the member is in equilibrium. The situation in Fig. 4-9 is not possible because only the summation of forces is zero. There is a net moment of magnitude *Td* with respect to point *A* or *B*, so the member cannot be in equilibrium.

The two-force member does not have to be straight or simple in shape. For example, the component in Fig. 4-10 must have collinear forces acting on it to be in equilibrium. These findings lead to the general statement that *a body subjected to forces applied at two points is in equilibrium only when the resultant forces at those points have the same magnitude the same line of action, and the opposite direction.*

The numerous possibilities of constraining a body with two-force members are illustrated with several two-dimensional models in Fig. 4-11. Assume that the two-force member is a pinned rod in each case. The situations in Fig. 4-11a, b, and c represent *partial constraint* because the

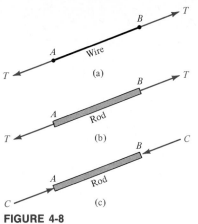

(a)

(b)

(c)

**FIGURE 4-8**

Two-force members

**FIGURE 4-9**

Two-force member not in equilibrium because the forces are not
collinear

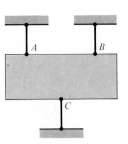

**FIGURE 4-10**

Member of arbitrary shape is in
equilibrium if the forces are collinear

(a) Partial constraint

(b) Partial constraint

(c) Partial constraint

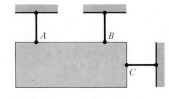

(d) Complete constraint

(e) Redundant constraint

**FIGURE 4-11**

Various constraints of a body by two-force rods

body could move in the $x$ direction (a and b) or rotate a little in the $xy$ plane (c). *Complete constraint* is achieved by a rearrangement of only three members, as in Fig. 4-11d. Naturally, it is easy to overdo the fixation of a member in theory, and have *redundant constraints* as in Fig. 4-11e. Certain aspects of such statically indeterminate bodies have been mentioned in Sec. 4-3.

Generally it is found that partial constraint (sometimes called *improper constraint*) exists when all the restraining links are parallel (Fig. 4-11a and b) or concurrent (Fig. 4-11c). For the latter, it is not necessary to have mutually perpendicular lines.

The concepts of constraint described above can be extended to bodies with simple roller or pivot joints (these can be considered as two-force links of zero length). In general, the adequacy of constraint is closely related to the number of ways a body could move, which is the number of independent coordinates that completely specify the position of the body. This is called the number of *degrees of freedom* of the body. A single degree of freedom exists for a body sliding along a fixed axis or rotating about a fixed axis. Plane motion of a single rigid body may be characterized by a maximum of three degrees of freedom (two linear coordinates and one angular coordinate). The constraints required for static equilibrium must prevent all possible motion of the body.

Finally, the practical difference between rods (rigid) and chains or wires (flexible) is worth remembering. Rigid rods are often the most desirable constraints since they transmit tension and compression (sometimes these forces alternate as in bars $A$, $B$, and $C$ in Fig. 4-12). Thus, two flexible members would be required to replace one rigid rod. Nevertheless, wires and chains are often used (such as wire $A$ in Fig. 4-7) because they are self-aligning and have simple end connections.

**FIGURE 4-12**

Rods A, B, and C apply forces to the right front axle of a car in a test. (Courtesy MTS Systems Corporation, Minneapolis, Minnesota.)

EXAMPLE 4-1

The selection of a reference point is arbitrary for the moment summation in the two-dimensional equilibrium equations of Eq. 4-4. A mistake made in another part of the analysis, however, may cause an incorrect determination of a moment reaction. For a simple example, imagine an attempt to analyze how the moment reaction is created at the supporting wall of a cantilever beam. Somebody proposes that an internal force **F** caused by the wall resists the moment of the applied load **W** as in Fig. 4-13a. Thus, taking moments about $O$,

$$F = \frac{WL}{d}$$

Is this the correct view of the situation?

## SOLUTION

It is not entirely correct. For one thing, the force **F** is the only force in the horizontal direction, so the beam could not be in equilibrium in that direction. For another, the beam is not in equilibrium with respect to moments, either. This is more subtle than the matter of force summation since it appears correct to take moments about $O$. However, it is also correct to take moments about any other point, for example, point $A$ in Fig. 4-13a. Now, it is clear that both **F** and **W** cause clockwise moments, so the beam could not be in equilibrium.

The proper construction of the free-body diagram could be initiated as in Fig. 4-13a because the moment of **W** must indeed have a moment reaction. But it is necessary to go further and make sure that equilibrium is attained in all respects. Figure 4-13b is a correct free-body diagram for the cantilever beam.

## JUDGMENT OF THE RESULTS

Figure 4-13b is acceptable because the force summations are zero (by inspection), and taking moments about *any* point gives the same result. Note that taking moments about two different points is often enough for checking the result, but it does not necessarily reveal a mistake. For example, taking moments about points $B$ and $O$ in Fig. 4-13a are equally inadequate.

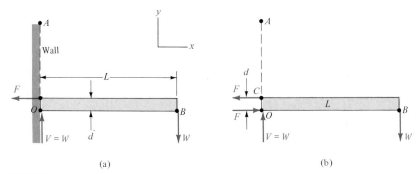

(a)                                        (b)

**FIGURE 4-13**

Analysis of the supporting reactions of a cantilever beam

EXAMPLE 4-1    135

# EXAMPLE 4-2

Determine the possible conditions for the equilibrium of a two-dimensional *three-force member*, which means that forces act on the body at only three points. Consider the concurrent, parallel, and general force systems shown in Fig. 4-14.

## SOLUTION

**CONCURRENT FORCES:** It is clear that equilibrium is possible, and the only equations that should be checked are for the summation of forces. With respect to Fig. 4-14a,

$$\sum F_x = 0 \quad \sum F_y = 0$$

There are no moments about point $O$. This condition of equilibrium is valid for more than three concurrent forces, as well.

**PARALLEL FORCES:** Again, equilibrium is possible. The only requirements with respect to Fig. 4-14b are

$$\sum F_y = 0$$

$$\sum M_O = 0 \qquad \text{where } O \text{ is an arbitrary point}$$

This condition is valid for more than three parallel forces, as well.

**GENERAL FORCES (Not Concurrent, Not Parallel):** Complete equilibrium is not possible, as seen after a simple change from Fig. 4-14a to c. Equilibrium of the forces is still conceivable, but not of moments. Taking moments about a point $O$ where the lines of action of any two forces intersect,

$$\sum M_O \neq 0$$

because the third force must have a moment about point $O$ (Fig. 4-14c). For more than three forces, equilibrium may be possible. In conclusion, *a two-dimensional body subjected to three forces may be in equilibrium only when the forces are concurrent or parallel.*

(a) Concurrent forces

(b) Parallel forces

(c) General forces

**FIGURE 4-14**

Analysis of a two-dimensional three-force member

EXAMPLE 4-3

Consider a weightless pulley (Fig. 4-15a) which has a frictionless axle at its center $O$. Prove that the magnitude of the tension in a rope or wire is the same on the two sides of the pulley using **(a)** point $O$ for reference, and **(b)** point $A$ for reference. Assume that tension $T$ is known.

## SOLUTION

**(a)** Figure 4-15b is the free-body diagram of the pulley. The required proof is obtained from the equilibrium equations. From Eq. 4-4, $\sum F_x = 0$, $\sum F_y = 0$, and $\sum M_O = 0$ are applicable. Note that the unknown reactions at point $O$ must be included in the free-body diagram. Thus, there are three unknowns, $T_1$, $O_x$, and $O_y$, and three independent equations.

| $\sum F_x = 0$ | $\sum F_y = 0$ | $\sum M_O = 0$ |
|---|---|---|
| $-T + T_1 \sin \theta + O_x = 0$ | $O_y - T_1 \cos \theta = 0$ | $Tr - T_1 r = 0$ |
| $O_x = T - T_1 \sin \theta$ | $O_y = T_1 \cos \theta$ | $T_1 = T$ |

Note that only the moment equation is necessary for the proof.

**(b)** The force summations are applicable as in part (a). Summing moments about point $A$,

$$\sum M_A = O_x r - O_y(2r) + T_1(\cos \theta)(2r - r \cos \theta)$$
$$+ T_1(\sin \theta)(r - r \sin \theta) = 0$$

Substituting for $O_x$ and $O_y$ as found in part (a),

$$(T - T_1 \sin \theta)r - (T_1 \cos \theta)(2r) + (T_1 \cos \theta)(2r) + (T_1 \sin \theta)r$$
$$- T_1 r(\cos^2 \theta + \sin^2 \theta) = 0$$
$$T = T_1$$

It is interesting that the tension $T_1$ is independent of the angle $\theta$ but the reaction at the axle $O$ depends on $\theta$.

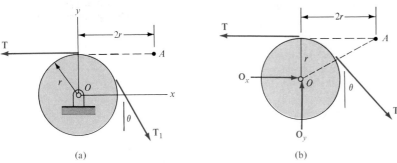

(a)    (b)

**FIGURE 4-15**

Equilibrium of a pulley

EXAMPLE 4-3    137

# EXAMPLE 4-4

Susie weighs 120 lb, and when she walks on her stilts (Fig. 4-16a) the whole weight is periodically applied to a triangular peg. Each peg is attached to the pole by a nail set 3 in. above corner $B$ of the peg. Determine the forces $\mathbf{F}_A$ and $\mathbf{V}$ in the nail and $\mathbf{F}_B$ on the peg, using the free-body diagram in Fig. 4-16b for the peg. Assume that the peg is touching the stilt leg only at points $A$ and $B$ and that force $\mathbf{F}_B$ must be horizontal (this implies that the peg is free to slide at $B$).

(a)

## SOLUTION

Using the free-body diagram, the applicable equations can be written at once from Eq. 4-4,

$$\sum F_x = 0$$
$$-F_A + F_B = 0 \qquad \textbf{(a)}$$
$$\sum F_y = 0$$
$$V - 120 \text{ lb} = 0 \qquad \textbf{(b)}$$
$$\sum M_A = 0$$
$$-(120 \text{ lb})(2 \text{ in.}) + F_B (3 \text{ in.}) = 0 \qquad \textbf{(c)}$$

Both Eqs. (b) and (c) can be solved directly, giving

$$V = 120 \text{ lb}$$
$$F_B = 80 \text{ lb}$$

From Eq. (a),

$$F_A = F_B = 80 \text{ lb}$$

The positive answers for $V$, $F_A$, and $F_B$ indicate that the directions were correct as assumed in Fig. 4-16b. Note that an equally valid set of equilibrium equations would be Eq. 4-5 with

$$\sum F_y = 0 \qquad \sum M_A = 0 \qquad \sum M_B = 0$$

## JUDGMENT OF THE RESULTS

It is seen by inspection of the free-body diagram that the clockwise couple formed by the 120-lb load and the force $\mathbf{V}$ is opposed by the counterclockwise couple formed by forces $\mathbf{F}_A$ and $\mathbf{F}_B$. Since forces $\mathbf{F}_A$ and $\mathbf{F}_B$ have a larger distance between them than the vertical forces do, the 80-lb force on the nail is reasonable. This also shows that increasing the vertical dimension of the peg may improve the reliability of the stilts by decreasing the magnitude of the force $\mathbf{F}_A$. Also, since the peg is a *three-force* member, a quick check reveals that all forces intersect at point $C$ in Fig. 4-16c.

(b)

(c)

**FIGURE 4-16**

Analysis of a supporting peg on a stilt

EXAMPLE 4-5

Consider the model of a crane in Fig. 4-17a. The weights of the column and boom are neglected. Assume that the reactions from the column to the boom are entirely horizontal at points $A$ and $B$. Determine the forces $\mathbf{F}_A$ and $\mathbf{F}_B$ at $A$ and $B$, respectively. Also calculate the tension $T$ in the cable which passes over a frictionless pulley at the top of the crane.

## SOLUTION

First draw the free-body diagram in Fig. 4-17b. The applicable equilibrium equations are written from Eq. 4-4,

$$\sum F_x = 0$$

$$T \cos 30° - T \cos 30° + F_A - F_B = 0 \qquad \textbf{(a)}$$

$$\sum F_y = 0$$

$$2T \sin 30° - 1 \text{ MN} = 0 \qquad \textbf{(b)}$$

$$\sum M_A = 0$$

$$(1 \text{ MN})(16 \text{ m}) - T(\sin 30°)(8.5 \text{ m}) + T(\sin 30°)(6.5 \text{ m}) \qquad \textbf{(c)}$$
$$- F_B(1.5 \text{ m}) = 0$$

Equation (b) can be solved first, giving

$$T = 1 \text{ MN}$$

From Eq. (c),

$$F_B = 10 \text{ MN}$$

From Eq. (a),

$$F_A = F_B = 10 \text{ MN}$$

## JUDGMENT OF THE RESULTS

The moment of the load about point $C$ is 15 MN·m. This is opposed by the couple acting at points $A$ and $B$, (10 MN)(1.5 m) = 15 MN·m, so the results appear to be correct. Note that the equal tension forces create no resultant moment since they are concurrent at the pulley.

(a)

**FIGURE 4-17**

Model of a crane

(b)

EXAMPLE 4-5    139

## EXAMPLE 4-6

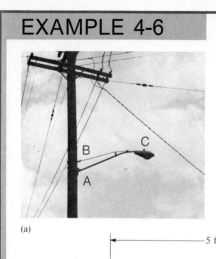

(a)

**FIGURE 4-18**

Street-light fixture

Some street-light fixtures are poorly designed. Such a fixture is shown and modeled in Fig. 4-18a and b, respectively. Find the tension $T$ in the cable $BC$ and the reactions at point $A$ in order to evaluate the fixture. Assume that $B$ is 0.75 ft above $A$, $C$ is 1 ft above $B$. Also assume that point $A$ is a pinned connection.

### SOLUTION

By the principle of transmissibility, a force can be applied anywhere along its line of action and still have the same effect on the rigid body. Consider the free-body diagram of the bar and cable together, Fig. 4-18b. By applying the tension at point $B$, moment summations about points $A$ and $B$ contain only one unknown each. With $\theta = \tan^{-1}\frac{1}{5} = 11.31°$,

$$\sum M_A = 0$$

$$(T\cos 11.31°)(0.75\text{ ft}) - (75\text{ lb})(5\text{ ft}) = 0 \qquad T = 510\text{ lb}$$

$$\sum M_B = 0$$

$$A_x(0.75\text{ ft}) - (75\text{ lb})(5\text{ ft}) = 0 \qquad A_x = 500\text{ lb}$$

The third equation can be either a force summation in the $y$ direction (Eq. 4-5) or a moment summation about point $C$ (Eq. 4-6). The most convenient is the moment equation,

$$\sum M_C = 0$$

$$-A_y(5\text{ ft}) + (500\text{ lb})(1.75\text{ ft}) = 0 \qquad A_y = 175\text{ lb}$$

### JUDGMENT OF THE RESULTS

Upon inspection it is seen that the bar in the model is a *two-force member* with all the external forces applied at points $A$ and $C$. A quick check of the results above shows that the resultant force at each point is directed along the bar $AC$, and is a compressive force of 530 lb. The design is poor because the cable is pulling *down* on the light in this arrangement, contrary to the purpose of having a cable. The result is an unnecessarily large compressive force (which could cause buckling) on the tube $AC$. For example, if the cable were mounted horizontally, the compressive force in the bar would be reduced to 227 lb.

## EXAMPLE 4-7

The bracket $ABC$ in Fig. 4-19 is part of a road simulator for automobiles (see Sec. 1-7 for a description of these). It is desired to apply a force $\mathbf{F}_C$ from a hydraulic actuator to hold the bracket when a force of magnitude $P = 6$ kN is applied to it as shown in Fig. 4-19a. **(a)** Determine the force $\mathbf{F}_C$ and the reactions $\mathbf{A}_x$ and $\mathbf{A}_y$ at the frictionless pin $A$. **(b)** Where should the actuator be attached if it could be anywhere between $C$ and $E$ in Fig. 4-19b?

### SOLUTION

**(a)** Figure 4-19a is the free-body diagram of the bracket. The necessary equilibrium equations are written from Eq. 4-4,

$$\sum F_x = 0$$

$$A_x - 6 \text{ kN} + \frac{4}{5}F_C = 0 \qquad \textbf{(a)}$$

$$\sum F_y = 0$$

$$A_y - \frac{3}{5}F_C = 0 \qquad \textbf{(b)}$$

$$\sum M_A = 0$$

$$(6 \text{ kN})(0.7 \text{ m}) - F_C(0.5 \text{ m}) = 0 \qquad \textbf{(c)}$$

Equation (c) can be solved first, giving

$$F_C = 8.4 \text{ kN}$$

From Eq. (a),

$$A_x = -0.72 \text{ kN}$$

The negative sign indicates that $\mathbf{A}_x$ is actually to the left in Fig. 4-10a. From Eq. (b),

$$A_y = 5.04 \text{ kN}$$

**(b)** For the possibility of changing the attachment of the actuator, it is enough to calculate the required force $\mathbf{F}_E$ in the extreme position at $E$. Here the actuator is not perpendicular to the line $AC$, so it is easiest to use the vector method of summing moments. From Eq. 4-1,

$$\sum \mathbf{F} = 0 \qquad \sum \mathbf{M}_A = \mathbf{r} \times \mathbf{F} = 0$$

Here it is sufficient to solve the moment equation, which has only one unknown, $\mathbf{F}_E$. The force summation should be checked by the readers for an exercise.

First the unknown force $\mathbf{F}_E$ is expressed in vector form using Eqs. 2-20 and 2-22.

$$\mathbf{F}_E = F\mathbf{n}_{ED} = F\left(\frac{0.95\mathbf{i} - 0.4\mathbf{j}}{1.031}\right)$$

$$= 0.92F\mathbf{i} - 0.388F\mathbf{j}$$

The moment summation is written as

$$\sum \mathbf{M}_A = \sum \mathbf{r} \times \mathbf{F} = \mathbf{r}_P \times \mathbf{P} + \mathbf{r}_E \times \mathbf{F}_E = 0$$

$$(0.7 \text{ m } \mathbf{j}) \times (-6 \text{ kN } \mathbf{i})$$
$$+ (0.15 \text{ m } \mathbf{i} + 0.2 \text{ m } \mathbf{j}) \times (0.92\mathbf{i} - 0.388\mathbf{j})F_E = 0$$

from which

$$F_E = \frac{4.2 \text{ kN·m}}{0.242 \text{ m}} = 17.34 \text{ kN}$$

Comparing the magnitudes of the forces $\mathbf{F}_C$ and $\mathbf{F}_E$, it is clear that attaching the actuator at $C$ is the most efficient.

## JUDGMENT OF THE RESULTS

The magnitudes of forces $\mathbf{F}_C$ and $\mathbf{F}_E$ appear correct in comparison with the magnitude of force $\mathbf{P}$ by inspection. The reaction $\mathbf{A} = A_x\mathbf{i} + A_y\mathbf{j}$ can be determined from the force triangle in Fig. 4-19c and from $A = \sqrt{A_x^2 + A_y^2}$. $A = 5.09$ kN using both methods, so $A_x$ and $A_y$ appear correct.

(a)

(b)

(c)

**FIGURE 4-19**

Bracket of a laboratory testing mechanism

EXAMPLE 4-7   141

In Probs. 4–1 to 4–9, determine whether the rectangular plate is completely or partially constrained. Compute the reactions where possible.

**4–1**    There are pin joints at *A*, *B*, *C*, and *D*. The force acts at the center of the plate.

Fig. P4-1

**4–2**    There are pin joints at *A*, *B*, *C*, and *D*. The force acts at the center of the plate.

Fig. P4-2

**4–3**    There are pin joints at *A*, *B*, *C*, *D*, *E*, and *F*.

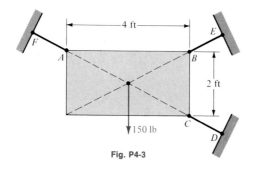

Fig. P4-3

**4–4**    There are pin joints at *A*, *B*, *C*, *D*, *E*, and *F*. The force acts at the center of the plate.

Fig. P4-4

**4–5**    There are pin joints at *A*, *B*, *D*, *E*, *F*, and *G*.

Fig. P4-5

**4–6**    There are pin joints at *A*, *B*, and *C*. The force acts at the center of the plate.

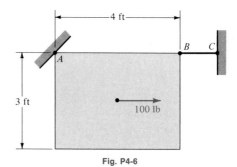

Fig. P4-6

**4–7**    There are pin joints at *A*, *B*, *C*, *D*, *E*, *F*, *G*, and *H*.

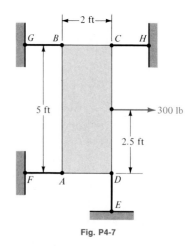

Fig. P4-7

**4–8**    There are pin joints at *A*, *B*, *C*, *E*, *F*, and *G*.

Fig. P4-13

**4-9** There are pin joints at $A$, $B$, $C$, $D$, $E$, and $F$. The force acts at the center of the plate.

**4-14** The magnitudes of the downward forces on the weightless beam are $F_1 = 1$ kN at $l_1 = d_1 = 0.4$ m, $F_2 = 3$ kN at $l_2 = 1$ m, and $F_3 = 2$ kN at $l_3 = 2$ m. Calculate the reactions at $A$ and $B$ if $d_2 = 2.5$ m.

Fig. P4-9

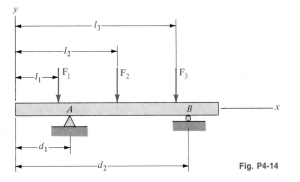

Fig. P4-14

**4-15** Assume that the largest allowable reactions $A_y = B_y = 5000$ lb on the beam in Fig. P4-14 are specified, and the dimensions are given as $d_1 = 0$, $d_2 = 10$ ft, $l_1 = 2$ ft, $l_2 = 5$ ft, and $l_3 = 9$ ft. Determine $F_1$, $F_2$, and $F_3$. If this is not possible, state the reason for the difficulty and determine the minimum additional information for solving the problem.

**4-10** Assume that tension $T_1$ and angles $\theta$ and $\phi$ are given for the pulley. Determine the tension $T_2$ and the reactions $C_x$ and $C_y$ at the center $C$. Neglect the weight of the pulley.

**4-16** The beam $AC$ is rigidly fixed at $A$. Determine the reaction acting on the beam at $A$ assuming that cable $BD$ is slack.

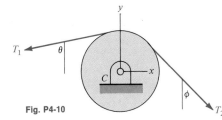

Fig. P4-10

**4-11** In Fig. P4-10, $T_1 = 5$ kN, $\theta = 70°$, and $\phi = 0$. Calculate the tension $T_2$ and the reactions on the pulley at point $C$.

**4-12** The largest allowable vertical reaction $C_y$ at center $C$ of the pulley in Fig. P4-10 is 1000 lb. Determine the maximum tension in the belt if $\theta = 60°$ and $\phi = 40°$.

**4-13** The magnitude of the force $\mathbf{F}_2$ on the weightless beam at $l_2 = 5$ ft is 2000 lb. The supports are at $d_1 = 2$ ft and $d_2 = 7$ ft. Determine the magnitude of the force $\mathbf{F}_1$ at $l_1 = 1$ ft that would make the vertical reactions at $A$ and $B$ equal.

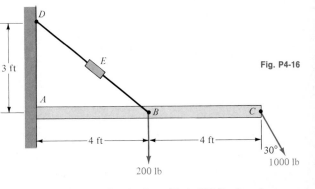

Fig. P4-16

**4-17** The tension in the cable is 300 lb after the turnbuckle $E$ is adjusted in Fig. P4-16. Determine the reactions acting on the beam at $A$.

**4–18** Calculate the reactions at hinge $A$ of the simple crane for $\theta = 70°$, $l = 3$ m, and $W = 3$ kN.

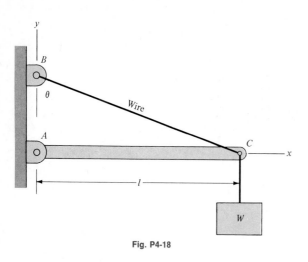

**Fig. P4-18**

**4–19** Calculate the reactions at hinge $A$ of the simple crane in Fig. P4-18 using $W = 1000$ lb, $l = 8$ ft, and $\theta = 45°$, $60°$, and $75°$. Plot $A_x$ and $A_y$ on the same diagram as a function of $\theta$. The wire is always attached to the end of the beam.

**4–20** (a) Prove that the beam which is hinged at $A$ is statically indeterminate for unknown forces $\mathbf{T}_B$, $\mathbf{T}_C$, $\mathbf{A}_x$, and $\mathbf{A}_y$. (b) Assume that the tension in wire $CD$ is measured with an instrument and found to be $T_C = 5$ kN. Calculate the tension $T_B$ and the reactions $A_x$ and $A_y$.

**Fig. P4-20**

**4–21** The forces $\mathbf{P}$ and $\mathbf{F}$ are holding the pinned, weightless bracket in the position shown. Use vector methods to determine the force $\mathbf{F}$ and the reactions at the pin support $O$ if $\mathbf{P} = (-5\mathbf{i} - 4\mathbf{j})$ kN. The coordinates are $(-0.2, 0.4, 0)$ m for point $A$, and $(0, 0.5, 0)$ m for point $B$.

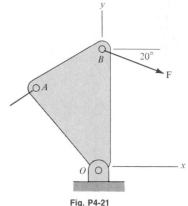

**Fig. P4-21**

**4–22** Calculate the reactions acting on the bracket at pin support $O$ and the force $\mathbf{F}$ in Fig. P4-21 if $\mathbf{P} = (-500\mathbf{i} - 350\mathbf{j})$ lb. The coordinates are $(-4, 9, 0)$ in. for $A$, and $(0, 12, 0)$ in. for $B$.

**4–23** Force $\mathbf{P} = -5$ kN $\mathbf{i}$ on the bracket. Consider different positions for point $C$ along the $x$ axis. Force $\mathbf{F}$ must always be directed to point $C$ (an actuator is located between $B$ and $C$). Plot the reactions $O_x$ and $O_y$ as functions of $d = 0.3$ m, $0.5$ m, and $1$ m. Plot both curves on the same diagram.

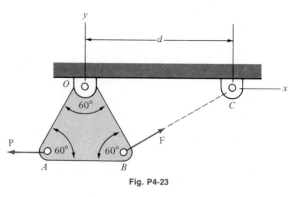

**Fig. P4-23**

**4–24** A highway sign is modeled as a 100-lb rigid member $ABDE$ and a weightless rigid bar $EF$. The 100-lb weight is assumed to act at $C$ of coordinates $(6, -1.5, 0)$ ft. Determine the forces acting at pin joints $A$ and $E$ on the sign.

144    EQUILIBRIUM OF RIGID BODIES

**Fig. P4-24**

**4-25** An athlete doing bench presses is modeled as follows. The 50-kg mass is concentrated at $B$, on a rigid arm $AB$ which is pivoted at $A$ and has a length of 0.6 m. Tabulate the magnitudes of the force $\mathbf{F}_A$ and the moment $\mathbf{M}_A$ acting at joint $A$ to maintain equilibrium as a function of the angle $\theta = 0$, $3°$, $6°$, and $9°$.

**Fig. P4-25**

**4-26** Consider the sketch of a gearbox. The external torques applied at the shafts are $\mathbf{C} = 200$ N·m $\mathbf{k}$ and $\mathbf{D} = -100$ N·m $\mathbf{k}$. Calculate the vertical forces at $A$ and $B$ necessary to resist the applied torques.

**Fig. P4-26**

**4-27** A lightweight aircraft is assumed to have a rigid wing of weight $W$, pivoted at joint $A$ and supported by wire $BD$. All members are assumed to be in the $xy$ plane. Tabulate the forces acting on the wing at support $A$ and the tension in wire $BD$ for connection $D$ at $2l/3$, $l$, and $4l/3$ from point $A$. The center of gravity is at point $C$.

**Fig. P4-27**

**4-28** The rigid bar $OB$ of negligible weight is pinned at $O$ and supported by a roller at $A$. (a) Determine the reactions at the supports. (b) Assume that an additional force $\mathbf{P} = P\mathbf{j}$ is applied at point $B$. What is the largest value that $P$ may have if bar $OB$ is to remain horizontal?

**Fig. P4-28**

**4-29** The mass of the truck bed is 10,000 kg. The bed is held in the tilted position by a hydraulic actuator $A$. Assume that there are single pivots at $O$, $B$, and $D$, all in the $xy$ plane. The coordinates are $B$: (1.3, 2.8, 0) m, $C$: (3.5, 2.4, 0) m, and $D$: (8, 0, 0) m. Determine the forces acting at joints $B$ and $D$ on the truck bed if $C$ is the location of the center of gravity.

**Fig. P4-29**

PROBLEMS **145**

**4–30** A member of a mechanism is supported by smooth pins $A$ and $B$ in vertical and horizontal tracks, respectively. Calculate the reactions on the pins.

$AB = 1.2$ m
$BD = 0.7$ m

End of track

**Fig. P4-30**

**4–31** Plot the reactions $A_x$, $A_y$, $B_x$, and $B_y$ in Prob. 4–30 for $\theta = 40°$, $50°$, and $60°$. Assume that force $\mathbf{P}$ is always in the given direction.

**4–32** Pins $A$ and $B$ can slide without friction in the narrow grooves. Determine the reactions at $A$ and $B$ if the bar is to be in equilibrium when a vertical force of magnitude $F$ is applied at $A$. Assume that the weight $\mathbf{W}$ of the bar acts at its midpoint. What is the complete system of forces required at $A$ and $B$ if the weight of the bar is neglected?

**Fig. P4-32**

**4–33** A jet engine is mounted on a test stand, with the vertical legs being rigidly attached to the 200-lb engine. The weight $\mathbf{W}$ and thrust $\mathbf{T} = -15{,}000$ lb $\mathbf{i}$ can be assumed to act at $A$ of coordinates $(-3, 4, 0)$ ft. Support $B$ on a frictionless surface has coordinates $(-5.5, 0, 0)$ ft. Determine the reactions at points $O$ and $B$.

**Fig. P4-33**

**4–34** The boom of a light crane is hinged at $A$ and is raised or lowered by a rope going over the pulley at $D$. Determine the reactions at $A$ and the tension in the rope for $\theta = 60°$.

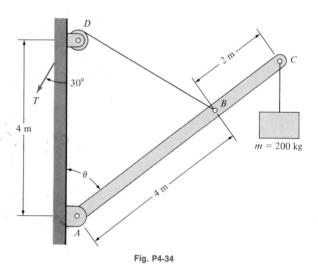

$m = 200$ kg

**Fig. P4-34**

**4–35** The boom of the light crane is supported by a smooth pin at $A$ and a roller at $B$. A small pulley at $C$ supports the load $W$. Calculate the reactions at $A$ and $B$ if the load weighs 45 kips and $\theta = 30°$.

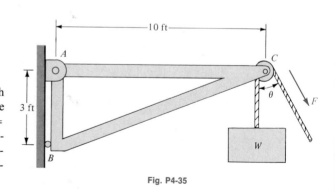

**Fig. P4-35**

**4–36** Consider the model of a tall crane where $ABC$ is assumed to be a single rigid member pivoted at $A$. Determine the tension $T$ in the supporting wire and the reactions at $A$. $\theta = 10°$.

(a)

(b)

Fig. P4-36

**4–37** Plot $T$, $A_x$, and $A_y$ as functions of $\theta = 6°$, $8°$, and $10°$ from Prob. 4–36. Use a single diagram.

**4–38** A ladder is modeled as a uniform bar of mass $m = 30$ kg and length $l = 4$ m. Calculate the reactions at $A$ and $B$ assuming that all contacting surfaces are frictionless, and that the weight acts at the midpoint of the bar.

Fig. P4-38

**4–39** A 20-ft ladder is set against a wall. It is on grass at $A$ and acts as if it were pinned. The wall is smooth and essentially frictionless at $B$. Calculate the changes in reactions at $A$ and $B$ when a 200-lb person moves from $D$ to $E$, if $AD = 7$ ft and $AE = 14$ ft.

Fig. P4-39

**4–40** The mass of a cart is $m = 800$ kg. The center of gravity is at point $C$, halfway between the wheels. Calculate the tension in the cable that holds the cart and the reactions on the frictionless wheels $A$ and $B$.

Fig. P4-40

**4–41** Plot the reactions on the frictionless wheels $A$ and $B$ for $d = 0.5$ m, $0.8$ m, and $1.1$ m of the cable in Prob. 4-40. Assume that $C$ remains at a distance of 0.8 m from the track.

**4–42** The 6000-lb cable car is suspended by two rigid, vertical bars on which the small, freely rotating wheels $A$ and $B$ are mounted. The cable is assumed to have a constant slope in the region shown. The coordinates of the center of gravity $C$ are $(5, -6, 0)$ ft. Determine the magnitude of the force $\mathbf{F}$ and the forces on the wheels from the cable when the gondola is in equilibrium.

Fig. P4-42

## CALCULATION OF UNKNOWN FORCES AND MOMENTS IN THREE DIMENSIONS

For the equilibrium of a rigid body in three dimensions, the six scalar equations of Eq. 4-2 are available. At most, six unknowns can be determined from these equations. If there are more than six unknowns, the problem is *statically indeterminate*.* As for problems in two dimensions, the unknowns are generally the constraints or reactions at the supports. As always, problem solutions rely on the drawing of proper free-body diagrams, so the table in Fig. 4-3 should be studied thoroughly. Equation 4-2 is repeated here for convenience.

$$\sum F_x = 0 \qquad \sum M_x = 0$$
$$\sum F_y = 0 \qquad \sum M_y = 0$$
$$\sum F_z = 0 \qquad \sum M_z = 0$$

The concepts of constraint described in Sec. 4-4 can be extended to the three-dimensional case. Three-dimensional problems may involve up to six degrees of freedom (linear motion in three directions and rotation about three axes). If there are less than six constraints, the rigid body is said to be *partially constrained* under general loading conditions, and there are more equations than unknowns. Under certain loading conditions, however, the rigid body may still be in equilibrium if the remaining equilibrium equations reduce to simple identities. If the body is *improperly constrained*, even six constraints will not ensure equilibrium. In three dimensions, improper constraint exists when all the restraining forces are either parallel or intersect the same axis.

The choice of a convenient point about which to sum moments was an important step for problem solving in two dimensions. In three-dimensional problems, the vector moment summation of Eq. 4-1 can also be made about a convenient point. The scalar moment summations are obtained by equating the components of the resulting vector equation. Moment summations, however, are not limited to those about the coordinate axes but can be made about any line in space. The choice of a convenient axis about which to sum moments can greatly simplify the solution of three-dimensional problems. Summing moments about a line that intersects two points at which the majority of unknowns act often results in an equation with only one unknown.

---

* The additional unknowns may be determined using the methods of mechanics of materials.

# EXAMPLE 4-8

An airplane of weight $W$ is resting on three wheels (Fig. 4-20), which are located at point $O$, point $A$ of coordinates $(a, 0, b)$ and point $B$ of coordinates $(a, 0, -b)$. The weight acts at point $C$ of coordinates $(0.7a, h, 0)$. Determine the forces acting on the wheels.

## SOLUTION

The airplane is only partially constrained, as there are only three reactions. For a loading such as a thrust from the engines, the plane would not remain at rest. For the particular loading of the weight alone, however, the plane will remain in equilibrium. Since all forces are in the $y$ direction, $\sum F_x$, $\sum F_z$, and $\sum M_y$ are all identically equal to zero.

The remaining three summations from Eq. 4-2 can be used to solve for the three unknowns, $O_y$, $A_y$, and $B_y$. By summing moments about the line through points $A$ and $B$ instead of about the $z$ axis, however, an equation is derived containing only one unknown.

$$\sum M_{AB} = O_y a - W(0.3a) = 0$$
$$O_y = 0.3W$$

$$\sum M_x = A_y b - B_y b = 0$$
$$A_y = B_y$$

$$\sum F_y = A_y + B_y + 0.3W - W = 0$$
$$2A_y = 0.7W$$
$$A_y = B_y = 0.35W$$

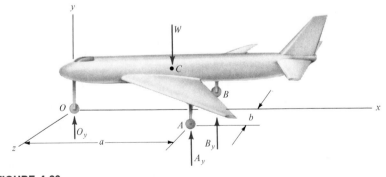

**FIGURE 4-20**

Model of an airplane resting on three wheels

EXAMPLE 4-8   149

## EXAMPLE 4-9

(a)

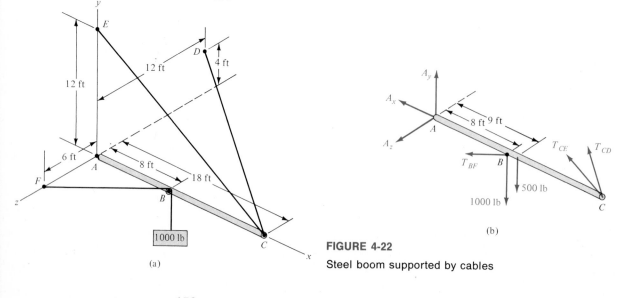

(b)

**FIGURE 4-21**

Model of a factory robot

The arm of a factory robot in Fig. 4-21a consists of three bars with coordinates $A$: $(0.6, -0.3, 0.4)$ m, $B$: $(1, -0.2, 0)$ m, and $C$: $(0.9, 0.1, -0.25)$ m. The weight of the arm is represented by $\mathbf{W}_A = -60$ N $\mathbf{j}$ at $A$, and $\mathbf{W}_B = -40$ N $\mathbf{j}$ at $B$. A moment $\mathbf{M}_C = (100\mathbf{i} - 20\mathbf{j} + 50\mathbf{k})$ N·m is applied to the arm at $C$. Determine the force and moment reactions at $O$, assuming that all joints are temporarily fixed.

### SOLUTION

To solve for the reactions at $O$, assume that the arm is rigid and consider equilibrium. The free-body diagram is drawn in Fig. 4-21b, showing the unknown force and moment reactions at $O$. From Eq. 4-1,

$$\sum \mathbf{F} = 0$$

$$\mathbf{F}_O + \mathbf{W}_A + \mathbf{W}_B = 0$$

$$\mathbf{F}_O - 60 \text{ N } \mathbf{j} - 40 \text{ N } \mathbf{j} = 0$$

$$\mathbf{F}_O = 100 \text{ N } \mathbf{j}$$

$$\sum \mathbf{M}_O = 0$$

$$\mathbf{M}_O + \mathbf{M}_C + (\mathbf{r}_{OA} \times \mathbf{W}_A) + (\mathbf{r}_{OB} \times \mathbf{W}_B) = 0$$

$$\mathbf{M}_O + (100\mathbf{i} - 20\mathbf{j} + 50\mathbf{k}) \text{ N·m} + (0.6\mathbf{i} - 0.3\mathbf{j} + 0.4\mathbf{k}) \text{ m} \times (-60 \text{ N } \mathbf{j})$$
$$+ (\mathbf{i} - 0.2\mathbf{j}) \text{ m} \times (-40 \text{ N } \mathbf{j}) = 0$$

$$\mathbf{M}_O + 100 \text{ N·m } \mathbf{i} - 20 \text{ N·m } \mathbf{j} + 50 \text{ N·m } \mathbf{k} - 36 \text{ N·m } \mathbf{k} + 24 \text{ N·m } \mathbf{i}$$
$$- 40 \text{ N·m } \mathbf{k} = 0$$

$$\mathbf{M}_O = (-124\mathbf{i} + 20\mathbf{j} + 26\mathbf{k}) \text{ N·m}$$

## EXAMPLE 4-10

The uniform steel boom in Fig. 4-22a weighs 500 lb. It is supported by two cables, $CE$ and $CD$, and a ball-and-socket joint at $A$. The cable supporting the 1000-lb load passes over a small pulley at $B$. Find the tension in cable $CE$ and in cable $CD$.

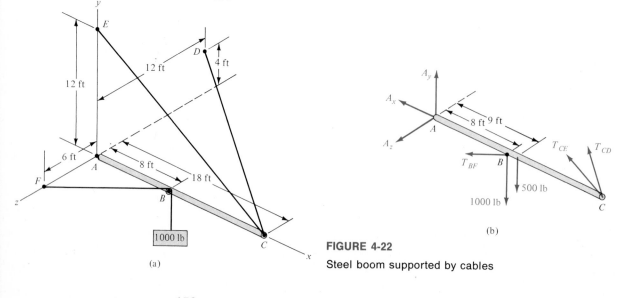

(a)

**FIGURE 4-22**

Steel boom supported by cables

(b)

EQUILIBRIUM OF RIGID BODIES

## SOLUTION

The free-body diagram of the boom is drawn in Fig. 4-22b. The six unknowns are the three force reactions at the ball-and-socket joint $A$ and the tensions in the three cables. The tension in cable $BF$ is determined by inspection to be 1000 lb, since it is supporting the weight and remains the same around the pulley. The problem statement does not ask for the reactions at $A$, so the easiest equilibrium equations to work with are moment summations about the $y$ and $z$ axes. The moment summation about the $y$ axis involves only the unknown $T_{CD}$. Once it is determined, summation of moments about the $z$ axis gives $T_{CE}$ directly.

First write all tension forces in vector form using Eqs. 2-20 and 2-23.

$$\mathbf{T}_{BF} = 1000 \text{ lb } \mathbf{n}_{BF}$$

$$= 1000 \text{ lb } \frac{-8\mathbf{i} + 6\mathbf{k}}{\sqrt{8^2 + 6^2}}$$

$$= (-800\mathbf{i} + 600\mathbf{k}) \text{ lb}$$

$$\mathbf{T}_{CD} = T_{CD}\mathbf{n}_{CD}$$

$$= T_{CD} \frac{-18\mathbf{i} + 4\mathbf{j} - 12\mathbf{k}}{\sqrt{18^2 + 4^2 + 12^2}}$$

$$= T_{CD}(0.8181\mathbf{i} + 0.1818\mathbf{j} - 0.545\mathbf{k})$$

$$\mathbf{T}_{CE} = T_{CE}\mathbf{n}_{CE}$$

$$= T_{CE} \frac{-18\mathbf{i} + 12\mathbf{j}}{\sqrt{18^2 + 12^2}}$$

$$= T_{CE}(-0.833\mathbf{i} + 0.556\mathbf{j})$$

Using determinants for $M_y$, as in Eq. 3-24,

$$\sum M_y = \begin{vmatrix} 0 & 1 & 0 \\ 8 \text{ ft} & 0 & 0 \\ -800 \text{ lb} & 0 & 600 \text{ lb} \end{vmatrix} + \begin{vmatrix} 0 & 1 & 0 \\ 18 \text{ ft} & 0 & 0 \\ -0.818 & 0.1818 & -0.545 \end{vmatrix} T_{CD} = 0$$

$$-4800 \text{ lb·ft} + (-18 \text{ ft})(-0.545)T_{CD} = 0$$

$$T_{CD} = 489 \text{ lb}$$

$$\mathbf{T}_{CD} = (-400\mathbf{i} + 88.9\mathbf{j} - 267\mathbf{k}) \text{ lb}$$

Solving for $T_{CE}$,

$$\sum M_z = \begin{vmatrix} 0 & 0 & 1 \\ 18 \text{ ft} & 0 & 0 \\ -400 \text{ lb} & 88.9 \text{ lb} & -267 \text{ lb} \end{vmatrix} + \begin{vmatrix} 0 & 0 & 1 \\ 18 \text{ ft} & 0 & 0 \\ -0.833 & 0.556 & 0 \end{vmatrix} T_{CE}$$

$$-(8 \text{ ft})(1000 \text{ lb}) - (9 \text{ ft})(500 \text{ lb}) = 0$$

$$(18 \text{ ft})(88.9 \text{ lb}) + (18 \text{ ft})(0.556 \text{ lb})T_{CE} - 8000 \text{ lb·ft} - 4500 \text{ lb·ft} = 0$$

$$T_{CE} = 1089 \text{ lb}$$

$$\mathbf{T}_{CE} = (-907\mathbf{i} + 605\mathbf{j}) \text{ lb}$$

Note that the scalar equations obtained by summing moments about the $y$ and $z$ axes are the same equations that would result from equating the $y$ and $z$ components of the vector equation $\sum \mathbf{M}_A = \mathbf{0}$.

EXAMPLE 4-10   151

**4–43** The aircraft rests on landing gears at $A$, $B$, and $D$. The total weight is represented by a concentrated force of magnitude $W = 50,000$ lb acting at $C$ (in the negative $z$ direction in the given coordinate system). Determine the reactions at wheels $A$, $B$, and $D$.

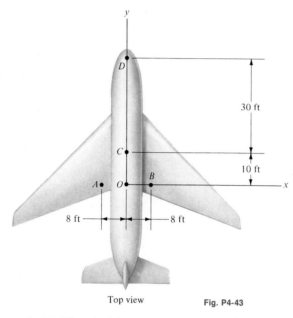

Top view      **Fig. P4-43**

**4–44** The wheel in a test has a radius $r = 0.35$ m and is rigidly connected to a shaft $AO = 0.5$ m. Determine the reactions at $A$ when a force $\mathbf{F} = 2$ kN $\mathbf{k}$ is applied at point $B$ of coordinates $(0, -0.35$ m, $0)$. Assume that the shaft is fixed at support $A$.

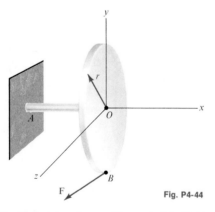

**Fig. P4-44**

**4–45** Determine the reactions at $A$ in Prob. 4–44 if $\mathbf{F} = (-1.5\mathbf{i} - 0.5\mathbf{j} + 2\mathbf{k})$ kN.

**4–46** Determine the reactions at $A$ and $O$ due to the three forces acting on the highway sign. $\mathbf{F} = -200$ lb $\mathbf{k}$, $\mathbf{P} = -130$ lb $\mathbf{j}$, and $\mathbf{W} = -300$ lb $\mathbf{j}$.

**Fig. P4-46**

**4–47** Consider the addition of a guy wire at point $A$ to the structure described in Prob. 4–46. The wire is also attached to point $C$ of coordinates $(-6$ ft, $0$, $0)$. The tension in the wire is adjusted to eliminate the moment of force $\mathbf{P}$ about point $O$. Calculate the resulting reactions at $O$.

**4–48** A force $\mathbf{F}_A = 30\mathbf{j}$ lb and a moment $\mathbf{M}_A = 70\mathbf{j}$ lb·ft are applied to the drill press at point $A$ of coordinates $(5, 4, 0)$ ft. Determine the reactions $\mathbf{F}_O$ and $\mathbf{M}_O$ at point $O$, neglecting the weight of the structure.

**Fig. P4-48**

**4-49** A tool is modeled as a rigid member $OA$. Determine the reactions at $O$ if $\mathbf{F} = -20$ N $\mathbf{k}$ and $\mathbf{M} = 10$ N·m $\mathbf{j}$ are applied at point $A$. Arm $AB$ is parallel to the $x$ axis.

Fig. P4-49

**4-50** Determine the reactions at $O$ in Prob. 4–49 if an additional force $\mathbf{P} = -10$ N $\mathbf{j}$ is acting at $A$.

**4-51** Consider the hand-crank hoist shown, where $A$ and $B$ are fixed sleeve bearings, and the wire holding the 100-lb weight is wound on the drum $C$. $\mathbf{F}$ is the minimum force required to hold the weight $W$. Calculate the reactions at $A$ and $B$ when the crank arm $DE$ is horizontal in position 1 (Fig. P4-51c).

(a) Top view ($ADE$ is horizontal)

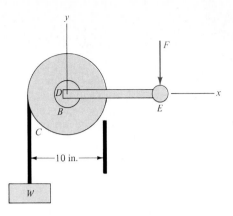

(b) End view from right side

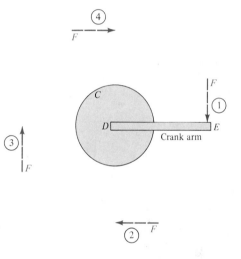

(c) End view from right side

Fig. P4-51

**4-52** Determine the reactions at $A$ and $B$ in Prob. 4–51 when the crank arm $DE$ is horizontal in position 3 (Fig. P4-51c).

**4-53** Determine the reactions at $A$ and $B$ in Prob. 4–51 when the crank arm $DE$ is vertical in position 2 (Fig. P4-51c).

**4-54** Determine the reactions at $A$ and $B$ in Prob. 4–51 when the crank arm is vertical in position 4 (Fig. P4-51c).

**4–55** A hand-crank hoist is designed to have a ratcheting crank arm $DE$ with a 60° range of motion. $A$ and $B$ are fixed sleeve bearings, and the wire holding the weight $W$ is wound on the drum $C$. Determine the reactions $A_z$ and $A_y$ when the crank arm $DE$ is parallel to the $y$ axis (position 2). Assume that $\mathbf{F}$ is always perpendicular to $DE$, and the ratchet applies no force.

(a) Side view

(b) End view

(c)

**Fig. P4-55**

**4–56** Calculate the reactions at $A$ and $B$ in Prob. 4–55 when arm $DE$ is in position 1 (Fig. P4-55c).

**4–57** Assume that in Prob. 4–55 the positions of $B$ and $C$ are fixed but bearing $A$ could be placed $\pm 50$ mm from the position indicated. If the bearing reactions should be minimized, where should bearing $A$ be placed on the $x$ axis?

**4–58** The weightless boom $OA$ is supported by a ball-and-socket joint at $O$ and wires $AB$ and $AC$. The mass of the 150-kg sign can be assumed to be evenly distributed between the two supports at $D$ and $E$. Determine the reaction at $O$ and the tension in the wires for $A$, $B$, and $C$ of coordinates $(0, 0, 2)$ m, $(-1, 1.5, 0)$ m, and $(1.2, 1.5, 0)$ m.

**Fig. P4-58**

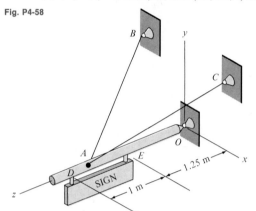

**4–59** The mast has a ball-and-socket joint at $O$ to allow rotation in any direction. The tensions in the wires are along $BA$ and $BC$. Calculate the tensions $T_A$ and $T_C$ if $\mathbf{F} = (-200\mathbf{i} - 100\mathbf{k})$ lb, $OA = OC = 8$ ft, $OB = 12$ ft, and $OD = 15$ ft.

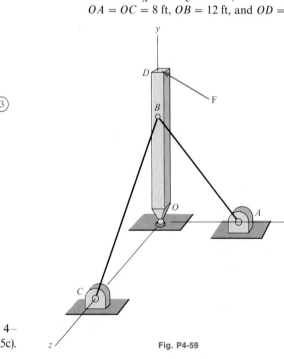

**Fig. P4-59**

**4–60** An antenna tower is modeled as a rigid bar with a ball-and-socket joint at $O$ and supported by three identical guy wires (top view). The effect of wind is represented by a concentrated force that causes wire $C$ to become slack. Determine the reaction at $O$ and the tension in wires $AD$ and $BD$.

(a) Top view

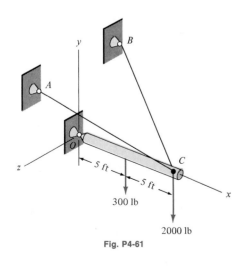

Fig. P4-60

(b) Side view

**4–61** The horizontal boom of the crane is hinged at $O$ and is held by cables at $A$ and $B$. Determine the reactions at $O$ for $A$ and $B$ of coordinates $A$: $(0, 6, 4)$ ft and $B$: $(1, 6, -3)$ ft.

Fig. P4-61

**4–62** Determine the reactions at point $O$ in Prob. 4–61 if the supports holding the cables have coordinates $A$: $(0, 4, 6)$ ft and $B$: $(0, 7, -4)$ ft.

**4–63** The hood of a car is modeled as a uniform rectangular plate of 15-kg mass (the weight acts at the center of the plate). Determine the reactions at hinges $A$ and $B$ and the force in bar $CD$, which is pinned at both ends. Assume there are no moment reactions at either hinge.

Fig. P4-63

**4–64** Solve Prob. 4–63 as stated and also with the hinges $A$ and $B$ moved to $x_A = 0.3$ m and $x_B = 1.1$ m, respectively. Which are the best positions for the hinges to minimize the supporting forces of the plate?

**4–65** What are the reactions at hinge $B$ necessary to maintain the equilibrium of the car hood in Prob. 4–63 if the hinge at $A$ rusts out and provides no support? Assume hinge $B$ can support moment reactions.

## INTERRELATED FREE-BODY DIAGRAMS.
## NEWTON'S THIRD LAW

An important concept in analyzing the equilibrium of rigid bodies is that parts of those rigid bodies are also in equilibrium. Consequently, complete free-body diagrams can be drawn for all parts that are of interest. Neighboring parts may cause forces and moments on each other, and these must be shown on the individual diagrams. The internal forces and moments are unknown when the whole body is first analyzed; of course, they are not shown on the free-body diagram of the whole body. The forces and moments caused by neighboring parts on each other are related according to Newton's Third Law: the forces are equal in magnitude and opposite in direction; thus, they cancel in pairs when the whole body is considered. These concepts are most easily appreciated with the aid of drawings.

Reconsider the two-piece wave board that was first discussed in Ch. 3. Assume that the two boards act as a single, weightless rigid body, with actuator $D$ "locked in position." Forces $\mathbf{F}_1$ and $\mathbf{F}_2$ in Fig. 4-23a are the net forces from the water on each board. They could be reduced to a single net force ($\mathbf{F} = \mathbf{F}_1 + \mathbf{F}_2$) when the board is considered as a single rigid body. The actuator $D$ is a two-force member between the two boards, so it applies two equal and opposite forces to the whole rigid body. Thus, the forces $\mathbf{D}$ and $-\mathbf{D}$ are not necessary in Fig. 4-23a, but at this stage it is helpful to see them in proceeding to the individual free-body diagrams.

The free-body diagram of the upper board is drawn in Fig. 4-23b. Here the forces $\mathbf{D}$ and $\mathbf{F}_1$ must be shown since they are external forces acting directly on this member. There is an unknown force at $B$, and this is conveniently shown by its rectangular components. The directions of $\mathbf{B}_x$ and $\mathbf{B}_y$ are arbitrarily assumed since they are not known.

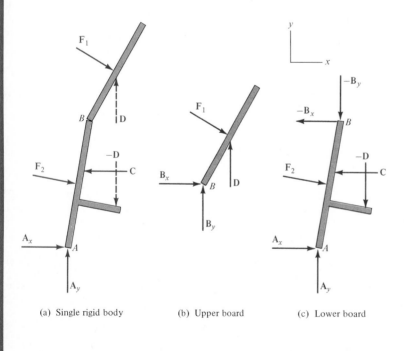

(a) Single rigid body      (b) Upper board      (c) Lower board

**FIGURE 4-23**

Interrelated free-body diagrams using vector notation

The free-body diagram of the lower board is readily drawn, but here the existence of the diagram for the upper board must be taken into account. The forces $-\mathbf{D}$ and $\mathbf{F}_2$ are external forces for the lower board, so they must be shown as in Fig. 4-23c. $\mathbf{A}_x$ and $\mathbf{A}_y$ are simply taken from Fig. 4-23a because at point $A$ it does not matter which free body is analyzed. The force at $B$ on the lower board is caused by the upper board; therefore, $-\mathbf{B}_x$ and $-\mathbf{B}_y$ in Fig. 4-23c must be equal in magnitude and opposite in direction to those in Fig. 4-23b. The concept of action and reaction (Newton's Third Law) must be used in each case when free-body diagrams are related to one another.

Frequently, a scalar notation is used with the arrows representing forces. In that case negative signs are not used to distinguish forces of action and reaction. The arrows in opposite directions represent such differences, while the symbols by the arrows show only the magnitudes of the forces. For example, Fig. 4-23b and c are repeated with vector magnitude notation in Fig. 4-24a and b.

For another example of interrelated free-body diagrams, consider the test setup for a seat belt in Fig. 4-25a. The system to analyze is the *load frame* (the part of the equipment that applies forces to test specimens) shown in Fig. 4-25a. The only external forces on the load frame are its weight $\mathbf{W}$ and the reaction $\mathbf{R}_1$ from the floor. Note that this diagram is completely independent of any force that may be applied to a test specimen $S$ in the load frame.

The test force in the load frame is applied by a hydraulic actuator $A$ that is not visible in Fig. 4-25a. The pump $P$ delivers oil under pressure to the upper chamber $C_1$ of the actuator shown in Fig. 4-25b. This forces the piston rod $B$ downward against the tension $\mathbf{T}$ developed in the specimen. The upward force $\mathbf{T}$ on the actuator is resisted by the reaction $\mathbf{R}_2$ at the place where the actuator is attached to the load frame. Naturally, if pressure is applied in chamber $C_2$, the rod $B$ is forced upward, applying compression to the specimen. In this case the upper force $\mathbf{T}$ and $\mathbf{R}_2$ would reverse directions to point toward each other.

The free-body diagram of the test specimen is drawn in Fig. 4-25c on the basis of part b of the figure. The lower force $\mathbf{T}_1$ is drawn first because it must be equal and opposite to the upward force $\mathbf{T}$ in Fig. 4-25b. The upper force $\mathbf{T}_2$ is readily drawn for equilibrium of the specimen.

Above the specimen there is a so-called *load cell L* for the measurement of forces using electronic equipment. The free-body diagram of the load cell is drawn in Fig. 4-25d. Force $\mathbf{T}_3$ on the bottom of the load cell is drawn first on the basis of the force $\mathbf{T}_2$ on top of the specimen. The upper force $\mathbf{T}_4$ on the load cell is again readily drawn for equilibrium. The dashed lines between Fig. 4-25b, c, and d show the progression in drawing interrelated free-body diagrams.

The load cell is attached to the horizontal *cross-head H* of the load frame. The free-body diagram of the cross-head is relatively complex, but advantage is taken of its obvious symmetry in Fig. 4-25e. The force $\mathbf{T}_5$ is from the load cell, and it must be opposed by vertical forces $\mathbf{V}$ where the cross-head is attached to the columns; for equilibrium, $T_5 = 2V$. The cross-head cannot rotate with respect to the columns, so it is necessary to show unknown moments $\mathbf{M}$ on the free-body diagram. Essentially, the

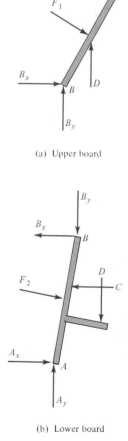

(a) Upper board

(b) Lower board

**FIGURE 4-24**

Interrelated free-body diagrams using vector magnitude notation

(a)

(b) Actuator $A$ applies tension to the specimen

(c) Specimen

(d) Load cell

(e) Cross head

(f) Left column of load frame

**FIGURE 4-25**
Test setup of a seat belt. (Photograph courtesy MTS Systems Corporation, Minneapolis, Minnesota.)

cross-head acts as a beam built in on both ends, which makes it a statically indeterminate member (one of the columns is redundant in theory).

For a last item in the sequence of major, related free-body diagrams (and many more are possible), consider one of the columns. $\mathbf{V}$ and $\mathbf{M}_1$ in Fig. 4-25f are caused by the cross-head on the column, while $\mathbf{R}_3$ and $\mathbf{M}_3$ are the reactions from the lower platen of the load frame.

Note that the whole load frame is quite heavy, so its weight is included in Fig. 4-25a. However, the weights of individual parts may be ignored in their free-body diagrams since often they have much larger external forces acting on them.

EXAMPLE 4-11

ONE-DIMENSIONAL PROBLEM

A heavy chain consists of 10 identical links, each of 3-kg mass. Draw the free-body diagrams of the top three links of the freely hanging chain, and determine the forces applied to link $A$ by its neighbors. See Fig. 4-26a.

SOLUTION

The free-body diagrams are drawn in Fig. 4-26b, c, and d by separating each link and applying Newton's Third Law for the forces between the links. The weight of each individual link is applied, and the reaction $O_y$ at the support $O$ is shown on the top link.

To determine the forces applied to link $A$ by its neighbors ($\mathbf{F}_2$ and $\mathbf{F}_3$), use Eq. 4-3, $\sum F_y = 0$, since all the forces are collinear vertically. This equation for link $A$ is an equation with two unknowns, so it is necessary to find a link about which more information is known.

Consider the top link and $O_y$. The reaction at the support is due to the weight of the total 10 links of the chain, $O_y = 10(3 \text{ kg})(9.81 \text{ m/s}^2) = 294 \text{ N}$. Using Eq. 4-3 with the $y$ direction positive upward:

Equilibrium for the first link:

$$\sum F_y = 0$$
$$O_y - mg - F_1 = 0$$
$$F_1 = O_y - mg = 294 \text{ N} - (3 \text{ kg})(9.81 \text{ m/s}^2)$$
$$F_1 = 265 \text{ N}$$

Equilibrium for the second link:

$$\sum F_y = 0$$
$$F_1 - mg - F_2 = 0$$
$$F_2 = F_1 - mg = 265 \text{ N} - 29.4 \text{ N}$$
$$F_2 = 236 \text{ N}$$

Equilibrium for link $A$:

$$\sum F_y = 0$$
$$F_2 - mg - F_3 = 0$$
$$F_3 = F_2 - mg = 236 \text{ N} - 29.4 \text{ N}$$
$$F_3 = 207 \text{ N}$$

The forces applied to link $A$ by its neighbor links are

$$\mathbf{F}_2 = 236 \text{ N} \, \mathbf{j} \quad \text{and} \quad \mathbf{F}_3 = -207 \text{ N} \, \mathbf{j}$$

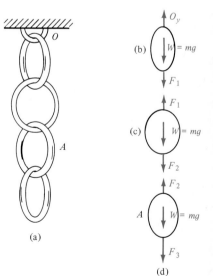

**FIGURE 4-26**

One-dimensional analysis of links in a chain

EXAMPLE 4-11    159

# EXAMPLE 4-12

(a)

## TWO-DIMENSIONAL PROBLEM

The boat winch of Fig. 4-27a is drawn schematically in Fig. 4-27b. The tension in the cable is up to 2000 lb. Determine the magnitude of the force **F** needed at the handle to start winding the cable up. Also determine the magnitude of the force **F**$_s$ that the stop must exert to keep the cable from unwinding. Handle $H$ is fixed on gear $B$.

## SOLUTION

The interrelated free-body diagrams of the gears are in Fig. 4-27c and d. The forces of magnitude $F_B$ are in a different direction on each gear. From the free-body diagrams and Eq. 4-6 (noting that only moment equations are useful since force directions are not given),

$$\sum M_A = 0$$

$$F_B(3 \text{ in.}) - (2000 \text{ lb})(1 \text{ in.}) = 0$$

$$F_B = 667 \text{ lb}$$

$$\sum M_B = 0$$

$$(667 \text{ lb})(1 \text{ in.}) - (F)(12 \text{ in.}) = 0$$

$$F = 56 \text{ lb}$$

The force on the stop is found from Fig. 4-27c with $F = 0$,

$$\sum M_B = 0$$

$$-(F_s)(1 \text{ in.}) + (667 \text{ lb})(1 \text{ in.}) = 0$$

$$F_s = 667 \text{ lb}$$

## JUDGMENT OF THE RESULTS

The magnitudes of **F**$_B$ and **F** are reasonable, and $F_s$ must equal $F_B$ when $F$ is zero (hands off). Note that $F$ can be reduced by decreasing the radius of gear $B$, and by increasing the length of the handle arm.

$R_A = 3$ in.

2000 lb

$B$

Stop

$r_B = r_A = 1$ in.

12 in.

$H$

$F$

(b)

(c)

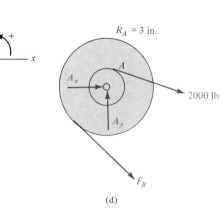

(d)

**FIGURE 4-27**

Two-dimensional analysis of a boat winch

# EXAMPLE 4-13

The mechanical hoe shown in Fig. 4-28a is assumed to have two identical, uniform bars $OA$ and $AB$, each of 200-lb weight and 8-ft length, with a force applied at $B$,

$$\mathbf{F} = (150\mathbf{i} + 250\mathbf{j} - 70\mathbf{k})\text{ lb}$$

Determine the forces and moments at points $O$ and $A$ necessary for equilibrium if $\theta = 30°$ and $\phi = 50°$. Bars $OA$ and $AB$ are in the $xy$ plane.

## SOLUTION

The interrelated free-body diagrams for arm $AB$ and arm $OA$ are drawn in Fig. 4-28b and c. Using Eq. 4-1 and Fig. 4-28b,

$$\sum \mathbf{F} = 0$$

$$(150\mathbf{i} + 250\mathbf{j} - 70\mathbf{k})\text{lb} - 200\text{ lb } \mathbf{j} + \mathbf{F}_A = 0$$

$$\mathbf{F}_A = (-150\mathbf{i} - 50\mathbf{j} + 70\mathbf{k})\text{ lb}$$

$$\sum \mathbf{M}_A = 0$$

$$\mathbf{M}_A + (4\sin 20°\,\mathbf{i} - 4\cos 20°\,\mathbf{j})\text{ ft} \times (-200\text{ lb}\,\mathbf{j})$$
$$+ (8\sin 20°\,\mathbf{i} - 8\cos 20°\,\mathbf{j})\text{ ft} \times (150\mathbf{i} + 250\mathbf{j} - 70\mathbf{k})\text{ lb} = 0$$

$$\mathbf{M}_A - 274\text{ lb·ft } \mathbf{k} + 684\text{ lb·ft } \mathbf{k} + 191.5\text{ lb·ft } \mathbf{j} + 1128\text{ lb·ft } \mathbf{k} + 526\text{ lb·ft } \mathbf{i} = 0$$

$$\mathbf{M}_A = (-526\mathbf{i} - 191.5\mathbf{j} - 1538\mathbf{k})\text{ lb·ft}$$

Note that $\mathbf{F}_A$ and $\mathbf{M}_A$ on bar $OA$ are equal and opposite to those found above on bar $AB$:

For bar $OA$,

$$\sum \mathbf{F} = 0$$

$$-(-150\mathbf{i} - 50\mathbf{j} + 70\mathbf{k})\text{ lb} - 200\text{ lb } \mathbf{j} + \mathbf{F}_O = 0$$

$$\mathbf{F}_O = (-150\mathbf{i} + 150\mathbf{j} + 70\mathbf{k})\text{ lb}$$

$$\sum \mathbf{M}_O = 0$$

$$\mathbf{M}_O - (-526\mathbf{i} - 191.5\mathbf{j} - 1538\mathbf{k})\text{ lb·ft} + (4\sin 30°\,\mathbf{i} + 4\cos 30°\,\mathbf{j})\text{ ft}$$
$$\times (-200\text{ lb } \mathbf{j}) + (8\sin 30°\,\mathbf{i} + 8\cos 30°\,\mathbf{j})\text{ ft}$$
$$\times [-(-150\mathbf{i} - 50\mathbf{j} + 70\mathbf{k})]\text{ lb} = 0$$

$$\mathbf{M}_O + 526\text{ lb·ft } \mathbf{i} + 191.5\text{ lb·ft } \mathbf{j} + 1538\text{ lb·ft } \mathbf{k} - 400\text{ lb·ft } \mathbf{k} + 200\text{ lb·ft } \mathbf{k}$$
$$+ 280\text{ lb·ft } \mathbf{j} - 1039\text{ lb·ft } \mathbf{k} - 485\text{ lb·ft } \mathbf{i} = 0$$

$$\mathbf{M}_O = (-41\mathbf{i} - 472\mathbf{j} - 299\mathbf{k})\text{ lb·ft}$$

(b)

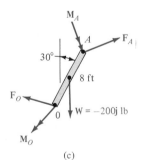

(c)

**FIGURE 4-28**

Three-dimensional analysis of a mechanical hoe

EXAMPLE 4-13    161

EXAMPLE 4-14

## TWO-DIMENSIONAL PROBLEM

A biaxial test setup of a shock absorber $S$ is shown in Fig. 4-29a. The magnitudes of the tensile forces from the testing machine are $F_y = 15$ kN and $F_x = 10$ kN when line $AB$ makes a 5° angle with the horizontal. Draw and analyze the free-body diagrams for the specimen $S$, load cell $L$, and member $ACD$. Calculate the reactions at $E$, the base of the horizontal actuator. Assume that the specimen is pinned at its ends, and that the weights of the individual parts are negligible compared to the applied forces.

### SOLUTION

Arbitrarily, start with the load cell in Fig. 4-29b. The vertical force of magnitude 15 kN is given (this is known quite accurately because it is measured by the load cell itself). The magnitude of the horizontal force $\mathbf{H}_1$ is not known precisely; it is caused by the specimen $S$ being pulled to the left by the horizontal actuator $AE$. The moment $M_1$ is expressed using Eq. 4-4, $\sum M_G = M_1 - H_1 l = 0$. Thus, $M_1 = H_1 l$.

The maximum total tension $T$ in cables $AB$ can be determined accurately because the magnitude of the force of the horizontal actuator is given as 10 kN. $\mathbf{T}$ alone resists this, but at a 5° angle. So, from Fig. 4-29d, and $\sum F_x = T \cos 5° - 10$ kN $= 0$,

$$T = \frac{10 \text{ kN}}{\cos 5°} = 10.04 \text{ kN}$$

From $\sum F_y = T \sin 5° - V_1 - V_2 = 0$, the magnitude of the vertical force at $C$ and $D$ is $V = V_1 + V_2 = T \sin 5°$. $V$ is shared by the two clamps unpredictably since the fixture is statically indeterminate here (either clamp alone could be enough for equilibrium). There are also horizontal forces at $C$ and $D$ that create a couple $\mathbf{M}_2$. From $\sum M_1 = M_2 - T(\sin 5°)(0.4 \text{ m}) = 0$,

$$M_2 = T(\sin 5°)(0.4 \text{ m}) = 0.35 \text{ kN·m}$$

The free-body diagram for the specimen $S$ is in Fig. 4-29c; $F$ and $T$ are known here. From $\sum F_y = 0$,

$$P = F - T \sin 5° = 15 - 0.88 = 14.12 \text{ kN}$$

For equilibrium in the horizontal direction, $\sum F_x = 0$, and

$$T \cos 5° = H_1 + H_2$$

Since $\sum M_B = 0$,

$$H_2 b - H_1 a = 0$$

The reactions at base $E$ of the horizontal actuator are obtained using Fig. 4-29e and Eq. 4-4,

$$\sum F_x = 0 \qquad H_3 = T \cos 5° = 10 \text{ kN}$$

$$\sum F_y = 0 \qquad V_3 = T \sin 5° = 0.88 \text{ kN}$$

$$\sum M_E = 0 \qquad M_3 = T(\sin 5°)(0.25 \text{ m}) = 0.22 \text{ kN·m}$$

Note that the reactions at $E$ could be obtained by drawing the free-body diagram of part $AE$ directly and realizing that $H_3 = 10$ kN is given.

AD horiz = 0.4 m
AE horiz = 0.25 m
CDvert = 0.1 m
∠AB/horiz = 5°

L

C

B

cable
A B

A          E          D          S

(a)

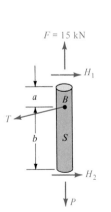

F = 15 kN

$M_1$

$H_1$

y

+

x

G
L

l

$H_1$

F = 15 kN
(given force for
starting the analysis)

(b)

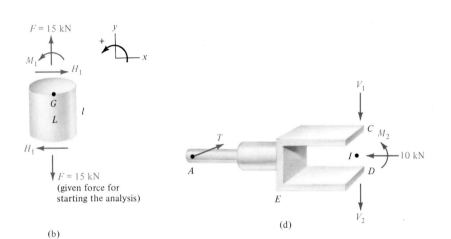

$V_1$

T

C          $M_2$

I          10 kN

A          D

E

$V_2$

(d)

F = 15 kN

$H_1$

a

B

T

b          S

$H_2$

P

(c)

T

$M_3$

E          $H_3$

A          $V_3$

(e)

**FIGURE 4-29**

Biaxial test setup of a shock absorber. (Photograph courtesy MTS Systems
Corporation, Minneapolis, Minnesota.)

EXAMPLE 4-14    163

**4-66**  Determine the reactions at $A$ and $B$ on the arm $AB$ of the adjustable lamp in order to hold it in equilibrium for the position shown.

**Fig. P4-66**

**4-67**  Determine the reactions on member $AB$ at $A$ and $B$ necessary to hold the excavator arm in the position shown.

**Fig. P4-67**

**4-68**  A mechanical hoe is modeled with forces $\mathbf{W}_1 = -1.2\,\text{kN}\,\mathbf{j}$ at $A$ of coordinates $(0.5, 1.5, 0)$ m, $\mathbf{W}_2 = -1.6\,\text{kN}\,\mathbf{j}$ at $C$ of coordinates $(1.4, 1.2, 0)$ m, and $\mathbf{F} = (800\mathbf{i} - 900\mathbf{j} + 400\mathbf{k})$ N at $E$ of coordinates $(1.7, -0.6, 0)$ m. Determine the forces on $OB$ and $BD$ necessary for equilibrium of the bucket. Point $B$ has coordinates $(1, 3, 0)$ m, and point $D$ has coordinates $(1.7, 0, 0)$ m.

**Fig. P4-68**

**4-69**  A clockwise moment $M = 600$ N·m is applied to gear $A$. Determine the moment acting on gear $C$ and the vertical reaction on its axle if the gears remain in equilibrium. Neglect the weights of the gears.

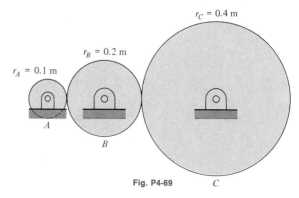

**Fig. P4-69**

**4-70**  In Fig. P4-69, $r_A = 2$ in., $r_B = 5$ in., and $r_C = 12$ in. Determine the largest moment that can be applied to gear $A$ if the vertical reaction should not exceed 1000 lb on any of the axles.

**4-71**  Determine the reactions at the wheels $A$, $B$, and $E$, and the force on axle $B$ due to the bar $BD$. The motorcycle weighs 250 lb, and the trailer weighs 50 lb. The bar $BD$ is hinged at $B$. The weights may be assumed to act at points $C$ and $F$, respectively.

**Fig. P4-71**

**4-72** A load of 7 kN may be placed anywhere within A and B in the trailer. Determine the reactions at the wheels at D, E, and F, and the force on the hitch H that is mounted on the car, for the extreme position A of the load. The mass of the car is 1500 kg.

**Fig. P4-72**

**4-73** Solve Prob. 4-72 assuming that the load in the trailer is at B.

**4-74** Pin C of the 0.25-m-long crank arm BC slides without friction in the slotted arm AD. Determine the vertical force **P** that maintains the mechanism in equilibrium if a couple of magnitude $M = 2$ kN·m is applied to the arm BC as shown.

**4-75** The piston B can slide without friction in the horizontal cylinder. Determine the force and moment reactions at point O if the pinned rod AB is to be in equilibrium when a horizontal force of magnitude $F = 300$ lb is applied.

**Fig. P4-75**

**Fig. P4-74**

**4-76** Determine the force and moment reactions at point O in Prob. 4-75 if a force of magnitude $F = 400$ lb is acting to the right on the piston.

# SUMMARY

## CONDITIONS OF EQUILIBRIUM

### VECTOR NOTATION:

$$\sum \mathbf{F} = 0$$
$$\sum \mathbf{M}_O = \sum (\mathbf{r} \times \mathbf{F}) = 0$$

where $O$ is an arbitrary point of reference.

### SCALAR NOTATION:

$$\sum F_x = 0 \qquad \sum M_x = 0$$
$$\sum F_y = 0 \qquad \sum M_y = 0$$
$$\sum F_z = 0 \qquad \sum M_z = 0$$

### MAXIMUM NUMBER OF EQUATIONS FOR ONE BODY:

1. One-dimensional problem:  $\sum F = 0$
2. Two-dimensional problem:

$$\sum F_x = 0, \qquad \sum F_y = 0, \qquad \sum M_A = 0$$
$$\sum F_x = 0, \qquad \sum M_A = 0, \qquad \sum M_B = 0 \quad (x \text{ axis } \not\perp AB)$$
$$\sum M_A = 0, \qquad \sum M_B = 0, \qquad \sum M_C = 0 \quad (AB \not\parallel BC)$$

3. Three dimensional problem:  $\sum F_x = 0, \quad \sum F_y = 0, \quad \sum F_z = 0$
$$\sum M_x = 0, \quad \sum M_y = 0, \quad \sum M_z = 0$$

**FREE-BODY DIAGRAMS:**   In solving for unknown forces or moments, always draw the free-body diagram first. The directions of unknowns may be assumed arbitrarily.

**CALCULATION OF UNKNOWNS IN TWO DIMENSIONS:**   There are three independent equilibrium equations available for a two-dimensional problem, so a maximum of three unknowns can be determined. Only statically determinate members, involving three unknowns, can be analyzed using the methods of statics.

## Section 4-4

**TWO-FORCE MEMBERS:**

1. Wires, chains, ropes: only tension may be applied to these
2. Rigid members: tension or compression may be applied to these

**GENERAL RULE:** A body subjected to forces applied at two points is in equilibrium only when the resultant forces at those points have the same magnitude, the same line of action, and the opposite direction.

## Section 4-5

**CALCULATION OF UNKNOWNS IN THREE DIMENSIONS:** The six independent equilibrium equations for a three-dimensional problem can be solved for six unknowns. If there are more than six unknowns, the problem is statically indeterminate and cannot be solved by the method of statics.

## Section 4-6

**INTERRELATED FREE-BODY DIAGRAMS:** When two or more bodies are in contact, separate free-body diagrams may be drawn for each body. The mutual forces and moments between the bodies are related according to Newton's Third Law (action and reaction). The directions of unknown forces and moments may be arbitrarily assumed in one diagram, but these initial choices affect the directions of unknowns in all other related diagrams. The number of unknowns and of usable equilibrium equations both increase with the number of related free-body diagrams.

**SCHEMATIC EXAMPLE IN TWO DIMENSIONS:**

Given: $F_1, F_2, F_3, M$

Unknowns: $P_1, P_2, P_3,$ and forces and moments
at joint $A$ (rigid connection)

**EQUILIBRIUM EQUATIONS:**

$$\sum F_x = -F_1 + P_3 = 0$$
$$\sum F_y = P_1 + P_2 - F_2 - F_3 = 0$$
$$\sum M_O = P_1 c + P_2(c + d + e) + M - F_2 a - F_3(a + b)$$
$$= 0$$

Three unknowns $(P_1, P_2, P_3)$ in three equations.

**RELATED FREE-BODY DIAGRAMS:**

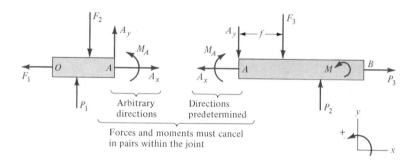

Dimensions $a$, $b$, $c$, $d$ and $e$ are also valid here.

**NEW SET OF EQUILIBRIUM EQUATIONS:**

Left part:   $\sum F_x = -F_1 + A_x = 0$
(*OA*)   $\sum F_y = P_1 + A_y - F_2 = 0$
$\sum M_O = P_1 c + A_y(c + d) + M_A - F_2 a = 0$

Right side:   $\sum F_x = -A_x + P_3 = 0$
(*AB*)   $\sum F_y = P_2 - A_y - F_3 = 0$
$\sum M_A = -M_A + P_2 e + M - F_3 f = 0$

Six unknowns $(P_1, P_2, P_3, A_x, A_y, M_A)$ in six equations.

NOTE: In the first diagram, the couple **M** may be moved anywhere from $O$ to $B$. $M$ is not shown in the second diagram because it is shown in the third diagram (in which it may be moved anywhere from $A$ to $B$).

**4-77** The crane's rigid boom $OC$ is supported by a cable $AB$. The boom and the cable have negligible weight. Determine the forces acting on the brackets at $O$ and $A$ for (a) the cable between $A$ and $B$, and (b) the cable between $A$ and $C$ instead of $A$ and $B$.

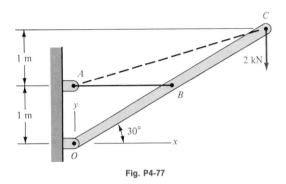

Fig. P4-77

**4-78** A mechanism includes two identical bars of negligible weight. Determine the magnitude of the reactions at pins $B$ and $C$ on bar $BC$ when the couple $\mathbf{M}$ is applied to it.

Fig. P4-79

**4-80** Determine the reactions at the supports $A$ and $D$ acting on member $AE$. Couple $\mathbf{M}_D$ acts directly on member $AE$ at pin $D$.

Fig. P4-78

**4-79** The 80-in.-diameter saw for road repair work rotates clockwise about its axis $A$. The whole assembly pivots about $B$ according to the positioning of the actuator $CD$. The equivalent weight at $A$ is 300 lb. Using the dimensions in the diagram, determine the magnitude of the force $\mathbf{F}$ from $D$ to $C$ which is large enough to lift the saw blade off the road.

Fig. P4-80

**4-81** A helicopter of weight $W$ in a test is attached to a fixture at $A$, which allows rotations about the $y$ axis with negligible friction but prevents other significant displacements. The main rotor applies a vertical lift force **L** while overcoming a moment $\mathbf{M}_D$ of air drag forces acting on the blades, which rotate clockwise when viewed from above. The tail rotor at $B$ is in the $xy$ plane and applies a thrust **P** to oppose the moment $\mathbf{M}_D$. Determine the reactions at $A$ if the helicopter's body should not move.

(a) Side view

**Fig. P4-81**    (b) Top view

**4-82** The door is hinged at $A$ and $D$, with the latter not supporting any vertical load. Determine the reactions at the two hinges if the door is ajar so that line $AB$ on the door makes an angle of $10°$ with the $x$ axis in the horizontal plane. Assume there are no moment reactions at either hinge.

**Fig. P4-82**

**4-83** Determine the applied moment **M** that must act on the shaker table leg $CD$ in order to hold the console in equilibrium for $\phi = 70°$. $BE = CD = 3$ ft and $BC = 5$ ft. The weights can be assumed to act at the center of the platform.

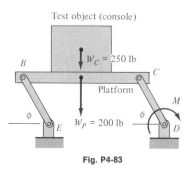

**Fig. P4-83**

**4-84** The medical X-ray apparatus has adjustable joints at the following points, with $O$ at the origin of the $xyz$ coordinates:

$A$: $(0, 0.1, 0.5)$ m
$B$: $(-0.1, 0.2, 1.0)$ m
$C$: $(-0.1, 0.6, 1.0)$ m
$D$: $(0.55, 1.1, 0.9)$ m
$E$: $(1.1, 0.85, 0.9)$ m
$F$: $(1.1, 0.4, 0.9)$ m

Assume weightless members and a 10-kg mass for the device at $F$. Determine the force and moment reactions at the joints of members $OA$, $BC$, and $DE$.

**Fig. P4-84**

# Forces and Moments in Beams

Basic Concepts in Modeling Beams

Beams are among the most common structural members. Frequently, beams are rather long in comparison with their lateral dimensions, and at least some of the loads acting on them cause bending of the beams. The bending of beams is often very important even if the deformations are not readily noticed. The following analyses of beams can be made using the methods of statics: (1) The external (supporting) reactions acting on a statically determinate beam can be calculated using the methods of Ch. 4; and (2) subsequently, the internal reactions (forces and moments) at any cross section in the beam can also be determined using the methods of Ch. 4. These analyses are essential background for investigating the geometric changes or safety aspects of real beams using the methods of mechanics of materials. Interestingly, the more advanced analysis of a beam (such as of its deformability) would not be possible without first assuming that it is perfectly rigid.

**SECTION 5-1** describes several common categories of supports. The reasonable modeling of the supports is indispensible in solving engineering problems. Quite frequently this can be done by inspection.

**SECTIONS 5-2 AND 5-3** extend the methods developed in Ch. 4 for determining internal forces and moments at any section in a beam. The methods are formalized to make them efficient and uniform for the analysis of many kinds of beams.

**SECTION 5-4** presents shear force and bending moment diagrams, which show at a glance how the internal reactions vary along a beam. These diagrams are important in locating the critical sections in a beam, and are used extensively in mechanics of materials.

**NOTE:** This chapter is devoted mainly to the analysis of beams on which concentrated forces are acting. Occasionally, couples acting on beams are also considered. The important but more complex problems of distributed loads on beams are analyzed in Ch. 7. This allows the readers to obtain a firm background in solving problems of beams and those relevant to beams.

## 5-1    CLASSIFICATION OF SUPPORTS

Several common supports for beams are shown in Fig. 5-1. The nature of the support determines the unknown reactions that must be shown at the appropriate place in the free-body diagram. The complete set of possible unknowns for two-dimensional loading of the beams in Fig. 5-1 is tabulated in Fig. 5-2. Various combinations of these supports are also common. For example, a roller may be placed under the right end of the cantilever beam in Fig. 5-1b.

**FIGURE 5-1**

Common supports for beams

(a) Simply supported

(b) Cantilevered

(c) Fixed-ended

(d) Continuous

| Beam | Possible reactions | | | | | |
|---|---|---|---|---|---|---|
| | $A_x$ | $A_y$ | $M_A$ | $B_x$ | $B_y$ | $M_B$ |
| | ✓ | ✓ | | | ✓ | |
| | ✓ | ✓ | ✓ | | | |
| | ✓ | ✓ | ✓ | ✓ | ✓ | ✓ |
| | ✓ | ✓ | | | ✓ similarly for C, D | |

**FIGURE 5-2**

Possible unknowns for common beams

The internal forces and moments (also called *internal reactions*) are essential for analyzing the strength of a beam. These reactions are determined using the concept that *if a beam is in equilibrium, any part of it is also in equilibrium.* Consider a cantilever beam with an external force **F** for a simple example, as in Fig. 5-3a. The free-body diagram is shown in part b of the figure. The reactions $V_O$ (called *vertical shear force,* or *transverse shear force,* in general) and $M_O$ are required for equilibrium, but their directions are arbitrarily assumed. From Eq. 4-2,

$$\sum F_y = 0$$

$$-F + V_O = 0, \quad V_O = F$$

$$\sum M_O = 0$$

$$M_O - FL = 0, \quad M_O = FL$$

Next, consider portion of the beam to the right of the cross section at point $A$. This part is also in equilibrium, and its free-body diagram is drawn in Fig. 5-3c. Note that $V_O$ and $M_O$ are not shown here because they are not acting directly on this part of the beam. Of course, there must be reactions at $A$ to keep the part in equilibrium. At this stage the internal reactions $V_A$ (vertical shear) and $M_A$ (often called *bending moment**) are unknown, but they are readily calculated from the equilibrium equations,

$$\sum F_y = 0$$

$$-F + V_A = 0, \quad V_A = F$$

$$\sum M_A = 0$$

$$M_A - Fl = 0, \quad M_A = Fl$$

Clearly, any other arbitrary length $x$ may be substituted for $l$, so the internal reactions as functions of $x$ are $V(x)$ and $M(x)$,

$$\boxed{V(x) = F \quad \text{and} \quad M(x) = Fx} \qquad \boxed{5\text{-}1}$$

which are valid for $0 < x < L$. Note that it is best to take moments about a point at the cross section of interest (point $A$ here) because the shear force on the section has no moment about that point.

The free-body diagram of the part to the left of $A$ is drawn in Fig. 5-3d. The internal reactions here are equal in magnitude and opposite in direction to those in Fig. 5-3c by the principle of action and reaction. One could say that $V_A$ and $M_A$ in part c are caused by the left part of the beam (in a sense, they are the remote effects of $V_O$ and $M_O$). Similarly, $V_A$ and $M_A$ in part d are caused by the right part of the beam (the remote effects of **F**).

---

* Sometimes it is necessary to distinguish a *bending moment* from a *twisting moment.* All bending moment vectors are perpendicular to the longitudinal axis of the beam, while twisting moment vectors are along that axis. The differences can be demonstrated by bending or twisting a pencil by hand.

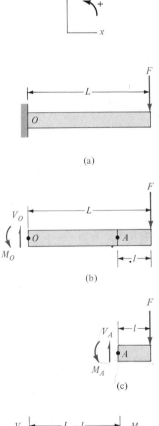

**FIGURE 5-3**

Internal forces and moments in a cantilever beam

It is reasonable to check if the piece $L - l$ is in equilibrium. Working with point $O$,

$$\sum F_y = V_O - V_A = F - F = 0$$
$$\sum M_O = M_O - M_A - V_A(L - l)$$
$$= FL - Fl - F(L - l) = 0$$

so the part of Fig. 5-3d is in equilibrium. It is equally reasonable to work with point $A$ on the piece $L - l$, which leads to the same results,

$$\sum F_y = V_O - V_A = F - F = 0$$
$$\sum M_A = M_O - M_A - V_O(L - l)$$
$$= FL - Fl - F(L - l) = 0$$

The cantilever beam is interesting since the internal reactions on any transverse section along the beam may be calculated directly, without first calculating the reactions at the fixed end. The simply supported beam is slightly more complicated in this respect. Its reactions must be calculated first; otherwise, there are too many unknowns in the equilibrium equations. For an example, consider the beam in Fig. 5-4a. From Fig. 5-4b and Eq. 4-2,

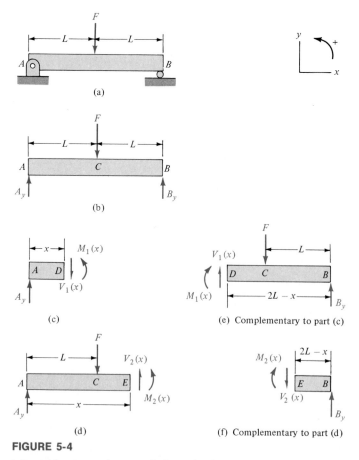

FIGURE 5-4

Internal forces and moments in a simply supported beam

$$\sum F_y = 0$$

$$A_y + B_y - F = 0$$

$$\sum M_A = 0$$

$$2LB_y - LF = 0, \quad B_y = \frac{F}{2}, \quad \text{so } A_y = B_y = \frac{F}{2}$$

Next consider an element of length $x$ at the left end of the beam. Its free-body diagram is shown in Fig. 5-4c. Here everything is unknown if the reaction $\mathbf{A}_y$ is not computed first, and nothing can be calculated. If $\mathbf{A}_y$ is known, $V_1(x)$ and $M_1(x)$ are readily calculated for any given $x$ up to the length $L$.

$$\sum F_y = 0$$

$$A_y - V_1(x) = 0, \qquad V_1(x) = A_y$$

$$\sum M_D = 0$$

$$-A_y x + M_1(x) = 0, \quad M_1(x) = A_y x$$

If $x$ were to get larger than $L$, a new free-body diagram and new equilibrium equations would be needed to include the force $\mathbf{F}$, as shown in Fig. 5-4d. Here $V_2(x)$ is clearly upward because $\mathbf{F}$ is larger than $\mathbf{A}_y$. The magnitudes of the internal reactions are obtained from the new equilibrium equations. Again using Eq. 4-2,

$$\sum F_y = 0$$

$$A_y + V_2(x) - F = 0, \qquad\qquad V_2(x) = F - A_y = \frac{F}{2}$$

$$\sum M_E = 0$$

$$-A_y x + F(x - L) + M_2(x) = 0, \quad M_2(x) = A_y x - F(x - L)$$

$$M_2(x) = -\frac{Fx}{2} + FL$$

It is worthwhile to draw the free-body diagrams of the parts adjoining those in Fig. 5-4c and d. These complementary parts are shown in Fig. 5-4e and f, respectively. Newton's Third Law must be obeyed when drawing the interrelated free-body diagrams.

Note that in Fig. 5-4 the dimension $x$ is always measured from point $A$ to the right. This is arbitrary; $x$ could be measured from point $B$ to the left without making any difference in the final results. However, there may be a notable difference in computational speed depending on which way a coordinate is chosen. This should be clear after thoughtful comparison of parts c, d, e, and f in Fig. 5-4. Experience with numerous free-body diagrams helps in choosing the simplest procedures.

Beams can also have axial forces acting on them, and these are easily added to the internal reactions already discussed. For example, an axial tension $\mathbf{T}$ is added to the beam in Fig. 5-4a, as shown in Fig. 5-5a. Now the free-body diagram of an element at the left end of the beam includes the tension, as shown in Fig. 5-5b. The presence of this axial tension does not affect any of the other external or internal reactions.

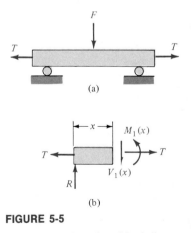

**FIGURE 5-5**

Internal reactions in a biaxially loaded beam

## SIGN CONVENTIONS

At the beginning of the analysis of any beam problem the positive directions for the internal forces and moments may be arbitrarily assumed and the vectors drawn on the free-body diagrams. It is only essential that the assumed directions obey Newton's Third Law if interrelated free-body diagrams are drawn. Nevertheless, most people prefer to follow a convention in analyzing beams so that the signs of the answers to the equilibrium equations are easily interpreted.

According to convention, a moment that makes a beam concave upward (not necessarily visible to the naked eye) is taken as positive. This is illustrated in Fig. 5-6. The element in part (a) of the figure is in equilibrium with the moments **M** acting on it. These moments are shown as force couples in Fig. 5-6b. The top fibers of the element are shortened because of the compressive force on them, and the bottom fibers are lengthened because they are in tension. A bending distortion of the element results as shown in Fig. 5-6c. This deformation is often negligible, so the assumption of ideal rigidity is reasonable. The magnitudes of such deformations are analyzed with the methods of mechanics of materials.

According to convention, a shear force acting downward on the right side of a section, or upward on the left side, is positive. Thus, both shear forces **V** are positive in Fig. 5-6d.

The signs of the actual moment and shear acting on a given section may or may not be the same for a given loading. Figure 5-7 shows the possible combinations of signs according to the common notation.

The positive directions as prescribed by these sign conventions should be used when drawing all free-body diagrams for beam analysis. In this way the signs of the answers to the equilibrium equations precisely give the correct directions of the shear force and bending moment. For example, by using the convention it is no longer necessary to see the free-body diagram to know exactly how a negative shear or bending moment affects the beam at any section.

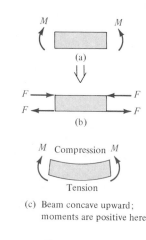

(a)

(b)

(c) Beam concave upward; moments are positive here

(d) Shear forces are positive here

**FIGURE 5-6**

Sign convention for shear force and bending moment

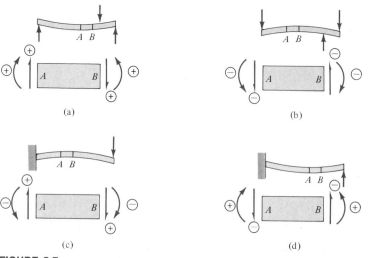

**FIGURE 5-7**

Possible combinations of signs of shear force and bending moment

EXAMPLE 5-1

Assume that a skyscraper with wind acting on it is modeled for convenience as in Fig. 5-8a. Determine the vertical shear and the bending moment at sections $A$, $B$, $C$, $D$, and $E$ in the cantilever beam.

## SOLUTION

Free-body diagrams are drawn for each section of interest. For each calculation, either the left or right portion of the beam can be considered. In this problem the right-hand portion makes for easiest calculations. Note that all internal shear forces and bending moments are assumed positive according to the sign conventions. Equation 4-2 is repeatedly used as follows:

**SECTION A:**   From Fig. 5-8b,

$$\sum F_y = 0 \qquad V_A - F = 0 \qquad V_A = F$$

$$\sum M_A = 0 \qquad -F\left(\frac{3L}{4}\right) - M_A = 0 \qquad M_A = -\frac{3}{4}FL$$

**SECTION B:**   From Fig. 5-8c,

$$\sum F_y = 0 \qquad V_B - F = 0 \qquad V_B = F$$

$$\sum M_B = 0 \qquad -F\left(\frac{L}{2}\right) - M_B = 0 \qquad M_B = -\frac{1}{2}FL$$

**SECTION C:**   From Fig. 5-8d,

$$\sum F_y = 0 \qquad V_C - F = 0 \qquad V_C = F$$

$$\sum M_C = 0 \qquad -F\left(\frac{L}{4}\right) - M_C = 0 \qquad M_C = -\frac{1}{4}FL$$

**SECTION D:**   From Fig. 5-8e,

$$\sum F_y = 0 \qquad$$ $V$ is undefined in the immediate neighborhood of a concentrated load (it is not known precisely how $\mathbf{V}$ and $\mathbf{F}$ act at the same point $D$)

$$\sum M_D = 0 \qquad -F(0) - M_D = 0 \qquad M_D = 0$$

**SECTION E:**   From Fig. 5-8f,

$$\sum F_y = 0 \qquad V_E - 0 = 0 \qquad V_E = 0$$

$$\sum M_E = 0 \qquad M_E + 0 = 0 \qquad M_E = 0$$

## JUDGMENT OF THE RESULTS

As expected from Fig. 5-8b, c, and d, $V_A = V_B = V_C = F$. The moment did vary by reasonable increments with the section considered, however. The obtained signs for $V$ and $M$ can be accepted as final since all internal reactions were assumed positive. It should also be noted that the unloaded end of a cantilevered beam will always have zero vertical shear and bending moment (Section E).

**FIGURE 5-8**

Analysis of a cantilever beam

EXAMPLE 5-1   177

EXAMPLE 5-2

Determine the vertical shear and bending moment at $B$, $C$, and $D$ in the simply supported beam shown in Fig. 5-9a.

## SOLUTION

First the reactions at supports $A$ and $E$ must be determined. From Fig. 5-9b for the whole beam, and using Eq. 4-2,

$$\sum F_x = 0 \qquad\qquad\qquad\qquad\qquad A_x = 0$$

$$\sum F_y = 0 \qquad A_y + E_y - 2F - F = 0$$

$$\sum M_A = 0 \qquad -2F\left(\frac{L}{3}\right) - F\left(\frac{2L}{3}\right) + E_y(L) = 0 \qquad E_y = \frac{4}{3}F$$

From $\sum F_y = 0$, $A_y = \frac{5}{3}F$.

**SECTION B:** Choosing the portion of the beam to the left of $A$ (Fig. 5-9c), and taking $V_B$ and $M_B$ as positive,

$$\sum F_y = 0 \qquad \frac{5}{3}F - V_B = 0 \qquad\qquad V_B = \frac{5}{3}F$$

$$\sum M_B = 0 \qquad M_B - \left(\frac{5}{3}F\right)\left(\frac{L}{4}\right) = 0 \qquad\qquad M_B = \frac{5}{12}FL$$

**SECTION C:** From Fig. 5-9d,

$$\sum F_y = 0 \qquad \frac{5}{3}F - 2F - V_C = 0 \qquad\qquad V_C = -\frac{1}{3}F$$

$$\sum M_C = 0 \qquad M_C - \left(\frac{5}{3}F\right)\left(\frac{L}{2}\right) + 2F\left(\frac{L}{6}\right) = 0 \qquad M_C = \frac{1}{2}FL$$

**SECTION D:** Here it is best to use the portion of the beam to the right of $D$ (this reduces the number of forces that must be considered). From Fig. 5-9e,

$$\sum F_y = 0 \qquad V_D + \frac{4}{3}F = 0 \qquad\qquad V_D = -\frac{4}{3}F$$

$$\sum M_D = 0 \qquad \frac{4}{3}F\left(\frac{L}{4}\right) - M_D = 0 \qquad\qquad M_D = \frac{1}{3}FL$$

## JUDGMENT OF THE RESULTS

The same results can be obtained by using complementary portions of the beam to those used here. To realize the facility of using either the left or right portion for calculating internal reactions, the student should compute the bending moments and shear forces at sections $A$ and $C$ using portions other than those used in the example. Remember to assume the appropriate positive shear force and bending moment at each section.

**FIGURE 5-9**

Analysis of a simply supported beam

EXAMPLE 5-3

Determine the vertical shear and bending moment at $A$, $B$, and $C$ in the cantilever beam shown in Fig. 5-10a. Determine the new reactions at $A$, $B$, and $C$ after adding a downward force of magnitude $F = 3$ kN acting at the middle of the beam.

## SOLUTION

**CASE 1. SECTION A:**   From Fig. 5-10b and using Eq. 4-2,

$$\sum F_y = 0 \qquad\qquad\qquad\qquad V_A = 0$$

$$\sum M_A = 0 \qquad M_A - 2\ \text{kN·m} = 0 \qquad\qquad M_A = 2\ \text{kN·m}$$

It can be realized that the shear force is zero throughout the beam since there are no vertical forces applied. Also, the internal bending moment is equal to 2 kN·m everywhere since the 2 kN·m couple, $M$, is applied at the end of the beam.

**CASE 2. SECTION A:**   From Fig. 5-10c, for the whole beam,

$$\sum F_y = 0 \qquad -3\ \text{kN} - V_A = 0 \qquad\qquad V_A = -3\ \text{kN}$$

$$\sum M_A = 0 \qquad M_A - 2\ \text{kN·m} + (3\ \text{kN})(0.75\ \text{m}) = 0 \qquad M_A = -0.25\ \text{kN·m}$$

**SECTION B:**   From Fig. 5-10d,

$$\sum F_y = 0 \qquad -3\ \text{kN} - V_B = 0 \qquad\qquad V_B = -3\ \text{kN}$$

$$\sum M_B = 0 \qquad M_B - 2\ \text{kN·m} + (3\ \text{kN})(0.25\ \text{m}) = 0 \qquad M_B = 1.25\ \text{kN·m}$$

**SECTION C:**   From Fig. 5-10e,

$$\sum F_y = 0 \qquad\qquad\qquad\qquad V_C = 0$$

$$\sum M_C = 0 \qquad M_C - 2\ \text{kN·m} = 0 \qquad\qquad M_C = 2\ \text{kN·m}$$

## JUDGMENT OF THE RESULTS

The same results are obtained when the complementary portions of the beam are used.

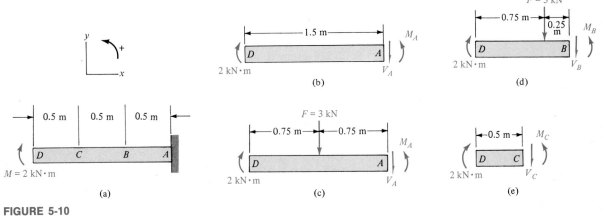

**FIGURE 5-10**

Analysis of a cantilever beam

EXAMPLE 5-3   179

# EXAMPLE 5-4

(a) Determine the internal forces and moments at $B$, $C$, and $D$ for the simply supported beam shown in Fig. 5-11a. (b) Determine the internal forces and moments at $B$, $C$, and $D$ after moving the force $\mathbf{F}$ to $B$.

## SOLUTION

(a) First the reactions must be determined from Fig. 5-11b for the whole beam. Using Eq. 4-2,

$$\sum F_x = 0 \qquad (600 \text{ lb}) \cos 60° - A_x = 0 \qquad A_x = 300 \text{ lb}$$

$$\sum F_y = 0 \qquad A_y + E_y - (600 \text{ lb}) \sin 60° = 0$$

$$\sum M_A = 0 \qquad -(600 \text{ lb})(\sin 60°)(30 \text{ in.})$$
$$+ E_y(60 \text{ in.}) = 0, \qquad E_y = 260 \text{ lb}$$

From $\sum F_y = 0$, $A_y = 260$ lb.

**SECTION A:** From Fig. 5-11c,

$$\sum F_x = 0 \qquad P_A - 300 \text{ lb} = 0 \qquad P_A = 300 \text{ lb}$$

$$\sum F_y = 0 \qquad V_A \text{ is undefined (see Ex. 5-1)}$$

$$\sum M_A = 0 \qquad (260 \text{ lb})(0) + M_A = 0 \qquad M_A = 0$$

**SECTION B:** From Fig. 5-11d,

$$\sum F_x = 0 \qquad P_B - 300 \text{ lb} = 0 \qquad P_B = 300 \text{ lb}$$

$$\sum F_y = 0 \qquad 260 \text{ lb} - V_B = 0 \qquad V_B = 260 \text{ lb}$$

$$\sum M_B = 0 \qquad M_B - (260 \text{ lb})(15 \text{ in.}) = 0 \qquad M_B = 3900 \text{ lb·in.}$$

**SECTION C:** From Fig. 5-11e,

$$\sum F_x = 0 \qquad P_C \text{ is undefined (see Ex. 5-1)}$$

$$\sum F_y = 0 \qquad V_C \text{ is undefined (see Ex. 5-1)}$$

$$\sum M_C = 0 \qquad M_C - (260 \text{ lb})(30 \text{ in.}) = 0 \qquad M_C = 7800 \text{ lb·in.}$$

**SECTION D:** From Fig. 5-11f,

$$\sum F_x = 0 \qquad\qquad\qquad\qquad P_D = 0$$

$$\sum F_y = 0 \qquad V_D + 260 \text{ lb} = 0 \qquad V_D = -260 \text{ lb}$$

$$\sum M_D = 0 \qquad (260 \text{ lb})(15 \text{ in.}) - M_D = 0 \qquad M_D = 3900 \text{ lb·in.}$$

(b) For the case when $F$ is applied at $B$, the reactions must be recalculated:

$$\sum F_x = 0 \qquad (600 \text{ lb}) \cos 60° - A_x = 0 \qquad A_x = 300 \text{ lb}$$

$$\sum F_y = 0 \qquad A_y + E_y - (600 \text{ lb}) \sin 60° = 0$$

$$\sum M_A = 0 \qquad -(600 \text{ lb})(\sin 60°)(15 \text{ in.})$$
$$+ E_y(60 \text{ in.}) = 0, \qquad E_y = 130 \text{ lb}$$

From $\sum F_y = 0$, $A_y = 390$ lb.

(a)   (b)   (c)

(d)   (e)   (f)

**FIGURE 5-11**

Analysis of a biaxially loaded, simply supported beam

**SECTION A:**

$$\sum F_x = 0 \qquad P_A - 300 \text{ lb} = 0 \qquad P_A = 300 \text{ lb}$$

$$\sum F_y = 0 \qquad 390 \text{ lb} - V_A = 0 \qquad V_A = 390 \text{ lb}$$

$$\sum M_A = 0 \qquad (390 \text{ lb})(0) + M_A = 0 \qquad M_A = 0$$

**SECTION B:**

$$\sum F_x = 0 \qquad P_B \text{ is undefined (see Ex. 5-1)}$$

$$\sum F_y = 0 \qquad V_B \text{ is undefined (see Ex. 5-1)}$$

$$\sum M_B = 0 \qquad M_B - (390 \text{ lb})(15 \text{ in.}) = 0 \qquad M_B = 5850 \text{ lb·in.}$$

**SECTION C (Using the Right-Hand Portion):**

$$\sum F_x = 0 \qquad\qquad\qquad\qquad P_C = 0$$

$$\sum F_y = 0 \qquad 130 \text{ lb} + V_C = 0 \qquad V_C = -130 \text{ lb}$$

$$\sum M_C = 0 \qquad (130 \text{ lb})(30 \text{ in.}) - M_C = 0 \qquad M_C = 3900 \text{ lb·in.}$$

**SECTION D:**

$$\sum F_x = 0 \qquad\qquad\qquad\qquad P_D = 0$$

$$\sum F_y = 0 \qquad 130 \text{ lb} + V_D = 0 \qquad V_D = -130 \text{ lb}$$

$$\sum M_D = 0 \qquad (130 \text{ lb})(15 \text{ in.}) - M_D = 0 \qquad M_D = 1950 \text{ lb·in.}$$

## JUDGMENT OF THE RESULTS

It is seen that by moving the force **F** from the midpoint to the left the internal forces become intensified on the left and diminished on the right as expected. The changes in reactions at $A$ and $E$ are acceptable by inspection.

EXAMPLE 5-4   181

# EXAMPLE 5-5

Show that the weightless beam in Fig. 5-12a is statically indeterminate for **F** in the range $0 \leq x \leq 1.2$ m. Next, assume that the beam is cut into two equal parts and that these are connected with a hinge at point $E$. Determine the reactions at $A$, $D$, and $E$ and the internal shear and bending moment at $B$ for the hinged beam, with **F** acting at $C$.

## SOLUTION

The free-body diagram of the beam in Fig. 5-12b shows there are four unknowns, $\mathbf{A}_x$, $\mathbf{A}_y$, $\mathbf{M}_A$, and $\mathbf{D}_y$. From Eq. 4-2, only three independent equations of equilibrium are available, $\sum F_x = 0$, $\sum F_y = 0$, and $\sum M_A = 0$. Hence, the beam is indeterminate. If the beam is hinged at $E$, there are two additional unknowns, $\mathbf{E}_x$ and $\mathbf{E}_y$ (Fig. 5-12c and d). Since the beam is actually two separate rigid bodies, however, there are six independent equations available and the beam is now determinate.

$F = 3$ kN

$A$    $B$   $\bullet E$   $C$     $D$

0.4 m    0.4 m    0.4 m

(a)

**PART E–D:** From Fig. 5-12d,

$$\sum F_x = 0 \qquad\qquad\qquad\qquad\qquad\qquad E_x = 0$$
$$\sum F_y = 0 \qquad D_y - E_y - 3\text{ kN} = 0 \qquad D_y = 1\text{ kN}$$
$$\sum M_E = 0 \qquad -(3\text{ kN})(0.2\text{ m}) + D_y(0.6\text{ m}) = 0 \qquad E_y = -2\text{ kN}$$

**PART A–E:** From Fig. 5-12c,

$$\sum F_x = 0 \qquad E_x - A_x = 0 \qquad\qquad A_x = 0$$
$$\sum F_y = 0 \qquad A_y + (-2\text{ kN}) = 0 \qquad A_y = 2\text{ kN}$$
$$\sum M_A = 0 \qquad -(2\text{ kN})(0.6\text{ m}) - M_A = 0 \qquad M_A = -1.2\text{ kN·m}$$

**SECTION B:** From Fig. 5-12e,

$$\sum F_y = 0 \qquad 2\text{ kN} - V_B = 0 \qquad\qquad\qquad V_B = 2\text{ kN}$$
$$\sum M_B = 0 \qquad -(-1.2\text{ kN·m}) - (2\text{ kN})(0.4\text{ m}) + M_B = 0 \qquad M_B = -400\text{ N·m}$$

Note that although the negative answers for $E_y$ and $M_A$ indicate that the wrong directions were assumed in the free-body diagrams, the negative signs were carried through in all the equations in order to arrive at the correct answers.

$F$

$M_A$

$A$       $C$    $D$

$A_x$

$A_y$            $D_y$

(b) Continuous beam from $A$ to $D$

$E_y$

$M_A$ $\leftarrow$ 0.6 m $\rightarrow$

$A$     $E$     $\rightarrow E_x$

$A_x$

$A_y$

(c) Assume that the beam is cut at $E$ and a hinge is installed there

$E_y$   $F$

     0.4 m

$E_x$

$E$      $D$

0.6 m $\rightarrow D_y$

(d) Part $ED$ with a hinge at $E$ and a roller at $D$

0.4 m      $M_B$

$A$   $B$

$M_A$      $V_B$

$A_y$

(e)

**FIGURE 5-12**

Statically indeterminate beam made into a determinate beam

**5–1** Determine the shear and bending moment at $x = 0$, 0.2 m, and 0.4 m.

Fig. P5-1

**5–2** Determine the shear and bending moment at $y = 0$, 1 ft, and 2 ft.

Fig. P5-2

**5–3** Determine the shear and bending moment at $x = 0$, 1 ft, and 2.5 ft.

**5–4** Calculate the shear and bending moment at $x = 0$, 0.5 m, and 1.5 m.

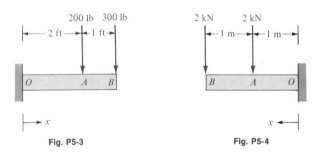

Fig. P5-3    Fig. P5-4

**5–5** Calculate the shear and bending moment at $y = 0$, 1 m, and 3 m.

Fig. P5-5

**5–6** Force **P** always acts at point $A$ on the beam, but force **F** has variable positions. Determine the shear and bending moment at support $O$ for $x_C = 1$ m, 2 m, and 4 m.

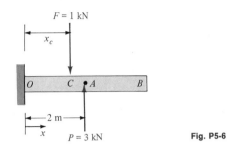

Fig. P5-6

**5–7** Calculate the shear and bending moment at $x = 0$, 0.5 ft, 1.2 ft, and 2.5 ft.

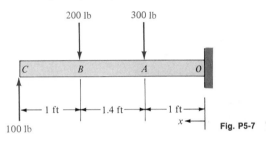

Fig. P5-7

**5–8** A simple crane is modeled as a weightless cantilever beam. The force **F** is applied through a symmetric carriage which has two wheels, $A$ and $B$. Calculate the shear and bending moment at support $O$ for $x_C = 3$ ft, 6 ft, and 9 ft.

Fig. P5-8

**5–9** Determine the reactions at $A$ and $B$ and calculate the shear and bending moment at $x = 1$ m and 2.5 m.

Fig. P5-9

**5–10**  Determine the reactions at $A$ and $B$ and calculate the shear and bending moment at $y = 2$ in. and 8 in.

10 in.

100 lb —

4 in.

**Fig. P5-10**

**5–11**  The force F acts at various places along the beam $A$–$B$. Calculate the bending moment at $C$ for $x_C = 1$ m, 2.5 m, and 4 m.

$F = 3$ kN

$x_c$

$L = 5$ m

**Fig. P5-11**

**5–12**  Let $F = 500$ lb and $L = 10$ ft in Fig. P5-11. Consider a cross section $D$ at $x_D = 5$ ft. Calculate the shear and bending moment at $D$ for **F** acting at $x_C = 1$ ft, 2.5 ft, and 4 ft (three separate cases).

**5–13**  Calculate the bending moments at $C$, $D$, and $E$.

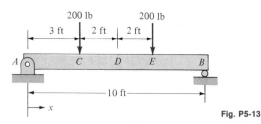

200 lb    200 lb

3 ft    2 ft    2 ft

10 ft

**Fig. P5-13**

**5–14**  Determine the bending moments at $C$ and $D$. Also calculate the shear and the bending moment at $E$, where $x_E = 4$ m.

2 kN    3 kN

1 m    1.5 m

5 m

**Fig. P5-14**

**5–15**  Determine the bending moments at $B$, $C$, and $D$, where $x_D = 2.5$ m. Also give the value of the shear at $D$.

2 kN    3 kN

1 m

0.8 m    2 m

**Fig. P5-15**

**5–16**  Calculate the shear and bending moment at cross section $C$ located at $x_C = 4$ ft, and at cross section $D$ located at $x_D = 6$ ft.

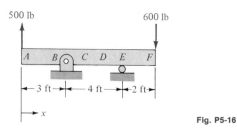

500 lb    600 lb

3 ft    4 ft    2 ft

**Fig. P5-16**

**5–17**  Calculate the shear and bending moment at $A$, $B$, and $C$.

$M_D = 300$ lb · ft

200 lb —

6 in.

6 in.

6 in.

$y$

**Fig. P5-17**

**5–18**  Determine the shear and bending moment at support $A$ and cross section $D$, where $x_D = 1$ m.

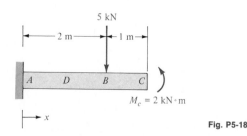

5 kN

2 m    1 m

$M_c = 2$ kN · m

**Fig. P5-18**

**5–19**  Determine the shear and bending moment at $C$.

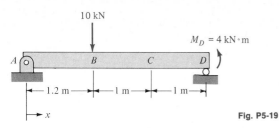

10 kN

$M_D = 4$ kN·m

A  B  C  D

|←1.2 m→|←1 m→|←1 m→|

x

**Fig. P5-19**

**5-20** When the two-wheeled carriage of a crane is positioned at C, a moment $M_E$ is applied to track AE by a neighboring structure. Calculate the bending moments at B, C, and D in the beam. Also give the shear at C assuming the load is evenly distributed between the carriage wheels.

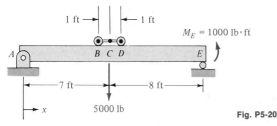

1 ft →  |  ← 1 ft

$M_E = 1000$ lb·ft

A  B C D  E

|←7 ft→|←8 ft→|

x

5000 lb

**Fig. P5-20**

**5-21** Consider the tall crane shown in the photograph and modeled below. The horizontal boom is pivoted at A, which is behind the operator's cab in the picture. The permanently attached cables at B make a 15° angle with the boom. The horizontal distance between A and B is 80 ft. The load carriage C can move along the 120-ft-long boom, which is assumed weightless. Calculate the reactions at A for a 10,000-lb vertical load on the carriage C at the position $x_C = 120$ ft. Also calculate the shear and bending moment in the boom at $x_D = 50$ ft assuming that the boom is a homogeneous beam.

(a)

$T_{AB}$

15°  B  C

A

$x_D$

|←80 ft→|

10,000 lb

$x_C$

(b)

**Fig. P5-21**

**5-22** It may be expected that the internal reactions in the boom described in Prob. 5–21 depend on the position $x_C$ of the carriage C. Determine the horizontal reaction, the vertical shear, and the bending moment at $x_E = 5$ ft when C is at $x_C = 50$ ft, 80 ft, and 110 ft, (three separate cases).

**5-23** Consider a proposed 9-m-long ladder. The possible internal reactions in the ladder must be known to minimize its weight. For an initial estimation, determine the transverse shear and bending moment at E and D for a 3-kN load applied at the center of the ladder, which is positioned at $d = 4$ m. Assume that the wall is frictionless at B, and that the ladder acts as if it were pinned at A.

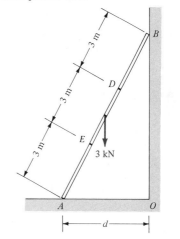

B

3 m

D

3 m

E

3 kN

3 m

A  O

|←d→|

**Fig. P5-23**

**5-24** It may be expected that the placement distance d makes a difference for the ladder described in Prob. 5–23. Plot the transverse shear and the bending moment at C and D as functions of $d = 2$ m, 4 m, and 6 m. Use one diagram for the shear and one for the bending moment.

**5-25** Consider the symmetrical spring BC of a railroad tank car. The wheel axle A applies a maximum upward force of 20,000 lb to the spring. The distance BC is 30 in. Determine the maximum bending moment and vertical shear in the spring for the angle $\theta = 30°$ that the links BB' and CC' make with the vertical. How does changing the angle $\theta$ affect the solution?

B

K14a

B'  A

θ

**Fig. P5-25**
Suspension system of a railroad car (Courtesy MTS Systems Corporation, Minneapolis, Minnesota)

**5–26** The light poles in the photograph are quite appealing esthetically. To investigate part of their technical merits, consider the bending moments in the curved poles and in poles made of straight members. To be specific, compare the bending moments at points $B$ and $B'$, $C$ and $C'$, and $D$ and $D'$ in Fig. P5-26b. Assume that a weight $\mathbf{W}$ acts at $A$, and that the pole is weightless for any geometry.

Fig. P5-28

**5–27** Consider the model of the large earth mover used in strip mining. The long arm $AD$ of the conveyor is pivoted at $A$ and has a roller at $B$. Determine the transverse shear and bending moment at points $E$ and $F$.

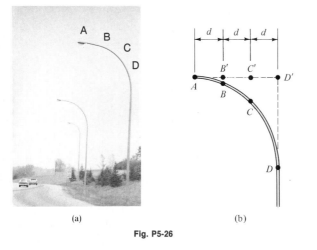

Fig. P5-26

**5–29** An amusement ride is to be modeled as a cantilever beam supported at $O$. The weights are assumed to be concentrated, $W_1$ acting at $x = R/2$, and $W_2$ acting at $A$. Furthermore, a clockwise couple of magnitude $M_A$ is acting at $A$ (as a reaction to initiating motion of the small arms). Determine the vertical shear and bending moment at $x = 0$ and $x = R/4$ in the main arm.

Fig. P5-29

Fig. P5-27

**5–28** An excavator is modeled with rigid arms $AB$ and $BC$, which can be locked in a variety of positions. Calculate the transverse shear and bending moment at point $F$.

**5–30** A drag racing car at the start is to be modeled as a weightless, horizontal beam pinned at $A$ and supported on a roller at $B$. The weight $\mathbf{W}$ acts at $C$, and a counterclockwise moment $\mathbf{M}_A$ is acting on $AB$ at point $A$ (an opposite moment is acting on the wheels at $A$; a large $M_A$ causes the occasional lifting of the front end). Determine the vertical shear and bending moment at $x = a/2$ and $x = a + b/2$, assuming a normal force of $0.1W$ acting on wheel $B$.

Fig. P5-30

A common problem in the design of beams is to find the critical sections where the shear and bending moment are the largest. The signs of the bending moment are also of great interest because they show where there is tension or compression in a beam. The signs of the shear forces are of smaller practical significance.*

The critical locations in some beams may be determined by inspection after a little experience. In any case, it is most satisfactory to plot shear and bending moment diagrams for the whole length of each beam. These are excellent pictorial views of the variations in shear and bending moment, and are essential when the loading is complex.

It is best to draw the shear and bending moment diagrams directly below the complete free-body diagram of the beam to obtain an overall view of the details of the problem. The construction of the diagrams is illustrated in Fig. 5-13. The free-body diagram of a simply supported beam is drawn in part a of the figure. From this, using Eq. 4-2, $P = R_1 + R_2$, $R_1 a = R_2 b$, and

$$R_1 = \frac{P}{1 + \dfrac{a}{b}}, \qquad R_2 = \frac{P}{1 + \dfrac{b}{a}}$$

Then vertical lines are drawn at the major points of interest such as where $R_1$, $R_2$, and $P$ act. Next, the coordinate axes of the shear ($V$) and bending moment ($M$) diagrams are drawn.

The plotting of the shear and bending moment diagrams is done after the calculation of shear and bending moment at a few judiciously selected sections in the beam. These are shown in Fig. 5-13d through j.

(a)

(b)

(c)

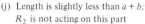

(d) Horizontal length of element is near zero

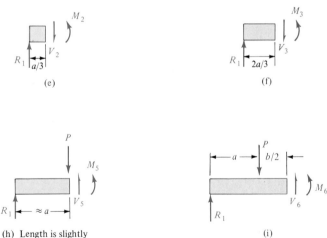

(e)

(f)

(g) Length is slightly less than $a$; $P$ is not acting on this part

(h) Length is slightly larger than $a$

(i)

(j) Length is slightly less than $a + b$; $R_2$ is not acting on this part

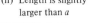

**FIGURE 5-13**

Shear and bending moment diagrams of a simply supported beam

* See B. I. Sandor, *Strength of Materials* (Englewood Cliffs, N.J.: Prentice-Hall, Inc., 1978).

If the beam is in equilibrium, each part of it is also in equilibrium. Thus, from the diagrams of the individual parts,

Fig. 5-13d: $\quad V_1 = R_1 = \dfrac{P}{1 + \dfrac{a}{b}}, \quad M_1 \approx 0.$

Fig. 5-13e: $\quad V_2 = R_1 \qquad\qquad M_2 = R_1 \dfrac{a}{3}$

Fig. 5-13f: $\quad V_3 = R_1 \qquad\qquad M_3 = R_1 \dfrac{2a}{3}$

Fig. 5-13g: $\quad V_4 = R_1 \qquad\qquad M_4 \approx R_1 a$

Fig. 5-13h: $\quad V_5 = P - R_1 \qquad M_5 \approx R_1 a$
$\qquad\qquad\qquad = R_2$

Fig. 5-13i: $\quad V_6 = R_2 \qquad\qquad M_6 = R_1\left(a + \dfrac{b}{2}\right) - P\dfrac{b}{2}$

Fig. 5-13j: $\quad V_7 = R_2 \qquad\qquad M_7 \approx R_1(a + b) - Pb$
$\qquad\qquad\qquad\qquad\qquad\quad \approx 0$

These values are plotted in Fig. 5-13b and c as points $d$ through $j$, respectively. The points are connected with straight lines; the areas between these lines and the $x$ coordinate are crosshatched to help in the visual evaluation of the plots. The shear and bending moment are zero in region $C$ (to the right of $\mathbf{R_2}$).

The sharp breaks in the plots are characteristic for sections where concentrated forces are acting. Such regions cannot be analyzed precisely. In theory, a force can act at a point, but in reality, even concentrated forces are distributed over small but finite areas or volumes. Thus, the shear and bending moment diagrams are not precise at $x \approx 0$, $x \approx a$, and $x \approx a + b$. However, the approximations with the straight lines as shown in the diagrams are reasonable in practice.

The procedure described above can be formalized by writing functions of the shear and bending moment in terms of the external loads and the distance $x$ along the beam. Each function is based on a single free-body diagram that is valid for a part of the beam. The beam in Fig. 5-13a has three unique parts, and there is a shear and bending moment function for each part as follows:

$$\left.\begin{array}{l} V_a(x) = R_1 \\ M_a(x) = R_1 x \end{array}\right\} \qquad 0 < x < a$$

$$\left.\begin{array}{l} V_b(x) = P - R_1 = R_2 \\ M_b(x) = R_1 x - P(x - a) \end{array}\right\} \qquad a < x < a + b$$

$$\left.\begin{array}{l} V_c(x) = 0 \\ M_c(x) = 0 \end{array}\right\} \qquad a + b < x < a + b + c$$

These functions can be evaluated at selected values of $x$. The points are plotted on the appropriate diagram and lines are drawn through the points as before. Much practice with these equations and diagrams is recommended.

EXAMPLE 5-6

Write the shear and bending moment equations and draw the shear and bending moment diagrams for the cantilever beam shown in Fig. 5-14a.

## SOLUTION

First the reactions at end $A$ are obtained (Fig. 5-14b) using Eq. 4-2,

$$\sum F_x = 0 \qquad\qquad A_x = 0$$

$$\sum F_y = 0 \qquad A_y - 5\text{ kN} = 0 \qquad A_y = 5\text{ kN}$$

$$\sum M_A = 0 \qquad M_A - (5\text{ kN})(2\text{ m}) = 0 \qquad M_A = 10\text{ kN·m}$$

To obtain the general shear and moment equations a section $K$ at a general distance $x$ from the left end is analyzed. Assuming positive shear and moment reactions at section $K$ in Fig. 5-14e according to the sign conventions,

**PART A–K:**

$$\sum F_y = 0 \qquad V_K = 5\text{ kN}$$

$$\sum M_K = 0 \qquad 10\text{ kN·m} - (5\text{ kN})(x) + M_K = 0$$

$$M_K = (5\text{ kN})(x) - 10\text{ kN·m}$$

Plotting the equations above, Fig. 14c and d are obtained.
    A check at the end of the beam shows (Fig. 5-14f):

**PART A–B (K is at B):**

$$\sum F_y = 0 \qquad 5\text{ kN} - 5\text{ kN} - V_K = 0,$$

$$\text{so } V_K = 0 \text{ as it should be}$$

$$\sum M_K = 0 \qquad 10\text{ kN·m} - (5\text{ kN})(2\text{ m}) + M_K = 0,$$

$$\text{so } M_K = 0 \text{ as it should be}$$

## JUDGMENT OF THE RESULTS

The results seem logical. Since there is no change in the applied load along the length of the beam, the shear, which resists the vertical load, should not change either. The linear moment distribution seems logical after studying the free-body diagram in Fig. 5-13e. Using only one general section is reasonable since no other sections would show a different loading situation.

**FIGURE 5-14**

Shear and bending moment diagrams of a cantilever beam

EXAMPLE 5-6    189

# EXAMPLE 5-7

Write the shear and bending moment equations and draw the shear and bending moment diagrams for the beam shown in Fig. 5-15a.

## SOLUTION

The reactions are given in Fig. 5-15b. These should be checked by the reader. Free-body diagrams are drawn to place a general section $K$ in each section between neighboring vertical forces (between $A$ and $B$, $B$ and $C$, and $C$ and $D$). Equations 4-2 are used for each part.

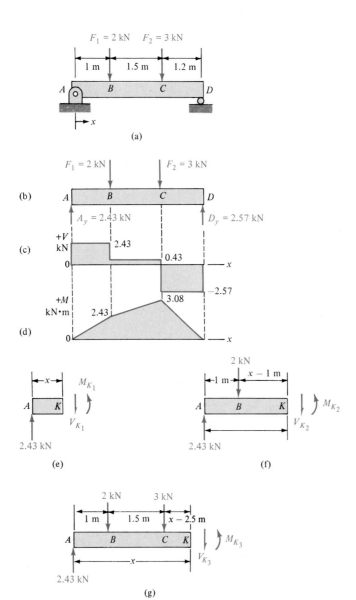

**FIGURE 5-15**

Shear and bending moment diagrams of a simply supported beam

**PART A–K (Fig. 5-15e):** For the region $0 \le x \le 1$ m,

$\sum F_y = 0 \qquad 2.43 \text{ kN} - V_{K_1} = 0 \qquad\qquad V_{K_1} = 2.43 \text{ kN}$

$\sum M_K = 0 \qquad M_{K_1} - (2.43 \text{ kN})(x) = 0 \qquad M_{K_1} = 2.43(x) \text{ kN·m}$

**PART A–B–K (Fig. 5-15f):** The equations are written for the region $1 \text{ m} \le x \le 2.5$ m,

$\sum F_y = 0 \qquad 2.43 \text{ kN} - 2 \text{ kN} - V_{K_2} = 0 \qquad V_{K_2} = 0.43 \text{ kN}$

$\sum M_K = 0 \qquad M_{K_2} - (2.43 \text{ kN})(x) + (2 \text{ kN})(x - 1 \text{ m}) = 0$

$\qquad\qquad\qquad M_{K_2} = 0.43(x) \text{ kN·m} + 2 \text{ kN·m}$

**PART A–B–C–K (Fig. 5-15g):** The equations are written for the region $2.5 \text{ m} \le x \le 3.7$ m,

$\sum F_y = 0$

$2.43 \text{ kN} - 2 \text{ kN} - 3 \text{ kN} - V_{K_3} = 0 \qquad\qquad V_{K_3} = -2.57 \text{ kN}$

$\sum M_K = 0$

$M_{K_3} - (2.43 \text{ kN})(x) + (2 \text{ kN})(x - 1 \text{ m}) + (3 \text{ kN})(x - 2.5 \text{ m}) = 0$

$M_{K_3} = (2.43 \text{ kN} - 2 \text{ kN} - 3 \text{ kN})(x) + (2 \text{ kN})(1 \text{ m}) + (3 \text{ kN})(2.5 \text{ m})$

$\qquad\quad = -2.57(x) \text{ kN·m} + 9.5 \text{ kN·m}$

These equations are plotted on the appropriate axes for the intervals for which they are valid (see Fig. 5-15c and d). Plotting is facilitated by evaluating each equation at the end points of the interval for which it is valid and plotting the values at those points. A straight line is drawn connecting the neighboring points. It is seen that two equations are valid where each force is acting, one for the portion of the beam to the left and one for the portion to the right of the point where the force is applied. If the reaction that is being sketched (shear or bending moment) is discontinuous at a section (for example, the shear at points B and C), then there are two different values plotted at that section and a vertical line is assumed to join the two points (see Fig. 5-15c). If the reaction being plotted is *not* discontinuous (for example, the bending moment throughout the beam), then the two equations should yield the same point at the interval end points. This helps to serve as a check of the obtained equations.

## JUDGMENT OF THE RESULTS

The shapes of the sketched plots appear reasonable when remembering the discussion above. Since the vertical loads are discontinuous at A, B, C, and D, the shear force diagram is expected to be discontinuous at these points, as well. When calculating the internal bending moments at B and C, the same values are obtained using either of the two equations which are valid for each point.

EXAMPLE 5-7    191

## EXAMPLE 5-8

Write the shear and bending moment equations and draw the shear and bending moment diagrams for the beam shown in Fig. 5-16a.

### SOLUTION

The reactions at $B$ and $D$ are given in Fig. 5-16b. The reader should check these. Free-body diagrams are drawn to place a general section $K$ in each section between neighboring vertical forces. Equations 4-2 are used for each part.

**PART A–K (Fig. 5-16e):**  For the region $0 \le x \le 2$ ft,

$$\sum F_y = 0 \qquad -400\text{ lb} - V_{K_1} = 0 \qquad V_{K_1} = -400\text{ lb}$$

$$\sum M_K = 0 \qquad M_{K_1} + (400\text{ lb})(x) = 0 \qquad M_{K_1} = -400(x)\text{ lb·ft}$$

**PART A–B–K (Fig. 5-16f):**  The equations are written for the region $2\text{ ft} \le x \le 3$ ft,

$$\sum F_y = 0 \qquad -400\text{ lb} + 750\text{ lb} - V_{K_2} = 0 \qquad V_{K_2} = 350\text{ lb}$$

$$\sum M_K = 0 \qquad M_{K_2} + (400\text{ lb})(x) - (750\text{ lb})(x - 2\text{ ft}) = 0$$

$$M_{K_2} = 350(x)\text{ lb·ft} - 1500\text{ lb·ft}$$

**PART A–B–C–K (Fig. 5-16g):**  The equations are written for the region $3\text{ ft} \le x \le 4$ ft,

$$\sum F_y = 0$$

$$-400\text{ lb} + 750\text{ lb} - 500\text{ lb} - V_{K_3} = 0 \qquad V_{K_3} = -150\text{ lb}$$

$$\sum M_K = 0$$

$$M_{K_3} + (400\text{ lb})(x) - (750\text{ lb})(x - 2\text{ ft}) + (500\text{ lb})(x - 3\text{ ft}) = 0$$

$$M_{K_3} = (750\text{ lb} - 400\text{ lb} - 500\text{ lb})(x) - 1500\text{ lb·ft} + 1500\text{ lb·ft}$$

$$= -150(x)\text{ lb·ft}$$

**PART A–B–C–D–K:**  For the region $4\text{ ft} \le x \le 6$ ft, it is seen that the calculation would become long and tedious because of the many forces involved. Remembering that the same internal reactions can be obtained using either the right or left portion of the beam it is advantageous to use the right-hand portion for region $DE$. However, the equations must remain consistent, with $x$ being measured from the left end (point $A$); hence the general distance shown in Fig. 5-16h is $(6\text{ ft} - x)$. It must also be remembered to assume the proper positive shear and bending moments for the left side of a section as shown. Thus, the last part used, instead of $A–B–C–D–K$, is

**PART K–E (Fig. 5-16h):**  For the region $4\text{ ft} \le x \le 6$ ft,

$$\sum F_y = 0 \qquad V_{K_4} - 300\text{ lb} = 0 \qquad V_{K_4} = 300\text{ lb}$$

$$\sum M_K = 0 \qquad -(300\text{ lb})(6\text{ ft} - x) - M_{K_4} = 0$$

$$M_{K_4} = 300(x)\text{ lb·ft} - 1800\text{ lb·ft}$$

These equations are plotted for the appropriate intervals (as outlined in Ex. 5-7) in Fig. 5-16c and d.

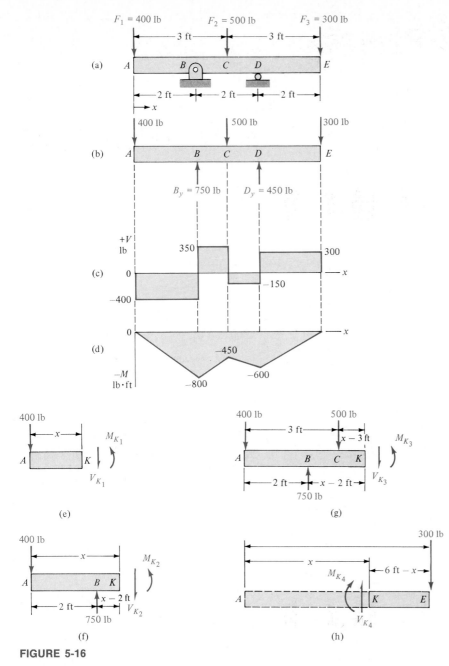

**FIGURE 5-16**

Shear and bending moment diagrams of a simply supported beam

## JUDGMENT OF THE RESULTS

The bending moments are checked at sections $B$, $C$, and $D$ by using both moment equations (for the part to the left and the part to the right from the point considered). It is seen that the same moments are obtained. Hence the results can be stated with confidence. The shear diagram is correct by inspection.

EXAMPLE 5-8   193

# EXAMPLE 5-9

Write the shear and bending moment equations and draw the shear and bending moment diagrams for the cantilever beam shown in Fig. 5-17a.

## SOLUTION

First the reactions at $A$ are obtained (Fig. 5-17b) using Eq. 4-2,

$$\sum F_x = 0 \qquad\qquad\qquad A_x = 0$$

$$\sum F_y = 0 \qquad\qquad\qquad A_y = 0$$

$$\sum M_A = 0 \quad 12,000 \text{ lb·in.} - M_A = 0$$

$$M_A = 12,000 \text{ lb·in.} = 1000 \text{ lb·ft}$$

Analyzing a general section at distance $x$, again assuming positive internal reactions, the shear and moment equations are obtained.

### PART A–K:

$$\sum F_y = 0 \qquad -V_k = 0 \qquad\qquad V_K = 0$$

$$\sum M_K = 0 \quad M_K - 1000 \text{ lb·ft} = 0 \quad M_K = 1000 \text{ lb·ft}$$

Plotting these equations, Fig. 5-17c and d are obtained.

## JUDGMENT OF THE RESULTS

The results appear logical. Since there are no applied vertical forces, no internal shear forces need exist to resist them. Note also that although the couple is a free vector with respect to the external reactions on the rigid beam, it is *not* a free vector with respect to the internal forces and moments. This concept will be illustrated in the next example.

(a)

(b)

(c)

(d)

$M = Fd = (600 \text{ lb})(20 \text{ in.})$
$= 12,000 \text{ lb·in.} = 1000 \text{ lb·ft}$

(e)

**FIGURE 5-17**

Shear and bending moment diagrams of a cantilever beam

EXAMPLE 5-10

Write the shear and bending moment equations and draw the shear and bending moment diagrams for the beam shown in Fig. 5-18a.

## SOLUTION

The reactions at $A$ are given in Fig. 5-18b, the free-body diagram of the whole beam. The reader should check these. There are two distinct parts of the beam, $A$–$K$ and $A$–$B$–$K$. Equations 4-2 are used for each of these parts.

**PART A–K (Fig. 5-18e):**  For the region $0 \leq x \leq 1$ m,

$$\sum F_y = 0 \qquad 2 \text{ kN} - V_{K_1} = 0 \qquad\qquad V_{K_1} = 2 \text{ kN}$$

$$\sum M_K = 0 \qquad M_{K_1} + 0.6 \text{ kN·m} - (2 \text{ kN})(x) = 0$$

$$M_{K_1} = 2(x) \text{ kN·m} - 0.6 \text{ kN·m}$$

**PART A–B–K (Fig. 5-18f):**  The equations are written for the region $1 \text{ m} \leq x \leq 2.3$ m,

$$\sum F_y = 0 \qquad 2 \text{ kN} - V_{K_2} = 0 \qquad\qquad V_{K_2} = 2 \text{ kN}$$

$$\sum M_k = 0 \qquad M_{K_2} + 0.6 \text{ kN·m} + 4 \text{ kN·m} - (2 \text{ kN})(x) = 0$$

$$M_{K_2} = 2(x) \text{ kN·m} - 4.6 \text{ kN·m}$$

The equations are plotted (by the method outlined in Ex. 5-7) in Fig. 5-18c and d. The discontinuity at $B$ in the bending moment curve is caused by the concentrated moment applied at this point. Plotting the moment equations near the end points of each part yields two different values at $B$, which are connected by a vertical line. It should be noted that the effect of a concentrated moment does not show up in the shear diagram. This result appears to be logical after reconsidering the free-body diagrams of the general parts.

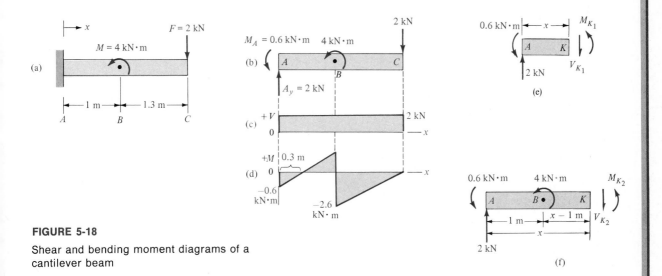

**FIGURE 5-18**

Shear and bending moment diagrams of a cantilever beam

EXAMPLE 5-10   195

# EXAMPLE 5-11

Write the shear and bending moment equations and draw the shear and bending moment diagrams for the beam shown in Fig. 5-19a.

## SOLUTION

The reactions at $A$ and $C$ are given in Fig. 5-19b, the free-body diagram of the whole beam. The reader should check these. There are two distinct parts of the beam, $A$–$K$ and $A$-$B$-$K$. Equations 4-2 are used for each of these parts.

**PART $A$–$K$ (Fig. 5-19e):**  For the region $0 \leq x \leq 3$ ft,

$$\sum F_y = 0 \qquad -V_{K_1} - 714 \text{ lb} = 0 \qquad V_{K_1} = -714 \text{ lb}$$

$$\sum M_K = 0 \qquad M_{K_1} + (714 \text{ lb})(x) = 0 \qquad M_{K_1} = -714(x) \text{ lb·ft}$$

**PART $A$–$B$–$K$ (Fig. 5-19f):**  The equations are written for the region $3$ ft $\leq x \leq 7$ ft,

$$\sum F_y = 0 \qquad -V_{K_2} - 714 \text{ lb} = 0 \qquad V_{K_2} = -714 \text{ lb}$$

$$\sum M_K = 0 \qquad M_{K_2} + (714 \text{ lb})(x) - 5000 \text{ lb·ft} = 0$$

$$M_{K_2} = 5000 \text{ lb·ft} - 714(x) \text{ lb·ft}$$

The equations are plotted (by the method outlined in Ex. 5-7) in Fig. 5-19c and d. Since there is a concentrated moment applied to the beam, the bending moment diagram is discontinuous. The shear diagram is unaffected at the point of the concentrated moment and would be the same regardless of the point at which the moment is applied.

## TIP

It can be noted above, as in all the examples of this section, that the shear is the first derivative of the moment equation with respect to $x$. This relationship is convenient to use for checking answers and the diagrams. A positive shear results in a positively sloped moment diagram. A negative value, as in this example, means that the moment diagram must have a negative slope. This relationship between the shear and bending moment will be discussed fully in Sec. 7-6.

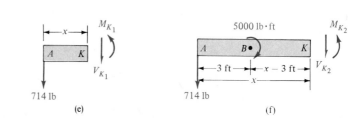

**FIGURE 5-19**

Shear and bending moment diagrams of a simply supported beam

# PROBLEMS

In Probs. 5–31 to 5–50, write the equations of transverse shear and bending moment for the beams shown. Sketch the complete shear and bending moment diagrams.

**5–31**

**5–32**

**5–33**

**5–34**

**5–35**

**5–36**

**5–37**

**5–38**

**5–39**

**5–40**

**5–41**

**5–46**

**5–42**

**5–47**

**5–43**

**5–48**

**5–44**

**5–49**

**5–45**

**5–50**

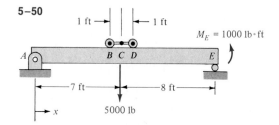

**5–51** Sketch the shear and bending moment diagrams for the boom of the crane described in Prob. 5–21. Make the plots for $0 < x < 120$ ft (the load is at $x_C = 120$ ft).

Fig. P5-51

**5–52** Derive the shear and bending moment equations and plot the shear and bending moment diagrams for the cantilever beam. Measure $x$ as shown.

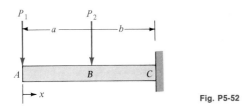

Fig. P5-52

**5–53** Sketch the transverse shear and bending moment diagrams for the ladder described in Prob. 5–23.

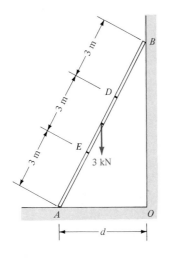

Fig. P5-53

**5–54** A 5-ft-long cantilever beam is modeled with five equal forces equally spaced along the beam. The forces represent the beam's weight and other loads. Draw the shear and bending moment diagrams.

Fig. P5-54

**5–55** In an assembly operation four bolts are tightened simultaneously as shown. Draw the shear and bending moment diagrams for the member $AF$.

Fig. P5-55

**5–56** Consider a light pole described in Prob. 5–26. Assume that section $AD$ is a quarter circle of radius $3d$. Calculate the bending moments at $A$, $B$, $C$, and $D$, and plot these as functions of the linear distance $l$ along the pole. Thus, make a standard bending moment diagram with $M$ on the vertical axis, and $l$ on the horizontal axis ($l$ increases from $A$ to $D$).

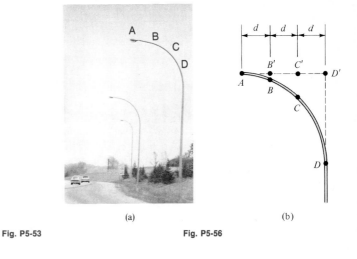

(a)                          (b)

Fig. P5-56

**5–57** Sketch the transverse shear and bending moment diagrams for the arm $AD$ of the excavator in Prob. 5–27.

**Fig. P5-57**

**5–58** Consider the excavator described in Prob. 5–28 and neglect the forces acting at $D$ and $E$. Sketch the bending moment diagrams for arms $AB$ and $BC$.

**Fig. P5-58**

**5–59** An amusement ride is modeled as a cantilever beam supported at $O$. The weight $W_1$ of arm $OA$ is represented by three forces, each of magnitude $P = W_1/3$. A force $P$ is acting at $x = R/4$, $x = R/2$, and $x = 3R/4$. The weight $W_2 = 2W_1$ of the small arms is acting at $A$. A clockwise couple of magnitude $M_A = PR$ is also acting at $A$ (as a reaction to initiating motion of the small arms). Sketch the bending moment diagram of arm $OA$.

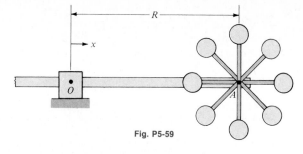

**Fig. P5-59**

**5–60** Draw the shear and bending moment diagrams for beam $AB$ of the drag racing car described in Prob. 5–30.

**Fig. P5-60**

**5–61** Write the equations of transverse shear and bending moment for the beam shown. Sketch the complete shear and bending moment diagrams.

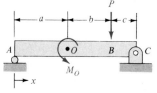

**Fig. P5-61**

**5–62** An axle considered in a design is modeled as a cantilever beam on which the following force and moments are acting: $\mathbf{F} = -F\mathbf{j}$, $\mathbf{M}_B = -M\mathbf{k}$, and $\mathbf{M}_C = 2M\mathbf{i}$. Draw the shear and bending moment diagrams.

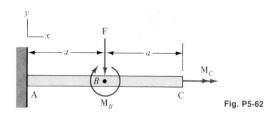

**Fig. P5-62**

**5–63** Assume that the end pipe of an agricultural sprinkler system is horizontal and is hinged at $A$. Its distributed weight is represented by nine concentrated forces as shown. Draw the shear and bending moment diagrams assuming that only the wire at $B$ is holding the pipe up ($T_C = T_D = 0$). For optional work, assume that either $T_B = T_D = 0$, or $T_B = T_C = 0$.

(a)

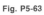

Fig. P5-63

(b)

**5-64** Each jet engine on the wing has a mass of 5000 kg. The magnitude of the resultant of the aerodynamic lift forces is $L = 2$ MN. Plot the moment diagram for the wing using these concentrated forces. Sketch on the same diagram your estimate of a more realistic moment variation. Assume that the wing is weightless and that **L** is the resultant of many parallel forces that are largest at $O$ and smallest at $D$ (the correct way of doing this is covered in Ch. 7; here, rely on simple techniques and intuition).

Fig. P5-64

(a)

(b)                    Fig. P5-65

**5-65** Consider the boom of the 100-metric-ton crane whose model is shown. The boom is hinged at $A$ and supported by wires at $B$ that make a 30° angle with it. Sketch on a single moment diagram two curves that show the bending moments in the boom with the load **P** at $C$ and later at $B$. $l = 15$ m, $P = 1$ MN.

**5-66** Solve Prob. 5–65 with the modification that the beam has a fixed end at point $A$ and that the tension in the wire $BD$ is given as $2P$.

Section 5-1

**COMMON SUPPORTS OF BEAMS AND POSSIBLE REACTIONS:**

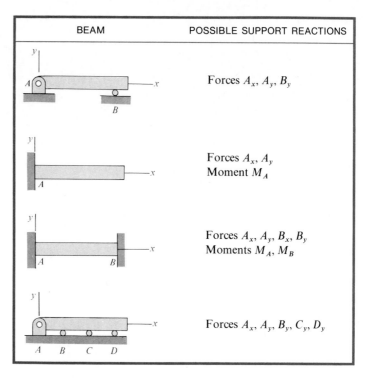

| BEAM | POSSIBLE SUPPORT REACTIONS |
|------|----------------------------|
| | Forces $A_x$, $A_y$, $B_y$ |
| | Forces $A_x$, $A_y$<br>Moment $M_A$ |
| | Forces $A_x$, $A_y$, $B_x$, $B_y$<br>Moments $M_A$, $M_B$ |
| | Forces $A_x$, $A_y$, $B_y$, $C_y$, $D_y$ |

Section 5-2

**INTERNAL FORCES AND MOMENTS—EXAMPLE:**
Given loading of a beam:

Free-body diagram of whole beam:

External reactions at support $O$:

External reactions at support $O$:

$$H_O = \text{horizontal reaction} = P$$
$$M_O = \text{bending moment} = FL$$
$$V_O = \text{vertical shear} = F$$

Free-body diagrams for parts $OA$ and $AB$:

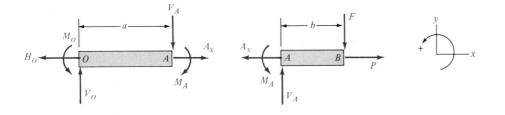

Internal reactions at arbitrary section $A$:

$$V_A = F \qquad M_A = \underbrace{M_O - Fa}_{\substack{\text{from} \\ \text{left piece}}} = \underbrace{Fb}_{\substack{\text{from} \\ \text{right piece}}}$$
$$A_x = P$$

$$\left( \begin{array}{l} \text{check:} \quad \text{as } a \to L,\ M_A \to 0 \\ \qquad\qquad \text{as } a \to 0,\ M_A \to M_O \end{array} \right)$$

## Section 5-3

**SIGN CONVENTIONS FOR SHEAR FORCE AND BENDING MOMENT:** According to convention, a moment that makes a beam concave upward is taken as positive. Thus, a clockwise moment is positive on the left side of a section, and a counterclockwise moment is positive on the right side. A shear force that acts upward on the left side of a section, or downward on the right side, is positive.

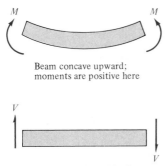

Beam concave upward;
moments are positive here

Shear forces are positive here

### SHEAR AND BENDING MOMENT DIAGRAMS—PROCEDURE:

**1.** Draw the free-body diagram of the whole beam and determine all reactions at the supports.

**2.** Draw the coordinate axes for the shear ($V$) and bending moment ($M$) diagrams directly below the free-body diagram.

**3.** Immediately plot those values of $V$ and $M$ that can be determined by inspection (especially where they are zero), observing the sign conventions.

**4.** Calculate and plot as many additional values of $V$ and $M$ as are necessary for drawing reasonably accurate curves through the plotted points.

**5.** Draw the complete curves of $V$ and $M$ and check a few points on them by using free-body elements that are complementary to those used in the original calculations.

### EXAMPLE:

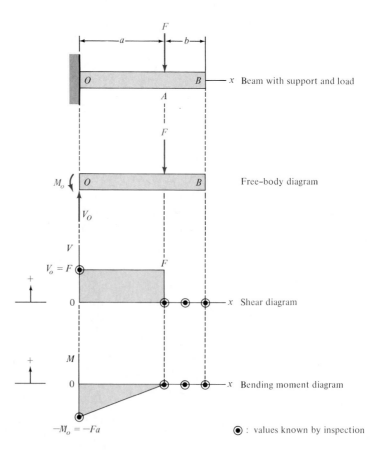

**5-67** Consider the model of a racing car mounted in a wind tunnel. The car is attached by two supports (each consisting of three parts, *A*, *B*, *C*) to the bottom of the tunnel. Each segment *A*, *B*, and *C* of the supports is in the *yz* plane and thus perpendicular to the direction of airflow. The supports *ABC* are instrumented to measure air drag and upward or downward lift on the car. For simplicity, assume that there is only one support *ABC*, with partial lengths *A* = 50 mm, *B* = 80 mm, and *C* = 400 mm. Ignore the weight of the car. Assume that a drag force $\mathbf{D} = -10\,\mathrm{N\,i}$ is acting at the top of rod *A* while the model is adjusted to have no lift force on it. Calculate the transverse shear and bending moment at the middle (*E*) and bottom (*F*) of rod *A*. Give the answers in vector form.

**5-68** Determine the transverse shear and bending moment at the top (*G*) and bottom (*H*) of rod *C* described in Prob. 5–67. Distinguish bending from twisting by using an equivalent force–couple system.

**5-69** Calculate the transverse shear and bending mo-

ment at the middle (*I*) of rod *B* described in Prob. 5–67. Assume that besides force **D** a lift force $\mathbf{L} = 2\mathrm{N}\,\mathbf{j}$ is also acting at the top of rod *A*. Distinguish bending from twisting by using an equivalent force–couple system.

**5-70** Determine the transverse shear and bending moment at the middle (*J*) of rod *C* described in Prob. 5–67. Assume that besides force **D** a downward aerodynamic force (negative lift) is also acting at the top of rod *A*, $\mathbf{L} = -4\mathrm{N}\,\mathbf{j}$. Distinguish bending from twisting by using an equivalent force-couple system.

**5-71** Consider Prob. 5–67 assuming that the drag force is $\mathbf{D} = -12\mathrm{N}\,\mathbf{i}$ while a negative lift force $\mathbf{L} = -3\mathrm{N}\,\mathbf{j}$ is also acting at the top of rod *A*. For purposes of instrumentation, the bending moments in the three rods should be compared. Calculate the bending moment at the middle of each rod: at point *E* in rod *A*, point *I* in rod *B*, and point *J* in rod *C*. (These points agree with those in the preceding related problems.) Distinguish bending from twisting by using an equivalent force–couple system.

(a)

(b)

**Fig. P5-67**

**5–72** A T-shaped fixture $ABC$ is tested in tension in a testing machine. An upward force of magnitude $F = 200$ lb is applied at $A$, while a downward force $\mathbf{F}$ is applied at $G$. Determine the internal bending moments at $B$, $C$, $D$, $E$, and $F$. $AB = GC = 2\ FC = 3\ AE = 8$ in. and $BD = DC = 5$ in.

**5–73** Draw the shear and bending moment diagrams for the simply supported beam.

**Fig. P5-73**

**5–74** Draw the shear and bending moment diagrams for the cantilever beam.

**Fig. P5-74**

# Analysis of Simple Structures and Machines

## Basic Concepts of Structures

Engineering structures are made of individual rigid members for the purpose of carrying loads safely and efficiently. For example, airplane wings must be strong to prevent failure but also minimum in weight for economical flying. Even stationary structures such as bridges and buildings must be designed to save materials.

Fortunately, it is possible to design beams and other members that are as strong but much lower in weight than the simplest solid members conceivable for the same purpose. The idea is to shape each member and the whole structure in such a way that most of the material carries reasonably high loads. This is the ultimate challenge in the engineering design of any structure: the amount of material with small loads in it should be minimized, but excessive loads are not permissible even in small regions of the whole structure. Knowledge of statics and strength of materials is important in such design work.

**SECTION 6-1** describes trusses, the most common structures, which have high strength and relatively low weight. Simple trusses are statically determinate and can be analyzed using the methods developed in Ch. 4. This section describes the simplifying assumptions necessary for solving basic problems.

**SECTION 6-2** presents the method of joints, which is based on equilibrium of a particle. This method provides substantial practice in the application of Newton's Third Law and interrelated free-body diagrams.

**SECTION 6-3** presents special cases of joints which are worth remembering for simplifying the analysis of trusses made of two-force members.

**SECTIONS 6-4 AND 6-5** will be omitted in most statics courses, but they should be of interest to those students who expect to work regularly with complex trusses. Section 6-4 describes a formal procedure for setting up problems, and Sec. 6-5 shows a graphic method of analyzing trusses on the basis of force polygons. This graphic method can be used with or without the formal notation described in Sec. 6-4.

**SECTION 6-6** presents the method of sections, which is another method for determining forces in a truss. This method is especially useful when it is required to calculate the forces only in a few members of a truss. It involves a simple procedure which is ideal for practicing basic skills with free-body diagrams and equilibrium equations.

**SECTION 6-7** extends the methods of analysis of plane trusses to three-dimensional trusses. This topic is often omitted in regular statics courses because, while it generalizes the basic methods, it requires rather laborious analyses.

**SECTION 6-8** presents the important method of analyzing frames and machines which involve multiforce members. This method is also based on concepts of equilibrium and interrelated free-body diagrams (Ch. 4). Proficiency in solving problems in this area is indicative of satisfactory knowledge of equilibrium principles.

**SECTION 6-9** describes a simple method of analysis for cables which are assumed to be weightless and have only concentrated forces acting on them. Such a cable can be analyzed as a set of two-force members without introducing any new concepts. This topic provides additional practice in solving for unknowns.

## 6-1    TRUSSES

*Trusses*, which are among the most common structures with high strength-to-weight ratios, are made in many different sizes and shapes, as can be seen in certain roof and bridge structures. They have straight, relatively slender members whose ends are connected at joints. Two-dimensional trusses (also called *plane trusses*) are designed to carry loads acting in their planes, and they are often connected to form three-dimensional trusses (also called *space trusses*).

The individual members of a truss are bolted, pinned, riveted, or welded at the connections. A major simplification in the analysis of any truss is to assume that all members are joined by frictionless pins at their ends. In this way each member can be treated as a two-force member, which of course can only have axial tension or compression in it. A consequence of this assumption is that in the analysis all forces must be applied at the joints. Thus, all members are assumed to be weightless, or their weights are divided and assumed to act at the joints.

Naturally, real structural frameworks are more complex than the models used in most analyses. For example, the boom of the crane in Fig. 6-1 is pinned only at sections *A* and *B*, and loads may be applied between the joints, laterally to some members. Still, the simplifying assumptions may be quite reasonable. The load on the boom is shared and

**FIGURE 6-1**

Part of a crane's boom

applied by eight small wheels, so the local lateral load on a member is reduced. The numerous welded joints between sections $A$ and $B$ make the structure somewhat more rigid than if it were pinned at every connection, so the assumption of pinned joints is a conservative one for practical purposes.

## Rigid Trusses

The basic element of a rigid plane truss is the triangle formed by three straight bars as modeled in Fig. 6-2a. This forms a rigid frame to the extent that the three bars and the connections themselves are rigid. In contrast, four or more bars connected to make polygons with as many sides are nonrigid frames. For example, the frame of four bars in Fig. 6-2b collapses to the right when a force $\mathbf{F}$ is applied at $D$. The addition of a single bar $BD$ or $AC$ would create triangular elements and make the frame rigid. This is called a *simple truss*. The addition of *both* bars $BD$ and $AC$ makes the frame rigid, but one of these members is redundant, so the truss is statically indeterminate in the analysis.

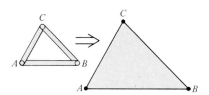

(a) Simplest drawing of truss element

(b)

**FIGURE 6-2**

Three pinned bars form the basic element of a rigid plane truss

## METHOD OF JOINTS

**6-2**

A truss consists only of pins and two-force members in the simplest cases, and is simply supported similar to beams. It is remarkable that all the forces on the pins and bars can be determined even when there are many such components. These calculations are based on the equilibrium of the whole truss and of its parts. The method is illustrated using the simple truss shown in Fig. 6-3a. The truss is pinned to a fixed bracket at $A$, while the roller at joint $C$ allows motion in the horizontal direction. This is a typical arrangement for the prevention of undesirable forces that would be caused by thermally or mechanically induced deformations in a redundantly supported truss.

The free-body diagram of the truss is in Fig. 6-3b. The external reactions $A_y$ and $C_y$ are obtained from the equilibrium equations of the truss. By inspection, an equation for moments can be written to solve for one unknown at once. Thus, from Eq. 4-2,

$$\sum M_A = 0 \qquad\qquad \sum F_y = 0$$
$$2.5lC_y - lP = 0 \qquad A_y + C_y - P = 0$$
$$C_y = 0.4P \qquad\qquad A_y = 0.6P$$

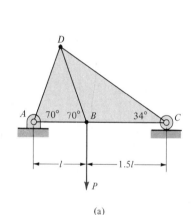

(a)

(b)

(c) (d) (e)

(f) (g)

**FIGURE 6-3**

Method of joints for the analysis of a simple truss

The solution of a truss problem should be started at a joint where only two unknown forces act. In this way the two equilibrium equations for forces at a joint ($\sum F_x = 0$, $\sum F_y = 0$) can be solved immediately for two unknowns. Joints $A$ and $C$ satisfy this criterion in the present problem. To start with $A$, draw the free-body diagram of this joint in Fig. 6-3c. The unknown forces $\mathbf{F}_{AB}$ and $\mathbf{F}_{AD}$ have known lines of action along members $AB$ and $AD$, respectively. They may be arbitrarily directed away from $A$ or toward $A$ in this diagram. To use a simple convention, assume that all members are in tension, which is indicated by arrows directed away from joint $A$ toward joints $B$ and $D$. The unknown forces are determined using the equilibrium equations for forces at the joint $A$. From Eq. 4-2,

$$\sum F_x = 0 \qquad\qquad\qquad \sum F_y = 0$$

$$F_{AB} + F_{AD} \cos 70° = 0 \qquad 0.6P + F_{AD} \sin 70° = 0$$

$$F_{AB} = -F_{AD} \cos 70° \qquad F_{AD} = -\frac{0.6P}{\sin 70°}$$

$$= -0.64P \text{ (the negative}$$
$$\text{sign indicates that}$$
$$\mathbf{F}_{AD} \text{ is actually toward}$$
$$\text{joint } A)(\text{compression})$$

Using the results for $AD$,

$$F_{AB} = 0.22P \text{ (the positive answer indicates that}$$
$$\text{the assumed direction of } \mathbf{F}_{AB} \text{ is correct) (tension)}$$

The same procedure is applied for joint $C$ with the aid of Fig. 6-3d. Again using Eq. 4-2,

$$\sum F_x = 0 \qquad\qquad\qquad \sum F_y = 0$$

$$-F_{CD} \cos 34° - F_{BC} = 0 \qquad 0.4P + F_{CD} \sin 34° = 0$$

$$F_{BC} = -F_{CD} \cos 34° \qquad F_{CD} = -\frac{0.4P}{\sin 34°}$$

$$= -0.72P$$
$$(\text{compression})$$

Using the result for $F_{CD}$,

$$F_{BC} = 0.6P \text{ (tension)}$$

The analysis of joint $B$ is made using the free-body diagram in Fig. 6-3e,

$$\sum F_x = 0 \qquad\qquad\qquad \sum F_y = 0$$

$$F_{BC} - F_{AB} - F_{BD} \cos 70° = 0 \qquad F_{BD} \cos 20° - P = 0$$

$$0.6P - 0.22P - F_{BD} \cos 70° = 0 \qquad F_{BD} = 1.06P \text{ (tension)}$$

$$F_{BD} = 1.11P \text{ (tension)}$$

The two answers for the same force $\mathbf{F}_{BD}$ are caused by round-off errors. Thus, one of the last two equilibrium equations can be used to check the result obtained from the other equation. Further checking can be done with the free-body diagram of joint $D$, where all of the forces are known at this stage. From Fig. 6-3f (noting the actual directions of the forces),

$$\sum F_x: \quad F_{AD} \sin 20° + F_{BD} \sin 20° - F_{CD} \cos 34°$$
$$= 0.64P(0.34) + 1.06P(0.34) - 0.72P(0.83)$$
$$\cong 0$$

$$\sum F_y: \quad F_{AD} \sin 70° - F_{BD} \sin 70° + F_{CD} \sin 34°$$
$$= 0.61P - 1.00P + 0.40P \cong 0$$

These results show that the forces were calculated correctly.

Note that Newton's Third Law is essential in using the method of joints. For example, the directions of forces in Fig. 6-3f are determined by the directions established in Fig. 6-3c, d, and e. Consider the joints $A$ and $D$ and the bar between them in detail. It is this bar that applies the force $\mathbf{F}_{AD}$ on joint $A$ as shown in Fig. 6-3g. In reaction, the joint $A$ applies the force $\mathbf{F}_{AD}$ at the bottom of this bar as shown. The bar is in equilibrium, so an equal and opposite force must be applied by the joint $D$ at the top of the bar as in Fig. 6-3g, putting the bar in compression. Again by Newton's law, the force $\mathbf{F}_{AD}$ acts on joint $D$ as shown in Fig. 6-3f and g.

The interrelated free-body diagrams must be drawn with complete consistency, but this is easily done after a little experience. Generally, it is sufficient to work only with the joints in the sketches, with a mental picture of the members between the joints such as $AD$ in Fig. 6-3g. It is helpful to draw the arrows at the joints on that side of the pin where the appropriate member is. For these, an arrow away from the pin indicates tension in the bar, and an arrow toward the pin indicates compression in the bar.

For a plane truss that contains $n$ pins it is possible to write $2n$ independent equilibrium equations (for summations in the $x$ and $y$ directions at each pin), so the maximum number of unknown forces that could be determined is $2n$. Another way to look at this is with respect to the total number of connecting members, $m$. In a simple truss $m = 2n - 3$ (to prove this, start with three members and three joints; there is one additional joint for every two new members that are added). Accordingly, the maximum number of unknown forces that can be determined with the method of joints is $m + 3$ for a truss having $m$ members. These considerations are valid for rigid plane trusses that are statically determinate externally and internally.

The procedure in conjunction with Fig. 6-3 typifies a simple and commonsense approach to understanding and solving truss problems. The emphasis in this approach is on using interrelated free-body diagrams of the joints, equilibrium equations for the diagrams, and on judgment of selecting the joints for efficient analysis. This procedure is ideal and sufficient for most students.

A few special arrangements of members at joints are worth remembering because they can save time and effort in the analysis. In all cases the special properties of the arrangement are the results of having only two-force members connected at the joints.

**1.** *Two members* connected at a joint must be collinear to transmit any force as in Fig. 6-4a. If there are forces in these members, the forces must be equal for equilibrium. Both tension and compression are possible in theory, but only tension is possible in reality. Since perfect alignment is difficult to achieve in practice, even small compressive loads would cause a collapse at a two-member joint. On the other hand, tension causes self alignment of the members. When the two members are not collinear, as in Fig. 6-4b, the forces in them must be zero for equilibrium.

**2.** *Three members* with two of them collinear are possible only if the third member has no force in it, as in Fig. 6-4c. The reason is that members $A$ and $B$ could not apply a lateral force to oppose the component of the force in $C$ that is perpendicular to $A$ and $B$. This is true for any angle $\theta$.

In that case, why have a member such as $C$ in Fig. 6-4c? The answer is again found after considering the difference between theory and practice. In theory, there is absolutely no reason to have the member $C$ when $A$ and $B$ are collinear. In practice, $A$ and $B$ may make a small angle, and member $C$ is important to maintain the geometry at the joint. This can be accomplished with a relatively small force in $C$. In other words, the so-called *zero-force member* $C$ may have a theoretically negligible force in it, which may even allow compressive loading of members $A$ and $B$. Of course, $C$ can be ignored in much of the analysis. An exception to this is when an external load is applied at the joint of $A$, $B$, and $C$, as in Fig. 6-4d. Another practical reason for having apparently unnecessary members is that most structures are subjected to changing loading conditions.

**3.** *Four members* that are collinear in pairs at a single joint must have equal and opposite forces in pairs, as shown in Fig. 6-4e. The force polygon for such a joint is a parallelogram, as in Fig. 6-4f.

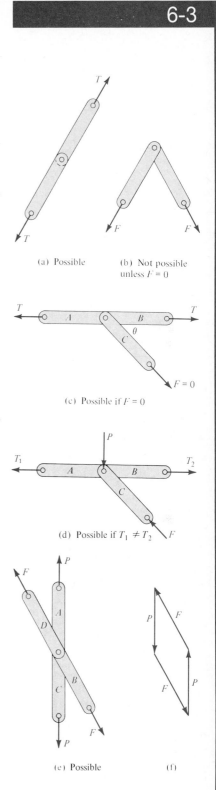

**FIGURE 6-4**

Special cases of joints

EXAMPLE 6-1

Determine the force in each member of the truss in Fig. 6-5a using the method of joints. Take advantage of symmetry.

## SOLUTION

Because of symmetry, it is sufficient to determine the forces at joints $A$, $B$, $C$, $G$, and $H$. The support reactions are determined first. By symmetry, $A_y = E_y = $ total load/2 = 75 kN. The joint-by-joint analysis is begun at a joint where there are only two unknown forces.

### JOINT $A$ (Fig. 6-5b):

$$\sum F_x = 0 \qquad\qquad \sum F_y = 0$$

$$F_{AH} + F_{AB} \cos 60° = 0 \qquad 75 \text{ kN} + F_{AB} \sin 60° = 0$$

$$F_{AB} = -86.6 \text{ kN (bar } AB$$
$$\text{is in compression)}$$

Using the result for $F_{AB}$,

$$F_{AH} = 43.3 \text{ kN} \quad \text{(tension, as assumed)}$$

(a)

(b)    (c)    (d)    (e)    (f)

**FIGURE 6-5**

Analysis of a symmetric plane truss

**JOINT B (Fig. 6-5c):**   A quick check shows there are three unknown forces and only two equations of equilibrium to use for solving. Hence this joint is not solvable yet.

**JOINT H (Fig. 6-5d):**

$$\sum F_x = 0 \qquad\qquad\qquad \sum F_y = 0$$

$$F_{GH} - F_{AH} = 0 \qquad\qquad F_{BH} - 50 \text{ kN} = 0$$

$$F_{GH} = 43.3 \text{ kN} \quad (\text{tension}) \quad F_{BH} = 50 \text{ kN} \quad (\text{tension})$$

**JOINT B:**   Joint B may now by analyzed. From Fig. 6-5c,

$$\sum F_x = 0$$

$$F_{BC} \cos 31° + F_{BG} \sin 40.9° - F_{AB} \sin 30° = 0$$

$$0.857 F_{BC} + 0.655 F_{BG} + 43.3 \text{ kN} = 0$$

$$\sum F_y = 0$$

$$F_{BC} \sin 31° - F_{AB} \cos 30° - F_{BH} - F_{BG} \cos 40.9° = 0$$

$$0.515 F_{BC} + 75 \text{ kN} - 50 \text{ kN} - 0.756 F_{BG} = 0$$

Solving the two equations simultaneously yields

$$F_{BG} = -0.88 \text{ kN} \quad (\text{compression})$$

$$F_{BC} = -49.8 \text{ kN} \quad (\text{compression})$$

**JOINT C (Figure 6-5e):**   From symmetry, $F_{CD} = F_{BC} = -49.8$ kN (compression),

$$\sum F_y = 0$$

$$-F_{CG} - F_{BC} \sin 31° - F_{CD} \sin 31° = 0$$

$$F_{CG} = -2 F_{BC} \sin 31° = -2(-49.8 \text{ kN})(0.515)$$

$$= 51.3 \text{ kN} \quad (\text{tension})$$

This completes the analysis of the symmetric truss.

## JUDGMENT OF THE RESULTS

A summation of the known forces in the $y$ direction at joint $G$ serves as a check on the other calculations. From Fig. 6-5f,

$$\sum F_y = F_{BG} \cos 40.9° + F_{CG} + F_{DG} \cos 40.9° - 50 \text{ kN}$$

With $F_{BG} = F_{DG}$ by symmetry,

$$-(0.88 \text{ kN})(0.756) + 51.3 \text{ kN} - (0.88 \text{ kN})(0.756) - 50 \text{ kN}$$

$$= -0.03 \text{ kN}$$

The deviation from zero in the final summation is caused by round-off errors. The results are accepted as correct.

EXAMPLE 6-1   215

EXAMPLE 6-2

Determine the forces acting at joints $A$, $B$, $C$, $H$, and $I$ of the truss in Fig. 6-6a. The angles are $\alpha = 56.3°$, $\beta = 38.7°$, $\phi = 39.8°$, and $\theta = 36.9°$.

## SOLUTION

First the reactions at the supports are determined and are shown in Fig. 6-6b. A joint at which only two unknown forces act is the best as the starting point for the solution. Joints $A$ and $F$ satisfy this criterion. Choosing joint $A$, the solution is progressively and thoughtfully developed, always seeking the next joint with only two unknowns. In each diagram circles indicate the quantities that are known from the preceding analysis.

(a)

(b)

(c)

(d)

(e)

(f)

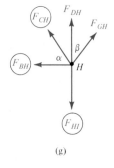

(g)

**FIGURE 6-6**

Analysis of a plane truss

**JOINT A (Fig. 6-6c):**

$\sum F_x = 0$

$F_{AI} = 0$

$\sum F_y = 0$

$F_{AB} - R_A = 0$

$F_{AB} - 50 \text{ kips} = 0$

$F_{AB} = 50 \text{ kips}$   (tension)

**JOINT I (Fig. 6-6d):**

$\sum F_x = 0$

$-F_{BI} \cos \alpha - F_{AI} = 0$

$F_{BI} \cos \alpha + 0 = 0$

$F_{BI} = 0$

$\sum F_y = 0$

$R_I + F_{HI} + F_{BI} \sin \alpha = 0$

$70 \text{ kips} + F_{HI} + 0 \sin \alpha = 0$

$F_{HI} = -70 \text{ kips}$   (compression)

**JOINT B (Fig. 6-6e):**

$\sum F_x = 0$

$F_{BH} + F_{BI} \cos \alpha = 0$

$F_{BH} + 0 \cos \alpha = 0$

$F_{BH} = 0$

$\sum F_y = 0$

$F_{BC} - F_{BI} \sin \alpha - F_{AB} = 0$

$F_{BC} - 0 \sin \alpha - 50 \text{ kips} = 0$

$F_{BC} = 50 \text{ kips}$   (tension)

**JOINT C (Fig. 6-6f):**

$\sum F_x = 0$

$F_{CD} + F_{CH} \cos \alpha = 0$

$F_{CD} - (60.1 \text{ kips})(0.555) = 0$

$F_{CD} = 33.3 \text{ kips}$
(tension)

$\sum F_y = 0$

$-F_{BC} - F_{CH} \sin \alpha = 0$

$-50 \text{ kips} - F_{CH}(0.832) = 0$

$F_{CH} = -60.1 \text{ kips}$
(compression)

**JOINT H (Fig. 6-6g):**

$\sum F_x = 0$

$F_{GH} \sin \beta - F_{CH} \cos \alpha$
$\quad - F_{BH} = 0$

$F_{GH}(0.625) + (60.1 \text{ kips})(0.555)$
$\quad - 0 = 0$

$F_{GH} = -53.4 \text{ kips}$
(compression)

$\sum F_y = 0$

$F_{CH} \sin \alpha + F_{DH}$
$\quad + F_{GH} \cos \beta - F_{HI} = 0$

$-(60.1 \text{ kips})(0.832) + F_{DH}$
$\quad - (53.4 \text{ kips})(0.780)$
$\quad + 70 \text{ kips} = 0$

$F_{DH} = 21.7 \text{ kips}$   (tension)

## JUDGMENT OF THE RESULTS

The magnitudes of the forces in the two-force members appear reasonable considering the applied force and the reactions at the supports. The senses of most of these forces, namely tension or compression in the members, are intuitively correct for the given situation. The fact that *AI*, *BI*, and *BH* are *zero-force members* should also be clear by inspection of the special loading conditions at joints *A*, *B*, and *I*.

EXAMPLE 6-2   217

**6-1** Determine the force in member $AC$ if $F = 1000$ lb, $\theta = 100°$, $AB = 8$ ft, and $BC = 12$ ft.

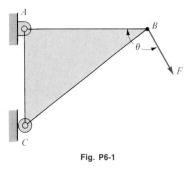

**Fig. P6-1**

**6-2** Determine the forces in the three members in Fig. P6-1 if $F = 3$ kN, $\theta = 70°$, $AB = 3$ m, and $AC = 1.4$ m.

**6-3** Calculate the forces in each member of the truss.

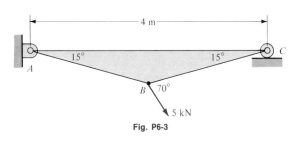

**Fig. P6-3**

**6-4** Determine the forces in the five members of the truss.

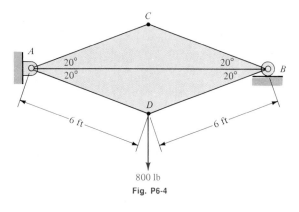

**800 lb**

**Fig. P6-4**

**6-5** The effect of wind on a structure is modeled as a concentrated force acting at $B$. Calculate the forces in the five members of the truss.

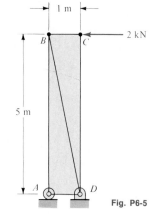

**Fig. P6-5**

**6-6** Consider Fig. P6-5. Replace member $BD$ with member $AC$ and determine the force in member $AC$.

**6-7** Calculate the forces in the five members of the truss as shown. Also solve after replacing member $AC$ with member $BD$. Which configuration appears to be the best to minimize the forces, especially to minimize compression in long members?

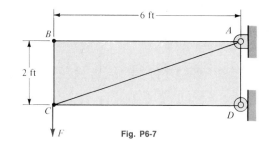

**Fig. P6-7**

**6-8** Calculate the forces in the five members of the truss.

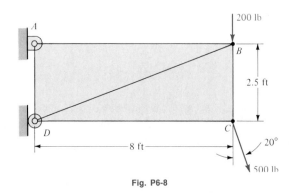

**Fig. P6-8**

**6–9**  Determine the forces in the five members if $F_1 = (2i - 1.5j)$ kN and $F_2 = 4i$ kN.

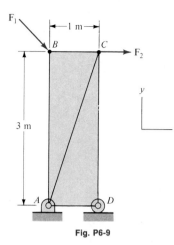

Fig. P6-9

**6–10**  Determine the forces in members $AB$, $BE$, and $EF$ if $P = 5$ kN and $l = h = 1.5$ m.

Fig. P6-10

**6–11**  Consider Fig. P6-10 with the following changes: replace member $BF$ with $AE$, and member $CE$ with $BD$. Calculate the forces in every member if $P = 1000$ lb, $l = 5$ ft, and $h = 4$ ft.

**6–12**  Determine the forces in members $AB$, $BD$, and $BE$ in the truss.

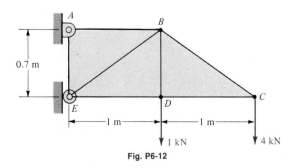

Fig. P6-12

**6–13**  Assume that members $AC$ and $BD$ of the truss have no common pin connection. List all the unknown forces and independent equilibrium equations for the truss.

Fig. P6-13

**6–14**  Calculate the forces acting at joint $F$ in the truss.

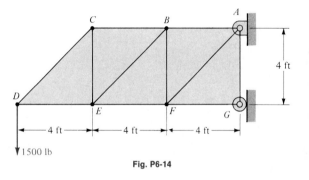

Fig. P6-14

**6–15**  The weight of the truss in Fig. P6-14 is modeled by a 200-lb force acting at $E$, and a 250-lb force acting at $F$. Calculate the other forces acting at $F$.

**6–16**  Determine the force in every member of the truss. Each member is 2 m long.

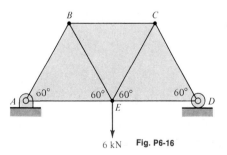

Fig. P6-16

**6–17**  Add 1-kN downward forces at joints $B$ and $C$ in Prob. 6–16, and determine the forces in the members.

**6–18**  The bridge truss is symmetric about line *DJ*. Each horizontal member is 8 ft long, and the structure is 10 ft high. Determine the forces acting on joints *L*, *K*, and *J*.

**6–20**  A common structure, the Pratt roof truss, is symmetric about line *DJ*. Calculate the forces in members *BC*, *LK*, and *KJ* if *AL* = *LK* = *KJ* = 5 ft, and *DJ* = 10 ft.

Fig. P6-18

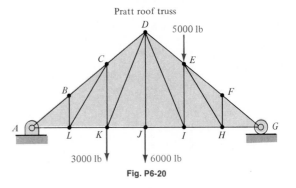

Pratt roof truss

Fig. P6-20

**6–19**  The Howe roof truss is symmetric about line *DJ*. Each horizontal member is 2 m long, and *DJ* = 3 m. Calculate the forces acting at joint *J*.

**6–21**  Determine the zero-force members of the trusses in (a) Fig. P6-19 and (b) Fig. P6-20 with the loading conditions given.

**6–22**  Determine the zero-force members for the symmetric truss if $F_2 = 2500$ lb, $F_4 = 4000$ lb, and $F_1 = F_3 = 0$.

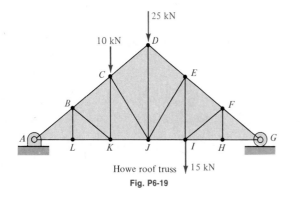

Howe roof truss

Fig. P6-19

**6–23**  Consider the truss in Prob. 6–22 with $F_1 = 2000$ lb, $F_2 = 2500$ lb, $F_3 = 3000$ lb, and $F_4 = 4000$ lb. Determine the forces acting at joints *E*, *N*, and *O*.

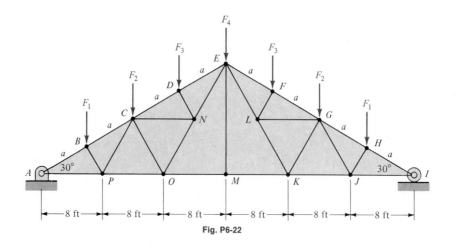

Fig. P6-22

Students who expect to work regularly with complex trusses may be interested in a formal procedure for setting up problems for analysis. The formal method of dealing with members, joints, and forces makes use of *Bow's notation*. This is illustrated in Fig. 6-7 for a simple plane truss.

First, every region between the external loads and reactions is denoted by a lowercase letter, going clockwise around the truss, as in Fig. 6-7b. The lettering is continued for the areas inside the truss, still going clockwise. These letters are used to denote the members, joints, and external and internal forces as follows.

**1.** *Members* are named by the letters of the two adjacent areas. For example, the diagonal member is *gh* or *hg*, the top member is *bh* or *hb*.

**2.** *Joints* are denoted by numbers or by listing the letters of areas at the joint in clockwise order. For example, the joint at $R_4$ is *bci*, or *cib*, or *ibc*.

**3.** *External forces* (loads and reactions) are named by the letters of the two areas adjacent to them, written in clockwise order with capital letters and vector notation with an arrow. For example, the force **P** is written as $\overrightarrow{AB}$, the reaction $R_4$ as $\overrightarrow{BC}$.

**4.** *Internal forces* (exerted by members on pins) are named by the letters of the two adjacent areas to the member. Capital letters with an arrow are used with respect to a particular joint. The arrows are omitted when the letters denoted only magnitudes of forces. For example, the force exerted by member *af* on joint 1 is $\overrightarrow{AF}$.

The whole scheme is illustrated in Fig. 6-8 for three joints of the truss in Fig. 6-7. Approximate force polygons are also given for the three joints to show an alternative method to the algebraic calculations.

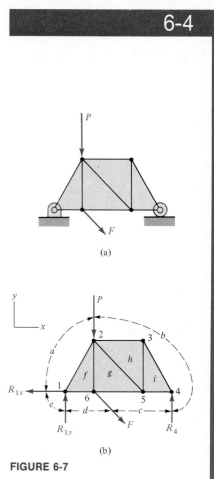

**FIGURE 6-7**

Formal procedure in joint analysis

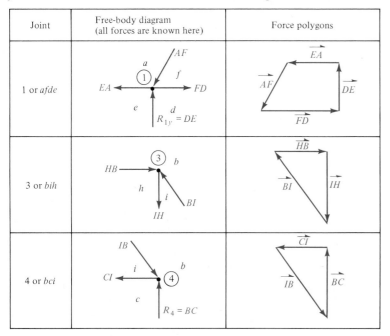

| Joint | Free-body diagram (all forces are known here) | Force polygons |
|---|---|---|
| 1 or *afde* | | |
| 3 or *bih* | | |
| 4 or *bci* | | |

**FIGURE 6-8**

Joint analysis schemes for Fig. 6-7

## GRAPHICAL ANALYSIS OF TRUSSES. MAXWELL DIAGRAM

The force polygons for a truss may be superimposed to make a composite figure called the *Maxwell diagram*. Construction of this diagram is based on the equilibrium of each joint and on the correct relationship of the individual free-body diagrams of the joints. The detailed, step-by-step procedure is shown in Fig. 6-9.

At first the external reactions must be determined from the equilibrium equations for the whole truss. Assume that the external load $P$ causes the reactions $R_1$ and $R_3$ shown in Fig. 6-9a. The graphic construction can be started and continued at any joint where the unknown forces can be determined directly from a force polygon (at most two unknown forces at a joint). For example, start with joint 1 in Fig. 6-9b. Here $\overrightarrow{CA}$ is known and must be drawn to scale. The force triangle is drawn knowing the lines of action of the other forces at the same joint. Note that the lettering sequence for the forces must be correct.

Consider joint 2 next, where force $\overrightarrow{DA}$ is known (it must be equal and opposite to force $\overrightarrow{AD}$) and there are only two unknown forces. Force $\overrightarrow{DA}$ is added as a dashed line in Fig. 6-9c. Successively, forces $\overrightarrow{AE}$ and $\overrightarrow{ED}$ are drawn in Fig. 6-9c since their lines of action are known.

The procedure is continued similarly in Fig. 6-9d and e. All previously drawn lines are retained at each step. New forces are added as solid arrows when first considered, and as dashed arrows when opposing already drawn forces. Figure 6-9e shows at a glance that each joint and the whole truss is in equilibrium since there is no net force acting anywhere.

The solid and dashed arrows are useful to illustrate the composite graphic construction of the Maxwell diagram, but they are not necessary in routine analysis. Simple lines are used in practice as in Fig. 6-9f to obtain a view of the force polygons. The magnitude of each force is measured from the length of the corresponding line. Tension or compression in a member is determined by considering whether the member is pushing or pulling a joint.

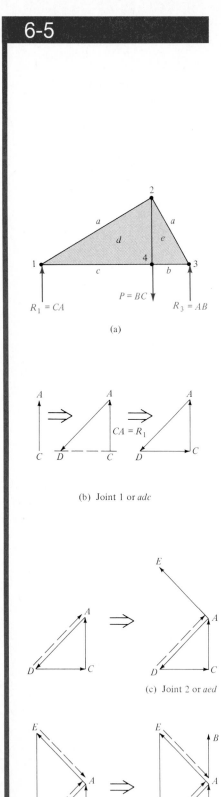

(a)

(b) Joint 1 or *adc*

**FIGURE 6-9**

Construction of a Maxwell diagram

(c) Joint 2 or *aed*

(d) Joint 3 or *abe*

(e) Joint 4 or *bcde*

(f) Maxwell diagram

EXAMPLE 6-3

Construct the Maxwell diagram for the truss in Fig. 6-10a. Use informal notation, solid and dashed arrows, and several diagrams to show the major steps in the construction. Each member is 3 m long.

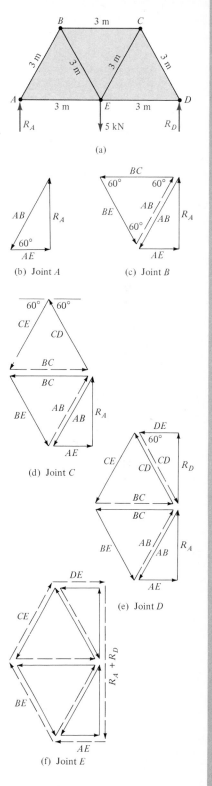

(a)

## SOLUTION

First the reactions are determined. Since there is symmetry, $R_A = R_D = 5$ kN/2 = 2.5 kN. Beginning with joint $A$, a force triangle is drawn.

**JOINT A (Fig. 6-10b):**

$$AB = \frac{R_A}{\sin 60°} = \frac{2.5 \text{ kN}}{0.866} = 2.89 \text{ kN}$$

$$AE = \frac{R_A}{\tan 60°} = \frac{2.5 \text{ kN}}{1.732} = 1.44 \text{ kN}$$

**JOINT B (Fig. 6-10c):**  Retaining the force triangle from joint $A$, the forces acting at joint $B$ are added. Forces $AB$, $BC$, and $BE$ form an equilateral force triangle, hence their magnitudes are all equal: $AB = BC = BE = 2.89$ kN.

(b) Joint $A$        (c) Joint $B$

**JOINT C (Fig. 6-10d):**  Continuing to add on to the previous diagram, solid arrows are shown for members being considered for the first time and dashed arrows are shown for members being considered the second time. Forces $CD$, $CE$, and $BC$ form an equilateral force triangle, hence their magnitudes are all equal: $CD = CE = BC = 2.89$ kN.

**JOINT D (Fig. 6-10e):**  Reaction $R_D$ is known already but may be used as a check on the other calculations.

$$R_D = CD \cos 30° = (2.89 \text{ kN})(0.866) = 2.5 \text{ kN} \quad \text{(check)}$$

$$DE = CD \sin 30° = (2.89 \text{ kN})(0.5) = 1.45 \text{ kN}$$

**JOINT E (Fig. 6-10f):**  Realizing that the applied force (5 kN) is equal and opposite to the sum of the reactions $R_A$ and $R_D$, it is seen that the Maxwell diagram is indeed closed.

(d) Joint $C$

(e) Joint $D$

## JUDGMENT OF THE RESULTS

There should be considerable confidence in the results obtained from force triangles. However, it should be noted that the sense (tension or compression) of the forces is not apparent from these diagrams. In order to obtain that extra information from the analysis, a combination of the methods discussed so far has to be used.

(f) Joint $E$

**FIGURE 6-10**

Construction of a Maxwell diagram using informal notation

EXAMPLE 6-3    223

# EXAMPLE 6-4

Construct the Maxwell diagram for the truss in Fig. 6-11a. Use Bow's notation, solid and dashed arrows, and several diagrams to show the major steps in the construction. The angles are $\theta = 53.1°$ and $\phi = 63.4°$.

(a)

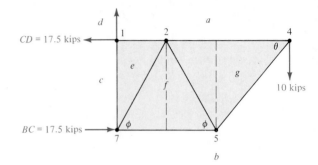

(b) Note: $a$, $b$, $c$, and $d$ denote regions between forces

(c) Joint 4

(d) Joint 5

(e) Joint 2

(f) Joint 7

(g) Joint 1

## SOLUTION

It is noted that the vertical members at joints 3 and 6 are zero-force members, so these joints can be ignored in the analysis. First the reactions are calculated by considering the truss as a whole (Fig. 6-11b). Choosing joint 4 as the starting point, a force triangle is drawn.

**JOINT 4 (Fig. 6-11c):**

$$BG = \frac{BA}{\sin\theta} = \frac{10 \text{ kips}}{\sin 53.1°} = 12.5 \text{ kips}$$

$$GA = BG \cos 53.1° = 7.5 \text{ kips}$$

**JOINT 5 (Fig. 6-11d):** Since member $BG$ is being considered for the second time, it is shown as a dashed arrow in the opposite direction originally shown.

$$FG = \frac{BA}{\sin\phi} = \frac{10 \text{ kips}}{\sin 63.4°} = 11.2 \text{ kips}$$

$$BF = GA + FG \cos 63.4° = 7.5 \text{ kips} + 11.2 \text{ kips} \cos 63.4°$$
$$= 12.51 \text{ kips}$$

**JOINT 2 (Fig. 6-11e):** From symmetry, $EF = FG = 11.2$ kips,

$$AE = BF + EF \cos\phi = 12.51 \text{ kips} + 11.2 \text{ kips} \cos 63.4°$$
$$= 17.5 \text{ kips}$$

**JOINT 7 (Fig. 6-11f):**

$$EC = BA = 10 \text{ kips}$$

$$CB = EA = 17.5 \text{ kips}$$

**JOINT 1 (Fig. 6-11g):** It is seen that the only way for the force diagram to be consistent with the lines of action of the reactions and members is for point $D$ to be the same as point $B$. It is also seen that the Maxwell diagram does indeed form a closed polygon, so the whole structure is in equilibrium.

**FIGURE 6-11**

Construction of a Maxwell diagram using Bow's notation

**6–24** Establish a formal notation for the cantilever truss starting with the upper support.

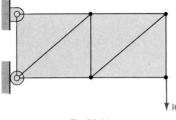

**Fig. P6-24**

**6–25** Establish a formal notation for the roof truss, starting with the left support.

**Fig. P6-25**

In Probs. 6–26 to 6–36, construct the Maxwell diagram for the truss or part of the truss indicated. Use solid and dashed arrows and several diagrams to show the major steps in the construction. Determine the magnitudes of the required forces from the diagrams.

**6–26** Prob. 6–2. $F_{AB} = ?$ $F_{BC} = ?$

**6–27** Prob. 6–3. $F_{AC} = ?$

**6–28** Prob. 6–4. $F_{AB} = ?$

**6–29** Prob. 6–5. $F_{AB} = ?$

**6–30** Prob. 6–6. $F_{AC} = ?$

**6–31** Prob. 6–7. $F_{AC} = ?$

**6–32** Prob. 6–8. $F_{AB} = ?$

**6–33** Prob. 6–9. $F_{AC} = ?$

**6–34** Prob. 6–10. Construct the Maxwell diagram for section *BCDE* of the truss. $F_{BC} = ?$

**6–35** Prob. 6-10. Give the force in every member.

**6–36** Prob. 6–12. Give the force in every member.

# METHOD OF SECTIONS

It is generally necessary to determine the forces in all members of a truss during its design. The method of joints and Maxwell's diagram are the most useful for such detailed analysis. There are other situations where only a few members must be analyzed regardless of the size and complexity of the whole structure (for example, checking the critical regions in a truss). In such cases the *method of sections* is the most efficient because it involves the members of interest at once without requiring many intermediate steps in the calculation.

The method of sections is based on the idea that any portion of a truss may be considered as a free body in equilibrium. The chosen portion may have any number of joints and members in it, but the number of unknown forces acting on it should not exceed three in most cases because only three equilibrium equations can be written for each portion of a plane truss.

To illustrate the possible choices of sections in a truss, assume that only the force in member *CD* in Fig. 6-12a must be known. A section at line *l* would involve the member *CD*, but both the upper and the lower portions of the truss have the unknown forces $\mathbf{F}_{BA}$, $\mathbf{F}_{BI}$, $\mathbf{F}_{CI}$, and $\mathbf{F}_{CH}$ also acting on them as shown in Fig. 6-12b. Thus, this particular imaginary sectioning is useless in the analysis because there are too many unknowns.

The useful sections for the truss in Fig. 6-12a are those that cut member *CD* and only two other members. Lines *m* or *n* are equally good choices, and either the left or the right portions of the truss can be used as shown in Fig. 6-12c and d for the section *m*. The unknown forces $\mathbf{F}_{CD}$, $\mathbf{F}_{CH}$, and $\mathbf{F}_{IH}$ have arbitrarily assumed directions at this stage, but these have to be consistent in the interrelated diagrams in Fig. 6-12c and d.

Judgment is again necessary to expedite the solution. Since only the force $\mathbf{F}_{CD}$ must be determined, it is best to take moments about point *H* in Fig. 6-12d. This gives an equation with only one unknown.

$$\sum M_H = 0$$

$$F_{CD}(b) + P(2a) = 0$$

$$F_{CD} = -\frac{2a}{b} P$$

(a)

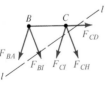

(b)

**FIGURE 6-12**

Method of sections for the analysis of selected members

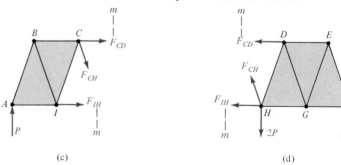

(c)  (d)

The negative sign indicates that $\mathbf{F}_{CD}$ is not a tensile force as assumed, but a compressive force (pushing joint $C$ to the left and joint $D$ to the right).

In general, equilibrium equations for horizontal and vertical forces can also be used, so three independent equations are available for a section of a plane truss. However, equilibrium equations involving only one unknown can usually be derived by using Eq. 4-5 with two moment summations and one force summation. To illustrate this convenient method, consider Fig. 6-12c with the three unknown forces $\mathbf{F}_{CD}$, $\mathbf{F}_{CH}$, and $\mathbf{F}_{IH}$. By summing moments about point $C$, the point of intersection of $\mathbf{F}_{CD}$ and $\mathbf{F}_{CH}$, the force $\mathbf{F}_{IH}$ can be directly determined. Similarly, summing moments about point $H$ involves only the force $\mathbf{F}_{CD}$. Since $\mathbf{F}_{CD}$ and $\mathbf{F}_{CH}$ do not intersect but are parallel, summing forces perpendicular to their lines of action results in an equation with the remaining unknown, $\mathbf{F}_{CH}$.

Note that in certain cases the sectioning is allowed to result in more than three unknowns if it is not necessary to solve for every unknown. For example, there may be any number of unknowns at point $H$ in Fig. 6-12d if only the force $\mathbf{F}_{CD}$ must be determined.

SPACE TRUSSES     **6-7**

The basic elements of the plane truss may be combined in three-dimensional structures called *space trusses*. The idealized, statically determinate space truss is made of rigid bars connected at their ends by ball-and-socket joints. These joints cannot apply moments to the bars, so each bar can be treated as a two-force member. The supports of the ideal space truss also apply only forces to the structure.

The basic rigid unit of space trusses has six bars connected to form the edges of a tetrahedron as shown in Fig. 6-13a. Each additional rigid unit requires three more members attached at different existing joints and connected to each other at a new joint as in Fig. 6-13b. Here $AE$, $BE$, and $DE$ are the new members connected at the old joints $A$, $B$, and $D$, and at the new joint $E$. It can be established from Fig. 6-13 that $m = 3n - 6$, where $m$ is the total number of members and $n$ is the total number of joints in a space truss.

A space truss can be analyzed with the method of joints or with the method of sections. For each joint, there are three scalar equilibrium equations, $\sum F_x = 0$, $\sum F_y = 0$, and $\sum F_z = 0$. The analysis must begin at a joint where there are at least one known force and no more than three unknown forces. The solution must progress to other joints in a similar fashion.

There are six scalar equations available when the method of sections is used,

$$\sum F_x = 0 \qquad \sum F_y = 0 \qquad \sum F_z = 0$$
$$\sum M_x = 0 \qquad \sum M_y = 0 \qquad \sum M_z = 0$$

Therefore, the number of unknowns at a section generally should not exceed six. This method for space trusses is less useful in practice than it appears at first. The reason is in the difficulty of finding moment axes involving only one unknown in each moment equation.

(a)

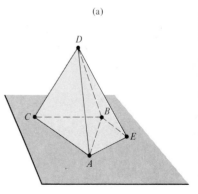

(b)

**FIGURE 6-13**

Six bars form the basic element, a tetrahedron, of a rigid space truss

EXAMPLE 6-5

Use the method of sections to determine the forces in members *CJ* and *DJ* of Fig. 6-14a.

## SOLUTION

First, the support reactions are found; because of the symmetry, $A_y = G_y = 1200$ lb. Next an appropriate section must be chosen such that forces $\mathbf{F}_{CJ}$ and $\mathbf{F}_{DJ}$

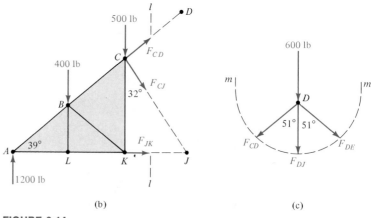

**FIGURE 6-14**

Analysis of a plane truss using the method of sections

are unknowns, and there are no more than three unknowns involved. It is seen that no one section fulfills all these criteria, hence two separate sections need to be analyzed.

Choosing the section to the left of line $l$–$l$, the force $\mathbf{F}_{CJ}$ can be solved for (Fig. 6-14b). Since the other unknown forces are both directed through joint $A$, this is a convenient point to sum moments about.

$$\sum M_A = 0$$

$$-(400 \text{ lb})(10 \text{ ft}) - (500 \text{ lb})(20 \text{ ft}) - F_{CJ}(\cos 32°)(20 \text{ ft})$$
$$- F_{CJ}(\sin 32°)(16 \text{ ft}) = 0$$

$$F_{CJ} = \frac{-14,000 \text{ lb·ft}}{(20 \cos 32° + 16 \sin 32°) \text{ ft}} = -550 \text{ lb} \quad \text{(compression)}$$

The calculation of force $\mathbf{F}_{DJ}$ is not as straightforward. The only section that cuts through only three members including $DJ$ is $m$–$m$. This, however, is essentially the same as considering joint $D$ alone (Fig. 6-14c). Since all forces at a joint are concurrent, the moment equation becomes trivial and there are only two equations of equilibrium available ($\sum F_x = 0$, $\sum F_y = 0$). Joint $D$ has three unknown forces acting on it, hence more information is needed before force $\mathbf{F}_{DJ}$ may be determined. It is seen that force $\mathbf{F}_{CD}$ can be obtained from the previous section (Fig. 6-14b). By summing moments about joint $J$ an equation involving $F_{CD}$ as the only unknown is obtained since forces $\mathbf{F}_{CJ}$ and $\mathbf{F}_{JK}$ are both directed through joint $J$.

$$\sum M_J = 0$$

$$-(1200 \text{ lb})(30 \text{ ft}) + (400 \text{ lb})(20 \text{ ft}) + (500 \text{ lb})(10 \text{ ft})$$
$$- F_{CD}(\cos 39°)(16 \text{ ft}) - F_{CD}(\sin 39°)(10 \text{ ft}) = 0$$

$$F_{CD} = \frac{-23,000 \text{ lb·ft}}{(16 \cos 39° + 10 \sin 39°) \text{ ft}}$$

$$= -1230 \text{ lb} \quad \text{(compression)}$$

Reconsidering Fig. 6-14c, by symmetry $F_{CD} = F_{DE} = -1230$ lb. Now force $\mathbf{F}_{DJ}$ is the only unknown at joint $D$. From Fig. 6-14c,

$$\sum F_y = 0$$

$$-F_{CD} \cos 51° - F_{DE} \cos 51° - 600 \text{ lb} - F_{DJ} = 0$$

$$F_{DJ} = 1230 \text{ lb} \cos 51° + 1230 \text{ lb} \cos 51° - 600 \text{ lb}$$
$$= 950 \text{ lb} \quad \text{(tension)}$$

It is seen that the method of sections allows for much faster calculation of forces if only a few forces are desired. While calculation of the force $\mathbf{F}_{DJ}$ might seem lengthy, it should be remembered that a solution by the method of joints would be even more complicated. In that case joints $A$, $L$, $B$, $K$, $C$, and $D$ would have to be analyzed.

EXAMPLE 6-5    229

**EXAMPLE 6-6**

Assume that the triangular structure on top of the tall crane in Fig. 6-15a consists of simple plane trusses as modeled in Fig. 6-15b. Determine the forces in members *BC*, *CJ*, and *IJ* using the method of sections.

## SOLUTION

First it is decided where to "cut" the truss for easiest analysis. Section *l–l* shown in Fig. 6-15b involves the three members of concern and no others. A decision must now be made whether to use the top section or the bottom section. By choosing the top section calculation of the support reactions is avoided, so a free-body diagram of this section is drawn (Fig. 6-15c).

From trigonometry the angle that member *CJ* makes with the horizontal is found to be 43.6°. By summing moments about joint *C* an equation involving only $F_{IJ}$ is obtained.

$$\sum M_C = 0$$

$$-(50 \text{ kN})(\cos 15°)(2.8 \text{ m}) + (50 \text{ kN})(\sin 15°)(6 \text{ m})$$
$$+ F_{IJ}(\sin 5°)(0 \text{ m}) - F_{IJ}(\cos 5°)(2.27 \text{ m}) = 0$$

$$F_{IJ} = \frac{-57.6 \text{ kN·m}}{(\cos 5°)(2.27 \text{ m})} = -25.5 \text{ kN} \quad \text{(compression)}$$

By summing moments about joint *J* an equation involving only $F_{BC}$ is obtained since $\mathbf{F}_{CJ}$ and $\mathbf{F}_{IJ}$ are both directed through this joint. Note that it is valid to sum moments about a point that is not part of the section being considered because moment equilibrium is valid about *any* point in space.

$$\sum M_J = 0$$

$$-(50 \text{ kN})(\cos 15°)(0.7 \text{ m}) + (50 \text{ kN})(\sin 15°)(8 \text{ m})$$
$$+ F_{BC}(\cos 25°)(2.1 \text{ m}) + F_{BC}(\sin 25°)(2 \text{ m}) = 0$$

$$F_{BC} = \frac{-69.7 \text{ kN·m}}{\cos 25° (2.1 \text{ m}) + \sin 25° (2 \text{ m})}$$

$$= -25.4 \text{ kN} \quad \text{(compression)}$$

By summing moments about joint *F* an equation involving only $F_{CJ}$ is obtained. It is seen that the external force acts through this joint as well as $\mathbf{F}_{IJ}$ and $\mathbf{F}_{BC}$; hence the moment equation is trivial and $F_{CJ} = 0$.

A check may be made by summing forces in the *x* and *y* directions for the whole section.

It should be noted that by choosing certain points to sum moments about, equations involving only one unknown are often available. Summing forces in certain directions can also reduce the number of unknowns. This has obvious advantages and should be used whenever possible.

(a)

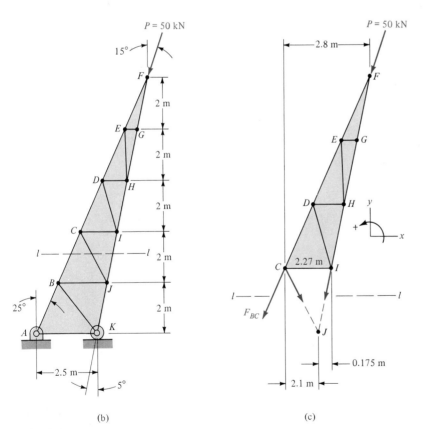

(b)                                    (c)

**FIGURE 6-15**

Analysis of a plane truss using the method of sections

EXAMPLE 6-6    231

## EXAMPLE 6-7

The boatyard crane shown in Fig. 6-16a has a basic space truss element as its stationary structure, $ABCD$ in Fig. 6-16b. The position of joint $A$ is fixed, while joints $B$ and $C$ (in the same horizontal plane as $A$) are only constrained to maintain their positions with respect to $A$ in the horizontal plane. Determine the forces in members $AD$ and $BD$ when $L = 6$ m and $P = 10$ kN. The other dimensions, coordinates, and angles are given as constants or quantities that are valid for the particular situation.

### SOLUTION

Since the supports at $B$ and $C$ resist motion in the $y$ direction only, $y$ component support reactions only are shown. From symmetry it is seen that $R_B = R_C = R$. These are determined by taking moments about the $x$ axis,

$$\sum M_x = 0$$

$$P(5.69 \text{ m}) - 2R(3 \text{ m} \cos 30°) = 0$$

$$R = \frac{(10 \text{ kN})(5.69 \text{ m})}{2(3 \text{ m})(\cos 30°)} = 10.95 \text{ kN} = R_B = R_C$$

The reactions at joint $A$ are determined by making summations of forces along $x$, $y$, and $z$.

$$\sum F_x = 0$$

$$A_x = 0$$

$$\sum F_y = 0$$

$$2R - A_y - P = 0$$

$$A_y = 2(10.95 \text{ kN}) - 10 \text{ kN} = 11.9 \text{ kN}$$

$$\sum F_z = 0$$

$$A_z = 0$$

The best joint to analyze next is joint $E$, which has only two unknowns. Here vector notation is initiated to facilitate the solution. The unknown forces are given as vectors of known directions using Eq. 2-23.

### JOINT $E$ (Fig. 6-16c):

$$\mathbf{F}_{ED} = F_{ED} \frac{(0\mathbf{i} - 1.9\mathbf{j} - 5.69\mathbf{k}) \text{ m}}{6 \text{ m}} = -0.32F_{ED}\mathbf{j} - 0.95F_{ED}\mathbf{k}$$

$$\mathbf{F}_{EA} = F_{EA} \frac{(0\mathbf{i} - 8.22\mathbf{j} - 5.69\mathbf{k}) \text{ m}}{10 \text{ m}} = -0.82F_{EA}\mathbf{j} - 0.57F_{EA}\mathbf{k}$$

Summing the scalar components of the forces acting at joint $E$,

| | |
|---|---|
| $\sum F_y = 0$ | $\sum F_z = 0$ |
| $-10 \text{ kN} - 0.32F_{ED} - 0.82F_{EA} = 0$ | $-0.95F_{ED} - 0.57F_{EA} = 0$ |
| $-10 \text{ kN} - 0.32(-0.6F_{EA})$ | $F_{ED} = -0.6F_{EA}$ |
| $\quad - 0.82F_{EA} = 0$ | |

$$F_{EA} = -15.9 \text{ kN} \quad \text{(compression)}$$

$$F_{ED} = 9.5 \text{ kN} \quad \text{(tension)}$$

(a)

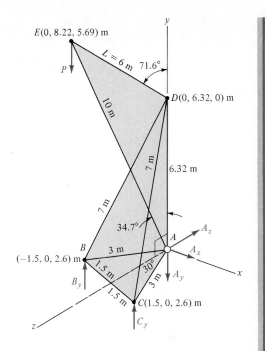

$E(0, 8.22, 5.69)$ m

$L = 6$ m   71.6°

$D(0, 6.32, 0)$ m

$P$

10 m

7 m

6.32 m

7 m

34.7°

$A_z$

$A$

$A_x$

$B$   3 m

$(-1.5, 0, 2.6)$ m

30°

$A_y$

$x$

$B_y$

1.5 m

3 m

1.5 m   $C(1.5, 0, 2.6)$ m

$z$   $C_y$

(b)

**JOINT D (Fig. 6-16d):**

$$\mathbf{F}_{DB} = F_{DB} \frac{(-1.5\mathbf{i} - 6.32\mathbf{j} + 2.6\mathbf{k})\ \text{m}}{7\ \text{m}}$$

$$= -0.21F_{DB}\mathbf{i} - 0.9F_{DB}\mathbf{j} + 0.37F_{DB}\mathbf{k}$$

$$\mathbf{F}_{DC} = F_{DC} \frac{(1.5\mathbf{i} - 6.32\mathbf{j} + 2.6\mathbf{k})\ \text{m}}{7\ \text{m}}$$

$$= 0.21F_{DC}\mathbf{i} - 0.9F_{DC}\mathbf{j} + 0.37F_{DC}\mathbf{k}$$

From symmetry, $F_{DB} = F_{DC}$. By Newton's law, $\mathbf{F}_{DE} = -\mathbf{F}_{ED} = 9.5(0.32\mathbf{j} + 0.95\mathbf{k})$ kN. Summing forces in the $z$ direction at joint $D$, only one unknown (force $F_{DB} = F_{DC}$) is involved,

$$\sum F_z = 0$$

$$9.5(0.95)\ \text{kN} + 2(0.37)F_{DB} = 0$$

$$F_{DB} = F_{DC} = -12.2\ \text{kN} \quad \text{(compression)}$$

Summing forces in the $y$ direction at joint $D$ gives the solution for the unknown force $\mathbf{F}_{DA}$,

$$\sum F_y = 0$$

$$0.32F_{DE} - 2(0.9)F_{DB} - F_{DA} = 0$$

$$0.32(9.5)\ \text{kN} - 2(0.9)(-12.2)\ \text{kN} - F_{DA} = 0$$

$$F_{DA} = 25\ \text{kN} \quad \text{(tension)}$$

## JUDGMENT OF THE RESULTS

All of the forces in the $y$ direction at joint $A$ are known, and a summation of these serves as a check. From Fig. 6-16e,

$$\sum F_y = F_{AD} - A_y + F_{AE}(0.82)$$
$$= 25\ \text{kN} - 11.9\ \text{kN} - 15.9(0.82)\ \text{kN} = 0.06\ \text{kN}$$

The small net force calculated is attributed to round-off errors. The results are accepted as correct.

$E$

$W = -10$ kN $\mathbf{j}$

$F_{ED}$

$F_{EA}$

(c)

$F_{DE}$

$D$

$F_{DB}$   $F_{DA}$

$F_{DC}$

(d)

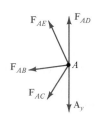

$F_{AE}$   $F_{AD}$

$F_{AB}$   $A$

$F_{AC}$

$A_y$

(e)

**FIGURE 6-16**

Analysis of a space truss

EXAMPLE 6-7   233

**6–37**   Determine the forces in members *AB*, *BD*, and *CD*.

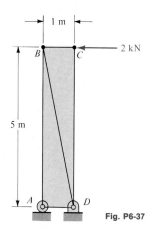

Fig. P6-37

**6–38**   In Prob. 6–37 replace member *BD* with a member *AC*. Determine the forces in members *AB*, *AC*, and *CD*.

**6–39**   Determine the forces in members *AB*, *AC*, and *CD*.

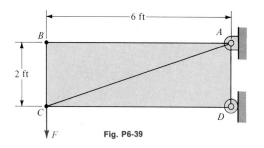

Fig. P6-39

**6–40**   Determine the forces in members *AB*, *BD*, and *CD*.

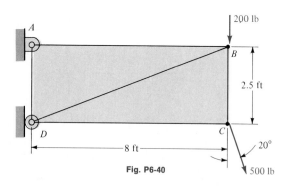

Fig. P6-40

**6–41**   Calculate the forces in members *AB*, *AC*, and *CD*. $F_1 = 2.5$ kN, $F_2 = 4$ kN.

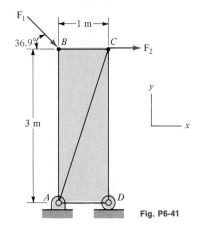

Fig. P6-41

**6–42**   Calculate the forces in members *AB*, *BF*, and *EF*. $P = 4$ kN, $l = 1.5$ m, and $h = 1$ m.

Fig. P6-42

**6–43**   Consider Fig. P6-42, assuming that $P = 600$ lb, $l = 7$ ft, and $h = 5$ ft. Calculate the forces in members *BC*, *BE*, and *EF*.

**6–44**   Determine the forces in members *AB*, *BE*, and *DE*.

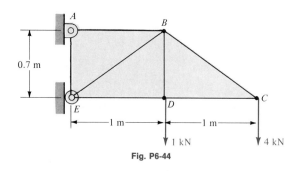

Fig. P6-44

**6–45**   Take a section through members *AB*, *AC*, *BD*, and *CD*. Show that there are not enough independent equations of equilibrium available to solve for the unknown forces in these members. Consider both the upper and lower sections.

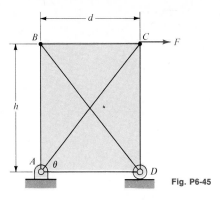

Fig. P6-45

**6-46** Calculate the forces in members BC, BE, and EF.

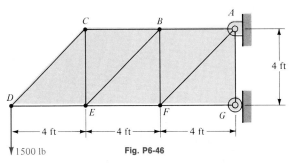

Fig. P6-46

**6-47** Consider Fig. P6-46 and determine the force in member AB using two different sections: (a) through AB, AF, and FG, and (b) through AB, BF, and EF.

**6-48** Calculate the forces in members BC, CE, and DE.

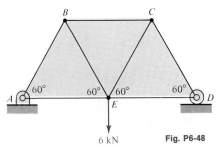

6 kN                    Fig. P6-48

**6-49** Determine the forces in members CD, DK, and JK. AL = LK = KJ = 5 ft and DJ = 10 ft. The truss is symmetric.

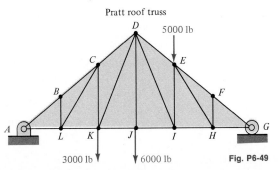

Fig. P6-49

**6-50** Solve Prob. 6–49 as stated, then replace member DK with CJ, and member DI with EJ. Calculate the forces in CD, CJ, and JK after the change.

**6-51** In the symmetric bridge truss AL = LK = KJ = 8 ft, and DJ = 10 ft. Determine the forces in members CD, CJ, and KJ. Check the force in member KJ by using a section through BC, CK, and KJ.

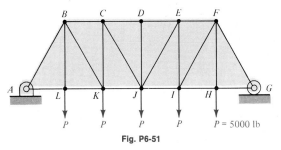

Fig. P6-51

**6-52** Calculate the forces in members DE, EJ, and IJ. AL = LK = KJ = 2 m, and DJ = 3 m. The truss is symmetric.

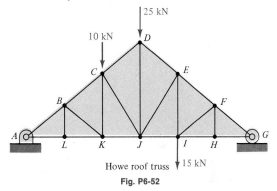

Howe roof truss     15 kN
Fig. P6-52

**6-53** Solve Prob. 6–52 as stated, then replace all internal slanted members with oppositely slanting bars (for example, BK is replaced by CL). Determine the forces in DE, DI, and IJ after the change.

**6-54** Calculate the forces in members BC, CH, and GH.

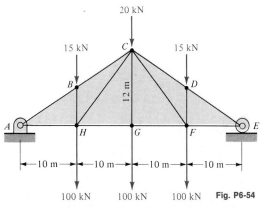

Fig. P6-54

**6-55**  Calculate the forces acting at joint $I$ if $F_1 = 20$ kN, $F_2 = 25$ kN, and $F_3 = 30$ kN.

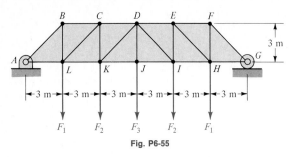

**Fig. P6-55**

**6-56**  Calculate the forces acting at joint $K$.

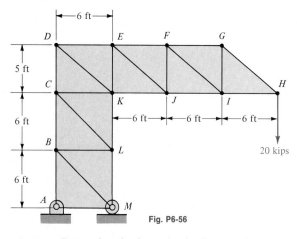

**Fig. P6-56**

**6-57**  Determine the forces in the four vertical members of the truss.

**Fig. P6-57**

**6-58**  Determine the forces acting at joint $R$.

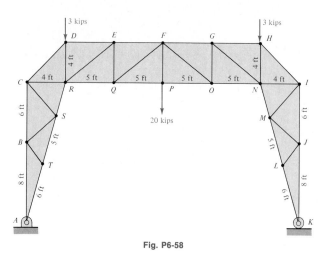

**Fig. P6-58**

**6-59**  Consider the container-loading derrick and assume that the supporting cables at joint $K$ become slack. Determine the forces acting at joints $B$ and $O$.

(a)

(b) Plane truss model of one side of boom

**Fig. P6-59**

*Multiforce* members (with three or more forces acting on each member) are common in structures. In these cases the forces are not directed along the members, so they are a little more complex to analyze than the two-force members in simple trusses. Multiforce members are used in two kinds of structure. *Frames* are usually stationary and fully constrained. *Machines* always having moving parts, so the forces acting on a member depend on the location and orientation of that member.

The analysis of multiforce members is based on the consistent use of interrelated free-body diagrams. The solution is often facilitated by representing forces by their rectangular components. Scalar equilibrium equations are the most convenient for two-dimensional problems, and vector notation is advantageous in three-dimensional situations.

To illustrate the analysis of frames and machines, consider the two-dimensional model of a robot arm supporting a weight $W$ in Fig. 6-17. Assume that $AE$ and $EG$ are rigid members held in position by two hydraulic actuators, $BC$ and $DF$, with smooth pin connections at $A$ and $B$. A rotary actuator controls the "hand" at $G$.

Suppose that it is necessary to determine all the forces and moments acting on members $AE$ and $EG$. First the fact that the actuators $BC$ and $DF$ are *two-force members* should be recognized. The free-body diagrams for $AE$ and $EG$ are then drawn in Fig. 6-17b and c; the *principle of action and reaction* must be carefully observed in this process. For example, in Fig. 6-17b, forces $A_x$, $A_y$, $E_x$, and $E_y$ are all unknown and may be drawn arbitrarily in the positive $x$ and $y$ directions and the hydraulic actuators are assumed to be in compression. After this drawing is made, $E_x$, $E_y$, and

**FIGURE 6-17**

Model of a robot arm

$\mathbf{F}_{DF}$ have fixed directions in Fig. 6-17c, as shown. Here the forces and moment at $G$ are also assumed arbitrarily. Note that the actual direction of each unknown remains to be determined from the sign of the result in the equilibrium equations.

The number of unknowns to be determined is nine, but the number of available equilibrium equations for the two free-body diagrams is six. One more free-body diagram with its accompanying three equilibrium equations is necessary for the complete solution. By using the free-body diagram for the whole rigid body, Fig. 6-17d, three more independent equations are acquired by which the external forces, $\mathbf{A}_x$, $\mathbf{A}_y$, and $\mathbf{F}_{BC}$, can be determined. The free-body diagram for member $GH$ can then be used as a check on the solution.

The next task is to write the equilibrium equations and determine the unknown forces. Any set of three independent force and moment summations, as given in Eq. 4-4, 4-5, or 4-6, can be used for each free-body diagram. The moment equations should be chosen to contain only one unknown, or as few as possible. Careful choice of these equations is necessary to reduce the need to solve simultaneous equations.

The equilibrium equations and the unknown to be determined from each are as follows:

$$
\begin{array}{llll}
\text{Fig. 6-17d:} & \sum M_A = 0 & \rightarrow & F_{BC} \\
& \sum F_x = 0 & \rightarrow & A_x \\
& \sum F_y = 0 & \rightarrow & A_y \\
\text{Fig. 6-17b:} & \sum M_B = 0 & \rightarrow & F_{DF} \\
& \sum F_x = 0 & \rightarrow & B_x \\
& \sum F_y = 0 & \rightarrow & B_y \\
\text{Fig. 6-17c:} & \sum M_G = 0 & \rightarrow & M_G \\
& \sum F_x = 0 & \rightarrow & G_x \\
& \sum F_y = 0 & \rightarrow & G_y \\
\end{array}
$$

These equations are available to calculate the unknown forces if the geometry of the robot arm and the loading are specified. A positive number for a result indicates that the direction of that force was assumed correctly in the beginning. A negative answer indicates that the force is acting in the direction opposite to the assumed direction. Generally, it is not necessary to redraw any of the forces correctly in the free-body diagrams since the signs are easy to interpret. If the answer is to be given without an accompanying diagram, the correct direction *and* the member on which the force acts must be specified.

Often, an applied force acts at a pin joining two or more members, or a support or connection may exist at a joint between two or more members. In these cases, a choice should be made of a single member at the joint on which to assume the external force to be acting. This decision should be stated in the analysis.

Special Cases of Frames

The number and arrangement of members and supports in a frame can be quite varied, so a few distinct possibilities are worth mentioning.

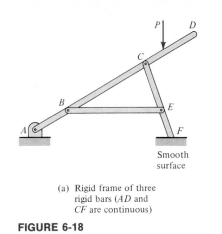

(a) Rigid frame of three rigid bars (*AD* and *CF* are continuous)

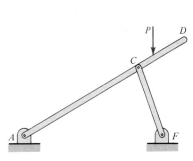

(b) Semirigid frame

(c) Statically indeterminate rigid frame

**FIGURE 6-18**

Special cases of frames

**1.** *Rigid frames* may be treated as rigid bodies whether or not they are attached to supports. Figure 6-18a shows such a frame that remains rigid in any orientation, under any loading (as long as the individual bars remain rigid), and with or without being attached to fixed supports.

**2.** *Semirigid frames* may be treated as rigid bodies only when they are attached to fixed supports. For example, the member *BE* in Fig. 6-18a may be removed only if ends *A* and *F* are attached to fixed supports as in Fig. 6-18b. An interesting aspect of this frame is that all the external reactions at *A* and *F* cannot be determined from the free-body diagram of the whole frame consisting of members *AD* and *CF*, since this frame is not rigid by itself. The individual free-body diagrams are needed to complete the analysis.

**3.** *Statically determinate and rigid frames* have an equal number of unknown force components and independent equilibrium equations. All the unknown forces can be determined, and all the equations can be satisfied (Fig. 6-18a). If there are more equations than unknowns, the frame is called *nonrigid*. This seems to be the case if equilibrium equations for the whole frame are added to the equations for the individual parts. However, these equations, although valid, would not be independent because the equations for the whole frame could be obtained by superimposing all of the individual free-body diagrams (this cancels all internal forces).

**4.** *Statically indeterminate and rigid frames* have more unknown forces than independent equilibrium equations. For example, member *DE* in Fig. 6-18c is redundant in theory, although it may be useful in practice. A statically indeterminate frame may also be nonrigid if there are more equations than unknown forces.

**5.** *Machines* are used to transmit and modify forces and motion. They consist of two or more main rigid parts that are relatively constrained. The simplest devices (hand tools such as pliers) and large and complex mechanisms can all be analyzed using the basic equilibrium conditions. Machines always have moving parts, so they are nonrigid on the whole even if they include some rigid structures.

EXAMPLE 6-8

The folding chair shown in Fig. 6-19a is represented by the two-dimensional model of the front or rear legs in Fig. 6-19b. The floor supporting the chair is assumed to be frictionless. Determine the reaction at the bottom of the leg and the forces at the joints $B$ and $C$ on member $DBCH$.

## SOLUTION

Consider the whole chair's free-body diagram (Fig. 6-19b) to determine the reaction $H_y$. Sum the moments about point $A$. This yields an equation with only one unknown.

$$\sum M_A = 0$$

$$-P(\cos 30°)(0.5\text{ m}) + P(\sin 30°)(0\text{ m}) + P(\cos 30°)(0.5\text{ m})$$

$$-P(\sin 30°)(0.714\text{ m}) + H_y(0.714\text{ m}) = 0$$

$$H_y = P \sin 30° = (600\text{ N})(0.5) = 300\text{ N}$$

It is now necessary to work with free-body diagrams of the individual members.

**MEMBER $DCBH$ (Fig. 6-19c):**

$$\sum F_x = 0$$

$$P \cos 30° + C_x + B_x = 0$$

$$520\text{ N} + C_x + B_x = 0 \qquad \textbf{(a)}$$

$$\sum F_y = 0$$

$$-P \sin 30° + C_y + B_y + H_y = 0$$

$$-300\text{ N} + C_y + B_y + 300\text{ N} = 0$$

$$C_y = -B_y \qquad \textbf{(b)}$$

It is best to take moments about point $B$ or $C$, where two unknown forces act. Choosing $C$,

$$\sum M_C = 0$$

$$-P(\cos 30°)(0.15\text{ m}) + P(\sin 30°)(0.214\text{ m}) + B_y(0.143\text{ m})$$
$$+ B_x(0.1\text{ m}) + H_y(0.5\text{ m}) = 0$$

$$-77.9\text{ N·m} + 64.2\text{ N·m} + B_y(0.143\text{ m})$$
$$+ B_x(0.1\text{ m}) + 150\text{ N·m} = 0 \qquad \textbf{(c)}$$

There are four unknowns ($B_x$, $B_y$, $C_x$, and $C_y$) and only three equations, so no unknowns can be determined. Equilibrium equations must be written for other members.

(a)

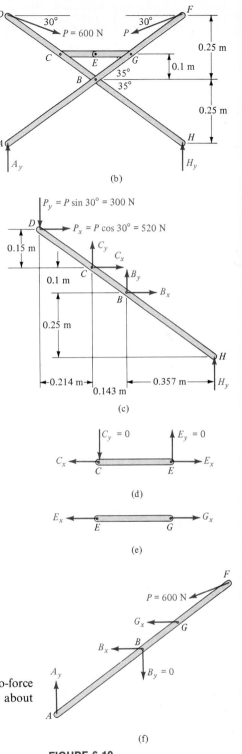

(b)

(c)

(d)

(e)

(f)

**FIGURE 6-19**

Analysis of a folding chair

**MEMBER *CE* (Fig. 6-19d):**

$$\sum M_C = \sum M_E = 0 \qquad \text{(member } CE \text{ is a two-force member)}$$

Hence,

$$C_y = E_y = 0 \quad \text{and} \quad C_x = E_x$$

By Eq. (b),

$$C_y = -B_y = 0$$

Equation (c) now has only one unknown, and this is readily calculated,

$$B_x = -\frac{136.3 \text{ N·m}}{0.1 \text{ m}} = -1360 \text{ N}$$

From Eq. (a), $C_x$ can now be determined,

$$C_x = -520 \text{ N} - B_x = -520 \text{ N} - (-1360 \text{ N}) = 840 \text{ N}$$

**SUMMARY:** On *DCBH*,

$$H_x = 0 \qquad\qquad H_y = 300 \text{ N}\uparrow$$

$$B_x = 1360 \text{ N}\leftarrow \qquad B_y = 0$$

$$C_x = 840 \text{ N}\rightarrow \qquad C_y = 0$$

## JUDGMENT OF THE RESULTS

The forces calculated can be checked using Fig. 6-19e, which is another two-force member, and Fig. 6-19f, where every force is known. Summing moments about point *F* is convenient.

$$\sum M_F = -(300 \text{ N})(0.714 \text{ m}) + (1360 \text{ N})(0.25 \text{ m}) - (840 \text{ N})(0.15 \text{ m})$$
$$= -0.2 \text{ N·m}$$

The net moment calculated is negligible, so the results are accepted as correct.

EXAMPLE 6-8    241

# EXAMPLE 6-9

A winch and mast-crutch frame of a boat trailer is shown in Fig. 6-20a. Its simplified model is shown in Fig. 6-20b, where the given forces are estimated for a braking situation. Calculate the forces at points $A$ and $B$ on member $ABC$ and the reaction at $D$.

## SOLUTION

First the reaction components at the supports $A$ and $D$ are sought by analysis of the frame as a whole (Fig. 6-20c).

$$\sum F_x = 0$$

$$- 50 \text{ lb} \cos 10° - 500 \text{ lb} \cos 20° + A_x + D_x = 0$$

$$A_x + D_x = 470 \text{ lb} + 49 \text{ lb} = 519 \text{ lb} \qquad \text{(a)}$$

$$\sum F_y = 0$$

$$- 50 \text{ lb} \sin 10° - 500 \text{ lb} \sin 20° + A_y + D_y = 0$$

$$A_y + D_y = 180 \text{ lb} \qquad \text{(b)}$$

$$\sum M_A = 0$$

$$D_y(0.91 \text{ ft}) - (500 \text{ lb} \sin 20°)(1 \text{ ft} \sin 20°)$$
$$+ (500 \text{ lb} \cos 20°)(1.66 \text{ ft} \cos 20°) + (50 \text{ lb} \cos 10°)(5 \text{ ft}) = 0$$

$$D_y = \frac{59 \text{ lb·ft} - 733 \text{ lb·ft} - 246 \text{ lb·ft}}{0.91 \text{ ft}} = -1011 \text{ lb}$$

From Eq. (b),

$$A_y = 180 \text{ lb} + 1011 \text{ lb} = 1191 \text{ lb}$$

After these results it is necessary to analyze the individual members. From Fig. 6-20d,

$$\sum F_x = 0$$

$$A_x + B_x - 50 \text{ lb} \cos 10° = 0 \qquad \text{(c)}$$

$$\sum F_y = 0$$

$$A_y + B_y - 50 \text{ lb} \sin 10° = 0$$

$$B_y = -1191 \text{ lb} + 9 \text{ lb} = -1182 \text{ lb}$$

$$\sum M_B = 0$$

$$A_x(2.5 \text{ ft}) + (50 \text{ lb} \cos 10°)(2.5 \text{ ft}) = 0$$

$$A_x = -49 \text{ lb}$$

From Eq. (c),

$$B_x = 50 \text{ lb} \cos 10° - A_x = 98 \text{ lb}$$

From Eq. (a),

$$D_x = 519 \text{ lb} - A_x = 568 \text{ lb}$$

(a)

(b)

(c)

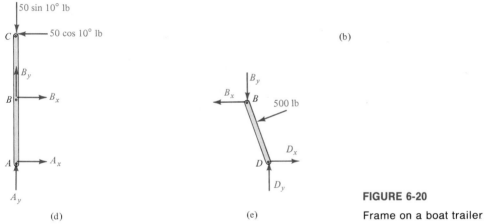

(d)                                    (e)

**FIGURE 6-20**

Frame on a boat trailer

**SUMMARY:** On member $ABC$,

$$A_x = 49 \text{ lb} \leftarrow \qquad A_y = 1191 \text{ lb} \uparrow$$

$$B_x = 98 \text{ lb} \rightarrow \qquad B_y = 1182 \text{ lb} \downarrow$$

Reaction at $D$, $\qquad D_x = 568 \text{ lb} \rightarrow \qquad D_y = 1011 \text{ lb} \downarrow$

## JUDGMENT OF THE RESULTS

The solution may be checked using Fig. 6-20e, where each force is known. Taking moments about point $B$,

$$\sum M_B = D_x(2.5 \text{ ft}) + D_y(0.91 \text{ ft}) - 500 \text{ lb}(1 \text{ ft}) = -0.01 \text{ lb·ft}$$

The net moment calculated is negligible, so the results are accepted as correct.

EXAMPLE 6-9    243

# EXAMPLE 6-10

The hood support mechanism of a car is shown in Fig. 6-21a, with the details of its extended position given in part b of the figure. **(a)** First determine the force $S$ in the spring required to hold the position of the linkage. **(b)** Also calculate the forces at pin $A$.

**FIGURE 6-21**

Hood support mechanism

(a)

(b)

(c)

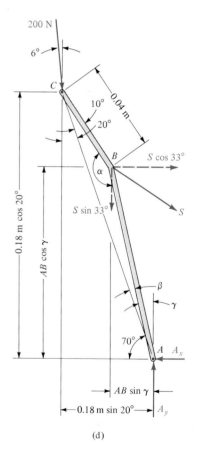

(d)

244   ANALYSIS OF SIMPLE STRUCTURES AND MACHINES

# SOLUTION

**(a)** The solution is best started by analyzing the member that has the known force acting on it.

**MEMBER CD (Fig. 6-21c):** This is a two-force member, hence the reaction $CD$ at $C$ is 200 N (by inspection) acting at an angle of 6° with the vertical as shown.

**MEMBER ABC (Fig. 6-21d; Using a Simplified, Equivalent Shape for the Bar):** First the geometry of this member must be determined.

$$|AB|^2 = (0.04 \text{ m})^2 + (0.18 \text{ m})^2 - 2(0.04 \text{ m})(0.18 \text{ m}) \cos 10°$$

$$|AB| = 0.141 \text{ m}$$

$$\frac{\sin \alpha}{0.18 \text{ m}} = \frac{\sin 10°}{0.141 \text{ m}} = \frac{\sin \beta}{0.04 \text{ m}}, \quad \alpha = 167.2°, \quad \beta = 2.8°$$

$$\gamma = 90° - 70° - 2.8° = 17.2°$$

By summing moments about point $A$, an equation is obtained involving the spring force, $S$, as the only unknown.

$$\sum M_A = 0$$

$$S (\sin 33°)(AB \sin \gamma) - S (\cos 33°)(AB \cos \gamma)$$
$$+ (200 \text{ N} \cos 6°)(0.18 \text{ m} \cos 20°)$$
$$- (200 \text{ N} \sin 6°)(0.18 \text{ m} \sin 20°) = 0$$

$$S (\sin 33°)(0.141 \text{ m} \sin 17.2°)$$
$$- S(\cos 33°)(0.141 \text{ m} \cos 17.2°) + 32.4 \text{ N·m} = 0$$

$$S = \frac{32.4 \text{ N·m}}{0.09 \text{ m}} = 360 \text{ N}$$

**(b)**
$$\sum F_x = 0$$

$$200 \text{ N} \sin 6° + S \cos 33° - A_x = 0$$

$$A_x = 20.9 + 360 \text{ N} \cos 33° = 323 \text{ N}$$

$$\sum F_y = 0$$

$$-200 \text{ N} \cos 6° - S \sin 33° + A_y = 0$$

$$A_y = 199 \text{ N} + 360 \text{ N} \sin 33° = 395 \text{ N}$$

## JUDGMENT OF THE RESULTS

The magnitudes and directions of the reaction forces appear reasonable. A check on the results may appear to be possible by making sure the whole system is in equilibrium, but this would not be a bona fide check. Since member $CD$ is a two-force member, it merely transmits the 200 N force along line $CD$. Hence, by the principle of transmissibility, analyzing the whole free body is the same as analyzing member $ABC$ alone.

EXAMPLE 6-10    245

## EXAMPLE 6-11

Determine the forces at points $B$ and $C$ on member $ABC$ and at points $D$ and $E$ on member $CDE$ of the fireplace tongs shown in Fig. 6-22a and b.

### SOLUTION

First a free-body diagram of member $ABC$ is drawn and analyzed.

**MEMBER ABC (Fig. 6-22c):**

$$\sum F_x = 0 \qquad \sum F_y = 0$$

$$B_x + C_x = 0 \qquad 30\ \text{lb} + B_y + C_y = 0$$

$$\sum M_B = 0$$

$$-(30\ \text{lb})(5\ \text{in.}) + C_y(7\ \text{in.}) - C_x(7\ \text{in. } \tan 15°) = 0$$

There are three equations and four unknowns ($B_x$, $B_y$, $C_x$, and $C_y$); hence no forces may be determined yet. Member $CDE$ is drawn next (Fig. 6-22d). It is noticed that the total number of unknowns increases by four, so analyzing this member does not help at this time. From symmetry, it is found that free-body diagrams of members $A'BC'$ and $C'DE'$ yield the same equations as for $ABC$ and $CDE$, thus still no forces can be determined. A free-body diagram of the whole system is then drawn (Fig. 6-22e). It is seen from symmetry considerations that $E_x = E'_x$, but from equilibrium of forces in the $x$ direction $E_x = -E'_x$. Hence, $E_x = 0 = E'_x$. Free-body diagrams of combinations of members should then be considered. Looking at member $ABCDE$ (Fig. 6-22f) it is seen there are three additional unknowns ($D_x$, $D_y$, and $E_y$), to make a total of seven unknowns and six equations of equilibrium when used in conjunction with Fig. 6-22c. The combination of $ABC$ and $A'BC'$ is then analyzed (Fig. 6-22g), and this turns out to be fruitful.

$$\sum F_x = 0$$

$$C_x + C'_x = 0$$

$$\sum M_{C'} = 0$$

$$(30\ \text{lb})(12\ \text{in.}) - (30\ \text{lb})(12\ \text{in.}) - C_x(2)(7\ \text{in. } \tan 15°) = 0$$

$$C_x = 0, \quad C'_x = 0$$

Now member $ABC$ may be solved (Fig. 6-22c),

$$\sum F_x = 0 \qquad \sum F_y = 0$$

$$B_x = 0 \qquad 30\ \text{lb} + B_y + C_y = 0$$

$$\sum M_B = 0$$

$$C_y(7\ \text{in.}) - (30\ \text{lb})(5\ \text{in.}) = 0$$

$$C_y = \frac{150\ \text{lb·in.}}{7\ \text{in.}} = 21.4\ \text{lb}$$

$$B_y = -C_y - 30\ \text{lb} = -21.4\ \text{lb} - 30\ \text{lb} = -51.4\ \text{lb}$$

**MEMBER *CDE* (Fig. 6-22d):**

$$\sum F_x = 0 \qquad \sum F_y = 0$$

$$D_x = 0 \qquad -C_y + D_y - E_y = 0$$

$$\sum M_D = 0$$

$$C_y(7 \text{ in.}) - E_y(7 \text{ in.}) = 0$$

$$E_y = C_y = 21.4 \text{ lb}$$

$$D_y = E_y + C_y = 21.4 \text{ lb} + 21.4 \text{ lb} = 42.8 \text{ lb}$$

**SUMMARY:** On *ABC*, $\quad B_x = 0 \qquad B_y = 51.4 \text{ lb}\downarrow$

$$C_x = 0 \qquad C_y = 21.4 \text{ lb}\uparrow$$

On *CDE*, $\quad D_x = 0 \qquad D_y = 42.8 \text{ lb}\uparrow$

$$E_x = 0 \qquad E_y = 21.4 \text{ lb}\downarrow$$

(a)

**FINAL COMMENTS:** Persistence pays off! It was known that this problem was solvable since there were 12 unknowns (two unknowns at each of *B*, *C*, *C'*, *D*, *E*, and *E'*) and there were 12 equations available (three equations for each member). The magnitudes of the forces appear reasonable from an intuitive appeal.

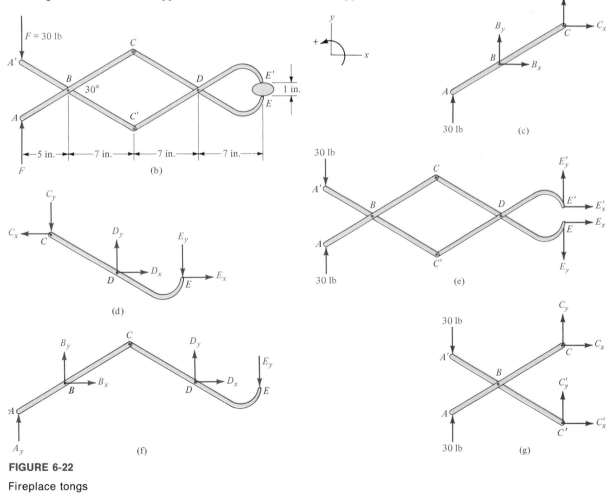

**FIGURE 6-22**

Fireplace tongs

EXAMPLE 6-11   247

**6–60**  The rigid boom *EC* of the crane is supported by three other bars as shown. Determine the reaction at support *A* and the forces acting at *C*, *D*, and *E* of the boom if $F = 5$ kN, $ED = DC = 1.3$ m, and $AE = 0.8$ m.

**Fig. P6-60**

**Fig. P6-62**

**6–61**  In Fig. P6-60, $F = 1500$ lb, $ED = 5$ ft, $DC = 7$ ft, and $BD = 4.5$ ft. The weight of the structure is represented by a force of magnitude $W = 250$ lb acting at *D*. Calculate the forces in members *BC* and *BD*.

**6–62**  A force $F = 6$ kN **i** is acting at joint *C* of the structure. Determine all forces acting on member *AC* if $AB = BE = 2$ m and $AC = 4$ m.

**6–63**  Consider Fig. P6-62 with a force $\mathbf{F}_C = 800$ lb **i** at *C*, and a force $\mathbf{F}_B = (-300\mathbf{i} - 350\mathbf{j})$ lb at *B*. Calculate the forces acting on member *DF* if $AF = 4$ ft, $AB = 6$ ft, and $AC = 15$ ft.

**6–64**  The rigid boom *AC* is pinned at *B* and *C* and supports a simple pulley at *A*. Determine the forces at *B* and *C* on the boom if $F = 3$ kN, $AB = BC = 1.5$ m, $CD = 1$ m, and $r = 0.5$ m.

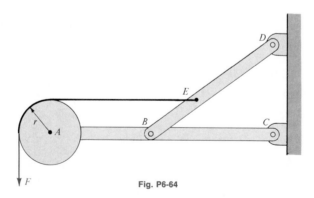

**Fig. P6-64**

**6-65** In Fig. P6-64, $F = 1200$ lb, $AB = 6$ ft, $BC = 5$ ft, $CD = 4$ ft, and $r = 2$ ft. The weight of the frame is represented by a force of magnitude $P = 200$ lb acting at $B$. Calculate the total forces acting at $B$ and $C$ on the boom.

**6-66** A symmetric bridge section is modeled as a frame consisting of rigid bars $AE$ and $BD$. Determine the forces acting on bars $BG$ and $CG$ if $P = 20$ kips.

**6-68** Consider the model of an ironing board. The two legs are pinned to the board at $A$ and $B$ and stand on a smooth, frictionless floor. $EC = DC = 32$ in., $CA = CB = 16$ in., and $ED = 37$ in. The weight of the board is $W = 15$ lb, assumed concentrated directly above joint $C$. Determine the forces acting at $A$, $C$, and $D$ on member $AD$.

Fig. P6-66

Fig. P6-68

**6-67** Determine all forces acting on member $AD$ of the pin-connected frame for the loading shown. $AB = GF = 2.5$ m, $BC = CD = 1.25$ m, $BF = AG = 1.5$ m, $P = 8$ kN, $F = 4$ kN, and $M = 2$ kN·m.

**6-69** Solve Prob. 6–68 assuming that an additional downward force of magnitude $F = 10$ lb is acting on the board directly above point $E$.

**6-70** An adjustable lamp arm has rigid bars $AJ$, $BC$, $CI$, $JD$, $DH$, $IE$, $EG$, and $HF$. A typical element $DH = 6$ in. is pinned at its center. Determine the reaction on the lamp at support $F$, the force in member $BC$, and the forces at joint $D$ on member $JD$ when $\theta = 90°$.

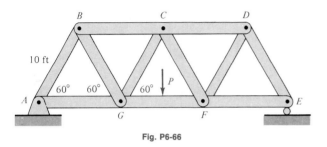

Fig. P6-67

**6-71** Solve Prob. 6–70 after adding three more downward forces, each of 0.5 lb magnitude, acting at $H$, $I$, and $J$.

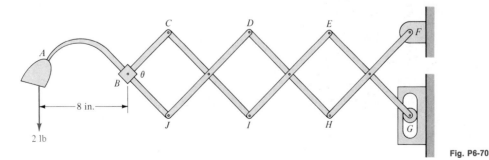

Fig. P6-70

**6–72** The "cherry-picker" mechanism of a utility truck consists of two rigid arms, *AB* and *BC*, each of 12 ft length and negligible weight. The hydraulic actuators *DG* and *EF* are pin connected at points *D* and *G* and *E* and *F*. *AD* = 6 ft, *BE* = 5 ft, *BF* = 3 ft, and *W* = 400 lb. Determine the magnitude of the force applied by each hydraulic actuator when $\theta = 20°$ and $\phi = 70°$.

**6–73** The hinged frame on the utility pole in Fig. P6-73 appears simple to analyze, but it requires a lot of work. Write down all the equations that are necessary to determine the force at pin *A*. Discuss the cause of any major difficulty if the solution were attempted.

**Fig. P6-72**

**Fig. P6-73**

(a)

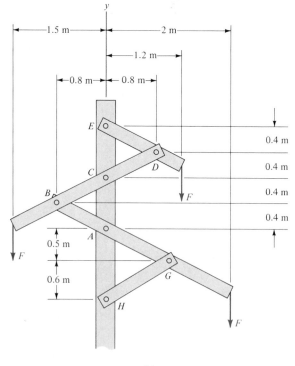

(b)

**6-74** Points $A$, $B$, $C$, and $D$ denote pin joints in the vise grip. Determine the magnitude of the gripping force at $F$ if the magnitude $P$ of the applied forces is 200 N.

(a)

(b)

**Fig. P6-74**

**6-75** Discuss the change in the force at $F$ in Prob. 6–74 that would be caused by shortening the distance $AD$ (this is done using the screw $E$).

**6-76** Determine the magnitude of the force applied by the bolt cutter to the bolt at $A$.

**6-77** It is assumed that the force on the bolt at $A$ in Fig. P6-76 depend on where the bolt is placed in the jaws. Plot the magnitude of the force acting on the bolt as a function of $d = 30$ mm, 40 mm, and 50 mm.

**6-78** The symmetric lifting tongs hold the 500-lb machine part as shown. Determine the forces acting at joints $B$ and $C$ on member $BCD$.

**Fig. P6-78**

**Fig. P6-76**

**6-79** Determine the forces acting on member *DF* of the front loader. Members *BC* and *EG* are hydraulic actuators which can be treated as two-force members of adjustable length. Their lengths are constant in Fig. P6-79b for maintaining the given configuration.

**6-80** Calculate the forces acting on member *AI* in Prob. 6-79.

**Fig. P6-79**

(a)

(b)

**6–81** Determine the forces acting on arm *BF* of the excavator. Joint *D* connects three rigid members as shown in Fig. P6-81c.

**6–82** Determine the forces acting on member *KG* in Prob. 6–81.

(a)

(b)

Fig. P6-81

(c)

# CABLES WITH CONCENTRATED LOADS

Cables are commonly used in certain structures and machines such as suspension bridges, cranes, transmission lines, and antenna towers. In the simplest situations cables can be analyzed as two-force members. It is only slightly more complicated to analyze a cable with concentrated loads acting on it. For example, the suspension cable in Fig. 6-23a can be modeled as in part b of the figure. Here the weight of the upper cable can be neglected in comparison with the large loads applied to it by the vertical cables. The more complex problem of including distributed loads on a cable (its own weight, ice, or wind) is discussed in Ch. 7.

Another assumption made in the analysis is that the cable is flexible and consequently in pure axial tension between neighboring points where the loads are applied. This way the cable is treated as a set of two-force members. The only unknown forces are at the external supports and the tensions in the individual segments of the cable. The solution for these unknowns is straightforward if the loads and the geometry are known, but it must be remembered that the cable is not a rigid body, so changes in the loading would change its shape.

The method of solution is illustrated using Fig. 6-23b, where the loading and the geometry are symmetric with respect to $D$ ($P_B = P_F$, $y_B = y_F$, etc.). First, the free-body diagram of the whole cable is drawn in Fig. 6-23c. The directions of the unknown reactions at $A$ and $G$ can be correctly drawn at once since there is only tension in the cable. From symmetry, $A_y = G_y = W/2$ where $W = P_B + P_C + P_D + P_E + P_F$. $\mathbf{A}_x$ and $\mathbf{G}_x$ are equal in magnitude by inspection, but they cannot be calculated from Fig. 6-23c. It is useful to draw another diagram such as that in Fig. 6-23d. If the coordinates of $A$, $B$, and $Q$ are known, $A_x$ can be calculated by taking moments about point $Q$, writing $\sum M_Q = 0$.

With the external reactions $\mathbf{A}_x$ and $\mathbf{A}_y$ known, the tension in each segment of the cable can be calculated using the force summations $\sum F_x = 0$ or $\sum F_y = 0$ for the appropriate portion of the cable. For example, $T_{CD}$ is determined from Fig. 6-23e using $\sum F_x = 0$,

$$T_{CD} = \frac{A_x}{\cos \theta_{CD}}$$

Since $A_x$ is a constant, the horizontal component of the tension is also constant for the entire cable, so

$$T_{AB} \cos \theta_{AB} = T_{BC} \cos \theta_{BC} = \ldots$$

Another useful observation from these results is that the tension in the cable is largest where $\cos \theta$ is smallest. This has to be at a support such as $A$ and $G$ in Fig. 6-23b.

(a)

**FIGURE 6-23**

Suspension cable with concentrated loads

(b)

(d)

(c)

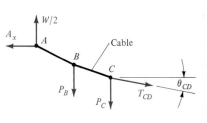

(e)

EXAMPLE 6-12

A cable supports three loads as shown in Fig. 6–24a. What is the required maximum tension in the cable if point $C$ should be 2 m below point $A$? What are the coordinates $y_B$ and $y_D$ in this configuration?

## SOLUTION

First a free-body diagram of the whole cable is drawn (Fig. 6-24b) and the $y$ components of the support reactions are determined.

$$\sum F_y = 0 \quad A_x = E_x \tag{a}$$

$$\sum F_y = 0$$

$$A_y - 2 \text{ kN} - 3 \text{ kN} - 1 \text{ kN} + E_y = 0$$

$$\sum M_E = 0$$

$$-A_y(43 \text{ m}) + (2 \text{ kN})(33 \text{ m}) + (3 \text{ kN})(21 \text{ m}) + (1 \text{ kN})(10 \text{ m}) = 0$$

$$A_y = \frac{(66 + 63 + 10) \text{ kN·m}}{43 \text{ m}} = 3.23 \text{ kN}$$

$$E_y = 6 \text{ kN} - A_y = 6 \text{ kN} - 3.23 \text{ kN} = 2.77 \text{ kN}$$

The $x$ components of the support reactions are obtained by considering a section of the cable and summing moments about a point that has known coordinates (point $C$ in this case). From Fig. 6-24c,

$$\sum M_C = 0$$

$$A_x(2 \text{ m}) - (3.23 \text{ kN})(22 \text{ m}) + (2 \text{ kN})(12 \text{ m}) = 0$$

$$A_x = \frac{(71.1 - 24) \text{ kN·m}}{2 \text{ m}} = 23.5 \text{ kN}$$

From Eq. (a), $E_x = A_x = 23.5$ kN. The largest tension in a cable occurs where the angle between the horizontal and the cable is largest (where $\cos \theta$ is smallest since $T_{\text{segment}} = A_x/\cos \theta$). For this case segments $AB$ and $DE$ must be considered.

**SEGMENT AB:** Since $AB$ is considered a two-force member, the force $\mathbf{A} = -A_x\mathbf{i} + A_y\mathbf{j}$ must be collinear with cable segment $AB$. Therefore, $\theta_{AB}$ equals the angle that $\mathbf{A}$ makes with the negative horizontal axis (Fig. 6-24d).

$$\tan \theta_{AB} = \frac{A_y}{A_x} = \frac{3.23 \text{ kN}}{23.5 \text{ kN}}, \quad \theta_{AB} = 7.8°$$

$$T_{AB} = A = \frac{A_x}{\cos \theta_{AB}} = 23.7 \text{ kN}$$

Also,

$$\frac{y_B}{10 \text{ m}} = \tan \theta_{AB} = 0.137, \quad y_B = (10 \text{ m})(0.137) = 1.37 \text{ m}$$

(a)

(b)

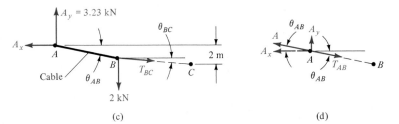

(c)                                           (d)

**FIGURE 6-24**

Suspension cable with three loads supported at the same elevation

**SEGMENT _DE_:**   By analogy to segment $AB$,

$$\tan \theta_{DE} = \frac{E_y}{E_x} = \frac{2.77 \text{ kN}}{23.5 \text{ kN}}, \quad \theta_{DE} = 6.7°$$

Since $\theta_{DE}$ is smaller than $\theta_{AB}$, the tension in segment $AB$ is the largest. Also,

$$\frac{y_D}{10 \text{ m}} = \tan \theta_{DE} = 0.118, \quad y_D = (10 \text{ m})(0.118) = 1.18 \text{ m}$$

## JUDGMENT OF THE RESULTS

The largest tension occurred in the expected segment. Since the largest loads were applied closer to the left end it was expected that this portion of the cable should make the largest angle with the horizontal and have the largest tension. The large tensions caused by relatively small loads also seem reasonable from an intuitive standpoint (the smaller the sag, the larger the tensions).

EXAMPLE 6-12   257

EXAMPLE 6-13

A cable supports three loads as shown in Fig. 6-25a. Determine the tension in each segment of the cable and the coordinates $y_B$ and $y_D$ of points $B$ and $D$.

### SOLUTION

Considering the whole cable as a free body, equations involving the support reaction components are obtained. From Fig. 6-25b,

$$\sum F_x = 0$$
$$A_x = E_x \tag{a}$$
$$\sum F_y = 0$$
$$A_y + E_y - 800 \text{ lb} - 500 \text{ lb} - 400 \text{ lb} = 0$$
$$A_y + E_y = 1.7 \text{ kips} \tag{b}$$
$$\sum M_E = 0$$
$$-A_x(20 \text{ ft}) - A_y(78 \text{ ft}) + (800 \text{ lb})(53 \text{ ft})$$
$$+ (500 \text{ lb})(33 \text{ ft}) + (400 \text{ lb})(15 \text{ ft}) = 0$$
$$20A_x + 78A_y = 64.9 \text{ kips} \tag{c}$$

Three equations involving four unknowns are obtained since none of the $x$ or $y$ components of the support reactions are collinear. Therefore, a section of the cable involving at most two additional unknowns must be analyzed. Cutting the cable through section $BC$ only two more unknowns are involved, the tension $\mathbf{T}_{BC}$ of unknown direction and magnitude (Fig. 6-25c).

$$\sum M_C = 0$$
$$-A_y(45 \text{ ft}) + A_x(2 \text{ ft}) + (800 \text{ lb})(20 \text{ ft}) = 0$$
$$-2A_x + 45A_y = 16 \text{ kips} \tag{d}$$

Solving Eqs. (c) and (d) simultaneously,

$$A_x = 1590 \text{ lb}$$
$$A_y = 426 \text{ lb}$$

From Eqs. (a) and (b),

$$E_x = 1590 \text{ lb}$$
$$E_y = 1270 \text{ lb}$$

Since each segment is considered a two-force member, the reaction force $\mathbf{A}$ and the tension $\mathbf{T}_{AB}$ must be equal in magnitude and opposite in direction (Fig. 6-25d).

$$T_{AB} = (A_x^2 + A_y^2)^{1/2} = 1650 \text{ lb}$$

$$\theta_{AB} = \tan^{-1} \frac{A_y}{A_x} = \tan^{-1} \left( \frac{426}{1590} \right) = 15°$$

$$\tan \theta_{AB} = \frac{y'_B}{25 \text{ ft}} = 0.27, \quad y'_B = (25 \text{ ft})(0.27) = 6.7 \text{ ft (Fig. 6-25c)}$$

$$y'_B = 20 \text{ ft} + 6.7 \text{ ft} = 26.7 \text{ ft}$$

(a)

(b)

For the tension in segment $BC$, Fig. 6-25c:

$$\tan \theta_{BC} = \frac{y'_B - 2 \text{ ft}}{20 \text{ ft}} = \frac{6.7 \text{ ft} - 2 \text{ ft}}{20 \text{ ft}} = 0.24, \quad \theta_{BC} = 13.2°$$

$$\sum F_x = 0$$

$$T_{BC} \cos \theta_{BC} - A_x = 0$$

$$T_{BC} = \frac{A_x}{\cos \theta_{BC}} = \frac{1590 \text{ lb}}{\cos 13.2°} = 1630 \text{ lb}$$

(c)

For the tension in segment $DE$, Fig. 6-25e:

$$T_{DE} = (E_x^2 + E_y^2)^{1/2} = 2030 \text{ lb}$$

$$\theta_{DE} = \tan^{-1} \frac{E_y}{E_x} = \tan^{-1}\left(\frac{1270}{1590}\right) = 38.6°$$

$$\frac{y_D}{15 \text{ ft}} = \tan \theta_{DE} = 0.8, \quad y_D = (15 \text{ ft})(0.8) = 12 \text{ ft}$$

(d)

For the tension in segment $CD$, Fig. 6-25f:

$$\tan \theta_{CD} = \frac{22 \text{ ft} - 12 \text{ ft}}{18 \text{ ft}} = 0.56, \quad \theta_{CD} = 29.1°$$

$$\sum F_x = 0$$

$$-T_{CD} \cos \theta_{CD} + E_x = 0$$

$$T_{CD} = \frac{E_x}{\cos \theta_{CD}} = \frac{1590 \text{ lb}}{\cos 29.1°} = 1820 \text{ lb}$$

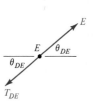

(e)

## JUDGMENT OF THE RESULTS

The largest tension ($T_{DE}$) occurs in a segment connected to a support as it should. It would be expected that $T_{DE}$ be greater than $T_{AB}$ since the angle that $DE$ makes with the horizontal is much larger than the angle of $AB$. The magnitudes of the other tensions appear reasonable for the given loading situation.

(f)

**FIGURE 6-25**

Suspension cable with three loads supported at different elevations

EXAMPLE 6-13   259

**6-83** Determine the tension in each segment of the cable if $F_B = F_C = 1$ kN, $d = 12$ m, and $y_B = 1.5$ m.

**6-87** In Fig. P6-85, $F_B = F_C = F_D = F$. Plot the maximum tension in the cable as a function of $y_C = 0.1d$, $0.2d$, and $0.3d$.

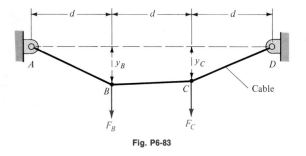

**Fig. P6-83**

**6-88** The forces acting on the cable are $F_B = 500$ N, $F_C = 600$ N, and $F_D = 900$ N. Calculate the maximum tension in the cable if $a = b = c = d = 5$ m, $y_D = 10$ m, and $y_E = 8$ m.

**6-84** In Fig. P6-83, $F_C = 2F_B = 400$ lb, $d = 15$ ft, and $y_C = 3$ ft. Calculate $y_B$ and the tension in each segment of the cable.

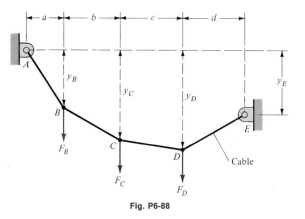

**Fig. P6-88**

**6-85** Calculate the tension in each segment of the cable if $F_C = 800$ N, $F_B = F_D = 600$ N, $d = 15$ m, and $y_C = 2$ m.

**6-89** In Fig. P6-88, $F_B = F_C = F_D = 200$ lb. Determine the tension in each segment of the cable if $a = 10$ ft, $b = 14$ ft, $c = 16$ ft, $d = 15$ ft, $y_C = 17$ ft, and $y_E = 13$ ft.

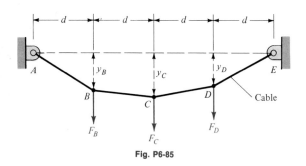

**Fig. P6-85**

**6-90** In Fig. P6-88, $F_B = 800$ N, $F_C = 700$ N, and $F_D = 500$ N; $a = 3$ m, $b = 5$ m, $c = 6$ m, $d = 5.5$ m, $y_B = 4$ m, and $y_E = 7$ m. Calculate $y_C$ and $y_D$.

**6-91** In Fig. P6-88, $F_C = 3F_B = 4F_D = 360$ lb; $a = 12$ ft, $b = 15$ ft, $c = 20$ ft, $d = 17$ ft, $y_C = 23$ ft, and $y_E = 18$ ft. Determine the tensions in $BC$ and $CD$.

**6-86** In Fig. P6-85, $F_B = 300$ lb, $F_C = 100$ lb, $F_D = 50$ lb, $d = 10$ ft, and $y_B = 4$ ft. Calculate $y_C$ and $y_D$.

## Section 6-1

**TRUSSES:** Assume that each member of a truss is a two-force member connected at frictionless pins at its ends. The members are assumed to be weightless, with forces applied at some of the joints. The basic element of a simple, rigid truss is the triangle formed by three bars.

## Section 6-2

**METHOD OF JOINTS:** Two equilibrium equations may be written for each joint of a two-dimensional truss:

$$\sum F_x = 0 \qquad \sum F_y = 0$$

**PROCEDURE FOR ANALYSIS:**

1. Determine the support reactions of the truss.
2. Start the detailed analysis at a joint where only one or two unknown forces act. Draw the free-body diagram of this joint, assuming that the unknown forces are tensions (arrows directed away from the joint).
3. Draw free-body diagrams for the other joints to be analyzed, using Newton's Third Law (action and reaction) consistently with respect to the first diagram.
4. Write the equilibrium equations for the forces acting at the joints and solve them. Attempt to progress from joint to joint in such a way that each equation contains only one unknown (to simplify the calculations). Positive answers indicate that the assumed directions of the unknown forces were correct, and vice versa.

## Section 6-3

**TWO MEMBERS AT A JOINT:**

(a) Possible      (b) Not possible unless $F = 0$

**THREE MEMBERS AT A JOINT:**

(c) Possible if $F = 0$

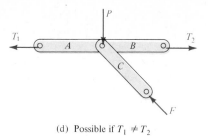

(d) Possible if $T_1 \neq T_2$

(e) Possible      (f)

**FOUR MEMBERS:** Collinear in pairs with equal and opposite forces in pairs:

Section 6-4

**FORMAL PROCEDURE USING BOW'S NOTATION:** For examples of lettering regions of a truss, its members, and external or internal forces, see Figs. 6-7 and 6-8.

Section 6-5

**PROCEDURE FOR GRAPHICAL CONSTRUCTION OF TRUSSES:**

1. Determine the support reactions for the truss.
2. Start the graphical construction at any joint where the unknown forces can be determined from a force polygon.
3. Add force triangles using the known lines of action at neighboring joints.
4. Each part of the diagram should have closed polygons showing equilibrium.
5. The magnitude of each force is measured from the length of the corresponding line, based on the scale established in the beginning for the known forces.

Section 6-6

**METHOD OF SECTIONS:** This method is most valuable when only a few forces must be determined in a complex truss. An imaginary cut is made through the members of interest. Three independent equations of equilibrium may be written for each portion of a plane truss (if the part has two or more joints in it), for example,

$$\sum F_x = 0 \qquad \sum M_A = 0 \qquad \sum M_B = 0$$

where line $AB$ is not perpendicular to the $x$ axis.

**PROCEDURE FOR ANALYSIS:**

1. Determine the support reactions if the section used in the analysis includes the joints supported.

2. Make the cut through the members of interest, preferably to cut only three members in which the forces are unknown (assume tensions).

3. Write the equations of equilibrium. Choose a convenient point of reference for moments to simplify the calculations such as the point of intersection of the lines of action for two of the unknown forces. If two unknown forces are parallel, sum forces perpendicular to their lines of action.

4. Solve the equations. If necessary, use more than one cut in the vicinity of interest to allow writing more equilibrium equations. Positive answers indicate that the assumed directions of the unknown forces were correct, and vice versa.

## Section 6-7

**SPACE TRUSSES:** For analyzing a three-dimensional truss, six scalar equations of equilibrium are available when using the method of sections:

$$\sum F_x = 0 \qquad \sum F_y = 0 \qquad \sum F_z = 0$$
$$\sum M_x = 0 \qquad \sum M_y = 0 \qquad \sum M_z = 0$$

Only the three force summations are available when using the method of joints.

## Section 6-8

**FRAMES AND MACHINES:** In general, a member may have two or more forces acting on it. Three independent equations of equilibrium are available for each member or combination of members in two-dimensional loading; for example,

$$\sum F_x = 0 \qquad \sum F_y = 0 \qquad \sum M_A = 0$$

where $A$ is an arbitrary point of reference.

**PROCEDURE FOR ANALYSIS:**

1. Determine the support reactions if necessary.

2. Determine all two-force members.

3. Draw the free-body diagram of the first member on which the unknown forces act assuming that the unknown forces are tensions.

4. Draw the free-body diagrams of the other members or groups of members using Newton's Third Law (action and reaction) consistently with respect to the first diagram. Proceed until the number of equilibrium equations available is no longer exceeded by the total number of unknowns.

5. Write the equilibrium equations for the members or combinations of members and solve them. Positive answers indicate that the assumed directions of the forces were correct, and vice versa.

## Section 6-9

**CABLES WITH CONCENTRATED LOADS:** These are to be analyzed as two-force members for which the method of joints is applicable. Note that cables are not rigid, so any change in the loading changes the configuration of a cable. The horizontal component of the tension in each segment of a given cable is a constant.

**6–92** Calculate the forces acting at joints $G$ and $H$ in the truss.

**6–95** Calculate the forces acting at joints $B$ and $I$.

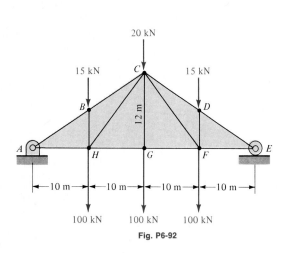

20 kN

15 kN   C   15 kN

B

12 m

D

A   H   G   F   E

|← 10 m →|← 10 m →|← 10 m →|← 10 m →|

100 kN   100 kN   100 kN

**Fig. P6-92**

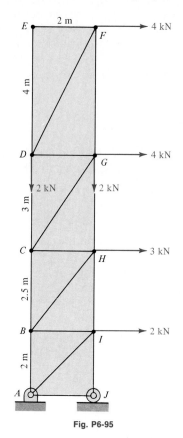

Fig. P6-95

**6–93** Construct the Maxwell diagram for the truss in Prob. 6–16. Determine the magnitudes of the forces $F_{AE}$ and $F_{BC}$.

**6–94** Determine the forces acting at joints $Q$, $R$, and $S$ in the truss.

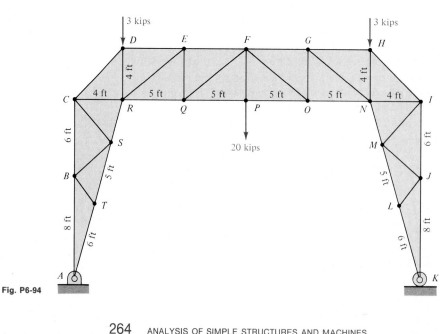

Fig. P6-94

**6-96** Determine the zero-force members in (a) Fig. P6-94 and (b) Fig. P6-95 for the loading shown.

**6-97** Two frictionless pulleys of 3-ft radius are attached to joints $E$ and $F$ of the plane truss. The cable supports a 20,000-lb load and is attached to joint $I$. Determine the forces acting at joints $A$, $B$, $C$, $H$, and $I$. The angles are $\alpha = 56.3°$, $\beta = 38.7°$, $\phi = 39.8°$, and $\theta = 36.9°$.

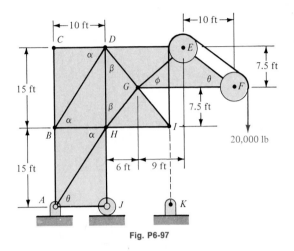

Fig. P6-97

**6-98** In Prob. 6-97 remove members $HI$ and $GI$ and assume that the cable is attached to point $K$. Determine all forces at $A$, $B$, $C$, $H$, and $J$.

**6-99** A ladder is modeled as a two-dimensional structure standing on a smooth, frictionless floor. Calculate the forces acting on the rigid leg $DG$ if a vertical force of magnitude $W = 200$ lb is applied to rung 5. Leg $AD$ is a rigid member.

Fig. P6-99

**6-100** Solve Prob. 6-99 with $\mathbf{W}$ acting on rung 3, and an additional downward force of magnitude $P = 50$ lb acting at joint $D$.

**6-101** Determine the distances $h_C$ and $h_D$ of the cable if $h_B = 15$ m. What is the maximum tension in the cable?

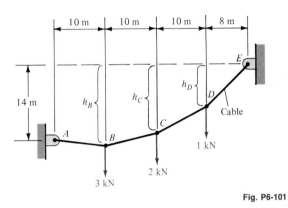

Fig. P6-101

**6-102** The same downward force $\mathbf{F}$ is applied at points $B$, $C$, and $D$; $a = b = c = d$, and $y_E = 2a$. Plot the maximum and minimum tensions in the cable as functions of $y_C = 2.5a$, $2.75a$, and $3a$. (The maximum tension is of interest to assure safety; the minimum tension is relevant to considerations of efficient use of the cable.)

Fig. P6-102

# 7 Distributed Forces

Basic Concepts of Distributed Forces

So far in this text forces have always been assumed to be concentrated and acting at distinct points on bodies. This idealization is very useful even in advanced analysis, but it is not immediately possible or adequate in certain practical situations. For example, the force of gravity acts as a distributed force on a rigid beam. The distribution of the force over the beam depends on the distribution of matter in the body. The distributed force (a parallel force system here) can be represented by an equivalent concentrated force which is useful in the determination of support reactions on the beam, but both the magnitude and location of the equivalent force must first be determined. In more advanced analyses, such as determining the load-carrying ability of the beam, the local magnitude of the distributed internal force is needed rather than the equivalent force for the given distribution. Several important special cases of distributed forces are presented in this chapter.

**SECTION 7-1** presents the concept and method of determination of the center of gravity. This topic is essential for engineers who will analyze dynamics of rigid bodies, stability of structures, or mechanics of materials.

**SECTIONS 7-2, 7-3, AND 7-4** describe purely geometrical approaches to determine centers of gravity. These methods are useful because most rigid bodies are sufficiently homogeneous so that the center of gravity coincides with the centroid, the geometric center of the body. The topics of Secs. 7-2 and 7-3 are used in several courses and in engineering practice, but the theorems of Sec. 7-4 are mainly for academic exercise.

**SECTION 7-5** presents the method for determining the external support reactions for structures on which distributed loads are acting. The method employs the concept of centroid in order to reduce the distributed loading to one of concentrated loads. This topic is important in that even the weight of a structure is a distributed load, and the analysis lays the foundation for determining the internal forces and moments.

**SECTION 7-6** presents the analysis of the internal reactions in beams under distributed loading. It is an extension of the methods used for beams under concentrated loading, but considerably more difficult for many students. This topic is frequently encountered in mechanics of materials and structural analysis, so it should be studied with emphasis. Students are advised to review the sign conventions in Sec. 5-3.

**SECTION 7-7** introduces the relatively complex analysis of cables under distributed loads. This topic is important in the design of certain bridges and power lines, but it is omitted from most statics courses.

**SECTIONS 7-8 AND 7-9** present the important basic concepts of fluid statics. These can be readily learned after Sec. 7-5. However, sometimes these topics are omitted from statics because they are also covered in fluid mechanics courses.

**SECTION 7-10** is a brief introduction to stress analysis to show another area involving distributed forces. This very important topic is often omitted from statics courses because it is covered in much more detail in mechanics of materials courses.

## CENTER OF GRAVITY                                                  7-1

The most common distributed force is the gravitational force of the earth acting on a body. It is essentially a parallel force system distributed all over the body. The resultant of this distributed force is called the *weight* of the body and it acts at a mathematically defined point called the *center of gravity*. The center of gravity is a point where the mass of the whole body may be assumed to be concentrated as far as gravitational attraction is concerned. The concept of *center of mass* is based on a similar theoretical concentration of mass, but independent of a particular gravitational field. The centers of mass and gravity are at the same point for most practical purposes. They are not precisely the same for extremely large bodies because the magnitude and direction of the gravitational attraction (toward the center of the earth) varies between parts of the body.

The concept of the center of gravity is illustrated with the aid of Fig. 7-1. Assume that two very small bodies $A$ and $B$ of different masses

**FIGURE 7-1**

Center of gravity $C$ of masses $A$ and $B$

are situated along the $x$ axis. These *point masses* may be separate or connected by a rigid rod of negligible mass. The center of gravity of these masses is the point where the equivalent resultant force caused by gravity is acting. In other words, the two weights balance each other with respect to their center of gravity. Clearly, the magnitude of the resultant force is $R = \sum W_i = 4W$. The location of its line of action is determined by setting up the moments for the given system of bodies in two different ways. Since the net moment about any chosen point must be the same regardless of how it is calculated in detail, the moments written in different ways can be equated (Varignon's Theorem, Sec. 3-4). For example, the summation of moments for bodies $A$ and $B$ with respect to point $O$ in Fig. 7-1a is

$$\sum M_O = \sum W_i x_i = Wa + 3W(a + b) = W(4a + 3b)$$

This is equivalent to working with the resultant force $\mathbf{R}$ in Fig. 7-1b where $\bar{x}$ is the unknown location of the line of action of $\mathbf{R}$. The moment of $\mathbf{R}$ is $\sum M_O = R\bar{x}$. Since $\sum W_i x_i$ must equal $R\bar{x}$,

$$W(4a + 3b) = 4W\bar{x}$$

and

$$\bar{x} = \frac{4a + 3b}{4} = a + \frac{3}{4}b$$

It makes sense that the center of gravity $C$ is closer to the larger of the masses. To check the details, take moments about point $A$ as in Fig. 7-1c. Here $3Wb$ is equivalent to $4W\bar{x}'$, where $\bar{x}' = \bar{x} - a$,

$$\bar{x}' = \frac{3}{4}b \quad \text{and} \quad \bar{x} = a + \bar{x}' = a + \frac{3}{4}b$$

The same procedure can be applied to more complex systems of point masses, and using other coordinate axes if necessary.

### Center of Gravity of a Three-Dimensional Body

The concepts used for point masses can be extended to deal with bodies of substantial dimensions. For an example, consider the homogeneous plate of constant thickness $t$ in Fig. 7-2a. Here every infinitesimal part $dA t$ of the total volume $At$ has the same force $d\mathbf{W}$ acting on it, but the moments caused by them with respect to the $y$ axis are not necessarily the same. The $x$ coordinate of the center of gravity is established as follows. First the resultant force is determined,

$$R = \int dW = W \tag{7-1}$$

the total weight of the plate. Next, the moments with respect to point $O$ are written in two ways using Fig. 7-2a and b, respectively,

$$\sum M_O = \int x\, dW \quad \text{and} \quad \sum M_O = \bar{x}W \tag{7-2}$$

The last two summations are equivalent, so

$$\boxed{\bar{x} = \frac{\int x\, dW}{W}} \tag{7-3}$$

**FIGURE 7-2**

Center of gravity $C$ of a uniform plate

To determine the $y$ coordinate of the center of gravity, imagine that the plate and the coordinate axes are rotated 90° to obtain Fig. 7-2c. The two ways of writing the moments with respect to point $O$ are now

$$\sum M_O = \int y \, dW \quad \text{and} \quad \sum M_O = \bar{y} W \qquad \boxed{7\text{-}4}$$

from which

$$\bar{y} = \frac{\int y \, dW}{W} \qquad \boxed{7\text{-}5}$$

In practice this can be written directly from Fig. 7-2a without drawing Fig. 7-2c.

Similarly, the method is applicable to the third dimension, and

$$\bar{z} = \frac{\int z \, dW}{W} \qquad \boxed{7\text{-}6}$$

Although Eqs. 7-3, 7-5, and 7-6 have been developed for a thin plate of constant thickness which lies in the $xy$ plane, these equations are valid for any three-dimensional homogeneous body.

The magnitude of the elemental force $d\mathbf{W}$ can be expressed as

$$dW = \gamma \, dV \qquad \boxed{7\text{-}7}$$

where $\gamma$ = specific weight of the material (weight per unit volume) and $dV$ = elemental volume. With this expression, Eqs. 7-3, 7-5, and 7-6 can be written as

$$\bar{x} = \frac{\int x\gamma \, dV}{\int \gamma \, dV} \qquad \bar{y} = \frac{\int y\gamma \, dV}{\int \gamma \, dV} \qquad \bar{z} = \frac{\int z\gamma \, dV}{\int \gamma \, dV} \qquad \boxed{7\text{-}8}$$

Having presented these derivations, the general meaning of the center of gravity can be stated: any body suspended or supported at the center of gravity would be perfectly balanced regardless of the orientation of the body. This is the basis for rapid and accurate experimental locations of the center of gravity. In general, the balance of a body is checked in three orthogonal orientations.

## CENTROIDS OF LINES, AREAS, AND VOLUMES 7-2

Equations 7-8 are generally valid even when the specific weight $\gamma$ varies throughout the body. Of course, these equations are most easily solved when $\gamma$ can be expressed as simple functions of the $x$, $y$, and $z$ coordinates. Fortunately, in most bodies $\gamma$ is a constant and can be canceled from Eq. 7-8. In such cases the center of gravity coincides with the *centroid*, which is a purely geometrical property of the body. The centroid is determined in a manner similar to that used for the center of gravity in homogeneous bodies.

## Centroids of Lines

Slender bodies of uniform cross-sectional areas $A$ can be treated simply as line segments. For illustration, consider the two-dimensional slender body of length $L$ in Fig. 7-3. It is good practice to guess and show the approximate location of the center of gravity $C$, which is also the centroid of the line when $\gamma$ and $A$ are constant over the whole length of the body. The coordinates of the centroid are obtained from Eqs. 7-3 and 7-5 using $dW = \gamma \, dV = \gamma A \, dL$ and $W = \gamma A L$,

$$\bar{x} = \frac{\int x \, dL}{L} \qquad \bar{y} = \frac{\int y \, dL}{L}$$

$\boxed{7\text{-}9}$

The integral $\int x \, dL$ is called the *first moment of the line $L$ with respect to the $y$ axis*. Similarly, $\int y \, dL$ is the first moment with respect to the $x$ axis. The first moment of the line can be positive or negative, depending on the position of the line with respect to the axis. The method is extended to deal with a three-dimensional line in Fig. 7-4. Equations 7-9 are also valid for this line, and a similar equation can be added for the $z$ coordinate,

$$\boxed{\bar{x} = \frac{\int x \, dL}{L} \qquad \bar{y} = \frac{\int y \, dL}{L} \qquad \bar{z} = \frac{\int z \, dL}{L}}$$

$\boxed{7\text{-}10}$

In this case the integral $\int x \, dL$ is the first moment of the line $L$ with respect to the $yz$ plane. Similar definitions can be made for the other integrals in Eq. 7-10. Note that the centroid $C$ seldom lies on the line $L$ unless the line is entirely straight.

**FIGURE 7-3**

Centroid $C$ of line $L$ in the $xy$ plane

**FIGURE 7-4**

Centroid $C$ of line $L$ in three dimensions

## Centroids of Areas

Bodies of small thickness $t$ can be treated as surface areas. The method of determining the centroid is illustrated with the three-dimensional sheet in Fig. 7-5. Assuming that $\gamma$ and $t$ are constant over the entire area $A$, and using $dW = \gamma \, dV = \gamma t \, dA$ and $W = \gamma t A$,

$$\bar{x} = \frac{\int x \, dA}{A} \qquad \bar{y} = \frac{\int y \, dA}{A} \qquad \bar{z} = \frac{\int z \, dA}{A}$$

7-11

Note again that the centroid seldom lies in the body except when the entire surface is a simple plane area. The integral $\int x \, dA$ is called the *first moment of the area A with respect to the yz plane*. Similarly, first moments can be defined with respect to any plane in three dimensions and with respect to any axis in two dimensions. The first moment of an area can be either positive or negative, depending on the position of the area with respect to the plane or the axis.

## Centroids of Volumes

The concepts described above can be readily extended to determine the coordinates of the centroid of a general body of volume $V$. Assuming that $\gamma$ is constant over the whole volume, and using $dW = \gamma \, dV$,

$$\bar{x} = \frac{\int x \, dV}{V} \qquad \bar{y} = \frac{\int y \, dV}{V} \qquad \bar{z} = \frac{\int z \, dV}{V}$$

7-12

The integral $\int x \, dV$ is called the *first moment of the volume V with respect to the yz plane*. Similarly, first moments can be defined with respect to the other planes.

Lines, areas, and volumes with a plane of symmetry have their centroids in that plane and have a zero first moment with respect to that plane. Lines, areas, and volumes with two planes of symmetry have their centroids on the line of intersection of those planes. Lines, areas, and volumes with three planes of symmetry that intersect at a single point have the centroid at that point. Similar statements can be made for two-dimensional lines and areas with one or two axes of symmetry. Since these

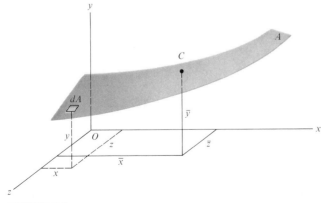

**FIGURE 7-5**
Centroid $C$ of a three-dimensional area

properties can be determined by inspection, the centroids are easily found for squares cubes, circles, spheres, and other symmetric bodies.

The centroidal coordinates of several common lines, areas, and volumes are given in Table 3, inside the back cover. In practice these tabulated coordinates are used in order to avoid calculating integrals for common shapes.

## COMPOSITE OBJECTS

Frequently, the center of gravity of a complex body is much easier and more accurate to locate experimentally (see the last paragraph in Sec. 7-1) than by calculations. Nevertheless, such calculations provide useful practice with the analytic methods and also may facilitate the experimentation by locating the center of gravity approximately before the actual tests. A method for such calculations is outlined here for the centroid of a composite object.

The centroids of complex objects are located in a manner similar to dealing with point masses. The object is broken down into simple parts, aiming to locate the centroid of each part by inspection. The method is illustrated for areas in which case the ideal simple parts are circles and rectangles. The composite centroid is located by equating the first moment of the whole object with the sum of the first moments of the individual parts, using the same reference system. For the centroidal coordinate $\bar{x}$ of an area $A$ consisting of discrete parts $A_i$,

$$\underbrace{\bar{x}A}_{\substack{\text{first moment} \\ \text{of whole area}}} = \underbrace{\sum \bar{x}_i A_i}_{\substack{\text{sum of first moments} \\ \text{of individual parts}}}$$ 

$\boxed{\text{7-13a}}$

where $\bar{x}_i$ are the centroidal coordinates of parts $A_i$ and $A = \sum A_i$. Similarly for the other coordinates,

$$\boxed{\bar{y}A = \sum \bar{y}_i A_i \quad \bar{z}A = \sum \bar{z}_i A_i}$$

$\boxed{\text{7-13b}}$

The method, which is valid for objects of any shape, is illustrated for the area $A$ in Fig. 7-6a.

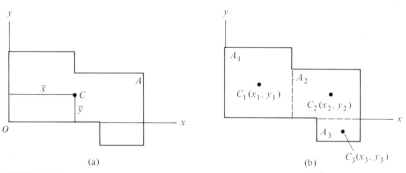

(a)  (b)

**FIGURE 7-6**

Centroids: $C_1$ of coordinates $x_1$ and $y_1$ is centroid of area $A_1$, and so on; $C$ of coordinates $\bar{x}$ and $\bar{y}$ is centroid of whole area $A = A_1 + A_2 + A_3$

First choose a convenient coordinate system along two sides of the area. Guess and show the approximate location of the centroid $C$ as in Fig. 7-6a. Next the area is divided into rectangles; one possibility is shown in Fig. 7-6b. The centroid of each rectangle ($C_1$, etc.) is at the intersection of its diagonals (this makes sense but can also be proved using integration). The first moment of area $A$ about point $O$ is written in two different ways for each axis of reference. Working with the $x$ coordinates, the resulting equation is

$$\underbrace{A\bar{x}}_{\substack{\text{first moment} \\ \text{of whole area}}} = \underbrace{A_1x_1 + A_2x_2 + A_3x_3}_{\substack{\text{sum of first moments} \\ \text{of individual parts}}} \qquad \boxed{\text{7-14a}}$$

$$\bar{x} = \frac{A_1x_1 + A_2x_2 + A_3x_3}{A}$$

Similarly,

$$\bar{y} = \frac{A_1y_1 + A_2y_2 + A_3y_3}{A} \qquad \boxed{\text{7-14b}}$$

where $A = \sum A_i = A_1 + A_2 + A_3$.

Whenever the complex body cannot be broken down into simple geometrical shapes, it is often reasonable to make approximations using rectangles and other simple shapes to facilitate the calculations. The error resulting from the approximation can be reduced by increasing the number of simple parts used. When the number of parts is large, it is advantageous to make a systematic table to list all areas (or lines, or volumes) and their centroidal distances with the proper magnitudes and signs. This is illustrated for Fig. 7-6 as follows.

| AREA PART $A_i$ | CENTROIDAL COORDINATE $\bar{x}_i$ | PRODUCT $\bar{x}_iA_i$ | CENTROIDAL COORDINATE $\bar{y}_i$ | PRODUCT $\bar{y}_iA_i$ |
|---|---|---|---|---|
| $+A_1$ | $x_1$ | $x_1A_1$ | $y_1$ | $y_1A_1$ |
| $+A_2$ | $x_2$ | $x_2A_2$ | $y_2$ | $y_2A_2$ |
| $+A_3$ | $x_3$ | $x_3A_3$ | $y_3$ | $y_3A_3$ |
| $\sum A_i = A$ | | $\sum \bar{x}_iA_i = \bar{x}A$ | | $\sum \bar{y}_iA_i = \bar{y}A$ |

The areas considered are positive if they are real parts of the body analyzed; they are negative for missing parts, such as a circular hole in a plate. Note that the product $y_3A_3$ for Fig. 7-6 is negative since the $y_3$ coordinate is negative.

This method is easily extended to locate the center of gravity of a complex body made of a homogeneous material, for then the centroid is concurrent with the center of gravity. For a three-dimensional body, the centroid can be located by replacing the areas in the analysis with the corresponding volumes. If the body is composed of parts made of different materials, the center of gravity can be located by replacing the areas in the analysis with the corresponding weights.

# EXAMPLE 7-1

Locate the centroid of the circular arc shown in Fig. 7-7 using Table 3, inside the back cover.

## SOLUTION

From symmetry it is seen that $\bar{y} = 0$. From Table 3,

$$\bar{x} = \frac{r \sin \theta}{\theta}$$

where $\theta$ must be stated in radians.
$\theta = 40° = 0.698$ rad, so

$$\bar{x} = \frac{(0.1 \text{ m})(\sin 0.698)}{0.698}$$

$$= 0.0921 \text{ m}$$

## JUDGMENT OF THE RESULTS

The answer is reasonable since $r \cos 40° < \bar{x} < r$, as should be expected.

**FIGURE 7-7**

Circular arc in the $xy$ plane

# EXAMPLE 7-2

The centroid $C$ of the rectangle $ab$ in Fig. 7-8 has coordinates $(a/2, b/2)$ by inspection. Prove this using integration.

## SOLUTION

From Eq. 7-11 and Fig. 7-8,

$$\bar{x} = \frac{\int x \, dA}{A} = \frac{\int xb \, dx}{ab} = \frac{1}{a}\left[\frac{x^2}{2}\right]_0^a = \frac{a}{2}$$

Similarly for the $y$ coordinate, but using a horizontal elemental area $dA = a \, dy$,

$$\bar{y} = \frac{\int y \, dA}{A} = \frac{\int ya \, dy}{ab} = \frac{1}{b}\left[\frac{y^2}{2}\right]_0^b = \frac{b}{2}$$

**TIP**

The solution by integration was facilitated by the choice of an elemental strip of area $b \, dx$ (and $a \, dy$) instead of the elemental area $dx \, dy$, which would have required a double integral for solution. Care must be taken with this method, however, to ensure that only like variables appear in the single integral.

**FIGURE 7-8**

Centroid $C$ of a rectangle in the $xy$ plane

EXAMPLE 7-3

(a) State the centroidal coordinates $\bar{x}$, $\bar{y}$ of the right triangle in Fig. 7-9a. Use Table 3, inside the back cover.

(b) Apply the results from part (a) to the oblique triangle in Fig. 7-9b using Eq. 7-13.

## SOLUTION

(a) Since the relation $\bar{y} = h/3$ in Table 3 is for a general triangle, it is valid for a right triangle. This result is applied to the triangle in Fig. 7-9a, using the $y$ and $x$ axes for reference to determine $\bar{x}$ and $\bar{y}$, respectively. The centroidal coordinates (which are worth remembering) are

$$\bar{x} = \frac{b}{3} \quad \bar{y} = \frac{h}{3}$$

(b) The results from part (a) are readily applied to triangles 1 and 2 in Fig. 7-9b after making adjustments (by inspection) for the particular configuration.

$$\bar{x}_1 = \frac{2}{3}\left(\frac{b}{5}\right) = \frac{2b}{15}$$

$$\bar{y}_1 = \frac{h}{3}$$

$$\bar{x}_2 = \frac{b}{5} + \frac{1}{3}\left(\frac{4}{5}b\right) = \frac{7b}{15}$$

$$\bar{y}_2 = \frac{h}{3}$$

The centroidal coordinates $\bar{x}_C$ and $\bar{y}_C$ of the composite triangle are obtained using Eq. 7-13,

$$\bar{x}_C = \frac{\bar{x}_1 A_1 + \bar{x}_2 A_2}{A_1 + A_2} \quad \bar{y}_C = \frac{\bar{y}_1 A_1 + \bar{y}_2 A_2}{A_1 + A_2}$$

where

$$A_1 = \frac{1}{2}\left(\frac{bh}{5}\right) = \frac{bh}{10} \quad \text{and} \quad A_2 = \frac{1}{2}\left(\frac{4b}{5}h\right) = \frac{2bh}{5}$$

Thus,

$$\bar{x}_C = \frac{\dfrac{2b}{15}\left(\dfrac{bh}{10}\right) + \dfrac{7b}{15}\left(\dfrac{2bh}{5}\right)}{\dfrac{bh}{10} + \dfrac{2bh}{5}} = \frac{2b}{5}$$

$$\bar{y}_C = \frac{\dfrac{h}{3}\left(\dfrac{bh}{10}\right) + \dfrac{h}{3}\left(\dfrac{2bh}{5}\right)}{\dfrac{bh}{10} + \dfrac{2bh}{5}} = \frac{h}{3}$$

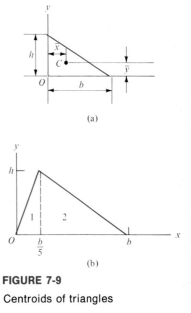

(a)

(b)

**FIGURE 7-9**

Centroids of triangles

EXAMPLE 7-3    275

## EXAMPLE 7-4

Determine the first moment $M_x$ of the rectangle $ab$ in Fig. 7-10 with respect to the $x$ axis. Use two methods: **(a)** Break the rectangle into four areas as divided by the $x$ and $y$ axes. **(b)** Work directly with the centroid of the whole area. **(c)** Use the simpler of the methods in parts (a) and (b) to calculate the first moment $M_y$ of the rectangle with respect to the $y$ axis.

### SOLUTION

**(a)** The first moment $M_x$ of the whole area is the sum of the first moments of the individual parts, $M_x = \sum \bar{y}_i A_i$, where $\bar{y}_i$ is the $y$ coordinate of the centroid of area $A_i$. The $y_i$ are easily determined by inspection.

$$M_x = \sum \bar{y}_i A_i = \bar{y}_1 A_1 + \bar{y}_2 A_2 + \bar{y}_3 A_3 + \bar{y}_4 A_4$$

$$= \left(\frac{1}{3}b\right)\left[\left(\frac{5}{6}a\right)\left(\frac{2}{3}b\right)\right] + \left(-\frac{1}{6}b\right)\left[\left(\frac{5}{6}a\right)\left(\frac{1}{3}b\right)\right]$$

$$+ \left(\frac{1}{3}b\right)\left[\left(\frac{1}{6}a\right)\left(\frac{2}{3}b\right)\right] + \left(-\frac{1}{6}b\right)\left[\left(\frac{1}{6}a\right)\left(\frac{1}{3}b\right)\right]$$

$$= \frac{18}{108}ab^2 = \frac{1}{6}ab^2$$

Note that the first moments of areas $A_2$ and $A_4$ are negative since they lie below the $x$ axis.

**(b)** The moment may also be calculated by considering the whole area $A$ and the vertical distance of its centroid from the $x$ axis:

$$M_x = \bar{y}A = \left(\frac{2}{3}b - \frac{1}{2}b\right)(ab) = \frac{1}{6}ab^2$$

**(c)** Realizing the facility of the method in part (b) the moment of the area about the $y$ axis is found to be

$$M_y = \bar{x}A = \left(\frac{5}{6}a - \frac{1}{2}a\right)(ab) = \frac{1}{3}a^2b$$

### JUDGMENT OF THE RESULTS

The identical results from parts (a) and (b) indicate that the calculations were performed correctly. The first moments of the total area with respect to the $x$ and $y$ axes would be expected to be positive since the centroid of the total area has positive $x$ and $y$ coordinates by inspection.

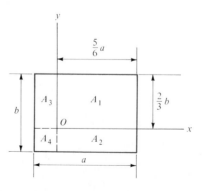

**FIGURE 7-10**

First moments of a rectangular area

EXAMPLE 7-5

Locate the centroid of area $A$ in Fig. 7-11 if it consists of three squares $A_1 = 5 \times 5$, $A_2 = 3.5 \times 3.5$, $A_3 = 1.4 \times 1.4$ units of length. Locate the new centroid after a circular hole of radius 0.8 is cut at the centroid of $A_1$.

## SOLUTION

It is helpful to set up a table for the three regions as discussed in Sec. 7-3.

| PART | $A_i$ | $\bar{x}_i$ | $\bar{x}_i A_i$ | $\bar{y}_i$ | $\bar{y}_i A_i$ |
|------|-------|-------------|-----------------|-------------|-----------------|
| $A_1$ | 25 | 2.5 | 62.5 | 2.5 | 62.5 |
| $A_2$ | 12.25 | 6.75 | 82.69 | 1.75 | 21.44 |
| $A_3$ | 1.96 | 7.8 | 15.29 | −0.7 | −1.37 |
| Summations: | 39.21 | | 160.48 | | 82.57 |

$$\bar{x}A = \sum \bar{x}_i A_i, \quad \bar{x} = \frac{\sum \bar{x}_i A_i}{A} = \frac{160.48}{39.21} = 4.09$$

$$\bar{y}A = \sum \bar{y}_i A_i, \quad \bar{y} = \frac{\sum \bar{y}_i A_i}{A} = \frac{82.57}{39.21} = 2.11$$

Next the area of the hole is analyzed. The hole may be considered a negative area, $A_4 = -2.01$ with centroid $(\bar{x}_4, \bar{y}_4) = (2.5, 2.5)$. The problem is now considered that of a composite area consisting of two regions, the original area, $A$, and $A_4$.

$$\bar{x}' = \frac{\bar{x}A + \bar{x}_4 A_4}{A + A_4} = \frac{(4.09)(39.21) + (2.5)(-2.01)}{39.21 + (-2.01)} = 4.18$$

$$\bar{y}' = \frac{\bar{y}A + \bar{y}_4 A_4}{A + A_4} = \frac{(2.11)(39.21) + (2.5)(-2.01)}{39.21 + (-2.01)} = 2.09$$

## JUDGMENT OF THE RESULTS

The results appear reasonable since the centroid of the final area is to the right and below the centroid of the original area. This was expected as a larger percent of the solid area is now to the right of the original centroid. Similarly, the centroid also moved down because a larger percent of the solid area is now below the original centroid.

**FIGURE 7-11**

Centroid of a composite area

EXAMPLE 7-5    277

# EXAMPLE 7-6

Locate the center of gravity of the composite solid hemisphere and attached hollow cylinder shown in Fig. 7-12. The cylinder is made of a steel alloy with a specific weight of $\gamma_S = 0.30$ lb/in$^3$, the hemisphere is aluminum with $\gamma_A = 0.10$ lb/in$^3$, and $r = 2$ in.

## SOLUTION

It is seen from symmetry that $\bar{x} = \bar{z} = 0$. Labeling the hemisphere body 1 and the cylinder body 2, the centroid of the composite piece is found using Eq. 7-13 with $W = \gamma V$ substituted for $A$ and the relations from Table 3, inside the back cover.

$$\bar{y} = \frac{\bar{y}_1 W_1 + \bar{y}_2 W_2}{W_1 + W_2} = \frac{\bar{y}_1 \gamma_1 V_1 + \bar{y}_2 \gamma_2 V_2}{\gamma_1 V_1 + \gamma_2 V_2}$$

$$= \frac{\gamma_A \left(\dfrac{3}{8} r\right)\left(\dfrac{2}{3} \pi r^3\right) + \gamma_S(-2r)(4r\pi)\left[r^2 - \left(\dfrac{3}{4} r\right)^2\right]}{\gamma_A \dfrac{2}{3} \pi r^3 + \gamma_S 4 r\pi\left[r^2 - \left(\dfrac{3}{4} r\right)^2\right]}$$

$$= \frac{\gamma_A \dfrac{\pi r^4}{4} - 8\gamma_S \pi r^4 + \dfrac{9}{2} \gamma_S \pi r^4}{\dfrac{2}{3} \gamma_A \pi r^3 + 4\gamma_S \pi r^3 - \dfrac{9}{4} \gamma_S \pi r^3}$$

$$= \frac{-3.22r}{1.859}$$

$$= -3.46 \text{ in.}$$

**FIGURE 7-12**

Centroid of a composite body

# PROBLEMS

**7–1** Two thin, uniform wires are soldered at $A$ to form the T shape. Determine the centroid of the line $OABC$.

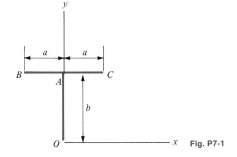

Fig. P7-1

**7–2** Determine the centroid of the thin, uniform wire $OAC$ in Fig. P7-1.

**7–3** Determine the centroid of the thin, uniform wire $AOBC$.

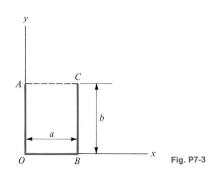

Fig. P7-3

**7–4** Solve Prob. 7–3 assuming that the segment $OA$ is doubled in length to $2b$.

**7–5** Determine the centroid of the thin, uniform wire $AOBC$. $a = 2$ in., $b = 4$ in., and $c = 6$ in.

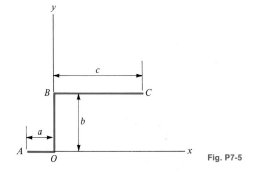

Fig. P7-5

**7–6** Determine the centroid of the thin, uniform wire $AOB$.

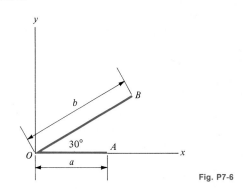

Fig. P7-6

**7–7** Determine the centroid of the circular arc $AB$ for $\theta = 40°$ using integration.

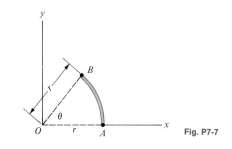

Fig. P7-7

**7–8** Determine the centroid of the circular arc $AB$ in Fig. P7-7 for $\theta = 90°$. Use Table 3, inside the back cover.

**7–9** Determine the centroid of the triangular area by integration.

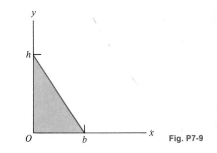

Fig. P7-9

**7–10** Assume that the triangle in Fig. P7-9 is truncated horizontally at $y = h/2$. Determine the centroid of the resulting trapezoid. Use the centroids and areas of the original triangle and of the smaller triangle that is removed.

**7–11**    Determine the centroid of the L-shaped area working with two rectangles: one of area $ac$, the other of area $(b - c)d$. $a = 200$ mm, $b = 250$ mm, $c = 110$ mm, and $d = 130$ mm.

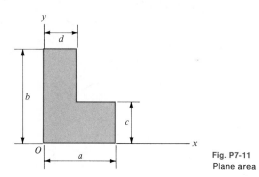

Fig. P7-11
Plane area

**7–12**    Solve Prob. 7–11 working with two imaginary rectangles: one of area $ab$, the other of area $(a - d)(b - c)$.

**7–13**    Determine the centroid of the triangular area.

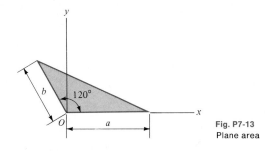

Fig. P7-13
Plane area

**7–14**    Determine the centroid of the trapezoidal area working with the rectangle of area $ab$ and the triangle above it. $a = 13$ in., $b = 11$ in., and $c = 15$ in.

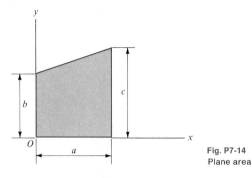

Fig. P7-14
Plane area

**7–15**    Solve Prob. 7–14 working with two imaginary areas: a rectangle of area $ac$ and a triangle of area $\frac{1}{2}a(c - b)$.

**7–16**    Determine the centroid of the shaded area.

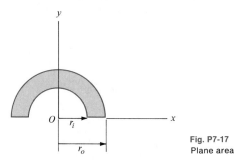

Fig. P7-16
Plane area

**7–17**    Determine the centroid of the shaded area if the inner radius is $r_i = r_o/2$.

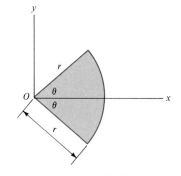

Fig. P7-17
Plane area

**7–18**    Assume that in Fig. P7-17 the inner radius is $r_i = 0.95r_o$. Determine the centroid of this area. For comparison, determine the centroid of a semicircular line of radius $r_o$.

**7–19**    Determine the centroid of the circular segment by integration.

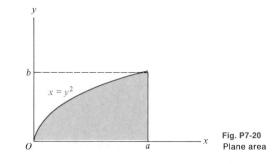

Fig. P7-19
Plane area

**7–20**    Determine the centroid of the shaded area.

Fig. P7-20
Plane area

**7–21** Determine the centroid of the area between the dashed line and the curve $x = y^2$ in Fig. P7-20.

**7–22** Determine the centroid of the shaded area.

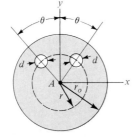

Fig. P7-22
Plane area

**7–23** Determine the centroid of the area between the dashed line and the curve $y = x^2/2$ in Fig. P7-22.

**7–24** Determine the centroid of a circular plate of radius $r_o$ after one hole of diameter $d = 0.2r_o$ is drilled at a distance $r = r_o/2$ from the center $A$. Consider the hole to the right. (The second hole is for Prob. 7–25.) $r_o = 150$ mm and $\theta = 30°$.

Fig. P7-24
Plane area

**7–25** Solve Prob. 7–24 assuming that two identical holes of diameter $d$ are drilled (Fig. P7-24).

**7–26** Determine the centroid of the area composed of a rectangle $a(2b)$ and a semicircle of radius $b$.

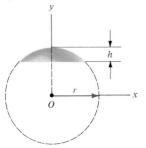

Fig. P7-26
Plane area

**7–27** Determine the centroid of the segment of a solid spherical volume. The height of the segment is $h = r/3$.

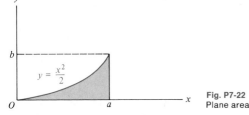

Fig. P7-27
Segment of spherical volume

**7–28** Determine the centroid of the truncated pyramid's volume. The height of the small pyramid that is cut off at the top is $h = a/4$.

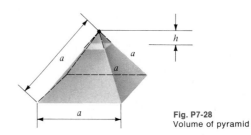

Fig. P7-28
Volume of pyramid

**7–29** Determine the centroid of the conical volume using integration. (The dimension $d$ is for Prob. 7–30.)

Fig. P7-29
Volume of circular cone

**7–30** Assume that the conical volume in Fig. P7-29 is truncated at $d = h/3$ from the top. Determine the centroid of the truncated cone using Table 3, inside the back cover.

**7–31** Determine the centroid of the composite volume.

Fig. P7-31
Volume of cylinder and cone

**7–32** A cylindrical piece is machined to produce an internal cone shape as shown in the cross-sectional view. Determine the centroid of the final volume if $r = 10$ mm, $d = 4$ mm, and $h = 15$ mm.

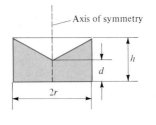

Fig. P7-32
Cross section of volume of cylinder minus cone

# THEOREMS OF PAPPUS

Bodies with *surfaces of revolution* are commonly used in engineering. A surface of revolution is defined as a surface generated by rotating a plane curve about a fixed axis. Conversely, the body may be rotated with respect to a fixed cutting edge as in a lathe. For example, consider the hourglass-shaped specimen $S$ and the grip $G$ in Fig. 7-13a. The specimen in Fig. 7-13b is rotated about the $x$ axis during machining while the cutting tool swings in a plane about a fixed point to generate the surface line $L-L$. The grip is made similarly.

An element of the surface of revolution generated is always a ring of width $dL$, as in Fig. 7-13b. The outer surface area of this ring is

$$dA = 2\pi y \, dL$$

The total area of the surface of revolution is

$$A = 2\pi \int y \, dL \qquad \boxed{7\text{-}15}$$

The volume $dV$ of an elemental ring of revolution with cross section $dA$ is obtained using Fig. 7-12c,

$$dV = 2\pi y \, dA$$

The total volume within the surface is

$$V = 2\pi \int y \, dA \qquad \boxed{7\text{-}16}$$

Simpler and more useful methods for calculating the surface areas and volumes of bodies of revolution were formulated by Pappus, a Greek geometer, in the third century A.D. The name of Guldinus, a Swiss mathematician of the seventeenth century, is also mentioned occasionally since he had restated these concepts. The two theorems of Pappus are based on the definitions of centroidal coordinates of lines and areas. These are given in concise form as follows and illustrated in Fig. 7-13c and d.

**THEOREM I.** Since $\int y \, dL = \bar{y}L$,

$$\boxed{A = 2\pi \bar{y} L} \qquad \boxed{7\text{-}17}$$

where $L$ = length of the generating curve
$\bar{y}$ = centroid of line $L$ with respect to the axis of revolution

**THEOREM II.** Since $\int y \, dA = \bar{y}A$,

$$\boxed{V = 2\pi \bar{y} A} \qquad \boxed{7\text{-}18}$$

where $A$ = area of the generating surface
$\bar{y}$ = centroid of area $A$ with respect to the axis of revolution

Naturally, Eqs. 7-17 and 7-18 can also be used to locate the centroids when the surface area $A$ and the volume $V$ of a body of revolution are known.

(a)

(b)

(c)     (d)

(e)

**FIGURE 7-13**

Hourglass-shaped specimen. (Photograph courtesy MTS
Systems Corporation, Minneapolis, Minnesota.)

EXAMPLE 7-7

Use the theorems of Pappus to locate the centroids of **(a)** a semicircular arc, and **(b)** a semicircular area of radius $r$. The well-known formulas for the surface area and volume of a sphere can be used.

### SOLUTION

**(a)** The semicircular arc in Fig. 7-14a would generate a spherical shell when rotated about the $x$ axis. The total surface area is $4\pi r^2$. From Eq. 7-17, $A = 2\pi\bar{y}L$,

$$4\pi r^2 = 2\pi\bar{y}(\pi r)$$

$$\bar{y} = \frac{2r}{\pi}$$

**(b)** The semicircular area in Fig. 7-14b would generate a solid sphere when rotated about the $x$ axis. The total volume of the sphere is $\frac{4}{3}\pi r^3$. From Eq. 7-18, $V = 2\pi\bar{y}A$,

$$\frac{4}{3}\pi r^3 = 2\pi\bar{y}\left(\frac{\pi r^2}{2}\right)$$

$$\bar{y} = \frac{4r}{3\pi}$$

### JUDGMENT OF THE RESULTS

Both results are reasonable by inspection. Note that the centroidal distances are different in the two cases even though the same symbol is used for them (this is common practice) and $r$ is the same in both cases.

(a)

(b)

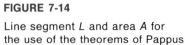

**FIGURE 7-14**

Line segment $L$ and area $A$ for the use of the theorems of Pappus

Determine the surface area and volume of the pulley whose cross section is shown in Fig. 7-15.

EXAMPLE 7-8

## SOLUTION

To find the surface area of the pulley the centroid of the boundary line *abcdef* must be found. Dividing the region into segments *ab, bc, cd, de, ef,* and *fa,* $\bar{y}$ is found for the composite area:

$$\bar{y} = \frac{\sum \bar{y}_i l_i}{\sum l_i}$$

$$= \frac{2(250 \text{ mm})(5 \text{ mm}) + \left[ 250 \text{ mm} - \dfrac{2(15 \text{ mm})}{\pi} \right](15\pi \text{ mm})}{2(5 \text{ mm}) + 15\pi \text{ mm} + 2(220 \text{ mm}) + 40 \text{ mm}}$$

$$+ \frac{2(140 \text{ mm})(220 \text{ mm}) + (30 \text{ mm})(40 \text{ mm})}{2(5 \text{ mm}) + 15\pi \text{ mm} + 2(220 \text{ mm}) + 40 \text{ mm}}$$

$$= \frac{76.6 \times 10^3 \text{ mm}^2}{537 \text{ mm}} = 143 \text{ mm} \quad \text{(relative to the axis of rotation)}$$

From Eq. 7-17,

$$A = 2\pi \bar{y} L = 2\pi(143 \text{ mm})(537 \text{ mm}) = 4.82 \times 10^5 \text{ mm}^2$$

Similarly, the centroid of the area *abcdef* is needed to calculate the volume of the pulley. Dividing the area into the rectangle *adef* and semicircular hole *bc* the centroid of the composite area is found.

$$\bar{y} = \frac{\sum \bar{y}_i A_i}{\sum A_i}$$

$$= \frac{(140 \text{ mm})(8800 \text{ mm}^2) + \left[ 250 \text{ mm} - \dfrac{4(15 \text{ mm})}{3\pi} \right]\left[ -(15 \text{ mm})^2 \dfrac{\pi}{2} \right]}{8800 \text{ mm}^2 - (15 \text{ mm})^2 \left( \dfrac{\pi}{2} \right)}$$

$$= \frac{11.46 \times 10^5 \text{ mm}^3}{8.45 \times 10^3 \text{ mm}^2} = 136 \text{ mm} \quad \text{(relative to the axis of rotation)}$$

From Eq. 7-18,

$$V = 2\pi \bar{y} A = 2\pi(136 \text{ mm})(8.45 \times 10^3 \text{ mm}^2) = 7.22 \times 10^6 \text{ mm}^3$$

**FIGURE 7-15**

Cross section of a pulley

EXAMPLE 7-8    285

## EXAMPLE 7-9

The removable seat of an O-ring consists of two halves as illustrated in Figure 7-16. Determine the surface area and the volume of the two halves put together to form the complete circular seat.

### SOLUTION

Using the relations of Table 3, inside the back cover, the centroid of the boundary of a cross-sectional area is found.

$$\bar{y}' = \frac{\sum \bar{y}_i l_i}{\sum l_i}$$

$$= \frac{(10 \text{ mm})(10 \text{ mm}) + (9 \text{ mm})(2 \text{ mm}) + (5 \text{ mm})(10 \text{ mm})}{10 \text{ mm} + 2 \text{ mm} + 10 \text{ mm} + 2 \text{ mm} + \dfrac{\pi(8 \text{ mm})}{2}}$$

$$+ \frac{(0 \text{ mm})(2 \text{ mm}) + \dfrac{2(8 \text{ mm})}{\pi}\left(\dfrac{8\pi}{2} \text{ mm}\right)}{10 \text{ mm} + 2 \text{ mm} + 10 \text{ mm} + 2 \text{ mm} + \dfrac{\pi(8 \text{ mm})}{2}}$$

$$= \frac{232 \text{ mm}^2}{36.6 \text{ mm}} = 6.3 \text{ mm}$$

$$\bar{y} = R + \bar{y}' = 100 \text{ mm} + 6.3 \text{ mm} = 106.3 \text{ mm}$$

From Eq. 7-17,

$$A = 2\pi\bar{y}L = 2\pi(106.3 \text{ mm})(36.6 \text{ mm})$$
$$= 2.44 \times 10^4 \text{ mm}^2$$

Similarly, by first finding the centroid of the cross-sectional area the volume of revolution can be found.

$$\bar{y}' = \frac{\sum \bar{y}_i A_i}{\sum A_i} = \frac{(5 \text{ mm})(100 \text{ mm}^2) + \dfrac{4(8 \text{ mm})}{3\pi}\left[-\dfrac{\pi(8 \text{ mm})^2}{4}\right]}{100 \text{ mm}^2 - \dfrac{\pi(8 \text{ mm})^2}{4}}$$

$$= \frac{330 \text{ mm}^3}{50 \text{ mm}^2} = 6.6 \text{ mm}$$

$$\bar{y} = R + \bar{y}' = 100 \text{ mm} + 6.6 \text{ mm} = 106.6 \text{ mm}$$

From Eq. 7-18,

$$V = 2\pi\bar{y}A = 2\pi(106.6 \text{ mm})(50 \text{ mm}^2) = 3.35 \times 10^4 \text{ mm}^3$$

**FIGURE 7-16**

Circular seat of an O-ring

**7–33** The solid ring has a circular cross section of radius $r$. Determine the surface area of the ring.

**Fig. P7-33**
Solid circular ring

**7–34** Determine the volume of the ring in Prob. 7–33.

**7–35** The segment of a solid circular ring is cut at an angle $\theta = 120°$. Determine the volume of the segment.

**Fig. P7-35**
Segment of circular ring

**7–36** Determine the total surface area of the ring segment described in Prob. 7–35.

**7–37** Determine the total surface area of the solid half ring.

**Fig. P7-37**
Solid half ring

**7–38** Determine the volume of the half ring described in Prob. 7–37.

**7–39** Determine the total surface area and volume of a solid hemisphere of radius $r$.

**7–40** Determine the total surface area and volume of a solid circular cone of base radius $r$ and height $h$.

**7–41** Determine the total surface area of the composite solid body.

**Fig. P7-41**
Surface area of cylinder and cone

**7–42** Determine the volume of the solid body for $r = 10$ mm, $d = 4$ mm. and $h = 15$ mm.

**Fig. P7-42**
Cross section of volume of cylinder minus cone

**7–43** Determine the surface area of the spherical cap for $\phi = 40°$.

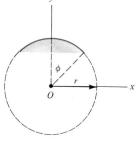

**Fig. P7-43**
Spherical cap

**7–44** Calculate the inside surface area (toward the $y$ axis) of the parabolic antenna if $a = 2$ m and $b = 0.5$ m. The antenna appears as a circle when viewed from the $y$ direction. Use integral tables.

**Fig. P7-44**
Parabolic dish antenna

**7–45** Calculate the volume of the parabolic antenna dish itself in Fig. P7-44 if $a = 8$ ft, $b = 2$ ft, and $t = 1$ in. The antenna appears as a circle when viewed from the $y$ direction. Use integral tables.

**7–46** Determine the total surface area of the tapered, hollow shaft if $2r = R$ and $h_1 = h_2 = h$.

**Fig. P7-47**
Hollow shaft

**7–47** Calculate the volume of the tapered, hollow shaft in Prob. 7–46 if $r = 10$ mm, $R = 16$ mm, $h_1 = 200$ mm, and $h_2 = 50$ mm.

**7–48** The concrete cooling tower of a power plant is modeled as an open-ended body, with the center of curvature at half height. Calculate the area of the external curved surface if $R = 80$ m, $r = 15$ m, and $h = 50$ m.

**Fig. P7-48**
Concrete cooling tower, open on ends

**7–49** Calculate the volume of the concrete wall of the cooling tower in Prob. 7–48. Let $R = 260$ ft, $r = 50$ ft, $h = 160$ ft, and $t = 1.7$ ft.

288    DISTRIBUTED FORCES

The analysis of structures often involves distributed loading. The most common such loading is from the weights of the members themselves. These and all other distributed loads acting simultaneously on a structure can be superimposed vectorially to determine the resultant force distribution. The concept of centroid can be applied to the distributed loading to simplify the calculations, as shown in the following.

Calculation of the external reactions on a member with a distributed load is illustrated for an airplane wing in Fig. 7-17. Assume that the airplane is stationary on the ground so that the distributed load on it is caused by weight alone. The loading $\mathbf{w}(x)$ (in terms of force per unit length) is continuous but irregular because of the numerous components in the wing. The support reactions at the fuselage are the shear force $\mathbf{V}_O$ and the moment $\mathbf{M}_O$. The shear $V_O$ is evaluated by integrating the load $w(x)$ on the element $dx$ over the whole length $l$ of the wing. For equilibrium of the wing in the vertical direction,

$$V_O = \int_0^l w(x)\, dx \qquad \boxed{7\text{-}19}$$

The moment $M_O$ is evaluated by integrating the elemental moment $xw(x)\, dx$ over the length $l$ of the wing. For the equilibrium of moments with respect to point $O$,

$$M_O = \int_0^l xw(x)\, dx \qquad \boxed{7\text{-}20}$$

Another way to state these is that $V_O$ is equal to the total area $A$ under the load distribution curve [since $A = \int dA = \int w(x)\, dx$], and $M_O$ is equal to the first moment of the area $A$ with respect to point $O$. In short, $V_O = A = W$, and

$$\boxed{M_O = \bar{x}A} \qquad \boxed{7\text{-}21}$$

where  $A = W$ = resultant of the distributed forces
$\quad\quad\quad \bar{x} = x$ coordinate of centroid of area $A$ under the load distribution curve

Accordingly, the distributed load may be replaced by a concentrated load whose magnitude is equal to the area under the load curve and whose line of action is at the centroid of that area. This result is very useful in calculating the support reactions for a structural member on which a distributed load is acting.

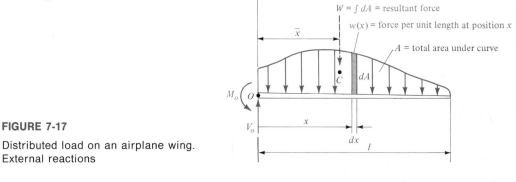

**FIGURE 7-17**

Distributed load on an airplane wing.
External reactions

## EXAMPLE 7-10

Determine the support reactions for the cantilevered beam with the distributed load shown in Fig. 7-18a.

### SOLUTION

The free-body diagram for the beam is drawn in Fig. 7-18b. The distributed load is represented by a composite area made up of a rectangle of length $l$ and height $w_B$ and a triangle of base $l/2$ and height $w_B$. The distributed load on the beam can be replaced by two concentrated loads with magnitudes equal to these areas, $W_1 = lw_B$ and $W_2 = \frac{1}{4}lw_B$. These concentrated loads must be applied at the centroids of the areas representing the distributed loads. In a beam problem with only transverse loading, however, only the $x$ coordinate of the centroid is needed. $\mathbf{W}_1$ must be applied at $x = l/2$ and $\mathbf{W}_2$ must be applied at $x = \frac{1}{3}(l/2) = \frac{1}{6}l$. The concentrated loads are shown as dashed lines in Fig. 7-18c. The reactions at the fixed support at $A$ are determined using Eq. 4-4,

$$\sum F_y = 0$$

$$A_y - W_1 - W_2 = 0$$

$$A_y = lw_B + \frac{1}{4}lw_B$$

$$A_y = \frac{5}{4}lw_B$$

$$\sum M_A = 0$$

$$M_A - \frac{l}{6}\left(\frac{1}{4}lw_B\right) - \frac{l}{2}(lw_B) = 0$$

$$M_A = \left(\frac{1}{24} + \frac{1}{2}\right)(l^2w_B)$$

$$M_A = \frac{13}{24}l^2w_B$$

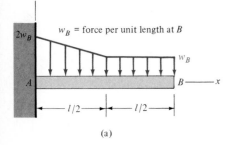

$w_B$ = force per unit length at $B$

(a)

(b)

(c)

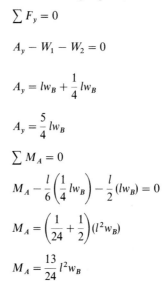

**FIGURE 7-18**

External reactions due to a distributed load on a cantilever beam

EXAMPLE 7-11

Determine the reactions at the supports for the simply supported beam with the parabolic load distribution shown in Fig. 7-19.

SOLUTION

The free-body diagram for the beam is drawn in Fig. 7-19b. The distributed load may be replaced by a concentrated load equal in magnitude to the area under the curve and applied at the centroid of the area. Using an elemental strip of area $y\, dx$,

$$A = \int dA = \int_0^3 (3 - x)^2\, dx$$

$$= \int_0^3 (9 - 6x + x^2)\, dx$$

$$= \left[ 9x - 3x^2 + \frac{1}{3} x^3 \right]_0^3$$

$$= 9\ \text{N}$$

$$\bar{x}A = \int x\, dA = \int_0^3 x(3 - x)^2\, dx$$

$$= \int_0^3 (9x - 6x^2 + x^3)\, dx$$

$$= \left[ \frac{9}{2} x^2 - 2x^3 + \frac{1}{4} x^4 \right]_0^3$$

$$= 6.75\ \text{N·m}$$

$$\bar{x} = \frac{\bar{x}A}{A} = 0.75\ \text{m}$$

The concentrated load is shown in Fig. 7-19c as a dashed arrow. The reactions are calculated using Eq. 4-4,

$$\sum M_A = 0$$

$$-(9\ \text{N})(0.75\ \text{m}) + B_y(3.0\ \text{m}) = 0$$

$$B_y = 2.25\ \text{N}$$

$$\sum F_y = 0$$

$$A_y - 9\ \text{N} + 2.25\ \text{N} = 0$$

$$A_y = 6.75\ \text{N}$$

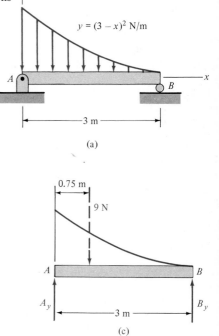

(a)

**FIGURE 7-19**

External reactions due to a distributed load on a simply supported beam

(b)

(c)

EXAMPLE 7-11    291

# DISTRIBUTED LOADS ON BEAMS.
# INTERNAL REACTIONS

The concepts described in Sec. 7-5 are also useful in determining internal forces and moments (such as $V_1$ and $M_1$ at section $B$ in Fig. 7-20) when applied with caution. This is illustrated in Fig. 7-20b, where $V_1$ and $M_1$ can be evaluated by considering only the part of the wing that is to the right of section $B$. Here the area $A_1$ under the load distribution curve, the centroidal dimension $\bar{x}_1$ of the area $A_1$, and the resultant force $W_1$ are only for part $DE$ of the wing and the load on it (Fig. 7-20b). Alternatively, the segment from $O$ to $D$ could be considered to calculate $V_1$ and $M_1$. In that case the distributed loading from $O$ to $D$ *and* the reactions $V_O$ and $M_O$ (which were obtained for the whole wing) must be used in the equilibrium equations. In any case, the correct free-body diagram for the part considered must be used.

The distributed external load on a beam causes variations in the internal shear $V$ and bending moment $M$ as functions of position along the beam. For illustration, consider an infinitesimal element of the beam (Fig. 7-21). It can be assumed that the load $w(x)$ is constant for an infinitesimal element $dx$ even if $w(x)$ varies in a macroscopic way. Assume that the shear $V$ and bending moment $M$ at section $B_1$ are known from calculations involving a large segment of the beam. There must be a different shear $V + dV$ and a different bending moment $M + dM$ at section $B_2$ since these are caused by a different load distribution than that used for section $B_1$. In other words, looking at the infinitesimal element alone, $dV$ and $dM$ represent the necessary difference in internal reactions between sections $B_1$ and $B_2$ because of the presence of $w(x)$ on the element. Note that the shear forces and bending moments in Fig. 7-21 are drawn in the positive directions according to the sign conventions in Sec. 5-3.

The element in Fig. 7-21 is in equilibrium so the summation of vertical forces on it is zero,

$$V - (V + dV) - w(x)\,dx = 0$$

and

$$w(x) = -\frac{dV}{dx}$$

7-22

Thus, the value of the slope $dV/dx$ is equal to the negative value of the load function at that location. It should be noted that the negative sign

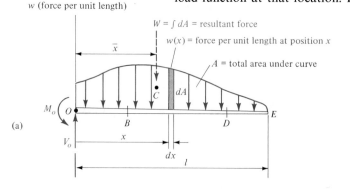

w (force per unit length)

$W = \int dA$ = resultant force

$w(x)$ = force per unit length at position $x$

$A$ = total area under curve

(a)

(b)

**FIGURE 7-20**

Distributed load on an airplane wing. Internal reactions

here results from the fact that a downward load is acting on the element. This sign convention should be remembered in using Eq. 7-22. It should also be noted that in the analysis above, $x$ is measured from the left end of the beam, and $V$ and $M$ are assumed to increase to the right. For $x$ increasing to the left, $w(x) = dV/dx$.

The difference in the shear at any two cross sections an arbitrary distance apart is equal in magnitude to the area under the load distribution curve between the two sections. Considering sections $B$ and $D$ in Fig. 7-20(a) and integrating Eq. 7-22,

**FIGURE 7-21**

Free-body element of a beam

$$\int_{V_B}^{V_D} dV = -\int_{x_B}^{x_D} w(x)\,dx$$

$$V_D - V_B = -\int_{x_B}^{x_D} w(x)\,dx,$$

or

$$V_B - V_D = \int_{x_B}^{x_D} w(x)\,dx \qquad \boxed{7\text{-}23}$$

$$= \text{area under } w(x)$$

The element in Fig. 7-21 is also in equilibrium with respect to moments. The resultant of $w(x)$ is put at the centroid of the $w(x)$ distribution, which is halfway between $B_1$ and $B_2$. Taking moments about $B_2$,

$$-M + (M + dM) + w(x)\,dx\,\frac{dx}{2} - V\,dx = 0$$

Ignoring the differential of higher order ($dx^2$), the following simple relation is obtained for $x$ increasing to the right,

$$\boxed{V = \frac{dM}{dx}} \qquad \boxed{7\text{-}24}$$

Thus, the shear at any section is equal to the slope of the moment curve. Conversely, the difference in moments at two cross sections $B$ and $D$ is the integral of the shear in that region,

$$\int_{M_B}^{M_D} dM = M_D - M_B = \int_{x_B}^{x_D} V\,dx \qquad \boxed{7\text{-}25}$$

where $M_B$ is the bending moment at the location $x_B$, and so on. Another way to state Eq. 7-25 is that

$$M_D - M_B = \text{area under shear diagram from } x_B \text{ to } x_D \quad \boxed{7\text{-}26}$$

Equation 7-26 provides a simple and practical way of constructing the bending moment diagram in some cases.

Another relationship involving the shear and the bending moment in a beam under continuous loading is obtained by differentiating $V$ with respect to $x$. From Eq. 7-24,

$$\boxed{\frac{dV}{dx} = \frac{d^2M}{dx^2}} \qquad \boxed{7\text{-}27}$$

The positive signs in Eqs. 7-24 through 7-27 result from the particular loading used in Fig. 7-21. Negative signs in these equations are possible for different loadings.

EXAMPLE 7-12

Determine the shear $V$ and bending moment $M$ as functions of $x$ for the cantilever beam shown in Fig. 7-22a. Draw the shear and bending moment diagrams.

## SOLUTION

The load function $w(x)$ varies linearly with $x$ and has a slope $\Delta w/\Delta x = w_B/l$, hence $w(x) = (w_B/l)x$. Consideration of a section $K$ at a general position $x$ (Fig. 7-22d) yields the shear and bending moment as functions of $x$. The effect of the distributed load is found by replacing that portion (and only that portion) of the load applied to part $AK$ with its equivalent concentrated load $\mathbf{W}$ applied at the centroid of its area. The area of the triangular load distribution is $\frac{1}{2}bh$, where $b = x$ and $h = (w_B/l)x$, and its centroid is at $x/3$ measured from section $K$. Since part $AK$ is in equilibrium,

$$\sum F_y = 0$$

$$-W - V = 0$$

$$V = -\frac{1}{2}\left(\frac{w_B}{l}x\right)x = -\frac{w_B x^2}{2l}$$

$$\sum M_K = 0$$

$$W\left(\frac{x}{3}\right) + M = 0$$

$$M = -\frac{w_B x^2}{2l}\left(\frac{x}{3}\right) = -\frac{w_B x^3}{6l}$$

It is seen that the shear curve is parabolic and the bending moment curve is cubic. The appropriate curves are sketched in Fig. 7-22b and c.

## ALTERNATE SOLUTION

Using Eq. 7-22, with $w(x) = (w_B/l)x$,

$$\frac{dV}{dx} = -\frac{w_B}{l}x$$

Integrating from 0 to $x$ where the shear is 0 and $V$ respectively,

$$\int_0^V dV = -\int_0^x \frac{w_B}{l}x\, dx$$

$$V(x) = -\frac{w_B}{2l}x^2$$

**FIGURE 7-22**

Shear and bending moment diagrams of a cantilever beam

Note that Eq. 7-23 gives the same results as follows,

$$V_K - V_A = -[\text{area under } w(x)]$$

$$V - 0 = -\frac{1}{2}\left(\frac{w_B}{l}x\right)(x)$$

$$V = V(x) = -\frac{1}{2}\frac{w_B}{l}x^2$$

At $x = l$, $V = -(w_B/2)l$.

Using Eq. 7-24,

$$\frac{dM}{dx} = V = -\frac{w_B}{2l}x^2$$

Since the bending moment is 0 at $x = 0$,

$$\int_0^M dM = -\int_0^x \frac{w_B}{2l}x^2\,dx$$

$$M(x) = -\frac{w_B}{6l}x^3$$

At $x = l$, $M = -(w_B/6)l^2$.

The equation for the shear is parabolic. Since $w(x)$ is zero at $x = 0$, the shear curve has zero slope at that point and an increasingly negative slope as $w(x)$ increases. The bending moment equation is cubic. Its curve also has an increasingly negative slope with zero slope at $x = 0$. These curves are sketched directly below the beam in Fig. 7-22b and c.

EXAMPLE 7-12   295

# EXAMPLE 7-13

Determine the shear $V$ and bending moment $M$ as functions of $x$ for the cantilever beam shown in Fig. 7-23a. Draw the shear and bending moment diagrams.

## SOLUTION

Since the right end of the beam is unsupported, consideration of sections measured from this free end avoids the necessity of calculating the reaction at the wall. Evaluating a general section, $0 \le x \le l/2$ (Fig. 7-23d), it is seen that only that portion of the distributed load up to the general location $K_1$ is considered. The effect of this portion of the load is handled by replacing the distributed load by its equivalent concentrated load equal to the area under the load curve located at the centroid of the area. Part $BK_1$ of the beam is in equilibrium, so

$$\sum F_y = 0$$

$$V_1 - w_B x = 0, \quad V_1 = w_B x$$

$$\sum M_{K_1} = 0$$

$$M_1 + w_B x \left(\frac{x}{2}\right) = 0, \quad M_1 = -w_B \left(\frac{x^2}{2}\right)$$

A quick check shows that $V_1 = -dM_1/dx$ and $w(x) = w_B = dV_1/dx$, which have the opposite signs than they should have by Eqs. 7-22 and 7-24. This is due to the fact that $x$ is being measured from the right end in this example rather than from the left, which was used for the derivation of these equations. For $l/2 \le x \le l$ (Fig. 7-23e), the load curve is considered as a composite area as shown. The load $W_2$ is a triangle of area $\frac{1}{2}bh$. The height of the triangle $h$ at a general position $x$ is determined by realizing that the load curve has a linear slope $\Delta w/\Delta x = w_B/(l/2)$ and that this portion of the load magnitude varies directly with $b = x - l/2$. Hence,

$$h = \frac{w_B}{l/2}(x - l/2) \quad \text{and} \quad W_2 = \frac{1}{2}bh = \frac{w_B}{l}(x - l/2)^2$$

Part $BK_2$ is in equilibrium, so

$$\sum F_y = 0$$

$$V_2 - w_B x - \frac{w_B}{l}(x - l/2)^2 = 0, \quad V_2 = \frac{w_B x^2}{l} + \frac{w_B l}{4}$$

$$\sum M_{K_2} = 0$$

$$-M_2 - \frac{w_B}{3l}(x - l/2)^2(x - l/2) - w_B x \left(\frac{x}{2}\right) = 0$$

$$M_2 = -\frac{w_B x^3}{3l} - \frac{w_B x l}{4} + \frac{w_B l^2}{24}$$

As a check it is found that $V_2 = -dM_2/dx$.

As a result, relations for both the internal shear $V$ and bending moment $M$ as they vary with $x$ have been determined for the total length of the beam. The $V$ and $M$ diagrams are drawn for the proper intervals as shown in Fig. 7-23b and c. Graphing is facilitated by plotting values at the end points of the intervals where each equation is valid and sketching in a curve of reasonable shape between points.

## JUDGMENT OF THE RESULTS

Substituting $x = l$ in the equations for the shear and bending moment gives the reactions at the wall. These calculated values agree with those found in Ex. 7-10. It is seen from the equations that both the shear and the moment at the free end are zero, as they should be. The following check is also useful.

## CHECK USING CALCULUS

The equations for the internal shear and bending moment can be calculated directly from the equation of the loading curve using Eqs. 7-22 and 7-24 with the appropriate change in signs for measuring $x$ from end $B$. Note that the equation for the shear must be determined first.

For $0 \leq x \leq l/2$:

$$w(x) = w_B$$

$$dV = w(x)\, dx$$

$$\int_0^V dV = \int_0^x w_B\, dx$$

$$V(x) = w_B x$$

$V(x)$ is a linear curve with a constant, positive slope. At $x = l/2$ its value is $w_B l/2$.

$$dM = -V(x)\, dx$$

$$\int_0^M dM = -\int_0^x w_B x\, dx$$

$$M(x) = -\frac{1}{2} w_B x^2$$

$M(x)$ is a parabolic curve with zero slope at $x = 0$ and an increasingly negative slope. At $x = l/2$, its value is $-w_B l^2/8$.

Similarly, for $l/2 \leq x \leq l$, $w(x) = (2w_B/l)x$

$$dV = w(x)\, dx$$

$$\int_{w_B(l/2)}^V dV = \int_{l/2}^x \frac{2w_B}{l} x\, dx$$

$$V - \frac{w_B l}{2} = \left[ w_B \frac{x^2}{l} \right]_{l/2}^x$$

$$V(x) = w_B \frac{x^2}{l} + w_B \frac{l}{4}$$

$$dM = V(x)\, dx$$

$$\int_{-w_B(l^2/8)}^M dM = -\int_{l/2}^x V(x)\, dx$$

$$M + w_B \frac{l^2}{8} = -\left[ w_B \frac{x^3}{3l} + w_B \frac{l}{4} x \right]_{l/2}^x$$

$$M(x) = -w_B \frac{x^3}{3l} - w_B \frac{l}{4} x + w_B \frac{l^2}{24}$$

$V(x)$ is parabolic with increasingly positive slope, and $M(x)$ is cubic with increasingly negative slope. These results agree with those found above.

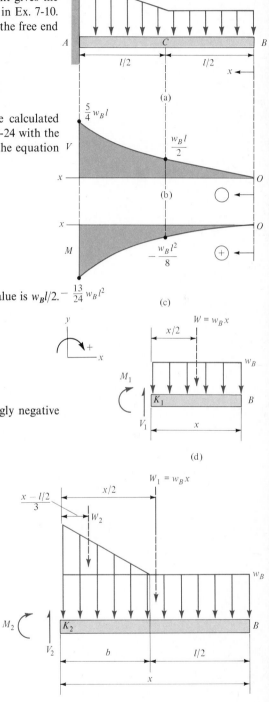

**FIGURE 7-23**

Shear and bending moment diagrams of a cantilever beam

EXAMPLE 7-13   297

# EXAMPLE 7-14

Determine the shear $V$ and bending moment $M$ as functions of $x$ for the simply supported beam shown in Fig. 7-24a. Draw the shear and bending moment diagrams.

## SOLUTION

Before a general section can be considered the support reactions must be determined. However, the total weight of the distributed load must be calculated before the support reactions can be solved. It is easiest to consider the load as a composite area as shown in Fig. 7-24d.

$$W = W_1 + W_2 = (400 \text{ lb/ft})(8 \text{ ft}) + \frac{1}{2}(800 - 400) \text{ lb/ft}(8 \text{ ft})$$

$$= 3200 \text{ lb} + 1600 \text{ lb} = 4800 \text{ lb}$$

**FIGURE 7-24**

Shear and bending moment diagrams of a simple supported beam

From symmetry,

$$A_y = B_y = \frac{1}{2} W = 2400 \text{ lb}$$

A general section $K$ at a position $x$, for $0 \le x \le 4$ ft, is considered, and part $AK$ is then drawn (Fig. 7-24e). The loads $W'_1$ and $W'_2$ on this section must be found first as functions of $x$. For the rectangular distribution, $W'_1 = 400x$ lb. The load $W'_2$ is the area of a triangle ($\frac{1}{2}bh$) with $b = x$ and

$$h = \frac{\Delta w}{\Delta x} x = \frac{(800 - 400) \text{ lb/ft}}{4 \text{ ft}} x$$

For equilibrium of part $AK$,

$$\sum F_y = 0$$

$$A_y - W'_1 - W'_2 - V = 0$$

$$2400 \text{ lb} - (400 \text{ lb/ft})x - \frac{1}{2}\left(\frac{400 \text{ lb/ft}}{4 \text{ ft}}\right)(x)(x) - V = 0$$

$$V = 2400 - 400x - 50x^2 \text{ (lb)}$$

$$\sum M_K = 0$$

$$-(A_y)(x) + W'_1\left(\frac{x}{2}\right) + W'_2\left(\frac{x}{3}\right) + M = 0$$

$$-2400(x) + 400\left(\frac{x^2}{2}\right) + 50\frac{x^3}{3} + M = 0$$

$$M = 2400x - 200x^2 - \frac{50}{3}x^3 \text{ (lb·ft)}$$

These equations are valid only for $0 \le x \le 4$ ft but due to the symmetry of loading only this half of the beam need be considered. The appropriate shear and bending moment equations are sketched in Fig. 7-24b and c.

## JUDGMENT OF THE RESULTS

A check to see that $V = dM/dx$ and $w(x) = -dV/dx$ lends confidence to the results. These relationships can be used further to check the curves directly as follows.

1. The values at the end points of the shear curve are not zero but are equal to the reactions at $A$ and $B$. The slope is not zero at either end point because $w(x)$ is not zero there.
2. The values at the end points of the bending moment diagram must be zero, since simple supports such as rollers cannot apply a moment. The slope is not zero, but equals the values of $V(x)$ at the end points.
3. Due to the symmetry of the loading, the maximum bending moment occurs at point $C$. Since this is a maximum, the slope of the bending moment curve at $C$ must be zero, so the value of the shear at $C$ is zero as shown on the diagrams.
4. Since $\mathbf{w}(x)$ acts downward on the top side of the beam, the slope of the shear curve is negative at all points—increasingly negative as $w(x)$ increases ($0 \le x \le 4$ ft), decreasingly negative as $w(x)$ decreases ($4$ ft $\le x \le 8$ ft).
5. The slope of the bending moment curve must be positive when $V(x)$ is positive ($0 \le x \le 4$ ft) and negative when $V(x)$ is negative ($4$ ft $\le x \le 8$ ft).

EXAMPLE 7-14    299

# EXAMPLE 7-15

Determine the shear $V$ and bending moment $M$ as functions of $x$ for the beam shown in Fig. 7-25a. Draw the shear and bending moment diagrams. $P = 2wa$, $M_O = 0.3Pa = 0.6wa^2$.

## SOLUTION

First the distributed load is replaced by its equivalent concentrated load $W$ at the centroid of the load area and the support reactions are calculated. From Fig. 7-25d,

$$\sum M_A = 0$$

$$-0.6wa^2 + 2wa(0.6a) + B_y a = 0$$

$$B_y = -0.6wa$$

$$\sum M_B = 0$$

$$-0.6wa^2 + 2wa(1.6a) + w(2a)(a) - A_y a = 0$$

$$A_y = 4.6wa \qquad \text{(for a check, } A_y + B_y = P + W\text{)}$$

**FIGURE 7-25**

Analysis of a beam under complex loading

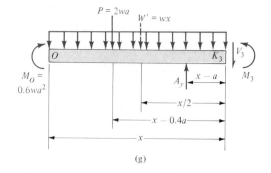

General sections between the concentrated forces are considered. The free-body diagrams are sketched and the equations of equilibrium are used to find the internal shear and bending moment as functions of $x$.

**REGION 1, $0 \leq x \leq 0.4a$ (Fig. 7-25e):**

$$\sum F_y = 0$$

$$-wx - V_1 = 0$$

$$V_1 = -wx$$

$$\sum M_{K_1} = 0$$

$$-0.6wa^2 + wx\left(\frac{x}{2}\right) + M_1 = 0$$

$$M_1 = 0.6wa^2 - w\left(\frac{x^2}{2}\right)$$

**REGION 2, $0.4a \leq x \leq a$ (Fig. 7-25f):**

$$\sum F_y = 0$$

$$-wx - 2wa - V_2 = 0$$

$$V_2 = -w(2a + x)$$

$$\sum M_{K_2} = 0$$

$$-0.6wa^2 + wx\left(\frac{x}{2}\right) + 2wa(x - 0.4a) + M_2 = 0$$

$$M_2 = -w\left(\frac{x^2}{2} + 2ax - 1.4a^2\right)$$

**REGION 3, $a \leq x \leq 2a$ (Fig. 7-25g):**

$$\sum F_y = 0$$

$$-2wa - wx + 4.6wa - V_3 = 0$$

$$V_3 = 2.6 - wx$$

$$\sum M_{K_3} = 0$$

$$-0.6wa^2 + 2wa(x - 0.4a) + wx\left(\frac{x}{2}\right) - 4.6wa(x - a) + M_3 = 0$$

$$M_3 = -w\left(\frac{x^2}{2} - 2.6ax + 3.2a^2\right)$$

The curves are drawn by plotting the shear and bending moment values at the end points of each interval for which each equation is valid and sketching a curve of appropriate order between the plotted points (Fig. 7-25b and c).

## JUDGMENT OF THE RESULTS

Several specific checks give one confidence in the results. The shear curve ends at $V = 0.6wa$, which is equal but opposite in sign to the reaction at $B$, as it should be. The moment curve goes to zero at the right end, as it should, since no moment can be applied by a pin. One further check would be to see if Eq. 7-26 holds. This is left as an exercise for the student.

EXAMPLE 7-15  301

# PROBLEMS

In Probs. 7–50 to 7–55, determine the reactions at the supports.

Fig. P7-50

Fig. P7-51

Fig. P7-52

Fig. P7-53

Fig. P7-54

Fig. P7-55

In Probs. 7–56 and 7–57, draw the shear and bending moment diagrams using the methods of calculus.

Fig. P7-56

Fig. P7-57

**7–58** A uniformly distributed force of 200-lb magnitude, caused by wind, is acting normal to the highway sign $BC$. Draw the bending moment diagram for the arm $AC$ assuming that the wind causes no force acting directly on part $AB$. Neglect weight.

**Fig. P7-58**

**7–59** The boom of a crane is modeled as a uniform beam of 120-kg mass. It is hinged at $A$ and supported by a wire at $B$. Draw the bending moment diagram for the beam.

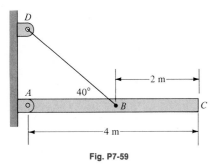

**Fig. P7-59**

**7–60** Solve Prob. 7–59 assuming that a concentrated vertical load of 3-kN magnitude is applied at point $C$.

**7–61** An airplane wing is modeled as a uniform beam with a linearly varying net upward force on it. Determine the shear $V$ and bending moment $M$ as functions of $x$.

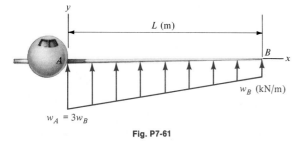

**Fig. P7-61**

**7–62** Draw the complete shear and bending moment diagrams for the wing in Prob. 7–61 if the weight $W = W_B L/8$ of the engine is applied at $2L/3$ (m).

**7–63** Draw the shear and bending moment diagrams for member $BF$ of the pin-connected frame.

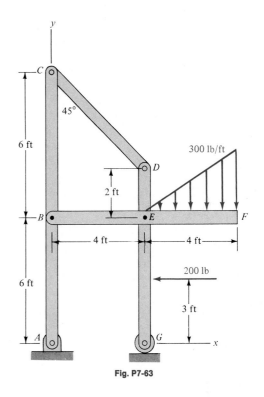

**Fig. P7-63**

**7–64** Determine the shear $V$ and bending moment $M$ as functions of $x$.

**Fig. P7-64**

**7–65** Draw the complete shear and bending moment diagrams for the beam in Fig. P7-64.

**7-66**    Draw the complete shear and bending moment diagrams for the beam.

**7-70**    Draw the shear and bending moment diagrams for the beam in Fig. P7-69.

**7-71**    Express the shear $V$ and bending moment $M$ as functions of $x$ for the region $AB$. $\mathbf{M}_B$ is an applied couple.

**7-67**    Express the shear $V$ and bending moment $M$ as functions of $x$.

**Fig. P7-71**

**7-72**    Draw the complete shear and bending moment diagrams for the beam in Prob. 7-71.

**7-73**    The simply supported beam's length is $l = 15$ ft, the distributed load has a maximum value of $w = 60$ lb/ft, and the concentrated force on the beam has a magnitude of $P = 1000$ lb. Draw the complete shear and bending moment diagrams for the beam.

**Fig. P7-67**

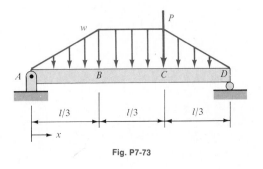

**7-68**    Draw the complete shear and bending moment diagrams for the beam in Fig. P7-67.

**7-69**    Determine the shear $V$ and bending moment $M$ as functions of $x$. $\mathbf{M}_A$ is an applied couple.

**Fig. P7-73**

The mechanics analysis of flexible cables with distributed loads is some-what complex, but it is important to understand the effects of self weight, ice, or wind on such members. The basic assumptions in the analysis are that a flexible cable has practically no resistance to bending and that the internal force at any point is tangent to the cable at that point. In the following discussions $\mathbf{w}(x)$ denotes a continuous, but possibly variable, distributed load applied to a cable. The units of $\mathbf{w}(x)$ are always force per unit length.

## General Considerations

Assuming that the whole cable is in equilibrium, an infinitesimal element of it, shown in Fig. 7-26, is also in equilibrium. Here $\mathbf{w}(x)\,dx$ is the resultant force of $\mathbf{w}(x)$ distributed over the distance $dx$ of the cable. On the basis of equilibrium in all directions, two independent equations can be written for the small element of the cable.

$$\sum F_x = 0$$
$$(T + dT) \cos (\theta + d\theta) - T \cos \theta = 0 \qquad \boxed{7\text{-}28}$$
$$\sum F_y = 0$$
$$(T + dT) \sin (\theta + d\theta) - T \sin \theta - w(x)\,dx = 0 \qquad \boxed{7\text{-}29}$$

These can be simplified using the trigonometric identities for the sum of two angles and the approximations $\sin d\theta = d\theta$ and $\cos d\theta = 1$ for the very small angle $d\theta$. The resulting two equations are, respectively,

$$d(T \cos \theta) = 0 \qquad \boxed{7\text{-}30}$$
$$d(T \sin \theta) = w(x)\,dx \qquad \boxed{7\text{-}31}$$

Equation 7-30 shows that the horizontal component of the tension is constant along the cable, as can be expected by inspection of the element. This component is often denoted by $T_o$. Using $T \cos \theta = T_o$, $\tan \theta = dy/dx$, and Eq. 7-31 allows one to write

$$\frac{d^2 y}{dx^2} = \frac{d}{dx}\left(\frac{\sin \theta}{\cos \theta}\right) = \frac{d}{dx}\left(\frac{T \sin \theta}{T_o}\right)$$

$$\boxed{\frac{d^2 y}{dx^2} = \frac{w(x)}{T_o}} \qquad \boxed{7\text{-}32}$$

FIGURE 7-26

Free-body element of a cable

The final result is the most useful differential equation of the cable. Integrating Eq. 7-32 gives the equation for the slope $dy/dx$ of the cable, and integrating the latter gives the equation for the position $y$ as a function of $x$. These results are illustrated for two important classes of cables in the following.

## Parabolic Cables

Sometimes interesting reversals are made in the modeling process. Clearly, a distributed load on a cable may be represented by numerous concentrated forces for all practical purposes. Conversely, it is advantageous to to substitute (in the analysis) a simple function of a distributed load for a large number of concentrated forces that would be cumbersome to use in detail. Common examples of such problems are suspension bridges (Fig. 6-23a).

In a suspension bridge the weight of the cable is negligible compared to the weight of the roadway. The latter is often a constant load $w$ per unit length measured horizontally in the $x$ direction. It is convenient to place the origin of the coordinate axes at the lowest point on the cable (Fig. 7-27a), which must be halfway between the supports if they are of the same height. Since $w$ and $T_o$ are constant, Eq. 7-32 is readily integrated to give

$$\frac{dy}{dx} = \frac{wx}{T_o} + C$$

With the slope of the cable equal to zero at the lowest point, $C$ is also zero, so the slope at any position $x$ is

$$\frac{dy}{dx} = \frac{wx}{T_o} \qquad \boxed{7\text{-}33}$$

Integration of Eq. 7-33 gives

$$y = \frac{wx^2}{2T_o} + C_1$$

where $C_1 = 0$ because $y = 0$ at $x = 0$. Thus,

$$\boxed{y = \frac{wx^2}{2T_o}} \qquad \boxed{7\text{-}34}$$

which is the equation of a parabola.

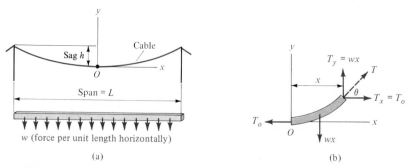

**FIGURE 7-27**

Parabolic cable of a suspension bridge

Designers are most interested in the magnitudes of the tension in the cable. A simple formula that gives the tension $T$ as a function of the horizontal position $x$ along the cable can be obtained with the aid of Fig. 7-27b. The segment of the cable to the right of the origin may be of any finite length $x$. The resultant of the distributed load on this segment is $wx$. Since $T_y = wx$,

$$T = \sqrt{T_o^2 + w^2x^2} \qquad \boxed{7\text{-}35}$$

Clearly, the minimum tension is at the lowest point of the cable ($x = 0$), and the maximum tension is at the supporting towers ($x = L/2$). Putting the values of $x = L/2$ and $y = h$ in Eq. 7-34,

$$T_{\min} = T_o = \frac{wL^2}{8h} \qquad \boxed{7\text{-}36}$$

Note that $T_o$ can be obtained by using Eq. 7-34 for any point on the curve except $x = y = 0$. The maximum tension is obtained from Eqs. 7-35 and 7-36,

$$T_{\max} = \sqrt{\left(\frac{wL^2}{8h}\right)^2 + w^2\left(\frac{L}{2}\right)^2} = \frac{wL}{2}\sqrt{1 + \frac{L^2}{16h^2}} \qquad \boxed{7\text{-}37}$$

The concepts and results discussed above are also relevant to unsymmetric cables where the supports are at different elevations, as in Fig. 7-28. Using the appropriate values of $x$ and $y$ for each support in Eqs. 7-34 and 7-35, the tensions at points $A$ and $B$ are

$$T_A = wa\sqrt{1 + \frac{a^2}{4h_A^2}} \quad \text{and} \quad T_B = wb\sqrt{1 + \frac{b^2}{4h_B^2}} \qquad \boxed{7\text{-}38}$$

The lengths $s_A$ or $s_B$ of the cable from the lowest point to the supports can be expressed as in elementary calculus for the length of a curve. For example, referring to Fig. 7-28,

$$s_A = \int_0^a \sqrt{1 + \left(\frac{dy}{dx}\right)^2}\, dx = \int_0^a \sqrt{1 + \frac{w^2x^2}{T_o^2}}\, dx$$

Substituting $T_o = wa^2/2h_A$ from Eq. 7-34,

$$s_A = \int_0^a \sqrt{1 + \frac{4h_A^2x^2}{a^4}}\, dx \qquad \boxed{7\text{-}39}$$

This is most easily evaluated by using the binomial theorem to get an infinite series whose first few terms can be integrated, resulting in

$$s_A = a\left(1 + \frac{2h_A^2}{3a^2} - \frac{2h_A^4}{5a^4} + \dots\right) \qquad \boxed{7\text{-}40}$$

This series converges for all values of $h_A/a < 0.5$, and usually the first two or three terms give a reasonable approximation to the cable length. The segment $s_B$ can be evaluated similarly.

$w$ (force per unit length horizontally)

**FIGURE 7-28**

Unsymmetric suspension cable

## Catenary Cables

Frequently, a cable has to support a weight that is uniformly distributed along the cable itself (self weight, ice). This is in contrast to the parabolic cable, whose own weight is ignored and which supports a load uniformly distributed in the horizontal direction, not along the cable. The shape of a cable under its own weight is called a *catenary*. For a tight cable (with a small ratio of sag to span) this shape is nearly the same as that of a parabolic cable and can be analyzed accordingly. More deeply sagging cables must be analyzed differently, as illustrated with the aid of Fig. 7-29.

In Fig. 7-29a the origin of the coordinate axes is placed at the lowest point on the cable, as before. A small part of the cable to the right of the origin is drawn in Fig. 7-29b. The length of this piece is $s$, so the resultant of the distributed load on it is $ws$. For equilibrium of any infinitesimal element $ds$ in the $y$ direction, Eq. 7-32 can be rewritten (after multiplying it by $dx/dx$) as

$$\frac{d^2y}{dx^2} = \frac{w}{T_o}\frac{ds}{dx} \qquad \boxed{7\text{-}41}$$

Since $s$ is a function of $x$ and $y$, $ds$ can be expressed in terms of $dx$ and $dy$, using $(ds)^2 = (dx)^2 + (dy)^2$. On this basis, Eq. 7-41 becomes

$w$ (force per unit length along cable)

(a)

(b)

**FIGURE 7-29**

Model for a catenary cable

$$\frac{d^2y}{dx^2} = \frac{w}{T_o} \sqrt{1 + \left(\frac{dy}{dx}\right)^2} \qquad \boxed{7\text{-}42}$$

which is the differential equation of the catenary curve. To solve this equation, let $dy/dx = f$. With this substitution,

$$\frac{df}{\sqrt{1 + f^2}} = \frac{w}{T_o} dx$$

This equation can be integrated at once, and

$$\ln(f + \sqrt{1 + f^2}) = \frac{w}{T_o} x + C_1 \qquad \boxed{7\text{-}43}$$

Since the slope $dy/dx = 0$ at $x = 0$, the constant $C_1 = 0$. Equation 7-43 can be further manipulated to advantage as follows:

$$f + \sqrt{1 + f^2} = e^{(w/T_o)x}$$

$$1 + f^2 = (e^{(w/T_o)x} - f)^2$$

$$= e^{2(w/T_o)x} - 2e^{(w/T_o)x}f + f^2$$

$$f = \frac{dy}{dx} = \frac{e^{(w/T_o)x} - e^{-(w/T_o)x}}{2} = \sinh \frac{wx}{T_o} \qquad \boxed{7\text{-}44}$$

The position $y$ of the cable is obtained by integrating Eq. 7-44,

$$y = \frac{T_o}{w} \cosh \frac{wx}{T_o} + C_2$$

Since $y = 0$ at $x = 0$, the constant $C_2 = -T_o/w$, so

$$y = \frac{T_o}{w} \left(\cosh \frac{wx}{T_o} - 1\right) \qquad \boxed{7\text{-}45}$$

This is the equation of the shape of the catenary curve.

The length $s$ of the cable in Fig. 7-29b (measured from point $O$) is obtained from Eq. 7-44. With $\tan \theta = dy/dx = ws/T_o$ from Fig. 7-29b,

$$s = \frac{T_o}{w} \sinh \frac{wx}{T_o} \qquad \boxed{7\text{-}46}$$

The tension $T$ in the cable is also obtained from Fig. 7-29b,

$$T^2 = T_x^2 + T_y^2 = T_o^2 + w^2 s^2$$

Using Eq. 7-46,

$$T^2 = T_o^2 \left(1 + \sinh^2 \frac{wx}{T_o}\right) = T_o^2 \cosh^2 \frac{wx}{T_o}$$

The result can be finally simplified using Eq. 7-45,

$$\boxed{T = T_o \cosh \frac{wx}{T_o} = T_o + wy} \qquad \boxed{7\text{-}47}$$

The equations for catenary cables given above can be solved graphically or numerically (where computers are helpful).

EXAMPLE 7-16

Determine the position of the cable shown in Fig. 7-30a at $x = 5$ m for two different loadings: **(a)** $w_1 = 100$ N/m = constant load per unit length measured horizontally, and **(b)** $w_2 = 100$ N/m = constant load per unit length measured along the cable. The sag $h = 5$ m and the span $L = 20$ m are the same for both loadings. Sketch the two cable shapes on the same diagram using the given and calculated few positions. Calculate the minimum and maximum tensions for both loadings.

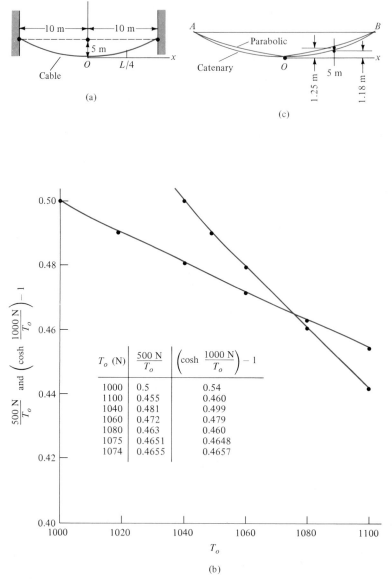

**FIGURE 7-30**

Cable shapes under two different loadings

# SOLUTION

(a) The cable is parabolic and Eqs. 7-33 to 7-40 apply. The minimum tension occurs at the midpoint of the cable. From Eq. 7-36,

$$T_o = \frac{wL^2}{8h} = \frac{(100 \text{ N/m})(20 \text{ m})^2}{8(5 \text{ m})} = 1000 \text{ N} \quad \text{(at midpoint)}$$

The maximum tension occurs at the supports (at $x = 10$ m). From Eq. 7-37,

$$T_{max} = \frac{wL}{2} \sqrt{1 + \frac{L^2}{16h^2}}$$

$$= \frac{(100 \text{ N/m})(20 \text{ m})}{2} \sqrt{1 + \frac{(20 \text{ m})^2}{16(5 \text{ m})^2}} = 1410 \text{ N}$$

The position $y$ at $x = 5$ m is obtained using Eq. 7-34,

$$y = \frac{wx^2}{2T_o} = \frac{(100 \text{ N/m})(5 \text{ m})^2}{2(1000 \text{ N})} = 1.25 \text{ m}$$

(b) The cable makes a catenary curve and Eqs. 7-41 to 7-47 apply. By knowing the coordinates of any point on the cable, Eq. 7-45 can be used to evaluate the magnitude of $T_o$. The point $(0, 0)$ is known but is of no help since Eq. 7-45 yields the trivial result that $0 = 0$ for this point. The point $(5, 10)$, however, does allow for determination of $T_o$, but it can only be solved by trial and error. Solution is facilitated by rearranging Eq. 7-45 as shown below and solving graphically in Fig. 7-30b.

$$\frac{yw}{T_o} = \left( \cosh \frac{wx}{T_o} \right) - 1 = \frac{(5 \text{ m})(100 \text{ N/m})}{T_o}$$

$$= \left( \cosh \frac{(100 \text{ N/m})(10 \text{ m})}{T_o} \right) - 1$$

$$\frac{500 \text{ N}}{T_o} = \left( \cosh \frac{1000 \text{ N}}{T_o} \right) - 1, \quad T_o = 1075 \text{ N}$$

From Eq. 7-47, $T = T_o + wy$.

Therefore,

$$T_{min} = T_o = 1075 \text{ N at } y = 0 \quad \text{(midpoint of cable)}$$

and

$$T_{max} = 1075 + \left( 100 \frac{\text{N}}{\text{m}} \right)(5 \text{ m}) = 1575 \text{ N} \quad \text{(at support)}$$

The height at $x = 5$ m is obtained from Eq. 7-45,

$$y = \frac{1075 \text{ N}}{100 \text{ N/m}} \left( \cosh \frac{(100 \text{ N/m})(5 \text{ m})}{1075 \text{ N}} - 1 \right) = 1.18 \text{ m}$$

Sketches of the cables are shown in Fig. 7-30c.

It is seen in this case that the maximum tensions in parabolic and catenary cables do not differ much. However, cables with larger sag ($h/L$) ratios or larger distributed loads will have larger differences between the maximum tensions in parabolic and catenary cables under similar loading.

EXAMPLE 7-16   311

**7-74** The cable supports a load of magnitude $w = 1$ kN/m (uniformly distributed horizontally). Calculate the tensions in the cable at the supports if $L = 50$ m and $h = 5$ m.

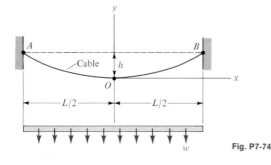

**Fig. P7-74**

**7-75** Determine the maximum tension in the cable of Fig. P7-74 and the position of the cable at $x = L/4$ if $w = 300$ lb/ft (uniformly distributed horizontally), $L = 200$ ft, and $h = 15$ ft.

**7-76** Determine the tensions at points $A$, $O$, and $B$ in the cable. Let $w = 200$ lb/ft (uniformly distributed horizontally), $a = 400$ ft, $b = 150$ ft, and $h_B = 10$ ft.

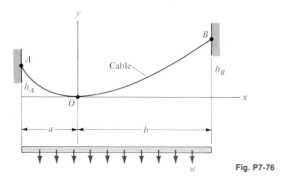

**Fig. P7-76**

**7-77** Plot the maximum tension in Prob. 7–76 as a function of $h_B = 10$ ft, 20 ft, and 30 ft.

**7-78** The cable in Prob. 7–76 supports a load of magnitude $w = 2$ kN/m (uniformly distributed horizontally). Determine the tensions in the cable at the supports if $a = 40$ m, $b = 60$ m, and $h_A = 5$ m.

**7-79** The cable supports a load of magnitude $w = 10$ N/m uniformly distributed along the cable. Calculate the tensions in the cable at the supports if $L = 70$ m and $h = 8$ m.

**Fig. P7-79**

**7-80** Determine the maximum and minimum tensions in the cable in Fig. P7-79. Also calculate the position of the cable at $x = L/4$. Let $w = 2$ lb/ft distributed uniformly along the cable, $L = 200$ ft, and $h = 18$ ft.

**7-81** Plot the maximum tension in Prob. 7–79 as a function of $h = 8$ m, 10 m, and 12 m.

**7-82** Calculate the maximum and minimum tensions in the cable. Let $w = 0.7$ lb/ft uniformly distributed along the cable, $a = 100$ ft, $b = 350$ ft, and $h_A = 15$ ft.

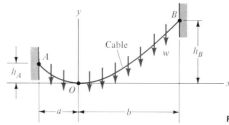

**Fig. P7-82**

**7-83** Plot the maximum tension in Prob. 7–82 for $h_A = 10$ ft, 15 ft, and 20 ft.

**7-84** Determine the maximum tension in the cable in Prob. 7–82 if $w = 5$ N/m uniformly distributed along the cable, $a = 30$ m, $b = 60$ m, and $h_A = 4$ m.

**7-85** A proposed pipeline must cross a ravine and is to be supported by a cable. The pipe weighs $w = 5$ kN/m (uniformly distributed horizontally). Calculate the maximum tension in the cable $AB$.

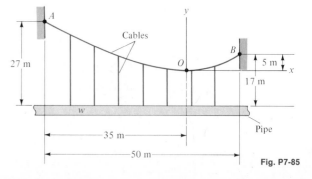

**Fig. P7-85**

A common problem area involving distributed forces is fluid mechanics. Fluids are classified as *liquids* and *gases*, or, on the basis of relative constancy of volume under the action of forces, as *incompressible* and *compressible fluids*. An important characteristic of all fluids is that at rest they can support only normal forces and not shear forces. This is the basis for Pascal's law, according to which *the pressure is the same in all directions at any given point in a fluid at rest*. Pressure is defined as the force per unit area, $p = \text{force/area}$, and it always implies compression in fluid statics. Pascal's law is intuitively obvious to many people (the pressure on the ear in deep water is independent of body orientation).

To illustrate Pascal's law, consider an infinitesimal element of fluid. The rectangle in Fig. 7-31a represents a three-dimensional element of dimensions $dx$, $dy$, $dz$, and weight $dW$. Pressures $p_1, \ldots$ act on the element, while it is assumed that no shear forces are acting on it. For equilibrium in the $x$ direction,

$$\sum F_x = 0$$

$$p_1 \underbrace{dy\, dz}_{\text{(pressure)(area)}} - p_2\, dy\, dz = 0, \quad \text{so } p_1 = p_2$$

$$\underbrace{\phantom{p_1\, dy\, dz}}_{\text{force}}$$

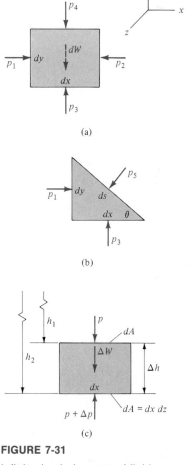

A similar statement can be made for the pressures acting in the $z$ direction. In the $y$ direction,

$$\sum F_y = 0$$

$$p_3\, dx\, dz - p_4\, dx\, dz - dW = 0$$

where $dW = \gamma\, dx\, dy\, dz$
$\gamma =$ specific weight (weight per unit volume)

Here the weight term $dW$ can be neglected since it is of higher order and approaches zero more rapidly than the terms involving pressures. Thus, $p_3 = p_4$.

To show that the pressure is the same on nonparallel planes, consider a wedge from the rectangular element in Fig. 7-31a. The pressures $p_1$, $p_3$, and $p_5$ in Fig. 7-31b are potentially different from one another at this stage in the analysis. For equilibrium in the $x$ and $y$ directions,

$$\sum F_x = 0$$

$$p_1\, dy\, dz - p_5\, ds \sin \theta\, dz = 0$$

$$\sum F_y = 0$$

$$p_3\, dx\, dz - p_5\, ds \cos \theta\, dz = 0$$

Since $ds \sin \theta = dy$, and $ds \cos \theta = dx$, $p_1 = p_5 = p_3 = p$, where $p$ is the pressure in any direction at the point considered.

For an important development based on the concepts discussed above, consider an element with an infinitesimal cross-sectional area ($dA = dx\, dz$) and a small height $\Delta h$ that makes the element's weight $\Delta W$

**FIGURE 7-31**

Infinitesimal elements of fluid

**FIGURE 7-32**

Pressure distribution in a liquid with a free surface

significant (Fig. 7-31c). In this case, for equilibrium of the element in the $y$ direction,

$$-p\,dx\,dz + (p + \Delta p)\,dx\,dz - \Delta W = 0$$

Finally, using $\Delta W = \gamma\,dA\,\Delta h$,

$$\boxed{\Delta p = \gamma\,\Delta h} \qquad \boxed{7\text{-}48}$$

This shows that the pressure varies with the depth $h$ in the fluid (downward $h$ is positive). The equation is valid for liquids and gases.

When $\gamma$ is a constant (as in liquids, which are essentially incompressible for most purposes), the reference for pressure and depth measurement is often taken at the open surface of the liquid, where the pressure is caused by the atmosphere above that surface. The pressure at such a surface is 1 atmosphere and the depth $h$ is zero. At any other depth the measured pressure shows only the increment above atmospheric pressure. This increment in pressure is called *gage pressure*. With these in mind, Eq. 7-48 can be written simply as

$$p = \gamma h \qquad \boxed{7\text{-}49}$$

where $p$ is the gage pressure and $h$ is the depth from the open surface of the liquid to the point where $p$ is measured. This is depicted in Fig. 7-32, where the horizontal lines show how the pressure (magnitude only) varies linearly with the depth $h$. The small triangle denotes the free surface of the liquid.

## Hydrostatic Forces on Plane Surfaces

It is frequently necessary to know the magnitude, direction, and point of action of the resultant force of *hydrostatic pressure* on a solid surface submerged in a liquid (the term *aerostatics* is used when a gas is involved). For example, consider a two-piece wave board (Fig. 3-1) in a stationary position shown in Fig. 7-33a. The pressure $p$ increases linearly to the bottom of the channel. The force of the water on each board is analyzed individually, keeping in mind the linear variation in pressure and that the pressure acts normal to any surface.

The upper board is drawn in Fig. 7-33b with the triangular pressure distribution; the reactions $\mathbf{R}_B$ and $\mathbf{M}_B$ at joint $B$ complete the free-body diagram. $\mathbf{P}_B$ is the resultant force caused by the distributed pressure; its magnitude equals the average pressure ($p_B/2$) times the total submerged area of board $BC$. Assume a board of constant width $b$. Using Eq. 7-49 and an elemental area $dA = b\,dh$,

$$P_B = \int p\,dA = \int_0^{l_B} \gamma h b\,dh$$

$$= \gamma b \frac{l_B^2}{2} = \underbrace{\gamma\left(\frac{l_B}{2}\right)}_{\substack{\text{average pressure,} \\ \text{occurs at } l_B/2}} l_B b$$

Note that a constant pressure $p$ is acting on the elemental strip of area $dA$.

The resultant force $\mathbf{P}_B$ acts at a distance $\bar{h}$ below the surface. This distance is determined as for a beam with distributed loading, by writing the moments about a point in two different ways and equating them. Working with point $C$ in Fig. 7-33b,

$$P_B\bar{h} = \int hp\, dA, \quad \text{so} \quad \bar{h} = \frac{\int hp\, dA}{\int p\, dA} \qquad \boxed{7\text{-}50}$$

After reviewing Eq. 7-11, it is found that $\mathbf{P}_B$ acts at the centroid of the pressure distribution, called the *center of pressure* (not at the centroid of the plate), which simplifies the calculations conveniently. For the triangular distribution, $\bar{h} = \frac{2}{3}l_B$.

A similar procedure is applied to the lower board with the aid of Fig. 7-33c, but this board has a more complex trapezoidal pressure distribution on it. For convenience, divide the area of the arrows into a rectangular distribution with the resultant force $\mathbf{P}_{A_1}$ acting halfway between points $A$ and $B$, and a triangular distribution with the resultant force $\mathbf{P}_{A_2}$ acting at $l_A/3$ from point $A$. Thus, the magnitudes of the resultant forces are

$$P_{A_1} = P_B (\text{area of board } AB)$$

$$P_{A_2} = \frac{1}{2}(p_A - p_B)(\text{area of board } AB)$$

$$P_A = P_{A_1} + P_{A_2}$$

The location $\bar{d}$ where $\mathbf{P}_A$ is acting is found by taking moments about a convenient point such as $B$,

$$P_A\bar{d} = P_{A_1}\frac{l_A}{2} + P_{A_2}\frac{2l_A}{3}$$

where $\bar{d}$ is the only unknown.

At this stage it is useful to restate and summarize several aspects of forces caused by hydrostatic pressure on vertical or inclined plane surfaces. The following are true for any plane area and are worth remembering:

**1.** *The magnitude of the resultant force* of hydrostatic pressure on a plane surface is equal to the area of the surface times the pressure at the centroid of the surface area. This force is independent of the inclination of the plane if the centroid of the plane area remains at a constant depth.

**2.** *The center of pressure* is where the resultant force of hydrostatic pressure is acting on the plane area. This point is the perpendicular projection of the centroid of the pressure distribution on the plane area. The center of pressure is always below the centroid of the area, and its location is independent of the specific weight of the liquid and of the inclination of the plane at a given depth.

## Hydrostatic Forces on Curved Surfaces

An algebraic summation of elemental hydrostatic forces on a curved surface is impractical because the direction of pressure is not constant on

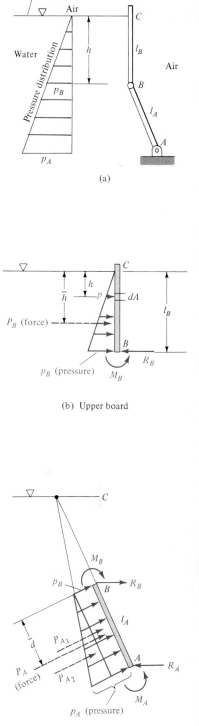

(a)

(b) Upper board

(c) Lower board

**FIGURE 7-33**

Pressure distribution on a two-piece wave board

**FIGURE 7-34**

Hydrostatic forces on a curved surface $S$

(a)

(b)

the whole area. Fortunately, the resultant forces in the horizontal and vertical directions can be relatively easily determined even for curved surfaces. For example, consider the surface $S$ in Fig. 7-34a. This area is projected as area $S'$ on an arbitrary vertical plane $V$. The pressure distribution and the resultant force $F'_x$ on the imaginary area $S'$ can be determined exactly as for any plane area. For horizontal equilibrium of the imaginary volume represented by $AA'BB'$, $F_x = F'_x$ with identical lines of action. Thus, *the horizontal force on a submerged surface of any shape is equal to the force on a projected vertical area at the same depth.*

The vertical force on a submerged surface of any shape is equal to the weight of the volume of fluid above that surface, extending to the free surface. The line of action of this force ($\mathbf{F}_y$ in Fig. 7-34b) must pass through the centroid $C$ of the volume represented by $AA''BB''$.

## BUOYANCY

The theory of flotation of bodies was first formulated by Archimedes. A simple statement of Archimedes' principle is that a submerged body is subjected to an upward buoyant force equal to the weight of the displaced fluid. The buoyant force $\mathbf{B}$ acts at the *center of buoyancy* (C.B. in Fig. 7-35), which is the center of gravity of the displaced fluid. Note that the center of gravity of the body (C.G. in Fig. 7-35) does not necessarily coincide with the center of buoyancy. The relative magnitudes and points of action of the weight $\mathbf{W}$ and the buoyant force $\mathbf{B}$ affect the position and stability of the submerged body in the following distinct ways.

(a) Stable position

(b) Unstable position

C.G. = center of gravity of solid body
C.B. = center of buoyancy
= center of gravity of fluid displaced by the solid body

**FIGURE 7-35**

Center of buoyancy

**1.** $B > W$: the body rises in the fluid.

**2.** $B = W$: the body floats in equilibrium.

**3.** $B < W$: the body sinks in the fluid.

**4.** C.B. is above C.G. (Fig. 7-35a): stable position with respect to rotation because $\mathbf{B}$ and $\mathbf{W}$ form a couple that tends to bring C.B. directly above C.G.; an increase in the angle $\theta$ increases the moment of the *righting couple* of $B$ and $W$, and this tends to return the body to its original position.

**5.** C.B. is below C.G. (Fig. 7-35b): unstable position with respect to rotation because $\mathbf{B}$ and $\mathbf{W}$ form a couple that tends to increase the angle $\theta$; as a result, the body overturns until C.B. is above C.G., where it becomes stable.

EXAMPLE 7-17

Determine the horizontal and vertical hydrostatic forces (magnitudes and locations) on each part of the two-piece wave board shown in Fig. 7-36a. The wave board is 6 m wide, and the density of water is $10^3$ kg/m$^3$.

SOLUTION

Since the density of the water is given in SI units, the specific weight must first be calculated by multiplying the density by the acceleration due to gravity,

$$\gamma = \rho g$$
$$= (10^3 \text{ kg/m}^3)(9.81 \text{ m/s}^2)$$
$$= 9.81 \text{ kN/m}^3$$

Top board: The maximum pressure, at $B$, is calculated first (Fig. 7-36b).

$$p_B = \gamma h_B = (9.81 \text{ kN/m}^3)(2.5\text{m} \sin 50°) = 18.78 \text{ kN/m}^2$$

The resultant force of the distributed load is equal to the area of the pressure distribution times the width of the plate,

$$P_B = \frac{1}{2}\left(18.78 \frac{\text{kN}}{\text{m}^2}\right)(2.5 \text{ m})(6 \text{ m}) = 141 \text{ kN}$$

Therefore,

$$P_{B_H} = 141 \text{ kN} \cos 40° = 108 \text{ kN}$$
$$P_{B_V} = -141 \text{ kN} \sin 40° = -91 \text{ kN}$$

The location of this resultant force is at $\frac{1}{3}$ the distance along the board:

$$\bar{h} = \frac{BC}{3} = \frac{2.5 \text{ m}}{3} = 0.83 \text{ m}$$

Bottom board: The pressure distribution is trapezoidal as shown in Fig. 7-36c. The force caused by the pressure is found by considering the pressure distribution in the two parts shown.

$$p_A = \gamma h_A = (9.81 \text{ kN/m}^3)(4.4 \text{ m}) = 43.2 \text{ kN/m}^2$$
$$P_{A_1} = \rho_B(A) = (18.78 \text{ kN/m}^2)(2.5 \text{ m})(6 \text{ m}) = 282 \text{ kN}$$
$$P_{A_2} = \frac{1}{2}(p_A - p_B)(A) = \frac{1}{2}(43.2 - 18.78)\frac{\text{kN}}{\text{m}^2}(2.5 \text{ m})(6 \text{ m})$$
$$= 183 \text{ kN}$$
$$P_A = P_{A_1} + P_{A_2} = 282 \text{ kN} + 183 \text{ kN} = 465 \text{ kN}$$

Therefore,

$$P_{A_H} = 465 \text{ kN} \cos 10° = 458 \text{ kN}$$
$$P_{A_V} = -465 \text{ kN} \sin 10° = -80.7 \text{ kN}$$

The location of this resultant force is at the centroid of the trapezoid,

$$\bar{d} = \frac{P_{A_1}\bar{h}_1 + P_{A_2}\bar{h}_2}{P_{A_1} + P_{A_2}}$$
$$= \frac{(282 \text{ kN})(1.25 \text{ m}) + (183 \text{ kN})(0.83 \text{ m})}{282 \text{ kN} + 183 \text{ kN}} = 1.08 \text{ m}$$

(a)

(b)

(c)

**FIGURE 7-36**

Hydrostatic forces on a two-piece wave board

EXAMPLE 7-17   317

# EXAMPLE 7-18

The buoy shown in Fig. 7-37a is a circular cylinder 5 ft in diameter and 6 ft in length. It is made of steel with a uniform wall thickness of 0.4 in. What is the tension in the cable at the bottom of the buoy if a 4-ft-long section of the buoy is submerged? Assume that no ship is tied at the buoy. Steel weighs 0.283 lb/in$^3$.

## SOLUTION

The free-body diagram of the buoy is drawn in Fig. 7-37c. Since the buoy is in equilibrium,

$$\sum F_y = 0$$
$$-W + B - T = 0$$

where $W = \gamma_{st} V_B = \gamma_{st}(2\pi rth + 2\pi r^2 t)$
$$= (0.283 \text{ lb/in}^3)[2\pi(30 \text{ in.})(0.4 \text{ in.})(72 \text{ in.}) + \pi(30 \text{ in.})^2(0.4 \text{ in.})2]$$
$$= 2180 \text{ lb}$$

$B$ = buoyant force = weight of water displaced
$$= \gamma_w V_w = \gamma_w(\pi r^2 h) = (62.4 \text{ lb/ft}^3)[\pi(2.5 \text{ ft})^2](4 \text{ ft})$$
$$= 4900 \text{ lb}$$

Therefore,

$$T = B - W = (4900 - 2180) \text{ lb} = 2720 \text{ lb}$$

## JUDGMENT OF THE RESULTS

The large magnitude of the results might cause one to reconsider the calculations. These magnitudes are reasonable, however, due to the large volume of water displaced by a relatively small amount of steel. A large tension is needed to maintain the buoy in this position.

(a)

(b)

(c)

**FIGURE 7-37**

Analysis of an anchored buoy

**7–86**  Consider a single-piece wave board in a stationary position (see Sec. 3-1 for an introduction to these devices). Determine the magnitudes and locations of the horizontal and vertical hydrostatic forces acting on the 3-m-wide board if $\theta = 70°$, $h = 2$ m, and the density $\rho = 10^3$ kg/m³.

Fig. P7-86

**7–87**  Plot the magnitude of the required supporting moment $\mathbf{M}_A$ in Prob. 7–8 as functions of $\theta = 50°$, $70°$, and $90°$.

**7–88**  A test setup for investigating flow in channels includes a rigid wall $AB$ on which $CD$ represents a square window. Determine the resultant force acting on the window from the water.

Fig. P7-88

**7–89**  The square plate is a model for a gate under water. Determine the magnitude and location of the hydrostatic force acting on one side of the plate if there is air on the other side. $\gamma = 62.4$ lb/ft³.

Fig. P7-89

**7–90**  Determine the magnitude of the hydrostatic force on the circular window of a small submarine if $h = 50$ m and the density $\rho = 10^3$ kg/m³.

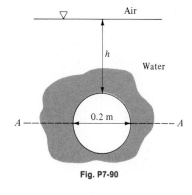

Fig. P7-90

**7–91**  Plot the distance $d$ of the center of pressure from the centroidal line $AA$ of the window in Prob. 7–90. Let $h = 0$, 50 m, and 100 m.

**7-92** Assume that the stationary rudder *ABCD* is under water. Determine the horizontal hydrostatic force and the center of pressure on the rudder. Discuss the possible effect of this force on (a) a solid metal rudder, and (b) a hollow, lightweight rudder. Salt water weighs 64 lb/ft³.

(a)

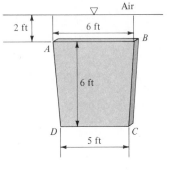

(b) Symmetric rudder     **Fig. P7-92**

**7-93** Consider a proposed curved wave board with a 10-m length in the *z* direction. Determine the magnitudes and locations of the horizontal and vertical hydrostatic forces on the board if the density $\rho = 10^3$ kg/m³.

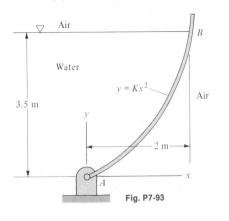

**Fig. P7-93**

**7-94** Estimate the forces at the hinged joints *A*, *B*, and *C* in the framed dam. There is a strut *BC* every 5 ft along the dam. Water weighs 62.4 lb/ft³.

**Fig. P7-94**

**7-95** The cross section of a proposed *gravity dam* is approximated by a triangle *ABC*. The *hydrostatic uplift* between *A* and *B* is caused by water seeping under the dam. The stability of the gravity dam depends on its weight (for concrete, $\rho = 2.4 \times 10^3$ kg/m³) and on the horizontal and vertical hydrostatic forces; the stability is independent of the total length of the dam. Is there any danger of this dam overturning? Water has a density of $10^3$ kg/m³.

**Fig. P7-95**

**7-96** An aquarium has a window *ABCD* of dimensions $AB = 1$ m, $BC = 0.5$ m, $CD = 0.6$ m, and a width (perpendicular to the plane of the paper) of 0.8 m. Determine the resultant force acting on the window from the water. $\rho = 10^3$ kg/m³.

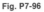

Fig. P7-96

**7–97** The gate of a spillway is pivoted at point $A$. Determine the total force and moment at $A$ for a 5-m-wide gate. $\rho = 10^3$ kg/m$^3$.

Fig. P7-97

**7–98** A balloon is designed to carry a load of 200 lb using no more than 20,000 ft$^3$ of helium. What is the maximum allowable magnitude of the balloon's own weight? Helium weighs 0.0112 lb/ft$^3$ and air weighs 0.0807 lb/ft$^3$ at the same pressure and temperature.

Fig. P7-98

**7–99** Consider solid cylinders whose specific gravity is half of that of water. Discuss why a short cylinder (a disk with $d \gg h$) is stable in water in the position shown, while a long rod (with $h \gg d$) is unstable in the same position. Assume small displacements from the equilibrium position where the cylinder's axis is vertical.

Fig. P7-99

**7–100** A small raft is designed with a square platform 6 ft × 6 ft in size. It has three buoyant elements, wooden boards 2 in. × 12 in. in cross section, 6 ft long, and weighing 0.02 lb/in$^3$. Determine the distance $d$ of the platform above the water.

Fig. P7-100

**7–101** Determine $h$ for the solid hemisphere shown in Fig. P7-101a. The specific weight of the solid is half of that of water.

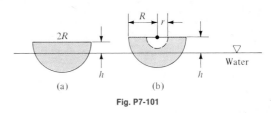

Fig. P7-101

**7–102** Determine $h$ for the hemispherical shell shown in Fig. P7-101b if $r = 0.6R$. The specific weight of the solid is half of that of water.

**7–103** Plot $h$ as a function of $r = 0$, $0.3R$, $0.6R$, and $0.9R$ for the solid in Fig. P7-101. $R$ is the same radius in each case, and the specific weight of the solid is half of that of water. Estimate $r$ to maximize h.

**7–104** A deep-sea drill rod is assembled from uniform pieces of steel pipe that has an 80-mm inside diameter and a 100-mm outside diameter. There is seawater ($\rho = 1019$ kg/m$^2$) inside and outside the pipe. Determine the force that supports the hanging pipe at the ship. Assume that the drill is not touching bottom at a depth of 5000 m, the length of the pipe. Steel has a density of 7830 kg/m$^3$.

**7–105** The thin-walled hulls of a catamaran are approximately triangular in cross section and 3 m in length. Assume that $d = h/2 = 150$ mm and $w = 200$ mm. Ignore any changes in the cross-sectional shape and area along the hull. Calculate the magnitude of the force **P**.

(a)

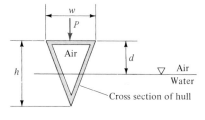

Fig. P7-105

**7–106** The concrete flotation chambers of the floating dock are 5 ft long, 1 ft high, and 2.5 ft wide. The uniform wall thickness is $t = 1$ in. of each box. What is the magnitude of the external force **P** that allows the box to extend 3 in. above the water? Concrete weighs 150 lb/ft$^3$, and seawater weighs 64 lb/ft$^3$.

(a) Floating dock

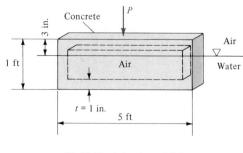

(b) Width of chamber = 2.5 ft

Fig. P7-106

In solid materials there are internal distributed forces somewhat analogous to the pressure in a fluid. The main difference is that a solid element may have normal and shear forces acting on it as shown in Fig. 7-38, while only normal forces act on a fluid at rest. Internal normal and shear forces can be generated by axial tension or compression, bending, and torsion of solid materials. These are studied in detail in the area of *mechanics of materials* (also called *strength of materials*). A brief introduction to this subject is given here to show the basic concepts and their similarities or differences to other kinds of distributed forces.

The magnitudes of the internal forces are of great practical interest because these forces can be directly related to the strength of any solid. This relationship is important in design to prevent failures of load-carrying members. The best way to determine the extreme values of internal forces is to calculate their local intensities.

The intensity of an internal force is called *stress* and is defined as the force per unit area. The common symbol for normal stress is the lowercase Greek letter $\sigma$, and for shear stress the lowercase Greek letter $\tau$. They are calculated by dividing the total force acting on an area by the size of that area,

$$\sigma = \frac{F_{normal}}{A} \quad \text{and} \quad \tau = \frac{F_{shear}}{A} \qquad \boxed{7\text{-}51}$$

The units for both $\sigma$ and $\tau$ are pascals (Pa = N/m$^2$) or pounds per square inch (psi).

The stresses calculated from Eq. 7-51 are the average stresses on the given area. The magnitudes of the stresses can vary from position to position in a member even if the external loading is constant. Thus, a more satisfactory definition of stress requires working with an elemental load $d\mathbf{F}$ acting on an elemental area $dA$,

$$\sigma = \frac{dF_{normal}}{dA} \quad \text{and} \quad \tau = \frac{dF_{shear}}{dA} \qquad \boxed{7\text{-}52}$$

If the functional relation for the variation of stress is known, the total force on an area is calculated by integration of Eq. 7-52,

$$F_{normal} = \int \sigma \, dA \quad \text{and} \quad F_{shear} = \int \tau \, dA \qquad \boxed{7\text{-}53}$$

In words, the total force acting on an area is equal to the stress integrated over the area it acts upon. It should be emphasized in this regard that *force and stress are not the same* even though they are directly related.

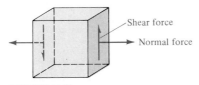

**FIGURE 7-38**
Element of a solid body

Normal and shear stresses on an infinitesimal element are commonly shown as in Fig. 7-39, where each arrow represents a distributed load. Three-dimensional states of stress (which have stresses acting in the $x$, $y$, and $z$ directions) frequently exist, but the two-dimensional case (called plane stress) is the most illustrative at this stage. The notations of the normal stresses (which may be tensile or compressive) are self-explanatory. For shear stress notation, the first subscript denotes the face on which the stress is acting: the subscript shows the coordinate axis normal to the face. The second subscript denotes the direction of the shear stress. For example, $\tau_{xy}$ is acting on a face normal to the $x$ axis, and in the $y$ direction (positive or negative).

There are several aspects of the magnitudes and directions of the stresses on a given element that are noteworthy, assuming a thickness of unity for the element. Also assume that $\sigma$ and $\tau$ may act simultaneously, or either one may act without the other. From the equilibrium of forces in Fig. 7-39,

$\sigma_x$ on face 1 is equal and opposite to $\sigma_x$ on face 3

$\sigma_y$ on face 2 is equal and opposite to $\sigma_y$ on face 4

$\tau_{xy}$ on face 1 is equal and opposite to $\tau_{xy}$ on face 3

$\tau_{yx}$ on face 2 is equal and opposite to $\tau_{yx}$ on face 4

From the equilibrium of moments taken about any corner of the element in Fig. 7-39, $\tau_{xy} = \tau_{yx}$ in magnitude, and they cause opposite moments (this is true even when $dx \neq dy$). When considering equilibrium of internal stresses, it is important to remember that *equilibrium equations can be written only for forces and moments of forces*, and not for stresses directly. For example, take moments about point $A$ in Fig. 7-39: the normal forces cancel each other, two shear forces have zero moment arms, so only the shear forces on faces 1 and 4 cause moments,

$$\sum M_A = \underbrace{\tau_{xy}\underbrace{(dy)(1)}_{\text{area}} \underbrace{(dx)}_{\text{moment arm}}}_{\text{force}} - \underbrace{\tau_{yx}\underbrace{(dx)(1)}_{\text{area}} \underbrace{(dy)}_{\text{moment arm}}}_{\text{force}} = 0 \qquad \boxed{7\text{-}54}$$

from which $\tau_{xy} = \tau_{yx}$.

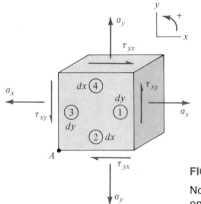

**FIGURE 7-39**

Normal stresses and shear stresses on an infinitesimal element

The equilibrium equations for forces based on known stresses are useful to determine the stresses on other planes in a given element. For example, consider the element in Fig. 7-40a, where the stresses on plane $AB$ must be found. Next draw only the triangular element $ABC$ on which the unknown stresses $\sigma_\theta$ and $\tau_\theta$ are acting. Note that $\sigma_x$ and $\tau_{xy}$ are redrawn with the original magnitudes since the concept of stress is independent of the area in a given small neighborhood. For convenience, assume that the plane $AB$ has an area of unity, and sum forces along axes $m$ and $n$ in Fig. 7-40b. For equilibrium in the $m$ direction,

$$\sum F_m = \underbrace{\underset{\text{area}}{\tau_\theta(1)}}_{\substack{\text{force in} \\ m \text{ direction}}} + \underbrace{\underset{\text{area}}{\sigma_x \sin\theta\cos\theta}}_{\substack{\text{force} \\ \text{component of} \\ \text{force of } \sigma_x \\ \text{in } m \text{ direction}}} + \underbrace{\underset{\text{area}}{\tau_{yx}\cos\theta\cos\theta}}_{\substack{\text{force} \\ \text{component of} \\ \text{force of } \tau_{yx} \\ \text{in } m \text{ direction}}} \qquad \boxed{7\text{-}55}$$

$$- \underbrace{\underset{\text{area}}{\tau_{xy}\sin\theta\sin\theta}}_{\substack{\text{force} \\ \text{component of} \\ \text{force of } \tau_{xy} \\ \text{in } m \text{ direction}}} = 0$$

For additional clarification of the components of stresses, consider Fig. 7-40c, d, and e.

For equilibrium in the $n$ direction,

$$\sum F_n = \sigma_\theta(1) - \sigma_x \sin\theta\sin\theta - \tau_{xy}\sin\theta\cos\theta \qquad \boxed{7\text{-}56}$$
$$- \tau_{yx}\cos\theta\sin\theta$$
$$= 0$$

Since the only unknowns are $\sigma_\theta$ and $\tau_\theta$ in these equations, they are readily determined. This method of stress analysis is called the *wedge method* because in each case a triangular element is used.

(a)

(b)

(c) Components of $\tau_{xy}$ on face $BC$ in $m$ and $n$ directions

(d) Face $BC$  (e) Face $AC$

**FIGURE 7-40**

Wedge method of stress analysis

# EXAMPLE 7-19

In a graphite–epoxy composite plate the fibers are oriented as shown in Fig. 7-41a. Determine the stresses parallel and normal to the fibers.

## SOLUTION

The stresses $\sigma_\theta$ and $\tau_\theta$ are found by realizing that the wedge in Fig. 7-41b is in equilibrium. It is most convenient to sum forces in the directions of the unknown stresses to avoid the necessity of solving simultaneous equations. Therefore, since each force is obtained by multiplying the stress by the area it acts upon,

$$\sum F_m = 0$$

$$\tau_\theta A - (30 \text{ MPa sin } 20°)(A \cos 20°) = 0$$

$$\tau_\theta = 30 \text{ MPa sin } 20° \cos 20° = 9.6 \text{ MPa}$$

$$\sum F_n = 0$$

$$\sigma_\theta A - (30 \text{ MPa cos } 20°)(A \cos 20°) = 0$$

$$\sigma_\theta = 30 \text{ MPa cos}^2 \, 20° = 26.5 \text{ MP}$$

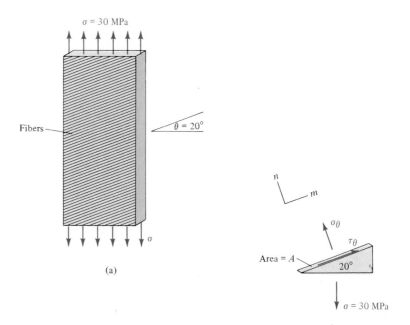

(a)

(b)

**FIGURE 7-41**

Stress analysis of a composite plate

EXAMPLE 7-20

The element in Fig. 7-42a is from a large glued plate. Determine the shear and normal stresses on plane $AB$ of the joint.

## SOLUTION

Resolving the known stress components into components in the $m$ and $n$ directions (Fig. 7-42b),

$$\sum F_m = 0$$

$$\tau_\theta A + (\tau \sin 30°)A \sin 30° - (\sigma_x \cos 30°)A \sin 30°$$
$$- (\tau \cos 30°)A \cos 30° - (\sigma_y \sin 30°)A \cos 30° = 0$$

$$\tau_\theta = \tau \cos^2 30° - \tau \sin^2 30° + \sigma_x \cos 30° \sin 30°$$
$$+ \sigma_y \sin 30° \cos 30°$$
$$= (500 \text{ psi})(\cos^2 30° - \sin^2 30°)$$
$$+ (1000 \text{ psi} + 2000 \text{ psi}) \sin 30° \cos 30°$$
$$= 1550 \text{ psi}$$

$$\sum F_n = 0$$

$$\sigma_\theta A + (\tau \cos 30°)A \sin 30° + (\sigma_x \sin 30°)A \sin 30°$$
$$+ (\tau \sin 30°)A \cos 30° - (\sigma_y \cos 30°)A \cos 30° = 0$$

$$\sigma_\theta = -\tau \cos 30° \sin 30° - \tau \sin 30° \cos 30°$$
$$- \sigma_x \sin^2 30° + \sigma_y \cos^2 30°$$
$$= -(500 \text{ psi})(2 \cos 30° \sin 30°) - 1000 \text{ psi} \sin^2 30°$$
$$+ 2000 \text{ psi} \cos^2 30°$$
$$= 820 \text{ psi}$$

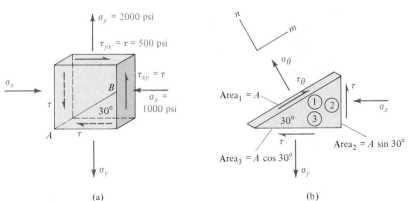

**FIGURE 7-42**

Stress analysis of a glued plate

EXAMPLE 7-20    327

# PROBLEMS

In Probs. 7–107 to 7–116, use the wedge method of analysis to determine the shear and normal stresses on plane $AB$ in the appropriate figure.

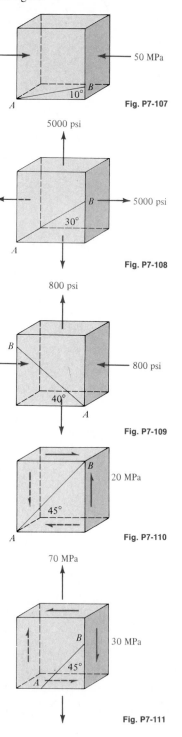

50 MPa

**Fig. P7-107**

5000 psi

5000 psi

30°

**Fig. P7-108**

800 psi

800 psi

40°

**Fig. P7-109**

20 MPa

45°

**Fig. P7-110**

70 MPa

30 MPa

45°

**Fig. P7-111**

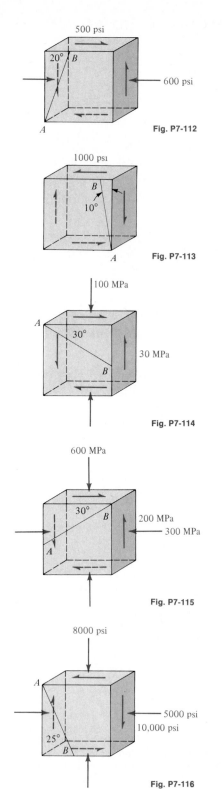

500 psi

20°

600 psi

**Fig. P7-112**

1000 psi

10°

**Fig. P7-113**

100 MPa

30°

30 MPa

**Fig. P7-114**

600 MPa

30°

200 MPa

300 MPa

**Fig. P7-115**

8000 psi

5000 psi

10,000 psi

25°

**Fig. P7-116**

Section 7-1

**COORDINATES OF CENTER OF GRAVITY:**

$$\bar{x} = \frac{\int x \, dW}{W} \quad \bar{y} = \frac{\int y \, dW}{W} \quad \bar{z} = \frac{\int z \, dW}{W}$$

where $dW = \gamma \, dV$

$\gamma$ = specific weight of the material (weight per unit volume)

$dV$ = volume element

$W = \int \gamma \, dV$ = total weight of the body

Section 7-2

**CENTROIDS:**

**LINES:** $\quad \bar{x} = \frac{\int x \, dL}{L} \quad \bar{y} = \frac{\int y \, dL}{L} \quad \bar{z} = \frac{\int z \, dL}{L}$

where $dL$ = line element

$L$ = total length of line

**AREAS:** $\quad \bar{x} = \frac{\int x \, dA}{A} \quad \bar{y} = \frac{\int y \, dA}{A} \quad \bar{z} = \frac{\int z \, dA}{A}$

where $dA$ = area element

$A$ = total area

**VOLUMES:** $\quad \bar{x} = \frac{\int x \, dV}{V} \quad \bar{y} = \frac{\int y \, dV}{V} \quad \bar{z} = \frac{\int z \, dV}{V}$

where $dV$ = volume element

$V$ = total volume

Section 7-3

**CENTROIDS OF COMPOSITE OBJECTS:** Equate the first moment of the whole object with the sum of the first moments of the individual parts, using the same reference system.

**LINES:** $\quad \bar{x}L = \sum \bar{x}_i L_i, \quad \bar{x} = \frac{\sum \bar{x}_i L_i}{L}$

$$\bar{y}L = \sum \bar{y}_i L_i, \quad \bar{y} = \frac{\sum \bar{y}_i L_i}{L}$$

$$\bar{z}L = \sum \bar{z}_i L_i, \quad \bar{z} = \frac{\sum \bar{z}_i L_i}{L}$$

where $\bar{x}, \bar{y}, \bar{z}$ are the centroidal coordinates of the whole line $L$, and $\bar{x}_i$, $\bar{y}_i, \bar{z}_i$ are the centroidal coordinates of the line element $L_i$; $L = \sum L_i$.

Similarly for areas and volumes,

**AREAS:** $\quad \bar{x} = \dfrac{\sum \bar{x}_i A_i}{A} \quad \bar{y} = \dfrac{\sum \bar{y}_i A_i}{A} \quad \bar{z} = \dfrac{\sum \bar{z}_i A_i}{A}$

**VOLUMES:** $\quad \bar{x} = \dfrac{\sum \bar{x}_i V_i}{V} \quad \bar{y} = \dfrac{\sum \bar{y}_i V_i}{V} \quad \bar{z} = \dfrac{\sum \bar{z}_i V_i}{V}$

**CENTER OF GRAVITY OF COMPOSITE BODY:**

$$\bar{x}W = \bar{x}_i W_i = \bar{x}_i \gamma_i V_i \quad \bar{x} = \dfrac{\sum \bar{x}_i W_i}{W}$$

$$\bar{y}W = \bar{y}_i W_i = \bar{y}_i \gamma_i V_i \quad \bar{y} = \dfrac{\sum \bar{y}_i W_i}{W}$$

$$\bar{z}W = \bar{z}_i W_i = \bar{z}_i \gamma_i V_i \quad \bar{z} = \dfrac{\sum \bar{z}_i W_i}{W}$$

Section 7-4

**THEOREMS OF PAPPUS:**

**THEOREM I:** $\quad A = 2\pi \bar{y} L$

where $A$ = total area of surface of revolution
   $L$ = length of the generating curve
   $\bar{y}$ = centroid of line $L$ with respect to the axis of revolution

**THEOREM II:** $\quad V = 2\pi \bar{y} A$

where $V$ = total volume within the generating surface of revolution
   $A$ = area of the generating surface
   $\bar{y}$ = centroid of area $A$ with respect to the axis of revolution

Section 7-5

**DISTRIBUTED LOADS ON STRUCTURES.**
**EXTERNAL REACTIONS:** Replace the distributed load with a concentrated force equal in magnitude to the area under the load distribution curve and applied at the centroid of that area.

Section 7-6

**DISTRIBUTED LOADS ON BEAMS.**
**INTERNAL REACTIONS:** Assume the following distributed load $\mathbf{w}(x)$ on the infinitesimal element $B_1 B_2$ of a beam:

The load $w(x)$, shear $V$, and bending moment $M$ are related as

$$w(x) = -\frac{dV}{dx} \qquad V = \frac{dM}{dx}$$

where the signs are for the situation shown here. Other useful expressions for any two cross sections $A$ and $B$ of a beam:

$$V_A - V_B = \int_{x_A}^{x_B} w(x)\,dx = \text{area under } w(x)$$

$$M_B - M_A = \int_{x_A}^{x_B} V\,dx = \text{area under shear diagram}$$

**EXAMPLE:**

Free-body diagram for $AB$:

where   $\mathbf{P}$ = resultant external force on part $AB$ (known from $w(x)$ and length of $AB$) = area $ABCD$

$x_P$ = location where force $\mathbf{P}$ acts to make $P$ equivalent to load $ABCD$ = centroid of area $ABCD$ as measured horizontally from point $B$

$\mathbf{V}_B$ = shear force at $B$ (unknown, assumed positive)

$V_B = -P$ from equilibrium of part $AB$

$\mathbf{M}_B$ = bending moment at $B$ (unknown, assumed positive)

$M_B = -Px_P$ from equilibrium of part $AB$

**DIFFERENTIAL EQUATION OF A CABLE:**

$$\frac{d^2y}{dx^2} = \frac{w(x)}{T_o}$$

where $w(x)$ = distributed load as a function of $x$ ($x$ is horizontal axis)

$T_o$ = constant = horizontal component of the tension in the cable

**PARABOLIC CABLES:** The cable supports a load $w$ which is uniformly distributed horizontally. The shape of the cable is given by

$$y = \frac{wx^2}{2T_o} \qquad (x = 0 \text{ at lowest point})$$

The tension in a symmetric cable is $T = \sqrt{T_o^2 + w^2 x^2}$

**CATENARY CABLES:** The cable supports a load $w$ which is uniformly distributed along the cable. The shape of the cable is given by

$$y = \frac{T_o}{w}\left(\cosh \frac{wx}{T_o} - 1\right)$$

The tension in the cable is $T = T_o + wy$.

**FLUID STATICS:**

$$\text{Pressure } p = \frac{\text{force}}{\text{area}}$$

**PASCAL'S LAW**

The pressure is the same in all directions at any given point in a fluid at rest.

The pressure varies linearly with the depth $h$ in a fluid,

$$\Delta p = \gamma \, \Delta h$$

where $\gamma$ is the specific weight of the fluid (weight per unit volume).

Gage pressure is measured from the open surface of a liquid,

$$p = \gamma h$$

Hydrostatic force = resultant of hydrostatic pressure on a solid surface

= area under pressure distribution curve

Center of pressure = location where the hydrostatic force acts

= centroid of pressure distribution (not at the centroid of the solid surface on which the pressure acts)

**ARCHIMEDES' PRINCIPLE**
A submerged body is subject to an upward buoyant force equal to the weight of the displaced fluid. The buoyant force acts at the center of buoyancy, which is the center of gravity of the displaced fluid (not necessarily the center of gravity of the body that displaces the fluid).

**GENERAL DEFINITIONS OF STRESS:**

$$\text{Stress} = \frac{\text{force}}{\text{area}}$$

$$\text{Normal stress } \sigma = \frac{\text{normal force}}{\text{area}}$$

$$\text{Shear stress } \tau = \frac{\text{shear force}}{\text{area}}$$

For precise definitions, use a differential force acting on a differential area,

$$\sigma = \frac{dF_{\text{normal}}}{dA} \qquad \tau = \frac{dF_{\text{shear}}}{dA}$$

To calculate force from stress,

$$F_{\text{normal}} = \int \sigma \, dA \qquad F_{\text{shear}} = \int \tau \, dA$$

**WEDGE METHOD OF STRESS ANALYSIS:** Draw a triangular element with all known $(\sigma_x, \dots)$ and unknown $(\sigma_\theta, \tau_\theta)$ stresses on it:

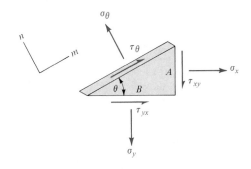

Write two equilibrium equations, one for the $m$ direction, the other for the $n$ direction. These equations are for *forces*, so each term must involve a stress times the area on which that stress is acting. Solve the equations for the unknown stresses.

**7–117** The probe of a mechanical testing instrument is modeled as a flat plate of negligible thickness. Locate its centroid assuming that the slot on the right end is not made.

**Fig. P7-117**
Plane area

**7–118** Solve Prob. 7–117 as stated, then make the correction for the 20-mm long slot as shown in Fig. P7-117.

**7–119** The shield of a small electronic measuring device is made of sheet metal of negligible thickness. Locate the centroid of the five-sided box shape.

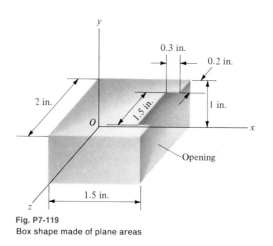

**Fig. P7-119**
Box shape made of plane areas

**7–120** Determine the external surface area of the bellows which appears as a circle when viewed from the right or left.

**Fig. P7-120**
Thin-walled bellows with ends closed

**7–121** Determine the volume enclosed by the bellows in Fig. P7-120.

**7–122** A bracket is made from a thin, uniform steel plate of 1 mm thickness. It is supported by a smooth pin at $C$ and a wire at $B$. Determine the reaction at support $C$ and the tension in the wire if the density of steel is 7830 kg/m$^3$, and the holes at $D$ and $E$ have a diameter of 0.04 m.

**Fig. P7-122**

**7–123** Determine the reactions on the beam at the fixed support at $A$ for the distributed load.

$$w = (3 - x)^2 \text{ kN/m}$$

**Fig. P7-123**

**7–124** Determine the shear $V$ and bending moment $M$ as functions of $x$ for the beam in Prob. 7–123.

**7–125** Draw the complete shear and bending moment diagrams for the simply supported beam.

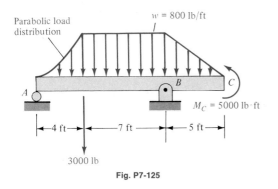

Fig. P7-125

**7–126** The cable supports a uniform pipe of total weight $W = 12$ kips. Determine the maximum tension in the cable.

Fig. P7-126

**7–127** The trough shown may carry muddy water whose specific weight varies linearly between the values given. Determine the internal reactions at corner $A$ per unit length of the trough.

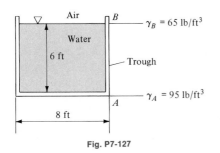

Fig. P7-127

**7–128** The pontoons of a seaplane are modeled with a uniformly varying circular cross section. What is the magnitude of the force $P$ and its location $d$ to place the centerline of the pontoon at the surface of the water? Assume that the pontoon itself is weightless.

Fig. P7-128

In Probs. 7–129 and 7–130 determine the shear and normal stresses on plane $AB$.

Fig. P7-129

Fig. P7-130

# 8 Friction

Basic Concepts in Analyzing the Effects of Friction

Some phenomena involving friction are frequently observed and are naturally understood by most people. For example, consider the heat generated by rubbing dry hands together, the skid marks made by automobile tires sliding on dry pavements, and the action of bicycle brakes. It appears from these and other phenomena that a *friction force* acts between contacting bodies when one slides relative to the other or when sliding tends to occur. The friction force is tangential to each body at the point of contact, as evidenced by the displacement of skin when the hands are slowly rubbed. It is also clear from the common examples that the friction force depends on the *normal force* pressing the bodies together and on the condition of the contacting surfaces.

These concepts can be extended to qualitatively explain more complex phenomena of friction. For example, there is *internal friction* in solid materials when deformed, as evidenced by the heat generated in the sidewalls of rolling tires or in repeatedly bent wires. Such friction involves atoms or molecules moving tangentially relative to one another in the material. *Fluid friction* or *drag* involves a fluid moving past a solid body. These phenomena can be briefly introduced in statics courses after an extensive analysis of friction forces between solid bodies.

**SECTION 8-1** presents the fundamental aspects of dry friction between solid bodies. These include the concepts of static and kinetic friction (the latter for moving bodies). These concepts are essential in solving many statics and dynamics problems and should be mastered.

**SECTION 8-2** describes the principles for experimentally determining the coefficient of static friction, which is assumed a constant in most problems. Engineers rarely do such experiments since approximate values are available from handbooks, but the method is also illustrative of friction problems in general.

**SECTION 8-3** presents a useful classification of problems involving dry friction and procedures for solving them.

**SECTION 8-4** extends the basic concepts of static friction to the analysis of wedges and screws, the simplest machines. These topics are very useful for the exercise of skills in modeling real situations with free-body diagrams and solving equilibrium equations.

**SECTIONS 8-5 and 8-6** present the application of the concepts of dry friction to disk bearings, disk clutches, disk brakes, and journal bearings for axles. These components of machines and vehicles provide for useful practice in statics analyses. They also build a background for more advanced courses in machine design.

**SECTION 8-7** introduces the topic of rolling resistance of wheels and balls. This is sufficiently specialized and different from those involving simple dry friction that it is often omitted from statics courses. The topic is presented after Sec. 8-6 because of similarities in the analytical approaches used.

**SECTION 8-8** presents the method of analysis and simple formulas for the friction problems of belts, ropes, and wires on pulleys or drums. This topic is relevant to machine design.

**SECTION 8-9** presents the elementary concepts of fluid friction. This topic is often omitted from statics courses since it is covered in fluid mechanics.

## DRY FRICTION

Assume that the driver of a moving vehicle on a steep, dry road applies enough braking force to lock every wheel. The vehicle slides on the road and may come to a complete stop without hitting anything. For analyzing these situations, both the sliding and the stationary vehicle (Fig. 8-1) can be modeled simply as a block on an inclined plane, as in Fig. 8-2a. The next logical step is to consider the free-body diagram of the block, which is shown in Fig. 8-2b. Here **W** is the weight of the block and for convenience it is resolved into components parallel and perpendicular to the inclined plane, $\mathbf{W}_m$ and $\mathbf{W}_n$, respectively.

If the body in Fig. 8-2b is stationary, there must be a reaction from the inclined plane acting on the body that is equal and opposite to $W$. This reaction can be resolved into convenient components $\mathscr{F}$ and **N** that act in opposite directions to $\mathbf{W}_m$ and $\mathbf{W}_n$, respectively. The normal force $N$ always acts perpendicular to the surface and prevents motion in that direction. Thus, $N$ is equal to $W\cos\theta$ whether or not there is motion parallel to the surface. The tangential force $\mathscr{F}$ is the friction force, which may or may not prevent motion. It is often denoted by a script letter to distinguish it from other forces. The friction force always acts at the sur-

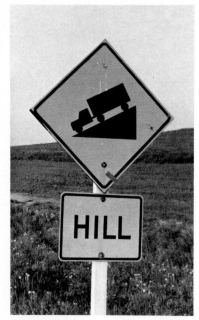

**FIGURE 8-1**

Friction may be involved in the analysis of both moving and stationary bodies

(a)

(c)

$N = W$

**FIGURE 8-2**

A vehicle on a sloping road is modeled as a block on an incline

(b)

$W_n = W \cos \theta$

$W_m = W \sin \theta$

$N = W \cos \theta$

faces of contact between bodies, parallel (or tangent) to the surfaces, and it acts in the direction opposite to the direction in which the body would tend to move. The friction force does not depend on the area of the contact patch if there are no large deformations there. However, the line of action of the resultant normal force is not known unless there is a distinct point of contact.

The magnitude of the friction force may have the following values with respect to $W \sin \theta$, the force that tends to cause motion along the incline.

1. $\mathcal{F} = W \sin \theta$, there is no motion, or there is motion at constant velocity.
2. $0 < \mathcal{F} < W \sin \theta$, the body accelerates downhill.
3. $\mathcal{F} > W \sin \theta > 0$, the body initially moving downhill has a decreasing velocity and eventually stops; after stopping, $\mathcal{F} = W \sin \theta$.

These generalizations should make sense even without much knowledge of dynamics. They can also be extended to situations such as that in Fig. 8-2c where $W \sin \theta = 0$, but another external force **P** tends to cause motion parallel to the supporting surface.

The magnitude of the friction force may vary from zero to a maximum value for any given pair of bodies. $\mathcal{F}$ is always zero when there is no force for the friction force to oppose, as in Fig. 8-2c if $P = 0$. The friction force can be equal to the applied force up to a limiting value $\mathcal{F}_{max}$ shown in Fig. 8-3. The maximum friction force depends on the materials involved, on the geometry and condition of the surfaces in contact, and on the normal force $N$ that presses the bodies together. The material and surface properties are lumped together and represented in analysis by the *coefficient of friction* $\mu$. It is found that under most conditions the limiting friction force is proportional to the normal force between the bodies,

$$\mathcal{F}_{max} = \mu N \quad (0 \le \mathcal{F} \le \mathcal{F}_{max})$$

8-1

The coefficient of friction is largest for a given pair of materials when motion is impending; the *coefficient of static friction* is denoted by $\mu$ or

FIGURE 8-3

Variation of the friction force between two bodies

$\mu_s$. The coefficient decreases when there is relative sliding of the bodies; this is called the *coefficient of kinetic friction* (or sliding friction, or dynamic friction) and is denoted by $\mu_k$. The magnitude of the kinetic friction force is given by $\mathscr{F}_k = \mu_k N$. The coefficient of kinetic friction ($\mu_k < \mu_s$) is approximately constant at moderate rubbing speeds. At higher speeds the coefficient of friction decreases further because of the heat generated by friction (Fig. 8-3). This is the reason for the phenomenon of brake fading at high speeds or during long descents. Clearly, locking the wheels in a moving vehicle immediately reduces the braking force at the road surface ($\mu_k < \mu_s$), and may cause lateral sliding of the vehicle since sliding in any direction facilitates sliding in other directions. The heat generated at the sliding contact patch of each tire may cause further changes in the friction force acting on a vehicle.

## ANGLES OF FRICTION

The experimental determination of the coefficient of static friction is normally an easy matter. To illustrate the principles used in this work, consider Fig. 8-4. Here a block of weight **W** is on a plane that is represented by a dashed line. In Fig. 8-4a the plane is horizontal, so the entire reaction **R** acting on the block is normal to the plane. Next assume that the plane is inclined at a small angle $\theta$ with the horizontal, as in Fig. 8-4b. The reaction **R** now is the resultant of the friction force $\mathscr{F}$ and the normal force **N**. As the angle $\theta$ is further increased, it reaches a critical value $\theta_c$, at which angle the block begins to slide (Fig. 8-4c). The reaction **R** is again the resultant of forces $\mathscr{F}$ and **N**. The largest possible magnitude of the friction force, $\mathscr{F}$, for the given bodies occurs when motion is impending, which is at the critical angle where motion begins, for all practical purposes. In

FIGURE 8-4

Angles of friction

(a)  No friction

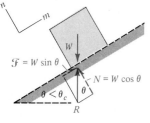

(b)  No motion

(c)  Motion impending or beginning

(d)  Motion

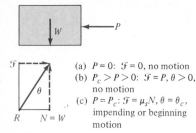

(a) $P = 0$: $\mathscr{F} = 0$, no motion
(b) $P_c > P > 0$: $\mathscr{F} = P$, $\theta > 0$, no motion
(c) $P = P_c$: $\mathscr{F} = \mu_s N$, $\theta = \theta_c$, impending or beginning motion

**FIGURE 8-5**

Critical value of a lateral force for causing relative motion between two bodies

fact, there is no way to prove that motion is impending without letting the motion begin. At angles larger than the critical angle, the block accelerates down the incline. In these cases the reaction **R** is no longer parallel to the weight **W** since $\mathscr{F} < W \sin \theta$.

The critical angle $\theta_c$ at which motion is impending is called the *angle of repose*. The friction force is at its maximum at this angle, so from Fig. 8-4c and Eq. 8-1,

$$\boxed{\tan \theta_c = \frac{\mathscr{F}}{N} = \frac{\mu_s N}{N} = \mu_s} \qquad \boxed{8\text{-}2}$$

Thus, measurement of the angle of repose is the only major step required in determining the coefficient of static friction.

The concepts discussed above can be extended to situations where the weight of a body does not tend to cause motion. For example, the block in Fig. 8-5 is on a horizontal plane, so it can move only if a large enough force **P** is applied. Here motion is impending or beginning when the lateral force reaches a critical magnitude $P_c$. At that instant the friction force is at its maximum, so from Fig. 8-5c and Eq. 8-1,

$$\tan \theta_c = \frac{\mathscr{F}}{N} = \frac{\mu_s N}{N} = \mu_s$$

which is Eq. 8-2. In this case the angle $\theta_c$ is called the *angle of static friction*, which is equivalent to the angle of repose.

The coefficients of friction in dry friction (also called Coulomb friction) are normally sensitive to humidity, temperature, dust, contamination, oxide films, surface finish, and chemical reactions. The contact area and the normal force affect the coefficients only when they cause large deformations of one or both bodies in contact. Published values of the coefficients of friction are all empirically determined, and most often they are for static friction. As a rule of thumb, high values of the coefficient of friction are between 0.6 and 1.0, and low values are about 0.1 or less. Approximate values of $\mu_s$ for many materials are given inside the front cover of this book for easy reference. For rough or sticky surfaces, or any other situation when there is a high surface interaction, $\mu_s$ may be larger than unity.

## 8-3    CLASSIFICATION OF PROBLEMS OF DRY FRICTION

There are numerous ways to classify problems involving dry friction. Generally, the first step is to decide whether Eq. 8-1 is applicable or not. It is essential to remember that $\mathscr{F} = \mu_s N$ *when there is motion or impending motion*. There are many situations where there is friction that is less than the maximum possible friction for the given conditions; in such cases $\mathscr{F} < \mu_s N$ and the equilibrium equations must be used to determine the friction force.

The major steps in solving problems of dry friction are listed in three categories in Table 8-1. Note that at least the directions of unknown friction forces are easily determined by inspection. The correct directions should be used in the solutions. The details of the procedures are illustrated in several examples.

**TABLE 8-1**

Procedures for Solving Friction Problems

| CLASS OF PROBLEM | GIVEN INFORMATION | PROCEDURE |
|---|---|---|
| A | Bodies, forces, coefficients of friction are known. Impending motion is not assured; $\mathscr{F} \neq \mu_s N$. | To determine if equilibrium is possible:<br>1. Construct the free-body diagram.<br>2. Assume that the system is in equilibrium.<br>3. Determine the friction and normal forces necessary for equilibrium.<br>4. Results: (a) $\mathscr{F} < \mu_s N$, the body is at rest. (b) $\mathscr{F} > \mu_s N$, motion is occurring, static equilibrium is not possible. Since there is motion, $\mathscr{F} = \mu_k N$. Complete solution requires principles of dynamics. |
| B | Bodies, forces, or coefficients of friction are given. Impending motion is specified, $\mathscr{F} = \mu_s N$ is valid. | To determine the unknowns:<br>1. Construct the free-body diagram.<br>2. Write $\mathscr{F} = \mu_s N$ for all surfaces where motion is impending.<br>3. Determine $\mu_s$ or the required forces from the equation of equilibrium. |
| C | Bodies, forces, coefficients of friction are known. Impending motion is specified, but the exact motion is not given. The possible motions may be sliding, tipping, or rolling, or relative motion if two or more bodies are involved. Alternatively, the forces or coefficients of friction may have to be determined to produce a particular motion from several possible motions. | To determine the exact motion that may occur, or unknown quantities required:<br>1. Construct the free-body diagram.<br>2. Assume that motion is impending in one of the two or more possible ways. Repeat this for each possible motion, and write the equation of equilibrium.<br>3. Compare the results for the possible motions and select the likely event. Determine the required unknowns for any preferred motion. |

# EXAMPLE 8-1

A car of 1000-kg mass is to be parked on the same $10°$ incline year round. The static coefficient of friction between the tires and the road varies between the extremes of 0.05 and 0.9. Is it always possible to park the car at this place? Assume that the car can be modeled as a particle.

## SOLUTION

The problem is in class A of Table 8-1. It is most reasonable to consider the lowest coefficient of friction first since the larger the coefficient of friction, the larger the resistance to sliding. The free-body diagram is drawn in Fig. 8-6. For equilibrium,

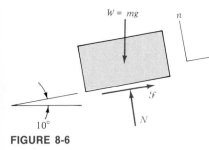

**FIGURE 8-6**

Model of car as a single block on an incline

$$\sum F_n = 0 \qquad\qquad \sum F_m = 0$$

$$N - mg \cos 10° = 0 \qquad \mathscr{F} - mg \sin 10° = 0$$

$$N = mg \cos 10° = 9.66 \text{ kN} \qquad \mathscr{F} = mg \sin 10° = 1.703 \text{ kN}$$
$$= \text{friction force necessary}$$
$$\text{for equilibrium}$$

The largest available friction force for the lowest coefficient of friction will be

$$\mathscr{F}_{\text{max}_1} = \mu_{s_1} N = 0.05(9.66 \text{ kN}) = 0.483 \text{ kN}$$

Since $\mathscr{F}_{\text{max}_1} < \mathscr{F}$, the car would slide and cannot be parked year round on the same incline.

## JUDGMENT OF THE RESULTS

The result can be checked by determining the angle of repose $\theta_c$ for the smallest coefficient of friction. From Eq. 8-2,

$$\theta_c = \tan^{-1} \mu_{s_1} = \tan^{-1} 0.05 = 3°$$

Since $\theta_c < 10°$, the car would slide.

# EXAMPLE 8-2

A car is to be parked with all wheels locked on an incline as in Fig. 8-7. Can the car remain at rest? If not, what is the magnitude of the additional force **P** required parallel to the incline for equilibrium? Assume that the forces are symmetric with respect to the longitudinal axis of the car.

## SOLUTION

The problem is in class A of Table 8-1. Figure 8-7 is the free-body diagram for the car as a rigid body. Assuming symmetry of the left and right sides of the car, the normal and friction forces at the two back tires and two front tires are represented

**FIGURE 8-7**

Car parked on an incline

by their sums $\mathbf{N}_A$, $\mathscr{F}_A$, and $\mathbf{N}_B$, $\mathscr{F}_B$, respectively (Fig. 8-7). The magnitudes of these four forces are unknown. First, it is necessary to calculate the normal forces. This is started by taking moments about either $A$ or $B$.

$$\sum M_A = 0$$

$$mg \, (\sin 10°)(0.3 \text{ m}) - mg \, (\cos 10°)(2 \text{ m}) + N_B(3.5 \text{ m}) = 0$$

$$N_B = \frac{(1000 \text{ kg})(9.81 \text{ m/s}^2)(2 \text{ m} \cos 10° - 0.3 \text{ m} \sin 10°)}{3.5 \text{ m}} = 5.37 \text{ kN}$$

$$\sum F_n = 0$$

$$N_A + N_B - mg \cos 10° = 0$$

$$N_A = 9.81 \text{ kN} \cos 10° - 5.37 \text{ kN} = 4.29 \text{ kN}$$

In investigating equilibrium parallel to the incline, the total friction force necessary for equilibrium, $\mathscr{F}_T = \mathscr{F}_A + \mathscr{F}_B$, must be determined. Summing forces along $m$,

$$\sum F_m = 0$$

$$\mathscr{F}_A + \mathscr{F}_B - mg \sin 10° = 0$$

$$\mathscr{F}_T = \mathscr{F}_A + \mathscr{F}_B = 9.81 \text{ kN} \sin 10° = 1.703 \text{ kN}$$

The available friction force is

$$\mathscr{F}_{max} = (\mathscr{F}_A + \mathscr{F}_B)_{max} = \mu N_A + \mu N_B = (0.1)(N_A + N_B)$$
$$= (0.1)(mg \cos 10°) = 966 \text{ N}$$

$\mathscr{F}_T > \mathscr{F}_{max}$; therefore, the car will slide. The magnitude of the additional force $\mathbf{P}$ required parallel to the incline for equilibrium is obtained from

$$\sum F_m = 0$$

$$P + \mathscr{F}_{max} - mg \sin 10° = 0$$

$$P = 1703 \text{ N} - 966 \text{ N} = 737 \text{ N}$$

Note that once the force $\mathbf{P}$ is applied, the relative magnitudes of $\mathbf{N}_A$ and $\mathbf{N}_B$ will change. The total normal force, however, will remain equal in magnitude to $mg \cos 10°$.

EXAMPLE 8-2   343

EXAMPLE 8-3

Consider a wheel at rest on an incline as in Fig. 8-8a. The forces of magnitude $F$ press the internal brake pads $B$ against the wheel. The brake pads are fixed with respect to the vehicle frame and always remain parallel to the inclined plane. **(a)** Determine $F$ that prevents the wheel from rolling downhill. **(b)** For this $F$, at what value of $\mu_A$ does the wheel slide down without rolling?

## SOLUTION

**(a)** The problem is in class C of Table 8-1. The free-body diagram is drawn in Fig. 8-8b. The weight of the wheel is $W = (1000 \text{ kg})(9.81 \text{ m/s}^2) = 9.81$ kN. By summing moments about point $A$, calculation of the unknown forces $\mathscr{F}_A$ and $\mathbf{N}_A$ is avoided. For impending motion, the magnitude of the friction forces at the brakes is $\mathscr{F}_B = \mathscr{F}_{max} = \mu_B F$. Realizing that the braking forces form a couple (a free moment vector),

$$\sum M_A = 0$$

$$W_m R - \mathscr{F}_B (2R) = 0$$

$$\mathscr{F}_B = \frac{1}{2} W_m = \frac{1}{2}(9.81 \text{ kN} \sin 20°) = 1.678 \text{ kN} = \mu_B F = 0.5F$$

Hence, $F = 1.678$ kN/0.5 = 3.36 kN. Note that this solution is independent of the radius of the wheel.

**(b)** For $F = 3.36$ kN, rolling is not a possible motion, so the problem is now in class B, Table 8-1. For impending sliding in Fig. 8-8c,

$$\mathscr{F}_A = \mathscr{F}_{max} = \mu_A N_A$$

For equilibrium,

$$\sum F_n = 0$$

$$N_A - W_n = 0$$

$$N_A = W \cos 20°$$

$$\sum F_m = 0$$

$$-W_m + \mathscr{F}_A = 0$$

$$-W \sin 20° + \mu_A W \cos 20° = 0$$

$$\mu_A = \tan 20° = 0.36$$

Note that this solution is independent of the weight of the wheel.

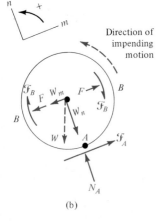

$R = 0.4$ m
$m = 1000$ kg
$\mu_B = 0.5$
$W = mg$
$\mu_A$
20°

(a)

Direction of impending motion

Direction of impending motion

(b)

(c)

**FIGURE 8-8**
Wheel at rest on an incline

EXAMPLE 8-4

Consider the crate shown in Fig. 8-9a. **(a)** Determine the largest force **F** that can be applied without the crate sliding or tipping. **(b)** What are the limiting values of $F$ and $\mu_s$ such that sliding occurs first? This case is preferable to tipping.

## SOLUTION

**(a)** This is a situation as discussed in Table 8-1, class C. The possible motions must be considered separately.

**MOTION 1, SLIDING (Motion Impending):** From Fig. 8-9b,

$$\sum F_y = 0 \qquad \sum F_x = 0$$
$$N - W = 0 \qquad F - \mathscr{F}_{max} = 0$$
$$N = W \qquad F = \mathscr{F}_{max} = \mu_s N = 0.3W = 0.3(800 \text{ lb}) = 240 \text{ lb}$$

**MOTION 2, TIPPING (Fig. 8-9c):** If tipping is impending, the only point of contact is corner $A$, hence the location of the normal force $N$ is known. Therefore,

$$\sum M_A = 0$$
$$(800 \text{ lb})(1 \text{ ft}) - F(4 \text{ ft}) = 0$$
$$F = 200 \text{ lb}$$

The smallest value of $F$ is the governing value, so

$$F_{max} = 200 \text{ lb}$$

which would cause the crate to tip rather than slide.

**(b)** If slipping is the desired first motion, then a smaller coefficient of friction is required. Now the problem is in class B, Table 8-1: determine $\mu_s$ for slipping and tipping to be equally likely; then for all $\mu < \mu_s$ sliding occurs first. From equilibrium in the $x$ direction when slipping is impending,

$$F = \mathscr{F} = \mathscr{F}_{max} = \mu_s N = \mu_s W$$

When tipping is impending, Fig. 8-9c is applicable:

$$\sum M_A = 0$$
$$W(1 \text{ ft}) - F(4 \text{ ft}) = 0$$
$$F = \frac{1}{4} W$$

Setting the two values of $F$ equal makes slipping or tipping equally likely,

$$\mu_s W = \frac{1}{4} W, \qquad \mu_s = 0.25$$

Hence, for slipping to occur first, the coefficient of static friction $\mu$ must be less than 0.25.

**FIGURE 8-9**

Sliding or tipping of a crate

EXAMPLE 8-4    345

## EXAMPLE 8-5

The 25-ft-long ladder is resting on a crate, $C$, and a smooth wall at $A$. See Fig. 8-10a. The crate weighs 100 lb. Determine the distance $h$ that a 200-lb person can safely walk up the ladder. The weight of the ladder can be neglected, and the crate can be treated as a particle due to the positioning of the ladder.

### SOLUTION

First the free-body diagram of the ladder–crate combination is considered in Fig. 8-10b. For impending sliding of the crate on the floor, $\mathscr{F}_{C\max} = \mu_C N_C$ and

$$\sum F_x = 0 \qquad\qquad \sum F_y = 0$$
$$N_A - \mathscr{F}_C = 0 \qquad\quad N_C - 200 - 100 = 0$$
$$N_A = \mathscr{F}_C \qquad\qquad N_C = 300$$

$$\sum M_B = 0$$
$$N_A(25\cos 30°) - 200(h\sin 30°) = 0$$
$$21.7 N_A = 100h$$

Then

$$h_{\max} = 0.217\mathscr{F}_{C\max} = 0.217\mu_C N_C = (0.217)(0.3)(300)$$
$$h = 19.53 \text{ ft}$$

The possibility that the ladder might slide on the crate must also be considered. The free-body diagrams of the ladder and crate are drawn in Fig. 8-10c and d. Forces $\mathscr{F}_B$ and $N_B$ are shown in equal and opposite pairs. The equilibrium equations give

$$\sum F_x = 0 \qquad\qquad \sum F_y = 0$$
$$N_A - \mathscr{F}_B = 0 \qquad\quad N_B - 200 = 0$$
$$N_A = \mathscr{F}_B \qquad\qquad N_B = 200$$

$$\sum M_B = 0$$
$$N_A(25\cos 30°) - 200(h\sin 30°) = 0$$
$$21.7 N_A = 100h$$

Then

$$h_{\max} = 0.217\mathscr{F}_{B\max} = (0.217)(0.4)(200) = 17.36 \text{ ft}$$

so the ladder would slide at point $B$ first. The maximum safe distance that the person can travel up the ladder is 17.36 ft.

### JUDGMENT OF THE RESULTS

A quick check of $\sum F_x$ for the crate in Fig. 8-10d shows that

$$\mathscr{F}_{B\max} = (0.4)(200\text{ lb}) < \mathscr{F}_{C\max} = (0.3)(300\text{ lb})$$

so the ladder will always slide on the crate before the crate slides on the floor.

(a)

(b)

(c)

(d)

**FIGURE 8-10**

Analysis of the slipping of a ladder on a crate

**8-1** Consider a block on an incline where $\mu_s = 0.2$. Plot the angle $\theta$ at which the block begins to slide as a function of the weight, $W = 10$ lb, 50 lb, and 100 lb.

Fig. P8-1

**8-2** Determine whether the block in Fig. P8-1 can remain at rest on the inclined plane if the mass $m = 30$ kg, $\theta = 10°$, and $\mu_s = 0.1$.

**8-3** The block weighs 100 lb, $P = 80$ lb, $\theta = 10°$, and $\mu_s = 0.25$. Does the block begin to slide when $\mathbf{P}$ is applied?

Fig. P8-3

**8-4** The block has a mass of 100 kg and $\mu_s = 0.4$. Determine the magnitude of the force $\mathbf{P}$ at $\theta = 20°$ to start moving the block if $\mathbf{P}$ acts in the opposite direction to that shown in Fig. P8-3.

**8-5** Determine the minimum magnitude of the force $\mathbf{P}$ (in terms of $W$) that causes the block to start moving up the incline if $\theta = 25°$ and $\mu_s = 0.3$.

Fig. P8-5

**8-6** Assume that in Prob. 8-5 the force $\mathbf{P}$ has a fixed direction and magnitude. Determine the weight $W$ (if any) to make the block slide down the incline in spite of the force $\mathbf{P}$.

**8-7** Determine if boxes $A$ and $B$ can remain in equilibrium as shown. The boxes have weights $W_A = 200$ lb and $W_B = 300$ lb.

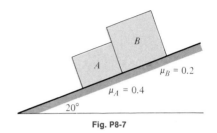

Fig. P8-7

**8-8** Calculate the magnitude of the required force $\mathbf{F}$ acting on box $A$ parallel to the incline in Prob. 8-7 to cause impending upward motion of the boxes.

**8-9** Determine the magnitude of the horizontal force $\mathbf{P}$ required to initiate motion of block $B$. The coefficient of friction between all surfaces is $\mu_s = 0.25$. Block $A$ weighs 10 lb, block $B$ weighs 15 lb, and block $C$ weighs 30 lb.

Fig. P8-9

**8-10** Block $A$ weighs 30 lb, and block $B$ weighs 20 lb. If no tipping occurs, determine the minimum magnitude of the horizontal force $\mathbf{P}$ required for impending motion. $\mu_s = 0.5$ between blocks $A$ and $B$ and $\mu_s = 0.2$ between block $B$ and the floor. Between which surfaces would slipping occur?

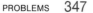

Fig. P8-10

**8–11** Solve Prob. 8–10 if an additional force of magnitude $P/2$ is acting to the left on block $B$.

**8–12** The ladder $AB$ is supported by a smooth wall at $B$. Determine whether the 5-m-long ladder can remain at rest with a person of mass $m = 80$ kg standing at $h = 4$ m if $\theta = 70°$ and $\mu_A = 0.4$. Neglect the mass of the ladder.

**Fig. P8-12**

**8–13** Determine the angle $\theta$ in Prob. 8–12 at which the ladder starts to slide.

**8–14** A uniform ladder $AB$ is supported by a smooth wall at $B$. It is to be adjusted by applying a horizontal force $F$ to the right at point $D$. Determine the magnitude of the force $F$ required to cause motion if the ladder has mass $m = 50$ kg, $L = 7$ m, $h = 1.3$ m, $\theta = 75°$, and $\mu_A = 0.3$.

**Fig. P8-14**

**8–15** Determine if the 60-lb, 8-ft-long beam can remain in equilibrium on the two steps. If yes, calculate the total friction force acting on the beam.

**Fig. P8-15**

**8–16** Repeat Prob. 8–15 for $\theta = 20°$.

**8–17** In Prob. 8–15 determine the magnitude of horizontal force $F$ at point $C$ that would cause the beam to start sliding upward.

**8–18** A 2000-lb lathe is to be slid into a new position. The coefficient of static friction is 0.5 at both supports $A$ and $B$. Point $C$ is the mass center. (a) Determine the required magnitude of the force $F$ to start the lathe moving if $\theta = 0$, and (b) determine $F$ if $\theta = 30°$.

**Fig. P8-18**

**8–19** A jet aircraft has a mass of 7000 kg. At the beginning of the takeoff run the aircraft is at rest with the brakes applied to the two main wheels $A$ but not to the nose wheel $B$. Determine the total horizontal thrust $T$ during the final check of the engines that would cause the locked wheels to slide. Point $C$ is the mass center of the plane.

**Fig. P8-19**

**8-20** Determine if the uniform rectangular console of 500-lb weight can remain in equilibrium as shown. If yes, calculate the friction force and the location of the normal force on the console. Point C is the location of the mass center.

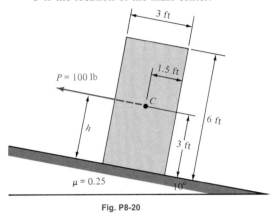

**Fig. P8-20**

**8-21** (a) Determine the minimum magnitude of the force **P**, parallel to the ramp, for impending motion of the console in Prob. 8–20. (b) What is the largest value of h possible for the console to slide rather than tip?

**8-22** A uniform crate weighs 800 lb. Workers on the truck apply a horizontal force **F** until the crate moves on the ramp. Determine the coefficient of friction to make the crate slide rather than tip.

**Fig. P8-22**

**8-23** Determine the magnitude of the counterclockwise couple that would start wheel C to rotate while a braking force of magnitude $P = 100$ lb is applied to arm AB of the brake. $r = 8$ in., $a = 10$ in., $b = 40$ in., and $c = 13$ in.

**Fig. P8-23**

**8-24** Determine the magnitude of the clockwise couple that would start wheel C to rotate while a braking force of magnitude $P = 500$ N is applied to arm AB of the brake in Fig. P8-23. $r = 0.2$ m, $a = 0.25$ m, $b = 0.75$ m, and $c = 0.3$ m.

**8-25** Solve Prob. 8–24 with $c = 0.3$ m and 0.4 m. Plot the clockwise and counterclockwise starting couples (magnitudes only) for the two values of c using the same diagram.

**8-26** Determine the magnitude of the counterclockwise couple that would start wheel C to rotate while a braking force of magnitude $P = 100$ lb is applied to arm AB of the brake. $r = 18$ in., $a = 12$ in., $b = 35$ in., and $c = d = 13$ in.

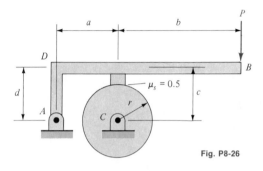

**Fig. P8-26**

**8-27** Solve Prob. 8–26 with $c = 13$ in., $d = 10$ in., 13 in., and 16 in. Plot the clockwise and counterclockwise starting couples (magnitudes only) for the three values of d. Make a single diagram.

**8-28** Consider a static test of brake pads B of a bicycle wheel. Determine the magnitude of the normal force, F, that must act on each pad to hold the wheel at rest on a horizontal road if a horizontal force of magnitude $P = 0.5W = 150$ N is applied to the axle of the wheel.

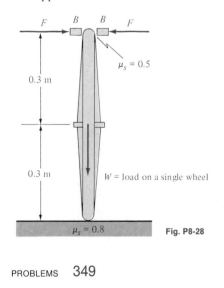

**Fig. P8-28**

**8-29** Consider a wheel of a car on an incline. A 1000-lb force is acting on the wheel spindle as shown. Determine the couple **M** that must be applied at the wheel spindle to hold the wheel at rest. Note that two different motions, sliding and spinning of the wheel, must be prevented.

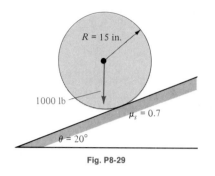

Fig. P8-29

**8-30** An exercise machine is modeled as a uniform wheel $A$ of weight $W$ and radius $r$. The wheel is supported by an idler wheel $B$ which rotates freely, and by the wall $C$. Determine the magnitude of the required couple **M** to start wheel $A$ rotating counterclockwise.

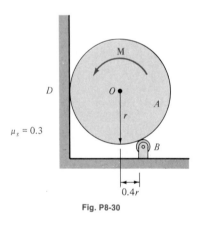

Fig. P8-30

**8-31** An exercise wheel is modeled as resting on a V-block. Write an expression for the magnitude of the couple, $M_C$, that would cause the wheel of weight $W$ to rotate.

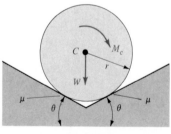

Fig. P8-31

**8-32** The launching ramp of a ship is modeled as a V-block, and the ship is assumed to have a circular cylindrical hull. Determine the magnitude of the force **P** necessary to start a ship of weight $W = 50,000$ lb to move if $r = 10$ ft, $\theta = 5°$, $\phi = 15°$, and $\mu = 0.1$.

(a)

(b) Cross section of hull and ramp

Fig. P8-32

**8-33** There is a so-called *skating force* acting on the tone arms of most record players. This force $\mathbf{F}_s$ appears to push the arm toward the center $C$ of the turntable. Explain this phenomenon by considering friction on the pick-up stylus.

Fig. P8-33

**8-4**

Wedges are among the oldest and simplest machines. They are still very useful in amplifying the effects of small forces or making precise, small adjustments in the positions of bodies. Friction plays an important role in the behavior of every wedge as illustrated in the following.

Figure 8-11a shows a body $A$ of weight $\mathbf{W}$ that is to be raised by driving the wedge $B$ of negligible weight to the left by the force $\mathbf{F}$. Assume that the coefficient of static friction is $\mu_s$ at all surfaces where motion is possible. Two directions of motion must be considered for the given arrangement, with the friction forces always opposing any impending or actual motion.

First assume that wedge $B$ is used to raise body $A$, and the force $\mathbf{F}$ that initiates motion must be determined. The free-body diagram of the wedge under this condition is shown in Fig. 8-11b. The corresponding diagram of body $A$ is shown in Fig. 8-11c. Note that force $\mathcal{F}_3$ is downward because $A$ would move upward if $B$ were driven to the left. Four equations of equilibrium can be written for the forces in the $x$ and $y$ directions from the two free-body diagrams. The magnitudes of the unknown forces $\mathbf{F}$, $\mathbf{N}_1$, $\mathbf{N}_2$, and $\mathbf{N}_3$ can be determined in terms of $W$, $\theta$, and $\mu_s$ from these equations. The problem can also be solved by drawing force polygons (these become force triangles if each pair of normal and friction forces is replaced by their resultant).

It is also worthwhile to check whether the arrangement in Fig. 8-11a is *self-locking*, which means that the wedge $B$ remains in place when the force $\mathbf{F}$ is removed. This situation is characterized by the free-body diagrams in Fig. 8-11d and e. Note the differences in the directions of the friction forces with respect to those in Fig. 8-11b and c. The solution of a specific problem is again straightforward using either the analytical or the graphical method.

**FIGURE 8-11**
Analysis of a wedge

(a)

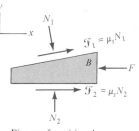

Diagram for raising $A$

(b)

(c)   Diagram for raising $A$

(d)   Diagram to check self-locking

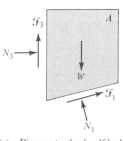

(e)   Diagram to check self-locking

*Screws* represent a most clever application of the concept of wedges: each screw thread may be considered a wedge wrapped on a circular shaft. Screw threads are made in a variety of shapes (square, trapezoidal, V-shape, etc.) for numerous purposes. Of these, square threads are the easiest to analyze since they are the most similar to wedges or to blocks moving on inclined planes. Square-threaded screws are commonly used in jacks, presses, and for transmitting power in machines.

For the general analysis of a square-threaded screw, consider Fig. 8-12a. Assume that a couple **M** turns the screw in its fixed frame so that the screw moves out of the frame against the axial load **P**. The situation is modeled in Fig. 8-12b, where the element represents the screw in the fixed frame. The angle $\alpha$ depends on the mean radius $r$ and of the *lead L* (advancement per revolution) of the screw. For single-threaded screws $L$ is equal to the *pitch p*, the distance between similar points on adjacent threads (multiple-threaded screws are also available; for them, $L = np$, where $n$ is the multiplicity of threads). Thus, from a right triangle formed by unwrapping a complete turn of the single thread,

$$\alpha = \tan^{-1} \frac{L}{2\pi r} = \tan^{-1} \frac{np}{2\pi r} \qquad \boxed{8\text{-}3}$$

In Fig. 8-12b the force **F** represents the turning effect of the moment **M** since $M = Fr$. The reaction of the fixed frame on the screw is the force **R**. If the screw could move without friction, the reaction would be **R'**, normal to the thread. The angle $\phi$ of **R** with the normal to the thread is the angle of friction, so $\tan \phi = \mu$, the appropriate coefficient of friction. For equilibrium of the element in the horizontal and vertical directions,

$$F = R \sin (\alpha + \phi) = \frac{M}{r}$$

$$P = R \cos (\alpha + \phi)$$

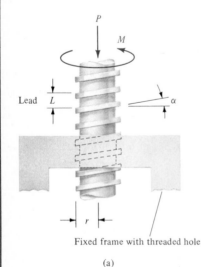

Lead

Fixed frame with threaded hole

(a)

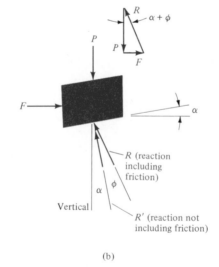

**FIGURE 8-12**

Square-threaded screw

(b)

From these two equations, the magnitude of the moment of the couple required to move the screw against the axial load is

$$M = Pr \tan(\alpha + \phi)$$

8-4

The relative values of $\alpha$ and $\phi$ control whether the screw is self-locking. Three possibilities are shown in Fig. 8-13. If $\phi > \alpha$, the screw can support an axial load without unwinding (Fig. 8-13a). In that case a moment must be applied causing a force **F** opposite to that in Fig. 8-12b if the screw should move in the direction of the force **P**. The magnitude of the couple required to move a self-locking screw in the direction of the axial load is $M_1 = Pr \tan(\phi - \alpha)$. At the opposite extreme, a couple of magnitude $M_2 = Pr \tan(\alpha - \phi)$ must be applied to the screw modeled in Fig. 8-13c to hold the screw stationary, making it equivalent to the model in Fig. 8-13b.

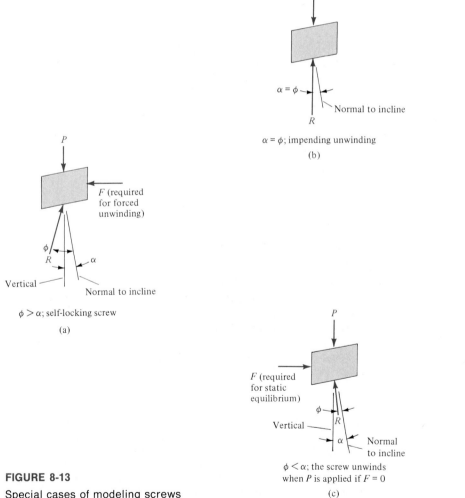

α = φ; impending unwinding

(b)

F (required for forced unwinding)

φ > α; self-locking screw

(a)

F (required for static equilibrium)

φ < α; the screw unwinds when P is applied if F = 0

(c)

**FIGURE 8-13**

Special cases of modeling screws

# EXAMPLE 8-6

In Fig. 8-14, $\theta = 5°$, $\mu_s = 0.3$ on all surfaces, and $m = 1000$ kg. Determine the magnitude of the horizontal force **F** and the force applied by the vertical wall during the process of slowly raising block $A$. What is the horizontal force **F** required for lowering block $A$?

## SOLUTION

Since the block is raised slowly, the equations for static equilibrium are used. From Fig. 8-14c,

$$\sum F_x = 0$$
$$-F + \mu_s N_2 + \mu_s N_1 \cos \theta + N_1 \sin \theta = 0 \qquad \textbf{(a)}$$
$$\sum F_y = 0$$
$$N_2 - N_1 \cos \theta + \mu_s N_1 \sin \theta = 0 \qquad \textbf{(b)}$$

From Fig. 8-14b,

$$\sum F_x = 0$$
$$N_3 - \mu_s N_1 \cos \theta - N_1 \sin \theta = 0 \qquad \textbf{(c)}$$
$$\sum F_y = 0$$
$$-\mu_s N_3 - mg - \mu_s N_1 \sin \theta + N_1 \cos \theta = 0 \qquad \textbf{(d)}$$

From Eq. (c),

$$N_3 = N_1(\mu_s \cos \theta + \sin \theta) = 0.39 N_1$$

Substituting into Eq. (d),

$$N_1 = \frac{mg}{\cos \theta - \mu_s \sin 5° - \mu_s(0.39)} = \frac{(1000 \text{ kg})(9.81 \text{ m/s}^2)}{0.85} = 11.54 \text{ kN}$$

Substituting into Eq. (b),

$$N_2 = 11.54 \text{ kN} \cos 5° - 0.3(11.54 \text{ kN}) \sin 5° = 11.19 \text{ kN}$$

Substituting into Eq. (a),

$$F = 0.3(11.19 \text{ kN}) + (11.54 \text{ kN})(0.3 \cos 5° + \sin 5°) = 7.81 \text{ kN}$$

For lowering the block, the directions of friction forces change. From Fig. 8-14d,

$$\sum F_x = 0$$
$$N_3 + \mu_s N_1 \cos \theta - N_1 \sin \theta = 0$$
$$N_3 = N_1(\sin \theta - \mu_s \cos \theta) = -0.21 N_1$$

A negative value for $N_3$ implies that the vertical wall must apply a pulling force, which it cannot do. Hence, it is seen that the block $A$ is self-locking. This result could have also been found by checking to see if block $A$ has a tendency to slip down the inclined plane of the wedge. A quick check shows that $\tan^{-1} \mu_s = \tan^{-1} 0.3 = 16.7° > 5°$. Thus, block $A$ does not tend to slip, even if the wall were not there, and cannot be lowered simply by changing **F** on block $B$.

(a)

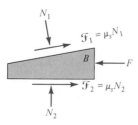

(b) Free-body diagram for raising $A$

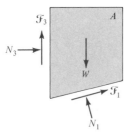

(c) Free-body diagram for raising $A$

(d) Free-body diagram for lowering $A$

**FIGURE 8-14**

Analysis of a wedge

EXAMPLE 8-7

In Fig. 8-15a $W = 3000$ lb and $\mu_s = 0.2$ for all contacting surfaces. Determine the range of the angle $\theta$ for which the wedge is self-locking.

## SOLUTION

The system is self-locking if it remains still after removal of the force $\mathbf{F}$ ($F = 0$). The upper limit on $\theta$ is obtained by considering the case of motion impending for both blocks. From Fig. 8-15c,

$$\sum F_x = 0$$
$$-\mu_s N_2 - \mu_s N_1 \cos\theta + N_1 \sin\theta = 0 \qquad \textbf{(a)}$$
$$\sum F_y = 0$$
$$N_2 - N_1 \cos\theta - \mu_s N_1 \sin\theta = 0 \qquad \textbf{(b)}$$

From Fig. 8-15b

$$\sum F_x = 0$$
$$N_3 - N_1 \sin\theta + \mu_s N_1 \cos\theta = 0 \qquad \textbf{(c)}$$
$$\sum F_y = 0$$
$$\mu_s N_3 - W + N_1 \cos\theta + \mu_s N_1 \sin\theta = 0 \qquad \textbf{(d)}$$

Adding Eq. (b) and (d),

$$N_2 + \mu_s N_3 - W = 0 \qquad \textbf{(e)}$$

Adding Eq. (a) and (c),

$$N_3 - \mu_s N_2 = 0 \qquad \textbf{(f)}$$

Solving (e) and (f) simultaneously yields

$$N_2 = 2885 \text{ lb} \qquad N_3 = 577 \text{ lb}$$

From Eq. (b), assuming $N_2 \simeq N_1 \cos\theta$ and substituting into Eq. (c),

$$N_3 - N_1 \sin\theta \left(\frac{\cos\theta}{\cos\theta}\right) + \mu_s N_2 = 0$$
$$N_3 - N_2 \tan\theta + \mu_s N_2 = 0$$
$$\tan\theta = \frac{N_3 + \mu_s N_2}{N_2} = \frac{577 + 0.2(2885)}{2885} = 0.4 \qquad \theta = 21.8°$$

The lower limit on $\theta$ is of course $\theta = 0$.

(a)

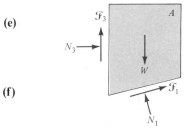

(b) Free-body diagram to check self-locking

(c) Free-body diagram to check self-locking

**FIGURE 8-15**

Analysis of a wedge

EXAMPLE 8-7   355

## EXAMPLE 8-8

(a)

Lead $L$

$\alpha$

Fixed frame with threaded hole

(b)

$P$

$F$

$\alpha$

$R$ (reaction including friction)

$\alpha$  $\phi$

Vertical

$R'$ (reaction not including friction)

A square-threaded screw has a mean diameter of 60 mm, a pitch of 7 mm, and a coefficient of static friction of 0.2. Determine the moment of the couple required to raise the load when the axial force on the screw is 5 kN (Fig. 8-16a).

### SOLUTION

Referring to Fig. 8-16b, the angle of inclination is

$$\alpha = \tan^{-1} \frac{np}{2\pi r} = \tan^{-1} \frac{(1)(7 \text{ mm})}{2\pi(30 \text{ mm})} = 2.13°$$

When motion is just beginning the reaction **R** is its maximum and

$$\phi = \phi_s = \tan^{-1} \mu_s = 11.3°$$

Using a force triangle such as in Fig. 8-16b, the necessary magnitude of the pushing force **F** is found,

$$F = P \tan(\phi + \alpha) = 5 \text{ kN} \tan(11.3° + 2.13°) = 1.19 \text{ kN}$$

The magnitude of the moment caused by **F** which must be overcome by the applied couple **M** is

$$Fr = M = (1.19 \text{ kN})(0.03 \text{ m}) = 35.7 \text{ N·m}$$

The solution shows that the necessary moment to raise the load decreases as the coefficient of friction decreases and also as the pitch decreases.

$R$

$\alpha + \phi$

$P$

$F$

**FIGURE 8-16**

Analysis of a square-threaded screw

## EXAMPLE 8-9

$P$

$\alpha = \phi$

Normal to incline

$R$

$\alpha = \phi$; impending unwinding

**FIGURE 8-17**

Analysis of a jackscrew

The jackscrew for a sports car is designed with a mean diameter of 0.6 in. The maximum load on the jack is 1200 lb. What is the required number of threads per inch on the single-threaded screw if unwinding is to be prevented even with the minimum coefficient of friction of 0.1?

### SOLUTION

From Fig. 8-17, a screw is just self-locking if the angle of inclination equals the angle of repose or when

$$\alpha = \tan^{-1} \frac{np}{2\pi r} = \tan^{-1} \mu_s = \tan^{-1}(0.1) = 5.71°$$

$$\frac{np}{2\pi r} = \frac{(1)p}{2\pi(0.3 \text{ in.})} = 0.1$$

$$p = 2\pi(0.3 \text{ in.})(0.1) = 0.188 \text{ in.}$$

Since $p$ is the distance between adjacent threads, the number of adjacent threads per inch equals $1/p = 5.31$ threads/in. It should be noted that the result was independent of the magnitude of the load.

**8–34** Determine whether the force **F** acting on the weightless wedge $B$ is sufficient to raise block $A$ if $\mu = 0.3$ at all surfaces and $\theta = 6°$.

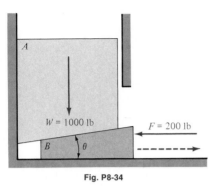

**Fig. P8-34**

**8–35** Plot the magnitude of the force **F** acting on wedge $B$ in Fig. P8-34 to lower block $A$. Use $W = 500$ lb, 1000 lb, and 2000 lb. $\mu = 0.3$ at all surfaces.

**8–36** A 3000-lb machine is modeled as a uniform block $A$ that is to be moved away from a wall using a wedge $B$. The coefficient of static friction is 0.3 at all surfaces. Determine the magnitude of the force **P** required to start block $A$ moving using (a) $\theta = 3°$, and (b) $\theta = 6°$.

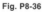

**Fig. P8-36**

**8–37** Block $A$ of 100-kg mass is to be raised using wedges $B$ and $C$, which have negligible weight. The coefficient of friction is 0.4 at all surfaces. Determine the magnitude of the horizontal force **F** that can move the block $A$ upward if $\theta = 8°$.

**Fig. P8-37**

**8–38** Determine the magnitude of the horizontal force **F** acting on wedge $C$ in Fig. P8-37 to lower block $A$. $\theta = 4°$, and $\mu = 0.4$ on all surfaces.

**8–39** Plot the magnitude of the force **F** as a function of $\theta = 4°$, $8°$, and $12°$ in Prob. 8–38. Estimate from the diagram the angle $\theta$ at which the wedge ceases to be self-locking.

**8–40** Determine the magnitude of the force **F** acting on the weightless wedge $B$ to raise block $A$ if block $A$ has a mass of 100 kg, $\theta = 5°$, $\mu_1 = 0.2$, $\mu_2 = 0.3$, and $\mu_3 = 0.5$.

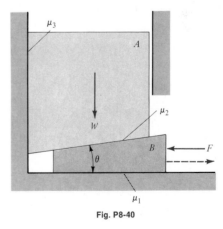

**Fig. P8-40**

**8–41** Determine the magnitude of the force **F** acting on wedge $B$ in Fig. P8-40 to lower block $A$ if block $A$ weighs 2000 lb, $\theta = 4°$, $\mu_1 = 0.2$, $\mu_2 = 0.4$, and $\mu_3 = 0.3$.

**8–42** Two equal blocks are to be separated by driving a wedge between them. Write an expression for the required magnitude of the force **P**, assuming that the coefficient of friction is $\mu$ at all surfaces, and that the two blocks start moving simultaneously.

**Fig. P8-42**

**8–43** A log-splitting machine is modeled in Fig. P8-43. Plot the magnitude of the required driving force **F** as a function of $\theta = 3°$, $6°$, and $9°$. Assume that in all cases the friction force from the wood on the wedge is 5 kN and the coefficient of friction is 0.5.

**Fig. P8-43**

**8–44** Consider the model of a mechanism in which wedges $A$ and $B$ have negligible weight. Determine the required coefficient of friction $\mu_{AB}$ to make the system self-locking (if possible) when a force $\mathbf{F}_A$ is applied.

**Fig. P8-44**

**8–45** The jack of a sports car is modeled as a threaded rod $C$ moving in the fixed base $A$. The rod is turned by applying a couple **M** to the arms $B$. The mean diameter of the screw is $d = 20$ mm. The pitch is $p = 1.4$ mm, and the coefficient of static friction is 0.3. Determine the magnitude of the couple necessary to (a) raise and (b) lower the jack when $P = 10$ kN.

**Fig. P8-45**

**8–46** Consider the two square-threaded screw columns of a mechanical testing machine. Each column has a mean diameter of 8 mm, a coefficient of friction of 0.25, and a maximum axial load of 100 kN. Determine the necessary magnitude of the couple $M$ for turning each column to raise the cross head if the pitch of the single thread is $p = 8$ mm.

**Fig. P8-46**

**8–47** Plot $M$ vs. the pitch $p = 8$ mm, $10$ mm, and $12$ mm in Prob. 8–46. Discuss, on the basis of the diagram, whether the pitch of the screws is an important factor in choosing a motor for turning the columns.

**8–48** One threaded rod has right-handed threads, the other left-handed threads in the turn-buckle. The mean diameter of the screws is $d = 1$ in., the pitch of the single threads is $p = 0.08$ in., and the coefficient of friction is 0.2. Determine the moments with which the turnbuckle could be slightly tightened or loosened if $T = 10{,}000$ lb. The two rods cannot rotate.

Fig. P8-48

**8–49** The tension in a guy wire is to be set approximately using a turnbuckle (Fig. P8-48). The mean diameter of the screws is $d = 24$ mm, the pitch of the single threads is $p = 2$ mm, and the coefficient of friction is 0.25. Determine the tension in the nonrotating rods that is set by applying a torque of 300 N·m to the turnbuckle.

**8–50** Shaft $B$ is to be pushed out of pulley $A$ by turning the double-threaded screw in the clamp $C$. The mean diameter of the screw is $d = 0.8$ in., the pitch is $p = 0.05$ in., and the coefficient of friction is 0.2. The screw has negligible friction at point $D$. Calculate the total force that the clamp applies to the pulley when the screw is tightened with a torque of 150 lb·ft.

Fig. P8-50

**8–51** Blocks $A$ and $B$ are pressed together in a single-threaded vise. The mean diameter of the screw is $d = 20$ mm, the pitch is $p = 1.9$ mm, and the coefficient of friction is 0.3. Determine the torque necessary to apply a compressive force of 3-kN magnitude to the blocks. There is negligible friction at $D$.

Fig. P8-51

**8–52** At what pitch would the screw in Prob. 8–51 not be able to compress the blocks if $M$ were zero?

**8–53** The large locking nuts $A$ of a mechanical testing machine are designed to be tightened by hand. The mean diameter of a single-threaded column is $d = 3$ in., the pitch is $p = 0.2$ in., the coefficient of friction is 0.25 at the threads and assumed negligible at the cross-head's surface. Calculate the axial force on a nut when a 100-lb·ft torque is applied to tighten it

Fig. P8-53

**8–54** Determine the magnitude of the smallest force at point $B$ in Fig. P8-53 that would loosen the locking nut $A$ if $l = 8$ in. Assume that the nut was tightened as described in Prob. 8–53.

**8–55** A tube-reaming device is designed to have a single-threaded screw of $d = 40$ mm mean diameter, pitch $p = 5$ mm, and coefficient of friction of 0.3. Estimate the axial force on the tube when the net torque driving the tool into the tube is 400 N·m.

Fig. P8-55

**8–56** Estimate the axial forces in Prob. 8–55 for single-, double-, and triple- threaded screws, with everything else being the same.

## DISK FRICTION

The friction between flat surfaces in relative rotary motion must be analyzed in a number of devices. These include end bearings and collar bearings of shafts for axial support, disk clutches, and disk brakes.

For a general analysis of disk friction, consider a flat-ended hollow shaft that is rotated by a couple **M** while being pressed against another flat surface as in Fig. 8-18. Assume that the same coefficient of friction $\mu$ is valid for the whole ring-shaped area (outer radius $R_o$, inner radius $R_i$), and that the pressure is constant over the whole area. The magnitude of the normal force $dN$ on an elemental area $dA$ is

$$dN = \frac{P}{A}\, dA = \frac{P}{\pi(R_o^2 - R_i^2)}\,(r\, d\theta\, dr)$$

The magnitude of the frictional force on the area $dA$ is $\mu\, dN$, and the magnitude of the moment of this frictional force about the axis of the shaft is approximately $dM = \mu\, dN\, r$. The total moment caused by friction on the whole area is

$$M = \int \mu\, dN\, r = \frac{\mu P}{\pi(R_o^2 - R_i^2)} \int_0^{2\pi} \int_{R_i}^{R_o} r^2\, dr\, d\theta \qquad \boxed{8\text{-}5}$$

$$= \frac{2}{3}\,\mu P\, \frac{R_o^3 - R_i^3}{R_o^2 - R_i^2}$$

For a solid disk with radius $R$, Eq. 8-5 becomes

$$M = \frac{2}{3}\,\mu P R \qquad \boxed{8\text{-}6}$$

Equations 8-5 and 8-6 may be used in static or dynamic situations with the appropriate coefficient of friction, $\mu_s$ or $\mu_k$. In the static case they give the largest torque that can be transmitted without slippage. In the case of relative motion, they give the moment required to maintain motion against the frictional resistance. It has been found that the frictional moment decreases for well-worn surfaces to about 75% of its original value for new surfaces (solid disks).

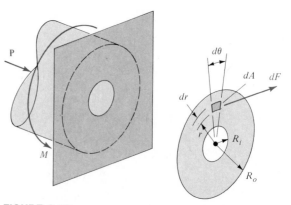

**FIGURE 8-18**
Model for analyzing disk friction

The lateral support of rotating axles is provided by bearings. The simplest of these are called *journal bearings* or *sliding bearings*. Such a bearing consists of a stationary member with a hole that is slightly larger than the shaft it has to support, as shown in Fig. 8-19. In dry or partially lubricated bearings the shaft makes contact with the bearing on a long and narrow area parallel to the centerline of the shaft. Such bearings can be analyzed using the principles of dry friction.

The rotating shaft in a journal bearing climbs a little as shown by the point of contact $A$ in the cross-sectional view in Fig. 8-19a. The height of the climb is limited by the inevitable (and desirable) slippage that occurs. The frictional resistance of the bearing can be determined from the climbing displacement of the shaft in the bearing.

Assume that the vertical load applied to a shaft is **P** in Fig. 8-19. The free-body diagram of the shaft is drawn in Fig. 8-19b. Here **M** is the torque required to maintain steady motion with slippage occurring at point $A$. For equilibrium of forces in the vertical direction, the reaction **R** acting on the shaft must be equal and opposite to **P**, with parallel lines of action. The applied torque **M** must be large enough to overcome the resisting couple formed by **P** and **R** with the distance $r_f$ between them. If **M** is reversed to act counterclockwise, the reaction **R'** acts at point $A'$ in Fig. 8-19b, at the same distance $r_f$ from the line of action of **P**. The circle that can be drawn with the radius $r_f$ is called the *circle of friction* which is a constant for a given journal bearing.

The reaction **R** of the bearing can be conveniently resolved into the normal for **N** and the friction force $\mathscr{F}$ as in Fig. 8-19b. The angle $\phi$ between **R** and **N** is related to the coefficient of kinetic friction $\mu_k$ according to Eq. 8-2, $\tan \phi = \mu_k$. For the equilibrium of moments about the center of the shaft $C$,

$$M = Rr_f = \mathscr{F}r = Rr \sin \phi \qquad \boxed{8\text{-}7}$$

Properly constructed bearings have low coefficients of friction, so $\sin \phi \simeq \tan \phi = \mu_k$. This leads to the following approximate formulas for the magnitude of the friction moment $M$ and the radius $r_f$ of the circle of friction,

$$M = Rr\mu_k = Pr\mu_k \qquad \boxed{8\text{-}8}$$

$$r_f = r\mu_k \qquad \boxed{8\text{-}9}$$

Thus, the friction moment depends on the applied load **P** and the properties of the bearing, whereas the circle of friction is independent of the load.

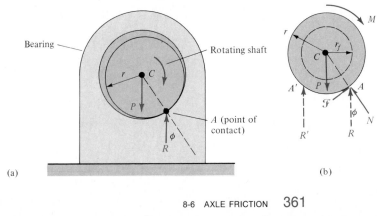

**FIGURE 8-19**

Cross section of an axle
in a sliding bearing    (a)

(b)

## ROLLING RESISTANCE

Wheels and balls are very useful in minimizing the resistance to motion of bodies. This does not mean, however, that the friction of a rolling wheel or ball can be zero. To give a qualitative example, the rolling resistance of tires is the major part of the total losses in cars traveling at speeds up to about 70 km/h (at higher speeds air drag becomes the largest resistance; the size and shape of the vehicle may greatly affect the speed at which the transition occurs). Steel rolling on steel is much more efficient than rubber tires, but it still cannot eliminate friction.

The rolling resistance of any wheel, not counting axle friction, is caused by deformations of the wheel and its track at the region of mutual contact. In the extreme, a soft wheel may deform considerably on a relatively rigid track (Fig. 8-20a), or a relatively rigid wheel may cause much deformation of a soft track (Fig. 8-20b). The deformations of rigid wheels and tracks may be extremely small and therefore difficult to measure. In any case, the internal friction of the deforming materials is the cause of the rolling resistance. Internal friction is somewhat similar to dry friction between bodies in contact. Both generate heat when displacement occurs as can be easily demonstrated by rapidly rubbing the hands together (surface friction) or bending soft wires back and forth (internal friction).

A wheel with rolling resistance can be modeled as in Fig. 8-21. Here **P** is the vertical load on the wheel, and **F** is the minimum horizontal force that can keep the wheel rolling at constant speed. The reaction **R** from the track is the resultant of all distributed forces between the track and the wheel. Assume that the length $l$ of the contact patch (Fig. 8-21) is much shorter than the radius $r$ of the wheel, and that, in Fig. 8-21, the moment arm of **F** is approximately $r$. Thus,

$$\sum M_A = 0, \quad Pa \simeq Fr \qquad \boxed{8\text{-}10}$$

The distance $a$ is taken as the *coefficient of rolling resistance;* it has units of length. The values of $a$, which must be experimentally determined, range upward from the low of about 0.005 mm for hardened steel on hardened steel.

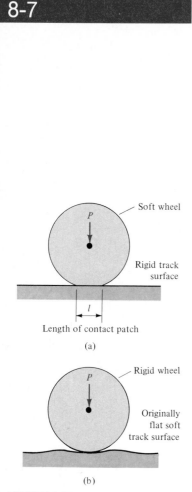

**Length of contact patch**

(a)

**Rigid wheel**

**Originally flat soft track surface**

(b)

**FIGURE 8-20**

Deformations of wheel or track cause resistance to rolling

**FIGURE 8-21**

Model for analyzing rolling resistance

EXAMPLE 8-10

Determine the maximum torque that can be transmitted without slippage by the disk clutch shown in Fig. 8-22. The disk plates are new and they are solid circles. How much would the maximum torque change if the disk plates were hollow circles with $R_o = 0.14$ m and $R_i = 0.07$ m?

## SOLUTION

The disks transmit the maximum torque when slipping is impending. From Eq. 8-6, for solid shafts, the torque that may be applied is

$$M_S = \frac{2}{3}\mu_s PR_o = \frac{2}{3}(0.7)(4 \text{ kN})(0.14 \text{ m}) = 0.26 \text{ kN·m}$$

For the hollow shaft, from Eq. 8-5,

$$M_H = \frac{2}{3}\mu_s P \frac{R_o^3 - R_i^3}{R_o^2 - R_i^2} = \frac{2}{3}(0.7)(4 \text{ kN})\frac{(0.14^3 - 0.07^3) \text{ m}^3}{(0.14^2 - 0.07^2) \text{ m}^2}$$

$$= 0.3 \text{ kN·m}$$

$$\text{Percent change} = \frac{M_H - M_S}{M_S} \times 100\%$$

$$= \frac{0.3 - 0.26}{0.26} \times 100\% = 15.4\%$$

## JUDGMENT OF THE RESULTS

It is surprising that the hollow rings transmit a larger torque than the solid circles. After checking the calculations, the results are still the same. Upon further thought, these results make sense. The torque caused by friction increases with the moment arm from the center and also with normal force exerted at each point on the surfaces in contact. By reducing the area in contact, the normal force at each point is increased. Hence, even though some resistance is lost due to loss of material, the maximum torque is increased due to the larger normal force at those points that also have the largest moment arms.

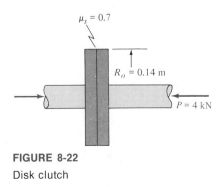

**FIGURE 8-22**
Disk clutch

EXAMPLE 8-10   363

## EXAMPLE 8-11

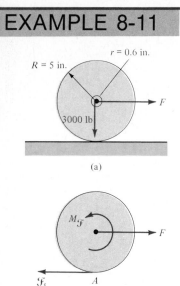

$r = 0.6$ in.

$R = 5$ in.

$F$

3000 lb

(a)

$M_{\mathscr{F}}$

$F$

$\mathscr{F}_s$       $A$

(b)

**FIGURE 8-23**

Analysis of axle friction

A small hoist is supported by two 10-in.-diameter wheels, with each wheel rotating on a 1.2-in.-diameter shaft. Assuming that the major friction is at the wheel axles with $\mu_s = 0.2$ and $\mu_k = 0.17$, determine the magnitude of the force **F** required to start the hoist moving horizontally (Fig. 8-23a).

### SOLUTION

From Fig. 8-23b it is seen that the friction moment $M_{\mathscr{F}}$ is overcome by the moment caused by **F** and $\mathscr{F}_s$, the friction force at the surface. Summing moments about point $A$ and using Eq. 8-8,

$$\sum M_A = 0 \qquad \text{(at impending motion)}$$

$$M_{\mathscr{F}} - FR = 0$$

$$Pr\mu_s - FR = 0$$

$$F = \frac{(3000 \text{ lb})(0.6 \text{ in.})(0.2)}{5 \text{ in.}} = 72 \text{ lb}$$

The actual force applied must be slightly greater than 72 lb.

## EXAMPLE 8-12

The uniform wheel shown in Fig. 8-24a has a rolling coefficient of friction $a = 2$ mm on a steel surface. Determine the largest angle $\theta$ with which the wheel can remain at rest.

### SOLUTION

Using Fig. 8-24b, the weight of the wheel can be broken into components that are analogous to forces **P** and **F** in Fig. 8-21.

$$\sum M_O = 0$$

$$W(\sin \theta)(R) - W(\cos \theta)(a) = 0$$

$$R \sin \theta = a \cos \theta$$

$$\frac{\sin \theta}{\cos \theta} = \tan \theta = \frac{a}{R} = \frac{2 \text{ mm}}{300 \text{ mm}}$$

$$\theta = 0.38°$$

$M = 50$ kg
$R = 0.3$ m

$\theta$

(a)

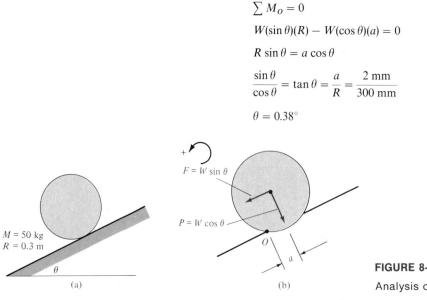

$+$

$F = W \sin \theta$

$P = W \cos \theta$

$O$

$a$

(b)

**FIGURE 8-24**

Analysis of rolling resistance

**8–57** Plot the calculated torque required to start the vertical shaft rotating vs. the diameter $d$ for $d = 20$ mm, 50 mm, and 100 mm. Extrapolate the curve and estimate the torque at $d = 0$. Make a calculation to check the estimation.

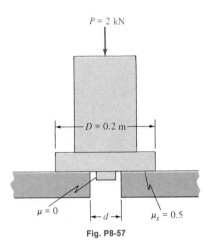

$P = 2$ kN

$D = 0.2$ m

$\mu = 0$   $d$   $\mu_s = 0.5$

**Fig. P8-57**

**8–58** In Fig. P8-57, $d = 2$ in. and $P = 3000$ lb. Determine the largest allowable value of $D$ if the starting torque should not exceed 350 lb·ft.

**8–59** The load-bearing face $A$ of a bolt has an outer diameter $D = 40$ mm, an inner diameter $d = 25$ mm, and a coefficient of friction of 0.3. Calculate the original tension in the bolt if it is loosened with a torque of 230 N·m.

$A$

**Fig. P8-59**

**8–60** In Prob. 8–50, the tip of the screw at $D$ is flat with a 0.3-in. diameter and a coefficient of friction of 0.25. Determine the additional torque necessary to overcome the friction at $D$, using the axial force calculated in Prob. 8–50.

$C$

$\mu = 0.2$

$A$

$B$

$D$

**Fig. P8-60**

**8–61** In Prob. 8–51, the tip of the screw at $C$ is flat with a 16 mm diameter and a coefficient of friction of 0.3. Determine the required torque in this case, with everything else being the same.

Glued joint

$\mu = 0.3$

$A$   $B$   $C$   $D$   $M$

**Fig. P8-61**

**8–62** The load-bearing flat face of the locking nut $A$ in Prob. 8–53 has an outer diameter $D_o = 4.4$ in., an inner diameter $D_i = 3.2$ in., and a coefficient of friction of 0.25. Determine the total torque necessary for applying an axial force of 1000-lb magnitude by tightening the nut.

$l$

$M$

$A$   $A$   $B$

Cross-head

**Fig. P8-62**

**8–63** A flat buffing disk of diameter $D = 0.12$ m is gripped in a small electric drill and rotated clockwise. The disk is pressed uniformly against a flat surface with a force of magnitude $P = 100$ N. Determine the magnitude of the couple necessary at the handle of the drill to hold it steady if the coefficient of friction at the surface being polished is 0.7.

Fig. P8-63

**8–64** Each brush of the street sweeper has an outer diameter $D = 4$ ft and an inner diameter $d = 1$ ft at the ground. The vertical load on a brush spindle is 50 lb, and the coefficient of friction at the ground is 0.5. Calculate the moment reaction on the spindle arm $A$ caused by rotation of the brush, assuming that it has uniform contact with the ground.

Fig. P8-64

**8–65** A disk brake is modeled as a flat disk $A$ sandwiched between circular pads $B$ and $C$. There are four clamping forces of magnitude $P = 100$ lb, but the normal force per unit area is uniform on each side of disk A. Determine the magnitude of the couple, $M$, that can turn disk $A$ relative to plates $B$ and $C$ if $D = 8$ in. and $d = 2$ in.

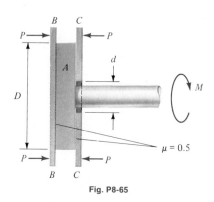

Fig. P8-65

**8–66** An inventor wishes to improve pointe shoes used in ballet. To analyze the frictional resistance of a shoe during a turn as shown, it is assumed to have a circular, flat tip at $A$ of diameter $D = 40$ mm and coefficient of friction of 0.3. Calculate the magnitude of the moment, $M$, of the couple that a dancer weighing 500 N must apply to start turning on this shoe.

Fig. P8-66

**8–67** Plot the required starting moment $M$ as a function of $D = 30$ mm, 40 mm, and 50 mm in Prob. 8–66. If possible, state the ideal diameter $D$ on the basis of the diagram, assuming that a small moment is advantageous for turning easily and gracefully, while a large diameter $D$ is desirable for stability in any elevated position on the toes.

**8–68** Determine the magnitude of the couple that must be applied to the bar $B$ to keep it rotating about the axle $A$, which has a diameter $d = 30$ mm.

$\mu_k = 0.25$

$P = 600$ N

**Fig. P8-68**

**8–69** Plot the moment as a function of $d = 20$ mm, 30 mm, and 40 mm in Prob. 8–68. Use two values of the lateral load, $P = 300$ N and 600 N, and show all results on the same diagram.

**8–70** The upper hinge of a heavy door has a 0.5-in.-diameter vertical shaft and a coefficient of friction of 0.3. This hinge is designed to support the door only horizontally. Determine the magnitude of the moment caused by friction in the hinge as the door is turned.

$P = 500$ lb

**Fig. P8-70**

**8–71** A pulley of diameter $D = 0.3$ m is supported by a fixed axle of diameter $d = 32$ mm. The coefficient of kinetic friction at the axle is 0.3. Determine the magnitude of the smallest force **P** at $\theta = 0$ for raising the load of mass $m = 200$ kg.

$m$

**Fig. P8-71**

**8–72** Determine the magnitude of the smallest force **P** in Prob. 8–71 if the load is to be slowly lowered. Assume that $\theta = 0$.

**8–73** Each pulley has a 12-in. diameter and rotates on a 0.75-in.-diameter shaft, where the coefficient of friction is 0.2. Determine the tension in segments $B$ and $C$ of the rope when the load $W = 400$ lb is slowly raised.

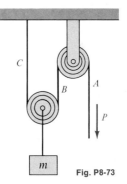

$m$

**Fig. P8-73**

**8–74** Determine the tension in segments $A$, $B$, and $C$ of the rope in Prob. 8–73 when the load is slowly lowered.

**8–75** In a remote region a hand-operated well includes a drum of radius $R = 0.15$ m on a steel shaft of radius $r = 13$ mm. The wooden bearings $A$ and $B$ are 1 m apart with the load of mass $m = 25$ kg midway between them. The coefficient of friction in each bearing is 0.3. Determine the magnitude of the couple required to slowly raise the load.

**Fig. P8-75**

**8-76** A stepped shaft loosely fits in a hole drilled through a flat plate. Axial and lateral forces act on the shaft as shown (assume that the shaft has no tendency to tip away from the vertical). Calculate the minimum torque required to keep the shaft turning at constant speed. $\mu_k = 0.3$ at all surfaces.

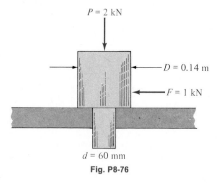

$P = 2$ kN

$D = 0.14$ m

$F = 1$ kN

$d = 60$ mm

**Fig. P8-76**

**8-77** The four wheels of a crane's carriage share equally in supporting the 40,000 lb load. The wheel diameters are 10 in., and the axles have 2 in. diameters with coefficient of kinetic friction of 0.2. Calculate the magnitude of the horizontal force **F** required to maintain steady motion of the carriage if rolling friction is negligible. Assume that $F$ is evenly distributed between all four wheels.

$D$ $d$

$F$

40 kips

**Fig. P8-77**

**8-78** In Prob. 8–77, the coefficient of rolling friction is $a = 0.02$ in. for each wheel. Determine the component of the horizontal force **F** required to overcome the rolling friction.

**8-79** A horizontal force of magnitude $F = 1$ kN is required to start a car of mass $m = 1800$ kg rolling on a horizontal road. The tires are 0.6 m in diameter. Determine the coefficient of rolling resistance assuming that other frictions are negligible and that the weight and the applied force are evenly distributed between all four wheels.

$F = 1$ kN

$d = 0.6$ m

**Fig. P8-79**

**8-80** A uniform wheel of radius $r = 0.1$ m and mass $m = 10$ kg is tested for its rolling resistance on a slightly inclined track. Determine the smallest angle $\theta$ of the incline with the horizontal on which the wheel rolls under the force **W** alone. The coefficient of rolling friction is constant, $a = 0.5$ mm.

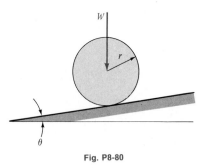

$W$

$r$

$\theta$

**Fig. P8-80**

**8-81** (a) Plot the angle $\theta$ vs. the radius $r$ in Prob. 8–80. Let $r = 0.1$ m, 0.15 m, and 0.2 m. (b) Plot the angle $\theta$ vs. the weight of the wheel in Prob. 8–80. Let $m = 5$ kg, 10 kg, and 15 kg.

The friction properties of belts, ropes, and wires on pulleys and drums are important in the design of belt drives, band brakes, and gripping devices for flexible members. The main features of belt friction are illustrated in Fig. 8-25. A belt under tensions $T_1$ and $T_2$ is laid on a drum making contact from $A$ to $B$. Generally, $T_1 \neq T_2$. A force $\mathbf{R}$ and a couple $\mathbf{M}$ act on the drum to oppose the resultant force and moment of $\mathbf{T}_1$ and $\mathbf{T}_2$ as shown in Fig. 8-25a. Part b of the figure shows the free-body diagram of the belt with distributed normal forces $\mathbf{N}$ and friction $\mathscr{F}$ caused by the drum. The friction forces $\mathscr{F}$ produce a counterclockwise moment here if $T_2 > T_1$.

For a detailed analysis of belt friction and the tensions $T_1$ and $T_2$, assume that the belt is on the verge of sliding to the right on the drum shown in Fig. 8-25a. The free-body diagram of an infinitesimal element $dl$ of the belt is drawn in Fig. 8-25c. Here the elemental friction force $d\mathscr{F}$ applies for impending motion under the obvious net tension to the right. For equilibrium in the $x$ direction, $\sum F_x = 0$,

$$(T + dT) \cos \frac{d\theta}{2} - T \cos \frac{d\theta}{2} - \mu_s \, dN = 0$$

This equation can be greatly simplified since $d\theta$ is infinitesimally small, so $\cos (d\theta/2) \simeq 1$. Thus,

$$dT = \mu_s \, dN \qquad \boxed{8\text{-}11}$$

(a)

(b)

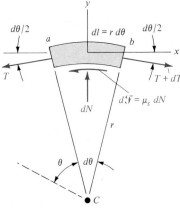

(c)

**FIGURE 8-25**

Analysis of belt friction

For equilibrium in the $y$ direction, $\sum F_y = 0$,

$$dN - T \sin \frac{d\theta}{2} - (T + dT) \sin \frac{d\theta}{2} = 0$$

This can be reduced using the following approximations that are valid for differential quantities:

$$\sin \frac{d\theta}{2} \simeq \frac{d\theta}{2} \quad \text{and} \quad dT \sin \frac{d\theta}{2} \simeq 0.$$

With these,

$$dN = T \, d\theta \qquad \boxed{8\text{-}12}$$

Equations 8-11 and 8-12 can be combined to eliminate the quantity $dN$, resulting in

$$\frac{dT}{T} = \mu_s \, d\theta \qquad \boxed{8\text{-}13}$$

The ultimate aim of relating the belt friction to the tensions $T_1$ and $T_2$ can be achieved by integrating Eq. 8-13 between the appropriate limits of tension and angle of contact,

$$\int_{T_1}^{T_2} \frac{dT}{T} = \int_0^\beta \mu_s \, d\theta \qquad \boxed{8\text{-}14}$$

$$\ln \frac{T_2}{T_1} = \mu_s \beta$$

or

$$\boxed{T_2 = T_1 e^{\mu_s \beta}} \qquad \boxed{8\text{-}15}$$

The total angle of belt contact $\beta$ must be expressed in radians. For a member wrapped around a drum $n$ times, the angle $\beta$ is $2\pi n$. It should be noted that the derivations above are valid for $T_2 > T_1$ and for impending slipping. If slipping occurs, $\mu_s$ may be replaced by the coefficient of kinetic friction $\mu_k$. The equations are not valid when there is no slipping or impending slipping.

The most common belts in belt drives are V belts, which make contact on two slanted sides with the pulleys as shown in cross section in Fig. 8-26. The relation of the high and low tensions in such a belt is derived in a

FIGURE 8-26

Cross section of a V-belt on a pulley

similar procedure to that used above. The details of the derivation are left as an exercise for the students. For a V belt of angle $2\phi$, the final result is

$$T_2 = T_1 e^{\mu_s \beta / \sin \phi} \qquad \boxed{\text{8-16}}$$

This equation has the same limitations as those for flat belts.

## FLUID FRICTION

There is always a *drag force* (or resistance) on a body when there is relative motion between the body and a surrounding fluid. The details of analyzing drag are somewhat complex and should be learned in fluid mechanics. It is useful, however, to learn the basic qualitative aspects of fluid friction at this stage since practical problems often involve this kind of friction along with others.

In simplistic form, the magnitude of the drag force **D** is expressed as

$$D = Cv^2 \qquad \boxed{\text{8-17}}$$

where $C$ is a constant depending mainly on the size and shape of the body and the properties of the fluid. $v$ is the relative velocity between the body and the fluid (either one may be stationary while the other is moving).

It is useful to distinguish between *form* (or *pressure*) *drag* and *friction* (or *surface*) *drag*. Form drag depends strongly on the projected area of the body normal to the direction of the flow. Friction drag depends on the friction of the surface in a given fluid and on the surface area parallel to the flow. In both cases the coefficients in the equation for the drag must be determined experimentally.

To illustrate the difference between the form and friction drags, consider the belt drive in Fig. 8-27. The spokes $A$ of the pulley "cut" through the air, so their projected areas normal to the flow should be minimized to reduce the drag. The same spokes also have surface friction, so their areas parallel to the flow should also be minimized. The outer rim $B$ of the pulley and the belt $C$ have only friction drag, which means that their shapes are much less important than their surface areas and the quality of their surfaces.

It is important that any kind of drag depends on the square of the velocity of relative fluid motion. In most automobiles this becomes a very important factor at speeds of about 70 km/h or higher. Clearly, even small reductions in the operating speeds of vehicles and other machines are highly effective in reducing the energy losses caused by drag.

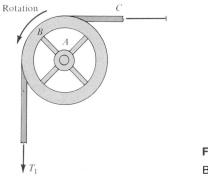

**FIGURE 8-27**

Belt drive

# EXAMPLE 8-13

Determine the resisting tension $T_1$ in the rope shown in Fig. 8-28 if the pulling tension is $T_2 = 500$ N. Assume that $\mu_s = 0.2$ and that slippage of the rope is impending. Also assume that both $T_1$ and $T_2$ are horizontal.

## SOLUTION

From Eq. 8-15, the angle $\beta$ is all that is needed to solve for $T_1$. The rope is wrapped $1\frac{1}{4}$ times around the post; hence $\beta = 2n\pi = 2\pi(1.25) = 7.85$ rad. Therefore,

**FIGURE 8-28**

Rope wrapped on a small post of a boat

$$T_1 = \frac{T_2}{e^{\mu_s\beta}} = \frac{500 \text{ N}}{e^{(0.2)(7.85)}} = 104 \text{ N}$$

# EXAMPLE 8-14

(a) Determine which arrangement shown in Fig. 8-29 is best to prevent clockwise slippage of the rope around the posts. Thus, the minimum resisting tension $T_1$ is sought. (b) What would $T_1$ have to be in each case to cause counterclockwise slippage of the rope (with $\mathbf{T}_1$ pulling and $\mathbf{T}_2$ opposing it)? $\mu_s = 0.3$.

## SOLUTION

(a) **ONE POST:** $\beta = 4\pi$ rad (since the rope is wrapped around the post twice),

$$T_1 = \frac{T_2}{e^{\mu_s\beta}} = \frac{2000 \text{ lb}}{e^{(0.3)(4\pi)}} = 46.1 \text{ lb}$$

**TWO POSTS:** This case should be split into two calculations, one for calculation of $T_1'$ on the left side of the right post and then using this as the pulling tension ($T_2$) for the left post.

$$\beta_1 = \beta_2 = 2\pi \text{ rad}$$

$$T_1' = \frac{T_2}{e^{\mu_s\beta_1}} = \frac{2000 \text{ lb}}{e^{(0.3)(2\pi)}} = 304 \text{ lb}$$

$$T_1 = \frac{T_1'}{e^{\mu_s\beta_2}} = \frac{304 \text{ lb}}{e^{(0.3)(2\pi)}} = 46.1 \text{ lb}$$

Both cases yield the same result.

(b) For the case of slippage to the left, let $T_1''$ be the tension to the left and remember that the larger tension must be in the numerator. Now Eq. 8-15 yields

$$T_1'' = T_2 e^{\mu_s\beta} = (2000 \text{ lb})e^{(0.3)(4\pi)} = 86{,}750 \text{ lb}$$

The same result is valid for both cases on the basis of the results in part (a).

## JUDGMENT OF THE RESULTS

$T_1$            $T_2 = 2000$ lb

(a) Rope is wrapped around post two times

The fact that using one or two posts yields the same result should not be too surprising. From a physical point of view the two cases are the same in that both involve a rope wrapped two full revolutions around the same material. From a mathematical point of view it is seen that the calculations for two posts reduce to

$T_1$            $T_2 = 2000$ lb

(b) Rope is wrapped around each post once

**FIGURE 8-29**

Cross sections of posts with ropes wrapped around them

$$T_1 = \frac{T_1'}{e^{\mu_s\beta_2}} = \frac{T_2/e^{\mu_s\beta_1}}{e^{\mu_s\beta_2}} = \frac{T_2}{e^{(\mu_s\beta_1 + \mu_s\beta_2)}} = \frac{T_2}{e^{\mu_s(\beta_1 + \beta_2)}}$$

**8–82** Determine the resisting tension $T_1$ if the pulling tension $T_2$ is 400 N, and $\mu_s = 0.2$. Assume that both $\mathbf{T}_1$ and $\mathbf{T}_2$ act horizontally and that slipping of the rope is impending. The angle of contact on the post is 270°.

Fig. P8-84

**8–85** Plot the tension $T_1$ vs. the number of turns $n = 0.75$, 2, and 3 in Prob. 8–84. Extrapolate the curve to $n = 0$ and $n = 10$. Discuss the practical significance of the results, assuming that a worker can easily apply a tension of $T_1 \simeq 500$ N.

Fig. P8-82

**8–86** A broken locking mechanism prevents the pulley from rotation. Determine the required tension $T$ to raise the load by sliding the rope over the pulley if $\theta = 90°$ and $\mu_k = 0.3$.

**8–83** Calculate the required pulling tension $T_2$ (before inserting the rope in the cam cleats) if the resisting tension $T_1$ is 50 lb. Assume that both $\mathbf{T}_1$ and $\mathbf{T}_2$ act horizontally, $\mu_k = 0.2$, and slippage is occurring. Estimate the angle of contact to the nearest quarter turn around the post.

$W = 200$ lb

Fig. P8-86

**8–87** Plot the tension $T$ as a function of $\theta = 0, 45°$, and 90° in Prob. 8–86.

**8–88** Consider a magnetic tape passing over a curved tape head. The minimum tension $T_1$ must be 0.2 lb. Determine the pulling tension $T_2$ if $\mu_k = 0.07$.

Fig. P8-83

Fig. P8-88

**8–84** A rope is to be wound around a mooring post. The pulling tension $T_2$ from the ship is 50 kN, and $\mu_s = 0.3$. Determine the required resisting tension $T_1$ (applied by hand of dock workers) for two complete turns ($n = 2$) of rope on the post if slippage is impending.

**8–89** Plot the tension $T_2$ vs. $\mu_k = 0.04, 0.07$, and 0.1 in Prob. 8–88. What is your conclusion from this diagram concerning the significance of friction in attempting to minimize the pulling tension $T_2$ in the machine?

**8–90** Prove that the tape tension in Prob. 8–88 can be analyzed the same way even if the tape head has a noncircular contour provided that $\beta$ is the angle as shown in Fig. P8-90 and $\mu_k$ is the same.

Noncircular contour

$T_1$        $\beta$    $T_2$

Tape head

**Fig. P8-90**

**8–91** Derive Eq. 8-16,

$$T_2 = T_1 e^{\mu_s \beta / \sin \phi}$$

**8–92** Consider a planned V-belt drive. The desired minimum tension is $T_1$, the angle of contact is $\beta = 130°$, and $\mu_s = 0.5$. Express the pulling tension $T_2$ if the belt angle is $2\phi = 50°$.

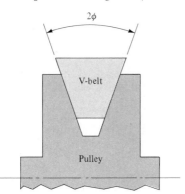

$2\phi$

V-belt

Pulley

**Fig. P8-92**

**8–93** Plot the tension $T_2$ vs. the belt angle $2\phi = 20°$, $50°$, and $90°$ in Prob. 8–92. Extrapolate the curve to $2\phi = 180°$ and discuss whether this extrapolation is reasonable or not.

**8–94** Determine the tensions at $A$ and $B$ in the band brake. Consider both clockwise and counterclockwise rotations of the drum.

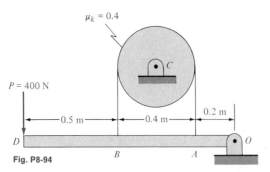

$\mu_k = 0.4$

$C$

$P = 400$ N

0.5 m    0.4 m    0.2 m

$D$      $B$      $A$      $O$

**Fig. P8-94**

**8–95** Motor $A$ must deliver a torque of 2 kN·m to pulley $B$. Determine the largest tension in the flat belt if the coefficient of static friction is 0.3 at both pulleys.

**Fig. P8-95**

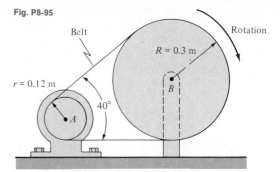

Belt          Rotation

$R = 0.3$ m

$r = 0.12$ m

$B$

$40°$

$A$

**8–96** Solve Prob. 8–95 assuming that the flat belt drive is replaced by a V-belt drive with a belt angle of $2\phi = 60°$.

**8–97** In Prob. 8–95 an idler pulley is proposed to make the belt's angle of contact $180°$ on pulley $A$ and $270°$ on pulley $B$. Determine the horizontal and vertical bearing reactions of the idler pulley if all friction at this pulley can be ignored.

**8–98** The magnitude of the horizontal force $\mathbf{F}$ on the drawbar of a trailer traveling at a constant speed of 40 mph is 500 lb. Sketch a reasonable curve of $F$ vs. the speed $v$ for speeds ranging from zero to 70 mph.

**Fig. P8-98**

$V$

Trailer

$F$

**8–99** A tugboat is towing two large container barges. Discuss whether the relative cable distances $a$ and $b$ are reasonable or not with respect to total fluid friction on the two barges. This is a complex but important practical problem. Base your discussion on concepts presented in this chapter and on your intuition.

**Fig. P8-99**

**DRY FRICTION:** A friction force $\mathscr{F}$ may act at the surfaces of contact between bodies. This force is always tangent to the surfaces in contact, and acts in the direction opposite to motion or impending relative motion. The magnitude of the friction force may vary from zero to a maximum value, depending on the external force system and the condition of the bodies in contact. $\mathscr{F}$ is always zero when there is no force for the friction force to oppose. The maximum value occurs when motion is impending and has the form

$$\mathscr{F}_{max} = \mu_s N$$

where $\mu_s$ = coefficient of static friction
$N$ = magnitude of the normal force between the bodies

Note that in problems involving static equilibrium, the magnitude of the *actual* friction force, $\mathscr{F}$, is frequently less than the value of the *possible* maximum friction force, $\mathscr{F}_{max}$, for the given conditions. When there is relative motion between the bodies,

$$\mathscr{F} = \mu_k N$$

where $\mu_k$ is the kinetic coefficient of friction.

Section 8-2

**ANGLES OF FRICTION:**

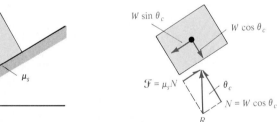

The angle of repose is the critical angle $\theta_c$ of the incline at which motion of the block is impending because of the weight of $W$.

$$\tan \theta_c = \frac{\mathscr{F}}{N} = \frac{\mu_s N}{N}$$

$$= \mu_s = \text{coefficient of static friction}$$

On a horizontal plane where force $\mathbf{P}_c$ causes impending motion,

$$\tan \theta_c = \frac{\mathscr{F}}{N} = \frac{\mu_s N}{N} = \mu_s$$

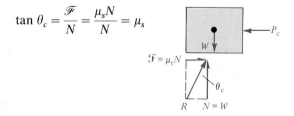

Here $\theta_c$ is the angle of static friction, which is equivalent to the angle of repose for inclines.

**PROCEDURES FOR SOLVING FRICTION PROBLEMS:**

| CLASS OF PROBLEM | GIVEN INFORMATION | PROCEDURE |
|---|---|---|
| A | Bodies, forces, coefficients of friction are known. Impending motion is not assured; $\mathscr{F} \neq \mu_s N$. | To determine if equilibrium is possible: <br> 1. Construct the free-body diagram. <br> 2. Assume that the system is in equilibrium. <br> 3. Determine the friction and normal forces necessary for equilibrium. <br> 4. Results: (a) $\mathscr{F} < \mu_s N$, the body is at rest. (b) $\mathscr{F} > \mu_s N$, motion is occurring, static equilibrium is not possible. Since there is motion, $\mathscr{F} = \mu_k N$. Complete solution requires principles of dynamics. |
| B | Bodies, forces, or coefficients of friction are given. Impending motion is specified, $\mathscr{F} = \mu_s N$ is valid. | To determine the unknowns: <br> 1. Construct the free-body diagram. <br> 2. Write $\mathscr{F} = \mu_s N$ for all surfaces where motion is impending. <br> 3. Determine $\mu_s$ or the required forces from the equation of equilibrium. |
| C | Bodies, forces, coefficients of friction are known. Impending motion is specified, but the exact motion is not given. The possible motions may be sliding, tipping, or rolling, or relative motion if two or more bodies are involved. Alternatively, the forces or coefficients of friction may have to be determined to produce a particular motion from several possible motions. | To determine the exact motion that may occur, or unknown quantities required: <br> 1. Construct the free-body diagram. <br> 2. Assume that motion is impending in one of the two or more possible ways. Repeat this for each possible motion, and write the equation of equilibrium. <br> 3. Compare the results for the possible motions and select the likely event. Determine the required unknowns for any preferred motion. |

**WEDGES:** Draw a free-body diagram for each wedge and body to be supported or moved by the wedge. Write equilibrium equations for each body and solve these for the unknown quantities. Note that all friction forces must oppose the impending relative motions.

**SCREWS (Square Threads):**

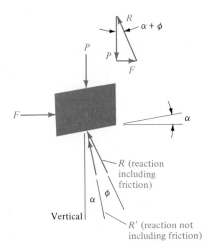

The magnitude of the couple **M** required to move the screw against the axial load **P** is

$$M = Pr \tan (\alpha + \phi)$$

where  $\alpha = \tan^{-1} L/2\pi r = \tan^{-1} np/2\pi r$
$L$ = lead (advancement per revolution)
$p$ = pitch (distance between similar points on adjacent threads)
$n$ = multiplicity of threads
$\phi = \tan^{-1} \mu$
$\mu$ = coefficient of friction ($\mu_s$ or $\mu_k$)

**DISK FRICTION:**

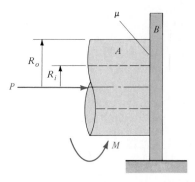

For the hollow member $A$ pressed against the flat member $B$, the magnitude of the couple $\mathbf{M}$ to cause relative rotation is

$$M = \frac{2}{3}\mu P \frac{R_o^3 - R_i^3}{R_o^2 - R_i^2}$$

For a solid shaft of radius $R$,

$$M = \frac{2}{3}\mu P R$$

where $P$ = magnitude of the normal force between the bodies
$\mu$ = coefficient of friction ($\mu_s$ or $\mu_k$)

**AXLE FRICTION:**

The magnitude of the couple $\mathbf{M}$ required to overcome friction in steady rotation of the axle is approximately

$$M = Pr\mu_k$$

where $P$ = magnitude of the lateral force on the axle
$\mu_k$ = coefficient of kinetic friction

**ROLLING RESISTANCE:**

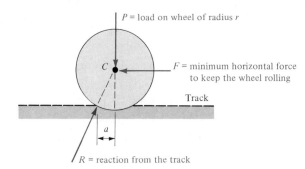

The coefficient of rolling resistance is

$$a \simeq \frac{Fr}{P} \text{ in units of length}$$

Section 8-8

**BELT FRICTION:**

For impending slipping to the right,

$$T_2 = T_1 e^{\mu_s \beta}$$

where  $\beta$ = angle of belt contact in radians
$\mu_s$ = coefficient of static friction (may be replaced by $\mu_k$ if there is slipping)

For a V belt of belt angle $2\phi$,

$$T_2 = T_1 e^{\mu_s \beta / \sin \phi}$$

Section 8-9

**FLUID FRICTION:**   The magnitude of the drag force **D** caused by fluid friction is approximately

$$D = Cv^2$$

where  $C$ = constant depending on size and shape of the body and the properties of the fluid
$v$ = relative velocity of the body and the fluid

**8–100** A uniform, 2500-lb concrete beam is resting on two rails. Calculate the magnitude of the horizontal force **P** parallel to the rails that would start the beam to slide if $\mu_A = \mu_B = 0.3$.

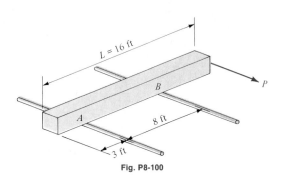

**Fig. P8-100**

**8–101** Consider the single-wheel model of a small electric truck used in a foundry to drag blocks of steel on the floor. Of course, a block may be so heavy that it will not slide while the wheel spins. Determine the smallest torque $M$ that would cause slipping of the block or of the wheel. The wheel has a mass of 400 kg, and the block has a mass of 300 kg.

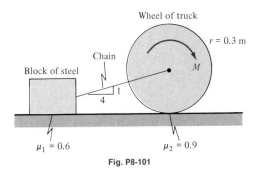

**Fig. P8-101**

**8–102** In Prob. 8–101 determine the smallest mass of the block that causes the wheel to slip first (this is where the method becomes useless with the given truck).

**8–103** A uniform crate has a mass of 200 kg. The coefficient of static friction is $\mu_s = 0.4$. Worker $D$ applies a constant force of magnitude $F = 500\,\text{N}$. Worker $E$ applies a gradually increasing force **P** until the crate moves either by sliding or tipping. Determine the magnitude of the force **P** when the crate starts moving. Point $C$ is the mass center of the crate.

**Fig. P8-103**

**8–104** The mean diameter of the single-threaded worm gear $A$ is 1.6 in., and of gear $B$ is 10 in. The coefficient of friction between the gear teeth is 0.25. Ignoring friction elsewhere in the mechanism, determine the required torque on gear $A$ to turn gear $B$ clockwise against a 500 lb·ft torque on gear $B$. The pitch is $p = 0.2\,\text{in.}$

**Fig. P8-104**

**8–105** Plot the required torque as a function of the pitch in Prob. 8–104 for $p = 0.2$ in., 0.25 in., and 0.3 in. Is the pitch an important factor is designing the machine that includes this worm gear?

**8–106** Determine the possibility of the unwinding of the mechanism in Prob. 8–104 if the maximum resisting torque acting on gear $A$ is 10 lb·ft while a 500 lb·ft torque is acting on gear $B$.

**8–107** For cooling purposes, a disk brake has radial slots in one disk which is pressed against a full circle of the same radius $R_o$. Determine the torque that can start one disk to slip on the other when they are pressed together by a force of magnitude $P = 1\,\text{kN}$. $R_o = 150\,\text{mm}$, $r_i = 20\,\text{mm}$, $r_o = 130\,\text{mm}$, and $\mu_s = 0.6$. The slots remove half of the area between $r_i$ and $r_o$.

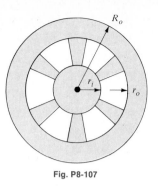

Fig. P8-107

**8–108** Derive a formula to determine the frictional resisting torque of the conical pivot.

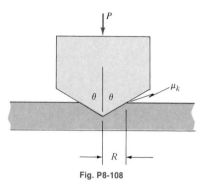

Fig. P8-108

**8–109** The 0.12-m-diameter uniform roller $A$ has a mass of 100 kg. It is supported by 20-mm-diameter shafts in vertical panels. Assume that the holes in the panels also have a 20-mm diameter but the shafts fit without interference. The coefficient of static friction at all surfaces is 0.5. Determine the torque $M$ required to start the roller spin-

ning if a force of magnitude $F = 500$ N is also applied during that time. The roller is supported against the horizontal force **F** at surface $D$. The surfaces at $D$ are flat and parallel.

Fig. P8-109

**8–110** Determine the tension in both segments of the band brake attached to the brake arm at $B$. The drum is rotating clockwise.

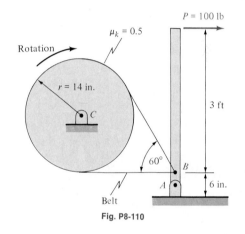

Fig. P8-110

# 9 Moments of Inertia

### Reasons for Studying the Topics of Inertia in Statics

The concepts and methods of inertia are used in the areas of dynamics and mechanics of materials. As far as subject matter is concerned, these topics have relatively little to do with statics since they do not involve the analysis of equilibrium. The reasons for presenting the topics of inertia in statics are twofold: First, the methods of first moments (Secs. 7-2 and 7-3) can be readily extended to analyze inertias. Second, in most academic situations it is easiest to schedule the detailed introduction of topics of inertia in statics, requiring only that they be rapidly reviewed in dynamics or mechanics of materials. This enhances the understanding and facility of using the methods of inertia even if they do not seem entirely natural in statics.

### Basic Concepts of Inertia

Superficially, there are two ways of viewing the concepts of inertia. From the mathematical point of view, they involve the methods of elementary calculus, and they can be readily used without trying to understand the relevant physical phenomena. A satisfactory physical view of the concepts of inertia is more difficult to attain because the concepts are rather abstract. Basically, the term *inertia* is used in dynamics to describe the tendency of matter to remain at rest or continue moving with a constant

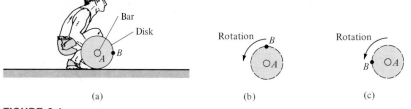

**FIGURE 9-1**

Weightlifting is used to illustrate the concepts of inertia. End view of a barbell

velocity. Inertia is also used in other areas to show resistance to any change. These concepts are illustrated in a qualitative discussion of lifting a barbell from the floor as follows.

Consider an end view of a barbell in Fig. 9-1. If the disks were fixed on the bar, each disk would rotate counterclockwise when the hands rotate the bar during a lift (Fig. 9-1b and c). This rotation would require an extra effort from the lifter, especially if the disks were large. The inertia, or resistance of a disk to rotation, depends on its mass and how it is distributed with respect to the axis of rotation $A$. In practice, the disks are relatively free to rotate on the bar, so the line $AB$ tends to remain horizontal because of inertia even though the bar rotates (this minimizes the effort of the lifter).

Next consider a side view of the barbell during a lift (Fig. 9-2). If the disks are heavy, the bar bends noticeably. The bending of the bar allows another rotation of each disk, that shown in Fig. 9-2b. The inertia of the disk in this rotation depends on its mass and how it is distributed with respect to the axis of rotation, which is perpendicular to the bar in this case. The two inertias discussed are not the same since the mass distributions with respect to the two axes of rotation are different.

A more abstract concept of inertia is relevant to the bending of the bar itself in Fig. 9-2. Here the resistance to change is taken as a property of the cross-sectional area of the bar (note that an originally vertical cross section becomes slanted as the bar bends). This property is used extensively in mechanics of materials.

Since the inertia of a mass is easier to understand intuitively than the inertia of an area, they are presented in that order. Mathematically they are similar, so in problems the difference between masses and areas is a difference of symbols and units.

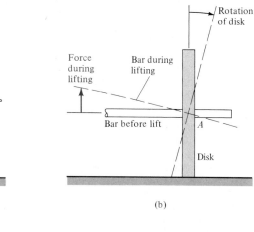

**FIGURE 9-2**

Side view of a barbell

**SECTION 9-1** presents the definition of moment of inertia of a mass with respect to an axis of reference. This topic is essential for studying the dynamics of rigid bodies.

**SECTION 9-2** presents the definition of moment of inertia (or second moment) of an area with respect to an axis of reference. This topic is essential for studying the mechanics of materials (bending of beams and columns).

**SECTION 9-3** extends the concept of moment of inertia of an area to the polar moment of inertia of the area with respect to a point of reference. This topic is essential for studying the mechanics of materials (torsion of cylindrical shafts).

**SECTION 9-4** presents the parallel-axis theorem for a mass or an area. This useful theorem facilitates the solving of problems in dynamics and mechanics of materials. The simple formulas for the transformation of moments of inertia from a centroidal axis to a parallel axis are worth remembering.

**SECTION 9-5** shows that the moment of inertia of a complex area or body about a specified axis may be determined in several ways using the parallel-axis theorem. The results of the different calculations should be the same in a given situation, but they are not necessarily equal in computational effort. It is demonstrated that an efficient approach can be found with a little experience and judgment.

**SECTIONS 9-6, 9-7, AND 9-8** present products of inertia, principal axes, and transformations of moments of inertia. These are advanced topics and are seldom discussed in statics courses because they are normally used only in advanced courses in dynamics and mechanics of materials. They are included in this chapter to facilitate studying them when required in a later course.

## 9-1    MOMENT OF INERTIA OF A MASS

The moment of inertia $dI_x$ of an elemental mass $dM$ about the $x$ axis (Fig. 9-3) is defined as

$$dI_x = r^2 \, dM = (y^2 + z^2) \, dM \qquad \boxed{9\text{-}1}$$

where $r$ is the nearest distance from the element of mass to the $x$ axis. The moment of inertia of an extended body is obtained by integrating Eq. 9-1 over the whole mass $M$ while always referring to the same axis,

$$\boxed{I_x = \int r^2 \, dM = \int (y^2 + z^2) \, dM} \qquad \boxed{9\text{-}2a}$$

Similarly, for using the $y$ or $z$ axis for reference in Fig. 9-3,

$$\boxed{I_y = \int (x^2 + z^2) \, dM} \qquad \boxed{9\text{-}2b}$$

$$\boxed{I_z = \int (x^2 + y^2) \, dM} \qquad \boxed{9\text{-}2c}$$

### Radius of Gyration

**FIGURE 9-3**

Mass element $dM$ at distance $r$ from the $x$ axis

Imagine that a body of mass $M$ is replaced by a point mass $M$ (same mass) at a distance $r_g$ from a given $x$ axis. If the moments of inertia $I_x$ are the same for the body and the point mass, $I_x = r_g^2 M$ from Eqs. 9-1 and 9-2, and

$$r_g = \sqrt{I_x/M} \qquad \boxed{9\text{-}3}$$

The distance $r_g$ is called the *radius of gyration* and it can be similarly defined for any other axis of reference. A point mass may be replaced by a thin strip or shell to obtain Eq. 9-3 if all mass is at a constant distance $r_g$ from the axis of reference.

Because the moment of inertia of a mass depends on $r^2$ of differential masses at distance $r$ from a common axis of reference, the geometry of a body strongly influences the moment of inertia. This is true for any axis of reference such as those in Figs. 9-1 and 9-2 for the same body. The moments of inertia of masses have units of mass times length squared.

## MOMENT OF INERTIA OF AN AREA

Consider an element of area in the $xy$ plane (Fig. 9-4). The moment of inertia $dI_x$ of the infinitesimal area $dA$ about the $x$ axis is defined as

$$dI_x = y^2\, dA \qquad \boxed{9\text{-}4}$$

where $y$ is the nearest distance from the element of area to the $x$ axis. The moment of inertia $I_x$ of an area $A$ is obtained by integrating Eq. 9-4 over the whole area $A$ while always referring to the same axis,

$$I_x = \int y^2\, dA \qquad \boxed{9\text{-}5}$$

The term *second moment* of the area $A$ is sometimes used in connection with Eq. 9-5. The radius of gyration of an area is defined the same way as it is for a mass in Eq. 9-3,

$$r_g = \sqrt{I_x/A} \qquad \boxed{9\text{-}6}$$

**FIGURE 9-4**

Area element $dA$ at distance $y$ from the $x$ axis

The moment of inertia of areas is an abstract concept. To obtain a feeling for what it means physically, assume that it represents the resistance of an area to rotational displacement from its original plane. For example, the bending of the bar in Fig. 9-2 would cause any vertical cross section in this part of the bar to rotate a little clockwise. The imaginary resistance of areas to such rotation is a useful concept in analyzing the strength of members even in static situations.

As in the case of masses, geometry strongly affects the moments of inertia of areas. Thus, the choice of the reference axis is important in any calculation of moments of inertia. For example, the area $A$ in Fig. 9-5 is likely to have different moments of inertia with respect to the $x$ and $y$ axes,

$$I_x = \int y^2\, dA, \quad I_y = \int x^2\, dA \qquad \boxed{9\text{-}7}$$

Note that in this case (Fig. 9-5) a moment of inertia $I_z$ cannot be defined in the same way as $I_x$ or $I_y$ but is defined in the following section.

**FIGURE 9-5**

Area $A$ in the $xy$ plane

## POLAR MOMENT OF INERTIA OF AN AREA

The *polar moment of inertia of an area*, $J_O$, is interpreted as the (imaginary) resistance of an area $A$ to rotational displacements in its own plane about point $O$. Using the concepts discussed above and the area $A$ in Fig. 9-5, $J_O$ is defined as

$$J_O = \int r^2 \, dA$$

9-8a

In other words, the polar moment of inertia is calculated with respect to an axis perpendicular to the area considered. Note that, since $r^2 = x^2 + y^2$ in Fig. 9-5, $J_O = \int (x^2 + y^2) \, dA$, so

$$J_O = I_x + I_y$$

9-8b

By dividing Eq. 9-8b by the total area, $A$, an equation relating the radii of gyration is obtained,

$$r_{gO}^2 = r_{g_x}^2 + r_{g_y}^2$$

9-9

A relationship similar to Eq. 9-8b is derived for the mass moments of inertia for thin plates lying in the $xy$ plane. From Fig. 9-6, $I_z = r^2 \, dM$ and

$$I_z = I_x + I_y$$

9-10

It should be noted that the moment of inertia is always positive regardless of the position of the body or area with respect to the reference axis. Also, the moments of inertia are additive; this is advantageous in working with bodies of complex shape. The moment of inertia of an area has units of length raised to the fourth power, $m^4$ or $in^4$.

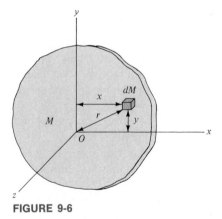

**FIGURE 9-6**

Thin plate of mass $M$ in the $xy$ plane

# EXAMPLE 9-1

Derive the equations for the moments of inertia of a rectangular area with sides $a$ and $b$ with respect to axes $x$, $x_1$, and $y$, $y_1$ shown in Fig. 9-7, and determine the polar moment of inertia, $J_O$.

## SOLUTION

The moments of inertia about the centroidal $xy$ axes are determined first. Using the elemental area of width $dy$ and length $a$, at a distance $y$ from the $x$ axis,

$$I_x = \int dI_x = \int y^2 \, dA = \int_{-b/2}^{b/2} y^2 a \, dy = \frac{1}{12} ab^3$$

Using the elemental area of width $dx$ and length $b$, at a distance $x$ from the $y$ axis,

$$I_y = \int dI_y = \int x^2 \, dA = \int_{-a/2}^{a/2} x^2 b \, dx = \frac{1}{12} ba^3$$

The same procedure is used for $I_{x1}$ and $I_{y1}$, but with different limits of integration,

$$I_{x1} = \int dI_{x1} = \int y^2 \, dA = \int_0^b y^2 a \, dy = \frac{1}{3} ab^3$$

$$I_{y1} = \int dI_{y1} = \int x^2 \, dA = \int_0^a x^2 b \, dx = \frac{1}{3} ba^3$$

To determine the polar moment of inertia, $J_O$, use Eq. 9-8b and the results above for $I_x$ and $I_y$.

$$J_O = I_x + I_y$$

$$= \frac{1}{12} ab^3 + \frac{1}{12} ba^3$$

$$= \frac{1}{12} ab(a^2 + b^2)$$

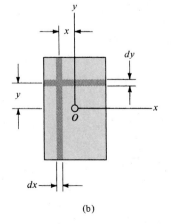

**FIGURE 9-7**

**Area *ab* in the *xy* plane**           (a)                                              (b)

EXAMPLE 9-1    387

## EXAMPLE 9-2

**(a)** Determine the moments of inertia $I_x$ and $I_y$ of the semicircular area shown in Fig. 9-8. **(b)** Calculate the mass moment of inertia $I_x$ of a thin plate with a semicircular area shown in Fig. 9-8. The plate has thickness $t = 0.1$ in. and density $\rho = 0.001$ slug/in$^3$.

### SOLUTION

**(a)** It is easiest to use polar coordinates for this problem with an elemental area $dA = R\,d\theta\,dR$ as shown in Fig. 9-8. From Eq. 9-7,

$$I_x = \int y^2\,dA = \int_{\theta=0}^{\theta=\pi}\int_{R=0}^{R=r}(R\sin\theta)^2\,R\,d\theta\,dR$$

$$= \int_{\theta=0}^{\theta=\pi}\left(\frac{r^4}{4}\sin^2\theta\right)d\theta = \frac{r^4}{4}\int_0^\pi\left(\frac{1}{2}-\frac{\cos 2\theta}{2}\right)d\theta$$

$$= \frac{r^4}{4}\left(\frac{\pi}{2}\right) = \frac{\pi r^4}{8}$$

$$= 245\ \text{in}^4$$

$$I_y = \int x^2\,dA = \int_{\theta=0}^{\theta=\pi}\int_{R=0}^{R=r}(R\cos\theta)^2\,R\,d\theta\,dR$$

$$= \frac{r^4}{4}\int_0^\pi(\cos^2\theta)\,d\theta = \frac{r^4}{4}\int_0^\pi\left(\frac{1}{2}+\frac{\cos 2\theta}{2}\right)d\theta$$

$$= \frac{r^4}{4}\left(\frac{\pi}{2}\right) = \frac{\pi r^4}{8}$$

$$= 245\ \text{in}^4$$

**FIGURE 9-8**

Semicircular area in the *xy* plane

**(b)** For the mass moment of inertia, Eq. 9-2 is used. Since the plate is so thin, however, assume that $I_x$ is not a function of $z$. Then

$$I_x = \int y^2\,dM$$

$dM = \rho\,dV$, where $dV$ is an elemental volume $dV = t\,dA$. Hence,

$$I_{x(mass)} = \int y^2\rho\,dV = \int y^2\rho t\,dA = \rho t\int y^2\,dA$$

$$= \rho t I_{x(area)}$$

$$= \left(0.001\ \frac{\text{slug}}{\text{in}^3}\right)(0.1\ \text{in.})(245\ \text{in}^4) = 0.0245\ \text{slug·in}^2$$

**TIP**

Note that the mass moment of inertia of a thin homogeneous plate (constant thickness and density) is the area moment of inertia multiplied by the thickness, $t$, and density, $\rho$.

EXAMPLE 9-3

Determine the radii of gyration of the area shown in Fig. 9-9a **(a)** about the $x$ axis and **(b)** about the $y$ axis.

## SOLUTION

**(a)** For a horizontal elemental area $dA = (a - x)\, dy$ with $x = y^{1/3}$ in Fig. 9-9b,

$$I_x = \int y^2\, dA = \int_0^b y^2(a - y^{1/3})\, dy$$

$$= a\frac{b^3}{3} - \frac{3}{10} b^{10/3} = b^3\left(\frac{a}{3} - \frac{3}{10} b^{1/3}\right)$$

and

$$A = \int dA = \int_0^b (a - y^{1/3})\, dy$$

$$A = ab - \frac{3}{4} b^{4/3} = b\left(a - \frac{3}{4} b^{1/3}\right)$$

Using Eq. 9-6,

$$r_{g_x} = \sqrt{\frac{I_x}{A}} = \sqrt{\frac{b^3\left(\dfrac{a}{3} - \dfrac{3}{10} b^{1/3}\right)}{b\left(a - \dfrac{3}{4} b^{1/3}\right)}} = b\left(\frac{\dfrac{a}{3} - \dfrac{3}{10} b^{1/3}}{a - \dfrac{3}{4} b^{1/3}}\right)^{1/2}$$

But for $y = x^3$, $b = a^3$ or $a = b^{1/3}$, so

$$r_{g_x} = \sqrt{\frac{I_x}{A}} = b\left(\frac{\dfrac{1}{3} - \dfrac{3}{10}}{1 - \dfrac{3}{4}}\right)^{1/2}$$

$$= 0.365b$$

**(b)** For a vertical elemental area $dA = y\, dx = x^3\, dx$ in Fig. 9-9c,

$$I_y = \int x^2\, dA = \int_0^a x^2(x^3)\, dx$$

$$= \frac{1}{6} a^6$$

Using Eq. 9-6,

$$r_{g_y} = \sqrt{\frac{I_y}{A}} = \sqrt{\frac{\dfrac{1}{6} a^6}{b\left(a - \dfrac{3}{4} b^{1/3}\right)}}$$

And for $b = a^3$,

$$r_{g_y} = a\left(\frac{\dfrac{1}{6}}{1 - \dfrac{3}{4}}\right)^{1/2}$$

$$= 0.816a$$

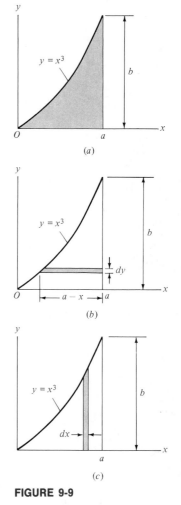

**FIGURE 9-9**

Area in the xy plane

EXAMPLE 9-3  389

# PROBLEMS

**9–1** Determine the mass moment of inertia $I_z$ of the uniform slender rod of mass $M$.

Fig. P9-1

**9–2** Determine the mass moment of inertia $I_{y'}$ of the uniform slender rod in Fig. P9-1. The axis $y'$ is parallel to $y$ at point $C$, $x = L/2$. The rod has mass $M$.

**9–3** A uniform slender bar of mass $M$ is parallel to the $y$ axis at $x = a$. Determine the mass moments of inertia $I_x$ and $I_y$, and the radii of gyration.

Fig. P9-3

**9–4** A thin hoop of radius $R$ and mass $M$ is in the $xz$ plane. Determine the mass moments of inertia $I_x$ and $I_y$ of the hoop.

Fig. P9-4

**9–5** A thin cylindrical shell of radius $R$ and mass $M$ is centered on the $x$ axis. Determine the mass moment of inertia $I_x$ and the radius of gyration.

Fig. P9-5

**9–6** The thin plate $ab$ has thickness $t$ and density $\rho$. Determine the mass moments of inertia $I_x$ and $I_y$ assuming that they are not functions of $z$ (since $t$ is small).

Fig. P9-6

**9–7** Determine the moments of inertia $I_x$ and $I_y$ of the area $ab$ in Fig. P9-6. Express the radii of gyration.

**9–8** Denote by $C$ the centroid of the area $ab$ in Fig. P9-6. Place centroidal axes $x'$ and $y'$ at $C$, parallel to $x$ and $y$, respectively. Determine the polar moment of inertia $J_C$ of the area $ab$.

**9–9** Calculate the mass moments of inertia $I_x$ and $I_y$ of the uniform slender rod of mass $M$.

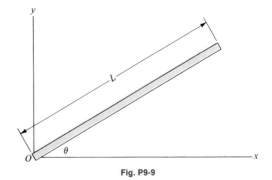

Fig. P9-9

**9–10** Determine the moment of inertia $I_y$ of the triangular area.

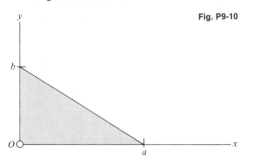

Fig. P9-10

**9–11** Determine the polar moment of inertia $J_O$ of the triangular area in Fig. P9-10.

**9–12** Determine the mass moment of inertia $I_x$ of the thin triangular plate of thickness $t$ and density $\rho$ in Fig. P9-10. Assume that $I_x$ is not a function of $z$ (since $t$ is small).

**9–13** Calculate the mass moment of inertia $I_y$ of the thin triangular plate of thickness $t$ and density $\rho$. Assume that $I_y$ is not a function of $z$ (since $t$ is small).

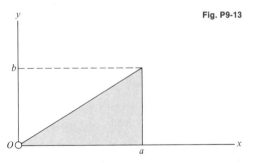

Fig. P9-13

**9–14** Calculate the moment of inertia $I_x$ of the triangular area in Fig. P9-13. Express the radius of gyration. Use $a = 10$ and $b = 5$.

**9–15** Determine the polar moment of inertia $J_O$ of the triangular area in Fig. P9-13.

**9–16** Determine the moments of inertia $I_x$ and $I_y$ of the semicircular area.

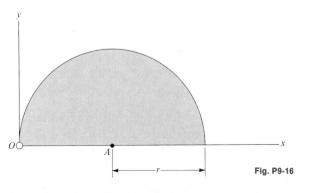

Fig. P9-16

**9–17** Determine the polar moment of inertia $J_A$ of the semicircular area in Fig. P9-16.

**9–18** The thin semicircular plate in Fig. P9-16 has thickness $t$ and density $\rho$. Calculate the mass moment of inertia $I_y$. Assume that $I_y$ is not a function of $z$ (since $t$ is small).

**9–19** Determine the mass moment of inertia $I_z$ of the thin circular plate of thickness $t$ and density $\rho$.

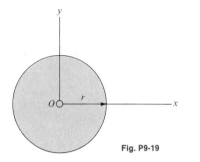

Fig. P9-19

**9–20** Calculate the moments of inertia $I_x$ and $I_y$ of the circular area in Fig. P9-19.

**9–21** Determine the polar moment of inertia $J_O$ of the circular area in Fig. P9-19 using two methods: (a) direct integration (Eq. 9-8a), and (b) using Eq. 9-8b.

**9–22** Determine the polar moment of inertia $J_O$ of the shaded area if $r = 50$ mm and $R = 80$ mm.

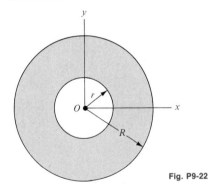

Fig. P9-22

**9–23** A thin, uniform wire of length $l = a + b$ and mass $m$ is bent into the shape shown. Determine (a) the mass moment of inertia $I_x$ of the piece and the corresponding radius of gyration, (b) the mass moment of inertia $I_y$, and (c) the mass moment of inertia $I_z$.

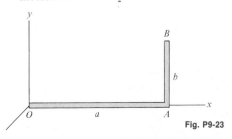

Fig. P9-23

## PARALLEL-AXIS THEOREM

In problems involving moments of inertia it is often convenient to first calculate the moment of inertia about a centroidal axis and then transform this to a moment of inertia about a required axis which is parallel to the given centroidal axis. The transformation can be done using a simple formula based on the *parallel-axis theorem*,

| | | |
|---|---|---|
| $I = I_C + Md^2$ | for a mass $M$ | 9-11 |
| $I = I_C + Ad^2$ | for an area $A$ | 9-12 |
| $J_O = J_C + Ad^2$ | for an area $A$ | 9-13 |

where $I$ or $J_O$ = moment of inertia of mass $M$ or area $A$ about any line $l$
$I_C$ or $J_C$ = moment of inertia of mass $M$ or area $A$ about a line through the mass center of $M$ or the centroid of $A$ and parallel to line $l$
$d$ = perpendicular distance between the two parallel lines

By dividing Eq. 9-11 by the mass $M$ or Eq. 9-12 by the area $A$, an equation relating the radii of gyration between two parallel axes is obtained.

$$r_g^2 = r_{gc}^2 + d^2 \qquad \boxed{9\text{-}14}$$

where $d$ is defined as above.

The parallel-axis theorem is proved in general terms for the rectangular area $A$ shown in Fig. 9-10. Here the axes $x'$ and $y'$ are centroidal axes of the area $A$. Assume that the moments of inertia with respect to the axes $x$ and $y$ must be determined. Considering the $x$ axis first, $dI_x = (y' + b)^2\, dA$ by definition of the moment of inertia (Eq. 9-4). For the whole area,

$$I_x = \int_A (y' + b)^2\, dA$$

$$= \int_A y'^2\, dA + 2b \int_A y'\, dA + b^2 \int_A dA$$

Note that $\int_A y'\, dA = 0$ since it is the first moment of the area $A$ about the $x'$ axis which is a centroidal axis (the first moment is always zero about any centroidal axis). Thus,

$$I_x = I_{x'} + Ab^2 \qquad \boxed{9\text{-}15}$$

Using the same procedure in reference to the $y$ axis,

$$I_y = I_{y'} + Aa^2 \qquad \boxed{9\text{-}16}$$

The method is also applicable to transformations of the polar moment of inertia. With respect to Fig. 9-10,

$$J_O = J_{O'} + Ac^2 = I_x + I_y \qquad \boxed{9\text{-}17}$$

The proof of this equation is left as an exercise for the reader.

**FIGURE 9-10**

Area $A$ in the $xy$ plane

**TIP**

> It is important to note that Eqs. 9-11 through 9-17 are valid only if one of the two axes in each equation is a centroidal axis. If the moment of inertia about a centroidal axis is to be determined and the moment of inertia about a noncentroidal axis is given, the equations must be rearranged; for example, $I_C = I - Md^2$. Transformations of moments of inertia between two noncentroidal axes cannot be made without an intermediate transformation involving the centroidal axis.

## MOMENTS OF INERTIA OF COMPLEX AREAS AND BODIES

**9-5**

In some problems it is necessary to determine the moments of inertia for masses or areas of complex shape. The procedure for these is reasonably easy if the object can be broken down into simple parts in the analysis. For each part, the location of the centroid should be obtained by inspection, and the centroidal moment of inertia should be calculated using a readily available formula (such as that for rectangles). Finally, the centroidal moments of inertia may have to be transformed to a common axis of interest using the parallel-axis theorem. The method is illustrated schematically for the moment of inertia of a complex area. Generally, the details of the calculations can be done in several different ways to obtain the same result.

Assume that it is required to determine the moments of inertia of the channel section about the $x$ and $y$ axes ($I_x$ and $I_y$) in Fig. 9-11. $I_x$ is calculated in two ways to illustrate that judgment is necessary in finding the most efficient procedure. First break the area down into three rectangles ($A$ and two $B$) as shown by the dashed lines in Fig. 9-11. For area $A$, $I_{x_A}$ is its centroidal moment of inertia. For each area $B$, using Eq. 9-12,

$$I_{x_B} = I_{x'_B} + (ad)\left(b + \frac{d}{2}\right)^2$$

For the whole channel section (since the moments of inertia are additive),

$$I_x = I_{x_A} + 2I_{x_B}$$

In an alternative procedure the channel section is imagined to be the difference of two rectangles, area $C = a(2b + 2d)$ and area $D = c(2b)$. Since the $x$ axis is centroidal to both of these rectangles, the moment of inertia can be calculated without using the parallel-axis theorem. Thus,

$$I_x = I_{x_C} - I_{x_D}$$

Clearly, the latter method is the simpler of the two since it avoids the calculation of $(ad)(b + d/2)^2$.

In the calculation of $I_y$ it is necessary to use the parallel-axis theorem for every part of the total area, so again it is best to minimize the number of parts used. Working with area $C = a(2b + 2d)$ and area $D = c(2b)$,

$$I_y = I_{y'_C} + a(2b + 2d)\left(e + \frac{a}{2}\right)^2 - \left[I_{y'_D} + c(2b)\left(e + d + \frac{c}{2}\right)^2\right]$$

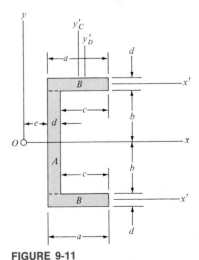

**FIGURE 9-11**

Channel cross section in the *xy* plane

EXAMPLE 9-4

Determine the moments of inertia of the rectangle $ab$ in Fig. 9-12 with respect to axes $x_1$, $y_1$, and $x_2$, $y_2$. Example 9-1 is relevant to this problem.

### SOLUTION

Using Eq. 9-12, the parallel-axis theorem for an area, and the results of Ex. 9-1 for the centroidal moments of inertia $I_x$ and $I_y$,

$$I_{x_1} = I_x + Ad^2 = \frac{1}{12} ab^3 + ab \left(\frac{b}{2}\right)^2$$

$$= \frac{1}{12} ab^3 + \frac{1}{4} ab^3 = \frac{1}{3} ab^3$$

$$I_{y_1} = I_y + Ad^2 = \frac{1}{12} ba^3 + ab \left(\frac{a}{2}\right)^2$$

$$= \frac{1}{12} ba^3 + \frac{1}{4} ba^3 = \frac{1}{3} ba^3$$

$$I_{x_2} = I_x + Ad^2 = \frac{1}{12} ab^3 + ab \left(\frac{b}{2}\right)^2 = \frac{1}{3} ab^3$$

$$I_{y_2} = I_y + Ad^2 = \frac{1}{12} ba^3 + ab \left(\frac{a}{2}\right)^2 = \frac{1}{3} ba^3$$

**TIP**

> Note that the moments of inertia about parallel axes equal distances away from the centroid, but in different directions, are equal. This is because the distance term is always squared, hence the sign of $d$ in Eq. 9-12 makes no difference.

### JUDGMENT OF THE RESULTS

The results obtained using the parallel-axis theorem for $I_{x1}$ and $I_{y1}$ are the same as those obtained by integration in Ex. 9-1.

**FIGURE 9-12**

Area $ab$ in the $xy$ plane

EXAMPLE 9-5

Determine the moments of inertia of the whole channel section area shown in Fig. 9-13 with respect to axes $x$, $x'$, and $y$. $a = 100$ mm, $b = 80$ mm, $c = 90$ mm, $d = 20$ mm, and $e = 40$ mm. Also compute the polar moment of inertia $J_O$.

## SOLUTION

Let $I_A$ and $I_B$ be centroidal moments of inertia of each of the areas $A$ and $B$.

$$I_x = I_{A_x} + A_A d_{A_y}^2 + 2(I_{B_x} + A_B d_{B_y}^2)$$

$$= \frac{1}{12}(2b)^3 d + 2bd(0) + 2\left[\frac{1}{12}d^3 a + ad\left(b + \frac{d}{2}\right)^2\right]$$

$$= \frac{1}{12}(160 \text{ mm})^3(20 \text{ mm}) + 0 + 2\left[\frac{1}{12}(20 \text{ mm})^3(100 \text{ mm})\right.$$

$$\left. + (100 \text{ mm})(20 \text{ mm})(80 \text{ mm} + 10 \text{ mm})^2\right]$$

$$= 683 \times 10^4 \text{ mm}^4 + 2(6.7 \times 10^4 + 1620 \times 10^4) \text{ mm}^4$$
$$= 3.94 \times 10^7 \text{ mm}^4$$

$$I_y = I_{A_y} + A_A d_{A_x}^2 + 2(I_{B_y} + A_B d_{B_x}^2)$$

$$= \frac{1}{12}(2b)d^3 + 2bd\left(e + \frac{d}{2}\right)^2 + 2\left[\frac{1}{12}da^3 + ad\left(e + \frac{a}{2}\right)^2\right]$$

$$= \frac{1}{12}(160 \text{ mm})(20 \text{ mm})^3 + 2(80 \text{ mm})(20 \text{ mm})(40 \text{ mm} + 10 \text{ mm})^2$$

$$+ 2\left[\frac{1}{12}(20 \text{ mm})(100 \text{ mm})^3\right.$$

$$\left. + (100 \text{ mm})(20 \text{ mm})(40 \text{ mm} + 50 \text{ mm})^2\right]$$

$$= 0.107 \times 10^6 \text{ mm}^4 + 8.0 \times 10^6 \text{ mm}^4$$
$$+ 2(1.667 \times 10^6 + 16.20 \times 10^6) \text{ mm}^4$$
$$= 4.38 \times 10^7 \text{ mm}^4$$

$$I_{x'} = I_{A_x} + A_A d_{A_y'}^2 + I_{B_{x''}} + A_B d_{B_{y'}}^2 + I_{B_{x'}}$$

$$= \frac{1}{12}(160 \text{ mm})^3(20 \text{ mm}) + (160 \text{ mm})(20 \text{ mm})(80 \text{ mm} + 10 \text{ mm})^2$$

$$+ \frac{1}{12}(100 \text{ mm})(20 \text{ mm})^3 + (100 \text{ mm})(20 \text{ mm})$$

$$\times (160 \text{ mm} + 20 \text{ mm})^2 + \frac{1}{12}(100 \text{ mm})(20 \text{ mm})^3$$

$$= (6.83 \times 10^6 + 2.59 \times 10^7 + 6.7 \times 10^4$$
$$+ 6.48 \times 10^7 + 6.7 \times 10^4) \text{ mm}^4 = 9.77 \times 10^7 \text{ mm}^4$$

$$J_O = I_x + I_y = 3.94 \times 10^7 \text{ mm}^4 + 4.38 \times 10^7 \text{ mm}^4$$
$$= 8.32 \times 10^7 \text{ mm}^4$$

**FIGURE 9-13**

Channel cross section in the *xy* plane

## JUDGMENT OF THE RESULTS

The large magnitude of $I_{x'}$ in relation to $I_x$ might be surprising. It means that rotating the whole channel cross-sectional area about the $x'$ axis would be much more difficult than rotating it about the $x$ axis. This is reasonable considering the $r^2$ dependence of moments of inertia.

EXAMPLE 9-5    395

## EXAMPLE 9-6

The thin, uniform, rectangular plate in Fig. 9-14a has mass $m$ before a circular hole of diameter $a$ is cut in it at point $A$. Determine the mass moment of inertia, $I_x$ of the plate with the hole and its radius of gyration, $r_g$, about the $x$ axis. Use tabulated formulas and the parallel-axis theorem.

### SOLUTION

Consider the plate as a composite of the whole rectangular plate, part 1, of unit thickness and mass per unit area of

$$\frac{m}{(3a)(2a)} = \frac{m}{6a^2}$$

less the circular hole, part 2, of mass

$$\pi r^2 \left(\frac{m}{6a^2}\right) = \frac{\pi m}{24}, \qquad \text{since } r = \frac{a}{2}.$$

Let $I_1$ and $I_2$ be the centroidal moments of inertia of parts 1 and 2. Using the parallel-axis theorem for each part, Eq. 9-11, and Fig. 9-14b,

$$I_x = (I_{1_x} + m_1 d_{1_y}^2) - (I_{2_x} + m_2 d_{2_y}^2)$$

$$= \left[\frac{1}{12} m(3a)^2 + m\left(\frac{3a}{2}\right)^2\right] - \left[\frac{1}{2}\left(\frac{\pi m}{24}\right)\left(\frac{a}{2}\right)^2 + \frac{\pi m}{24}(2a)^2\right]$$

$$= 0.75ma^2 + 2.25ma^2 - (0.01636ma^2 + 0.524ma^2)$$

$$= 2.46ma^2$$

To determine the radius of gyration, first determine the total mass, $M$.

$$M = m - \frac{\pi m}{24} = 0.869m$$

The radius of gyration with respect to the $x$ axis is determined using Eq. 9-3.

$$r_g = \sqrt{\frac{I_x}{M}} = \sqrt{\frac{2.46ma^2}{0.869m}} = 1.683a$$

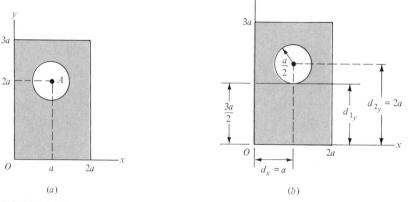

(a)

(b)

**FIGURE 9-14**

Rectangular plate with a circular hole

Use the parallel-axis theorem in Probs. 9–24 to 9–46. Take advantage of tabulated values of moments of inertia when appropriate.

**9–24** Determine the mass moments of inertia $I_y$ and $I_z$ of the uniform slender rod of mass $M$.

**Fig. P9-24**

**9–25** Calculate the mass moments of inertia $I_x$, $I_y$, and $I_z$ of the uniform slender rod of mass $M$.

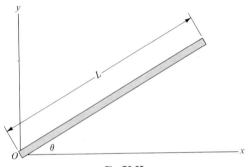

**Fig. P9-25**

**9–26** Calculate the moments of inertia $I_x$ and $I_y$ of the rectangular area $ab$.

**Fig. P9-26**

**9–27** Determine the polar moment of inertia $J_O$ of the rectangular area $ab$ in Fig. P9-26. Use the centroidal polar moment of inertia $J_C$.

**9–28** A rectangular prism has edges of lengths $a$, $b$, and $c$, with $a = 3b = 4c$. Calculate the mass moment of inertia $I_l$ of the prism if $l$ coincides with one of the edges of length $c$.

**9–29** Determine the moments of inertia $I_x$ and $I_y$ of the triangular area.

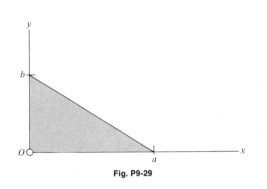

**Fig. P9-29**

**9–30** Calculate the polar moment of inertia $J_O$ of the triangular area in Fig. P9-29. Determine and use the centroidal polar moment of inertia $J_C$.

**9–31** A thin hoop of radius $R$ and mass $M$ is in the $xz$ plane. Determine the mass moment of inertia $I_{y_1}$ where $y_1$ is parallel to $y$ at $x = R$.

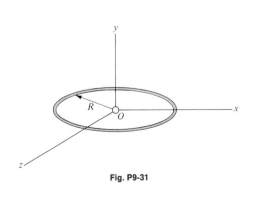

**Fig. P9-31**

**9–32** The thin cylindrical shell of radius $R$ and mass $M$ is centered on the $x$ axis. Calculate the mass moment of inertia $I_{x_1}$ where $x_1$ is parallel to $x$ at $y = R$.

Fig. P9-32

**9–33** Consider a sphere of radius $r$ and mass $M$. Calculate the mass moment of inertia $I_l$ where $l$ is any line tangent to the sphere.

**9–34** Calculate the moment of inertia $I_{x'}$ of the semicircular area. Axis $x'$ is parallel to $x$ through the centroid $C$ of the area.

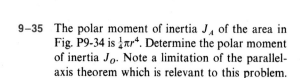

Fig. P9-34

**9–35** The polar moment of inertia $J_A$ of the area in Fig. P9-34 is $\frac{1}{4}\pi r^4$. Determine the polar moment of inertia $J_O$. Note a limitation of the parallel-axis theorem which is relevant to this problem.

**9–36** Consider a disk of mass $M$, radius $r$, and negligible thickness. Calculate the mass moment of inertia $I_{y_1}$, where $y_1$ is parallel to $y$ at $x = 2r$.

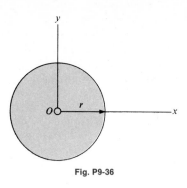

Fig. P9-36

**9–37** Determine the moment of inertia $I_x$ of the area in Fig. P9-36. Axis $x_1$ is parallel to $x$ at $y = r/2$.

**9–38** Determine the moments of inertia $I_x$ and $I_y$ of the shaded area.

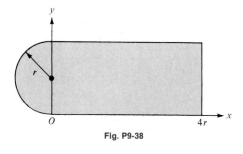

Fig. P9-38

**9–39** A thin, uniform wire of total mass $m$ is used to make the T-shaped piece. Determine the mass moments of inertia $I_x$ and $I_y$.

Fig. P9-39

**9–40** Determine the moment of inertia $I_x$ of the shaded area. Work with a rectangle $ac$, from which a triangular piece is removed. $a = 6$ in., $b = 4$ in., and $c = 7$ in.

Fig. P9-40

**9–41** Determine the moment of inertia $I_y$ of the shaded area in Fig. P9-40. Work with a rectangle $ac$ from which a triangular piece is removed. $a = 15$ mm, $b = 20$ mm, and $c = 25$ mm.

**9–42** A thin, uniform, rectangular plate has mass $m$ before a circular hole of diameter $d = a$ is cut in it at point $A$. Determine the mass moment of inertia $I_y$ and the radius of gyration about the $y$ axis of the plate with the hole.

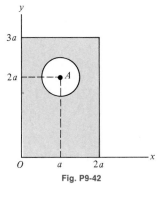

Fig. P9-42

**9–43** Calculate the moment of inertia $I_y$ of the shaded area.

Fig. P9-43

**9–44** Assume that the shaded area in Fig. P9-43 represents a thin plate of density $\rho$. The thickness is $t$ in the lower part $bd$, and $2t$ in the upper part $he$. Determine the mass moment of inertia $I_x$.

**9–45** Determine the moment of inertia $I_x$ of the shaded area in terms of the dimensions given without units.

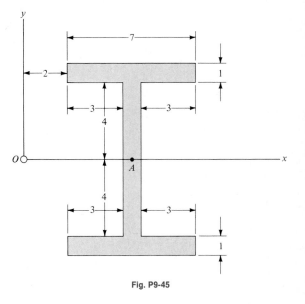

Fig. P9-45

**9–46** Calculate the moment of inertia $I_y$ in Prob. 9–45.

## PRODUCTS OF INERTIA

In certain problems of mechanics of materials and dynamics it is necessary to determine the maximum and minimum moments of inertia for areas or masses and the orientations of the corresponding reference axes. For this purpose it is useful to define the *product of inertia* $I_{xy}$, which is calculated with respect to the $x$ and $y$ axes simultaneously. This is illustrated using areas. For the element of area $dA$ in Fig. 9-15, the product of inertia is defined as $dI_{xy} = xy\, dA$. For the entire area $A$,

FIGURE 9-15

Area $A$ in the $xy$ plane

$$I_{xy} = \int xy\, dA \qquad \boxed{\text{9-18a}}$$

Similarly for the other pairs of axes,

$$I_{yz} = \int yz\, dA \qquad \boxed{\text{9-18b}}$$

$$I_{xz} = \int xz\, dA \qquad \boxed{\text{9-18c}}$$

The product of inertia has the same units as the moment of inertia of an area, length raised to the fourth power, $m^4$ or $in^4$. In contrast to the moment of inertia, which is always positive, the product of inertia may be positive, negative, or zero, since $x$ and $y$ may be zero or of various signs.

The product of inertia $I_{xy}$ is zero if either the $x$ or $y$ axis (or both) is an axis of symmetry of the area. For example, consider the rectangle in Fig. 9-16 that has the $y$ axis as an axis of symmetry. Here any element $dA$ with coordinates $+x$ and $\pm y$ has a corresponding image element $dA'$ with coordinates $-x$ and $\pm y$. The products of inertia for the elements shown in Fig. 9-16 are $xy\, dA$ and $-xy\, dA'$, where the areas $dA$ and $dA'$ are equal. Thus, the sum of the products of inertia for the pair of elements, $dA$ and $dA'$ is zero. On this basis the product of inertia for the whole area $A$ in Fig. 9-16 is zero.

### Parallel-Axis Theorem

A parallel-axis theorem that is valid for products of inertia clearly must take into account the simultaneous reference to *two* new axes. To derive the appropriate formula, consider the area $A$ in Fig. 9-17. Here the $x'$ and $y'$ axes are centroidal axes of the nonsymmetric area $A$, and it is desired to find the product of inertia with respect to the set of parallel axes $x$ and $y$.

The product of inertia of the element $dA$ in Fig. 9-17 can be defined with respect to both the $x'y'$ and the $xy$ system of axes,

$$dI_{x'y'} = x'y'\, dA$$

$$dI_{xy} = (x' + d_x)(y' + d_y)\, dA$$

FIGURE 9-16

Area $A$ in the $xy$ plane

These equations can be integrated to obtain the products of inertia for the whole area $A$,

$$I_{x'y'} = \int x'y'\, dA$$

$$I_{xy} = \int (x' + d_x)(y' + d_y)\, dA$$

$$= \int x'y'\, dA + d_x \int y'\, dA + d_y \int x'\, dA + d_x d_y \int dA$$

The second and third terms in the last equation are zero because the first moment of any area about a centroidal axis is zero. The fourth integral is equal to the area $A$, so

$$\boxed{I_{xy} = I_{x'y'} + A d_x d_y} \qquad \text{9-19a}$$

Similarly for the other pairs of axes,

$$\boxed{I_{yz} = I_{y'z'} + A d_y d_z} \qquad \text{9-19b}$$

$$\boxed{I_{xz} = I_{x'z'} + A d_x d_z} \qquad \text{9-19c}$$

where $d_x$, $d_y$, and $d_z$ have signs depending on their locations with respect to the $xy$ reference system. This equation is similar to the parallel-axis theorem for moments of inertia and it is convenient to use in dealing with composite areas.

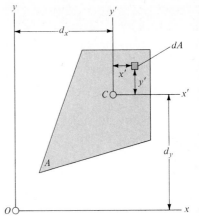

**FIGURE 9-17**

Area $A$ in the $xy$ plane

## Mass Products of Inertia

Mass products of inertia are defined similarly to those for areas. For a body of mass $M$ three products of inertia may be calculated with respect to axes $x$, $y$, and $z$,

$$\boxed{I_{xy} = \int xy\, dM \qquad I_{yz} = \int yz\, dM \\ I_{xz} = \int xz\, dM} \qquad \text{9-20}$$

Mass products of inertia may be zero when the body has symmetry, as in the case of areas. The parallel-axis theorem for mass products of inertia is also similar to that for areas, resulting in

$$\boxed{I_{xy} = I_{x'y'} + M d_x d_y} \qquad \text{9-21a}$$

$$\boxed{I_{yz} = I_{y'z'} + M d_y d_z} \qquad \text{9-21b}$$

$$\boxed{I_{xz} = I_{x'z'} + M d_x d_z} \qquad \text{9-21c}$$

where $x'$, $y'$, and $z'$ are axes through the center of mass, and $x$, $y$, and $z$ are parallel to them, respectively. $d_x$, $d_y$, and $d_z$ are the apparent distances between the pairs of parallel axes as viewed from another coordinate direction (see Ex. 9-8). $I_{x'y'}$, $I_{y'z'}$, and $I_{x'z'}$ are products of inertia with respect to the indicated centroidal axes. As before, Eqs. 9-21 are convenient in dealing with composite bodies.

# PRINCIPAL AXES AND MOMENTS OF INERTIA OF AN AREA

It is sometimes necessary to transform the known moments and product of inertia $I_x$, $I_y$, and $I_{xy}$ to refer them to other axes $u$ and $v$ that are inclined to the $x$ and $y$ axes. The transformation can be performed with the aid of Fig. 9-18. The $u$ and $v$ coordinates can be expressed in terms of $x$, $y$, and $\theta$,

$$u = x \cos \theta + y \sin \theta$$

$$v = y \cos \theta - x \sin \theta$$

Using these transformation equations, the moments and product of inertia of the area $A$ can be written as

$$I_u = \int v^2 \, dA = \int (y \cos \theta - x \sin \theta)^2 \, dA$$

$$= \cos^2 \theta \int y^2 \, dA - 2 \sin \theta \cos \theta \int xy \, dA$$

$$+ \sin^2 \theta \int x^2 \, dA$$

Finally, using the fundamental definitions for inertias,

$$I_u = I_x \cos^2 \theta - 2I_{xy} \sin \theta \cos \theta + I_y \sin^2 \theta \qquad \boxed{9\text{-}22}$$

Following a similar procedure for $I_v$ and $I_{uv}$ yields

$$I_v = I_x \sin^2 \theta + 2I_{xy} \sin \theta \cos \theta + I_y \cos^2 \theta \qquad \boxed{9\text{-}23}$$

$$I_{uv} = I_x \sin \theta \cos \theta + I_{xy}(\cos^2 \theta - \sin^2 \theta) \qquad \boxed{9\text{-}24}$$
$$- I_y \sin \theta \cos \theta$$

After adding Eqs. 9-22 and 9-23 it is found that

$$I_u + I_v = I_x + I_y = J_O \qquad \boxed{9\text{-}25}$$

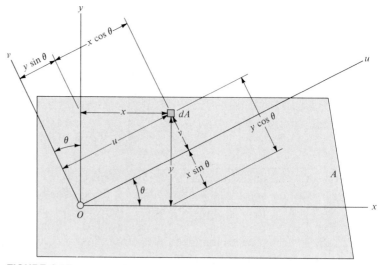

**FIGURE 9-18**

Area $A$ in the $xy$ plane

This is reasonable since the polar moment of inertia $J_O$ for a given area should depend only on its position with respect to the reference point $O$ (Eq. 9-8) and not on the orientation of any axes through that point.

It is worthwhile to change Eqs. 9-22, 9-23, and 9-24 using the trigonometric identities $\sin 2\theta = 2 \sin \theta \cos \theta$ and $\cos 2\theta = \cos^2 \theta - \sin^2 \theta$. The resulting equations are

$$I_u = \frac{I_x + I_y}{2} + \frac{I_x - I_y}{2} \cos 2\theta - I_{xy} \sin 2\theta \qquad \boxed{9\text{-}26}$$

$$I_v = \frac{I_x + I_y}{2} - \frac{I_x - I_y}{2} \cos 2\theta + I_{xy} \sin 2\theta \qquad \boxed{9\text{-}27}$$

$$I_{uv} = \frac{I_x - I_y}{2} \sin 2\theta + I_{xy} \cos 2\theta \qquad \boxed{9\text{-}28}$$

These can be further manipulated to advantage if the functions of $\theta$ are eliminated. A simple but less than obvious procedure involving Eqs. 9-26 and 9-28 involves the following major steps. From Eq. 9-26,

$$\left(I_u - \frac{I_x + I_y}{2}\right)^2 = \left(\frac{I_x - I_y}{2} \cos 2\theta - I_{xy} \sin 2\theta\right)^2$$

Adding this equation to the square of Eq. 9-28 and noting that $\sin^2 2\theta + \cos^2 2\theta = 1$,

$$\left(I_u - \frac{I_x + I_y}{2}\right)^2 + I_{uv}^2 = \left(\frac{I_x - I_y}{2}\right)^2 + I_{xy}^2 \qquad \boxed{9\text{-}29}$$

Here $I_x$, $I_y$, and $I_{xy}$ may be assumed known constants or relatively easily calculated quantities. It is required to determine the unknown quantities $I_u$ and $I_{uv}$. This task is facilitated by substituting in Eq. 9-29

$$\frac{I_x + I_y}{2} = I_{\text{ave}} \quad \text{and} \quad \left(\frac{I_x - I_y}{2}\right)^2 + I_{xy}^2 = R^2 \qquad \boxed{9\text{-}30}$$

Here $I_{\text{ave}}$ is the average of $I_x$ and $I_y$. With these, Eq. 9-29 becomes

$$\boxed{(I_u - I_{\text{ave}})^2 + I_{uv}^2 = R^2} \qquad \boxed{9\text{-}31}$$

which is the equation of a circle of radius $R$ in the $I_u$ and $I_{uv}$ coordinate system as shown in Fig. 9-19. This is called *Mohr's circle* after the German engineer Otto Mohr (1835–1918). Since $I_x$, $I_u$, and so on, are related and their units are all the same, these quantities may be plotted in the same diagram. $I_{\text{ave}}$ is the location of the center of the circle on the $I_u$ axis, and it is always on that axis according to Eq. 9-31.* The moments of inertia labeled $I_{\max}$ and $I_{\min}$ in Fig. 9-19 are called the *principal moments of inertia* of the area $A$ with respect to point $O$ in Fig. 9-19. Their magnitudes can be obtained from Eq. 9-31 after setting $I_{uv} = 0$. In detail, the extreme values of $I_u$ are

$$\boxed{I_{\substack{\max \\ \min}} = \frac{I_x + I_y}{2} \pm \sqrt{\left(\frac{I_x - I_y}{2}\right)^2 + I_{xy}^2}} \qquad \boxed{9\text{-}32}$$

* The equation of a circle in the $xy$ plane with radius $R$ and center at coordinates $(a, b)$ is $(x - a)^2 + (x - b)^2 = R^2$.

**FIGURE 9-19**

Mohr's circle in the plane of $I_u$ and $I_{uv}$ axes

Another item of interest is the location of the principal axes $x'$ and $y'$, which are the reference axes for $I_{max}$ and $I_{min}$. The simplest case is when the area has at least one axis of symmetry. Since $I_{uv} = 0$ was used in deriving Eq. 9-32, and since the product of inertia is zero with respect to any axis of symmetry, *any axis of symmetry is a principal axis of the area.* The reverse is not necessarily true, so a principal axis may or may not be an axis of symmetry.

The problem of locating the principal axes is more complex when the area has no symmetry. For the analysis of such a case, assume that $I_x$, $I_y$, and $I_{xy}$ are relatively easy to calculate for the area $A$ in Fig. 9-20a, so these are considered as known quantities. The circle representing Eq. 9-31 can be drawn with the following considerations involving $I_x$, $I_y$, and $I_{xy}$.

(a)

(b)

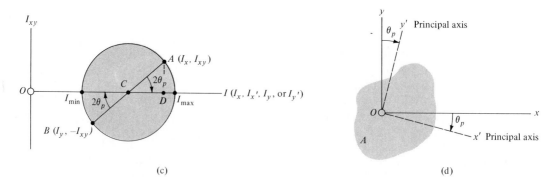

(c)

(d)

**FIGURE 9-20**

Principal axes and moments of inertia of area $A$ in the $xy$ plane

**1.** The center of the circle must be on the horizontal $I$ axis, and its location is at $C = (I_x + I_y)/2$ in Fig. 9-20b.

**2.** $I_x$ and $I_{xy}$ must be plotted at a single point (assume point $A$ in Fig. 9-20b) since they must simultaneously satisfy Eq. 9-31 as a point on the circle.

**3.** $I_y$ and $I_{xy}$ must also be at a single point, and on the same circle as $I_x$ and $I_{xy}$, since $I_x$, $I_y$, and $I_{xy}$ refer to the same area. It is reasonable to assume at this stage that point $B'$ in Fig. 9-20b could represent $I_y$ and $I_{xy}$. However, careful thought leads to the finding that point $B$ with coordinates $I_y$ and $-I_{xy}$ is the correct point on the circle. The reasons are outlined as follows.

**a.** $I_y$ and $-I_{xy}$ satisfy Eq. 9-31 just as $I_y$ and $I_{xy}$ do since the product of inertia is squared.

**b.** The $x$ and $y$ axes are 90° apart in Fig. 9-20a, but after careful consideration they "appear" to be 180° apart in the Mohr's circle representation of Fig. 9-20b. Clearly, a rotation through 360° would be required in Fig. 9-20b to go along the circle from $A$ and return to $A$. However, a 180° rotation of the $x$ axis in Fig. 9-20a produces the same reference axis for $I_x$ and $I_{xy}$. In other words, $I_x$ and $I_{xy}$ are not changed if the reference axis $x$ is rotated 180° about the origin of the coordinate system. Thus, any rotation $\theta$ in Fig. 9-20a must be represented by a rotation $2\theta$ in the construction of Mohr's circle (Fig. 9-20b). With these considerations, point $B$ is the logical choice for $I_y$ and $I_{xy}$ in Fig. 9-20b.

The location of the principal axes with respect to the known axes $x$ and $y$ is now straightforward using Fig. 9-20c. The principal axes are reached by a clockwise rotation through an angle $2\theta_p$ in Mohr's circle. From Fig. 9-20c,

$$\tan 2\theta_p = \frac{AD}{CD} = \frac{2I_{xy}}{I_x - I_y} \qquad \boxed{\text{9-33}}$$

The locations of the principal axes with respect to area $A$ are obtained in Fig. 9-20d by a rotation from the $x$ and $y$ axes through an angle $\theta_p$. The direction of this rotation is the same as in Mohr's circle.

## TRANSFORMATION OF MOMENTS OF INERTIA OF A MASS

**9-8**

The dynamics analysis of bodies in some situations involves the concept of *principal moments of inertia of a body*. This concept is similar to the principal moments of inertia for an area, and it is reasonable to use here in preparation for a course in dynamics. The basic problem again is to transform known (or relatively easily calculated) moments of inertia to refer them to other axes of interest.

The transformation procedure is explained with the aid of Fig. 9-21. A body of total mass $M$ is outlined here with respect to the $xyz$ coordinate system. Assume that the moment of inertia of the body should be determined with respect to the line $l$. By definition, $I_l = \int r^2 \, dM$, where $r$ is the shortest distance from the elemental mass $dM$ to the line $l$. It is advantageous to replace $r$ by the vector product $\mathbf{n} \times \mathbf{q}$, where $\mathbf{n}$ is a unit vector along the line $l$ (this is allowed since $r = q \sin \theta$, the magnitude of $\mathbf{n} \times \mathbf{q}$). In terms of the rectangular components $n_x$, $n_y$, and $n_z$ of $\mathbf{n}$, and $x$, $y$, and $z$ of the mass $dM$,

$$I_l = \int \left[ (n_x y - n_y x)^2 + (n_y z - n_z y)^2 + (n_z x - n_x z)^2 \right] dM \qquad \boxed{9\text{-}34a}$$

This equation can be written as

$$I_l = n_x^2 \int (y^2 + z^2) \, dM + n_y^2 \int (z^2 + x^2) \, dM$$

$$+ \, n_z^2 \int (x^2 + y^2) \, dM - 2n_x n_y \int xy \, dM$$

$$- \, 2n_y n_z \int yz \, dM - 2n_z n_x \int zx \, dM$$

With $\int (y^2 + z^2) \, dM = I_x$, $\int xy \, dM = I_{xy}$, and so on,

$$I_l = I_x n_x^2 + I_y n_y^2 + I_z n_z^2 - 2I_{xy} n_x n_y - 2I_{yz} n_y n_z - 2I_{zx} n_z n_x \qquad \boxed{9\text{-}34b}$$

For the case where principal moments of inertia about principal axes $x'$, $y'$, and $z'$ are used, $I_{x'y'} = I_{y'z'} = I_{z'x'} = 0$. Thus, Eq. 9-34b reduces to

$$\boxed{I_l = I_{x'} n_{x'}^2 + I_{y'} n_{y'}^2 + I_{z'} n_{z'}^2} \qquad \boxed{9\text{-}35}$$

This equation is valid for bodies of any shape since even irregular bodies have maximum and minimum values of their moments of inertia. Naturally, it is easiest to work with symmetric bodies for which the principal axes can be established by inspection. These equations are also valid for the moment of inertia of an area about a line $l$ where the appropriate quantities for the area are used. Note that the principal moments of inertia for objects of simple shape are available in Table 4, inside the back cover.

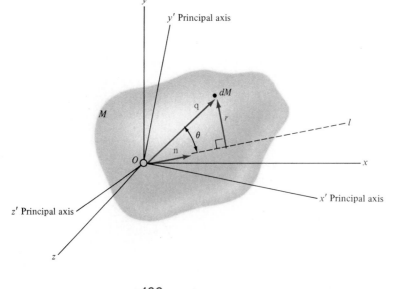

**FIGURE 9-21**

Transformation of moments of inertia of mass $M$ in the $xyz$ frame

EXAMPLE 9-7

Determine the product of inertia $I_{xy}$ of the area in Fig. 9-22a.

## SOLUTION

Choosing an elemental area of width $dx$ in Fig. 9-22b,

$$I_{xy} = \int xy \, dA = \int x\left(\frac{y}{2}\right) y \, dx$$

For $y = kx^2$, the value of $k$ is determined by substituting in known values for a point on the curve:

$$b = ka^2$$

$$k = \frac{b}{a^2}$$

Using $y = (b/a^2) x^2$ in the equation for $I_{xy}$,

$$I_{xy} = \frac{1}{2} \int_0^a x \left(\frac{b}{a^2} x^2\right)^2 dx = \frac{b^2}{2a^4} \int_0^a x^5 \, dx = \frac{b^2}{2a^4}\left(\frac{a^6}{6}\right)$$

$$= \frac{1}{12} a^2 b^2$$

(a)

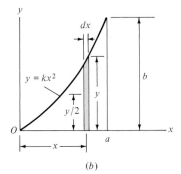

(b)

**FIGURE 9-22**

Area in the $xy$ plane

EXAMPLE 9-7   407

EXAMPLE 9-8

Determine the principal axes and moments of inertia about the centroid of the area shown in Fig. 9-23. Also determine the moments and product of inertia with respect to the $uv$ axes. Use the equations of Mohr's circle.

## SOLUTION

It is necessary to determine the moments and product of inertia about the $x'$ and $y'$ axes (parallel to the $x$ and $y$ axes) with the origin at $C$. First $C$ must be located, however.

$$\bar{x}_C = \frac{\sum \bar{x}_i A_i}{\sum A_i} = \frac{(1 \text{ in.})(1 \text{ in}^2) + (0.25 \text{ in.})(2.5 \text{ in}^2) + (2 \text{ in.})(2 \text{ in}^2)}{1 \text{ in}^2 + 2.5 \text{ in}^2 + 2 \text{ in}^2}$$

$$= 1.02 \text{ in.}$$

$$\bar{y}_C = \frac{\sum \bar{y}_i A_i}{\sum A_i} = \frac{(5.75 \text{ in.})(1 \text{ in}^2) + (3 \text{ in.})(2.5 \text{ in}^2) + (0.25 \text{ in.})(2 \text{ in}^2)}{1 \text{ in}^2 + 2.5 \text{ in}^2 + 2 \text{ in}^2}$$

$$= 2.5 \text{ in.}$$

Using the parallel-axis theorem (with $I_{x_i}$ as centroidal moments of inertia of sub-areas $A_i$),

$$I_{x'} = \sum (I_{x_i} + A_i d_{y_i}^2)$$

$$= \frac{1}{12} (0.5 \text{ in.})^3 (2 \text{ in.}) + (1 \text{ in}^2)(5.75 - 2.5)^2 \text{ in}^2$$

$$+ \frac{1}{12} (5 \text{ in.})^3 (0.5 \text{ in.}) + (2.5 \text{ in}^2)(3 - 2.5)^2 \text{ in}^2$$

$$+ \frac{1}{12} (0.5 \text{ in.})^3 (4 \text{ in.}) + (2 \text{ in}^2)(0.25 - 2.5)^2 \text{ in}^2$$

$$= 26.6 \text{ in}^4$$

Similarly,

$$I_{y'} = \sum (I_{y_i} + A_i d_{x_i}^2)$$

$$= \frac{1}{12} (0.5 \text{ in.})(2 \text{ in.})^3 + (1 \text{ in}^2)(1 \text{ in.} - 1.02 \text{ in.})^2$$

$$+ \frac{1}{12} (5 \text{ in.})(0.5 \text{ in.})^3 + (2.5 \text{ in}^2)(0.25 \text{ in.} - 1.02 \text{ in.})^2$$

$$+ \frac{1}{12} (0.5 \text{ in.})(4 \text{ in.})^3 + (2 \text{ in}^2)(2 \text{ in.} - 1.02 \text{ in.})^2$$

$$= 6.46 \text{ in}^4$$

$$I_{x'y'} = \sum (I_{xy_i} + A_i d_{y_i} d_{x_i})$$

$$= 0 + (1 \text{ in}^2)(5.75 \text{ in.} - 2.5 \text{ in.})(1 \text{ in.} - 1.02 \text{ in.})$$

$$+ 0 + (2.5 \text{ in}^2)(3 \text{ in.} - 2.5 \text{ in.})(0.25 \text{ in.} - 1.02 \text{ in.})$$

$$+ 0 + (2 \text{ in}^2)(0.25 \text{ in.} - 2.5 \text{ in.})(2 \text{ in.} - 1.02 \text{ in.})$$

$$= -0.07 \text{ in}^4 + (-0.96 \text{ in}^4) + (-4.41 \text{ in}^4)$$

$$= -5.44 \text{ in}^4$$

**FIGURE 9-23**

Area in the $xy$ plane

The principal moments of inertia are obtained by Eq. 9-32,

$$I_{\substack{max \\ min}} = \frac{26.6 + 6.46}{2} \pm \sqrt{\left(\frac{26.6 - 6.46}{2}\right)^2 + (-5.44)^2}$$

$$I_{max} = 28.0 \text{ in}^4, \quad I_{min} = 5.08 \text{ in}^4$$

From Eq. 9-29,

$$\tan 2\theta = \frac{2I_{x'y'}}{I_{x'} - I_{y'}} = \frac{2(-5.44)}{26.6 - 6.46} = -0.54$$

$$2\theta = 151.6° \quad \text{or} \quad -28.4°$$

$$\theta = 75.8° \quad \text{or} \quad -14.2$$

From Mohr's circle it is seen that the axis of maximum moment of inertia is obtained by rotating the $x'$ axis counterclockwise 16.3°. Using Eqs. 9-26, 9-27, and 9-28, for $\theta = -30°$ (a clockwise rotation from axis $x'$ for the transformation equations),

$$I_u = \frac{26.6 + 6.46}{2} + \frac{26.6 - 6.46}{2} \cos(-60°) - (-5.44) \sin(-60°)$$

$$= 16.9 \text{ in}^4$$

$$I_v = \frac{26.6 + 6.46}{2} - \frac{26.6 - 6.46}{2} \cos(-60°) + (-5.44) \sin(-60°)$$

$$= 16.2 \text{ in}^4$$

$$I_{uv} = \frac{26.6 - 6.46}{2} \sin(-60°) + (-5.44) \cos(-60°)$$

$$= -11.4 \text{ in}^4$$

**TIP**

It is important to be careful with the signs of $d_x$ and $d_y$ when computing products of inertia. The correct signs are obtained by going *from* the centroidal axes to the other axes. The correct sign of $\theta$ is obtained by remembering that positive $\theta$ is measured clockwise from the $x'$ axis going *from* the original axis *to* the new axis (this $\theta$ is for the transformation equations).

## JUDGMENT OF THE RESULTS

*Check:*

$$I_u + I_v = (16.9 + 16.2) \text{ in}^4 = 33.1 \text{ in}^4$$

$$= (26.6 + 6.46) \text{ in}^4 = I_{x'} + I_{y'}$$

EXAMPLE 9-8    409

# EXAMPLE 9-9

A railroad freight car on a test track (Fig. 9-24a) is modeled as a rectangular body of uniform density $\rho$ in Fig. 9-24b. Determine the mass moments of inertia of the body with respect to the $x$ axis, line $AB$, and the diagonal $OG$. Use Table 4, inside the back cover.

## SOLUTION

From Table 4 (inside back cover), the moment of inertia about the $x'$ axis through the centroid of the body is $I_{x'} = (M/12)(b^2 + c^2)$. Using the parallel axis theorem, and $d_1$ as the nearest distance between the $x$ and $x'$ axes, the mass moment of inertia with respect to the $x$ axis is

$$I_x = I_{x'} + Md_1^2 = \frac{M}{12}(b^2 + c^2) + M\left[\left(\frac{b}{2}\right)^2 + \left(\frac{c}{2}\right)^2\right]$$

$$= \frac{M}{3}(b^2 + c^2)$$

where $M = \rho abc$. Similarly,

$$I_{AB} = I_{y'} + Md_2^2 = \frac{M}{12}(a^2 + c^2) + M\left[\left(\frac{a}{2}\right)^2 + \left(\frac{c}{2}\right)^2\right]$$

$$= \frac{M}{3}(a^2 + c^2)$$

Equation 9-34b is used to calculate the moment of inertia with respect to line $OG$,

$$I_{OG} = I_x n_x^2 + I_y n_y^2 + I_z n_z^2 - 2I_{xy}n_x n_y - 2I_{yz}n_y n_z - 2I_{xz}n_x n_z$$

Several of these terms must be calculated yet.

$$I_y = I_{AB} = \frac{M}{3}(a^2 + c^2)$$

$$I_z = I_{z'} + Md_3^2 = \frac{M}{12}(a^2 + b^2) + M\left[\left(\frac{a}{2}\right)^2 + \left(\frac{b}{2}\right)^2\right]$$

$$= \frac{M}{3}(a^2 + b^2)$$

$$I_{xy} = I_{x'y'} + Md_x d_y = 0 + M\left(-\frac{b}{2}\right)\left(-\frac{a}{2}\right) = \frac{M}{4}ab$$

where $d_x$ is the vertical distance in going from the $x'$ to the $x$ axis, as viewed from the $z$ axis, and so on (the sign is taken going *from* the centroidal axes).

$$I_{yz} = I_{y'z'} + Md_y d_z = 0 + M\left(-\frac{c}{2}\right)\left(-\frac{b}{2}\right) = \frac{M}{4}bc$$

$$I_{xz} = I_{x'z'} + Md_x d_z = 0 + M\left(-\frac{c}{2}\right)\left(-\frac{a}{2}\right) = \frac{M}{4}ac$$

$$I_{x'y'} = I_{y'z'} = I_{x'z'} = 0$$

because the body is symmetric with respect to the centroidal $x'y'z'$ axes. The unit vector in the direction of $OG$ is

(a)

**FIGURE 9-24**

Modeling of a railroad freight car for determining its moments of inertia.
(Photograph courtesy MTS Systems Corporation, Minneapolis, Minnesota.)

(b)

$$\mathbf{n}_{OG} = \frac{\overrightarrow{OG}}{OG} = \frac{a\mathbf{i} + c\mathbf{k}}{(a^2 + c^2)^{1/2}}, \quad \text{so} \quad n_x = \frac{a}{(a^2 + c^2)^{1/2}},$$

$$n_y = 0, \quad n_z = \frac{c}{(a^2 + c^2)^{1/2}}$$

Finally,

$$I_{OG} = \frac{M}{3}(b^2 + c^2)\frac{a^2}{a^2 + c^2} + \frac{M}{3}(c^2 + a^2)(0)$$

$$+ \frac{M}{3}(a^2 + b^2)\frac{c^2}{a^2 + c^2} - 2\left(\frac{M}{4}ab\right)(0)$$

$$- 2\left(\frac{M}{4}bc\right)(0) - 2\left(\frac{M}{4}ac\right)\left(\frac{ac}{a^2 + c^2}\right)$$

$$= \frac{M}{3}\left(\frac{a^2b^2 + b^2c^2}{a^2 + c^2}\right) + \frac{M}{6}\left(\frac{a^2c^2}{a^2 + c^2}\right)$$

$$= \frac{M}{6}\left(2b^2 + \frac{a^2c^2}{a^2 + c^2}\right)$$

## JUDGMENT OF THE RESULTS

The units obtained are correct.

EXAMPLE 9-9    411

# PROBLEMS

In Probs. 9–47 to 9–65, calculate the indicated moments of inertia and products of inertia of the area $A$ or mass $M$. The masses are equal to the volume times the density $\delta$. The thickness of each body is $t$.

**9–47**  Area $A = ab$. $I_{xy} = ?$ $I_{uv} = ?$

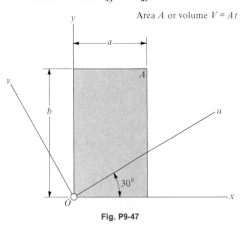

**Fig. P9-47**

**9–48**  Body $V = abt$ in Fig. P9-47. $I_{xy} = ?$ $I_{uv} = ?$

**9–49**  Area $A = ab$ in Fig. P9-47. Plot Mohr's circle showing $I_x, I_y, I_u, I_v, I_{xy}, I_{max}, I_{min}$. Use simple formulas to calculate any of these quantities when possible.

**9–50**  Area $A$. $I_{xy} = ?$ $I_{uv} = ?$

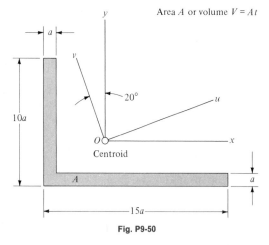

**Fig. P9-50**

**9–51**  Body $V$ in Fig. P9-50. $I_{xy} = ?$ $I_{uv} = ?$

**9–52**  Area $A$ in Fig. P9-50. Plot Mohr's circle showing $I_x, I_y, I_u, I_v, I_{xy}, I_{uv}, I_{max}$, and $I_{min}$. Use simple formulas to calculate any of these quantities when possible.

**9–53**  Area $A$; $\theta = 40°$. $I_{uv} = ?$

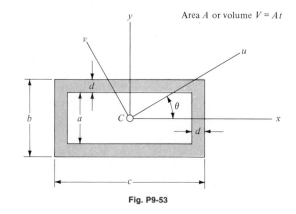

**Fig. P9-53**

**9–54**  Body $V$ in Fig. P9-53; $\theta = 30°$. $I_{uv} = ?$

**9–55**  Area $A$. $I_{xy} = ?$

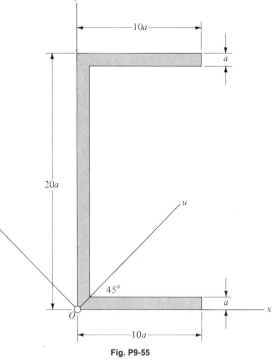

**Fig. P9-55**

**9–56**  Body $V$ in Fig. P9-55. $I_{uv} = ?$

**9–57**  Area $A$ in Fig. P9-55. Plot Mohr's circle showing $I_x, I_y, I_u, I_v, I_{xy}, I_{uv}, I_{max}$, and $I_{min}$. Use simple formulas to calculate any of these quantities when possible.

**9–58**   Area $A$. $I_{xy} = ?$

**Fig. P9-58**

**9–59**   Body $V$ in Fig. P9-58. $I_{uv} = ?$

**9–60**   Body $V$. $I_{xy} = ?$

Area $A$ or volume $V = At$

**Fig. P9-60**

**9–61**   Area $A$ in Fig. P9-60. $I_{uv} = ?$

**9–62**   Calculate the principal moments of inertia of the area. Determine the orientation of the principal axes.

**Fig. P9-62**

**9–63**   The thin hoop in the $xz$ plane and the two pegs on it are used as a simple gear. The material has a mass of 1 kg/m, and $r$ is 30 mm. Determine the principal moments of inertia of this gear.

**Fig. P9-63**

**9–64**   The thin-walled cylindrical shell is made of steel and weighs 0.283 lb/in³. Determine the principal moments of inertia of the shell. The origin of the coordinates is at the centroid of the shell.

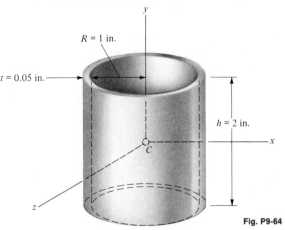

**Fig. P9-64**

**9–65**   A hole of radius $r = 0.2$ in. is drilled at the positive $x$ axis in the shell shown in Fig. P9-64. Calculate the mass products of inertia of the new member. Estimate the principal moments of inertia. Discuss the assumptions and limitations of your analysis.

## Section 9-1

### MOMENT OF INERTIA OF A MASS $M$:

$$dI_x = r^2\, dM = (y^2 + z^2)\, dM$$

$$I_x = \int r^2\, dM = \int (y^2 + z^2)\, dM$$

$$I_y = \int (x^2 + z^2)\, dM$$

$$I_z = \int (x^2 + y^2)\, dM$$

**RADIUS OF GYRATION:**  $r_g = \sqrt{I/M}$, where $r_g$ and $I$ are with respect to the same axis.

## Section 9-2

### MOMENT OF INERTIA OF AN AREA $A$ IN THE $XY$ PLANE:

$$dI_x = y^2\, dA$$

$$I_x = \int y^2\, dA \quad I_y = \int x^2\, dA$$

**RADIUS OF GYRATION:**  $r_g = \sqrt{I/A}$, where $r_g$ and $I$ are with respect to the same axis.

## Section 9-3

### POLAR MOMENT OF INERTIA OF AN AREA $A$:

$$J_O = \int r^2\, dA = I_x + I_y$$

**RADII OF GYRATION:**  $r_{g_O}^2 = r_{g_x}^2 + r_{g_y}^2$

### THIN PLATES IN THE $XY$ PLANE:

$$I_z = I_x + I_y$$

$$I_{\text{mass}} = I_{\text{area}}(t)(\rho)$$

where   $t$ = constant thickness of the plate
   $\rho$ = constant density of the plate

## Section 9-4

### PARALLEL-AXIS THEOREM:

For a mass $M$:    $I = I_C + Md^2$

For an area $A$:    $I = I_C + Ad^2$

$$J_O = J_C + Ad^2$$

where $I$ or $J_O$ = moment of inertia of mass $M$ or area $A$ about any
appropriate line $l$

$I_C$ or $J_C$ = moment of inertia of mass $M$ or area $A$ about a line
through the mass center of $M$ or the centroid of $A$ and
parallel to line $l$

$d$ = perpendicular distance between the two parallel lines

For radii of gyration: $\quad r_g^2 = r_{gc}^2 + d^2$

*Note:* Each equation of the parallel-axis theorem is valid only if one of
the two axes is a centroidal axis.

Section 9-5

**PROCEDURE TO DETERMINE MOMENTS OF INERTIA OF
COMPLEX AREAS AND BODIES:**

1. Break the object down into simple parts for which the centroids can be
   determined by inspection.
2. Calculate the centroidal moments of inertia of each part. Use tabulated
   formulas when possible.
3. Transform the individual centroidal moments of inertia to the required
   common axis using the parallel-axis theorem.

*Note:* Use judgment in the initial choosing of parts to simplify the calculations throughout the solution.

Section 9-6

**PRODUCTS OF INERTIA:**

**FOR AN AREA $A$:**

$$I_{xy} = \int xy \, dA$$

$$I_{yz} = \int yz \, dA$$

$$I_{xz} = \int xz \, dA$$

Parallel-axis theorem:

$$I_{xy} = I_{x'y'} + A d_x d_y$$

$$I_{yz} = I_{y'z'} + A d_y d_z$$

$$I_{xz} = I_{x'z'} + A d_x d_z$$

**FOR A MASS $M$:**

$$I_{xy} = \int xy \, dM$$

$$I_{yz} = \int yz \, dM$$

$$I_{xz} = \int xz \, dM$$

Parallel-axis theorem:

$$I_{xy} = I_{x'y'} + M d_x d_y$$

$$I_{yz} = I_{y'z'} + M d_y d_z$$

$$I_{xz} = I_{x'z'} + M d_x d_z$$

where $x'y'z'$ are centroidal axes, and $xyz$ are arbitrary axes parallel to $x'y'z'$, respectively. $d_x$, $d_y$, and $d_z$ are the apparent distances between the pairs of parallel axes as viewed from another coordinate direction (see Ex. 9-8).

Section 9-7

**PRINCIPAL AXES AND MOMENTS OF INERTIA OF AN AREA:**  The principal axes $x'$ and $y'$ are the reference axes for $I_{max}$ and $I_{min}$. The angle $\theta_p$ between the principal axes $x'y'$ and given axes $xy$ is obtained from

$$\tan 2\theta_p = \frac{2I_{xy}}{I_x - I_y}$$

The principal moments of inertia $I_{max}$ and $I_{min}$ are obtained from

$$I_{\substack{max \\ min}} = \frac{I_x + I_y}{2} \pm \sqrt{\left(\frac{I_x - I_y}{2}\right)^2 + I_{xy}^2}$$

Section 9-8

**TRANSFORMATION OF MOMENTS OF INERTIA OF A MASS:**

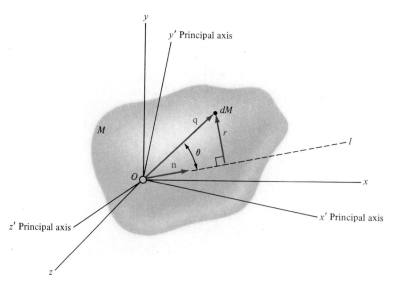

For an arbitrary line $l$,

$$I_l = I_x n_x^2 + I_y n_y^2 + I_z n_z^2 - 2I_{xy} n_x n_y - 2I_{yz} n_y n_z - 2I_{zx} n_z n_x$$

When principal moments of inertia about principal axes $x'y'z'$ are used,

$$I_l = I_{x'} n_{x'}^2 + I_{y'} n_{y'}^2 + I_{z'} n_{z'}^2$$

**9–66** The thin plate has thickness $t$ and density $\rho$. Determine the mass moment of inertia $I_y$, assuming that it is not a function of $z$ (since $t$ is small).

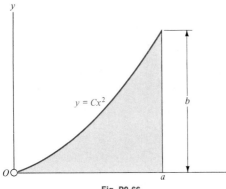

Fig. P9-66

**9–67** Calculate the moment of inertia $I_{x'}$ of the shaded area in Fig. P9-66. Axis $x'$ is parallel to $x$ through the centroid of the area.

**9–68** Calculate the moment of inertia $I_x$ of the shaded area in Fig. P9-66. Express the radius of gyration.

**9–69** Determine the polar moment of inertia $J_O$ of the shaded area in Fig. P9-66.

**9–70** The cross section of an I-beam is approximated by the shaded area. Determine the moments of inertia, $I_x$ and $I_y$.

**9–71** Determine the product of inertia $I_{xy}$ of the shaded area in Fig. P9-70.

**9–72** Determine the product of inertia $I_{xy}$ of the triangular area.

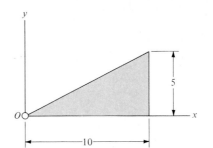

Fig. P9-72

**9–73** Determine the principal moments of inertia and the orientation of the principal axes of the shaded area.

Fig. P9-73

Fig. P9-70

# 10 Work and Potential Energy

Basic Concepts of Work and Energy

The term *work* has an exact, quantitative meaning when used in the context of mechanics. Quantitatively, *work is the product of the force and the corresponding displacement*. This simple definition has to be carefully interpreted and used. The term *energy* is used to describe the capacity of a system to do work on another system.

These concepts are advantageous for analyzing the equilibrium of complex systems in statics, a variety of motions in dynamics, and the behavior of engineering members in mechanics of materials. Several practical problem areas of work and energy are introduced here using the flight simulation of aircraft structures. In the old days flight simulation was done by putting people on the wings or applying dead weights at the regions of interest. The modern method is to use numerous hydraulic actuators controlled by electronic equipment to produce a wide range of realistic forces that can vary in magnitude and repetitions. This requires an elaborate reaction-frame structure around the aircraft, as shown in Fig. 10-1.

The strength and performance characteristics of the test equipment must be determined in the design almost as carefully as those of the aircraft to be tested. For example, some wings are highly flexible, as shown in Fig. 10-2. The minimum requirement of the test equipment is to apply static forces to the wing over the full range of expected deformations. A

**FIGURE 10-1**

Boeing 747 aircraft in a test setup. (The Boeing Company, Seattle, Washington.)

**FIGURE 10-2**

Static deflections of a long wing in a test. The maximum upward deflection is 22 ft and the downward deflection is 10 ft (double exposure) from the undeflected position. (The Boeing Company, Seattle, Washington.)

more severe requirement is to apply the forces dynamically, even repeatedly (since the wing may vibrate in service). Thus, the concepts of work (force times displacement) and the rate of doing work are needed in designing such tests. In all cases the strength properties of the cantilever wing are also involved in the design of the test equipment.

**SECTIONS 10-1 AND 10-2** define the work of forces and couples. The basic definitions and several useful expressions of work should be learned in preparation for other courses in engineering mechanics.

**SECTION 10-3** presents the concept and method of virtual work. In this analysis, displacements do not actually take place. The use of imaginary small displacements is very appealing philosophically and powerful practically for investigating the equilibrium of complex systems. However, this topic is sometimes omitted from regular statics courses because it is not essential for most subsequent courses.

**SECTION 10-4** defines the work of friction forces when relative motion between bodies occurs. The distinction of useful work and frictional losses is used in the concept of mechanical efficiency. This concept is important in designing machines.

**SECTION 10-5** presents the concept of potential energy (the capacity to do work) of a mechanical system because of the effect of gravity on the system or elastic deformability of its members. This topic is especially useful in the study of dynamics.

**SECTIONS 10-6 AND 10-7** apply the relationship of work and potential energy to investigate the equilibrium and the stability of a system. These topics are often omitted from standard statics courses because they are not used in subsequent elementary courses.

**STUDY GOALS**

## WORK OF A FORCE        10-1

By definition, work is done by a force on a body when it causes a corresponding displacement of that body. There are several aspects of this concept that must be distinguished in engineering analyses.

(a)

(b)

**FIGURE 10-3**

Displacement of a block along an incline

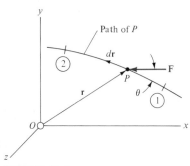

**FIGURE 10-4**

Infinitesimal displacements of a particle

**FIGURE 10-5**

The work of internal forces $T_A$ and $T_B$ cancel each other if the member displaced is rigid

A simple case is illustrated in Fig. 10-3a where the constant force **F** slowly moves the block from point 1 to 2 along a frictionless inclined plane. If **F** is parallel to the incline, the work $U$ of the force is defined as

$$U = Fs \qquad \boxed{10\text{-}1}$$

where $s$ is the displacement of the block. If **F** is at an angle $\theta$ with the plane as in Fig. 10-3b,

$$\boxed{U = F \cos \theta \, s} \qquad \boxed{10\text{-}2}$$

The reason for Eq. 10-2 is that only the component $F \cos \theta$ is associated with the displacement of the block. The other component, $F \sin \theta$, presses the block to the inclined plane but is not involved in the displacement (assuming that the plane is rigidly fixed) and therefore does no work on the block according to the definition stated above.

A more generally useful form of Eq. 10-2 is necessary because both the magnitude of **F** and its direction with respect to the path of the displacement may vary during the displacement. This can be obtained by considering the infinitesimal work $dU$ of a force **F** during an infinitesimal displacement. Work on a particle (such as $P$ in Fig. 10-4) can be defined vectorially as the scalar product of two vectors, which agrees in principle with Eq. 10-2,

$$dU = \mathbf{F} \cdot d\mathbf{r} = F \, dr \cos \theta \qquad \boxed{10\text{-}3}$$

The total work $U_{12}$ in displacing particle $P$ from point 1 to 2 is obtained by integrating Eq. 10-3. Using the rectangular components of **F** and $d\mathbf{r}$,

$$\boxed{\begin{aligned} U_{12} &= \int_1^2 \mathbf{F} \cdot d\mathbf{r} = \int_1^2 F \cos \theta \, dr \\ &= \int_1^2 (F_x \, dx + F_y \, dy + F_z \, dz) \end{aligned}} \qquad \boxed{10\text{-}4}$$

Work is a scalar quantity with units of force times distance, N·m or ft·lb.* Work may be positive or negative according to the cosine function in Eq. 10-3. Forces that do no work are common. For example, a force applied to a fixed particle causes no displacement, so its work is zero. Similarly, a force acting perpendicular to the displacement has no component in the direction of the displacement, and its work is also zero.

There are several special aspects of total work that are worth remembering. It is easy to prove that the total work of concurrent forces is equal to the work of the resultant of those forces. This proof is left as an exercise for the reader. Another aspect that should be noted is that the work of internal forces in an inextensible member cancels out. For example, assume that the segment $dl$ of a rigid rod in Fig. 10-5 is displaced to the right through a distance $s$. If $T_A = T_B$, the positive work of $\mathbf{T}_B$ is canceled by the negative work of $\mathbf{T}_A$ since they are equal in magnitude.

---

* In the U.S. customary system, the units for moment are lb·ft and the units for work are ft·lb. These are dimensionally the same units, but in common usage they are written differently. In the SI system, this difference does not exist.

An interesting and important aspect of total work must be considered when the displacements go through one or more complete cycles. In such a case the total work is zero when each cycle is completed, with no frictional losses during the displacements. For example, the total work done on an object is zero in the complete cycle of lifting it up from the floor and putting it back on the floor. Getting tired in the process is caused by internal losses in the human body and not by doing net work on the object that was moved.

## Qualitative Example of Major Concepts of Work

For a synthesis of the concepts discussed above, consider several test loads on the wing of an airplane. Two views of the wing are shown in Fig. 10-6, where the coordinate system is fixed at the root of the wing, a region of relatively high rigidity. The forces are applied by the hydraulic actuators $A$ and $B$, whose piston rods move axially, rod $A$ in the $y$ direction and rod $B$ in the $z$ direction.

Assume at first that actuator $B$ is not connected to the wing. An aerodynamic lift force is simulated by rod $A$ moving upward. The force

**FIGURE 10-6**
Model of a statically tested wing

$F_A$ required for this increases in magnitude with increasing deflection of the wing. Actuator $A$ does positive work $\int F_A \, dy_A$ as the wing is raised. Note that this work is simple to calculate since $F_A$ is only a function of $y_A$, no matter how the wing deflects. As the wing is slowly lowered back to its neutral position, actuator $A$ resists the motion and does negative work $\int F_A \, dy_A$ on the wing. It is equally valid to say that now the wing does positive work $\int F_A \, dy_A$ on the actuator. The net work of the actuator on the wing (also of the wing on the actuator) in the complete cycle is zero, neglecting frictional effects.

Assume next that actuator $B$ is connected to the wing to simulate drag forces, but actuator $A$ is not connected. Since the force $\mathbf{F}_B$ is a function of the displacement $z_B$, the work of $\mathbf{F}_B$ is calculated using $\int F_B \, dz_B$ and the same consideration for signs as discussed above for $\mathbf{F}_A$.

The situation is more complex when actuators $A$ and $B$ cause simultaneous displacements $y_A$ and $z_B$. If $B$ applies only a horizontal force to the wing (which can be done approximately if necessary), the work of force $\mathbf{F}_B$ does not affect the work of force $\mathbf{F}_A$, and vice versa. Otherwise, each actuator can do work not only on the wing but also on the other actuator. In all cases, the work of internal forces cancel out (as was discussed for Fig. 10-5) in the relatively rigid piston rods, connecting linkages, and pin joints.

## 10-2

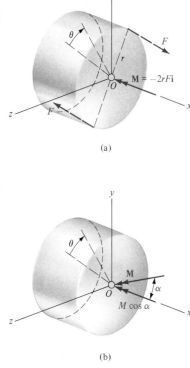

(a)

(b)

**FIGURE 10-7**

Work of a couple

## WORK OF A COUPLE

The work of a couple of moment $\mathbf{M}$ is defined in a similar way as that of a force. The only essential difference is that a couple does work on a body when there is a corresponding angular displacement. In the simplest case, the work of a constant couple of moment $\mathbf{M}$ acting on a rigid body is defined with respect to Fig. 10-7a as

$$U = M\theta \qquad \boxed{10\text{-}5}$$

where the magnitude of the moment $M = 2rF$ (in units of force times distance) and the angular displacement $\theta$ (in radians) are measured in the same plane (the moment vector $\mathbf{M}$ is perpendicular to the plane of the displacement). The derivation of Eq. 10-5 using the work of the two forces of magnitude $F$ is left as an exercise for the reader.

In the case where $\mathbf{M}$ is not perpendicular to the plane of the angular displacement, only the perpendicular component of $\mathbf{M}$ does work on the body. Thus, referring to Fig. 10-7b, the work of the constant couple of moment $\mathbf{M}$ is defined as

$$U = M\theta \cos \alpha \qquad \boxed{10\text{-}6}$$

In the general case the magnitude of the couple and its orientation with respect to the angular displacement may vary, so it is useful to consider infinitesimal quantities. Expressing both the moment $\mathbf{M}$ and the infinitesimal rotation $d\boldsymbol{\theta}$ with respect to the same coordinate system as in Fig. 10-8 (where the body is not shown), the incremental work is

$$dU = \mathbf{M} \cdot d\boldsymbol{\theta} \qquad \boxed{10\text{-}7}$$

which has the units of moment since the rotations are measured in radians.

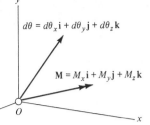

The total work $U$ in a rotation from angular position $\theta_1$ to $\theta_2$ can be expressed using Eqs. 10-6 and 10-7,

$$U = \int_{\theta_1}^{\theta_2} M \cos \alpha \, d\theta = \int_{\theta_1}^{\theta_2} \mathbf{M} \cdot d\boldsymbol{\theta} \qquad \boxed{\text{10-8}}$$

Equation 10-8 is analogous to the expressions defining the work of a force. The work of a couple is a scalar, and it may be positive or negative.

The work of a couple should be computed using Eq. 10-8 rather than Eq. 10-5 when successive rotations about different axes occur. The reason is that *finite rotations are not commutative*, but *infinitesimal rotations are commutative*. This concept is illustrated in Fig. 10-9. First consider $\Delta\theta_x = -90°$ and $\Delta\theta_z = +90°$ rotations about the $x$ and $z$ axes, respectively. These rotations are shown according to the right-hand rule in Fig. 10-9a. Point $P$ rotates to different positions $P_1$ and $P_2$ depending on the order of the $\Delta\theta_x$ and $\Delta\theta_z$ rotations. $P_1$ is obtained when $\Delta\theta_x$ is followed by $\Delta\theta_z$ (the latter does not move the point to a new position). $P_2$ is the final location when $\Delta\theta_z$ is followed by $\Delta\theta_x$ (the latter does not move the point to a new position).

The infinitesimal rotations are shown in Fig. 10-9b. Clearly, $P$ moves to the same point $P_2$ whether the sequence of rotations is $d\theta_x + d\theta_z$ or $d\theta_z + d\theta_x$. Thus, these rotations are commutative. Consequently, an infinitesimal angular displacement $d\boldsymbol{\theta}$ is a vector, but a finite rotation, $\Delta\theta$, is not a vector. The infinitesimal angular displacements $d\boldsymbol{\theta}$ add according to the parallelogram law.

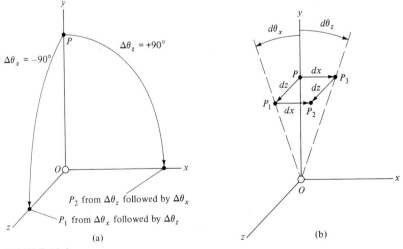

(a)

(b)

**FIGURE 10-9**

Distinction of infinitesimal and finite rotations

# EXAMPLE 10-1

Assume that a block of weight $W$ has to be moved from position $A$ to $B$, and two different paths are reasonable to consider, as shown in Fig. 10-10. Evaluate the total input work in moving the block from $A$ to $B$ for the two paths. **(a)** First, assume that all paths are frictionless. **(b)** Next, assume the same coefficient of friction $\mu$ for all surfaces of contact. Ignore the differences in angular orientation of the block.

## SOLUTION

**(a)** Frictionless ($\mu = 0$):

**STRAIGHT PATH:** The force **F** necessary to begin motion has magnitude $F = W \sin \alpha$, hence the work done on the block is

$$U_{AB} = Fs = Ws \sin \alpha$$

**LONG PATH:** From $A$ to $B'$ the force $\mathbf{F}_1$ necessary to begin motion has magnitude $F_1 = W \cos 10°$, and the distance $\mathbf{F}_1$ acts through is

$$\frac{\Delta y}{\cos 10°} = \frac{s \sin \alpha}{\cos 10°}$$

$$U_{AB'} = F_1 \frac{s \sin \alpha}{\cos 10°} = W (\cos 10°) \left( \frac{s \sin \alpha}{\cos 10°} \right) = Ws \sin \alpha$$

From $B'$ to $B$, $F_2 \simeq 0$ since the surface is frictionless, hence $U_{B'B} = 0$ and the total work done is $U_{AB} = Ws \sin \alpha$.

In the frictionless situation the work done is independent of the path.

**(b)** With coefficient of friction $\mu$:

**STRAIGHT PATH:** The magnitude of the force **F** necessary to cause motion is

$$F = W \sin \alpha + \mu N = W \sin \alpha + \mu W \cos \alpha$$

**FIGURE 10-10**

Work in moving a block to a given elevation using different paths

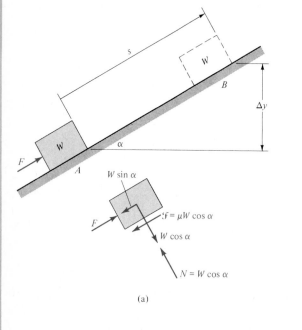

(a)

(b)

Hence, the work done is $U_{AB} = Fs = (W \sin \alpha + \mu W \cos \alpha)s$.

**LONG PATH:** From $A$ to $B'$, the magnitude of the force $\mathbf{F}_1$ necessary to cause motion is $F_1 = W \cos 10° + \mu N_{AB'} = W \cos 10° + \mu W \sin 10°$.

Thus, the work done is $U_{AB'} = (W \cos 10° + \mu W \sin 10°) \dfrac{s \sin \alpha}{\cos 10°}$

$$U_{AB'} = Ws \sin \alpha + \mu Ws \tan 10° \sin \alpha$$

The force $\mathbf{F}_2$ necessary to cause motion from $B'$ to $B$ has magnitude $F_2 = \mathscr{F} = \mu W$ and the work done is $U_{B'B} = \mu W \Delta x \approx \mu Ws \cos \alpha$.

The total work on the body is $U_{AB} = Ws \sin \alpha + \mu Ws \tan 10° \sin \alpha + \mu Ws \cos \alpha$.

## JUDGMENT OF THE RESULTS

The form and magnitude of the results seem reasonable. It is seen that for small angles $\beta$ ($\tan \beta \approx 0$) and small coefficients of friction, the work done is almost the same for either case.

---

# EXAMPLE 10-2

**(a)** Calculate the work of the couple $\mathbf{C}$ in Fig. 10-11 for raising a block of mass $m = 100$ kg through a vertical distance $h = 50$ m. Compare the results for two drum radii, $r_1 = 0.2$ m and $r_2 = 0.4$ m. Ignore frictional losses. **(b)** What is the total work of $\mathbf{C}$ in a complete cycle of raising and lowering the cable if the mass raised is always $m$ and the one lowered is always $0.1 \, m$?

## SOLUTION

**(a)** From Eq. 10-5, the work done by the couple $\mathbf{C}$ is $C\theta$, where $\theta$ is the angle through which the drum rotates. In a frictionless system the work $U_C$ of the couple is equal in magnitude and opposite in sign to the work $U_W$ of the weight, so

$$C\theta = Wh = mgh = (100 \text{ kg})(9.81 \text{ m/s}^2)(50 \text{ m}) = 49.1 \text{ kN·m}$$

It is seen that the work done by $\mathbf{C}$ does not depend on the radius of the drum.

**(b)** The total work $U_T$ of the couple $\mathbf{C}$ in a complete cycle is

$$U_T = U_{up} + U_{down} = mgh - 0.1 \, mgh = 0.9 \, mgh$$
$$= (0.9)(100 \text{ kg})(9.81 \text{ m/s}^2)(50 \text{ m}) = 44.1 \text{ kN·m}$$

## JUDGMENT OF THE RESULTS

The result of part (a) can be checked using a different approach. Summing moments about point $0$ for a static situation yields

$$\sum M_O = 0$$
$$C - mgr = 0$$
$$C = mgr$$

The angle $\theta$ through which $\mathbf{C}$ acts can be determined. Since $h = r\theta$,

$$\theta = \frac{h}{r} \quad \text{and} \quad U_C = C\theta = (mgr)\left(\frac{h}{r}\right) = mgh = 49.1 \text{ kN·m}$$

**FIGURE 10-11**

Work of a couple $\mathbf{C}$ in raising a block

EXAMPLE 10-2    425

# PROBLEMS

**10-1** A block of weight $W$ is slowly moved by a machine from $A$ to $B$. Determine the total work done by the machine if friction is negligible.

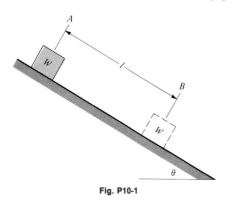

**Fig. P10-1**

**10-2** Solve Prob. 10-1 using $W = 500$ lb, $l = 10$ ft, $\theta = 30°$, and a coefficient of friction $\mu_k = 0.1$.

**10-3** The block in Fig. P10-1 weighs $W = 100$ lb. A machine slowly moves the block through a complete cycle $A$ to $B$ to $A$. Calculate the total work done by the machine in the cycle of displacement $A - B - A$ if $l = 8$ ft, $\theta = 25°$, and the coefficient of friction is $\mu_k = 0.3$.

**10-4** A block of mass $m = 10$ kg is slowly moved by a machine from $A$ to $B$. Calculate the total work done by the machine if friction is negligible.

**Fig. P10-4**

**10-5** Solve Prob. 10-4 assuming that the coefficient of friction is $\mu_k = 0.3$ everywhere.

**10-6** The block in Fig. P10-4 has mass $m = 100$ kg. It is slowly moved by a machine from $A$ to $B$ and back to $A$. Determine the total work done by the machine in the complete cycle of displacement $A-B-A$ if the coefficient of friction is $\mu_k = 0.4$ everywhere.

**10-7** A machine slowly moves the block of weight $W$ from $A$ to $B$. Write an expression for the total work done by the machine if the coefficient of kinetic friction is $\mu_1$ on the incline and $\mu_2$ on the horizontal path. The block could rest on the incline.

**Fig. P10-7**

**10-8** Assume that in Prob. 10-7, $\mu_1 = \mu_2 = \mu$. Express the ratio of the work $U_{AB}$ to $U_{BA}$ done by the machine during a complete cycle $A - B - A$ of the displacements.

**10-9** A block of weight $W = 200$ lb is slowly moved by a machine from $A$ to $B$. Calculate the total work done by the machine if friction is negligible.

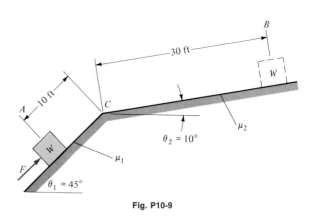

**Fig. P10-9**

**10-10** Solve Prob. 10-9 assuming that the coefficients of kinetic friction are $\mu_1 = 0.2$ and $\mu_2 = 0.3$.

**10–11** The block in Fig. P10-9 represents a container used in a mine. Its total weight is 3000 lb as it moves from $A$ to $B$, and 400 lb as it moves from $B$ to $A$. Determine the total work done by the slowly moving hoisting machine during a complete cycle $A - B - A$ of the motion if $\mu_1 = \mu_2 = 0.1$.

**10–12** A machine slowly moves the block of mass $m = 200$ kg from $A$ to $B$. Calculate the total work done by the machine if the coefficient of friction is $\mu_k = 0.3$ and $l = 10$ m.

**Fig. P10-12**

**10–13** Consider the model of a dump truck. Two hydraulic actuators $A$ are mounted on the sides of the truck bed as shown. Assume that the location of the center of gravity $C$ of the truck bed is fixed. Calculate the total work of the actuators when the truck bed is raised from $\theta = 0$ to $30°$. The mass of the truck bed is $m = 10,000$ kg.

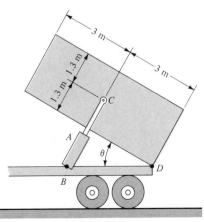

**Fig. P10-13**

**10–14** A uniform slender bar of weight $W$ is hinged at point $O$. Determine the work of the couple $\mathbf{M}_O$ when the bar is slowly raised from $\theta = 0$ to $90°$.

**Fig. P10-14**

**10–15** An amusement ride is modeled as a cage of weight $W_1 = 800$ lb fixed on a uniform arm of weight $W_2 = 300$ lb, which is hinged at $A$. Determine the work of the couple $\mathbf{M}_A$ when the arm is slowly raised from $\theta = 0$ to $60°$. $AB = 20$ ft and $BC = 3$ ft.

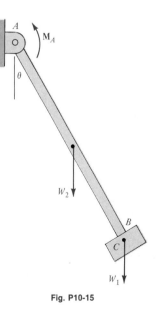

**Fig. P10-15**

**10–16** A polishing machine is modeled as a small block $A$ of mass $m = 10$ kg mounted on a weightless arm $OA$. The arm is hinged at $O$ to freely swing vertically. A couple $\mathbf{M}_O$ slowly moves the arm in the horizontal plane. The motion is opposed by friction acting on the block. Calculate the work of $\mathbf{M}_O$ in a complete revolution.

Fig. P10-16

**10–17** In Prob. 10–16, a spring $S$ is added to the arm at hinge $O$. The spring is in the vertical plane and doubles the normal force between block $A$ and the supporting surface. Determine the work of $\mathbf{M}_O$ and the work of the spring in a complete revolution.

Fig. P10-17

**10–18** Determine the work of the couple $\mathbf{M}_B$ which slowly rotates the wheel through a complete revolution.

Fig. P10-18

**10–19** Assume that the wheel in Fig. P10-18 is alternately moving in the clockwise and counterclockwise directions. Determine the total work of the driving couple $\mathbf{M}_B$ in two complete but opposite turns of the wheel.

**10–20** Part of a machine is modeled as block $A$ of mass $m = 300$ kg, which is raised using the weightless wedge $B$. Calculate the total work of force $\mathbf{F}$ in raising block $A$ through a distance of 50 mm. The coefficient of friction on all surfaces is $\mu_k = 0.3$.

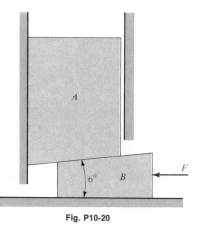

Fig. P10-20

**10–21** Determine the work of force $\mathbf{F}$ in Prob. 10–20 if block $A$ is to be lowered through a distance of 50 mm.

**10–22** A car is stuck with one of its driving wheels slowly spinning. Calculate the work required to make one complete turn of the wheel if its radius is 15 in. and the axle load is 600 lb.

Fig. P10-22

The concept of work is useful in the analysis of certain problems even when displacements do not actually take place. The idea is to imagine that infinitesimal displacements occur within the constraints of a given system (that is, the displacements should be possible), and examine the resulting work done by the external forces acting on the system. The method can be used to investigate the equilibrium of the system. This is somewhat analogous to testing the strength of ice on a pond by gently pressing it with an extended foot. It is not necessary to break the ice to prove that it is unsafe; one can imagine from its response to small loads that the ice would break under the full body weight.

The imaginary small displacement is called a *virtual displacement* and is denoted by $\delta\mathbf{r}$, which is to be distinguished from a similarly small but real displacement $d\mathbf{r}$. The imaginary work done by a force $\mathbf{F}$ during a virtual displacement of a point where the force is acting is called *virtual work* and defined as

$$\delta U = \mathbf{F} \cdot \delta\mathbf{r}$$ 　10-9

Similarly for the virtual work of a couple of moment $\mathbf{M}$,

$$\delta U = \mathbf{M} \cdot \delta\boldsymbol{\theta}$$ 　10-10

where $\delta\boldsymbol{\theta}$ is a virtual angular displacement of infinitesimal magnitude.

It is reasonable to assume that both forces and couples remain constant during the virtual displacements. This is because any change in a force or a couple caused by an infinitesimal displacement must itself be infinitesimal. The resulting difference in work is a second-order differential and can be neglected.

The concept of virtual work can be used to analyze the equilibrium of particles and rigid bodies because the sum of all virtual work done by a system of forces or couples is zero when the system is in equilibrium. This is true for any system, and it is easiest to prove for a particle or single rigid body under the action of two forces. For example, consider the particle $A$ in Fig. 10-12. If the particle is in equilibrium, the forces $\mathbf{F}_1$ and $\mathbf{F}_2$ must be equal in magnitude, collinear, and opposite in direction. Thus, $\mathbf{F}_1$ does positive work during a virtual displacement to the right, while $\mathbf{F}_2$ does negative work of equal magnitude. Formally, with $F_1 = F_2$,

$$\delta U = \sum_{i=1}^{n=2} \mathbf{F}_i \cdot \delta\mathbf{r}_i = F_1 \, \delta r + (-F_2 \, \delta r) = 0$$

This procedure can be extended to deal with any number of forces or couples acting in any direction. The general statement of equilibrium on the basis of virtual work defined by Eqs. 10-9 and 10-10 is

$$\delta U = \sum_{i=1}^{m} \mathbf{F}_i \cdot \delta\mathbf{r}_i + \sum_{j=1}^{n} \mathbf{M}_j \cdot \delta\boldsymbol{\theta}_j = 0$$ 　10-11

where the subscript $i$ refers to an individual force and the corresponding virtual displacement, while $j$ refers to an individual couple and the corresponding virtual rotation. Equation 10-11 is valid for particles (with the moment terms naturally absent) or for rigid bodies. This equation includes the work of all forces and moments that act through a distance or rotation under a virtual displacement. The work may be of applied forces or mo-

**FIGURE 10-12**

Virtual work is done in the imaginary displacement $\delta\mathbf{r}$ of a particle ments, including body forces (such as weight or magnetic force), and friction forces. Equation 10-11 is not advantageous in comparison with the fundamental equations of equilibrium unless the system consists of at least several interconnected rigid bodies (see Ex. 10-3). A particular advantage of the method of virtual work is in the determination of the equilibrium configuration for a system of rigid bodies.

There are several items of practical interest in solving a problem using the method of virtual work. First, it is not necessary to draw the interrelated free-body diagrams for the members of the given system of rigid bodies. The required diagram has to show only those *active forces and couples* that have virtual displacements associated with them. Thus, the reactions at any fixed joint do not have to be shown The number of equations that has to be solved in a given problem equals the number of degrees of freedom $n$ for that system. The reason is that the virtual work according to Eq. 10-11 can be written in $n$ different ways, one for each particular possible virtual displacement.

## MECHANICAL EFFICIENCY OF REAL SYSTEMS

The concepts of work and virtual work are extremely useful because of their simplicity when frictional effects are ignored. However, real mechanical systems always operate with frictional losses, so a more precise analysis may have to be done. This is simple if the magnitude of the friction force is known. The work of the friction force is always taken as negative for a sliding body because the direction of the friction force is always opposite to the direction of the displacement of the body.

To illustrate the work of friction in a simple case, consider the block of weight $W$ on the inclined plane in Fig. 10-13a. The active forces are shown as solid lines in Fig. 10-13b. The force $F$ required to push the block up the incline can be determined using Eq. 10-11,

$$\delta U = \underbrace{F\,\delta s}_{\substack{\text{work by} \\ \text{applied} \\ \text{force}}} - \underbrace{W \sin \theta\, \delta s}_{\substack{\text{work by} \\ \text{weight}}} - \underbrace{\mu W \cos \theta\, \delta s}_{\substack{\text{work by} \\ \text{friction force}}} = 0$$

$$F > W(\sin \theta + \mu \cos \theta) \qquad \text{to cause upward motion}$$

This answer could also be found without using the concept of work.

It is useful to rewrite the expression for the virtual work as follows,

$$F\,\delta s = W \sin \theta\, \delta s + \mu W \cos \theta\, \delta s$$

which is stated in words as

$$\text{input work} = \text{useful work} + \text{work of friction} \qquad \boxed{10\text{-}12}$$

The useful work is also called *output work;* in the case of a machine this is the work available to be done on another body or system of bodies.

The ratio of output work to input work is defined as the *mechanical efficiency $\eta$* of a machine,

$$\boxed{\eta = \frac{\text{output work}}{\text{input work}} = \frac{\text{useful work}}{\text{total work required}}} \qquad \boxed{10\text{-}13}$$

The mechanical efficiency of an ideal machine is $\eta = 1$ and of any real machine is $\eta < 1$.

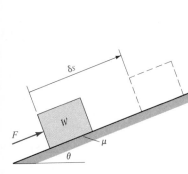

(a)

(b)

**FIGURE 10-13**

Model of a real system with friction

EXAMPLE 10-3

The two uniform booms $A$–$B$ of the electric trolley bus shown in Fig. 10-14a are 18 ft long, weigh 30 lb each, and have essentially frictionless hinges at $A$. The normal force at $B$ needed for proper contact is 10 lb and the coefficient of friction is 0.6 for each contact slider. Use the method of virtual work to determine the couple $M$ required at joint $A$ of each boom to maintain proper contact at $B$. The nearly vertical strings on the booms are slack during normal operation. $\theta = 30°$.

## SOLUTION

A reference system that moves with the trolley car with its origin at $A$ is considered as shown in Fig. 10-14b. The position of the boom is totally defined by the angle $\theta$.

$$y_B = (18\ \text{ft}) \sin\theta$$

$$\frac{dy_B}{d\theta} = (18\ \text{ft}) \cos\theta$$

$$\delta y_B = (18\ \text{ft}) \cos\theta\ \delta\theta$$

$$x_B = (18\ \text{ft}) \cos\theta$$

$$\frac{dx_B}{d\theta} = -(18\ \text{ft}) \sin\theta$$

For the center of gravity $C$,

$$\bar{y} = (9\ \text{ft}) \sin\theta$$

$$\frac{d\bar{y}}{d\theta} = (9\ \text{ft}) \cos\theta$$

$$\delta\bar{y} = (9\ \text{ft}) \cos\theta\ \delta\theta$$

(a)

A virtual displacement $\delta\theta$ is imagined as shown in Fig. 10-14b, and Eq. 10-11 is used. Note that the signs are obtained by deciding if the force or moment being considered acts in the direction of its displacement (positive work) or against it (negative work).

$$
\begin{aligned}
\delta U &= M\ \delta\theta - N\ \delta y_B + \mathscr{F}\ \delta x_B - W\ \delta\bar{y} = 0 \\
&= M\ \delta\theta - (10\ \text{lb})(18\ \text{ft} \cos\theta)\ \delta\theta + (6\ \text{lb})\ (18\ \text{ft} \sin\theta)\ \delta\theta \\
&\quad - (30\ \text{lb})(9\ \text{ft} \cos\theta)\ \delta\theta = 0 \\
M &= 180\ \text{lb·ft} \cos 30° - 108\ \text{lb·ft} \sin 30° + 270\ \text{lb·ft} \cos 30° \\
&= 234\ \text{lb·ft}
\end{aligned}
$$

## JUDGMENT OF THE RESULTS

A check using the equations of equilibrium confirms this result. Note that the check is much faster than using the method of virtual work. The example is worked out here for practice with Eq. 10-11.

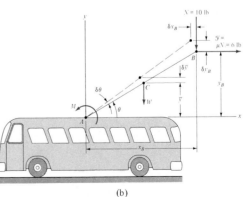

(b)

**FIGURE 10-14**

Analysis of the booms of an electric trolley bus using virtual displacements

# EXAMPLE 10-4

Consider the wing-loading mechanism with the hydraulic actuator $A$ shown in Fig. 10-15a. Use the method of virtual work to determine the magnitude of the actuator force $\mathbf{F}$ required to apply a vertical force $\mathbf{P}$ to the wing first upward, then downward. Assume that all joints are frictionless and that the mechanism is weightless.

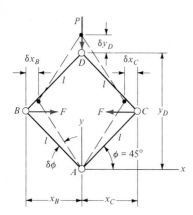

(a)

## SOLUTION

First the mechanism is sketched showing virtual displacements. All of the virtual displacements of interest are determined in terms of a single parameter, the angle $\phi$. For $\mathbf{P}$ acting upward on the wing (Fig. 10-15b),

$$x_B = -l\cos\phi \qquad \delta x_B = l\sin\phi\,\delta\phi$$
$$x_C = l\cos\phi \qquad \delta x_C = -l\sin\phi\,\delta\phi$$
$$y_D = 2l\sin\phi \qquad \delta y_D = 2l\cos\phi\,\delta\phi$$

Equation 10-11 is written after carefully checking the resultant signs in each term,

$$\delta U = F\,\delta x_B + (-F)\,\delta x_C - P\,\delta y_D = 0$$
$$= 2F(l\sin\phi)\,\delta\phi - P(2l\cos\phi)\,\delta\phi = 0$$
$$2Fl\sin\phi = 2Pl\cos\phi$$
$$F = P\cot\phi = P\cot 45° = P$$

For $\mathbf{P}$ acting down on the wing, Fig. 10-15c,

$$\delta U = (-F)\,\delta x_B + F\,\delta x_C + P\,\delta y_D = 0$$
$$= -2Fl\sin\phi + 2Pl\cos\phi = 0$$
$$F = P\cot\phi = P\cot 45° = P$$

It should be noted that the signs of the work terms are checked by deciding if the force being considered acts in the direction of its displacement (positive work) or against it (negative work). Also note that the vertical deflections of the forces $F$ are never considered since the forces always act horizontally, hence $F$ does not work through the vertical displacement.

## JUDGMENT OF THE RESULTS

This problem could be solved using the equations of equilibrium but it would be a much longer solution. In this case, using the method of virtual work is fastest. The results appear reasonable because of the geometry ($|\delta AD| = |\delta BC|$), but they are valid only for $\phi = 45°$.

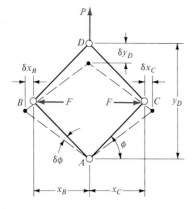

(b)

(c)

**FIGURE 10-15**

Virtual displacements of a mechanism

**10-23** Determine the magnitude of the force **P** required to maintain equilibrium of the weightless bar supported by a frictionless hinge.

Fig. P10-23

**10-24** Assume that in Fig. P10-23 $P = 4F$ and the uniform bar's weight is $W = 2F$. Determine the couple $\mathbf{M}_O$ required to maintain equilibrium.

**10-25** Assume that in Fig. P10-23 $F = 3P$ and the uniform bar's weight is $W = 0.5F$. Determine the couple $\mathbf{M}_O$ required to maintain equilibrium.

**10-26** Assume that the rigid bar $AB$ is weightless and has a frictionless pin at $A$. Derive an expression for the magnitude of the force **P** required to maintain equilibrium.

Fig. P10-26

**10-27** A uniform bar of weight $W$ is supported by frictionless surfaces and a force **F**. Determine the magnitude of the force **F** required to maintain equilibrium.

Fig. P10-27

**10-28** Solve Prob. 10-27 after moving force **F** from point $A$ to point $B$, where it acts in the vertical direction.

**10-29** Determine the magnitude of the force **P** required to maintain equilibrium of the weightless bars, which have frictionless connections.

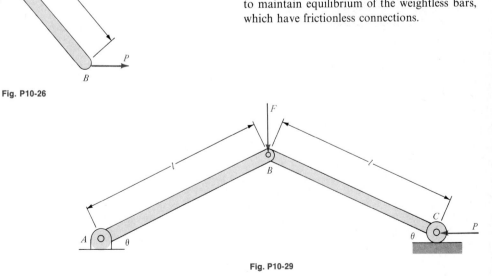

Fig. P10-29

**10–30** Solve Prob. 10–29 assuming that each bar is uniform and has weight $W = 0.5F$.

**10–31** Calculate the magnitude of the horizontal force **F** that the toggle vise applies to block $A$ if friction is negligible and $\theta = 5°$, $10°$, and $15°$. Plot $F$ vs. the angle $\theta$.

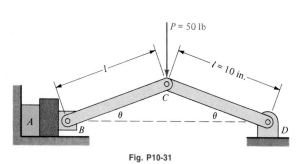

**Fig. P10-31**

**10–32** Determine the mechanical efficiency of the system which slowly moves the block from $A$ to $B$. $m = 60$ kg, $l = 10$ m, $\theta = 20°$, and $\mu_k = 0.3$. Other frictional losses are ignored.

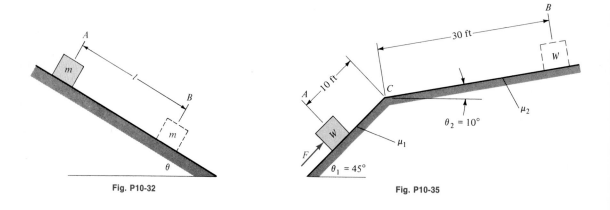

**Fig. P10-32**

**10–33** The block in Prob. 10–32 is moved through a complete cycle of displacements $A–B–A$. Calculate the mechanical efficiencies in the distinct motions $A–B$ and $B–A$.

**10–34** Write an expression for the mechanical efficiency of the system which moves the block from $A$ to $B$. The coefficient of friction is $\mu_k$ everywhere. The block could rest on the incline.

**Fig. P10-34**

**10–35** The block represents a container used in a mine. Its total weight is 3000 lb as it moves from $A$ to $B$ and 400 lb as it moves from $B$ to $A$. Calculate the mechanical efficiency of the slowly moving container system in a complete cycle of displacements $A–B–A$ if $\mu_1 = \mu_2 = 0.1$.

**Fig. P10-35**

**10–36** The mechanical efficiency is 0.85 in the tilting action of the dump truck described in Prob. 10–13. Calculate the work of friction in a tilt from $\theta = 0$ to $30°$.

434    WORK AND POTENTIAL ENERGY

The method of potential energy in engineering has an interesting historical background. The majority of inventors of perpetuum mobiles have tried for centuries to find clever ways to get continuously available and cheap work from the force of gravity. The basic concept is illustrated in Fig. 10-16. When the weight $W$ is allowed to go from level 1 to a lower level 2 (with respect to the center of the earth), the weight does useful work on the system that lowers it (drives a clock, pumps water, etc.). In an ideal system this work, according to Eq. 10-4, is

$$U_{12} = \int_1^2 W \, dy = W(y_2 - y_1) = Wh$$

The potential of a weight $W$ to do work because of its relative height $h$ is defined as its *potential energy*. The term "potential energy" comes from the perception that a weight situated at any level could do useful work if allowed to descend to a lower level. Potential energy thus depends on the levels chosen. The weight $W$ has potential energy $U_{23}$ at level 2 with respect to level 3 in Fig. 10-16,

$$U_{23} = \int_2^3 W \, dy = Wl$$

Naturally,

$$U_{13} = \int_1^3 W \, dy = W(h + l)$$

Many inventors have tried to make a machine that does work in lowering a weight from level 1 to level 2 (Fig. 10-16), then uses part of this work to return the weight to level 1 so that the cycle can be repeated. The problem, of course, is that the work required to move the weight from 2 to level 1 in an ideal system is

$$U_{21} = \int_2^1 -W \, dy = -Wh$$

The minus sign is present since gravity acts in opposing the motion upward. Thus, in a complete cycle from 1 to 2 to 1, the net work is

$$U_{12} + U_{21} = Wh - Wh = 0$$

In a real machine some work is dissipated by friction (in the form of heat), so the work available in the descent from 1 to 2 is never enough to completely raise the weight from 2 to 1.

The concept of potential energy is very useful in judging most of the proposed perpetuum mobiles (but not all of them). The idea is to ignore the paths of descent and ascent of all members of the machine and concentrate on the changes in their elevations. If the net change in potential energy in a complete cycle is zero when friction is ignored, the machine is proved to be unable to run perpetually on its own. Often there are other methods also for proving the same result, but the method of potential energy is generally the simplest.

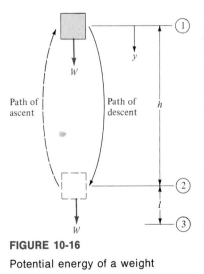

**FIGURE 10-16**

Potential energy of a weight

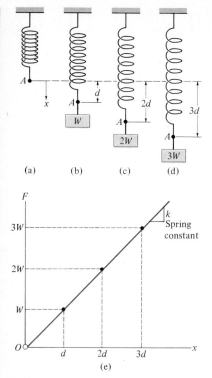

**FIGURE 10-17**

Theory of springs for an ideally elastic (linearly deforming) spring

## Elastic Potential Energy

There are many other forms of potential energy besides that associated with the force of gravity. The potential energy of elastic members is the most important of these in elementary mechanics.

The background for analyzing any elastic member is the *theory of springs* discovered by Robert Hooke in 1642. This important concept is easy to understand with the aid of Fig. 10-17. Consider a helical spring which is loaded with progressively larger weights $W$ as in Fig. 10-17a through d. If the deformations $x$ are not excessive, they are proportional to the applied force. Consequently, a plot of force magnitude $F$ vs. deformation $x$ is a straight line as in Fig. 10-17e. The slope $k$ of this line is called the *spring constant*. It has units of N/m or lb/ft, and it is the property of each particular spring (depending on material, length of spring, wire diameter, and coil diameter). The formal relation of the force and deflection in a spring, most commonly given in terms of the spring constant, $k$, is

$$F = kx$$

$\boxed{\text{10-14}}$

The work of a force $\mathbf{F}$ on a spring is obtained from Eq. 10-4. With respect to Fig. 10-18, where the force $\mathbf{F}_1$ causes a deformation $x_1$ from the undeflected position,

$$U_1 = \int_0^{x_1} kx \, dx = \frac{1}{2} kx_1^2$$

$\boxed{\text{10-15a}}$

which is also equal to the shaded area under the force vs. deflection line in Fig. 10-18. Care must be taken to use consistent units for $k$ and $x$. In a more general case, as in going from $x_1$ to $x_2$ in Fig. 10-18,

$$\boxed{U = \int_{x_1}^{x_2} kx \, dx = \frac{1}{2} k(x_2^2 - x_1^2)}$$

$\boxed{\text{10-15b}}$

The result is true for a linearly deforming spring, whether the spring is being stretched or compressed. The work $U$ of a force $\mathbf{F}$ on the spring is

$$U = \frac{1}{2} kx^2$$

$\boxed{\text{10-16}}$

where $x$ is the change in length (from the undeformed length of the spring) caused by the force. Once deformed, the spring has the ability to do useful work if allowed to return to its undeformed length. Thus, Eq. 10-16 is the *elastic potential energy* of the spring.

## Torsion Springs

The concepts described above are readily extended to analyze the deformation and energy in a member which is twisted through an angle $\theta$ (measured in radians) when a couple of moment $\mathbf{M}$ is applied to it. In this case Eq. 10-14 becomes

$$M = k\theta$$

where $k$ is the torsional spring constant in units of N·m/rad or lb·ft/rad.

**FIGURE 10-18**

Potential energy of a deformed spring

## Notation for Potential Energy

The units of potential energy are the units of work, N·m or ft·lb, but it is useful to have a special notation for potential energy as shown in Secs. 10-6 and 10-7. One of the most common symbols for potential energy is $V$. The relation of work $U$ done on a mechanical system by an external force and the *change* in the potential energy $V$ of the system is generally expressed as

$$U = -\Delta V \qquad \boxed{10\text{-}17}$$

This equation shows that negative work is done by a system while its own potential energy is increased through the action of an external force or moment. Of course, the external agent does positive work in this process since it acts in the same direction as that of the resulting motion.

Equation 10-17 is valid for any ideal system based on gravity, springs, and so on. These ideal systems are called *conservative*, and the forces involved are called *conservative forces*. In reality there is always some friction, so Eq. 10-17 is not strictly correct. Even in the best case, there is internal friction in the spring material. Nevertheless, Eq. 10-17 is quite useful in practice when idealized systems are analyzed.

## POTENTIAL ENERGY AT EQUILIBRIUM

10-6

It was shown in Sec. 10-3 that the concept of virtual work can be used to determine the equilibrium of forces and moments in a system. This method can be altered advantageously using the relation of work and potential energy. Assume that Eq. 10-17 is valid for virtual displacements, so $\delta U = -\delta V$. For equilibrium, $\delta U = 0$, and consequently, $\delta V = 0$. The physical interpretation of this is that small variations $\delta$ in the configuration of a system do not change the potential energy $V$ of that system. Assuming that the virtual variations are equivalent to taking the differential with respect to an independent parameter $q$, the condition of equilibrium on the basis of total potential energy $V$ of the system is

$$\frac{dV}{dq} = 0 \qquad \boxed{10\text{-}18}$$

The independent parameter $q$ should be an independent coordinate along which displacement of the system is possible. There is only one such coordinate for a system with one degree of freedom. For a system with $n$ degrees of freedom, the partial derivatives of the potential energy with respect to each independent parameter $q_i$ must be zero for the system to be in equilibrium. The mathematical statement of this is

$$\frac{\partial V}{\partial q_i} = 0 \qquad \boxed{10\text{-}19}$$

where $i = 1, 2, \ldots, n$. The resulting $n$ equations are independent, and they can be used to check the equilibrium of the system.

## STABILITY OF EQUILIBRIUM

An important aspect of the condition of equilibrium is the stability of the configuration. Various possibilities of equilibrium are illustrated in Fig. 10-19. A ball in a bowl (Fig. 10-19a) is in *stable equilibrium* because when it is displaced from its resting position, it tends to return there. In the position of *neutral equilibrium* (Fig. 10-19b) the ball has no tendency to move away from or toward any particular position. In unstable equilibrium (Fig. 10-19c and d) a small displacement can cause the ball to move even farther from its resting position.

The condition of stability can be analyzed on the basis of potential energy of the system as outlined in the following.

1. **Stable:**   Displacement causes an increase in potential energy.
2. **Neutral:**   Displacement causes no change in potential energy.
3. **Unstable:**   Displacement causes a decrease in potential energy.

A graphical illustration of each of these is given in Fig. 10-20, where the potential energy $V$ is sketched as a function of $q$, the independent coordinate along which displacement of the system is possible. Since the maximum or minimum values of a function can be determined by considering its second derivative, the conditions of equilibrium and stability can be summarized for a system with one degree of freedom as follows.

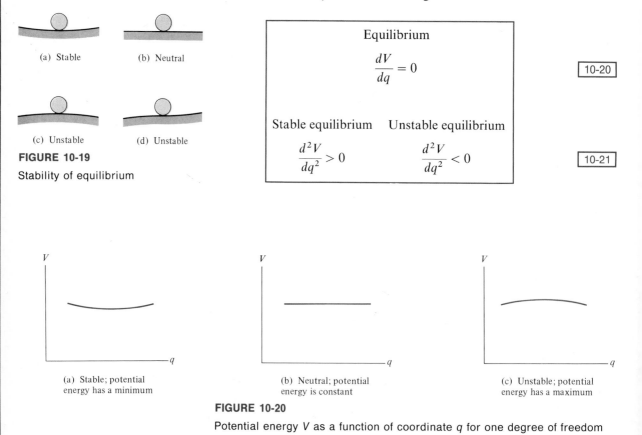

(a) Stable          (b) Neutral

(c) Unstable          (d) Unstable

**FIGURE 10-19**

Stability of equilibrium

Equilibrium

$$\frac{dV}{dq} = 0$$

$\boxed{10\text{-}20}$

Stable equilibrium     Unstable equilibrium

$$\frac{d^2V}{dq^2} > 0 \qquad \frac{d^2V}{dq^2} < 0$$

$\boxed{10\text{-}21}$

(a) Stable; potential energy has a minimum

(b) Neutral; potential energy is constant

(c) Unstable; potential energy has a maximum

**FIGURE 10-20**

Potential energy $V$ as a function of coordinate $q$ for one degree of freedom

The possible result that $dV/dq = 0$ and $d^2V/dq^2 = 0$ does not indicate neutral equilibrium for certain because derivatives of higher order may be positive, negative, or zero. The higher-order derivatives, while small in magnitude, may affect the variation of the potential energy near the equilibrium position. This variation is the basis of the criterion for stability. In general (without giving proof, because of the subtle details):

1. **Neutral equilibrium:** All derivatives are zero.
2. **Stable equilibrium:** The first nonzero derivative is positive and of even order.
3. **Unstable equilibrium:**
   a. The first nonzero derivative is of odd order (positive or negative), or
   b. The first nonzero derivative is negative and of even order.

The complexities of these statements indicate that in some problems both the qualitative aspects of the stability (stable, neutral, or unstable) and the degree of extent of that condition must be considered. To illustrate this, imagine a small dip at the peak of the curve in Fig. 10-20c. The dip indicates stability, but it is small and is superimposed on a large-scale instability. Naturally, many important problems can be solved without having to consider many derivatives of the potential energy function.

In systems with $n$ degrees of freedom the potential energy depends on $n$ variables, and it may be difficult to visualize or analytically determine the extreme values of the potential energy function. For example, in the case of a system with two degrees of freedom the variation of potential energy is represented as three-dimensional surfaces of $V$ vs. $x$ and $y$ (analogous to Fig. 10-20). The analysis of such a system would involve checking the following:

For *equilibrium* (stable or unstable),

$$\frac{\partial V}{\partial x} = \frac{\partial V}{\partial y} = 0 \qquad \boxed{10\text{-}22}$$

For *stability* (minimum potential energy),

$$\left(\frac{\partial^2 V}{\partial x\, \partial y}\right)^2 - \left(\frac{\partial^2 V}{\partial x^2}\right)\left(\frac{\partial^2 V}{\partial y^2}\right) < 0 \qquad \boxed{10\text{-}23}$$

and

$$\frac{\partial^2 V}{\partial x^2} + \frac{\partial^2 V}{\partial y^2} > 0 \qquad \boxed{10\text{-}24}$$

For *instability* (maximum potential energy),

$$\left(\frac{\partial^2 V}{\partial x\, \partial y}\right)^2 - \left(\frac{\partial^2 V}{\partial x^2}\right)\left(\frac{\partial^2 V}{\partial y^2}\right) < 0 \qquad \boxed{10\text{-}25}$$

and

$$\frac{\partial^2 V}{\partial x^2} + \frac{\partial^2 V}{\partial y^2} < 0 \qquad \boxed{10\text{-}26}$$

The proof of these criteria are beyond the scope of this text but can be found in texts on advanced calculus.

# EXAMPLE 10-5

Show that the scale mechanism shown in Fig. 10-21a is always in equilibrium regardless of the angle $\theta$ and the positions of the weights $W$ on the horizontal arms. The bars $AB$, $CD$, $EF$, $CE$, and $DF$ are rigid members. $B$, $C$, $D$, $E$, $F$, and $G$ are smooth pin joints. $EG = GF = l$. What reaction moment is necessary at point $A$?

## SOLUTION

The potential energy of the system arises from the elevation of the weights only. An expression for the potential energy of the system with respect to point $A$ (arbitrarily chosen datum) is written in terms of the parameter $\theta$,

$$V = W_{y_1} + W_{y_2} = W(h + l\sin\theta + a) + W(h - l\sin\theta + a)$$

Using Eq. 10-18,

$$\frac{dV}{d\theta} = Wl\cos\theta - Wl\cos\theta = 0$$

$$\frac{d^2V}{d\theta^2} = 0$$

All derivatives of $V$ with respect to $\theta$ are zero, hence the mechanism is always in neutral equilibrium and no position is preferred to another. Summing moments about point $A$ (Fig. 10-21b) yields

$$\sum M_A = 0$$

$$M_A + W(l\cos\theta + d_1) - W(l\cos\theta + d_2) = 0$$

$$M_A = W(d_2 - d_1)$$

The resultant moment caused by the two weights is counteracted by the reaction moment $\mathbf{M}_A$ at the base. This moment depends only on $d_1$ and $d_2$, not on $\theta$. This moment reaction is transferred to the mechanism through the horizontal reactions $\mathbf{B}_x$ and $\mathbf{G}_x$ (not shown), which form a couple.

(a)                    (b)

**FIGURE 10-21**

Scale mechanism with two horizontal arms for the movable weights $W$

EXAMPLE 10-6

The double-decker bus shown in Fig. 10-22a weighs 100 kN when empty. The wheels have a radius of 0.4 m and the track width of the wheels is 2.2 m. The center of gravity is halfway between the wheels, 1.2 m above the ground. **(a)** Determine the angle $\theta$ (Fig. 10-22b) where the bus is in equilibrium in a tipped position. Check for stability in this position. **(b)** Determine whether or not the bus would be stable on a 0.2-m high curb.

## SOLUTION

**(a)** First the bus is sketched in the tipped position, as shown in Fig. 10-22b. The potential energy of the system is written in terms of the displacement parameter, in this case $\theta$.

$$V = Wy = W(1.2 \text{ m} \cos \theta + 1.1 \text{ m} \sin \theta)$$

For equilibrium, using Eq. 10-18,

$$\frac{dV}{d\theta} = 0 = W(-1.2 \text{ m} \sin \theta + 1.1 \text{ m} \cos \theta)$$

$$\frac{\sin \theta}{\cos \theta} = \tan \theta = \frac{1.1 \text{ m}}{1.2 \text{ m}}$$

$$\theta = 42.5°$$

Check for stability for $\theta = 42.5°$,

$$\frac{d^2V}{d\theta^2} = W(-1.2 \text{ m} \cos \theta - 1.1 \text{ m} \sin \theta)$$

$$= W(-1.2 \text{ m} \cos 42.5° - 1.1 \text{ m} \sin 42.5°) = -1.63 \text{ m } W$$

Since $d^2V/d\theta^2 < 0$, the position $\theta = 42.5°$ is an unstable equilibrium, as would be expected.

**(b)** With the bus on a 0.2-m high curb,

$$\theta = \sin^{-1} \frac{0.2 \text{ m}}{2.2 \text{ m}} = 5.2° < 42.5°$$

Hence the bus is stable when tipped up on the curb.

(a)

(b)

**FIGURE 10-22**

Equilibrium and stability of a tilted bus

EXAMPLE 10-6    441

**10–37** A block of mass $m = 10$ kg is placed (between frictionless guides) on a helical spring whose original length is $L_o$. Calculate the potential energy of the spring and the position of the block.

**Fig. P10-37**

**10–38** Assume that a second block of mass $m$ is placed on top of the first block in Prob. 10–37. Calculate the change in potential energy of the spring caused by the second block.

**10–39** A torsion spring of spring constant $k$ is attached to the hinged bar $AB$. The spring is undeformed at $\theta = 0$. Determine $\theta$ and the potential energy in the spring when $F = 60$ lb.

**Fig. P10-39**

**10–40** Assume that the magnitude of the force $\mathbf{F}$ in Prob. 10–39 increases. Calculate the change in potential energy of the spring as $F$ changes from 60 lb to 150 lb.

**10–41** A shaft is fixed at $A$, and a couple is applied to it at end $B$. The torsional spring constant of the shaft is $k = 5$ kN·m/rad (the angular displacement $\theta$ is that of end $B$ relative to end $A$). Calculate the magnitude $F$ required to cause a rotation of $\theta = 1°$.

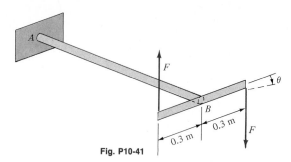

**Fig. P10-41**

**10–42** Determine the change in potential energy of the shaft in Prob. 10–41 as the angle $\theta$ increases from 1° to 2°.

**10–43** The uniform bar of weight $W$ and length $l$ is pulled to the right by a force of magnitude $P = 3W$ at end $B$. Determine the angle $\theta$ at equilibrium.

**Fig. P10-43** $B$

**10–44** Each of the uniform bars has weight $W$ and length $l$. Determine the angles $\theta$ and $\phi$ at equilibrium after a force of magnitude $F = W$ is applied at point $C$.

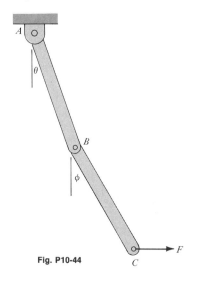

**Fig. P10-44** $C$

**10–45** Each of the uniform bars has weight $W$ and length $l$. Determine the angle $\theta$ at equilibrium after a force of magnitude $P = 2W$ is applied at point $C$. Is this a stable position?

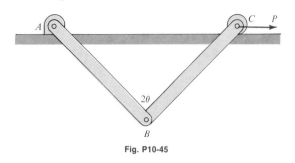

**Fig. P10-45**

**10–46** The rope supporting the mass $m = 5$ kg is attached to the unstretched spring when $A$ is at $x = 0.2$ m. Calculate the position of $A$ after the spring is deformed and check the system for stability.

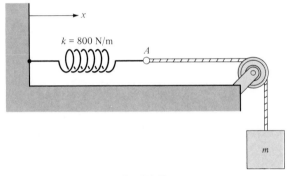

**Fig. P10-46**

**10–47** The weightless bar of length $l = 6$ in. is supported by a spring at $d = 2$ in. The bar is horizontal when $P = 0$. Determine the angle $\theta$ of the bar when $P = 3$ lb. Is this a stable position?

**10–48** Consider the model of a compact scale mechanism. The dimensions $a$ and $b$ are measured when $P$ and the force in the spring $AB$ are both zero. Determine the angle $\theta$ for $P = 100$ N and check the stability of the system. Solve by trial and error if necessary.

**Fig. P10-48**

**10–49** An engine throttle mechanism is modeled as a hinged bar. The torsional spring has a constant $k = 200$ lb·in./rad; it is unstrained when $\theta = 30°$ counterclockwise from the horizontal. Determine the angle $\theta$ for $P = 5$ lb (always vertical) and check the stability of the system. Solve by trial and error if necessary.

**Fig. P10-47**                          **Fig. P10-49**

**10–50** Consider the boom of an electric trolley (Fig. 10-14). Assume that the mass of each boom is uniformly distributed, 4 kg/m. Determine the spring constant $k$ (there are two springs on each boom) for $\theta = 30°$ and a spring elongation of 0.02 m (from 0.48 m to 0.5 m). Check the equilibrium using the potential energy in the given configuration. Use statics, small deflections, and assume that the spring is parallel to the boom in the deformed state.

Fig. P10-50

**10–51** A proposed shake table for vibration testing is modeled as a heavy bar supported by two hinged arms of negligible weight. A torsion spring of stiffness $k = 500$ lb·ft/rad is attached to arm $AB$. Determine the angle $\theta$ at equilibrium if the spring applies no moment to the arm at $\theta = 90°$. Solve by trial and error if necessary.

**10–52** A uniform bar of mass $m = 20$ kg is placed on two springs of identical initial height $h = 0.1$ m. Calculate the angle $\theta$ of the bar as it rests on the springs if $l = 1$ m, $k_1 = 4$ kN/m, and $k_2 = 6$ kN/m.

Fig. P10-52

**10–53** A weightless bar is placed on two identical springs. Determine the angle $\theta$ of the bar after the force $\mathbf{P}$ is applied at $l = 8$ in.

Fig. P10-53

**10–54** Consider a rocking chair of simple construction. Determine if there is stable equilibrium for a person of weight $W = 200$ lb with center of gravity at $h = 2$ ft.

Fig. P10-54

Fig. P10-51

**WORK OF A FORCE:**

$$U = Fs$$

where  $U$ = work

  $F$ = magnitude of constant force

  $s$ = displacement of a body in the same direction in which **F** is acting

For an arbitrary path from point 1 to 2,

$$U = \int_1^2 \mathbf{F} \cdot d\mathbf{r} = \int_1^2 (F_x \, dx + F_y \, dy + F_z \, dz)$$

There is no work when:

**1.** A force is applied to a fixed, rigid body

$$(dr = 0, \text{ so } dU = 0)$$

**2.** A force acts perpendicular to the displacement ($\mathbf{F} \cdot d\mathbf{r} = 0$)

Section 10-2

**WORK OF A COUPLE:**

$$U = M\theta$$

where  $M$ = magnitude of the moment of the couple

  $\theta$ = angular displacement (in radians) in the same plane in which the couple is acting

In a rotation from angular position $\alpha$ to $\beta$,

$$U = \int_\alpha^\beta \mathbf{M} \cdot d\boldsymbol{\theta} = \int_\alpha^\beta (M_x \, d\theta_x + M_y \, d\theta_y + M_z \, d\theta_z)$$

Section 10-3

**VIRTUAL WORK $\delta U$ OF FORCE F OR MOMENT M:**

$$\delta U = \mathbf{F} \cdot \delta\mathbf{r}$$

$$\delta U = \mathbf{M} \cdot \delta\boldsymbol{\theta}$$

where $\delta\mathbf{r}$ and $\delta\boldsymbol{\theta}$ are imaginary infinitesimal displacements.

A general statement of equilibrium on the basis of virtual work is

$$\delta U = \sum_{i=1}^m \mathbf{F}_i \cdot \delta\mathbf{r}_i + \sum_{j=1}^n \mathbf{M}_j \cdot \delta\boldsymbol{\theta}_j = 0$$

where $i$ refers to an individual force and the corresponding virtual displacement, and $j$ refers to an individual couple and the corresponding virtual rotation.

**MECHANICAL EFFICIENCY $\eta$ OF REAL SYSTEMS:**

input work = useful work + work of friction

$$\eta = \frac{\text{output work}}{\text{input work}} = \frac{\text{useful work}}{\text{total work required}}$$

For an ideal machine:  $\eta = 1$

For any real machine:  $\eta < 1$

Section 10-5

**GRAVITATIONAL WORK AND POTENTIAL ENERGY:**  The work of weight $W$ as the block moves from level 1 to 2 is

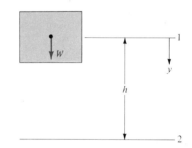

$$U_{12} = \int_1^2 W \, dy = Wh$$

= potential energy of the block at level 1
with respect to 2

The work of weight $W$ as the block moves from level 2 to 1 is

$$U_{21} = \int_2^1 -W \, dy = -Wh$$

= potential energy of the block at level 2
with respect to 1

**RELATION OF WORK $U$ AND RESULTING POTENTIAL ENERGY $V$:**  The work $U$ done on a mechanical system by an external force and the resulting change in the potential energy $V$ of the system are expressed as

$$U = -\Delta V$$

The change in potential energy as the block moves from level 2 to 1 is

$$\Delta V = V_1 - V_2 = -U_{21} = Wh$$

This is an alternative way to express the potential energy of the block at level 1 with respect to 2.

**ELASTIC POTENTIAL ENERGY:** The work of a force $(F = kx)$ on a spring is

$$U = \frac{1}{2} kx^2$$

= potential energy of the deformed spring

where  $k$ = spring constant of the spring (N/m or lb/ft)
  $x$ = change in length (from the length of the undeformed spring)

In deforming a spring from length $x_1$ to $x_2$,

$$U = \int_{x_1}^{x_2} kx\, dx = \frac{1}{2} k(x_2^2 - x_1^2)$$

= change in potential energy from length $x_2$
  to length $x_1$

Section 10-6

**POTENTIAL ENERGY AT EQUILIBRIUM:**  For equilibrium,

$$\frac{dV}{dq} = 0$$

where  $q$ = an independent coordinate along which displacement of the
  system is possible

For a system with $n$ degrees of freedom,

$$\frac{\partial V}{\partial q_i} = 0 \qquad \text{where} \quad i = 1, 2, \ldots, n$$

Section 10-7

**STABILITY OF EQUILIBRIUM:**

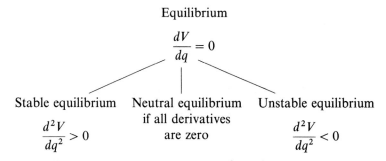

Equilibrium

$$\frac{dV}{dq} = 0$$

Stable equilibrium     Neutral equilibrium     Unstable equilibrium
            if all derivatives
$$\frac{d^2V}{dq^2} > 0$$      are zero      $$\frac{d^2V}{dq^2} < 0$$

Note that the conditions $dV/dq = 0$ and $d^2V/dq^2 = 0$ do not necessarily indicate neutral equilibrium because derivatives of higher order may be other than zero. In complex configurations the derivatives of higher order and their signs must be considered as described in Sec. 10-7 to determine the stability of the system.

**10–55** A block of weight $W = 200$ lb is raised to a height of 8 ft by a rope which passes over the fixed member $A$. Calculate the total work done by the tension $\mathbf{T}$ in raising the block.

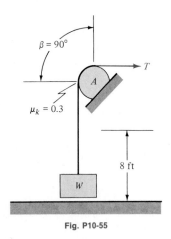

**Fig. P10-55**

**10–56** Four identical bars of length $l = 2$ ft and negligible weight are pinned at $B$, $C$, $D$, and $E$. A hydraulic actuator $A$ (two-force member) applies a force $\mathbf{F}$ at $B$ and $D$. Determine the magnitude of the force $\mathbf{F}$ of the actuator required to maintain equilibrium if $P = 10,000$ lb, $l = 3$ ft, and $\theta = 30°$.

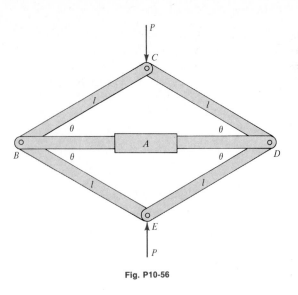

**Fig. P10-56**

**10–57** Plot the magnitude of the force $\mathbf{F}$ vs. the angle $\theta = 30°$, $45°$, and $60°$ in Prob. 10–56. State the range of angles $\theta$ for which $F \leq P$.

**10–58** Determine the mechanical efficiency of the polishing machine described in Prob. 10–16.

**10–59** The mechanism shown is a model for a typical office chair. Determine the required spring constant $k$ for equilibrium at $\theta = 30°$ and check the stability of equilibrium. The spring applies no force when $\theta = 0$.

**Fig. P10-59**

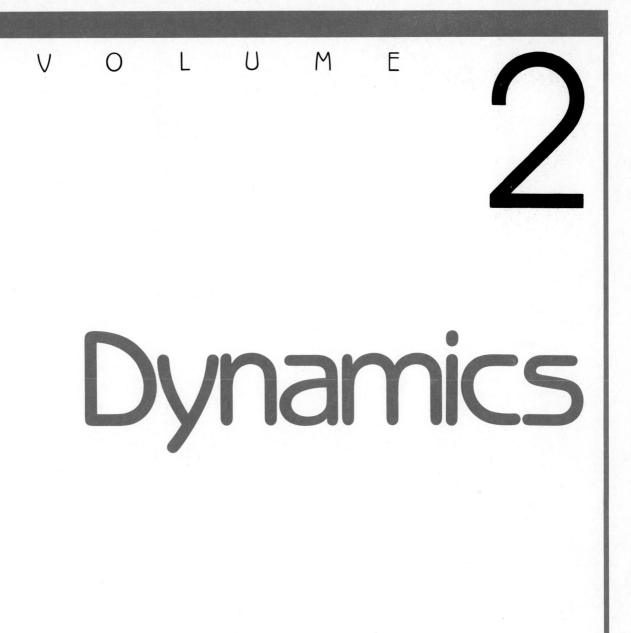

VOLUME

2

Dynamics

# Kinematics of Particles: Scalar Analysis

**11**

Dynamics is the analysis of moving bodies. The fundamental concepts of this area were first investigated and stated by Galileo Galilei (1564–1642) and Isaac Newton (1642–1727). Since that time, dynamics has been applied in many areas of engineering work. It is indispensable in the analysis of all moving vehicles and high-speed machinery. Dynamics is frequently used in the design of massive and apparently static structures such as bridges and high-rise buildings. There are also applications of dynamics in biomechanics work for aerospace travel, and in medical and sports investigations.

Dynamics consists of two major areas. *Kinematics* deals with the geometry and the time-dependent aspects of motion without considering the forces causing motion. In these studies forces may or may not be associated with the motion. The parameters of interest in kinematics are position, displacement, velocity, acceleration, and time. *Kinetics* is based on kinematics and it includes the effects of forces on masses.

Further divisions of dynamics are made according to the sizes and rigidities of the masses whose motions are analyzed. *Particles* include those masses whose dimensions can be ignored when considering their motion. Bodies that are not rotating may be considered as particles in many cases. The criterion for this assumption is that only the motion of the center of mass should be important. On this basis, particles may range in size from submicroscopic to astronomical. Often a mass cannot be treated as a particle. In that case, it should be analyzed either as a *rigid body*, a *deformable solid*, or a *fluid*. In this text the emphasis is on the dynamics of particles and rigid bodies.

**OVERVIEW**

451

In simple problems of dynamics, a scalar method of analysis is easily understood by the average student and is adequate to obtain solutions. Many scalar methods are special cases of vector methods, which are advantageous for solving complex problems. There are pedagogical advantages and disadvantages of starting a study of dynamics with a scalar analysis. Those who prefer a heavy emphasis on vector analysis should rapidly progress through this chapter and devote the time gained to Ch. 12. The topics of this chapter are introduced in the following overview.

**SECTION 11-1** presents definitions and mathematical relationships for the position, velocity, and acceleration of a particle that moves linearly in a rectangular coordinate system. These statements are part of the basic vocabulary in dynamics.

**SECTION 11-2** has simple examples in which either the position, the velocity, or the acceleration are given as a function of time, and the other two quantities are determined by differentiation or integration.

**SECTION 11-3** presents several useful expressions for common problems where the acceleration is given as a function of another quantity. Some of the equations derived in this section are frequently used in solving problems where initial conditions of position and velocity are given.

**SECTION 11-4** introduces the concept of relative motion of two particles, which may be separate or belong to the same body. Here the scalar method of analysis is limited to the case where both particles move along the same path or when modeling with a single path is reasonable.

## 11-1

## RECTILINEAR POSITION, VELOCITY, AND ACCELERATION

The motion of a particle can be analyzed by considering its position and the rates of change of its position. These are most conveniently described in terms of rectangular coordinates. The simplest motion of a particle is along a straight line which is also conveniently specified by rectangular coordinates. Such motion is called *rectilinear motion*. To understand all aspects of this simple motion it is sufficient to consider motion along one coordinate axis.

Consider the motion shown in Fig. 11-1. Assume that a particle is at point $A$ at time $t = 0$. The particle is at position $B$ at time $t$, and at position $C$ at time $t + \Delta t$. A positive direction for the position coordinate $x$ is chosen in order to describe the direction of the displacement from the original point of reference, here taken as the origin.

The next item of interest concerning the motion of the particle is the time rate of change of its position. The *average velocity* $v_{ave}$, is the quotient of the net displacement to the total time interval. For the particle moving from point $B$ to $C$ in Fig. 11-1,

$$\text{average velocity} = v_{ave_{BC}} = \frac{\Delta x}{\Delta t}$$

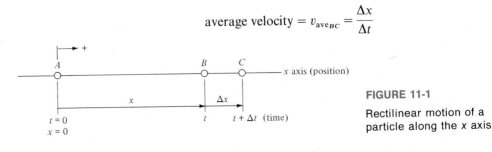

**FIGURE 11-1**

Rectilinear motion of a particle along the $x$ axis

The quotient of displacement and time interval at the limit is called the *instantaneous velocity v.* For the particle at point *B* in Fig. 11-1,

$$\text{instantaneous velocity} = v = \lim_{\Delta t \to 0} \frac{\Delta x}{\Delta t}$$

$$= \frac{dx}{dt} = \dot{x}$$

11-1

The units of velocity are always in terms of distance divided by time. Typically, they are meters per second (m/s) in the SI terminology, and feet per second (ft/s) in the U.S. customary system. Velocity may have a positive or negative sign, denoting the direction of the motion in the chosen coordinate system. For example, if the particle is moving away from the origin in the positive direction, $\Delta x$ is positive and the velocity is positive. The same is true for motion toward the origin from a negative position. If the position of the particle is decreasingly positive or increasingly negative, the velocity will be negative. The magnitude of the velocity is known as the *speed* of the particle. Velocity is actually a vector quantity, as shown in Ch. 12, but the scalar analysis of magnitude and sense of direction (sign) is adequate for rectilinear motion.

The instantaneous velocity is usually not constant during a displacement. The variation of velocity over time is called *acceleration*. The *average acceleration* is determined as the quotient of the net change in velocity $\Delta v$ to the corresponding total time interval $\Delta t$,

$$\text{average acceleration} = a_{\text{ave}} = \frac{\Delta v}{\Delta t}$$

The change in instantaneous velocity divided by the corresponding infinitesimal time interval is defined as the

$$\text{instantaneous acceleration} = a = \lim_{\Delta t \to 0} \frac{\Delta v}{\Delta t}$$

$$= \frac{dv}{dt} = \dot{v}$$

11-2

Also, from Eq. 11-1,

$$a = \frac{dv}{dt} = \frac{d(dx/dt)}{dt} = \frac{d^2 x}{dt^2} = \ddot{x}$$

The units of acceleration are always in terms of distance divided by time squared. Typically, they are $m/s^2$ in SI units, and $ft/s^2$ in the U.S. customary system. Acceleration may be positive or negative, depending on its direction with reference to the chosen positive direction along the coordinate axis. If the signs of the velocity and the acceleration are the same, the speed of the particle is increasing and the particle is accelerating. If the signs of the velocity and acceleration are opposite, the speed of the particle is decreasing. Sometimes this situation is called deceleration. The different possible cases are illustrated in Fig. 11-2. Acceleration is a vector quantity, as shown in Ch. 12. The scalar analysis of acceleration must be limited to rectilinear motion.

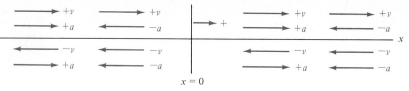

**FIGURE 11-2**

In many problems it is advantageous to eliminate the time $dt$ from the equations. Using the chain rule,

$$a = \frac{dv}{dt} = \frac{dv}{dx}\frac{dx}{dt} = \frac{dv}{dx}v$$

gives

$$\boxed{v\,dv = a\,dx}$$ 11-3

Equations 11-1, 11-2, and 11-3 are the differential equations of motion of a particle. Similar equations can be written using the $y$ or $z$ coordinates when the motion is rectilinear.

## 11-2 · MEASURED AND DERIVED QUANTITIES AS FUNCTIONS OF TIME

In practice it is possible to measure either position, velocity, or acceleration as a function of time. From any of these three, the other two quantities may be calculated. The three basic possibilities in rectilinear motion are discussed in the following.

**Position measured, $x = f(t)$.**   When the position of a particle in rectilinear motion is measured continuously as a function of time, the position from the origin can be expressed as a mathematical function. For example, consider the position $x$ given by

$$x = t^3 - 4t^2 + 5 \tag{a}$$

It is seen from Eq. 11-1 that the velocity of the particle can be obtained by differentiating Eq. (a) with respect to time $t$,

$$v = \frac{dx}{dt} = 3t^2 - 8t \tag{b}$$

The velocity can be calculated directly for chosen times $t$, or for positions $x$ using Eq. (a).

It is seen from Eq. 11-2 that acceleration of the particle is determined by differentiating Eq. (b) with respect to time $t$,

$$a = \frac{dv}{dt} = \frac{d^2x}{dt^2} = 6t - 8 \tag{c}$$

The position, velocity, and acceleration of the particle can be plotted on separate graphs using the same time base for a coordinate, as shown in Fig. 11-3. Such graphs are valuable in showing the various aspects of the motion at a glance.

The three graphs in Fig. 11-3 have several notable features that are typical for all such plots. The slope of the position vs. time curve at any

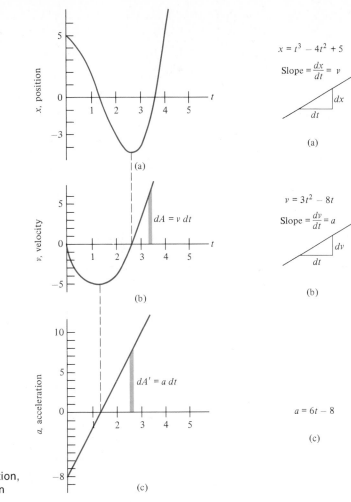

**FIGURE 11-3**

Plots of a particle's position, velocity, and acceleration

time $t$ is the velocity $v$ of the particle at that particular time. The slope of the velocity vs. time curve at any time $t$ is the acceleration $a$ at that time. The infinitesimal element of area under the velocity curve (Fig. 11-3b) is $dA = v\,dt$, which is equal to the infinitesimal displacement $dx$. Integrating the velocity between two arbitrary times $t_1$ and $t_2$ is thus equivalent to integrating $dx$ between the corresponding positions $x_1$ and $x_2$. The area $A$ under the velocity curve from time $t_1$ to $t_2$ is expressed as

$$A = \int_{t_1}^{t_2} v\,dt = \int_{x_1}^{x_2} dx = x_2 - x_1 \qquad \boxed{11\text{-}4}$$

This means that the area under the velocity–time curve in a specified time interval is equal to the net displacement during that time interval.

A distinction should be made here between the position, net displacement, and total distance traveled. The solution of $x = f(t)$ gives the position of the particle with respect to the origin at a specific time $t$. The position may be positive or negative depending on whether the particle is to the right or left of the origin (assuming positive direction to the right). The net displacement during a time interval $\Delta t = t_2 - t_1$ is the difference $\Delta x = x_2 - x_1$, which depends only on the position of the particle at the beginning and end points of its motion. The net displacement may be positive or negative depending on whether $x_2$ is more positive, or less negative, than $x_1$. The total distance traveled during a time interval $\Delta t = t_2 - t_1$ is the total of the absolute values of the net displacements between points at which the particle changes direction. These "turning-around points" are the extrema of the position curve; the points where the slope, and hence the velocity, are zero.

For example, in Fig. 11-3 the position of the particle at $t = 0$ and at $t = 4$ is 5. The net displacement during the time interval from $t = 0$ to $t = 4$ is $\Delta x = 5 - 5 = 0$, so the negative area below the $t$ axis of the velocity curve must equal the positive area above the axis. Since the motion of the particle changes direction when the velocity is zero at $t = \frac{8}{3}$, the total distance traveled during the interval from $t = 0$ to $t = 4$ is $|x_{t=4} - x_{t=8/3}| + |x_{t=8/3} - x_{t=0}| = |5 - (-4.48)| + |-4.48 - 5| = 18.96$.

The velocities and accelerations can be related in a similar procedure. From Fig. 11-3c, $dA' = a\,dt = dv$, and the area $A'$ under the acceleration curve is expressed as

$$A' = \int_{t_1}^{t_2} a\,dt = \int_{v_1}^{v_2} dv = v_2 - v_1 \qquad \boxed{11\text{-}5}$$

This means that the area under the acceleration–time curve during a specified time interval is equal to the net change in velocity during that time.

**Velocity measured, $v = f(t)$.** Sometimes the velocity is determined directly as a function of time. For example, assume that the velocity of a particle in rectilinear motion is estimated from a plot of data in the time interval from $t_0 = 0$ to $t$ as

$$v = \sin 2t + 3 \qquad \textbf{(d)}$$

From this function the position $x$ and the acceleration $a$ at time $t$ can be determined by integration and differentiation, respectively. From Eqs. 11-1 and 11-2,

$$x = \int_{t_0}^{t} v\,dt = \left[\frac{-1}{2}\cos 2t + 3t\right]_0^t = -\frac{\cos 2t}{2} + 3t + \frac{1}{2} \qquad \textbf{(e)}$$

$$a = \frac{dv}{dt} = 2\cos 2t \qquad \textbf{(f)}$$

**Acceleration measured, $a = f(t)$.** Acceleration as a function of time is frequently the most convenient to measure in modern experimental analysis of motion. For example, assume that the acceleration of a particle in rectilinear motion is measured from time $t_0 = 0$ to $t$, and it is expressed as

$$a = 2t - 3 \qquad \textbf{(g)}$$

with $x = v = 0$ at $t = 0$. The velocity and position at time $t$ can be determined from this by successive integration. From Eqs. 11-2 and 11-1,

$$v = \int_{t_0}^{t} a\,dt = [t^2 - 3t]_0^t = t^2 - 3t \qquad \textbf{(h)}$$

$$x = \int_{t_0}^{t} v\,dt = \left[\frac{1}{3}t^3 - \frac{3}{2}t^2\right]_0^t = \frac{t^3}{3} - \frac{3t^2}{2} \qquad \textbf{(i)}$$

which satisfy the given initial conditions.

Acceleration is often the basic quantity in the analysis of motion because it is easily measured or is specified in the problem. Therefore, it is useful to learn to work with expressions based on functions of acceleration. These expressions involve time, velocity, displacement, or a combination of these quantities, as shown in the following. It should be noted that there are a few basic expressions (these are the most important to remember) which can be applied in several special cases.

### Basic Expressions

When time is the specified variable in a problem, $a = dv/dt$ (Eq. 11-2) is rearranged and integrated:

$$\int_{v_0}^{v} dv = \int_{0}^{t} a\, dt \qquad \boxed{11\text{-}6}$$

When displacement is the specified variable in a problem, the expression $v\, dv = a\, dx$ (Eq. 11-3) is integrated:

$$\int_{v_0}^{v} v\, dv = \int_{x_0}^{x} a\, dx \qquad \boxed{11\text{-}7}$$

In these equations the initial conditions $t = 0$, $x = x_0$, and $v = v_0$ are used as the lower limits of integration. The upper limits are the arbitrary time, position, and velocity of interest. Note the relationship between Eqs. 11-6 and 11-7 since Eq. 11-3 was obtained from Eq. 11-2 using the chain rule.

### Special Cases

**Constant acceleration.** The simplest and also one of the most common cases is when the acceleration is constant. From Eq. 11-6,

$$\int_{v_0}^{v} dv = a \int_{0}^{t} dt$$

Thus, at an arbitrary time $t$,

$$\boxed{v = at + v_0} \qquad \boxed{11\text{-}8}$$

The velocity can also be obtained as a function of the distance $x$. From Eq. 11-7,

$$\int_{v_0}^{v} v\, dv = a \int_{x_0}^{x} dx$$

From this, at an arbitrary displacement $x$,

$$\boxed{v^2 = 2a(x - x_0) + v_0^2} \qquad \boxed{11\text{-}9}$$

A useful expression for the position $x$ is derived from $v = dx/dt$ (Eq. 11-1) and Eq. 11-8,

$$\int_{x_0}^{x} dx = \int_{0}^{t} (at + v_0)\, dt$$

The result is in terms of constant acceleration $a$, initial velocity $v_0$, and arbitrary time $t$,

$$x = \frac{1}{2} at^2 + v_0 t + x_0$$

11-10

Note that Eqs. 11-8, 11-9, and 11-10 are valid *only* for constant acceleration; this includes the possibility of $a = 0$. One of the most common applications is that for a freely falling body, in which case $a = g$, the acceleration due to gravity.

**COMMON ERROR**

Students often make the mistake of memorizing Eqs. 11-8, 11-9, and 11-10 and using them even when the acceleration is not constant. Errors can be avoided by using Eqs. 11-6 and 11-7 or by making certain that the special cases of constant acceleration or zero acceleration are satisfied before applying Eq. 11-8, 11-9, or 11-10.

**Variable acceleration given as a function of time.** Assume that $a = f(t)$. From Eqs. 11-6 and 11-1,

$$\int_{v_0}^{v} dv = \int_{0}^{t} f(t)\, dt \quad v = \int_{0}^{t} f(t)\, dt + v_0$$

$$\int_{x_0}^{x} dx = \int_{0}^{t} v\, dt \quad x = \int_{0}^{t} v\, dt + x_0$$

These expressions can be further evaluated only when $f(t)$ is specified.

**Variable acceleration given as a function of displacement.** Assume that $a = f(x)$. From Eq. 11-7,

$$\int_{v_0}^{v} v\, dv = \int_{x_0}^{x} f(x)\, dx \quad v^2 = 2 \int_{x_0}^{x} f(x)\, dx + v_0^2$$

This can be solved for $x$ as a function of $t$ when $f(x)$ is specified and $v$ is replaced by $dx/dt$.

**Variable acceleration given as a function of velocity.** Assume that $a = f(v)$. From $a = dv/dt$,

$$dt = \frac{dv}{f(v)}$$

$$t = \int_{0}^{t} dt = \int_{v_0}^{v} \frac{dv}{f(v)}$$

After integrating this expression, $v$ is obtained as a function of $t$. Another integration results in the relation for the displacement $x$. Equation 11-3 may also be used, giving

$$a = f(v) = v\frac{dv}{dx} \quad dx = \frac{v\, dv}{f(v)}$$

$$\int_{0}^{x} dx = x = \int_{v_0}^{v} \frac{v\, dv}{f(v)}$$

EXAMPLE 11-1

A particle $P$ moves along a straight line and its position is given by $x = 2t^3 - 4t^2 + 3$, where $x$ is in meters and $t$ is in seconds. Calculate **(a)** the times when the velocity and the acceleration are zero, **(b)** the net displacement from $t = 0$ to $t = 2$ s, and **(c)** the total distance traveled in part (b). Sketch $x$ vs. $t$, $v$ vs. $t$, and $a$ vs. $t$, and check the numerical answers using the graphs.

## SOLUTION

**(a)** A relation for the velocity as a function of time is determined using Eq. 11-1,

$$v = \frac{dx}{dt} = 6t^2 - 8t = t(6t - 8)$$

Setting this equation to zero yields the result that the velocity is zero at times $t = 0$ and $t = \frac{4}{3}$ s.

The acceleration is found using Eq. 11-2,

$$a = \frac{dv}{dt} = 12t - 8$$

Setting this equation to zero yields $a = 0$ at $t = \frac{2}{3}$ s.

**(b)** The displacement depends only on the initial position $x_{t=0}$ and final position $x_{t=2\,s}$,

$$x_{t=0} = 2(0)^3 - 4(0)^2 + 3 = 3\,\text{m}$$

$$x_{t=2\,s} = 2(2)^3 - 4(2)^2 + 3 = 3\,\text{m}$$

$$\Delta x = 3\,\text{m} - 3\,\text{m} = 0$$

**(c)** The total distance traveled, $D$, depends on the path of the particle from $t = 0$ to $t = 2$ s. The particle changes direction when the velocity is zero at $t = \frac{4}{3}$ s. At $t = \frac{4}{3}$ s,

$$x = 2\left(\frac{4}{3}\right)^3 - 4\left(\frac{4}{3}\right)^2 + 3 = 0.63\,\text{m}$$

Then the distance traveled is

$$D = |0.63 - 3| + |3 - 0.63| = 4.74\,\text{m}$$

Figure 11-4e is helpful in understanding the motion of the particle. Although the particle moves in a straight line, the plot of the position vs. time is shown as the curve in Fig. 11-4b. The velocity curve should be a sketch of the slope of the $x(t)$ curve at each point and hence the velocity should be zero when the position is at a local maximum or minimum (when the slope of the tangent to the curve is zero). Similarly, the acceleration curve should be a graph of the slope of the velocity curve and should be zero when the velocity is a local maximum or minimum. Using these principles and calculations with several values of $t$, the three curves are sketched in Fig. 11-4.

**FIGURE 11-4**

Plots of a particle's position, velocity, and acceleration

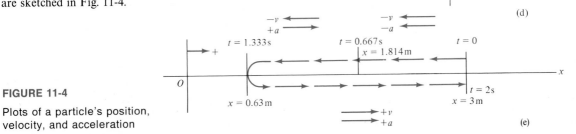

EXAMPLE 11-1    459

# EXAMPLE 11-2

The velocity of a large-displacement piston (Fig. 11-5a) is measured from $t = 0$ as $v = \sin(t/2) + 1/\pi$, where $v$ is in feet per second and $t$ is in seconds. The piston starts at the position $x_0 = 0$ at $t = 0$. Calculate the maximum acceleration and sketch $x$ vs. $t$, $v$ vs. $t$, and $a$ vs. $t$ for a complete cycle after which the motion is repeated.

## SOLUTION

The position $x$ as a function of time is determined first. From Eq. 11-1,

$$\int_{x_0=0}^{x} dx = \int_{t=0}^{t} v\, dt = \int_{0}^{t} \left( \sin\frac{t}{2} + \frac{1}{\pi} \right) dt$$

$$x = -2\cos\frac{t}{2} + \frac{t}{\pi} + 2$$

The acceleration $a$ is determined from Eq. 11-2,

$$a = \frac{dv}{dt} = \frac{1}{2}\cos\frac{t}{2}$$

The maximum acceleration occurs when $\cos(t/2) = 1$ or when $t = 0$, $4\pi$, $8\pi$, and so on. The acceleration at these times is $a = \frac{1}{2}\,\text{ft/s}^2$.

In order to sketch the $x$ vs. $t$, $v$ vs. $t$, and $a$ vs. $t$ curves, other extreme values are calculated. The acceleration is a minimum for $\cos(t/2) = -1$, when $t = 2\pi$, $6\pi$, and so on. Its value is $-\frac{1}{2}\,\text{ft/s}^2$. The acceleration is zero for $\cos(t/2) = 0$, when $t = \pi$, $3\pi$, and so on.

The maximum and minimum values for the velocity occur when $a = 0$, at $t = \pi$ and $3\pi$. Then

$$v_{\text{max}} = \sin\frac{\pi}{2} + \frac{1}{\pi} = \left( 1 + \frac{1}{\pi} \right) = 1.32\,\text{ft/s}$$

$$v_{\text{min}} = \sin\frac{3\pi}{2} + \frac{1}{\pi} = \left( -1 + \frac{1}{\pi} \right) = -0.68\,\text{ft/s}$$

When $t = 0$, $2\pi$, $4\pi$, $v = 1/\pi$. The local maximum and minimum values of $x$ and the times at which they occur are found by solving for the times when the velocity is zero and substituting these times into the expression for $x$.

$$v = \sin\frac{t}{2} + \frac{1}{\pi} = 0 \quad \text{when} \quad \sin\frac{t}{2} = -\frac{1}{\pi}$$

Therefore, $v = 0$ at $t = 2\sin^{-1}(-1/\pi)$,

$$t = 2(-0.324) = -0.648\,\text{s}$$

$$t = 2(\pi + 0.324) = 6.93\,\text{s}$$

$$t = 2(-0.324 + 2\pi) = 11.92\,\text{s}$$

Discarding the negative value due to the initial conditions of the problem,

$$x_{\text{max}} = 6.10\,\text{ft}$$

$$x_{\text{min}} = 3.90\,\text{ft}$$

The other values for $x$ are shown in the table in Fig. 11-5e, and the sketches of $x$ vs. $t$, $v$ vs. $t$, and $a$ vs. $t$ are shown in Fig. 11-5b, c, and d. Note that the oscillatory component of the position $x$ is found to be small relative to the total displacement when time is large in the equation for $x$.

Piston rod

Piston

(a)

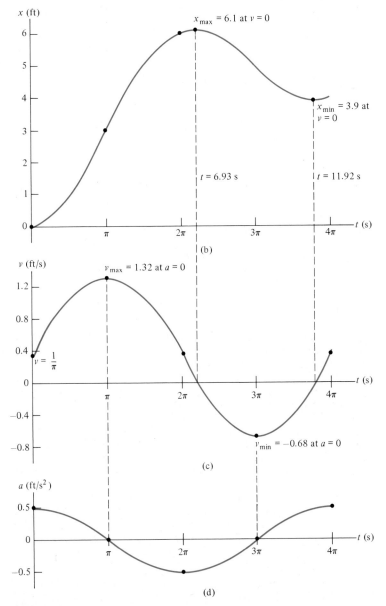

$x$ (ft)

$x_{max} = 6.1$ at $v = 0$

$x_{min} = 3.9$ at $v = 0$

$t = 6.93$ s

$t = 11.92$ s

$t$ (s)

(b)

$v$ (ft/s)

$v_{max} = 1.32$ at $a = 0$

$v = \dfrac{1}{\pi}$

$v_{min} = -0.68$ at $a = 0$

$t$ (s)

(c)

$a$ (ft/s$^2$)

$t$ (s)

(d)

| $t$ (s) | $x$ (ft) |
|---------|----------|
| 0       | 0        |
| $\pi$   | 3        |
| $2\pi$  | 6        |
| $3\pi$  | 5        |
| $4\pi$  | 4        |

(e)

**FIGURE 11-5**

Plots of a piston's position, velocity, and acceleration

EXAMPLE 11-2    461

# EXAMPLE 11-3

Worker $A$ throws a small tool vertically toward worker $B$ (Fig. 11-6a). The tool's initial velocity is $v_0 = 50$ ft/s upward at the initial height of $y_0 = 6$ ft. The acceleration of the tool is constant, $a = 32.2$ ft/s$^2$ downward. Determine **(a)** the highest elevation $y_{max}$ that the tool could reach, and **(b)** the velocity of the tool at the ground if it falls back without interference. Use integration of basic equations and draw the $y$ vs. $t$ and $v$ vs. $t$ curves.

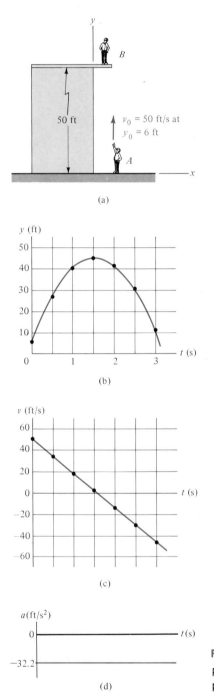

**FIGURE 11-6**

Plots of a vertically thrown object's position, velocity, and acceleration

## SOLUTION

Using the given reference system, $a = -32.2 \, \text{ft/s}^2$, and at $t = 0$, $y_0 = 6 \, \text{ft}$ and $v_0 = 50 \, \text{ft/s}$.

**(a)** First it is necessary to obtain expressions for the velocity and elevation at any time $t$. From Eq. 11-6,

$$\int_{v_0=50}^{v} v = -\int_{0}^{t} 32.2 \, dt$$

$$[v]_{50}^{v} = -[32.2t]_{0}^{t}$$

$$v = 50 - 32.2t \qquad \text{(a)}$$

From Eq. 11-1 and using $y_0 = 6 \, \text{ft}$ at $t = 0$,

$$v = \frac{dy}{dt} = 50 - 32.2t$$

$$\int_{y_0=6}^{y} dy = \int_{0}^{t} (50 - 32.2t) \, dt$$

$$[y]_{6}^{y} = [50t - 16.1t^2]_{0}^{t}$$

$$y = 6 + 50t - 16.1t^2 \qquad \text{(b)}$$

At the highest elevation, $v = 0$. Substituting into Eq. (a), the time $t_1$ of upward flight is obtained,

$$50 - 32.2t_1 = 0$$

$$t_1 = 1.553 \, \text{s} \qquad \text{(c)}$$

Substituting Eq. (c) into Eq. (b) yields

$$y_{\text{max}} = 6 + (50)(1.553) - (16.1)(1.553)^2$$
$$= 44.8 \, \text{ft}$$

**(b)** When the tool hits the ground, $y = 0$. Substituting into Eq. (b), the total time $t_2$ of upward and downward flight is obtained,

$$6 + 50t_2 - 16.1t_2^2 = 0$$

$$t_2 = 1.553 \pm 1.668$$

Using the positive root for the total flight measured from $t = 0$ yields

$$t_2 = 3.221 \, \text{s} \qquad \text{(d)}$$

The velocity of the tool at the ground is obtained from Eqs. (a) and (d),

$$v = 50 - (32.2)(3.221) = -53.7 \, \text{ft/s}$$

The $y$ vs. $t$ and $v$ vs. $t$ plots are obtained from Eqs. (b) and (a), respectively, and are drawn in Fig. 11-6b and c.

## JUDGMENT OF THE RESULTS

The final velocity is reasonable considering the initial velocity, the initial and final elevations, and the magnitude of the acceleration. Since the acceleration has a constant negative value, the velocity curve must be linear with a constant negative slope. The position curve must be parabolic with a positive slope when the velocity is positive and a negative slope when the velocity is negative. The calculated results agree with these conditions.

EXAMPLE 11-3   463

## EXAMPLE 11-4

A car moves along the highway (Fig. 11-7a) with an initial velocity of $v_0 = 25$ m/s to the right and is at point $O$ when $t = 0$. At that instant the driver notices a large rock ahead, and after a reaction time of 1 s the brakes are applied, producing a constant rate of deceleration of $a = -5$ m/s². Determine (a) the time elapsed from point $O$ until the vehicle is brought to rest at point $P$, and (b) the distance $D$ measured from point $O$ to $P$. Sketch the $x$–$t$, $v$–$t$, and $a$–$t$ curves for the motion of the car and check the answers by using these motion curves.

### SOLUTION

(a) From $t = 0$ to $t = 1$ s, $v = 25$ m/s (constant). At $t = 1$ s, $x = (25 \text{ m/s})(1 \text{ s}) = 25$ m (right of $O$). From $t = 1$ s to $t = t_P$, $a = -5$ m/s² (constant) until $v = 0$, which occurs at $t = t_P$ (unknown). Since $a = dv/dt = -5$ m/s²,

$$\int_{25}^{0} dv = -5 \int_{1}^{t_P} dt$$

Integration yields

$$0 - 25 = -5(t_P - 1)$$

$$t_P = 6 \text{ s}$$

(a)

$t = 1\text{s}$        $t = t_P$

(b)

**FIGURE 11-7**

Plots of a car's acceleration, velocity, and position

464    KINEMATICS OF PARTICLES: SCALAR ANALYSIS

**ALTERNATIVE METHOD:** In the *definite integral* above, the limits of integration on time were matched to the corresponding limits of integration on velocity. Specifically, $v = 25\,\text{m/s}$ at $t = 1\,\text{s}$ and $v = 0$ at $t = t_P$. It is equally acceptable to integrate by using an *indefinite integral* plus a constant of integration as follows:

$$a = \frac{dv}{dt} = -5\,\text{m/s}^2$$

$$\int dv = -5 \int dt + C_1$$

$$v = -5t + C_1 \qquad\qquad \textbf{(a)}$$

The constant of integration must be determined by considering the initial condition that $v = 25\,\text{m/s}$ at $t = 1\,\text{s}$. Substituting in Eq. (a),

$$25 = -5(1) + C_1 \quad \text{so} \quad C_1 = 30$$

Then Eq. (a) becomes

$$v = -5t + 30 \qquad (\text{valid for } 1 \le t \le t_P) \qquad \textbf{(b)}$$

Solving for the time $t = t_P$ at which $v = 0$, we obtain from Eq. (b),

$$0 = -5t_P + 30$$

$$t_P = 6\,\text{s} \qquad (\text{as above})$$

   **(b)** The brakes are applied after the automobile has traveled 25 m to the right of point $O$ and when it has a speed of 25 m/s. After braking,

$$a = -5\,\text{m/s}^2 = \frac{dv}{dt} = v\frac{dv}{dx} \qquad\qquad \textbf{(c)}$$

The latter form is often preferred for relating a change in position ($dx$) to a change in velocity ($dv$), although $a = dv/dt$ would also work. By integrating Eq. (c) with the appropriate limits on the integrals, the distance $D$ is obtained,

$$-5 = v\frac{dv}{dx}$$

$$-5 \int_{25\,\text{m}}^{D} dx = \int_{25\,\text{m/s}}^{0} v\,dv$$

$$-5(D - 25) = -\frac{1}{2}(25)^2$$

$$D = 87.5\,\text{m}$$

The easiest motion curve to sketch in this example is the $a$–$t$ curve, since it is a constant $a = 0$ for $0 \le t \le 1\,\text{s}$ and a constant $a = -5\,\text{m/s}^2$ for $t > 1\,\text{s}$, as shown in Fig. 11-7b. From the $a$–$t$ curve,

$$\text{area } A_1 = \Delta v = (-5)(t_P - 1)$$

From the $v$–$t$ curve,

$$\Delta v = -25 = A_1$$

$$-25 = -5(t_P - 1)$$

$$t_P = 6\,\text{s}$$

Also, since the area under the $v$–$t$ curve equals the change in position $\Delta x$,

$$\Delta x = \Delta x_1 + \Delta x_2 = A_2 + A_3 = D$$

where $A_2 = 25(1) = 25\,\text{m}$ and $A_3 = \frac{1}{2}(t_p - 1)(25) = 62.5\,\text{m}$. Hence,

$$D = 25\,\text{m} + 62.5\,\text{m}$$

$$= 87.5\,\text{m}$$

EXAMPLE 11-4    465

# EXAMPLE 11-5

An airplane approaches a stationary aircraft carrier at a speed of 300 ft/s (Fig. 11-8). Once the wheels make contact with the landing surface and the braking forces are applied, it is observed that the deceleration of the airplane is proportional to the speed, written mathematically as $a = -0.5v$. Determine **(a)** the minimum length $L$ of runway required for landing, **(b)** the position on the runway where the speed is 50 ft/s (for placement of a "catching" mechanism), and **(c)** the time elapsed after touchdown for the speed to reach 50 ft/s.

## SOLUTION

**(a)** For rectilinear motion,

$$a = \frac{dv}{dt} = v\frac{dv}{dx}$$

Since the problem statements for parts (a) and (b) involve a "change in velocity related to a change in position," the latter form, $a = v(dv/dx)$, is best suited for integration.

Given that $a = -0.5v = v(dv/dx)$, we obtain by separating variables and integrating,

$$dx = -2\,dv$$

$$x = -2v + C_1$$

At $x = 0$, $v = 300$, so $C_1 = 600$. Therefore,

$$x = -2v + 600 \tag{a}$$

The velocity $v$ must equal 0 at $x = L$, so from Eq. (a) we have

$$L = 600\text{ ft} \qquad \text{(minimum length of runway)}$$

**(b)** From Eq. (a), setting $v = 50$ ft/s,

$$x = -100\text{ ft} + 600\text{ ft}$$
$$= 500\text{ ft}$$

**(c)** The statement for part (c) requires a relationship between "a change in velocity and a change in time," so the form $a = dv/dt$ is now best suited for this integration. Since $a = -0.5v = dv/dt$, we may separate variables to yield

$$dt = -2\frac{dv}{v}$$

and obtain by integration,

$$t = -2\ln v + C_2$$

At $t = 0$, $v = 300$, so $C_2 = 2\ln 300$. Therefore,

$$t = 2\ln\frac{300}{v}$$

For $v = 50$ ft/s,

$$t = 2\ln\frac{300}{50}$$

$$= 3.58\text{ s}$$

$v = 300$ ft/s

**FIGURE 11-8**

Airplane landing on the runway of an aircraft carrier

EXAMPLE 11-6

The rocket sled in Fig. 11-9 starts from rest and accelerates at a rate of $a = 9x\,\text{ft/s}^2$ to a speed of 88 ft/s at point $B$ along a launching track of length $D$. After leaving the launching track at point $B$, the rocket sled begins to decelerate at a rate of $a = -0.2t\,\text{ft/s}^2$ until it comes to rest at point $C$ (where $t_e$ is the time elapsed since passing point $B$). Determine **(a)** the necessary length $D$ of the launching track, **(b)** the time required for the rocket sled to go from $B$ to $C$ where it stops, and **(c)** the necessary length of track $L$ from $B$ to $C$.

## SOLUTION

**(a)** The problem statement involves a "change in velocity related to a change in position." Therefore, use

$$a = v\frac{dv}{dx} = 9x \quad \text{or} \quad 9x\,dx = v\,dv \qquad \text{(a)}$$

Integrating Eq. (a) yields

$$\frac{1}{2}(9)x^2 = \frac{1}{2}v^2 + C_1$$

Since $v = 0$ at $x = 0$, $C_1 = 0$ and

$$v = 3x\,\text{ft} \qquad \text{(b)}$$

Therefore, solving Eq. (b) for $v = 88\,\text{ft/s}$,

$$x = D = 29.3\,\text{ft}$$

**(b)** From $B$ to $C$ the acceleration $a = -0.2t_e$, which must be integrated by using the form $a = dv/dt$. Integration yields

$$v = -0.1(t_e)^2 + C_2$$

At $t_e = 0$, $v = 88\,\text{ft/s}$, so $C_2 = 88$ and

$$v = -\frac{1}{10}(t_e)^2 + 88\,\text{ft/s} \qquad \text{(c)}$$

Integrating Eq. (c) to determine $x_e$, we have

$$x_e = -\frac{1}{30}(t_e)^3 + 88t_e + C_3\,\text{(ft)} \qquad \text{(d)}$$

where $C_3 = 0$ because $x_e = 0$ at $t_e = 0$. According to Eq. (c), the sled comes to rest at

$$0 = -\frac{1}{10}(t_e)^2 + 88$$

$$t_e = 29.7\,\text{s}$$

**(c)** From Eq. (d), the position $x_e = L$, reached 29.7 s after passing point $B$, is calculated,

$$L = -\frac{1}{30}(29.7)^3 + 88(29.7)$$

$$= 1740\,\text{ft}$$

**FIGURE 11-9**

Rocket sled

EXAMPLE 11-6    467

**11-1** The position $x$ of a particle is defined by the expression $x = 3t + 4$, where $x$ is in meters and $t$ is in seconds. Determine the velocity and acceleration when $t = 4\,\text{s}$.

**11-2** The position $z$ of a particle is defined by the expression $z = t^3 - 2t$, where $z$ is in meters and $t$ is in seconds. Determine the position and acceleration when the velocity is zero.

**11-3** The position $y$ of a particle is defined by the relationship $y = 2t^2 - 5$, where $y$ is in feet and $t$ is in seconds. Calculate the velocity and acceleration at $t = 3\,\text{s}$.

**11-4** The position $x$ of a particle is defined by the relationship $x = 5 \sin 2t + 4$, where $x$ is in inches and $t$ is in seconds. Calculate the velocity and acceleration at $t_1 = 1\,\text{s}$ and $t_2 = 2\,\text{s}$.

**11-5** The position $y$ of a particle is defined by the expression $y = t^4 - 4t^2 + t + 2$, where $y$ is in meters and $t$ is in seconds. Determine the maximum speed attained by the particle.

**11-6** The position $x$ of a particle is defined by the expression $x = t^3 - 4t^2 + 5$, where $x$ is in meters and $t$ is in seconds. Determine (a) the velocity and acceleration when $t = 2\,\text{s}$, and (b) the maximum speed of the particle.

**11-7** Determine the total distance traveled during the first 4 s of motion by the particle in Prob. 11-6.

**11-8** The velocity of a particle is defined by the expression $v = 3 \cos t + 1$, where $v$ is in meters per second and $t$ is in seconds. The particle is at $y_0 = 2\,\text{m}$ when $t = 0$. Determine the position and acceleration at $t = 10\,\text{s}$.

**11-9** The velocity of a particle is defined by the expression $v = 2t$, where $v$ is in feet per second and $t$ is in seconds. The particle is at $x_0 = 2\,\text{ft}$ when $t = 0$. Calculate the position and acceleration at $t = 3\,\text{s}$.

**11-10** The velocity of a particle is estimated from data and expressed as $v = t^2 + 3t - 2$, where $v$ is in inches per second and $t$ is in seconds. The particle is at $x_0 = -2\,\text{in.}$ at $t = 0$. Determine the position and acceleration at $t = 5\,\text{s}$.

**11-11** The velocity of a particle is defined by the relationship $v = 3x$, where $v$ is in meters per second and $x$ is in meters. Determine the position, velocity, and acceleration at $t = 0.5\,\text{s}$ if initially $v_0 = 0$ at $x_0 = 0.1\,\text{m}$.

**11-12** The velocity of a particle is defined by the expression $v = ky^2$, where $v$ is in feet per second, $y$ is in feet, and $k$ is a constant. Determine the acceleration at $y = 100\,\text{ft}$ if initially $v_0 = 2\,\text{ft/s}$ at $y_0 = 5\,\text{ft}$.

**11-13** The acceleration of a particle is given as $a = 6\,\text{m/s}^2$. At time $t = 0$ the position is $x_0 = 0$ and the velocity is $v_0 = 0$. Calculate the position and velocity at time $t = 5\,\text{s}$.

**11-14** A particle moves along the $x$ axis with a constant acceleration of $6\,\text{m/s}^2$ to the right. At time $t = 0$ the particle is located 4 m to the right of the origin and has a velocity of 18 m/s to the left. Determine the equations for the position and velocity as functions of time.

**11-15** The acceleration of a particle is given as $a = 4t\,\text{ft/s}^2$, where $t$ is in seconds. At time $t = 0$ the position is $x_0 = 0$ and the velocity is $v_0 = 5\,\text{ft/s}$. Determine the position and velocity at $t = 2\,\text{s}$.

**11-16** The acceleration of a particle is defined by the relationship $a = kx\,\text{in./s}^2$, where $k$ is a constant and $x$ is in inches. At $x_0 = 0$, $v_0 = 0$, and at $x_1 = 2\,\text{in.}$, $v_1 = 10\,\text{in./s}$. Determine the velocity $v_2$ at $x_2 = 5\,\text{in.}$

**11-17** The acceleration of a particle is defined by the expression $a = 10 - x^2$, where $a$ is in m/s$^2$ and $x$ is in meters. At $x_0 = 0$, $v_0 = 0$. Determine the position $x$ where the velocity is maximum.

**11–18** For the particle in Prob. 11–17, determine (a) the velocity at $x = 4\,\text{m}$, and (b) the position where the velocity returns to zero.

**11–19** The acceleration of a particle is defined by the expression $a = -kx^{-2}\,\text{ft/s}^2$, where $x$ is in feet and $k$ is a constant. At $x_0 = 0.1\,\text{ft}$, $v_0 = 0$, and at $x_1 = 1\,\text{ft}$, $v_1 = 10\,\text{ft/s}$. Determine the velocity $v_2$ at $x_2 = 5\,\text{ft}$.

**11–20** The acceleration of a particle is defined by the relationship $a = -v$, where $a$ is in m/s$^2$ and $v$ is in m/s. At $t = 0$, $v_0 = 500\,\text{m/s}$ and $x_0 = 0$. Determine the distance traveled by the particle as it comes to rest from the initial velocity.

**11–21** For the particle in Prob. 11–20, write the equations for the velocity as (a) a function of the position $x$ and (b) a function of the time $t$.

**11–22** The acceleration of a particle is defined by the expression $a = 400 - 5v$, where $a$ is in ft/s$^2$ and $v$ is in ft/s. At $t = 0$, $v_0 = 200\,\text{ft/s}$. Determine the time required to reduce the velocity (a) to half its initial value, and (b) to zero.

**11–23** The acceleration of a particle is given as $a = -v^2$, where $a$ is in m/s$^2$ and $v$ is in m/s. Determine the distance $d$ traveled by the particle as it nearly comes to rest from its initial velocity $v_0$. Plot $v/v_0$ vs. $d$ and calculate $d$ when $v = 5\%$, $1\%$, and $0.1\%$ of $v_0$.

**11–24** The acceleration of a particle falling through the atmosphere or a liquid is defined by the expression $a = g(1 - k^2v^2)$, where $g = 32.2\,\text{ft/s}^2$, $k$ is a constant, and $v$ is velocity in ft/s. Initially, the position is $y_0 = 0$ and the velocity is $v_0 = 0$. Determine the velocity $v$ as a function of the position $y$.

**11–25** A projectile is shot in the vertical direction for atmospheric research. Determine the highest elevation reached by the projectile if its initial velocity is $v_0 = 500\,\text{m/s}$ and its downward acceleration is constant, $a = 9.81\,\text{m/s}^2$.

**11–26** In Prob. 11–25, determine the time of flight before the projectile hits the ground, and the velocity just before impact.

**11–27** In the laboratory simulation of impacts on hard hats, a hammer is dropped from a height of 3 m. Determine the hammer's velocity just before impact if during the free fall the hammer is subject to a constant acceleration of 9.81 m/s$^2$ downward. What is the required initial velocity of the hammer at the height of 3 m to simulate a free fall from a height of 20 m?

**Fig. P11-27**

**11–28** A rock is thrown vertically upward leaving the child's hand at $O$ with a speed $v_0 = 30\,\text{m/s}$ at $t = 0$. The acceleration of the rock due to gravity is constant at 9.81 m/s$^2$ downward. Determine (a) the equations for the position $y$ and velocity $v$ of the rock, (b) its maximum height $y_{\text{max}}$, (c) its velocity when it returns to $O$, and (d) the velocity at which it strikes the water 10 m below $O$.

**Fig. P11-28**

# RELATIVE MOTION OF TWO PARTICLES ALONG THE SAME LINE

The motion of a body may be evaluated with respect to another moving body. There are numerous practical reasons for analyzing such relative motion. The most common cases are of vehicles whose motion may be monitored with respect to other vehicles (e.g., police using radar equipment). Some crashes between moving vehicles are in the same category. Another important area of relative motion is in structural mechanics. The relative motion of various parts of a given structure (which on the whole may or may not have a fixed location) is indicative of deformations and possible damage in the structural members. In simple cases the relative motions may be modeled by two particles moving along the same line. The particles may represent two separate bodies, or two distinct points of a deformable body.

Assume that two particles $A$ and $B$ are moving along the same straight line $x$ in Fig. 11-10a, but that they are independent of each other. The positions, velocities, and accelerations of $A$ and $B$ are given in reference to a fixed point $O$ as $x_A$, $v_A$, $a_A$, and $x_B$, $v_B$, $a_B$, respectively.

The *scalar relative position*, *velocity*, and *acceleration* of particle $B$ with respect to $A$ are defined using Fig. 11-10a as follows.

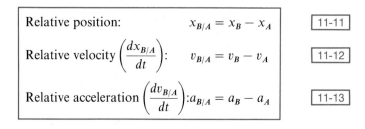

| | | |
|---|---|---|
| Relative position: | $x_{B/A} = x_B - x_A$ | 11-11 |
| Relative velocity $\left(\dfrac{dx_{B/A}}{dt}\right)$: | $v_{B/A} = v_B - v_A$ | 11-12 |
| Relative acceleration $\left(\dfrac{dv_{B/A}}{dt}\right)$: | $a_{B/A} = a_B - a_A$ | 11-13 |

Note that according to Fig. 11-10 there are signs to consider. For example, $x_{A/B} = -x_{B/A}$, $v_{A/B} = -v_{B/A}$, and so on.

**TIP**

In analyzing the relative motion of the two cars in Fig. 11-10, it is helpful to remember that the notation "$B/A$" means viewing the position, velocity, and acceleration of car $B$ as if one were a passenger in car $A$.

The significance of Eqs. 11-11, 11-12, and 11-13 varies depending on the particular problem. For example, in vehicle crashes the relative velocity is ultimately the most important, while in structural deformations the permanent changes in relative positions of chosen points indicate the extent of damage. The more general vector analyses of relative motion are presented in Ch. 12.

(a)

for $v_B > v_A$:

for $v_A > v_B$:

(b)

**FIGURE 11-10**

Relative motion of two cars

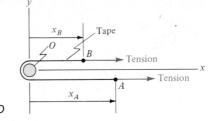

**FIGURE 11-11**

Tape wrapped around a fixed peg $O$

**Dependent relative motions. Degrees of freedom.** In some mechanical systems the position of a particle depends on the position of another particle or of several other particles. For example, a flexible, inextensible tape is situated about a fixed peg $O$ as in Fig. 11-11. In this case only one of the two coordinates $x_A$ and $x_B$ (of particles $A$ and $B$) may be chosen arbitrarily, so by inspection

$$x_A + x_B = \text{constant}$$

Any system in which only one coordinate is chosen arbitrarily is said to have *one degree of freedom*.

Consider a different tape transport system as in Fig. 11-12, where the tape is wrapped about fixed pegs $O$ and $O'$ and a peg $P$ which is movable in the $x$ direction. Here the coordinate $x_A$ depends on coordinates $x_B$ and $x_P$, both of which may be arbitrarily chosen, so this system has *two degrees of freedom*. To determine the relationship of the coordinates, consider the following facts: the tape $AB$ has a constant length, and the lengths of tape in contact with the pegs are constant. Thus,

$$x_A + x_B + 2x_P = \text{constant}$$

The relationship of the particles' velocities or of their accelerations is obtained from the expression of their positions by differentiation with respect to time. In the case of the tape in Fig. 11-12,

$$v_A + v_B + 2v_P = 0$$

$$a_A + a_B + 2a_P = 0$$

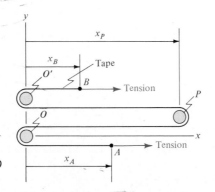

**FIGURE 11-12**

Tape wrapped around fixed pegs $O$ and $O'$ and movable peg $P$

# EXAMPLE 11-7

Assume that police car $P$ in Fig. 11-13 is equipped with an advanced radar that allows simultaneous tracking of two vehicles. At a particular instant $v_P = 40\,\text{mph}$, $a_P = 10\,\text{ft/s}^2$, $v_{A/P} = 25\,\text{mph}$, $a_{A/P} = -8\,\text{ft/s}^2$, $v_{B/P} = 20\,\text{mph}$, $a_{B/P} = -13\,\text{ft/s}^2$. Determine the absolute velocities, $v_A$ and $v_B$, and absolute accelerations, $a_A$ and $a_B$, of vehicles $A$ and $B$.

**FIGURE 11-13**

Relative motion of three vehicles on the same path

## SOLUTION

From Eq. 11-12,

$$v_{A/P} = v_A - v_P$$

$$v_A = v_P + v_{A/P} = 40\,\text{mph} + 25\,\text{mph} = 65\,\text{mph}$$

$$v_B = v_P + v_{B/P} = 40\,\text{mph} + 20\,\text{mph} = 60\,\text{mph}$$

From Eq. 11-13,

$$a_{A/P} = a_A - a_P$$

$$a_A = a_P + a_{A/P} = 10\,\text{ft/s}^2 - 8\,\text{ft/s}^2$$
$$= 2\,\text{ft/s}^2 \quad \text{(increasing velocity)}$$

$$a_B = a_P + a_{B/P} = 10\,\text{ft/s}^2 - 13\,\text{ft/s}^2$$
$$= -3\,\text{ft/s}^2 \quad \text{(decreasing velocity)}$$

# EXAMPLE 11-8

In an automobile road race the vehicles are started at 1-minute intervals. Assume that car $A$ precedes car $B$, and that their accelerations are defined by $6e^{-t/\tau}\,\text{m/s}^2$, where $t$ is time in seconds and $\tau$ is a constant. The maximum velocity of car $A$ is $165\,\text{km/h}$, of car $B$ it is $170\,\text{km/h}$. Determine the relative position and velocity of car $B$ with respect to $A$ at $t = 3600\,\text{s}$. Measure time from the start of car $A$.

## SOLUTION

First, put the maximum velocities in units of meters and seconds. The velocity of car $A$ is 45.83 m/s, of car $B$ it is 47.22 m/s. The velocity and acceleration are determined using indefinite integrals according to Eqs. 11-1 and 11-2,

$$\int dv = \int a\, dt$$
$$= \int 6e^{-t/\tau}\, dt$$
$$v = -6\tau e^{-t/\tau} + C_1$$

and

$$\int dx = \int v\, dt$$
$$x = \int (-6\tau e^{-t/\tau} + C_1)\, dt$$
$$= 6\tau^2 e^{-t/\tau} + C_1 t + C_2$$

where $C_1$ and $C_2$ are constants of integration. They are determined using the initial conditions that at $t = 0$, $v = x = 0$. Thus,

$$C_1 = 6\tau$$
$$C_2 = -6\tau^2$$

From $v_{A_{max}} = 45.83$ m/s $= 6\tau_A$, and $v_{B_{max}} = 47.22$ m/s $= 6\tau_B$,*

$$\tau_A = 7.638\,\text{s} \quad\text{and}\quad \tau_B = 7.870\,\text{s}$$

The equations for velocity and displacement are

$$v_A = 45.83(1 - e^{-t/7.638})\ \text{m/s}$$
$$v_B = 47.22(1 - e^{-t/7.87})\ \text{m/s}$$
$$x_A = 45.83t + 350.0(e^{-t/7.638} - 1)\ \text{m}$$
$$x_B = 47.22t + 371.6(e^{-t/7.87} - 1)\ \text{m}$$

Using Eqs. 11-11 and 11-12, the relative position and velocity can be found. At $t = 3600$ s,

$$t_A = 3600\,\text{s} \qquad\qquad t_B = 3540\,\text{s}$$
$$v_A(3600) = 45.83\,\text{m/s} \qquad v_B(3540) = 47.22\,\text{m/s}$$
$$x_A(3600) = 164.64\,\text{km} \qquad x_B(3540) = 166.79\,\text{km}$$
$$v_{B/A} = 47.22 - 45.83 = 1.39\,\text{m/s}$$
$$x_{B/A} = 166.79\,\text{km} - 164.64\,\text{km} = 2.15\,\text{km}$$

## JUDGMENT OF THE RESULTS

Car $A$ covered about 2.5 km before car $B$ was allowed to start. With a difference between the cars of 5 km/h, it is reasonable that car $B$ should be a little over 2 km ahead of car $A$ after an hour of racing. Note that the effect of the exponential decreases rapidly with increasing time.

* Note that these are approximations. Usually, $v_{max}$ is determined using $dv/dt = 0$, but here $dv/dt = a = 6e^{-t/\tau}$, which only gradually approaches zero as $t$ becomes large. Thus, $v_{max} = -6\tau e^{-t/\tau} + 6\tau \simeq 6\tau$.

EXAMPLE 11-8    473

# EXAMPLE 11-9

Truck $A$ and car $B$ are approaching one another as shown in Fig. 11-14 and at time $t = 0$ they are 2 km apart. Truck $A$ is accelerating to the right at a rate $a_A = 0.2 \, \text{m/s}^2$ and at $t = 0$ its speed $v_A = 15 \, \text{m/s}$. Car $B$ is accelerating to the left at a constant rate $a_B = 0.5 \, \text{m/s}^2$ and at $t = 0$ its speed $v_B = 20 \, \text{m/s}$. Determine **(a)** the position $x_m$ where the vehicles meet one another along the highway, and **(b)** the velocity of truck $A$ relative to car $B$ at that instant.

## SOLUTION

**(a)** Choose the origin of the coordinate system as the position of the truck $A$ at $t = 0$. Positive is taken to the right and negative to the left.

For truck $A$:

$$a_A = 0.2 \, \text{m/s}^2 \tag{a}$$

$$v_A = \int 0.2 \, dt + C_1$$

$$= 0.2t + C_1$$

$v_A = 15 \, \text{m/s}$ at $t = 0$, so $C_1 = 15$ and

$$v_A = (0.2t + 15) \, \text{m/s} \tag{b}$$

Also,

$$x_A = \int (0.2t + 15) \, dt + C_2$$

$$= 0.1t^2 + 15t + C_2$$

$x_A = 0$ at $t = 0$, so $C_2 = 0$ and

$$x_A = (0.1t^2 + 15t) \, \text{m} \tag{c}$$

For car $B$:

$$a_B = -0.5 \, \text{m/s}^2 \tag{d}$$

Therefore,

$$v_B = \int (-0.5) \, dt + C_3$$

$$= -0.5t + C_3$$

$v_B = -20 \, \text{m/s}$ at $t = 0$, so $C_3 = -20$ and

$$v_B = (-0.5t - 20) \, \text{m/s} \tag{e}$$

**FIGURE 11-14**

Relative motion of two approaching vehicles

Also,

$$x_B = -\int (0.5t + 20)\, dt + C_4$$
$$= -0.25t^2 - 20t + C_4$$

$x_B = +2000\,\text{m}$ at $t = 0$, so $C_4 = 2000$ and

$$x_B = (-0.25t^2 - 20t + 2000)\,\text{m} \qquad \textbf{(f)}$$

Setting $x_A = x_B = x_m$ from Eqs. (c) and (f) yields an equation for the time $t_m$ when the vehicles meet, given by

$$0.35t_m^2 + 35t_m - 2000 = 0 \qquad \textbf{(g)}$$

Solving Eq. (g) we obtain

$$t_m = 40.7\,\text{s}$$
$$x_m = 772\,\text{m}$$

**(b)** At $t = t_m = 40.7\,\text{s}$,

$$v_A = 0.2(40.7) + 15 = 23.1\,\text{m/s}$$

and

$$v_B = -0.5(40.7) - 20 = -40.4\,\text{m/s}$$

Then

$$v_{A/B} = v_A - v_B = 23.1\,\text{m/s} - (-40.4\,\text{m/s})$$
$$v_{A/B} = 63.5\,\text{m/s}$$

EXAMPLE 11-9   475

# EXAMPLE 11-10

The truck shown in Fig. 11-15 is used for hoisting crate $B$ from a ship by a pulley system. The cable of length $L$ is inextensible and all pulleys have a radius $r$. The truck starts from rest at $x = 0$ and accelerates to the right at a constant rate of $6\,\text{ft/s}^2$ until it has traveled a distance of 75 ft to the right. For the instant the truck reaches $x = 75\,\text{ft}$, determine **(a)** the velocity of the truck, and **(b)** the velocity of crate $B$.

## SOLUTION

**(a)** For the truck,

$$a = 6\,\text{ft/s}^2 \tag{a}$$

Since $v = \int a\,dt + C_1$,

$$v = 6t + C_1$$

At $t = 0$, $v = 0$, so $C_1 = 0$. In general,

$$v = 6t\,\text{ft/s} \tag{b}$$

Integrating again,

$$x = \int v\,dt + C_2 = \int 6t\,dt + C_2$$
$$= 3t^2 + C_2$$

At $t = 0$, $x = 0$, so $C_2 = 0$. In general,

$$x = 3t^2\,\text{ft} \tag{c}$$

Denoting the time for which $x = 75\,\text{ft}$ as $t = t_f$, Eq. (c) yields

$$75 = 3t_f^2 \quad \text{or} \quad t_f = 5\,\text{s}$$

The truck's velocity according to Eq. (b) is

$$v = 6(5)\,\text{ft/s}$$
$$= 30\,\text{ft/s}$$

**(b)** The cable's length (starting from the truck and working through the pulley system toward point $P$) is given by

$$L = x + \frac{1}{2}\pi r + h + \pi r + (x_B - 1) + \pi r + (x_B - 1) + \pi r + x_B$$

or simply,

$$3x_B + x = \text{constant}$$

Differentiating this expression yields

$$3v_B + v = 0 \quad \text{or} \quad v_B = -\frac{1}{3}v \tag{d}$$

Note that, since $x_B$ is shown positive downward, velocity $v_B$ is also positive downward because $v_B = +dx_B/dt$. Therefore, when the truck reaches $x = 75\,\text{ft}$,

$$v_B = -\frac{30\,\text{ft/s}}{3}$$
$$= -10\,\text{ft/s}$$

**FIGURE 11-15**

Truck hoisting a crate with pulley system

**11–29** The positions of two particles $A$ and $B$ are defined by $x_A = 3t^2 + 4$ and $x_B = 3t - 2$, where $x$ is in meters and $t$ is in seconds. Determine the relative velocity $v_{B/A}$ and relative acceleration $a_{B/A}$ at $t = 3$ s.

**11–30** The positions of two particles $A$ and $B$ are defined by $x_A = t^3 - 2t$ and $x_B = 2t^3 + t^2 - 5$, where $x$ is in feet and $t$ is in seconds. Determine the relative velocity $v_{A/B}$ and relative acceleration $a_{A/B}$ at $t = 2$ s.

**11–31** The motions of two particles $A$ and $B$ are defined by $x_A = 4t + 2$ and $v_B = 3t^2 + t$, where $x$ is in feet, $v$ is in ft/s, and $t$ is in seconds. Determine $x_{B/A}$, $v_{B/A}$, and $a_{B/A}$ at $t = 10$ s. When $t = 0$, $x_B = 1$ ft.

**11–32** The velocities of two particles $A$ and $B$ are defined by $v_A = 4t^2 + t$ and $v_B = 3t - 4$, where $v$ is in m/s and $t$ is in seconds. Determine the relative position $x_{B/A}$ and relative acceleration $a_{B/A}$ at $t = 5$ s. When $t = 0$, $x_A = x_B$.

**11–33** The motions of two particles $A$ and $B$ are defined by $a_A = 20$ and $a_B = t + 2$, where $a$ is in m/s$^2$ and $t$ is in seconds. Determine $x_{A/B}$ and $v_{A/B}$ at $t = 5$ s. At $t = 0$, $x_A = x_B = v_A = v_B = 0$.

**11–34** The motions of two particles $A$ and $B$ are defined by $a_A = 3t$ and $v_B = 2t^2$, where $a$ is in ft/s$^2$, $v$ is in ft/s, and $t$ is in seconds. Determine $x_{B/A}$ and $a_{B/A}$ at $t = 3$ s. At $t = 0$, $x_A = 1$ ft, $x_B = 3$ ft, $v_A = 2v_B = 2$ ft/s.

**11–35** Cars $A$ and $B$ are traveling toward each other at constant velocities as shown. Initially, $x_A = 0$ and $x_B = 1000$ ft. Determine when and where the cars would meet with respect to the initial configuration.

**11–36** Cars $A$ and $B$ are traveling in the same direction at constant velocities as shown. Determine the distance between the cars 10 s after the instant when they were 500 m apart.

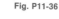

Fig. P11-36

**11–37** The police car $P$ starts from rest when car $A$ passes it at a constant speed of 40 m/s. Car $P$ accelerates at a constant rate of 1 m/s$^2$. Determine the speed of car $P$ and the distance it has traveled when it overtakes car $A$.

Fig. P11-37

**11–38** The police car $P$ is parked on the shoulder of a highway and starts from rest when car $A$ passes it at a constant speed of 60 mph (Fig. P11-37). The police car accelerates at a constant rate of 8 ft/s$^2$. Determine (a) the speed of car $P$ and the distance traveled before it overtakes car $A$, and (b) the position, velocity, and acceleration of car $P$ relative to car $A$ at $t = 10$ s.

Fig. P11-35

**11–39**  A motorist in car $A$ passes an unmarked police car $P$ at $x = 0$. At that instant car $P$ accelerates at a constant rate of $3\,\text{ft/s}^2$ and $A$ decelerates at a constant rate of $4\,\text{ft/s}^2$. Determine the time required for car $P$ to overtake car $A$.

**Fig. P11-39**

**11–40**  Airplanes $A$ and $B$ are $2\,\text{km}$ apart and traveling in the same direction at the given velocities when $B$ accelerates at a constant rate of $15\,\text{m/s}^2$. Determine the time required for $B$ to overtake $A$.

**Fig. P11-40**

**11–41**  In a simulation of automobile crashes two cars $A$ and $B$ are started from rest toward each other from a distance of $400\,\text{m}$. The cars have constant accelerations, $6\,\text{m/s}^2$ and $5\,\text{m/s}^2$, respectively. Determine the relative velocity and the position of the cars with respect to the origin just before the impact.

**Fig. P11-41**

**11–42**  Two drag racing cars $A$ and $B$ start side by side in a race to cover a quarter mile ($1320\,\text{ft}$). Assume that each car can accelerate constantly over this distance. Their accelerations are $a_A = 50.00\,\text{ft/s}^2$ and $a_B = 50.02\,\text{ft/s}^2$. Determine the relative velocity and distance of the cars when car $B$ crosses the finish line. Assume that there is no lateral distance between the cars.

**11–43**  Blocks $A$ and $B$ are initially at rest when $y_A = 3\,\text{m}$ and $y_B = 2\,\text{m}$. Block $B$ is accelerated downward at a constant rate of $0.5\,\text{m/s}^2$. Determine the relative vertical position and velocity of the blocks $3\,\text{s}$ after the motion begins.

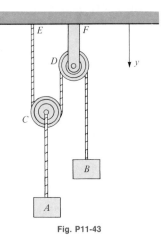

**Fig. P11-43**

**11–44**  The three blocks are initially at rest when $y_A = 5\,\text{ft}$, $y_B = 6\,\text{ft}$, and $y_C = 8\,\text{ft}$. Blocks $B$ and $C$ are both accelerated upward at a constant rate of $3\,\text{ft/s}^2$. Determine the relative vertical position and velocity of blocks $A$ and $B$ at a time $2\,\text{s}$ after the motion begins.

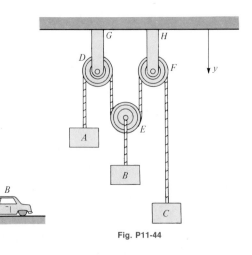

**Fig. P11-44**

**Rectilinear position, velocity, and acceleration**

## AVERAGE VELOCITY:

$$v_{ave} = \frac{\Delta x}{\Delta t}$$

where $\Delta x$ = net displacement
$\Delta t$ = total time interval

## INSTANTANEOUS VELOCITY:

$$v = \lim_{\Delta t \to 0} \frac{\Delta x}{\Delta t} = \frac{dx}{dt} = \dot{x}$$

where $x$ = position coordinate
$t$ = time

## AVERAGE ACCELERATION:

$$a_{ave} = \frac{\Delta v}{\Delta t}$$

where $\Delta v$ = net change in velocity
$\Delta t$ = total time interval

## INSTANTANEOUS ACCELERATION:

$$a = \lim_{\Delta t \to 0} \frac{\Delta v}{\Delta t} = \frac{dv}{dt} = \dot{v} = \ddot{x}$$

Section 11-2

**Measured and derived quantities as functions of time**
See representative example in the figure on page 480.

## TYPICAL FEATURES

**1.** $v = 0$ is at a local maximum or minimum of the displacement–time curve.
**2.** $a = 0$ is at a local maximum or minimum of the velocity–time curve.
**3.** The area under the velocity–time curve in a specified time interval equals the net displacement in that interval.
**4.** The area under the acceleration–time curve in a specified time interval equals the net change in velocity in that interval.

**EXAMPLE:**

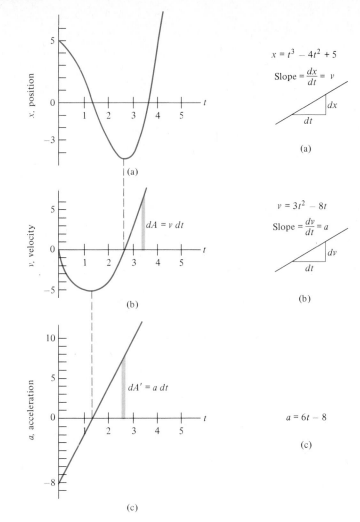

(a)

(b)

(c)

$x = t^3 - 4t^2 + 5$

$\text{Slope} = \dfrac{dx}{dt} = v$

(a)

$v = 3t^2 - 8t$

$\text{Slope} = \dfrac{dv}{dt} = a$

(b)

$a = 6t - 8$

(c)

Section 11-3

**Useful expressions based on acceleration**

**FOR PROBLEMS INVOLVING VELOCITY, ACCELERATION, AND TIME:**

$$a = \frac{dv}{dt} \quad \Rightarrow \quad \int_{v_0}^{v} dv = \int_{0}^{t} a\, dt$$

**FOR PROBLEMS INVOLVING VELOCITY, ACCELERATION, AND DISPLACEMENT:**

$$v\, dv = a\, dx \quad \Rightarrow \quad \int_{v_0}^{v} v\, dv = \int_{x_0}^{x} a\, dx$$

**Special cases**

1. For constant acceleration $a$:

$$\int_{v_0}^{v} dv = a \int_{0}^{t} dt \quad \Rightarrow \quad v = at + v_0$$

$$\int_{v_0}^{v} dv = a \int_{x_0}^{x} dx \quad \Rightarrow \quad v^2 = 2a(x - x_0) + v_0^2$$

Position $x = \frac{1}{2}at^2 + v_0 t + x_0$.

2. For variable acceleration $a = f(t)$:

$$\int_{v_0}^{v} dv = \int_{0}^{t} f(t)\, dt \quad \Rightarrow \quad v = \int_{0}^{t} f(t)\, dt + v_0$$

$$\int_{x_0}^{x} dx = \int_{0}^{t} v\, dt \quad \Rightarrow \quad x = \int_{0}^{t} v\, dt + x_0$$

3. For variable acceleration $a = f(x)$:

$$\int_{v_0}^{v} v\, dv = \int_{x_0}^{x} f(x)\, dx \quad \Rightarrow \quad v^2 = 2 \int_{x_0}^{x} f(x)\, dx + v_0^2$$

4. For variable acceleration $a = f(v)$:

$$a = \frac{dv}{dt} \quad \Rightarrow \quad dt = \frac{dv}{f(v)} \quad \Rightarrow \quad t = \int_{0}^{t} dt = \int_{v_0}^{v} \frac{dv}{f(v)}$$

$$a = f(v) = v \frac{dv}{dx} \quad \Rightarrow \quad x = \int_{0}^{x} dx = \int_{v_0}^{v} \frac{v\, dv}{f(v)}$$

## Section 11-4

**Relative motion of particles $A$ and $B$**
The positions, velocities, and accelerations of $A$ and $B$ are given with respect to a fixed point on the line of the particles as $x_A$, $v_A$, $a_A$, and $x_B$, $v_B$, $a_B$, respectively.

**RELATIVE POSITION:**

$$x_{B/A} = x_B - x_A$$

**RELATIVE VELOCITY:**

$$v_{B/A} = v_B - v_A$$

**RELATIVE ACCELERATION:**

$$a_{B/A} = a_B - a_A$$

*Note:* The following equations refer to position, velocity and accelerations.

$$x_{A/B} = -x_{B/A}$$

$$v_{A/B} = -v_{B/A}$$

etc.

The degrees of freedom of a mechanical system is the number of independent coordinates required to define the position of the system.

**11–45** The position of a particle $P$ moving along a straight line is given by $x = 3t^2 - 18t + 4$, where $x$ is in meters and $t$ in seconds. Determine (a) the position, velocity, and acceleration of $P$ at $t = 0$ and $t = 4$ s, (b) the displacement from $t = 0$ to $t = 4$ s, and (c) the total distance traveled in part (b).

**11–46** The rectilinear motion of a particle is defined by $y = v^2/25$, where $y$ is in ft and $v$ is in ft/s. At time $t = 0$, $v_0 = 4$ ft/s. Determine the position, velocity, and acceleration of the particle when $t = 4$ s.

**Fig. P11-50**

Test sled of crash simulator. Courtesy MTS Systems Corporation, Minneapolis, Minnesota

**11–47** The acceleration of a particle is defined by the expression $a = 4 \sin t$, where $a$ is in m/s² and $t$ is in seconds. At time $t = 0$ the position is $x_0 = 0$ and the velocity is $v_0 = 8$ m/s. Determine the first position past the origin, $x_1$, where the velocity is maximum.

**11–48** A package is released from a hovering helicopter. The downward acceleration of the package is estimated as $(32.2 - 0.02 v^2)$ ft/s² where $v$ is in ft/s. Determine the velocity of the package after it has fallen (a) 200 ft, and (b) 400 ft.

**11–49** A particle falling toward a planet is accelerated by the gravity of that planet according to $a = -g_p(R^2/r^2)$, where $g_p$ is the acceleration of gravity, measured at the surface, $R$ is the radius of the planet, and $r$ is the distance from the center of the spherical planet to the particle. Derive an equation for the velocity $v$ at which a particle freely falling from a height $h$ above the surface would hit the surface.

**11–50** Consider the crash simulator shown in Fig. P11–50. The track is horizontal, and the sled is uniformly accelerated ($a = $ const.) over a distance of 4 m from $v_0 = 0$ to $v = 100$ km/h. The sled is then uniformly decelerated and stoppped in a distance of 1 m. Calculate the total time elapsed during the motion of the sled.

**11–51** Two accelerometers $A$ and $B$ are mounted on an airplane fuselage for measurements in a structural test (two such transducers are shown in proximity in Fig. P11–51). $A$ is located at $x_A = 10$ ft, and $B$ at $x_B = 15$ ft. Thus, $x_{B/A} = 5$ ft initially. Starting at the same time, the acceleration outputs are $a_A = (\pi^2/5000) \sin (\pi t/10)$ ft/s² and $a_B = -(\pi^2/6000) \sin (\pi t/12)$ ft/s². Determine the value of $x_{B/A}$ at $t = 10$ s. Assume that all motions are along the $x$ axis and that $v_A = v_B = 0$ at $t = 0$.

**Fig. P11-51**

Two accelerometers on an airplane fuselage. The Boeing Company, Seattle, Washington

# Kinematics of Particles: Vector Analysis

**12**

Velocity and acceleration are vector quantities because they have magnitudes and directions. The vector methods that are developed in this chapter are useful in the analysis of a wide range of dynamics problems. These methods are most advantageous when solving complex problems in either two-dimensional or three-dimensional motion. Students are strongly advised to thoroughly study the basic principles and methods of vector analysis.

## OVERVIEW

## STUDY GOALS

**SECTION 12-1** introduces the mathematical method of determining the velocity and acceleration vectors from the position vector of a particle which moves along a curved path. This analysis is the basis for studying all motions that are more complex than rectilinear motion. The concepts in Secs. 11-1 and 12-1 are readily correlated.

**SECTION 12-2** presents an analysis of the physical aspects of a particle's motion along a curved path. Several concepts and definitions given here are essential for most work in dynamics.

**SECTION 12-3** presents the concepts and analysis of tangential and normal components of the velocity and acceleration of a particle traveling on a curved path. These components are frequently used in dynamics, so they should be studied thoroughly. The vector expressions of the velocity and acceleration in curvilinear motion (Eqs. 12-8 and 12-11) and their implications are especially worthwhile to remember.

**SECTION 12-4** extends the concepts of normal and tangential components of the velocity and acceleration to define radial and transverse components in polar coordinates. These coordinates are often convenient in solving problems of plane motion. The analysis in this section also provides useful practice with time derivatives of unit vectors.

**SECTION 12-5** presents cylindrical coordinates which are used for the analysis of three-dimensional motion. The concept and application of these coordinates are based on polar coordinates. The additional mathematical complexity is minimal.

**SECTION 12-6** presents the derivation of the velocity and acceleration of a particle using spherical coordinates. This derivation provides useful experience in three-dimensional vector mechanics, but it is quite difficult and is omitted from many courses.

**SECTION 12-7** describes the vector analysis of absolute and relative motions of two particles using rectangular coordinate systems in translation. The radial and transverse components of velocity and acceleration studied in Sec. 12-4 using polar coordinates are applied in solving some problems of relative motion.

(a)

(b)

**FIGURE 12-1**

Position vector **r** in rectangular coordinates

## CURVILINEAR MOTION. RECTANGULAR CARTESIAN COORDINATES

The motion of a particle along a curved path is classified as *plane* (two-dimensional) or as *space* (three-dimensional) *curvilinear motion*. The basic concepts for analyzing all such motions can be established by considering space motion of a particle. The mathematical method developed here is simple and immediately allows the solution of some problems of curvilinear motion. A more detailed analysis is presented in Sec. 12-2 to show the physical concepts of curvilinear motion that must be understood for solving a larger variety of problems.

### Rectangular Cartesian Coordinates

The fixed rectangular Cartesian coordinate system is useful for analyzing curvilinear motion. Consider that the position vector **r** of a particle $P$ is given in terms of rectangular components $x$, $y$, and $z$ as in Fig. 12-1a. With the unit vectors **i**, **j**, and **k**,

$$\mathbf{r} = x\mathbf{i} + y\mathbf{j} + z\mathbf{k} \qquad \boxed{12\text{-}1}$$

Assume that $x$, $y$ and $z$ are differentiable functions of time $t$. In that case the velocity and the acceleration can be written as

$$\mathbf{v} = \frac{d\mathbf{r}}{dt} = \frac{dx}{dt}\mathbf{i} + \frac{dy}{dt}\mathbf{j} + \frac{dz}{dt}\mathbf{k} = \dot{x}\mathbf{i} + \dot{y}\mathbf{j} + \dot{z}\mathbf{k} \qquad \boxed{12\text{-}2}$$

$$\mathbf{a} = \frac{d\mathbf{v}}{dt} = \frac{dv_x}{dt}\mathbf{i} + \frac{dv_y}{dt}\mathbf{j} + \frac{dv_z}{dt}\mathbf{k} = \ddot{x}\mathbf{i} + \ddot{y}\mathbf{j} + \ddot{z}\mathbf{k} \qquad \boxed{12\text{-}3}$$

since the time derivatives of the constant unit vectors in the fixed $xyz$ frame are zero; $v_x = \dot{x}$, $a_x = \ddot{x}$, and so on, are the scalar components of the corresponding vectors. The signs of these quantities denote the directions of the components in the positive or negative coordinate directions. Thus, curvilinear motion in general is the vector sum of orthogonal motions occurring simultaneously in the $x$, $y$, and $z$ directions. For example, the absolute velocity **v** of particle $P$ in Fig. 12-1b is the vector sum of velocities $v_x\mathbf{i}$, $v_y\mathbf{j}$, and $v_z\mathbf{k}$.

In a reverse procedure, the acceleration or velocity components may be individually integrated with respect to time to obtain velocity or position components, respectively.

## PHYSICAL ASPECTS OF VELOCITY AND ACCELERATION IN CURVILINEAR MOTION

To understand the physical aspects of curvilinear motion, it is necessary to analyze the changes of the position and velocity vectors in detail. The effects of these changes on the acceleration of a particle are complex, but it is essential to study them in order to solve many problems in dynamics.

Consider a particle that is at position $P$ at time $t$ and at position $P'$ at time $t + \Delta t$ as in Fig. 12-2a. The two positions are defined by vectors $\mathbf{r}$ and $\mathbf{r}'$ with respect to the origin of a fixed reference frame. The vector $\Delta \mathbf{r}$ defines the change of the position vector $\mathbf{r}$ to $\mathbf{r}'$, while $\Delta s$ represents the actual displacement of the particle along the path of motion.

The *average velocity* of the particle over the time interval $\Delta t$ is defined as the quotient $\Delta \mathbf{r} / \Delta t$. This is a vector in the direction of $\Delta \mathbf{r}$ since $\Delta t$ is a scalar. The *instantaneous velocity* $\mathbf{v}$ of the particle at time $t$ is defined at the limit as

$$\mathbf{v} = \lim_{\Delta t \to 0} \frac{\Delta \mathbf{r}}{\Delta t} = \frac{d\mathbf{r}}{dt}$$

12-4

As $\Delta t$ approaches zero, point $P'$ approaches point $P$, and the vector $\Delta \mathbf{r}$ becomes tangent to the path of the particle. Consequently, *the velocity vector $\mathbf{v}$ at any time $t$ is tangent to the path of the particle* as in Fig. 12-2b. This fact will be very useful in the analysis of curvilinear motion.

The magnitude $v$ of the vector $\mathbf{v}$ is called the *speed* of the particle. It may also be defined by the quotient of the scalar quantities $\Delta s$ and $\Delta t$ at the limit,

$$v = \lim_{\Delta t \to 0} \frac{\Delta s}{\Delta t} = \frac{ds}{dt}$$

12-5

where $s$ is a length along the path of the particle.

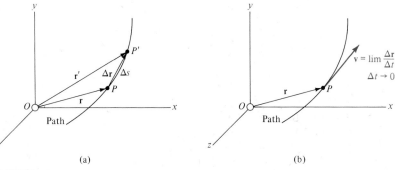

(a)                                    (b)

**FIGURE 12-2**

Definition of a particle's velocity

(a)

(b)

(c)

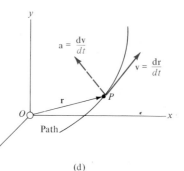

(d)

**FIGURE 12-3**

Definition of a particle's acceleration

The acceleration of the particle is analyzed by considering the velocity vectors at two nearby positions along the path as in Fig. 12-3a. For convenience, the vectors **v** and **v'** are redrawn with a common origin $O'$ in Fig. 12-3b. The vector $\Delta\mathbf{v}$ defines the change of vector **v** to **v'**.

The *average acceleration* of the particle over the time interval $\Delta t$ is defined as the quotient $\Delta\mathbf{v}/\Delta t$. This is a vector in the direction of $\Delta\mathbf{v}$ since $\Delta t$ is a scalar. The *instantaneous acceleration* **a** of the particle at time $t$ is defined at the limit as

$$\mathbf{a} = \lim_{\Delta t \to 0} \frac{\Delta\mathbf{v}}{\Delta t} = \frac{d\mathbf{v}}{dt}$$

$$\boxed{12\text{-}6}$$

The acceleration vector is tangent to the curve described by the tip of the velocity vector **v**. This curve, called the *hodograph* of the motion, is illustrated in Fig. 12-3c. Thus, *the acceleration is not tangent to the path of the particle and is not collinear with the velocity vector in curvilinear motion.* A general case is shown schematically in Fig. 12-3d.

## 12-3 TANGENTIAL AND NORMAL COMPONENTS

Since the velocity vector is always tangent to the path of the particle, it is often convenient to define the velocity and acceleration of a particle using components along the tangent and the normal to the curved path and in the plane of that path. The concepts of this approach are established using plane motion for clarity, but they are valid for space motion as well.

# Curvilinear Plane Motion

The tangential and normal dir~~ections~~ ... l by unit
vectors $\mathbf{n}_t$ and $\mathbf{n}_n$ ... direction
of travel, and $\mathbf{n}_n$ ... curvature
of the path. The ... particle
moves along the ...

The tangen... on origin
$O'$ in Fig. 12-4b. ... r $\mathbf{n}_t$ to $\mathbf{n}_t'$
in an infinitesima... n for the
magnitude of $d\mathbf{n}_t$ ... that $d\mathbf{n}_t$
and $\mathbf{n}_n$ are in th... n. Thus,
$d\mathbf{n}_t = d\theta\,\mathbf{n}_n$, or

12-7

This result is usef... ngential
and normal comp...

The velocity ... th (Sec.
12-2), so it can be

12-8

The acceleration i... pect to
time $t$,*

12-9

In an effort to elir... ~~unit vector~~ $\mathbf{n}_t$, the following substitutions are found useful:

$$\frac{d\mathbf{n}_t}{dt} = \frac{d\mathbf{n}_t}{d\theta}\frac{d\theta}{ds}\frac{ds}{dt}$$

where $\dfrac{d\mathbf{n}_t}{d\theta} = \mathbf{n}_n$ from Eq. 12-7

$\dfrac{d\theta}{ds} = \dfrac{1}{\rho} =$ curvature of the path from Fig. 12-5
$\qquad (\rho =$ radius of curvature$)$

$\dfrac{ds}{dt} = v$ from Eq. 12-5

Thus,

$$\frac{d\mathbf{n}_t}{dt} = \frac{v}{\rho}\,\mathbf{n}_n \qquad \boxed{12\text{-}10}$$

* The standard rule for the differentiation of products of scalar functions is extended
to the product of a scalar function $A(t)$ and a vector function $\mathbf{B}(t)$ of the same variable $t$:

$$\frac{d(A\mathbf{B})}{dt} = \lim_{\Delta t \to 0} \frac{(A + \Delta A)(\mathbf{B} + \Delta\mathbf{B}) - A\mathbf{B}}{\Delta t}$$

$$= \lim_{\Delta t \to 0}\left(\frac{\Delta A}{\Delta t}\mathbf{B} + A\frac{\Delta\mathbf{B}}{\Delta t}\right) = \frac{dA}{dt}\mathbf{B} + A\frac{d\mathbf{B}}{dt}$$

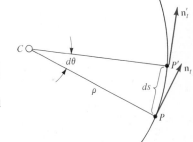

**FIGURE 12-4**

Unit vectors in curvilinear plane
motion

**FIGURE 12-5**

Quantities used in defining the
curvature of the path

**FIGURE 12-6**

Normal and tangential accelerations

and the acceleration can be expressed from Eq. 12-9 as

$$\mathbf{a} = \frac{dv}{dt}\mathbf{n}_t + \frac{v^2}{\rho}\mathbf{n}_n$$

12-11

where the tangential and normal components of the acceleration are

$$a_t = \frac{dv}{dt} \qquad a_n = \frac{v^2}{\rho}$$

12-12

Note that $a_t$ is equal to the time rate of change of the speed of the particle and may be positive or negative, depending on whether the speed along the path is increasing or decreasing, while $a_n$ is always positive and $\mathbf{a}_n$ is directed toward the center of curvature of the path. These components are illustrated in Fig. 12-6.

## Curvilinear Space Motion

The analysis of motion in space is made as an extension of the concepts used for motion in a plane. The connection between the two-dimensional and three-dimensional situations is based on the fact that at each point of interest on the path of the particle, a plane can be drawn to include at least a small part of the path in that plane. Thus, the motion of the particle in that neighborhood can be analyzed using the methods established for plane motion.

It is necessary to define the directions of the unit vectors $\mathbf{n}_t$ and $\mathbf{n}_n$. As in plane motion, $\mathbf{n}_t$ is tangent to the path of the particle and is in the direction of travel; $\mathbf{n}_n$ is normal to $\mathbf{n}_t$ and is in the direction of the center of curvature of the small segment of path in the neighborhood of the particle. The plane containing the unit vectors $\mathbf{n}_t$ and $\mathbf{n}_n$ at a given position of the particle is called the *osculating plane*. Naturally, there is an infinite number of osculating planes as a particle moves along a three-dimensional path. Such a plane is illustrated in Fig. 12-7 for a particle moving upward along a helical path. Here the path segment and the unit vectors at $P$ are drawn in the same color as the parallelogram representing the osculating plane.

Osculating plane contains short segment of path at $P$ and the unit vectors $\mathbf{n}_t$ and $\mathbf{n}_n$ at that point

**FIGURE 12-7**

Definition of osculating plane

EXAMPLE 12-1

The recording pen $P$ is being used to draw the line $QP$ on graph paper by an automatic $x$-$y$ recorder shown in Figure 12-8a. The velocity of carriage $AB$ is given as $\dot{x} = (2t + 4)$ ft/s and the velocity of the pen relative to carriage $AB$ is $\dot{y} = (2/y)$ ft/s. At time $t = 0$ the pen is at the position $(x, y) = (1\text{ ft}, 0)$. Determine (a) the equation for the shape of the graph, (b) the velocity and acceleration of point $P$ at $t = 2$ s, and (c) the slope of the graph at $t = 2$ s.

## SOLUTION

(a) Since $\dot{x} = 2t + 4$, integration yields

$$x = t^2 + 4t + C_1$$

At $t = 0$, $x = 1$, so $C_1 = 1$. Then

$$x = t^2 + 4t + 1 \qquad \text{(a)}$$

Also, $\dot{y} = 2/y = dy/dt$, which can be written as $dt = \frac{1}{2}y\, dy$. Then

$$t = \frac{1}{4}y^2 + C_2 \qquad \text{(b)}$$

where $C_2 = 0$ because $y = 0$ at $t = 0$. Therefore, eliminating time as a variable from Eqs. (a) and (b), we obtain the shape of the graph as

$$x = \left(\frac{1}{16}y^4 + y^2 + 1\right) \text{ft}$$

(b) Since $\dot{x} = 2t + 4$, differentiation yields

$$\ddot{x} = 2$$

At $t = 2$ s, $\qquad\qquad \dot{x} = 2(2) + 4 = 8$ ft/s

$$\ddot{x} = 2 \text{ ft/s}^2$$

Also, since $y = 2t^{1/2}$, $\qquad\qquad \dot{y} = t^{-1/2}$

$$\ddot{y} = -\frac{1}{2}t^{-3/2}$$

At $t = 2$ s,

$$\dot{y} = \frac{1}{\sqrt{2}} = 0.707 \text{ ft/s}$$

$$\ddot{y} = -\frac{1}{2\sqrt{8}} = -0.177 \text{ ft/s}^2$$

Then the velocity and acceleration vectors are

$$\mathbf{v} = (8\mathbf{i} + 0.707\mathbf{j}) \text{ ft/s}$$
$$\text{(at } t = 2 \text{ s)}$$
$$\mathbf{a} = (2\mathbf{i} - 0.177\mathbf{j}) \text{ ft/s}^2$$

(c) Since the velocity vector $\mathbf{v}$ is tangent to the curve, its direction defines the slope (Fig. 12-8b). Using the chain rule,

$$\frac{dy}{dx} = \frac{dy/dt}{dx/dt} = \frac{\dot{y}}{\dot{x}}$$

At $t = 2$ s,

$$\frac{dy}{dx} = \frac{0.707}{8}$$

$$= 0.0884$$

(a)

(b)

**FIGURE 12-8**

Automatic $x$-$y$ recorder

EXAMPLE 12-1    489

## EXAMPLE 12-2

Supplies are dropped from an airplane to land at point $A$ in Fig. 12-9. The velocity of the package at release from the airplane is the speed of the airplane, $\mathbf{v}_0 = 200\mathbf{i}$ ft/s. The acceleration of the package due to gravity is $\mathbf{a} = -32.2\mathbf{j}$ ft/s². (a) Determine the required distance $x$ for the release of the package neglecting any effect of air on the package. (b) Next, change the initial velocity to $\mathbf{v}_0 = (200\mathbf{i} + 10\mathbf{j})$ ft/s.

### SOLUTION

(a) The velocity in the $x$ direction is constant throughout the fall since there is no acceleration in that direction. By finding the time for a vertical fall alone and multiplying it by the velocity in the $x$ direction, the displacement in the $x$ direction can be found. Using Eq. 11-10 for constant acceleration in the $y$ direction,

$$y - y_0 = v_{0_y}t + \frac{1}{2}a_y t^2$$

$$0 - 300 = 0t + \frac{1}{2}(-32.2)t^2$$

$$\left(\frac{300}{16.1}\right)^{1/2} = t$$

$$4.32\,\text{s} = t$$

Then

$$x - x_0 = v_{0_x}t + \frac{1}{2}a_x t^2$$

$$x - 0 = v_{0_x}t + 0$$

$$x = (200\,\text{ft/s})(4.32\,\text{s}) = 864\,\text{ft}$$

(b) For $v_{0_y} = 10$ ft/s,

$$y - y_0 = v_{0_y}t + \frac{1}{2}a_y t^2$$

$$0 - 300 = 10t + \frac{1}{2}(-32.2)t^2$$

$$16.1t^2 - 10t - 300 = 0$$

$$t = \frac{10 \pm \sqrt{100 + 19{,}300}}{32.2}$$

$$= \frac{10 \pm 139}{32.2}$$

$$= 4.63\,\text{s} \quad \text{and} \quad -4.01\,\text{s}$$

Using the positive root,   $x = (4.63\,\text{s})(200\,\text{ft/s}) = 926\,\text{ft}$

The negative root is not in the domain of the definition of the problem.

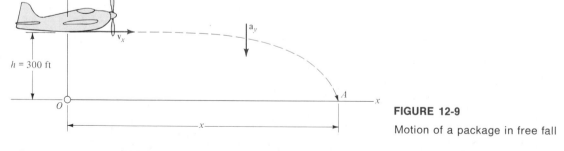

**FIGURE 12-9**

Motion of a package in free fall

This package is in *projectile motion*, for which rectangular coordinates are most useful. By neglecting air resistance, the projectile is assumed to have constant velocity in the horizontal direction ($x$) and constant acceleration in the vertical direction ($y$). Using axes as shown in Fig. 12-9, the equations of projectile motion are derived from Eqs. 11-8, 11-9, and 11-10 as follows:

$$v_y = v_{0y} - gt$$

$$v_y^2 = v_{0y}^2 - 2g(y - y_0)$$

$$y = y_0 + v_{0y}t - \tfrac{1}{2}gt^2$$

$$x = v_{0x}t$$

Note that the acceleration due to gravity is downward, so it is negative in the equations.

# EXAMPLE 12-3

A racing car is traveling at a speed of 50 m/s on a circular track (Fig. 12-10). The driver, anticipating a straight section of the track, is accelerating the car at the rate of 2 m/s². Determine the total acceleration of the car at this instant in the coordinate system shown.

## SOLUTION

**TANGENTIAL ACCELERATION:**   The magnitude of this component is given as $a_t = 2$ m/s². The vector expression is obtained by inspection,

$$\mathbf{a}_t = (2\text{ m/s}^2)(-\cos 20° \,\mathbf{i} + \sin 20° \,\mathbf{j})$$

$$= (-1.88\mathbf{i} + 0.68\mathbf{j})\text{ m/s}^2$$

**NORMAL ACCELERATION:**   From Eq. 12-12, and by inspection of the direction of $\mathbf{n}_n$,

$$\mathbf{a}_n = \frac{v^2}{\rho}\,\mathbf{n}_n = \frac{50^2\text{ (m/s)}^2}{200\text{ m}}(-\cos 70° \,\mathbf{i} - \sin 70° \,\mathbf{j})$$

$$= (-4.28\mathbf{i} - 11.75\mathbf{j})\text{ m/s}^2$$

**TOTAL ACCELERATION:**   The net acceleration is

$$\mathbf{a} = \mathbf{a}_t + \mathbf{a}_n = (-6.16\mathbf{i} - 11.07\mathbf{j})\text{ m/s}^2$$

The results are shown in Fig. 12-10b.

**COMMON ERROR**

A common error in working problems of curvilinear motion is to neglect the normal component of the acceleration. This mistake occurs most often when the particle is traveling with a *constant speed* along a curved path. Note in this example that even if the car were not accelerating along the path, the normal component of the acceleration would still exist due to the curvature of the track.

(a)

(b)

**FIGURE 12-10**

Motion of a car on a circular track

EXAMPLE 12-3   491

# EXAMPLE 12-4

The position of a charged particle moving in a horizontal plane is measured electronically. This information is fed into a computer which employs curve-fitting techniques to generate an analytical expression for its position given by $\mathbf{r} = t^3\mathbf{i} + t^4\mathbf{j}$, where $\mathbf{r}$ is in meters and $t$ in seconds. For $t = 1$ s, determine **(a)** the acceleration of the particle expressed in rectangular components, **(b)** its normal and tangential components of acceleration, and **(c)** the radius of curvature of the path.

## SOLUTION

**(a)** For a general time $t$,

$$\mathbf{r} = (t^3\mathbf{i} + t^4\mathbf{j}) \text{ m}$$

$$\dot{\mathbf{r}} = \mathbf{v} = (3t^2\mathbf{i} + 4t^3\mathbf{j}) \text{ m/s}$$

$$\ddot{\mathbf{r}} = \mathbf{a} = (6t\mathbf{i} + 12t^2\mathbf{j}) \text{ m/s}^2$$

For $t = 1$ s,

$$\mathbf{r} = (\mathbf{i} + \mathbf{j}) \text{ m}$$

$$\mathbf{v} = (3\mathbf{i} + 4\mathbf{j}) \text{ m/s}$$

$$\mathbf{a} = (6\mathbf{i} + 12\mathbf{j}) \text{ m/s}^2$$

The position, velocity, and acceleration vectors at $t = 1$ s are shown in Fig. 12-11.

**(b)** Since the velocity vector is always tangent to the path of a particle, its direction defines the tangential direction. The unit vector in the tangential direction is given by

$$\mathbf{n}_t = \frac{\mathbf{v}}{v}$$

and the tangential component of acceleration is

$$\mathbf{a}_t = a_t\mathbf{n}_t$$

where $a_t = \mathbf{a} \cdot \mathbf{n}_t$. At $t = 1$ s,

$$\mathbf{n}_t = \frac{\mathbf{v}}{v} = \frac{3\mathbf{i} + 4\mathbf{j}}{\sqrt{3^2 + 4^2}} = \frac{3}{5}\mathbf{i} + \frac{4}{5}\mathbf{j}$$

Then

$$a_t = (6\mathbf{i} + 12\mathbf{j}) \cdot \left(\frac{3}{5}\mathbf{i} + \frac{4}{5}\mathbf{j}\right)$$

and

$$\mathbf{a}_t = 13.2\,\mathbf{n}_t \text{ m/s}^2$$

Also, since $a^2 = a_t^2 + a_n^2$,

$$a_n^2 = a^2 - a_t^2 = [(6 \text{ m/s}^2)^2 + (12 \text{ m/s}^2)^2] - (13.2 \text{ m/s}^2)^2$$

Then

$$\mathbf{a}_n = 2.4\,\mathbf{n}_n \text{ m/s}^2$$

**(c)** At $t = 1$ s, $a_n = 2.4 \text{ m/s}^2 = v^2/\rho$ where $v = 5$ m/s, so

$$\rho = \frac{v^2}{a_n} = \frac{(5 \text{ m/s})^2}{2.4 \text{ m/s}^2}$$

$$= 10.4 \text{ m}$$

**FIGURE 12-11**

Analysis of the motion of a charged particle

EXAMPLE 12-5

In a packaging assembly operation small containers are deflected by a parabolic plate which is parallel to the ground (Fig. 12-12). What is the maximum speed $v$ of the containers if their total acceleration caused by the deflector plate should not exceed 200 in./s²? The plate does not change the speed of the containers.

## SOLUTION

The magnitude of the total acceleration vector from Eq. 12-11 is

$$a = \left[\left(\frac{dv}{dt}\right)^2 + \left(\frac{v^2}{\rho}\right)^2\right]^{1/2}$$

Since all motion is occurring in a plane horizontal to the ground, the speed of each package is constant as it moves along the path, and $dv/dt = 0$. Therefore,

$$a_{max} = \frac{v^2}{\rho} = 200 \text{ in./s}^2$$

Since $v$ is constant, $a$ has its largest magnitude when the radius of curvature $\rho$ is minimum. The minimum radius of curvature for a parabola is at its vertex (point $O$ in Fig. 12-12). The relation for the radius of curvature at a point is known from calculus,*

$$\rho = \frac{[1 + (dy/dx)^2]^{3/2}}{d^2y/dx^2}$$

Evaluating the appropriate derivatives at the point $O$, the minimum radius of curvature is

$$\rho = \frac{[1 + 0]^{3/2}}{1/5} = 5 \text{ in.}$$

and the maximum speed is

$$v^2 = a_{max}(\rho)$$
$$= (200 \text{ in./s}^2)(5 \text{ in.}) = 1000 \text{ in}^2/\text{s}^2$$
$$v = 31.6 \text{ in./s} = 2.64 \text{ ft/s}$$

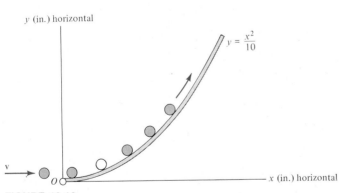

$y$ (in.) horizontal

$y = \dfrac{x^2}{10}$

$x$ (in.) horizontal

**FIGURE 12-12**

Motion of particles along a parabolic deflector

* George B. Thomas, Jr., *Calculus and Analytic Geometry*, Alternate Edition (Reading, Mass.: Addison-Wesley Publishing Co., 1972), p. 601.

EXAMPLE 12-5    493

**12–1** The position vector of a particle in plane curvilinear motion is $\mathbf{r} = [(t^2 + 3t)\mathbf{i} + (3t - 2)\mathbf{j}]$ m, where $t$ is in seconds. Determine the velocity and acceleration at $t = 3$ s.

**12–2** The position of a particle is defined by $x = 2t^3$ and $y = 4t^2 + 3$, where $x$ and $y$ are in feet and $t$ is in seconds. Determine the velocity and acceleration vectors at $t = 5$ s.

**12–3** The velocity vector of a particle is $\mathbf{v} = [20t\mathbf{i} - (8t + 3)\mathbf{j}]$ ft/s, where $t$ is in seconds. Determine the position and acceleration vectors at $t = 10$ s if initially $x_0 = 10$ ft and $y_0 = 15$ ft.

**12–4** The velocity of a particle is defined by $v_x = 2t$ and $v_y = 3t^2 + 2$, where $v_x$ and $v_y$ are in m/s and $t$ is in seconds. Determine the position and acceleration at $t = 2$ s if initially $x_0 = 2$ m and $y_0 = 5$ m

**12–5** The acceleration of a particle is defined by $a_x = 5$ m/s$^2$ and $a_y = -9$ m/s$^2$ at time $t_0 = 0$ when $x_0 = 0$, $y_0 = 100$ m, and $v_{x_0} = v_{y_0} = 0$. Determine the position of the particle at $t = 2$ s.

**12–6** The acceleration vector of a particle is $\mathbf{a} = (10t\mathbf{i} - 30\mathbf{j})$ ft/s$^2$, where $t$ is in seconds. Determine the velocity vector at $t = 3$ s if initially $x_0 = y_0 = 0$, and at $t = 1$ s, $x_1 = 40$ ft and $y_1 = 25$ ft.

**12–7** The velocity of a particle in three-dimensional motion is defined by $\mathbf{v} = (2t\mathbf{i} + t^2\mathbf{j} - t\mathbf{k})$ m/s, where $t$ is in seconds. Determine the position vector at $t = 10$ s if initially $x_0 = 0$, $y_0 = 200$ m, and $z_0 = 0$.

**12–8** A projectile has an initial speed $v_0 = 100$ ft/s at $\theta = 30°$ at point $O$. Determine the maximum elevation $h$ and horizontal distance $d$ reached by the projectile, neglecting air resistance.

Fig. P12-8

**12–9** For the same initial speed $v_0$ in Prob. 12–8, plot $h$ vs. $\theta$ and $d$ vs. $\theta$ as functions of $\theta = 30°$, 45°, and 60°.

**12–10** A projectile is fired at an angle $\theta = 35°$ toward the target $A$ as shown in Fig. P12-8. Determine the required speed $v_0$ if $d = 1000$ m. Neglect air resistance.

**12–11** A helicopter at $x = 0$ and $y = 300$ ft has speed $v_{x_0} = 60$ ft/s when a package is ejected from it with a downward velocity of 5 ft/s. Determine the horizontal distance $OA$ traveled by the package. Neglect air resistance.

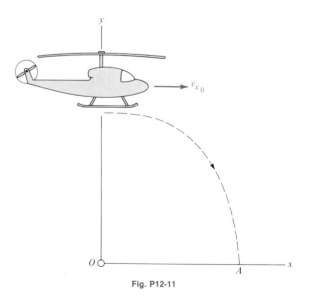

Fig. P12-11

**12–12**  A person on a bluff horizontally throws a rock aimed at a target at point $A$. What should the initial speed $v_0$ of the rock be to hit the target if air resistance can be neglected, $h = 40\,\text{m}$, and $d = 30\,\text{m}$?

**Fig. P12-12**

**12–13**  A person in Fig. P12-12 horizontally throws a rock with an initial speed $v_0 = 30\,\text{ft/s}$ from a height $h = 100\,\text{ft}$. Determine the horizontal distance $d$ and the velocity vector of the rock just before it lands at $A$. Neglect air reistance.

**12–14**  A ski jumper leaves the track at point $O$ with a horizontal speed $v_0$. Determine $v_0$ if the skier touches down at $s = 150\,\text{ft}$, neglecting air resistance.

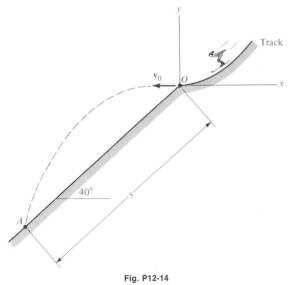

**Fig. P12-14**

**12–15**  Assume that the takeoff velocity of the ski jumper in Prob. 12–14 is $\mathbf{v}_0 = (-20\mathbf{i} - 2\mathbf{j})\,\text{m/s}$. Determine the distance $s$ to the touchdown point at $A$.

**12–16**  A person releases a ball $2\,\text{m}$ above the ground and $7\,\text{m}$ away from a wall. The ball has an initial velocity $v_0 = 20\,\text{m/s}$ at an angle $\theta = 70°$. Determine the height $h$ where the ball hits the wall. Neglect air resistance.

**Fig. P12-16**

**12–17**  In Prob. 12–16, determine the smallest speed $v_0$ with which the ball could land on the flat roof $18\,\text{m}$ above the ground.

**12–18**  A projectile is fired with an initial speed $v_0 = 3000\,\text{ft/s}$ toward a target at point $A$. Determine the two firing angles $\alpha_1$ and $\alpha_2$ with which the projectile can reach the target. Also calculate the two times of flight. Neglect air resistance.

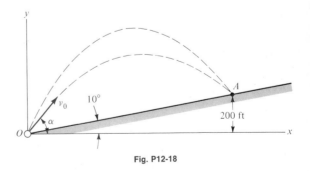

**Fig. P12-18**

**12–19** A nozzle at point $A$ discharges water which strikes the incline at point $B$. Determine the initial speed $v_A$ of the water if air resistance can be neglected.

Fig. P12-19

**12–20** A projectile is thrown with an initial velocity $\mathbf{v}_0 = a\mathbf{i} + b\mathbf{j}$ from point $A$. The elevation $h_{\max}$ is known to be twice the elevation of the upper plateau with respect to $A$. (a) Determine the final velocity $\mathbf{v}_f$ of the projectile just before impact at point $B$. The acceleration of gravity is $g$. (b) Assume that the same projectile is thrown from point $B$ with an initial velocity $-\mathbf{v}_f$. Determine the final velocity $\mathbf{v}_f'$ of the projectile just before impact at the lower level. Is this impact at point $A$?

Fig. P12-20

**12–21** A particle $P$ is moving with a constant speed $v = 10$ m/s on a circular track of radius $r = 0.2$ m. Determine the acceleration of the particle at $\theta = 0$ and $90°$.

Fig. P12-21

**12–22** A particle $P$ is moving clockwise on a circular track of radius $r = 1$ ft (Fig. P12-21). The speed of $v = 10$ ft/s is increasing at the rate of $2$ ft/s$^2$ when $\theta = 30°$. Determine the acceleration vector of the particle in the $xy$ coordinate system shown.

**12–23** A particle $P$ starts from rest at $\theta = 0$ and travels counterclockwise on a circular track of radius $r = 0.5$ m (Fig. P12-21). With the speed increasing at a constant rate, the particle is at $\theta = 90°$ at time $t = 0.1$ s after the motion began. Determine the acceleration vectors in the $xy$ coordinate system at $\theta = 0$ and $90°$.

**12–24** A racing car travels at a constant speed $v$ on a circular part of the track which has a radius of $100$ m. Determine the maximum allowable speed assuming that the normal component of the acceleration must not exceed $7$ m/s$^2$.

**12–25** The car in Prob. 12–24 is traveling at a speed of $v = 20$ m/s when the driver wishes to accelerate forward. Determine the maximum rate at which $v$ may be increased if the maximum total acceleration must not exceed $7$ m/s$^2$.

**12–26** An amusement ride has seats arranged in a circular pattern of $20$ ft diameter and rotates in a horizontal plane. The speed of each person is $18$ ft/s when the operator wishes to decrease the speed. Determine the maximum allowable rate of decrease of the speed if the total acceleration of each rider must not exceed $40$ ft/s$^2$.

Fig. P12-26

**12–27** A roller coaster travels at speed $v_A = 60$ ft/s at point $A$ where the radius of curvature is $r_A = 30$ ft. At this point $v_A$ is decreasing at the rate of $3$ ft/s$^2$. Determine the total acceleration of a person at point $A$.

Fig. P12-27

**12-28** A roller coaster travels at speed $v_B = 5\,\text{m/s}$ at point $B$ in Fig. P12-27 where the radius of curvature is $r_B = 8\,\text{m}$. At this point $v_B$ is decreasing at the rate of $1\,\text{m/s}^2$. Determine the total acceleration of a person at point $B$, noting that the train and its passengers are constrained to follow the track.

**12-29** A prototype car moves on a complex test track. Recordings of the motion show that at time $t$ the car's speed $v = 40\,\text{ft/s}$ was increasing at the rate of $3\,\text{ft/s}^2$, and its total acceleration was $25\,\text{ft/s}^2$. Determine the radius of curvature of the car's path at that instant.

**12-30** An airplane in a diving test has a speed $v_A = 200\,\text{m/s}$, decreasing at the rate of $15\,\text{m/s}^2$, at the bottom of its trajectory. Determine the radius of curvature of the trajectory at point $A$ if the total acceleration of the airplane is $80\,\text{m/s}^2$ at point $A$.

Fig. P12-30

**12-31** A particle $P$ moves with a constant speed $v = 10\,\text{ft/s}$ on a path defined by $y = 4x^2$ ft. Determine the radius of curvature $\rho$ and the acceleration $\mathbf{a}$ of the particle at $x = 0$ and $2$ ft.

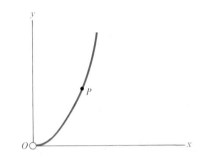

Fig. P12-31

**12-32** A particle $P$ starts from rest at point $O$ in Fig. P12-31. The path of motion is defined by $y = 3x^2$ m. The particle's speed increases at a constant rate and it reaches $x = 1$ m at time $t = 0.2$ s after the motion began. Determine the acceleration vectors at $x = 0$ and $1$ m.

**12-33** The position vector of a particle in plane curvilinear motion is $\mathbf{r} = [(4t^3 + 5)\mathbf{i} - t^2\mathbf{j}]$ m, with $t$ in seconds. Calculate the radius of curvature $\rho$ and the acceleration $\mathbf{a}$ of the particle at $t_1 = 1$ s and $t_2 = 2$ s.

**12-34** The position vector of a particle is defined by $\mathbf{r} = [t^3\mathbf{i} + (5t^3 - t)\mathbf{j}]$ ft, where $t$ is in seconds. Calculate the radius of curvature $\rho$ and the acceleration $\mathbf{a}$ of the particle at $t_0 = 0$ and $t_1 = 5$ s.

**12-35** A projectile is fired with an initial speed $v_0 = 1000\,\text{m/s}$. Determine the radius of curvature of the path of motion (a) immediately after the projectile leaves the muzzle, and (b) at the maximum elevation.

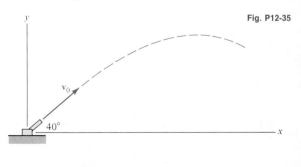

Fig. P12-35

# MOTION OF A PARTICLE IN POLAR COORDINATES

Sometimes the plane motion of a particle is convenient to analyze using polar coordinates. This is done with the aid of the methods developed in Sec. 12-3. According to common notation, the position, velocity, and acceleration of the particle are given in terms of the radial coordinate $r$ and the angular coordinate $\theta$. Of course, these can be readily converted to a rectangular system as shown in Fig. 12-13a, where $x = r \cos \theta$ and $y = r \sin \theta$.

The infinitesimal displacement $d\mathbf{r}$ is resolved into *radial* and *transverse components* $d\mathbf{r}_r$ and $d\mathbf{r}_\theta$, respectively, as in Fig. 12-13b. The magnitude of $d\mathbf{r}_r$ is $dr$, the infinitesimal change in length of $\mathbf{r}$. The approximation is made that the magnitude $dr_\theta$ is equal to the arc length $r\,d\theta$. With these, the radial and transverse scalar components of the velocity are

$$v_r = \frac{dr}{dt} = \dot{r}$$

12-13

$$v_\theta = r\frac{d\theta}{dt} = r\dot{\theta} = r\omega$$

12-14

**FIGURE 12-13**

Analysis of motion using polar coordinates

where the symbols $\dot{\theta}$ and $\omega$ denote the *angular speed* of a radial line.* This scalar quantity is the time rate of change of the angle $\theta$. The angular speed $\dot{\theta}$ can be positive or negative depending on whether the angle $\theta$ is increasing or decreasing (Fig. 12-14). Equations 12-13 and 12-14 can be combined vectorially using the unit vectors $\mathbf{n}_r$ and $\mathbf{n}_\theta$ (Fig. 12-13b),

$$\mathbf{v} = \dot{r}\mathbf{n}_r + r\dot{\theta}\mathbf{n}_\theta \qquad \boxed{12\text{-}15}$$

Equation 12-15 is the expression for the velocity in polar coordinates. The units of angular speed are radians per second. The commonly used units of revolutions per minute (rpm) must be converted to rad/s for use in Eq. 12-14 or 12-15.

**FIGURE 12-14**

The positive angular speed of a radial line

The same result can be obtained by working directly with the derivative of the vector $\mathbf{r} = r\mathbf{n}_r$. This approach is lengthier than the pictorially advantageous method above, but it is so useful in some analyses that its details are presented here.

First express the orthogonal unit vectors $\mathbf{n}_r$ and $\mathbf{n}_\theta$ from Fig. 12-11 as

$$\mathbf{n}_r = \cos\theta\mathbf{i} + \sin\theta\mathbf{j} \quad \mathbf{n}_\theta = -\sin\theta\mathbf{i} + \cos\theta\mathbf{j} \qquad \boxed{12\text{-}16}$$

It is important to realize that the unit vectors $\mathbf{n}_r$ and $\mathbf{n}_\theta$ are constant only in magnitude but change directions as the particle moves to a new, nearby location. These changes are modeled as the rotations from $\mathbf{n}_r$ to $\mathbf{n}_r'$ and from $\mathbf{n}_\theta$ to $\mathbf{n}_\theta'$ in Fig. 12-11c. The infinitesimal changes of the unit vectors can be determined from this diagram using the arc-length approximations that $dn_r = n_r\,d\theta = d\theta$ and $dn_\theta = n_\theta\,d\theta = d\theta$. The time derivatives of the unit vectors can be written from these equations or directly from the expressions for $\mathbf{n}_r$ and $\mathbf{n}_\theta$. For example,

$$\dot{\mathbf{n}}_r = \frac{d\mathbf{n}_r}{dt} = \frac{d\mathbf{n}_r}{d\theta}\frac{d\theta}{dt} = (-\sin\theta\mathbf{i} + \cos\theta\mathbf{j})\dot{\theta} = \dot{\theta}\mathbf{n}_\theta \qquad \boxed{12\text{-}17}$$

The derivation of the expression for $\dot{\mathbf{n}}_\theta$ is left as an exercise for the reader. In summary, the infinitesimal changes and time derivatives of the unit vectors are tabulated as follows.

Magnitudes:
$$dn_r = n_r\,d\theta \quad dn_\theta = n_\theta\,d\theta \qquad \boxed{12\text{-}18}$$
$$= d\theta \qquad\quad = d\theta$$

Vector derivatives:
$$\dot{\mathbf{n}}_r = \frac{d\mathbf{n}_r}{dt} \qquad \dot{\mathbf{n}}_\theta = \frac{d\mathbf{n}_\theta}{dt} \qquad \boxed{12\text{-}19}$$
$$= \dot{\theta}\mathbf{n}_\theta \qquad\quad = -\dot{\theta}\mathbf{n}_r$$

With these vectors, the velocity of the particle can be obtained by differentiating $\mathbf{r} = r\mathbf{n}_r$,

$$\mathbf{v} = \frac{d\mathbf{r}}{dt} = \frac{dr}{dt}\mathbf{n}_r + r\frac{d\mathbf{n}_r}{dt} = \dot{r}\mathbf{n}_r + r\dot{\theta}\mathbf{n}_\theta$$

which is identical to the result in Eq. 12-15.

* The *angular velocity* vector $\omega$ is defined in App. A.

The acceleration of the particle is obtained by differentiating Eq. 12-15 with respect to time,

$$\mathbf{a} = \dot{\mathbf{v}} = \ddot{r}\mathbf{n}_r + \dot{r}\dot{\mathbf{n}}_r + \dot{r}\dot{\theta}\mathbf{n}_\theta + r\ddot{\theta}\mathbf{n}_\theta + r\dot{\theta}\dot{\mathbf{n}}_\theta$$

After substituting the expressions for $\dot{\mathbf{n}}_r$ and $\dot{\mathbf{n}}_\theta$ and grouping the terms, the acceleration in polar coordinates becomes

$$\boxed{\mathbf{a} = (\ddot{r} - r\dot{\theta}^2)\mathbf{n}_r + (r\ddot{\theta} + 2\dot{r}\dot{\theta})\mathbf{n}_\theta}$$

12-20

Here the radial and transverse scalar components of the acceleration are

$$a_r = \ddot{r} - r\dot{\theta}^2$$

12-21

$$a_\theta = r\ddot{\theta} + 2\dot{r}\dot{\theta}$$

12-22

It should be noted that $a_r \neq \dot{v}_r$ and $a_\theta \neq \dot{v}_\theta$ because the derivative of the radial velocity $\mathbf{v}_r$ has components in both the radial and transverse directions, and similarly for the derivative of the velocity $\mathbf{v}_\theta$. The quantity $\ddot{\theta} = \dot{\omega} = \alpha$ is called the *angular acceleration* of a radial line, and has units of rad/s².* This scalar quantity is the time rate of change of the angular speed $\dot{\theta}$. The sign of the angular acceleration $\ddot{\theta}$ depends on its direction in the chosen coordinate system (Fig. 12-15). If the angular speed and angular acceleration have the same sign, the angular speed is increasing.

An important special case of curvilinear motion is *circular motion* for which $r$ = constant, $\dot{r} = \ddot{r} = 0$. In that case Eqs. 12-15 and 12-20 for velocity and acceleration become

$$\mathbf{v} = r\dot{\theta}\mathbf{n}_\theta \quad \mathbf{a} = -r\dot{\theta}^2\mathbf{n}_r + r\ddot{\theta}\mathbf{n}_\theta$$

12-23

These agree, as they should, with Eqs. 12-8 and 12-11 when the differences in the sense of the vectors $\mathbf{n}_n$ and $\mathbf{n}_r$ are noted. Thus, for circular motion

$$v = r\dot{\theta} = r\omega$$

12-24

$$a_t = \frac{dv}{dt} = r\ddot{\theta} = r\alpha$$

12-25

$$a_n = \frac{v^2}{r} = r\dot{\theta}^2 = r\omega^2$$

12-26

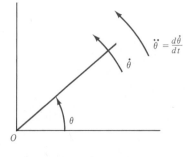

**FIGURE 12-15**

The positive angular speed and angular acceleration of a radial line (the angular speed is increasing)

* The angular acceleration vector $\boldsymbol{\alpha}$ is defined in Sec. 17-2.

The analysis of a particle's motion in three dimensions can be done in any one of three major ways: with the use of rectangular coordinates as introduced in Sec. 12-1, with polar coordinates adapted to space motion as described in this section, or with the application of spherical coordinates as presented in Sec. 12-6.

The components of velocity and acceleration that were derived in Sec. 12-4 are also valid in three dimensions. It is only necessary to consider the additional motion which is perpendicular to the plane of the $r$ and $\theta$ coordinates. When $r$ and $\theta$ are defined in terms of the rectangular coordinates $x$ and $y$ as in Fig. 12-16, the third dimension is the $z$ coordinate. The position vector $\mathbf{R}$ of particle $P$ in cylindrical coordinates is defined according to Fig. 12-16 as

$$\mathbf{R} = r\mathbf{n}_r + z\mathbf{k} \qquad \boxed{12\text{-}27}$$

The expressions for velocity and acceleration in cylindrical coordinates are as follows. From Eqs. 12-27 and 12-15, the velocity vector is

$$\boxed{\mathbf{v} = \dot{r}\mathbf{n}_r + r\dot{\theta}\mathbf{n}_\theta + \dot{z}\mathbf{k}} \qquad \boxed{12\text{-}28}$$

where $\dot{z}$ is the scalar component of the velocity in the $z$ direction, and $\mathbf{k}$ is the corresponding unit vector. The other terms were described in Sec. 12-4.

Similarly, the acceleration vector is obtained from Eq. 12-28 and 12-20,

$$\boxed{\mathbf{a} = (\ddot{r} - r\dot{\theta}^2)\mathbf{n}_r + (r\ddot{\theta} + 2\dot{r}\dot{\theta})\mathbf{n}_\theta + \ddot{z}\mathbf{k}} \qquad \boxed{12\text{-}29}$$

Note that the time derivatives of $z\mathbf{k}$ are $\dot{z}\mathbf{k}$ and $\ddot{z}\mathbf{k}$ because $\mathbf{k}$ is displaced only to parallel positions in cylindrical coordinates. Accordingly, $\dot{\mathbf{k}}$ is always zero.

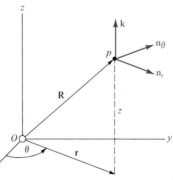

**FIGURE 12-16**

Unit vectors in cyclindrical coordinates. Note: $\mathbf{n}_r$ is parallel to $r$; $\mathbf{n}_\theta$ is parallel to $xy$ plane; $\mathbf{k}$ is parallel to $z$.

# SPHERICAL COORDINATES

The derivation of the equations of motion in spherical coordinates is very difficult for many people. The major concepts and steps of such a derivation are presented here for those who wish to obtain further experience in three-dimensional vector mechanics. A review of the derivatives of unit vectors in Sec. 12-2 and App. A is useful for this analysis. The concepts of polar coordinates presented in Sec. 12-5 are also relevant.

(a)

(b)

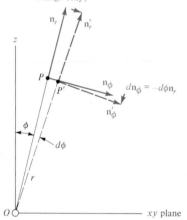

$dn_r = d\phi n_\phi$
(rotational change only)

$dn_\phi = -d\phi n_r$

(c) View $A$ from (b) ($\perp$ to $n_r$ and $n_\phi$)

**FIGURE 12-17**

Unit vectors in spherical coordinates

The position of particle $P$ on a spherical surface (Fig. 12-17a) is defined by the coordinates $r$, $\phi$, and $\theta$. Note that $\theta$ is not the angle to the line $OP$ but to its projection $OQ$ in the $xy$ plane (thus, $\theta$ is not the same angle as that defined for polar coordinates). The significance of this distinction can be appreciated by considering an arbitrary large rotation through an angle $\Delta\theta$. This rotation causes a large displacement of point $P$ if it is located in the $xy$ plane. The same rotation causes no displacement of point $P$ if it is located on the $z$ axis.

The derivation of the particle's velocity and acceleration is based on a triad of orthogonal unit vectors $\mathbf{n}_r$, $\mathbf{n}_\phi$, and $\mathbf{n}_\theta$ which satisfy the right-hand rule according to the vector product $\mathbf{n}_r \times \mathbf{n}_\phi = \mathbf{n}_\theta$. These unit vectors may change in directions, so their time derivatives must be included in the analysis. Pictorial views of these changing unit vectors are shown in Fig. 12-17c, which is typical of three orthogonal views $A$, $B$, $C$ from Fig. 12-17b. Each of these views shows only two unit vectors as they change.

The complex rotations that are possible for the triad of unit vectors necessitate the use of vector notation for the angular velocities. To illustrate a convenient scheme, assume that the $y$ and $z$ axes rotate counterclockwise about the $x$ axis with a common angular velocity $\omega_x$. This angular velocity can be expressed as a vector $\boldsymbol{\omega}_x$ in the $x$ direction, with its sense determined by the right-hand rule, $\boldsymbol{\omega}_x = \omega_x\mathbf{i}$. Similarly, the scheme can be applied to angular velocities about the $y$ and $z$ axes. A general rotational velocity $\boldsymbol{\omega}$ of a line such as $OP$ in Fig. 12-17 can be expressed as the vector sum of three angular velocities concurrent at point $O$,

$$\boldsymbol{\omega} = \omega_x\mathbf{i} + \omega_y\mathbf{j} + \omega_z\mathbf{k} \qquad \boxed{12\text{-}30}$$

In seeking relationships of unit vectors and angular velocities, a useful property is obtained from App. A. For any unit vector $\mathbf{n}$,

$$\dot{\mathbf{n}} = \boldsymbol{\omega}_n \times \mathbf{n} \qquad \boxed{12\text{-}31}$$

where $\omega_n$ is the angular velocity of that unit vector. For example, $\dot{\mathbf{n}}_r = \dot{\phi}\mathbf{n}_\phi$ from Fig. 12-17c. Writing the vector product according to Eq. 12-31, with $\dot{\boldsymbol{\phi}} = \dot{\phi}\mathbf{n}_\theta$ (using the right-hand rule for $\dot{\boldsymbol{\phi}}$),

$$\dot{\boldsymbol{\phi}} \times \mathbf{n}_r = \begin{vmatrix} \mathbf{n}_r & \mathbf{n}_\phi & \mathbf{n}_\theta \\ 0 & 0 & \dot{\phi} \\ 1 & 0 & 0 \end{vmatrix} = \dot{\phi}\mathbf{n}_\phi = \dot{\mathbf{n}}_r$$

For the complete analysis of the particle's velocity and acceleration in the given spherical system, assume that its position $\mathbf{r}$ and the angles $\phi$ and $\theta$ are given as functions of time. In unit vector notation, the position is

$$\mathbf{r} = r\mathbf{n}_r \qquad \boxed{12\text{-}32}$$

The total angular velocity $\omega$ of vector $\mathbf{r}$ can be expressed in terms of components,

$$\omega = \dot{\phi}\mathbf{n}_\theta + \dot{\theta}\mathbf{n}_z = \dot{\phi}\mathbf{n}_\theta + \dot{\theta}\cos\phi\,\mathbf{n}_r - \dot{\theta}\sin\phi\,\mathbf{n}_\phi \qquad \boxed{12\text{-}33}$$

where Fig. 12-13c was used to convert from $\mathbf{n}_z$ to $\mathbf{n}_r$ and $\mathbf{n}_\phi$.

The velocity of the particle is obtained by differentiating Eq. 12-32 with respect to time,

$$\mathbf{v} = \dot{\mathbf{r}} = \dot{r}\mathbf{n}_r + r\dot{\mathbf{n}}_r \qquad \boxed{12\text{-}34}$$

where $\dot{\mathbf{n}}_r$ can be written as

$$\dot{\mathbf{n}}_r = \omega \times \mathbf{n}_r = \dot{\phi}\mathbf{n}_\phi + \dot{\theta}(\sin\phi)\mathbf{n}_\theta \qquad \boxed{12\text{-}35}$$

Substituting Eq. 12-35 into Eq. 12-34, the general expression for the velocity is obtained,

$$\boxed{\mathbf{v} = \dot{r}\mathbf{n}_r + r\dot{\phi}\mathbf{n}_\phi + r\dot{\theta}(\sin\phi)\mathbf{n}_\theta} \qquad \boxed{12\text{-}36}$$

The acceleration of the particle is derived by differentiating Eq. 12-36 with respect to time,

$$\begin{aligned} \mathbf{a} = \dot{\mathbf{v}} = \ddot{r}\mathbf{n}_r &+ \dot{r}\dot{\mathbf{n}}_r + \dot{r}\dot{\phi}\mathbf{n}_\phi + r\ddot{\phi}\mathbf{n}_\phi \\ &+ r\dot{\phi}\dot{\mathbf{n}}_\phi + \dot{r}\dot{\theta}(\sin\phi)\mathbf{n}_\theta + r\ddot{\theta}(\sin\phi)\mathbf{n}_\theta \\ &+ r\dot{\theta}(\cos\phi)\dot{\phi}\mathbf{n}_\theta + r\dot{\theta}(\sin\phi)\dot{\mathbf{n}}_\theta \end{aligned} \qquad \boxed{12\text{-}37}$$

This can be simplified by the following substitutions:

$$\dot{\mathbf{n}}_\phi = \omega \times \mathbf{n}_\phi = -\dot{\phi}\mathbf{n}_r + \dot{\theta}(\cos\phi)\mathbf{n}_\theta \qquad \boxed{12\text{-}38}$$

$$\dot{\mathbf{n}}_\theta = \omega \times \mathbf{n}_\theta = -\dot{\theta}(\sin\phi)\mathbf{n}_r - \dot{\theta}(\cos\phi)\mathbf{n}_\phi \qquad \boxed{12\text{-}39}$$

The resulting general expression for the acceleration is

$$\boxed{\begin{aligned} \mathbf{a} = (\ddot{r} &- r\dot{\phi}^2 - r\dot{\theta}^2\sin^2\phi)\mathbf{n}_r \\ &+ (r\ddot{\phi} + 2\dot{r}\dot{\phi} - r\dot{\theta}^2\sin\phi\cos\phi)\mathbf{n}_\phi \\ &+ (r\ddot{\theta}\sin\phi + 2\dot{r}\dot{\theta}\sin\phi + 2r\dot{\phi}\dot{\theta}\cos\phi)\mathbf{n}_\theta \end{aligned}} \qquad \boxed{12\text{-}40}$$

where each parenthetical expression is a component of the total acceleration in the direction of the appropriate unit vector.

EXAMPLE 12-6

Experiments show that package $P$ will slide off the rotating disk shown when the magnitude of its acceleration equals $30 \, \text{m/s}^2$. The table starts from rest at $\theta = 0$ and accelerates counterclockwise at a constant angular acceleration of $\ddot{\theta} = 2 \, \text{rad/s}^2$ (Fig. 12-18a). Determine the angle $\theta$ at which the package slides. The $xy$ axes define the horizontal plane.

## SOLUTION

The package $P$ is in circular motion (until sliding occurs) at an angular rate of $\ddot{\theta} = 2 \, \text{rad/s}^2$ (constant). Since

$$\frac{d\dot{\theta}}{dt} = \ddot{\theta} \quad \text{and} \quad \frac{d\theta}{dt} = \dot{\theta}$$

$$d\dot{\theta} = \ddot{\theta} \, dt \quad \text{and} \quad d\theta = \dot{\theta} \, dt$$

Substitution and integration yield

$$\dot{\theta} = 2t + C_1 \quad \text{and} \quad \theta = t^2 + C_2$$

where $C_1 = C_2 = 0$ since $\theta = \dot{\theta} = 0$ when $t = 0$. Since $a^2 = a_n^2 + a_t^2 = a_r^2 + a_\theta^2$ for circular motion,

$$a^2 = (r\dot{\theta}^2)^2 + (r\ddot{\theta})^2$$

Then

$$(30 \, \text{m/s}^2)^2 = [(3 \, \text{m})(2t \, \text{rad/s})^2]^2 + [(3 \, \text{m})(2 \, \text{rad/s}^2)]^2$$

Solving for $t$, $t = 1.565 \, \text{s}$. Substituting this value of $t$ in the equation for $\theta$,

$$\theta = (1.565)^2 \, \text{rad} = 2.45 \, \text{rad}$$

(a)　　　　　　　　　(b)

**FIGURE 12-18**

Circular motion of a package on a rotating disk

EXAMPLE 12-7

A radar station is tracking a rocket shortly after launch (Fig. 12-19). The data obtained at $\theta = 40°$ are $\dot{\theta} = 0.1\,\text{rad/s}$, $\ddot{\theta} = 0.05\,\text{rad/s}^2$, $r = 500\,\text{ft}$, $\dot{r} = 40\,\text{ft/s}$, and $\ddot{r} = 0$. **(a)** Determine the magnitudes of the velocity and acceleration of the rocket. **(b)** Obtain an expression for $r$ in terms of $\theta$, assuming that $\dot{r} = $ constant and $\dot{\theta} = $ constant for $0 < \theta < 40°$.

## SOLUTION

**(a)** The magnitude of the velocity is obtained using Eq. 12-15,

$$v = \sqrt{\dot{r}^2 + r^2\dot{\theta}^2} = \sqrt{40^2 + (500^2)(0.1^2)}$$
$$= 64.0\,\text{ft/s}$$

The magnitude of the acceleration is obtained using Eq. 12-20,

$$a = \sqrt{(\ddot{r} - r\dot{\theta}^2)^2 + (r\ddot{\theta} + 2\dot{r}\dot{\theta})^2}$$
$$= \sqrt{[0 - (500)(0.01)^2]^2 + [(500)(0.05) + 2(40)(0.1)]^2}$$
$$= 33.0\,\text{ft/s}^2$$

Note that quantities with units of rad/s and rad/s² required no conversion in this solution.

**(b)** At all times $t$ for motion in the given range of angles (since $\dot{r} = $ constant),

$$r = d + \dot{r}t$$

and (since $\dot{\theta} = $ constant),

$$\theta = \dot{\theta}t$$

Thus,

$$r = d + \frac{\dot{r}\theta}{\dot{\theta}}$$

**FIGURE 12-19**

Tracking a rocket by radar

EXAMPLE 12-7    505

# EXAMPLE 12-8

An automobile is traveling uphill along a mountain road at a speed of 20 m/s which is increasing at a rate of 5 m/s². The slope of the highway is 10° to the horizontal and its radius of curvature with respect to a vertical axis is constant at 50 m. Using cylindrical coordinates, determine the acceleration of the automobile at this instant.

## SOLUTION

The motion of the automobile is shown schematically in Fig. 12-20, with the automobile at point $P$, corresponding to $\theta = 0$. Since $r = 50$ m (constant), $\dot{r} = \ddot{r} = 0$. Using Eq. 12-28,

$$\mathbf{v} = r\dot{\theta}\mathbf{n}_\theta + \dot{z}\mathbf{k} = v\cos 10° \, \mathbf{n}_\theta + v\sin 10° \, \mathbf{k}$$
$$(50\,\text{m})\dot{\theta}\mathbf{n}_\theta + \dot{z}\mathbf{k} = (20\,\text{m/s})\cos 10° \, \mathbf{n}_\theta + (20\,\text{m/s})\sin 10° \, \mathbf{k}$$

Then $\dot{z} = 3.47$ m/s and $\dot{\theta} = 0.394$ rad/s. Similarly, breaking the vector $\dot{\mathbf{v}} = \mathbf{a}_t = 5\mathbf{n}_t$ m/s² (shown) into vertical and horizontal components yields

$$\ddot{z} = (5\,\text{m/s}^2)\sin 10° = 0.868\,\text{m/s}^2$$

and

$$\ddot{\theta} = \frac{(5\,\text{m/s}^2)\cos 10°}{50\,\text{m}} = 0.0985\,\text{rad/s}^2$$

Using Eq. 12-29,

$$\mathbf{a} = -r\dot{\theta}^2\mathbf{n}_r + r\ddot{\theta}\mathbf{n}_\theta + \ddot{z}\mathbf{k}$$
$$= -(50\,\text{m})(0.394\,\text{rad/s})^2\mathbf{n}_r + (50\,\text{m})(0.0985\,\text{rad/s}^2)\mathbf{n}_\theta + 0.868\mathbf{k}\,\text{m/s}^2$$
$$= (-7.76\mathbf{n}_r + 4.93\mathbf{n}_\theta + 0.868\mathbf{k})\,\text{m/s}^2$$

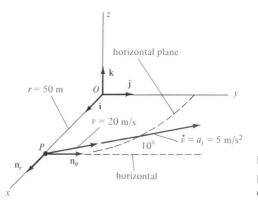

**FIGURE 12-20**

Motion of an automobile on a cylindrical path

# EXAMPLE 12-9

A satellite tracking antenna system includes a hemispherical body which rotates about its vertical axis at 50 rpm as illustrated in cross-sectional view in Fig. 12-21a. **(a)** Determine the acceleration of $P$ for a constant angle $\phi = 50°$ using spherical coordinates. Convert the result to rectangular coordinates. **(b)** Also determine the acceleration of $P$ for the instant shown with the angle $\phi$ changing at a constant rate of $\dot{\phi} = -2$ rad/s.

## SOLUTION

**(a)** Applying Eq. 12-40 with $\dot{r} = \ddot{r} = \dot{\phi} = \ddot{\phi} = \ddot{\theta} = 0$, and

$$\dot{\theta} = \left(50 \, \frac{\text{rev}}{\text{min}}\right)\left(\frac{2\pi \, \text{rad}}{\text{rev}}\right)\left(\frac{1 \, \text{min}}{60 \, \text{s}}\right)$$

$$= 5.24 \, \text{rad/s}$$

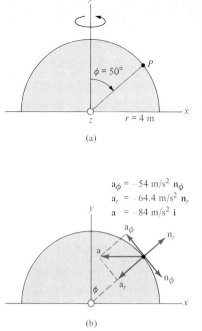

(a)

yields

$$\mathbf{a} = -r\dot{\theta}^2(\sin^2\phi)\mathbf{n}_r - r\dot{\theta}^2(\sin\phi)(\cos\phi)\mathbf{n}_\phi + 0\mathbf{n}_\theta$$
$$= -(4\,\text{m})(5.24\,\text{rad/s})^2(\sin^2 50°)\mathbf{n}_r$$
$$\quad -(4\,\text{m})(5.24\,\text{rad/s})^2(\sin 50°)(\cos 50°)\mathbf{n}_\phi$$
$$= -64.4\,\text{m/s}^2\,\mathbf{n}_r - 54.0\,\text{m/s}^2\,\mathbf{n}_\phi$$

The spherical and rectangular components are sketched in Fig. 12-21, where

$$a_x = a_\phi \cos\phi + a_r \sin\phi$$
$$= (-54\,\text{m/s}^2)\cos 50° - (64.4\,\text{m/s}^2)\sin 50°$$

$$\mathbf{a}_x = -84\,\text{m/s}^2\,\mathbf{i}$$

$$a_y = -a_\phi \sin\phi + a_r \cos\phi$$
$$= (54\,\text{m/s}^2)\sin 50° - (64.4\,\text{m/s}^2)\cos 50°$$

$$\mathbf{a}_y = 0\mathbf{j}$$

**(b)** $\dot{\phi} = -2\,\text{rad/s}$. From Eq. 12-40,

$$\mathbf{a} = (-r\dot{\phi}^2 - r\dot{\theta}^2\sin^2\phi)\mathbf{n}_r - r\dot{\theta}^2(\sin\phi)(\cos\phi)\mathbf{n}_\phi + 2r\dot{\phi}\dot{\theta}(\cos\phi)\mathbf{n}_\theta$$

Using the results that have already been obtained yields

$$\mathbf{a} = [-(4\,\text{m})(-2\,\text{rad/s})^2 - 64.4\,\text{m/s}^2]\mathbf{n}_r - 54.0\,\text{m/s}^2\,\mathbf{n}_\phi$$
$$\quad + 2(4\,\text{m})(-2\,\text{rad/s})(5.24\,\text{rad/s})(\cos 50°)\mathbf{n}_\theta$$
$$= -80.4\,\text{m/s}^2\,\mathbf{n}_r - 54.0\,\text{m/s}^2\,\mathbf{n}_\phi - 53.8\,\text{m/s}^2\,\mathbf{n}_\theta$$

Realizing the relations between $\mathbf{a}_x$, $\mathbf{a}_y$ and $\mathbf{a}_r$, $\mathbf{a}_\phi$ stated above, and also that $\mathbf{n}_\phi = -\mathbf{k}$ (since $\mathbf{n}_\phi = \mathbf{n}_r \times \mathbf{n}_\phi$), $\mathbf{a}_z = -\mathbf{a}_\phi$ for the position shown,

$$\mathbf{a} = (a_\phi \cos\phi + a_r \sin\phi)\mathbf{i} + (-a_\phi \sin\phi + a_r \cos\phi)\mathbf{j} + a_\theta\mathbf{k}$$
$$= (-54.0 \cos 50° - 80.4 \sin 50°)\,\text{m/s}^2\,\mathbf{i} + (54.0 \sin 50°$$
$$\quad - 80.4 \cos 50°)\,\text{m/s}^2\,\mathbf{j} + 53.8\,\text{m/s}^2\,\mathbf{k}$$
$$= (-96.3\mathbf{i} - 10.3\mathbf{j} + 53.8\mathbf{k})\,\text{m/s}^2$$

**FIGURE 12-21**

Accelerations of a particle using spherical coordinates

## JUDGMENT OF THE RESULTS IN PART (a)

Since the path of particle $P$ is a circle in a horizontal plane, and the speed of $P$ is constant, the only component of acceleration is the normal component parallel to the $xz$ plane and directed toward the $y$ axis. Thus,

$$\mathbf{a} = \mathbf{a}_n = \rho\omega^2\mathbf{n}_n = (4\,\text{m})(\sin 50°)(5.24\,\text{rad/s})^2\,\mathbf{n}_n = 84\,\text{m/s}^2\,\mathbf{n}_n$$

which agrees with the result obtained using spherical coordinates.

EXAMPLE 12-9    507

**12–36** A particle $P$ travels with a speed $v = 5\,\text{m/s}$ at $\theta = 30°$ on a circular path of radius $r = 2\,\text{m}$. The speed is increasing at the rate of $3\,\text{m/s}^2$ at the given position. Determine the total acceleration of the particle using polar coordinates.

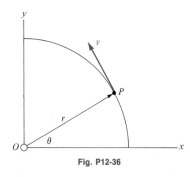

**Fig. P12-36**

**12–37** A particle $P$ travels with a constant speed $v$ on a circular track of radius $r = 2\,\text{ft}$ as in Fig. P12-36. Determine the maximum speed $v$ if the total acceleration of the particle must not exceed $300\,\text{ft/s}^2$. Use polar coordinates.

**12–38** The following data are obtained by radar for the rocket: $\theta = 60°$, $\dot{\theta} = 0.03\,\text{rad/s}$ (increasing $\theta$), $\ddot{\theta} = -0.001\,\text{rad/s}^2$ (decreasing $\dot{\theta}$), $r = 7000\,\text{ft}$, $\dot{r} = 800\,\text{ft/s}$, and $\ddot{r} = 50\,\text{ft/s}^2$. Determine the magnitude of the velocity and acceleration of the rocket.

**Fig. P12-38**

**12–39** The rocket in Prob. 12–38 is tracked by radar giving the data $\theta = 30°$, $\dot{\theta} = 0.02\,\text{rad/s}$ (increasing $\theta$), $\ddot{\theta} = 0$, $r = 2\,\text{km}$, $\dot{r} = 30\,\text{m/s}$, and $\ddot{r} = 15\,\text{m/s}^2$. Determine the vertical components of the velocity and the acceleration of the rocket.

**12–40** An aircraft is tracked by radar and the following data are obtained: $\theta = 80°$, $\dot{\theta} = 0.01\,\text{rad/s}$ (increasing $\theta$), $\ddot{\theta} = 0$, $r = 3\,\text{km}$, $\dot{r} = -100\,\text{m/s}$, and $\ddot{r} = -5\,\text{m/s}^2$. Determine the horizontal components of the velocity and acceleration of the aircraft.

**Fig. P12-40**

**12–41** The following data are obtained by radar for the aircraft in Fig. P12-40: $\theta = 60°$, $\dot{\theta} = 0.03\,\text{rad/s}$ (increasing $\theta$), $\ddot{\theta} = 0.01\,\text{rad/s}^2$ (increasing $\dot{\theta}$), $r = 10{,}000\,\text{ft}$, $\dot{r} = -400\,\text{ft/s}$, and $\ddot{r} = -30\,\text{ft/s}^2$. Determine the horizontal components of the velocity and acceleration of the aircraft.

**12–42** The following data are obtained from an aircraft by radio: $\mathbf{v} = (200\mathbf{i} + 40\mathbf{j})\,\text{m/s}$ and $\mathbf{a} = 0$ at $x = 800\,\text{m}$, $y = 3\,\text{km}$. Determine $\dot{r}$, $\ddot{r}$, $\dot{\theta}$, and $\ddot{\theta}$ measured at that instant at a radar station located at point $O$.

**Fig. P12-42**

**12-43** The following data are reported from an aircraft by radio: $\mathbf{v} = 400\mathbf{i}$ ft/s and $\mathbf{a} = 100\mathbf{i}$ ft/s$^2$ at $x = y = 6000$ ft. Determine $\dot{r}$, $\ddot{r}$, $\dot{\theta}$, and $\ddot{\theta}$ measured by radar at point $O$ in Fig. P12-42.

**12-44** The data obtained from radar tracking of a meteor are as follows: $\theta = 60°$, $\dot{\theta} = -3$ rad/s (decreasing $\theta$), $\ddot{\theta} = 0.3$ rad/s$^2$ (decreasing $\dot{\theta}$), $r = 28$ km, $\dot{r} = -40$ m/s, and $\ddot{r} = 130$ m/s$^2$. Determine the velocity and acceleration vectors of the meteor.

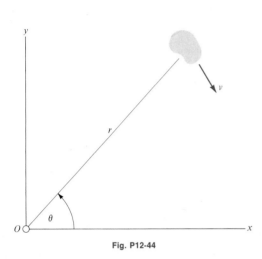

**Fig. P12-44**

**12-45** Peg $P$ moves in the slotted bar with a constant speed $v_r$ while the bar rotates with a constant angular speed $\omega$. Write an expression for the total acceleration of the peg at position $r$ using polar coordinates. Prove that peg $P$ and a particle $Q$ of the bar $OA$ have different accelerations at the instant when $P$ and $Q$ are momentarily in contact.

**Fig. P12-45**

**12-46** Block $B$ slides along bar $OA$ with a constant speed $v = 6$ ft/s while the bar is rotating clockwise with a constant speed of 120 rpm. Determine the velocity and acceleration of block $B$ at $r = 2$ ft and $\theta = 40°$.

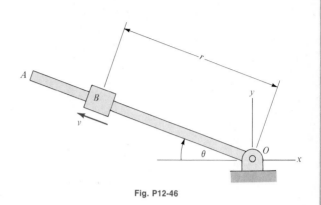

**Fig. P12-46**

**12-47** Arm $A$ in Fig. P12-46 starts from rest at $\theta = 0$; its position is given by $\theta = 3t^2$, where $\theta$ is in radians and $t$ is in seconds. The position of block $B$ is given by $r = t^2/2 + 0.1$ m. Determine the total velocity $\mathbf{v}_B$ and acceleration $\mathbf{a}_B$ of the block at $t = 2$ s.

**12-48** A particle $P$ travels with a constant speed $v = 3$ m/s on a path defined by $y = x^2$. Determine the angular speed and angular acceleration of the line $OP$ at $x = 2$ m with the angle $\theta$ measured from the $x$ axis.

**Fig. P12-48**

**12-49** A particle $P$ travels with a constant speed $v = 10$ ft/s as in Fig. P12-48. Determine the total acceleration of the particle at $x = 5$ ft using polar coordinates. The angle $\theta$ of line $OP$ is measured from the $x$ axis.

**12–50** Peg $P$ in a mechanism is to be pushed to the right with a constant speed $v = 8\,\text{m/s}$. Determine the required angular speed and angular acceleration of the arm $OA$ at $\theta = 60°$ if $l = 0.5\,\text{m}$.

**Fig. P12-50**

**12–51** The acceleration of particle $P$ in the horizontal groove in Fig. P12-50 must not exceed $100\,\text{ft/s}^2$. What are the limitations on the rotational motion of arm $OA$, which pushes the particle, if $l = 1\,\text{ft}$ and $\theta = 70°$? The arm starts from rest at $\theta = 90°$.

**12–52** A car travels with constant speed $v = 80\,\text{ft/s}$ on a circular part of a test track. The car's motion is monitored with radar located at point $O$. Determine the total acceleration of the car at $\theta = 70°$ if $R = 300\,\text{ft}$. Use polar coordinates.

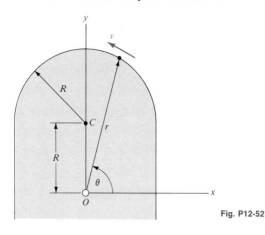

**Fig. P12-52**

**12–53** An amusement ride consists of a rotating cage with people standing in it by the wall. Calculate the velocity and acceleration of a person if $r = 4\,\text{m}$, $v = 0$, $a = 3\,\text{m/s}^2$, and the cage is rotating at a constant speed of 8 rpm.

**Fig. P12-53**

**12–54** Calculate the velocity and acceleration of a person in the cage described in Prob. 12–53 if $r = 15\,\text{ft}$, $\omega = 10\,\text{rad/s}$, $\alpha = 2\,\text{rad/s}^2$, $v = 5\,\text{ft/s}$, and $a = 10\,\text{ft/s}^2$.

**12–55** The rotary elevator on the outside of the tall tower has an angular speed $\omega = 0.5\,\text{rad/s}$ which increases at the rate $\alpha = 0.1\,\text{rad/s}^2$. The elevator is moving downward at this instant with $v = 2\,\text{ft/s}$ and $a = 1\,\text{ft/s}^2$. Calculate the acceleration of a person who is 12 ft from the center of the tower.

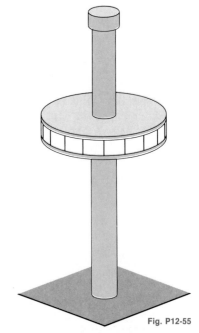

**Fig. P12-55**

**12–56** Small containers on a tray are lowered as the supporting screw rotates. The lead of the screw (advancement per revolution) is 0.3 in., and $r$ is 10 in. Calculate the acceleration of a container if $\omega$ is 10 rad/s and decreasing at the rate of 1 rad/s².

$r = 10$ in.

$\omega$

$d = 2$ in.

$L = 0.3$ in.

**Fig. P12-56**

**12-57** A car is descending on a curved road with a speed $v = 15$ m/s which is decreasing at the rate of $4$ m/s$^2$. Calculate the acceleration of the car if $r = 200$ m and the road makes a $10°$ angle with the horizontal.

$r$

Path of car

**Fig. P12-57**

**12-58** Small containers are moving on the helical path of an automated packaging machine. Calculate the acceleration of a particle at $P$ if it has a constant speed $v = 2$ m/s, and $D = 0.6$ m.

$v$

$P$

$\alpha$

$v$

$\alpha = 10°$ (angle of path with the horizontal at point $P$)

$D$

**Fig. P12-58**

**12-59** Calculate the maximum allowable speed $v$ of a container on the helical path shown in Fig. P12-58. The speed is constant, $D = 2$ ft, and the acceleration should not exceed $20$ ft/s$^2$.

**12-60** The train of the amusement ride has a helical track. The passengers travel on a 28-ft-diameter helix which makes a complete turn in a 50-ft horizontal distance. Calculate the acceleration of the riders when their speed along the track is $40$ ft/s, increasing at the rate of $5$ ft/s.

**Fig. P12-60**

Fig. P12-61

**12–61** The loop of the amusement ride is approximately circular with a diameter of 74 ft. The lower parts $A$ and $B$ of the helical track are horizontally separated by a distance of 6 ft. Calculate the acceleration of the train at point $C$, where its speed along the track is 50 ft/s and increasing at the rate of 4 ft/s$^2$.

**12–62** An airplane is descending in a helical pattern. It has a speed of 80 m/s horizontally, a downward speed of 5 m/s, and a downward acceleration of 2 m/s$^2$. Calculate the airplane's acceleration. Assume that the helix has a constant pitch ($\ddot{\theta} = 0$).

Path

|—————400 m—————|

Fig. P12-62

**12–63** In a proposed amusement ride people sit in cages at a distance $r$ from the pivot $O$. Hydraulic actuators $A_1$ and $A_2$ control the arm length $r$ and the angle $\phi$. Calculate the acceleration of a rider if the ride rotates about the $y$ axis at an angular speed of 5 rpm, $r = 15$ ft, $\dot{r} = -2$ ft/s ($r$ is decreasing), $\phi = 30°$, $\dot{\phi} = 0$, and $\ddot{\phi} = 0.1$ rad/s$^2$.

Fig. P12-63

**12–64** Particles of fluid inside the nozzle move at a constant speed $v = 5$ m/s relative to the nozzle. Calculate the acceleration of the fluid at $r = 0.2$ m when the nozzle rotates at a constant speed of 20 rpm. The nozzle makes a 20° angle with a horizontal plane.

Fig. P12-64

Fig. P12-66

**12–65** A fluid particle at point $P$ in the rotating nozzle is moving with a constant radial speed $\dot{r} = 3$ m/s at $r = 0.15$ m. Calculate the acceleration of the particle if $\phi = 60°$, $\dot{\phi} = 0.2$ rad/s, and $\ddot{\phi} = 0.2$ rad/s².

**12–67** The hemispherical structure of a radar antenna is rotating about the $y$ axis with $\omega_y = 2$ rad/s, which is increasing at the rate of 0.1 rad/s². Calculate the acceleration of a particle at $P_1$ if $R = 5$ m, $\phi_1 = 30°$, and $\dot{\phi}_1 = 3$ rad/s.

Fig. P12-65

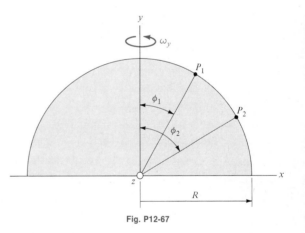

Fig. P12-67

**12–66** Consider the model of the large, high-speed earth mover used in strip mining. The long arm rotates about a vertical line with $\omega_1 = 0.01$ rad/s, while it also rotates about a horizontal line with $\omega_2 = 0.005$ rad/s. Calculate the total acceleration of a particle $P$ at the end of the arm.

**12–68** The radar antenna described in Prob. 12–67 is rotating with $\omega_y = 5$ rad/s, which is decreasing at the rate of 0.2 rad/s². Calculate the acceleration of a particle at $P_2$ if $R = 15$ ft, $\phi_2 = 60$, $\dot{\phi}_2 = 0$, and $\ddot{\phi}_2 = -0.3$ rad/s² ($P_2$ starts moving toward the $y$ axis).

# RELATIVE MOTION OF TWO PARTICLES. TRANSLATION OF COORDINATES

The preceding analyses of curvilinear motion were made with respect to fixed coordinate systems. In some practical problems it is desirable to work with a moving coordinate system. This can be done by extending the concepts of relative motion that were introduced for rectilinear motion. The approach is to define the motion of a particle with respect to a convenient moving reference system, and define the motion of this system with respect to a fixed reference frame. For example, the motion of a component in an aircraft engine can be analyzed with respect to a coordinate system fixed in the aircraft. The motion of this system can be described with respect to a coordinate system fixed on earth. Clearly, no reference system is fixed in an absolute sense. The coordinates fixed on earth move with respect to the sun, and so forth. In most engineering problems the fixed coordinate system is taken to be stationary with respect to the earth. The moving reference frame may or may not be rotating with respect to the fixed frame. The following analysis is for those cases where the moving reference frame is not rotating.

## Translating Coordinate Systems

Consider Fig. 12-22 and assume that $XYZ$ reference frame is fixed and the $xyz$ system is moving. This motion is *translation* when the respective

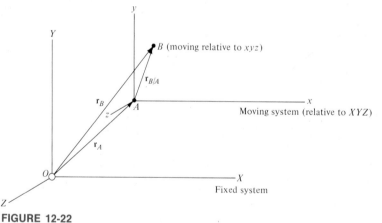

**FIGURE 12-22**

Translation of the *xyz* frame with respect to the *XYZ* frame coordinate axes always remain parallel to each other. The moving co-ordinate system's position is defined by the vector $\mathbf{r}_A$, its velocity by $\mathbf{v}_A = \dot{\mathbf{r}}_A$, and its acceleration by $\mathbf{a}_A = \ddot{\mathbf{r}}_A$ with respect to the fixed coordinates $X$, $Y$, and $Z$.

Next assume that a particle $B$ is moving with respect to the $xyz$ coordinates. Its position vector with respect to point $A$ is $\mathbf{r}_{B/A}$. It follows by vector addition that the position vector of particle $B$ with respect to the fixed coordinates $X$, $Y$, and $Z$ is completely specified by

$$\mathbf{r}_B = \mathbf{r}_A + \mathbf{r}_{B/A} \qquad \boxed{12\text{-}41}$$

where the same set of unit vectors $\mathbf{i}$, $\mathbf{j}$, $\mathbf{k}$ is used in both reference frames. The velocity and acceleration of particle $B$ in the $XYZ$ frame is obtained by differentiating the position vector with respect to time,

$$\mathbf{v}_B = \dot{\mathbf{r}}_B = \dot{\mathbf{r}}_A + \dot{\mathbf{r}}_{B/A} = \mathbf{v}_A + \mathbf{v}_{B/A} \qquad \boxed{12\text{-}42}$$
$$\mathbf{a}_B = \ddot{\mathbf{r}}_B = \ddot{\mathbf{r}}_A + \ddot{\mathbf{r}}_{B/A} = \mathbf{a}_A + \mathbf{a}_{B/A} \qquad \boxed{12\text{-}43}$$

It should be noted in this derivation that the time derivatives of $\mathbf{i}$, $\mathbf{j}$, and $\mathbf{k}$ are zero since their magnitudes and directions are constant when the motion is translation.

The motion of particle $B$ with respect to the moving reference frame at point $A$ is called the *relative motion* of $B$ and is denoted by the sub-script $B/A$. The motion of particle $B$ with respect to the fixed coordinate system at point $O$ is called the *absolute motion* of $B$. This is denoted by the subscript $B$, the symbol for the particle. The choice of the subscripts is arbitrary in setting up a problem, but consistency is necessary in using the subscripts in Eqs. 12-41, 12-42, and 12-43. For example, interchange $A$ and $B$ in Fig. 12-22. Now Eq. 12-41 becomes $\mathbf{r}_A = \mathbf{r}_B + \mathbf{r}_{A/B}$, from which $\mathbf{r}_B = \mathbf{r}_A - \mathbf{r}_{A/B}$. Therefore, $\mathbf{r}_{B/A} = -\mathbf{r}_{A/B}$. Note that the position vector $\mathbf{r}_{B/A}$ is the vector *from* point $A$ to point $B$, whereas $\mathbf{r}_{A/B}$ is the vector *from* point $B$ to point $A$. The same is true for velocities and accelerations. Note that radial and transverse components are often useful in analyzing relative motion as illustrated in the following example.

# EXAMPLE 12-10

Ship $A$ is monitoring the position and velocity of ship $B$ by radar. The information is received at $A$ in the form of the distance $r_{B/A}$ between the ships, the angle $\theta$ that the information beam makes with the north direction, and the rates at which these parameters change (Fig. 12-23). For the instant shown the absolute velocity of $A$ is $\mathbf{v}_A = (-10\mathbf{i} + 40\mathbf{j})$ km/h, $\theta = 30°$, $\dot{\theta} = -0.001$ rad/s, $r_{B/A} = 3$ km, and $\dot{r}_{B/A} = 2$ m/s. Determine the absolute velocity of $B$ in units of m/s.

## SOLUTION

From Eq. 12-42, $\mathbf{v}_B = \mathbf{v}_A + \mathbf{v}_{B/A}$. The velocity $\mathbf{v}_{B/A}$ is calculated using radial and transverse components with respect to a translating reference system with its origin at $A$. From Eq. 12-15,

$$\begin{aligned}
\mathbf{v}_{B/A} &= \dot{r}_{B/A}\mathbf{n}_r + r_{B/A}\dot{\theta}\mathbf{n}_\theta \\
&= 2\,\text{m/s}\,\mathbf{n}_r + (3000\,\text{m})(-0.001\,\text{rad/s})\,\mathbf{n}_\theta \\
&= 2\,\text{m/s}\,\mathbf{n}_r - 3\,\text{m/s}\,\mathbf{n}_\theta
\end{aligned}$$

Transforming from radial and transverse components to rectangular coordinates with the aid of Fig. 12-23b gives

$$\begin{aligned}
(v_{B/A})_x &= -v_r \sin\theta + v_\theta \cos\theta \\
&= -(2\,\text{m/s}) \sin 30° + (3\,\text{m/s}) \cos 30° \\[6pt]
(v_{B/A})_y &= v_r \cos\theta + v_\theta \sin\theta \\
&= (2\,\text{m/s}) \cos 30° + (3\,\text{m/s}) \sin 30° \\[6pt]
\mathbf{v}_{B/A} &= (1.6\mathbf{i} + 3.2\mathbf{j})\,\text{m/s}
\end{aligned}$$

Converting $\mathbf{v}_A$ to units of m/s yields

$$\begin{aligned}
\mathbf{v}_A &= (-10\mathbf{i} + 40\mathbf{j})\,\text{km/h}\left(\frac{1000\,\text{m}}{\text{km}}\right)\left(\frac{1\,\text{h}}{3600\,\text{s}}\right) \\
&= (-2.8\mathbf{i} + 11.1\mathbf{j})\,\text{m/s}
\end{aligned}$$

Then

$$\begin{aligned}
\mathbf{v}_B = \mathbf{v}_A + \mathbf{v}_{B/A} &= [(-2.8 + 1.6)\mathbf{i} + (11.1 + 3.2)\mathbf{j}]\,\text{m/s} \\
&= (-1.2\mathbf{i} + 14.3\mathbf{j})\,\text{m/s}
\end{aligned}$$

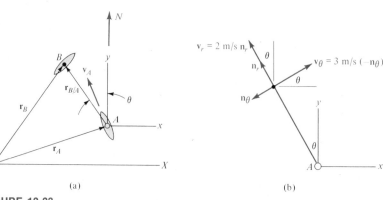

(a)            (b)

**FIGURE 12-23**

Relative motion of two ships

EXAMPLE 12-11

The radar equipment in a police aircraft during a calibration procedure is tracking a car as in Fig. 12-24a. Both the car and the airplane are traveling horizontally, the airplane in a transverse direction to the road. Calculate $r$ and $\dot{r}$ for the given conditions.

## SOLUTION

A translating coordinate system with its origin at $P$ and its $xyz$ axes parallel to the fixed $XYZ$ axes is used as shown in Fig. 12-24b. The information requested is more accurately stated as $r_{C/P}$ and $v_{C/P}$. These scalars must be determined from vector quantities.

$$\mathbf{r}_{C/P} = \mathbf{r}_C - \mathbf{r}_P = (300\mathbf{i}\,\text{m} - 150\mathbf{k}\,\text{m}) - 200\mathbf{j}\,\text{m}$$

$$= 300\mathbf{i}\,\text{m} - 200\mathbf{j}\,\text{m} - 150\mathbf{k}\,\text{m}$$

$$r_{C/P} = \sqrt{300^2 + 200^2 + 150^2} = 391\,\text{m}$$

To calculate $\mathbf{v}_{C/P} = \dot{\mathbf{r}}_{C/P}$, $v_C$ and $v_P$ must be expressed in m/s.

$$v_C = \frac{100\,\text{km}}{\text{h}}\left(\frac{1000\,\text{m}}{\text{km}}\right)\left(\frac{1\,\text{h}}{3600\,\text{s}}\right) = 27.8\,\text{m/s}$$

$$v_P = \frac{150\,\text{km}}{\text{h}}\left(\frac{1000\,\text{m/km}}{3600\,\text{s/h}}\right) = 41.7\,\text{m/s}$$

Then

$$\mathbf{v}_{C/P} = \dot{\mathbf{r}}_{C/P} = \mathbf{v}_C - \mathbf{v}_P = -27.8\,\mathbf{k}\,\text{m/s} - 41.7\,\mathbf{i}\,\text{m/s}$$

$$= (-41.7\mathbf{i} - 27.8\mathbf{k})\,\text{m/s}$$

$$v_{C/P} = \sqrt{-41.7^2 + 27.8^2} = 50\,\text{m/s}$$

(a)

(b)

**FIGURE 12-24**

Relative motion of a car and an airplane

EXAMPLE 12-11    517

**12–69** A coordinate system $xy$ is in translational motion at a constant velocity with respect to fixed axes $XY$. Prove that a particle's acceleration is the same with respect to the $xy$ and $XY$ coordinate systems.

**12–70** Consider three particles $A$, $B$, and $C$ in plane motion. Derive an equation for the relative velocity $\mathbf{v}_{A/C}$ in terms of $\mathbf{v}_{A/B}$ and $\mathbf{v}_{C/B}$.

**12–71** Three particles $A$, $B$, and $C$ are in plane motion. Derive an equation for the relative acceleration $\mathbf{a}_{B/A}$ in terms of $\mathbf{a}_{C/A}$ and $\mathbf{a}_{B/C}$.

**12–72** The positions of particles $A$ and $B$ are defined in a fixed $XY$ frame by $\mathbf{r}_A = 4t^2\mathbf{i} + 3t\mathbf{j}$ and $\mathbf{r}_B = 4t^2\mathbf{i} + 5t^2\mathbf{j}$, where $r_A$ and $r_B$ are in meters and $t$ is in seconds. Determine the relative velocity $\mathbf{v}_{B/A}$ and the relative acceleration $\mathbf{a}_{B/A}$ at $t = 2\,\text{s}$.

**12–73** The positions of particles $A$ and $B$ are defined in a fixed $XY$ frame by $\mathbf{r}_A = -3t^3\mathbf{i} - 2t^3\mathbf{j}$ and $\mathbf{r}_B = -4t^2\mathbf{i} + 3t^3\mathbf{j}$, where $r_A$ and $r_B$ are in feet and $t$ is in seconds. Determine the relative velocity $\mathbf{v}_{A/B}$ and the relative acceleration $\mathbf{a}_{A/B}$ at $t = 4\,\text{s}$.

**12–74** A police car $P$ is traveling at velocity $\mathbf{v}_P = 100\mathbf{j}\,\text{km/h}$. The data from the police car's radar indicate that $r = 200\,\text{m}$, $\dot{r} = 30\,\text{km/h}$, $\ddot{r} \approx 0$, $\theta = 0.1\,\text{rad}$, $\dot{\theta} = -10^{-4}\,\text{rad/s}$ ($\theta$ decreasing), and $\ddot{\theta} \approx 0$. Determine $\mathbf{v}_A$.

Fig. P12-74

**12–75** In Prob. 12–74, $\mathbf{v}_P = 50\mathbf{j}\,\text{mph}$, $r = 400\,\text{ft}$, $\dot{r} = 30\,\text{ft/s}$, $\ddot{r} = 6\,\text{ft/s}^2$, $\theta = 0.12\,\text{rad}$, $\dot{\theta} = -10^{-5}\,\text{rad/s}$ ($\theta$ decreasing), and $\ddot{\theta} \approx 0$. Determine the magnitudes of the velocity and acceleration of car $A$.

**12–76** A police car $P$ is traveling at velocity $\mathbf{v}_P = 20\mathbf{j}\,\text{km/h}$ when its radar gives the following data about car $A$: $r = 100\,\text{m}$, $\dot{r} = -60\,\text{km/h}$ ($r$ decreasing), $\ddot{r} \approx 0$, $\theta = 0.5\,\text{rad}$, $\dot{\theta} = -10^{-3}\,\text{rad/s}$ ($\theta$ decreasing), and $\ddot{\theta} \approx 0$. Determine the velocity $\mathbf{v}_A$.

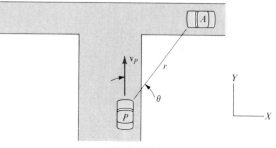

Fig. P12-76

**12–77** A police car $P$ is traveling at velocity $\mathbf{v}_P = (30\mathbf{i} + 10\mathbf{j})\,\text{mph}$ when its radar shows that $r = 400\,\text{ft}$, $\dot{r} = -100\,\text{ft/s}$ ($r$ decreasing), $\ddot{r} \approx 0$, $\theta = 0.11\,\text{rad}$, $\dot{\theta} = -10^{-5}\,\text{rad/s}$ ($\theta$ decreasing), and $\ddot{\theta} \approx 0$. Determine the velocity of car $A$.

Fig. P12-77

**12–78** A police car's velocity is $\mathbf{v}_P = 40\mathbf{j}\,\text{km/h}$ when its radar gives the following data about car $A$: $r = 80\,\text{m}$, $\dot{r} = -35\,\text{m/s}$ ($r$ decreasing), $\ddot{r} = -3\,\text{m/s}^2$, $\theta = 0.9\,\text{rad}$, $\dot{\theta} = -0.2\,\text{rad/s}$ ($\theta$ decreasing), and $\ddot{\theta} \approx 0$. Determine the velocity and acceleration of car $A$.

**Fig. P12-78**

**Fig. P12-81**

**12-79** A police aircraft is directly over point $O$ at an altitude of 400 ft, flying with a velocity $\mathbf{v}_P = (200\mathbf{i} + 30\mathbf{k})$ ft/s. The car is at $X = 500$ ft and traveling at a velocity of $\mathbf{v}_A = 95\mathbf{i}$ ft/s. Determine the relative velocity $v_{P/A}$.

**12-82** The hang glider's acceleration in Prob. 12–81 is defined by $\mathbf{a}_A = (2\mathbf{i} + \mathbf{j})$ m/s² while the boat's acceleration is $\mathbf{a}_B = (1.5\mathbf{i} + 2\mathbf{k})$ m/s². Determine the angular acceleration of the rope at that instant.

**12-83** A mechanism consists of two rigid bars $AB$ and $BC$ which are pinned at $A$ and $B$. $AB = BC = 1$ m, $\omega_1 = 0.3$ rad/s, $\alpha_1 = 0$ (in the $XY$ frame), $\omega_2 = 0.2$ rad/s, and $\alpha_2 = 0$ (in the $xy$ frame). Calculate and sketch the velocity $\mathbf{v}_C$ and acceleration $\mathbf{a}_C$ of point $C$ in the $XY$ and the $xy$ frames.

**Fig. P12-79**

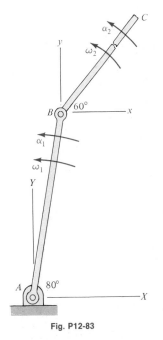

**Fig. P12-83**

**12-80** At the given instant in Prob. 12–79 the aircraft's acceleration is $\mathbf{a}_P = (15\mathbf{i} - 10\mathbf{j})$ ft/s² and the car's acceleration is $\mathbf{a}_A = -8\mathbf{i}$ ft/s². Calculate the relative acceleration $a_{A/P}$.

**12-81** A hang glider $A$ is being towed by boat $B$. The rope has a constant length $l = 30$ m. At a particular instant the velocities are defined by $\mathbf{v}_A = (-12\mathbf{i} + 4\mathbf{j} + 9\mathbf{k})$ m/s and $\mathbf{v}_B = (-10\mathbf{i} + 12\mathbf{k})$ m/s. Determine the angular speed $\omega$ of the rope.

**12-84** Solve Prob. 12–83 after changing the angular accelerations to $\alpha_1 = 0.3$ rad/s² and $\alpha_2 = -0.1$ rad/s² (decreasing $\omega_2$).

Section 12-1

**Curvilinear motion in rectangular Cartesian coordinates**

**POSITION VECTOR:**

$$\mathbf{r} = x\mathbf{i} + y\mathbf{j} + z\mathbf{k}$$

**VELOCITY VECTOR:**

$$\mathbf{v} = \frac{d\mathbf{r}}{dt} = \frac{dx}{dt}\mathbf{i} + \frac{dy}{dt}\mathbf{j} + \frac{dz}{dt}\mathbf{k} = \dot{x}\mathbf{i} + \dot{y}\mathbf{j} + \dot{z}\mathbf{k}$$

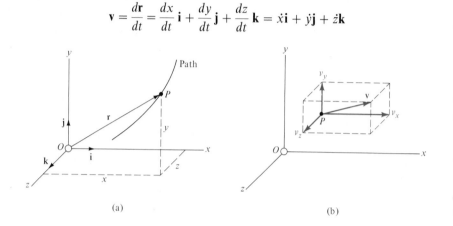

(a)

(b)

**ACCELERATION VECTOR:**

$$\mathbf{a} = \frac{d\mathbf{v}}{dt} = \frac{dv_x}{dt}\mathbf{i} + \frac{dv_y}{dt}\mathbf{j} + \frac{dv_z}{dt}\mathbf{k} = \ddot{x}\mathbf{i} + \ddot{y}\mathbf{j} + \ddot{z}\mathbf{k}$$

The time derivatives of the unit vectors are zero when the $xyz$ frame is fixed. The scalar components $\dot{x}$, $\ddot{x}$, and so on, may be analyzed by using appropriate equations from Sec. 11-1, 11-2, or 11-3. Each equation must include only those quantities which are relevant to the coordinate direction being considered.

Section 12-2

**Physical aspects of velocity and acceleration vectors**

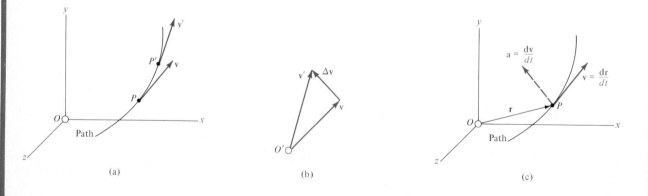

(a)

(b)

(c)

The instantaneous velocity **v** is always tangent to the path of the particle. Speed = magnitude $v$ of vector **v**. The instantaneous acceleration **a** is in the direction of the infinitesimal change of the velocity vector.

The vector **a** is not tangent to the path of the particle and is not collinear with the velocity vector in curvilinear motion.

Section 12-3

**Tangential and normal components**

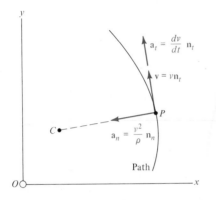

**UNIT VECTORS:**

$\mathbf{n}_t$    in direction of travel
$\mathbf{n}_n$    in direction of center of curvature of path

**VELOCITY:**

$\mathbf{v} = v\mathbf{n}_t$    always tangent to the path

**ACCELERATION:**

$$\mathbf{a} = a_t\mathbf{n}_t + a_n\mathbf{n}_n$$

$$a_t = \frac{dv}{dt} \quad a_n = \frac{v^2}{\rho}$$

$$\rho = \frac{[1 + (dy/dx)^2]^{3/2}}{d^2y/dx^2}$$

For a circular path, $\rho = r =$ radius of circle.

## Polar coordinates

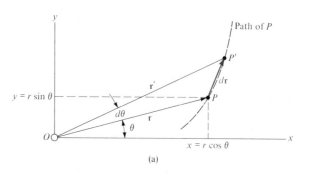

(a)

**UNIT VECTORS:**

$\mathbf{n}_r$      in radial direction (of position vector $\mathbf{r}$)

$\mathbf{n}_\theta$      in transverse direction to $\mathbf{r}$, in plane of path

**VELOCITY:**

$$\mathbf{v} = \dot{r}\mathbf{n}_r + r\dot{\theta}\mathbf{n}_\theta \qquad \text{always tangent to the path}$$

**ANGULAR SPEED OF A RADIAL LINE:**

$$\frac{d\theta}{dt} = \dot{\theta} = \omega, \text{rad/s}$$

**ACCELERATION:**

$$\mathbf{a} = (\ddot{r} - r\dot{\theta}^2)\mathbf{n}_r + (r\ddot{\theta} + 2\dot{r}\dot{\theta})\mathbf{n}_\theta$$

**ANGULAR ACCELERATION OF A RADIAL LINE:**

$$\frac{d\dot{\theta}}{dt} = \ddot{\theta} = \dot{\omega} = \alpha, \ \text{rad/s}^2$$

**CIRCULAR MOTION:**

$$\mathbf{v} = r\dot{\theta}\mathbf{n}_\theta$$

$$\mathbf{a} = -r\dot{\theta}^2\mathbf{n}_r + r\ddot{\theta}\mathbf{n}_\theta$$

**Conversion to Tangential and Normal Components:**

$$v = r\dot{\theta} = r\omega$$

$$a_t = \frac{dv}{dt} = r\ddot{\theta} = r\alpha$$

$$a_n = \frac{v^2}{r} = r\dot{\theta}^2 = r\omega^2$$

## Section 12-5

### Cylindrical coordinates

**UNIT VECTORS:**

$\mathbf{n}_r$      in radial direction in $xy$ plane

$\mathbf{n}_\theta$      in transverse direction to radial line in $xy$ plane

$\mathbf{k}$      in $z$ direction of $xyz$ frame

**VELOCITY:**

$$\mathbf{v} = \dot{r}\mathbf{n}_r + r\dot{\theta}\mathbf{n}_\theta + \dot{z}\mathbf{k}$$

**ACCELERATION:**

$$\mathbf{a} = (\ddot{r} - r\dot{\theta}^2)\mathbf{n}_r + r\ddot{\theta} + 2\dot{r}\dot{\theta})\mathbf{n}_\theta + \ddot{z}\mathbf{k}$$

These are based on the expressions for polar coordinates; $r$ and $\theta$ are defined in the $xy$ plane or in any plane parallel to $xy$.

## Section 12-6

### Spherical coordinates

**UNIT VECTORS:**

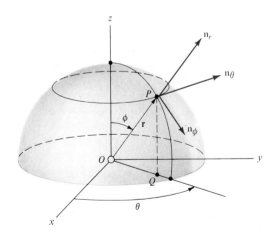

**VELOCITY:**

$$\mathbf{v} = \dot{r}\mathbf{n}_r + r\dot{\phi}\mathbf{n}_\phi + r\dot{\theta}\sin\phi\,\mathbf{n}_\theta$$

**ACCELERATION:**

$$\mathbf{a} = (\ddot{r} - r\dot{\phi}^2 - r\dot{\theta}^2 \sin^2 \phi)\mathbf{n}_r$$
$$+ (r\ddot{\phi} + 2\dot{r}\dot{\phi} - r\dot{\theta}^2 \sin \phi \cos \phi)\mathbf{n}_\phi$$
$$+ (r\ddot{\theta} \sin \phi + 2\dot{r}\dot{\theta} \sin \phi + 2r\dot{\phi}\dot{\theta} \cos \phi)\mathbf{n}_\theta$$

Section 12-7

### Relative motion in translating coordinates

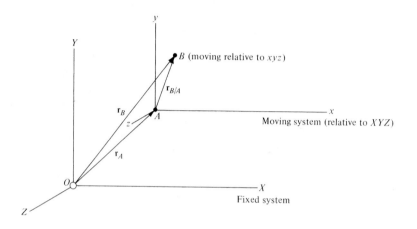

The motion of the *xyz* frame is translation when the respective coordinate axes of *xyz* and *XYZ* are always parallel to each other. The same unit vectors **i**, **j**, **k** are used in both reference frames. Absolute motions of particle *B* are defined with respect to point *O* of *XYZ*. Relative motions are defined with respect to point *A* of *xyz*.

**POSITION:**

$$\mathbf{r}_B = \mathbf{r}_A + \mathbf{r}_{B/A}$$

**VELOCITY:**

$$\mathbf{v}_B = \mathbf{v}_A + \mathbf{v}_{B/A}$$

**ACCELERATION:**

$$\mathbf{a}_B = \mathbf{a}_A + \mathbf{a}_{B/A}$$

Radial and transverse components (polar coordinates) are useful for determining the relative velocity $\mathbf{v}_{B/A}$ and relative acceleration $\mathbf{a}_{B/A}$. Subscripts are arbitrary but must be used consistently since

$$\mathbf{r}_{B/A} = -\mathbf{r}_{A/B} \quad \mathbf{v}_{B/A} = -\mathbf{v}_{A/B} \quad \mathbf{a}_{B/A} = -\mathbf{a}_{A/B}$$

**12-85** A skier starts at time $t = 0$ from the top of a hill, which is approximately parabolic in shape given by $y = x^2/50$. After numerous experiments it is determined that the skier's speed $\dot{s}$ is given approximately by $\dot{s} = 10t$, where $t$ is in seconds and $\dot{s}$ in ft/s. (a) If it takes 3 s to reach point $A$ shown, determine the instantaneous velocity of the skier at $A$ expressed in terms of $x$ and $y$ components. (b) If it takes 5 s to reach point $O$ at the bottom of the hill, determine the skier's acceleration at point $O$ expressed in terms of $x$ and $y$ components.

**Fig. P12-86**

**Fig. P12-85**

**12-86** In preparation for the Winter Olympics, an engineer is asked to help design a ski jump high enough so that the skier will attain the appropriate velocity $v_0$ at point $A$, which will result in a "soft landing" at point $B$ (i.e., a flight path that is tangent to the hill at point $B$). The slope of the hill is 45° at point $B$, which is shown below along with the location of point $B$ from the origin $O$. The ski jump is designed so that the skier's mass center leaves the jump at $A$ at a speed $v_0$ at an angle of 10° from the horizontal as shown. Neglect air resistance. (a) Determine the required velocity $v_0$ for a flight path that is tangent to the hill at $B$ and (b) determine the required height $h$ at the end of the ski jump for the conditions in part (a).

**12-87** A water skier travels up the ramp shown at a speed of 25 ft/s. The rescue boat is parked a distance $d$ from the end of the jump. Neglecting air resistance and considering the water skier to be a particle in free flight until landing on the water, determine (a) the minimum allowable distance $d$, assuming that the height of boat and man equals 4 ft as shown, and (b) the maximum height of the skier's path above the water level.

**Fig. P12-87**

**12-88** In Prob. 12-87, determine the radius of curvature of the skier's path (a) at the time $t = 0.466$ s when maximum height is achieved, and (b) at the instant immediately after leaving the ramp.

**12-89** A resourceful civil engineer, who is not particularly interested in surveying in the swampland shown, is given the job of finding the approximate radius of curvature $\rho$ of an existing set of railroad tracks. The original maps have been lost. The supervisor says it is sufficient to know $\rho$ as a function of $s$, where $s$ is the distance along the tracks from the train station at $O$ to point $A$ located at $s = 800$ ft. The engineer recalls that if one rides on a railroad car moving at constant speed $v$, the acceleration normal to the path, $a_n$, is dependent on $\rho$. An accelerometer, which is an electromechanical device for measuring a component of acceleration, may be mounted perpendicular to the path to measure $a_n$. In carrying out the experiment the railroad car maintains a constant speed $v = 40$ ft/s. The data, which are plotted as shown, can be approximated by a straight line described by $a_n = 4t$, where $t$ is in seconds and $t = 0$ for $s = 0$. Determine the radius of curvature $\rho$ as a function of $s$ from the experimental data.

**Fig. P12-90**

**Fig. P12-89**

**12-90** The rocket is launched vertically at $P$ and is tracked by radar at $O$. Assume that the values of $\theta$, $\dot{\theta}$, and $\ddot{\theta}$ are accurately recorded at $O$. (a) Determine general expressions for the speed $v$ and acceleration $a$ of the rocket in terms of $\theta$, $\dot{\theta}$, and $\ddot{\theta}$ only. (b) What is the speed $v$ and acceleration $a$ at $\theta = 30°$ for $\dot{\theta} = 1$ rad/s and $\ddot{\theta} = 0.5$ rad/s²?

**12-91** The position of a particle $P$ is given in polar coordinates by $r = 2\theta$ and $\theta = 2t^2$, where $r$ is in meters, $t$ in seconds, and $\theta$ in radians. (a) Determine the equations for velocity and acceleration as general functions of time in polar coordinates, and (b) determine the components of acceleration tangent and normal to the path at $t = 1$ s.

**12-92** Particles of fluid are flowing outward in a rotating nozzle that has a constant angular velocity $\omega = 10$ rad/s. At the instant before leaving the nozzle the speed of the fluid particles is $v_r = 8$ m/s, increasing at a rate of $a_r = 2$ m/s², both relative to the nozzle. Using cylindrical coordinates, determine the velocity and acceleration of the fluid particles just before leaving the nozzle.

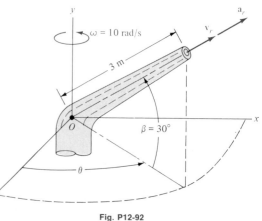

**Fig. P12-92**

**12–93** In Prob. 12–92 assume that in addition to the motion described, the angle $\beta = 30°$ is instantaneously increasing at a constant rate of 0.1 rad/s about a horizontal pivot at 0. Using spherical coordinates, determine the velocity and acceleration of the fluid particles at the instant before leaving the nozzle.

**12–94** In an amusement ride people sit in cages mounted on arms of constant length $r = 6.5$ m. The arms rotate at a constant angular speed $\omega_y = 2$ rad/s. A hydraulic actuator $A$ controls the angle $\phi$. Calculate the acceleration of a rider at $R = 8$ m from the pivot $O$ if $\phi = 30°$, $\dot\phi = 0.2$ rad/s.

Fig. P12-94

**12–95** Consider the air traffic situation in the illustration. The controlling aircraft $A$ is cruising at a steady speed $v_A$ in a circular pattern at height $h = 15,000$ ft while another aircraft $B$ is at height $l = 8000$ ft. Assume that $B$ and $C$ are in the $XY$ plane. Determine $r$ and $\dot r$ for $\mathbf{v}_A = -300\mathbf{i}$ ft/s, $\mathbf{v}_B = -600\mathbf{k}$ ft/s, $R = 5000$ ft, and $d = 9000$ ft.

Fig. P12-95

**12–96** Determine the relative acceleration $\mathbf{a}_{B/A}$ in Prob. 12–95 assuming that the pilot of $A$ is increasing the speed at the rate of 15 ft/s² while remaining in the same circular pattern.

# 13

# Kinetics of Particles: Newton's Second Law

## OVERVIEW

The methods developed for the kinematics analysis of particles are readily applied in the important area of kinetics of particles. In this area the relationship of forces, masses, and the acceleration of masses is investigated. The fundamental principles of kinetics were established by Isaac Newton. These principles are used in a wide range of kinetics analyses of particles, rigid bodies, deformable solids, and fluids.

## STUDY GOALS

**SECTION 13-1** defines Newton's second law of motion, the relationship of force, mass, and acceleration. The concepts and limitations of using this relationship should be remembered for future work in kinetics.

**SECTION 13-2** presents definitions and important distinctions of engineering units that are used in all areas of kinetics. The student should note that the units and terminologies given in Sec. 13-2 are the correct ones for use in engineering work. Incorrect units are not acceptable even if they are commonly used in everyday life (these are mentioned in App. B to remove them from the mainstream of kinetics analyses).

**SECTION 13-3** presents vector and scalar expressions of the equations of motion for the practical application of Newton's second law. These equations are indispensable for solving problems in kinetics.

**SECTION 13-4** outlines a procedure for solving problems in particle kinetics. The student is advised to review the method for drawing proper free-body diagrams learned in statics. Free-body diagrams are as important for applying the equations of motion as they were for the equations of equilibrium.

528

It is known from common experience that a body accelerates when a net force is acting on it. The acceleration is in the same direction as the direction of the resultant force. The magnitude of the acceleration depends on the magnitude of the force and the mass of the body. The role of the mass is not clear to many people, as evidenced by those who believe erroneously that all heavy objects fall faster than light objects. Galileo was the first to demonstrate that the apparent differences caused by mass are only differences in the air friction of the objects; otherwise, heavy and light objects are accelerated at the same rate in the same gravitational field (e.g., in a given region of the earth's atmosphere).

The formal relation of force, mass, and acceleration of a particle is Newton's second law of motion, which may be stated as follows:

> The magnitude of the acceleration of a particle is directly proportional to the magnitude of the resultant force acting on it, and inversely proportional to its mass. The direction of the acceleration is the same as the direction of the resultant force.

This is a postulated law which can be proved only by experiment. For example, if a particle is subject to different forces $\mathbf{F}_1, \mathbf{F}_2, \mathbf{F}_3, \ldots$ at different times, the corresponding accelerations $\mathbf{a}_1, \mathbf{a}_2, \mathbf{a}_3, \ldots$ are determined using data of positions and times and kinematics relationships. Each acceleration is in the same direction as the force causing it, as shown in Fig. 13-1. Calculating the ratios of the magnitudes $F_1/a_1$, and so on, it is found that these ratios are constant for a given particle,

$$\frac{F_1}{a_1} = \frac{F_2}{a_2} = \frac{F_3}{a_3} = \cdots = \text{constant} = m$$

These constant ratios are defined as the *mass* of the particle, which is denoted by $m$. Mass is a property of the particle and is a measure of its *inertia*, or its *resistance to changes in velocity*. Newton's second law is expressed most concisely as

$$\boxed{\mathbf{F} = m\mathbf{a}}$$ 

$$\boxed{13\text{-}1}$$

This equation accounts both for the directions of the force and acceleration vectors, and the proportions of the quantities involved. Equation 13-1 is easy to use, but caution is required in several particular situations as follows.

(a)  (b)  (c)

**FIGURE 13-1**

Acceleration of a particle

1. **F** must be the *resultant force* when two or more forces are acting on the particle.

2. Equation 13-1 is defined for a single particle. It may be used for systems of particles if **a** is taken as the acceleration of the mass center of the system (see Ch. 17).

3. The motion of the particle is determined with respect to a "fixed," nonaccelerating reference frame. In most engineering problems the frame used is attached to the surface of the earth, which is not truly fixed. However, the results obtained are accurate enough for most applications. In some cases, other primary reference systems must be used (for example, one located at the sun). An ideal, fixed reference system is called a *Newtonian reference frame* or *inertial reference frame* (generally it is taken with respect to distant stars).

4. According to Einstein's theory of relativity, Eq. 13-1 is increasingly in error as the particle's velocity approaches the speed of light.

## 13-2    SYSTEMS OF UNITS

It is necessary to use a correct, internally consistent, system of units when applying Eq. 13-1 to solve problems. The following brief statements concern the two major systems in use, and are relevant to all areas of kinetics. These systems are the SI, which is adopted by international agreements, and the U.S. customary system, which is used by tradition.

### SI Units

There are four units in Eq. 13-1, those of force, mass, length, and time. Any three of these may be arbitrarily chosen (they are independent of one another), while the fourth is dependent on the other three in order to satisfy $F = ma$. The base units of SI are the *kilogram* (kg) for mass, the *meter* (m) for length, and the *second* (s) for time. These three units are arbitrarily chosen and are dimensionally independent. The SI units form an absolute system, so the three base units are also independent of location in space. The unit of force is the *newton* (N), which is defined as the force that accelerates a 1-kg mass at the rate of $1 \, \text{m/s}^2$. Its units are obtained from $F = ma$,

$$1 \, \text{N} = (1 \, \text{kg})(1 \, \text{m/s}^2) = 1 \, \text{kg} \cdot \text{m/s}^2$$

It is important to distinguish between the mass and the weight of a body whose motion is to be analyzed. Mass is a property of the body independent of its location or motion. Weight is a force that is caused by a particular gravitational field on the body. For any body near the earth, the weight of the body is the force of attraction of the earth on that body. The magnitude of this force **W** is calculated as

$$\boxed{W = mg} \qquad \text{13-2}$$

where $m$ is the mass of the body and $g$ is the acceleration of gravity. $g$ is approximately equal to $9.81 \, \text{m/s}^2$ near the surface of the earth.

## U.S. Customary Units

The base units in the U.S. customary system are the *pound* (lb) for force, the *foot* (ft) for length, and the *second* (s) for time. The pound is defined as the weight of a certain platinum cylinder that has a mass of about $0.453592 \, \text{kg}$. The foot is defined as $0.3048 \, \text{m}$. The units of time are the same in SI and in the U.S. customary system.

Since the pound is not an absolute unit (the weight slightly depends on location due to differences in elevation and the oblate shape of the earth), the U.S. customary units do not form an absolute system. It is called a *gravitational* system of units.

The mass is a derived quantity in the U.S. system of units, and is called a *slug*. From Eq. 13-1, $m = F/a$, so

$$1 \text{ slug} = \frac{1 \text{ lb}}{1 \text{ ft/s}^2} = 1 \, \frac{\text{lb} \cdot \text{s}^2}{\text{ft}}$$

In general, the mass $m$ in slugs is calculated from

$$m = \frac{W}{g} \qquad \boxed{13\text{-}3}$$

where $W$ is in pounds and the acceleration of gravity $g$ is about $32.2 \, \text{ft/s}^2$.

Further comments about the applications of units are made in App. B.

## EQUATIONS OF MOTION

In most situations there are several forces acting on a particle. Equation 13-1 can be expressed for an arbitrary number of concurrent forces $\mathbf{F}_1$, $\mathbf{F}_2, \mathbf{F}_3, \ldots$, as

$$\boxed{\sum \mathbf{F}_i = m\mathbf{a}} \qquad \boxed{13\text{-}4}$$

This is called the *equation of motion*. It is the formal statement of the fact that the particle has only one acceleration no matter how many forces are acting on it simultaneously. The acceleration is caused by the resultant of the forces.

For solving problems it is convenient to express Eq. 13-4 in terms of scalar component equations. This is done using the most important coordinate systems in the following.

**Rectangular components.** The forces and the acceleration can be resolved into $x$, $y$, and $z$ components. In vector notation,

$$\sum (F_x\mathbf{i} + F_y\mathbf{j} + F_z\mathbf{k}) = m(a_x\mathbf{i} + a_y\mathbf{j} + a_z\mathbf{k})$$

The equations of motion in scalar form are

$$\boxed{\sum F_x = ma_x \quad \sum F_y = ma_y \quad \sum F_z = ma_z} \qquad \text{13-5a}$$

These may also be stated as differential equations of motion,

$$\boxed{\sum F_x = m\ddot{x} \quad \sum F_y = m\ddot{y} \quad \sum F_z = m\ddot{z}} \qquad \text{13-5b}$$

which can be integrated successively to express the velocity and position of the particle as functions of time.

**Normal and tangential components.** For plane curvilinear motion, the components of Eq. 13-4 may also be written using normal and tangential coordinates. This is similar to using polar coordinates, and is particularly advantageous in circular motion where the radius of curvature $\rho$ is constant. The scalar components of Eq. 13-4 are written on the basis of Eq. 12-12,

$$\boxed{\sum F_n = ma_n = m\frac{v^2}{\rho}} \qquad \text{13-6a}$$

$$\boxed{\sum F_t = ma_t = m\dot{v} = mv\frac{dv}{ds}} \qquad \text{13-6b}$$

where $\dot{v}$ is the time derivative of the magnitude of the velocity and $s$ is distance along the path. The directions of these components should be noted. As shown in Fig. 13-2, the normal component $\sum F_n$ is always toward the center of curvature of the particle's path, while the tangential component $\sum F_t$, if not zero, is collinear with the velocity. There is always a normal component when the particle moves on a curved path, but the

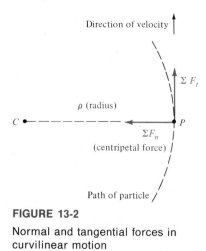

**FIGURE 13-2**

Normal and tangential forces in curvilinear motion

tangential component may be absent. The normal component $\sum F_n$ is called *centripetal force.*\*

**Radial and transverse components.** For plane curvilinear motion, the components of Eq. 13-4 are written in terms of $r$ and $\theta$ according to Eqs. 12-21 and 12-22,

$$\sum F_r = ma_r = m(\ddot{r} - r\dot{\theta}^2) \qquad \boxed{\text{13-7a}}$$

$$\sum F_\theta = ma_\theta = m(r\ddot{\theta} + 2\dot{r}\dot{\theta}) \qquad \boxed{\text{13-7b}}$$

## PROBLEM SOLVING IN PARTICLE KINETICS

The first and most important step in solving problems in kinetics is the drawing of a correct free-body diagram. Once a particle is isolated and sketched, all the known and unknown forces acting on it are drawn. The force vectors must be shown in the correct direction, if known, and the directions of the unknown force vectors must be assumed. Since a force by definition is the action of one body on another, the free-body diagram for a kinetics problem will be just the same as if one were considering equilibrium. In other words, *there are no extra forces in kinetics due to the motion or acceleration of the particle.* The student should become accustomed to the unbalanced look of a free-body diagram for an accelerating particle. Sideline designations of velocity or acceleration vectors are often useful in the analysis, but these vectors are *not* to be made part of any free-body diagram.

Once a complete free-body diagram is drawn, a convenient coordinate system should be chosen. As was pointed out in Sec. 12-1, the acceleration must be determined with respect to a nonaccelerating reference frame. The equations of motion which are appropriate for the chosen coordinate system are then applied. In plane motion there will be two equations of motion for each free-body diagram; in space motion there will be three. The number of unknowns that can be determined is limited to the number of equations. The unknowns to be determined in kinetics problems are either forces or accelerations. If information about the velocity or position of the particle is given or required in the analysis, the equations of motion can be supplemented with the equations of kinematics from Chs. 11 and 12.

If a negative answer appears in the solution, the assumed direction for the force or the acceleration was incorrect. It is especially important to be consistent with the assumed direction of the acceleration when it is used in both the equations of kinematics and the equations of motion. The correct direction of all forces and accelerations, along with the correct units, should be included in the answer.

---

\* A related term is *centrifugal force* as described in App. C. The concept of centrifugal force is from D'Alembert's definition of dynamic equilibrium for analyzing problems of dynamics as problems of statics. This method may be of interest to some people, but it should be omitted from the mainstream of kinetics topics at the elementary level for the reasons given in App. C.

# EXAMPLE 13-1

The mass of each apple in a laboratory test (Fig. 13-3a) is 0.2 kg. Calculate the net force acting on the bottom of an apple which is accelerated upward at $a_y = 20\,\text{m/s}^2$ and horizontally at $a_x = 3\,\text{m/s}^2$, simultaneously.

## SOLUTION

First the free-body diagram of an apple is sketched in Fig. 13-3b. The required force acting on the fruit is the resultant of the force components $\mathbf{R}_x$ and $\mathbf{R}_y$. For rectangular coordinates the equations of motion are from Eq. 13-5a,

$$\sum F_x = ma_x$$
$$R_x = ma_x$$
$$= (0.2\,\text{kg})(3\,\text{m/s}^2)$$
$$= 0.6\,\text{kg}\cdot\text{m/s}^2$$
$$= 0.6\,\text{N}$$

$$\sum F_y = ma_y$$
$$R_y - W = (0.2\,\text{kg})(20\,\text{m/s}^2)$$
$$R_y = 4\,\text{kg}\cdot\text{m/s}^2 + (0.2\,\text{kg})(9.81\,\text{m/s}^2)$$
$$= 5.96\,\text{kg}\cdot\text{m/s}^2$$
$$= 5.96\,\text{N}$$

The net force at the bottom of the apple is

$$R = (R_x^2 + R_y^2)^{1/2} = [(0.6\,\text{N})^2 + (5.96\,\text{N})^2]^{1/2}$$
$$= 5.99\,\text{N} \quad \text{(at an angle of 84° from the horizontal)}$$

## JUDGMENT OF THE RESULT

In static equilibrium the magnitude of the net force at the bottom of the fruit would be $R = R_y = W = 1.96\,\text{N}$. Thus, the result is reasonable for the given accelerations $a_x$ and $a_y$.

(a)

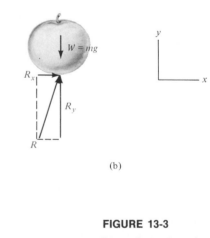

(b)

**FIGURE 13-3**

Dynamic testing of packaged fruit (Photograph courtesy MTS Systems Corporation, Minneapolis, Minnesota.)

EXAMPLE 13-2

The total weight $W$ of the elevator shown in Fig. 13-4a at rest is 5000 lb. Determine the tension $T$ in the cable **(a)** during an upward acceleration $a_1 = 3\,\text{ft/s}^2$, and **(b)** during a downward acceleration of $a_2 = 4\,\text{ft/s}^2$.

## SOLUTION

**(a)** The free-body diagram of the elevator is drawn in Fig. 13-4b. Next the mass of the elevator is calculated from Eq. 13-3,

$$m = \frac{W}{g} = \frac{5000\,\text{lb}}{32.2\,\text{ft/s}^2} = 155\,\frac{\text{lb}\cdot\text{s}^2}{\text{ft}} = 155\,\text{slug}$$

The equation of motion in the $y$ direction is, from Eq. 13-5a,

$$\sum F_y = ma_y$$

$$T - W = ma_y$$

$$T = W + ma_y = 5000\,\text{lb} + \left(155\,\frac{\text{lb}\cdot\text{s}^2}{\text{ft}}\right)\left(3\,\frac{\text{ft}}{\text{s}^2}\right)$$

$$= 5000\,\text{lb} + 465\,\text{lb} = 5465\,\text{lb}$$

**(b)** The free-body diagram of the elevator is drawn in Fig. 13-4c. The equation of motion is from Eq. 13-5a, as above,

$$\sum F_y = ma_y$$

$$T - W = ma_y$$

$$T = W + ma_y = 5000\,\text{lb} + \left(155\,\frac{\text{lb}\cdot\text{s}^2}{\text{ft}}\right)\left(-4\,\frac{\text{ft}}{\text{s}^2}\right)$$

$$= 4380\,\text{lb}$$

## JUDGMENT OF THE RESULTS

In static equilibrium the tension in the cable is 5000 lb. Thus, the calculated small changes in this tension for the given small accelerations (in comparison with the acceleration of gravity) are reasonable.

**COMMON ERROR**

A common error of beginning students in kinetics is the assumption that the tension in a cable holding a weight is equal to that weight, as it would be in a static situation. As problems become more complicated, it is always essential to check to see if the mass is accelerating.

(a) Elevator at rest

(b) Elevator accelerating upward

(c) Elevator accelerating downward

**FIGURE 13-4**

Analysis of an elevator

EXAMPLE 13-2    535

# EXAMPLE 13-3

An automobile of 1000-kg mass is traveling along a level road (Fig. 13-5a) at $v_0 = 100$ km/h when its brakes are applied locking the wheels. Calculate the stopping distance if the kinetic coefficient of friction of the wheels is 0.5. Neglect air resistance.

## SOLUTION

The car is modeled as a block and its free-body diagram is drawn in Fig. 13-5b. Assuming that the wheels of the car do not leave the pavement during the braking process, there is no acceleration in the $y$ direction and consequently no net force, so $N = W = mg$. Using Eq. 13-5a and b for the $x$ direction,

$$-\mathscr{F} = ma_x = m\frac{dv_x}{dt} = m\ddot{x}$$

where the magnitude of the friction force $\mathscr{F} = \mu_k N = \mu_k mg$. Since the distance traveled is of interest rather than time, $x$ is the desired independent variable, not $t$. For the conversion, use

$$\frac{dv_x}{dt} = \frac{dv_x}{dx}\frac{dx}{dt} = v_x\frac{dv_x}{dx}$$

Thus, with $\mathscr{F}$ constant (since $\mu$ and $N$ are constant),

$$-\mathscr{F} = mv_x\frac{dv_x}{dx}$$

After integration,

$$-\mathscr{F}x = \frac{1}{2}mv_x^2 + C$$

Taking $x = 0$ at $v_x = v_0$, $C = -\frac{1}{2}mv_0^2$. The stopping distance is at $x = x_s$, where $v_x = 0$, so

$$x_s = \frac{mv_0^2}{2\mathscr{F}} = \frac{mv_0^2}{2mg\mu_k} = \frac{(100\text{ km/h})^2(1000\text{ m/km})^2}{2(3600\text{ s/h})^2(9.81\text{ m/s}^2)(0.5)}$$

$$= 78.7\text{ m}$$

## JUDGMENT OF THE RESULTS

The result is reasonable on the basis of experience in driving cars. Note that the car's mass cancels out.

(a)

(b)

**FIGURE 13-5**

Analysis of a car's motion

EXAMPLE 13-4

A crate weighing 96.6 lb is being pushed up the 30° incline by a worker who pushes *horizontally* with a force of magnitude $P = 200$ lb (Fig. 13-6a). The coefficient of kinetic friction between the inclined plane and the crate is $\mu = 0.1$. Determine **(a)** the acceleration of the crate, and **(b)** the acceleration and the magnitude of the force between the crates if the worker decides to push two identical crates at once. (Assume negligible friction between the two crates.)

## SOLUTION

**(a)** The free-body diagram of the single crate $A$ is drawn in Fig. 13-6b. For the axes shown, the total acceleration of the crate would be along the incline in the $x$ direction. Using Eq. 13-5a,

$$\sum F_y = ma_y = 0$$

$$N_A - (200\,\text{lb})\sin 30° - (96.6\,\text{lb})\cos 30° = 0$$

$$N_A = 184\,\text{lb}$$

$$\sum F_x = ma_x$$

$$(200\,\text{lb})\cos 30° - (96.6\,\text{lb})\sin 30° - 0.1(184\,\text{lb}) = \frac{96.6\,\text{lb}}{32.2\,\text{ft/s}^2}\,a$$

$$a = 35.5\,\text{ft/s}^2 \;\nearrow$$

**(b)** Although one could consider both crates as one particle for calculating the acceleration because there is no relative motion, it is necessary to draw a free-body diagram of each to determine the magnitude of the force reaction, $R$, between them (Fig. 13-6c). Note: $a_A = a_B = a$.

For $A$:

$$\sum F_y = ma_y = 0$$

$$N_A = 184\,\text{lb}$$

$$\sum F_x = ma_x$$

$$(200\,\text{lb})(0.866) - (96.6\,\text{lb})(0.5) - 0.1(184\,\text{lb}) - R = \frac{96.6\,\text{lb}}{32.2\,\text{ft/s}^2}\,a$$

$$106.5\,\text{lb} - R = 3\,\frac{\text{lb}\cdot\text{s}^2}{\text{ft}}\,a \qquad \textbf{(a)}$$

For $B$:

$$F_y = ma_y = 0$$

$$N_B - (96.6\,\text{lb})(0.866) = 0$$

$$N_B = 83.7\,\text{lb}$$

$$\sum F_x = ma_x$$

$$R - (96.6\,\text{lb})(0.5) - 0.1(83.7\,\text{lb}) = \frac{96.6\,\text{lb}}{32.2\,\text{ft/s}^2}\,a$$

$$R - 56.7\,\text{lb} = 3\,\frac{\text{lb}\cdot\text{s}^2}{\text{ft}}\,a \qquad \textbf{(b)}$$

Solving Eqs. (a) and (b) simultaneously,

$$a = 8.3\,\text{ft/s}^2 \;\nearrow \quad \text{and} \quad R = 81.6\,\text{lb}$$

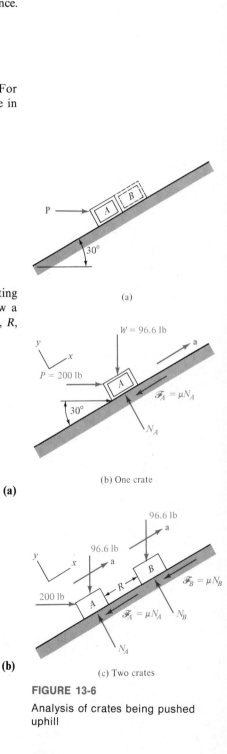

(a)

(b) One crate

(c) Two crates

**FIGURE 13-6**
Analysis of crates being pushed uphill

EXAMPLE 13-4   537

# EXAMPLE 13-5

A ski-jump track has a circular part of radius $r = 40\,\text{m}$ at point $A$, where the skier has a horizontal speed $v = 20\,\text{m/s}$ (Fig. 13-7a). The 70-kg skier's speed is decreasing at point $A$ at the rate of $1.3\,\text{m/s}^2$ because of air resistance. Determine the resultant force vector acting on the skier just before takeoff at $A$.

## SOLUTION

The free-body diagram of the skier is drawn in Fig. 13-7b. The weight $W$ is $mg = (70\,\text{kg})(9.81\,\text{m/s}^2) = 687\,\text{N}$. The unknown forces $R_x$ and $R_y$ are determined using normal and tangential components according to Fig. 13-7b. The equations of motion are from Eq. 13-6,

$$\sum F_t = ma_t = m\dot{v}$$

$$-R_x = (70\,\text{kg})(-1.3\,\text{m/s}^2)$$

$$R_x = 91\,\text{N}$$

$$\sum F_n = ma_n = m\frac{v^2}{\rho}$$

$$R_y - W = m\frac{v^2}{\rho}$$

$$R_y = 687\,\text{N} + (70\,\text{kg})\left(\frac{400\,\text{m}^2/\text{s}^2}{40\,\text{m}}\right)$$

$$= 1387\,\text{N}$$

The positive signs of $R_x$ and $R_y$ show that their directions were assumed correctly. The resultant force on the skier is

$$\mathbf{R} = R_x\mathbf{i} + (R_y - W)\mathbf{j} = (91\mathbf{i} + 700\mathbf{j})\text{N}$$

## JUDGMENT OF THE RESULTS

In static equilibrium the magnitude of the upward force on the skier would be $R_y = W = 687\,\text{N}$. Since the center of curvature is above the skier, the normal acceleration caused by moving on a curved path is upward, so the increase in $R_y$ is reasonable when compared with static equilibrium. The direction of $\mathbf{R}$ is reasonable by inspection.

(a)

(b)

FIGURE 13-7

Analysis of a ski jumper

EXAMPLE 13-6

Consider the amusement ride in Fig. 13-8a. A train of small cars with passengers travels inside the loop of the tracks, on a circular path of 70 ft diameter. Assume that the tracks can apply only normal forces $P_n$ to the cars. The tracks and wheels are so constructed that the train cannot fall away from the tracks. Each car has a weight of $W = 700$ lb. **(a)** Determine the speed $v_A$ of the train at point $A$ at which there is no force from the tracks on a car at $A$. **(b)** At point $B$ the train's speed is $v_B = 50$ ft/s. Determine the normal force on a car at point $B$.

## SOLUTION

**(a)** The free-body diagram of a car at $A$ is drawn in Fig. 13-8b. Since the path has a constant radius of curvature and only normal forces have to be considered, the equation of motion is most convenient in terms of normal components. From Eq. 13-6a,

$$\sum F_n = m \frac{v_A^2}{r}$$

From Fig. 13-8b,

$$P_n + W = m \frac{v_A^2}{r}$$

where $P_n$ must be zero according to the statement of the problem. Using $W = mg$, the mass cancels out, and

$$v_A = (gr)^{1/2} = [(32.2 \, \text{ft/s}^2)(35 \, \text{ft})]^{1/2} = 33.57 \, \text{ft/s}$$

**(b)** It is seen from the free-body diagram in Fig. 13-8c that $W$ cannot affect the normal force in that particular position of the car. From Eq. 13-6a,

$$P_n = m \frac{v_B^2}{r} = \left(\frac{W}{g}\right) \frac{v_B^2}{r}$$

$$= \left(\frac{700 \, \text{lb}}{32.2 \, \text{ft/s}^2}\right) \frac{(50 \, \text{ft/s})^2}{35 \, \text{ft}} = 1553 \, \text{lb}$$

(a)

**FIGURE 13-8**

Analysis of an amusement ride with motion in a vertical plane

(b) Free-body diagram of a car at $A$

(c) Free-body diagram of a car at $B$

EXAMPLE 13-6   539

## EXAMPLE 13-7

At the instant the bullet $P$ of mass $m$ is at position $r$, the hunter is rotating the rifle about a vertical axis through point $O$ at an angular velocity $\dot{\theta}$ and angular acceleration $\ddot{\theta}$ (Fig. 13-9a). The speed of the bullet relative to the gun barrel, $v_r = \dot{r}$, is increasing in magnitude at a rate $a_r = \ddot{r}$. Determine **(a)** the general expression for the total normal force $\mathbf{F}_B$ exerted by the bullet on the inside of the gun barrel, and **(b)** the magnitude of this force for $\dot{\theta} = 3 \text{ rad/s}$, $\ddot{\theta} = 5 \text{ rad/s}^2$, $r = 0.5 \text{ m}$, $v_r = 400 \text{ m/s}$, and $a_r = 50 \text{ m/s}^2$ for a bullet of mass $0.03 \text{ kg}$.

### SOLUTION

**(a)** Polar coordinates are well suited for this problem where $\dot{r} = v_r$ and $\ddot{r} = a_r$ in the polar coordinate equations, Eq. 13-7. Let the normal force *on* the bullet due to the gun barrel be called $\mathbf{F}$ as shown in Fig. 13-9b. Using Eq. 13-7b for the transverse direction,

$$\sum F_\theta = ma_\theta = m(r\ddot{\theta} + 2\dot{r}\dot{\theta}) = F$$

The total normal force $\mathbf{F}_B$ exerted *by* the bullet on the side wall of the gun barrel is due to gravity and the dynamic reaction $-\mathbf{F}$ (Newton's third law),

$$\mathbf{F}_B = -m(r\ddot{\theta} + 2\dot{r}\dot{\theta})\mathbf{n}_\theta - mg\mathbf{k}$$

**(b)** For $\dot{\theta} = 3 \text{ rad/s}$, $\ddot{\theta} = 5 \text{ rad/s}^2$, $r = 0.5 \text{ m}$, $\dot{r} = v_r = 400 \text{ m/s}$, $m = 0.03 \text{ kg}$, and $g = 9.81 \text{ m/s}^2$,

$$\mathbf{F}_B = -(0.03 \text{ kg})[(0.5 \text{ m})(5 \text{ rad/s}^2) + 2(400 \text{ m/s})(3 \text{ rad/s})]\mathbf{n}_\theta - (0.03 \text{ kg})(9.81 \text{ m/s}^2)\mathbf{k}$$
$$= (-72.1\mathbf{n}_\theta - 0.29\mathbf{k}) \text{ N}$$

Note that the dominant part of the force $\mathbf{F}_B$ is due to the acceleration term $2\dot{r}\dot{\theta}$. This term is called the "Coriolis acceleration." Its magnitude depends on twice the product of a velocity relative to a rotating body multiplied by the body's angular velocity. In this example it is twice the velocity of the bullet, $v_r$, relative to the rotating gun barrel times the angular velocity $\dot{\theta}$ of the rotating gun. Coriolis acceleration terms can become very large in high-speed rotating machinery, thus resulting in large forces. This example demonstrates the physical nature of these Coriolis effects (see App. D).

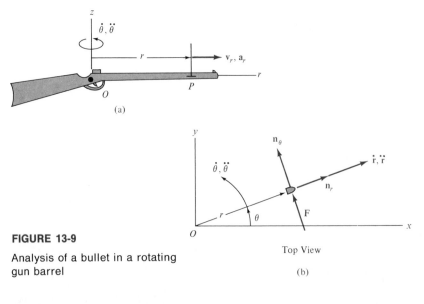

**FIGURE 13-9**

Analysis of a bullet in a rotating gun barrel

EXAMPLE 13-8

In designing the angle of bank $\theta$ for highway curves (Fig. 13-10a) it is possible to design each section of the curve so that at a given design speed a car can stay on the banked curve even if $\mathscr{F} = \mu N = 0$ (glare ice). If $\mathscr{F} \neq 0$, it simply means that the frictional force makes it possible to stay on the curve for a wider range of speeds. Determine the design relationship between $\rho$, $v$, and $\theta$ for a banked curve for $\mathscr{F} = 0$.

## SOLUTION

The free-body diagram for the car in the vertical plane at cross section A–A is drawn in Fig. 13-10b. Since the car is traveling on a circular path, it has a normal acceleration directed toward the center of curvature, $C$. The orthogonal axis for which another equilibrium equation can be written is the $y$ axis, along which the acceleration is zero. Any increase or decrease in speed is an acceleration in the tangential direction, which is into the paper in Fig. 13-10b. Using Eqs. 13-5a and 13-6a with $\mathscr{F} = 0$,

$$\sum F_y = ma_y$$

$$N \cos \theta - W = 0$$

$$N \cos \theta = W \qquad \textbf{(a)}$$

$$\sum F_n = ma_n = m\frac{v^2}{\rho}$$

$$N \sin \theta = \frac{W}{g}\left(\frac{v^2}{\rho}\right) \qquad \textbf{(b)}$$

Combining Eqs. (a) and (b),

$$\tan \theta = \frac{v^2}{g\rho} \qquad \text{for } \mathscr{F} = 0$$

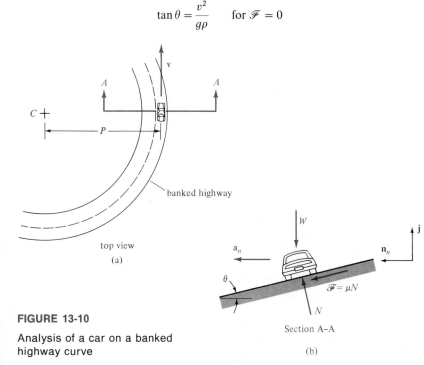

top view
(a)

banked highway

Section A–A

(b)

**FIGURE 13-10**

Analysis of a car on a banked highway curve

EXAMPLE 13-8    541

**13-1** Express Newton's second law of motion in terms of cylindrical component equations.

**13-2** Write the scalar equations of motion using spherical coordinates.

**13-3** Determine the weight of a 70-kg person at sea level where $g = 9.81 \text{ m/s}^2$, and at an elevation $h_1$ where $g_1 = 8 \text{ m/s}^2$. Draw a free-body diagram of this person in a free fall at the elevation $h_1$.

**13-4** An electronic instrument weighs 15 lb on the earth. Determine the mass of this device and its weight on the moon where the acceleration of gravity is $5.31 \text{ ft/s}^2$.

**13-5** A rocket is accelerating vertically at the rate of $80 \text{ ft/s}^2$ shortly after takeoff. Determine the net force from the seat on a 170-lb astronaut.

**Fig. P13-5**

**13-6** The acceleration of the rocket in Fig. P13-5 is $\mathbf{a} = (3\mathbf{i} + 25\mathbf{j}) \text{ m/s}^2$ shortly after takeoff. Determine the vertical and horizontal forces from the seat on a 70-kg astronaut.

**13-7** In a test of packaging materials a 10-kg electronic instrument is dropped on a cushioning pad. The speed of the device is changed by the pad from 15 m/s to zero in the time interval of $\Delta t = 0.002 \text{ s}$. Calculate the force on the electronic device during its deceleration.

**Fig. P13-7**

**13-8** Assume that in Prob. 13-7 the time interval of deceleration is not known, but the deformation of the pad is measured. Determine the force on the device if the maximum deformation of the pad is 40 mm.

**13-9** In a laboratory test of restraining belts for automobile passengers the initial horizontal speed of a dummy is $v_0 = 120 \text{ ft/s}$. Its final speed is zero after a time interval of $\Delta t = 0.01 \text{ s}$. Determine the net force that a 200-lb dummy applies to the belt in the crash test. All motion is horizontal.

**Fig. P13-9**

**13-10** A truck travels horizontally at speed $v_0 = 60 \text{ ft/s}$ when its brakes are applied locking the wheels. Determine the coefficient of kinetic friction if the truck stops after sliding 350 ft.

**Fig. P13-10**

**13-11** A car is traveling horizontally at speed $v_0 = 30 \text{ m/s}$ when its brakes are applied locking the wheels. The car hits a structure after sliding 40 m. Determine the car's velocity just before impact if the coefficient of kinetic friction is 0.4.

Fig. P13-11

**13–12** For a car involved in an accident, it is estimated that its speed just before impact was $v_f = 40$ ft/s. Skid marks over a distance of $d = 150$ ft indicate the wheels were locked. Determine the initial speed $v_0$ of the car when the brakes were applied if the coefficient of kinetic friction was 0.5.

**13–13** A 5-kg block is moved horizontally by a force $\mathbf{P} = (200\mathbf{i} + 25\mathbf{j})$ N. Determine the acceleration vector of the block if the coefficient of kinetic friction is 0.3.

Fig. P13-13

**13–14** Solve Prob. 13–13 assuming that $\mathbf{P} = (-200\mathbf{i} - 25\mathbf{j})$ N.

**13–15** The coefficient of static friction between the 10-lb block and the conveyor belt is 0.6. Determine the maximum upward acceleration of the block along the incline that can be achieved without slippage of the block on the belt.

Fig. P13-15

**13–16** An experimental racing car of 2500-lb weight has initial velocity $v_0 = 150\mathbf{i}$ ft/s when its brakes are applied locking the wheels. (a) Determine the stopping distance if the coefficient of kinetic friction is 0.6. (b) To assist in the braking, a jet applies a downward thrust $\mathbf{P} = -300\mathbf{j}$ lb throughout the deceleration. Calculate the stopping distance for this case.

Fig. P13-16

**13–17** An experimental car is equipped with a jet of thrust $\mathbf{P}$ (Fig. P13-16) to shorten the stopping distance of the car from its initial velocity $\mathbf{v}_0$. For any angle $\theta$, $P = 0.2\ W$. Plot the normalized stopping distance $d/(v_0^2/g)$ as a function of $\theta = 0°, 45°,$ and $90°$ for a coefficient of friction $\mu = 0.2$.

**13–18** A 1000-kg car has initial speed $v_0 = 25$ m/s down the incline when its brakes are applied locking the wheels. Determine the stopping distance if the coefficient of kinetic friction is 0.7.

Fig. P13-18

**13–19** A 150-kg box on the horizontal forks of the lift truck is not to shift during motion of the truck. The forks are not moving vertically when the truck is slowing down from a speed of 15 km/h. What is the shortest stopping distance allowed if the coefficient of static friction is 0.3?

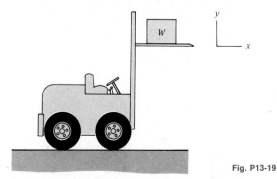

Fig. P13-19

**13–20** The fork lift truck in Prob. 13–19 is lowering a 300-lb box at a downward acceleration of 4 ft/s². What is the maximum horizontal acceleration of the truck if the box is not to shift on the horizontal forks where the coefficient of static friction is 0.4?

**13–21** A truck is decelerating at the rate of 20 ft/s² when the 500-lb crate begins to shift forward on the sloping truck bed. Determine the coefficient of friction between the crate and the truck bed.

Fig. P13-21

**13–22** The truck in Prob. 13–21 travels at a speed of $v_0 = 90$ km/h when its brakes are applied. What is the shortest stopping distance allowed if the 200-kg crate is not to shift on the sloping truck bed? The coefficient of static friction at the crate is 0.25.

**13–23** The locomotive $A$ weighs $2 \times 10^5$ lb and the freight car $B$ weighs $1.5 \times 10^5$ lb. They travel at a speed of $v_0 = 50$ mph when a 6000-lb braking force is applied to $A$ and a 4000-lb braking force is applied to $B$. Determine the stopping distance and the force in the coupling between $A$ and $B$.

**Fig. P13-23**

**13–24** In Prob. 13–23, an identical freight car $C$ is coupled to $B$. The braking forces on $B$ and $C$ are 4000 lb each. Determine the stopping distance and the forces in the two couplings ($A$–$B$ and $B$–$C$) if everything else is as in Prob. 13–23.

**13–25** A small elevator $A$ is connected to a counterweight $B$ through a pulley of negligible weight and axle friction. (a) Determine the acceleration of $A$ starting from rest. (b) Assume that the counterweight $B$ is replaced by a constant-force spring which applies the same force $\mathbf{F}$ over the full range of displacements. Determine the acceleration of $A$ in this case.

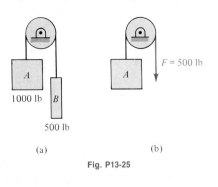

**Fig. P13-25**

**13–26** The blocks are released from rest. Determine the tensions in rope segments $a$, $b$, $c$, and $d$ assuming that the pulleys have negligible mass and axle friction.

**Fig. P13-26**

**13–27** Consider an $X$-$Y$ plotter which records an electric signal $Y$ as a function of another signal $X$. Bar $A$ always remains vertical as shown but moves horizontally according to the $X$ signal. Carriage $B$ with the pen moves along the bar as controlled by the $Y$ signal. The 0.2-kg bar and the 0.04-kg carriage move in the same vertical plane. Determine the forces $\mathbf{F}_A$ on the bar and $\mathbf{F}_B$ on the carriage at an instant when $\mathbf{a}_A = 10\mathbf{i}$ m/s$^2$ and $\mathbf{a}_{B/A} = 25\mathbf{j}$ m/s$^2$.

**Fig. P13-27**

**13–28** Consider the steam-driven pile driver. The 200-kg hammer is raised by steam pressure before each drop through a height of 2 m. The steam pressure can also be applied to cause a downward force on the falling hammer. Determine the magnitude of the constant downward force which would make the 2-m forced fall equivalent to a 4-m free fall for driving a pile.

**Fig. P13-28**

**13–29** A speed skater travels at a constant speed of 15 m/s on a curved path of 30 m radius. Calculate the magnitude of the horizontal force on the 70-kg skater.

Fig. P13-29

**13–30** A speed boat travels at $v = 35$ mph on a circular path of radius $r = 50$ ft. Determine the centripetal force on the 400-lb boat.

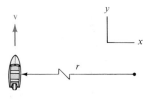

Top view                    Fig. P13-30

**13–31** The road segment is horizontal and has a radius of curvature of 150 ft. The 2000-lb car travels at a constant speed of 40 mph. Determine the centripetal force on the car and the coefficient of friction necessary to keep the car on the circular path.

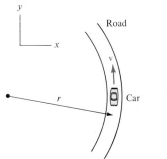

Fig. P13-31

**13–32** A 1000-kg car on a horizontal road of radius $r = 100$ m (Fig. P13-31) travels at speed $v = 30$ m/s, which is decreasing at the rate of 5 m/s$^2$. Determine the total horizontal force on the car.

**13–33** A speed boat travels at a speed of 60 km/h on a circular path of 30 m radius. Determine the required angle $\theta$ of the boat with the horizontal if the passengers must not have any tendency to slide on the seats. For this, it is necessary to assume a coefficient of static friction of zero.

Front view                  Fig. P13-33

**13–34** A speed boat travels at a speed of 40 mph on a circular path of 50 ft radius. Determine the required angle $\theta$ of the boat with the horizontal (Fig. P13-33) if the passengers must not slide on the seats, where the coefficient of static friction is 0.3.

**13–35** The centerline of a road is in the horizontal plane and has a radius of curvature $r = 300$ ft. The legal speed at this location is planned to be 40 mph. Determine the appropriate banking angle $\theta$ of the road if a car traveling at the legal speed should not skid on the road where the coefficient of static friction is 0.2.

Rear view

Fig. P13-35

**13–36** A car travels on a road of radius $r = 100$ m and banking angle $\theta = 10°$ (Fig. P13-35). Determine the maximum speed $v$ of the car if it is not to skid on the road, where the coefficient of static friction is 0.4.

**13–37** The track of the amusement ride has a radius of curvature $r = 40$ ft and a banking angle $\theta = 35°$ with the horizontal. Determine the speed $v$ of the train if the passengers must not have any tendency to slide on the seats. Assume that the track is horizontal and the coefficient of friction at the seats is zero.

Fig. P13-37

**13–38** In Prob. 13–37, $v = 40$ ft/s and the coefficient of static friction at the seats is 0.4. Determine whether the passengers would slide on the seats.

**13–39** A high-speed train is proposed on a monorail track. Calculate the force on the track when a 50,000-lb coach travels at a speed of 200 ft/s on a horizontal track of 1000 ft radius (Fig. P13-39a).

(a)  **Fig. P13-39**  (b)

**13–40** Determine the required banking angle $\alpha$ of the seats in the coach in Prob. 13–39 if the passengers are not to slide on the seats, where the coefficient of static friction is 0.5 (Fig P13-39b).

**13–41** The airplane is flying on a horizontal circular path of 400 m radius at a speed of 100 m/s which is decreasing at the rate of 5 m/s². Determine whether the pilot would slide on the seat where the coefficient of static friction is 0.5 if the banking angle is $\theta = 20°$. Assume that the seat is flat.

**Fig. P13-41**

**13–42** An amusement ride consists of a rotating cage with people standing in it by the wall. When the cage is rotating at a constant angular velocity $\omega_0$ about the vertical axis, the floor drops out from underneath the passengers. Determine the minimum angular velocity $\omega_0$ for which a passenger of weight $W$ will not slide down the wall when the floor is removed. The coefficient of friction between the passenger and the wall is $\mu = 0.2$, and the radius of the cage is $r = 7$ m.

**Fig. P13-42**

**13–43** A 70-kg person on roller skates is doing a stunt. At the bottom of the circular track of 6 m radius the skater's speed is $v_A = 10$ m/s. Determine the normal force on the skater at $A$.

**Fig. P13-43**

**13–44** A gymnast swinging on a high bar is modeled as a 160-lb particle at a distance of 3.5 ft from bar $B$. Determine the force of the bar on the gymnast's arms when he is at point $A$ below the bar, the "particle" moving at a horizontal speed of $v_A = 15$ ft/s.

**Fig. P13-44**

**13–45** A 160-lb pilot is belted in the seat of the airplane which makes a vertical loop of 800-ft diameter. Determine the net force of the seat on the pilot at point $A$ if the speed of the airplane is $v_A = 100$ mph, decreasing at the rate of 20 ft/s².

**Fig. P13-45**

**13–46** The airplane in Fig. P13-45 is flying at a speed of $v_B = 200\,\text{m/s}$ and accelerating at the rate of $15\,\text{m/s}^2$ at the low point $B$ of a vertical loop of 400-m diameter. Determine the net force of the seat on the 65-kg pilot.

**13–47** A child is twirling a 1-kg stone at the end of a 1-m-long rope at the rate of 200 rpm. Determine the maximum and minimum tension in the rope, assuming that the path is in the vertical plane and its center $C$ is fixed.

Fig. P13-47

**13–48** Solve Prob. 13–47 assuming that the plane of the stone's path makes a 45° angle with the horizontal.

**13–49** Passengers are belted in a roller coaster which travels in a vertical plane. At a high point $A$ the speed of the train is $v_A = 8\,\text{ft/s}$, decreasing at the rate of $10\,\text{ft/s}^2$, and $r = 25\,\text{ft}$. Calculate the net force of the seat on a 200-lb passenger.

(a)

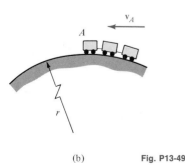

(b)          Fig. P13-49

**13–50** Determine the constant speed $v_A$ in Prob. 13–49 at which the net force of the seat on the passenger is zero (the passenger experiences weightlessness).

**13–51** An experimental car traveling on a horizontal road is tracked by radar located at point $O$. At a particular instant $r = 100\,\text{m}$, $\dot{r} = 3\,\text{m/s}$, $\ddot{r} = 0$, $\theta = 60°$, $\dot{\theta} = 0.3\,\text{rad/s}$, and $\ddot{\theta} = 0$. Calculate the net horizontal force on the 1300-kg car.

Fig. P13-51

**13–52** Radar data of a car traveling on a horizontal road (Fig. P13-51) indicate that $r = 250\,\text{ft}$, $\dot{r} = 0$, $\ddot{r} = 10\,\text{ft/s}$, $\theta = 45°$, $\dot{\theta} = 0.6\,\text{rad/s}$, $\ddot{\theta} = 0.01\,\text{rad/s}^2$. Calculate the net horizontal force on the 2600-lb car.

**13–53** Consider a wire-controlled model airplane which flies in a horizontal plane. Calculate the net horizontal force on the 10-lb airplane if $r = 30\,\text{ft}$ (in the plane of the path), $\dot{r} = 0$, $\ddot{r} = 0$, $\dot{\theta} = 1\,\text{rad/s}$, and $\ddot{\theta} = 0.3\,\text{rad/s}^2$.

Fig. P13-53

**13–54** The 2-kg block $A$ on the rotating arm is at $r = 0.4\,\text{m}$ when $\dot{r} = 2\,\text{m/s}$, $\ddot{r} = 0$, $\dot{\theta} = 3\,\text{rad/s}$, $\ddot{\theta} = 0.7\,\text{rad/s}^2$. Determine the $x$ and $y$ components of the net force on the block for $\theta = 30°$.

Fig. P13-54

Section 13-1

**Newton's second law of motion**

$$\mathbf{F} = m\mathbf{a}$$

The magnitude of the acceleration **a** of a particle is directly proportional to the magnitude of the resultant force **F** acting on it, and inversely proportional to its mass $m$. The direction of the acceleration is the same as the direction of the resultant force. The motion of the particle is determined with respect to a non-accelerating (Newtonian) reference frame.

Section 13-2

**SI units**

**FORCE:**

$$F = ma$$

where  $F$ = newtons
      $m$ = kilograms
      $a$ = meters per second squared

**WEIGHT:**

$$W = mg$$

where  $W$ = newtons
      $m$ = kilograms
      $g$ = acceleration of gravity $\simeq 9.81\,\text{m/s}^2$

**U.S. customary units**

**FORCE:**

$$F = ma$$

where  $F$ = pounds
      $m$ = slug
      $a$ = feet per second squared

**WEIGHT:**

$$W = mg$$

where  $W$ = pounds
      $m$ = slug
      $g$ = acceleration of gravity $\simeq 32.2\,\text{ft/s}^2$

**Equations of motion**   Basic equation of motion for any number of forces $\mathbf{F}_1, \mathbf{F}_2, \mathbf{F}_3, \ldots$, acting on one particle.

$$\sum \mathbf{F}_i = m\mathbf{a}$$

**Scalar component equations of motion**

**RECTANGULAR COMPONENTS:**

$$\sum F_x = ma_x \quad \sum F_y = ma_y \quad \sum F_z = ma_z$$

**NORMAL AND TANGENTIAL COMPONENTS:**

$$\sum F_n = ma_n = m\frac{v^2}{\rho} \quad \text{(centripetal force)}$$

$$\sum F_t = ma_t = m\dot{v} = mv\frac{dv}{ds}$$

**RADIAL AND TRANSVERSE COMPONENTS:**

$$\sum F_r = ma_r = m(\ddot{r} - r\dot{\theta}^2)$$
$$\sum F_\theta = ma_\theta = m(r\ddot{\theta} + 2\dot{r}\dot{\theta})$$

Section 13-4

**Procedure for solving problems**

**1.** Draw the free-body diagram of the particle showing known and unknown forces.

**2.** Establish a convenient nonaccelerating reference frame.

**3.** Apply the equations of motion which are appropriate for the chosen coordinate system to determine forces or accelerations.

**4.** If the velocity or position of the particle must be calculated, use the equations of kinematics after the acceleration has been determined.

**13–55** The truck weighs 9660 lb and is dragging a crate weighing 1610 lb by means of a cable. The truck is able to develop a maximum traction of $F_D = 4000$ lb between the tires and the horizontal surface. Assume that both the truck and the crate can be approximated as particles. Determine (a) the maximum acceleration of the crate, and (b) the corresponding tension in the cable.

**Fig. P13-55**

**13–56** Consider a wrecker's ball suspended from a crane. Assume that point $A$ is fixed. A proposed safety device would measure the tension in the cable at $A$ and compute the velocity and acceleration of the ball as a function of $\theta$. The ball swings in the vertical plane. Calculate the velocity and acceleration for $L = 10$ m and $\theta = 20°$ if the tension in the cable is twice the weight of the ball.

**Fig. P13-56**

**13–57** The 1000-kg wrecker's ball swings in the vertical plane (Fig. P13-56). Determine the tension in the wire rope supporting the ball if point $A$ is fixed, $L = 10$ m, $\dot{L} = 1$ m/s, $\ddot{L} = -0.5$ m/s² ($\dot{L}$ is decreasing), $\theta = 30°$, $\dot{\theta} = 0.8$ rad/s. Also determine $\ddot{\theta}$ at this instant.

**13–58** A speed-governing mechanism is modeled with two 0.2-kg balls $A$ suspended by weightless rods hinged at $B$. Calculate the angular speed $\omega$ and the tension in the rods if $L = 0.1$ m, and $\theta = 60°$. The hinges allow rotation of the rods only in a vertical plane. The given angle $\theta$ is constant in the present case.

**Fig. P13-58**

**13–59** In the speed-governing mechanism of Prob. 13–58, $m = 0.2$ kg, $L = 0.1$ m (rods $AB$ have negligible mass), $\theta = 40°$, $\dot{\theta} = 0$. Determine $\ddot{\theta}$ at that instant and the tension in the rod if the mechanism is rotating at 500 rpm.

# Kinetics of Particles: Work and Energy Methods

# 14

The fundamental method of analyzing the kinetics of particles uses Newton's second law, $\mathbf{F} = m\mathbf{a}$. This method enables one to analyze the velocities and displacements of a particle as well as the acceleration of the particle, but such analysis is not always convenient in practice. For example, consider the mechanized hay-baling and transporting system shown in Fig. 14-1. The baler takes hay from the ground, makes bales of hay, and throws the bales to a higher level in the trailer. On the basis of kinematics, the bales require a certain initial velocity as they are launched in order to attain the level of the trailer. The acceleration of the bale in the launching phase of the motion is predetermined depending on the physical constraints of the machine. However, although the displacement of the launching arm of the baler is known, the exact magnitude and direction of the force of the arm on each bale is not easily measured. For such a case, it is useful to develop supplementary methods for the analysis of forces, velocities, and displacements of a particle, without the need to consider accelerations. The concepts of work and energy enable one to make such a simplification of the basic methods of kinetics.

**FIGURE 14-1**

Hay-baling machine in operation

**SECTION 14-1** presents the definition of work of a force acting on a particle. Work is a scalar quantity and is frequently used in engineering mechanics because it simplifies the solution of certain problems. A prior study of work in a statics course is useful but not essential for studying this material.

**SECTION 14-2** describes the concept of potential energy for gravitational and elastic systems. Potential energy is frequently used in the analysis of moving bodies. The calculation of potential energy in most mechanical systems should be a routine procedure.

**SECTION 14-3** presents the definition of kinetic energy of a particle on the basis of Newton's second law. All moving bodies have kinetic energy, and this quantity is very useful in the analysis of motion. The relationship of work and energy is especially important in dynamics analyses.

**SECTION 14-4** introduces the concepts of power and efficiency which are important in the design of machines.

**SECTION 14-5** illustrates the advantages and limitations of using energy methods in kinetics. Overall, the advantages are substantial, and it is important to acquire a working knowledge of these methods.

**SECTION 14-6** defines the motion of particles in conservative systems in which friction does not oppose the motion. The elementary concepts of conservative systems are important for all students of dynamics. The advanced analysis at the end of Sec. 14-6 requires the use of partial derivatives, and for that reason it is often omitted from elementary dynamics courses.

**SECTION 14-7** presents the principle of conservation of mechanical energy. The relationship of kinetic and potential energies in a conservative system is one of the most important and useful expressions in dynamics, and it should be studied with emphasis.

## 14-1  WORK OF A FORCE

The term *work* has an exact, quantitative meaning in the area of mechanics. Quantitatively, *work is the product of the force and the corresponding displacement.* For example, the constant force **F** slowly moves the block from position 1 to position 2 along a smooth plane in Fig. 14-2a. If **F** is parallel to the incline, the work $U$ of the force is

$$U = Fs \qquad \boxed{14\text{-}1}$$

If **F** is at an angle $\theta$ with the plane as in Fig. 14-2b, only the component $F \cos \theta$ is associated with the displacement $s$ of the block if the inclined plane is fixed. The component $F \sin \theta$ does no work because there is no displacement of the block perpendicular to the fixed plane. Alternately,

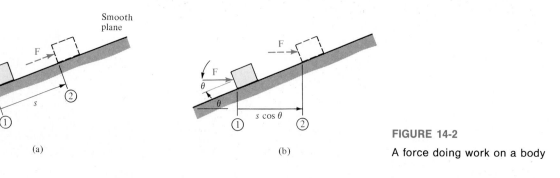

**FIGURE 14-2**

A force doing work on a body

the displacement corresponding to the horizontal force **F** in Fig. 14-2b is $s \cos\theta$. The work $U$ of the force **F** in Fig. 14-2b is

$$U = (F \cos\theta)s = F(s \cos\theta)$$ 14-2

In general, the force **F** may change in magnitude and direction during the displacement. The work in that case is determined by considering the infinitesimal work $dU$ of a force **F** during an infinitesimal displacement $d\mathbf{r}$ of particle $P$ as in Fig. 14-3. The work $dU$ is defined as the scalar product of the vectors **F** and $d\mathbf{r}$,

$$dU = \mathbf{F} \cdot d\mathbf{r} = F \, dr \cos\theta$$ 14-3

The total work $U_{12}$ in displacing particle $P$ from point 1 to 2 is obtained by integrating Eq. 14-3. Using the rectangular components of **F** and $d\mathbf{r}$ gives

$$U_{12} = \int_1^2 \mathbf{F} \cdot d\mathbf{r} = \int_1^2 F \cos\theta \, dr$$ 14-4
$$= \int_1^2 (F_x \, dx + F_y \, dy + F_z \, dz)$$

Work is a scalar quantity with units of force times distance, N·m or ft·lb (dimensionally the same as for moments). Work may be positive or negative according to the cosine function in Eq. 14-3.

Forces that do no work are common. For example, a force applied to a fixed particle causes no displacement, so its work is zero. Similarly, a force acting perpendicular to the displacement has no component in the direction of the displacement, and its work is also zero.

The definition of work can be alternately stated using Newton's second law, $\mathbf{F} = m\mathbf{a}$. From Eq. 14-4,

$$U_{12} = \int_1^2 \mathbf{F} \cdot d\mathbf{r} = \int_1^2 m\mathbf{a} \cdot d\mathbf{r}$$ 14-5

Note that $\mathbf{a} \cdot d\mathbf{r} = a_t \, dr$, where $a_t$ is the tangential acceleration of the particle and $dr$ is approximately the infinitesimal displacement along the path.

The following special aspects of total work are easy to prove and are worth remembering:

1. The total work of concurrent forces is equal to the work of the resultant of those forces.
2. The total work of internal forces in a rigid member is zero.

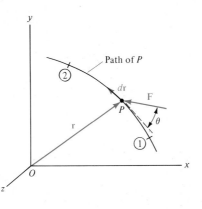

**FIGURE 14-3**

Analysis of a general displacement for the definition of work

## POTENTIAL ENERGY AND WORK

Historically, the term *potential energy* comes from the perception that a weight $W$ situated at any level could do useful work (drive a clock, pump water, etc.) if allowed to descend to a lower level with respect to the center of the earth. The formal expression of this concept is based on Fig. 14-4, where a body of weight $W$ moves without friction between levels 1 and 2. In going from level 1 to 2, the total work of the constant force $\mathbf{W}$ according to Eq. 14-4 is

$$\underbrace{U_{12}}_{\substack{\text{work of } W \\ \text{from 1 to 2}}} = \int_1^2 W \, dy = W(y_2 - y_1) = \underbrace{Wh}_{\substack{\text{potential energy} \\ \text{of } W \text{ at 1} \\ \text{with respect to 2}}} \qquad \boxed{14\text{-}6}$$

The potential of the weight $\mathbf{W}$ to do work because of its relative height $h$ is defined as its *gravitational potential energy*. For further illustration of this concept, consider the potential energies at level 2 with respect to 3, at 1 with respect to 3, and at 2 with respect to 1 in Fig. 14-4. The results are, respectively,

$$U_{23} = \int_2^3 W \, dy = Wl \qquad \boxed{14\text{-}7\text{a}}$$

$$U_{13} = \int_1^3 W \, dy = W(h + l) \qquad \boxed{14\text{-}7\text{b}}$$

$$U_{21} = \int_2^1 -W \, dy = -Wh = -U_{12} \qquad \boxed{14\text{-}7\text{c}}$$

The minus sign in the last equation shows that gravity opposes the motion up.

It should be noted that Eq. 14-6 is valid only if $W = mg$ is essentially a constant over the displacement $h$.

### Elastic Potential Energy

The potential energy of elastic members* is frequently important in mechanical systems. The background for analyzing any elastic member is the *theory of springs* discovered by Robert Hooke in 1642 and illustrated in Fig. 14-5. A helical spring is loaded with progressively larger weights $W$, $2W$, and so on (Fig. 14-5a–d). The deformations $x$ are proportional to the applied force $F$ for small deformations as shown by the straight-line plot in Fig. 14-5e. The slope $k$ of this line is called the *spring constant* or *spring stiffness*. It has units of N/m or lb/in., and it is the property of each particular spring (depends on material, wire diameter, coil length, and coil diameter). From Fig. 14-5e,

$$F = kx \qquad \boxed{14\text{-}8}$$

**FIGURE 14-4**

Model for defining gravitational potential energy

* Ideal elastic members return to their original dimensions after the forces acting on them are removed. Elastic members are classified as linear or nonlinear depending on their diagrams of force vs. deformation. The term *inelastic behavior* means that the member does not return to its original shape after the forces on it are removed. Such behavior is analyzed in mechanics of materials.

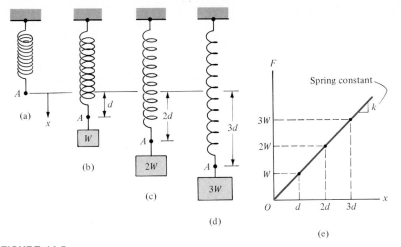

(a) (b) (c) (d) (e)

**FIGURE 14-5**

Model for the theory of elastic springs

which is used in Eq. 14-4 to obtain the work of the external force $\mathbf{F}$ in stretching the spring from $O$ to $x_1$,

$$U_1 = \int_O^{x_1} kx\, dx = \frac{1}{2} kx_1^2$$

This is equal to the shaded area in Fig. 14-6. In general, the work $U$ of an *external force* $\mathbf{F}$ in *causing* a deformation $x$ from the undeformed length of a spring is

$$\underline{U} \qquad = \qquad \frac{1}{2} kx^2$$

work of $F$    potential energy of spring
from $O$ to $x$    at $x$ with respect to $O$

14-9a

The potential of a spring to do positive work because of its deformation $x$ is defined as its *elastic potential energy*. In the same deformation $x$, the work $U$ of the *force exerted by the spring* is

$$U = -\frac{1}{2} kx^2$$

14-9b

since the spring opposes the deformation which is caused by the external force $\mathbf{F}$.

In general, the work done by an *external force* in deforming a linear, elastic spring from $x_1$ to $x_2$ (Fig. 14-6) is

$$U = \int_{x_1}^{x_2} kx\, dx = \frac{1}{2} k(x_2^2 - x_1^2)$$

14-10a

**FIGURE 14-6**

Definition of elastic potential energy

while the work of the *force exerted by the spring* is

$$U = -\frac{1}{2} k(x_2^2 - x_1^2)$$  14-10b

To avoid confusion concerning the signs, recall that a force does positive work when it acts in the direction of the displacement, and vice versa. Use free-body diagrams to determine the correct directions of forces and displacements. Note that a deformed spring does positive work on a body when the spring is returning to its undeformed position. This is true for either tensile or compressive deformations of the spring since in both cases the displacement and the force of the spring are in the same direction.

## Notation for Potential Energy

A commonly used symbol for potential energy is $V$. The *gravitational potential energy* is denoted by $V_g$ and is a quantity that depends on the weight $W$ of the particle and on the relative vertical position $h$ of the particle with respect to a defined horizontal datum plane,

$$V_g = Wh$$  14-11a

This energy must be taken as positive when $h$ is above the datum plane and negative for positions below the datum plane.

The *elastic potential energy*, $V_e$, of an elastic spring in linear deformation is given by

$$V_e = \frac{1}{2} kx^2$$  14-11b

where $k$ is the spring constant and $x$ is the change in the length of the spring from its unstretched position. The elastic potential energy is always positive.

Sometimes it is useful to consider the *total potential energy $V$* of a system consisting of a particle in a gravitational field and a spring acting on the particle. The total potential energy in this case is expressed as the algebraic sum of the gravitational and elastic potential energies,

$$\boxed{V = V_g + V_e}$$  14-11c

Although the units of potential energy are the units of work, N·m or ft·lb, work is defined for a displacement of a particle over a specific path, whereas potential energy is defined for a particular configuration. When the potential energy is evaluated at the end points of the path, the following equation applies:

$$\boxed{U = -\Delta V}$$  14-12

Equation 14-12 is valid for any ideal system based on gravity, springs, and so on, in which the work of a force depends only on the net change of position and not on the path of reaching that position. Such ideal systems are called *conservative*, and the forces involved are called *conservative forces*. These concepts are discussed in more detail in Sec. 14-6.

When a particle is in motion it has *kinetic energy* because of its velocity. To analyze this form of energy, consider a particle of mass $m$ which moves on a curved path as in Fig. 14-7.

The kinetic energy associated with the velocity of the particle is defined on the basis of Newton's second law as follows. Denoting the displacement by $s$, and the magnitude of the velocity by $v$,

$$F_t = ma_t = m\frac{dv}{dt} = m\frac{dv}{ds}\frac{ds}{dt} = mv\frac{dv}{ds}$$

$$F_t\,ds = mv\,dv$$

This equation is integrated from position 1 to 2 where the position and velocity have values $s_1$, $v_1$, and $s_2$, $v_2$, respectively. For a constant mass $m$,

$$\int_{s_1}^{s_2} F_t\,ds = m\int_{v_1}^{v_2} v\,dv = \frac{1}{2}mv_2^2 - \frac{1}{2}mv_1^2 \qquad \boxed{14\text{-}13}$$

The integral $\int F\,ds$, a scalar quantity, is the work of the force on a particle between the positions indicated in Fig. 14-7. The scalar quantity $\frac{1}{2}mv^2$ is defined as the *kinetic energy* of the particle at the particular speed $v$. Kinetic energy is generally denoted by the symbol $T$,

$$\boxed{T = \frac{1}{2}mv^2} \qquad \boxed{14\text{-}14}$$

*Kinetic energy is always positive*, regardless of the direction of the velocity vector, since mass is positive and speed is squared. The units of kinetic energy are the units of work,

$$\frac{\text{kg} \cdot \text{m}^2}{\text{s}^2} = \left(\frac{\text{kg} \cdot \text{m}}{\text{s}^2}\right)\text{m} = \text{N} \cdot \text{m (joule)}$$

in SI units and

$$\frac{\text{slug} \cdot \text{ft}^2}{\text{s}^2} = \left(\frac{\text{slug} \cdot \text{ft}}{\text{s}^2}\right)\text{ft} = \text{lb} \cdot \text{ft}$$

in U.S. customary units.

**FIGURE 14-7**

Quantities considered in the definition of kinetic energy

Path of particle

The relationship of work and kinetic energy as expressed by Eq. 14-13 can be written concisely as

$$\boxed{U_{12} = T_2 - T_1} \qquad \text{14-15}$$

where $U_{12}$ is the work of the force on the particle as it moves from position 1 to 2, and $T_1$ and $T_2$ are the corresponding initial and final kinetic energies of the particle. Equation 14-15 is called the *principle of work and energy*. Since work may be positive or negative, a force may increase or decrease the kinetic energy of a particle as outlined in the following.

$$\textit{Positive work:} \quad T_1 + U_{12} = T_2, \quad T_2 > T_1 \qquad \text{14-16}$$

In words, positive work always increases the kinetic energy.

$$\textit{Negative work:} \quad T_1 + U_{12} = T_2, \quad 0 \le T_2 < T_1 \qquad \text{14-17}$$

where it is noted that kinetic energy is always positive.

## 14-4

## POWER AND EFFICIENCY

In the design of machines it is often necessary to consider how much work must be done by the whole machine in a given time. The *time rate* at which work is done is defined as *power*:

$$\text{power} = \frac{dU}{dt} \qquad \text{14-18}$$

This can be manipulated to obtain an expression in terms of force and velocity which are relatively easily measured or known in advance in the design. Since $dU = \mathbf{F} \cdot d\mathbf{r}$ and $d\mathbf{r}/dt = \mathbf{v}$,

$$\boxed{\text{power} = \frac{dU}{dt} = \frac{\mathbf{F} \cdot d\mathbf{r}}{dt} = \mathbf{F} \cdot \mathbf{v}} \qquad \text{14-19}$$

This gives a scalar quantity since both work and time are scalars. The units of power are the watt (W) in SI units, and ft·lb/s or *horsepower* (hp) in U.S. customary units. The details and relations of these units are as follows.

$$1\,\text{W} = 1\,\text{J/s} = 1\,\text{N·m/s}$$

$$1\,\text{hp} = 550\,\text{ft·lb/s} = 33{,}000\,\text{ft·lb/min}$$

$$1\,\text{ft·lb/s} = 1.356\,\text{J/s} = 1.356\,\text{W}$$

$$1\,\text{hp} = 550(1.356\,\text{W}) = 746\,\text{W} = 0.746\,\text{kW}$$

### Mechanical Efficiency

Real mechanical systems always operate with frictional losses, so the useful work done by a machine is always less than the total work done by that machine. The *mechanical efficiency* $\eta$ is defined as

$$\boxed{\eta = \frac{\text{output work}}{\text{input work}} = \frac{\text{useful work}}{\text{total work required}}} \qquad \text{14-20}$$

Since both the output work and the input work are done in the same time interval, the efficiency can be defined in terms of power,

$$\eta = \frac{\text{power output}}{\text{power input}} = \frac{\text{useful power}}{\text{total power required}} \qquad \boxed{14\text{-}21}$$

The mechanical efficiency of an ideal machine is $\eta = 1$ and of any real machine is $\eta < 1$.

## Overall Efficiency

Equation 14-21 is also valid when a system transforms energy from one form to another. The meaning of overall efficiency is that all losses of energy (electric, frictional, thermal, etc.) can be expressed in one consistent system of units and treated as loss in general. Of course, energy is conserved in a physical sense, and loss is simply the energy that must be expended in the process of doing useful work but not available for useful work.

## OPPORTUNITIES AND LIMITATIONS OF USING ENERGY METHODS

The principle of work and energy (Eq. 14-15) is extremely useful in solving certain problems. For example, consider the initial steps in the design of the amusement ride shown in Fig. 14-8. The train with passengers is to start from rest at point $O$. The train should accelerate over a short distance $OA$ and make an inside loop at a height $h$ with a radius of curvature $r_B$. The speed $v_B$ at the top of the loop is specified. The known friction force $\mathscr{F}$ is constant along the track of given length $s$. It is required to determine the work $U_{OA}$ of the starter machine acting on the motorless train so that it can roll from point $A$ to $B$ and have a speed $v_B$ at $B$. This requirement leads to the following major steps in the analysis of an arbitrary mass $m$ traveling from point $O$ to $B$:

**FIGURE 14-8**

Illustration for using the principle of work and energy

1. The required work of the starter machine is $U_{OA} = T_A - T_O$ from Eq. 14-15.
2. The kinetic energy at $A$ is $T_A = \frac{1}{2}mv_A^2$ from Eq. 14-14. $T_O = 0$.
3. The kinetic energy at $A$ should be sufficient for the train to rise to $B$ and have kinetic energy $T_B = \frac{1}{2}mv_B^2$ at $B$.
4. The kinetic energies $T_A$ and $T_B$ are related according to Eq. 14-15, $U_{AB} = T_B - T_A$ where $U_{AB}$ is the work done on the train from $A$ to $B$.
5. The work $U_{AB}$ involves the work of gravity and the work of friction from $A$ to $B$, so $U_{AB} = -mgh - \int_A^B \mathscr{F}\, ds$, where $ds$ is measured along the path of the train.
6. Thus, the required work of the starter machine on the train is expressed in terms of the given quantities as

$$U_{OA} = T_A = T_B - U_{AB} = \frac{1}{2}mv_B^2 + mgh + \mathscr{F}s_{AB}$$

This schematic example shows the following advantages and limitations of using energy methods.

## Advantages

1. The design can be made without determining any accelerations of the train.
2. Additional parts of the path are easy to include in the analysis. For example, making the train to continue from $B$ to $D$ and back all the way to $O$ requires a simple modification of the procedure described above.
3. Only a summation of scalar quantities is necessary even if the path of the motion is complex.
4. Forces that do no work need not be considered. For example, the force of wind perpendicular to the vertical plane of the path $O$ to $D$ can do no work on the train and is ignored.

## Limitations

1. The acceleration cannot be calculated directly from the quantities of work and energy. This may be necessary for other aspects of the problem besides getting a rudimentary solution. For example, the maximum initial ($O$ to $A$) and final ($A$ to $O$) accelerations of the train in Fig. 14-8 are important with respect to the comfort and safety of the passengers. The accelerations can be calculated but Newton's second law must be used directly to do this.*
2. Forces that do no work cannot be calculated from the quantities of work and energy. For example, the normal forces between the train and its track anywhere from $O$ to $D$ can be calculated only by using Newton's second law.*

Overall, it is worth noting that the limitations of using energy methods are not serious because a problem solution can always be augmented by Newton's second law if necessary. The substantial advantages of these methods and their limitations are illustrated with the following examples.

* It should be remembered that the work–energy method involves no new principles, but is a reformulation of Newton's second law for convenience in solving certain problems.

# EXAMPLE 14-1

Redo Ex. 13-3 using energy methods. To repeat the problem, a car of 1000 kg mass is traveling at $v_0 = 100$ km/h when its brakes are applied. Calculate the stopping distance if $\mu = 0.5$ and compare the solution with that of Ex. 13-3.

## SOLUTION

By inspection of the free-body diagram in Fig. 14-9b, it is clear that the vertical forces, **W** and **N**, do no work on the car because the car is moving horizontally. From Eq. 14-15, $U_{12} = T_2 - T_1$. For this case,

$$U_{12} = -(\mathscr{F})\,\Delta x = -\mu N_{car}\,\Delta x = -\mu W_{car}\,\Delta x$$
$$= -(0.5)(m_{car}g)\,\Delta x = -(0.5)(9.81\ \text{m/s}^2)\,m_{car}\,\Delta x$$
$$= -\left(4.9\,\frac{\text{m}}{\text{s}^2}\right)m_{car}\,\Delta x$$

$$T_2 = \frac{1}{2}mv_2^2 = \frac{1}{2}m_{car}(0)^2 = 0$$

$$T_1 = \frac{1}{2}mv_1^2 = \frac{1}{2}m_{car}\left[\left(\frac{100\ \text{km}}{\text{h}}\right)\left(\frac{1000\ \text{m}}{\text{km}}\right)\left(\frac{1\ \text{h}}{3600\ \text{s}}\right)\right]^2$$

$$T_1 = (385.8\ \text{m}^2/\text{s}^2)\,m_{car}$$

Hence,

$$-(4.9\ \text{m/s}^2)\,m_{car}\,\Delta x = 0 - (385.8\ \text{m}^2/\text{s}^2)\,m_{car}$$

Dividing through by $m_{car}$,

$$\Delta x = \frac{385.8\ \text{m}^2/\text{s}^2}{4.9\ \text{m/s}^2} = 78.7\ \text{m}$$

which agrees with the previous result obtained in Ex. 13-3. Note that the work done by friction on the car was negative because it acted opposite to the direction of displacement of the car.

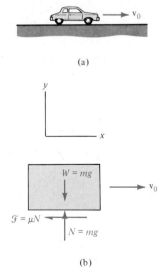

(a)

(b)

**FIGURE 14-9**

Analysis of a car's motion

EXAMPLE 14-1    561

# EXAMPLE 14-2

A mass of 10 kg is dropped from rest at a height of 0.8 m onto a very lightweight (massless) platform which is supported by a spring of spring constant $k = 1000 \text{ N/m}$ (Fig. 14-10a). The spring has a natural length of 1.0 m and is initially compressed 0.4 m by the cables $c$ shown. Determine **(a)** the maximum displacement of the platform, and **(b)** the magnitude of the maximum compressive force transmitted to the ground.

## SOLUTION

**(a)** The spring is initially compressed 0.4 m at $x = 0$. Therefore, the magnitude of the force in the spring for a general platform displacement $x$ is given by $F_s = 400 \text{ N} + (1000 \text{ N/m}) x$. Let the initial position correspond to the instant the weight is released and let the final position be at maximum deflection of the spring after impact for which $v = 0$. Therefore,

$$T_1 = 0 \quad \text{and} \quad T_2 = 0$$

so

$$U_{12} = T_2 - T_1 = 0 \qquad \textbf{(a)}$$

From the free-body diagram in Fig. 14-10b, work is done on the mass by both the spring force $\mathbf{F}_s$ and the weight $\mathbf{W}$.

$$U_{\text{spring}} = -\int_0^{x_{\max}} F_s \, dx = -\int_0^{x_{\max}} (400 + 1000x) \, dx$$

$$= \text{area under diagram shown in Fig. 14-10c}$$

$$= -(400 \text{ N}) \, x_{\max} - (500 \text{ N/m}) \, x_{\max}^2$$

$$U_{\text{weight}} = (10 \text{ kg})(9.81 \text{ m/s}^2)(0.8 \text{ m} + x_{\max}) = 78.5 \text{ N·m} + (98.1 \text{ N}) \, x_{\max}$$

Substituting these values into Eq. (a),

$$-(400 \text{ N}) \, x_{\max} - (500 \text{ N/m}) \, x_{\max}^2 + 78.5 \text{ N·m} + (98.1 \text{ N}) \, x_{\max} = 0$$

$$(500 \text{ N/m}) \, x_{\max}^2 + (302 \text{ N}) \, x_{\max} - 78.5 \text{ N·m} = 0$$

$$x_{\max} = 0.196 \text{ m}$$

**(b)** The magnitude of the maximum compressive force transmitted to the ground occurs at maximum deflection,

$$(F_s)_{\max} = 400 \text{ N} + (1000 \text{ N/m}) \, x_{\max}$$

$$= 400 \text{ N} + (1000 \text{ N/m})(0.196 \text{ m})$$

$$= 596 \text{ N}$$

(a)

$W = 10(9.81) \text{ N}$

$F_s = (400 + 1000 \, x) \text{ N}$

(b)

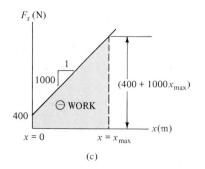

$F_s \text{ (N)}$

$1000$

$1$

$(400 + 1000x_{\max})$

$\ominus$ WORK

$400$

$x = 0$

$x = x_{\max}$

$x(\text{m})$

(c)

**FIGURE 14-10**

Analysis of a mass falling on a pre-compressed spring

EXAMPLE 14-3

A precompressed spring is initially held by a latch mechanism $OA$ (Fig. 14-11a). When the latch is released the spring propels a 30-kg machine part, which is being heat treated at $A$, up the inclined plane onto a conveyor belt at $B$. The spring is precompressed 0.2 m, the coefficient of friction between the machine part and the 30° incline is $\mu = 0.1$, and the desired speed of the machine part when it reaches the top of the incline at $B$ is 5 m/s. This is the same speed as the speed of the conveyor belt. Determine the spring constant $k$ (N/m) that the engineer must use in this design.

## SOLUTION

From Eq. 14-15, $U_{AB} = T_B - T_A$. Since the machine part starts from rest, $T_A = 0$. The total work done on the machine part during its motion from $A$ to $B$ is expressed using the free-body diagram in Fig. 14-11b,

$$U_{AB} = \frac{1}{2}kx^2 - mgh - \mathscr{F}s$$

where $x = 0.2$ m, $h = 4$ m sin 30°, and $s = 4$ m. To determine $\mathscr{F} = \mu N$,

$$\sum F_y = 0,$$

$$N - mg\cos 30° = 0$$

$$N = 255\,\text{N}$$

So $\mathscr{F} = 0.1(255\,\text{N}) = 25.5\,\text{N}$. Then

$$U_{AB} = \frac{1}{2}k(0.2\,\text{m})^2 - (294\,\text{N})(4\,\text{m})(\sin 30°) - (25.5\,\text{N})(4\,\text{m})$$

$$= -690\,\text{N·m} + (0.02\,\text{m}^2)\,k$$

The final kinetic energy is $T_B = \frac{1}{2}mv_B^2$:

$$T_B = \frac{1}{2}(30\,\text{kg})(5\,\text{m/s})^2 = 375\,\text{N·m}$$

Using Eq. 14-15,

$$-690\,\text{N·m} + (0.02\,\text{m}^2)\,k = 375\,\text{N·m}$$

Solving for $k$,

$$k = 53.3\,\text{kN/m}$$

(a)

**FIGURE 14-11**

Propelling of a machine part onto a conveyor belt

EXAMPLE 14-3    563

## EXAMPLE 14-4

The arresting mechanism of an aircraft carrier is modeled in Fig. 14-12a. The tensioning devices in the mechanisms at $A$ and $B$ (which let out the cable) apply a constant tension $T$ to the cable throughout the braking process. Consider an airplane of weight $W = 15{,}000$ lb with a landing speed of $v_0 = 180$ ft/s. What must the tension in the cable be if the airplane is to stop after rolling 700 ft horizontally?

### SOLUTION

From the free-body diagram in Fig. 14-12b and the fact that the plane is being brought to rest, it is clear that the cables are doing negative work on the plane. Using Eq. 14-17, $T_1 + U_{12} = T_2$,

$$\frac{1}{2}\left(\frac{W}{g}\right)v_0^2 - 2\int_0^{700\,\text{ft}} T\cos\theta\,dx = 0$$

Since $\cos\theta$ varies with $x$, a relationship for it in terms of $x$ must be found. From Fig. 14-12c,

$$\cos\theta = \frac{x}{(x^2 + 35^2)^{1/2}}$$

Therefore,

$$\frac{1}{2}\left(\frac{15{,}000\,\text{lb}}{32.2\,\text{ft/s}^2}\right)(180\,\text{ft/s})^2 - 2\int_0^{700\,\text{ft}} T\left[\frac{x}{(x^2 + 35^2)^{1/2}}\right]dx = 0$$

Evaluating the integral yields, since $T$ is constant,

$$7.55 \times 10^6\,\text{ft·lb} - 2T(x^2 + 35^2)^{1/2}\Big|_0^{700\,\text{ft}} = 0$$

$$7.55 \times 10^6\,\text{ft·lb} - 2T\{[(700\,\text{ft})^2 + (35\,\text{ft})^2]^{1/2}$$
$$- [0^2 + (35\,\text{ft})^2]^{1/2}\} = 0$$

$$7.55 \times 10^6\,\text{ft·lb} - 2T(666\,\text{ft}) = 0$$

$$T = 5670\,\text{lb}$$

### JUDGMENT OF THE RESULTS

For a check, assume that $\cos\theta = 1$ during the whole motion. This is reasonable since $\theta$ is small for most of the displacement (for example, at $x = 200$ ft, $\cos\theta = 0.98$). Thus, assuming that $(x^2 + 35^2)^{1/2} \simeq 700$ ft,

$$7.55 \times 10^6\,\text{ft·lb} - 2T(700\,\text{ft}) = 0$$

$$T = 5400\,\text{lb}$$

so the more detailed solution appears correct.

**FIGURE 14-12**

Arresting of an aircraft by a cable

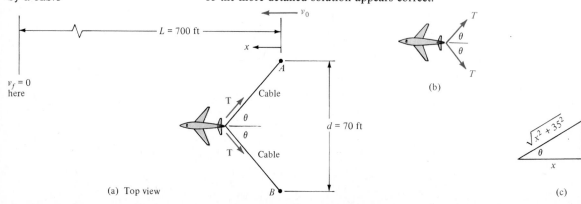

(a) Top view

(b)

(c)

EXAMPLE 14-5

Consider the amusement ride shown in Fig. 14-8. Assume that the mass of the train is $m = 3000\,kg$, the height $h$ is $25\,m$, and the minimum speed in the loop is $v_B = 2\,m/s$ (Fig. 14-13a). Calculate the required takeoff speed $v_A$ and the maximum power input if the train is accelerated from rest at $O$ to $v_A$ at $A$ over a distance of $15\,m$ by a constant horizontal force $\mathbf{F}$. **(a)** Solve the problem ignoring friction. **(b)** Assume that the system operates with 75% efficiency.

## SOLUTION

**(a)** From Eq. 14-15, $U_{AB} = T_B - T_A$.
Using a free-body diagram of the train as it moves up (Fig. 14-13b),

$$U_{AB} = -mgh = -(3000\,kg)(9.81\,m/s^2)(25\,m)$$
$$= -7.36 \times 10^5\,kg{\cdot}m^2/s^2$$

$$T_B = \frac{1}{2}mv_B^2 = \frac{1}{2}(3000\,kg)(2\,m/s)^2 = 6000\,kg{\cdot}m^2/s^2$$

$$T_A = \frac{1}{2}mv_A^2 = \frac{1}{2}(3000\,kg)(v_A)^2 = 1500\,kg\,v_A^2$$

Hence,

$$-7.36 \times 10^5\,kg{\cdot}m^2/s^2 = 6000\,kg{\cdot}m^2/s^2 - 1500\,kg\,v_A^2$$

$$v_A^2 = \frac{742\,000\,kg{\cdot}m^2/s^2}{1500\,kg} = 494.7\,m^2/s^2$$

$$v_A = 22.2\,m/s$$

To calculate the maximum power input, Eq. 14-18 is used. Since $F$ is constant, and $\mathbf{F}$ is in the same direction as $\mathbf{v}$,

$$P_{\max} = Fv_{\max}$$

$v_{\max}$ occurs at point $A$, so the maximum power input must be at $A$. To find $F$, the principle of work and energy is used again,

$$U_{OA} = T_A - T_O$$

$$Fd = \frac{1}{2}mv_A^2 - 0$$

$$F(15\,m) = \frac{1}{2}(3000\,kg)(22.2\,m/s)^2$$

$$= 739 \times 10^3\,\frac{kg{\cdot}m^2}{s^2} = 739\,kN{\cdot}m$$

$$F = \frac{739\,kN{\cdot}m}{15\,m} = 49.3\,kN$$

Hence,

$$P_{\max} = Fv_{\max} = (49.3\,kN)(22.2\,m/s) = 1094\,kN{\cdot}m/s$$
$$= 1094\,kW = 1466\,hp$$

**(b)** From Eq. 14-21,

$$\text{total power required} = \frac{\text{useful power}}{\eta} = \frac{1094\,kW}{0.75}$$

$$= 1459\,kW = 1956\,hp$$

(a)

(b)

**FIGURE 14-13**

Analysis of an amusement ride

EXAMPLE 14-5    565

# EXAMPLE 14-6

Consider the hay baler shown in Fig. 14-1 and assume that the machine and the trailer are stationary. Each bale weighs 30 lb and must travel a maximum of $h = 10$ ft vertically and $d = 15$ ft horizontally from the launching arm (Fig. 14-14). This arm starts from rest and moves over a maximum range of $\Delta y = 1.5$ ft vertically and $\Delta x = 1$ ft horizontally. Calculate the required launching velocity $\mathbf{v}$ and the maximum power input $P$ if the system operates with a $\eta = 60\%$ efficiency, and air resistance is negligible.

## SOLUTION

The bale's aerial velocity in the $x$ direction is assumed constant since there are no retarding forces in that direction. Gravity is the only force doing work on the bale in the air. In the analysis it is necessary to consider the launching phase and the flight phase of the motion separately.

**LAUNCHING PHASE:** From Eq. 14-15, $U_{12} = T_2 - T_1$, where 2 denotes the end of the launching phase. The force $\mathbf{F}_{\text{arm}}$ in the free-body diagram of Fig. 14-14c has components $\mathbf{F}_x$ and $\mathbf{F}_y$. Thus,

$$F_x \, \Delta x = \frac{1}{2} m v_{x_2}^2 \qquad \text{and} \qquad -W \, \Delta y + F_y \, \Delta y = \frac{1}{2} m v_{y_2}^2 \qquad \textbf{(a)}$$

These equations cannot be solved yet since $F_x$, $F_y$, $v_x$, and $v_y$ are all unknown.

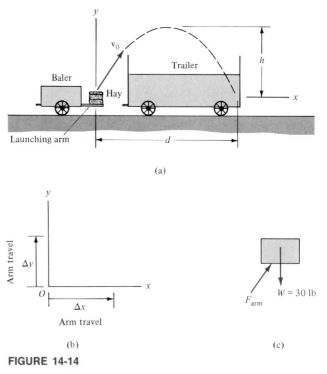

(a)

(b)                                                                 (c)

**FIGURE 14-14**

Analysis of the launching and flight of bales of hay

**FLIGHT PHASE:** From Eq. 14-15, $U_{23} = T_3 - T_2$, where 3 denotes the peak height of the flight.

$$-Wh = \frac{1}{2} mv_{x_3}^2 - \frac{1}{2} m(v_x^2 + v_y^2)_2$$

Note that $v_{y_3} = 0$. Also, $v_{x_2} = v_{x_3}$, so the two $v_x$ terms can be eliminated, yielding

$$-mgh = 0 - \frac{1}{2} mv_{y_2}^2$$

$$v_{y_2} = \sqrt{2gh} = \sqrt{2(32.2 \text{ ft/s}^2)(10 \text{ ft})} = 25.4 \text{ ft/s}$$

Next, the time of flight $t$ to the peak of the trajectory is obtained from Eq. 11-8,

$$v_y = v_{0_y} - gt$$
$$0 = 25.4 \text{ ft/s} - (32.2 \text{ ft/s}^2)t$$
$$t = 0.789 \text{ s}$$

In this time the bale moves 7.50 ft in the $x$ direction. Thus, $v_x$ can be calculated from Eq. 11-10,

$$x = x_0 + v_{0_x}t$$
$$15 \text{ ft} = 0 + v_{0_x}(0.789 \text{ s})(2)$$
$$v_{0_x} = v_x = 9.51 \text{ ft/s}$$
$$\mathbf{v} = (9.51\mathbf{i} + 25.4\mathbf{j}) \text{ ft/s at takeoff}$$

**LAUNCHING PHASE:** The equations of work and energy for the $x$ and $y$ directions [Eq. (a)] can now be solved.

$$F_x \, \Delta x = \frac{1}{2} mv_{x_2}^2$$

$$F_x(1 \text{ ft}) = \frac{1}{2}\left(\frac{30 \text{ lb}}{32.2 \text{ ft/s}^2}\right)(9.51 \text{ ft/s})^2 = 42.1 \text{ ft} \cdot \text{lb}$$

$$F_x = 42.1 \text{ lb}$$

$$-W \, \Delta y + F_y \, \Delta y = \frac{1}{2} mv_{y_2}^2$$

$$F_y(1.5 \text{ ft}) = \frac{1}{2}\left(\frac{30 \text{ lb}}{32.2 \text{ ft/s}^2}\right)(25.4 \text{ ft/s})^2 + (30 \text{ lb})(1.5 \text{ ft}) = 345 \text{ ft} \cdot \text{lb}$$

$$F_y = 230 \text{ lb}$$

The maximum power is required when the speed is maximum (at the end of the launching phase). From Eq. 14-19,

$$\begin{aligned}
P_{\max} = \mathbf{F}_{\text{arm}} \cdot \mathbf{v}_{\max} &= F_x(v_x) + F_y(v_y) \\
&= (42.1 \text{ lb})(9.51 \text{ ft/s}) + (230 \text{ lb})(25.4 \text{ ft/s}) \\
&= 6242 \text{ lb} \cdot \text{ft/s}
\end{aligned}$$

From Eq. 14-21,

$$P_{\text{required}} = \frac{P_{\text{useful}}}{\eta} = \frac{6242 \text{ ft} \cdot \text{lb}}{0.6 \quad \text{s}} = 18.9 \text{ hp}$$

EXAMPLE 14-6   567

**14-1** A particle of mass $m$ freely falls from rest at a height $h = 20\,\text{m}$. Calculate the speed $v_f$ of the particle just before it hits the ground.

**14-2** A 3-lb particle is dropped with no initial velocity from a height $h$. Its final speed just before impact is $v_f = 40\,\text{ft/s}$. Determine $h$.

**14-3** A ball of mass $m$ is released at height $h$ with an initial velocity $\mathbf{v}_0$. Determine the final velocity $\mathbf{v}_f$ at the ground for (a) the case when $\mathbf{v}_0$ is downward, and (b) the case when $\mathbf{v}_0$ is upward.

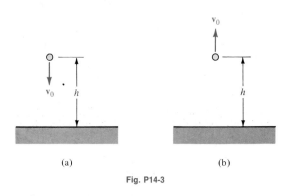

(a)                    (b)

**Fig. P14-3**

**14-4** A 0.5-kg particle is projected at height $h = 40\,\text{m}$ with an initial velocity $\mathbf{v}_0 = (20\mathbf{i} + 15\mathbf{j})\,\text{m/s}$. Use energy methods to determine its final velocity $\mathbf{v}_f$ just before impact at the ground.

**Fig. P14-4**

**14-5** A 2-lb particle is released at height $h = 50\,\text{ft}$ with an initial velocity $\mathbf{v}_0 = (30\mathbf{i} - 10\mathbf{j})\,\text{ft/s}$. Use energy methods to determine its final velocity $\mathbf{v}_f$ just before impact at the ground.

**Fig. P14-5**

**14-6** A 10-kg package has a speed $v_A = 8\,\text{m/s}$ at point $A$ of the sloping board. Determine its speed at point $B$ if the coefficient of kinetic friction is 0.2.

**Fig. P14-6**

**14-7** A 5-kg block is sliding on a horizontal surface on which the coefficient of kinetic friction is 0.3. Determine the required initial speed $v_0$ to stop the block by friction after a travel of 4 m.

**14-8** A 20-lb box is pushed along the incline by a constant force $\mathbf{F}$. The box starts from rest at point $A$ and must have a speed of $v_B = 5\,\text{ft/s}$ after traveling 10 ft to point $B$. Determine the magnitude of the force $\mathbf{F}$ neglecting friction.

**Fig. P14-8**

**14-9** Determine the required magnitude of the force $\mathbf{F}$ in Prob. 14–8 assuming a coefficient of kinetic friction of 0.4.

**14–10** A ski slope has a straight section on which the coefficient of kinetic friction is 0.1. Where should the starting point *A* be located with respect to *B* if a 150-lb skier's speed at *B* should not exceed 30 ft/s? Assume that the poles are not used.

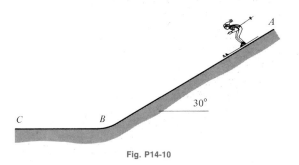

Fig. P14-10

**14–11** In Prob. 14–10, the path of the skiers from *B* to *C* is horizontal. Determine the maximum distance *BC* that a skier could glide without using the poles.

**14–12** Skiers start from rest at point *A*, which has an elevation of 200 m above point *B*. The total path length is 800 m, and the net friction force is 10 N over the path. Determine the speed $v_B$ of a 70-kg skier who is not using the poles for propulsion.

Fig. P14-12

**14–13** A 1000-kg car is traveling downhill at a speed of $v_0 = 20$ m/s when its brakes are applied locking the wheels. The coefficient of kinetic friction on the road is 0.3. Plot the stopping distance *s* as a function of $\theta = 5°$, $10°$, and $15°$. Estimate from the diagram the smallest incline on which the car would not stop by using the brakes.

Fig. P14-13

**14–14** Determine the stopping distance of a 3000-lb car traveling up the incline on which the coefficient of kinetic friction is 0.5. The initial speed is $v_0 = 50$ mph when the brakes are applied locking the wheels.

Fig. P14-14

**14–15** A 3000-lb experimental car can apply a downward thrust **P** in an emergency braking situation. Calculate the stopping distance *d* on a level road if the initial speed is $v_0 = 60$ mph, $P = 500$ lb, and $\mu_k = 0.5$.

Fig. P14-15

**14–16** An interesting question concerning the vehicle in Prob. 14–15 is the direction of the thrust **P** for the shortest stopping distance. To obtain some insight, solve Prob. 14–15 as stated, and also with **P** acting not vertically but horizontally to oppose the motion directly.

**14–17** A mass *m* on the end of a cable of length *l* and of negligible mass is released from rest at $\theta_0$. The mass *m* strikes the side of a building in a demolition operation at a speed $v_f$. Using the work–energy principle, determine (a) the general expression for speed *v* at any position $\theta$ before impact, and (b) the impact speed $v_f$.

Fig. P14-17

**14–18** A 2-kg mass is suspended on a rigid rod of negligible mass and length $l = 0.5$ m. The rod has a smooth hinge at $O$. The pendulum is released from rest at $\theta = 0$. Determine the speed of the mass at $\theta = 90°$ and $135°$ neglecting friction.

Fig. P14-18

**14–19** A bicycle racer and his bicycle have a total mass of 100 kg. He can develop 0.2 hp of power in a sustained effort. Determine his speed up the incline neglecting frictional losses.

Fig. P14-19

**14–20** A motor applies a constant, 200-lb force after the cart starts from rest. Determine the velocity of the cart after it has moved 30 ft up the incline. Also calculate the maximum power required of the motor for this motion.

Fig. P14-20

**14–21** A power winch is pulling a 100-lb crate up on the incline, where the coefficient of kinetic friction is 0.4. The crate starts from rest at $A$ and has a speed of 5 ft/s at $B$. Calculate the required maximum power of the winch assuming a constant tension in the rope.

Fig. P14-21

**14–22** A 100-kg crate is to be raised by an electric motor which is rated at 60% efficiency and requires an electrical input of 1 kW. Calculate the constant speed $v$ of the crate moving upward.

Fig. P14-22

**14–23** A 5000-lb elevator is being raised by a motor (Fig. P14-22) at a constant speed $v = 5$ ft/s. Determine the required power of the motor.

**14–24** The elevator cab has a mass $m_A = 2000$ kg and the counterweight has a mass $m_B = 1500$ kg. The system is moved by an electric motor at point $O$. Determine the required power of the motor to maintain a velocity $v_A = 2$ m/s upward.

Fig. P14-24

**14-25** A machine partially shown in the photograph is designed to test electronic equipment $E$ of mass $m = 10$ kg. The stroke (maximum travel) of the constant-force actuator $A$ is $\Delta y = 50$ mm. Calculate the required maximum power of the actuator to impart an upward velocity of 1 m/s to the equipment $E$.

**Fig. P14-25**

Dynamic testing of a portable electronic instrument (Courtesy MTS Systems Corporation, Minneapolis, Minnesota)

**14-26** A 5-lb block is suspended from a spring which has a spring constant $k = 0.5$ lb/in. When not stretched, the spring has length $l_0 = 20$ in. The block is released from rest at $y = 0$ (the spring is not stretched initially). Determine (a) the speed of the spring at $y = 5$ in., and (b) the maximum displacement of the block.

**Fig. P14-26**

**14-27** The block in Prob. 14-26 is displaced upward and released from rest with the spring at half its original length. Determine the lowest position of the block.

**14-28** A 1200-kg car has a light bumper supported horizontally by two springs of stiffness $k = 15$ kN/m. Determine the initial speed $v$ of impact with a fixed wall that causes a 0.2-m compression of the springs.

**Fig. P14-28**

**14-29** At an amusement park people may jump by parachutes which are guided to the ground by vertical wires. The jumpers land on a weightless spring-supported platform at a speed of 6 ft/s. The effective spring constant (as if one spring supported each platform) is $k = 30$ lb/in. How much space should be allowed for the vertical displacement of each platform assuming that a 250-lb person may land on it?

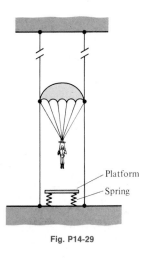

**Fig. P14-29**

**14–30** A spring of stiffness $k = 2\,\text{kN/m}$ is compressed through a distance $d = 50\,\text{mm}$ by a mechanism. The spring propels a 1-kg block along the incline when the mechanism is released. Determine the maximum distance $l$ traveled by the block if friction is negligible. The block and the spring are not attached.

Fig. P14-30

**14–31** Determine the coefficient of kinetic friction in Prob. 14–30 if the block stops after traveling $l = 0.4\,\text{m}$.

**14–32** The bumper system of a tuned mass damper for a skyscraper (for a description of these, see Sec. 1-7 of the associated statics volume) is modeled as a mass $m$ sliding at speed $v_0$ just before touching the bumper spring. Determine the rebound speed $v_f$ of the block if the coefficient of kinetic friction is $\mu_k$.

Fig. P14-32

**14–33** In Prob. 14–32, $m = 3 \times 10^5\,\text{kg}$, $\mu_k = 0.05$, and $v_0 = 2\,\text{m/s}$. Plot the ratio $v_f/v_0$ as a function of the spring stiffness, using $k = 200\,\text{kN/m}$, $400\,\text{kN/m}$, and $600\,\text{kN/m}$.

**14–34** A 5000-lb aircraft is launched from a catapult which has a spring of stiffness $k = 100\,\text{lb/in}$. The spring is compressed through a distance of 4 ft, then it is released to push the aircraft, whose engine applies a constant force of magnitude $P = 1200\,\text{lb}$ during the launch. Determine the takeoff speed $v$ of the aircraft at the instant when the spring regains its original length. Neglect friction and assume that the aircraft is propelled by horizontal forces.

Fig. P14-34

**14–35** In Prob. 14–34, the launch angle $\theta$ is also important besides the takeoff speed. Plot the speed $v$ as a function of $\theta = 0$, $15°$, and $30°$, assuming that the two propelling forces on the aircraft are acting parallel to the launching ramp.

Fig. P14-35

**14–36** The device shown is for measuring coefficients of friction. The fixture is slowly rotated counterclockwise until the 4-lb block starts sliding at $\theta = 10°$. Determine the coefficients of static and kinetic friction if the spring of stiffness $k = 5\,\text{lb/in}$. deflects a maximum of 1.5 in.

Fig. P14-36

The work of a force involved in a displacement of a particle between two specified points is independent of the path when the force is conservative. This concept is illustrated in Fig. 14-15, where the work of gravity is $U_{AB} = -Wh$ regardless of what path is followed.

The work of the force of an elastic spring is also independent of path for most practical purposes. For example, the work of the spring in the elongation from $A$ to $B$ in Fig. 14-16 is $U_{AB} = -\frac{1}{2}k(x_B^2 - x_A^2)$. This is independent of the orientation $\theta$ of the spring.

In general, a conservative system may involve motion in three dimensions. For example, a particle may move from point $P_1$ to $P_2$ while a conservative force $\mathbf{F}$ is acting on it as in Fig. 14-17. The potential energy $V(x, y, z)$ at any point $x, y, z$ along a path is known as the *potential function* of the conservative force $\mathbf{F}$. The work $U_{12}$ of the force $\mathbf{F}$ between arbitrary points $P_1$ and $P_2$ can be written in terms of the corresponding potential functions,

$$U_{12} = V_1(x_1, y_1, z_1) - V_2(x_2, y_2, z_2) \qquad \boxed{14\text{-}22}$$

or, simply,

$$U_{12} = V_1 - V_2 \qquad \boxed{14\text{-}23}$$

which agrees with the definition in Eq. 14-12.

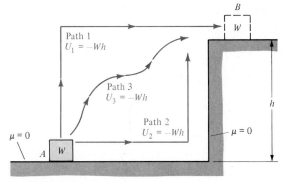

**FIGURE 14-15**

Work in a conservative system is independent of the path

**FIGURE 14-16**

Conservative system with an elastic spring

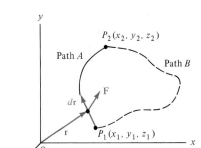

**FIGURE 14-17**

Analysis of a general conservative system

An important special case of conservative systems is when the particle moves on a *closed path*, such as leaving from $P_1$ and finishing at $P_1$. One closed path starting from any point is called a *cycle* of displacement. In that case $V_1 = V_2$ in Eq. 14-23, and the total work in the cycle is zero. The detailed mathematical expression of this condition is made by using the line integral $\oint$ for the closed path,

$$U = \oint dU = \oint \mathbf{F} \cdot d\mathbf{r}$$

$$= \oint (F_x\, dx + F_y\, dy + F_z\, dz) = 0$$

14-24

### Advanced Analysis

In some cases it is useful to consider infinitesimal variations of the potential function $V$. Since the difference in potential energies at two positions is equal to the work of the force between these positions,

$$dU = V(x, y, z) - V(x + dx, y + dy, z + dz)$$

$$dU = -V(dx, dy, dz) = -dV(x, y, z)$$

14-25

where $dU$ is the elementary work of a conservative force* and $-dV$ is the corresponding change in potential energy. The differential change in $V$ can be expressed with partial derivatives as the sum of independent changes along $x$, $y$, and $z$ (the total derivative of $V$),

$$dV = \frac{\partial V}{\partial x}\, dx + \frac{\partial V}{\partial y}\, dy + \frac{\partial V}{\partial z}\, dz$$

From Eqs. 14-24 and 14-25,

$$F_x\, dx + F_y\, dy + F_z\, dz = -\frac{\partial V}{\partial x}\, dx - \frac{\partial V}{\partial y}\, dy - \frac{\partial V}{\partial z}\, dz$$

14-26

Equating the terms that refer to the same axes,

$$F_x = -\frac{\partial V}{\partial x} \qquad F_y = -\frac{\partial V}{\partial y} \qquad F_z = -\frac{\partial V}{\partial z}$$

14-27

These expressions can be written more concisely using vector notation. From Eq. 14-26,

$$\mathbf{F} = F_x\mathbf{i} + F_y\mathbf{j} + F_z\mathbf{k} = -\left(\frac{\partial V}{\partial x}\mathbf{i} + \frac{\partial V}{\partial y}\mathbf{j} + \frac{\partial V}{\partial z}\mathbf{k}\right)$$

The vector quantity on the right side is called the *gradient of the scalar function V* and is denoted by **grad** $V$. Another common notation uses the vector operator $\nabla$ (called del), which is written as

$$\nabla = \frac{\partial}{\partial x}\mathbf{i} + \frac{\partial}{\partial y}\mathbf{j} + \frac{\partial}{\partial z}\mathbf{k}$$

---

\* Equation 14-25 implies that $dU$ is an *exact differential*.

With these two notations, the force and the potential function can be related as

$$\mathbf{F} = -\mathbf{grad}\ V = -\nabla V \qquad \boxed{14\text{-}28}$$

Thus, a force is conservative when all of its components can be derived from a potential function based on position coordinates alone and when it can be expressed according to Eq. 14-28. This implies that the work done by a conservative force is independent of the path used in reaching a given position and the potential energy associated with it. Conversely, if a force cannot be expressed as the gradient of a potential function, the force is *nonconservative*. For example, when the force depends not only on position but also on frictional resistance (i.e., $\mu$ and length of path), a potential function does not exist, and the force is nonconservative. A force that is dependent on velocity (such as the drag force in a fluid) is also nonconservative.

## CONSERVATION OF ENERGY

In any conservative system the work of a force on a particle can be converted to kinetic or potential energies of that particle. The resulting possibilities are illustrated using a particle of mass $m$ in Fig. 14-18 and are explained as follows (these concepts are also valid for systems containing elastic members):

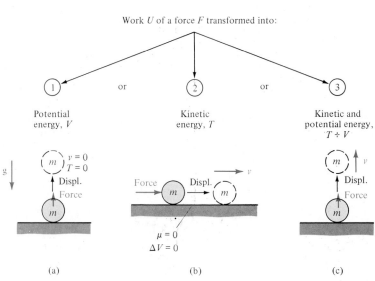

Work $U$ of a force $F$ transformed into:

1    or    2    or    3

Potential energy, $V$

Kinetic energy, $T$

Kinetic and potential energy, $T \div V$

$v = 0$
$T = 0$
Displ.
Force
$m$

Force   Displ.
$m \longrightarrow m$
$v$
$\mu = 0$
$\Delta V = 0$

$m$   $v$
Displ.
Force
$m$

(a)       (b)       (c)

**FIGURE 14-18**

Transformation of work into potential energy or kinetic energy

1. All of the work of the force is transformed into potential energy at a new elevation where the velocity of the particle is zero (Fig. 14-18a).

2. The motion is in a horizontal plane. All of the work of the force is transformed into kinetic energy while there is no change in potential energy (Fig. 14-18b).

3. Some of the work of the force is transformed into potential energy, the rest of it is transformed into kinetic energy (Fig. 14-18c).

The work of conservative forces can be expressed entirely as a change in potential energy ($U_{12} = V_1 - V_2$ in Eq. 14-23) or as a change in kinetic energy ($U_{12} = T_2 - T_1$ in Eq. 14-15). Therefore, a relationship involving only potential and kinetic energies can be obtained for the general case illustrated in Fig. 14-18c. From Eqs. 14-23 and 14-15,

$$V_1 - V_2 = T_2 - T_1$$

where $T$ and $V$ have the same units. Rearranging yields

$$\boxed{T_1 + V_1 = T_2 + V_2} \qquad \text{14-29}$$

which means that the sum of the kinetic and potential energies is constant in any position of a particle that moves under the action of conservative forces. Equation 14-29 is a statement of the *conservation of mechanical energy*. In general, the *total mechanical energy E* of a particle at a particular position is $E = T + V$. At positions 1, 2, 3, ... , $n$,

$$E_1 = E_2 = \cdots = E_n = \text{constant} \qquad \text{14-30}$$

In many systems there are other energies besides the two mechanical energies that should be considered. These include chemical, electrical, and thermal energies, among others. In a broad sense, total energy is conserved but its parts are constantly being transformed. Some of these energies can be relatively easily considered in modified versions of Eq. 14-29, but their experimental determination may be quite difficult. For example, velocity and position ($T$ and $V$) are easily measured while the work of friction forces (dissipated as heat) is very difficult to measure.

**TIP**

Equation 14-29 is commonly used to determine the velocity of a particle, the elevation of a particle, or the deformation of a spring before or after a particular displacement has occurred. Since the reference level (or datum plane) from which to determine the gravitational potential energy is arbitrary, the calculations are simplified if the datum plane is chosen at either position 1 or position 2 in Eq. 14-29. In this way the gravitational potential energy is zero on one side of the equation.

EXAMPLE 14-7

Consider the swing shown in Fig. 14-19. Assume that the mass $m$ of the child is 40 kg at the distance $L = 3$ m from point $O$. Calculate the maximum speed of the child if the maximum angle of the supporting chains with the vertical is $\theta = 70°$. Neglect frictional losses.

## SOLUTION

A free-body diagram and a sketch of two extreme positions of the mass are drawn in Fig. 14-19b and c. The tension force does no work since it is always perpendicular to the path of the particle. The only other force acting on the particle is its weight, which is a conservative force, so the conservation of mechanical energy equation applies. Since $T + V = E = $ constant, $T$ is a maximum when $V$ is a minimum. A datum from which to measure height $h$ is defined at the lowest position of the swing. Therefore, $V_{\min}$ occurs at the datum $h = 0$. Using Eq. 14-29,

$$T_1 + V_1 = T_2 + V_2$$

$$T_1 = 0 \qquad (v_1 = 0 \text{ at the top of the path by inspection})$$

$$\begin{aligned} V_1 = V_{g_1} = mgh_1 &= mg(L - L\cos 70°) \\ &= (40\,\text{kg})(9.81\,\text{m/s}^2)(3\,\text{m})(1 - \cos 70°) \\ &= 775\,\text{kg·m}^2/\text{s}^2 \end{aligned}$$

$$T_2 = \frac{1}{2}mv_2^2$$

$$V_2 = V_{g_2} = mgh_2 = mg(0) = 0$$

Therefore,

$$0 + 775\,\text{kg·m}^2/\text{s}^2 = \frac{1}{2}(40\,\text{kg})(v_2)^2$$

$$v_2 = v_{\max} = 6.22\,\text{m/s}$$

(a)

(b)

(c)

**FIGURE 14-19**

Analysis of a swing

**TIP**

Note that in this example of a particle starting from rest with its weight the only force doing work,

$$v = \sqrt{2gh}$$

This is a useful result to remember in order to simplify many problems.

EXAMPLE 14-7    577

EXAMPLE 14-8

A 3-kg collar slides with negligible friction along a vertical rod as shown in Fig. 14-20a. The spring is undeformed when the collar $A$ is at the same elevation as point $O$. The collar is released from rest at $y_1 = 0.4$ m. Determine the velocities of the collar as it first passes $y_2 = 0$ and $y_3 = -0.4$ m.

$A$

$k = 500$ N/m

$y$

0.3 m

(a)

$A$

0.5 m

$y_1 = 0.4$ m

$O$

0.3 m

(b)

0.3 m

$O$

$y_3 = -0.4$ m

0.5 m

$A$

(c)

**FIGURE 14-20**

Sliding collar and spring mechanism

## SOLUTION

For convenience, first the terms of Eq. 14-29 are evaluated at each position.

**POSITION 1** (Fig. 14-20b):   Potential  energy:   $V_1 = V_{g_1} + V_{e_1} = mgy_1 + \frac{1}{2}kd_1^2$, where $d$ is the elongation of the spring.

$$V_1 = (3\,\text{kg})(9.81\,\text{m/s}^2)(0.4\,\text{m}) + \frac{1}{2}(500\,\text{N/m})(0.5\,\text{m} - 0.3\,\text{m})^2$$

$$= 11.8\,\text{N}\cdot\text{m} + 10\,\text{N}\cdot\text{m} = 21.8\,\text{N}\cdot\text{m}$$

Kinetic energy:  $T_1 = 0$ since $v_1 = 0$.

**POSITION 2:**   Potential energy:

$$V_2 = mgy_2 + \frac{1}{2}kd_2^2$$

$$= (3\,\text{kg})(9.81\,\text{m/s}^2)(0) + \frac{1}{2}(500\,\text{N/m})(0)^2 = 0$$

Kinetic energy:  $T_2 = \frac{1}{2}mv_2^2 = (1.5\,\text{kg})v_2^2$.

**POSITION 3**   (Fig. 14-20c):   Potential energy:

$$V_3 = mgy_3 + \frac{1}{2}kd_3^2$$

$$V_3 = (3\,\text{kg})(9.81\,\text{m/s}^2)(-0.4\,\text{m}) + \frac{1}{2}(500\,\text{N/m})(0.5\,\text{m} - 0.3\,\text{m})^2$$

$$= -11.8\,\text{N}\cdot\text{m} + 10\,\text{N}\cdot\text{m} = -1.8\,\text{N}\cdot\text{m}$$

Kinetic energy:  $T_3 = \frac{1}{2}mv_3^2 = (1.5\,\text{kg})v_3^2$.

The required speeds are calculated by applying the principle of conservation of energy (Eq. 14-29) first for positions 1 and 2, then for positions 1 and 3. To obtain the speed $v_2$,

$$T_1 + V_1 = T_2 + V_2$$

$$0 + 21.8\,\text{N}\cdot\text{m} = (1.5\,\text{kg})v_2^2 + 0$$

$$v_2 = \pm 3.8\,\text{m/s}$$

By inspection, $\mathbf{v}_2$ is downward.
   To obtain the speed $v_3$,

$$T_1 + V_1 = T_3 + V_3$$

$$0 + 21.8\,\text{N}\cdot\text{m} = (1.5\,\text{kg})v_3^2 - 1.8\,\text{N}\cdot\text{m}$$

$$v_3 = \pm 4\,\text{m/s}$$

The velocity $\mathbf{v}_3$ is still downward. The spring has not yet reached its maximum deformation at which point the speed of the collar will be decreased to zero. Afterwards, the collar will have an upward velocity.

EXAMPLE 14-8   579

## EXAMPLE 14-9

Consider the sport of pole vaulting, in which tremendous improvements have been made by using poles that act as springs. Determine the additional height that a jumper can reach by using an elastic pole. To be specific, use the following data about a world-class athlete: weight = 170 lb, and maximum running speed = 30 ft/s. Assume that the speed of the vaulter in the horizontal direction is 7 ft/s during and after takeoff (Fig. 14-21a). To simplify this complex problem, assume that the vaulter is a particle that travels horizontally and compresses a coil spring, the spring rotates to the vertical position with the particle on top of it, and the spring propels the particle upward (Fig. 14-21b–f).

(a)

$$V_1 = 0$$
$$T_1 = \frac{1}{2} m v_1^2$$

$$v_1 = 30 \text{ ft/s}$$

(b)

$$V_2 = T_1 - T_2$$
$$T_2 = \frac{1}{2} m v_2^2$$

$$v_2 = 7 \text{ ft/s}$$

(c) Maximum compression of spring

$$v_3 = v_2$$
$$v_y = 0$$

$$T_3 = T_2$$
$$V_3 = V_2$$

(d) Maximum compression of spring

$$v_3 = v_2$$

(e)

$$v_4 = v_3 = v_2$$
$$v_y = 0$$

$$V_4 = mgh$$
$$T_4 = T_3$$

$$h$$

(f) Maximum elevation

**FIGURE 14-21**

Modeling the system configurations of pole vaulting

## SOLUTION

Assuming that the frictional loss in the pole is negligible, the energy stored in the pole can be calculated. Using the model of a mass and helical spring in Fig. 14-21b and c, and applying Eq. 14-29,

$$T_1 + V_1 = T_2 + V_2$$

where

$$T_1 = \frac{1}{2} mv_1^2 = \frac{1}{2}\left(\frac{170\,\text{lb}}{32.2\,\text{ft/s}^2}\right)(30\,\text{ft/s})^2 = 2376\,\text{ft·lb}$$

$$V_1 = 0$$

$$T_2 = \frac{1}{2} mv_2^2 = \frac{1}{2}\left(\frac{170\,\text{lb}}{32.2\,\text{ft/s}^2}\right)(7\,\text{ft/s})^2$$

$\qquad = 129\,\text{ft·lb}$     (kinetic energy of the particle at maximum deformation of the spring)

$V_2 =$ potential energy in the spring at its maximum deformation

Then

$$2376\,\text{ft·lb} + 0 = 129\,\text{ft·lb} + V_2$$

$V_2 = 2247\,\text{ft·lb}$     (maximum potential energy in the pole from the kinetic energy of running horizontally)

Eventually, the energy $V_2$ is released and the vaulter is projected upward by the pole, as modeled in Fig. 14-21d–f. The height reached by the mass $m$ is calculated by using Eq. 14-29 again,

$$T_3 + V_3 = T_4 + V_4$$

where the subscripts 3 denote the initial conditions of Fig. 14-21d, and the subscripts 4 denote the quantities at the maximum elevation reached. Since the horizontal speed is 7 ft/s at positions 3 and 4, and the vertical velocity is zero at both 3 and 4, $T_3 = T_4$. Thus, with $V_3 = V_2$,

$$V_2 = V_4$$

$$2247\,\text{ft·lb} = mgh = (170\,\text{lb})h$$

$$h = 13.2\,\text{ft}$$

This is the additional height that can be reached, considering the pole as a spring and the vaulter as a particle during a jump. The actual height reached is greater than $h$ because the athlete's center of gravity is initially above the ground by about 3.5 ft, he also jumps upward at takeoff, and he also uses his arms to push upward from the finally straightened pole. The performances of pole vaulters who clear bars set at heights over 18 ft can be explained using these considerations.

EXAMPLE 14-9    581

Problems 14–37 to 14–44 are intended only for those students who have worked with partial derivatives.

**14–37** Prove that a force $\mathbf{F}(x, y, z) = F_x\mathbf{i} + F_y\mathbf{j} + F_z\mathbf{k}$ is conservative if

$$\frac{\partial F_x}{\partial y} = \frac{\partial F_y}{\partial x} \qquad \frac{\partial F_y}{\partial z} = \frac{\partial F_z}{\partial y} \qquad \frac{\partial F_z}{\partial x} = \frac{\partial F_x}{\partial z}$$

Show that these are necessary and sufficient conditions.

In Probs. 14–38 to 14–44, determine whether $\mathbf{F}$ acting on a particle is a conservative force. Also determine the associated potential function if it exists. Use the appropriate relationships given in Prob. 14–37.

**14–38** $\mathbf{F} = x\mathbf{i} + 2y\mathbf{j}$

**14–39** $\mathbf{F} = 3x^2\mathbf{i}$

**14–40** $\mathbf{F} = -2x^2\mathbf{i} + 5y^3\mathbf{j}$

**14–41** $\mathbf{F} = xy\mathbf{i} + 3z^2\mathbf{k}$

**14–42** $\mathbf{F} = \dot{x}\mathbf{i} - z\mathbf{k}$

**14–43** $\mathbf{F} = x^2y\mathbf{i} - xy^2\mathbf{j}$

**14–44** $\mathbf{F} = y\dot{y}\mathbf{j} + 3z^2\dot{z}\mathbf{k}$

**14–45** A 1-lb ball is thrown vertically upward at an initial speed of 30 ft/s. Determine the maximum elevation reached by the ball from its point of release.

**14–46** In the laboratory simulation of impacts on hard hats, a 4-lb plunger falls without resistance through a height of 6 ft. Determine the required initial downward speed of the plunger if its final speed should be 30 ft/s.

Fig. P14-46

**14–47** A 1-kg block starts from rest at point $A$ and slides along a frictionless incline. Determine the speed of the block at point $B$, at a distance of 2 m from $A$.

Fig. P14-47

**14–48** A 3-kg slider is moving up the smooth incline at a speed $v_A = 5$ m/s at point $A$. Calculate the speed at point $B$.

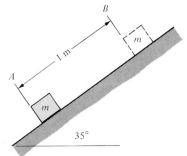

Fig. P14-48

**14–49** A 20-lb package is to be projected upward along a smooth incline toward a conveyor. Determine the required speed $v_A$ at point $A$ if the package should have a speed $v_B = 2$ ft/s at point $B$, 6 ft above $A$.

Fig. P14-49

**14–50** A 160-lb gymnast is doing giant circles on a high bar, moving with negligible velocity in the position shown. Assume that his body may be modeled as a particle at the center of gravity, attached to a rigid, weightless rod of length $l = 3.5$ ft from the horizontal steel bar, and rotating without frictional resistance. Determine the speed of the gymnast's center of gravity at position $A$.

Fig. P14-50

**14–51** An impact tester is modeled as a pendulum of mass $m = 5$ kg on a thin wire of length $l = 0.8$ m. Determine the angle $\theta$ for releasing the ball from rest if its speed at point $A$ should be 2.3 m/s.

Fig. P14-51

**14–52** A person on roller skates is doing a stunt on a curved track. Determine the skater's speed at point $A$ if the height reached at $B$ is 5 m above $A$. Assume that the skater is a particle sliding without friction from $A$ to $B$. $v_B = 0$.

Fig. P14-52

**14–53** A high jumper's center of gravity is 1.05 m above the ground. The jumper's maximum vertical speed at takeoff is 4.8 m/s. Estimate the elevation for setting the bar, which must be 0.15 m below the maximum height reached by the jumper's center of gravity.

**14–54** A ski jumper of mass $m = 70$ kg starts at the top of a tower that has a nearly straight track of 30 m vertical height and a 40 slope. The jumper acquires an initial speed of $v_0 = 2$ m/s by pulling on the railing. How much does this pulling improve the takeoff speed at the bottom of the track, assuming a frictionless slide?

Fig. P14-54

**14–55** A ski jumper of weight $W = 160$ lb starts at the top of a tower that has a nearly straight track of 100 ft elevation at a $40°$ slope. Instead of pulling himself for an initial speed (as in Prob. 14–54), he decides to jump onto the track with a high, slightly forward jump. Assume that he can elevate his center of gravity 2 ft from a standing

position, and that the track is frictionless. How much does this initial jump improve his takeoff speed at the bottom of the track?

**14–56** A 10-lb block is dropped from a height $h = 2$ ft on a spring of length $l = 10$ in. and spring constant $k$. Determine the required value of $k$ if the spring should be compressed by the block to $y = 6$ in.

Fig. P14-56

**14–57** The block in Prob. 14–56 has an initial speed $v_0 = 7$ ft/s downward at height $h$. Determine the required stiffness in this case. Discuss the expected result if the same initial speed is applied upward.

**14–58** A 5-kg block is attached to a spring of stiffness $k = 800$ N/m whose unstretched length is 0.3 m. The block is released from rest at (a) $y_1 = 0.3$ m, and (b) $y_2 = 0.25$ m. Determine the maximum deformation of the spring in each case.

Fig. P14-58

**14–59** Assume that the block in Prob. 14–58 is released at an initial speed $v_0$ at $y = 0.4$ m. Calculate $v_0$ if the spring should be compressed to a length of 0.2 m by the moving block.

**14–60** The suspension system of a vehicle is modeled with springs of stiffness $k = 10$ kN/m and free lengths of $l = 0.3$ m. The maximum compression of each spring is to a length of 0.15 m. Determine the mass $m$ that can be allowed to fall from rest on the spring from a height of $h = 0.2$ m so that the spring just bottoms out (its final length is 0.15 m).

Fig. P14-60

**14–61** Determine the maximum force transmitted to the axle in Prob. 14–60 if a mass of 30 kg falls on the spring from a height of 0.15 m.

**14–62** A high-speed tester for leaf springs has a falling mass $m_A = 5$ kg which has a downward speed of 5 m/s when it contacts the undeformed spring. Assume that the spring constant is $k = 200$ kN/m. Determine (a) the speed of $A$ when the spring's deflection is 0.01 m, and (b) the reactions at supports $B$ and $C$ at that instant. The supports can freely move horizontally. Neglect friction and the mass of the spring.

**Fig. P14-62**

Leaf-spring-testing equipment (Photograph courtesy MTS Systems Corporation, Minneapolis, Minnesota)

**14–63** Assume that a rigid bar of negligible mass is supported by two identical springs of length $l = 0.2$ m and stiffness $k$. A 15-kg block is dropped on the bar from a height of $h = 0.5$ m so that the springs deform identically. Determine the required stiffness $k$ of each spring if the lowest position of the bar is to be at $y = 0.1$ m.

**Fig. P14-63**

**14–64** A 160-lb athlete exercises with a spring device to decrease the time between "explosive" jumps. A rigid bar $AB$ of negligible weight is held on the shoulders and is connected by identical springs of stiffness $k = 2$ lb/in. to the ground.

The jumper's vertical speed at takeoff is $v_0 = 13$ ft/s. At that instant $y = 5.3$ ft and the springs are not deformed. Determine the maximum force that the bar applies to the athlete.

**Fig. P14-64**

**14–65** Assume that in Prob. 14–64 the maximum force of the bar on the athlete is specified as 200 lb. Determine the stiffness $k$ of each spring to achieve this force for the same speed $v_0$.

**14–66** A rigid bar $AB$ of negligible mass is used with flexible strings of constant length at $A$ and $B$ to compress a spring of stiffness $k = 10$ lb/in. and initial length $l = 14$ in. to a length $d = 12$ in. A 6-lb block is dropped on the bar from a height $h$. Determine $h$ if the spring is compressed to $y = 9$ in. by the block.

**Fig. P14-66**

**14–67** A 75-kg mountain climber slips at point $A$ and freely falls 10 m before the safety rope becomes taut. The rope's spring constant is approximately $k = 2$ kN/m when its length is 10 m. Determine the maximum extension of the rope.

**Fig. P14-67**

### Work of a force

Total work of force **F** in displacing particle $P$ from position 1 to position 2,

$$U_{12} = \int_1^2 \mathbf{F} \cdot d\mathbf{r} = \int_1^2 F \cos \theta \, dr$$

$$= \int_1^2 (F_x \, dx + F_y \, dy + F_z \, dz)$$

Section 14-2

### Potential energy and work

**GRAVITATIONAL SYSTEM:**

$$\underset{\substack{\text{work of } \mathbf{W} \\ \text{from 1 to 2}}}{\underline{U_{12}}} = \int_1^2 W \, dy = Wh = \underset{\substack{\text{potential energy} \\ \text{of } W \text{ at 1} \\ \text{with respect to 2}}}{\underline{V_g}}$$

$$U_{23} = Wl$$

$$U_{13} = W(h + l)$$

$$U_{21} = -Wh = -U_{12}$$

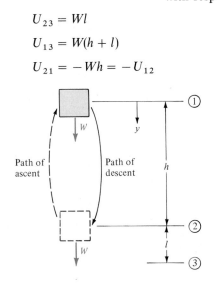

**ELASTIC SYSTEM:**
at $x = 0$, spring is undeformed

$$F = kx$$

$$\underset{\substack{\text{work of } \mathbf{F} \\ \text{from } O \text{ to } x}}{\underline{U}} = \int_0^x kx\,dx = \frac{1}{2}kx^2 = \underset{\substack{\text{potential energy} \\ \text{of spring at } x}}{\underline{V_e}}$$

Section 14-3

**Kinetic energy. Principle of work and energy**

**KINETIC ENERGY OF MASS $m$ AT SPEED $v$:**

$$T = \frac{1}{2}mv^2$$

**PRINCIPLE OF WORK AND ENERGY:**

$$U_{12} = T_2 - T_1$$

where  $U_{12}$ = work of a force on the particle from position 1 to 2
$\phantom{where}$  $T_1$ = kinetic energy of the particle at position 1 (initial kinetic energy)
$\phantom{where}$  $T_2$ = kinetic energy of the particle at position 2 (final kinetic energy)

Section 14-4

**Power and efficiency**

**POWER:**

$$\frac{dU}{dt} = \mathbf{F} \cdot \mathbf{v}$$

**MECHANICAL EFFICIENCY:**

$$\eta = \frac{\text{output work}}{\text{input work}} = \frac{\text{power output}}{\text{power input}}$$

Section 14-5

**Opportunities and limitations using energy methods**

**ADVANTAGES:**

1. Accelerations need not be determined.
2. Modifications of problems are easy to include in the analysis.

**3.** Scalars are summed even if the path of motion is complex.

**4.** Forces that do no work are ignored.

**LIMITATIONS:**

The quantities of work and energy cannot be used to determine accelerations or forces that do no work. In both cases Newton's second law must be used directly.

Section 14-6

**Conservative systems. Potential functions**

The work of a conservative force is independent of the path of displacement of a particle.

Potential energy $V(x, y, z) =$ potential function of a conservative force **F**

$$\underbrace{U_{12}}_{\substack{\text{work of } \mathbf{F} \\ \text{from 1 to 2}}} = \underbrace{-\Delta V}_{\substack{\text{difference of potential} \\ \text{energies at 1 and 2}}}$$

For a closed path, $U = \oint \mathbf{F} \cdot d\mathbf{r} = 0$

Analysis using partial derivatives:

$$\mathbf{F} = F_x \mathbf{i} + F_y \mathbf{j} + F_z \mathbf{k} = -\left( \frac{\partial V}{\partial x} \mathbf{i} + \frac{\partial V}{\partial y} \mathbf{j} + \frac{\partial V}{\partial z} \mathbf{k} \right)$$

Section 14-7

**Conservation of mechanical energy**

$$T_1 + V_1 = T_2 + V_2$$

where $T =$ kinetic energy of a particle
$V =$ potential energy of a particle or elastic member of the system

    1 and 2 refer to two different positions of a particle. The equation is useful to determine the speed or elevation of a particle or the deformation of a spring. Choose reference levels to minimize the number of terms in the equation.

**14-68** The horizontally moving shake table $T$ has a mass of 300 kg. Determine the maximum power required to move the table with negligible friction if the horizontal acceleration is defined by $a = 0.5 \sin 5\pi t$ m/s$^2$ where $t$ is in seconds.

Fig. P14-68

**14-69** In Prob. 14–68 the total weight of the moving parts is 500 lb. Determine the maximum power if the horizontal displacement is defined by $d = 0.6 \sin 10\pi t$ in. where $t$ is in seconds.

**14-70** A 10,000-kg aircraft lands at $x = 0$ where its horizontal speed is $v_0 = 60$ m/s. The magnitude of the total retarding force on the aircraft is constant, $P = 120$ kN (the sum of braking force, reversed engine thrust, and force from the arresting cable). At $x = 50$ m the arresting cable breaks, and the magnitude of the retarding force becomes a constant $P_1 = 50$ kN. Determine the total stopping distance of the airplane.

Fig. P14-70

**14-71** Assume that in Prob. 14–70 the brakes are automatically released when the cable breaks, and the reversed engine thrust is instantaneously changed to a constant forward thrust **T**. Determine the magnitude of the required force **T** if the aircraft must reach its takeoff speed of $v_t = 80$ m/s at $x = 200$ m.

**14-72** A toboggan slide is being planned for a community park. The maximum length of the runway is limited to 600 ft, and the slope of the slide is designed to be 45°. The coefficient of friction $\mu$ for the slide and runway is determined by experiment to be $\mu = 0.05$. (a) Determine the maximum allowable height $h$ of the toboggan slide. Neglect air resistance. (b) What is the maximum speed attained during the toboggan ride? Use the height $h$ from part (a).

Fig. P14-72

**14-73** A 5-kg block is attached to a rigid bar of negligible mass which is pivoted at point $O$. The spring of stiffness $k = 700$ N/m is attached to the middle of the bar and is undeformed when the bar is released from rest in the horizontal position. Calculate the speed of the block at $\theta = 30°$.

Fig. P14-73

**14-74** The 50-lb block slides horizontally with negligible friction. The springs are undeformed in the position shown. The block is displaced to $x = 1$ ft and is released from rest. Determine the speed at $x = 0$ and the maximum displacement to the left.

$k_1 = 20$ lb/in.  $k_2 = 30$ lb/in.

$l_1 = 4$ ft    1 ft    $l_2 = 3$ ft

$x$

Fig. P14-74

**14–75** A vibration damper is modeled as a 10-kg block which slides horizontally with negligible friction. The springs apply no force to the block when $A$, $B$, $C$, and $D$ are lined up. At that instant the speed of the block is $v = 5$ m/s. Determine the stiffness $k$ of each spring if the maximum displacement should be $d = 0.6$ m.

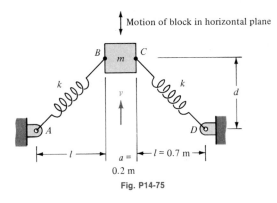

Motion of block in horizontal plane

$B$  $m$  $C$

$k$  $v$  $k$

$d$

$A$  $D$

$l$  $a = 0.2$ m  $l = 0.7$ m

**Fig. P14-75**

**14–76** A mass weighing 50 lb is used for shearing the rough edges off manufactured parts that are successively placed at $C$ by an automatic feeder mechanism. The spring of spring constant $k = 100$ lb/in. has an unstretched length of 12 in. When latch $OA$ is released the mass moves downward, striking the part at $C$. Determine (a) the impact speed at $C$, and (b) the maximum distance $d$ that the mass would fall if the automatic feeder fails to place a part at $C$.

$O$    20 in.

$B$

$A$    $k = 100$ lb/in.

15 in.

$C$

$d$

$D$

**Fig. P14-76**

# 15

# Kinetics of Particles: Momentum Methods

## OVERVIEW

The foundation of all work in the area of kinetics is Newton's second law of motion (Eq. 13-1). It was demonstrated in Ch. 14 that special definitions and methods are useful in reformulating this law to facilitate the solution of problems where it is not essential to consider accelerations. In this chapter another approach is taken to reformulate Newton's second law and develop special methods that are advantageous in solving problems. The quantities of interest in this part of kinetics are force, mass, velocity, and time.

## STUDY GOALS

**SECTION 15-1** defines linear momentum by reformulating Newton's second law. The method of momentum is useful in solving common problems in kinetics, and it should be learned by all students of dynamics.

**SECTIONS 15-2 AND 15-3** present the application of the method of momentum to curvilinear motion of a particle. The resulting concept of angular momentum is frequently used in dynamics. The method of angular momentum is convenient to use as a scalar method for analyzing circular motion in a plane, but it is particularly advantageous as a vector method for solving complex problems.

**SECTION 15-4** presents Newton's law of universal gravitation, the mathematical expression of the gravitational force acting between two masses. This expression is used to define the acceleration of gravity g. The law of universal gravitation is most useful in aerospace engineering and scientific work.

**SECTION 15-5** states Kepler's laws of planetary motion and the proofs of these laws based on Newton's second law. The method of momentum is used in the derivation. The initial steps in the analysis are of general interest, but the subsequent details are not important to the majority of students.

**SECTIONS 15-6 AND 15-7** present the method of impulse and momentum which is useful when time is involved in the analysis of motion of one or more particles. The definitions and major formulas of these sections are often used in dynamics analyses.

**SECTIONS 15-8 THROUGH 15-10** extend the method of impulse and momentum to the analysis of deformable bodies in collisions. These sections introduce a few practical considerations and provide additional practice in solving problems of impulsive motion.

## LINEAR MOMENTUM. TIME RATE OF CHANGE OF LINEAR MOMENTUM

**15-1**

Equation 13-1 can be expressed in a different form by noting that acceleration is the time rate of change of velocity. With this,

$$\mathbf{F} = m\frac{d\mathbf{v}}{dt} \qquad \text{15-1}$$

If it can be assumed that the mass is constant during the motion,

$$\mathbf{F} = \frac{d}{dt}(m\mathbf{v}) \qquad \text{15-2}$$

The vector $m\mathbf{v}$ (which is sometimes denoted by $\mathbf{G}$) is defined as the *momentum*, or, more specifically, the *linear momentum* of the particle. The direction of the momentum is in the direction of the velocity. Its magnitude is equal to the mass times the speed $v$ of the particle. Thus, Newton's second law may be stated as

force = time rate of change of linear momentum

If several forces act simultaneously on a particle, the resultant force must be used in determining the change of linear momentum of the particle,

$$\sum \mathbf{F} = \dot{\mathbf{G}} = \frac{d}{dt}(m\mathbf{v}) \qquad \text{15-3}$$

where the mass $m$ is assumed constant. When the resultant force on a particle is zero, the momentum vector $\mathbf{G}$ is constant in magnitude and direction. This is called *conservation of linear momentum*, which is a special definition of Newton's first law. The units of linear momentum are kg·m/s or N·s in SI, and lb·s in U.S. customary units.

## ANGULAR MOMENTUM

A highly useful analytical method can be developed by considering a particle's momentum with respect to a fixed center of rotation. First assume that a particle $P$ is moving at a constant speed $v$ in a circular path as in Fig. 15-1. In circular motion the vector $m\mathbf{v}$ is perpendicular to the position vector $\mathbf{r}$. It is evident from Sec. 15-1 and Ch. 13 that both $m\mathbf{v}$ and $\mathbf{r}$ are important vectors for the complete description of the motion. Therefore, it is useful to consider their cross product, the moment of the linear momentum vector about the fixed point $O$. This *moment of linear momentum* or *angular momentum* about point $O$ is commonly denoted by $\mathbf{H}_O$,

$$\mathbf{H}_O = \mathbf{r} \times m\mathbf{v} \qquad \boxed{15\text{-}4}$$

According to the definition of the cross product of two vectors, $\mathbf{H}_O$ is perpendicular to the plane of motion (the $xy$ axes in Fig. 15-1). $\mathbf{H}_O$ has a magnitude $rmv$ for circular motion. The direction of $\mathbf{H}_O$ is determined by applying the right-hand rule. In Fig. 15-1, $\mathbf{H}_O$ is out of the paper in the positive $z$ direction. The units of angular momentum are $\text{kg} \cdot \text{m}^2/\text{s}$ in SI, and $\text{ft} \cdot \text{lb} \cdot \text{s}$ in U.S. customary units.

### General Plane Motion

The concept of angular momentum is also applicable to general curvilinear motion. Consider Fig. 15-2, which shows the plane curvilinear motion of a particle $P$. The vector $m\mathbf{v}$ is tangent to the path, and it can be resolved into radial and transverse components $m\mathbf{v}_r$ and $m\mathbf{v}_\theta$, respectively. Equation 15-4 is applicable to this configuration also. The magnitude of the cross product $\mathbf{r} \times m\mathbf{v}$ is $H_O = rmv \sin \phi$. (With $\phi = 90°$, this agrees with the result for circular motion.) It is seen from Fig. 15-2 that $v \sin \phi = v_\theta$; hence the magnitude of the angular momentum is $H_O = rmv_\theta$. This result makes sense since the component $m\mathbf{v}_r$ does not create a moment about point $O$.

$$H_O = \mathbf{r} \times m\mathbf{v} = rmv\mathbf{k}$$

**FIGURE 15-1**

Definition of angular momentum of a particle

$$H_O = \mathbf{r} \times m\mathbf{v} = rmv_\theta\mathbf{k}$$

**FIGURE 15-2**

Angular momentum in general plane motion

Therefore, the magnitude of the angular momentum vector may be stated in several ways,

$$H_O = rmv \sin \phi = rmv_\theta = mr^2\dot\theta \qquad \boxed{15\text{-}5}$$

The direction of $\mathbf{H}_O$ is again determined by applying the right-hand rule. $\mathbf{H}_O$ is in the positive $z$ direction (out of the paper) in Fig. 15-2.

## General Space Motion

The use of the determinant form for the calculation of the vector cross product is useful for the analysis of three-dimensional motion of a particle. Using rectangular coordinates, the position vector is $\mathbf{r} = x\mathbf{i} + y\mathbf{j} + z\mathbf{k}$ and the momentum vector is $m\mathbf{v} = mv_x\mathbf{i} + mv_y\mathbf{j} + mv_z\mathbf{k}$. With these, the most general form of the angular momentum can be written as follows:

$$\mathbf{H}_O = \mathbf{r} \times m\mathbf{v} = \begin{vmatrix} \mathbf{i} & \mathbf{j} & \mathbf{k} \\ x & y & z \\ mv_x & mv_y & mv_z \end{vmatrix} = m\begin{vmatrix} \mathbf{i} & \mathbf{j} & \mathbf{k} \\ x & y & z \\ v_x & v_y & v_z \end{vmatrix} \qquad \boxed{15\text{-}6}$$

The angular momentum vector can also be expressed in terms of its rectangular components,

$$\mathbf{H}_O = H_x\mathbf{i} + H_y\mathbf{j} + H_z\mathbf{k} \qquad \boxed{15\text{-}7}$$

where $H_x$ is the angular momentum about the $x$ axis, and so on. The magnitudes of the components in Eq. 15-7 can be expressed after evaluating the determinant in Eq. 15-6,

$$H_x = m(yv_z - zv_y)$$
$$H_y = m(zv_x - xv_z) \qquad \boxed{15\text{-}8}$$
$$H_z = m(xv_y - yv_x)$$

In any plane motion two of the components of Eq. 15-7 are zero. For example, if the particle moves in the $xz$ plane, $H_x = H_z = 0$ because $y = v_y = 0$.

## TIME RATE OF CHANGE OF ANGULAR MOMENTUM

**15-3**

In the general case a net force is acting on the particle which changes the particle's angular momentum. The net force and the resultant change in angular momentum can be related after evaluating the time derivative of the angular momentum vector,

$$\dot{\mathbf{H}}_O = \frac{d}{dt}(\mathbf{r} \times m\mathbf{v}) = \dot{\mathbf{r}} \times m\mathbf{v} + \mathbf{r} \times m\dot{\mathbf{v}}$$

$$= \mathbf{v} \times m\mathbf{v} + \mathbf{r} \times m\mathbf{a}$$

$$= \mathbf{r} \times m\mathbf{a}$$

The last step is valid because the cross product of the two collinear vectors $\mathbf{v}$ and $m\mathbf{v}$ is zero. In the final term, $m\mathbf{a}$ is equal to the resultant of all the forces acting on the particle. Hence,

$$\dot{\mathbf{H}}_O = \mathbf{r} \times m\mathbf{a} = \mathbf{r} \times \sum \mathbf{F} = \sum \mathbf{M}_O$$

Thus, the term $\mathbf{r} \times m\mathbf{a}$ is equivalent to the moment $\mathbf{M}_O$ caused by the net force acting on the particle. It is concluded that *the time rate of change of angular momentum of a particle is equal to the sum of the moments of the forces acting on the particle.* This is expressed mathematically as

$$\boxed{\sum \mathbf{M}_O = \dot{\mathbf{H}}_O} \qquad \boxed{15\text{-}9}$$

It is often convenient to write Eq. 15-9 in terms of rectangular components,

$$\sum M_x = \dot{H}_x$$
$$\sum M_y = \dot{H}_y \qquad \boxed{15\text{-}10}$$
$$\sum M_z = \dot{H}_z$$

where $M_x$ is the moment of all forces about the $x$ axis, and so on. Equations 15-9 and 15-10 are very useful in more advanced dynamics analyses, especially in dealing with various systems of particles. In the special case when the moment $\mathbf{M}_O$ is zero, the angular momentum of the particle is constant. This case is called *conservation of angular momentum.* Note that if there are no external forces acting on a particle so that the resultant force is zero, both the angular momentum and the linear momentum of the particle will be conserved. There are many cases, however, for which linear momentum is not conserved but angular momentum is conserved. Consider the case in which there is a resultant force acting on a particle but all the forces are either parallel to or intersect a particular axis. The angular momentum of the particle will be conserved about the axis, as in Ex. 15-3. Another interesting case follows.

## Motion under a Central Force

The net force acting on a particle that moves on a curved path is sometimes directed toward the same point $O$ during a time interval. Such a force is called a *central force.* It is convenient to use the point $O$ also as the reference point for the moments and angular momentum, as expressed in Eq. 15-9. In that case the line of action of the central force is through point $O$, $\sum \mathbf{M}_O = 0$, and

$$\dot{\mathbf{H}}_O = 0 \qquad \boxed{15\text{-}11}$$

This equation can be integrated with respect to time, resulting in

$$\mathbf{H}_O = \mathbf{r} \times m\mathbf{v} = \textbf{constant} \qquad \boxed{15\text{-}12}$$

Conservation of the angular momentum means that the particle's velocity and distance from the reference point tend to have an inverse relationship since the magnitude $H_O = r_1 m v_1 \sin \phi_1 = r_2 m v_2 \sin \phi_2 = $ constant. If the distance increases, the velocity must decrease, and vice versa.

# EXAMPLE 15-1

A 10-kg particle moves in the horizontal $xy$ plane under the action of force $\mathbf{F}$. The position of the particle is desired to be $\mathbf{r} = 3t\mathbf{i} + 2t^2\mathbf{j}$ measured in meters. Determine at $t = 5\,s$ (a) the linear momentum $\mathbf{G}$, (b) the angular momentum $\mathbf{H}_O$, (c) the required force $\mathbf{F}$ acting on the particle, and (d) the moment $\mathbf{M}_O$.

## SOLUTION

(a) $\mathbf{G} = m\mathbf{v} = m\dot{\mathbf{r}}$

$$= (10\,\text{kg})(3\mathbf{i} + 4t\mathbf{j})\,\text{m/s} = (30\mathbf{i} + 40t\mathbf{j})\frac{\text{kg}\cdot\text{m}}{\text{s}}$$

At $t = 5\,s$,

$$\mathbf{G} = (30\mathbf{i} + 200\mathbf{j})\,\text{kg}\cdot\text{m/s}$$

(b) $\mathbf{H}_O = \mathbf{r} \times m\mathbf{v} = \mathbf{r} \times \mathbf{G}$

$$= (3t\mathbf{i} + 2t^2\mathbf{j})\,\text{m} \times (30\mathbf{i} + 40t\mathbf{j})\frac{\text{kg}\cdot\text{m}}{\text{s}} = 60t^2\mathbf{k}\,\frac{\text{kg}\cdot\text{m}^2}{\text{s}}$$

At $t = 5\,s$,

$$\mathbf{H}_O = 1500\mathbf{k}\,\text{kg}\cdot\text{m}^2/\text{s}$$

(c) $\mathbf{F} = \dfrac{d}{dt}(\mathbf{G}) = \dot{\mathbf{G}}$

$$= 40\mathbf{j}\,\text{N}$$

At $t = 5\,s$,

$$\mathbf{F} = 40\mathbf{j}\,\text{N}$$

(d) $\mathbf{M}_O = \dot{\mathbf{H}}_O$

$$= 120t\mathbf{k}\,\text{N}\cdot\text{m}$$

At $t = 5\,s$,

$$\mathbf{M}_O = 600\mathbf{k}\,\text{N}\cdot\text{m}$$

# EXAMPLE 15-2

A pendulum consists of a particle of mass $m$ suspended by an inextensible wire of negligible weight (Fig. 15-3a). Write an expression for $\ddot{\theta}$ using the concept of angular momentum.

## SOLUTION

The free-body diagram of the pendulum is drawn in Fig. 15-3b for the general case when $\theta \neq 0$. Since the tension force $\mathbf{T}$ always intersects point $O$, it creates no moment about point $O$. The only moment is due to the force of gravity. Assuming that the pendulum moves in the $xy$ plane, the moment of force $\mathbf{W}$ is about the $z$ axis, so the last of Eq. 15-10 is used,

$$\sum M_z = \dot{H}_z$$

By inspection,     $\sum M_z = -mgL \sin \theta$

Also,     $H_z = mrv$

where $r = L$ and $v = L\dot{\theta}$ since the pendulum moves in circular motion. Thus,

$$\dot{H}_z = mL^2\ddot{\theta}$$

$$\ddot{\theta} = -\frac{g}{L}\sin\theta$$

## JUDGMENT OF THE RESULTS

The angular acceleration $\ddot{\theta}$ is negative, which means that it is in the opposite direction to that shown for the angular velocity $\dot{\theta}$. This result makes sense since the speed of the pendulum decreases as it rises (see also Ex. 14-7).

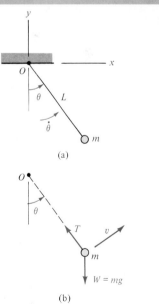

**FIGURE 15-3**

Analysis of a pendulum using the method of angular momentum

EXAMPLE 15-2     595

# EXAMPLE 15-3

A proposed amusement ride is modeled as two particles of weight $W = 600 \, \text{lb}$ mounted on a weightless rod $AB$ which rotates in a horizontal plane (Fig. 15-4a). The rod is rotating at an angular speed $\omega_1 = 1 \, \text{rad/s}$ when $r_1 = 20 \, \text{ft}$. An internal mechanism pulls both particles toward point $O$, to positions where $r_2 = 10 \, \text{ft}$. Determine the final angular speed $\omega_2$. Assume that rod $AB$ is freely rotating about the $y$ axis while the particles are moved from one position to another.

## SOLUTION

The free-body diagram for each particle is drawn in Fig. 15-4b. Since all forces are parallel to or intersect the $y$ axis, the total angular momentum must be conserved about the $y$ axis or point $O$ in the horizontal plane of Fig. 15-4c. Since all motion is in the same plane, the angular momentum does not change direction. Thus, the problem is solved using scalar quantities. The initial total angular momentum is

$$H_O = rmv = 2\left(r_1 \frac{W}{g} r_1 \omega_1\right)$$

which must be equal to the final angular momentum

$$H_O = 2\left(r_2 \frac{W}{g} r_2 \omega_2\right)$$

Thus,

$$r_1^2 \omega_1 = r_2^2 \omega_2$$

The final angular speed is

$$\omega_2 = \frac{r_1^2}{r_2^2} \omega_1 = \frac{20^2}{10^2} (1) = 4 \, \text{rad/s}$$

(a)

Top view

(c)

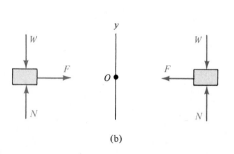

(b)

**FIGURE 15-4**

Conservation of angular momentum in an amusement ride

**15–1** Bar $OA$ of negligible mass is rotating in a horizontal plane about point $O$ at a speed of 50 rpm. A 3-kg block is at a constant distance $r = 0.4$ m from point $O$. Determine the linear momentum $\mathbf{G}$ and the angular momentum $\mathbf{H}_O$ of the block in the coordinate system shown.

**Fig. P15-4**

**Fig. P15-1**

**15–5** The rod of negligible mass in Fig. P15-4 is rotating in a vertical plane at $\dot{\theta} = 6$ rad/s. The 5-kg block $B$ is at $r = 1$ m and moving toward point $O$ at a speed of $\dot{r} = 2$ m/s. Determine the couple $\mathbf{M}$ that must act about point $O$ to maintain the rod's angular speed. Neglect friction.

**15–6** The unstretched length of the spring is 0.3 m, and its stiffness is 400 N/m. The 2-kg block attached to the spring moves on a smooth horizontal plane at velocity $\mathbf{v} = (3\mathbf{i} - 6\mathbf{j})$ m/s when $\theta = 10°$, $\mathbf{F} = (100\mathbf{i} - 70\mathbf{j})$ N, and $l = 0.5$ m. Determine the angular momentum $\mathbf{H}_O$ of the block and the time rate of change of $\mathbf{H}_O$.

**15–2** Solve Prob. 15–1 assuming that the block is moving away from the pin $O$ at a constant speed of $\dot{r} = 6$ m/s along the rod.

**15–3** A 4-lb particle is moving in a horizontal $xy$ plane under the action of force $\mathbf{F}$. The position of the particle is defined by $\mathbf{r} = (5\mathbf{i} + 2t^3\mathbf{j})$ ft, where $t$ is in seconds. Determine the linear momentum $\mathbf{G}$ and the angular momentum $\mathbf{H}_O$ of the particle at $t = 3$ s. Also determine the force $\mathbf{F}$.

**Fig. P15-6**

**15–7** A bicycle wheel in a test is supported at point $O$ and is initially at rest. A couple of magnitude $M_O = 100$ lb·ft is applied to the wheel. Determine the angular acceleration $\ddot{\theta}$ using momentum methods. Assume that the thin rim weighs 4 lb, the spokes are weightless, and $r = 13$ in. Model the rim as a particle at the distance $r$ from point $O$.

**Fig. P15-3**

**15–4** Rod $OA$ is assumed weightless and that it rotates without friction in a horizontal plane. The rod is rotating at 100 rpm. The 10-lb block $B$ is at $r = 2$ ft and moving away from point $O$ at a speed of $\dot{r} = 4$ ft/s. Determine the couple $\mathbf{M}$ that must act on the rod to maintain its angular speed.

**Fig. P15-7**

**15–8** A proposed amusement ride consists of a flat disk rotating in a horizontal plane at angular speed $\dot{\theta} = 0.1$ rad/s. A 70-kg person $P$ walks between safety rails from point $O$ to $A$ with a constant speed $v = 1$ m/s relative to the disk. Determine the couple $\mathbf{M}_O$ that must be applied to the disk to maintain its angular speed when the person is at $r_1 = 2$ m and at $r_2 = 3$ m, still walking toward point $A$. Neglect friction.

Top view

**Fig. P15-8**

**15–9** Determine the required couple $\mathbf{M}_O$ in Prob. 15–8 if $\dot{\theta} = 0.1$ rad/s must increase at the rate of 0.2 rad/s$^2$ when the person is at $r_1 = 2$ m. Consider only this position and neglect the mass of the disk.

**15–10** A 12-lb block $A$ is attached to a rigid bar of length $l = 4$ ft and negligible weight. A spring applies an upward force of magnitude $F = 5$ lb at the middle of the bar when it starts from rest. The bar swings without frictional resistance in a vertical plane. Determine the initial acceleration of the block using momentum methods.

**Fig. P15-10**

**15–11** Solve Prob. 15–10 assuming that at the position shown block $A$ is moving at a constant speed $v = 2$ ft/s toward point $O$.

**15–12** The amusement park ride shown rotates freely about $O$ when released from rest at $\theta = 90°$. It may be modeled as a concentrated mass $m$ at the end of a massless arm of length $l$. Determine the angular acceleration $\ddot{\theta}$ at the positions $\theta = 0$ and $30°$ using momentum methods for $l = 10$ m.

**Fig. P15-12**

**15–13** A 2-kg mass is suspended by a wire of length $l = 1$ m and negligible mass. At $\theta = 30°$ the angle is decreasing at the rate of 0.3 rad/s. Determine the linear momentum $\mathbf{G}$ and the angular momentum $\mathbf{H}_O$ of the mass using rectangular coordinates.

**Fig. P15-13**

**15–14** Calculate the tension in the wire in Prob. 15–13 using momentum methods.

**15-15** An impact tester is modeled as a particle on the end of a weightless, rigid rod. At $\theta = 60°$ the angle is increasing at the rate of 0.2 rad/s. Determine the axial force in the bar using momentum methods.

Fig. P15-15

**15-16** In Prob. 15-15, a couple $\mathbf{M}_O$ is applied to the bar by a mechanism to make the angular acceleration $\ddot{\theta} = 2$ rad/s² clockwise at $\theta = 60°$. Calculate $M_O$.

**15-17** Assume a 160-lb gymnast on the flying rings is modeled as a particle $P$ which can move along the supporting rope. At $\theta = 0$, $l = 10$ ft and $P$ has a horizontal speed $v = 8$ ft/s. At that instant $l$ is increased to 15 ft by the gymnast's own action. Calculate the new speed of $P$.

Fig. P15-17

**15-18** An astronaut during a space walk is on a tether of length $L = 20$ m which has an angular speed $\omega_y = 0.01$ rad/s (the motion is in the $xz$ plane). The astronaut is pulled closer to the space vehicle by shortening the tether to $d = 4$ m. Determine the linear speed of the astronaut in the final position in the $xyz$ frame neglecting the effects of gravity.

Fig. P15-18

**15-19** The astronaut shown in Fig. P15-18 pushes off the space vehicle at $d = 5$ ft at an initial velocity $\mathbf{v}_0 = (2\mathbf{i} - 3\mathbf{j})$ ft/s while the tether is loose. The tether becomes taut at length $L = 40$ ft. Determine the final linear velocity of the astronaut in the $xyz$ frame neglecting the effects of gravity.

**15-20** A ball of mass $m$ is attached to a rigid rod of negligible mass and length $r$. The position $\phi$ of the rod is controlled by an internal mechanism. At $\phi = 0$, the rod rotates in a horizontal plane at angular speed $\dot{\theta}$. The mechanism adjusts the angle $\phi$ to 30° below the horizontal. Determine the new angular speed $\dot{\theta}_1$ assuming that the system is freely rotating about the vertical axis.

Fig. P15-20

**15-21** Assume that in Prob. 15-20 the original angular speed $\dot{\theta}$ must be maintained for all positions of the rod $r$. Determine the required couple to achieve this if the angle $\phi$ changes at a constant rate of $\dot{\phi} = 0.1\dot{\theta}$.

According to Newton's law of action and reaction, any force acting on a particle must give rise to a force on the body that causes the force on the particle. These two forces must be equal in magnitude and opposite in direction, as illustrated in Fig. 15-5, where $r$ is the distance between two masses $M$ and $m$. Thus, the mass $m$ moves under the force **F**, while $M$ moves under the force $-$**F**, so that the forces are collinear. The mutual forces have the same magnitude, but the resulting accelerations of the two bodies depend on their masses. Clearly, if $M$ is much larger than $m$, the latter appears to move entirely with respect to $M$, while the force $-$**F** centered at $m$ has a negligible effect on the larger mass. These differences depend on the relative masses and not on the magnitudes of the forces between them. Of course, the magnitudes of the forces are important in analyzing the motion of each body in detail.

The mutual force between two masses that are not in contact results from gravitational, electric, or magnetic effects. Of these, the gravitational force was first investigated theoretically and experimentally. Newton recognized that the magnitude of the mutual force, $F$, between any two masses $M$ and $m$ is directly proportional to the product of the two masses and inversely proportional to the square of the distance $r$ between them. The mathematical expression of Newton's *law of universal gravitation is*

$$F = G\frac{Mm}{r^2}$$

15-13

where $G$ is a universal constant, the *constant of gravitation*. The value of this constant can be determine only experimentally. It was first measured by Cavendish in 1797, and many times since then to obtain a reliable value. This value in SI and U.S. customary units is

$$G = 6.673 \times 10^{-11}\,\text{m}^3/\text{kg}\cdot\text{s}^2$$

$$G = 3.442 \times 10^{-8}\,\text{ft}^4/\text{lb}\cdot\text{s}^4$$

The gravitational force $F$ is significant only if at least one of the masses is very large because $G$ is extremely small.

Equation 15-13 can be used to define the acceleration of gravity $g$ and to partially explain why $g$ is not a constant over the surface of the earth. Denoting the mass of the earth by $M$, its radius by $R$, and the mass of a body near or on the surface by $m$,

$$W = mg = G\frac{Mm}{R^2} \qquad g = \frac{GM}{R^2}$$

15-14

**FIGURE 15-5**

Mutual attraction of two masses

Here $G$ and $M$ are constant but $R$ is not since the earth is not a perfect sphere. Other variations in $g$ are caused by the spin of the earth and the fact that $M$ is constant but it is not uniformly distributed. Nevertheless, the mass of the earth is much larger than the variations in the other quantities, so $g = 9.81\,\text{m/s}^2 = 32.2\,\text{ft/s}^2$ can be assumed in most engineering problems.

## KEPLER'S LAWS OF PLANETARY MOTION     15-5

The methods of analyzing the motion of a particle under the action of a central force are ideally suitable in the area of celestial mechanics. When studying the motion of our solar system, the sun is often assumed fixed because its mass is 700 times the total mass of all its planets. The gravitational attraction between different planets is also assumed negligible when compared to that between the sun and a planet. Thus, the mutual gravitational force between the sun and a planet is analyzed as a central force on the planet.

The first specific statements about planetary motion were made by the German astronomer Johannes Kepler (1571–1630), who analyzed astronomical data. The following two laws that Kepler discovered empirically are relevant to the concepts presented in this chapter.

**KEPLER'S FIRST LAW (OR LAW OF SECTORIAL AREAS)**
The radius drawn from a planet to the sun sweeps equal areas in equal times.

**KEPLER'S SECOND LAW**
The orbit of each planet is an ellipse, with the sun at one of its foci.

The mathematical proofs of these laws were provided by Newton. For a derivation consider the orbit of a planet as in Fig. 15-6. It is most

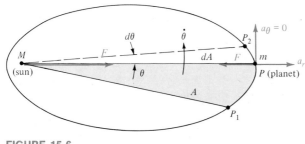

**FIGURE 15-6**

Illustration for proving Kepler's laws of planetary motion

convenient to use polar coordinates $r$ and $\theta$ to describe the position and motion of the planet. The angle $\theta$ is measured from an arbitrary line of reference. Since the only force acting on the planet is the central force $F$ directed toward the sun, the equations of motion are

$$F = -ma_r \quad \text{and} \quad a_\theta = 0 \quad\quad (1)$$

From Eqs. 13-7 and 15-13,

$$F = G\frac{Mm}{r^2} = -m(\ddot{r} - r\dot{\theta}^2) \quad\quad (2)$$

Since the angular momentum is constant, Eq. 15-5 can be written as

$$\dot{H}_O = \frac{d}{dt}(mr^2\dot{\theta}) = 0 \quad\quad (3)$$

Thus, after integration of Eq. (3) with respect to time,

$$r^2\dot{\theta} = C \quad\quad (4) \quad \boxed{15\text{-}15}$$

where $C$ is a constant.

Equation (4) is the mathematical statement of Kepler's first law as can be seen from the following considerations:

1. A finite sectorial area $A$ swept by the radius $r$ is defined in Fig. 15-6 as the planet moves from point $P_1$ to $P$.
2. An infinitesimal sectorial area $dA$ is swept by the radius as the planet moves from point $P$ to $P_2$. Since the area $dA$ is essentially a triangle of two sides $r$ and a third side $r\,d\theta$, $dA \simeq \frac{1}{2}r^2\,d\theta$ for any orbital path.
3. Dividing $dA$ by $dt$, and realizing that $r$ is essentially constant during the time interval $dt$, $\dot{A} = \frac{1}{2}r^2(d\theta/dt)$. On the basis of Eq. (4), this is equivalent to the verbal statement of Kepler's first law.

Kepler's second law can be proved by manipulating Eq. (2). Since the aim is to describe the shape of the orbit, time derivatives should be eliminated, and $r$ should be expressed as a function of $\theta$. Using Eq. (4), $\dot{\theta}$ is eliminated from Eq. (2),

$$\ddot{r} + \frac{GM}{r^2} - \frac{C^2}{r^3} = 0 \quad\quad (5)$$

The following expressions can be used to eliminate $\ddot{r}$,

$$\dot{r} = \frac{dr}{dt} = \frac{dr}{d\theta}\frac{d\theta}{dt} = \frac{dr}{d\theta}\dot{\theta}$$

$$\ddot{r} = \frac{d\dot{r}}{dt} = \frac{d}{dt}\left(\frac{dr}{d\theta}\dot{\theta}\right) = \frac{d}{d\theta}\left(\frac{dr}{d\theta}\dot{\theta}\right)\frac{d\theta}{dt} = \dot{\theta}\frac{d}{d\theta}\left(\dot{\theta}\frac{dr}{d\theta}\right)$$

Once again, $\dot{\theta}$ can be eliminated by using Eq. (4), and

$$\ddot{r} = \frac{C^2}{r^2}\frac{d}{d\theta}\left(\frac{1}{r^2}\frac{dr}{d\theta}\right)$$

Thus, Eq. (5) becomes

$$\frac{C^2}{r^2}\frac{d}{d\theta}\left(\frac{1}{r^2}\frac{dr}{d\theta}\right) - \frac{C^2}{r^3} + \frac{GM}{r^2} = 0 \qquad (6)$$

This differential equation can be significantly simplified by making the substitution $q = 1/r$. The result is

$$\frac{d^2q}{d\theta^2} + q = \frac{GM}{C^2} \qquad (7)$$

The general solution of this equation is

$$q = \frac{GM}{C^2}[1 - K\cos(\theta - \alpha)] \qquad (8)$$

where $K$ and $\alpha$ are constants. Now the two quantities of interest, $r$ and $\theta$, can be related in a usable form,

$$r = \frac{C^2/GM}{1 - K\cos(\theta - \alpha)} \qquad \boxed{15\text{-}16}$$

For repetitive orbital motion, $r$ must be finite. Since $-1 \le \cos(\theta - \alpha) \le 1$, the constant $K$ must be in the range of values $-1 < K < 1$ to obtain positive, noninfinite values of $r$. It can be shown that under these conditions, Eq. 15-16 is the general equation of an ellipse. This proves Kepler's second law. In addition, three special cases of applying Eq. 15-16 are worth mentioning. For $K = 0$, $r =$ constant, and the orbit is circular. For $K = 1$, the orbit is parabolic. For $K > 1$, the orbit is hyperbolic. There are angles for which the radius $r$ is infinite in the cases of parabolic and hyperbolic paths. The four distinct motions under a central force are shown schematically in Fig. 15-7.

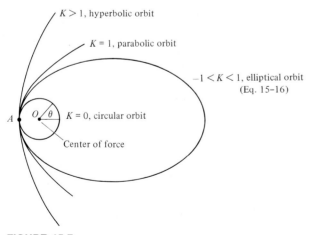

**FIGURE 15-7**

Four distinct planetary orbits

The foregoing considerations are also relevant to the flight of spacecraft. An important special aspect of such flights is the velocity required to allow a vehicle to escape from a celestial body. For a vehicle to escape from its dominant center of force, it must be moving on a parabolic or hyperbolic path with respect to that center of force. To understand the escape problem of spacecraft, consider a particle at point $A$ in Fig. 15-7. The particle must have a minimum angular momentum with respect to point $O$ to maintain a circular or elliptical orbit; otherwise, it will fall toward the center of force at point $O$. For escape, the angular momentum (thus, velocity) of the particle must be high enough to obtain a parabolic or hyperbolic path. Therefore, $K \geq 1$ in Eq. 15-16. Since it is most convenient to work with the particle at its closest approach to the center of force, $r$ in Eq. 15-16 should be minimized. Letting $K = 1$ and $\cos(\theta - \alpha) = -1$,

$$2r_A = \frac{C^2}{GM}$$

The velocity at point $A$ is $v_A = r_A \dot{\theta}_A$. From Eq. 15-15, $C = r_A^2 \dot{\theta}_A$; hence,

$$2r_A = \frac{v_A^2 r_A^2}{GM}$$

$$v_A = \sqrt{\frac{2GM}{r_A}} = v_{\text{esc}} \qquad \boxed{15\text{-}17}$$

This is the *escape velocity* since it allows the particle to move on a parabolic path away from the center of force. In the case of any celestial body the escape velocity can be expressed in two ways using Eq. 15-14, $GM = gR^2$,

$$v_{\text{esc}} = \sqrt{\frac{2GM_{\text{cel}}}{r_{\text{esc}}}} = \sqrt{\frac{2g_{\text{cel}}R_{\text{cel}}^2}{r_{\text{esc}}}} \qquad \boxed{15\text{-}18}$$

where  $G$ = universal constant of gravitation
$M_{\text{cel}}$ = mass of celestial body from which the escape is made
$g_{\text{cel}}$ = acceleration of gravity of celestial body
$R_{\text{cel}}$ = radius of celestial body
$r_{\text{esc}}$ = distance (from center of celestial body) at which the escape velocity is computed

Naturally, any particular escape is with respect to only one particular celestial body. For example, a vehicle that escapes from the earth may still remain a satellite of the sun.

## Orbital Geometries of Planets and Satellites

The relationships of focal positions and characteristic dimensions of elliptic orbits are established using Fig. 15-8. The orbit has semi-major axis $a$ and semi-minor axis $b$, and the foci are points $B$ and $C$ (of course, the major mass $M$ is only at one of the foci). Consider a satellite of mass $m$ at point $A$, the perigee of the orbit. It is seen that $2a = r_{min} + r_{max}$, so

$$a = \frac{1}{2}(r_{min} + r_{max})$$

$\boxed{15\text{-}19}$

Next, it should be recalled that the sum of the distances from any point on an ellipse to the two foci is a constant for that ellipse. Therefore,

$$BD + CD = AB + AC = 2a$$

From the symmetry of the right triangles,

$$BD = CD = a = OC + r_{min}$$
$$b^2 = (OD)^2 = (CD)^2 - (OC)^2 = a^2 - (a - r_{min})^2$$
$$= r_{min}(2a - r_{min}) = r_{min}r_{max}$$

and

$$b = \sqrt{r_{min}r_{max}}$$

$\boxed{15\text{-}20}$

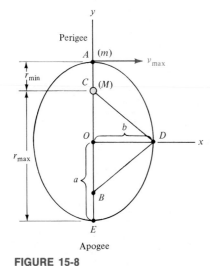

**FIGURE 15-8**

Geometry of elliptic orbits

## EXAMPLE 15-4

The radar record of a comet indicates that it is moving at a speed of 26,000 mph in a direction perpendicular to the radial line $r$, and that $r$ is 5000 mi from the center of the earth. Will this comet be captured by the earth?

### SOLUTION

Since the velocity is perpendicular to the radial vector, the velocity is a maximum or a minimum at this point. Assume that the velocity is a maximum and compare it to the escape velocity. From Eq. 15-18,

$$v_{esc} = \sqrt{\frac{2g_{cel}R_{cel}^2}{r_{esc}}} = \sqrt{\frac{2(32.2\,\text{ft/s}^2)[(3960\,\text{mi})(5280\,\text{ft/mi})]^2}{5000\,\text{mi}\,(5280\,\text{ft/mi})}}$$

$$= 32,660\,\text{ft/s} = 22,270\,\text{mph}$$

The comet will escape.

## EXAMPLE 15-5

An earth satellite of mass $m = 1000\,\text{kg}$ is planned to move in an elliptic orbit with semimajor axis $a = 3 \times 10^7\,\text{m}$ and semiminor axis $b = 2 \times 10^7\,\text{m}$. Calculate the maximum and minimum velocities of the satellite. The earth is at focus $C$ of the ellipse in Fig. 15-9a.

### SOLUTION

Since the satellite is moving under a central force, from Eq. 15-5,

$$r_1 m v_1 \sin \phi_1 = r_2 m v_2 \sin \phi_2$$

The maximum and minimum velocities occur when $\sin \phi = \sin 90° = 1$, at points $A$ and $E$ in Fig. 15-9a. The mass remains constant, hence the angular momentum relation can be written as

$$r_{max} v_{min} = r_{min} v_{max} \tag{a}$$

This equation involves two unknown velocities. Therefore, one of the velocities must be calculated using another relationship. Note that at points $A$ and $E$ the radial direction (through the center of the earth) coincides with the normal direction for the path of the satellite. Considering $v_{max}$, the following equation is written using Fig. 15-9, Eq. 13-6, and Eq. 15-13,

$$\sum F_n = ma_n = m\frac{v^2}{\rho}$$

$$\frac{GMm}{r_{min}^2} = m\frac{v_{max}^2}{\rho} \tag{b}$$

The radius of curvature $\rho$ at point $A$ in Fig. 15-9 is expressed according to a well-known formula as

$$\rho = \frac{[1 + (dx/dy)^2]^{3/2}}{|d^2x/dy^2|} \tag{c}$$

The general equation of an ellipse centered at the origin of the $xy$ axes is

$$\frac{x^2}{a^2} + \frac{y^2}{b^2} = 1 = \frac{x^2}{(3 \times 10^7\,\text{m})^2} + \frac{y^2}{(2 \times 10^7\,\text{m})^2}$$

Taking the derivative twice with respect to $y$ yields

$$\frac{2x}{a^2}\left(\frac{dx}{dy}\right) + \frac{2y}{b^2} = 0$$

$$\frac{dx}{dy} = -\frac{a^2}{b^2}\left(\frac{y}{x}\right)$$

$$\frac{d^2x}{dy^2} = -\frac{a^2}{b^2}\left(\frac{1}{x}\right)$$

At point $A$, $x = a$ and $y = 0$, so

$$\frac{dx}{dy} = 0 \quad \text{and} \quad \frac{d^2x}{dy^2} = -\frac{a}{b^2}$$

Then

$$\rho = \frac{[1 + (0)^2]^{3/2}}{a/b^2} = \frac{(2 \times 10^7 \text{ m})^2}{3 \times 10^7 \text{ m}} = 1.33 \times 10^7 \text{ m}$$

Note the difference between radius of curvature ($\rho$) and distance from the center of the earth ($r$) at any given point on the elliptical orbit. Note that $\rho = r$ only for a circular orbit. Solving Eq. (b) for $v_{max}$ and substituting from Eq. 15-14 gives

$$v_{max} = \left(\frac{\rho GM}{r_{min}^2}\right)^{1/2} = \left(\frac{\rho g R^2}{r_{min}^2}\right)^{1/2}$$

$$= \left[\frac{(1.33 \times 10^7 \text{ m})(9.81 \text{ m/s}^2)(6.37 \times 10^6 \text{ m})^2}{(7.64 \times 10^6 \text{ m})^2}\right]^{1/2}$$

where $r_{min} = 7.64 \times 10^6$ m was determined from Eqs. 15-19 and 15-20. The result is

$$v_{max} = 9524 \text{ m/s} \quad \text{at} \quad x = 3 \times 10^7 \text{ m} \quad \text{and} \quad y = 0$$

From Eq. (a),

$$v_{min} = \frac{r_{min}}{r_{max}} v_{max} = \left(\frac{7.64 \times 10^6 \text{ m}}{5.236 \times 10^7 \text{ m}}\right)(9524 \text{ m/s})$$

where $r_{max} = 5.236 \times 10^7$ m was determined from Eq. 15-19. Finally,

$$v_{min} = 1390 \text{ m/s} \quad \text{at} \quad x = -3 \times 10^7 \text{ m} \quad \text{and} \quad y = 0$$

## JUDGMENT OF THE RESULTS

The value of $\rho$ is acceptable because it is less than $a$, the radius of a circle that could be drawn through point $A$ while centered at $O$. The value of $r_{min}$ might appear small, but it is reasonable for an elliptical orbital, and it allows the satellite to pass about 1300 km above the surface of the earth. The large velocities are possible in celestial motion.

(a)

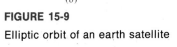

(b)

**FIGURE 15-9**

Elliptic orbit of an earth satellite

EXAMPLE 15-5  607

**15-22** Assume that the moon travels in a circular orbit of radius $r = 239{,}000$ mi around the earth. Calculate the time of a complete revolution of the moon.

**15-23** Two satellites revolve around the earth in concentric circular orbits. Determine the orbital speeds $v_1$ and $v_2$ for $r_1 = 2 \times 10^7$ m and $r_2 = 3 \times 10^7$ m, respectively.

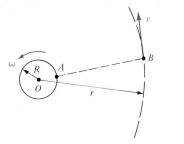

Fig. P15-23

**15-24** Satellites in synchronous orbit always remain above the same point on the earth, as $O$, $A$, and $B$ in Fig. P15-23 remain lined up. Calculate the radius $r$ of the synchronous orbit and the speed $v$ of the satellite. Use $R = 6370$ km.

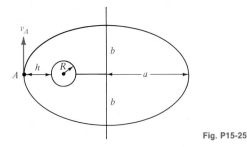

Fig. P15-24

**15-25** An earth satellite is planned to travel in an elliptical orbit whose major axis is $a = 10{,}000$ mi. Determine the required speed $v_A$ of the satellite at point $A$ if $h$ must be 300 mi and $R = 3960$ mi.

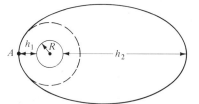

Fig. P15-25

**15-26** An earth satellite is moving in an elliptical orbit (Fig. P15-25). The closest approach is at point $A$, at $h = 1000$ km. Determine the dimension $a$ of the orbit if $v_A = 30\,000$ km/h and $R = 6370$ km.

**15-27** The radar record of an earth satellite's orbit shows that $h = 700$ km and $a = 10\,000$ km (Fig. P15-24). Determine the maximum and minimum velocities of the satellite using $R = 6370$ km.

**15-28** An earth satellite is in circular orbit at an altitude $h_1 = 500$ km and traveling at a speed $v_1$. At point $A$ the original speed is increased by 25%. Determine the maximum altitude $h_2$ of the new orbit. $R = 6370$ km.

Fig. P15-28

**15-29** An earth satellite is in circular orbit at an altitude $h_1 = 400$ mi. Calculate the velocity in this orbit. Next, calculate the required change in velocity at point $A$ to obtain a new orbit with a maximum altitude $h_2 = 30{,}000$ mi as shown in Fig. P15-28. $R = 3960$ mi.

**15-30** An earth satellite is in elliptical orbit with minimum altitude $h_1 = 300$ mi and maximum altitude $h_2 = 21{,}000$ mi (Fig. P15-28). Determine the velocity at point $A$.

**15-31** In Prob. 15–30, determine the required change in velocity at point $A$ to obtain a circular orbit of radius $r = h_1 + R$, where $R = 3960$ mi.

**15-32** A space vehicle is in circular orbit of radius $r = 10\,000$ km around the earth. Calculate the escape velocity from the earth starting from this orbit.

**15-33** Sketch and discuss the orbits of a particle about a center of force for the borderline case of nearly identical maximum velocities but different orbits, one elliptical, the other parabolic.

**15-34** A space vehicle is in circular orbit of radius $r = 2200$ mi around a planet whose radius is $R = 2000$ mi. Determine the escape velocity from this planet starting from the given orbit if the acceleration of gravity is 10 ft/s² at the surface of the planet.

The concept of momentum is especially useful for analyzing motion where time must be considered. For example, a force may act on a particle during a specified time interval. The final momentum in that case depends on the initial momentum, the force, and the time interval of the action of the force. The analysis of such a problem is facilitated by using a special method based on Newton's second law for what is called *impulsive motion* of a particle. This method is developed in the following.

Assume that a force $\mathbf{F}$ is acting on a particle of constant mass $m$. Newton's second law is written as

$$\mathbf{F} = \frac{d}{dt}(m\mathbf{v}) = \dot{\mathbf{G}}$$

where $\mathbf{G} = m\mathbf{v}$ is the linear momentum of the particle. Multiplying this equation by $dt$ and integrating it from time $t_1$ to time $t_2$ yields

$$\mathbf{F}\,dt = d(m\mathbf{v})$$
$$\int_{t_1}^{t_2} \mathbf{F}\,dt = m\mathbf{v}_2 - m\mathbf{v}_1 \qquad \boxed{15\text{-}21}$$

The integral on the left side in Eq. 15-21 is called the *linear impulse*, or simply the *impulse*. The impulse is said to be *of the force*, and acting *on the mass*. In simple terms, impulse is equal to the change in linear momentum of the particle. The force $\mathbf{F}$ in Eq. 15-21 is sometimes called the *impulsive force*.

In general, the force may change both in magnitude and direction, even during a very short time interval. To analyze such motion it is necessary to evaluate the integral of $\mathbf{F}\,dt$. It is often most convenient to use rectangular components of the force. For convenience, the change in linear momentum is written as the vector $\Delta\mathbf{G}$,

$$\Delta\mathbf{G} = \mathbf{G}_2 - \mathbf{G}_1 = m\mathbf{v}_2 - m\mathbf{v}_1$$

where the subscripts refer to time $t_2$ and time $t_1$. With this,

$$\Delta\mathbf{G} = \int_{t_1}^{t_2} \mathbf{F}\,dt \qquad \boxed{15\text{-}22}$$
$$= \mathbf{i}\int_{t_1}^{t_2} F_x\,dt + \mathbf{j}\int_{t_1}^{t_2} F_y\,dt + \mathbf{k}\int_{t_1}^{t_2} F_z\,dt$$

The $x$, $y$, and $z$ components of the change in linear momentum are set equal to the $x$, $y$, and $z$ components of the impulse,

$$\Delta G_x = mv_{x_2} - mv_{x_1} = \int_{t_1}^{t_2} F_x\,dt$$
$$\Delta G_y = mv_{y_2} - mv_{y_1} = \int_{t_1}^{t_2} F_y\,dt \qquad \boxed{15\text{-}23}$$
$$\Delta G_z = mv_{z_2} - mv_{z_1} = \int_{t_1}^{t_2} F_z\,dt$$

It must be kept in mind that impulse and momentum are vectors, and the scalar components are the most convenient to use in solving problems. A few major possibilities involving impulsive motion in a plane are shown in Fig. 15-10. Assume that in each case the initial momentum $m\mathbf{v}_1$ is known, a given force $\mathbf{F}$ acts on the particle, and the final momentum $m\mathbf{v}_2$ is to be determined. From Eq. 15-21,

$$\underbrace{m\mathbf{v}_1}_{\substack{\text{initial}\\\text{momentum}}} + \underbrace{\int_{t_1}^{t_2} \mathbf{F}\, dt}_{\text{impulse}} = \underbrace{m\mathbf{v}_2}_{\substack{\text{final}\\\text{momentum}}} \qquad \boxed{15\text{-}24}$$

The scalar or vector calculations with this equation are illustrated in Fig. 15-10. When the initial momentum and the impulse are collinear, their magnitudes can be summed as scalars to obtain the final momentum. In

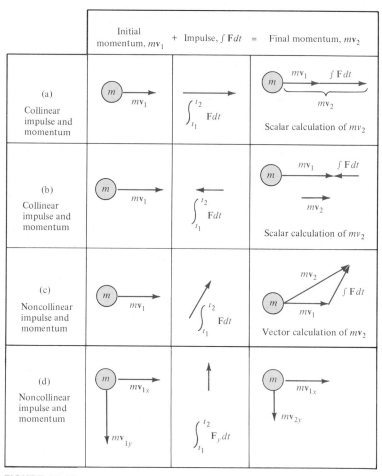

**FIGURE 15-10**

Major possibilities of impulsive motion of a particle

all other cases they must be summed as vectors. Very often the impulse acts in only one coordinate direction and the momentum in the other direction is conserved (Fig. 15-10d). In this case the scalar component equations of Eq. 15-23 are the easiest to use. These concepts are valid for all bodies that can be treated as particles that cannot deform or rotate for all practical purposes.

The units of impulse of a force are the same as those for linear momentum. In SI it is $N \cdot s = kg \cdot m/s$, and in U.S. customary units it is $lb \cdot s = slug \cdot ft/s$.

## Approximations of Impulses

The impulsive force that is easiest to analyze has a square-wave shape as in Fig. 15-11a. In that case the impulse can be expressed as

$$\int_{t_1}^{t_2} F \, dt = F \, \Delta t \qquad \boxed{15\text{-}25}$$

In a real mechanical system the force varies relatively smoothly such as $F(t)$ in Fig. 15-11b. Nevertheless, in some cases the area $\int F \, dt$ under the curve $F(t)$ may be approximated by inspection using the equivalent quantities $F_{eq}$ and $\Delta t_{eq}$ in Eq. 15-25.

(a)

(b)

**FIGURE 15-11**
Ideal and realistic impulsive forces

# CONSERVATION OF TOTAL MOMENTUM OF PARTICLES

Consider the motion of two particles that temporarily interact. The mutual forces between the particles can be described using Newton's law of action and reaction. The force acting on each particle is equal in magnitude and opposite in direction to the force acting on the other particle. Both forces act in the same time interval. The resulting change in velocity of each particle depends on its mass. To analyze these changes in velocity it is useful to consider the changes in momentum of two interacting particles.

Assume that two particles of mass $m_A$ and $m_B$ are at rest and touching a compressed spring as in Fig. 15-12a. The spring is suddenly released and the particles are propelled apart (Fig. 15-12b). The forces acting on the two particles during their accelerations are equal in magnitude and opposite in direction so the changes in the particles' momentum can be written as

$$m_A \mathbf{v}_{A_2} - m_A \mathbf{v}_{A_1} = \int_{t_1}^{t_2} \mathbf{F}\, dt$$

$$m_B \mathbf{v}_{B_2} - m_B \mathbf{v}_{B_1} = -\int_{t_1}^{t_2} \mathbf{F}\, dt$$

(a) Particles are at rest, spring is compressed

(b) Spring is released, particles are propelled apart

**FIGURE 15-12**

Conservation of total momentum of two interacting particles

Adding these two equations gives

$$m_A \mathbf{v}_{A_2} - m_A \mathbf{v}_{A_1} + m_B \mathbf{v}_{B_2} - m_B \mathbf{v}_{B_1} = 0$$

$$m_A \mathbf{v}_{A_1} + m_B \mathbf{v}_{B_1} = m_A \mathbf{v}_{A_2} + m_B \mathbf{v}_{B_2}$$

where the individual changes occur in the same time interval $\Delta t$. Note that it does not matter how the force varies during $\Delta t$. It is concluded that the sum of the momenta of the particles $A$ and $B$ at any time is a constant,

$$m_A \mathbf{v}_A + m_B \mathbf{v}_B = \text{constant} \qquad \boxed{15\text{-}26}$$

which means that *the total momentum of the interacting particles is conserved* when only a mutual force is acting on them.

The same result can be obtained by applying Eq. 15-24 to analyze the impulsive motion of two or more particles simultaneously. For each particle $i$ the initial momentum is $m_i \mathbf{v}_{i_1}$, and the final momentum is $m_i \mathbf{v}_{i_2}$. After vectorially adding the momenta for all the particles and the impulses for all the forces acting on them, Eq. 15-24 can be written for $n$ particles as

$$\sum_i^n (m_i \mathbf{v}_i)_1 + \sum_i^n \int_{t_1}^{t_2} \mathbf{F}_i \, dt = \sum_i^n (m_i \mathbf{v}_i)_2 \qquad \boxed{15\text{-}27}$$

Here the integral represents the impulses of external forces acting on the particles *and* of interactive forces between the particles. The interactive forces always cancel out in pairs. Thus, in the absence of external forces, the final and initial momenta are the same,

$$\underbrace{\sum_i^n (m_i \mathbf{v}_i)_1}_{\substack{\text{total initial} \\ \text{momentum at} \\ \text{time } t_1}} = \underbrace{\sum_i^n (m_i \mathbf{v}_i)_2}_{\substack{\text{total final} \\ \text{momentum at} \\ \text{time } t_2}} \qquad \boxed{15\text{-}28}$$

Equations 15-26 and 15-28 both express the conservation of total momentum for particles that have only mutually interactive forces acting on them. It is worth noting that the change in velocity of a particle in mutual impulsive motion with another particle is inversely proportional to its mass.

## EXAMPLE 15-6

A ballistics test consists of the setup shown in Fig. 15-13. The coefficient of friction between the 100-kg gun–platform system and the support is $\mu_k = 0.7$. Determine **(a)** the speed $(v_P)_1$ of the platform just after the gun fires a 0.05-kg bullet at a speed $v_B = 1000$ m/s, and **(b)** the distance $\Delta x_P$ traveled by the platform before the bullet's impact at $A$.

### SOLUTION

**(a)** The initial speed of the platform $(v_P)_1$ is found using Eq. 15-26,

$$\underbrace{\sum (m_i v_i)_0}_{0} + \underbrace{\sum \int F\, dt}_{0} = \sum (m_i v_i)_1$$

$$0 = (0.05\,\text{kg})(-1000\,\text{m/s}) + (100\,\text{kg})(v_P)_1$$

$$(v_P)_1 = 0.5\,\text{m/s}$$

**(b)** The distance $\Delta x_P$ traveled by the platform is equal to $x - x_0$ in Eq. 11-9.

$$v^2 = v_0^2 + 2a(x - x_0)$$

Here $v_0 = (v_P)_1 = 0.5\,\text{m/s}$, but $v = (v_P)_2{}^*$ and $a$ must be calculated yet. This involves several intermediate steps as follows. The relative velocity of the bullet to the platform just after firing is

$$(v_{B/P})_1 = v_B - (v_P)_1 = -1000\,\text{m/s} - (0.5\,\text{m/s}) = -1000.5\,\text{m/s}$$

Neglecting air resistance, the time $\Delta t$ the bullet takes to travel the fixed distance of 2 m is

$$x_{B/P} = (x_{B/P})_0 + (v_{B/P})_1\, \Delta t$$

$$-2\,\text{m} = 0 + (-1000.5\,\text{m/s})\, \Delta t$$

$$\Delta t = 0.002\,\text{s}$$

---

\* The different notations from those in Eq. 11-9 are useful to distinguish the various phases of the impulsive motion.

**FIGURE 15-13**

Gun–platform system for a ballistics test

Looking at the motion of the platform alone during this time interval, the friction force ($\mathscr{F} = \mu N = \mu W$) is an impulsive force opposing the direction of motion. Thus,

$$mv_1 - \mathscr{F}\,\Delta t = mv_2$$

$$(100\,\text{kg})(0.5\,\text{m/s}) - (0.7)(100\,\text{kg})(9.81\,\text{m/s}^2)(0.002\,\text{s})$$
$$= (100\,\text{kg})(v_P)_2$$

$$(v_P)_2 = 0.486\,\text{m/s} \qquad \text{(velocity of platform to the right}$$
$$\text{just prior to impact)}$$

From Newton's second law, $-\mathscr{F} = ma$ ($\mathscr{F}$ is negative since friction opposes the motion). Thus, the acceleration $a$ of the platform is

$$a = -\frac{\mathscr{F}}{m} = \frac{-(0.7)(100\,\text{kg})(9.81\,\text{m/s}^2)}{(100\,\text{kg})}$$

$$= -6.87\,\text{m/s}^2$$

The distance traveled is calculated using Eq. 11–9,

$$v^2 = v_0^2 + 2a(x - x_0)$$

$$(0.486\,\text{m/s})^2 = (0.5\,\text{m/s})^2 + 2(-6.87\,\text{m/s}^2)(\Delta x_p)$$

$$\Delta x_p = 0.001\,\text{m}$$

## JUDGMENT OF THE RESULTS

The small initial speed of the platform is reasonable considering the masses of the bullet and the platform. The small distance traveled by the platform prior to the bullet's impact at $A$ is reasonable considering the small average velocity of the platform and the short time of travel of the bullet in the air.

EXAMPLE 15-6    615

# EXAMPLE 15-7

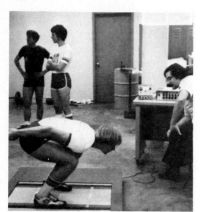

(a)

Consider an athlete making vertical jumps on a force platform which electronically measures force (Fig. 15-14a). Figure 15-14b is the recording of force vs. time in a single jump from start to finish. Zero force corresponds to the person standing still, and the minimum force represents the person's weight (while the jumper is airborne). The takeoff velocity is the important factor in jumping. Determine the approximate takeoff velocity and height of the jump by estimating the area under the force vs. time plot in the relevant part of the motion. Each division of the vertical axis is 20 lb. The force recording is a graph of the magnitude of the *resultant* force $R_y = N - W$ acting on the person ($N$ = normal force, $W$ = weight).

## SOLUTION

First, it is useful to analyze the major parts of the motion. While the person is standing still, the magnitude of the normal force $N$ exactly equals the weight $W$, so

$$R_y = \sum F = N - W = ma_y = 0$$

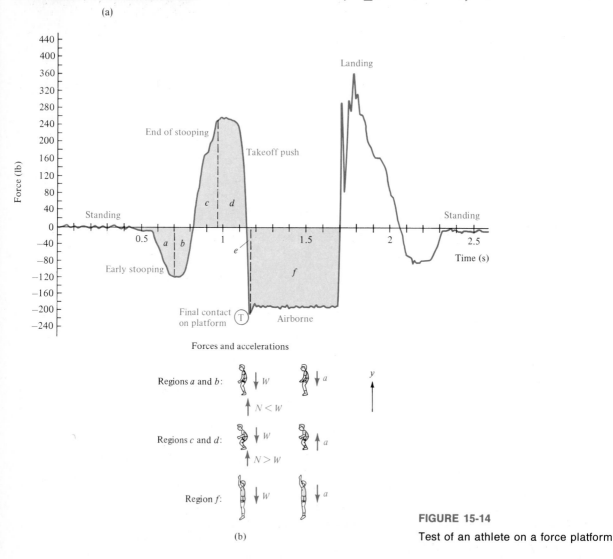

**FIGURE 15-14**

Test of an athlete on a force platform

(b)

At the beginning of the stooping process the acceleration of the center of gravity $C$ is negative, hence a negative resultant force is registered,

$$R_y = ma_y = -ma_C \qquad (\text{i.e., } N < W)$$

Nearing the end of the stooping process, the person is decreasing a negative velocity which requires a positive force larger than the body weight,

$$R_y = ma_y = ma_C \qquad (\text{i.e., } N > W)$$

During the push-off process the person accelerates his center of gravity up (increasing a positive velocity), so the force registered is still positive. While in the air, the resultant force acting on the person is his weight (in the negative direction) and hence the force platform records a negative force of

$$R_y = \sum F = -W = -mg \qquad (N = 0)$$

Therefore, the weight of the jumper is seen to be about 180 lb. Considering the original problem, Eq. 15-24 is applicable:

$$mv_1 + \int R_y \, dt = mv_2$$

The integral $\int R_y \, dt$ is estimated as the area under the force vs. time curve. "Takeoff" occurs at point $T$. Breaking the area up into several component areas and approximating them as triangles and a rectangle,

$$\int R_y \, dA = A_a + A_b + A_c + A_d + A_e$$

$$= \frac{1}{2}(0.12\,\text{s})(-120\,\text{lb}) + \frac{1}{2}(0.11\,\text{s})(-120\,\text{lb})$$

$$+ \frac{1}{2}(0.14\,\text{s})(240\,\text{lb}) + (0.16\,\text{s})(240\,\text{lb})$$

$$+ \frac{1}{2}(0.02\,\text{s})(-180\,\text{lb})$$

$$= 39.6\,\text{lb·s}$$

Therefore,

$$0 + 39.6\,\text{lb·s} = \left(\frac{180\,\text{lb}}{32.2\,\text{ft/s}^2}\right) v_2$$

$$v_2 = 7.08\,\text{ft/s}$$

The height of the jump is calculated from the work–energy equation (Eq. 14-15),

$$T_1 + U_{12} = T_2$$

$$\frac{1}{2}(m)(7.08\,\text{ft/s})^2 - mgh = 0$$

$$h = \frac{(7.08\,\text{ft/s})^2}{2g} = 0.778\,\text{ft (elevation of center of gravity)}$$

A more accurate analysis of the jump using the same plot is left as an exercise for the student. Note the negative signs of the areas $A_a$, $A_b$, and $A_e$ in the summation for $\int R_y \, dA$. These negative signs represent the fact that the resultant force allows the body to fall in those intervals.

EXAMPLE 15-7   617

**15–35** A 1-kg particle is initially at rest. A force $\mathbf{F} = 10\mathbf{i}$ N acts on the particle from time $t_1 = 0$ to $t_2 = 2$ s. Calculate the particle's final velocity.

**15–36** A 4-lb particle is traveling at a velocity $\mathbf{v}_1 = 10$ ft/s $\mathbf{j}$ when a force $\mathbf{F} = 50$ lb $\mathbf{j}$ starts acting on it. Determine the particle's velocity $\mathbf{v}_2$ after a time interval of 0.5 s during which the force has acted.

**15–37** A 10-kg block is traveling at a velocity $\mathbf{v}_1 = 20\mathbf{i}$ m/s at time $t_1 = 0$. A force $\mathbf{F} = -2\mathbf{i}$ kN acts on the block from time $t_1$ to $t_2 = 1$ s. Calculate the block's velocity $\mathbf{v}_2$ at times $t > t_2$.

**15–38** A 0.5-kg particle is initially at rest. A force $\mathbf{F} = (50 + 10t)\mathbf{i}$ N acts on the particle in a time interval of 3 s. Calculate the final velocity of the particle if $t$ is in seconds.

**15–39** A 5-lb block is traveling at a velocity $\mathbf{v}_1 = (-3\mathbf{j} + 5\mathbf{k})$ ft/s at time $t_1 = 2$ s. A force $\mathbf{F} = (400\mathbf{j} - 400\mathbf{k})$ lb acts on the block from time $t_1$ to $t_2 = 4$ s. Determine the block's velocity at time $t_2$.

**15–40** A 0.2-lb particle is traveling at a velocity $\mathbf{v}_1 = 20\mathbf{i}$ ft/s at time $t_1 = 1$ s. A force $\mathbf{F} = (30t + 40t^2)\mathbf{j}$ lb acts on the particle from time $t_1$ to $t_2 = 1.2$ s. Determine the particle's velocity at times $t > t_2$.

**15–41** A new rifle is to be fired while firmly attached to a heavy bench. Calculate the magnitude of the average force applied to the vise $A$ if a 0.05-lb bullet traverses the barrel in a time of 2 ms and exits at a speed of 2000 ft/s.

Fig. P15-41

**15–42** A highway divider is to be protected from crashing automobiles by a large, resilient bumper $A$. What is the magnitude of the average force acting on the bumper when it arrests a 2000-kg car traveling initially at 40 m/s? The car is stopped in a time interval of 0.2 s.

Fig. P15-42

**15–43** A 2000-kg car is traveling on a horizontal road at a speed of 100 km/h. Calculate the time required for the car to skid to a stop with the brakes applied if $\mu_k = 0.9$. Neglect air resistance.

**15–44** The seat belt for a car is assumed to be in the same plane as the velocity vector of the car. The belt should not break in restraining a 100-kg passenger when the car traveling at 100 km/h is stopped in 0.05 s. Calculate the required minimum strength of the belt.

Fig. P15-44

**15–45** A 1500-kg car is traveling down a 15% grade at a speed of 70 km/h when the brakes are applied. Determine the time required for the car to skid to a stop if $\mu_k = 0.7$, neglecting air resistance.

**15–46** A 3000-lb car is traveling up a 20% grade at a speed of 60 mph when the brakes are applied. Calculate the time required for the car to skid to a stop if $\mu_k = 0.5$, neglecting air resistance.

**15–47** A 2500-lb car is traveling down a 5% grade at 50 mph when the brakes are applied. Because of heat generated at the skidding tires, $\mu_k = 0.9 - 0.03t$, where $t$ is in seconds. Calculate the time required to stop the car neglecting air resistance.

**15–48** A 2000-kg car is pulling a 1000-kg trailer at a speed of 90 km/h. The brakes on the car (and none on the trailer) are applied causing skidding of its wheels. Calculate the magnitude of the horizontal force, $F$, between the car and the trailer during the initial time interval $\Delta t = 1\,\text{s}$ of the skidding if $\mu_k = 0.5$. Treat the car and the trailer as particles.

**Fig. P15-48**

**15–49** Consider the recording of force vs. time for a single vertical jump on a force platform. Estimate the initial impulse and check if it should make the jumper airborne for the length of time measured. Zero force corresponds to the person standing (or crouching) still on the platform.

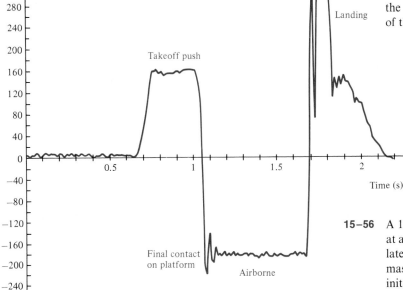

**Fig. P15-49**

**15–50** Reconsider Ex. 15-7 and Fig. 15-14. Compare and discuss the approximate total impulse of the jumper during takeoff with the total impulse during landing on the force platform.

**15–51** Because of improper programming of the automatic recording equipment of a force platform, part of the data during landing after a jump was "clipped" as shown in Fig. P15-49. Estimate the part of the recording that was lost.

**15–52** A 130-lb high jumper elevates her center of gravity by 2.5 ft. What is the magnitude of the average force exerted by her leg if it touches the ground for a time of 0.14 s during takeoff?

**15–53** An important training method for athletes who need explosive movement by the legs is called *depth jumping*. In this exercise a person jumps from a bench to the floor and immediately rebounds vertically. Consider a 70-kg person who freely falls from a height of 0.6 m and rebounds to the same elevation. Calculate the magnitude of the average normal force acting on this person if the time interval of contact with the floor is 0.5 s.

**15–54** A 0.05-kg bullet is fired from a 4-kg gun. The bullet leaves the barrel at a speed of 700 m/s. Calculate the recoil speed of the gun.

**15–55** A box that weighs 20 lb must be given an initial upward velocity of 20 ft/s by a spring-loaded mechanism. Calculate the recoil speed of the 150-lb base structure. Assume that the box and the base structure are particles on opposite sides of the spring.

**15–56** A 10 000-kg truck leaves a loosely moored ferry at a speed of 10 km/h relative to the ferry. Calculate the velocity imparted to the 20 times more massive ferry, assuming that both were at rest initially.

**15–57** Two particles of mass $m_A$ and $m_B$ are at rest next to a compressed spring. Prove that the mass center remains fixed at any time after the spring is released sending the particles apart. Assume there is no friction.

**Fig. P15-57**

## COLLISION OF DEFORMABLE BODIES

The principles of impulse and momentum were introduced in Secs. 15-6 and 15-7 for particles and for bodies that could be treated as particles in situations where the deformations and rotations of them could be ignored. At this stage the concepts developed for particles and particle-like bodies are extended to deal with deformable bodies that can be assumed to interact without rotations. The additional analysis of rotations is in the area of rigid body dynamics (Ch. 20).

Before considering the details of impulsive motion with deformations of the bodies, it is necessary to define the various possible relative motions at the instant of impact. These relative motions are described in reference to the line that is perpendicular to the plane of contacting surfaces of the bodies. This line is called the *line of impact* and is shown in Fig. 15-15. The relative motion is a *central impact* when the mass centers of the two bodies are on the line of impact during the impact. Otherwise, the motion involves *eccentric impact*. Only central impact is discussed in this chapter; it may occur in two ways as shown in Fig. 15-15. These are discussed in the following two sections.

(a) Direct central impact

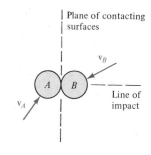

(b) Oblique central impact

**FIGURE 15-15**

Direct and oblique central impacts

## DIRECT CENTRAL IMPACT

In the simple case when two masses are in collinear impact, as in Fig. 15-15a, the only matter of interest is the rebound of the bodies as it is affected by their deformations. It should be remembered that all bodies deform during an impact since all forces cause deformations. Each deformation has two components, an *elastic or recoverable deformation*, and a *plastic or permanent deformation*. Generally, both of these occur simultaneously, but either one may be dominant. They can also be equal.

The plastic deformation mainly dissipates energy in the form of heat. During a collision, energy is also dissipated by the generation of sound waves and elastic waves. It is the loss of energy that reduces the rebound velocities.

The ability of a body to deform plastically depends on the material, on the temperature, and on the rate of deformation. In the impact of two bodies, their deformations may be quite different, both overall and in the proportions of elastic and plastic deformations. It is useful to understand this, but it is not necessary to analyze all of these details for each impact. What is important is the overall changes of velocities, regardless of which body deformed more than the other. If one body deforms plastically and the other one does not, it still affects the motion of both bodies.

The unified treatment of the impact problem is based on the principles of impulse and momentum. Assume that the bodies in impact are spheres of different masses, deformation characteristics, and velocities. It is worthwhile to consider four successive instants of the impact event as shown in Fig. 15-16.

1. The bodies before impact are traveling on a collision course. Their velocities may be in the same direction (pursuit), or in opposite directions heading toward each other. The momenta in this situation are the initial information for the analysis.

2. Contact is made, and both bodies deform. Their velocities rapidly change in this phase but are somewhat similar as the two bodies travel together for an instant. At maximum deformation $v'_A = v'_B = v_c$, where the common speed $v_c$ is at the end of the phase.

3. The bodies are still in contact but are rebounding. At the beginning of this phase, the particles still have the common speed $v''_A = v''_B = v_c$. The deformation is decreasing throughout this phase, which is called *restitution*.

4. If there is separation, the bodies travel at new velocities. The final speeds $v_{Af}$ and $v_{Bf}$ depend on the initial momenta and the energy lost during impact. $v_{Af}$ may be positive or negative even if both $v_A$ and $v_B$ were positive.

Assume that there are no other forces acting on the two bodies besides the mutual force between them. In this case the total momentum of the two bodies is conserved. Using Eq. 15-26 gives

$$m_A v_A + m_B v_B = m_A v_{Af} + m_B v_{Bf} \qquad \boxed{15\text{-}29}$$

(a) Before impact (bodies approach)

(b) Increasing deformation during contact

(c) Decreasing deformation or restitution during contact

(d) After impact (bodies move apart; $v_{Af}$ may be to the left)

**FIGURE 15-16**

Four successive events of a direct central impact

This equation contains two unknowns, $v_{Af}$ and $v_{Bf}$. An additional relation can be obtained by considering the mutual impulses during contact. These impulses are expressed using Fig. 15-17. The impulses during increasing and decreasing deformations are caused by the mutual contact forces of magnitude $F_D$ and $F_R$, as shown in the figure. Using the equation of impulse and momentum for body $A$ yields

$$\underbrace{m_A v_c}_{\substack{\text{momentum at} \\ \text{max. deformation}}} = \underbrace{m_A v_A}_{\substack{\text{initial} \\ \text{momentum}}} - \underbrace{\int_0^{t_1} F_D\, dt}_{\substack{\text{impulse of} \\ \text{contact force} \\ \text{during} \\ \text{increasing} \\ \text{deformation}}} \qquad \boxed{15\text{-}30a}$$

and similarly for the restitution,

$$\underbrace{m_A v_{Af}}_{\substack{\text{final} \\ \text{momentum}}} = \underbrace{m_A v_c}_{\substack{\text{momentum} \\ \text{at max.} \\ \text{deformation}}} - \underbrace{\int_{t_1}^{t_2} F_R\, dt}_{\substack{\text{impulse of} \\ \text{contact force} \\ \text{during restitution}}} \qquad \boxed{15\text{-}30b}$$

The ratio $e$ of the restoration impulse and of the deformation impulse is written to represent the ability of the bodies to recover from the deformations.

$$e = \frac{\int_{t_1}^{t_2} F_R\, dt}{\int_0^{t_1} F_D\, dt} = \frac{m_A(v_c - v_{Af})}{m_A(v_A - v_c)} \qquad \boxed{15\text{-}31}$$

For body $B$,

$$m_B v_c = m_B v_B + \int_0^{t_1} F_D\, dt \qquad \boxed{15\text{-}32a}$$

$$m_B v_{Bf} = m_B v_c + \int_{t_1}^{t_2} F_R\, dt \qquad \boxed{15\text{-}32b}$$

The ratio $e$ for body $B$ becomes

$$e = \frac{\int_{t_1}^{t_2} F_R\, dt}{\int_0^{t_1} F_D\, dt} = \frac{m_B(v_{Bf} - v_c)}{m_B(v_c - v_B)} \qquad \boxed{15\text{-}33}$$

Since the mass cancels out in each equation, the common speed $v_c$ at maximum deformation can be eliminated from the two expressions for $e$. Note that a quotient remains unchanged if an equal quotient is added to it by summing the numerators and denominators [for example, $a/b = (a + a)/(b + b)$]. Thus, from Eqs. 15-31 and 15-33,

$$\boxed{\begin{aligned} e &= \frac{v_{Bf} - v_c + v_c - v_{Af}}{v_c - v_B + v_A - v_c} = \frac{v_{Bf} - v_{Af}}{v_A - v_B} \qquad 15\text{-}34 \\[4pt] &= \frac{|\text{relative velocity of separation}|}{|\text{relative velocity of approach}|} \end{aligned}}$$

(a) Increasing deformation during contact from time 0 to $t_1$; velocity of $A$ changes from $v_A$ to $v_c$; $v_A' \approx v_B'$

(b) Decreasing deformation or restitution during contact from time $t_1$ to $t_2$; velocity of $A$ changes from $v_c$ to $v_{Af}$; $v_A'' \approx v_B''$

**FIGURE 15-17**

Mutual contact forces during direct central impact

Students are advised to be careful of signs when applying Eq. 15-34. Note that the form

$$\frac{v_{B_f} - v_{A_f}}{v_A - v_B}$$

is derived for the specific example shown in Fig. 15-17. Particle $A$ is to the left of particle $B$ and they are both moving to the right. Other conditions will affect the signs in the expression above. For example, if particle $A$ is moving to the right and particle $B$ is moving to the left, toward $A$, the denominator will be $v_A + v_B$. Instead of just memorizing the form for $e$ given above, the safest check on the results is to remember the form

$$e = \frac{|\text{relative velocity of separation}|}{|\text{relative velocity of approach}|}$$

The ratio $e$ is called the *coefficient of restitution*. Two extreme values of this coefficient are worth remembering.

**1. *Perfectly elastic impact, e = 1.*** Ideally, elastic materials sustain no permanent deformations during impact, and the generation of sound and elastic waves is negligible. The impulse during the increasing deformation equals the impulse during the rebound from maximum deformation. Consequently, $e = 1$, and the bodies separate with the same relative velocity at which they were approaching each other (Eq. 15-34). *The total kinetic energy of the bodies is conserved in this case.*

**2. *Perfectly plastic impact, e = 0.*** Ideally, plastic materials are permanently deformed during impact and have no restoring force after the maximum deformation occurs. This means that $F_R = 0$ in Fig. 15-17b and $e = 0$ for both bodies when the ratio of the impulses is considered for the two phases of contact. Note that only one of the bodies has to be ideally plastic to obtain this condition. It is interesting that *the total momentum is conserved even though energy is lost* during the permanent deformation. The reason is that all forces are internal and cancel out in the isolated system of the two bodies. The conservation of momentum and having no rebound lead to the conclusion that the bodies move together after the impact with a common speed $v_c$. For initial speeds $v_A$ and $v_B$,

$$m_A v_A + m_B v_B = (m_A + m_B)v_c \qquad \boxed{15\text{-}35}$$

It can be shown analytically that some of the total initial kinetic energy is lost during plastic impact. Most of this energy is converted into heat.

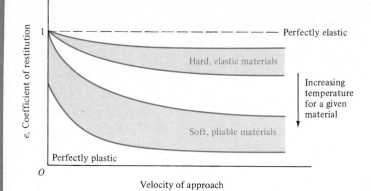

FIGURE 15-18

Schematic variation of coefficient of restitution for various materials and conditions

**Real materials in impact.** For real materials $0 < e < 1$ because forces cause elastic and plastic deformations simultaneously, even if one or the other is practically immeasurable. A complicating factor in this respect occasionally is the adhesive quality of the surfaces in contact, which may cause the bodies to stick together. Other effects on the coefficient of restitution are caused by the velocity of approach and temperature of the bodies. It is not possible to accurately determine $e$ without experiments under the conditions of interest. Figure 15-18 shows the general tendencies of behavior that can be expected on the basis of experimental results.

## 15-10

## OBLIQUE CENTRAL IMPACT

The concepts of direct central impact can be readily extended to the case where the initial and final velocities are not collinear. This case is called *oblique impact* and is illustrated in Fig. 15-19, where a coordinate system is conveniently positioned at the place of impact. The normally available initial information consists of the masses $m_A$ and $m_B$, and their initial velocities $\mathbf{v}_A$ and $\mathbf{v}_B$. The final velocities $\mathbf{v}_{Af}$ and $\mathbf{v}_{Bf}$ are unknown in magnitude and direction, so four independent equations are required to determine the two components of each velocity.

The problem of oblique central impact is solved using the superposition of two motions if there is no mutual friction force between the bodies

(a)

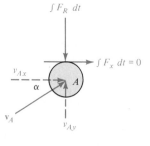

(b)

FIGURE 15-19

Oblique central impact

during contact. The two distinct motions are those along the $x$ and $y$ axes in Fig. 15-19. Motion along the $x$ axis is equivalent to two bodies moving along parallel lines without any chance of collision. Motion along the $y$ axis is equivalent to direct central motion, which was analyzed in Sec. 15-9. With this approach the following four equations can be written:

1. First consider mass $A$ by itself (Fig. 15-19b). From Eq. 15-24,

$$m_A(v_A)_x + \int F_x \, dt = m_A(v_{Af})_x$$

Since friction is assumed to be negligible, $F_x = 0$, and the $x$ component of the momentum of body $A$ is conserved,

$$(v_A)_x = (v_{Af})_x \qquad \boxed{15\text{-}36}$$

From Fig. 15-19a, $v_A \cos \alpha = v_{Af} \cos \alpha_f$.

2. Similarly, the $x$ component of the momentum of body $B$ is conserved,

$$(v_B)_x = (v_{Bf})_x \qquad \boxed{15\text{-}37}$$

From Fig. 15-19a,

$$v_B \cos \beta = v_{Bf} \cos \beta_f$$

3. The total momentum of the bodies in the $y$ direction is conserved as in direct central impact,

$$m_A(v_A)_y + m_B(v_B)_y = m_A(v_{Af})_y + m_B(v_{Bf})_y \qquad \boxed{15\text{-}38}$$

From Fig. 15-19a,

$$m_A v_A \sin \alpha - m_B v_B \sin \beta = -m_A v_{Af} \sin \alpha_f + m_B v_{Bf} \sin \beta_f$$

4. The coefficient of restitution is applicable to the motion along the $y$ axis (as in direct central impact). From Eq. 15-34,

$$e = \left( \frac{|\text{relative velocity of separation}|}{|\text{relative velocity of approach}|} \right)_{y \text{ direction}} \qquad \boxed{15\text{-}39}$$

From Fig. 15-19a,

$$e = \frac{v_{Bf} \sin \beta_f + v_{Af} \sin \alpha_f}{v_A \sin \alpha + v_B \sin \beta}$$

Equations 15-36 to 15-39 can be used to solve problems where the assumption of no friction during the collision is reasonable. Note that the detailed equations based on Fig. 15-19 are only for that example.

TIP

The specific steps given above are summarized as follows: Consider the motion along the common normal (line of impact) and along the common tangent line between the two particles by finding the components of each particle's velocity along these lines. The tangential component of velocity is conserved for *each* particle, while the *total* momentum of the system is conserved in the normal direction. The equation for $e$ applies only in the normal direction.

# EXAMPLE 15-8

The energy-absorbing bumper of a 1500-kg car should be designed to allow for low-velocity impacts. Assume that the car travels at 10 km/h when it hits a stationary 2000-kg car which has a rigid bumper (Fig. 15-20) and is free to roll. Calculate the required coefficient of restitution of the bumper if car A is to continue at a speed of 3 km/h. Also calculate the energy that must be dissipated.

## SOLUTION

Since no resultant external forces act on the cars, momentum of the system is conserved. A 3-km/h final speed means that the 1500-kg car travels in the same direction after impact. The final velocity of car $B$ must be determined. From Eq. 15-29 and with the positive direction taken to the left in Fig. 15-20,

$$m_A v_{A_1} + m_B v_{B_1} = m_A v_{A_2} + m_B v_{B_2}$$

$$(1500\,\text{kg})(10\,\text{km/h}) + 0 = (1500\,\text{kg})(3\,\text{km/h}) + (2000\,\text{kg})v_{B_2}$$

$$v_{B_2} = 5.25\,\text{km/h}$$

From Eq. 15-34,

$$e = \frac{v_{B_2} - v_{A_2}}{v_{A_1} - v_{B_1}} = \frac{5.25\,\text{km/h} - 3\,\text{km/h}}{10\,\text{km/h} - 0} = 0.225$$

The energy dissipated is calculated using Eq. 14-17,

$$T_1 + U_{12} = T_2$$

where $U_{12} =$ energy dissipated (negative work).

$$\frac{1}{2} m_A v_{A_1}^2 + \frac{1}{2} m_B v_{B_1}^2 + U_{12} = \frac{1}{2} m_A v_{A_2}^2 + \frac{1}{2} m_B v_{B_2}^2$$

Converting the speeds to m/s (with 10 km/h = 2.78 m/s, 3 km/h = 0.833 m/s, and 5.25 km/h = 1.46 m/s),

$$\frac{1}{2}(1500\,\text{kg})(2.78\,\text{m/s})^2 + U_{12} = \frac{1}{2}(1500\,\text{kg})(0.833\,\text{m/s})^2$$

$$+ \frac{1}{2}(2000\,\text{kg})(1.46\,\text{m/s})^2$$

$$U_{12} = -3140\,\text{N·m}$$

**FIGURE 15-20**

Collision of two cars

EXAMPLE 15-9

A ball is thrown with a speed of 10 m/s at an angle of 30° onto a smooth floor as shown in Fig. 15-21. Determine the angle of rebound $\beta_r$ and the rebound speed $v_r$ using a coefficient of restitution of $e = 0.7$.

## SOLUTION

Since the floor is smooth (i.e., $\mu = 0$) the component of linear momentum in the $x$ direction remains constant during impact. Using Eq. 15-29,

$$m(10 \text{ m/s})(\cos 30°) = m(v_r \cos \beta_r)$$

Canceling the mass,

$$8.66 \text{ m/s} = v_r \cos \beta_r \qquad \text{(a)}$$

The definition of coefficient of restitution $e$ applies *only* to the components of velocity along the line of impact ($y$ axis for this case).

$$e = \left( \frac{|\text{relative velocity of separation}|}{|\text{relative velocity of approach}|} \right)_y$$

$$0.7 = \frac{v_r \sin \beta_r - 0}{(10 \text{ m/s})(\sin 30°) - 0}$$

$$3.5 \text{ m/s} = v_r \sin \beta_r \qquad \text{(b)}$$

Dividing Eq. (a) by Eq. (b) yields

$$0.404 = \tan \beta_r$$

Then

$$\beta_r = 22°$$

$$v_r = 9.34 \text{ m/s}$$

Note that in this problem the conservation of total momentum equation is not used and the velocity of the floor is taken as zero. This is done because the collision occurs between a ball of relatively negligible mass and the floor, which is attached to the earth, of relatively infinite mass. For a collision of a particle with an infinite mass, the total momentum equation is indeterminate, and the velocity of the infinite mass must be zero, or it would have an infinite change in momentum.

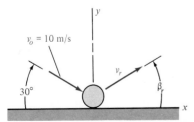

**FIGURE 15-21**

Rebound of a ball on a smooth floor

EXAMPLE 15-9    627

EXAMPLE 15-10

It is desired to hit pool ball $B$, which is initially at rest, so it will move toward pocket $P$ after being hit by the cue ball $A$ (Fig. 15-22a). The coefficient of restitution is $e = 0.9$, and both $A$ and $B$ are of equal mass and frictionless. Cue ball $A$ has a speed $v_A = 20\,\text{ft/s}$ before impact in the direction shown. Determine the velocity of each ball after impact.

## SOLUTION

Because the impact is frictionless, ball $B$ can have a change of momentum along the line of impact only. Therefore, the line of impact must be directed toward pocket $P$. Using the coordinate system shown in Fig. 15-22b, after impact,

$$\mathbf{v}_{B_2} = v_{B_2}\mathbf{j}$$

Also, assume that after impact ball $A$ has velocity components

$$\mathbf{v}_{A_2} = -(v_{A_2})_x\mathbf{i} + (v_{A_2})_y\mathbf{j}$$

For a frictionless impact,

$$(v_{A_2})_x = (v_A)_x = (20\,\text{ft/s})\cos 45°$$
$$= 14.1\,\text{ft/s}$$

Along the line of impact ($y$ axis), the total momentum of the system is conserved. Using Eq. 15-26,

$$m_A(14.1\,\text{ft/s}) + 0 = m_A(v_{A_2})_y + m_B(v_{B_2})$$
$$(v_{A_2})_y + v_{B_2} = 14.1\,\text{ft/s} \tag{a}$$

Using Eq. 15-34, the definition of the coefficient of restitution (applies to velocity components along line of impact *only*),

$$e = \frac{v_{B_2} - (v_{A_2})_y}{14.1\,\text{ft/s} - 0} = 0.9$$
$$v_{B_2} - (v_{A_2})_y = 12.7\,\text{ft/s} \tag{b}$$

Solving Eqs. (a) and (b) simultaneously,

$$\mathbf{v}_{B_2} = 13.4\mathbf{j}\,\text{ft/s}$$
$$\mathbf{v}_{A_2} = (-14.1\mathbf{i} + 0.7\mathbf{j})\,\text{ft/s}$$

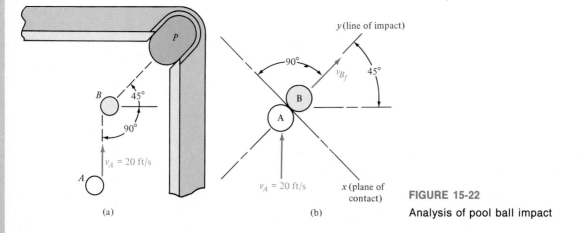

**FIGURE 15-22**

Analysis of pool ball impact

**15–58** The $3 \times 10^5$-kg block of a tuned mass damper in a skyscraper (see Sec. 1-7 in the associated statics volume) is moving at $v_0 = 2\,\text{m/s}$ when it hits the overtravel bumpers. Calculate the rebound speed if the coefficient of restitution is $e = 0.5$. Assume that friction is negligible and the bumpers are fixed.

**15–63** Railroad car $A$ of mass $m_A = 30\,000\,\text{kg}$ is moving at velocity $\mathbf{v}_A = 2\mathbf{i}\,\text{km/h}$ toward car $B$ of mass $m_B = 15,000\,\text{kg}$, which moves at velocity $\mathbf{v}_B = 0.3\mathbf{i}\,\text{km/h}$. The cars are coupled on contact and move together at velocity $\mathbf{v}_f$. Determine $\mathbf{v}_f$.

**Fig. P15-63**

Top view

Bumpers

**Fig. P15-58**

**15–59** Calculate the energy that the bumpers must dissipate during a single impact in the system described in Prob. 15–58.

**15–60** Calculate the rebound speed of the block in Prob. 15–58 after a total of three impacts between the left and right bumpers, starting with $v_0$. Plot the rebound speed as a function of successive impacts. Estimate the number of bumps after which the rebound speed would be about $0.01v_0$.

**15–64** A railroad car $A$ of mass $20\,000\,\text{kg}$ is moving at a velocity of $\mathbf{v}_A = 3\mathbf{i}$ (km/h) toward an identical car $B$ which is initially at rest. Cars $A$ and $B$ become coupled on contact and move to the right, striking and coupling with a third identical car $C$, which is also initially at rest. Determine the final velocity of the system as they move together along frictionless tracks.

**Fig. P15-64**

**15–61** Freight car $A$ is moving at $v_A = 1\,\text{mph}$, and car $B$ at $v_B = 2\,\text{mph}$, as shown. Calculate their new velocities after an impact where the coefficient of restitution is $e = 0.4$.

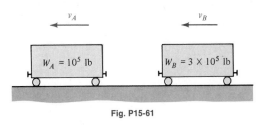

**Fig. P15-61**

**15–65** Particle $A$ of mass $m_A = 1\,\text{kg}$ moves at speed $v_A$ before hitting the stationary particle $B$ of mass $m_B = 2\,\text{kg}$. The speed of particle $B$ after the impact is $5\,\text{m/s}$. Calculate the speed $v_A$ if $e = 0.6$ and friction is negligible. Determine the energy lost during the collision.

**15–62** Assume that cars $A$ and $B$ in Prob. 15–61 are coupled together at impact. Calculate their common speed and the energy lost during the coupling.

**Fig. P15-65**

**15-66** Consider the rear-end collision of two cars whose initial velocities are collinear. Determine the final velocities if $m_A = 1000\,\text{kg}$, $m_B = 2000\,\text{kg}$, $v_A = 40\,\text{km/h}$, $v_B = 80\,\text{km/h}$, and $e = 0.5$.

**Fig. P15-66**

**15-67** A 3000-lb car travels at $v_B = 60\,\text{mph}$ when it collides with a 2200-lb car that is moving in the same direction (Fig. 15-66). The speed of car $B$ is 40 mph just after the impact. Determine the initial and final speeds of car $A$ if $e = 0.4$.

**15-68** An easy way to measure the coefficient of restitution between two materials is to drop a small sphere of material $A$ from a height $h_i$ onto a large flat surface of material $B$ and measure the rebound height $h_f$. Determine the coefficient of restitution, $e$, in terms of $h_i$ and $h_f$.

**Fig. P15-68**

**15-69** The 300-kg ram of a pile driver is dropped on a 700-kg pile from a height of 2 m and drives it into soft soil. Determine the velocities of the ram and the pile after impact if the coefficient of restitution is $e = 0.5$.

(a)

(b)

**Fig. P15-69**

**15-70** It is known that the ram of a pile driver (Fig. 15-69) stops momentarily after hitting an 800-lb pile in a free fall from a height of 6 ft. The coefficient of restitution is $e = 0.5$. Determine the weight of the ram.

**15–71** An 80 000-kg airplane lands at a horizontal speed of 60 m/s and a vertical descent of 2 m/s. Calculate the height $h$ of the rebound after the first and second bounce on the runway if the coefficient of restitution is $e = 0.3$. Assume that there is no friction, aerodynamic lift, or engine thrust involved. Treat the airplane as a particle.

Fig. P15-71

**15–72** The block of a biaxial tuned mass damper of a skyscraper (see Sec. 1-7 in the associated statics volume) moves in the horizontal plane and may hit overtravel bumpers. Assume that this motion can be modeled by using a small particle bouncing from walls as shown. Determine $v_f$ and $\phi$ in terms of the given initial quantities, assuming that the coefficient of restitution is $e$ for each impact. Neglect friction.

Fig. P15-72

**15–73** A ping-pong ball hits a table at an angle as shown. Calculate $v_f$ and $\theta$ neglecting friction if the coefficient of restitution is $e = 0.7$.

Fig. P15-73

**15–74** A 20-ton aircraft is to land on a 60 000-ton aircraft carrier. Their initial velocities are $v_A = -60\mathbf{i}$ km/h, and $v_B = (-200\mathbf{i} - 3\mathbf{j})$ km/h. Calculate the rebound velocities of the aircraft assuming that (a) the ship cannot move up or down, and (b) the ship has no impulsive support reaction. The coefficient of restitution is $e = 0.4$. Neglect friction, engine thrust, aerodynamic lift, and treat the ship and aircraft as particles.

Fig. P15-74

**15–75** Two cars are colliding as they travel at initial velocities $v_A = 50\mathbf{j}$ mph and $v_B =$ unknown. The cars weigh $W_A = 2000$ lb and $W_B = 3000$ lb, and they become entangled during the impact. Their common velocity just after impact is estimated from skid marks as $v_C = (10\mathbf{i} + 55\mathbf{j})$ mph. Determine the initial velocity $v_B$.

Fig. P15-75

Section 15-1

**Linear momentum**

$$\mathbf{G} = m\mathbf{v}$$

**Time rate of change of linear momentum**

$$\sum \mathbf{F} = \dot{\mathbf{G}} = \frac{d}{dt}(m\mathbf{v})$$

**SPECIAL CASE:**

$$\sum \mathbf{F} = 0 \qquad m\mathbf{v} = \text{constant} \qquad \text{(conservation of linear momentum)}$$

Section 15-2

**Angular momentum**

$$\mathbf{H}_O = \mathbf{r} \times m\mathbf{v} = rmv_\theta \mathbf{k}$$

**Magnitude of angular momentum**

**GENERAL PLANE MOTION:**

$$H_O = rmv \sin \phi = rmv_\theta = mr^2 \dot{\theta}$$

**GENERAL SPACE MOTION:**

$$H_x = m(yv_z - zv_y)$$
$$H_y = m(zv_x - xv_z)$$
$$H_z = m(xv_y - yv_x)$$

Section 15-3

**Time rate of change of angular momentum**

$$\dot{\mathbf{H}}_O = \frac{d}{dt}(\mathbf{r} \times m\mathbf{v}) = \mathbf{r} \times \sum \mathbf{F} = \sum \mathbf{M}_O$$

$$\sum \mathbf{M}_O = \dot{\mathbf{H}}_O$$
$$\sum M_x = \dot{H}_x \qquad \sum M_y = \dot{H}_y \qquad \sum M_z = \dot{H}_z$$

**SPECIAL CASE:**

$$\sum \mathbf{M}_O = 0$$
$$\mathbf{H}_O = \mathbf{r} \times m\mathbf{v} = \text{constant} \qquad \text{(conservation of angular momentum)}$$

Section 15-4

## Newton's law of universal gravitation

$$F = G\frac{Mm}{r^2}$$ for masses $M$ and $m$ at a distance $r$ between them

**UNIVERSAL CONSTANT OF GRAVITATION:**

$$G = 6.673 \times 10^{-11}\,\text{m}^3/\text{kg}\cdot\text{s}^2$$

$$G = 3.442 \times 10^{-8}\,\text{ft}^4/\text{lb}\cdot\text{s}^4$$

**ACCELERATION OF GRAVITY AT THE SURFACE OF THE EARTH (MASS $M$, RADIUS $R$):**

$$g = \frac{GM}{R^2}$$

Section 15-5

## Kepler's laws of planetary motion

**FIRST LAW (LAW OF SECTORIAL AREAS)**
The radius drawn from a planet to the sun sweeps equal areas in equal times.

**SECOND LAW**
The orbit of each planet is an ellipse, with the sun at one of its foci.

These laws are valid in any situation where the motion of a body is similar to planetary motion under the action of a central force.

Section 15-6

## Impulse and momentum

**LINEAR IMPULSE:**

$$\int_{t_1}^{t_2} \mathbf{F}\, dt = \underbrace{m\mathbf{v}_2}_{\substack{\text{final} \\ \text{momentum}}} - \underbrace{m\mathbf{v}_1}_{\substack{\text{initial} \\ \text{momentum}}}$$

Section 15-7

## Conservation of total momentum of particles
For $n$ particles which move in such a way that only interactive forces are applied to them,

$$\underbrace{\sum_{i}^{n}(m_i\mathbf{v}_i)_1}_{\substack{\text{total initial} \\ \text{momentum at time } t_1}} = \underbrace{\sum_{i}^{n}(m_i\mathbf{v}_i)_2}_{\substack{\text{total final momentum} \\ \text{at time } t_2}}$$

Sections 15-8 and 15-9

## Direct central impact of deformable bodies

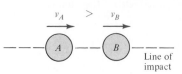

(a) Before impact (bodies approach)

(b) Increasing deformation during contact

(c) Decreasing deformation or restitution during contact

(d) After impact (bodies move apart; $v_{A_f}$ may be to the left)

**COEFFICIENT OF RESTITUTION:**

$$e = \frac{v_{Bf} - v_{Af}}{v_A - v_B} = \frac{|\text{relative velocity of separation}|}{|\text{relative velocity of approach}|}$$

For real materials, $0 < e < 1$.

Section 15-10

## Oblique central impact of deformable bodies

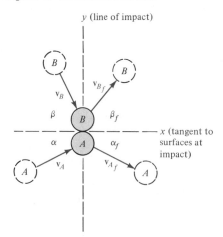

Assume that friction is negligible.

1. The $x$ component of the momentum of body $A$ is conserved.
2. The $x$ component of the momentum of body $B$ is conserved.
3. The total momentum of bodies $A$ and $B$ in the $y$ direction is conserved.
4. The coefficient of restitution is applicable to the motion along the $y$ axis.

**15-76** An amusement ride is modeled as a 300-kg particle at point $A$, attached to a rigid rod of negligible mass and length $l = 8\,m$. The arm $OA$ rotates about the vertical column at an angular speed $\dot{\theta} = 1.4\,rad/s$. At the angle $\phi = 50°$ force $\mathbf{F}$ causes $\phi$ to decrease at the constant rate of $0.2\,rad/s$. Determine the required net moment to maintain $\dot{\theta}$ constant, neglecting friction.

**Fig. P15-76**

**15-77** In Prob. 15–76, $\dot{\theta} = 1.4\,rad/s$ when $\phi = 50°$. Determine the new angular speed $\dot{\theta}_1$ after the angle $\phi$ is changed to 25°, assuming that the system is rotating about the vertical axis without any interference.

**15-78** It is reported on television that a satellite orbiting Earth has a maximum distance above Earth's surface of 26,400 mi and a minimum distance of 160 mi. It is also reported that (1) the maximum speed along the orbital path is 23,000 mi/h, and (2) the orbital period $T$ is 12.8 hs. From the information given, determine:
  (a) the minimum speed along the orbital path.
  (b) the rate at which area is swept out by the radius vector $\mathbf{r}$ (called the areal rate $dA/dt$).
  (c) the radius of curvature of the orbit at perigee $r_p$,
  (d) the semiminor axis $b$ and the semimajor axis $a$ of the orbital elliptic path given that the area of an ellipse $A = \pi ab$.

**15-79** In an assembly operation a 20-lb empty tray is sliding on rollers with $v_t = 5\,ft/s$. Three 5-lb objects are dropped on the tray in succession. Determine the new speed of the tray after each object comes to rest on it. What would be the final speed if all three objects were dropped simultaneously?

**Fig. P15-79**

**15-80** Consider two billiard balls of equal mass $m$ and radius $r$. Ball $A$ moves at velocity $\mathbf{v}_A$ and strikes ball $B$ which is at rest. Determine the velocities of the balls after impact if $d = 0.7r$ and the coefficient of restitution is $e = 0.6$. Neglect friction.

**Fig. P15-80**

**15-81** Solve Prob. 15–80 for $\mathbf{v}_B$ with $d = 0.5r$, $1.2r$, and $1.8r$. Draw the vectors $\mathbf{v}_B$ on the same $xy$ diagram with a common origin for the three vectors. Estimate from the diagram the vector $\mathbf{v}_B$ relative to $\mathbf{v}_A$ for the situation when the two balls just graze each other.

**15–82** In an automated packaging operation a 20-lb box is dropped on a 40-lb tray that can freely roll horizontally. Calculate the velocity of the box and of the tray just after the first impact if initially $\mathbf{v}_A = (3\mathbf{i} - 4\mathbf{j})$ ft/s, $\mathbf{v}_B = 0$, and the coefficient of restitution is $e = 0.2$.

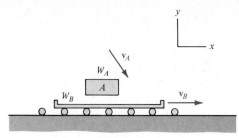

**Fig. P15-82**

# Kinetics of Particles: Compilation of Methods. Strategies for Solving Problems

# 16

## OVERVIEW

Since both the energy methods and the momentum methods available for solving problems in particle kinetics are derived from Newton's second law of motion, it follows that more than one way of solving a particular problem may exist. In this chapter the different methods are outlined and compared, and the procedure for choosing the best method for solving a particular problem is explored.

## STUDY GOALS

In the previous three chapters all the problems at the end of a section were solved using the method presented in that section. In this chapter the problems requiring different methods appear randomly so that the student gains insight for choosing the most efficient method. Many problems involve different steps for which only one method of solution works well. Understanding the best strategy for solving these problems is important for the development of engineering proficiency.

# OVERVIEW OF KINETICS ANALYSIS OF PARTICLES

### Newton's Second Law

The fundamental principle in kinetics is Newton's second law of motion. For a particle,

$$\sum \mathbf{F} = m\mathbf{a}$$

Newton's second law is used in all problems requiring the analysis of accelerations. Other methods are based on this equation and are developed for convenience in solving problems for which other variables, such as velocity, displacement, and time, are the main concern.

### Energy Methods

Energy methods are employed when the problem requires an analysis of forces related to changes in speeds and displacements without an evaluation of the acceleration. The principle of work and energy must be used when the problem includes nonconservative forces such as friction. The conservation of energy equation is often used when the forces involved are due to gravity and springs.

### Momentum Methods

When a problem involves a change in velocity due to external forces applied to a mass over an interval of time, the principle of impulse and momentum is often the most efficient equation to use. If no external force acts on a particle, or if the only force acting on two particles is a mutual force between them, the conservation of linear momentum equation is used. Conservation of linear momentum, in conjunction with the coefficient of restitution equation, is the only practical method to employ for impact problems involving a collision between two bodies.

When the only forces acting on a particle either intersect an axis or are parallel to that axis, the conservation of angular momentum equation with respect to that axis is used. In plane motion problems, the axis is a point about which angular momentum is conserved. Central force motion, such as the motion of orbiting satellites and planets, is best analyzed using the principle of conservation of angular momentum. Figure 16-1 summarizes the various methods.

| | | |
|---|---|---|
| FUNDAMENTAL LAW | Newton's second law (forces, mass, acceleration) | $\sum \mathbf{F} = m\mathbf{a}$ |

| | | | |
|---|---|---|---|
| METHODS DERIVED FROM NEWTON'S LAW | Energy methods | Work | $U_{12} = \int_1^2 \mathbf{F} \cdot d\mathbf{r}$ |
| | | Kinetic energy | $T = \dfrac{1}{2} mv^2$ |
| | | Principle of work and energy (forces, mass, speed, displacement) | $U_{12} = T_2 - T_1$ |
| | | Gravitational potential energy | $V_g = Wh$ |
| | | Elastic potential energy | $V_e = \dfrac{1}{2} kx^2$ |
| | | Principle of conservation of energy (weight, springs, speed, position) | $T_1 + V_1 = T_2 + V_2$ |
| | Momentum methods | Linear momentum | $\mathbf{G} = m\mathbf{v}$ |
| | | | $\sum \mathbf{F} = \dot{\mathbf{G}} = \dfrac{d}{dt}(m\mathbf{v})$ |
| | | Principle of impulse and momentum (forces, mass, velocity, time) | $\int_{t_1}^{t_2} \mathbf{F}\, dt = m\mathbf{v}_2 - m\mathbf{v}_1$ |
| | | Conservation of linear momentum ($\sum \mathbf{F} = 0$, impact) | $\sum (m_i\mathbf{v}_i)_1 = \sum (m_i\mathbf{v}_i)_2$ |
| | | Angular momentum | $\mathbf{H}_O = \mathbf{r} \times m\mathbf{v}$ |
| | | | $\sum \mathbf{M}_O = \dot{\mathbf{H}}_O = \dfrac{d}{dt}(\mathbf{r} \times m\mathbf{v})$ |
| | | Conservation of angular momentum ($\sum \mathbf{M}_O = 0$, central force) | $\mathbf{H}_{O_1} = \mathbf{H}_{O_2}$ |

**FIGURE 16-1**

Table of equations for the methods of particle kinetics

# EXAMPLE 16-1

The track of a proposed ski-jump tower is in the shape of a circular arc as shown in Fig. 16-2. The jumpers start from rest at $A$ and become airborne at $B$. Determine the takeoff speed $v$ if $r = 60$ m and $\theta = 40°$, for a skier of weight $W$. Assume that the track is frictionless.

## METHODS OF SOLUTION

The unknown quantity in this problem is the speed, and the known quantities are the force of gravity and the distance traveled. Two methods are available for the solution: (1) the speed can be determined by integration of the tangential acceleration obtained through kinetics analysis, and (2) the speed can be determined directly by the work–energy method. Both methods are given below for comparison.

## KINETICS AND KINEMATICS SOLUTION

The free-body diagram of the skier is drawn in Fig. 16-2b. Summing forces in the tangential direction for a general position,

$$F_t = ma_t$$

$$mg \sin \theta = ma_t$$

$$g \sin \theta = a_t$$

Note that this acceleration is independent of the weight of the skier. Using Eq. 12-12,

$$a_t = g \sin \theta = \frac{dv}{dt}$$

Separating variables yields

$$g \sin \theta \, dt = dv$$

A problem arises here since it is not known how $\theta$ varies with time, so the equation cannot be integrated. Since the path is circular, however, the tangential and transverse directions are the same. Using transverse components from Eq. 12-22 gives

$$a_t = a_\theta = g \sin \theta = r\ddot{\theta} + 2\dot{r}\dot{\theta}$$

$$g \sin \theta = r \frac{d\dot{\theta}}{dt} + 0 \qquad \text{since } \dot{r} = 0$$

(a)

(b) Free-body diagram

(c)

**FIGURE 16-2**

Analysis of a skier using alternate methods

The chain rule is applied to the right side of this equation to allow for separation of variables,

$$g \sin \theta = r \frac{d\dot{\theta}}{dt} = r \frac{d\dot{\theta}}{d\theta} \frac{d\theta}{dt} = r \frac{d\omega}{d\theta} \omega$$

$$\omega \frac{d\omega}{d\theta} - \frac{g}{r} \sin \theta = 0$$

$$\omega \, d\omega - \frac{g}{r} \sin \theta \, d\theta = 0$$

Integrating yields

$$\frac{\omega^2}{2} + \frac{g}{r} \cos \theta = C = \text{constant}$$

Initial conditions require that $\omega = 0$ at $\theta = 0°$. Hence

$$C = \frac{g}{r} \quad \text{and} \quad \omega = \sqrt{\frac{2g}{r}(1 - \cos \theta)}$$

For $\theta = 40°$,

$$\omega = \sqrt{\frac{2(9.81 \text{ m/s}^2)}{60 \text{ m}}(1 - \cos 40°)} = 0.277 \text{ rad/s}$$

From Eq. 12-14,

$$v = r\omega = (60 \text{ m})(0.277 \text{ rad/s}) = 16.6 \text{ m/s}$$

## WORK–ENERGY SOLUTION

Using Fig. 16-2c and Eq. 14-15, $U_{12} = T_2 - T_1$. Since the skier starts from rest, $T_1 = \frac{1}{2}mv_1^2 = 0$. The work done on the skier is that done by gravity, $U_{12} = Wh = mgh$. This is positive according to the free-body diagram in Fig. 14-9b. The final kinetic energy is $T_2 = \frac{1}{2}mv_2^2$, and

$$mgh = \frac{1}{2}mv_2^2 - 0$$

$$v_2 = \sqrt{2gh} = \sqrt{2g(r - r \cos 40°)}$$
$$= \sqrt{2(9.81 \text{ m/s}^2)[60 \text{ m} - (60 \text{ m}) \cos 40°]}$$
$$= 16.6 \text{ m/s}$$

which agrees with the previous result. Note that Eq. 14-15 allowed a much faster solution.

## JUDGMENT OF THE METHODS AND TIPS

It is clear from the analysis given above that the work–energy method is the most efficient. When problems involve a change in speed, forces, and displacement, the work–energy method should always be tried first. In cases where the force is not constant or not a function of displacement, however, kinetics and kinematics must be used. In general, one should begin by reading the problem statement carefully for these clues.

EXAMPLE 16-1    641

# EXAMPLE 16-2

A car of weight $W$ skids downhill until coming to rest at $B$ (Fig. 16-3a). The speed of the car at $A$ is $v_0$ and the coefficient of kinetic friction $\mu$ is constant along the slope. Determine **(a)** the time $t_s$ required for the car to skid to rest, and **(b)** the distance $D$ through which the car skids before coming to rest. (Charts based on this equation for $D$ are useful to police in accident investigations.)

**(a) KINETICS AND KINEMATICS METHOD:**  The free-body diagram is drawn in Fig. 16-3b. Using Eq. 13-5b,

$$\sum F_y = m\ddot{y} = 0$$

$$N - W\cos\theta = 0$$

$$N = W\cos\theta \tag{a}$$

$$\sum F_x = m\ddot{x}$$

$$W\sin\theta - \mu N = \frac{W}{g}a$$

$$W\sin\theta - \mu W\cos\theta = \frac{W}{g}a \tag{b}$$

From Eqs. (a) and (b),

$$a = -g(\mu\cos\theta - \sin\theta) \quad \text{(constant)} \tag{c}$$

Since $v = \int a\,dt + C_1$,

$$v = -g(\mu\cos\theta - \sin\theta)t + v_0$$

Therefore,

$$v = 0 \quad \text{at} \quad t_s = \frac{v_0}{g(\mu\cos\theta - \sin\theta)} \quad \text{(time to skid to rest)} \tag{d}$$

**IMPULSE–MOMENTUM METHOD:**  Since this problem involves velocity, forces, and time, the impulse–momentum method is also valid. Using the free-body diagram in Fig. 16-3b and Eq. 15-23 for the velocity component normal to the incline,

$$\int_{t_1}^{t_2} F_y\,dt = mv_{y_2} - mv_{y_1} = 0$$

$$N - W\cos\theta = 0$$

$$N = W\cos\theta \quad \text{and} \quad \mathscr{F} = \mu W\cos\theta$$

Note that the equation $\sum F_y = m\ddot{y} = 0$ leading to Eq. (a) could have been used here just as well. The implication of this is that in special cases like this, one can mix methods to simplify the solution. For the component of the velocity parallel to the incline,

$$\int_{t_1}^{t_2} F_x\,dt = mv_{x_2} - mv_{x_1}$$

$$\int_0^{t_2} (W\sin\theta - \mu W\cos\theta)\,dt = 0 - \frac{W}{g}v_0$$

Since the integrand is constant,

$$W(\sin\theta - \mu\cos\theta)t_s = -\frac{W}{g}v_0$$

Therefore,

$$t_s = \frac{v_0}{g(\mu\cos\theta - \sin\theta)}$$

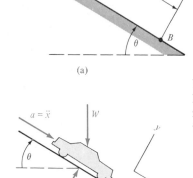

(a)

(b)

**FIGURE 16-3**

Analysis of a car's motion using alternate methods

Note that $t_s$ is independent of $W$, and $t_s$ is greater for skidding to rest downhill $(\sin \theta > 0)$ than uphill $(\sin \theta < 0)$.

**(b) KINETICS AND KINEMATICS METHOD:** When the car comes to rest at $t = t_s$, $x = D$. Since $x = \int v \, dt + C_2$,

$$x = -\frac{g}{2} (\mu \cos \theta - \sin \theta) t^2 + v_0 t \qquad \text{(e)}$$

Substituting Eq. (d) for $t_s$ into (e) yields

$$D = \frac{v_0^2}{2g(\mu \cos \theta - \sin \theta)}$$

*Alternate solution:* The statement of part (b) seeks a relationship between "change in position" and "change in speed" for which the form $a = v(dv/dx)$ is often preferable over $a = dv/dt$. The latter is best suited for relating a "change in speed" to time. Using Eq. (c) above,

$$a = -g(\mu \cos \theta - \sin \theta) = v \frac{dv}{dx}$$

$$-g(\mu \cos \theta - \sin \theta) \int_0^D dx = \int_{v_0}^0 v \, dv$$

$$D = \frac{v_0^2}{2g(\mu \cos \theta - \sin \theta)}$$

Note that $D$ is also independent of the weight $W$, and that for a given initial speed $v_0$, one skids farther downhill $(\sin \theta > 0)$ than uphill $(\sin \theta < 0)$, as expected from experience.

**WORK–ENERGY METHOD:** From $A$ to $B$ work is done by the weight $W$ and friction $\mathscr{F} = \mu N$. The normal force $\mathbf{N}$ does no work because it is perpendicular to the displacement. Using Eq. 14-15,

$$U_{AB} = T_B - T_A$$

$$W(D \sin \theta) - \mu(W \cos \theta)D = 0 - \frac{1}{2} \frac{W}{g} v_0^2$$

Solving for $D$ yields

$$D = \frac{v_0^2}{2g(\mu \cos \theta - \sin \theta)}$$

Note that the work–energy solution is very similar to the "Alternate Solution" given above in which $a = v(dv/dx)$ is used in the integration. This is not an accident since the work–energy method was derived by separating variables in much the same way.

## JUDGMENT OF THE METHODS

Neither part of this problem required a determination of acceleration, so the work–energy and the momentum methods are more direct. Since the problem statement for part (a) clearly asks to relate a change in velocity to a time interval, the work–energy method will not work well and the impulse–momentum method is the easiest to use. Likewise, the statement for part (b) asks to relate a change in speed to a change in position, a situation for which the work–energy method is ideal. Note again that by reading the problem statement carefully, one could readily select the most efficient method of solution.

EXAMPLE 16-1    643

# EXAMPLE 16-3

A skier of weight $W$ starts from rest at the top of a hill and slides with negligible friction down the hill (Fig. 16-4a). The shape of the upper portion of the hill from $A$ to $C$ is circular of radius $R$ (from $\theta = 0$ to $60°$). The skier leaves contact with the hill at point $B$ and crashes at point $D$. Neglecting friction and air resistance, determine **(a)** the speed of the skier at a general position $\theta$ before reaching $B$, **(b)** the angle $\theta_B$ at which the skier leaves the hill at $B$, and **(c)** the speed at which the skier crashes at point $D$ for a height $h = 60\,\text{ft}$.

## SOLUTION

Parts (a) and (c) of this problem ask for a determination of speed for a given weight and displacement, so the work–energy method is the logical choice. From the free-body diagram in Fig. 16-4b, the normal force does no work and the principle of conservation of energy applies. Part (b), however, requires a solution for the point at which the normal force becomes zero. Since the normal force is not included in a work–energy analysis, Newton's second law must be used. The most convenient force summation is in the normal direction for the circular motion described. The work–energy analysis works well with a force summation in the normal direction because the normal acceleration is a function of the speed.

    **(a)** Let $P$ be the position for a general angle $\theta$. From Eq. 14-29, with the datum at point $A$,

$$T_A + V_A = T_P + V_P$$

where $T_A = 0$

$\qquad V_A = 0$

$\qquad T_P = \frac{1}{2}mv^2$

$\qquad V_P = -Wh = -WR(1 - \cos\theta)$

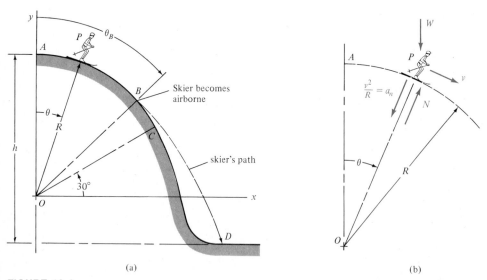

(a)

**FIGURE 16-4**

(b)

Then

$$0 = \frac{1}{2} mv^2 - mgR(1 - \cos\theta)$$

$$v = \sqrt{2gR(1 - \cos\theta)} \qquad \text{for a general position } \theta$$

**(b)** Using Eq. 13-6a with the free-body diagram in Fig. 16-4b,

$$F_n = ma_n$$

$$W \cos\theta - N = \frac{W}{g} \frac{v^2}{R}$$

Substituting for $v^2$ from part (a) yields a general expression for the magnitude of the normal force $N$ at any position $\theta$ given by

$$N = 3W \cos\theta - 2W$$

At the position $\theta = \theta_B$ where $N = 0$ (skier becomes airborne),

$$3W \cos\theta_B = 2W$$

$$\cos\theta_B = \frac{2}{3}$$

$$\theta_B = 48.2°$$

**(c)** The principle of conservation of energy also applies from point $A$ to point $D$,

$$T_A + V_A = T_D + V_D$$

$$0 = \frac{1}{2} mv^2 - mgh$$

$$v^2 = 2(32.2 \, \text{ft/s}^2)(60 \, \text{ft})$$

$$v = 62.2 \, \text{ft/s}$$

**TIP #1**

Note that the principle of conservation of energy could have been applied (1) from $A$ to $B$ to determine the speed at $B$, and then (2) from $B$ to $D$ to determine the speed at $D$. However, it involves less calculation to treat the motion in a single stage from $A$ to $D$ as done here.

**TIP #2**

Note that in part (b) the normal force is set equal to zero in order to solve for the limiting position of contact between the skier and the slope. Such limiting cases are often encountered in dynamics problems.

EXAMPLE 16-3   645

EXAMPLE 16-4

A bullet of mass 0.07 kg is fired into a bag of sand of mass 10 kg at a speed $v_1$ (Fig. 16-5a). The bullet is embedded in the bag of sand, which is suspended by a massless cord. The maximum height $h_{max}$ is used as a means for measuring the velocity of the bullet. Determine the muzzle velocity $v_1$ if the maximum angle of travel is $\theta = 25°$.

(a)

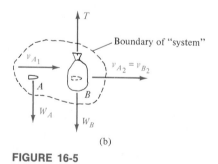

(b)

**FIGURE 16-5**

## SOLUTION

Consider this problem in two phases:

**PHASE I:** During the impact time (measured from the instant the bullet strikes the bag of sand until it becomes embedded in it) conservation of linear momentum applies in the horizontal direction. This follows because no external forces act on the "system" (Fig. 16-5b) in the horizontal direction. The impact forces between $A$ and $B$ are internal to the system and are equal and opposite.

$$m_A v_{A_1} + m_B(0) = m_A v_{A_2} + m_B v_{B_2} = (m_A + m_B) v_{B_2} \qquad \textbf{(a)}$$

In Eq. (a), $v_{A_2} = v_{B_2}$ because $e = 0$ (perfectly plastic impact for which the "relative velocity of separation" is zero). From Eq. (a),

$$0.07 v_{A_1} = 10.07 v_{B_2} \qquad \textbf{(b)}$$

**PHASE II:** After impact, the system, comprising the mass of the bag of sand plus bullet (Fig. 16-5b), starts from a speed $v_{B_2}$ and swings to rest at an angular position $\theta = 25°$ along a circular arc of 2 m radius. The work–energy principle is ideal for relating this "change in speed" to "change in position." The only work done is that by the weights in moving vertically upward through the distance $h_{max} = 2(1 - \cos 25°)$. Using Eq. 14-15,

$$U_{23} = T_3 - T_2$$

$$-W h_{max} = 0 - \frac{1}{2} \frac{W}{g} v_{B_2}^2$$

$$h_{max} = (2\,\text{m})(1 - \cos 25°) = \frac{v_{B_2}^2}{2(9.81\,\text{m/s}^2)}$$

$$v_{B_2} = 1.92\,\text{m/s}$$

Substituting this result into Eq. (b) above yields

$$v_{A_1} = 276\,\text{m/s}$$

**TIP**

Note that in Phase I of this problem not only was the linear momentum conserved, but the angular momentum about point $O$ was also conserved. The angular momentum about point $O$ is *not* conserved in Phase II, because the weight of the bullet and bag creates a moment about point $O$.

EXAMPLE 16-4    647

## EXAMPLE 16-5

The hammer $A$ of a drop forge system weighs 1000 lb and is dropped from rest from a height of 2 ft onto an anvil $B$ weighing 8000 lb (Fig. 16-6a). The anvil $B$ is supported by a system of springs with a combined stiffness of $k = 20,000$ lb/ft. The coefficient of restitution is $e = 0.5$ between the hammer and anvil. Determine the height $h$ to which the hammer $A$ rebounds after impacting with the anvil $B$ and the maximum deflection $y$ of the springs.

### SOLUTION

This problem involves an impact between two bodies, so the only practical method is that of impulse–momentum with the equation for the coefficient of restitution. Energy is lost due to stress waves, sound waves, and heat generated during the collision. For the motion directly preceding and following the collision, however, only the forces of gravity and of the springs are involved, and only velocities need to be calculated, so the work–energy method with the conservation of energy equation is ideal. This problem is most easily evaluated by separating it into three phases of motion: (I) before impact, (II) during impact, and (III) after impact as follows.

(a)

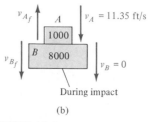

During impact

(b)

**FIGURE 16-6**

**PHASE I:** Using $T_1 + V_1 = T_2 + V_2$ during the free fall of hammer $A$,

$$(1000\,\text{lb})(2\,\text{ft}) = \frac{1}{2}\frac{1000\,\text{lb}}{32.2\,\text{ft/s}^2}\,v_A^2 - 0$$

$$v_A = 11.35\,\text{ft/s (just before impact)}$$

**PHASE II:** Use (a) conservation of linear momentum during the impact time $\Delta t$ plus (b) the definition of coefficient of restitution.

$$\frac{1000\,\text{lb}}{g}(11.35\,\text{ft/s}) + 0 = -\frac{1000\,\text{lb}}{g}\,v_{Af} + \frac{8000\,\text{lb}}{g}\,v_{Bf}$$

$$8v_{Bf} - v_{Af} = 11.35\,\text{ft/s} \qquad \text{(a)}$$

$$e = \frac{|\text{relative velocity of separation}|}{|\text{relative velocity of approach}|}$$

$$0.5 = \frac{v_{Bf} + v_{Af}}{11.35\,\text{ft/s}} \qquad \text{(b)}$$

Solving Eqs. (a) and (b) yields

$$v_{Bf} = 1.892\,\text{ft/s}\,(\downarrow)$$

$$v_{Af} = 3.79\,\text{ft/s}\,(\uparrow)$$

**PHASE III:** Use $T_1 + V_1 = T_2 + V_2$ to determine the height $h$ for the rebound of $A$ and the maximum compression $y$ of the springs by $B$ after impact.

For $A$:

$$\frac{1}{2}mv_{A_f}^2 = W_A h$$

$$\frac{1}{2}\frac{1000\,\text{lb}}{32.2\,\text{ft/s}^2}(3.79\,\text{ft/s})^2 = (1000\,\text{lb})\,h$$

$$h = 0.233\,\text{ft}$$

For $B$:

$$T_1 + V_{e1} + V_{g1} = 0 + V_{e2} + V_{g2} \qquad \text{with}$$

$$V_{e1} = \frac{1}{2}(20{,}000\,\text{lb/ft})\,\delta^2 \qquad \left(\delta = \frac{8000\,\text{lb}}{20{,}000\,\text{lb/ft}} = 0.4\,\text{ft}\right)$$

$$= 1600\,\text{ft}\cdot\text{lb}$$

$$V_{g1} = 0 \text{ (take datum at static equilibrium position)}$$

$$V_{e2} = \frac{1}{2}(20{,}000\,\text{lb/ft})(y + \delta)^2$$

$$= (10{,}000\,\text{lb/ft})\,y^2 + (8000\,\text{lb})\,y + 1600\,\text{ft}\cdot\text{lb}$$

$$V_{g2} = -W_B y = -(8000\,\text{lb})\,y$$

Then

$$\frac{1}{2}\left(\frac{8000\,\text{lb}}{32.2\,\text{ft/s}^2}\right)(1.892\,\text{ft/s})^2 + 1600\,\text{ft}\cdot\text{lb}$$

$$= (10{,}000\,\text{lb/ft})\,y^2 + (8000\,\text{lb})\,y + 1600\,\text{ft}\cdot\text{lb} - (8000\,\text{lb})\,y$$

$$445\,\text{ft}\cdot\text{lb} = (10{,}000\,\text{lb/ft})\,y^2$$

$$y = 0.211\,\text{ft}$$

EXAMPLE 16-5    649

# EXAMPLE 16-6

The 96.6-lb circus acrobat shown is hanging by her heels at the end of a rope which is supported by a swivel at $O$. Initially the acrobat is moving in a circular path of radius 5 ft at the end of a rope measuring 30 ft from $O$, as shown in Fig. 16-7a, at an initial speed $v_1$. A circus hand then very gradually pulls on the rope $OP$ until the angle $\theta$ increases from $\theta_1$ to 30°. Now the acrobat moves in a new circular path of radius $r_2$ at a speed $v_2$ at a rope angle of 30°. Determine (a) the speed $v_1$ and tension in the rope $T_1$ for the initial motion, and (b) the radius $r_2$, speed $v_2$, and tension in the rope $T_2$ for the final circular path of the acrobat.

## SOLUTION

By inspection of the free-body diagram in Fig. 16-7b, the forces on the acrobat either intersect or are parallel to the $y$ axis. Thus, throughout the motion, the angular momentum of the acrobat about the $y$ axis is conserved. For position 1 and position 2,

$$r_1 m v_1 = r_2 m v_2$$

$$r_1 v_1 = r_2 v_2$$

$$(5\,\text{ft})(v_1) = r_2 v_2 \tag{a}$$

There are three unknowns in this equation, so two more equations are needed for a solution. Since the acrobat can be assumed to travel in circular motion at each position, a summation of the forces in the normal direction will also involve the radius $r$ and the speed $v$. From the free-body diagram in Fig. 16-7b,

(a)

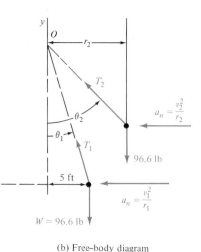

(b) Free-body diagram

**FIGURE 16-7**

Circus acrobat

$$\sum F_n = ma_n = m\frac{v^2}{r}$$

$$T \sin \theta = m\frac{v^2}{r} \qquad \textbf{(b)}$$

For motion in the horizontal plane,

$$\sum F_y = 0$$

$$T \cos \theta - mg = 0$$

$$T = \frac{mg}{\cos \theta} \qquad \textbf{(c)}$$

Substitution of Eq. (c) into Eq. (b) yields

$$g \tan \theta = \frac{v^2}{r}$$

For position 1 with $r = 5\,\text{ft}$,

$$\sin \theta_1 = \frac{5}{30}$$

$$\theta_1 = 9.59°$$

Then

$$T_1 = 96.6\,\text{lb}/\cos 9.59° = 98\,\text{lb}$$

and

$$v_1^2 = (5\,\text{ft})(32.2\,\text{ft/s}^2)(\tan 9.59°)$$
$$v_1 = 5.22\,\text{ft/s}$$

For position 2 with $\theta = 30°$,

$$T_2 = \frac{96.6\,\text{lb}}{\cos 30°} = 111.5\,\text{lb}$$

$$r_2 = \frac{v^2}{g \tan 30°} = \frac{v^2}{18.59\,\text{ft/s}^2} \qquad \textbf{(d)}$$

Using Eqs. (a) and (d),

$$(5\,\text{ft})(5.22\,\text{ft/s}) = \frac{v_2^2}{18.59\,\text{ft/s}}\,v_2$$

$$v_2^3 = 485\,\text{ft}^3/\text{s}^3$$

$$v_2 = 7.86\,\text{ft/s}$$

$$r_2 = 3.32\,\text{ft}$$

**TIP**

This example is typical of conical pendulum-type problems (circular motion in the horizontal plane) for which angular momentum is conserved. Note that in order to determine the forces in such problems, a summation of forces in the normal and vertical directions is necessary.

EXAMPLE 16-6    651

**16–1** In designing the carnival ride shown, the engineer wants the height $h$ to be high enough so the carriage will push upward on the track with a force of 5000 N when at the top of the loop, point $B$. Neglecting friction, determine the required height $h$ if the radius of the loop is $r = 10$ m, and the mass of the carriage plus passengers is 500 kg. The carriages start from rest at point $A$.

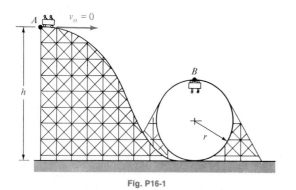

**Fig. P16-1**

**16–2** Crates $A$ and $B$ each weigh 64.4 lb and the coefficient of restitution between them is $e = 0.6$. Crate $B$ is at rest and is in contact with the spring shown in its initially undeformed state. The spring constant $k = 600$ lb/ft. Crate $A$ starts from rest and slides down the frictionless ramp until it strikes crate $B$. Determine the maximum deflection of the spring immediately after impact.

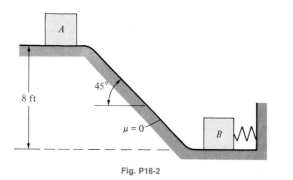

**Fig. P16-2**

**16–3** A simple device for the measurement of velocities of thrown balls or other projectiles is called a *ballistic pendulum*. This catches the ball while hanging at rest and swings up with the ball as shown. The mass of the pendulum is 30 kg, and of the ball it is 0.5 kg. Calculate the initial velocity of the ball if the pendulum on a light rod swings to a height $h = 0.15$ m with the ball.

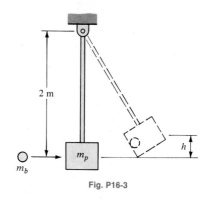

**Fig. P16-3**

**16–4** A spring–mass system is used for crushing material at $C$ by an impact load at $B$. The weight of 161 lb is released from rest at $A$ and pulled by the spring of spring constant $k = 100$ lb/ft. The spring's natural length is 8 ft. Neglecting friction, determine the normal force exerted on the shaft $AB$ at $B$ just before impact.

**Fig. P16-4**

**16–5** In order to cut construction costs, a civil engineer decides to clear the top of a mountain for a roadway by simply pushing the rocks down the side of the mountain. A bulldozer weighing 6440 lb starts from rest and pushes a rock weighing 644 lb a distance $d$ from point $O$ to $A$. At point $A$ the bulldozer suddenly stops and the rock continues sliding for 5 ft from $A$ to $B$. The bulldozer creates a constant ground traction force of 4000 lb forward while pushing the rock. The coefficient of friction for the rock as it slides from $O$ to $B$ is $\mu = 0.5$. Determine (a) the distance $d$ the rock must be pushed, and (b) the magnitude of the force $\mathbf{F}$ exerted by the bulldozer on the rock as it accelerates from $O$ to $A$ in order for the rock to clear the roadway at $C$.

Fig. P16-5

**16-8** A manufacturer of tennis balls sorts the balls according to their rebounding quality as shown. Calculate the distance $d$ of horizontal travel if the coefficient of restitution with the fixed racket is $e = 0.5$. Neglect friction.

Fig. P16-8

**16-6** A 70-kg gymnast on the flying rings is modeled as a particle $P$. At $\theta = 60°$, $l = 6$ m, and $P$ has no velocity. Assume that at $\theta = 0$ the gymnast applies a large force to suddenly make the length $l = 5.5$ m. Calculate the final speed of $P$ at $\theta = 0$.

**16-9** Solve Prob. 16-8 using $e = 0.4, 0.5$, and $0.6$. Plot $d$ as a function of $e$ and discuss the general tendency observed.

Fig. P16-6

**16-10** At an amusement park people may jump by parachutes which are guided to the ground by vertical wires. The jumpers land on a platform supported by four springs at a speed of 6 ft/s. The spring constant of each spring is $k = 7.5$ lb/in, and the platform weighs 50 lb. How much space should be allowed for the vertical displacement of each platform assuming that a 250 lb person may land on it?

**16-7** In the sorting of ball-bearing balls the rebound height $h$ and horizontal distance $d$ are to be electronically measured. Calculate $h$ and $d$ for the approximate positioning of the sensors if the platen is fixed and the coefficient of restitution is $e = 0.75$.

Fig. P16-7

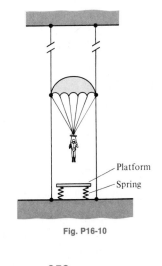

Fig. P16-10

**16–11** In an amusement ride people sit in cages mounted on arms of constant length $r = 6.5$ m. The arms are rotating freely at an angular speed $\omega_z = 2$ rad/s when $\phi = 30°$ for both arms. Determine the work done by the force exerted by each actuator $A$ in changing the angle $\phi$ of both arms to $60°$. Assume that a person of 70-kg mass sits in each cage at a distance $R = 8$ m and neglect the mass of the arms and cages.

**Fig. P16-11**

**16–12** The 300-kg ram of a pile driver is dropped on a 700-kg pile from a height of 2 m and drives it into soft soil. The coefficient of restitution between the ram and pile is $e = 0.5$. Determine (a) the height above the point of impact to which the ram rebounds, and (b) the magnitude of the average force exerted by the soil on the pile if it is observed to stop moving in 2 s.

**Fig. P16-12**

**16–13** In an automated packaging operation a 20-lb box is dropped on a 40-lb tray that can freely roll horizontally. The coefficient of kinetic friction between the box and the tray is 0.3. Determine how far the box slides on the tray before coming to rest, assuming no rebound of the box upon landing on the tray with a velocity $\mathbf{v}_A = (3\mathbf{i} - 4\mathbf{j})$ ft/s.

**Fig. P16-13**

**16–14** A person releases a ball 2 m above the ground and 7 m away from a wall. The ball has an initial velocity $v_0 = 20$ m/s at an angle $\theta = 70°$. Determine the horizontal distance from the wall where the ball will land on the floor after rebounding off of the wall. The coefficient of resitution between the ball and the wall is 0.7.

**Fig. P16-14**

**16–15** An engineer designs a ramp and bumper mechanism so that workers can slide crates down the ramp into an excavation site. The crates are stopped by a spring which is initially unstretched. In order that the crates do not fall off the end of the ramp, the maximum allowable deflection of the spring is limited to 1 ft. Assume that frictional effects are negligible. (a) Determine the required spring constant $k$ for a 322-lb crate started from rest at $A$. (b) There is always the possibility that the supporting structure of the ramp will collapse due to dynamic loads. Therefore, determine the magnitude of the maximum normal force acting on the ramp and the position where it occurs for the 322-lb crate.

Fig. P16-15

**16–16** Automobile $A$ is traveling at a speed of 45 mi/h and strikes automobile $B$, which is parked with its brakes locked. The coefficient of restitution between $A$ and $B$ is $e = 0.1$, and the coefficient of friction between $B$ and the pavement is $\mu = 0.5$. $A$ weighs 4000 lb and $B$ weighs 3000 lb. Determine (a) the distance $D$ required for $B$ to skid to rest after impact, and (b) the time $t_s$ for $B$ to skid to rest after impact.

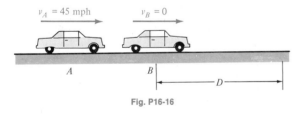

Fig. P16-16

**16–17** Car $A$ weighing 3000 lb was traveling east through an intersection when struck broadside by car $B$ weighing 4000 lb which was traveling northward. After impact car $A$ skids to a stop in 40 ft on a surface with a coefficient of friction $\mu = 0.6$ in the direction shown, and car $B$ skids 25 ft down the pavement for which $\mu = 0.5$. From these measurements (a) determine the approximate velocity of each car before impact by neglecting friction during impact, and (b) determine the approximate coefficient of restitution between the two vehicles. Assume that the $y$ axis is the approximate line of impact and the $x$ axis the plane of contact.

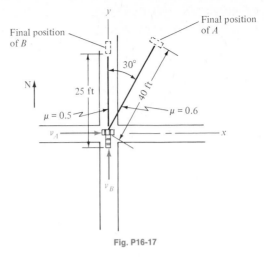

Fig. P16-17

**16–18** Reconsider Ex. 15-6 and Fig. 15-13. Determine the distance $d$ that the platform travels after the bullet's impact at $A$. Assume that the bullet is stopped instantaneously upon impact.

**16–19** The 1000-kg wrecker's ball swings in the vertical plane. Determine the maximum tension in the cable supporting the ball if it is released from rest at $\theta = 30°$. $L = 10$ m.

Fig. P16-19

**16–20** Reconsider Prob. 15–78 and determine the impulse that must be applied to the satellite of mass $m$ at apogee $A$ to cause it to transfer into a circular orbit of radius $r_A = 30{,}360$ mi.

**16–21** A rocket sled is moving on a frictionless track at a speed of 100 ft/s. The aerodynamic drag resisting the motion of the sled can be approximated by $F_D = 0.004v^2$. If the weight of the sled is 5000 lb, what distance will it travel before its speed is reduced to 50 ft/s?

# 17 Kinetics of Systems of Particles

Simple systems involving only a few particles have been analyzed on numerous occasions in previous sections. It is necessary to review and generalize the methods already developed to deal with a larger variety of problems than before. For this purpose it is useful to distinguish systems consisting of discrete particles from systems of continuously distributed particles.

**1. *Discrete particles.*** An example of this is the solar system. In certain studies it is reasonable to work with only two major masses such as the sun and the earth, but accurate analysis requires the consideration of numerous other interactions, as well. In this kind of a situation even the analysis of three interacting bodies is much more difficult than that of two bodies. Other discrete systems, such as certain exploding bodies, can be relatively easily analyzed.

**2. *Continuous particles in fluids.*** Even in small volumes of fluids the number of particles is enormous, so they cannot be individually analyzed. However, some macroscopic aspects of the motion of fluids can be modeled by lumping many particles in a simple system.

**3. *Continuous particles in rigid bodies.*** These bodies also contain innumerable particles for individual analysis, but the methods developed for the motion of particles can be extended to rigid bodies. Such generalizations of the methods are possible because in rigid bodies the geometric configuration of all particles is fixed.

656

The fundamentals of discrete systems and fluids are presented in this chapter. Much more information is available in other texts such as those dealing entirely with fluid mechanics. The motions of rigid bodies are more extensively covered in this text starting with Ch. 18. Chapter 17 serves, in a large part, as a bridge between the kinetics of particles and of rigid bodies.

**SECTION 17-1** extends Newton's second law of motion to a system of particles. The analysis of external and internal forces acting on the system and the resulting governing equation of motion are relevant to further work in kinetics.

**SECTION 17-2** presents important equations for the position, velocity, and acceleration of the center of mass of a system of particles. These equations must be understood by all students of dynamics.

**SECTIONS 17-3 THROUGH 17-5** present the potential energy, kinetic energy, and work–energy relationship for systems of particles. The major equations derived are particularly useful in the dynamics analysis of rigid bodies.

**SECTIONS 17-6 AND 17-7** are important for distinguishing and analyzing the linear and rotational motions of systems of particles.

**SECTION 17-8** extends the results of Sec. 17-7 to the special case of using the center of mass for reference. Several useful expressions are derived, but the details of these derivations are not necessary to study in order to make substantial progress in elementary dynamics.

**SECTION 17-9** presents the equations of conservation of linear and angular momentum for a system as a whole. These equations are useful for solving many problems and are worth remembering.

**SECTION 17-10** describes the methods of analyzing impulsive motion of systems of particles. This special topic provides for additional practice in the application of momentum methods.

**SECTIONS 17-11 THROUGH 17-13** present the concepts and basic methods of analyzing the flow of mass using the momentum methods developed for particles. These special topics are not essential for making progress in elementary dynamics, and may be omitted from the course syllabus. However, they are useful for practicing the solution of momentum problems. The students may also note that a large variety of interesting problems are available in this area.

## STUDY GOALS

## NEWTON'S SECOND LAW APPLIED TO A SYSTEM OF PARTICLES

### 17-1

The analysis of a system consisting of many particles can be based on a system of only two particles. Assume that the forces acting on such a system are completely described in Fig. 17-1. The masses of particles $P_i$ and $P_j$ are $m_i$ and $m_j$, respectively. The forces $\mathbf{f}_{ij}$ and $\mathbf{f}_{ji}$ are the mutual central forces between the particles. They are always collinear, equal in magnitude, and opposite in direction (attraction in the case of gravity, attraction or repulsion in the cases of electricity and magnetism). It is tempting to cancel the *internal forces* $\mathbf{f}_{ij}$ and $\mathbf{f}_{ji}$ at once, but it is best to consider them individually in trying to relate the motion of each particle to the motion of the system as a whole. The forces $\mathbf{F}_i$ and $\mathbf{F}_j$ are *external forces* whether the particles are considered individually or together as a

**FIGURE 17-1**

System of two interacting particles

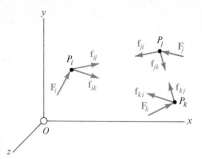

**FIGURE 17-2**

System of three interacting particles

system. If there are only two particles in the system, Newton's second law can be written as

$$\mathbf{F}_i + \mathbf{f}_{ij} = m_i\mathbf{a}_i \qquad \text{for particle } P_i \qquad \boxed{\text{17-1a}}$$

$$\mathbf{F}_j + \mathbf{f}_{ji} = m_j\mathbf{a}_j \qquad \text{for particle } P_j \qquad \boxed{\text{17-1b}}$$

This can be extended to include a third particle $P_k$, which exerts mutual internal forces on each of the other particles in the system (Fig. 17-2). Summing all the forces acting on particle $P_i$, there are two internal forces now, and the external force $\mathbf{F}_i$. Therefore,

$$\mathbf{F}_i + \mathbf{f}_{ij} + \mathbf{f}_{ik} = m_i\mathbf{a}_i$$

Generalizing this to $n$ particles in the system, the equation of motion for particle $P_i$ is

$$\mathbf{F}_i + \sum_{j=1}^{n} \mathbf{f}_{ij} = m_i\mathbf{a}_i \qquad \boxed{\text{17-2}}$$

The similar equation for particle $P_j$ is

$$\mathbf{F}_j + \sum_{i=1}^{n} \mathbf{f}_{ji} = m_j\mathbf{a}_j$$

In both cases the forces $\mathbf{f}_{ii}$ and $\mathbf{f}_{jj}$ have no meaning.

The general situation is that each particle of $n$ particles in the system has a resultant external force $\mathbf{F}$ on it, and also $n - 1$ internal forces $\mathbf{f}$ on it. Newton's second law for the whole system can be concisely written as

$$\sum_{i=1}^{n} \mathbf{F}_i + \sum_{\substack{i=1 \\ i \ne j}}^{n} \sum_{j=1}^{n} \mathbf{f}_{ij} = \sum_{i=1}^{n} m_i\mathbf{a}_i \qquad \boxed{\text{17-3}}$$

The second term on the left side can be eliminated since each force $\mathbf{f}_{ij}$, and so on, has an equal and opposite force $\mathbf{f}_{ji}$, and so on, in the system. The resulting governing equation for the motion of the whole system is

$$\boxed{\sum_{i=1}^{n} \mathbf{F}_i = \sum_{i=1}^{n} m_i\mathbf{a}_i} \qquad \boxed{\text{17-4}}$$

where $\sum \mathbf{F}_i$ is the vector sum of all forces acting on the system.

## 17-2    MOTION OF THE CENTER OF MASS

For a complete description of the motion of a system, it is useful to determine the position, velocity, and acceleration of the center $C$ of the total mass. First, assume that $\mathbf{r}_i$ is the position vector of particle $P_i$ in the chosen reference frame, and $\mathbf{r}_C$ is the position vector of the mass center of the system of particles in the same reference frame. For a system of $n$ particles with a total mass $m$,

$$\boxed{m\mathbf{r}_C = \sum_{i=1}^{n} m_i\mathbf{r}_i} \qquad \boxed{\text{17-5}}$$

by definition of the center of mass. This equation is valid regardless of any velocities and accelerations of the particles at a particular instant.

The velocities of the system as a whole and of the particles are obtained by differentiating Eq. 17-5 with respect to time,

$$m\mathbf{v}_C = \sum_{i=1}^{n} m_i\mathbf{v}_i \qquad \boxed{\text{17-6}}$$

It is assumed in this differentiation that the mass of each particle is constant. Note that the mass is not necessarily constant (e.g., when fuel is burned).

The accelerations of the system as a whole and of the particles are obtained by differentiating Eq. 17-6 with respect to time,

$$m\mathbf{a}_C = \sum_{i=1}^{n} m_i\mathbf{a}_i \qquad \boxed{\text{17-7}}$$

From Eqs. 17-4 and 17-7,

$$\sum \mathbf{F} = m\mathbf{a}_C \qquad \boxed{\text{17-8}}$$

where $\sum \mathbf{F}$ is the resultant of all forces acting on the system, $m$ is the total mass of the system, and $\mathbf{a}_C$ is the acceleration of the center of mass.

On the basis of Eqs. 17-4 through 17-8, the following important remarks are made concerning the motion of a system of particles:

*If there are only mutual forces acting among the particles* (there are forces **f** but no forces **F** in Fig. 17-2), the center of mass of the system has no acceleration. Therefore, the center of mass of the system remains at rest if initially at rest, or it moves with a constant velocity regardless of the velocities of the constituent particles. These results agree with Newton's first and second laws for a single particle. In summary,

> The center of mass of a system of particles moves as if the total mass of the system were concentrated at the center of mass if there is no resultant force acting on the system.

These statements can be generalized for the situation where internal and external forces act on the particles (both forces **f** and **F** in Fig. 17-2). In that case,

> The center of mass of a system of particles moves under the action of internal and external forces as if the total mass of the system and all the external forces were at the center of mass.

Note that the last statement does not mean that the resultant of all the external forces must act at the center of mass, causing only a translation of the system. In general, the resultant force would not act through the center of mass, causing both a translation and a rotation (because of a moment about $C$) of the system. The derivations and statements above concern only the translation part of the total motion that is possible. The rotation will be considered in Sec. 17-7.

# EXAMPLE 17-1

In the delicate docking maneuver of two spacecraft, even small changes in their relative positions are important. Assume that the total weight of a space vehicle with crew is 4000 lb. What is the allowable displacement of a 160-lb astronaut in this vehicle to limit the resulting displacement of the vehicle's frame to 0.2 in.? The astronaut moves away from the center of mass of the vehicle as modeled in Fig. 17-3. The $xy$ frame is fixed with respect to the center of mass $C$ of the whole system, and the astronaut moves in the $xy$ plane.

## SOLUTION

To determine the allowable displacement, apply Eq. 17-5,

$$m_T \mathbf{r}_C = \sum m_i \mathbf{r}_i = m_V \mathbf{r}_V + m_A \mathbf{r}_A$$

where $m_T$ is the total mass of the system, $m_A$ is the mass of the astronaut, and $m_V$ is the mass of the vehicle without the moving astronaut, $m_V = m_T - m_A$. The corresponding position vectors $\mathbf{r}$ are defined with respect to the $xy$ reference frame. Since the reference frame's origin is at the mass center, $\mathbf{r}_C = 0$. Thus,

$$0 = m_V \mathbf{r}_V + m_A \mathbf{r}_A$$

For a displacement $\mathbf{r}_A$ in any direction with respect to the origin, $\mathbf{r}_A$ and $\mathbf{r}_V$ are collinear by inspection. Thus, a scalar solution can be used,

Frame of vehicle

Displacement of frame of vehicle with respect to $C$

**FIGURE 17-3**

Motion of an astronaut in a space-craft

$$-\frac{3840\,\text{lb}}{32.2\,\text{ft/s}^2}(0.2\,\text{in.}) = \frac{160\,\text{lb}}{32.2\,\text{ft/s}^2}(r_A)$$

The allowable displacement $r_A$ of the astronaut is

$$r_A = -4.8\,\text{in.}$$

Note the inverse proportionality of the displacements to the masses involved. The negative sign indicates that the astronaut and the vehicle move in opposite directions.

# EXAMPLE 17-2

A spacecraft consists of a main cabin of mass $M = 6000\,\text{kg}$ and a nuclear power source of mass $m = 300\,\text{kg}$ (Fig. 17-4). The two main parts are treated as particles which are connected by a tube of negligible mass. An attitude-controlling jet applies a force $\mathbf{F} = (40\mathbf{i} + 20\mathbf{j})\,\text{N}$ to the main cabin. Determine the acceleration of the mass center.

## SOLUTION

The acceleration is obtained from Eq. 17-8,

$$\sum \mathbf{F} = m_T \mathbf{a}_C$$

where $m_T = M + m = 6300\,\text{kg}$. Then

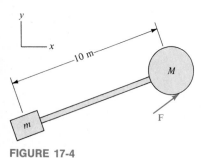

**FIGURE 17-4**

Model of a spacecraft

$$\mathbf{a}_C = \frac{40\,\text{N}}{6300\,\text{kg}}\mathbf{i} + \frac{20\,\text{N}}{6300\,\text{kg}}\mathbf{j} = (0.00635\mathbf{i} + 0.00317\mathbf{j})\,\text{m/s}^2$$

EXAMPLE 17-3

A meteor breaks into two pieces at a height of 500 m as it approaches Earth (Fig. 17-5). Radar tracking records its approach velocity $v_0$ as 1000 m/s in the $xy$ plane at an angle of 45° to the vertical. One piece is observed to be twice as large as the other and the larger piece is found later at $A$ (300 m, 0, 150 m). Where would you expect to find the smaller piece, neglecting air resistance?

## SOLUTION

Neglecting air resistance, the resultant force acting on the system is due to gravity. Therefore, the system's mass center moves under free-fall conditions. Applying Eq. 17-8,

$$\Sigma \mathbf{F} = -W\mathbf{j} = -mg\mathbf{j} = m\mathbf{a}_C$$

$$\mathbf{a}_C = -9.81\mathbf{j}\,\text{m/s}^2 \qquad \textbf{(a)}$$

Integrating Eq. (a), $\mathbf{v}_C = (-9.81t\mathbf{j} + \mathbf{C}_1)$ m/s. At $t = 0$,

$$\mathbf{v} = (707\mathbf{i} - 707\mathbf{j})\,\text{m/s}, \qquad \text{so } \mathbf{C}_1 = (707\mathbf{i} - 707\mathbf{j})\,\text{m/s}$$

Therefore,

$$\mathbf{v}_C = 707\mathbf{i} - (9.81t + 707)\mathbf{j}\,\text{m/s} \qquad \textbf{(b)}$$

Similarly,

$$\mathbf{r}_C = (707t\mathbf{i} - (4.905t^2 + 707t - 500)\mathbf{j})\,\text{m} \qquad \textbf{(c)}$$

Since each piece falls freely, they both strike the ground at the same time $t_g$ given by Eq. (c) with $y_C = 0$.

$$4.905t_g^2 + 707t_g - 500 = 0, \qquad t_g = 0.704\,\text{s} \qquad \textbf{(d)}$$

Also, at $t = 0.704$ s, $\mathbf{r}_C = 498\mathbf{i}$ m.
From the definition of mass center applied at $t = 0.704$ s,

$$m(498\mathbf{i}) = \frac{2m}{3}(300\mathbf{i} + 150\mathbf{k}) + \frac{m}{3}(x\mathbf{i} - z\mathbf{k}) \qquad \textbf{(e)}$$

Solving Eq. (e) for $x$ and $z$ yields $x = 894$ m and $z = 300$ m. Therefore, the smaller piece lands at

$$B(894\,\text{m}, 0, -300\,\text{m})$$

**FIGURE 17-5**

Pieces of a falling meteor

EXAMPLE 17-3    661

**17–1** The masses and position vectors of three particles are $m_1 = 1\,\text{kg}$, $\mathbf{r}_1 = (0.5\mathbf{i} + 0.8\mathbf{j})$ m, $m_2 = 2\,\text{kg}$, $\mathbf{r}_2 = (0.9\mathbf{i} - 0.4\mathbf{j})$ m, $m_3 = 1.5\,\text{kg}$, $\mathbf{r}_3 = (2\mathbf{i} + 0.7\mathbf{j})$ m. $\mathbf{v}_3 = 6\mathbf{i}$ m/s. Determine the position vector and the velocity of the center of mass.

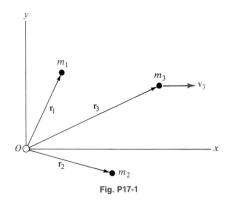

Fig. P17-1

**17–2** The position vectors of the three particles are $\mathbf{r}_1 = (2\mathbf{j} + 2\mathbf{k})$ ft, $\mathbf{r}_2 = (3\mathbf{i} + 2.5\mathbf{j} + 2\mathbf{k})$ ft, and $\mathbf{r}_3 = (4\mathbf{i} - 0.6\mathbf{j} - 2\mathbf{k})$ ft. Particle 2 has an acceleration $\mathbf{a}_2 = -8\mathbf{j}$ ft/s². Determine the position vector and the acceleration of the center of mass.

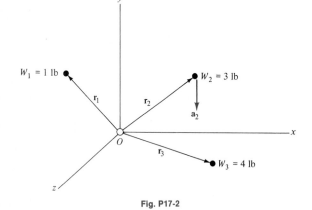

Fig. P17-2

**17–3** The three particles have identical mass $m = 2\,\text{kg}$. Calculate the velocity of the center of mass.

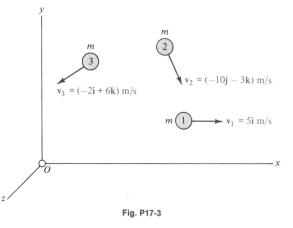

Fig. P17-3

**17–4** The speed of each particle is increasing at the rate of $1\,\text{m/s}^2$ in Prob. 17–3. Determine the resultant force acting on the system, assuming rectilinear motion.

**17–5** The three particles are released from rest and they all fall in a vertical plane. Force **F** acts horizontally on particle 3. Determine the acceleration of the center of mass of the system.

Fig. P17-5

**17–6** A 10-ton truck starts from the back of a 200-ton ferry. What would be the distance of the ferry from the road with the truck at $A$ if the ferry initially touched land but was not tied? $C$ is the center of mass of the ferry. Assume that there is no friction when the ferry moves.

Fig. P17-6

**17–7** A 70-kg person steers a 120-kg boat to gently touch a dock. The person is 6 m away from the dock, gets up, and moves to the front of the boat to climb out. How far is the person from the dock after having moved to the front of the boat? $C$ is the center of mass of the boat. Assume that there is no friction when the boat moves.

Fig. P17-7

**17–8** A 10-kg projectile traveling at $v_0 = 200\,\text{m/s}$ breaks into two parts at point $A$. Calculate $v_1$ and $v_2$ if $m_1 = 2\,\text{kg}$, $\theta = 35°$, and $\phi = 30°$.

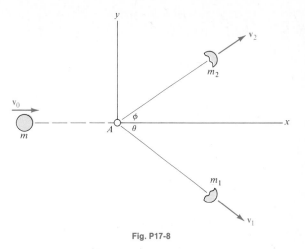

Fig. P17-8

**17–9** A meteor (tracked with radar) is traveling at $v_0 = 4000\,\text{mph}$. It breaks into two parts at point $A$ in Fig. P17-8. Calculate the weight $W$ of the meteor if $\theta = 30°$, $W_1 = 400\,\text{lb}$, $v_2 = 5110\,\text{mph}$, and $\phi = 40°$. The weight $W_1$ is measured at the surface of the earth after the fragment is recovered.

**17–10** A rocket (tracked with radar) is traveling at $v_0 = 8000\,\text{km/h}$. It breaks into three parts at point $A$. The fragments move in the $xy$ plane at $v_1 = 9000\,\text{km/h}$, $\theta_1 = 45°$, $v_2 = 10,000\,\text{km/h}$, $\theta_2 = 0°$, $v_3 = 11,000\,\text{km/h}$, and $\theta_3 = 45°$. $m_1$ is recovered and found to be 1600 kg. Calculate $m_2$, $m_3$, and the total mass of the rocket.

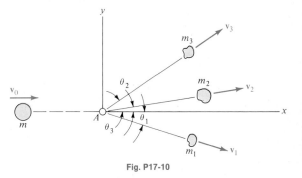

Fig. P17-10

**17–11** A rocket (tracked with radar) is traveling at $v_0 = 8000\,\text{km/h}$. It breaks into three parts at point $A$ as in Fig. P17-10. The fragments move in the $xy$ plane at $v_1 = 9000\,\text{km/h}$, $\theta_1 = 40°$, $v_2 = 10,500\,\text{km/h}$, $\theta_2 = 10°$, $v_3 = 14,000\,\text{km/h}$, and $\theta_3 = 45°$. $m_1$ is recovered and found to be 800 kg. Try to determine the masses $m_2$ and $m_3$ and show that the measured data are likely to be in error.

## 17-3

**FIGURE 17-6**

Schematic for the gravitational potential energy of a system of particles

# GRAVITATIONAL POTENTIAL ENERGY OF A SYSTEM OF PARTICLES

The gravitational potential energy of a system of particles is defined as the sum of the potential energies of the individual particles of the system. The individual potential energies may be measured with respect to a common reference level (for convenience) such as the $x$ axis in Fig. 17-6. The potential energy $V_{gi}$ of a particle $P_i$ is $m_i g y_i$, and the same factor $g$ appears in every individual term of the total potential energy. Thus, the potential energy $V_g$ of a system of $n$ particles is expressed as

$$V_g = g \sum_{i=1}^{n} m_i y_i = \sum_{i=1}^{n} W_i y_i$$

17-9

where $m_i$ and $W_i$ are the mass and weight of particle $P_i$, respectively. Equation 17-9 is valid regardless of any motion of the particles.

The potential energy can also be expressed as the product of the total mass $m$ of the system, the position $y_C$ of its center of mass, and $g$,

$$V_g = mgy_C = Wy_C$$

17-10

where $W$ is the total weight of the system. In common engineering problems $C$ represents both the center of mass and the center of gravity of the system.

## 17-4

# KINETIC ENERGY OF A SYSTEM OF PARTICLES

The kinetic energy of a system of particles is defined as the sum of the kinetic energies of the individual particles of the system. The kinetic energy $T_i$ of a particle $P_i$ is $\frac{1}{2}m_i v_i^2$, where $v_i$ is the magnitude of the particle's velocity in a fixed reference frame as in Fig. 17-7. The kinetic energy $T$ of a system of $n$ particles is given by

$$T = \frac{1}{2} \sum_{i=1}^{n} m_i v_i^2$$

17-11

**FIGURE 17-7**

Schematic for the kinetic energy of a system of particles

**FIGURE 17-8**

Two particles with identical kinetic energies in different motions

It should be remembered that the kinetic energy is always positive if there is motion, regardless of the number of particles and the directions of their velocities. For example, the two particles of equal mass in Fig. 17-8a have velocities that are equal in magnitude and opposite in direction. Their total kinetic energy according to Eq. 17-11 is $T = 2(\frac{1}{2}mv^2) = mv^2$. The same is true for the system of two particles in Fig. 17-8b.

### Reference Frame Translating with the Mass Center

There are many situations where it is desirable to determine the kinetic energy of a system of particles with respect to a given fixed frame, but it is not convenient to make the direct summation using Eq. 17-11. For example, consider the model of an amusement ride in Fig. 17-9. Each cage of mass $m$ may be considered a particle in circular motion with respect to the carriage which translates with the $xy$ frame. In such a case it is most convenient to superimpose the kinetic energy of the system with respect to its own center of mass (point $C$ for the 4 cages in Fig. 17-9) and the kinetic energy of the total mass assumed to be concentrated at the center of mass. This concept is developed for the common special case of a translating reference frame as follows.

Assume that the $xyz$ frame is at the center $C$ of the total mass of a system of particles in Fig. 17-10. This frame is in *translation only* with respect to the $XYZ$ frame at velocity $\mathbf{v}_C$. The particle $P_i$ has velocity $\mathbf{v}_i$ and $\mathbf{v}_i'$ in the $XYZ$ and $xyz$ frames, respectively. $\mathbf{v}_i$ is the vector sum of $\mathbf{v}_C$ and $\mathbf{v}_i'$ as shown in Fig. 17-10,

$$\mathbf{v}_i = \mathbf{v}_C + \mathbf{v}_i' \qquad \boxed{17\text{-}12}$$

**FIGURE 17-9**

Amusement ride

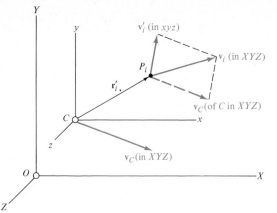

**FIGURE 17-10**

Schematic for using a translating reference frame
located at the mass center

The total kinetic energy $T$ of a system of $n$ particles in the $XYZ$ frame
can be expressed using the scalar product $\mathbf{v}_i \cdot \mathbf{v}_i = v_i^2$,

$$T = \frac{1}{2} \sum_{i=1}^{n} m_i v_i^2 = \frac{1}{2} \sum_{i=1}^{n} (m_i \mathbf{v}_i \cdot \mathbf{v}_i)$$

This expression is expanded using Eq. 17-12,

$$T = \frac{1}{2} \sum_{i=1}^{n} \left[ m_i (\mathbf{v}_C + \mathbf{v}_i') \cdot (\mathbf{v}_C + \mathbf{v}_i') \right]$$

$$= \frac{1}{2} \left( \sum_{i=1}^{n} m_i \right) v_C^2 + \frac{1}{2} \sum_{i=1}^{n} m_i v_i'^2 + \sum_{i=1}^{n} (m_i \mathbf{v}_C \cdot \mathbf{v}_i')$$

The first term on the right side equals $\frac{1}{2} m v_C^2$, where $m$ is the total mass
of the system. Recognizing that $\mathbf{v}_C$ is the same for every term in the
summation in the last term, it can be rewritten as

$$\sum_{i=1}^{n} (m_i \mathbf{v}_C \cdot \mathbf{v}_i') = \mathbf{v}_C \cdot \sum_{i=1}^{n} m_i \mathbf{v}_i' = \mathbf{v}_C \cdot \frac{d}{dt} \sum_{i=1}^{n} m_i \mathbf{r}_i' = 0$$

since $\sum_{i=1}^{n} m_i \mathbf{r}_i' = 0$ by definition of the center of mass. Thus, the total
kinetic energy of the system in a fixed frame is

$$T = \underbrace{\frac{1}{2} m v_C^2}_{\substack{\text{motion of total} \\ \text{mass imagined to} \\ \text{be concentrated} \\ \text{at } C \text{ moving in} \\ \text{fixed } XYZ \text{ frame}}} + \underbrace{\frac{1}{2} \sum_{i=1}^{n} m_i v_i'^2}_{\substack{\text{motion of all} \\ \text{particles relative} \\ \text{to mass center } C \text{ at} \\ \text{origin of translating} \\ xyz \text{ frame}}} \qquad \boxed{17\text{-}13}$$

Equation 17-13 is particularly useful in the dynamics analysis of rigid
bodies. It is emphasized that Eq. 17-13 was obtained using a translating
reference frame, and is not valid for a rotating frame.

The work–energy relationship for a single particle $P_i$ can be expressed as the sum of changes in the potential energy $V_i$ and kinetic energy $T_i$,

$$U'_i = \Delta V_i + \Delta T_i \qquad \boxed{17\text{-}14}$$

where $U'_i$ is the work of all nonconservative forces acting on the particle.

To evaluate the work done on a system of particles it is necessary to consider the work of the net external force $\mathbf{F}_i$ and of all internal forces $\mathbf{f}_{ij}$ on each particle $P_i$. These forces were discussed in detail in Sec. 17-1. The work of the internal forces $\mathbf{f}_{ij}$ and $\mathbf{f}_{ji}$ may or may not cancel out depending on the displacements of the particles $P_i$ and $P_j$. This can be evaluated by inspection in many situations. Frequently, the pairs of displacements are identical in magnitude and only the external forces do work. The total work on a system of $n$ particles and the resulting changes in potential and kinetic energies are expressed by

$$\sum_{i=1}^{n} U'_i = \sum_{i=1}^{n} \Delta V_i + \sum_{i=1}^{n} \Delta T_i$$

which is commonly written in the concise form

$$\boxed{U' = \Delta V + \Delta T} \qquad \boxed{17\text{-}15}$$

The concept of a particle's potential energy can be generalized to include the internal elastic energy resulting from deformations. In that case the term $\Delta V$ may have two components, a gravitational and an elastic potential energy.

## Conservation of Energy

There are many situations where it is reasonable to assume that all forces acting on a system of particles are conservative. This means that the work of external or internal friction forces must be zero or negligible in the entire system of particles. In that case Eq. 17-15 becomes

$$\boxed{\Delta V + \Delta T = 0} \qquad \boxed{17\text{-}16}$$

which is the statement of the *conservation of total mechanical energy* in a system of particles. Equation 17-16 is frequently used to express the same total energy at two different configurations of a system and thus solve for an unknown quantity. This method is stated schematically for configurations 1 and 2,

$$\boxed{V_1 + T_1 = V_2 + T_2} \qquad \boxed{17\text{-}17}$$

Equations 17-16 and 17-17 express the same principle of conservation of total mechanical energy. The applicability of the method is generally determined by judging the expected frictional losses.

# EXAMPLE 17-4

The track of a roller coaster has circular sections both at $A$ and $B$ for which the radius $R = 30$ m to the center of mass of each 300 kg car. The speed of the train at the instant the mass center of the middle car of a seven-car train passes point $A$ is $v_A = 5$ m/s. The cars are all identical and the distance between neighboring centers of mass is $d = 3.14$ m. Determine **(a)** the maximum speed of the train at the instant the mass center of the middle car is at point $B$, neglecting friction, and **(b)** the resultant force exerted by the track on the train at position $B$.

(a)

## SOLUTION

(b)

**(a)** The centers of mass of neighboring cars are separated by a distance $d$ defined in Fig. 17-11b where

$$\frac{\theta_d}{360} \simeq \frac{d}{2\pi R} \quad \text{or} \quad \theta_d \simeq \frac{(3.14\,\text{m})(360°)}{6.28(30\,\text{m})} = 6°$$

Therefore the distance of the train's center of mass, $C$, above $O$ (at point $A$) and below $O'$ (at point $B$) is given by

$$7(300\,\text{kg})y_C = (300\,\text{kg})[30\,\text{m} + 2(30\,\text{m})\cos 6° + 2(30\,\text{m})\cos 12° + 2(30\,\text{m})\cos 18°]$$

$$y_C = 29.3\,\text{m}$$

The work done by gravity is equal to the weight of the system multiplied by the change in height of the system's mass center. The work-energy principle, Eq. 14-15, is used to determine the speed at $B$,

$$U_{AB} = T_B - T_A$$

$$W(\Delta h_C) = \frac{1}{2}m(v_B^2 - v_A^2)$$

$$7(300\,\text{kg})(9.81\,\text{m/s}^2)(2)(29.3\,\text{m}) = \frac{1}{2}(7)(300\,\text{kg})\{v_B^2 - (5\,\text{m/s})^2\}$$

$$v_B = 34.3\,\text{m/s}$$

(c)

**FIGURE 17-11**

Analysis of a roller coaster

**(b)** At point $B$, where the positive normal direction is vertically upward (Fig. 17-11c),

$$\sum F_n = m(a_C)_n$$

$$F_B - W = m\frac{v_B^2}{y_C}$$

$$F_B = 7(300\,\text{kg})(9.81\,\text{m/s}^2) + 7(300\,\text{kg})\frac{(34.3\,\text{m/s}^2)^2}{29.3\,\text{m}}$$

$$\mathbf{F}_B = 105\mathbf{j}\,\text{kN}$$

EXAMPLE 17-5

Consider the model of an amusement ride in Fig. 17-12a. The carriage weighs 1000 lb and moves at a constant speed of 15 ft/s on a horizontal track. The rigid arms of 12-ft length and negligible weight rotate at a constant angular speed of 15 rpm. Each cage can carry 3 to 4 passengers and can be modeled as an 800-lb particle. Determine the total kinetic energy of the amusement ride.

## SOLUTION

The problem is solved using Eq. 16-13,

$$T = \frac{1}{2} m v_C^2 + \frac{1}{2} \sum_{i=1}^{n} m_i v_i'^2$$

A reference frame $xy$, translating with resect to the fixed $XY$ frame, is attached to the carriage at point $C$ as in Fig. 17-12b. First the total mass of the carriage and the two cages is imagined to be concentrated at point $C$, the center of mass of the system. The contribution to the total kinetic energy is

$$\frac{1}{2} m v_C^2 = \frac{1}{2} \left\{ \frac{1000 \text{ lb}}{32.2 \text{ ft/s}^2} + 2 \frac{800 \text{ lb}}{32.2 \text{ ft/s}^2} \right\} (15 \text{ ft/s})^2$$

Next the motion of both cages relative to the mass center $C$ at the origin of the translating $xy$ frame is analyzed. Both cages are in circular motion with respect to point $C$ with a relative velocity of $v_i' = r\dot{\theta}$, where $\dot{\theta} = (15 \text{ rpm})(2\pi \text{ rad/rev})(\frac{1}{60} \text{ min/s}) = 1.571 \text{ rad/s}$. The kinetic energy due to this motion is

$$\frac{1}{2} \sum_{i=1}^{2} m_i v_i'^2 = \frac{1}{2} \left\{ 2 \frac{800 \text{ lb}}{32.2 \text{ ft/s}} [(12 \text{ ft})(1.571 \text{ rad/s})]^2 \right\}$$

The total kinetic energy is the sum,

$$T = 9080 \text{ ft} \cdot \text{lb} + 8830 \text{ ft} \cdot \text{lb}$$
$$= 17,910 \text{ ft} \cdot \text{lb}$$

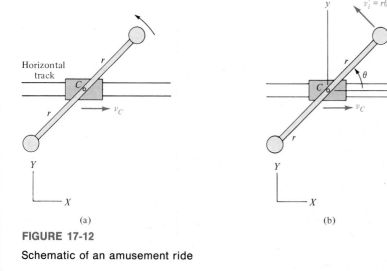

**FIGURE 17-12**

Schematic of an amusement ride

EXAMPLE 17-5    669

**17–12** Each car of the roller coaster weighs 400 lb including passengers. The front car reaches an elevation $h = 50$ ft. Determine the speed $v_A$ of the train to make it coast without friction to that height if $l = 20$ ft.

**Fig. P17-12**

**17–13** A roller coaster consists of five identical cars, each of 200 kg mass including passengers (Fig. P17-12). The train of length $l = 8$ m has speed $v_A = 15$ m/s on a horizontal part of the track. Determine the elevation $h$ reached by the front car if friction is negligible.

**17–14** The track of a roller coaster has a circular part of radius $R = 50$ m to the center of mass of each 250-kg car. The maximum speed of the train is $v_A = 20$ m/s. Determine the speed $v_B$ at a position $\theta = 40°$ of the front car. A typical distance between cars is $d = 2$ m. Neglect friction.

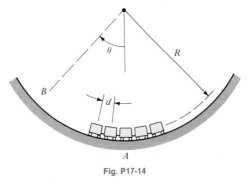

**Fig. P17-14**

**17–15** The front car of a roller coaster reaches a maximum displacement $\theta = 45°$ in Fig. P17-14. The radius of curvature to the center of mass of each car is $R = 180$ ft, each car weighs 400 lb, and a typical distance between cars is $d = 5$ ft. Determine the maximum speed $v_A$ of the train to make it coast without friction to point $B$.

**17–16** The train of the amusement ride has a speed $v_A = 6$ ft/s when the train's center is at $A$. Each of the seven cars in the 40-ft-long train weighs 800 lb with riders. Calculate the maximum elevation $h_B$ of the train's front if $h_A = 70$ ft and friction is negligible as the train moves from $A$ to $B$.

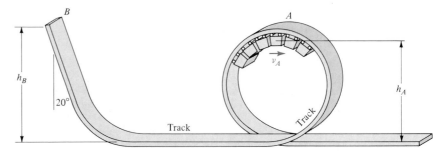

**Fig. P17-16**

**17–17** The chain-drive mechanism has 100 identical links. The weight of each link is 0.05 lb and $R = 0.5$ ft. Determine the average power requirement of the machine to start the chain from rest and bring it to a speed $v = 20$ ft/s in a time of 7 s.

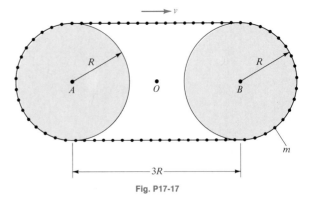

**Fig. P17-17**

**17–18** A turbine disk has 50 identical 0.15-kg blades of negligible dimensions. Assume that the disk itself has no mass and is rotating at 5000 rpm. Determine the work of the braking mechanism that stops the rotation of the disk.

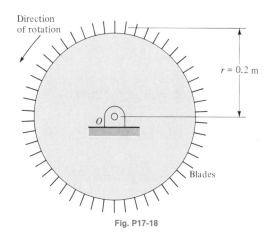

**Fig. P17-18**

**17–19** A Ferris wheel has three evenly spaced cages, each weighing 1000 lb with riders. Calculate the total kinetic energy of the cages if $\omega_A = 0.2$ rad/s and $\omega_Y = 0$.

**Fig. P17-19**

**17–20** Calculate the total kinetic energy of the three cages in Prob. 17–19 assuming an additional motion at angular speed $\omega_Y = 0.15$ rad/s for the position shown. Note that with this additional motion, the cages can be assumed to be in circular motion about the $Y$ axis at any instant. The total motion of a cage can be obtained as a superposition of circular motion in the vertical plane about point $A$ and circular motion about the $Y$ axis in the position shown.

**17–21** Determine the total work required to start the Ferris wheel of Prob. 17–19 from rest and bring it to the angular speeds $\omega_A = 0.2$ rad/s and $\omega_y = 0.15$ rad/s. Neglect friction and changes in elevation.

**17–22** An amusement ride is designed to rotate arm $AOB$ about point $O$ in the horizontal plane at 5 rpm. The two cages with two riders in each are to rotate at 6 rpm relative to the arm $AOB$. The average rider has a 70-kg mass and sits at $r = 1$ m from the center of the cage. Calculate the total kinetic energy of (a) riders 1 and 2, and (b) riders 3 and 4 for the position shown.

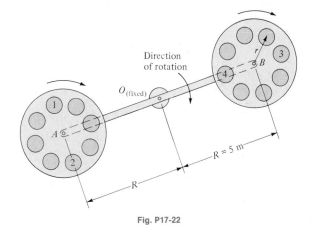

**Fig. P17-22**

**17–23** A proposed mechanism would change $r$ from 1 m to 1.5 m during operation of the ride described in Prob. 17–22. Calculate the change in total kinetic energy and the required work to make the change. Ignore the masses of structural members. Assume that the change occurs instantaneously in the position shown.

## MOMENTS OF FORCES ON A SYSTEM OF PARTICLES

The analysis of forces in Sec. 17-1 can be readily extended to consider the moments of the forces about an arbitrary point of reference. This is done using Fig. 17-13. Assume that the moments of all forces acting on the system of particles should be evaluated with respect to point $O$. For generality, the moments of external and internal forces are included in the analysis. They are written for particle $P_i$ as

$$\mathbf{r}_i \times \mathbf{F}_i + \mathbf{r}_i \times \mathbf{f}_{ij} = \mathbf{r}_i \times m_i \mathbf{a}_i$$

The moments of forces for a whole system of $n$ particles can be written from this equation following the procedure used in Sec. 17-1,

$$\sum_{i=1}^{n} (\mathbf{r}_i \times \mathbf{F}_i) + \sum_{i=1}^{n} \sum_{\substack{j=1 \\ i \neq j}}^{n} (\mathbf{r}_i \times \mathbf{f}_{ij}) = \sum_{i=1}^{n} (\mathbf{r}_i \times m_i \mathbf{a}_i) \qquad \boxed{17\text{-}18}$$

The second term on the left side can be eliminated since the moments of the internal forces cancel in pairs for any number of interacting particles. The resulting governing equation for the motion of the system about the arbitrary point $O$ caused by moments of forces is

$$\sum_{i=1}^{n} (\mathbf{r}_i \times \mathbf{F}_i) = \sum_{i=1}^{n} \mathbf{M}_{io} = \sum_{i=1}^{n} (\mathbf{r}_i \times m_i \mathbf{a}_i) \qquad \boxed{17\text{-}19}$$

**FIGURE 17-13**
System of two interacting particles

## LINEAR AND ANGULAR MOMENTA OF A SYSTEM OF PARTICLES

The linear momentum of a particle $P_i$ is $\mathbf{G}_i = m_i \mathbf{v}_i$ by definition. The linear momentum of a system of $n$ particles is, by definition,

$$\mathbf{G} = \sum_{i=1}^{n} m_i \mathbf{v}_i$$

Substituting Eq. 17-6 gives

$$\mathbf{G} = m \mathbf{v}_C \qquad \boxed{17\text{-}20}$$

which means that the linear momentum of a system of particles is equal to the total mass of the particles times the velocity of the center of mass. Equation 17-20 is valid if the mass of the system is constant.

Differentiating Eq. 17-20 with respect to time yields

$$\dot{\mathbf{G}} = m \dot{\mathbf{v}}_C$$

where $m\dot{\mathbf{v}}_C = m\mathbf{a}_C$. Using a simplified notation for the result obtained from Eqs. 17-4 and 17-7,

$$\boxed{\sum \mathbf{F} = \dot{\mathbf{G}}} \qquad \boxed{17\text{-}21}$$

which means that *the resultant of the external forces on a system of particles equals the time rate of change of linear momentum of that system.* Equation 17-21 is an alternative expression of Eq. 17-4. These two equations are the generalized forms of Newton's second law of motion. It is assumed that the mass of the system is constant.

The angular momentum **H** of a system of particles is defined with respect to an arbitrary, fixed point of reference. Referring to point $O$ in Fig. 17-1,

$$\mathbf{H}_O = \sum_{i=1}^{n} (\mathbf{r}_i \times m_i \mathbf{v}_i) \qquad \boxed{17\text{-}22}$$

Differentiating this equation with respect to time gives

$$\dot{\mathbf{H}}_O = \sum_{i=1}^{n} (\dot{\mathbf{r}}_i \times m_i \mathbf{v}_i) + \sum_{i=1}^{n} (\mathbf{r}_i \times m_i \dot{\mathbf{v}}_i)$$

Since $\dot{\mathbf{r}}_i = \mathbf{v}_i$, $\dot{\mathbf{r}}_i \times m_i \mathbf{v}_i = \mathbf{v}_i \times m_i \mathbf{v}_i = 0$ (recall that the cross product of two collinear vectors is zero). Writing $\dot{\mathbf{v}}_i = \mathbf{a}_i$ yields

$$\dot{\mathbf{H}}_O = \sum_{i=1}^{n} (\mathbf{r}_i \times m_i \mathbf{a}_i) \qquad \boxed{17\text{-}23}$$

It is seen from Eq. 17-19 that the right side of Eq. 17-23 is the sum of the moments of the external forces about point $O$ for the whole system of particles. Denoting this sum by $\sum \mathbf{M}_O$, the resulting equation is

$$\boxed{\sum \mathbf{M}_O = \dot{\mathbf{H}}_O = \sum (\mathbf{r}_i \times m_i \mathbf{a}_i)} \qquad \boxed{17\text{-}24}$$

which means that *the resultant of the moments of the external forces on a system of particles equals the time rate of change of angular momentum of that system.* The moments and the angular momentum are defined with respect to a fixed point $O$. The mass of the system is assumed constant.

## ANGULAR MOMENTUM ABOUT THE CENTER OF MASS

17-8

An important special case in using Eq. 17-24 is when the center of mass is taken as the point of reference for the moments of forces and the angular momentum. In this case it must be assumed that the center $C$ of the mass may be moving with respect to the fixed coordinate system whose origin $O$ was used in obtaining Eq. 17-24. The analysis of angular momentum is done using Fig. 17-14. The center $C$ of the total mass of the

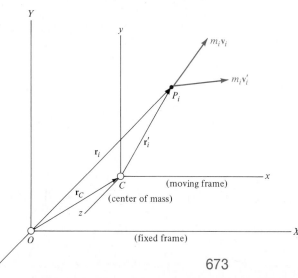

**FIGURE 17-14**

Motion of a particle with respect to the center of mass and a fixed reference frame

673

particles moves in an arbitrary way with respect to the $XYZ$ frame. A translating but not rotating reference frame $xyz$ is moving with point $C$. The position vectors, velocities, and accelerations of a particle $P_i$ with mass $m_i$ are represented according to the following scheme:

|  | FIXED $XYZ$ FRAME | MOVING $xyz$ FRAME |
|---|---|---|
| Position vector | $\mathbf{r}_i$ | $\mathbf{r}_i'$ |
| Velocity vector | $\mathbf{v}_i$ | $\mathbf{v}_i'$ |
| Acceleration vector | $\mathbf{a}_i$ | $\mathbf{a}_i'$ |

The corresponding quantities for the mass center $C$ moving in the $XYZ$ frame are denoted by $\mathbf{r}_C$, $\mathbf{v}_C$, and $\mathbf{a}_C$.

The *absolute angular momentum* of the system of particles is defined with respect to point $O$ of the $XYZ$ frame using Eq. 17-22,

$$\mathbf{H}_O = \sum_{i=1}^{n} (\mathbf{r}_i \times m_i\mathbf{v}_i)$$

The angular momentum of the same system with respect to its own mass center $C$ is

$$\mathbf{H}_C = \sum_{i=1}^{n} (\mathbf{r}_i' \times m_i\mathbf{v}_i) \qquad \boxed{17\text{-}25}$$

The same velocity $\mathbf{v}_i$ is used in Eqs. 17-22 and 17-25 because it is the same linear momentum $m_i\mathbf{v}_i$ that is being considered with two different position vectors to obtain the angular momenta in the two systems. From the expression of relative velocities,

$$\mathbf{v}_i = \mathbf{v}_C + \mathbf{v}_i'$$

Substituting for $\mathbf{v}_i$ in Eq. 17-25 gives

$$\mathbf{H}_C = \sum_{i=1}^{n} \left[\mathbf{r}_i' \times m_i(\mathbf{v}_C + \mathbf{v}_i')\right]$$

$$= \sum_{i=1}^{n} (\mathbf{r}_i' \times m_i\mathbf{v}_C) + \sum_{i=1}^{n} (\mathbf{r}_i' \times m_i\mathbf{v}_i') \qquad \boxed{17\text{-}26}$$

This equation can be simplified if it is recognized that the first term on the right side can be written as

$$\left(\sum_{i=1}^{n} \mathbf{r}_i' m_i\right) \times \mathbf{v}_C$$

since $\mathbf{v}_C$ is the same in each term of the summation and thus it can be factored out, and $m_i$ is constant for each particle. Note that $\mathbf{v}_i'$ cannot be factored out since it is the velocity of only one particle in the most general case. By definition of the center of mass, $\sum_{i=1}^{n} \mathbf{r}_i' m_i = 0$ (this is the location of the center of mass with respect to itself). Thus,

$$\mathbf{H}_C = \sum_{i=1}^{n} (\mathbf{r}_i' \times m_i\mathbf{v}_i') \qquad \boxed{17\text{-}27}$$

The last result also could be obtained if the *xyz* system were not moving and Eq. 17-22 were applied directly with respect to the center of mass. It is concluded that Eqs. 17-25 and 17-27 are the same. This means that *the angular momentum of a system of particles about its center of mass is the same whether it is observed from a newtonian reference frame or from the centroidal frame which may be translating but not rotating.* This result is not true for points other than the mass center.

The rate of change of angular momentum of a system of particles about its mass center depends on the moments of the external forces about the same point. This is shown by differentiating Eq. 17-27 with respect to time,

$$\dot{\mathbf{H}}_C = \sum_{i=1}^{n} (\dot{\mathbf{r}}_i' \times m_i \mathbf{v}_i') + \sum_{i=1}^{n} (\mathbf{r}_i' \times m_i \dot{\mathbf{v}}_i') \qquad \boxed{17\text{-}28}$$

With $\mathbf{v}_i' = \dot{\mathbf{r}}_i'$, the first term on the right side is zero since the cross product of two collinear vectors is zero. In the second term, $\dot{\mathbf{v}}_i' = \mathbf{a}_i'$. From the expression of relative accelerations,

$$\mathbf{a}_i = \mathbf{a}_C + \mathbf{a}_i'$$

The second term on the right side of Eq. 17-28 is expanded as

$$\sum_{i=1}^{n} [\mathbf{r}_i' \times m_i (\mathbf{a}_i - \mathbf{a}_C)]$$

$$= \sum_{i=1}^{n} (\mathbf{r}_i' \times m_i \mathbf{a}_i) - \sum_{i=1}^{n} (\mathbf{r}_i' \times m_i \mathbf{a}_C)$$

where

$$\sum_{i=1}^{n} (\mathbf{r}_i' \times m_i \mathbf{a}_i) = \sum_{i=1}^{n} (\mathbf{r}_i' \times \mathbf{F}_i) = \sum_{i=1}^{n} \mathbf{M}_{iC}$$

from Eq. 17-4. Also,

$$\sum_{i=1}^{n} (\mathbf{r}_i' \times m_i \mathbf{a}_C) = \left( \sum_{i=1}^{n} \mathbf{r}_i' m_i \right) \times \mathbf{a}_C = 0$$

as was done with $\mathbf{v}_C$ in deriving Eq. 17-27. The resulting equation in the same concise form as that of Eq. 17-24 is

$$\boxed{\sum \mathbf{M}_C = \dot{\mathbf{H}}_C = \sum (\mathbf{r}_i' \times m_i \mathbf{a}_i) = \sum (\mathbf{r}_i' \times m_i \mathbf{a}_i')} \qquad \boxed{17\text{-}29}$$

This means that *the resultant of the moments of the external forces about the mass center of a system of particles equals the time rate of change of the angular momentum about the mass center.* It is assumed that the total mass of the system is constant. Note that Eqs. 17-24 and 17-29 are similar in form, and the only difference between them is the point of reference. Both of these equations are useful in dynamics analysis. Remember in using Eq. 17-24 or 17-29 that moments may be taken only about a fixed point or the center of mass. If another point is used for reference, the equation of motion for rotation is in a more complicated form.

## Useful Expressions

1. The angular momenta determined with respect to a fixed point $O$ and the center of mass $C$ are related as follows. From Eq. 17-22 and Fig. 17-14,

$$\mathbf{H}_O = \sum (\mathbf{r}_i \times m_i\mathbf{v}_i) = \sum (\mathbf{r}_C + \mathbf{r}'_i) \times m_i\mathbf{v}_i$$

$$= \sum (\mathbf{r}_C \times m_i\mathbf{v}_i) + \sum (\mathbf{r}'_i \times m_i\mathbf{v}_i)$$

where $\mathbf{r}_C$ is the same for considering any particle of the system, and $\sum m_i\mathbf{v}_i = m\mathbf{v}_C$ from Eq. 17-6. The last term is $\mathbf{H}_C$ from Eq. 17-25. Thus,

$$\boxed{\mathbf{H}_O = \mathbf{H}_C + \mathbf{r}_C \times m\mathbf{v}_C} \qquad \boxed{17\text{-}30}$$

2. The moments about a fixed point $O$ are related to the time rate of change of angular momentum with respect to the center of mass $C$ as follows. From Eqs. 17-24 and 17-30,

$$\sum \mathbf{M}_O = \dot{\mathbf{H}}_O = \frac{d}{dt}(\mathbf{H}_C + \mathbf{r}_C \times m\mathbf{v}_C)$$

$$= \dot{\mathbf{H}}_C + \dot{\mathbf{r}}_C \times m\mathbf{v}_C + \mathbf{r}_C \times m\dot{\mathbf{v}}_C$$

Since $\dot{\mathbf{r}}_C = \mathbf{v}_C, \dot{\mathbf{r}}_C \times m\mathbf{v}_C = 0$. Writing $\dot{\mathbf{v}}_C = \mathbf{a}_C$ for the acceleration of the center of mass gives

$$\boxed{\sum \mathbf{M}_O = \dot{\mathbf{H}}_C + \mathbf{r}_C \times m\mathbf{a}_C} \qquad \boxed{17\text{-}31}$$

## 17-9 CONSERVATION OF MOMENTUM OF A SYSTEM OF PARTICLES

During any period of time when the resultant of the external forces acting on a system of particles is zero, $\dot{\mathbf{G}} = 0$ in Eq. 17-21. In that case,

$$\boxed{\mathbf{G} = \text{constant}} \qquad \boxed{17\text{-}32}$$

which means that the total linear momentum of the system is conserved. This does not mean that the linear momentum of any particle of the system must be conserved. For example, some particles could lose momentum while others gain an identical amount, thus leaving the total unchanged.

Similarly, the resultant moment of all external forces acting on a system of particles may be zero, either in reference to a fixed point $O$ or to the mass center $C$. In the first case, $\dot{\mathbf{H}}_O = 0$ from Eq. 17-24, and

$$\boxed{\mathbf{H}_O = \text{constant}} \qquad \boxed{17\text{-}33}$$

In the second case, $\dot{\mathbf{H}}_C = 0$ from Eq. 17-29, and

$$\boxed{\mathbf{H}_C = \text{constant}} \qquad \boxed{17\text{-}34}$$

Equations 17-33 and 17-34 express the conservation of angular momentum. Note that the two constants may be different. Furthermore, either $\mathbf{H}_O$ or $\mathbf{H}_C$ may be constant while the other one is not constant.

Equations 17-21 and 17-24 are the fundamental expressions relating forces and moments to the time rates of change of linear and angular momenta for systems of particles. However, these equations are not the best to analyze impulsive motion when the relationship of force and time is known. The concept of impulse which was developed for individual particles is also valid for systems of particles.

The time rate of change of linear momentum is expressed from Eq. 17-21 as

$$\sum \mathbf{F} = \frac{d}{dt} \mathbf{G}$$

This is integrated between time $t_1$ and $t_2$ during which the forces act and change the momentum of the system of particles from $\mathbf{G}_1$ to $\mathbf{G}_2$.

$$\boxed{\sum_{i=1}^{n} \int_{t_1}^{t_2} \mathbf{F}_i \, dt = \mathbf{G}_2 - \mathbf{G}_1 = m\mathbf{v}_{C_2} - m\mathbf{v}_{C_1}} \qquad \boxed{17\text{-}35}$$

where the integral is the linear impulse of an external force $F_i$ on a particle $P_i$ of the system. The summation of these impulses over all particles equals the change in total linear momentum of the system of $n$ particles.

A similar procedure is followed to express the *angular impulse* using Eqs. 17-24 and 17-22,

$$\boxed{\begin{aligned} \sum \mathbf{M}_O &= \frac{d}{dt} \mathbf{H}_O \\ \sum_{i=1}^{n} \int_{t_1}^{t_2} \mathbf{M}_{i_O} \, dt &= \mathbf{H}_{O_2} - \mathbf{H}_{O_1} \\ &= \sum_{i=1}^{n} (\mathbf{r}_i \times m_i \mathbf{v}_i)_2 - \sum_{i=1}^{n} (\mathbf{r}_i \times m_i \mathbf{v}_i)_1 \end{aligned}} \qquad \boxed{17\text{-}36}$$

where the integral is the *angular impulse of the moment* $\mathbf{M}_{i_O} = \mathbf{r}_i \times \mathbf{F}_i$ *on a particle $P_i$ of the system*. The summation of these impulses over all particles equals the change in total angular momentum of the system of $n$ particles about the fixed reference point $O$. A similar expression can be written for the angular momentum about the center of mass.

EXAMPLE 17-6

A 1000-kg space vehicle is moving at a velocity $\mathbf{v}_0 = 2000\mathbf{k}$ km/h when spring-loaded mechanisms simultaneously eject two small probes in a plane normal to the parent vehicle's original velocity. The masses of the probes and their directions of motion are shown in Fig. 17-15. Determine the velocities of the three parts after the separation if $\mathbf{v}_f = (-0.56\mathbf{i} - 0.3\mathbf{j} + 0.77\mathbf{k})v_f$. In other words, the direction of the parent vehicle's final velocity is given, but its magnitude $v_f$ and the magnitudes $v_1$ and $v_2$ are unknown.

SOLUTION

Since there are no external forces acting on this system of particles, the total linear momentum of the system is conserved,

$$\mathbf{G} = m_0\mathbf{v}_0 = m\mathbf{v}_f + m_1\mathbf{v}_1 + m_2\mathbf{v}_2$$

where $m_0$ = total mass of the system = $m + m_1 + m_2 = 1000$ kg
  $m$ = mass of parent vehicle = 650 kg
  $\mathbf{v}_0$ = original velocity of space vehicle
  $\mathbf{v}_f$ = velocity of parent vehicle after firing

Substituting the given quantities,

$$(1000\,\text{kg})(2000\,\text{km/h})\,\mathbf{k} = (650\,\text{kg})(-0.56\mathbf{i} - 0.3\mathbf{j} + 0.77\mathbf{k})v_f$$
$$+ (200\,\text{kg})(\cos 45°\,\mathbf{i} + \sin 45°\,\mathbf{j})v_1$$
$$+ (150\,\text{kg})\,v_2\mathbf{i}$$

Equating vector components gives

x direction: $\quad 0 = (650)(-0.56)v_f + (200)(\cos 45°)v_1 + (150)v_2$

y direction: $\quad 0 = (650)(-0.3)v_f + (200)(\sin 45°)v_1$

z direction: $\quad (1000)(2000) = (650)(0.77)v_f$

$$v_f = \frac{(1000)(2000)}{(650)(0.77)} = 4000\,\text{km/h}$$

$$v_1 = \frac{(650)(0.3)(4000)}{(200)(\sin 45°)} = 5515\,\text{km/h}$$

$$v_2 = \frac{(650)(0.56)(4000) - (200)(\cos 45°)(5515)}{150} = 4507\,\text{km/h}$$

and

$$\mathbf{v}_f = (-2240\mathbf{i} - 1200\mathbf{j} + 3080\mathbf{k})\,\text{km/h}$$
$$\mathbf{v}_1 = (3900\mathbf{i} + 3900\mathbf{j})\,\text{km/h}$$
$$\mathbf{v}_2 = (4507\mathbf{i})\,\text{km/h}$$

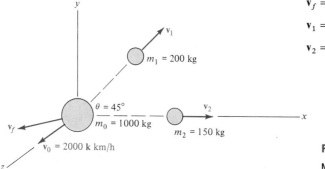

**FIGURE 17-15**

Motions of a spacecraft and its separated parts

EXAMPLE 17-7

Each carriage of an amusement ride (Fig. 17-16a) weighs 1200 lb (when loaded with its passengers) and is a distance of 10 ft from the carriage frame's mass center $C$. The carriage frame has a constant angular speed of $\omega_r = 3$ rad/s *relative* to the main frame $OC$ which rotates at a constant angular speed of $\omega_0 = 1$ rad/s, both counterclockwise as shown. Neglect the mass of the structural frame and approximate the carriages as particles. Determine **(a)** the absolute angular momentum $\mathbf{H}_C$ about mass center $C$, and **(b)** the angular momentum $\mathbf{H}_O$ about fixed point $O$.

## SOLUTION

**(a)** According to the definition of absolute angular momentum $\mathbf{H}_C$ given by Eq. 17-27, one has a choice of working either with relative velocities $\mathbf{v}'_i$ or absolute velocities $\mathbf{v}_i$. For this problem the relative velocities $\mathbf{v}'_i$ of each carriage relative to mass center $C$ are easier to describe because each carriage simply has circular motion about $C$. So use

$$\mathbf{H}_C = \sum_{i=1}^{n} (\mathbf{r}'_i \times m_i \mathbf{v}'_i)$$

For a typical mass $m_i$ at $P$, in Fig. 17-16b,

$$\mathbf{r}'_i = 10\mathbf{n}_r \text{ ft}$$

$$\mathbf{v}'_i = \mathbf{v}_i - \mathbf{v}_C = \mathbf{v}_{P/C} = (10\text{ ft})(1 \text{ rad/s} + 3 \text{ rad/s})\,\mathbf{n}_\theta$$
$$= 40\,\mathbf{n}_\theta \text{ ft/s}$$

$$\mathbf{r}'_1 \times m_i \mathbf{v}'_i = \frac{1200\,\text{lb}}{32.2\,\text{ft/s}^2}\,(10\mathbf{n}_r \text{ ft} \times 40\mathbf{n}_\theta \text{ ft/s})$$
$$= 14{,}907\mathbf{k}\,\text{ft} \cdot \text{lb} \cdot \text{s}$$

Note that $\mathbf{r}'_1 \times m_i \mathbf{v}'_i$ is the same for all carriages, so the total angular momentum is

$$\mathbf{H}_C = 6(14{,}907\mathbf{k}\,\text{ft} \cdot \text{lb} \cdot \text{s})$$

$$\mathbf{H}_C = 8.94 \times 10^4 \mathbf{k}\,\text{ft} \cdot \text{lb} \cdot \text{s}$$

**(b)** From Eq. 17-30, $\mathbf{H}_O = \mathbf{H}_C + \mathbf{r}_C \times m\mathbf{v}_C$. From Fig. 17-16b, $\mathbf{r}_C = \mathbf{R} = 16\mathbf{n}_R$ ft and

$$m\mathbf{v}_C = \frac{6(1200\,\text{lb})}{32.2\,\text{ft/s}^2}\,(16\,\text{ft})(1 \text{ rad/s})\,\mathbf{n}_\phi$$
$$= 3{,}580\mathbf{n}_\phi\,\text{lb} \cdot \text{s} \qquad \text{(always normal to } OC \text{ in the direction of motion)}$$

Using the results for $\mathbf{H}_C$ from part (a),

$$\mathbf{H}_O = 8.94 \times 10^4\,\mathbf{k}\,\text{ft} \cdot \text{lb} \cdot \text{s} + (16\mathbf{n}_R\,\text{ft}) \times (3580\mathbf{n}_\phi\,\text{lb} \cdot \text{s})$$
$$= 1.467 \times 10^5\mathbf{k}\,\text{ft} \cdot \text{lb} \cdot \text{s}$$

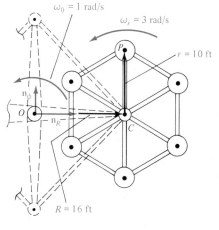

**FIGURE 17-16**

Model of an amusement ride

EXAMPLE 17-7    679

# EXAMPLE 17-8

Referring to Ex. 17-7, assume that it is desired to apply a constant external resultant force $\mathbf{F}_C$ perpendicular to line $OC$ of the carriage frame at $C$ and a constant torque about $C$ to bring the carriage frame up to its steady operating speed in 20 s. Determine **(a)** the required resultant force $\mathbf{F}_C$, and **(b)** the resultant torque $\mathbf{M}_C$.

## SOLUTION

**(a)** At steady-state operation, the linear momentum of the carriage frame is $\mathbf{G} = m\mathbf{v}_C = 3{,}580\mathbf{n}_\phi$ lb·s, as determined in Ex. 17-7(b). Applying the impulse–momentum principle given by Eq. 17-35,

$$\sum_{i=1}^{n} \int_{t_1}^{t_2} \mathbf{F}_i \, dt = \int_{t_1}^{t_2} \mathbf{F}_C \, dt = \mathbf{G}_2 - \mathbf{G}_1$$

$$\int_0^{20} \mathbf{F}_C \, dt = \mathbf{F}_C \int_0^{20} dt = 20\mathbf{F}_C = 3580\mathbf{n}_\phi - 0$$

$$\mathbf{F}_C = 179\mathbf{n}_\phi \, \text{lb} \qquad \text{(always normal to } OC\text{)}$$

**(b)** At steady-state operation, the absolute angular momentum of the carriage frame about its mass center $C$ is given in Ex. 17-7 as $\mathbf{H}_C = 8.944 \times 10^4 \mathbf{k}$ ft·lb·s. Applying the angular impulse–momentum principle given by the second of Eqs. 17-36,

$$\sum_{i=1}^{n} \int_{t_1}^{t_2} \mathbf{M}_{i_C} \, dt = \int_{t_1}^{t_2} \mathbf{M}_C \, dt = \mathbf{H}_2 - \mathbf{H}_1$$

$$\int_0^{20} \mathbf{M}_C \, dt = \mathbf{M}_C \int_0^{20} dt = 20\mathbf{M}_C = 8.944 \times 10^4 \mathbf{k} - 0$$

$$\mathbf{M}_C = 4470\mathbf{k} \, \text{lb·ft}$$

Note that the magnitudes of both $\mathbf{F}_C$ and $\mathbf{M}_C$ are inversely proportional to the required time for bringing the system up to speed or stopping it (for constant values of $\mathbf{F}_C$ and $\mathbf{M}_C$).

**17–24** Two particles of mass $m_1 = 2\,\text{kg}$ and $m_2 = 5\,\text{kg}$ move in the fixed $XY$ frame. Calculate the total angular momentum about point $O$ and about the center of mass if $\mathbf{r}_1 = (2\mathbf{i} + \mathbf{j})\,\text{m}$, $\mathbf{v}_1 = (0.4\mathbf{i} + 6\mathbf{j})\,\text{m/s}$, $r_2 = (\mathbf{i} + 2.5\mathbf{j})\,\text{m}$, and $\mathbf{v}_2 = (4\mathbf{i} + 3\mathbf{j})\,\text{m/s}$.

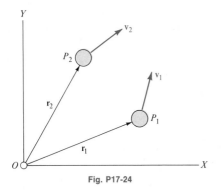

Fig. P17-24

**17–25** Two particles of weight $W_1 = 3\,\text{lb}$ and $W_2 = 4\,\text{lb}$ move in the fixed $XY$ frame as shown in Fig. P17-24. Calculate the total angular momentum about point $O$ and about the center of mass if $\mathbf{r}_1 = (5\mathbf{i} + 2\mathbf{j})\,\text{ft}$, $\mathbf{v}_1 = (30\mathbf{i} + 30\mathbf{j})\,\text{ft/s}$, $\mathbf{r}_2 = (3\mathbf{i} + 6\mathbf{j})\,\text{ft}$, and $\mathbf{v}_2 = (-2\mathbf{i} + 10\mathbf{j})\,\text{ft/s}$.

**17–26** Two particles of weight $W_1 = 2\,\text{lb}$ and $W_2 = 5\,\text{lb}$ move in the vertical $XY$ frame. (a) Calculate the total linear momentum and its time rate of change. (b) Calculate the total angular momentum and its time rate of change about point $O$ and about the center of mass. $\mathbf{r}_1 = (5\mathbf{i} + 8\mathbf{j})\,\text{ft}$, $\mathbf{v}_1 = (20\mathbf{i} + 25\mathbf{j})\,\text{ft/s}$, $\mathbf{F}_1 = (100\mathbf{i} - 150\mathbf{j})\,\text{lb}$, $\mathbf{r}_2 = (-3\mathbf{i} + 7\mathbf{j})\,\text{ft}$, $\mathbf{v}_2 = 0$, and $\mathbf{F}_2 = (-100\mathbf{i} + 150\mathbf{j})\,\text{lb}$. The particles are located near the surface of the earth.

Fig. P17-26

**17–27** Two particles of mass $m_1 = 3\,\text{kg}$ and $m_2 = 1\,\text{kg}$ move in the fixed $XY$ frame as shown in Fig. P17-26. (a) Determine the total linear momentum and its time rate of change. (b) Calculate the total angular momentum and its time rate of change about point O and about the center of mass if $\mathbf{r}_1 = (2\mathbf{i} + 3\mathbf{j})\,\text{m}$, $\mathbf{v}_1 = (4\mathbf{i} + 5\mathbf{j})\,\text{m/s}$, $\mathbf{F}_1 = (-20\mathbf{i} + 30\mathbf{j})\,\text{N}$, $\mathbf{r}_2 = (\mathbf{i} + 10\mathbf{j})\,\text{m}$, $\mathbf{v}_2 = (6\mathbf{i} + 2\mathbf{j})\,\text{m/s}$, $\mathbf{F}_2 = (10\mathbf{i} + 25\mathbf{j})$.

**17–28** The masses, positions, and velocities of three particles are given as $m_1 = 80\,\text{kg}$, $m_2 = 70\,\text{kg}$, $m_3 = 100\,\text{kg}$. $\mathbf{r}_1 = (4\mathbf{i} + 3\mathbf{j})\,\text{m}$, $\mathbf{r}_2 = (2\mathbf{i} + \mathbf{j} + 3\mathbf{k})\,\text{m}$, $\mathbf{r}_3 = (3\mathbf{j} + 5\mathbf{k})\,\text{m}$, $\mathbf{v}_1 = (2\mathbf{j} - 0.5\mathbf{k})\,\text{m/s}$, $\mathbf{v}_2 = (0.5\mathbf{i} + 2\mathbf{j} + \mathbf{k})\,\text{m/s}$, and $\mathbf{v}_3 = (5\mathbf{i} + \mathbf{j})\,\text{m/s}$. Determine the angular momentum of the system about its center of mass and about point $O$.

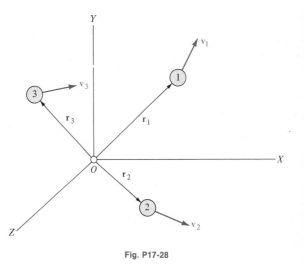

Fig. P17-28

**17–29** The three particles of Prob. 17–28 are located near the surface of the earth. Calculate the time rate of change of the angular momentum $\mathbf{H}_O$ of the system assuming that $Y$ is a vertical axis.

**17–30** A turbine disk has 50 identical 0.15-kg blades of negligible dimensions. Assume that the disk itself has no mass and is rotating at 5000 rpm. Assume that one blade breaks off. Determine the total angular momentum $\mathbf{H}_O$ of the 50 blades just before and just after the fracture of the blade.

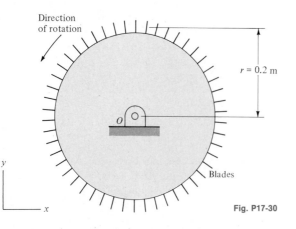

Fig. P17-30

**17–31** Determine the required average couple $\mathbf{M}_O$ to stop the rotation of the turbine disk described in Prob. 17–30. The braking of the intact disk is to be done in a time of 10 s. What is the time of stopping the disk after the fracture if the same couple $\mathbf{M}_O$ is applied? Assume that a constant couple of magnitude $M_W = 10$ N·m caused by the displaced weight opposes $\mathbf{M}_O$.

**17–32** A small airplane is flying horizontally at $v_0 = 120$ mph when two parachutists jump out horizontally. Parachutist $A$ weighs 180 lb and pushes against the airplane with a 200-lb force applied for 0.3 s. Parachutist $B$ weighs 190 lb and jumps shortly after $A$, pushing with a 150-lb force for 0.2 s. Determine the final linear momentum of the airplane, which weighs 6500 lb without the two parachutists. What would be the difference in the final linear momentum if the parachutists jump simultaneously, applying their respective forces for 0.3 s?

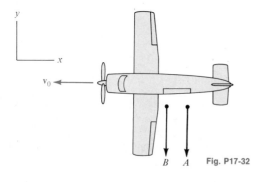

Fig. P17-32

**17–33** Coordinates $xyz$ are fixed at point $A$ in a large space vehicle which has a translational velocity $\mathbf{v}_A = (10\,000\mathbf{i} + 2000\mathbf{j})$ km/h with respect to the fixed $XYZ$ frame. Three small probes of mass $m_1 = 200$ kg, $m_2 = 300$ kg, and $m_3 = 250$ kg are moving relative to the $xyz$ frame with velocities $\mathbf{v}_{1/A} = (500\mathbf{i} + 100\mathbf{j} - 50\mathbf{k})$ km/h, $\mathbf{v}_{2/A} = (600\mathbf{i} + 100\mathbf{k})$ km/h, and $\mathbf{v}_{3/A} = (700\mathbf{i} - 100\mathbf{j})$ km/h. (a) Calculate the total linear momentum of the three probes with respect to the moving and the fixed frames. (b) Determine the total angular momenta $\mathbf{H}_A$ and $\mathbf{H}_O$ of the three probes. The position vectors at the instant considered are $\mathbf{r}_{1/A} = (3\mathbf{i} - 100\mathbf{j} + 8\mathbf{k})$ m, $\mathbf{r}_{2/A} = (-400\mathbf{i} + 2\mathbf{k})$ m, $\mathbf{r}_{3/A} = -600$ m$\mathbf{k}$, and $\mathbf{r}_A = (1800\mathbf{i} + 100\mathbf{k})$ m.

**17–34** Generalize from Prob. 17–33 and derive a simple equation to express the total linear momentum $\mathbf{G}_O$ (in the fixed system) of any number of particles with total mass $m$. Write the equation in terms of the total momentum $\mathbf{G}_A$ (in the moving frame), the velocity $\mathbf{v}_A$ of the moving frame, and the total mass $m$.

**17–35** An amusement ride consists of 12 cages (2 are shown) of mass $m = 500$ kg with riders. The cages are on rigid arms at distance $R = 5$ m from points $B$. The arms are pivoted at points $B$ on a rigid horizontal platform of $r = 4$ m which rotates about the vertical axis at 10 rpm. The angle $\theta$ of each arm is controlled by a hydraulic actuator $A$. Calculate the total angular momentum $\mathbf{H}_O$ if $\theta = 20°$ for every arm.

Fig. P17-35

**17–36** Determine the required couple $\mathbf{M}_C$ about the center of mass of the 12 cages in Prob. 17–35 if a 50% increase in the angular speed $\omega_Y$ must be made in a time of 10 s. Neglect the mass of the supporting structure.

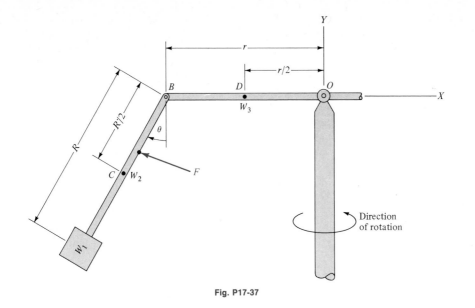

Fig. P17-37

**17-37** Part of an amusement ride is modeled as a particle of weight $W_1 = 1000$ lb on a rigid arm which is pivoted at $B$ to rotate in a vertical plane. This arm is supported by a rigid member $OB$ which rotates about the vertical axis at 12 rpm. For an approximation, the two rigid members are assumed to be particles of $W_2 = 150$ lb at point $C$, and $W_3 = 200$ lb at point $D$. The angle $\theta$ is controlled by force $\mathbf{F}$. Calculate the magnitude of the angular momentum $\mathbf{H}_O$ of the three particles if $R = 16$ ft, $r = 12$ ft, and $\theta = 40°$.

**17-38** Assume that a force $\mathbf{F} = (-500\mathbf{i} + 100\mathbf{j})$ lb is acting on the arm shown in Prob. 17–37. Determine the time rate of change of angular momentum about point $O$ at that instant if $\omega_Y$ is constant and $\mathbf{F}$ acts 5 ft from point $B$.

**17-39** An amusement ride is designed to rotate arm $AOB$ about point $O$ in the horizontal plane at 5 rpm. The two cages with two riders in each are to rotate at 6 rpm relative to the arm $AOB$. The average rider has a 70-kg mass and sits at $r = 1$ m from the center of the cage. Calculate the total angular momentum of four riders with respect to point $O$ for the position shown.

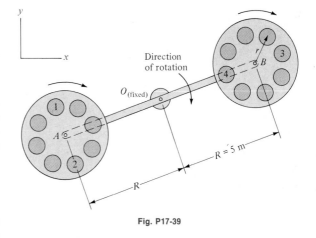

Fig. P17-39

**17-40** Assume that all rotations of the system described in Prob. 17–39 must be stopped in a time of 0.05 s. Determine the required braking couples $\mathbf{M}_A$, $\mathbf{M}_B$, and $\mathbf{M}_O$ to stop the motion of the riders.

**17–41** An amusement ride moves up to eight 200-lb riders on a horizontal track with an angular speed of 7 rpm about the vertical $Y$ axis. The weight of the rigid arm and cage, $W = 1000$ lb, is assumed to be concentrated at $R/2$. Once in each revolution a hydraulic actuator applies a vertical force of magnitude $F = 2500$ lb for a time of 1 s to the cage. Determine the magnitude of the maximum angular momentum $\mathbf{H}_O$ of the system if the arm is free to move up the track. $R = 12$ ft, and $r = 5$ ft.

**17–43** In Prob. 17–42, $m = 0.03$ kg, $R = 0.15$ m, and $v = 4$ m/s. Assume that a force of magnitude $F = 100$ N is applied tangentially to the chain and evenly around the edge for a time of 5 s. Determine the new speed.

**17–44** A Ferris wheel has three evenly spaced cages which weigh 1000 lb with riders. Calculate the total angular momentum of the cages if $\omega_A = 0.2$ rad/s and $\omega_Y = 0.15$ rad/s for the position shown. Consider the motion of the cages as a superposition of circular motion in the vertical plane about point $A$ and circular motion about the $Y$ axis at the instant shown.

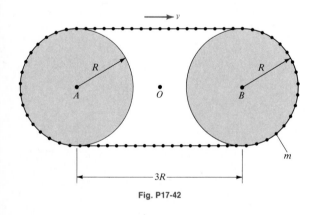

Top view of cage

Side view

**Fig. P17-41**

**17–42** The chain-drive mechanism has 100 identical links of mass $m$. Determine the total angular momentum of the chain with respect to point $O$.

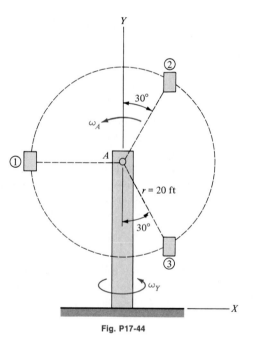

**Fig. P17-44**

**Fig. P17-42**

**17–45** Consider the rotating system in Prob. 17–44. Moments $\mathbf{M}_A = -12{,}000\mathbf{k}$ lb·ft and $\mathbf{M}_Y = -5000\mathbf{j}$ lb·ft are applied for a time of 0.03 s. Determine the new $\omega_A$ and $\omega_Y$, assuming that the system is still in the same position shown.

The flow of fluids is readily analyzed at the elementary level using methods developed for particles. Two categories of fluid flow are distinguished. *Steady flow of mass* means that either no mass enters or leaves the system, or that the mass entering the system equals the mass leaving it in the same period of time. Typical examples of this are the flow of fluids in closed pipelines. *Variable mass* means that the system gains or loses mass. Rockets are examples of variable-mass systems.

The fluids involved in either steady-flow or variable-mass systems may be compressible or incompressible. Strictly speaking, all fluids are *compressible*, which means that they change volume without changing mass under compressive forces. However, liquids are much less compressible than gases, so they are considered *incompressible* in many engineering problems. Even gases may be assumed incompressible in some cases such as the flow of air around cars and low-speed aircraft (up to speeds of about 120 m/s).

The method of analyzing any flow of mass is based on the concept of free-body diagrams. A system of interest, which is commonly called the *control volume*, is imagined to be isolated from its surroundings, and the masses entering and leaving the system are accounted for. The general case for steady flow of mass is illustrated in Fig. 17-17. Fluid flows at velocity $v_1$ through area $A_1$ into the control volume, while fluid leaves the control volume at velocity $v_2$ through area $A_2$. A net force $\sum F$ acts on the fluid as the momentum of the fluid changes in passing through the control volume. The flow of mass is described in more detail as follows.

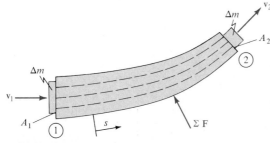

**FIGURE 17-17**

Control volume of a flowing fluid. Velocity and linear momentum analysis

## Velocity Change of the Fluid

A useful approximation in the analysis of the flow is that at any cross section $A$ all particles have the same velocity vector $\mathbf{v}$ ($\mathbf{v}$ is actually the average velocity of a nonuniform velocity distribution). Considering Fig. 17-17, assume that mass $m_1$ is entering the control volume at area $A_1$ in a unit of time. The exiting mass in the same time interval is $m_2 = m_1$ since mass is neither gained nor lost in steady flow. The density (mass per unit volume) of the fluid is denoted by $\rho$, and distance in the direction of flow is defined as the quantity $s$. An element of mass $\Delta m$ moving through the control volume can be expressed as $\Delta m = m_1 = m_2$,

$$\Delta m = \rho_1 A_1 \Delta s_1 = \rho_2 A_2 \Delta s_2$$

Dividing by the time interval $\Delta t$, letting $\Delta t$ approach zero at the limit, and substituting the speed $v$ for $ds/dt$ gives

$$\frac{dm}{dt} = \rho_1 A_1 v_1 = \rho_2 A_2 v_2 \qquad \boxed{17\text{-}37}$$

This is simplified in the case of incompressible fluids where $\rho_1 = \rho_2 = \rho$,

$$\boxed{\frac{1}{\rho}\frac{dm}{dt} = A_1 v_1 = A_2 v_2} \qquad \boxed{17\text{-}38}$$

The result $A_1 v_1 = A_2 v_2$ is useful in solving many common problems. The velocity vector is in the direction of the average flow at each cross section.

## Momentum Change of the Fluid

The change in linear momentum $\Delta \mathbf{G}$ between sections 1 and 2 in Fig. 17-17 is expressed

$$\Delta \mathbf{G} = \Delta m \mathbf{v}_2 - \Delta m \mathbf{v}_1 = \Delta m\, \Delta \mathbf{v} \qquad \boxed{17\text{-}39}$$

The force causing the change in momentum is determined by considering the interval of time $\Delta t$ during which an element of mass $\Delta m$ of velocity $\mathbf{v}_1$ enters at section 1 while an equal mass $\Delta m$ of velocity $\mathbf{v}_2$ exits at section 2. Dividing Eq. 17-39 by $\Delta t$ and letting $\Delta t$ approach zero at the limit yields

$$\dot{\mathbf{G}} = \frac{dm}{dt}\,\Delta \mathbf{v}$$

The time rate of change of momentum equals the resultant of all forces acting on the system of mass $m$ in a unit of time, so from Eq. 17-21,

$$\boxed{\sum \mathbf{F} = \frac{dm}{dt}\,\Delta \mathbf{v} = \frac{dm}{dt}(\mathbf{v}_2 - \mathbf{v}_1)} \qquad \boxed{17\text{-}40}$$

Equation 17-40 has consistent SI units with forces expressed in newtons, $dm/dt$ in kg/s, and the velocities in m/s. The corresponding units in the U.S. customary system are the lb, slug/s, and ft/s.

The methods described above can be extended to analyze the angular momentum of a steadily flowing mass. This is also illustrated using Fig. 17-18. Identical elements of mass $\Delta m$ are considered at sections 1 and

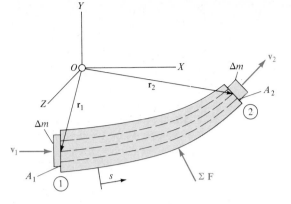

**FIGURE 17-18**

Control volume of a flowing fluid.
Angular momentum analysis

2, and the angular momenta of these are evaluated with respect to an arbitrary fixed point $O$. For a general flow,

$$\mathbf{H}_1 = \Delta m(\mathbf{r}_1 \times \mathbf{v}_1) \qquad \mathbf{H}_2 = \Delta m(\mathbf{r}_2 \times \mathbf{v}_2)$$

The net moment of the external forces about point $O$ equals the time rate of change of angular momentum. The equation of this is obtained in a way similar to that for linear momentum. Letting $\Delta t$ approach zero at the limit gives

$$\sum \mathbf{M}_O = \dot{\mathbf{H}}_O = \frac{d}{dt}(\mathbf{H}_2 - \mathbf{H}_1) = \frac{dm}{dt}(\mathbf{r}_2 \times \mathbf{v}_2 - \mathbf{r}_1 \times \mathbf{v}_1) \qquad \boxed{17\text{-}41}$$

where $dm/dt$ is the time rate of flow of mass through the system. Note that the angular momenta and the moments of the external forces are taken about the same point $O$.

## Pressures in Flowing Fluids

It should be remembered that in Eqs. 17-40 and 17-41 all forces acting on the chosen element of fluid must be considered. This includes the distributed forces caused by pressure within the fluid and the forces applied by solid objects in contact with the fluid. In some cases the pressure can be ignored if the element is small and the pressure is constant over the element. In the general case such as the free-body diagram in Fig. 17-19, the pressures $p_1$ and $p_2$ are not the same and they act on different areas. The forces caused by these must be evaluated by integrating the pressure over each appropriate area. With a constant pressure, force = (pressure) · (area).

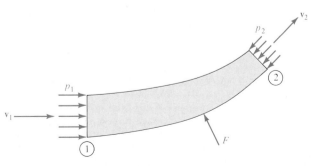

**FIGURE 17-19**

Pressure at two cross sections in a flowing fluid

## Common Applications of Steady Flow

The principles of steady flow of mass are applied in many areas of engineering work. These are presented with a few comments to help in the solution of simple problems.

**1. *Flow through pipes.*** The major forces acting on fluids are caused by changing cross sections of pipes and by changing directions of flow. In long segments of a pipe the wall friction has to be considered in evaluating the change in pressure in the fluid.

**2. *Flow through nozzles.*** The major changes in flow velocity and fluid pressure occur within the nozzle. The free stream away from the nozzle maintains its linear momentum in the absence of external forces.

**3. *Streams of fluid striking solid objects.*** A stream of fluid striking a solid object exerts a force on that object. As a result, the linear momentum of the free stream changes during the contact with the solid object.

**4. *Flow through propellers.*** Air or water moving through fans and propellers is accelerated from one side of the blades to the other side. For fans it is often assumed that the air has no initial velocity as it enters the region of the blades, as from the left in Fig. 17-20. At exit, the *slipstream* moves at a high velocity. The major external force on the stream of air is the thrust of the blades. The air is considered incompressible (with atmospheric pressure throughout) in these situations.

The flow of fluid through the propellers of airplanes and boats is similar to the flow through fans. The major difference is that the fluid has an initial relative velocity $v_0$ entering the region of the blades if the vehicle is in motion. The initial velocity is the velocity of the vehicle in the medium, and the velocity at exit is the velocity of the slipstream relative to the vehicle.

**5. *Flow through jet engines.*** Air entering a jet engine with a relative initial velocity $v_0$ is accelerated by the burning of fuel in the engine, as modeled in Fig. 17-21. This is similar to the flow of air through propellers

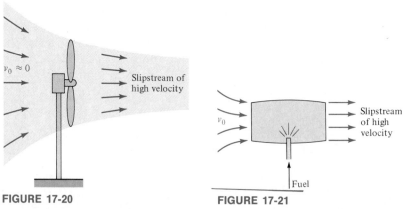

**FIGURE 17-20**

Fluid flow through a propeller

**FIGURE 17-21**

Airflow through a jet engine

and can be analyzed in three ways. As a first assumption, the relatively small mass of the fuel added to the stream of air may be neglected. This allows treating the jet engine as a propeller system. Second, the mass of the fuel can be considered by isolating the engine from the fuel tanks. In this case air and fuel are taken as entering the system at right angles to each other and leaving the system in a common stream of the same total mass. In the third method the engine and its fuel supply are considered as the system of interest. Of course, this system loses mass, and it must be analyzed using the method presented in Sec. 17-13.

## SYSTEMS WITH VARIABLE MASS 17-13

A system may continuously gain or lose mass. Examples are rockets and jet aircraft. The analysis of such systems is based on a model that gains mass as in Fig. 17-22.

Assume that the system of mass $m$ is moving with absolute velocity $\mathbf{v}$ at time $t$ and is about to overtake and absorb mass $\Delta m$, which is moving with absolute velocity $\mathbf{u}$ in the same direction. The original relative velocity of $\Delta m$ with respect to $m$ is $\mathbf{v}_{rel} = \mathbf{u} - \mathbf{v}$. The mass $m$ absorbs $\Delta m$ in time $\Delta t$ while external forces $\mathbf{F}$ may be acting on the system. The forces are included in the analysis for generality, but they exclude the mutual forces between $m$ and $\Delta m$ because these are internal to the new system of $m + \Delta m$. According to the principle of impulse and momentum, the final momentum equals the sum of the initial momentum and the impulse of the external forces,

$$(m + \Delta m)(\mathbf{v} + \Delta \mathbf{v}) = m\mathbf{v} + \Delta m\mathbf{u} + \sum \mathbf{F}\,\Delta t \qquad \boxed{17\text{-}42}$$

Expanding and rearranging gives

$$\sum \mathbf{F}\,\Delta t = m\,\Delta \mathbf{v} + \Delta m(\mathbf{v} - \mathbf{u}) + \Delta m\,(\Delta \mathbf{v})$$

**FIGURE 17-22**

Moving system gaining mass

This is simplified by substituting $-\mathbf{v}_{\text{rel}}$ for $\mathbf{v} - \mathbf{u}$ and neglecting the second-order term $\Delta m \, (\Delta \mathbf{v})$,

$$\sum \mathbf{F} \, \Delta t = m \, \Delta \mathbf{v} - \Delta m \mathbf{v}_{\text{rel}}$$

Dividing this equation by $\Delta t$ and letting $\Delta t$ approach zero at the limit yields

$$\sum \mathbf{F} = m \frac{d\mathbf{v}}{dt} - \frac{dm}{dt} \mathbf{v}_{\text{rel}} \qquad \boxed{\text{17-43}}$$

where $dm/dt$ is the time rate at which mass is absorbed by the original system of mass $m$.

Equation 17-43 is readily modified to obtain the governing equation for the motion of a system that loses mass. $dm/dt$ is negative in that case, and

$$\sum \mathbf{F} = m \frac{d\mathbf{v}}{dt} + \frac{dm}{dt} \mathbf{v}_{\text{rel}} \qquad \boxed{\text{17-44}}$$

It is interesting to compare Eqs. 17-43 and 17-44 with Newton's second law of motion for constant mass, $\sum \mathbf{F} = m(d\mathbf{v}/dt)$. The gain of mass is equivalent to an external force of magnitude $(v_{\text{rel}}) \, dm/dt$ that opposes the original motion of $m$ because $\mathbf{v}_{\text{rel}}$ and $\mathbf{v}$ are opposite in direction. Conversely, the expulsion of mass is equivalent to an external force in the direction of the velocity $\mathbf{v}$ of the system. This is the mechanism of propulsion by rockets.

## Special Cases of Systems with Variable Mass

**1. *All external forces are zero.*** A rocket propelling itself in space encounters no air resistance. If gravity acting on it is also negligible, $\sum \mathbf{F} = 0$, and Eq. 17-44 becomes

$$m \frac{d\mathbf{v}}{dt} = \frac{dm}{dt} \mathbf{v}_{\text{rel}} \qquad \boxed{\text{17-45}}$$

**2. *The system absorbs mass whose absolute initial velocity is zero.*** In this case $\mathbf{u} = 0$, $\mathbf{v}_{\text{rel}} = -\mathbf{v}$, so Newton's second law and Eq. 17-43 give the same result,

$$\sum \mathbf{F} = \frac{d}{dt}(m\mathbf{v}) = m \frac{d\mathbf{v}}{dt} + \frac{dm}{dt} \mathbf{v}$$

Newton's second law is not applicable to other systems that gain mass.

**3. *The system loses mass whose absolute final velocity is zero.*** In this case $\mathbf{u} = 0$, $\mathbf{v}_{\text{rel}} = \mathbf{v}$, so Newton's second law and Eq. 17-44 give the same result as in item 2 above. Newton's second law is not applicable to other systems that expel mass.

**4. *The system and the expelled mass move in the same direction.*** When the velocity $\mathbf{v}$ of the system is high, $v$ may be larger than $v_{\text{rel}}$, so $\mathbf{v}$ and $\mathbf{u}$ have the same direction.

**EXAMPLE 17-9**

Water flows into a horizontal pipe of 0.4 m diameter at the rate of 50 kg/s. The flow is diverted into a temporary pipe of 0.2 m diameter as shown in Fig. 17-23a. **(a)** Calculate the net horizontal and vertical forces acting on the water in the elbow region. **(b)** What are these forces after the temporary pipe is replaced by a 0.4-m-diameter pipe? The flow is in the horizontal $xy$ plane.

## SOLUTION

The control volume of the water is shown in Fig. 17-23b with a general unknown force **F** acting on it.

**(a)** The speed $v_1$ can be determined from Eq. 17-37,

$$\frac{dm}{dt} = A_1 v_1 \rho$$

$$v_1 = \frac{50 \text{ kg/s}}{(0.1257 \text{ m}^2)(1000 \text{ kg/m}^3)} = 0.4 \text{ m/s}$$

From Eq. 17-38,

$$v_2 = \frac{A_1}{A_2} v_1 = 1.6 \text{ m/s}$$

The unknown force is assumed to have positive rectangular components $F_x$ and $F_y$. From Eq. 17-40,

$$F_x = \frac{dm}{dt}(v_{2_x} - v_{1_x})$$

$$= (50 \text{ kg/s})(1.6 \cos 50° - 0.4) \text{ m/s} = 31.5 \text{ N}$$

$$F_y = \frac{dm}{dt}(v_{2_y} - v_{1_y})$$

$$= (50 \text{ kg/s})(1.6 \sin 50° - 0) \text{ m/s} = 61.3 \text{ N}$$

**(b)** If the pipes are the same,

$$A_2 = A_1 \qquad v_2 = v_1$$

$$F_x = (50 \text{ kg/s})(0.4 \cos 50° - 0.4) \text{ m/s} = -7.1 \text{ N}$$

$$F_y = (50 \text{ kg/s})(0.4 \sin 50° - 0) \text{ m/s} = 15.3 \text{ N}$$

The signs of the force components should be interpreted remembering that the unknown forces $F_x$ and $F_y$ were assumed to be acting in the positive $x$ and $y$ directions. The forces calculated are the resultants of forces from adjoining pipes, fixtures of the elbow joint, and pressure in the water.

Temporary, $d_2$ = 0.2 m
$A_2$ = 0.0314 m²
$v_2$
$\theta = 50°$
Final
$v_1$
$d_1$ = 0.4 m
$A_1$ = 0.1257 m²
$F$

(a)

$v_2$
$v_1$
$F$

(b)

**FIGURE 17-23**

Water flowing through a bent pipe

EXAMPLE 17-9   691

## EXAMPLE 17-10

A fixed nozzle of 0.05 m diameter emits a horizontal stream of water at the rate of 10 kg/s. The stream is diverted by a horizontally moving smooth vane whose absolute velocity is **u**. **(a)** Calculate the horizontal and vertical components of the force **F** that acts on the vane in this situation (Fig. 17-24a). **(b)** Determine the power transmitted to the vane.

### SOLUTION

The control volume of the water at the vane is drawn in Fig. 17-24b. **F** is the force of the vane on the water.

**(a)** The flow speed at the nozzle is obtained from Eq. 17-37,

$$v = \frac{dm/dt}{\rho A} = \frac{10 \text{ kg/s}}{1000 \text{ kg/m}^3 \, (\pi)(0.025 \text{ m})^2} = 5.1 \text{ m/s}$$

The relative speed of the water striking the vane at section 1 is

$$v_1 = v - u = \frac{v}{2}$$

and the water leaves the vane at section 2 without any change in the magnitude of the relative speed, so $v_2 = v_1$.

The forces are determined using Fig. 17-23b and Eq. 17-40, where the relative flow rate $dm_1/dt$ must be used rather than the flow rate at the nozzle. Thus,

$$F_x = \frac{dm_1}{dt}(v_{2_x} - v_{1_x}) = (\rho A v_1)(v_1 \cos 60° - v_1)$$

$$= (5 \text{ kg/s})(2.55 \text{ m/s})(\cos 60° - 1) = -6.4 \text{ N}$$

This is the horizontal force acting on the water. The same force in the opposite direction is acting on the vane. Similarly, for the $y$ direction,

$$F_y = \frac{dm_1}{dt}(v_{2_y} - v_{1_y}) = (\rho A v_1)(v_2 \sin 60° - 0)$$

$$= (5 \text{ kg/s})(2.55 \text{ m/s})(\sin 60°) = 11 \text{ N}$$

acting upward on the water and downward on the vane. The force of the water on the vane is

$$\mathbf{F} = (6.4\mathbf{i} - 11\mathbf{j}) \text{ N}$$

**(b)** Since the vane is moving only in the horizontal direction,

$$\text{power} = (\text{speed of vane})(F_x) = \frac{v}{2}F_x$$

$$= (2.55 \text{ m/s})(6.4 \text{ N}) = 16.3 \text{ W}$$

**FIGURE 17-24**

Water striking a moving vane

EXAMPLE 17-11

An escalator (Fig. 17-25) is designed to move people at a speed of 5 ft/s. **(a)** Assume that a group of people starts stepping on the empty escalator with a negligible relative speed. Calculate the magnitude of the required force to keep the escalator moving at its constant speed if weight is added on it at the rate of 200 lb/s. **(b)** What is the required force if the absolute initial velocity of each person is 3 ft/s parallel to the escalator? Neglect friction, the change in elevation of the people, and any changes in the direction of their velocity as they step on the escalator (i.e., their initial velocity is parallel to **v**).

## SOLUTION

The system is gaining mass, so Eq. 17-43 can be used.

**(a)** In this case $dv/dt = 0$ and $v_{rel} = 0$; therefore, the system maintains its initial speed without requiring any force (noting the assumptions).

**(b)** Here $v_{rel} = u - v = 3 - 5 = -2$ ft/s, while $dv/dt = 0$. From Eq. 17-43,

$$F = \frac{-dm}{dt} v_{rel} = -\left(\frac{200 \text{ lb/s}}{32.2 \text{ ft/s}^2}\right)(-2 \text{ ft/s})$$

$$= 12.4 \text{ lb}$$

**FIGURE 17-25**

People stepping on a moving escalator

EXAMPLE 17-12

A 5000-kg space vehicle is at zero velocity in empty space (no air, no gravity) when its ion-propulsion engine is started (Fig. 17-26). The engine emits mass at a constant rate of $dm/dt = 10^{-6}$ kg/s with a relative speed of 6000 km/h. Calculate the speeds of the vehicle after the engine has operated for 1 h and 100 h.

## SOLUTION

Since the system is losing mass and there are no external forces, Eq. 17-45 is applicable,

$$m\frac{dv}{dt} = -\frac{dm}{dt} v_{rel}$$

With $v_{rel}$ in the opposite direction as $v$,

$$(5000 \text{ kg})\left(\frac{dv}{dt}\right) = -(10^{-6} \text{ kg/s})(-6000 \text{ km/h})\left(\frac{3600 \text{ s}}{\text{h}}\right)$$

$$\frac{dv}{dt} = 0.00432 \text{ km/h}^2 = a$$

Because the space vehicle has zero initial velocity, the speed at time $t$ is

$$v_t = at$$

$$v_{1h} = (0.00432 \text{ km/h}^2)(1 \text{ h}) = 0.00432 \text{ km/h}$$

$$v_{100h} = (0.00432 \text{ km/h}^2)(100 \text{ h}) = 0.432 \text{ km/h}$$

Note that the total mass emitted in 100 h is only 0.0036 kg.

**FIGURE 17-26**

Propulsion of a spacecraft by ion emission

EXAMPLE 17-12    693

# PROBLEMS

**17–46** The jet engine on the test stand draws in air at the rate of 100 lb/s and expels it at a speed of 2000 ft/s. Calculate the thrust of the engine.

$v_0 \approx 0$      $v_1 = 2000$ ft/s

**Fig. P17-46**

**17–47** A small helicopter accelerates air (specific weight = 0.076 lb/ft³) to a maximum speed of 50 ft/s in its downward slipstream of 20 ft diameter. Determine the gross weight of the helicopter, which is hovering.

$v = 50$ ft/s

$d = 20$ ft

**Fig. P17-47**

**17–48** A fan accelerates air (density = 1.21 kg/m³) to a speed of 8 m/s in its slipstream. Calculate the force required to hold the fan.

$v_0 \approx 0$      $v_1 = 8$ m/s      $d = 0.6$ m

**Fig. P17-48**

**17–49** A research submarine should have a propeller with a downward thrust of 10 kN for floating at a given depth (the vehicle should rise to the surface if the motor fails). Assume that the proposed motor and blade can accelerate water (density = 1000 kg/m³) from zero speed to 6 m/s in the slipstream. Estimate the diameter of the propeller circle.

$v = 6$ m/s

**Fig. P17-49**

**17–50** The wind approaching a wind turbine is reduced from a speed of 30 ft/s to a speed of $v_2$ behind the blades. The radius of the stream of air influenced by the turbine is 3.27 ft in front of the blades and 5.66 ft behind the blades. (a) Determine the thrust, $\mathbf{F}_T$, acting on the blades of the wind turbine due to the slowing down of this column of air (specific weight = 0.076 lb/ft³). (b) What is the change in thrust if the approach wind speed is doubled?

$r_2 = 5.66$ ft      $r_1 = 3.27$ ft

$v_2$      $v_1 = 30$ ft/s

**Fig. P17-50**

**17–51** A stream of water is directed normal to a smooth, flat plate. The water flows away parallel to the plate. Calculate the required force to hold the plate.

Fig. P17-51

Fig. P17-54

**17-52** A jet engine draws in air at the rate of 70 kg/s and expels it at a speed of 700 m/s. The engine burns fuel at the rate of 1.2 kg/s. The fuel flows vertically to the engine at a speed of 5 m/s. Determine the $x$ and $y$ components of the force acting on the engine caused by the flowing fluids.

**17-55** Oil flows in a 1-m-diameter pipe at a speed of 2 m/s. The symmetric expansion pipe is in the horizontal plane. (a) Calculate the force caused by the flowing oil on the first elbow. (b) Is there a net force acting on the system of the four elbows? The density of the oil is 970 kg/m³.

Fig. P17-52

Fig. P17-55

**17-53** A stream of water flowing at the rate of 20 lb/s is deflected by a smooth, fixed vane. Determine the force required ($x$ and $y$ components) to hold the vane if $\theta = 60°$.

**17-56** Water flows through the horizontal pipe at the rate of 30 kg/s. Calculate the total horizontal force caused by the flowing water acting on flange 2 if $A_1 = 0.1\,\text{m}^2$, $A_2 = 0.06\,\text{m}^2$, $p_1 = 1\,\text{kPa}$, and $p_2 = 920\,\text{Pa}$.

Fig. P17-53

**17-54** A smooth, fixed blade diverts a stream of water (density = 1000 kg/m³) flowing at a speed $v_0 = 10$ m/s. Determine (a) the general expression for the force **F** required ($x$ and $y$ components) to hold the vane fixed, and (b) tabulate the magnitudes of **F** for $\theta = 45°$, $90°$, and $135°$.

Fig. P17-56

**17–57** Water flows into a branching pipe of equal cross sections. One branch carries water in the same direction as the original flow. Calculate the force acting on the control volume of water. The specific weight of water is $62.4 \, \text{lb/ft}^3$.

**Fig. P17-57**

**17–58** A jet engine in flight has an absolute speed of $1000 \, \text{km/h}$. The engine draws in air at the rate of $100 \, \text{kg/s}$ and expels it at a relative speed of $750 \, \text{m/s}$. Determine the thrust of the engine.

**Fig. P17-58**

**17–59** A fixed nozzle of $0.1 \, \text{m}$ diameter emits a horizontal stream of water at the rate of $60 \, \text{kg/s}$. The stream is to be diverted by a horizontally moving smooth vane whose absolute speed is $u = 5 \, \text{m/s}$. (a) Determine the horizontal component of the force acting on the vane if $\theta = 45°$. (b) Calculate the power transmitted to the vane.

**Fig. P17-59**

**17–60** A jet engine is mounted at the tail section of the fuselage, with the air intake parallel to the exhaust. The engine draws in air at the rate of $180 \, \text{lb/s}$ and discharges it at a relative speed of $2200 \, \text{ft/s}$. Determine the force acting on the tail section caused by the deflected and accelerated airflow and the position of the line of action where this force appears to be acting.

**Fig. P17-60**

**17–61** Each side scoop of the jet aircraft draws in air at the rate of $100 \, \text{kg/s}$. The single engine burns fuel at the rate of $3 \, \text{kg/s}$ and discharges the gases at a relative speed of $900 \, \text{m/s}$. Determine the thrust and the horsepower of the engine.

**Fig. P17-61**

**17–62** A 10,000-lb helicopter is designed to ascend at the maximum speed of $6 \, \text{ft/s}$. What is the required relative speed of air in the slipstream if its diameter is estimated to be 36 ft? Air weighs $0.076 \, \text{lb/ft}^3$.

**Fig. P17-62**

**17-63** An airplane lands at a speed of 150 km/h when the airflow from the propellers is reversed for braking. The relative speed of the air with respect to the airplane is $v_1 = 200$ m/s in the 2-m-diameter slipstream. Calculate the reversed thrust of each engine for the given conditions. The density of air is 1.21 kg/m$^3$.

Fig. P17-63

**17-64** A jet aircraft lands at a speed of 130 mi/h. Each engine draws in air at the rate of 200 lb/s and exhausts it at a relative speed of 1800 ft/s. The thrust is reversed by baffles for purposes of braking the aircraft. Calculate the braking force of each engine for the given configuration.

Fig. P17-64

**17-65** A rocket designed for atmospheric research is to hover briefly at its maximum elevation. The rocket's total mass at the beginning of this period is 4000 kg. Fuel is expelled at a constant speed of 2500 m/s. Find an expression for the rate $dm/dt$ of the fuel consumption. Sketch this rate for a time of 20 s of operation. The acceleration of gravity is 8 m/s$^2$.

Fig. P17-65

**17-66** An empty railroad car has a mass of $m_0 = 12{,}000$ kg. The car is rolling at a constant speed of $v = 1$ km/h while it is loaded with grain at the rate of 200 kg/s. Calculate the magnitude of the force **F** required to maintain the constant speed at time $t = 1$ min after the loading begins. Neglect friction.

Fig. P17-67

**17-67** The railroad car described in Prob. 17-66 is rolling at an initial speed of $v_0 = 2$ km/h when the loading begins. Calculate its speed 20 s later assuming that no horizontal forces are acting during this time.

**17-68** A pipeline-scrubber plug weighs 100 lb when empty. It accumulates scrapings at the rate of 0.02 lb/s while moving at a speed of 6 ft/s. Determine the magnitude of the force required to maintain the constant speed after 1 h of operation. Neglect friction.

Fig. P17-68

**17–69** A balloon used for atmospheric research is to be tethered on a long wire rope that weighs 0.3 lb/ft. Plot the magnitude of the force between the balloon and the wire at point $A$ as a function of time $t = 0$, 1 min, and 2 min. The balloon rises at a constant speed of $v = 5$ ft/s, and the wire uncoils at the ground without resistance.

**Fig. P17-69**

**17–70** A moving sidewalk is designed to carry people horizontally at a speed of 5 ft/s. A group of people starts stepping on the empty mover, adding weight at the rate of 300 lb/s. (a) Calculate the force required to keep the machine moving at its constant speed if $v_{rel} = 0$. Neglect friction. (b) Calculate the force assuming that the people do not stand on the moving strip but walk forward on it at a speed of 4 ft/s relative to the strip.

**Fig. P17-70**

**17–71** A 4000-lb rocket is fired vertically. Fuel is burned at the rate of 40 lb/s and ejected at a relative speed of 10,000 ft/s. Sketch the magnitudes of the acceleration and velocity of the rocket from time $t = 0$ to 50 s.

**17–72** A spacecraft is designed with a maximum total thrust of 2 MN, which is to be obtained by ejecting propellant at a relative speed of 3000 m/s. Determine the required rate of fuel consumption.

**17–73** Determine the total initial mass of a rocket that **begins** to hover ($v_0 = 0$) in the vertical position immediately after ignition (time $t = 0$) if it burns fuel at the rate of 100 kg/s, ejecting it with a relative speed of 3400 m/s. Calculate the magnitudes of the acceleration and velocity of this rocket at $t = 20$ s.

**Fig. P17-73**

**17–74** A 5000-lb space vehicle is moving at a speed of 6000 ft/s when its retrorocket is fired to reduce the speed. The engine consumes fuel at the rate of 10 lb/s and expels it at a relative speed of 9000 ft/s. Determine the final speed of the vehicle if 100 lb of fuel can be used.

**Fig. P17-74**

**17–75** A 9000-kg spacecraft is moving in empty space at a speed of 2 km/s. The engine is fired for 10 s, burning fuel at the rate of 20 kg/s and ejecting it at a relative speed of 3500 m/s. Determine the magnitudes of the maximum acceleration and velocity of the vehicle during the 10-s period of using the engine.

**Newton's second law for a system of $n$ particles**

$$\sum_{i=1}^{n} \mathbf{F}_i = \sum_{i=1}^{n} m_i \mathbf{a}_i$$

Section 17-2

**Motion of the center of mass**

**POSITION:**

$$m\mathbf{r}_C = \sum_{i=1}^{n} m_i \mathbf{r}_i$$

**VELOCITY:**

$$m\mathbf{v}_C = \sum_{i=1}^{n} m_i \mathbf{v}_i$$

**ACCELERATION:**

$$m\mathbf{a}_C = \sum_{i=1}^{n} m_i \mathbf{a}_i$$

$$\sum \mathbf{F} = m\mathbf{a}_C$$

Section 17-3

**Gravitational potential energy**

$$V = mgy_C = Wy_C$$

Section 17-4

**Kinetic energy**

$$T = \frac{1}{2} \sum_{i=1}^{n} m_i v_i^2 \qquad (v_i \text{ are with respect to a fixed frame})$$

$$T = \underbrace{\frac{1}{2} mv_C^2}_{\substack{\text{motion of total} \\ \text{mass imagined to} \\ \text{be concentrated at } C}} + \underbrace{\frac{1}{2} \sum_{i=1}^{n} m_i v_i'^2}_{\substack{\text{motion of all} \\ \text{particles relative} \\ \text{to mass center } C}} \qquad \substack{(v' \text{ are with} \\ \text{respect to a} \\ \text{translating} \\ \text{frame})}$$

Section 17-5

**Work and energy**

$$\sum_{i=1}^{n} U_i' = \sum_{i=1}^{n} \Delta V_i + \sum_{i=1}^{n} \Delta T_i$$

$$U' = \Delta V + \Delta T$$

**Conservation of energy**

$$\Delta V + \Delta T = 0$$

For configurations 1 and 2: $V_1 + T_1 = V_2 + T_2$

Section 17-6

**Moments of forces about point $O$**

$$\sum_{i=1}^{n} (\mathbf{r}_i \times \mathbf{F}_i) = \sum_{i=1}^{n} \mathbf{M}_{io} = \sum_{i=1}^{n} (\mathbf{r}_i \times m_i \mathbf{a}_i)$$

Section 17-7

**Linear and angular momenta**

**TRANSLATION:**

$$\mathbf{G} = \sum_{i=1}^{n} m_i \mathbf{v}_i \qquad \sum \mathbf{F} = \dot{\mathbf{G}}$$

**ROTATION:**

$$\mathbf{H}_O = \sum_{i=1}^{n} (\mathbf{r}_i \times m_i \mathbf{v}_i) \qquad \text{(point } O \text{ is fixed)}$$

$$\sum \mathbf{M}_O = \dot{\mathbf{H}}_O = \sum (\mathbf{r}_i \times m_i \mathbf{a}_i)$$

Section 17-8

**Angular momentum about the center of mass**   The angular momentum of a system of particles about its center of mass is the same whether it is observed from a fixed frame or from the centroidal frame which may be translating but not rotating.

**USEFUL EXPRESSIONS:**

$$\mathbf{H}_O = \mathbf{H}_C + \mathbf{r}_C \times m\mathbf{v}_C$$

$$\sum \mathbf{M}_O = \dot{\mathbf{H}}_C + \mathbf{r}_C \times m\mathbf{a}_C$$

where point $O$ is fixed, $C$ is the center of mass, and $\mathbf{r}_C$ is from $O$ to $C$.

## Section 17-9

**Conservation of momentum**

$$
\left.
\begin{aligned}
\mathbf{G} &= \text{constant} \\
\mathbf{H}_O &= \text{constant} \\
\mathbf{H}_C &= \text{constant}
\end{aligned}
\right\} \quad \text{not the same constants in general}
$$

## Section 17-10

**Impulse and momentum**

**TRANSLATION:**

$$
\sum_{i=1}^{n} \int_{t_1}^{t_2} \mathbf{F}_i \, dt = \mathbf{G}_2 - \mathbf{G}_1 = m\mathbf{v}_{C_2} - m\mathbf{v}_{C_1}
$$

**ROTATION:**

$$
\sum_{i=1}^{n} \int_{t_1}^{t_2} \mathbf{M}_{i_O} \, dt = \mathbf{H}_{O_2} - \mathbf{H}_{O_1}
$$

## Sections 17-11 and 17-12

**Steady flow of mass**

$$
\frac{dm}{dt} = \rho_1 A_1 v_1 = \rho_2 A_2 v_2
$$

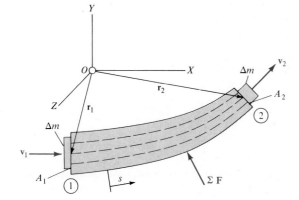

For an incompressible fluid, $\rho_1 = \rho_2 = \rho$,

$$\frac{1}{\rho}\frac{dm}{dt} = A_1 v_1 = A_2 v_2$$

$$\sum \mathbf{F} = \frac{dm}{dt}(\mathbf{v}_2 - \mathbf{v}_1)$$

$$\sum \mathbf{M}_O = \dot{\mathbf{H}}_O = \frac{dm}{dt}(\mathbf{r}_2 \times \mathbf{v}_2 - \mathbf{r}_1 \times \mathbf{v}_1)$$

Section 17-13

## Systems with variable mass

System at time $t$

(a)

System at
time $t + \Delta t$

(b)

**SYSTEM GAINING MASS:**

$$\sum \mathbf{F} = m\frac{d\mathbf{v}}{dt} - \frac{dm}{dt}\mathbf{v}_{\text{rel}}$$

**SYSTEM LOSING MASS:**

$$\sum \mathbf{F} = m\frac{d\mathbf{v}}{dt} + \frac{dm}{dt}\mathbf{v}_{\text{rel}}$$

$$\frac{dm}{dt} = \text{time rate at which mass is absorbed or lost}$$

**17–76** Three 80-kg skydivers link up as shown. Assume they are falling vertically and that atmospheric forces are negligible. Diver $C$ pushes away from the other two, acquiring a horizontal speed $v_C = 2\,\text{m/s}$. Determine the distance between diver $C$ and the others at $t = 5\,\text{s}$ after the separation if their common center of mass moves only vertically.

**Fig. P17-76**

**17–77** A 1-kg mass $m_1$ has a velocity $\mathbf{v}_1 = \mathbf{i} + 4\mathbf{j} - 2\mathbf{k}$ at the position $\mathbf{r}_1 = 5\mathbf{i} + \mathbf{k}$. At the same instant a second particle $m_2$ of mass 2 kg has a velocity $\mathbf{v}_2 = 3\mathbf{i} - 2\mathbf{j}$ at the position $\mathbf{r}_2 = 2\mathbf{i} + 3\mathbf{j} - 2\mathbf{k}$. Distances are in meters and velocities are in m/s. At this instant, determine: (a) the position and velocity of the system's mass center, (b) the linear momentum of the system, (c) the kinetic energy of the system, (d) the angular momentum $\mathbf{H}_O$ about point $O$, and (e) the angular momentum $\mathbf{H}_C$ about the system's mass center $C$.

**Fig. P17-77**

**17–78** The amusement ride has 16 carriages of mass $m = 200\,\text{kg}$ which rotate about point $A$ at 5 rpm with respect to the arm $OA$. The carriages with riders move in the horizontal plane when $\theta = 0$. A hydraulic actuator at point $A$ raises the arm with the rotating assembly. Calculate the kinetic energy of only two carriages with respect to point $O$ when $\theta = 0$ and the angular speed of arm $OA$

is 0.3 rpm. One carriage is nearest to $O$, the other is farthest from $O$ at the same instant.

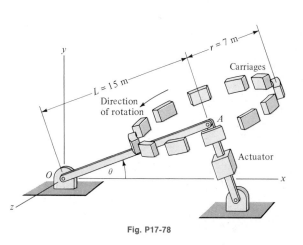

**Fig. P17-78**

**17–79** Solve Prob. 17–78 for the pair of carriages that are equidistant from $O$ at the same instant.

**17–80** The amusement ride in Fig. P17-78 has 16 carriages of mass $m = 600\,\text{kg}$ which rotate at 10 rpm with respect to the arm $OA$. The carriages with riders move in the horizontal plane when $\theta = 0$. A hydraulic actuator at point $A$ raises the arm with the rotating assembly. Calculate the angular momentum of two carriages with respect to point $O$ when $\theta = 0$ and the angular speed of arm $OA$ is 0.3 rpm. One carriage is nearest to $O$, the other is farthest from $O$ at the same instant. The carriages and arm $OA$ are always in the same plane for the purposes of this analysis.

**17–81** Consider the rotating system of two masses in Prob. 17-80. Moments $\mathbf{M}_A = -80\mathbf{j}\,\text{kN·m}$ and $\mathbf{M}_O = 12\mathbf{k}\,\text{kN·m}$ are applied for a time of 0.01 s. Determine the new angular speed of the carriages about point $A$ and of the arm $OA$ assuming that $\theta$ is still zero.

**17–82** In a game of billiards, ball $A$ is initially moving with a speed of 4 m/s as shown before impacting with balls $B$ and $C$, which are at rest side by side. All balls have the same mass and the impact is frictionless and perfectly elastic. The directions of balls $A$, $B$, and $C$ after impact are observed as shown. Determine the speeds of all balls after impact if ball $A$ strikes $B$ to the left of center.

**Fig. P17-82**

**17–83** The die of a machine that extrudes plastic rods receives the liquid polymer at the rate of 5 lb/s. Calculate the exit speed $v_2$ of the rods and the force for holding the die if $A_1 = 10\,\text{in}^2$, $A_2 = 0.5\,\text{in}^2$, $\gamma = 124.8\,\text{lb/ft}^3$, $p_1 = 10\,\text{psi}$ and $p_2 = 8.21\,\text{psi}$.

Die

**Fig. P17-83**

**17–84** The fireboat discharges water from two nozzles at the rates of $Q_1 = 6000\,\text{kg/min}$ at $v_1 = 30\,\text{m/s}$, and $Q_2 = 7000\,\text{kg/min}$ at $v_2 = 40\,\text{m/s}$. Deter-

mine the required propeller thrust (horizontal) to hold the boat in a fixed position.

**Fig. P17-84**

**17–85** An orbiting 3000-lb spacecraft has a speed of 15,000 mi/h which must be increased to 20,000 mi/h to change the orbit. How much fuel is required if the engine burns fuel at the rate of 30 lb/s and expels it at a relative speed of 10,000 ft/s?

**17–86** A 6000-kg vehicle is moving in space at a speed of $v = 30\,\text{km/h}$. The vehicle collects interstellar particles at the rate of $10^{-9}\,\text{kg/s}$. Determine the required rate of fuel consumption to keep the speed $v$ constant if fuel is expelled at a relative speed of 10 km/h. The speed $v$ is in a fixed reference frame in which the particles have negligible speed.

**Fig. P17-86**

# Kinematics of Rigid Bodies

A frequently used simplification in analyzing the motion of a body which cannot be treated as a particle is to assume that the body is a system of particles that are rigidly connected. By definition, all particles of a rigid body are in fixed positions with respect to each other. The basic problem in kinematics of rigid bodies is to relate the motions at specified locations in the body to the resulting motions at other arbitrary points. The concept of rigid body must be used with caution since nothing is perfectly rigid. Thus, there is always a difference between the actual and theoretical motions of a body when the assumption of rigid body is used in the analysis. The error depends on the deformability of the body. For example, pneumatic tires and springs are much less rigid than the frame of a car, which itself can bend or twist a great amount in comparison with a solid block of steel of the same overall dimensions. Nevertheless, the assumption of dealing with rigid bodies is a powerful concept in the analysis of numerous motions.

## OVERVIEW

It is convenient to distinguish two motions of rigid bodies, translation and rotation. These may occur in pure form or in combination, which is considered as general motion. The various possibilities are described in the following categories and illustrated in Fig. 18-1.

**1. *Translation.*** A rigid body is in translational motion when every fixed straight line in the body remains parallel to its original position throughout the motion. This implies that all particles of the body have

| Motion | Schematic representation of motion | Example |
|---|---|---|
| (a) Rectilinear translation | | Body and frame of car |
| (b) Curvilinear translation | | Test object and platform of shake table |
| (c) Rotation about a fixed axis intersecting the body | | Supporting arms of shake table |
| (d) Rotation about a fixed axis outside of the body | | Body and frame of a vehicle on a curved track |
| (e) General plane motion (translation and simultaneous rotation) | | Rolling wheel |

**FIGURE 18-1**

Categories of plane motion of rigid bodies

parallel paths during translation. The motion of one point in the body completely specifies the motion of the whole body. There are two subcategories of translation.

**a.** In *rectilinear translation* the paths of all particles of the body are parallel straight lines as in Fig. 18-1a.

**b.** In *curvilinear translation* the paths of the particles of the body are curved lines as in Fig. 18-1b.

**2. *Rotation about a fixed axis.*** A rigid body is in rotational motion about a fixed axis when all particles of the body move along circular paths whose centers are on the same fixed axis. The *axis of rotation* may or may not intersect the rigid body. If the axis intersects the body, particles on that axis have no velocity or acceleration (Fig. 18-1c). If the axis is outside the body, all particles of the body have velocity and acceleration (Fig. 18-1d). The orientation of a fixed line in a body that is not parallel to the axis of rotation of that body changes during rotational motion. This fact distinguishes rotation from curvilinear translation.

**3. General plane motion.**   Any plane motion of a rigid body that is a combination of translation and rotation is general plane motion. The most common example of this is a rolling wheel (Fig. 18-1e).

**4. General motion.**   All motions of rigid bodies which are not in one of the preceding three categories are called general motion. For example, the rolling wheel in Fig. 18-1e may be moving along a curved path on the ground. This means that the plane of rotation changes, and the motion is three-dimensional.

**Motion about a fixed point.** A special category of three-dimensional motion is considered for rigid bodies that are attached at a fixed point. For example, the center $A$ of the wheel in Fig. 18-1e may be fixed. This allows rotation of the wheel in a plane and also tilting of that plane about any line through point $A$.

## 18-1

## TRANSLATION

Assume that $A$ and $B$ are fixed particles in a rigid body which moves in rectilinear or curvilinear translation as in Fig. 18-2. The relative position vector $\mathbf{r}_{B/A}$ and the absolute position vectors $\mathbf{r}_A$ and $\mathbf{r}_B$ are related by the expression

$$\mathbf{r}_B = \mathbf{r}_A + \mathbf{r}_{B/A}$$  18-1

The relationship for the absolute and relative velocities is obtained by differentiating Eq. 18-1 with respect to time. By the definition of a rigid body and of translation, the vector $\mathbf{r}_{B/A}$ always has a constant magnitude and direction. Thus, $\dot{\mathbf{r}}_B = \dot{\mathbf{r}}_A + 0$, or

$$\mathbf{v}_B = \mathbf{v}_A$$  18-2

The relationship for the accelerations is obtained by differentiating Eq. 18-2 with respect to time,

$$\mathbf{a}_B = \mathbf{a}_A$$  18-3

Equations 18-2 and 18-3 are the mathematical statements of the fact that *when a rigid body is in translation, the motion of a single point completely specifies the motion of the whole body.* In rectilinear translation the velocity and acceleration have constant lines of action during the motion. In the case of curvilinear translation the directions of the velocity and acceleration vectors change throughout the motion depending on the curved path.

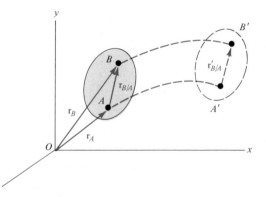

**FIGURE 18-2**
Translation of a rigid body

Assume that a rigid body of general shape is rotating about the fixed $y$ axis as in Fig. 18-3. Note that in any situation a coordinate axis may be chosen to coincide with the axis of rotation for convenience. Consider an arbitrary point $P$ defined by its position vector $\mathbf{r}$ at an angle $\phi$ with the $y$ axis. Point $P$ moves in circular motion about the axis of rotation and in a plane that is perpendicular to the axis of rotation. The position of the body is defined by the *angular coordinate* $\theta$ in the plane of rotation of point $P$. $\theta$ is measured from an arbitrary line of reference such as $z'$ to the line $AP$, where $A$ is the projection of point $P$ on the axis of rotation. The distance $\rho = r \sin \phi$ between points $A$ and $P$ is constant in a rigid body, and line $AP$ is perpendicular to the axis of rotation.

The angular velocity of line $AP$, as defined in App. A, is represented by the *angular velocity vector* $\boldsymbol{\omega} = \omega \mathbf{j} = \dot{\theta} \mathbf{j}$ directed along the fixed axis of rotation as shown in Fig. 18-3. The velocity $\mathbf{v}$ of point $P$ is obtained by differentiating the position vector $\mathbf{r}$ with respect to time,

$$\mathbf{v} = \frac{d\mathbf{r}}{dt}$$

Since the vector $\mathbf{r}$ has a constant magnitude for any given point $P$, the results of App. A can be applied,

$$\boxed{\mathbf{v} = \dot{\mathbf{r}} = \boldsymbol{\omega} \times \mathbf{r}} \qquad \boxed{\text{18-4}}$$

The velocity of point $P$ is a vector that is perpendicular to line $AP$ in the plane of rotation of point $P$ and has a direction determined by the angle $\theta$ and the right-hand rule for the cross product of two vectors. This is exactly the result one would expect, since point $P$ is in circular motion about the $y$ axis. Equation 18-4 can be used to determine the velocity of any point on the rigid body since the angular velocity vector $\boldsymbol{\omega}$ is the same for *any* line perpendicular to the axis of rotation in the rigid body. Thus, $\boldsymbol{\omega}$ is the *angular velocity of the rigid body*.

Writing the position vector as $\mathbf{r} = x\mathbf{i} + y\mathbf{j} + z\mathbf{k}$, Eq. 18-4 can be evaluated,

$$\mathbf{v} = \boldsymbol{\omega} \times \mathbf{r} = \omega z \mathbf{i} - \omega x \mathbf{k}$$

The magnitude of the velocity is expressed as

$$v = \omega \sqrt{z^2 + x^2} = \omega \rho = \omega r \sin \phi \qquad \boxed{\text{18-5}}$$

according to Fig. 18-3. Note that the magnitude of the velocity of any point on a rigid body in rotation about a fixed axis depends only on its distance from that fixed axis, since $\omega$ is a property of the rigid body and is independent of position in the body. For a thin slab parallel to the $xz$ plane at point $A$, the velocity of any point $P$ is obtained either from Eq. 18-4 or from the scalar expression $v = \rho \omega = \rho \dot{\theta}$ and the direction of $\omega$. The vector equation is most advantageous for analyzing three-dimensional motion.

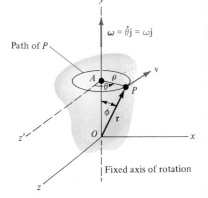

**FIGURE 18-3**

Analysis of velocities of a rigid body rotating about a fixed axis

The acceleration **a** of a particle $P$ in a rigid body is obtained by differentiating Eq. 18-4 with respect to time,

$$\mathbf{a} = \frac{d\mathbf{v}}{dt} = \frac{d}{dt}(\boldsymbol{\omega} \times \mathbf{r})$$

Using the rule for the differentiation of a vector product gives

$$\mathbf{a} = \frac{d\boldsymbol{\omega}}{dt} \times \mathbf{r} + \boldsymbol{\omega} \times \frac{d\mathbf{r}}{dt}$$

$$\mathbf{a} = \boldsymbol{\alpha} \times \mathbf{r} + \boldsymbol{\omega} \times \mathbf{v}$$

Since $\mathbf{v} = \boldsymbol{\omega} \times \mathbf{r}$,

$$\boxed{\mathbf{a} = \boldsymbol{\alpha} \times \mathbf{r} + \boldsymbol{\omega} \times (\boldsymbol{\omega} \times \mathbf{r})} \qquad \boxed{18\text{-}6}$$

where $\boldsymbol{\alpha} = d\boldsymbol{\omega}/dt$ is the *angular acceleration vector* of the rigid body. For a body rotating about a fixed axis, $\boldsymbol{\alpha}$ is along that axis and is the same for any line perpendicular to the axis of rotation in the rigid body. The magnitude of $\boldsymbol{\alpha}$ is $\dot{\omega}$, so its direction along the axis is determined by whether the magnitude of the angular velocity of the body is increasing or decreasing. In Fig. 18-4, $\boldsymbol{\alpha}$ is in the same direction as $\boldsymbol{\omega}$, so the angular velocity of the body is increasing. The term $\boldsymbol{\alpha} \times \mathbf{r}$ in Eq. 18-6 is the tangential acceleration $\mathbf{a}_t$ of the particle $P$. The term $\boldsymbol{\omega} \times (\boldsymbol{\omega} \times \mathbf{r})$ is a vector triple product whose direction is always from point $P$ to $A$ on the axis of rotation (since $\boldsymbol{\omega}$ is along the axis of rotation and $\boldsymbol{\omega} \times \mathbf{r} = \mathbf{v}$ is tangent to the path of $P$). The term $\boldsymbol{\omega} \times (\boldsymbol{\omega} \times \mathbf{r})$ is the normal acceleration $\mathbf{a}_n$ of the particle $P$. The two components of the acceleration of $P$ are shown in Fig. 18-4. Their magnitudes are

$$a_t = \rho\alpha \qquad a_n = \rho\omega^2 \qquad \boxed{18\text{-}7}$$

where $\rho$ is applicable to a specific point $P$. For a thin slab parallel to the $xz$ plane at point $A$, the acceleration **a** of any particle $P$ is obtained either from Eq. 18-6 or from the scalar expressions of Eq. 18-7 and the direction of $\boldsymbol{\alpha}$. These are important and useful results for the kinematics analysis of rigid bodies.

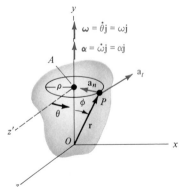

**FIGURE 18-4**

Analysis of accelerations of a rigid body rotating about a fixed axis

The plane rotational motion of a rigid body (with or without a fixed axis) can be completely specified in magnitude if one variable of that motion is known as a function of time. This variable may be the angular coordinate $\theta$, or the angular speed $\dot{\theta} = \omega$, or the angular acceleration $\ddot{\theta} = \alpha$. The time derivatives of the position coordinate are commonly written in one of the following scalar forms:

$$\omega = \frac{d\theta}{dt} = \dot{\theta}$$

18-8

$$\alpha = \frac{d\omega}{dt} = \frac{d^2\theta}{dt^2} = \ddot{\theta}$$

18-9

Sometimes it is convenient to eliminate time as a variable. Using the chain rule yields

$$\alpha = \frac{d\omega}{dt} = \frac{d\omega}{d\theta}\frac{d\theta}{dt} = \omega\frac{d\omega}{d\theta}$$

18-10

These three equations are analogous to those presented for rectilinear motion of a particle in Sec. 11-1, with $\theta$, $\omega$, and $\alpha$ replacing $x$, $v$, and $a$, respectively. They can be manipulated in the same way. In particular, two important special cases of rotation may be noted.

1. *Constant angular speed, $\omega$.* In this case the angular acceleration is zero, and the angular position $\theta$ is given by

$$\theta = \theta_0 + \omega t$$

18-11

where $\theta_0$ is the initial coordinate for reference at time $t = 0$.

2. *Constant angular acceleration, $\alpha$.* Integrating Eqs. 18-8 and 18-10 with respect to time $t$ and position $\theta$, respectively, yields relationships analogous to those in Sec. 11-3,

$$\omega = \omega_0 + \alpha t$$

18-12a

$$\theta = \theta_0 + \omega_0 t + \frac{1}{2}\alpha t^2$$

18-12b

$$\omega^2 = \omega_0^2 + 2\alpha(\theta - \theta_0)$$

18-12c

where $\omega_0$ is the initial angular speed corresponding to the initial position $\theta_0$ at time $t = 0$.

If the angular acceleration is neither zero nor constant as in these special cases, Eqs. 18-8 through 18-10 must be used. It should be noted that in rotations about a fixed axis the positive direction is the same for $\theta$, $\omega$, and $\alpha$.

EXAMPLE 18-1

The model of a tire tester in Fig. 18-5 shows a motor $M$ which drives a large flywheel $W$. The diameter of the pulley on the motor is $d_1 = 1.1$ ft and on the flywheel it is $d_2 = 4$ ft. The flywheel has an outside diameter of $d_3 = 8$ ft. The motor starts from rest and rotates at a constant angular acceleration of $\alpha = 1$ rad/s$^2$. Calculate the maximum speed at time $t = 20$ s for a point on a 3-ft-diameter tire that is pressed against the flywheel. Assume that there is no slippage anywhere in the system, and that consequently points in contact (of different members) have the same absolute velocity.

SOLUTION

The motion can be analyzed as planar motion. From Eq. 18-12a, using the subscript 1 for pulley 1,

$$\omega_1 = \omega_0 + \alpha_1 t$$

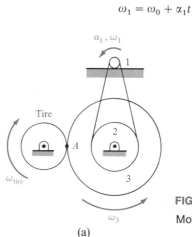

**FIGURE 18-5**

Model of a tire-testing system

(a)

(b)

712    KINEMATICS OF RIGID BODIES

If the motor starts from rest, $\omega_0 = 0$. The angular speed at $t = 20\,\text{s}$ is

$$\omega_1 = \alpha_1 t = (1\,\text{rad/s}^2)t = 1\,\text{rad/s}^2(20\,\text{s}) = 20\,\text{rad/s}$$

If there is no slipping, all points on the perimeters of pulleys 1 and 2 have the same speed since all parts of the belt have the same speed. Thus, $v_1 = v_2$. From Eq. 18-4, $\mathbf{v} = \boldsymbol{\omega} \times \mathbf{r}$, and in two dimensions $v = \omega r$. Consequently, the angular speed of pulley 2 is

$$\omega_2 = \frac{\omega_1 r_1}{r_2} = \frac{(20\,\text{rad/s})(0.55\,\text{ft})}{2\,\text{ft}} = 5.5\,\text{rad/s}$$

Since the pulley 2 is connected to the flywheel 3,

$$\omega_3 = \omega_2 \qquad \text{and} \qquad v_3 = \omega_3 r_3 = 22\,\text{ft/s}$$

If there is no slipping, the speed $v_3$ is the same as the speed at the outside edge of the tire, which is the required maximum speed. Therefore, $v_A = v_3$, and

$$v_{\text{tire}} = v_3 = 22\,\text{ft/s}$$

---

EXAMPLE 18-2

Determine the magnitude of the maximum total acceleration of a point on the tire described in Ex. 18-1 at $t = 20\,\text{s}$. Assume there is no slipping.

## SOLUTION

Both the tangential and normal components of the acceleration are maximum for particles at the outside edge of the tire according to Eq. 18-7. Since the belt does not slip, the tangential accelerations for points on the perimeter of pulleys 1 and 2 are obtained using $a_t = r\alpha$ from Eq. 18-7,

$$a_{t_2} = a_{t_1} = r_1\alpha_1 = \left(\frac{1.1}{2}\,\text{ft}\right)(1\,\text{rad/s}^2) = 0.55\,\text{ft/s}^2$$

Thus, because pulley 2 is attached to the flywheel,

$$\alpha_3 = \alpha_2 = \frac{a_{t_2}}{r_2} = \frac{0.55\,\text{ft/s}^2}{2\,\text{ft}} = 0.275\,\text{rad/s}^2$$

$$a_{t_3} = \alpha_3 r_3 = 1.1\,\text{ft/s}^2$$

If there is no slipping, points of the tire and the flywheel in contact at $A$ have the same tangential acceleration, so

$$a_{t_{\text{tire}}} = 1.1\,\text{ft/s}^2$$

The normal acceleration of a particle on the tire at point $A$ is

$$a_n = \omega^2 r = \frac{v^2}{r} = \frac{(22\,\text{ft/s})^2}{1.5\,\text{ft}} = 323\,\text{ft/s}^2$$

The total acceleration for any point on the tire is obtained from Eq. 18-6,

$$\mathbf{a}_{\text{total}} = \mathbf{a}_t + \mathbf{a}_n$$

Both $a_n$ and $a_t$ are magnitudes of perpendicular vector quantities, so

$$a_{\text{total}} = \sqrt{a_t^2 + a_n^2} = \sqrt{(1.1)^2 + (323)^2} = 323\,\text{ft/s}^2$$

EXAMPLE 18-2    713

## EXAMPLE 18-3

The rotor of an electric motor (Fig. 18-6a) has a speed of 3450 rpm when the power is shut off. It is observed to coast to rest in 40 s at a uniformly decelerating rate. Determine (a) the number of revolutions required for the rotor to coast to rest, and (b) the velocity and acceleration of point $P$ the instant after the power is shut off.

(a)

(b)

(c)

**FIGURE 18-6**

Rotor analysis

# SOLUTION

(a) The initial angular velocity is counterclockwise, as shown in Fig. 18-6a, $\omega_0 = \omega_0 \mathbf{k}$. Since the angular speed is decreasing, the angular acceleration may be written as $\alpha = -\alpha \mathbf{k}$. The number of revolutions required for the rotor to come to rest can be determined using Eq. 18-12b since the angular acceleration is constant, but the angular acceleration must first be determined using Eq. 18-12a. In order to use Eq. 18-12a, the angular speed must be converted to rad/s,

$$\omega_0 = 3450 \left( \frac{\text{rev}}{\text{min}} \right) \left( \frac{2\pi \, \text{rad}}{\text{rev}} \right) \left( \frac{\text{min}}{60 \, \text{sec}} \right) = 361 \, \text{rad/s}$$

Using Eq. 18-12a, $\omega = 361 \, \text{rad/s} - \alpha t$. At $t = 40 \, \text{s}$,

$$0 = 361 \, \text{rad/s} - \alpha(40 \, \text{s})$$

$$\alpha = 9.03 \, \text{rad/s}^2$$

Using Eq. 18-12b,

$$\Delta\theta = \theta - \theta_0 = \omega_0 t - \tfrac{1}{2}\alpha t^2$$
$$= (361 \, \text{rad/s})(40 \, \text{s}) - \tfrac{1}{2}(9.03 \, \text{rad/s}^2)(40 \, \text{s})^2$$
$$= 7220 \, \text{rad}$$

$$\text{number of revolutions} = \frac{7220 \, \text{rad}}{2\pi \, \text{rad/rev}} = 1150 \, \text{rev}$$

(b) The velocity and acceleration of point $P$ are calculated using two different methods. For the velocity, using Fig. 18-6b:

1. *Vector method:*  Using Eq. 18-4,
$$\mathbf{v}_P = \omega \times \mathbf{r} = (361\mathbf{k} \, \text{rad/s}) \times (0.2\mathbf{j} \, \text{m})$$
$$= -72.2\mathbf{i} \, \text{m/s}$$

2. *Elementary circular motion:*  Using Eq. 18-5,
$$\mathbf{v}_P = -r\omega_0\mathbf{i} = -(0.2 \, \text{m})(361 \, \text{rad/s})\mathbf{i} \quad \text{(direction by inspection)}$$
$$= -72.2\mathbf{i} \, \text{m/s}$$

For the acceleration, using Fig. 18-6c,

1. *Vector method:*
$$\mathbf{a}_P = \alpha \times \mathbf{r} + \omega \times (\omega \times \mathbf{r})$$

where $\omega \times \mathbf{r} = \mathbf{v}_P = -72.2\mathbf{i} \, \text{m/s}$. From the above,
$$\mathbf{a}_P = (-9.03\mathbf{k} \, \text{rad/s}^2) \times (0.2\mathbf{j} \, \text{m}) + (361\mathbf{k} \, \text{rad/s}) \times (-72.2\mathbf{i} \, \text{m/s})$$
$$= (1.81\mathbf{i} - 26{,}060\mathbf{j}) \, \text{m/s}^2$$

2. *Elementary circular motion:*
$$\mathbf{a}_P = r\alpha\mathbf{i} - r\omega^2\mathbf{j} \quad \text{(directions by inspection)}$$
$$= (0.2 \, \text{m})(9.03 \, \text{rad/s}^2)\mathbf{i} - (0.2 \, \text{m})(361 \, \text{rad/s})^2\mathbf{j}$$
$$= (1.81\mathbf{i} - 26{,}060\mathbf{j}) \, \text{m/s}^2$$

EXAMPLE 18-3  715

**18–1** The rigid bar of $2l = 1$ m is rotating about pin $O$ at a constant rate of $\omega = 8$ rad/s. Determine the velocities and accelerations of points $A$ and $B$ for the position shown.

**Fig. P18-1**

**18–2** The angular speed of the bar described in Prob. 18–1 is increasing at the rate of $\alpha = 2$ rad/s². Determine the acceleration vectors for points $A$ and $B$.

**18–3** The triangular bracket is rotating about pin $O$ at a constant rate of $\omega = 15$ rad/s. Determine the velocities and accelerations of pins $A$ and $B$ for the position shown.

$OA = 12$ in.
$OB = 9$ in.

**Fig. P18-3**

**18–4** The angular speed of the bracket described in Prob. 18–3 is decreasing at the rate of $\alpha = 3$ rad/s². Determine the acceleration vectors for pins $A$ and $B$.

**18–5** A car in a laboratory test is supported by hinges at the rear axles. The front wheels are simultaneously accelerated upward at the rate of 14 m/s². Calculate the angular acceleration of *the car and the vertical acceleration of point A* midway between the wheels for the given position. Assume that the whole car is rigid.

$a = 14$ m/s²

**Fig. P18-5**

**18–6** The car described in Prob. 18–5 is rotating about hinges $O$ at the rate of $\omega = 1$ rad/s counterclockwise when the front end is accelerated upwards as shown. Determine the total acceleration of point $A$, midway between the wheels.

**18–7** A 6-ft-tall electric console is tested on a shake table whose four supporting arms are 3 ft long. The arms are rotating at constant speeds of $\omega = 2$ rad/s, clockwise as viewed in Fig. P18-7(b), when the arms make 60° angles with the horizontal. Calculate the accelerations of points $A$ and $B$ at the bottom and at the top of the console, respectively. Assume that the platform is thin.

**Fig. P18-7**
Electric console on a shake table (Courtesy MTS Systems Corporation, Minneapolis, Minnesota)

**18–8** The angular velocities of the arms described in Prob. 18–7 are decreasing at the rate of 1 rad/s² in the given position. Determine the acceleration vectors at points $A$ and $B$ in the plane of rotation ($xy$ plane).

**18-9** A radial line $OA$ on a wheel has its position given by $\theta = 2t^3 - t^2 + 5$, where $\theta$ is in radians and $t$ is in seconds. Determine the angular velocity, acceleration, and number of turns of the wheel at a time $t = 10\,\text{s}$ after starting from rest. How much time is required for the wheel to reach 5000 rpm?

Fig. P18-9

**18-10** Determine the $x$ and $y$ components of the velocity and acceleration vectors of point $A$ in Prob. 18-9 at time $t = 5\,\text{s}$ if the wheel's radius is 0.3 m.

**18-11** A wheel is rotating at an angular velocity $\omega$. It is desirable to stop the wheel in one complete revolution of a radial line on it. Derive expressions for the required angular acceleration $\alpha$ and time $t$ for such a braking.

**18-12** A turbine disk is rotating at a speed of 30,000 rpm. Calculate the angular acceleration required to stop the disk in a time of 100 s.

**18-13** The rotor of a centrifugal pump starts from rest and reaches a speed of 1000 rpm in 300 revolutions. Calculate the time required for the 300 revolutions if the acceleration is at a constant rate.

**18-14** A flywheel is rotating at a speed of 800 rpm when it begins to decelerate at the constant rate of 0.02 rad/s. Calculate the time required to stop the wheel and the number of revolutions during the slowing process.

**18-15** A drill rod of geologic exploration is 20,000 ft long, and it lies radially in the earth's crust. Calculate the linear velocities and accelerations of the ends $A$ and $B$ of the rod caused by rotation of the earth. The rod is located at 40° north latitude. Assume that the radius of the earth is 3960 mi, and that the earth makes one complete revolution about its axis in 24 h.

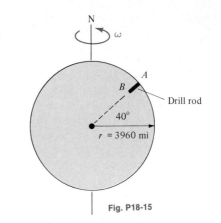

Fig. P18-15

**18-16** The large drum for drying paper moves the paper at a speed of 50 ft/s. Calculate the required angular speed of the small driving roller.

Fig. P18-16

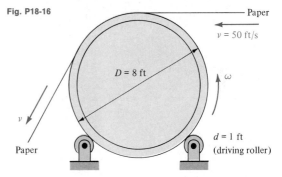

**18-17** The turntable of a record player is to have a friction drive by a smaller wheel. What is the required angular acceleration of the small wheel $A$ if the large disk $B$ is to reach a speed of 33.3 rpm in a time of 5 s? How many revolutions does the large disk make in this time?

Fig. P18-17

**18-18** The sprockets of a chain drive have diameters $d_1 = 0.1\,\text{m}$ and $d_2 = 0.2\,\text{m}$. The small sprocket is designed to reach a speed of 100 rpm in a time of 7 s. Determine the required angular acceleration of the large sprocket and the number of revolutions of both wheels during the acceleration.

Fig. P18-18

**18–19**  The large wheel in Fig. P18-18 is rotating at a speed of 200 rpm. Determine the angular acceleration of the small wheel to reduce the large wheel's speed to 100 rpm in a time of 5 s. $d_1 = 6$ in., $d_2 = 15$ in. There is no slipping.

**18–20**  The circular saw for road repairs has a diameter of 6 ft. Its chain drive has a 0.5 ft diameter sprocket at $A$, and a 0.8 ft diameter sprocket at $B$. Calculate the angular acceleration of sprocket $A$ if the cutting teeth should reach a speed of 8 ft/s in a time of 5 s.

Fig. P18-20

**18–21**  In a gearbox gear $A$ drives $B$, and $C$, which is firmly attached to $B$, drives gear $D$. Determine the ratios of the angular speeds $\omega_A/\omega_D$, of the angular accelerations $\alpha_A/\alpha_D$, and of the number of revolutions $n_A/n_D$ in terms of the diameters $d_A$, $d_B$, $d_C$, and $d_D$.

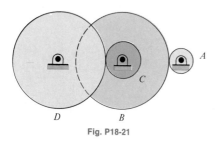

Fig. P18-21

**18–22**  A chain drive consists of sprocket $A$ with diameter $d_A = 0.26$ m, and a composite sprocket $B$ with diameters $d_{B_1} = 0.1$ m, $d_{B_2} = 0.08$ m, and $d_{B_3} = 0.06$ m. The angular acceleration of $A$ is 5 rad/s$^2$, starting from zero angular speed. Determine the angular speed of sprocket $B$ and its number of turns after a complete revolution of sprocket $A$ using each of the three diameters of $B$.

Fig. P18-22

**18–23**  A turntable has an inner wheel for friction drive as shown in the cross-sectional view. The small wheel has a diameter of 3 in., the large one of 12 in. (inside). Calculate the angular acceleration of the small wheel to increase the speed of the large wheel from 50 rpm to 100 rpm in a time of 3 s. Also calculate the number of revolutions of the large wheel during this time.

Fig. P18-23

**18–24**  A hoist mechanism consists of wheel $A$ ($r = 1$ m), gear $B$ ($r = 0.5$ m), which is firmly attached to wheel $A$, and pinion gear $C$ ($r = 0.1$ m), which drives gear $B$. The maximum allowed vertical acceleration of the load is 5 m/s$^2$. Its maximum speed is 8 m/s. Determine the maximum angular speed and angular acceleration of gear $C$, and the number of revolutions in which it reaches the maximum speed starting from rest.

Fig. P18-24

**18–25**  A failure in the hoist mechanism described in Prob. 18–24 allows the load to fall from rest at the rate of 7 m/s$^2$. Determine the angular acceleration of the gear $C$, and its angular speed after the load has descended 20 m.

A general plane motion of a rigid body can be considered a *simultaneous translation and rotation*. An example of this motion is a car moving on a plane, curved road as in Fig. 18-7a. The motion of the frame can be described as a combination of a pure curvilinear translation and a pure rotation shown in Fig. 18-7b and c. The absolute velocity of an arbitrary particle in such a rigid body is determined using the concept of relative motion of two particles.

Assume that the body in Fig. 18-8a is in general plane motion with respect to the fixed $XY$ frame. Also assume that the absolute velocity $\mathbf{v}_A$ of particle $A$ and the angular velocity $\omega$ of the body are known, and the absolute velocity $\mathbf{v}_B$ of a given particle $B$ is to be determined. The velocity $\mathbf{v}_B$ can be expressed (as for any two particles $A$ and $B$) as

$$\mathbf{v}_B = \mathbf{v}_A + \mathbf{v}_{B/A} \qquad \boxed{18\text{-}13}$$

where $\mathbf{v}_{B/A}$ is the relative velocity of $B$ with respect to $A$. Using the position vector $\mathbf{r}_{B/A}$ from point $A$ to $B$, $\mathbf{v}_{B/A} = \dot{\mathbf{r}}_{B/A}$. From Eq. 18-4,

$$\mathbf{v}_{B/A} = \omega \times \mathbf{r}_{B/A} \qquad v_{B/A} = r_{B/A}\omega \qquad \boxed{18\text{-}14}$$

$$\mathbf{v}_B = \mathbf{v}_A + \omega \times \mathbf{r}_{B/A} \qquad \boxed{18\text{-}15}$$

| (a) General motion, top view | (b) Pure curvilinear translation | (c) Pure rotation |

**FIGURE 18-7**

Schematic of general plane motion

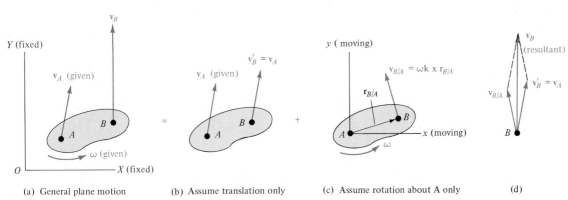

(a) General plane motion     (b) Assume translation only     (c) Assume rotation about A only     (d)

**FIGURE 18-8**

Analysis of velocities in general plane motion

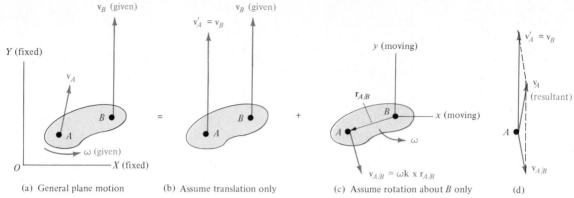

**FIGURE 18-9**

Alternative reference point for analyzing the motion illustrated in Fig. 18-7

since $\mathbf{r}_{B/A}$ has a constant magnitude in a rigid body. The velocity $\mathbf{v}_B$ is obtained as the vector sum of $\mathbf{v}_A$ and $\omega \times \mathbf{r}_{B/A}$, where each quantity is known. This procedure is illustrated in Fig. 18-8 as the superposition of translation and rotation. In Fig. 18-8b all particles have the same velocity $\mathbf{v}_A$. The relative velocity $\mathbf{v}_{B/A}$ would be seen by a nonrotating observer located at point $A$ (and moving with the body) as a pure rotation of $B$ about point $A$ as in Fig. 18-8c. The graphic vector summation of $\mathbf{v}_A$ and $\mathbf{v}_{B/A}$ gives the resultant $\mathbf{v}_B$ in Fig. 18-8d. Note that $\omega$ is the angular velocity of line $AB$ and of the body as a whole.

It is instructive to interchange the roles assigned to the points $A$ and $B$ in the analysis, to determine the velocity of $A$ in terms of $\mathbf{v}_B$. Assume that the rigid body in Fig. 18-9a has the same general plane motion as that in Fig. 18-8a. If the absolute velocity of $B$ and the body's angular velocity $\omega$ are known, the analysis is made by using point $B$ as an intermediate reference point, the origin of the moving reference frame. The translational component of the motion again causes points $A$ and $B$ to have identical velocities both in magnitude and in direction. Therefore, the translational velocity of point $A$ is $\mathbf{v}'_A = \mathbf{v}_B$, as shown in Fig. 18-9b. Here the velocities are different from those in Fig. 18-8b even though the same body and total motion are analyzed in the two figures.

To consider rotation alone, point $B$ is now assumed fixed, with point $A$ moving on a circular path with respect to $B$ as in Fig. 18-9c. Note that the angular velocities shown in Figs. 18-8c and 18-9c represent the same rotation of the body.* The velocity vector of $A$ caused by its rotation about

* To check that $\omega$ is the same using $A$ or $B$ for a moving reference point, assume $\omega_A$ for the angular velocity in Fig. 18-8c, and $\omega_B$ in Fig. 18-9c. The two figures refer to the same body at the same instant. Thus,

$$\mathbf{v}_B = \mathbf{v}_A + \omega_A \times \mathbf{r}_{B/A} \qquad \text{(Fig. 18-8c)}$$

$$\mathbf{v}_A = \mathbf{v}_B + \omega_B \times \mathbf{r}_{A/B} \qquad \text{(Fig. 18-9c)}$$

With $\mathbf{r}_{B/A} = -\mathbf{r}_{A/B}$,

$$\mathbf{v}_B = \mathbf{v}_A + \omega_A \times (-\mathbf{r}_{A/B}) = \mathbf{v}_A - \omega_A \times \mathbf{r}_{A/B}$$

$$\mathbf{v}_A = \mathbf{v}_B + \omega_A \times \mathbf{r}_{A/B}$$

$$\mathbf{v}_A = \mathbf{v}_A - \omega_A \times \mathbf{r}_{A/B} + \omega_B \times \mathbf{r}_{A/B}, \qquad \text{hence } \omega_A = \omega_B = \omega$$

Consequently, the angular velocity $\omega$ of a rigid body in plane motion is independent of the reference point and is a free vector.

point $B$ is

$$\mathbf{v}_{A/B} = \boldsymbol{\omega} \times \mathbf{r}_{A/B} \qquad v_{A/B} = r_{A/B}\omega \qquad \boxed{18\text{-}16}$$

It is seen from Eqs. 18-14 and 18-16 and Figs. 18-8c and 18-9c that $\mathbf{v}_{A/B}$ and $\mathbf{v}_{B/A}$ have parallel lines of action, equal magnitudes, and opposite directions.

  The absolute velocity of point $A$ is the vector sum of its translational velocity ($\mathbf{v}'_A = \mathbf{v}_B$) and its rotational velocity with respect to $B$, $\mathbf{v}_{A/B}$. This is shown in Fig. 18-9d, where $\mathbf{v}_A = \mathbf{v}_B + \boldsymbol{\omega} \times \mathbf{r}_{A/B}$. The resultant vector $\mathbf{v}_A$ is identical to $\mathbf{v}_A$ in Fig. 18-8, as it should be since the same total motion was analyzed in different ways.

  The same procedure can be applied using any pair of points in a rigid body. These concepts can be applied for analyzing the motion of individual or interconnected rigid bodies. In solving problems it is useful to remember the following:

1. The angular velocity $\omega$ of a rigid body in plane motion is independent of the reference point.

2. The common point of two or more pin-jointed members must have the same absolute velocity even though the individual members may have different angular velocities. For example, $\mathbf{v}_O = \mathbf{v}_{O_A} = \mathbf{v}_{O_B}$ in Fig. 18-10.

3. The points of contact in members that are in temporary contact may or may not have the same absolute velocity. If there is sliding between the members, the points in contact have different absolute velocities such as in Fig. 18-11. The absolute velocities of the contacting particles are always the same if there is no sliding (no slip condition) such as in Fig. 18-12.

4. Frequently, the angular velocity of a member is not known, but some points of the member move along defined paths (e.g., the end points of a piston rod). The geometric constraints of motion define the directions of the velocity vectors and are useful in the solution.

5. The geometric center of a wheel rolling on a flat surface moves in rectilinear motion. If there is no slipping at the point of contact, the linear distance the center point travels is equal to that portion of the rim circumference that has rolled along the flat surface (Fig. 18-13). For an angular displacement $\theta$ of any radial line $r$ on the wheel, the distance the center moves is given by the arc length formula, $s = r\theta$. Differentiating this equation once with respect to time gives the velocity of the geometric center of a wheel of radius $r$ rolling without slip, $v = r\omega$.

**FIGURE 18-10**

Velocity of a common point of two members

Slipping at the point of contact

**FIGURE 18-11**

Two identical wheels slipping relative to one another

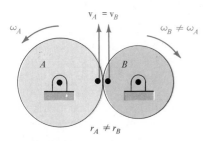

No slipping at the point of contact

**FIGURE 18-12**

Two different wheels rotating without slipping relative to one another

**FIGURE 18-13**

Wheel rolling without slipping on a flat surface

# INSTANTANEOUS CENTER OF ROTATION

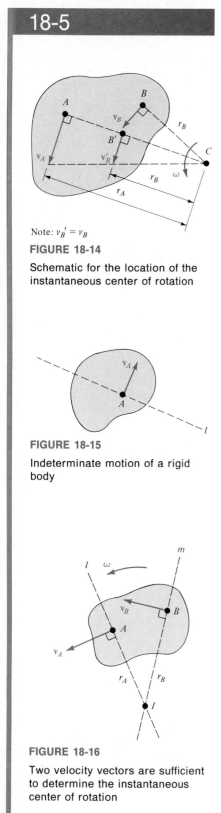

Note: $v_B' = v_B$

**FIGURE 18-14**

Schematic for the location of the instantaneous center of rotation

**FIGURE 18-15**

Indeterminate motion of a rigid body

**FIGURE 18-16**

Two velocity vectors are sufficient to determine the instantaneous center of rotation

All lines in the plane of motion in a rigid body have the same angular velocity at a given instant, and this fact is very useful in the analysis of velocities of the body. It is worthwhile to reconsider the common case of a body that is rotating about a fixed axis. In general this axis may or may not intersect the body.

For example, assume that the body in Fig. 18-14 is rotating at an angular velocity $\omega$ about the fixed point $C$. This is enough information to determine the velocity vector of any point such as $A$ or $B$ in the body if its distance from the axis of rotation is given. Each of these velocity vectors is normal to the line between the axis of rotation and the point of interest; $\mathbf{v}_A = \boldsymbol{\omega} \times \mathbf{r}_A$ and $\mathbf{v}_B = \boldsymbol{\omega} \times \mathbf{r}_B$. Their magnitudes are proportional to the angular velocity and the particular distances to the center of rotation, $C$. The proportionality of the magnitudes $v_A$ and $v_B$ can be illustrated by considering points $A$ and $B'$, with $B'$ on line $AC$, at a distance $r_B$ from point $C$ (Fig. 18-14). Here $v_A = r_A\omega$ and $v_{B'} = v_B = r_B\omega$. The ratio of these speeds involves only the radial distances from the center of rotation, $v_A/v_B = r_A/r_B$, since $v_A/r_A = v_B/r_B = \omega$. These concepts are useful in analyzing the rotational motion of bodies that do not rotate about fixed axes.

The *general plane motion* (translation *and* rotation) of a rigid body may be analyzed if there are at least two points in the body whose velocity vectors are known. The need for knowing two velocity vectors is explained as follows.

**1. *One velocity vector is specified.*** Assume that a rigid body is in plane motion and that the velocity of one point is given as in Fig. 18-15. The motion of the rigid body is indeterminate because the angular velocity of the body cannot be calculated from $\mathbf{v}_A$. The only statement that can be made about the total motion is that the body *may be* rotating at the given instant about a point on line $l$ which is perpendicular to $\mathbf{v}_A$. Pure translation may be considered a rotation about a point on line $l$ at an infinite distance from particle $A$. However, this motion cannot be analyzed with sufficient certainty.

**2. *Two velocity vectors are specified.*** Consider a rigid body for which the velocities of two points are given as in Fig. 18-16. The different magnitudes and directions of these vectors indicate that the body is rotating. Assume that the rotation is about a point that is fixed at the instant considered. The velocity vector of any point in the body must be perpendicular to its radial line drawn from the instantaneously fixed point, the center of rotation, as in rotation about a fixed axis. Referring to Fig. 18-16, the center of rotation must be somewhere on line $l$ for particle $A$, and somewhere on line $m$ for particle $B$. To simultaneously satisfy both of these conditions, the body must be rotating instantaneously about point $I$, the intersection of lines that are normal to the given velocity vectors. This means that at the instant when $v_A$ and $v_B$ are specified, the body is rotating about point $I$ which is called the *instantaneous center of rotation*. There are several properties of the instantaneous center of rotation that are useful or restrictive in the analysis of motion:

**a.** The angular velocity of the body can be determined from the velocity and radial distance of a single point when the instantaneous center is located. From Fig. 18-16, $\omega = v_A/r_A$, or $\omega = v_B/r_B$. Since $\omega$ applies to all lines in the plane of motion of the rigid body, the velocity of all other points in the body can be readily determined.

**b.** The instantaneous center of rotation has zero velocity at the given instant, and it may or may not be a point on the body.

**c.** The instantaneous center may be determined when the two given velocity vectors are parallel but of different magnitude. In that case the instantaneous center is obtained as point $C$ in Fig. 18-14, where a triangular construction can be made using points $A$ and $B'$ and the parallel vectors $\mathbf{v}_A$ and $\mathbf{v}'_B$. Alternatively, the method shown in Fig. 18-16 may be used.

**d.** Both the absolute position of the instantaneous center and its position relative to the body generally change in time. The instantaneous center can be imagined to be a fixed center of rotation only for the instant when the velocity vectors are valid. The locus of the absolute positions of the instantaneous center is called the *space centrode*. The locus of the positions of the instantaneous center relative to the body is called the *body centrode*. As shown schematically in Fig. 18-17, these curves are tangent at point $I$ for the instant considered. The body centrode appears to roll on the space centrode as the body moves. For example, consider a wheel rolling on a flat surface as in Fig. 18-18.

**e.** The instantaneous center of rotation has zero velocity, but in general motion it has an acceleration. Thus, the instantaneous center of velocity is *not* an instantaneous center of zero acceleration, and cannot be used for convenient analyses of accelerations. This limitation of the concept of instantaneous center of rotation is important to remember. It should also be noted that the instantaneous center of rotation is introduced for convenience, but it implies no new relationship in kinematics.

Body centrode
(Locus of instantaneous centers
relative to the body)

*I*

Space centrode
(Locus of instantaneous
centers, fixed in space)

**FIGURE 18-17**

Body and space centrodes

Body centrode (locus of point $I$,
relative to the wheel)

Space centrode (locus of point $C$,
fixed in space)

**FIGURE 18-18**

Body and space centrodes of a
rolling wheel

# EXAMPLE 18-4

The chord line $OB$ of an airfoil pivoted at point $O$ is controlled to rotate about $O$ by the link $AB$ (Fig. 18-19a). Point $A$ is the pivot on a slider which has an instantaneous velocity of 2.6 m/s, as shown. For the instantaneous position shown, **(a)** determine the angular velocities of $AB$ and $BO$ by the *instantaneous center method*, and **(b)** repeat part (a) by the *vector method*.

## SOLUTION

It is seen by inspection of Fig. 18-19a that point $A$ is constrained to move in rectilinear motion, rigid body $OB$ is constrained to rotate about point $O$, and rigid body $AB$ moves in general plane motion.

**(a)** Locate the instantaneous center $I$ for link $AB$ by constructing normals to the velocity vectors at $A$ and $B$ as shown in Fig. 18-19b. $A$ and $B$ move in circular paths about $I$ at the instant considered,

$$\omega_{AB}\,(\overline{IA}) = \omega_{AB}\,(1.3\,\text{m}) = v_A = 2.6\,\text{m/s}$$

$$\omega_{AB} = \frac{2.6\,\text{m/s}}{1.3\,\text{m}} = 2\,\text{rad/s} \qquad \text{(counterclockwise)}$$

Also,

$$v_B = \omega_{AB}\,(\overline{IB}) = (2\,\text{rad/s})(1.4\,\text{m}) = 1.2\omega_{OB}\,\text{m/s}$$

$$\omega_{OB} = 2.33\,\text{rad/s} \qquad \text{(clockwise)}$$

**(b)** $\mathbf{v}_B = \mathbf{v}_A + \mathbf{v}_{B/A} = \mathbf{v}_A + \boldsymbol{\omega}_{AB} \times \mathbf{r}_{B/A}$

where $\quad \mathbf{v}_B = \boldsymbol{\omega}_{OB} \times \mathbf{r}_{B/O} = (-\omega_{OB}\mathbf{k}) \times (-1.2\mathbf{i}\,\text{m}) = 1.2\omega_{OB}\mathbf{j}\,\text{m/s}$   *(assumed clockwise)*

$\mathbf{v}_A = (2.4\mathbf{i} + 1.0\mathbf{j})\,\text{m/s}$

$\mathbf{v}_{B/A} = (\omega_{AB}\mathbf{k}) \times (0.9\mathbf{i} + 1.2\mathbf{j})\,\text{m} = (-1.2\omega_{AB}\mathbf{i} + 0.9\omega_{AB}\mathbf{j})\,\text{m/s}$   *(assumed counterclockwise)*

Then

$$(1.2\,\text{m})\,\omega_{OB}\mathbf{j} = (2.4\mathbf{i} + 1.0\mathbf{j})\,\text{m/s} - (1.2\,\text{m})\,\omega_{AB}\mathbf{i} + (0.9\,\text{m})\,\omega_{AB}\mathbf{j}$$

Equating $\mathbf{i}$ and $\mathbf{j}$ components yields $\omega_{AB} = 2\,\text{rad/s}$ and $\omega_{OB} = 2.33\,\text{rad/s}$. Positive answers indicate that both assumed directions were correct. Using vector notation,

$$\boldsymbol{\omega}_{AB} = 2\mathbf{k}\,\text{rad/s} \qquad \text{and} \qquad \boldsymbol{\omega}_{OB} = -2.33\mathbf{k}\,\text{rad/s}$$

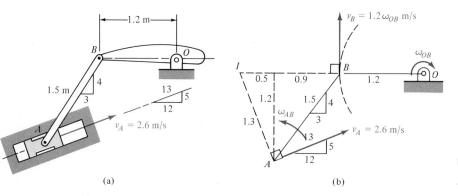

(a)        (b)

**FIGURE 18-19**

Analysis of pivoted airfoil

EXAMPLE 18-5

The wheel of radius $r$ is rolling without slipping along a horizontal surface at an angular velocity of $\omega$ as shown in Fig. 18-20a. Using the method of instantaneous center, determine the velocities of points $D$, $A$, $O$, and $B$.

## SOLUTION

Since there is no slipping, the point of contact $I$ has zero velocity at the instant shown. Therefore, $I$ is the instantaneous center and all points move along circular paths about $I$ at the instant shown. Using Fig. 18-20b,

$$v_D = d\omega \qquad \text{(velocity varies linearly along line } IA)$$

$$\mathbf{v}_D = d\omega\mathbf{i}, \qquad \mathbf{v}_O = r\omega\mathbf{i}, \qquad \text{and} \qquad \mathbf{v}_A = 2r\omega\mathbf{i}$$

Similarly, since $IB = \sqrt{2}\,r$,

$$v_B = \sqrt{2}\,r\omega \qquad \text{(direction of } v_B \text{ shown)}$$

In vector form,

$$\mathbf{v}_B = r\omega\mathbf{i} + r\omega\mathbf{j}$$

(a)

(b)

**FIGURE 18-20**

Analysis of wheel rolling without slipping

EXAMPLE 18-5    725

EXAMPLE 18-6

The ladder shown in Fig. 18-21a is sliding down the wall. At the instant shown. end $A$ has a velocity of 16 m/s to the right. Determine **(a)** the angular velocity of the ladder, and **(b)** the velocity of end $B$ at this instant. Use the instantaneous center method.

SOLUTION

End $B$ is constrained to vertical motion by the wall. Construct normals to $\mathbf{v}_A$ and $\mathbf{v}_B$ to show that the instantaneous center is at $I$ in Fig. 18-21b. Considering the ladder $AB$ as instantaneously rotating about $I$ at the rate $\omega_{AB}$,

$$\omega_{AB} (8\,\text{m}) = v_A = 16\,\text{m/s} \quad \text{and} \quad \omega_{AB} (6\,\text{m}) = v_B$$

Therefore,

$$\omega_{AB} = \frac{16\,\text{m/s}}{8\,\text{m}} = 2\,\text{rad/s} \quad \text{and} \quad v_B = (2\,\text{rad/s})(6\,\text{m}) = 12\,\text{m/s}$$

In vector form,

$$\omega_{AB} = 2\mathbf{k}\,\text{rad/s} \quad \text{and} \quad \mathbf{v}_B = -12\mathbf{j}\,\text{m/s}$$

(a)

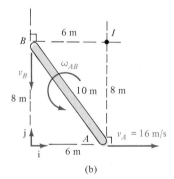

(b)

**FIGURE 18-21**

Analysis of ladder sliding down a wall

**18–26** In the design of a motorcycle the absolute velocities of particles $A$ and $B$ on the tire are to be determined. Assume that $r = 0.4\,\text{m}$, $v_0 = 100\,\text{km/h}$, and the wheel rolls without slipping. Determine the velocities of $A$ and $B$ using the vector equations 18-14 and 18-15.

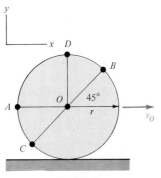

Fig. P18-26

**18–27** Solve Prob. 18–26 for the velocities of points $C$ and $D$, with $r = 18\,\text{in.}$ and $v_0 = 60\,\text{mi/h}$ using the instantaneous center of rotation.

**18–28** A cable drum of radius $R = 1\,\text{ft}$ is attached to rollers of radius $r = 3\,\text{in.}$ which roll on two horizontal tracks. Calculate the absolute speed of the cable if $\omega = 0.4\,\text{rad/s}$.

Fig. P18-28

**18–29** The wheel rolling to the left winds up a cable on a 0.2-m-diameter drum. The diameter of the wheel is 0.4 m. Calculate the absolute speed $v$ of the cable if the wheel's angular speed is 2 rad/s. Use vector methods and assume the wheel does not slip.

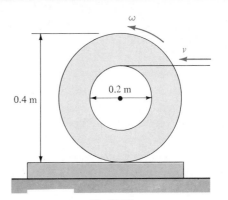

Fig. P18-29

**18–30** Solve Prob. 18–29 using the instantaneous center of rotation.

**18–31** The bicycle is traveling at a speed of 30 km/h. It has wheels of 0.7 m diameter, sprocket $A$ of 0.2 m diameter, and a pedal circle of 0.3 m diameter. Plot the angular speed of sprocket $A$ as a function of the sprocket diameter of $B$. Let $d_B = 0.5\,\text{m}$, $0.7\,\text{m}$, and $0.9\,\text{m}$.

Fig. P18-31

**18–32** The airspeed of a helicopter is $v = 100\,\text{ft/s}$. Determine the maximum and minimum air speed at the tips of the large blades if they rotate at an angular speed $\omega = 160\,\text{rad/s}$.

**18–33** A helicopter shown in top view is moving horizontally at $v_0 = 100$ mi/h. Determine the instantaneous centers of rotation and the maximum airspeeds for the blades $A$, $B$, and $C$ in the positions shown when the blades are rotating clockwise at 150 rpm.

*(helicopter top-view figure)*

**Fig. P18-33**

**18–34** The outer case of a ball bearing is fixed. The shaft has a diameter of $D = 0.1$ m, and the diameter of the balls is $d = 0.01$ m. Calculate the angular speed of the shaft for which the speed $v_B$ of the balls relative to the outer case is 10 m/s. The balls do not slip.

**Fig. P18-34**

**18–35** Determine the instantaneous center of rotation and the angular speed of a ball in the bearing described in Prob. 18–34.

**18–36** In an aerodynamic investigation of tennis balls a ball is given speeds $v$ and $\omega$ as shown. Determine the maximum and minimum speeds of points at the surface of a ball.

$\omega = 31.4$ rad/s

$v = 50$ m/s

$d = 65$ mm

**Fig. P18-36**

**18–37** Joint $A$ of the link is sliding down the incline at a speed of $v_A = 2$ m/s. Determine the speed of joint $B$ and angular velocity of the link $AB$ for the position $\theta = 20°$. Use the instantaneous center of rotation.

$A$  0.3 m

$60°$  $\theta$

$B$

**Fig. P18-37**

**18–38** The four identical links of a toggle mechanism are moved by the hydraulic actuator $A$. Calculate the relative velocity of joints $C$ and $E$ if $B$ and $D$ are separating at the rate of 4 ft/s.

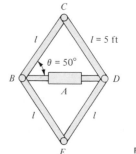

**Fig. P18-38**

**18–39** Determine the instantaneous center of rotation and the angular velocity of link *CD* in Prob. 18–38.

**18–40** A horizontal platform is supported by 4-m-long bars that are pin-jointed at their ends and centers. Calculate the vertical speed of the platform if the actuator *A* moves wheels *B* and *C* apart at a relative velocity of $v_{B/C} = 1.4\,\text{m/s}$.

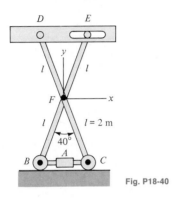

Fig. P18-40

**18–41** Determine the instantaneous center of rotation and the angular velocity of member *CD* in Prob. 18–40.

**18–42** A mechanism consists of two rigid bars *AB* and *BC* of length 2.4 m each. Bar *AB* rotates about pin *A* while pin *C* moves in a horizontal guide. Determine the velocity of pin *C* and the angular speed $\dot\phi$ of bar *BC* if $\theta = 20°$ to the right of the vertical and $\dot\theta = 0.8\,\text{rad/s}$ clockwise.

Fig. P18-42

**18–43** Determine the velocity of joint *B* in the three-bar linkage if $\theta$ is zero and bar *BC* is horizontal. Bar *CD* is rotating clockwise at a speed of 200 rpm. Use vector methods.

Fig. P18-43

**18–44** Solve Prob. 18–43 using the instantaneous center of rotation.

**18–45** Determine the velocity of piston *C* for which the angular speed of the crank *AB* is maintained at 800 rpm at $\theta = 30°$. Use vector methods.

Fig. P18-45

**18–46** Solve Prob. 18–45 using the instantaneous center of rotation.

**18–47** Calculate the angular velocity of the slotted arm *CD* for $\theta = 30°$. Arm *AB* rotates at 300 rpm clockwise. Use vector methods.

Fig. P18-47

**18–48** Solve Prob. 18–47 using the instantaneous center of rotation.

**18–49** Wheel $A$ rotates counterclockwise at 200 rpm. Determine the angular speed of wheel $B$ for which the angular speed of the arm $AB$ is 50 rpm clockwise.

$r_B = 0.08$ m

$r_A = 0.20$ m

**Fig. P18-49**

**18–50** A gear system is modeled by a small wheel rolling inside a larger cylinder. Calculate the angular velocity of the small wheel if the connecting arm $AB$ rotates clockwise at 150 rpm.

$R = 6$ in.

$r = 1$ in.

**Fig. P18-50**

**18–51** Determine the maximum and minimum absolute speeds of particles on the small wheel in Fig. P18-50. How do these speeds depend on the angular velocity of the connecting arm?

**18–52** An automobile has 0.60-m-diameter wheels and is traveling at $v_0 = 80$ km/h. Determine the instantaneous center of zero velocity and the maximum absolute velocity of a wheel that has a counterclockwise angular speed of 30 rpm.

$v_0$

**Fig. P18-52**

**18–53** A street sweeper is moving at speed $v = 6$ km/h while its 0.8-m diameter brush is rotating clockwise at 100 rpm. Determine the instantaneous center of rotation of the brush and the bristle's speed against the ground at point $A$.

$v$

$\omega$

$A$

**Fig. P18-53**

**18–54** A linear bearing has 0.4-in.-diameter rollers. Track $A$ is moving to the left at $v_A = 4$ ft/s while track $B$ is moving to the right at $v_B = 5$ ft/s. Determine the instantaneous center of rotation of a roller that moves without slipping. Also calculate the velocities of points 1 and 2. All linear velocities are with respect to the fixed $xy$ frame.

$y$

$x$

$B$ 2

$v_B$

$O$ 1

$v_A$

$A$

**Fig. P18-54**

The method developed for analyzing absolute and relative velocities of a rigid body in Sec. 18-4 can be extended for the analysis of accelerations. Differentiating Eq. 18-13 with respect to time results in the vector expression

$$\mathbf{a}_B = \mathbf{a}_A + \mathbf{a}_{B/A} \qquad \boxed{18\text{-}17}$$

where $\mathbf{a}_B$ is the absolute acceleration of an arbitrary point $B$ in the body, $\mathbf{a}_A$ is the absolute acceleration of a given reference point $A$ in the body, and $\mathbf{a}_{B/A}$ is the relative acceleration of $B$ with respect to $A$. This is the same expression as that for any two particles.

It is necessary to consider the relative acceleration in detail. Using Eqs. 18-14 and 18-6 gives

$$\mathbf{a}_{B/A} = \frac{d}{dt}(\omega \times \mathbf{r}_{B/A}) \qquad \boxed{18\text{-}18}$$

$$= \dot{\omega} \times \mathbf{r}_{B/A} + \omega \times \dot{\mathbf{r}}_{B/A}$$

$$= \alpha \times \mathbf{r}_{B/A} + \omega \times (\omega \times \mathbf{r}_{B/A})$$

Equation 18-17 is written in detail as

$$\mathbf{a}_B = \mathbf{a}_A + \alpha \times \mathbf{r}_{B/A} + \omega \times (\omega \times \mathbf{r}_{B/A}) \qquad \boxed{18\text{-}19}$$

Since terms in the form of $\alpha \times \mathbf{r}$ and $\omega \times (\omega \times \mathbf{r})$ represent tangential and normal accelerations, respectively, Eq. 18-19 can be expressed as

$$\mathbf{a}_B = \underbrace{\mathbf{a}_A}_{\text{translation}} + \underbrace{(\mathbf{a}_{B/A})_t + (\mathbf{a}_{B/A})_n}_{\text{rotation}} \qquad \boxed{18\text{-}20}$$

The magnitudes of the rotational components are written as

$$(a_{B/A})_t = r_{B/A}\alpha \qquad \boxed{18\text{-}21\text{a}}$$

$$(a_{B/A})_n = r_{B/A}\omega^2 \qquad \boxed{18\text{-}21\text{b}}$$

where the subscripts $B/A$ are frequently omitted if the meaning of each quantity is clear. It should be noted that in these expressions $\omega$ and $\alpha$ are the angular velocity and acceleration of the body, and they are common to all lines parallel to the plane of motion in that body. On the other hand, the position vector $\mathbf{r}_{B/A}$ is valid only for the two points $A$ and $B$.

Equation 18-20 can be appreciated from the schematic diagram in Fig. 18-22. Assume that a rigid body is in general plane motion and that the absolute accelerations of points $A$ and $B$ are known. One way to look at the details of this motion is to consider it as a simultaneous translation with acceleration $\mathbf{a}_A$ and a rotation about point $A$. The major complication in this analysis is that point $B$ may have two components of acceleration as it rotates with respect to point $A$. The tangential component $(\mathbf{a}_{B/A})_t$ exists whenever $\omega$ is not constant. The normal component $(\mathbf{a}_{B/A})_n$ always exists when $\omega$ is other than zero, and it is always directed toward $A$. The vector sum of the translational and rotational components must equal the absolute acceleration of point $B$ as in Fig. 18-22e.

(a) General plane motion

(b) Assume translation only

(c) Assume rotation about $A$ only

(d) Velocity and accelerations of point $B$
from part (c) (not to scale)

(e)

**FIGURE 18-22**

Analysis of accelerations in general plane motion

It is worth noting that in general the absolute acceleration of the moving reference point (such as $A$ in Fig. 18-22) is itself the resultant of normal and tangential components of acceleration. The motions of points $A$ and $B$ are still correctly related according to Eq. 18-18. In solving problems it is useful to remember the following:

1. The angular velocity $\omega$ and the angular acceleration $\alpha$ of a rigid body in plane motion are independent of the reference point.
2. The common points of pin-jointed members must have the same absolute acceleration even though the individual members may have different angular velocities and angular accelerations such as in Fig. 18-23.

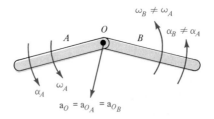

**FIGURE 18-23**

Acceleration of the common point of two members

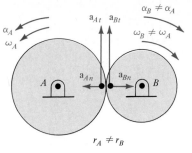

(a) No slipping at the point of contact

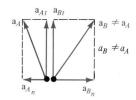

(b) Accelerations of points in contact on the two wheels

**FIGURE 18-24**

Accelerations of points in contact on two wheels

3. The points of contact in members that are in temporary contact generally do not have the same absolute acceleration. Even when there is no sliding between the members as in Fig. 18-24, only the tangential accelerations of the points in contact are the same, while the normal accelerations are frequently different in magnitude and direction.

4. The instantaneous center of zero velocity in general has an acceleration and should *not* be used as a reference point for accelerations unless its acceleration is known and included in the analysis.

5. Frequently, the angular acceleration of a member is not known, but some points of the member move along defined paths (e.g., the end points of a piston rod). The geometric constraints of motion can be used to define the directions of normal and tangential acceleration vectors and are useful in the solution.

6. The geometric center of a wheel rolling on a flat surface moves in rectilinear motion. If there is no slipping at the point of contact, the linear acceleration of the center point is parallel to the flat surface and equal to $r\alpha$ for a wheel of radius $r$ and angular acceleration $\alpha$ (Fig. 18-25).

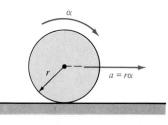

**FIGURE 18-25**

Acceleration of the center of a wheel rolling without slipping on a flat surface

EXAMPLE 18-7

The mechanical hoe in Fig. 18-26 has the following dimensions: $OA = 2.3$ m, $AB = 1.7$ m. Assume that $\theta = 25°$, $\omega_{OA} = 0.5$ rad/s counterclockwise, $\alpha_{OA} = 0$, $\phi = 30°$, $\omega_{AB} = 0.8$ rad/s clockwise, and $\alpha_{AB} = 2$ rad/s² counterclockwise. Determine the acceleration of the bucket joint $B$ for the given configuration.

## SOLUTION

Arm $OA$ is constrained to rotate about point $O$ with a constant angular velocity $\omega_{OA} = 0.5\mathbf{k}$ rad/s. Arm $AB$ is moving in general plane motion with $\omega_{AB} = -0.8\mathbf{k}$ rad/s and $\boldsymbol{\alpha}_{AB} = 2\mathbf{k}$ rad/s².

From Eq. 18-17,

$$\mathbf{a}_B = \mathbf{a}_A + \mathbf{a}_{B/A}$$

$$\begin{aligned}
\mathbf{a}_A &= \boldsymbol{\omega}_{OA} \times (\boldsymbol{\omega}_{OA} \times \mathbf{r}_{A/O}) + \boldsymbol{\alpha}_{OA} \times \mathbf{r}_{A/O} \\
&= 0.5\mathbf{k} \times [0.5\mathbf{k} \times 2.3(\sin 25°\,\mathbf{i} + \cos 25°\,\mathbf{j})] \\
&= (-0.24\mathbf{i} - 0.52\mathbf{j}) \text{ m/s}^2
\end{aligned}$$

$$\begin{aligned}
\mathbf{a}_{B/A} &= \boldsymbol{\omega}_{AB} \times (\boldsymbol{\omega}_{AB} \times \mathbf{r}_{B/A}) + \boldsymbol{\alpha}_{AB} \times \mathbf{r}_{B/A} \\
&= -0.8\mathbf{k} \times [(-0.8\mathbf{k}) \times 1.7(\sin 30°\,\mathbf{i} - \cos 30°\,\mathbf{j})] \\
&\quad + 2\mathbf{k} \times 1.7(\sin 30°\,\mathbf{i} - \cos 30°\,\mathbf{j}) \\
&= (-0.54\mathbf{i} + 0.94\mathbf{j} + 2.94\mathbf{i} + 1.7\mathbf{j}) \text{ m/s}^2 \\
&= (2.40\mathbf{i} + 2.64\mathbf{j}) \text{ m/s}^2
\end{aligned}$$

Combining the results for $\mathbf{a}_A$ and $\mathbf{a}_{B/A}$ gives

$$\begin{aligned}
\mathbf{a}_B &= -0.24\mathbf{i} - 0.52\mathbf{j} + 2.40\mathbf{i} + 2.64\mathbf{j} \\
&= (2.16\mathbf{i} + 2.12\mathbf{j}) \text{ m/s}^2
\end{aligned}$$

## JUDGMENT OF THE RESULTS

Since arm $OA$ is rotating with a constant angular velocity, point $A$ should have only a normal component of acceleration directed toward point $O$ of magnitude equal to $r_{OA}(\omega_{OA})^2 = (2.3 \text{ m})(0.5 \text{ rad/s})^2 = 0.575 \text{ m/s}^2$. A quick check of vector $\mathbf{a}_A$ calculated above reveals that

$$a_A = \sqrt{(0.24 \text{ m/s}^2)^2 + (0.52 \text{ m/s}^2)^2} = 0.573 \text{ m/s}^2$$

and

$$\tan^{-1}\frac{0.24}{0.52} = 24.8° \approx \theta$$

The slight discrepancies are due to round-off error.

(a)

(b)

**FIGURE 18-26**

Analysis of a mechanical hoe

EXAMPLE 18-8

An automobile tire of $r = 0.3\,$m is rolling without slipping. Determine the acceleration of points $A$, $B$ and $I$ in Fig. 18-27 if $v_O = 10\,$m/s and $a_O = 4\,$m/s$^2$, both to the right.

## SOLUTION

Since the wheel is rolling without slipping, the geometric center $O$ moves in rectilinear translation with $v_O = r\omega$ and $a_O = r\alpha$. Thus

$$\omega = \frac{v_O}{r} = \frac{10\,\text{m/s}}{0.3\,\text{m}} = 33.3\,\text{rad/s} \qquad \text{(clockwise)}$$

$$\boldsymbol{\omega} = -33.3\mathbf{k}\,\text{rad/s}$$

and

$$\alpha = \frac{a_O}{r} = \frac{4\,\text{m/s}^2}{0.3\,\text{m}} = 13.3\,\text{rad/s}^2 \qquad \text{(clockwise)}$$

$$\boldsymbol{\alpha} = -13.3\mathbf{k}\,\text{rad/s}^2$$

From Eq. 18-17,

$$\mathbf{a}_B = \mathbf{a}_O + \mathbf{a}_{B/O}$$

Using Eqs. 18-18 through 18-21b,

$$\mathbf{a}_{B/O} = (\mathbf{a}_{B/O})_n + (\mathbf{a}_{B/O})_t$$

$$(\mathbf{a}_{B/O})_n = \boldsymbol{\omega} \times (\boldsymbol{\omega} \times \mathbf{r}_{B/O}) = -33.3\mathbf{k} \times [-33.3\mathbf{k} \times (-0.3)\mathbf{i}]$$
$$= 333.3\mathbf{i}\,\text{m/s}^2$$

$$(\mathbf{a}_{B/O})_t = \boldsymbol{\alpha} \times \mathbf{r}_{B/O} = -13.3\mathbf{k} \times (-0.3)\mathbf{i}$$
$$= 4\mathbf{j}\,\text{m/s}^2$$

Combining these equations gives

$$\mathbf{a}_B = \mathbf{a}_O + \mathbf{a}_{B/O} = \mathbf{a}_O + (\mathbf{a}_{B/O})_n + (\mathbf{a}_{B/O})_t$$
$$= (4\mathbf{i} + 333.3\mathbf{i} + 4\mathbf{j})\,\text{m/s}^2$$
$$= (337.3\mathbf{i} + 4\mathbf{j})\,\text{m/s}^2$$

Solving for point $A$ similarly yields

$$\mathbf{a}_A = \mathbf{a}_O + \mathbf{a}_{A/O} = \mathbf{a}_O + (\mathbf{a}_{A/O})_n + (\mathbf{a}_{A/O})_t$$
$$= 4\mathbf{i} + \boldsymbol{\omega} \times (\boldsymbol{\omega} \times \mathbf{r}_{A/O}) + \boldsymbol{\alpha} \times \mathbf{r}_{A/O}$$
$$= 4\mathbf{i} + (-33.3\mathbf{k}) \times (-33.3\mathbf{k} \times 0.3\mathbf{j}) + (-13.3\mathbf{k} \times 0.3\mathbf{j})$$
$$= (8\mathbf{i} - 333.3\mathbf{j})\,\text{m/s}^2$$

Similarly for point $I$, the instantaneous center,

$$\mathbf{a}_I = \mathbf{a}_O + \mathbf{a}_{I/O}$$
$$= 4\mathbf{i} + (-33.3\mathbf{k}) \times [-33.3\mathbf{k} \times (-0.3\mathbf{j})] + [-13.3\mathbf{k} \times (-0.3\mathbf{j})]$$
$$= 333.3\mathbf{j}\,\text{m/s}^2$$

Note that the acceleration of this instantaneous center is *not* zero.

**FIGURE 18-27**

Analysis of accelerations on a rolling wheel

EXAMPLE 18-8    735

# EXAMPLE 18-9

A tire of radius 1 ft is rolling to the right with its geometric center $O$ moving with $\mathbf{v}_O = 16\mathbf{i}$ and $\mathbf{a}_O = 10\mathbf{i}$ at the instant shown. The angular velocity of the tire at the instant shown is $\omega = 20$ rad/s clockwise and the angular acceleration is $\alpha = 16$ rad/s² counterclockwise (Fig. 18-28a). Determine (a) the velocity of point $P$ by both the *vector method* and the *method of instantaneous center*, and (b) the acceleration of point $P$.

## SOLUTION

Since $v_O \neq r\omega$ and $a_O \neq r\alpha$, the wheel must be slipping at point $P$.

**(a) Vector Method:**

$$\mathbf{v}_P = \mathbf{v}_O + \boldsymbol{\omega} \times \mathbf{r}_{P/O}$$
$$= 16\mathbf{i} + (-20\mathbf{k}) \times (-\mathbf{j})$$
$$= 16\mathbf{i} - 20\mathbf{i}$$
$$\mathbf{v}_P = -4\mathbf{i} \, \text{ft/s} \qquad \text{(slipping to the left)}$$

**Method of Instantaneous Center:** Since the instantaneous center $I$ has zero velocity at this instant,

$$0 = \mathbf{v}_I = \mathbf{v}_O + \boldsymbol{\omega} \times (\mathbf{r}_{I/O})$$
$$= 16\mathbf{i} + (-20\mathbf{k}) \times (-p\mathbf{j})$$
$$0 = 16\mathbf{i} - 20p\mathbf{i}$$
$$p = 0.8 \, \text{ft} \qquad \text{(below 0)}$$

Consequently, point $P$ may be thought of as instantaneously moving in a circular path of radius $1 - 0.8 = 0.2$ ft about $I$ (Fig. 18-27b),

$$\mathbf{v}_P = -(0.2)(\omega)\mathbf{i} = -(0.2)(20)\mathbf{i} = -4\mathbf{i} \, \text{ft/s} \qquad \text{(slipping to the left)}$$

**(b)** $\mathbf{a}_P = \mathbf{a}_O + \boldsymbol{\alpha} \times \mathbf{r}_{P/O} + \boldsymbol{\omega} \times (\boldsymbol{\omega} \times \mathbf{r}_{P/O})$
$$= 10\mathbf{i} + (16\mathbf{k}) \times (-\mathbf{j}) + (-20\mathbf{k}) \times [(-20\mathbf{k}) \times (-\mathbf{j})]$$
$$= (26\mathbf{i} + 400\mathbf{j}) \, \text{ft/s}^2$$

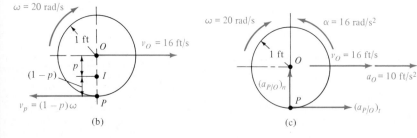

(a)

(b)

(c)

**FIGURE 18-28**

Analysis of a tire rolling and slipping

EXAMPLE 18-10

Determine the instantaneous acceleration $\mathbf{a}_A$ of the slider $A$ so the chord line $OB$ of the airfoil will have an angular acceleration of $2\,\text{rad/s}^2$ counterclockwise for the position shown in Fig. 18-29a. Refer to the results of Ex. 18-4 for the angular velocities of $AB$ and $OB$, which are $\omega_{AB} = 2\mathbf{k}\,\text{rad/s}$ and $\omega_{OB} = -2.33\mathbf{k}\,\text{rad/s}$. Also determine the angular acceleration of link $AB$ at this instant.

## SOLUTION

From Fig. 18-29b, realizing that $B$ moves in a circular path about $O$,

$$\mathbf{a}_B = 1.2(2.33)^2\mathbf{i} - 1.2(2)\mathbf{j}$$
$$= (6.51\mathbf{i} - 2.4\mathbf{j})\,\text{m/s}^2$$

Referring to Fig. 18-29c,

$$\mathbf{a}_B = \mathbf{a}_A + \mathbf{a}_{B/A} = \mathbf{a}_A + \boldsymbol{\alpha}_{AB} \times \mathbf{r}_{B/A} + \boldsymbol{\omega}_{AB} \times (\boldsymbol{\omega}_{AB} \times \mathbf{r}_{B/A})$$

where

$$(\mathbf{a}_{B/A})_t = \boldsymbol{\alpha}_{AB} \times \mathbf{r}_{B/A} = 1.5\alpha_{AB}\mathbf{n}_t$$

or

$$(\mathbf{a}_{B/A})_t = -\frac{4}{5}(1.5\alpha_{AB})\mathbf{i} + \frac{3}{5}(1.5\alpha_{AB})\mathbf{j} = (-1.2\alpha_{AB}\mathbf{i} + 0.9\alpha_{AB}\mathbf{j})\,\text{m/s}^2$$

Also,

$$(\mathbf{a}_{B/A})_n = \boldsymbol{\omega}_{AB} \times (\boldsymbol{\omega}_{AB} \times \mathbf{r}_{B/A}) = 6.0\mathbf{n}_n$$

or

$$(\mathbf{a}_{B/A})_n = -\frac{3}{5}(6.0)\mathbf{i} - \frac{4}{5}(6.0)\mathbf{j} = (-3.6\mathbf{i} - 4.8\mathbf{j})\,\text{m/s}^2$$

Therefore, since $\mathbf{a}_B = \mathbf{a}_A + \mathbf{a}_{B/A}$,

$$6.51\mathbf{i} - 2.4\mathbf{j} = \frac{12}{13}a_A\mathbf{i} + \frac{5}{13}a_A\mathbf{j} - 1.2\alpha_{AB}\mathbf{i} + 0.9\alpha_{AB}\mathbf{j} - 3.6\mathbf{i} - 4.8\mathbf{j}$$

Solving for $a_A$ and $\alpha_{AB}$ by equating the $\mathbf{i}$ and $\mathbf{j}$ components yields $a_A = 9.24$ and $\alpha_{AB} = -1.31$. Therefore, in vector notation,

$$\mathbf{a}_A = (8.53\mathbf{i} + 3.55\mathbf{j})\,\text{m/s}^2 \qquad \text{and} \qquad \boldsymbol{\alpha}_{AB} = -1.31\mathbf{k}\,\text{rad/s}^2$$

(a)

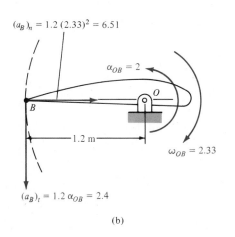

**FIGURE 18-29**

Acceleration analysis of pivoted airfoil

(b)

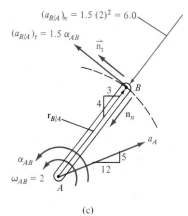

(c)

EXAMPLE 18-10   737

# PROBLEMS

**18–55** A wheel of radius $r = 0.5$ m is rolling without slipping. Determine the accelerations of points $A$ and $B$ for $\mathbf{v}_O = 20\mathbf{i}$ m/s and $\mathbf{a}_O = 3\mathbf{i}$ m/s$^2$.

**Fig. P18-55**

**18–56** A wheel of radius $r = 2$ ft is rolling without slipping. Determine the accelerations of points $C$ and $D$ in Fig. P18-55 for $\mathbf{v}_O = 100\mathbf{i}$ ft/s and $\mathbf{a}_O = -20\mathbf{i}$ ft/s$^2$.

**18–57** Block $A$ has an upward velocity of 7 ft/s and a downward acceleration of 15 ft/s$^2$ when $\theta = 30°$. Calculate the acceleration of block $B$ if $l = 18$ in.

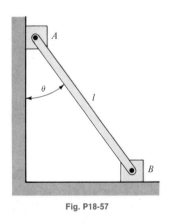

**Fig. P18-57**

**18–58** Block $A$ in Fig. P18-57 is sliding down at a speed of $v_A = 2$ m/s and an acceleration of $a_A = 6$ m/s$^2$. Determine the acceleration of block $B$ for $\theta = 40°$ and $l = 0.4$ m.

**18–59** The tracks of the linear bearing move at $v_A = 1$ m/s and $v_B = 1.5$ m/s (both constant). Determine the accelerations of points 1 and 2 on the 20-mm-diameter roller.

**Fig. P18-59**

**18–60** The outer case of the roller bearing is fixed. The shaft of diameter $D = 2$ in. is rotating with an angular speed of 5000 rpm counterclockwise. Determine the accelerations of both contacting points (at the shaft and at the outer case) of a 0.2-in.-diameter roller.

**Fig. P18-60**

**18–61** Gears $A$ and $B$ are rigidly attached to one another and roll along the horizontal rack with an angular velocity of 8 rad/s clockwise at the instant shown. The instantaneous acceleration of

The three-dimensional motion of a rigid body is analyzed using the methods developed for the analysis of two-dimensional motion in Secs. 18-4 and 18-6. The difference between plane motion and space motion is in the relationship of the angular velocity and angular acceleration vectors, as illustrated in Fig. 18-30. Plane motion implies that the angular velocity vector $\omega$ and the angular acceleration vector $\alpha$ are collinear (Fig. 18-30a). In other words, $\omega$ may change in magnitude but its direction is fixed in space, and all particles of the body move in fixed planes that are parallel to one another. In space motion both the magnitude and the direction of the angular velocity $\omega$ may change, which means that the points of the body do not move in fixed parallel planes, and $\omega$ and $\alpha$ are not collinear in general (Fig. 18-30b).

For the analysis of the velocity and acceleration of an arbitrary point $B$ in Fig. 18-30b, assume that the motion of point $A$ and the vectors $\omega$ and $\alpha$ are known at a particular instant. The $xyz$ frame is attached to point $A$ and is translating with respect to the $XYZ$ frame. As in Eq. 18-15,

$$\mathbf{v}_B = \mathbf{v}_A + \boldsymbol{\omega} \times \mathbf{r}_{B/A}$$

18-22

(a)

Plane motion

(b)

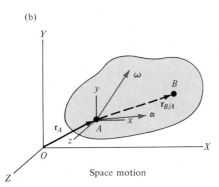

Space motion

**FIGURE 18-30**

Schematic for general motion of a rigid body

since $\mathbf{r}_{B/A}$ has a constant magnitude in a rigid body. From Eq. 18-19,

$$\mathbf{a}_B = \mathbf{a}_A + \boldsymbol{\alpha} \times \mathbf{r}_{B/A} + \boldsymbol{\omega} \times (\boldsymbol{\omega} \times \mathbf{r}_{B/A})$$

18-23

Equations 18-22 and 18-23 are valid for plane motion and for space motion since they are restatements of results from Secs. 18-4 and 18-6. The terms involving vector products are more complex to evaluate in the case of space motion than for plane motion. In both cases, the angular velocity and angular acceleration of a rigid body may vary with time, but they do not depend on what moving reference point is chosen.

**Motion about a fixed point.** A special case of three-dimensional motion is when a rigid body has a fixed point. This is in contrast to motion about a fixed axis, which is always two-dimensional motion. Assuming that the fixed point is $A$ in Fig. 18-28b, $\mathbf{v}_A = 0$ in Eq. 18-22, and $\mathbf{a}_A = 0$ in Eq. 18-23.

## Angular Acceleration

By definition, angular acceleration is the time derivative of angular velocity of a rigid body,

$$\boldsymbol{\alpha} = \frac{d\boldsymbol{\omega}}{dt}$$

18-24

This equation is valid in all situations, but it is useful to distinguish the following cases.

1. The direction of $\boldsymbol{\omega}$ is constant. This is plane motion and $\alpha = \dot{\omega}$ can be used in scalar solutions of problems.
2. Both the magnitude and direction of $\boldsymbol{\omega}$ change. This is space motion since all or some points of the rigid body have three-dimensional paths. An example is the wheel of a car which accelerates on a curved path. Equation 18-24 must be used in this case.
3. The magnitude of $\boldsymbol{\omega}$ is constant but its direction changes. This is space motion since all or some points of the rigid body have three-dimensional paths. An example is the wheel of a car which travels at a constant speed on a curved path. A useful expression for the angular acceleration in this case can be obtained as follows.

Assume that a rigid body is fixed at point $O$ in Fig. 18-31, and the body has an angular velocity $\boldsymbol{\omega}$ of constant magnitude. Also assume that the tip $A$ of the vector $\boldsymbol{\omega}$ moves on a circular path centered on the $Y$ axis, and the motion of $\boldsymbol{\omega}$ about the $Y$ axis is defined by the angular velocity vector $\boldsymbol{\Omega}$. Thus, $d\boldsymbol{\omega}/dt$ is tangent to the path of $A$. As for any vector from point $O$ to $A$,

$$\frac{d\boldsymbol{\omega}}{dt} = \boldsymbol{\alpha} = \boldsymbol{\Omega} \times \boldsymbol{\omega}$$

18-25

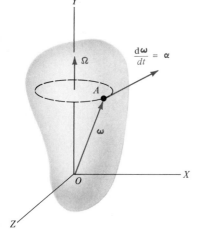

Schematic for the definition of angular acceleration when the magnitude $\omega$ of the angular velocity is constant

This result is analogous in vector operation to the expression for the linear velocity of a particle: Assume that $\mathbf{r}$ is the position vector from point $O$ to $A$ in Fig. 18-31; $\mathbf{v}_A = \dot{\mathbf{r}}_A = \boldsymbol{\Omega} \times \mathbf{r}$ from Eq. 18-4. For a review of these concepts, see App. A.

An automobile is traveling at a velocity $\mathbf{v} = 100\mathbf{i}$ km/h on a horizontal curved road of radius $R = 500$ m. Determine the velocity and acceleration of a particle $B$ on the 0.6-m-diameter wheel (Fig. 18-32a). Assume that the car is turning left as it travels on the curved road at a constant speed.

# EXAMPLE 18-11

## SOLUTION

Use the center of curvature $O$ of the road for the fixed frame $XYZ$, and the axle $A$ of the wheel for the translating frame $xyz$ (Fig. 18-32b). Thus,

$$\mathbf{r}_{A/O} = (0.3\mathbf{j} + 500\mathbf{k}) \text{ m}$$

$$\mathbf{r}_{B/A} = 0.3\mathbf{i} \text{ m}, \qquad \mathbf{r}_{A/C} = 0.3\mathbf{j} \text{ m}$$

From Eq. 18-22,

$$\mathbf{v}_B = \mathbf{v}_A + \boldsymbol{\omega} \times \mathbf{r}_{B/A}$$

where $\mathbf{v}_A = 27.8\mathbf{i}$ m/s by inspection. Note that the total angular velocity $\boldsymbol{\omega}$ of the wheel has two components: $\omega_A\mathbf{k}$ is the angular velocity of the wheel rotating about its own axis and $\omega_O\mathbf{j}$ is the angular velocity of the wheel about point $O$. $\omega_A$ and $\omega_O$ are determined first.

$$\mathbf{v}_A = \boldsymbol{\omega}_A \times \mathbf{r}_{A/C} \qquad\qquad \mathbf{v}_A = \boldsymbol{\omega}_O \times \mathbf{r}_{A/O}$$

$$27.8\mathbf{i} = \omega_A\mathbf{k} \times 0.3\mathbf{j} \qquad 27.8\mathbf{i} = \omega_O\mathbf{j} \times 500\mathbf{k}$$

$$27.8 = -\omega_A(0.3) \qquad\qquad 27.8 = 500\,\omega_O$$

$$\omega_A = -92.6\mathbf{k} \text{ rad/s} \qquad \omega_O = 0.056\mathbf{j} \text{ rad/s}$$

These vector components can be summed to obtain the total angular velocity $\boldsymbol{\omega}$ of the wheel. Thus,

$$\boldsymbol{\omega} \times \mathbf{r}_{B/A} = (0.056\mathbf{j} - 92.6\mathbf{k}) \times 0.3\mathbf{i} = (-27.8\mathbf{j} - 0.0168\mathbf{k}) \text{ m/s}$$

Summing $\mathbf{v}_A$ and $\boldsymbol{\omega} \times \mathbf{r}_{B/A}$ gives

$$\mathbf{v}_B = (27.8\mathbf{i} - 27.8\mathbf{j} - 0.0168\mathbf{k}) \text{ m/s}$$

From Eq. 18-23,

$$\mathbf{a}_B = \mathbf{a}_A + \dot{\boldsymbol{\omega}} \times \mathbf{r}_{B/A} + \boldsymbol{\omega} \times (\boldsymbol{\omega} \times \mathbf{r}_{B/A})$$

where $\boldsymbol{\omega} = \boldsymbol{\omega}_A + \boldsymbol{\omega}_O$ and

$$\mathbf{a}_A = \dot{\boldsymbol{\omega}}_O \times \mathbf{r}_{A/O} + \boldsymbol{\omega}_O \times (\boldsymbol{\omega}_O \times \mathbf{r}_{A/O})$$

$$\dot{\omega}_O = 0, \qquad \dot{\boldsymbol{\omega}} = \dot{\boldsymbol{\omega}}_A + \dot{\boldsymbol{\omega}}_O = \boldsymbol{\omega}_O \times \boldsymbol{\omega}_A$$

$$\mathbf{a}_A = 0.056\mathbf{j} \times [0.056\mathbf{j} \times (0.3\mathbf{j} + 500\mathbf{k})]$$

$$= -1.57\mathbf{k} \text{ m/s}^2$$

$$\dot{\boldsymbol{\omega}} \times \mathbf{r}_{B/A} = (\boldsymbol{\omega}_O \times \boldsymbol{\omega}_A) \times \mathbf{r}_{B/A} = [0.056\mathbf{j} \times (-92.6\mathbf{k})] \times 0.3\mathbf{i} = 0$$

$$\boldsymbol{\omega} \times (\boldsymbol{\omega} \times \mathbf{r}_{B/A}) = (0.056\mathbf{j} - 92.6\mathbf{k}) \times [(0.056\mathbf{j} - 92.6\mathbf{k}) \times 0.3\mathbf{i}]$$

$$= -2572\mathbf{i} \text{ m/s}^2$$

Combining these results gives

$$\mathbf{a}_B = (-2572\mathbf{i} - 1.57\mathbf{k}) \text{ m/s}^2$$

(a) Side view

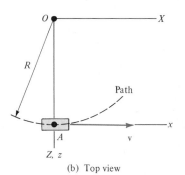

(b) Top view

**FIGURE 18-32**

Analysis of a particle's motion on a rolling wheel

EXAMPLE 18-11    743

## EXAMPLE 18-12

A helicopter is accelerating horizontally at the rate of $5\mathbf{i}\,\text{ft/s}^2$. At $\theta = 0$ the blades are rotating at $\omega_O = 12\,\text{rad/s}$ and $\alpha_O = 2\,\text{rad/s}^2$ (Fig. 18-33). A turbine disk of 4 in. diameter is rotating essentially in the vertical plane in the engine at the tip of each blade. At $\theta = 0$ the disk in engine $A$ is rotating about its own axis at an angular speed of 500 rad/s clockwise, as viewed from the rear of the helicopter, and increasing at $50\,\text{rad/s}^2$. Determine the acceleration $\mathbf{a}_P$ of a particle $P$ on the rim of the disk at 16 ft from the rotor $O$ (this is the outermost particle on the disk with respect to point $O$).

### SOLUTION

In this case point $O$ accelerates with respect to the $XYZ$ frame, point $C$ with respect to point $O$, and particle $P$ with respect to point $C$. Thus, it is necessary to use a set of equations based on Eqs. 18-17 and 18-23 as follows.

$$\mathbf{a}_P = \mathbf{a}_O + \mathbf{a}_{P/O} \qquad \mathbf{a}_{P/O} = \mathbf{a}_{C/O} + \mathbf{a}_{P/C}$$

where $\mathbf{a}_{C/O} = \boldsymbol{\alpha}_O \times \mathbf{r}_{C/O} + \boldsymbol{\omega}_O \times (\boldsymbol{\omega}_O \times \mathbf{r}_{C/O})$

$$= -2\mathbf{j} \times \left(-16 - \frac{2}{12}\right)\mathbf{k} + (-12\mathbf{j}) \times \left[-12\mathbf{j} \times \left(-16 - \frac{2}{12}\right)\mathbf{k}\right]$$

$$= (37.3\mathbf{i} + 2328\mathbf{k})\,\text{ft/s}^2$$

$$\mathbf{a}_{P/C} = \boldsymbol{\alpha} \times \mathbf{r}_{P/C} + \boldsymbol{\omega} \times (\boldsymbol{\omega} \times \mathbf{r}_{P/C})$$

$$\boldsymbol{\alpha} = \boldsymbol{\alpha}_C + \boldsymbol{\alpha}_O + \boldsymbol{\omega}_O \times \boldsymbol{\omega}_C = 50\mathbf{i} - 2\mathbf{j} + 6000\mathbf{k}$$

$$\mathbf{a}_{P/C} = (50\mathbf{i} - 2\mathbf{j} + 6000\mathbf{k}) \times \left(-\frac{2}{12}\right)\mathbf{k} + (500\mathbf{i} - 12\mathbf{j})$$

$$\times \left[(500\mathbf{i} - 12\mathbf{j}) \times \left(-\frac{2}{12}\right)\mathbf{k}\right]$$

$$= (0.333\mathbf{i} + 8.33\mathbf{j} + 41{,}691\mathbf{k})\,\text{ft/s}^2$$

$$\mathbf{a}_O = 5\mathbf{i}\,\text{ft/s}^2$$

Combining the intermediate results gives

$$\mathbf{a}_P = (37.6\mathbf{i} + 8.33\mathbf{j} + 44{,}019\mathbf{k})\,\text{ft/s}^2$$

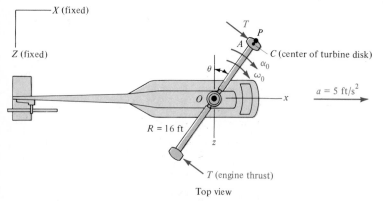

**FIGURE 18-33**

Analysis of a particle's motion in a helicopter

**18–73** A wheel of radius $r = 0.4$ m is always in a vertical plane as it travels on a horizontal track of radius $R = 100$ m at a constant speed $v = 20$ m/s. Determine the velocities of points $B$ and $D$. The center of the track is on the $Z$ axis.

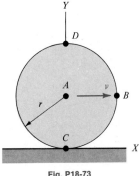

Fig. P18-73

**18–74** Determine the acceleration of point $D$ of the wheel described in Prob. 18–73 for a track of radius $R = 50$ m.

**18–75** The track of a roller coaster has a radius of curvature $r = 40$ ft in a vertical plane, and a radius of curvature $R = 70$ ft in the horizontal plane at the same point. The center of $R$ is on the $Z$ axis. Determine the angular acceleration of a car at point $O$ if it has a constant speed $v_0 = 50$ ft/s.

Fig. P18-75

**18–76** Determine the angular acceleration of a car in Prob. 18–75 if $v_0$ is decreasing at the rate of $15$ ft/s$^2$.

**18–77** The 40-m-long boom of the crane rotates about the vertical axis through point $A$ at an angular speed $\omega_Y = 0.2$ rad/s. At the same time the boom is raised vertically at an angular speed $\omega_Z = -0.4$ rad/s when $AB$ makes a $30°$ angle with the horizontal. Determine the velocity of point $B$.

Fig. P18-77

**18–78** Determine the acceleration of point $B$ in Prob. 18–77 assuming that the angular speeds are constant.

**18–79** Determine the acceleration of point $B$ in Prob. 18–77 assuming that $\omega_Y$ is increasing at the rate of $0.2$ rad/s$^2$, and that $\omega_Z$ is decreasing at the rate of $0.1$ rad/s$^2$.

**18–80**  The rotor shaft of an alternator in a car is in the horizontal plane. It rotates at a constant angular speed of 1500 rpm while the car travels at $v = 60$ ft/s on a horizontal road of 400 ft radius. Determine the angular acceleration of the rotor shaft if $v$ increases at the rate of 8 ft/s².

v = 60 ft/s

Car

R = 400 ft

Top view

**Fig. P18-80**

**18–81**  A wheel of 0.5 m radius is rolling at a constant speed of 30 m/s. Calculate the angular acceleration of the wheel if the wheel axle is turning at a constant angular velocity $\omega_Y = -0.2\mathbf{j}$ rad/s.

**Fig. P18-81**

**18–82**  Solve Prob. 18–81 with the additional angular velocity $\omega_Z = -0.1\mathbf{k}$ rad/s of the wheel axle when the axle lies along the $X$ axis.

**18–83**  An airplane is flying horizontally at a constant velocity $\mathbf{v} = 300\mathbf{i}$ ft/s when the angular velocity of the propeller is a constant $\omega_X = 210$ rad/s. Calculate the angular acceleration of the propeller immediately when the airplane begins to turn right on a horizontal path of 1500 ft radius.

Top view

**Fig. P18-83**

**18–84**  Determine the angular acceleration of the propeller in Prob. 18–83 if the airplane is simultaneously diving at the rate of $\omega_Y = 0.8\mathbf{j}$ rad/s.

**18–85**  The tire rotates at a constant angular velocity $\omega_Z = 100\mathbf{k}$ rad/s during a test. Determine the angular acceleration of the tire if the wheel axle also rotates at angular velocity $\omega_Y = 0.5\mathbf{j}$ rad/s.

**Fig. P18-85**
Model of a tire-testing system. (Courtesy MTS System Corporation, Minneapolis, Minnesota)

**18–86**  Determine the angular acceleration of the tire in Prob. 18–85 for the simultaneous angular velocities of $\omega_X = -0.3\mathbf{i}$ rad/s, $\omega_Y = 0.4\mathbf{j}$ rad/s, and $\omega_Z = 120\mathbf{k}$ rad/s. Calculate the acceleration of point $P$ the instant when it is at the bottom of the tire. The radius of the tire is 1.4 ft.

**18–87**  A two-engine aircraft is flying horizontally at a constant speed. The two propellers are rotating at the same rate $\omega$ but in opposite directions. What are the directions of the angular accelerations of the propellers if the airplane begins a nosedive?

Fig. P18-89

Top view

Fig. P18-87

**18–88**  In Prob. 18–87, $\omega_1 = 240\mathbf{j}$ rad/s and $\omega_2 = -240\mathbf{j}$ rad/s. The airplane flies at $\mathbf{v} = 300\mathbf{j}$ km/h and starts turning to the left on a horizontal path of 1000 m radius. Determine the angular acceleration of the propellers caused by this maneuver. Calculate the acceleration of particle $A$ for the position shown. The length of the propeller is 1.5 m.

**18–89**  A food processor has wheels $A$ and $B$ of 1 ft radius mounted on a common arm which rotates in the horizontal plane at a constant $\omega_Y = 15\mathbf{j}$ rad/s. The wheels rotate without slipping on the base platen. Determine the angular accelerations of the wheels caused by the rotation and calculate the acceleration of particle $P$ when the wheel axle lies along the $X$ axis.

**18–90**  The large amusement ride has two angular velocities at a given instant: the 12 small arms have $\omega_B = (0.2\mathbf{i} + 0.1\mathbf{k})$ rad/s, and the arm $AB$ has $\omega_Y = 0.05\mathbf{j}$ rad/s. Determine the angular acceleration of a small arm that rotates about $B$.

Fig. P18-90

**18–91**  Solve Prob. 18–90 with the additional angular velocity of $\omega_X = 0.03\mathbf{i}$ rad/s.

In some cases the motion of a particle or a rigid body is defined in a reference frame which is translating and rotating with respect to a fixed frame. For example, the motion of a turbine blade in a jet engine is conveniently defined with respect to the fuselage of the aircraft which translates and rotates with respect to the earth. The analysis of the absolute motion of the body in such a case is straightforward concerning translation of the moving frame, but the rotation of that frame requires additional considerations. These are discussed for an arbitrary vector $\mathbf{Q}$ as follows.

Assume that a fixed reference frame $XYZ$ and a rotating frame $xyz$ have a common origin $O$ as in Fig. 18-34. Frame $xyz$ is rotating at an angular velocity $\mathbf{\Omega}$ with respect to the $XYZ$ frame. A vector $\mathbf{Q}$ is defined in terms of its rectangular components in the $xyz$ frame,

$$\mathbf{Q} = Q_x\mathbf{i} + Q_y\mathbf{j} + Q_z\mathbf{k} \qquad \boxed{18\text{-}26}$$

The time rate of change of vector $\mathbf{Q}$ with respect to the $xyz$ frame is obtained by differentiating Eq. 18-26 with respect to time. This is expressed as

$$(\dot{\mathbf{Q}})_{xyz} = \dot{Q}_x\mathbf{i} + \dot{Q}_y\mathbf{j} + \dot{Q}_z\mathbf{k} \qquad \boxed{18\text{-}27}$$

since $d\mathbf{i}/dt$, $d\mathbf{j}/dt$, and $d\mathbf{k}/dt$ are all zero ($\mathbf{i}$, $\mathbf{j}$, and $\mathbf{k}$ are constant in magnitude and have fixed directions in the $xyz$ frame). The time rate of change of vector $\mathbf{Q}$ with respect to the $XYZ$ frame is written from Eq. 18-26 as

$$(\dot{\mathbf{Q}})_{XYZ} = \dot{Q}_x\mathbf{i} + \dot{Q}_y\mathbf{j} + \dot{Q}_z\mathbf{k} \qquad \boxed{18\text{-}28}$$
$$+ Q_x\frac{d\mathbf{i}}{dt} + Q_y\frac{d\mathbf{j}}{dt} + Q_z\frac{d\mathbf{k}}{dt}$$

where the time derivatives of the unit vectors are not zero since their directions change with respect to $XYZ$. The last three terms in Eq. 18-28 represent the velocity of a particle located at the tip of vector $\mathbf{Q}$ if the magnitude $Q$ is constant. Thus, from Eq. 18-25,

$$Q_x\frac{d\mathbf{i}}{dt} + Q_y\frac{d\mathbf{j}}{dt} + Q_z\frac{d\mathbf{k}}{dt} = \boldsymbol{\omega} \times \mathbf{Q} \qquad \boxed{18\text{-}29}$$

Using Eqs. 18-27 and 18-29, Eq. 18-28 is written as

$$\boxed{(\dot{\mathbf{Q}})_{XYZ} = (\dot{\mathbf{Q}})_{xyz} + \mathbf{\Omega} \times \mathbf{Q}} \qquad \boxed{18\text{-}30}$$

**FIGURE 18-34**

Fixed and rotating reference frames with a common origin

where $(\dot{\mathbf{Q}})_{xyz}$ can be thought of as the time rate of change of the magnitude of vector $\mathbf{Q}$, while $\mathbf{\Omega} \times \mathbf{Q}$ is the time rate of change of the direction of $\mathbf{Q}$, both in the fixed frame. Equation 18-30 is useful in the analysis of velocities and accelerations when a rotating reference frame must be considered.

Assume that the position of a particle $P$ is defined by the vector $\mathbf{r}$ with respect to the $xyz$ frame in Fig. 18-35. The $xyz$ frame is rotating at angular velocity $\mathbf{\Omega}$ with respect to the $XYZ$ frame. The absolute velocity $\mathbf{v}_P$ of the particle in the $XYZ$ frame is the time derivative of the position vector $\mathbf{r}$ with respect to the fixed frame. Using Eq. 18-30 and substituting the vector $\mathbf{r}$ for the arbitrary vector $\mathbf{Q}$ gives

$$\mathbf{v}_P = (\dot{\mathbf{r}})_{XYZ} = (\dot{\mathbf{r}})_{xyz} + \mathbf{\Omega} \times \mathbf{r} \qquad \boxed{18\text{-}31}$$

where $(\dot{\mathbf{r}})_{xyz} = \mathbf{v}_{xyz}$, the velocity of the particle with respect to the rotating $xyz$ frame. The term $\mathbf{\Omega} \times \mathbf{r}$ represents the absolute velocity $\mathbf{v}_{P'}$ of a particle $P'$ which is fixed in the $xyz$ frame and whose position coincides with the position of particle $P$ at the given instant. Thus, Eq. 18-31 is written more concisely as

$$\mathbf{v}_P = \mathbf{v}_{xyz} + \mathbf{v}_{P'} \qquad \boxed{18\text{-}32}$$

where  $\mathbf{v}_P$ = absolutely velocity of particle $P$ in the $XYZ$ frame
  $\mathbf{v}_{xyz}$ = velocity of $P$ with respect to the moving $xyz$ frame
  $\mathbf{v}_{P'}$ = absolute velocity of particle $P'$ fixed in the $xyz$ frame; $P$ and $P'$ coincide at the given instant

Note that particle $P$ has velocity $\mathbf{v}_{xyz}$ relative to $P'$.

The absolute acceleration $\mathbf{a}_P$ of particle $P$ is obtained by differentiating Eq. 18-31 with respect to time,

$$\mathbf{a}_P = \dot{\mathbf{v}}_P = \frac{d}{dt}\left[(\dot{\mathbf{r}})_{xyz}\right] + \dot{\mathbf{\Omega}} \times \mathbf{r} + \mathbf{\Omega} \times \dot{\mathbf{r}} \qquad \boxed{18\text{-}33}$$

**FIGURE 18-35**

Analysis of a particle's motion with respect to a fixed frame and a rotating frame

where $(\dot{\mathbf{r}})_{xyz}$ is with respect to the $xyz$ frame, while $(d/dt)[(\dot{\mathbf{r}})_{xyz}]$, $\dot{\mathbf{\Omega}} \times \mathbf{r}$, and $\mathbf{\Omega} \times \dot{\mathbf{r}}$ are all defined with respect to the $XYZ$ frame. Using Eq. 18-30 and substituting the vector $(\dot{\mathbf{r}})_{xyz}$ for the arbitrary vector $\mathbf{Q}$ gives

$$\frac{d}{dt}\left[(\dot{\mathbf{r}})_{xyz}\right] = (\ddot{\mathbf{r}})_{xyz} + \mathbf{\Omega} \times (\dot{\mathbf{r}})_{xyz} \qquad \boxed{18\text{-}34}$$

The quantity $\dot{\mathbf{r}}$ in Eq. 18-33 is the absolute velocity $\mathbf{v}_P$ of particle $P$. Thus, from Eq. 18-31,

$$\mathbf{\Omega} \times \dot{\mathbf{r}} = \mathbf{\Omega} \times (\dot{\mathbf{r}})_{xyz} + \mathbf{\Omega} \times (\mathbf{\Omega} \times \mathbf{r}) \qquad \boxed{18\text{-}35}$$

Using Eqs. 18-34 and 18-35, Eq. 18-33 can be written as

$$\mathbf{a}_P = \mathbf{a}_{xyz} + \dot{\mathbf{\Omega}} \times \mathbf{r} + \mathbf{\Omega} \times (\mathbf{\Omega} \times \mathbf{r}) + 2\mathbf{\Omega} \times \mathbf{v}_{xyz} \qquad \boxed{18\text{-}36}$$

where   $\mathbf{a}_{xyz} = (\ddot{\mathbf{r}})_{xyz}$ = acceleration of particle $P$ with respect to the moving $xyz$ frame
  $\dot{\mathbf{\Omega}} \times \mathbf{r}$ = tangential acceleration of $P'$ because of angular acceleration of the $xyz$ frame; it is always normal to $\mathbf{r}$

$\boldsymbol{\Omega} \times (\boldsymbol{\Omega} \times \mathbf{r}) =$ normal acceleration of $P'$ because of angular velocity of the $xyz$ frame; it is always directed toward the origin of the vector $\mathbf{r}$ in the $xyz$ frame

$\mathbf{v}_{xyz} =$ velocity of $P$ with respect to the moving $xyz$ frame

$2\boldsymbol{\Omega} \times \mathbf{v}_{xyz} = $ *Coriolis acceleration** = acceleration of particle $P$ with respect to $P'$, which is fixed in the $xyz$ frame; it is always normal to $\mathbf{v}_{xyz}$ and thus to the path of $P$ in $xyz$

Equation 18-36 is written more concisely as

$$\mathbf{a}_P = \mathbf{a}_{xyz} + \mathbf{a}_{P'} + \mathbf{a}_{\mathrm{Cor}} \qquad \boxed{18\text{-}37}$$

where 
$\mathbf{a}_P =$ absolute acceleration of particle $P$ in the $XYZ$ frame
$\mathbf{a}_{xyz} =$ acceleration of $P$ with respect to the moving $xyz$ frame
$\mathbf{a}_{P'} = \dot{\boldsymbol{\Omega}} \times \mathbf{r} + \boldsymbol{\Omega} \times (\boldsymbol{\Omega} \times \mathbf{r}) =$ absolute acceleration of particle $P'$ fixed in the $xyz$ frame; $P$ and $P'$ coincide at the given instant
$\mathbf{a}_{\mathrm{Cor}} = 2\boldsymbol{\Omega} \times \mathbf{v}_{xyz} =$ Coriolis acceleration of $P$ with respect to $P'$

## Translation and Rotation of the Moving Reference Frame

The general case of a moving reference frame is illustrated in Fig. 18-36a. The $xyz$ frame rotates at angular velocity $\boldsymbol{\Omega}$ while its origin $A$ translates with respect to the $XYZ$ frame. The absolute position, velocity, and acceleration of particle $P$ are defined, respectively, as

$$\mathbf{r}_P = \mathbf{r}_A + \mathbf{r}_{P/A} \qquad \boxed{18\text{-}38}$$

$$\mathbf{v}_P = \dot{\mathbf{r}}_P = \mathbf{v}_A + \mathbf{v}_{P/A} \qquad \boxed{18\text{-}39}$$

$$\mathbf{a}_P = \dot{\mathbf{v}}_P = \mathbf{a}_A + \mathbf{a}_{P/A} \qquad \boxed{18\text{-}40}$$

The translation and rotation of frame $xyz$ are distinguished using Fig. 18-36b, where $X'Y'Z'$ are parallel to $XYZ$. Since the velocity $\mathbf{v}_{P/A}$ can be determined by applying Eq. 18-31 to the $X'Y'Z'$ frame, Eq. 18-39 is written as

$$\mathbf{v}_P = \mathbf{v}_A + (\dot{\mathbf{r}}_{P/A})_{xyz} + \boldsymbol{\Omega} \times \mathbf{r}_{P/A} \qquad \boxed{18\text{-}41}$$

where $\mathbf{v}_P$ and $\mathbf{v}_A$ are with respect to the $XYZ$ frame.

The absolute acceleration $\mathbf{a}_P$ of particle $P$ is obtained by differentiating Eq. 18-41 with respect to time. Using Eqs. 18-36 and 18-40 gives

$$\mathbf{a}_P = \mathbf{a}_A + \mathbf{a}_{xyz} + \dot{\boldsymbol{\Omega}} \times \mathbf{r}_{P/A} + \boldsymbol{\Omega} \times (\boldsymbol{\Omega} \times \mathbf{r}_{P/A}) \\ + 2\boldsymbol{\Omega} \times \mathbf{v}_{xyz} \qquad \boxed{18\text{-}42}$$

---

* After the French mathematician DeCoriolis (1792–1843). The Coriolis acceleration is difficult to visualize, but the inclusion of this term (which may be significant) in the solution of problems causes little additional work. Those who wish to obtain further understanding of this acceleration should see App. D.

(a)

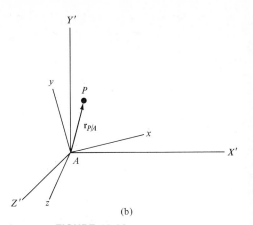

(b)

**FIGURE 18-36**

General analysis using a moving reference frame

where $\mathbf{a}_P$ and $\mathbf{a}_A$ are with respect to the $XYZ$ frame, and the other terms are defined as for Eq. 18-36. Note that the vector $\mathbf{r}_{P/A}$ in Eqs. 18-41 and 18-42 is analogous to the vector $\mathbf{r}$ in Eqs. 18-31 and 18-36; it is a position vector to point $P'$, which is fixed in the moving $xyz$ frame and is coincident with the particle $P$ at the instant considered. The fact that the temporarily coincident particles $P$ and $P'$ have different motions is especially important in the derivation of the kinematics relationship. In solving problems it is useful to remember the following:

1. Equations 18-31 through 18-42 are valid for individual particles or for points on rigid bodies.

2. The terms $\mathbf{v}_{xyz}$ and $\mathbf{a}_{xyz}$ are both zero when the rotating $xyz$ frame is fixed in the rigid body whose motion is analyzed. The reason is that $\mathbf{v}_{xyz}$ and $\mathbf{a}_{xyz}$ represent the motion of a particle $P$ relative to the $xyz$ frame (and to point $P'$ which is fixed in $xyz$). If $xyz$ is fixed in the body, $P$ and $P'$ are always at the same location with respect to the $xyz$ frame.

3. The common points of two or more pin-jointed members must have the same absolute velocity and the same absolute acceleration.

4. The points of contact in members that are in temporary contact generally have different absolute velocities and different absolute accelerations.

5. The geometric constraints of motion in a rotating frame are useful to define the directions of several components of the absolute velocity and acceleration. For an illustration of these, consider the plane motion of particle $P$ with respect to the $xy$ frame which rotates in the $XY$ frame in Fig. 18-37.

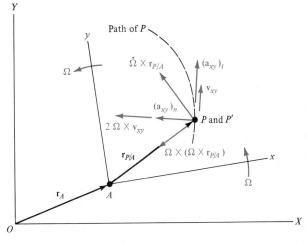

**FIGURE 18-37**

Analysis of a particle's plane motion using a rotating reference frame

EXAMPLE 18-13

Pin $B$ of the crank $r$ is moving in the slotted arm in Fig. 18-38. Assume that $l = 12$ in., $r = 5$ in., $\omega_1 = 10$ rad/s, and $\dot{\omega}_1 = 0$. Determine the angular velocity $\omega_2$ of arm $OC$ and the relative velocity $\mathbf{v}_{xyz}$ of pin $B$ in the arm for $\theta = 20°$, $\phi = 55°$.

## SOLUTION

The absolute velocity $\mathbf{v}_B$ of pin $B$ can be expressed in two ways, using first point $A$, then point $O$ as the fixed reference point. From Eq. 18-31,

$$\mathbf{v}_B = \omega_1 \times \mathbf{r}_{B/A} \qquad \text{(reference point } A\text{)} \qquad \text{(a)}$$

$$\mathbf{v}_B = \mathbf{v}_{xyz} + \omega_2 \times \mathbf{r}_{B/O} \qquad \text{(reference point } O\text{)} \qquad \text{(b)}$$

Equating these two expressions and using the two scalar equations for the $\mathbf{i}$ and $\mathbf{j}$ components gives two equations with two unknowns that can be determined. In the configuration shown,

$$\mathbf{r}_{B/A} = \frac{5}{12}(\cos 55° \, \mathbf{i} + \sin 55° \, \mathbf{j}) \text{ ft}$$

$$= (0.24\mathbf{i} + 0.34\mathbf{j}) \text{ ft}$$

Using $\omega_1 = -10\mathbf{k}$ rad/s, Eq. (a) becomes

$$\mathbf{v}_B = \omega_1 \times \mathbf{r}_{B/A} = (3.4\mathbf{i} - 2.4\mathbf{j}) \text{ ft/s} \qquad \text{(c)}$$

Using the law of sines to determine $\mathbf{r}_{B/O}$ gives

$$\mathbf{r}_{B/O} = \frac{8.4}{12}(\sin 20° \, \mathbf{i} - \cos 20° \, \mathbf{j}) \text{ ft}$$

$$= (0.24\mathbf{i} - 0.66\mathbf{j}) \text{ ft}$$

$$\mathbf{v}_{xyz} = v_{xyz}(0.34\mathbf{i} - 0.94\mathbf{j}) \text{ ft/s}$$

$$\omega_2 \times \mathbf{r}_{B/O} = \omega_2\mathbf{k} \times (0.24\mathbf{i} - 0.66\mathbf{j}) \text{ ft/s}$$

Thus, Eq. (b) becomes

$$\mathbf{v}_B = v_{xyz}(0.34\mathbf{i} - 0.94\mathbf{j}) \text{ ft/s} + \omega_2(0.66\mathbf{i} + 0.24\mathbf{j}) \text{ ft/s} \qquad \text{(d)}$$

Equating the $\mathbf{i}$ components from Eqs. (c) and (d) yields

$$3.4 = v_{xyz}(0.34) + \omega_2(0.66) \qquad \text{(e)}$$

Equating the $\mathbf{j}$ components from Eqs. (c) and (d) gives

$$-2.4 = v_{xyz}(-0.94) + \omega_2(0.24) \qquad \text{(f)}$$

Solving Eqs. (e) and (f),

$$v_{xyz} = 3.4 \text{ ft/s} \qquad \mathbf{v}_{xyz} = (1.2\mathbf{i} - 3.2\mathbf{j}) \text{ ft/s}$$

$$\omega_2 = 3.4 \text{ rad/s} \qquad \omega_2 = 3.4\mathbf{k} \text{ rad/s}$$

**FIGURE 18-38**

Analysis of a crank-slotted arm mechanism

EXAMPLE 18-14

The mechanism in Fig. 18-38 (Ex. 18-13) has dimensions $l = 0.3$ m and $r = 0.12$ m. Determine the angular acceleration $\boldsymbol{\alpha}_2$ of arm $OC$ and the relative acceleration $\mathbf{a}_{xyz}$ of pin $B$ in the arm for $\theta = 20°$, $\phi = 51.2°$, and a constant angular speed of $\omega_1 = 50$ rad/s.

## SOLUTION

The absolute acceleration $\mathbf{a}_B$ of pin $B$ can be expressed in two ways, using first point $A$, then point $O$ as the fixed reference point. From Eq. 18-36,

$$\mathbf{a}_B = \boldsymbol{\omega}_1 \times (\boldsymbol{\omega}_1 \times \mathbf{r}_{B/A}) \qquad \text{(reference point } A\text{)} \qquad \textbf{(a)}$$

$$\mathbf{a}_B = \mathbf{a}_{xyz} + \boldsymbol{\alpha}_2 \times \mathbf{r}_{B/O} + \boldsymbol{\omega}_2$$
$$\times (\boldsymbol{\omega}_2 \times \mathbf{r}_{B/O}) + 2\boldsymbol{\omega}_2 \times \mathbf{v}_{xyz} \qquad \text{(reference point } O\text{)} \qquad \textbf{(b)}$$

Using a method of solution as in Ex. 18-13, $\boldsymbol{\omega}_2$ and $\mathbf{v}_{B/O}$ are found to be

$$\boldsymbol{\omega}_2 = 13.95\mathbf{k} \text{ rad/s}$$

$$\mathbf{v}_{xyz} = (1.75\mathbf{i} - 4.81\mathbf{j}) \text{ m/s}$$

For the given configuration, $\mathbf{r}_{B/A} = 0.12(\cos 51.2° \, \mathbf{i} + \sin 51.2° \, \mathbf{j})$, so Eq. (a) becomes

$$\mathbf{a}_B = -50\mathbf{k} \times [-50\mathbf{k} \times 0.12(\cos 51.2° \, \mathbf{i} + \sin 51.2° \, \mathbf{j})]$$
$$= (-188\mathbf{i} - 234\mathbf{j}) \text{ m/s}^2$$

The components of Eq. (b) are determined as follows:

$$\mathbf{a}_{xyz} = a_{xyz} (\sin 20° \, \mathbf{i} - \cos 20° \, \mathbf{j})$$
$$= (0.34a_{xyz}\mathbf{i} - 0.94a_{xyz}\mathbf{j}) \text{ m/s}^2$$

For the given configuration, $\mathbf{r}_{B/O} = 0.22(\sin 20° \, \mathbf{i} - \cos 20° \, \mathbf{j})$, and

$$\boldsymbol{\alpha}_2 \times \mathbf{r}_{B/O} = \alpha_2\mathbf{k} \times [0.22(\sin 20° \, \mathbf{i} - \cos 20° \, \mathbf{j})]$$
$$= (0.075\alpha_2\mathbf{j} + 0.21\alpha_2\mathbf{i}) \text{ m/s}^2$$

$$\boldsymbol{\omega}_2 \times (\boldsymbol{\omega}_2 \times \mathbf{r}_{B/O}) = 13.95\mathbf{k} \times [13.95\mathbf{k} \times 0.22(\sin 20° \, \mathbf{i} - \cos 20° \, \mathbf{j})]$$
$$= (-14.6\mathbf{i} + 40.2\mathbf{j}) \text{ m/s}^2$$

$$2(\boldsymbol{\omega}_2 \times \mathbf{v}_{xyz}) = 2[13.95\mathbf{k} \times (1.75\mathbf{i} - 4.81\mathbf{j})]$$
$$= (134.2\mathbf{i} + 48.8\mathbf{j}) \text{ m/s}^2$$

Using these results provides two equations with two unknowns. Equating the $\mathbf{i}$ components from Eqs. (a) and (b) gives

$$-188 = 0.34a_{xyz} + 0.21\alpha_2 - 14.6 + 134.2$$

Equating the $\mathbf{j}$ components from Eqs. (a) and (b) yields

$$-234 = -0.94a_{xyz} + 0.075\alpha_2 + 40.2 + 38.8$$

These equations reduce to

$$-307.6 = 0.34a_{xyz} + 0.21\alpha_2$$
$$-323 = -0.94a_{xyz} + 0.075\alpha_2$$

Solving these equations simultaneously yields

$$a_{xyz} = 201 \text{ m/s}^2 \qquad \mathbf{a}_{xyz} = (68.3\mathbf{i} - 188.9\mathbf{j}) \text{ m/s}^2$$

$$\alpha_2 = -1790 \text{ rad/s}^2 \qquad \boldsymbol{\alpha}_2 = -1790\mathbf{k} \text{ rad/s}^2$$

EXAMPLE 18-14    753

EXAMPLE 18-15

An amusement ride is modeled in Fig. 18-39. People sit in two circular cages that are mounted on a rigid rotating arm at $A$ and $B$. The horizontal arm $AB$ is rotating about point $O$ at a constant speed of 10 rpm, and the cages rotate with respect to the arm $AB$ at constant speed of 3 rpm. Calculate the absolute velocity and acceleration of a person at position 2 for the position shown in Fig. 18-39b.

## SOLUTION

The first step is to choose coordinate systems, as shown in the figure. The absolute velocity of point 2 is determined using Eq. 18-41,

$$\mathbf{v}_2 = \mathbf{v}_A + \mathbf{v}_{2_{xyz}} + \boldsymbol{\omega}_1 \times \mathbf{r}_{2/A}$$

where $\mathbf{v}_{2_{xyz}} = \boldsymbol{\omega}_2 \times \mathbf{r}_{2/A}$. The velocity of the center of the cage is expressed as

$$\mathbf{v}_A = \mathbf{v}_{A/D} + \boldsymbol{\omega}_1 \times \mathbf{r}_{A/O}$$

where $\mathbf{v}_{A/D} = 0$ because the center of the cage is fixed on the rotating arm. Consequently,

$$\mathbf{v}_2 = \boldsymbol{\omega}_1 \times \mathbf{r}_{A/O} + \boldsymbol{\omega}_2 \times \mathbf{r}_{2/A} + \boldsymbol{\omega}_1 \times \mathbf{r}_{2/A}$$

where

$$\mathbf{r}_{A/O} = (5\mathbf{i} + 8.66\mathbf{j}) \text{ ft}$$

$$\mathbf{r}_{2/A} = (-4.33\mathbf{i} + 2.5\mathbf{j}) \text{ ft}$$

The units of angular velocity should be in rad/s,

$$\boldsymbol{\omega}_1 = -10 \frac{\text{rev}}{\text{min}} \left( 2\pi \frac{\text{rad}}{\text{rev}} \right) \left( \frac{1 \text{ min}}{60 \text{ s}} \right) \mathbf{k}$$

$$= -1.047\mathbf{k} \text{ rad/s}$$

$$\boldsymbol{\omega}_2 = -0.314\mathbf{k} \text{ rad/s}$$

Thus,

$$\mathbf{v}_2 = (-1.047\mathbf{k}) \times (5\mathbf{i} + 8.66\mathbf{j}) + (-0.314\mathbf{k}) \times (-4.33\mathbf{i} + 2.5\mathbf{j})$$
$$+ (-1.047\mathbf{k}) \times (-4.33\mathbf{i} + 2.5\mathbf{j})$$
$$= (12.47\mathbf{i} + 0.659\mathbf{j}) \text{ ft/s}$$

Solving for the acceleration using Eq. 18-42 gives

$$\mathbf{a}_2 = \mathbf{a}_A + \mathbf{a}_{2_{xyz}} + \dot{\boldsymbol{\omega}}_1 \times \mathbf{r}_{2/A} + \boldsymbol{\omega}_1$$
$$\times (\boldsymbol{\omega}_1 \times \mathbf{r}_{2/A}) + 2\boldsymbol{\omega}_1 \times \mathbf{v}_{2_{xyz}}$$

where $\mathbf{a}_A$ is recognized to be from pure rotation of $A$ about the fixed point $O$. According to Eq. 18-36,

$$\mathbf{a}_A = \mathbf{a}_{A/D} + \dot{\boldsymbol{\omega}}_1 \times \mathbf{r}_{A/O} + \boldsymbol{\omega}_1 \times (\boldsymbol{\omega}_1 \times \mathbf{r}_{A/O}) + 2\boldsymbol{\omega}_1 \times \mathbf{v}_{A/D}$$

Here $\mathbf{a}_{A/D} = \mathbf{v}_{A/D} = \dot{\boldsymbol{\omega}}_1 = 0$ since $A$ is fixed on the arm, which rotates at a constant angular velocity. Thus,

$$\mathbf{a}_A = \boldsymbol{\omega}_1 \times (\boldsymbol{\omega}_1 \times \mathbf{r}_{A/O})$$
$$= -1.047\mathbf{k} \times [-1.047\mathbf{k} \times (5\mathbf{i} + 8.66\mathbf{j})]$$
$$= (-5.48\mathbf{i} - 9.49\mathbf{j}) \text{ ft/s}^2$$

(a)

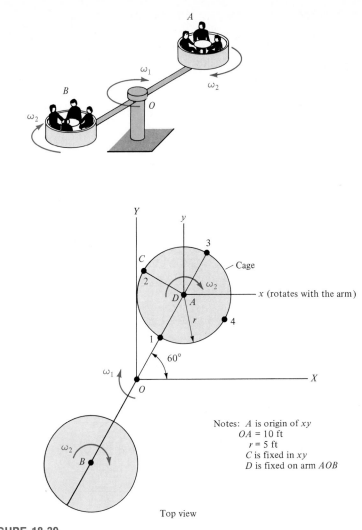

(b)

Notes: $A$ is origin of $xy$
$OA = 10$ ft
$r = 5$ ft
$C$ is fixed in $xy$
$D$ is fixed on arm $AOB$

Top view

**FIGURE 18-39**

Schematic of an amusement ride

Also, $\dot{\omega}_2 = 0$. Therefore,

$$
\begin{aligned}
\mathbf{a}_{2_{xyz}} &= \dot{\omega}_2 \times \mathbf{r}_{2/A} + \omega_2 \times (\omega_2 \times \mathbf{r}_{2/A}) \\
&= \omega_2 \times (\omega_2 \times \mathbf{r}_{2/A}) \\
&= -0.314\mathbf{k} \times [-0.314\mathbf{k} \times (-4.33\mathbf{i} + 2.5\mathbf{j})] \\
&= (0.416\mathbf{i} - 0.246\mathbf{j}) \text{ ft/s}^2
\end{aligned}
$$

$$
\begin{aligned}
\omega_1 \times (\omega_1 \times \mathbf{r}_{2/A}) &= -1.047\mathbf{k} \times [-1.047\mathbf{k} \times (-4.33\mathbf{i} + 2.5\mathbf{j})] \\
&= (4.75\mathbf{i} - 2.74\mathbf{j}) \text{ ft/s}^2
\end{aligned}
$$

$$
\begin{aligned}
2\omega_1 \times \mathbf{v}_{2_{xyz}} &= 2(-1.047\mathbf{k}) \times [-0.314\mathbf{k} \times (-4.33\mathbf{i} + 2.5\mathbf{j})] \\
&= (2.85\mathbf{i} - 1.64\mathbf{j}) \text{ ft/s}^2
\end{aligned}
$$

Combining these results, the person at point 2 has absolute acceleration

$$
\mathbf{a}_2 = (2.54\mathbf{i} - 14.12\mathbf{j}) \text{ ft/s}^2
$$

EXAMPLE 18-15    755

**18–92** The two rigid bars of length $AB = BC = 1.5\,\text{m}$ are pinned at $A$ and $B$. Determine the velocity $\mathbf{v}_C$ and acceleration $\mathbf{a}_C$ if $\omega_1 = 0.3\,\text{rad/s}$, $\alpha_1 = 0$, $\omega_2 = 0.2\,\text{rad/s}$, and $\alpha_2 = 0$.

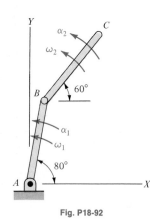

**Fig. P18-92**

**18–93** $AB = 3\,\text{ft}$ and $BC = 2\,\text{ft}$ in Fig. P18-92. $\omega_1 = 0.4\,\text{rad/s}$, $\alpha_1 = 0.2\,\text{rad/s}^2$ (increasing $\omega_1$), $\omega_2 = 0.3\,\text{rad/s}$, and $\alpha_2 = -0.2\,\text{rad/s}^2$ (decreasing $\omega_2$). Calculate the velocity $\mathbf{v}_C$ and acceleration $\mathbf{a}_C$.

**18–94** A train is traveling on a curved track at 70 mph when its brakes are applied causing a deceleration of $8\,\text{ft/s}^2$. At this instant persons $A$ and $B$ are walking in the train as shown. Calculate the acceleration of each person.

**Fig. P18-94**

**18–95** A flywheel is rotating in the horizontal plane in an experimental car. The car is traveling at a constant speed $v = 100\,\text{km/h}$ on a road of radius $R = 400\,\text{m}$. Calculate the acceleration of points 1 and 2 on the flywheel. The points are 0.2 m from the center of the wheel, which rotates at 5000 rpm.

**Fig. P18-95**

**18–96** A proposed merry-go-round is modeled as shown. The two superimposed rotations about points $O$ and $A$ are in the horizontal plane. Calculate the maximum allowed angular speed $\omega_A$ if the maximum absolute acceleration of a person rotating 2 m away from $A$ is $12\,\text{m/s}^2$. The angular speeds are constant, and the merry-go-round rotates at 5 rpm.

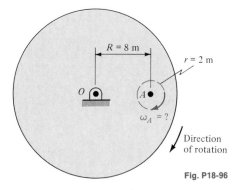

**Fig. P18-96**

**18–97** Determine the absolute velocities $\mathbf{v}_1$ and $\mathbf{v}_3$ of persons at positions 1 and 3 in Fig. 18-37 (Ex. 18-15).

**18–98** Calculate the absolute acceleration $\mathbf{a}_4$ of a person at position 4 in Fig. 18-37 (Ex. 18-15).

**18–99** Particle $P$ slides in the radial groove of the rotating disk. Determine the magnitudes of the absolute velocity and acceleration of the particle using a rotating frame for $\omega = 10\,\text{rad/s}$, $\alpha = 3\,\text{rad/s}^2$, $r = 4\,\text{in.}$, $\dot{r} = 6\,\text{in./s}$, and $\ddot{r} = 50\,\text{ft/s}^2$.

**Fig. P18-99**

**18–100** Assume that $\ddot{r} = 0$ in Prob. 18–99. What is the maximum velocity $\dot{r}$ of the particle if its acceleration normal to the groove should not exceed $60\,\text{ft/s}^2$?

**18–101** Particle $P$ slides in the straight groove of the rotating disk. Determine the absolute velocity and acceleration of the particle for $\omega = 5\,\text{rad/s}$, $\alpha = 4\,\text{rad/s}^2$, $x = -0.2\,\text{m}$, $\dot{x} = 2\,\text{m/s}$ (increasing $x$), and $\ddot{x} = 1\,\text{m/s}^2$ (increasing $\dot{x}$).

**Fig. P18-101**

**18–102** Assume that $\ddot{x} = 0$ in Prob. 18–101. What is the maximum velocity $\dot{x}$ of the particle if its acceleration normal to the groove should not exceed $20\,\text{m/s}^2$?

**18–103** A slender body of length $l = 1.5\,\text{ft}$ is sliding in the radial groove of the rotating disk. Determine the components radial and normal to the groove of the accelerations of the ends $A$ and $B$ of the body for $\omega = 20\,\text{rad/s}$, $r = 2\,\text{ft}$, and $\dot{r} = 100\,\text{ft/s}$ (increasing $r$).

**Fig. P18-103**

**18–104** A slender body of length $l = 0.3\,\text{m}$ is sliding in the straight groove of the rotating disk. Determine the $x$ and $y$ components of the accelerations of the ends $A$ and $B$ of the body for $\omega = 15\,\text{rad/s}$, $\alpha = 2\,\text{rad/s}^2$, $x_A = 1\,\text{m}$, and $\dot{x}_A = 50\,\text{m/s}$ (increasing $x_A$).

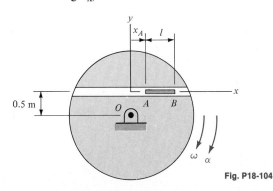

**Fig. P18-104**

**18–105** An amusement ride consists of arm $OA$ rotating at $\Omega = 0.8\,\text{rad/s}$, and the circular cage at $A$ rotating at $\omega = 0.5\,\text{rad/s}$ relative to the arm $OA$. A person is moving away from $A$ along the rail $ABC$ at a speed of $0.5\,\text{m/s}$. Determine the person's acceleration at point $B$ for $AB = 2\,\text{m}$, $\theta = 60°$, and $\phi = 90°$. The angular speeds are constant.

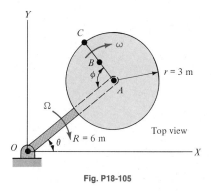

**Fig. P18-105**

**18–106** The crank arm $BC$ rotates clockwise at $200\,\text{rpm}$, and pin $C$ on the end of the crank slides in the slotted arm $AD$. Determine $\dot{\phi}$ and $\ddot{\phi}$ for $\theta = 30°$.

**Fig. P18-106**

**18–107** The Geneva mechanism of a mechanical counter converts the constant rotational motion of wheel $A$ to intermittent rotation of wheel $B$. Pin $C$ is mounted on $A$ and slides in the slots of wheel $B$. Determine the angular velocity and angular acceleration of wheel $B$ for $\theta = 30°$ and wheel $A$ rotating at $100\,\text{rpm}$ clockwise.

**Fig. P18-107**

Section 18-1

**Translation**

$$\mathbf{v}_B = \mathbf{v}_A \qquad \mathbf{a}_B = \mathbf{a}_A$$

Section 18-2

**Rotation about a fixed axis**

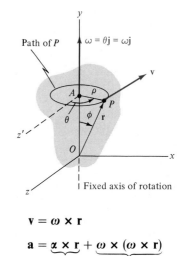

$$\mathbf{v} = \boldsymbol{\omega} \times \mathbf{r}$$

$$\mathbf{a} = \underbrace{\boldsymbol{\alpha} \times \mathbf{r}}_{\mathbf{a}_t} + \underbrace{\boldsymbol{\omega} \times (\boldsymbol{\omega} \times \mathbf{r})}_{\mathbf{a}_n}$$

$$a_t = \rho\alpha \qquad a_n = \rho\omega^2$$

$\omega$ *and* $\alpha$ are valid for any line parallel to the plane of motion of the rigid body at a given instant.

Section 18-3

**Kinematics equations for rotating rigid bodies.** If $\theta$ (angular coordinate) is known as a function of time,

$$\omega = \frac{d\theta}{dt}$$

$$\alpha = \frac{d\omega}{dt} = \omega\,\frac{d\omega}{d\theta}$$

**SPECIAL CASES:**

**1.** $\omega = $ constant

$\theta = \theta_0 + \omega t$      $(\theta = \theta_0 \text{ at } t = 0)$

**2.**   $\alpha = $ constant

$\omega = \omega_0 + \alpha t$      $(\omega = \omega_0 \text{ at } t = 0)$

$\theta = \theta_0 + \omega_0 t + \frac{1}{2}\alpha t$

$\omega^2 = \omega_0^2 + 2\alpha(\theta - \theta_0)$

## Section 18-4

### Velocities in general plane motion

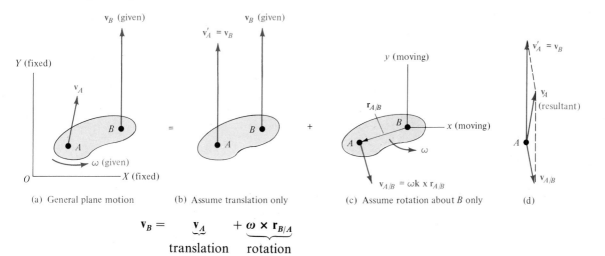

(a) General plane motion      (b) Assume translation only      (c) Assume rotation about $B$ only      (d)

$$\mathbf{v}_B = \underbrace{\mathbf{v}_A}_{\text{translation}} + \underbrace{\boldsymbol{\omega} \times \mathbf{r}_{B/A}}_{\text{rotation}}$$

## Section 18-5

### Instantaneous center of rotation

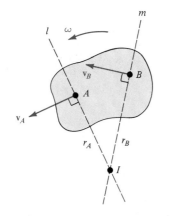

Velocities $\mathbf{v}_A$ and $\mathbf{v}_B$ are given; the body is rotating about point $I$ at that instant. Point $I$ has zero velocity at that instant, but in general it has an acceleration.

Section 18-6

## Accelerations in general plane motion

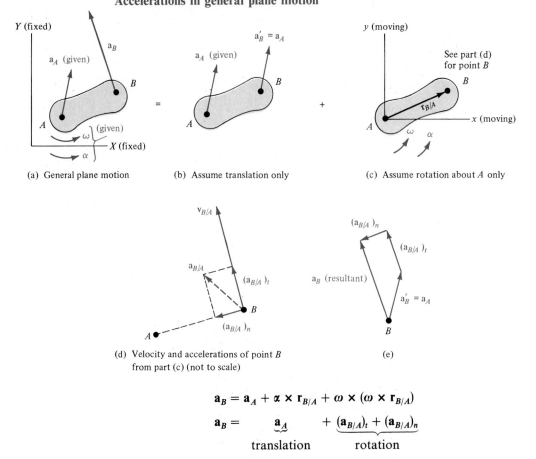

(a)  General plane motion

(b)  Assume translation only

(c)  Assume rotation about $A$ only

(d)  Velocity and accelerations of point $B$
from part (c) (not to scale)

(e)

$$\mathbf{a}_B = \mathbf{a}_A + \boldsymbol{\alpha} \times \mathbf{r}_{B/A} + \boldsymbol{\omega} \times (\boldsymbol{\omega} \times \mathbf{r}_{B/A})$$

$$\mathbf{a}_B = \underset{\text{translation}}{\underbrace{\mathbf{a}_A}} + \underset{\text{rotation}}{\underbrace{(\mathbf{a}_{B/A})_t + (\mathbf{a}_{B/A})_n}}$$

$$(a_{B/A})_t = r_{B/A}\alpha \qquad (a_{B/A})_n = r_{B/A}\omega^2$$

Section 18-7

## General motion

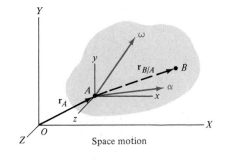

Space motion

$$\mathbf{v}_B = \underset{\text{translation}}{\underbrace{\mathbf{v}_A}} + \underset{\text{rotation}}{\underbrace{\boldsymbol{\omega} \times \mathbf{r}_{B/A}}}$$

$$\mathbf{a}_B = \underset{\text{translation}}{\underbrace{\mathbf{a}_A}} + \underset{\text{rotation}}{\underbrace{\boldsymbol{\alpha} \times \mathbf{r}_{B/A} + \boldsymbol{\omega} \times (\boldsymbol{\omega} \times \mathbf{r}_{B/A})}}$$

**Angular acceleration.** If $\omega$ has a constant magnitude,

$$\alpha = \frac{d\omega}{dt} = \Omega \times \omega$$

where $\Omega$ is the angular velocity of $\omega$.

## Section 18-8

**Time derivative of a vector using a rotating frame**

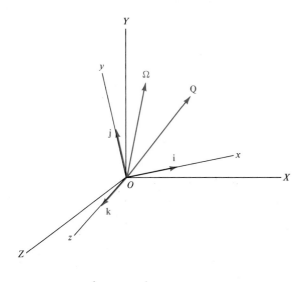

$$(\dot{\mathbf{Q}})_{XYZ} = (\dot{\mathbf{Q}})_{xyz} + \Omega \times \mathbf{Q}$$

where $\mathbf{Q}$ is an arbitrary vector and $\Omega$ is the angular velocity of the $xyz$ frame with respect to $XYZ$.

## Section 18-9

**Analysis of velocities and accelerations using a rotating frame**

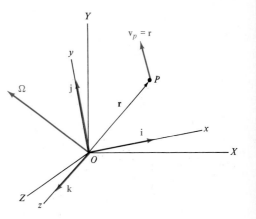

For common origin of $XYZ$ and $xyz$:

$$\mathbf{v}_P = \mathbf{v}_{xyz} + \Omega \times \mathbf{r}$$

$$\mathbf{a}_P = \mathbf{a}_{xyz} + \dot{\Omega} \times \mathbf{r} + \Omega \times (\Omega \times \mathbf{r}) + 2\Omega \times \mathbf{v}_{xyz}$$

For origin $A$ of $xyz$ translating with respect to $XYZ$:

$$\mathbf{v}_P = \mathbf{v}_A + (\dot{\mathbf{r}}_{P/A})_{xyz} + \Omega \times \mathbf{r}_{P/A}$$

$$\mathbf{a}_P = \mathbf{a}_A + \mathbf{a}_{xyz} + \dot{\Omega} \times \mathbf{r}_{P/A} + \Omega \times (\Omega \times \mathbf{r}_{P/A}) + 2\Omega \times \mathbf{v}_{xyz}$$

where $\Omega$ is the angular velocity of the $xyz$ frame with respect to $XYZ$.

**18-108** A bucket of tar weighing 966 lb is hoisted to the top of a building by an electric rotor $A$ whose angular speed is given by $\omega = (t^2 - t^3/60)$ rad/s for $0 \le t \le 60$ s. Determine (a) the maximum angular acceleration of the rotor $A$ and the time when it occurs, and (b) the corresponding maximum tension in the cable, neglecting the cable and pulley masses.

Tar

**Fig. P18-108**

**18-109** Gear $A$ starts from rest and rotates with constant acceleration of $\alpha = 4$ rad/s counterclockwise as shown. There is no slipping between the gear teeth (not shown) on the outer edges of gears $A$ and $B$. Gear $B$ has a pulley attached which lowers the mass $C$ by means of an inextensible cable. Determine (a) the angular velocity and acceleration of gear $A$ and of gear $B$, (b) the velocity and acceleration of point $P$ on gear $B$, and (c) the velocity and acceleration of the cable and mass $C$ at $t = 5$ s. The system starts from rest.

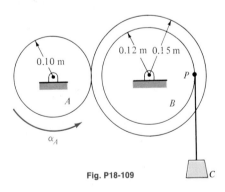

**Fig. P18-109**

**18-110** The machine element $ADCB$ rotates at an angular speed of $\omega = 6.5$ rad/s about sleeve bearings at $A$ and $B$ as shown. Determine the velocity and acceleration of points $D$ and $C$ for the position shown considering that the angular speed increases at 13 rad/s².

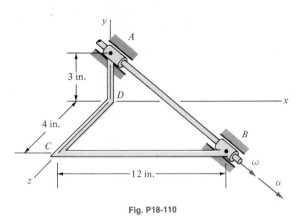

**Fig. P18-110**

**18-111** The large drum on a fishing trawler is designed to take the rope up at a constant speed of $v = 6$ ft/s. Determine the angular velocity of the drum when the core of wound rope has a diameter of 2 ft. Calculate the required angular deceleration of the drum at that instant if the core diameter increases at the rate of 0.001 ft/s.

**Fig. P18-111**

**18-112** An amusement ride consists of circular cages rotating at $\omega_2 = 3$ rad/s with respect to arm $AB$ which rotates about point $O$ at $\omega_1 = 2$ rad/s. Determine the velocity of point $A_1$ for $\theta = 30°$ and $\phi = 45°$. Locate the instantaneous center of rotation of line $AA_1$ and determine its angular velocity in the given position.

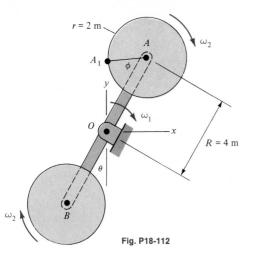

Fig. P18-112

**18–113** The 3-in.-radius disk rotates with a constant angular velocity of 10 rad/s clockwise. For the instantaneous position shown, determine (a) the velocity of point $B$, (b) the angular velocity of member $CB$, and (c) the angular velocity of member $AB$.

Fig. P18-113

**18–114** The hydraulic slider $D$ has an instantaneous velocity $\mathbf{v}_D = 0.4\mathbf{i}$ m/s and an acceleration $\mathbf{a}_D = -0.1\mathbf{i}$ m/s². Member $OBC$ is a rigid arm of a store windown model. For the instant shown, determine the angular velocity and angular acceleration of member $OBC$ and member $AD$.

Fig. P18-114

**18–115** The crank $OA$ is rotating clockwise at a constant angular speed of 1500 rpm. Determine the acceleration of the piston $B$ and the angular acceleration of the piston rod $AB$ for the position shown.

Fig. P18-115

**18–116** The center $O$ of the gear shown has an instantaneous velocity of $v_O = 5$ in./s and an instantaneous acceleration $a_O = 10$ in./s², both to the right. Determine the instantaneous acceleration of slider $A$ for the position shown.

Fig. P18-116

**18–117** An airplane takes off at a constant velocity $\mathbf{v} = 220\mathbf{k}$ km/h in an arc of radius $R = 1000$ m. The wheels are 0.4 m in diameter, and they are retracted soon after leaving the ground. The landing gear $A$ has a constant angular velocity $\boldsymbol{\omega}_2 = 1\mathbf{k}$ rad/s. Determine the angular acceleration of the landing gear during retraction.

**Fig. P18-117**

**18–118** Determine the angular acceleration of both wheels $A$ and $B$ in Prob. 18–117. Calculate the acceleration of point $P$ assuming that it is directly below the pivot in the wing, at a distance $d = 1$ m from the pivot.

**18–119** A 20-ft-long projectile for atmospheric research is to be launched from a long vertical tube. $R = 3960$ mi, $r \approx R$, $\dot{r} = 2000$ ft/s, $\ddot{r} = 0$, $\omega = 0.729 \times 10^{-4}$ rad/s. Determine the accelerations of the projectile's head $B$ and tail $A$ in the tube, in the direction normal to the tube's axis.

**Fig. P18-119**

**18–120** Determine the maximum value of $\dot{r}$ in Prob. 18–119 if the acceleration normal to the tube's axis should not exceed 100 ft/s$^2$.

# Kinetics of Rigid Bodies in Plane Motion

19

## OVERVIEW

The kinetics analysis of rigid bodies is based on Newton's second law of motion and the methods developed in Ch. 17. This chapter on plane motion is presented separately before space motion because the latter is quite complex and is often omitted from elementary courses. Planar kinetics analysis is very important in engineering practice because many components of systems of rigid bodies have two-dimensional motions.

## STUDY GOALS

**SECTION 19-1** presents the equation of motion for the translation of a rigid body. This equation is fundamental in analyzing the motion of rigid bodies and should be remembered for frequent application in dynamics.

**SECTION 19-2** presents the derivation of the equation of motion for the rotation of a rigid body. This equation is also fundamental in analyzing the motion of rigid bodies and should be remembered. Memorizing the derivation is not necessary, but the resulting equation of motion is very important in dynamics.

**SECTION 19-3** describes several practical aspects of applying the basic vector equations of motion, one for translation and another for rotation. Three scalar equations of motion are written from these for the convenient analysis of plane motion. Students should acquire skill in using these equations.

SECTION 19-4 presents the analysis of rigid bodies which rotate about fixed axes. This topic is important in machine design. The subtopic titled *center of percussion* is an interesting special case of rotation about a fixed axis, but it is not essential to make further progress in elementary dynamics.

SECTION 19-5 presents the methods of analyzing common problems of general plane motion, such as the rolling *and* sliding of wheels. These special problems provide for useful practice in applying the equations of motion.

## 19-1

## EQUATION OF TRANSLATIONAL MOTION

To establish the methods of analysis, assume that a rigid body is a thin slab of arbitrary shape and is moving in such a way that all of its constituent particles move parallel to the fixed $xy$ frame (Fig. 19-1a). The body consists of $n$ particles of mass $m_i$ with $i = 1, 2, \ldots, n$. Assume that the mass center $C$ and all significant forces acting on the body are in the $xy$ plane and are completely specified. In this case the system of forces can be replaced by an equivalent force–couple system consisting of the resultant force $\sum \mathbf{F}$ acting at $C$ and the resultant moment $\sum \mathbf{M}_C$ acting about $C$ as in Fig. 19-1b.

The resultant moment is a couple, so it does not affect the translational motion of the body in any way. The translational motion depends only on the net force $\sum \mathbf{F}$ and the total mass $m$ of the body. This can be expressed using Eq. 17-7, since the rigid body is a system of particles,

$$\sum \mathbf{F} = \sum m_i \mathbf{a}_i = m\mathbf{a}_C \qquad \boxed{19\text{-}1}$$

where $\mathbf{a}_C$ is the acceleration of the mass center. The resultant force and the acceleration vectors are collinear.

Equation 19-1 is fundamental in analyzing the motion of rigid bodies. It is simple to use, but certain aspects of it are worth noting as follows.

**1.** Newton's second law $\mathbf{F} = m\mathbf{a}$ is applicable only to individual particles having mass but no dimensions, or to the mass center of a system of discrete particles where $\mathbf{F}$ is the resultant of individual forces acting on individual particles. In the case of a rigid body it is assumed that the force acting on any one particle of the body is instantaneously acting on

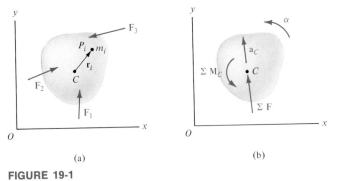

(a)                                        (b)

**FIGURE 19-1**

Plane motion of a rigid body

**FIGURE 19-2**

A rigid body modeled as a system of six identical particles

all particles of that body. This assumption is modeled in Fig. 19-2, where a single force is acting on a body of six equal particles. The force is directly acting on particle 2, but the six particles are rigidly connected. According to Eq. 19-1, this result is equivalent to having a single particle of mass $6\,m$, so the acceleration of the mass center is $\mathbf{a}_C = \mathbf{F}/6\,m$.

**2.** The acceleration of the mass center depends only on the total mass of the body and the resultant force acting on the body. Thus, in Fig. 19-2 the force $\mathbf{F}$ may be moved to any particle without affecting the magnitude or direction of $\mathbf{a}_C$. This is independent of any rotation of the body, and consequently it is extremely useful in the analysis of motion.

## EQUATION OF ROTATIONAL MOTION

**19-2**

The resultant force $\sum \mathbf{F}$ in Fig. 19-1b fully accounts for the translational acceleration of a given body caused by the system of forces. The resultant moment of these forces about the center of mass $C$ causes the body to rotate at an angular acceleration $\boldsymbol{\alpha}$ (independent of the linear acceleration $\mathbf{a}_C$). Using Eqs. 17-24 and 17-29 for a system of particles gives

$$\sum \mathbf{M}_C = \sum (\mathbf{r}_i \times m_i \mathbf{a}) = \dot{\mathbf{H}}_C \qquad \boxed{19\text{-}2}$$

where $\dot{\mathbf{H}}_C$ is the time rate of change of the angular momentum $\mathbf{H}_C$ of the system of particles about its mass center. The angular momentum for the body in Fig. 19-1 is expressed according to Eq. 17-27 as

$$\mathbf{H}_C = \sum_{i=1}^{n} (\mathbf{r}_i \times m_i \mathbf{v}_i) \qquad \boxed{19\text{-}3a}$$

where $\mathbf{r}_i$ is the position vector* and $m_i \mathbf{v}_i = m_i \dot{\mathbf{r}}_i$ is the linear momentum of particle $P_i$ with respect to the mass center of the body. With $\mathbf{v}_i = \boldsymbol{\omega} \times \mathbf{r}_i$ for any rigid body, where $\omega$ is the angular velocity of the body,

$$\mathbf{H}_C = \sum_{i=1}^{n} \left[ \mathbf{r}_i \times (\boldsymbol{\omega} \times \mathbf{r}_i) m_i \right] \qquad \boxed{19\text{-}3b}$$

Since $\mathbf{r}_i$ and the velocity $\mathbf{v}_i$ are both parallel to the $xy$ plane for every particle in a thin slab, $\mathbf{r}_i \times \mathbf{v}_i$ and hence the angular momentum vector $\mathbf{H}_C$ are perpendicular to the $xy$ plane. Thus, for plane motion, $\mathbf{H}_C$ is in the same direction and has the same sense as $\omega$ as found by using the right-hand rule. The magnitude of the angular momentum from Eq. 19-3b is $H_C = \omega \sum r_i^2 m_i$. Here each term $r_i^2 m_i$ is the moment of inertia of particle

---

\* $\mathbf{r}_i$ here is equivalent to $\mathbf{r}_i'$ in Eq. 17-27. The change in notation is advantageous in the following analysis.

$P_i$ about the center of mass. The summation of the moments of inertia for all infinitesimal particles of a rigid body (as $n \to \infty$ and $m_i \to 0$) is equivalent to writing $I_C = \int r^2 \, dm$, the mass moment of inertia of the body about the centroidal axis perpendicular to the $xy$ plane. Therefore, the concise expression for the angular momentum (for plane motion only) is

$$\mathbf{H}_C = I_C \omega \qquad \boxed{\text{19-4}}$$

Since $I_C$ is a constant property of the body in plane motion, differentiating Eq. 19-4 with respect to time gives

$$\dot{\mathbf{H}}_C = I_C \dot{\omega} = I_C \alpha \qquad \boxed{\text{19-5}}$$

where $\alpha$ is the angular acceleration of the body. Substituting Eq. 19-5 in Eq. 19-2 yields

$$\sum \mathbf{M}_C = I_C \alpha \qquad \boxed{\text{19-6}}$$

This equation is equally important to Eq. 19-1 in analyzing the motion of rigid bodies. The following comments are worth noting.

**1.** Equation 19-2 is strictly applicable only to a system of discrete particles where individual forces are acting on individual particles. In the case of a rigid body it is assumed that the moment of a force acting on any one particle is instantaneously changing the angular momentum of all particles of the body. This assumption is modeled in Fig. 19-3, where a single force is acting on six rigidly connected particles that rotate about point $C$. Here $M_C = Fd = I_C \alpha$ and the particles respond to the moment in unison, which means that $\omega$ and $\alpha$ are unique to the whole body at a given instant. This result is equivalent to having a single particle of mass $6m$ located on a massless link at a distance $r_g = \sqrt{I_C/6m}$ ($r_g$ = radius of gyration) from point $C$. It appears that the angular motion of particle 5 is retarded by the other five particles. This effect depends on the square of the particles' distance from the mass center, so particle 6 has a much greater effect than particle 4 does on the motion of particle 5, even though they have the same mass.

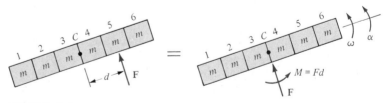

**FIGURE 19-3**

Rotation of a rigid body which is modeled as a system of six particles

**2.** The angular acceleration of the body depends only on its mass moment of inertia and the resultant moment acting on it. The rotational motion is independent of any translational motion of the body.

**3.** An important restriction of Eq. 19-6 is that the body is assumed to be a thin slab with $\mathbf{r}_i$ and $\mathbf{v}_i$ parallel to the $xy$ plane. This equation is also used for other bodies in special cases, when the $xy$ plane is a plane of symmetry (with products of inertia being zero), or when all $\mathbf{v}_i$ are parallel to the $xy$ plane because of constraints.

## APPLICATION OF EQUATIONS OF MOTION **19-3**

The fundamental equations (Eqs. 19-1 and 19-6) that govern the translation and rotation of a rigid body in plane motion are most convenient to use in scalar form. There are a maximum of three scalar equations for plane motion. Defining the $xy$ plane as the plane of motion,

$$\sum F_x = ma_{C_x} \quad \sum F_y = ma_{C_y} \quad \sum M_C = I_C \alpha \qquad \boxed{19\text{-}7}$$

These equations can be integrated to obtain the velocities and positions of a body in terms of initial conditions.

To apply the equations of motion (Eq. 19-7), it is necessary to use proper free-body diagrams of the body or system of bodies. It is essential that the external force system acting on each body be represented by its equivalent resultant force and moment. Equations 19-7 reduce to the following special cases.

**1. *Pure translation.*** The resultant force in each coordinate direction is applied to the total mass as if the rigid body were a particle. There is no rotation of the body, so from Eq. 19-7,

$$\sum F_x = ma_{C_x} \quad \sum F_y = ma_{C_y} \quad \sum M_C = 0 \qquad \boxed{19\text{-}8}$$

**2. *Pure rotation about the center of mass.*** The resultant moment about the center of mass causes rotation about that point. There is no translation of the center of mass, so from Eq. 19-7,

$$\sum F_x = 0 \quad \sum F_y = 0 \quad \sum M_C = I_C \alpha \qquad \boxed{19\text{-}9}$$

Note that the equation of rotational motion was obtained using the center of mass for reference.

### Constraints of Motion

The motions of rigid bodies are categorized according to the constraints affecting the motion:

**1.** *Unconstrained motion* is the general case which is exemplified by the motion of aircraft and rockets. In the case of unconstrained plane motion Eq. 19-7 can be directly applied with the three scalar equations being independent of one another.

**2.** *Constrained motion* is most commonly encountered in engineering work since the components of machines generally move in fixed patterns. When there is a physical constraint to the motion, a *kinematics analysis* has to be made first to determine the relation of linear and angular accelerations of the body. This is necessary because in constrained plane motion the three accelerations $a_{C_x}$, $a_{C_y}$, and $\alpha$ are interrelated, so the three equations of Eq. 19-7 are not independent of one another. The force and moment equations can be solved after the kinematics analysis. Two special cases of constrained motion are distinguished:

**a.** *Point constraint* is typical of rotation about a fixed point or axis as in Fig. 19-4.

**b.** *Line constraint* exists when the body moves along a fixed line or plane. An example of this motion is a wheel rolling along a road surface as in Fig. 19-5.

**FIGURE 19-4**

Axis of rotation is the point of constraint

**FIGURE 19-5**

Line of constraint is on the supporting surface of the wheel

## Systems of Rigid Bodies

Frequently, a member's motion is constrained by other members that are also in motion. Such motions can be readily analyzed if the motions of the interconnected bodies can be determined kinematically. After the kinematics analysis, the force and moment equations can be solved using individual but interrelated free-body diagrams or using the free-body diagram for the entire system if convenient. It should be remembered that the required number of free-body diagrams depends on the number of unknowns in the problem. At most three equations of motion (Eq. 19-7) are available from a given free-body diagram for plane motion to solve for three unknowns. The equations of motion are first applied in examples of translational motion in the following.

EXAMPLE 19-1

The combined mass of the motorcycle and its rider in Fig. 19-6a is 250 kg. The mass center is at $C$. The wheels are locked when the brakes are applied at a high speed. Determine the linear acceleration and the normal and tangential forces acting on each wheel assuming that the coefficient of kinetic friction is 0.5, and that the angular acceleration is zero.

## SOLUTION

First the free-body diagram is sketched in Fig. 19-6b, showing the known and unknown forces externally applied to the body. The constraints of motion are that $a_{C_y} = 0$ and $\alpha = 0$. Since the brakes have been applied, the motorcycle is slowing down. Thus, the acceleration is assumed in the positive $x$ direction, opposite the direction of motion. Using the three equations from Eq. 19-8 for pure translation,

$$\sum F_x = \mathscr{F}_A + \mathscr{F}_B = ma_{C_x} \tag{a}$$

$$\sum F_y = N_A + N_B - W = ma_{C_y} = 0 \tag{b}$$

$$\sum M_C = \mathscr{F}_A\,(0.7\,\text{m}) - N_A\,(0.8\,\text{m}) + \mathscr{F}_B\,(0.7\,\text{m}) + N_B\,(0.5\,\text{m})$$
$$= I_C\alpha = 0 \tag{c}$$

Since the wheels are locked and sliding, the magnitudes of the friction forces are $\mathscr{F}_A = \mu_k N_A$ and $\mathscr{F}_B = \mu_k N_B$. Hence, from Eqs. (b) and (c),

$$N_A + N_B = W = (250\,\text{kg})(9.81\,\text{m/s}^2) = 2453\,\text{N}$$

$$\mu_k N_A\,(0.7\,\text{m}) - N_A\,(0.8\,\text{m}) + \mu_k N_B\,(0.7\,\text{m}) + N_B\,(0.5\,\text{m}) = 0$$

$$N_A\,(0.35 - 0.8)\,\text{m} + N_B\,(0.35 + 0.5)\,\text{m} = 0$$

Solving these simultaneously yields

$$N_A = 1604\,\text{N} \qquad\qquad N_B = 849\,\text{N}$$

$$\mathscr{F}_A = \mu_k N_A = 802\,\text{N} \qquad \mathscr{F}_B = \mu_k N_B = 425\,\text{N}$$

Hence, from Eq. (a),

$$a_{C_x} = \frac{1}{250\,\text{kg}}\,(802 + 425)\,\text{N} = 4.91\,\text{m/s}^2$$

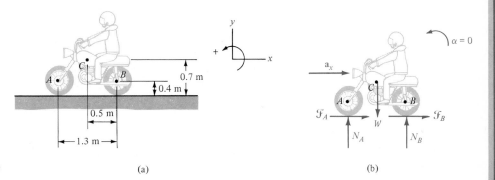

**FIGURE 19-6**

Motorcycle and its rider modeled as a rigid body

(a)

(b)

EXAMPLE 19-1    771

EXAMPLE 19-2

Assume that a small trailer of 70 kg mass is attached to the motorcycle described in Ex. 19-1. The drawbar is firmly attached to the trailer and is hinged at axle $B$ (Fig. 19-7a). The trailer has a single wheel of negligible mass and no brake. Model the trailer as a rigid body which has no angular acceleration and which moves without friction. Determine the linear acceleration $a_x$ of the system when the wheels of the motorcycle slide during braking.

## SOLUTION

Free-body diagrams are drawn to show the external forces on the whole system (Fig. 19-7b), on the motorcycle alone (Fig. 19-7c), and on the trailer alone (Fig. 19-7d). The subscripts $t$ and $m$ are used for the trailer and motorcycle, respectively. Considering the system as a whole, there are four unknowns, $N_A$, $N_B$, $N_t$, and $a_x$, but only three equations of motion. Analyzing the trailer and the motorcycle separately, there are six independent equations of motion and six unknowns, $N_A$, $N_B$, $N_t$, $a_x$, $R_x$, and $R_y$. From the free-body diagrams:

Trailer:

$$\sum F_x = R_x = m_t a_x \tag{a}$$

$$\sum F_y = R_y + N_D - W_t = m_t a_{t_y} = 0 \quad \Rightarrow \quad R_y = W_t - N_D \tag{b}$$

$$\sum M_{C_t} = I_{C_t} \alpha_t = 0$$

$$R_x (0.2\,\text{m}) - R_y (1.0\,\text{m}) = 0 \quad \Rightarrow \quad R_x = 5R_y \tag{c}$$

Motorcycle:

$$\sum F_x = \mathcal{F}_A + \mathcal{F}_B - R_x = m_m a_x \tag{d}$$

$$\sum F_y = N_A + N_B - W_m - R_y = m_m a_{m_y} = 0 \tag{e}$$

$$\sum M_{C_m} = \mathcal{F}_A (0.7\,\text{m}) + \mathcal{F}_B (0.7\,\text{m}) - N_A (0.8\,\text{m}) + N_B (0.5\,\text{m})$$
$$- (0.3\,\text{m})\, R_x - (0.5\,\text{m})\, R_y = I_{C_m} \alpha_m = 0 \tag{f}$$

There are several alternative procedures to obtain $a_x$ from these equations. Two alternatives are presented here.

(a)

(b)

(c)

(d)

FIGURE 19-7

Analysis of a motorcycle and the attached trailer

## SOLUTION A

Using Eqs. (a), (b), and (c),

$$m_t a_x = 5(W_t - N_D) \tag{g}$$

From Eqs. (a) and (d),

$$\mathscr{F}_A + \mathscr{F}_B = (m_m + m_t)a_x \tag{h}$$

From Eqs. (b), (c), and (e),

$$N_A + N_B = W_m + (0.2)m_t a_x \tag{i}$$

From Eqs. (a), (b), (g), and (f),

$$(0.7\,\text{m}) (\mathscr{F}_A + \mathscr{F}_B) - N_A (0.8\,\text{m}) + N_B (0.5\,\text{m})$$
$$- (0.3\,\text{m}) \, m_t a_x - (0.5\,\text{m})(0.2) \, m_t a_x = 0 \tag{j}$$

Using $\mathscr{F}_A = \mu_k N_A$ and $\mathscr{F}_B = \mu_k N_B$, Eq. (h), (i), and (j) can be solved giving

$$a_x = 3.91 \text{ m/s}^2$$

## SOLUTION B

The moment equations may be simplified if, instead of using the mass centers, the reference points are carefully chosen to minimize the number of terms. Such a point for the motorcycle is point $P$ at the intersection of forces $\mathscr{F}_B$ and $N_B$, eliminating the moments of $\mathscr{F}_A$, $\mathscr{F}_B$, $N_B$, and $\mathbf{R}_y$. The moment equation using Eq. 17-31 and 19-5 is

$$\sum \mathbf{M}_P = I_C \alpha + \mathbf{r}_{CP} \times m_m \mathbf{a}_C$$

where $\alpha = 0$. Thus,

$$- N_A(1.3) + W_m(0.5) + R_x(0.4) = -(0.7)(m_m)(a_x)$$

For the trailer, point $Q$ at the intersection of forces $\mathbf{R}_x$ and $N_D$ is advantageous for reference of moments, giving (again from Eq. 17-31)

$$- R_y(1) = -(0.2)(m_t)(a_x)$$

The solution using the original force summations and the latter moment equations also yields

$$a_x = 3.91 \text{ m/s}^2$$

## JUDGMENT OF THE RESULT

The answer is also reasonable in comparison with $a_x = 4.91 \text{ m/s}^2$ in Ex. 19-1 since the mass of the trailer moves without friction and its center of mass is directly above the trailer wheel (thus it has a small effect on the forces $N_A$ and $N_B$). For another check, solve for $N_A$ and $N_B$ using $a_x = 3.91 \text{ m/s}^2$ and prove that this value of $a_x$ is acceptable since it satisfies the equations of motion.

EXAMPLE 19-2   773

# EXAMPLE 19-3

The combined mass of the crate and platform $AB$ is 400 kg with a mass center at point $C$. Members $DA$ and $EB$ supporting the platform $AB$ have negligible mass. For the position shown member $DA$ is being raised by a motor at an angular velocity $\omega = 2$ rad/s at an increasing rate of $\alpha = 4$ rad/s² (Fig. 19-8a). Determine the instantaneous forces exerted on the platform at $A$ and $B$ for this motion.

## SOLUTION

The combined crate and platform $AB$ move in curvilinear translation, so all points have similar paths of motion and the same acceleration. Using Fig. 19-8b,

$$\mathbf{a}_C = \mathbf{a}_A = (5\,\text{m})(4\,\text{rad/s}^2)\,\mathbf{n}_t + (5\,\text{m})(2\,\text{rad/s})^2\,\mathbf{n}_n$$

or

$$\mathbf{a}_C = (-28\mathbf{i} + 4\mathbf{j})\,\text{m/s}^2$$

Member $EB$ is a *two-force member* because it is (1) massless (weightless) and (2) does not have a couple applied to its end as does member $DA$. Consequently, $\mathbf{F}_{EB}$ is in the direction of member $EB$ at all times, as shown in the free-body diagram in Fig. 19-8b. Using Eq. 19-8,

$$\sum F_x = ma_{C_x} \qquad A_x + \frac{4}{5} F_{EB} = (400\,\text{kg})(-28\,\text{m/s}^2) \tag{a}$$

$$\sum F_y = ma_{C_y} \qquad A_y + \frac{3}{5}(F_{EB}) - 3920\,\text{N} = (400\,\text{kg})(4\,\text{m/s}^2) \tag{b}$$

$$\sum M_C = 0 \qquad A_x\,(1\,\text{m}) - A_y\,(1.5\,\text{m}) + \frac{4}{5} F_{EB}\,(1\,\text{m}) + \frac{3}{5} F_{EB}\,(1.5\,\text{m}) = 0 \tag{c}$$

Solving Eqs. (a), (b), and (c) simultaneously yields $A_x = -19\,860$ N, $A_y = -973$ N, and $F_{EB} = 10\,820$ N. The minus signs on $A_x$ and $A_y$ simply mean that the wrong directions were assumed on the free-body diagram. Consequently, $A_x$ acts to the left, $A_y$ downward, and $F_{EB}$ as shown.

$$A_x = 19\,860\,\text{N} \quad (\leftarrow)$$

$$A_y = 973\,\text{N} \quad (\downarrow)$$

$$F_{EB} = 10\,820\,\text{N} \quad \left(\begin{array}{c}\text{\tiny 3}\\\text{\tiny 4}\end{array}\right)$$

**FIGURE 19-8**

Curvilinear translation of a crate and platform

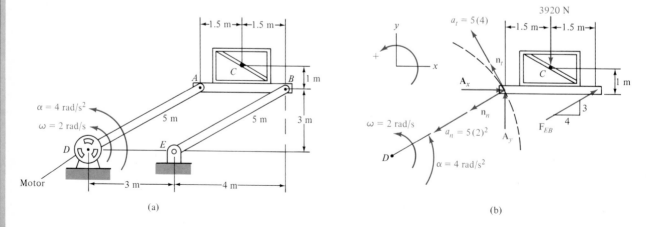

(a)

(b)

**19–1** The mass of a motorcycle and its rider is 300 kg. It starts from rest with the driving wheel spinning on the ground. Determine the linear acceleration if $a = 1.2$ m, $b = 0.5$ m, $c = 0.6$ m, and $\mu_k = 0.8$. Assume that the wheels have negligible mass.

**19–4** The weight of the motorcycle and its rider is 500 lb. It starts from rest going up on a $10°$ incline with the driving wheel spinning on the ground. Determine the normal and tangential forces on the wheels if $a = 4$ ft, $b = 1.7$ ft, $c = 2$ ft, and $\mu_k = 0.9$. Assume that the wheels have negligible mass.

**Fig. P19-1**

**Fig. P19-4**

**19–2** For the motorcycle in Prob. 19–1, determine the magnitude of the normal force on the front wheel.

**19–5** A drag racer weighs 2000 lb. Calculate the maximum acceleration of the vehicle and the friction force at the driving wheels $A$ that will allow at least minimal contact of wheels $B$ with the road. Assume that the wheels have negligible mass.

**19–3** A 2000-kg car is traveling downhill when the brakes are applied locking all wheels. Calculate the normal and tangential forces on the wheels for $\mu_k = 0.5$, assuming symmetry between the left and right wheels.

**Fig. P19-5**

**Fig. P19-3**

**19–6** Determine the required coefficient of kinetic friction in Prob. 19–5 to attain the maximum allowable acceleration. Note that $\mu_k > 1$ is possible in practice depending on the conditions of the tires and the road.

**19–7** An 1800-kg car is pulling a two-wheel trailer of 500 kg mass. The drawbar is rigidly attached to the trailer and is hinged at hitch $H$ on the car. The front-wheel-drive car accelerates uphill with the wheels spinning on the ground. Determine the linear acceleration of the car assuming that the six wheels have negligible mass and have symmetry of the left and right sides.

**Fig. P19-7**

**19–8** The car and trailer described in Prob. 19–7 travel downhill on a 10° slope when the brakes are applied locking all four wheels on the car. Determine the linear acceleration of the car.

**19–9** A uniform crate weighs 8000 lb and its coefficient of static friction with the truck is 0.2. What is the maximum acceleration of the truck that does not cause the crate to slide assuming that it does not tip?

**Fig. P19-9**

**19–10** The crate in Prob. 19–9 is tied down with two cables that are in the same vertical plane with the center of mass. The truck is decelerating at the rate of 20 ft/s² and the forward cable becomes slack. Calculate the tension in the rear cable.

**Fig. P19-10**

**19–11** A 322-lb window frame whose mass center is at $C$ is transported as shown, leaning against the truck at $D$ and secured by rope $AB$ to prevent slipping. All surfaces are smooth and the window frame is thin compared to other important dimensions. If the rope would break at a tension of 300 lb, determine the maximum allowable acceleration of the truck.

**Fig. P19-11**

**19–12** The structure on the seismic simulator is modeled as a two-dimensional rigid body of weightless members and four concentrated masses $m$. The hinge support at $A$ is accelerated to the right at the rate of $a_x$. The leg at $B$ is on a frictionless surface. Determine the forces acting on the legs at $A$ and $B$.

(a)

**19–14** The crate shown has a mass of 100 kg, and its mass center is at $C$. A worker pushes the crate with a force of magnitude $P$ of 500 N. The coefficient of friction between the crate and the surface is $\mu = 0.2$. Determine the magnitude and location of the resultant normal force acting on the bottom of the crate as it slides.

Fig. P19-14

**19–15** The fixture moving along horizontal tracks is modeled as three parts, $A$, $B$, and $C$, which remain in their vertical configuration during motion. Assume that $A$ and $C$ have concentrated masses as shown, and $W_A = 50$ lb, $W_C = 20$ lb. A force of 500-lb magnitude is applied horizontally to $A$ and the fixture slides without friction. Calculate the maximum bending moment in the uniform rod $B$ which has a weight of 40 lb.

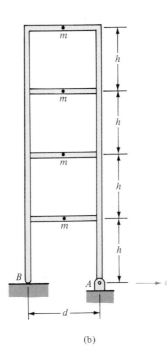

(b)

Fig. P19-12
Model of a building frame tested on a seismic simulator
(Photograph courtesy MTS Systems Corporation, Minneapolis, Minnesota)

(b)

Fig. P19-15
Fixture *ABC* moving along horizontal tracks (Photograph courtesy MTS Systems Corporation, Minneapolis, Minnesota)

**19–13** Determine the acceleration $a_x$ in Prob. 19–12 at which leg $B$ separates from its supporting surface.

**19–16** A 100-kg console can be moved on casters whose friction may be neglected. Calculate the range of the height $h$ at which a force of magnitude $F = 400$ N may be applied without causing the console to overturn.

**Fig. P19-16**

**19–17** Determine the maximum magnitude of force **F** in Prob. 19–16 that can be applied at a height of $h = 1.7$ m without causing the console to overturn. Assume that the casters have a coefficient of kinetic friction $\mu_k = 0.2$.

**19–18** A 3000-lb vehicle may be driven either by its front or rear wheels. Calculate the maximum acceleration in each mode of operation without allowing slippage of the wheels if $\mu_s = 0.5$, the wheels have negligible mass, and $d = 4$ ft.

**Fig. P19-18**

**19–19** Determine the position $d$ of the center of mass in Prob. 19–18 in order for the driving force to be equally distributed between the four wheels at maximum acceleration without slippage of the wheels.

**19–20** A 1000-kg trailer is pulled with a slanted drawbar. Determine the tension in the bar and the magnitudes of the normal forces on wheels $A$ and $B$ if the truck accelerates at the rate of $3 \text{ m/s}^2$. The wheels have negligible mass.

**Fig. P19-20**

**19–21** Calculate the force in the drawbar in Prob. 19–20 for a deceleration of the truck at the rate of $6 \text{ m/s}^2$. The wheels of the trailer are locked by the brakes and slide; $\mu_k = 0.3$.

**19–22** Consider a 3000-lb car traveling on a curved, banked road at a speed of 80 ft/s. Determine the required bank angle $\theta$ for equalizing the normal forces on all wheels.

**Fig. P19-22**

**19–23** Determine the maximum speed of the car in Prob. 19–22 so that it does not slip or tip on the road with $\theta = 10°$.

**19–24** A motorcycle and its rider have a total mass of 200 kg. Determine the maximum safe speed $v$ and the required angle $\theta$ of leaning on a horizontal, curved road to prevent slipping or tipping ($\ddot{\theta}$ should be zero).

$r = 50$ m

$0.8$ m

$\theta$

$\mu_s = 0.5$

Fig. P19-24

**19-25** The hydraulic actuator $A$ makes an angle $\theta$ with the horizontal and is attached to the horizontal platform in the plane of joints $B$ and $C$. The uniform platform weighs 200 lb, is 8 in. thick, and 5 ft long between $B$ and $C$. Each rocker arm $(BE)$ is 2.5 ft long, is essentially weightless, and makes an angle $\phi$ with the horizontal. The test object weighs 250 lb and its center of mass is 2 ft above the platform. Determine the magnitude of the actuator force that would cause slipping of the test object while raising the platform starting at rest from $\theta = 45°$, $\phi = 60°$, if $\mu_s = 0.2$. Assume that the force is applied instantaneously.

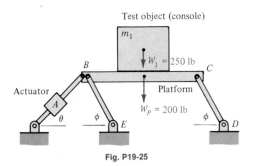

Test object (console)

$m_1$

$B$

$W_1 = 250$ lb

$C$

Actuator

$A$

$\theta$

$\phi$

$E$

Platform

$W_P = 200$ lb

$\phi$

$D$

Fig. P19-25

**19-26** Consider the shake table described in Prob. 19–25. Assume that the magnitude of the actuator force is rapidly decreased to zero when $\phi = 70°$ and $\dot{\phi} = 0$. Determine the accelerations of the test object and the platform assuming $\mu_s = 0$ between them.

**19-27** A person walking on stilts tries to stay upright even during a fall. This is modeled as a weightless bar $AB$ (the stilts) starting from rest and rotating about $A$. A uniform bar representing the person is supported at $D$ and remains vertical as $AB$ rotates. Write an expression for the horizontal acceleration of the center of mass $C$ assuming that the stilts start from rest with a small initial displacement $d$. What can be concluded concerning the stability of stilts as a function of their length $l$ and the mass of the person?

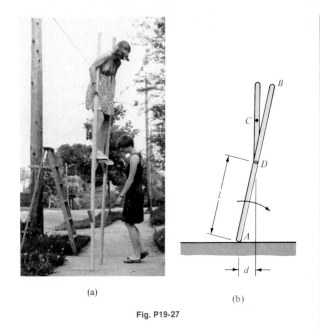

(a)

(b)

Fig. P19-27

**19-28** The uniform bars moving in the horizontal $xy$ plane have length $l = 10$ in. and weight $W = 2$ lb each. $AB$ is rotating at a constant angular speed $\omega = 50$ rad/s. Determine the required couple at $B$ to constrain bar $BC$ to remain parallel to the $y$ axis as shown.

Fig. P19-28

**19-29** Calculate the required couple in Prob. 19–28 if the angular speed is decreasing at the rate of 10 rad/s².

# ROTATION ABOUT A FIXED AXIS

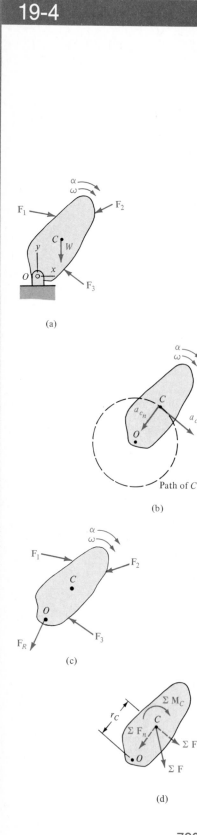

(a)

(b)

(c)

(d)

The methods developed in the preceding three sections can be applied in the analysis of motion of bodies that rotate about fixed axes. This is an important area in plane motion of rigid bodies because of the numerous rotating shafts, wheels, gears, and linkages in essentially two-dimensional motion in almost every machine. For a general model of such rotational motions, consider the rigid body in Fig. 19-9a. This body is pinned at point $O$ and moves in the vertical $xy$ plane. Thus, it can only rotate about the fixed $z$ axis at point $O$.

The mass center $C$ moves in circular motion about the fixed point $O$, so it is convenient to consider the normal and tangential components of the acceleration of the center of mass as shown in Fig. 19-9b. For an angular velocity $\omega$ and an angular acceleration $\alpha$, the components are

$$a_{C_n} = r_C\omega^2 \quad \text{and} \quad a_{C_t} = r_C\alpha \qquad \boxed{19\text{-}10}$$

where $r_C$ is the nearest distance between the fixed axis and the center of mass.

The free-body diagram of the rigid body is shown in Fig. 19-9c, where $\mathbf{F}_R$ is the reaction of the pin on the body. This system of forces can be replaced by the resultant force $\mathbf{F}$ acting at the center of mass $C$ and by the resultant moment $\mathbf{M}_C$ acting about $C$. In Fig. 19-9d the resultant force is shown broken into its normal and tangential components where the normal and tangential directions have been defined by the components of the acceleration of the center of mass. For this case of rotation about a fixed axis, the equations of motion are most conveniently written as

$$\sum F_n = ma_{C_n} = mr_C\omega^2 = m\frac{v_C^2}{r_C} \qquad \boxed{19\text{-}11}$$

$$\sum F_t = ma_{C_t} = mr_C\alpha \qquad \boxed{19\text{-}12}$$

$$\sum M_C = I_C\alpha \qquad \boxed{19\text{-}13}$$

An alternative moment equation can be written by summing the moments about the fixed point $O$. As can be seen in Fig. 19-9d, the normal component of the resultant force intersects point $O$ and creates no moment about that point. The moment summation is written as

$$\sum M_O = M_C + r_C F_t = I_C\alpha + r_C m a_{C_t}$$

$$= I_C\alpha + r_C m (r_C\alpha)$$

$$\sum M_O = (I_C + mr_C^2)\alpha$$

where the quantity in parentheses is equal to $I_O$, the mass moment of inertia about point $O$ by the parallel-axis theorem, as shown in Ch. 9 in the statics volume. Thus,

$$\sum M_O = I_O\alpha \qquad \boxed{19\text{-}14}$$

**FIGURE 19-9**

Rotation of a rigid body about a fixed axis

where the angular acceleration $\alpha$ is independent of the point of reference. By using Eq. 19-14, the angular acceleration of the body can often be determined without having first to determine the reactions at the fixed axis of rotation.

In summary, the equations of motion for a rigid body rotating about a fixed axis $O$ are

$$\sum F_n = mr_C\omega^2$$
$$\sum F_t = mr_C\alpha \qquad \boxed{19\text{-}15}$$
$$\sum M_O = I_O\alpha$$

where $r_C$ is the distance between the fixed axis at $O$ and the center of mass. The two force equations must include the normal and tangential components of the reaction at the fixed axis of rotation.

**Rotation about the centroidal axis.** Equation 19-15 may be used for rigid bodies that rotate about centroidal axes. In that special case $r_C = 0$, and $I_O = I_C$, so the equations of motion become

$$\sum F_n = 0$$
$$\sum F_t = 0 \qquad \boxed{19\text{-}16}$$
$$\sum M_C = I_C\alpha$$

where $n$ and $t$ may be taken as any pair of orthogonal directions in the plane of motion. This result agrees with Eq. 19-9 for pure rotation of a rigid body about the center of mass.

Center of Percussion

Another special case of rotation about a fixed axis occurs when the equations of motion can be written as two force summations and a moment summation which is equal to zero about a particular point $P$. The location of this point $P$ is found by considering the rotating body in Fig. 19-10, where again the fixed axis is at $O$ and the center of mass is at $C$. Assume

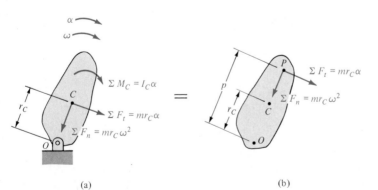

**FIGURE 19-10**

Schematic location of the center of percussion

that the resultants $\sum \mathbf{F}_n$, $\sum \mathbf{F}_t$, and $\sum \mathbf{M}_C$ have been determined from the system of external forces acting on the body, as shown in Fig. 19-10a. It is possible to find a point $P$ on the line $OC$ where the forces $\sum \mathbf{F}_n$ and $\sum \mathbf{F}_t$ are equivalent to the original force–couple system at $C$, so $\sum \mathbf{M}_C$ can be eliminated from consideration. The location of point $P$ is such that the moment about $C$ remains the same. The new location of the line of action of $\sum \mathbf{F}_t$ at point $P$ is at a distance $p$ from the axis of rotation as shown in Fig. 19-10b. The value of $p$ is calculated from the equality of moments in Fig. 19-10a and b, using point $O$ for reference,

$$\sum M_{O_{(b)}} = \sum M_{O_{(a)}}$$
$$p \sum F_t = r_C \sum F_t + \sum M_C$$

From this equation of moments,

$$mr_C\alpha p = mr_C^2\alpha + I_C\alpha \qquad \boxed{19\text{-}17}$$

This equation can be simplified by using the radius of gyration $r_g$ (with respect to the center of mass),

$$I_C = r_g^2 m$$

With this substitution, and canceling $m\alpha$, Eq. 19-17 reduces to

$$p = r_C + \frac{r_g^2}{r_C}$$

Note that $r_C^2 + r_g^2 = r_O^2$, where $r_O$ is the radius of gyration of the body about point $O$;

$$r_O^2 = \frac{I_O}{m} = \frac{I_C + r_C^2 m}{m} = \frac{(r_g^2 + r_C^2)m}{m}$$

Thus,

$$p = \frac{r_O^2}{r_C} \qquad \boxed{19\text{-}18}$$

The point $P$ located in this manner is called the *center of percussion*. Any system of forces acting on the body can be represented by a single resultant force (or by its components $\sum \mathbf{F}_n$ and $\sum \mathbf{F}_t$) acting at the center of percussion.

The scalar equations of motion can be written as

$$\sum F_n = mr_C\omega^2$$
$$\sum F_t = mr_C\alpha \qquad \boxed{19\text{-}19}$$
$$\sum M_O = 0$$

An interesting and important aspect of this last equation is that there can be no tangential component of reaction at the supporting hinge of a body when the only external force with a tangential component is applied at the center of percussion. The "sweet spot" of a tennis racket or baseball bat is located at the center of percussion. Thus, by hitting the ball at the sweet spot, no sting or jarring is felt in the grip. The concept of center of percussion is also useful in the design of impact devices such as hammers and testing machines.

EXAMPLE 19-4

The final step in a cable-making process is modeled in Fig. 19-11a. Assume that the cable drum is a homogeneous disk of 1 m radius and 1000 kg mass. The drum starts from rest and accelerates at a constant rate until the cable attains a velocity $\mathbf{v} = -2\mathbf{i}$ m/s in a period of 5 s. The drum winds up cable whose mass moving in the horizontal direction is 100 kg. Assume that the radius of the drum is constant in the given time interval and that the die applies a 5-kN friction force horizontally to the cable. Determine the magnitude of the required horizontal force and the moment at axle $O$.

## SOLUTION

A sketch (Fig. 19-11b) of the cable alone allows for the calculation of the horizontal component $T_{D_x}$ of the tension at the point of tangency to the disk. The acceleration of the cable is

$$a_{cable} = \frac{\Delta v}{\Delta t} = \frac{2\,\text{m/s}}{5\,\text{s}} = 0.4\,\text{m/s}^2 \quad \text{(to the left)}$$

$$\sum F_X = 5\,\text{kN} - T_{D_x} = m_{cable}a_{cable} = (100\,\text{kg})(-0.4\,\text{m/s}^2)$$

$$T_{D_x} = 5000\,\text{N} + 40\,\text{N} = 5040\,\text{N}$$

The free-body diagram of the drum alone is sketched in Fig. 19-11c. The angular acceleration $\alpha$ is calculated from the acceleration of the point of tangency $D$, $a_D = a_{cable} = r_{OD}\alpha$. Hence,

$$\alpha = \frac{a_{cable}}{r_{OD}} = \frac{0.4\,\text{m/s}^2}{1\,\text{m}} = 0.4\,\text{rad/s}^2 \quad \text{(counterclockwise)}$$

$$\sum F_x = 5040\,\text{N} - R_x = m_{drum}a_{drum} = 0$$

$$R_x = 5040\,\text{N}$$

$$\sum M_O = M_1 - (5040\,\text{N})(1\,\text{m}) = I_O\alpha = \frac{m_{drum}}{2}r_{drum}^2\alpha$$

$$M_1 = \left(\frac{1000\,\text{kg}}{2}\right)(1\,\text{m})^2(0.4\,\text{rad/s}^2) + 5040\,\text{N}\cdot\text{m}$$

$$= 5240\,\text{N}\cdot\text{m} \quad \text{(counterclockwise)}$$

(a)

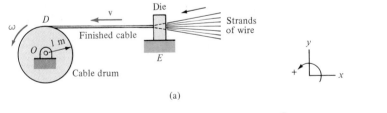

(b)                                                          (c)

**FIGURE 19-11**

Schematic of cable manufacturing

EXAMPLE 19-4    783

EXAMPLE 19-5

The composite body shown in Fig. 19-12a consists of a uniform rod $CD$ weighing 32.2 lb welded at $C$ to a uniform disk weighing 64.4 lb. A weight $W$ is supported by a massless cable attached at $D$, and no slipping occurs at the pulley. The weight of the pulley is 16.1 lb and its radius of gyration about its axis $B$ is 0.5 ft. Determine the weight $W$ which will accelerate downward at a rate of 12 ft/s² if the system is released from rest in the position shown.

## SOLUTION

The interrelated free-body diagrams are drawn in Fig. 19-12b.

**Kinematics:**

$$r_B \alpha_B = (1 \text{ ft})(\alpha_B) = 12 \text{ ft/s}^2$$

$$\alpha_B = 12 \text{ rad/s}^2 \quad \text{(clockwise)}$$

Also,

$$\alpha_{CD} (6 \text{ ft}) = \alpha_B (1 \text{ ft}) = 12 \text{ ft/s}^2$$

$$\alpha_{CD} = 2 \text{ rad/s}^2 \quad \text{(clockwise)}$$

**Kinetics:** For the weight $W$,

$$\sum F_y = ma$$

$$-W + T_2 = \frac{W}{32.2 \text{ ft/s}^2} (-12) \tag{a}$$

For the pulley at $B$,

$$\sum M_B = I_B \alpha_B = m_B r_g^2 \alpha_B$$

$$T_2(1) - T_1(1) = \left[ \frac{16.1 \text{ lb}}{32.2 \text{ ft/s}^2} (0.5 \text{ ft})^2 \right] (12 \text{ rad/s}^2)$$

$$T_2 - T_1 = 1.5 \text{ lb} \tag{b}$$

For the composite body,

$$\sum M_O = I_O \alpha$$

where

$$I_O = \left[ \frac{1}{2} \frac{64.4 \text{ lb}}{32.2 \text{ ft/s}^2} (1 \text{ ft})^2 + \frac{64.4 \text{ lb}}{32.2 \text{ ft/s}^2} (1 \text{ ft})^2 \right]$$

$$+ \left[ \frac{1}{12} \frac{32.2 \text{ lb}}{32.2 \text{ ft/s}^2} (4 \text{ ft})^2 + \frac{32.2 \text{ lb}}{32.2 \text{ ft/s}^2} (4 \text{ ft})^2 \right]$$

$$I_O = 20.3 \text{ ft} \cdot \text{lb} \cdot \text{s}^2$$

Then

$$-T_1 (6 \text{ ft}) + (32.2 \text{ lb})(4 \text{ ft}) + (64.4 \text{ lb})(1 \text{ ft}) = (20.3 \text{ ft} \cdot \text{lb} \cdot \text{s}^2)(-2 \text{ rad/s}^2)$$

$$T_1 = 39.0 \text{ lb}$$

Solving Eqs. (a) and (b) yields

$$T_2 = 40.5 \text{ lb}$$

$$W = 64.6 \text{ lb}$$

(a)

(b)

**FIGURE 19-12**

Analysis of a composite body

**19-30** A 40-kg gear is represented as a 1-m-diameter uniform disk. Calculate the magnitude of the tangential force **F** for accelerating the gear at the rate of $2\,\text{rad/s}^2$.

Fig. P19-30

**19-31** The small weight $w$ is attached to the uniform disk of weight $W$. Calculate the required couple to accelerate the disk at the rate of $3\,\text{rad/s}^2$.

$w = 2\,\text{lb}$

$2\,\text{ft}$

$W = 50\,\text{lb}$   Fig. P19-31

**19-32** The turntable of a record player is a 2-kg, 0.3 m-diameter, uniform disk. Calculate the required couple to bring the turntable from rest to a speed of 33.33 rpm in a time of 5 s.

**19-33** An amusement ride is a cylindrical ring of 20 ft outside diameter and 7 ft inside diameter. Assume that its 12,000-lb weight is uniformly distributed. Calculate the magnitude of the required tangential force at the inside circle to accelerate the ring at the rate of $0.5\,\text{rad/s}^2$.

**19-34** The pulley of an elevator is modeled as a 30-kg, 1-m-diameter, uniform disk. The elevator and its counterweight have masses $m_1 = 2000\,\text{kg}$ and $m_2 = 1800\,\text{kg}$, respectively. Calculate the required torque on the pulley for accelerating the elevator upward at the rate of $1\,\text{m/s}^2$.

$m_1$     $m_2$

Fig. P19-34

**19-35** Consider Prob. 19-34 with the counterweight replaced with a constant force. This force has the same magnitude as the static weight of $m_2$. Discuss the performances of the two systems for "assisting" the drive motor.

**19-36** The cable unwinds from a 2-ft-radius spool as the 500-lb weight is allowed to freely fall. The weight of the spool and cable is 1610 lb and the radius of gyration is $r_g = 1.5\,\text{ft}$. Neglecting friction, determine the angular acceleration of the spool as it unwinds and the tension in the cable.

$2\,\text{ft}$

$W$     $500\,\text{lb}$

Fig. P19-36

**19-37** Determine the required magnitude of the force **F** to accelerate a thin sheet of metal at the rate of $5\,\text{ft/s}^2$ through the rollers that are free to rotate about their axes. Each uniform roller has a weight of 200 lb and a diameter of 1 ft. The sheet does not slip between the rollers.

$F$

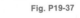

Fig. P19-37

**19–38**  The force **F** applied to an exercise treadmill is transmitted through a belt to 20 rollers that spin with negligible friction. Each roller has a mass of 2 kg and a diameter $d$ of 0.08 m. Calculate the tangential acceleration $a$ of the belt for $F = 1$ kN and $\theta = 60°$. Assume that only the top part of the belt is in contact with the rollers and does not slip.

(a)                               (b)

**Fig. P19-38**

**19–39**  The uniform, 10-kg wheel is rotating at an angular speed of 500 rpm when a braking force of magnitude $F = 800$ N is applied to the hinged bar. The friction at $O$ and $A$ is negligible, while $\mu_k = 0.5$ at $B$. Calculate the time required to stop the wheel.

**Fig. P19-39**

**19–40**  The driving wheels of a 3000-lb car are on the freely turning drum of a dynamometer. The drum is modeled as a thin-walled, open-ended cylinder of 300 lb weight and $D = 5$ ft diameter. What is the maximum angular acceleration of the dynamometer drum if the driving wheels have $\mu_s = 0.9$ on the drum?

**Fig. P19-40**

**19–41**  The base structure of a rotating restaurant is modeled as three uniform disks that rotate as a unit about the vertical axis $y$. The mass $m$ is 8000 kg, and the diameter $d$ is 7 m. Calculate the magnitude of the required couple to stop the disks in a time of 10 s from an angular speed of 0.1 rpm.

**Fig. P19-41**

**19–42**  Each wing of a small airplane is modeled as a uniform, slender bar of 150-kg mass. Determine the angular acceleration of the airplane caused by the two forces of magnitude $F$. Neglect the mass of the fuselage and air drag opposing the rotation.

**Fig. P19-42**

**19–43**  Solve Prob. 19–42 assuming that the fuselage is a uniform cylinder of 600-kg mass and 0.8 m diameter that is rigidly attached to the top of the wings. Also assume that the center of the fuselage is fixed.

**19–44**  Assume the following for the vertical arm $AC$ of of the excavator: $AB = 3$ ft, $BC = 10.5$ ft; weight of member $AC$: 300 lb (uniformly distributed); weight of load at $C$: 350 lb. Determine the initial acceleration of the load at $C$ caused by a force $\mathbf{F} = (-400\mathbf{i} - 70\mathbf{j})$ lb applied at joint $A$. Calculate the components of the force at joint $B$, about which arm $AC$ rotates.

Fig. P19-44

**19–45** The uniform bar of mass 300 kg is supported by a cable at $A$ and a frictionless pin at $O$. A clockwise couple of 400 N·m magnitude acts at $O$ and a force of 200 N magnitude is applied at $B$ as shown. Determine (a) the angular acceleration of bar $AB$ the instant after the cable at $A$ breaks, and (b) the corresponding reactions on the bar at $O$.

Fig. P19-45

**19–46** The uniform, slender bar has a mass of 4 kg. It is released from rest at $\theta = 30°$. Calculate the angular acceleration of the bar and the $x$ and $y$ reactions at point $O$ for the instant after release.

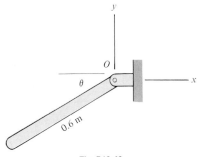

Fig. P19-46

**19–47** Solve Prob. 19–46 assuming that at $\theta = 30°$ the bar has an initial counterclockwise angular speed of $\dot\theta = 10$ rad/s.

**19–48** Consider a gymnast doing giant circles on the high bar. Assume that his body can be modeled as a uniform rigid bar $AB$ of 160 lb weight, firmly attached to a weightless rigid member $OA$. Assume that the angular speed is $\omega = 1$ rad/s at $\theta = 30°$. Determine the $x$ and $y$ reactions at the pivot $O$. Neglect friction of the hands on the bar.

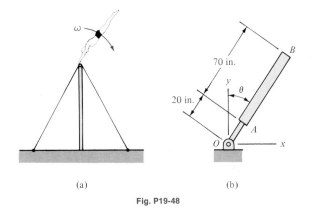

(a)　　　　　(b)

Fig. P19-48

**19–49** A 50-m-tall radio antenna tower is modeled as a uniform, rigid rod of 2000-kg mass, in a ball-and-socket joint at point $O$, and supported by four guy wires that are 90° apart (top view). The initial tension in each wire is 8 kN. Assume that one of these wires breaks. Calculate the initial angular acceleration of the tower and the reaction on the joint at $O$, assuming that the tension is constant in the other three wires for a short time after the breakage.

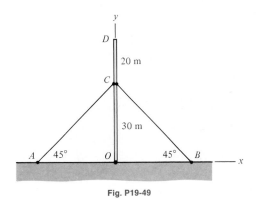

Fig. P19-49

**19–50** The 20-ft-long swinging arm of an amusement ride is a uniform rod of 200 lb weight. The 500-lb weight of the riders and the cab may be assumed as concentrated at $B$, the end of the rod. At $\theta = 30°$, $\dot{\theta} = 0$. Determine the magnitude of the force **F** at $\phi = 45°$ and the reaction at $O$ for causing an angular acceleration of 1 rad/s² of the rod.

Fig. P19-50

**19–51** The landing gear of an aircraft is modeled as a uniform rod $AB$ of 100 kg mass, and two uniform disks of 30 kg mass each. A force of magnitude $F = 3$ kN is applied at $C$ to retract the landing gear. Determine the angular acceleration of $AB$ and the reaction at $A$ caused by **F**, assuming that the wheels are thin disks positioned at 0.25 m on either side of point $B$.

**19–52** Solve Prob. 19–51 by treating the wheels as uniform cylinders of 0.3-m width located as in Fig. P19–51b.

**19–53** A uniform rectangular plate of mass $m$ is hinged at point $O$ and is released from rest when side $a$ is horizontal. Determine the angular acceleration of the plate and the $x$ and $y$ reactions at $O$ when the plate begins to swing.

Fig. P19-53

**19–54** A locomotive of 100 000-kg mass is tested in a laboratory. Assume that the mass is uniformly distributed in a rectangular volume 2.4 m × 2.4 m in cross section. Horizontal forces **F** are applied to the locomotive by hydraulic actuators until the wheels are slightly off track $B$. The locomotive acts as if it were hinged at $A$ when the magnitudes of the actuator forces are suddenly decreased to zero. Determine the angular acceleration of the locomotive and the linear accelerations measured by the horizontal and vertical accelerometers $H$ and $V$, respectively.

Fig. P19-51
Landing gear of an aircraft (Photograph courtesy The Boeing Company, Seattle, Washington)

(a)

(b)

(a)

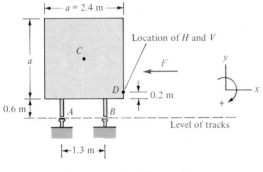

$a = 2.4$ m

Location of $H$ and $V$

$a$

$C$

$F$

$y$

$x$

$D$

0.2 m

0.6 m

$A$   $B$

Level of tracks

1.3 m

(b) Cross section of locomotive

**Fig. P19-54**

**19-55** The composite body shown rotates without friction about a horizontal axis at $O$. The rigid body consists of two slender rods of 64.4-lb weight each welded to a homogeneous cylinder of 193.2-lb weight. For the position shown the velocity of the mass center $C$ has an instantaneous velocity of 10 ft/s downward. Determine the instantaneous angular acceleration of the composite body for the position shown and the bearing reactions on the cylinder at $O$.

100 lb

4 ft

$O$   $C$

3 ft

2 ft

4 ft

200 lb

**Fig. P19-55**

**19-56** A uniform ring of thickness $t$ and density $\rho$ is hinged at point $O$ and is released from rest in the position shown. Determine the angular acceleration of the ring and the $x$ and $y$ reactions at $O$ when the ring begins to swing. $r = 0.9R$.

$y$

$r$

$O$

$x$

$R$

**Fig. P19-56**

**19-57** Plot the magnitude of the angular acceleration of the rigid body in Prob. 19-56 as a function of $r/R = 0$ (solid circular plate), 0.5, and 0.9.

**19-58** A propeller blade $OA$ is modeled as a uniform, slender bar of 20-lb weight. Determine the magnitudes of the axial and transverse forces in the blade at $d = 1$ ft for $\omega = 30$ rad/s and $\alpha = 50$ rad/s$^2$.

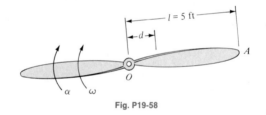

$l = 5$ ft

$d$

$A$

$O$

$\alpha$   $\omega$

**Fig. P19-58**

**19-59** A thin-walled cylinder of mass $m$ used for drying paper is rotating at a constant angular speed $\omega$. $R \gg t$. Write an expression for the tension $T$ in the cylinder caused by its rotation. This tension is normal to any radial plane through the axis of the cylinder.

$T$

Element of cylinder wall

$R$

$T$

$l$

$R$

$\omega$

$t$

**Fig. P19-59**

## GENERAL PLANE MOTION

A rigid body is said to be in general plane motion when it is translating and rotating. This combined motion is analyzed using Eq. 19-7. For a review, consider the free-body diagram of an arbitrary thin slab as in Fig. 19-13a. The basic set of resultants is shown in Fig. 19-13b, where $\sum \mathbf{F}$ is the resultant force and $\sum \mathbf{M}_C$ is the resultant moment about the center of mass $C$. The scalar equations of motion using Fig. 19-13b are, as in Sec. 19-3,

$$\sum F_x = ma_{C_x} \quad \sum F_y = ma_{C_y} \quad \sum M_C = I_C\alpha \qquad \boxed{19\text{-}7}$$

Note that the $xy$ frame and the forces are parallel to the plane of the thin slab.

The resultant moment may also be determined with respect to an arbitrary, noncentroidal reference point as in Sec. 19-3. Assume that the desired reference point is point $A$ in Fig. 19-13c. The general form of the resultant moment $\sum \mathbf{M}_A$ is

$$\sum \mathbf{M}_A = \sum \mathbf{M}_C + \mathbf{r} \times \sum \mathbf{F} \qquad \boxed{19\text{-}20}$$

where $\sum \mathbf{M}_C$ is the resultant moment of the external forces about the center of mass $C$, and $\mathbf{r} \times \sum \mathbf{F}$ is the moment of the resultant force about the reference point $A$. Substitution of Eqs. 19-1 and 19-6 into Eq. 19-20 gives

$$\boxed{\sum \mathbf{M}_A = I_C\alpha + \mathbf{r} \times m\mathbf{a}_C} \qquad \boxed{19\text{-}21}$$

This result is also obtained by considering the time rate of change of angular momentum about point $A$. Using Eq. 17-31 with 19-5.

$$\sum \mathbf{M}_A = \mathbf{H}_C + \mathbf{r} \times m\mathbf{a}_C$$
$$= I_C\alpha + \mathbf{r} \times m\mathbf{a}_C$$

The reference point $A$ for moments should be selected to simplify the solution.

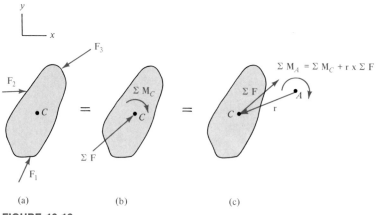

(a)          (b)          (c)

**FIGURE 19-13**

Resultant forces and moments with respect to the mass center of an arbitrary point

**COMMON ERROR**

Although summing moments about a convenient point simplifies a problem, students often make the error of neglecting the $\mathbf{r} \times m\mathbf{a}_C$ term in the calculation. A reduction of Eq. 19-21 occurs only if $A$ is a fixed axis of revolution, as in Sec. 19-4.

## Constraints

In many problems of general plane motion the solution is facilitated if there is some constraint to the motion as shown in the following special cases of considerable practical significance.

**Pure rolling of wheels.**  Assume that a uniform wheel rolls without slipping on a flat surface at angular speed $\omega$ and angular acceleration $\alpha$ as in Fig. 19-14a. Here the center of mass $C$ moves along a straight line and has acceleration $a_{C_x} = r\alpha$. (Note that the problem is more complex if the mass center is not at the geometric center since then the mass center is not in rectilinear translation.) The equation of motion for pure rolling is written using the free-body diagram in Fig. 19-14b. Summing moments about point $A$ and using Eq. 19-21,

$$\sum M_A = \mathrm{P}r = I_C\alpha + ma_{C_x}r = I_C\alpha + m(r\alpha)r$$
$$= (I_C + mr^2)\alpha$$

With $I_A = I_C + mr^2$,

$$\sum M_A = I_A\alpha \qquad \boxed{\text{19-22}}$$

where $I_A$ is the mass moment of inertia of the wheel about the point of contact. Equation 19-22 is valid only if the wheel rolls without slipping. Note that the friction force in this case has magnitude $\mathscr{F} \le \mu_s N$, where $\mu_s$ is the coefficient of static friction.

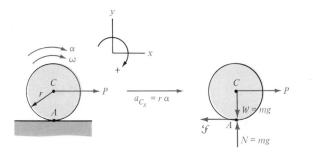

(a) No slipping at $A$                    (b)

**FIGURE 19-14**

Pure rolling of a wheel

**FIGURE 19-15**

Rolling motion of an unbalanced wheel

**UNBALANCED WHEELS:** A special problem of practical interest is the plane motion of unbalanced wheels that roll without slipping. In such a case the mass center $C$ and the geometric center $G$ do not coincide, as in Fig. 19-15, and the relationship $a_C = r\alpha$ is not valid. Instead, one must use the expression

$$a_G = r\alpha \qquad \boxed{19\text{-}23}$$

The acceleration of the mass center $C$ is the vector sum of the acceleration of the geometric center $G$, and of the relative acceleration of $C$ with respect to $G$,

$$\mathbf{a}_C = \mathbf{a}_G + \mathbf{a}_{C/G} = \mathbf{a}_G + (\mathbf{a}_{C/G})_n + (\mathbf{a}_{C/G})_t \qquad \boxed{19\text{-}24}$$

where $(a_{C/G})_n = (CG)\omega^2$ and $(a_{C/G})_t = (CG)\alpha$.

**Rolling and sliding of wheels.** In the general case a wheel may be rolling while slipping at the point of contact. This means that the magnitude of the friction force at point $A$ in Fig. 19-16 is known, $\mathscr{F} = \mu_k N$, where $\mu_k$ is the coefficient of kinetic friction. In this case, however, $a_{C_x} \neq \alpha r$ since the linear and angular accelerations are independent of one another.

## Analysis of Rolling or Sliding Balanced Wheels

In general, there are four equations available for determining four unknowns in problems involving wheels in either pure rolling or in rolling *and* sliding. The equations are from force and moment summations and kinematic relationships. The method of solution is illustrated for the common case of a balanced wheel moving on a horizontal track under the action of a horizontal, centroidal force **P** as in Fig. 19-16.

(a) Slipping at $A$ is possible

$$\sum F_x = ma_{C_x} \qquad P - \mathscr{F} = ma_{C_x} \qquad \boxed{19\text{-}25}$$

$$\sum F_y = ma_{C_y} \qquad N - mg = 0 \qquad \boxed{19\text{-}26}$$

$$\sum M_C = I_C\alpha \qquad \mathscr{F}r = I_C\alpha \qquad \boxed{19\text{-}27}$$

$$a_{C_x} = \alpha r \qquad \boxed{19\text{-}28}$$

$$\mathscr{F} = \mu_k N \qquad \boxed{19\text{-}29}$$

for no slipping   for slipping

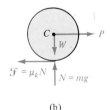

(b)

**FIGURE 19-16**

Schematic for a rolling and sliding wheel

If it is not certain whether the wheel is slipping or not, the condition for no slipping should be assumed and checked first. When the friction force has magnitude $\mathscr{F} \leq \mu_s N$, there is no slipping, and the assumption was correct. On the other hand, $\mathscr{F} > \mu_s N$ indicates that there is sliding, so the solution must be started over with $\mathscr{F} = \mu_k N$ now known but the accelerations $a_{C_x}$ and $\alpha$ unrelated. Note that Eqs. 19-25 through 19-29 can be used with appropriate modifications for problems involving inclined tracks, or a force **P** which is not parallel to the track or not acting at the center of mass.

EXAMPLE 19-6

A weight $W$ of negligible dimensions is attached at point $B$ to a uniform disk of radius $r$ and weight $10W$(Fig. 19-17a). The disk is released from rest in the position shown. Determine the acceleration of $B$ if the disk does not slip.

## SOLUTION

The free-body diagram is drawn in Fig. 19-17b. Since there is no slipping,

$$a_G = r\alpha$$

The acceleration of point $B$ is

$$\mathbf{a}_B = \mathbf{a}_G + (\mathbf{a}_{B/G})_n + (\mathbf{a}_{B/G})_t$$
$$= -\alpha\mathbf{k} \times r\mathbf{j} + 0 + (-\alpha\mathbf{k}) \times r\mathbf{i}$$
$$= r\alpha\mathbf{i} - r\alpha\mathbf{j}$$

From Eqs. 19-20 and 19-21,

$$\sum M_A = (I_A\alpha)_{\text{disk}} + m_W r(a_{B_x} - a_{B_y})$$

where $m_W r(a_{B_x} - a_{B_y}) = 2m_w r^2\alpha$. Then

$$10Wr \sin 30° + W(r \sin 30° + r \cos 30°)$$

$$= \left[\frac{1}{2}\frac{10W}{g}r^2 + \frac{10W}{g}r^2 + \frac{W}{g}(AB)^2\right]\alpha$$

Dividing by $W$ and $r$,

$$5 + 0.5 + 0.866 = \left(\frac{5r}{g} + \frac{10r}{g} + \frac{2r}{g}\right)\alpha$$

Solving for $\alpha$,

$$\alpha = 0.37\frac{g}{r} \qquad \boldsymbol{\alpha} = -0.37\frac{g}{r}\mathbf{k}$$

The acceleration of point $B$ is

$$\mathbf{a}_B = 0.37g\mathbf{i} - 0.37g\mathbf{j}$$

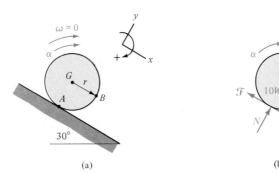

(a)        (b)

**FIGURE 19-17**

Analysis of a rolling, unbalanced wheel

EXAMPLE 19-6   793

EXAMPLE 19-7

A 20-kg uniform disk is at rest on a horizontal surface (Fig. 19-18a) when a horizontal force of magnitude $F = 400$ N is applied at the center of the disk. Determine the acceleration of the mass center.

## SOLUTION

The free-body diagram is drawn in Fig. 19-18b. It is not known whether the disk rolls without slipping or not. Hence, rolling without slipping is assumed and, the assumption will be checked later.

$$\sum F_x = F - \mathscr{F} = ma_{C_x}$$

$$400\,\text{N} - \mathscr{F} = m(r\alpha) = (20\,\text{kg})(0.3\,\text{m})\,\alpha \qquad \textbf{(a)}$$

$$\sum F_y = N - W = ma_{C_y} = 0$$

$$N = W = (20\,\text{kg})(9.81\,\text{m/s}^2) = 196.2\,\text{N}$$

$$\sum M_A = F(r) = I_A\alpha$$

$$(400\,\text{N})(0.3\,\text{m}) = \left[\frac{1}{2}(20\,\text{kg})(0.3\,\text{m})^2 + (20\,\text{kg})(0.3\,\text{m})^2\right]\alpha$$

$$120\,\text{N} \cdot \text{m} = (2.70\,\text{kg} \cdot \text{m}^2)\alpha$$

$$\frac{44.4\,\text{N} \cdot \text{m}}{\text{kg} \cdot \text{m}^2} = 44.4\,\text{rad/s}^2 = \alpha$$

Substituting the value obtained for $\alpha$ into Eq. (a) yields

$$\mathscr{F} = 400\,\text{N} - (20\,\text{kg})(0.3\,\text{m})(44.4\,\text{rad/s}^2) = 133\,\text{N}$$

This is the friction force (to the left) necessary if rolling without slipping were to occur. A check of the maximum possible magnitude of the friction force shows that

$$\mathscr{F}_{\text{max}} = \mu_S N = (0.4)(196.2\,\text{N}) = 78.5\,\text{N}$$

Since $\mathscr{F}$ cannot exceed $\mathscr{F}_{\text{max}}$, the wheel must roll and slip. The solution must be started over, using the kinetic friction force of magnitude $\mathscr{F}_k = (0.3)(196.2\,\text{N}) = 58.9\,\text{N}$.

$$\sum F_x = F - \mathscr{F}_k = ma_{C_x}$$

$$400\,\text{N} - 58.9\,\text{N} = (20\,\text{kg})(a_{C_x})$$

$$a_{C_x} = 17.1\,\text{m/s}^2$$

$$\sum M_C = \mathscr{F}_k(r) = I_C\alpha$$

$$(58.9\,\text{N})(0.3\,\text{m}) = \frac{1}{2}(20\,\text{kg})(0.3\,\text{m})^2\,\alpha$$

$$\alpha = 19.3\,\text{rad/s}^2$$

Note that if rolling and slipping had been assumed initially, there would have been no convenient check on the assumption available. Also notice that the sense of force $\mathscr{F}$ and angular acceleration $\boldsymbol{\alpha}$ are not even known at first. Directions must be assumed and the solution will yield the correct signs.

(a)

(b)

**FIGURE 19-18**

Analysis of a rolling wheel which may also be slipping

**19-60** A thin hoop of mass $m$ and radius $r$ starts from rest on an inclined plane. The surfaces are rough and prevent slipping. Determine the hoop's linear acceleration for $m = 2\,\text{kg}$, $r = 0.4\,\text{m}$, and $\theta = 20°$.

Fig. P19-60

**19-61** Plot the acceleration of the hoop shown in Fig. P19-60 for the following conditions:
  (a) $m = $ constant, $\theta = $ constant, radius $= r$, $2r$, $3r$
  (b) $m = $ constant, $r = $ constant, angle $= \theta$, $2\theta$, $3\theta$

**19-62** A 20-lb cylinder is released from rest. Determine the cylinder's linear acceleration if there is no slippage, $r = 6\,\text{in.}$, $R = 12\,\text{in.}$, and $\theta = 30°$.

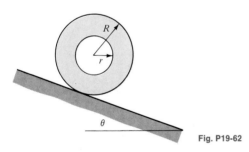

Fig. P19-62

**19-63** Calculate the acceleration of the cylinder in Fig. P19-62 for $r = 3\,\text{in.}$ and $r = 9\,\text{in.}$ All other quantities are the same as in Prob. 19-62.

**19-64** A uniform 5-lb cylinder is used in a simple experiment to demonstrate the acceleration of gravity. Assume that the cylinder rolls without slipping. Determine the angle $\theta$ for which the cylinder's linear acceleration along the incline is $20\,\text{ft/s}^2$.

$W = 5$ lb

$r = 2$ in.

Fig. P19-64

**19-65** A force of magnitude $P = 200\,\text{N}$ is acting at the mass center of a uniform disk of 10 kg mass and 0.4 m radius. Determine the linear acceleration of the disk if there is no slippage and $\theta = 25°$.

Fig. P19-65

Fig. P19-66

**19-66** A uniform cylinder of 20 kg mass and 0.3 m radius is supported by an inextensible band $AB$. Calculate the acceleration of the mass center and the tension in $A$ for the instant immediately after part $B$ of the band breaks. Assume that there is no slipping between the band and the cylinder.

**19-67** The gear of radius $r = 0.5\,\text{ft}$ weighing 10 lb is acted on by a force of magnitude $P = 30\,\text{lb}$ and a couple of magnitude $M = 15\,\text{lb·ft}$. The radius of gyration of the gear is $r_g = 0.4\,\text{ft}$. Determine the magnitude of the shearing force acting on the gear teeth and the angular acceleration of the gear.

Fig. P19-67

**19-68** A 50-lb uniform wheel of 1 ft radius rolls downhill. Determine the wheel's linear acceleration, and the minimum value of $\mu_s$ that is sufficient to prevent slipping.

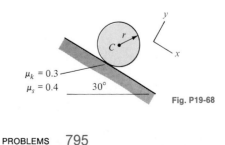

$\mu_k = 0.3$
$\mu_s = 0.4$

$30°$

Fig. P19-68

**19–69**  Add a force $\mathbf{P} = (20\mathbf{i} - 10\mathbf{j})$ lb acting at point $C$ to the wheel in Prob. 19–68. Calculate the linear acceleration in this case.

**19–70**  A force of magnitude $P = 500$ N is applied to a uniform disk of 30-kg mass and 0.5 m radius. Calculate the acceleration of the disk.

$\mu_s = 0.4$, $\mu_k = 0.35$

**Fig. P19-70**

**19–71**  A 100-lb uniform wheel of 2-ft radius is initially at rest on an inclined plane. Calculate the linear acceleration of the disk caused by a force of magnitude $P = 500$ lb.

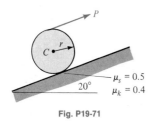

$\mu_s = 0.5$
$20°$
$\mu_k = 0.4$

**Fig. P19-71**

**19–72**  What is the maximum magnitude of force $\mathbf{P}$ if the wheel in Prob. 19–71 is to be moved uphill without slipping?

**19–73**  The console has a mass of 250 kg which may be assumed as uniformly distributed. It rests on casters that roll freely in any direction. Determine the acceleration of the mass center caused by a force $\mathbf{F} = 800\,\mathbf{k}$ N applied at point $A$.

**Fig. P19-73**

**19–74**  The barge carrying an empty spherical container for liquid natural gas is modeled as a rigid system in Fig. 19-74b. The thin-walled spherical shell has a mass of 800 t, and the barge is assumed to be a 40 m × 60 m × 4 m solid plate of 2000-t mass (1 t = 907.2 kg). Assume that the system is stable in the position shown and that center $A$ of the barge always moves horizontally without frictional resistance. Determine the angular acceleration caused by a gust of wind that has a resultant of magnitude $F = 50$ kN.

(a)

$D = 36$ m

$F$ —

$B$

$A$

4 m

40 m

(b)

**Fig. P19-74**
Spherical container for liquefied natural gas
(Photograph courtesy General Dynamics
Quincy Shipbuilding Division)

**19–75**  A 20-ft I-beam weighing 644 lb is being raised by two cables, $A$ and $B$. At the instant motion begins, cable $A$ accelerates at 5 ft/s² and cable $B$ at 9 ft/s², both upward. Determine the tension in each cable at this instant.

$a_A$     $a_B$

$A$     $B$

20 ft

**Fig. P19-75**

**19–76** On April 20, 1979, cable $A$ broke on a scaffolding used for window washing in Richmond, Virginia. Calculate the tension in cable $B$ and the angular acceleration of the scaffolding immediately after the failure of cable $A$. Assume that the scaffolding is a uniform, slender bar.

Fig. P19-76

**19–77** Consider the dry test of a two-piece waveboard (see Ch. 3 of the Statics volume) which consists of two slender, uniform boards $AB$ and $BC$, each of 2 m width and 300 kg mass. The lower board is stationary ($\omega_1 = 0$) when it is in the vertical position. At the same instant the upper board makes an angle of 30° with the vertical and has an angular speed of $\omega_2 = 0.3$ rad/s, counterclockwise. $\omega_2$ is increasing at the rate of 0.2 rad/s². Determine the moment about hinge $B$ and the force acting on this hinge on board $BC$ at the given instant.

(a)

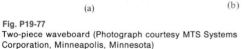

(b)

Fig. P19-77
Two-piece waveboard (Photograph courtesy MTS Systems Corporation, Minneapolis, Minnesota)

**19–78** Solve Prob. 19–17 assuming that $\omega_1 = 0.2$ rad/s, constant.

**19–79** A uniform rod of 40 lb weight is hanging from the axle of a weightless, frictionless roller of negligible size. Determine the angular acceleration of the rod and the acceleration of the roller immediately after a horizontal force of magnitude $F = 20$ lb is applied at $d = 1$ ft.

Fig. P19-79

**19–80** A 60-lb ladder is assumed to be a uniform bar which starts sliding without frictional resistance. Determine the reactions at $A$ and $B$.

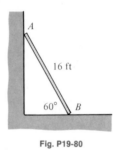

Fig. P19-80

**19–81** Solve Prob. 19–80 assuming that a 200-lb person is at the center of the ladder. Neglect the dimensions of the person.

**19–82** Pins $A$ and $B$ of the 5-kg, uniform bar have negligible mass and size, and they slide freely in their tracks. Calculate the acceleration of $B$ and the reactions of the guides on the pins immediately after the force $\mathbf{F}$ is applied.

Fig. P19-82

**19–83** The 2-lb crank arm $OA$ is rotating clockwise at a constant speed of 1500 rpm. The uniform piston rod $AB$ weighs 2.2 lb, and the piston $B$ weighs 2.4 lb. Calculate the force acting on crank arm $OA$ at pin $A$ for $\theta = 90°$, neglecting friction.

**Fig. P19-83**

**19–84** Determine the $x$ and $y$ components of the force acting on the pin on piston $B$ in Prob. 19–83 for $\theta = 0$.

**19–85** The toggle mechanism $OAB$ moves in the horizontal $xy$ plane. Each uniform bar has a mass of 12 kg, and block $B$ has a mass of 6 kg. Calculate the $x$ and $y$ components of the force acting on the pin on block $B$ immediately after a force of magnitude $F = 700$ N is applied when $\theta = 30°$. Friction is negligible.

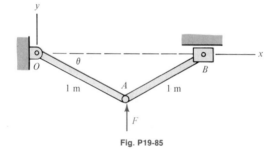

**Fig. P19-85**

**19–86** A uniform, 20-kg disk of 0.6 m radius has a 2-kg mass attached at point $A$. The disk is initially at rest when a force $\mathbf{F} = 50\mathbf{i}$ N is applied at its geometric center $O$. Determine the acceleration of axis $O$ for $\theta = 30°$ if there is no slippage.

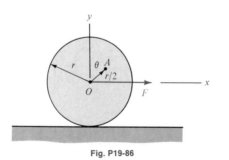

**Fig. P19-86**

**19–87** Determine the acceleration of the mass center of the system in Prob. 19–86.

**19–88** A linear bearing is modeled as a thin 5-lb bar on which a 10-lb force from a weightless spring mechanism is acting. The steel ball bearings have a coefficient of friction $\mu_s = 0.1$. Calculate the magnitude of the largest force $\mathbf{F}$ that can be applied without causing the balls to slip. Each bearing weighs $\frac{1}{16}$ lb.

**Fig. P19-88**

**19–89** The test sled is sliding on four linear bearings on two horizontal tracks. The dummy is restrained by a seat belt. Assume that the rigid torso (above the belt) has a 40-kg mass, a 0.35 m radius of gyration about its own mass center, which is 0.30 m from the pivot of the torso (at the seat belt). Determine the angular acceleration of an initially vertical torso if the sled is decelerating at the rate of 60 m/s$^2$.

**Fig. P19-89**

**19–90** $\theta$ is the angle of the torso's initial recline (backward) from the vertical in Fig. P19-89. Calculate the angular acceleration of the torso in Prob. 19–89 for $\theta = 30°$.

**19–91** A 200-kg spacecraft has two small attitude jets in the $xz$ plane, the plane perpendicular to the vertical centerline ($y$ direction) of the craft. The jets have the following thrusts and locations: $\mathbf{T}_1 = -0.5\,\mathbf{k}\,\text{N}$ at $x = 0.8\,\text{m}$, and $\mathbf{T}_2 = 0.5\,\mathbf{k}\,\text{N}$ at $x = -0.8\,\text{m}$. The spacecraft has an angular acceleration $\boldsymbol{\alpha} = 0.1\,\mathbf{j}\,\text{rad/s}^2$ when both jets are operating simultaneously. Determine the radius of gyration about the $y$ axis. Assume that $\mathbf{T}_1$, $\mathbf{T}_2$, and the center of mass are in the same plane, and that the 20-m-long supporting wire applies only a vertical force to the spacecraft.

**Fig. P19-91**
Spacecraft before a test in a vacuum chamber (Courtesy The Boeing Company, Seattle, Washington)

**19–92** Assume that only $\mathbf{T}_1$ is acting of the two jets in Prob. 19–91. Calculate the linear and angular accelerations of the spacecraft.

**19–93** Sometimes a rigid body is too complex for a direct calculation of its mass moment of inertia. An example of this is a small rocket which has numerous components in it. Describe a procedure for using experiments and calculations to determine the mass moment of inertia of the rocket about the transverse axis $y$ through the centroid $C$. Assume that forces and accelerations can be accurately measured using electronic equipment.

**Fig. P19-93**

**19–94** The mass of a rocket is 6000 kg, and its radius of gyration about the mass center $C$ is 3.5 m. Determine the thrust $T$ that can cause an angular acceleration of 0.1 rad/s² when applied at $\theta = 5°$. Calculate the absolute acceleration of point $A$ caused by this thrust.

**Fig. P19-94**

**19–95** Assume that the wings and the engines of the model aircraft are the major parts that have mass moment of inertia about the longitudinal centerline of the fuselage. This centerline is a fixed axis in a test. Assume that each wing is a uniform, slender bar of length $L$ and mass $m$, and each pair of engines is a concentrated mass $m_1$ at a distance $l_1$ and $l_2$ from the center. Write an expression for the angular acceleration $\alpha$ of the model caused by a moment $M$ about the centerline. Neglect the differences in elevation of the various parts. The angle of sweep of the wings with the fuselage is $\theta$.

(a)

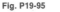

Note: Dimensions are in plane of paper, not along the swept wing

(b)

**Fig. P19-95**
Model aircraft in a laboratory test (Photograph courtesy The Boeing Company, Seattle, Washington)

## Section 19-1

**Equation of translational motion**

$$\sum \mathbf{F} = m\mathbf{a}_C$$

where $\mathbf{a}_C$ is the acceleration of mass center $C$ regardless of where the resultant force acts on the rigid body; this is independent of any rotation of the body.

## Section 19-2

**Equation of rotational motion**

$$\sum \mathbf{M}_C = I_C\boldsymbol{\alpha} \qquad \text{(for a thin slab perpendicular to } \sum \mathbf{M}_C \text{, or special cases of symmetry or constraints)}$$

where $\boldsymbol{\alpha}$ is the angular acceleration of the rigid body; this is independent of any translation of the body.

## Section 19-3

**Application of equations of motion.** Three scalar equations of motion can be written for each rigid body in plane motion:

$$\sum F_x = ma_{C_x} \quad \sum F_y = ma_{C_y} \quad \sum M_C = I_C\alpha$$

**PURE TRANSLATION:**

$$\sum F_x = ma_{C_x} \quad \sum F_y = ma_{C_y} \quad \sum M_C = 0$$

**PURE ROTATION ABOUT THE CENTER OF MASS:**

$$\sum F_x = 0 \quad \sum F_y = 0 \quad \sum M_C = I_C\alpha$$

**SYSTEMS OF RIGID BODIES:** At most three equations of motion are available from each free-body diagram for plane motion to solve for three unknowns. The motion of interconnected bodies must be analyzed using interrelated free-body diagrams.

## Section 19-4

**Rotation about a fixed axis not through the center of mass.** The scalar equations of motion using normal and tangential components are

$$\sum F_n = mr_C\omega^2 \quad \sum F_t = mr_C\alpha \quad \sum M_O = I_O\alpha$$

where $r_C$ is the nearest distance between the fixed axis $O$ and the mass center $C$. $\sum F_n$ and $\sum F_t$ must include the reaction on the rigid body at the axis of rotation.

**General plane motion.** The scalar equations of motion are

$$\sum F_x = ma_{C_x} \quad \sum F_y = ma_{C_y} \quad \sum M_C = I_C\alpha$$

**PURE ROLLING OF WHEELS:**

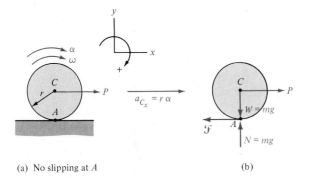

(a) No slipping at $A$                                 (b)

If the mass center $C$ coincides with the geometric center $G$ (balanced wheel),

$$a_{C_x} = r\alpha \quad \sum M_A = I_A\alpha$$

If $C$ is not at $G$ (unbalanced wheel),

$$a_C \neq r\alpha \quad a_G = r\alpha$$

$$\mathbf{a}_C = \mathbf{a}_G + \mathbf{a}_{C/G} = \mathbf{a}_G + (\mathbf{a}_{C/G})_n + (\mathbf{a}_{C/G})_t$$

**ROLLING AND SLIDING OF BALANCED WHEELS:**

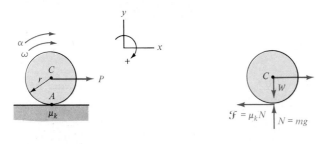

(a) Slipping at $A$ is possible                              (b)

$$\sum F_x = ma_{C_x} \qquad P - \mathscr{F} = ma_{C_x}$$

$$\sum F_y = ma_{C_y} \qquad N - mg = 0$$

$$\sum M_C = I_C\alpha \qquad \mathscr{F}r = I_C\alpha \qquad \text{for no slipping} \qquad \text{for slipping}$$

$$a_{C_x} = \alpha r$$

$$\mathscr{F} = \mu_k N$$

If slipping is not certain, assume no slipping and check whether $\mathscr{F} \leq \mu_s N$. $\mathscr{F} > \mu_s N$ is not possible; start the solution over, using $\mathscr{F} = \mu_k N$ without using $a_{C_x} = \alpha r$.

**19–96** The 400-lb crate, whose mass center is at $C$, is propped up against the frictionless surface $AB$. Edge $D$ rests on a rough surface. Determine (a) the maximum allowable acceleration of the truck to avoid tipping of the crate, and (b) the minimum coefficient of friction to avoid sliding at edge $D$ for part (a).

**Fig. P19-96**

**19–97** A 500-lb refrigerator with mass center at $C$ is being pushed up a 20° incline. The coefficient of friction $\mu$ is 0.1. Determine (a) the normal forces at $A$ and $B$ and the acceleration of the refrigerator for $P = 250$ lb, and (b) the maximum allowable magnitude of the force **P** for the refrigerator sliding up the incline without tipping.

**Fig. P19-97**

**19–98** A sliding door of mass 400 kg is supported by frictionless rollers at $D$ and $E$. The counterweight $B$ of mass 80 kg is connected to the door by a massless cable. Determine (a) the acceleration of the door, and (b) the reactions of the track on each roller. The mass center of the door is at $C$.

**Fig. P19-98**

**19–99** A playground swing is modeled as a 10-kg uniform bar on two wires swinging in the $xy$ plane. Determine the acceleration of the bar and the tension in the wires when the swing is released from $\theta = 60°$.

**Fig. P19-99**

**19–100** A 30-kg child sits on the swing of Prob. 19–99. The coefficient of static friction between the bar and the child is 0.3. Determine the accelerations of the bar and the child when the swing is released from rest at $\theta = 60°$.

**19–101** A uniform slender rod $AB$ weighing 96.6 lb is welded to a homogeneous disk $D$ weighing 64.4 lb as shown. A massless cable is attached to the center of disk $D$ and to a weight $E$ of 32.2 lb. Friction is negligible. If the support for $E$ is suddenly removed, determine the instantaneous acceleration of weight $E$ and the tension in the cable.

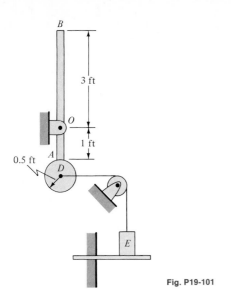

Fig. P19-101

**19–102** The unbalanced 10-kg cylinder, whose mass center is at $C$, has a radius of gyration of $r_g = 0.3\,\text{m}$, rotates about a horizontal pin $E$, and is driven by the 50-N·m couple. For the position shown, the velocity of point $O$ is 2 m/s downward. Determine (a) the angular acceleration of the cylinder and (b) the bearing reactions at point $E$ acting on the cylinder.

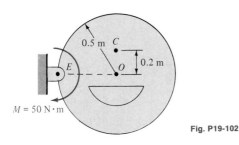

Fig. P19-102

**19–103** A proposed impact tester consists of a 2-m-long uniform bar $OA$ of 16-kg mass. The position $h$ of the 20-kg striker $B$ can be adjusted before a test. Determine $h$ so that there is no reaction at the pivot $O$ caused by $B$ striking the specimen $S$. The pendulum swings down with only its weight acting on it. Neglect the transverse dimensions of $OA$ and $B$.

Fig. P19-103

**19–104** A 5-kg uniform disk is rotating at a constant speed $\omega = 30\,\text{rad/s}$ when it is lowered onto a horizontal surface and released. Determine the time elapsed until the disk begins to roll without slipping if $\mu_k = 0.3$.

Fig. P19-104

**19–105** Determine the distance traveled by the disk in Prob. 19–104 during its motion involving skidding.

**19–106** A conveyor belt starts from rest with an acceleration of $1\,\text{m/s}^2$. Determine the angular acceleration of the 2-kg sphere and the 3-kg cylinder. For both bodies, $\mu_s = 0.4$, and $\mu_k = 0.3$.

Fig. P19-106

**19–107** The unbalanced wheel of 100 kg mass has its mass center at $C$ and has a radius of gyration of $r_g = 0.25\,\text{m}$. For the instantaneous position shown the velocity of $O$ is 2 m/s and its acceleration is 4 m/s, both to the right. A force $\mathbf{P}$ and a couple of magnitude $M = 60\,\text{N·m}$ counterclockwise are applied to the wheel. Determine all unknown forces acting on the wheel given that it rolls without slipping. What is the minimum allowable value of the coefficient of friction?

Fig. P19-107

**19-108** The eccentric wheel with mass center $C$ weighs 322 lb and has a radius of gyration about a horizontal axis through point $C$ of $r_g = 1$ ft. The wheel rolls and slips on the horizontal plane. In the position shown the velocity of point $O$ is 6 ft/s to the right, the angular velocity of the wheel is 5 rad/s clockwise, and the angular acceleration is 3 rad/s² clockwise. The coefficient of friction between the wheel and the horizontal surface is 0.3. Determine (a) the magnitude of the couple **M**, and (b) the instantaneous acceleration of point $O$.

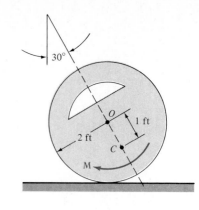

Fig. P19-108

# 20

# Energy and Momentum Methods for Rigid Bodies in Plane Motion

It was shown in Ch. 14 that the methods of work-energy are advantageous in solving a variety of problems in dynamics. These methods can be readily extended to analyze the motion of rigid bodies. Flywheel-powered cars, currently being investigated as an economical means of urban transportation, are an example of the usefulness of energy methods in the analysis of rigid body motion. A common feature of several proposed designs is a flywheel that can rotate at high speed about a vertical axis (Fig. 20-1). This disk may receive power periodically from an external source or an auxiliary engine in the car. The flywheel can be used to propel the car at the expense of gradually decreasing its own rotational speed. In turn, regenerative braking of the car can recharge the flywheel energy storage system. In such a case, there are repeated conversions of energy of the various components of the vehicle, and some energy is dissipated by fric-

**FIGURE 20-1**

Schematic of a flywheel-powered car

tion. The speed and acceleration of the vehicle depend on the quantities of work and energy associated with the motion.

Momentum methods are also useful in analyzing the motion of rigid bodies. The major results of Ch. 15 are readily applied to rigid bodies. Since both the energy and momentum methods can be described concisely at this stage, they are presented together in this chapter.

## STUDY GOALS

**SECTION 20-1** presents the definition of the work of a force acting on a rigid body. The definition of work is also extended to systems of forces. This topic should be thoroughly studied because it is essential for applying the principle of work and energy to rigid bodies (Sec. 20-3).

**SECTION 20-2** extends the concepts of kinetic energy of systems of particles to rigid bodies. The equation representing the kinetic energy due to translation and rotation of a body is frequently applied in dynamics analyses and is worth remembering.

**SECTION 20-3** presents the principle of work and energy for analyzing the motion of a rigid body or system of rigid bodies. This method is particularly advantageous in solving problems where velocities and displacements are of interest.

**SECTION 20-4** defines the conservation of energy in a conservative mechanical system. Even though an idealization, this is a very useful method for solving certain practical problems.

**SECTION 20-5** extends the definition of power to translating and rotating rigid bodies. These concepts are used in machine design.

**SECTIONS 20-6 AND 20-7** present the method of impulse and momentum of a rigid body or system of rigid bodies. This special method is useful in the analysis of motion when time and velocities are of interest.

**SECTION 20-8** defines the conservation of momentum for rigid bodies that have no external forces or moments acting on them. This method is useful in analyzing a mechanical system that changes configuration by internal means.

## 20-1

## WORK OF A FORCE ON A RIGID BODY

The work of a force acting on a rigid body is defined as it is for a particle: work equals the magnitude of the force times the displacement along the line of action of the force. Formally, the work $U_{12}$ of a force **F** during a displacement of its point of application from 1 to 2 is

$$U_{12} = \int_1^2 \mathbf{F} \cdot d\mathbf{r} \qquad \boxed{\text{20-1}}$$

where $d\mathbf{r}$ is the infinitesimal vector displacement of **F** along the path from 1 to 2 (see Sec. 14-1). In this equation the vectors **F** and $d\mathbf{r}$ are at any angle to each other since the dot product automatically accounts for the collinear components, the only ones that are involved in computing work. It is also useful to express the work of a force in terms of velocity. Since $\mathbf{v} = d\mathbf{r}/dt$,

$$U_{12} = \int_1^2 \mathbf{F} \cdot \mathbf{v} \, dt \qquad \boxed{\text{20-2}}$$

where **v** is the velocity of the instantaneous point of application of **F**.

The force $\mathbf{F}$ in Eqs. 20-1 and 20-2 may be the resultant of a system of forces. The work of these forces may be considered individually for a given rigid body. In that case the total work $U$ of $n$ forces is given by

$$U = \int \mathbf{F}_1 \cdot d\mathbf{r}_1 + \int \mathbf{F}_2 \cdot d\mathbf{r}_2 + \cdots + \int \mathbf{F}_n \cdot d\mathbf{r}_n \qquad \boxed{20\text{-}3}$$

where it must be observed that no relative motion may occur between particles of a rigid body. In other words, the work of internal forces (pairs of forces, equal in magnitude but opposite in direction) cancel for rigid bodies.

**Work of a moment.** The work of an applied moment acting on a rigid body may be developed from Eq. 20-3. For this it is desirable to establish a relationship involving the angular displacement caused by the moment. Consider an arbitrary rigid body with forces $\mathbf{F}$ and $-\mathbf{F}$ acting on it in the $xy$ plane as in Fig. 20-2. For convenience, a reference frame is placed at point $O$ such that line $AB$ coincides with the $x$ axis, and the $y$ axis bisects the distance $AB$. For any translational motion of the body, the net work of forces $\mathbf{F}$ and $-\mathbf{F}$ is zero by inspection. In a rotational displacement $d\theta$, however, the resultant work of the two forces is positive. Working with points $A$ and $B$, the total infinitesimal work $dU$ can be expressed (for convenience, in scalar form) as

$$dU = dU_A + dU_B = -F(-r\,d\theta) + F(r\,d\theta)$$

$$dU = F(2r)\,d\theta = M\,d\theta$$

where $M$ is the magnitude of the moment of the couple. The work $U_{12}$ of the couple during a finite rotation of the body is the integral of this expression. The result is

$$\boxed{U_{12} = \int_{\theta_1}^{\theta_2} M\,d\theta} \qquad \boxed{20\text{-}4}$$

which is limited to the common case where the moment vector $M$ is perpendicular to the plane of motion.

The equations above for the work of a force or a moment are useful for solving many problems. There are several special cases that are worth noting as follows.

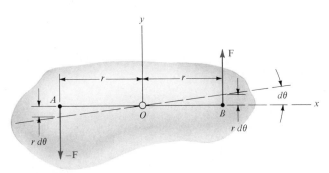

**FIGURE 20-2**

Schematic for defining the work of a couple

**Work of a constant force.** The force acting on the rigid body may have a constant magnitude $F$ and a constant direction $\theta$ with respect to the path $s$ of the body. The work $U_{12}$ of the force $F$ in moving the body from position $s_1$ to $s_2$ is written from Eq. 20-1 as

$$U_{12} = F(\cos\theta)(s_2 - s_1) \qquad \boxed{20\text{-}5}$$

**Work of a constant moment.** When the moment $M$ of the couple is constant, Eq. 20-4 becomes

$$U_{12} = M(\theta_2 - \theta_1) \qquad \boxed{20\text{-}6}$$

### Forces That Do No Work

Frequently, there are forces acting on moving rigid bodies without doing work on those bodies. The following list gives several such forces that are common.

1. Forces acting at fixed points on the body do no work. For example, the reaction at a fixed, frictionless pin does no work on the body that rotates about that pin.
2. A force which is always perpendicular to the direction of the motion does no work.
3. The weight of a body does no work when the body's center of gravity moves in a horizontal plane.
4. The friction force $\mathscr{F}$ at the point of contact on a body that rolls without slipping does no work. This is because the point of contact $I$ is the instantaneous center of zero velocity ($v_I = 0$). From Eq. 20-2, $U = \int \mathscr{F} \cdot \mathbf{v}_I \, dt = 0$

## 20-2

### KINETIC ENERGY OF A RIGID BODY

The system of particles comprising a rigid body has kinetic energy as a result of rotational motion as well as translational motion. These parts of the total kinetic energy are expressed using Fig. 20-3, where $P_i$ is a representative particle of the body. This particle has mass $m_i$, absolute velocity $\mathbf{v}_i$ in the fixed $XY$ frame, and relative velocity $\mathbf{v}_i'$ in the moving

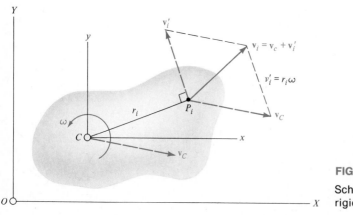

**FIGURE 20-3**

Schematic for defining the total kinetic energy of a rigid body

$xy$ frame, which is attached at the center of mass $C$. Frame $xy$ is translating with velocity $\mathbf{v}_C$ with respect to the fixed frame. The kinetic energy $T_i$ of the particle $P_i$ is written according to Sec. 17-4 as

$$T_i = \frac{1}{2} m_i v_i^2 = \frac{1}{2} m_i (v_C^2 + v_i'^2) \qquad \boxed{20\text{-}7}$$

where $v_i' = r_i \omega$ may be substituted to represent the velocity of the particle relative to the $xy$ frame. The total kinetic energy $T$ is obtained by summing $T_i$ over the $n$ particles of the body.

$$T = \frac{1}{2} mv_C^2 + \frac{1}{2} \sum_{i=1}^{n} m_i v_i'^2$$

$$T = \frac{1}{2} mv_C^2 + \frac{1}{2} \left( \sum_{i=1}^{n} m_i r_i^2 \right) \omega^2 \qquad \boxed{20\text{-}8}$$

Using an infinitesimal element of mass $dm_i$, and integrating over the whole body, the term in parentheses becomes the mass moment of inertia $I_C$ about the centroidal axis through point $C$ and normal to the plane of motion. This gives a concise expression for the kinetic energy,

$$\boxed{T = T_{\text{transl}} + T_{\text{rot}} = \frac{1}{2} mv_C^2 + \frac{1}{2} I_C \omega^2} \qquad \boxed{20\text{-}9}$$

where $\frac{1}{2}mv_C^2$ represents the kinetic energy due to *translation,* and $\frac{1}{2}I_C\omega^2$ represents the kinetic energy due to *rotation* of the body, even though the two motions may be simultaneous. Note that the kinetic energy is always a positive quantity if the body has any motion. Equation 20-9 is valid for the general plane motion of any rigid body of arbitrary shape.

## Rotation about an Arbitrary Fixed Axis

Equation 20-9 is valid for a body rotating about any fixed axis, but it can be further simplified for such a special case. Consider a body whose mass center is at $C$ rotating about an axis at $O$ as in Fig. 20-4. Here $v_C = r_C \omega$, and Eq. 20-9 can be written as

$$T = \frac{1}{2} (I_C + mr_C^2) \omega^2$$

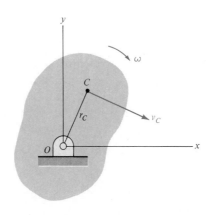

**FIGURE 20-4**

Rigid body rotating about an arbitrary fixed axis

where $I_c + mr_c^2 = I_O$, the mass moment of inertia about the axis $O$, by the parallel-axis theorem. Thus,

$$T = \frac{1}{2} I_O \omega^2$$

20-10

which is equivalent to Eq. 20-9 for a rigid body in rotation about a fixed axis $O$.

## PRINCIPLE OF WORK AND ENERGY

In a mechanical system work and energy are related for a rigid body as they are for a particle or a generalized system of particles. From Eq. 14-15,

$$T_2 = T_1 + U_{12}$$

20-11

where   $T_1$ = initial kinetic energy of the body
$T_2$ = final kinetic energy of the body
$U_{12}$ = work of all external forces and moments acting on the body during its motion from position 1 to 2

Equation 20-11 is particularly advantageous in solving problems of dynamics because it involves only scalar quantities and allows for direct calculations of velocities, distances, and rotations.

### Systems of Rigid Bodies

Most mechanical systems consist of interconnected parts that do work on one another while energy is transferred among the parts. If such a system can be assumed to be internally conservative (i.e., the internal friction forces do no net work), the principle of work and energy expressed by Eq. 20-11 may be applied to the system as a whole. In that case

$T_1$ = sum of initial kinetic energies of all parts of the system

$T_2$ = sum of final kinetic energies of all parts of the system

$U_{12}$ = work of all external forces and moments acting on the system as a whole; the internal forces (including friction forces) acting between various joints and parts of the system occur as pairs of forces equal in magnitude and opposite in direction, so the sum of their work is zero and need not be considered; there are no internal dissipative friction forces in the system

In analyzing an interconnected system of bodies, a free-body diagram of the whole system should be drawn to determine all significant external forces that are involved. All summations of kinetic energies and work are done arithmetically since these are scalar quantities.

The work of conservative forces (such as weight of a body or force of a spring) cause changes in the potential energy of a mechanical system. In the case of a conservative system the work of the external forces may be expressed as a difference in potential energies, $U_{12} = V_1 - V_2$, at the initial and final positions. Thus, Eq. 20-11 can be written as

$$T_1 + V_1 = T_2 + V_2 \qquad \boxed{20\text{-}12}$$

In words, *the sum of the kinetic and potential energies of a conservative system is constant.*

Note that in the general case the kinetic energy has translational and rotational parts. This possibility presents an interesting implication: a change from kinetic energy to potential energy, or vice versa, is not the only change possible in a conservative system of rigid bodies; it is also possible to convert translational kinetic energy to rotational kinetic energy, and vice versa. An example of the latter mechanism of energy conversion is a flywheel-powered automobile whose linear speed increases while the angular speed of its flywheel decreases.

The potential energy $V$ of the system is most commonly in the form of gravitational potential energy or elastic potential energy. These quantities are evaluated as in the case of particles.

The *gravitational potential energy* $V_g$ is a quantity that depends on the weight $W$ of a system of bodies and on the relative vertical position $h$ of the *center of gravity* with respect to a defined datum plane,

$$V_g = Wh \qquad \boxed{20\text{-}13}$$

This energy must be taken as positive when $h$ is above the datum plane (larger distance from the center of the earth), and negative for positions below that plane.

The *elastic potential energy* $V_e$ of an elastic spring in linear deformation is given by

$$V_e = \frac{1}{2} kx^2 \qquad \boxed{20\text{-}14}$$

where $k$ is the spring constant and $x$ is the change in length of the spring from its unstretched position. Note that a deformed spring does positive work on a body when the spring is returning to its undeformed position. This is true for either tensile or compressive deformations of the spring since in both cases the displacement and the force of the spring are in the same direction.

In general, the *total potential energy* $V$ of a system in a particular configuration is expressed as the algebraic sum of the gravitational and elastic potential energies,

$$V = V_g + V_e \qquad \boxed{20\text{-}15}$$

Note that conversions of gravitational potential energy to elastic potential energy, or vice versa, may occur as a system's configuration changes.

**Limitations.** The methods of work and energy described above are convenient to determine the speed of a body but not the reactions at joints and supporting surfaces. The equations of motion must be used to calculate forces.

## 20-5    POWER

The concept of power presented for particles in Sec. 14-4 is also valid for rigid bodies. However, it is necessary to distinguish analytically the power associated with translational and rotational motions. In both cases, power is the time rate at which work is done. For translation of a body under the action of a force $\mathbf{F}$,

$$\text{power} = \frac{dU}{dt} = \mathbf{F} \cdot \mathbf{v} = m\mathbf{a} \cdot \mathbf{v} \qquad \boxed{20\text{-}16}$$

where $\mathbf{v}$ is the velocity of any point in a translating body. Power can also be expressed as the time rate of change of kinetic energy,

$$\text{power} = \dot{T} = \frac{d}{dt} \left( \frac{1}{2} m\mathbf{v} \cdot \mathbf{v} \right)$$

$$= \frac{1}{2} m(\mathbf{a} \cdot \mathbf{v} + \mathbf{v} \cdot \mathbf{a})$$

$$= m\mathbf{a} \cdot \mathbf{v}$$

which is equivalent to Eq. 20-16. For rotation of a body under the action of a constant moment $\mathbf{M}$ whose direction is parallel to the axis of rotation (a common case),

$$\text{power} = \frac{dU}{dt} = \frac{d}{dt}(M\,d\theta) = M\frac{d\theta}{dt} = M\omega \qquad \boxed{20\text{-}17}$$

Equation 20-17 can also be obtained by taking the time derivative of the rotational kinetic energy; this is left as an exercise for the student.

The net power supplied to a system (or taken from it) is the algebraic sum of the powers required to change the kinetic and potential energies of the system in the same time interval. These can be expressed using the individual terms in concise notation as

$$\text{power} = \dot{T}_{\text{transl}} + \dot{T}_{\text{rot}} + \dot{V}_g + \dot{V}_e \qquad \boxed{20\text{-}18}$$

The terms on the right side of this equation are obtained from Eqs. 20-9, 20-13, and 20-14. Each quantity is considered positive when it represents power supplied to the system, and negative when it is for power taken from the system. The units, watts or horsepower, must be consistent in the equation.

EXAMPLE 20-1

Each wheel of a vehicle is assumed to be a uniform disk of 15 kg mass and 0.8 m diameter. Determine the total kinetic energy of one wheel at a vehicle speed of 100 km/h (Fig. 20-5).

## SOLUTION

A rotating wheel has kinetic energy due to both its translation and rotation. The angular speed of the wheel is calculated assuming that the wheel is rolling without slipping,

$$\omega = \frac{v_C}{R} = \frac{100 \, \text{km}}{\text{h}} \left( \frac{1000 \, \text{m/km}}{3600 \, \text{s/h}} \right) \left( \frac{1}{0.4 \, \text{m}} \right) = \frac{27.8 \, \text{m/s}}{0.4 \, \text{m}}$$

$$= 69.5 \, \text{rad/s}$$

From Eq. 20-9,

$$T = \frac{1}{2} m v_C^2 + \frac{1}{2} I_C \omega^2$$

$$= \frac{1}{2} (15 \, \text{kg})(27.8 \, \text{m/s})^2$$

$$+ \frac{1}{2} \left[ \frac{1}{2} (15 \, \text{kg})(0.4 \, \text{m})^2 \right] (69.5 \, \text{rad/s})^2$$

$$= 8694 \, \text{N·m}$$

## JUDGMENT OF THE RESULT

Since the point in contact with the ground is an instantaneous center of rotation, $v_C = R\omega$, so Eq. 20-9 can be written as

$$T = \frac{1}{2} (I_C + mR^2) \omega^2$$

$$= \frac{1}{2} \left( \frac{1}{2} mR^2 + mR^2 \right) \omega^2$$

$$= \frac{3}{4} mR^2 \omega^2 = \frac{3}{4} m v_C^2 = \left( \frac{3}{4} \right) (15 \, \text{kg})(27.8 \, \text{m/s})^2$$

$$= 8694 \, \text{N·m}$$

**TIP**

Since $I_C + mR^2 = I_O$ when point $O$ is an instantaneous center, Eq. 20-10, $T = \frac{1}{2} I_O \omega^2$, is also valid in this case.

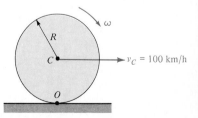

**FIGURE 20-5**

Analysis of a rolling wheel

EXAMPLE 20-1    813

# EXAMPLE 20-2

Flywheel *A* of the tire tester in Fig. 20-6a drives two wheels *B* and *C* simultaneously. The flywheel has a weight of 500 lb, a 6-ft diameter, and a 2.5-ft radius of gyration. Wheels *B* and *C* weigh 35 lb each, have a 2.5-ft diameter, and may be considered as uniform disks. The wheels are at rest when a 2000-lb·ft counterclockwise couple is applied to wheel *A*. Determine the number of revolutions of wheel *A* before the angular speed of wheel *C* is what it would be in a car traveling at 150 mi/h. Neglect frictional losses at axles *A*, *B*, and *C*.

(a)

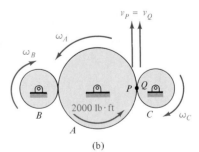

(b)

**FIGURE 20-6**

Tire-testing equipment (Courtesy MTS Systems Corporation, Minneapolis, Minnesota)

# SOLUTION

The three wheels are considered as a system of rigid bodies. Since there is no slipping, there is no dissipative friction force in the system. Applying Eq. 20-11 gives

$$T_1 = 0 \quad \text{(initially at rest)}$$

$$T_2 = \frac{1}{2} I_A \omega_A^2 + \frac{1}{2} m_A v_A^2 + \frac{1}{2} I_B \omega_B^2$$

$$+ \frac{1}{2} m_B v_B^2 + \frac{1}{2} I_C \omega_C^2 + \frac{1}{2} m_C v_C^2$$

$$U_{12} = M_A \theta_A$$

For wheel $C$ traveling (rolling without slipping) at 150 mi/h,

$$\omega_C = \frac{v_Q}{r_C} = \frac{150\,\text{mi}}{\text{h}} \left( \frac{5280\,\text{ft/mi}}{3600\,\text{s/h}} \right) \left( \frac{1}{1.25\,\text{ft}} \right)$$

$$= 176\,\text{rad/s}$$

The velocities of the points $P$ and $Q$ in contact between $A$ and $C$ must be the same if no slipping is occurring (Fig. 20-6b); hence,

$$v_Q = r_C \omega_C = r_A \omega_A$$

$$\omega_A = \left( \frac{r_C}{r_A} \right) \omega_C = \frac{1.25\,\text{ft}}{3\,\text{ft}} (176\,\text{rad/s})$$

$$= 73.3\,\text{rad/s}$$

Since wheel $B$ is in contact with wheel $A$ at the same radii as for wheels $A$ and $C$, $\omega_B = \omega_C = 176\,\text{rad/s}$ (opposite direction). The translational velocities of the mass centers are always zero.

The only unknown now left in the work–energy equation is the angle through which the applied moment acts, $\theta_A$. Hence,

$$T_1 + U_{12} = T_2$$

$$0 + (2000\,\text{lb}\cdot\text{ft})(\theta_A) = \frac{1}{2} \left[ \frac{500\,\text{lb}}{32.2\,\text{ft/s}^2} (2.5\,\text{ft})^2 \right] \left( 73.3\,\frac{\text{rad}}{\text{s}} \right)^2 + \frac{1}{2} m_A(0)^2$$

$$+ \frac{1}{2} \left[ \frac{1}{2} \left( \frac{35\,\text{lb}}{32.2\,\text{ft/s}^2} \right) (1.25\,\text{ft})^2 \right] \left( 176\,\frac{\text{rad}}{\text{s}} \right)^2 + \frac{1}{2} m_B(0)^2$$

$$+ \frac{1}{2} \left[ \frac{1}{2} \left( \frac{35\,\text{lb}}{32.2\,\text{ft/s}^2} \right) (1.25\,\text{ft})^2 \right] \left( 176\,\frac{\text{rad}}{\text{s}} \right)^2 + \frac{1}{2} m_C(0)^2$$

$$\theta_A = \frac{(260{,}700 + 13{,}150 + 13{,}150)\,\text{ft}\cdot\text{lb}}{2000\,\text{lb}\cdot\text{ft}}$$

$$= 143.5\,\text{rad}$$

The number of revolutions is

$$143.5\,\text{rad} \left( \frac{1\,\text{rev}}{2\pi\,\text{rad}} \right) = 22.8\,\text{rev}$$

EXAMPLE 20-2    815

# EXAMPLE 20-3

Consider the system in Ex. 20-2 and determine the magnitude of the required normal force between the wheels to prevent slipping if $\mu_s = 0.7$.

## SOLUTION

To calculate the minimum required normal force between the wheels, the friction force needed for the motion must be calculated. Considering wheel $A$ alone (Fig. 20-7), it is realized that the friction forces $\mathscr{F}_B$ and $\mathscr{F}_C$ do work *against* wheel $A$. The work of these friction forces was not included when the whole system was considered since the work of the friction forces acting on $B$ and $C$ canceled the work of $\mathscr{F}_B$ and $\mathscr{F}_C$ on $A$. The normal forces do no work since they have no component in the direction of the displacement. Hence, using the work energy equation for wheel $A$ alone,

$$T_1 + U_{12} = T_2$$

$$0 + M_A \theta_A - \mathscr{F}_B(s_B) - \mathscr{F}_C(s_C) = \frac{1}{2} I_A(\omega_2)_A^2 + \frac{1}{2} m_A(v_2)_A^2 \qquad \textbf{(a)}$$

The arc lengths $s_B$ and $s_C$ through which $\mathscr{F}_B$ and $\mathscr{F}_C$ act are

$$s_B = s_C = R_A \theta_A$$

From symmetry,

$$\mathscr{F}_B = \mathscr{F}_C = \mathscr{F}$$

Substituting the values found in Ex. 20-2 into Eq. (a),

$$(2000\,\text{ft})(143.5\,\text{rad}) - 2\mathscr{F}(3\,\text{ft})(143.5\,\text{rad})$$

$$= \frac{1}{2}\left(\frac{500\,\text{lb}}{32.2\,\text{ft/s}^2}\right)(2.5\,\text{ft})^2\left(73.3\,\frac{\text{rad}}{\text{s}}\right)^2 + \frac{1}{2} m_A(0)^2$$

Solving for $\mathscr{F}$ gives

$$\mathscr{F} = 30.5\,\text{lb}$$

The magnitude of the minimum normal force $\mathbf{N}$ necessary is the value for the condition of impending slipping, $\mathscr{F} = \mathscr{F}_{\text{max}} = \mu_s N$.

$$N = \frac{\mathscr{F}}{\mu_s} = \frac{30.5\,\text{lb}}{0.7} = 43.6\,\text{lb}$$

**FIGURE 20-7**

Flywheel $A$ from Fig. 20-6

# EXAMPLE 20-4

A uniform, 70-lb bar is released from rest with no force initially acting on it from the spring (Fig. 20-8). The ends of the bar slide along smooth surfaces. Calculate the speeds of ends $A$ and $B$ just before the bar strikes the horizontal surface.

## SOLUTION

The bar undergoes both translation and rotation. The two unknowns in this problem are the final linear speed $v_C$ and angular speed $\omega$. The work–energy principle allows for one equation. Another equation is obtained since $v_C$ and $\omega$ are related. This relationship varies with the angle of inclination, but the relationship for the bar's nearly horizontal position is readily determined. At this instant, by inspection, $v_B = 0$, $B$ is the instantaneous center of rotation, and $v_C = (\ell/2)\omega$. Work is done on the rod by the force of gravity and the spring force. Both of these forces are conservative forces, so the principle of conservation of energy applies (Eq. 20-12). The kinetic and potential energies at the beginning (Fig. 20-8a) and end (Fig. 20-8b) positions of the bar are

$$T_1 = 0 \text{ (released from rest)}$$

$$V_{1g} = W_{\text{rod}}(y_C) \qquad V_{1e} = 0 \text{ (unstretched spring)}$$

$$V_{2e} = \frac{1}{2} k y_A^2 \qquad V_{2g} = 0 \text{ (at datum plane)}$$

$$T_2 = \frac{1}{2} I_C \omega^2 + \frac{1}{2} m v_C^2$$

Using Eq. 20-12,

$$T_1 + V_1 = T_2 + V_2$$

$$0 + (70 \,\text{lb})(20 \,\text{in. sin } 45°) = \frac{1}{2}\left[ \frac{1}{12}\left( \frac{70 \,\text{lb}}{32.2 \,\text{ft/s}^2} \right) \frac{1 \,\text{ft}}{12 \,\text{in.}} (40 \,\text{in.})^2 \right] \omega^2$$

$$+ \frac{1}{2}\left( \frac{70 \,\text{lb}}{32.2 \,\text{ft/s}^2} \right) \frac{1 \,\text{ft}}{12 \,\text{in.}} \left[ \left( \frac{40 \,\text{in.}}{2} \right) \omega \right]^2$$

$$+ \frac{1}{2}(2 \,\text{lb/in.})[40 \,\text{in. sin } 45°]^2$$

$$990 \,\text{in.·lb} = (12.08 \,\text{in.·lb·s}^2)\omega^2$$
$$+ (36.2 \,\text{in.·lb·s}^2)\omega^2 + 800 \,\text{in.·lb}$$

$$\omega^2 = \frac{190 \,\text{in.·lb}}{48.3 \,\text{in.·lb·s}^2} = 3.93 \,\frac{1}{\text{s}^2}$$

$$\omega = 1.98 \,\text{rad/s}$$

Since point B is the instantaneous center of rotation,

$$v_A = l\omega = (40 \,\text{in.})\left( 1.98 \,\frac{\text{rad}}{\text{s}} \right)$$

$$= 79.3 \,\text{in./s}$$

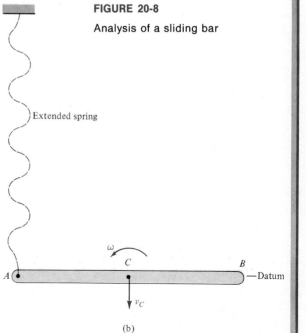

**FIGURE 20-8**

Analysis of a sliding bar

k = 2 lb/in.

y

A

y_A

l = 40 in.

C

y_C

45°

B

x

(a)

Extended spring

ω

C

A

B

Datum

v_C

(b)

EXAMPLE 20-5

The hammer consisting of a 322-lb disk $A$ welded to the end of a 6-ft slender rod weighing 161 lb is used to form sheet metal by impacting it at $EF$ (Fig. 20-9a). A spring of spring constant $k = 500\,\text{lb/ft}$ is attached at $B$; its unstretched length is 1 ft. Determine **(a)** the velocity of the center of the disk $A$ immediately before impact if it is released from rest as shown, and **(b)** the horizontal component of the bearing reaction at $O$ immediately before impact.

(a)

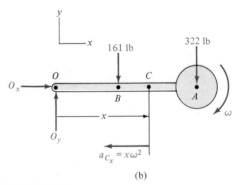

(b)

**FIGURE 20-9**

Model of a sheet metal hammer

## SOLUTION

**(a)** Applying the conservation of energy principle from the initial position (1) to the final position (2),

$$T_1 + V_1 = T_2 + V_2 = \text{constant} \qquad \text{(Eq. 20-12)}$$

Selecting the reference position for the spring as its unstretched length and for the weights as the horizontal position immediately before impact,

$$V_1 = V_{g1(\text{disk})} + V_{g1(\text{rod})} + V_{e1}$$

$$= (322 \text{ lb})(7 \text{ ft}) + (161 \text{ lb})(3 \text{ ft}) + \frac{1}{2}(500 \text{ lb/ft})(4 \text{ ft})^2 = 6737 \text{ ft} \cdot \text{lb}$$

$$V_2 = T_1 = 0$$

and

$$T_2 = \frac{1}{2} I_O \omega^2 = \frac{1}{2} \left[ \frac{1}{2}(10 \text{ slugs})(1 \text{ ft})^2 + (10 \text{ slugs})(7 \text{ ft})^2 + \frac{1}{3}(5 \text{ slugs})(6 \text{ ft})^2 \right] \omega^2$$

$$= \frac{1}{2}(555 \text{ ft} \cdot \text{lb} \cdot \text{s}^2) \omega^2$$

Therefore, from conservation of energy,

$$6737 \text{ ft} \cdot \text{lb} = (277.5 \text{ ft} \cdot \text{lb} \cdot \text{s}^2) \omega^2$$

$$\omega = 4.93 \text{ rad/s}$$

Since $v_A = (7 \text{ ft})\omega$,

$$v_A = 34.5 \text{ ft/s (before impact)}$$

**(b)** First the center of mass $C$ of the composite body must be located,

$$\bar{x}(483 \text{ lb}) = (3 \text{ ft})(161 \text{ lb}) + (7 \text{ ft})(322 \text{ lb})$$

$$\bar{x} = 5.67 \text{ ft}$$

From the free-body diagram in Fig. 20-9b,

$$\sum F_x = m_T a_{C_x} = m_T(-\bar{x}\omega^2)$$

$$O_x = (10 \text{ slugs} + 5 \text{ slugs})(-5.67 \text{ ft})(4.93 \text{ rad/s})^2$$

$$= -2070 \text{ lb}$$

$$\mathbf{O}_x = -2070 \text{ lb } \mathbf{i}$$

EXAMPLE 20-5   819

**20-1** A 10-kg uniform slender rod is released from the vertical position. It starts to fall with negligible initial speed and rotates without friction about hinge $A$. Calculate the rod's angular speed in the horizontal position.

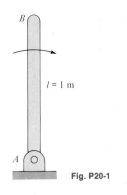

**Fig. P20-1**

**20-2** Solve Prob. 20–1 assuming that an additional mass $m_B = 3$ kg of negligible dimensions is attached at point $B$, and that $B$ has an initial speed $v_B = 1$ m/s when the rod is vertical.

**20-3** Two uniform slender rods are hinged at $A$ and $B$, and part $C$ slides on a smooth surface. The 10-lb rods are released from rest when $\theta = 60°$. Calculate the angular speed of $BC$ when it strikes the surface.

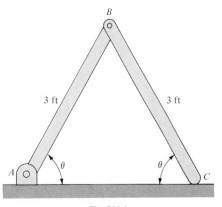

**Fig. P20-3**

**20-4** Solve Prob. 20–3 assuming that $\dot{\theta} = -0.5$ rad/s at $\theta = 60°$ (decreasing angle).

**20-5** A uniform, rectangular bar of 10-kg mass is released from rest at $\theta = 60°$. Determine the speed $v_A$ of point $A$ at $\theta = 90°$.

**Fig. P20-5**

**20-6** The bar in Prob. 20–5 has angular speed $\omega_o = 1$ rad/s at $\theta = 0$. Determine the speed $v_A$ at $\theta = 180°$.

**20-7** A long slender log weighing 1610 lb is used as a battering ram to knock down a stone wall. The log is suspended by two cables of length 10 ft attached to the ends of the log. Determine the speed at which the log strikes the wall at $P$ if it is released from rest at $30°$ as shown.

**Fig. P20-7**

**20-8** The test platform weighs 200 lb, and the console weighs 250 lb. There are four uniform rocker arms such as $BE$; each weighs 10 lb and makes an angle $\phi$ with the horizontal. The platform is released from rest at $\phi = 85°$. Calculate its speed when $\phi = 40°$. Assume that the actuator $A$ has no mass and applies no force to the platform during this motion. $BC = 5$ ft, $BE = CD = 2.5$ ft.

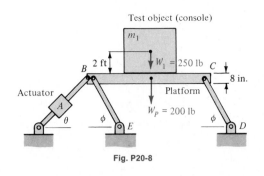

**Fig. P20-8**

**20–9**   A torsion spring attached at joint $E$ in Fig. P20-8 applies no moment to arm $BE$ when $\phi = 90°$. The platform slowly starts to swing downward at this angle, its motion only opposed by the spring. Calculate the platform's speed at $\phi = 50°$ if $k = 300\,\text{lb·ft/rad}$ of the single spring used.

**20–10**   A 30-kg flywheel is a uniform disk with a 0.2-m radius. Determine the magnitude of the couple that can increase the wheel's rotational speed from zero to 10,000 rpm in 100 revolutions.

**20–11**   A uniform, rectangular plate of 80 lb weight is released from rest with edge $OA$ horizontal. Determine the angular speed of the plate and the reaction at $O$ when $OA$ is vertical.

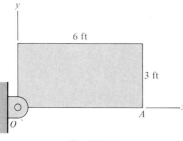

**Fig. P20-11**

**20–12**   Solve Prob. 20–11 if $\mathbf{v}_A = 3\,\text{ft/s}\,\mathbf{j}$ when point $A$ is on the $x$ axis.

**20–13**   An elevator system is modeled as $m_1 = 2000\,\text{kg}$, $m_2 = 1800\,\text{kg}$, and a uniform disk of $m_3 = 200\,\text{kg}$. Assume that the system starts from rest and moves under the action of gravity, with negligible friction. Determine the speed of $m_1$ after it moves to $y = 30\,\text{m}$.

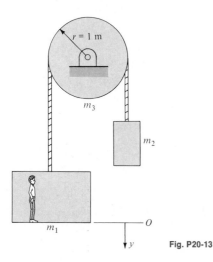

**Fig. P20-13**

**20–14**   Solve Prob. 20–13 assuming that $m_1$ has an initial upward speed of 2 m/s at $y = 0$.

**20–15**   A uniform, 30-lb bar is hinged at point $O$. It is released from rest in the horizontal position. Calculate its angular speed in the vertical position if the spring applies no force in that position.

**Fig. P20-15**

**20–16**   The 4-lb uniform bar pivots about point $O$. The spring is 13 in. long when not deformed. The bar is released from rest in the vertical position. Calculate the speed of point $C$ when $\theta = 0$.

**Fig. P20-16**

**20–17**   Determine the $x$ and $y$ components of the reaction at point $O$ in Prob. 20–16 when $\theta = 0$.

**20–18** An impact tester is modeled as a pendulum of 110 lb weight and 2.5 ft radius of gyration about point $O$. It is released from rest at $\theta = 0$ to strike a specimen $S$ at $\theta = 90°$. The specimen breaks absorbing 20 ft·lb of energy, and allows the pendulum to swing past the vertical position. Calculate the maximum angle $\theta$ from the horizontal position of the arm.

**Fig. P20-18**

**20–19** A 30-lb uniform disk is released from rest and rolls without slipping down the incline. Calculate $v_C$ at position $B$.

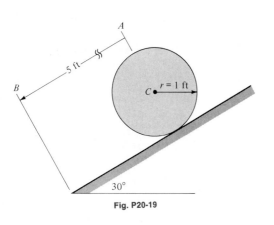

**Fig. P20-19**

**20–20** Solve Prob. 20–19 assuming that the disk has an initial speed $v_C = 4$ ft/s up the incline at position $A$.

**20–21** A proposed conveyor system is modeled as a plate of mass $M$ moving without slipping on small cylindrical rollers of mass $m$ and radius $r$. Determine the speed of the plate after it has moved a distance $l$, starting from rest, along an inclined plane. Consider (a) free rollers, and (b) rollers freely rotating about their fixed axes. Assume the plate is on three rollers at all times as it moves a distance $l$.

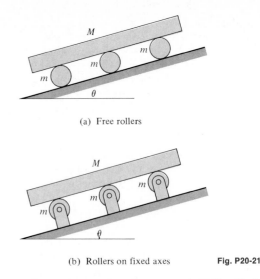

(a) Free rollers

(b) Rollers on fixed axes      **Fig. P20-21**

**20–22** The total mass of a roller coaster is 300 kg. Each of its four wheels has a 3-kg mass, a 0.1-m radius, and a 0.07-m radius of gyration. Determine the linear speed of the coaster at $A$. It starts from rest at $O$. The wheels roll without slipping.

**Fig. P20-22**

**20–23** Assume that each wheel of a 1000-kg automobile has a 15-kg mass, 0.4-m radius, and 0.2-m radius of gyration. Plot the total kinetic energy of the four wheels and of the rest of the vehicle as functions of speed, using 20 km/h, 60 km/h, and 100 km/h.

**20–24** A child rides a homemade Soapbox Derby car down a hill starting from rest at a height of 20 ft above the level ground. The child weighs 100 lb and the car weighs 222 lb, *including* the four wheels which each weigh 16.1 lb. The radius of each wheel is 1.0 ft, and each wheel has a centroidal radius of gyration of 0.5 ft. The child has designed a braking system which locks all four wheels, causing the car to skid from point $A$ until it comes to rest at $B$. The coefficient of kinetic friction is 0.25. Neglecting air resistance, determine (a) the speed of the car at point $A$, and (b) the distance $d$ required for the car to skid to rest.

**20–25** Pins $A$ and $B$ of the uniform, 40-lb bar slide without friction in their guides. The spring applies no force to the bar when it is released from rest at $\theta = 90°$. Calculate the speed of pin $A$ when $\theta = 45°$.

Fig. P20-25

**20–26** The bar in Prob. 20–25 is released from rest at $\theta = 135°$. Determine the smallest value of $\theta$ as the bar swings down.

**20–27** Show that in rotational motion power equals the time derivative of the rotational kinetic energy.

**20–28** The base structure of a revolving restaurant is modeled as three uniform disks that rotate as a unit about the vertical axis. The mass $m$ is 8000 kg, and the diameter $d$ is 7 m. Determine the power input to accelerate the three disks from rest to a rotational speed of 0.1 rpm in a time of 10 s.

Fig. P20-28

**20–29** The shaft of a motor is designed to transmit a torque of 2000 lb·ft. Plot the maximum power that can be developed by the motor as a function of its rotational speed at 1000 rpm, 1500 rpm, and 2000 rpm.

**20–30** The uniform, 10-kg disk is rotating at an angular speed of 500 rpm when a braking force $\mathbf{F}$ is applied to the hinged bar. The friction at $O$ and $A$ is negligible, while $\mu_k = 0.5$ at $B$. Determine the magnitude of the force $\mathbf{F}$ for stopping the disk in 10 revolutions.

Fig. P20-30

**20–31** A moment is applied to the disk in Prob. 20–30 to accelerate it from rest, while the force $\mathbf{F}$ on the bar has a magnitude of 400 N. Determine the power input if the disk attains an angular speed of 500 rpm in a time of 20 s.

**20–32** The double pulley has a total weight of 50 lb and a radius of gyration of 1.6 ft. The system starts from rest. Calculate the speed of $W_1$ after a travel of 5 ft.

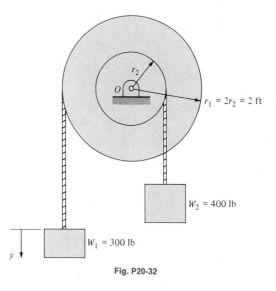

Fig. P20-32

**20–33** Determine the power of the system in Prob. 20–32 at a time of 3 s after the start.

**20–34** The amusement ride of 12 cages is modeled as 12 concentrated masses of 700 kg each, on the ends of uniform, straight bars of 8 m length and 250 kg mass. The system is started from rest and moves in a plane. Calculate the required work input for the system to attain a speed of 4 rpm.

Fig. P20-34

**20–35** Determine the power requirement of the system in Prob. 20–34 if it must attain the speed of 4 rpm in a time of 5 s.

**20–36** A 20,000-kg railroad car is modeled as a uniform bar of 3 m × 3 m cross section. It is firmly attached to the rigid rail bed $OA$, which is raised while pivoting about $O$ for rapid unloading. Determine the maximum power required to elevate the car and the angle at which this occurs if $\dot{\theta} = 0.3$ rad/s (constant) from $\theta = 0$ to $\theta = 60°$.

Fig. P20-36

**20–37** A concentrated mass $m_1 = 50$ kg is at the end of a uniform, hinged rod of mass $m_2 = 30$ kg. A couple $\mathbf{M}_0$ is applied about pivot $O$ to raise the rod from $\theta = 0$ to 60° at a constant angular speed of 0.5 rad/s. Calculate the maximum power required during this motion.

Fig. P20-37

**20–38** A flywheel-powered, 1200-kg car begins regenerative braking when traveling at a speed of 70 km/h. The 15-kg flywheel of 0.2-m radius of gyration is rotating about the vertical $y$ axis at a rate of 1000 rpm at this instant. What is the final angular speed of the flywheel when the car is stopped if 30% of the car's translational kinetic energy can be used for recharging the power plant? Neglect the rotation of the car's wheels.

Fig. P20-38

**20–39** A front end loader has a total weight of 10,000 lb. Each of its four wheels has a 350-lb weight, a 5-ft diameter, and a 1.5-ft radius of gyration. Calculate the required power to accelerate this machine from rest to a speed of 10 ft/s in a time of 5 s.

Fig. P20-39

The principles of impulse and momentum are particularly useful in the analysis of motion when time and velocities are involved in the problem as was demonstrated in Ch. 15. These concepts are applicable to the systems of particles that constitute rigid bodies, and they are presented here for two-dimensional motion.

## Linear Impulse and Momentum

The linear momentum **G** of a body is the vector sum of the linear momenta of all of its particles $P_i$ of mass $m_i$ and velocity $\mathbf{v}_i$. For $n$ particles, from Eq. 17-20,

$$\mathbf{G} = \sum_{i=1}^{n} m_i \mathbf{v}_i = m\mathbf{v}_C \qquad \boxed{20\text{-}19}$$

where $m$ is the total mass and $\mathbf{v}_C$ is the velocity of the mass center.

Newton's equation of motion can be written as

$$\sum \mathbf{F} = m\mathbf{a}_C = m\frac{d\mathbf{v}_C}{dt} = \frac{d}{dt}(m\mathbf{v}_C) \qquad \boxed{20\text{-}20}$$

since $m$ is a constant. Multiplying this by $dt$ and integrating from time $t_1$ to $t_2$,

$$\boxed{\int_{t_1}^{t_2} \sum \mathbf{F}\, dt = m(\mathbf{v}_{C_2} - \mathbf{v}_{C_1})} \qquad \boxed{20\text{-}21}$$

where $\mathbf{v}_{C_2}$ and $\mathbf{v}_{C_1}$ are the velocities of the mass center at $t_2$ and $t_1$, respectively. The left side of this equation is the *impulse of the external forces* acting on the body from time $t_1$ to $t_2$. This impulse is equal to the change in linear momentum of the body in the given time interval. It should be noted that Eq. 20-21 is a vector equation and can be used in terms of scalar components for motion along conveniently selected orthogonal $x$ and $y$ axes. In any case, the impulse term must include *all* external forces acting on the body. Unlike the work-energy equation in which some external forces could be neglected because they do no work, *all* external forces act over an interval of time and must be included in the impulse and momentum equation.

## Angular Impulse and Momentum

The angular momentum **H** of a body is the vector sum of the angular momenta of all its particles. Angular momentum of a particle $P_i$ is defined as the moment of its linear momentum with respect to a chosen reference point. Using the center of mass $C$ in Fig. 20-10 for reference, and recalling Eq. 19-4,

$$\mathbf{H}_C = I_C\omega \qquad \boxed{20\text{-}22}$$

where $I_C$ is the centroidal mass moment of inertia about an axis perpendicular to the plane of motion, and $\omega$ is the angular velocity of the body.

**FIGURE 20-10**

Schematic for defining angular momentum of a rigid body

The equation of motion for rotation relates the external moments $\sum \mathbf{M}$ to the time rate of change of angular momentum,

$$\sum \mathbf{M}_C = I_C \boldsymbol{\alpha} = I_C \frac{d\omega}{dt} = \frac{d}{dt}(I_C \omega) = \dot{\mathbf{H}}_C \qquad \boxed{20\text{-}23}$$

since $I_C$ is a constant. Multiplying this by $dt$ and integrating from time $t_1$ to $t_2$ gives

$$\boxed{\int_{t_1}^{t_2} \sum \mathbf{M}_C \, dt = I_C \omega_2 - I_C \omega_1 = \mathbf{H}_{C_2} - \mathbf{H}_{C_1}} \qquad \boxed{20\text{-}24}$$

where $\omega_2$ and $\omega_1$ are the angular velocities at $t_2$ and $t_1$, respectively. The left side of Eq. 20-24 is the *impulse of the external moments* acting on the body from time $t_1$ to $t_2$. This impulse equals the change in angular momentum of the body in the given time interval. The impulse term must include all external moments. For plane motion, Eq. 20-24 can be written in simplified scalar form if $\sum \mathbf{M}_C$ is parallel to $\boldsymbol{\omega}$,

$$\int_{t_1}^{t_2} \sum M_C \, dt = I_C(\omega_2 - \omega_1) \qquad \boxed{20\text{-}25}$$

Using the results of Eqs. 17-30 and 17-36 for any fixed point of reference $O$ (such as in Fig. 20-11), Eqs. 20-22 through 20-24 become

$$\mathbf{H}_O = I_C \boldsymbol{\omega} + \mathbf{r}_C \times m\mathbf{v}_C \qquad \boxed{20\text{-}26}$$

$$\sum \mathbf{M}_O = \dot{\mathbf{H}}_O \qquad \boxed{20\text{-}27}$$

$$\int_{t_1}^{t_2} \sum \mathbf{M}_O \, dt = \mathbf{H}_{O_2} - \mathbf{H}_{O_1} \qquad \boxed{20\text{-}28}$$

where $\mathbf{r}_C$ is the position vector from the reference point $O$ to the center of mass $C$, and $\mathbf{v}_C$ is the velocity of the center of mass. An important special case occurs when the point $O$ is the fixed axis of rotation for the rigid body.

In the case of plane motion about a fixed point $O$, the velocity of the center of mass is perpendicular to $r_C$ and related to the angular velocity of the rigid body by $v_C = r_C \omega$. Equation 20-26 can be written in scalar form as $H_O = I_C \omega + r_C(mr_C\omega)$ and substituted into Eq. 20-28,

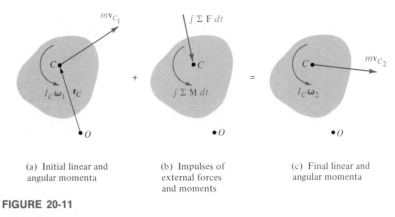

(a) Initial linear and angular momenta

(b) Impulses of external forces and moments

(c) Final linear and angular momenta

**FIGURE 20-11**

Analysis of impulse and momentum

$$\int_{t_1}^{t_2} \sum M_O \, dt = [I_C \omega + r_C(mr_C\omega)]_{t_1}^{t_2} \qquad \boxed{20\text{-}29}$$

$$= (I_C + mr_C^2)(\omega_2 - \omega_1)$$

$$= I_O(\omega_2 - \omega_1)$$

The results of Eq. 20-29 for the rotation of a rigid body about a fixed point enable one to sum moments about the fixed point by evaluating the moment of inertia with respect to that point. Thus, in Eq. 20-28 the reactions at the fixed point are eliminated from the moment calculation and Eq. 20-26 reduces to $\mathbf{H}_O = I_O\omega$.

## IMPULSE AND MOMENTUM OF A SYSTEM OF RIGID BODIES

The methods developed in Sec. 20-6 can be applied to a system of rigid bodies (frequently a system of interconnected bodies) in two ways. First, Eqs. 20-21 and 20-24 (or 20-28) can be used to analyze the motion of each rigid member of the system separately. In this case the mutual forces acting between the members must be included in the formulation of the solution. The second approach is to apply the principle of impulse and momentum to the entire system of bodies. The advantage of this method is that the mutual forces acting between members of the system do not have to be included in the solution since they cancel in pairs.

### Formulation of Problem Solutions

For a schematic illustration of common problems and the procedure for their solution, assume that the initial total linear and angular momenta are known for a given system of rigid bodies. The system may consist of one or more bodies, and the momenta may be entirely linear momentum, angular momentum, or both simultaneously. A system of external impulses is applied to the system of bodies, which thereby acquires a new, or final, set of momenta. This is shown schematically for a single rigid body in the *momentum and impulse diagrams* in Fig. 20-11.

The same procedure is used when the system consists of several bodies, but it is necessary to include in the formulation the linear and angular momenta and the external impulses for each moving member of the system. This is shown schematically for two interconnected bodies in Fig. 20-12.

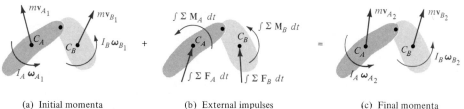

   (a) Initial momenta            (b) External impulses           (c) Final momenta

**FIGURE 20-12**

Impulse and momentum for two interconnected bodies

The choice between the above two approaches is made according to the number of unknowns that must be determined. For any system of bodies in plane motion there are three scalar equations of impulse and momentum that can be written for a given configuration. These are obtained from Eqs. 20-21 and 20-28. Schematically,

$$\sum \begin{pmatrix} \text{initial} \\ \text{linear} \\ \text{momentum} \end{pmatrix}_x + \sum \begin{pmatrix} \text{linear} \\ \text{impulse} \end{pmatrix}_x \qquad \boxed{20\text{-}30a}$$

$$= \sum \begin{pmatrix} \text{final} \\ \text{linear} \\ \text{momentum} \end{pmatrix}_x$$

$$\sum \begin{pmatrix} \text{initial} \\ \text{linear} \\ \text{momentum} \end{pmatrix}_y + \sum \begin{pmatrix} \text{linear} \\ \text{impulse} \end{pmatrix}_y \qquad \boxed{20\text{-}30b}$$

$$= \sum \begin{pmatrix} \text{final} \\ \text{linear} \\ \text{momentum} \end{pmatrix}_y$$

$$\sum \begin{pmatrix} \text{initial} \\ \text{angular} \\ \text{momentum} \end{pmatrix}_O + \sum \begin{pmatrix} \text{angular} \\ \text{impulse} \end{pmatrix}_O \qquad \boxed{20\text{-}30c}$$

$$= \sum \begin{pmatrix} \text{final} \\ \text{angular} \\ \text{momentum} \end{pmatrix}_O$$

where $x$ and $y$ are the chosen coordinate directions and $O$ is the reference point for the analysis of angular impulse and momentum. Point $O$ must be the same *fixed reference point* for all parts of a given system to allow using one equation for the rotational motion of the entire system. Equations 20-30 make it possible to determine up to three unknowns working with the system as a whole. For more than three unknowns it is necessary to apply the principle of impulse and momentum to each member separately. In this case, three scalar equations can be written for each member but the forces between members must be included in the external impulses.

## 20-8    CONSERVATION OF MOMENTUM

The principles of conservation of momentum that were developed for particles (Sec. 17-9) are valid for rigid bodies that have no external forces or moments acting on them. The formal statements of the conservation principle are considered separately for linear and rotational motions.

### Conservation of Linear Momentum

There is no linear impulse when $\sum \mathbf{F} = 0$, so Eqs. 20-19 and 20-21 become, respectively,

$$\Delta \mathbf{G} = 0 \qquad \boxed{\text{20-31a}}$$

$$m\mathbf{v}_{C_1} = m\mathbf{v}_{C_2} \qquad \mathbf{v}_{C_1} = \mathbf{v}_{C_2} \qquad \boxed{\text{20-31b}}$$

which express that the linear momentum is constant in the time interval $t_1$ to $t_2$ so that the center of mass moves with constant speed in a constant direction. These are readily generalized for a system of interconnected bodies where the total linear momentum is conserved if there is no resultant linear impulse. This implies that some parts of the system may have changes in linear momentum while there is no net change for the whole system.

## Conservation of Angular Momentum

The angular impulse is zero when $\sum \mathbf{M} = 0$, so Eqs. 20-22 and 20-24 become, respectively,

$$\Delta \mathbf{H}_C = 0 \qquad \boxed{\text{20-32a}}$$

$$I_C \omega_1 = I_C \omega_2 \qquad \boxed{\text{20-32b}}$$

which express that the angular momentum is constant in the time interval $t_1$ to $t_2$. If the moment of inertia about the center of mass of the body or system of bodies is constant during the time interval, the angular velocity must also be constant. If the moment of inertia changes during the time interval, the angular velocity will change in inverse proportion. For example, consider the rapid, reversible changes in the rate of rotation of a figure skater who is spinning about a vertical axis. The rotation is slow when the arms are extended horizontally and becomes rapid when the arms are pulled close to the body. As the skater's moment of inertia about the vertical axis decreases, the angular velocity increases.

If there are no external moments about an arbitrary reference point $O$ during a time interval $\Delta t$, Eq. 20-28 becomes $\Delta \mathbf{H}_O = 0$. If the reference point $O$ is a fixed point of rotation for a rigid body in plane motion. Eq. 20-29 becomes $I_O \omega_1 = I_O \omega_2$.

There are two noteworthy aspects of conservation of angular momentum. First, $\Delta \mathbf{H}_C = 0$ does not imply that $\Delta \mathbf{H}_O = 0$, or vice versa. The reason is that the angular impulse is defined with respect to a specific point as in Eqs. 20-24 and 20-28. Second, conservation of angular momentum does not require a simultaneous conservation of linear momentum. This means that there may be a linear impulse while the angular impulse is zero.

The net angular momentum of a system of interconnected bodies is conserved when there is no resultant angular impulse acting on the system. Again, some parts of the system may have changes in angular momentum while the system as a whole experiences no change. The same *fixed reference point* $O$ must be used for all parts of the system.

EXAMPLE 20-6

A new pole for street lights is designed to shear off by a relatively small force when an automobile crashes into it (Fig. 20-13a). For modeling such an event, assume that the car is a uniform rectangular plate of 1200 kg mass. The car is traveling at $\mathbf{v}_1 = 100\mathbf{j}$ km/h prior to the impact. Assume that the force applied to the car by the pole is $\mathbf{F} = -5\mathbf{j}$ kN acting for a time of 0.1 s. Determine the motion of the car after the impact.

## SOLUTION

The impulse and momentum equations for a rigid body can be applied directly here. From Eq. 20-30 and Fig. 20-13b,

$$\Sigma\begin{pmatrix} \text{initial linear} \\ \text{momentum} \end{pmatrix}_x + \Sigma\begin{pmatrix} \text{linear} \\ \text{impulse} \end{pmatrix}_x = \Sigma\begin{pmatrix} \text{final linear} \\ \text{momentum} \end{pmatrix}_x$$

$$0 \qquad + \qquad 0 \qquad = \qquad 0$$

$$\Sigma(\text{initial linear momentum})_y + \Sigma(\text{linear impulse})_y$$
$$= \Sigma(\text{final linear momentum})_y$$

$$mv_{1y_1} + \Sigma F_y\,\Delta t = mv_{y_2}$$

$$(1200\,\text{kg})\left(100\,\frac{\text{km}}{\text{h}}\right)\left(\frac{1000\,\text{m/km}}{3600\,\text{s/h}}\right)$$

$$+ (-5000\,\text{N})(0.1\,\text{s}) = (1200\,\text{kg})v_{y_2}$$

$$27.78\,\text{m/s} - \frac{500\,\text{N}\cdot\text{s}}{1200\,\text{kg}}\left(\frac{\text{kg}\cdot\text{m/s}^2}{\text{N}}\right) = v_2$$

$$v_2 = 27.78\,\text{m/s} - 0.42\,\text{m/s} = 27.36\,\text{m/s}$$
$$= 98.5\,\text{km/h}$$

$$\Sigma(\text{initial angular momentum})_C + \Sigma(\text{angular impulse})_C$$
$$= \Sigma(\text{final angular momentum})_C$$

$$I_C\omega_1 + \Sigma M_C\,\Delta t = I_C\omega_2$$

$$0 + (5000\,\text{N})(1\,\text{m})(0.1\,\text{s}) = \frac{1}{12}(1200\,\text{kg})[(2\,\text{m})^2 + (4\,\text{m})^2]\omega_2$$

$$\omega_2 = 0.25\,\text{rad/s} \qquad (\text{counterclockwise})$$

(a)

(b) Top view of car during impact

**FIGURE 20-13**

Motion of a car after an impact

# EXAMPLE 20-7

The 50-lb wheel of an aircraft has a 1.5-ft radius and a 1-ft radius of gyration. Just prior to touchdown the wheel has a horizontal speed $v = 200\,\text{ft/s}$ and does not rotate (Fig. 20-14a). Assume that during touchdown a constant load $W = 20{,}000\,\text{lb}$ is acting on the wheel, $v$ is constant, and there is no deformation or bouncing. Determine the distance traveled by the aircraft from the initial touchdown until the wheel can roll without sliding.

## SOLUTION

Equations 20-30 are illustrated for this body in Fig. 20-14b during the interval of interest. Some qualitative observations must be made first, however. When considering motion in the $x$ direction it must be remembered that the translational velocity of the wheel is the same as that of the aircraft. The horizontal velocity of the airplane will remain essentially constant until the brakes are applied (they are assumed "off" here). Therefore, when considering the linear momentum of the wheel it must be remembered that momentum of the whole aircraft is constant. On the other hand, the rotational velocity of the wheel may be considered separately from the rest of the aircraft since the wheel and aircraft rotate independently. Any change in motion in the $y$ direction is ignored here; hence,

$$\sum (mv_y)_1 + N\,\Delta t - W\,\Delta t = \sum (mv_y)_2$$

$$N = W = 20{,}000\,\text{lb}$$

The problem then is to determine the time $\Delta t$ after which the wheel rolls without slipping, when $v_C = r\omega = 200\,\text{ft/s}$. During slipping, the friction force is of magnitude $\mathscr{F} = \mathscr{F}_{\text{max}} = \mu_k N$. From Eq. 20-30c,

$$\sum (\text{initial angular momenta})_C + \sum (\text{angular impulse})_C$$
$$= \sum (\text{final angular momenta})_C$$

$$I_C(0) + \mathscr{F}r(\Delta t) = I_C\omega_2 = mr_g^2\left(\frac{v_C}{r}\right)$$

$$(0.4)(20{,}000\,\text{lb})(1.5\,\text{ft})(\Delta t) = \frac{50\,\text{lb}}{32.2\,\text{ft/s}^2}(1\,\text{ft})^2\left(\frac{200\,\text{ft/s}}{1.5\,\text{ft}}\right)$$

$$\Delta t = \frac{207\,\text{lb} \cdot \text{ft} \cdot \text{s}}{12{,}000\,\text{lb} \cdot \text{ft}} = 0.017\,\text{s}$$

The distance $x$ traveled by the aircraft until the wheel rolls without slipping is

$$x = v\,\Delta t = (200\,\text{ft/s})(0.017\,\text{s})$$
$$= 3.45\,\text{ft}$$

(a)

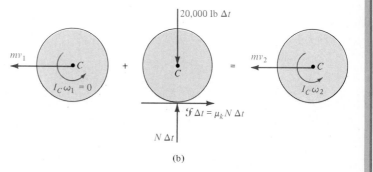

(b)

**FIGURE 20-14**

Motion of an airplane's wheel during landing

EXAMPLE 20-7   831

EXAMPLE 20-8

A system of two disks (Fig. 20-15a) is rotating without slipping. Disk $A$ has a 2-kg mass, a 0.2-m radius, a 0.16-m radius of gyration, and is rotating clockwise at the rate of 2000 rpm. The uniform disk $B$ has a 10-kg mass and a 0.4-m radius. Determine the required couple $\mathbf{M}_A$ on disk $A$ to stop the rotation of the disks in a time of 20 s.

## SOLUTION

The only unknown, at first glance, is the magnitude of the applied couple $\mathbf{M}_A$. Therefore, one might be tempted to write the angular impulse-momentum equation for the system to simply solve for $M_A$ (Fig. 20-15b). However, it must be remembered that stationary reaction impulsive forces, even though they do no work, must be included for impulse-momentum solution. Also, it must be recalled that one of the criteria for applying the angular impulse and momentum method was that moments must be taken *about the same point for the entire system*. Choosing point $A$ as the reference point, it is seen that the vertical reaction component at $B$ does cause an impulsive moment about $A$ and therefore must be included. Therefore, each disk must be analyzed separately.

Looking first at disk $A$ (Fig. 20-15c), the angular impulse-momentum equation is written,

$$I_A\omega_A + \mathscr{F}r_A\,(\Delta t) - M_A\,\Delta t = I_A\,(0) \tag{a}$$

The temptation to use $\mathscr{F} = \mathscr{F}_{\text{max}} = \mu N$ must be overcome here because slipping is not occurring and is not necessarily impending. Hence, both $\mathscr{F}$ and $M_A$ are unknown and cannot be determined yet.

Looking at disk $B$ (Fig. 20-15d), the angular impulse-momentum equation is written,

$$I_B\omega_B + \mathscr{F}r_B\,(\Delta t) = I_B\,(0) \tag{b}$$

The initial angular speed $\omega_B$ can be determined from

$$r_A\omega_A = r_B\omega_B$$

$$\omega_B = (r_A/r_B)\omega_A = \left(\frac{0.2\,\text{m}}{0.4\,\text{m}}\right)\left(2000\,\frac{\text{rev}}{\text{min}}\right)\left(\frac{2\pi\,\text{rad/rev}}{60\,\text{s/min}}\right)$$

$$= 105\,\text{rad/s}$$

Solving Eq. (b) for $\mathscr{F}$ (note the sign convention used) yields

$$\frac{1}{2}(10\,\text{kg})(0.4\,\text{m})^2(-105\,\text{rad/s}) + \mathscr{F}\,(0.4\,\text{m})(20\,\text{s}) = 0$$

$$\mathscr{F} = 10.5\,\text{N}$$

Substituting this into Eq. (a) and using $I_A = m_A r_g^2$ gives

$$(2\,\text{kg})(0.16\,\text{m})^2(210\,\text{rad/s}) + (10.5\,\text{N})(0.2\,\text{m})(20\,\text{s}) - M_A\,(20\,\text{s}) = 0$$

$$M_A = 2.64\,\text{N·m} \qquad \text{(counterclockwise)}$$

(a)

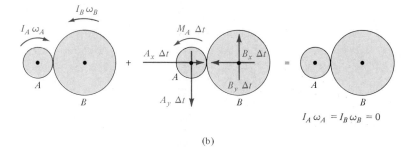

$$I_A \omega_A = I_B \omega_B = 0$$

(b)

(c)

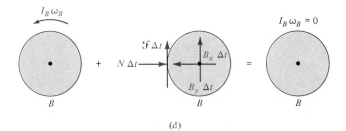

(d)

**FIGURE 20-15**

Analysis of a system of two wheels

EXAMPLE 20-8     833

EXAMPLE 20-9

It is proposed to control the spin of a satellite of mass $M$ and radius of gyration $r_g$ (about its spin axis $z$) by simultaneously extending four equal masses $m$. Assume that each mass $m$ is supported by a massless arm of length $l$ as shown in Fig. 20-16. The angle $\theta$ may be controlled from within the satellite. At $\theta = 0$ the satellite's spin velocity is $\omega_0$ about its spin axis $z$. Determine an equation for the satellite's spin velocity $\omega$ as a function of $\theta$.

### SOLUTION

There are no external moments about the spin axis $z$, so angular momentum is conserved about that axis.

$$[(H_C)_z]_1 = [(H_C)_z]_2$$

$$(I_C)_1\omega_0 = (I_C)_2\omega$$

where

$$(I_C)_1 = Mr_g^2 + 4[m(R + l\sin 0°)^2] = Mr_g^2 + 4mR^2$$

$$(I_C)_2 = Mr_g^2 + 4[m(R + l\sin\theta)^2]$$

Therefore,

$$\omega = \frac{(Mr_g^2 + 4mR^2)\omega_0}{Mr_g^2 + 4m(R + l\sin\theta)^2}$$

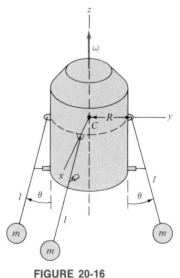

**FIGURE 20-16**

Analysis of a satellite

# EXAMPLE 20-10

A bullet weighing 0.125 lb is fired with a muzzle velocity $v_B$ into a wooden beam weighing 30 lb, which is initially at rest and serves as a ballistic pendulum (Fig. 20-17a). Its angular velocity after being struck by the bullet is a way of measuring the speed of the bullet before impact. Consider that the beam's angular velocity immediately after the bullet is fully embedded is electronically measured to be 1.1 rad/s. Determine (a) the speed $v_B$ of the bullet, and (b) the average horizontal impulsive reaction at hinge $O$ assuming that the bullet becomes fully embedded in 0.02 s.

## SOLUTION

(a) Consider the system to be comprised of the bullet and the beam so that the large impact force between them is an internal force. During the impact time of 0.02 s the moments about point $O$ of the weights of the beam and the bullet are zero (or at least very nearly so). Therefore, angular momentum is conserved about point $O$. Note, however, that linear momentum is not conserved because $O_x$ and $O_y$ depend on the impact force (Fig. 20-17b). Since $H_{O_1} = H_{O_2}$ for the entire system about the fixed point $O$,

$$\underbrace{\left(\frac{0.125\,\text{lb}}{32.2\,\text{ft/s}^2}\right)v_B}_{\substack{\text{linear momentum}\\\text{of bullet}}}\underbrace{\left[2\,\text{ft}\left(\frac{4}{5}\right)\right]}_{\substack{\text{moment arm}\\\text{of bullet}}} = I_O\omega_2 = I_O(1.1\,\text{rad/s})$$

$$I_O = \frac{1}{3}\frac{30\,\text{lb}}{32.2\,\text{ft/s}^2}(4\,\text{ft})^2 + \frac{0.125\,\text{lb}}{32.2\,\text{ft/s}^2}(2\,\text{ft})^2 = 4.98\,\text{lb}\cdot\text{ft}\cdot\text{s}^2$$

Solving for $v_B$ yields $v_B = 882\,\text{ft/s}$.

(b) Apply the impulse and momentum principle in the $x$ direction, using the motion of the entire system's mass center for computing $G_x$,

$$\int_{t_1}^{t_2} F_x\,dt = -\int_0^{0.02\,\text{s}} O_x\,dt = -(O_x)_{\text{ave}}(0.02\,\text{s})$$

Also,

$$\Delta G_x = G_{x_2} - G_{x_1} = \left(\frac{30\,\text{lb} + 0.125\,\text{lb}}{32.2\,\text{ft/s}^2}\right)2(1.1\,\text{rad/s}) - \frac{4}{5}\left(\frac{0.125\,\text{lb}}{32.2\,\text{ft/s}^2}\right)(882\,\text{ft/s})$$

$$-(0.02\,\text{s})(O_x)_{\text{ave}} = \Delta G_x = 2.06\,\text{lb}\cdot\text{s} - 2.74\,\text{lb}\cdot\text{s}$$

$$(O_x)_{\text{ave}} = 34.0\,\text{lb}\ (\leftarrow)$$

(a)

(b)

FIGURE 20-17
Imbedding of a bullet in a wooden beam

EXAMPLE 20-10    835

**20-40** A 50-lb flywheel has a radius of gyration of 15 in. The wheel starts to idle in a vacuum chamber at 15,000 rpm and comes to rest in 30 days. Determine the couple caused by friction in the bearings.

**20-41** A 30-kg flywheel is a uniform disk with a 0.25-m radius. Determine the couple that can increase the wheel's rotational speed from 1000 to 5000 rpm in a time of 100 s.

**20-42** An electric motor is shut off when its rotor is rotating at 3600 rpm, and it is observed to coast to rest in 55 s. The mass of the rotor is 30 kg and its radius of gyration is 60 mm. Determine the average magnitude of the couple due to bearing friction acting on the rotor while it is coasting to rest.

Initial rotor speed = 3600 rpm

Electric motor

**Fig. P20-42**

**20-43** A thin hoop of 1 kg mass and 0.4-m radius starts to roll without slipping on an incline of $\theta = 30°$. Determine its speed after a time of 10 s and the minimum coefficient of friction to prevent slipping.

$\theta$

**Fig. P20-43**

**20-44** Solve Prob. 20-43 for a uniform disk of 10 lb weight and 1 ft radius.

**20-45** A 40-lb uniform wheel is at rest when a horizontal force **F** is applied at its axis. Calculate the horizontal speed of the wheel after a time of 5 s assuming that it rolls without slipping. Also calculate the minimum coefficient of friction to prevent slipping.

$r = 1$ ft

$F = 50$ lb

**Fig. P20-45**

**20-46** A uniform 4-kg bar is released from rest at $\theta = 60°$. Determine its angular speed after a time of 0.05 s if the hinge at $O$ is frictionless. Assume that $\theta$ is constant during $\Delta t$.

$y$

$A$

$l = 1$ m

$O$

$\theta$

$x$

**Fig. P20-46**

**20-47** Determine the angular velocity of the bar in Prob. 20-46 assuming that it has an initial angular velocity $\omega_z = 0.5\,\mathbf{k}$ rad/s at $\theta = 60°$.

**20-48** Solve Prob. 20-46 assuming that a spring applies a constant downward force of 100-N magnitude at point $A$ on the bar during the initial 0.05 s of the motion.

**20-49** A uniform plate of 40-kg is released from rest with edge $OA$ horizontal. Determine the angular speed of the plate 0.05 s after its release.

Fig. P20-49

**20-54** Solve Prob. 20–53 for a uniform sphere of 4-kg mass and 0.12-m radius. Also determine the horizontal speed of the sphere when it starts rolling without slipping.

**20-50** Solve Prob. 20–49 assuming that the plate has an initial angular velocity $\omega_z = -0.3\,\mathbf{k}\,\text{rad/s}$ when $OA$ is horizontal.

**20-51** Two uniform, 10-lb rods are hinged at $A$ and $B$, and part $C$ slides on a smooth surface. The rods are released from rest when $\theta = 60°$. Calculate the vertical velocity of $B$ after a time of $0.02\,\text{s}$.

**20-55** The amusement ride of 12 cages is modeled as 12 concentrated masses of 700 kg each, on the ends of uniform, straight bars of 8 m length and 250 kg mass. The system is started from rest and moves in a plane. Calculate the required couple to attain an angular speed of 4 rpm in a time of 5 s.

Fig. P20-55

Fig. P20-51

**20-56** The test platform weighs 200 lb, and the console weighs 250 lb. There are four uniform rocker arms such as $BE$; each weighs 10 lb and makes an angle $\phi$ with the horizontal. The platform is released from rest at $\phi = 85°$. Calculate its speed 0.1 s after it is released from rest. Assume that the actuator $A$ has no mass and applies no force to the platform during this motion. $BC = 5\,\text{ft}$, $BE = CD = 2.5\,\text{ft}$.

**20-52** Solve Prob. 20–51 assuming that a spring applies a constant horizontal force of 50-lb magnitude acting to the right at point $C$ during the initial $0.02\,\text{s}$ of the motion.

**20-53** A thin hoop of 3 lb weight and 1 ft radius has an initial angular speed of $\omega_0 = 20\,\text{rad/s}$ when it is placed on a horizontal surface. The hoop does not bounce as it slides and rolls. Determine the time $t$ when the hoop starts rolling without slipping if $\mu_k = 0.3$.

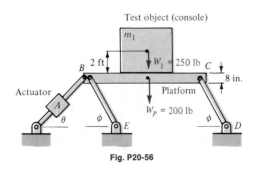

Fig. P20-56

**20–57** The flywheel $A$ of a tire tester has a 500-lb weight, a 6-ft diameter, and a 2.5-ft radius of gyration. It is freely rotating (no power input) at 100 rpm when a nonrotating wheel $C$ is pressed against it with a normal force of magnitude 500 lb acting for a time $t = 1.5$ s. Wheel $C$ is modeled as a uniform disk of 30 lb weight and 2.4 ft diameter. Determine the angular velocities of the two wheels after contact for the time of 1.5 s if $\mu_s = 0.7$ and $\mu_k = 0.6$.

**Fig. P20-57**
Tire-testing equipment (Courtesy MTS Systems Corporation, Minneapolis, Minnesota)

**20–58** An elevator system is modeled as $m_1 = 2000$ kg, $m_2 = 1800$ kg, and a uniform disk of $m_3 = 200$ kg. Calculate the velocity of $m_1$ at time $t = 10$ s after the system starts from rest.

**Fig. P20-58**

**20–59** Determine the required braking moment acting on $m_3$ in Prob. 20–58 to bring it to a stop in a time of 3 s from an angular speed of 3 rad/s.

**20–60** The double pulley has a total weight of 50 lb and a radius of gyration of 1.6 ft. The system starts from rest. Calculate the required moment at $O$ to give $W_1$ an upward velocity of 8 ft/s in a time of 2 s.

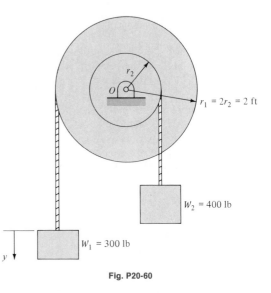

**Fig. P20-60**

**20–61** A front end loader has a total weight of 10,000 lb. Each of its four wheels has a 350-lb weight, a 5-ft diameter, and a 1.5-ft radius of gyration. Calculate the required torque applied to the two driving wheels in order to accelerate this machine from rest to a speed of 10 ft/s in a time of 5 s.

**Fig. P20-61**

**20–62**  The pulley has a 0.15-m radius of gyration. Calculate the tension $T$ to start the system from rest and attain an upward velocity of 8 m/s in a time of 5 s.

**Fig. P20-62**

**20–63**  A 20,000-kg railroad car is modeled as a uniform bar of $3\,m \times 3\,m$ cross section. It is firmly attached to the rigid rail bed $OA$, which is raised while pivoting about $O$ for rapid unloading. Determine the required moment of the force $\mathbf{F}$ to attain $\dot{\theta} = 0.3\,rad/s$ in a time of 2 s starting from rest at $\theta = 0$.

**Fig. P20-63**

**20–64**  A proposed conveyor system is modeled as a plate of mass $M$ moving without slipping on small cylindrical rollers of mass $m$ and radius $r$. Determine the speed of the plate at a time $t$ after it has moved along the incline starting from rest. Consider (a) free rollers, and (b) rollers freely rotating about their fixed axes. Assume the plate is on three rollers at all times as it moves.

(a) Free rollers

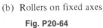

(b) Rollers on fixed axes

**Fig. P20-64**

**20–65**  The 4-lb uniform bar pivots about point $O$. The spring is 13 in. long when not deformed. The bar is released from rest in the vertical position. Calculate the speed of point $C$ at time $t = 0.01\,s$ after the bar is released.

**Fig. P20-65**

**20–66** Pulleys $A$ and $B$ of the wire rope tester are essentially uniform disks of 300 lb weight and 2 ft radius. The wire rope makes a complete loop around the pulleys, which are pushed apart by the hydraulic actuator $C$ applying a horizontal force of 100,000 lb magnitude. Determine the required moment at pulley $B$ to attain a clockwise angular speed of 10 rad/s in a time of 0.1 s starting from rest. Also calculate the tension in the upper and lower parts of the wire rope assuming no slipping. Pulley $A$ is free to rotate.

Fig. P20-66

**20–67** A 5-kg uniform bar is hinged at $A$. The bar is at rest when a 0.04-kg bullet strikes it horizontally and comes to rest in the bar. Calculate the initial angular speed of the bar with the bullet resting in it after the impact. Ignore direction $B$ which is used in Prob. 20–68.

Fig. P20-67

**20–68** Solve Prob. 20–67 with the bullet coming from the direction of $B$ at $\theta = 30°$. Also calculate the horizontal and vertical reactions at hinge $A$ caused by the bullet during its arrest, which occurs in a time of 0.0015 s.

**20–69** A 7-lb turntable has a 12-in. diameter and a 4-in. radius of gyration. The turntable is rotating at a speed of 33.33 rpm when a 0.08-lb, 12-in.-diameter, uniform, nonrotating disk is dropped on it. The disk stops slipping on the turntable in a time of 0.2 s. Determine the additional moment that the drive motor applies to the turntable during this 0.2-s interval to maintain the original speed $\omega_0$.

Fig. P20-69

**20–70** A twin-engine jet aircraft has a total weight of $10^5$ lb and a radius of gyration $r_z = 15$ ft about the $z$ axis. Assume that the aircraft is in level flight at a constant velocity $\mathbf{v}_0$ in the $y$ direction when engine $B$ fails. It takes a time of 3 s for an automatic correction to cause an effect. Determine the change in $v_0$ and the angular speed $\omega_z$ that the airplane acquires in that time of 3 s. Assume that air drag is negligible for the rotation.

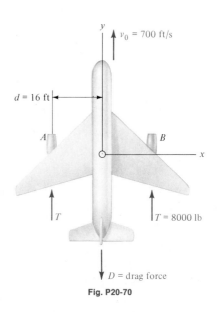

Fig. P20-70

**Work of a force on a rigid body**

$$U_{12} = \int_1^2 \mathbf{F} \cdot d\mathbf{r} = \int_1^2 \mathbf{F} \cdot \mathbf{v} \, dt$$

where  $d\mathbf{r}$ = infinitesimal displacement of force $\mathbf{F}$ along the path
from 1 to 2

$$\mathbf{v} = \frac{d\mathbf{r}}{dt}$$

**Work of a moment**

$$U_{12} = \int_{\theta_1}^{\theta_2} M \, d\theta$$

where $d\theta$ is the infinitesimal angular displacement.

**Kinetic energy of a rigid body**

$$T = T_{\text{transl}} + T_{\text{rot}} = \frac{1}{2} m v_C^2 + \frac{1}{2} I_C \omega^2$$

where $C$ denotes the center of mass.

**Rotation about an arbitrary fixed axis at point $O$**

$$T = \frac{1}{2} I_O \omega^2$$

**Principle of work and energy**

$$T_2 = T_1 + U_{12}$$

where   $T_1$ = initial kinetic energy of the body
$T_2$ = final kinetic energy of the body
$U_{12}$ = work of all external forces and moments acting on the
body from position 1 to 2

The equation is also valid for an internally conservative system of rigid
bodies.

Section 20-4

**Conservation of energy**
In a conservative system,

$$T_1 + V_1 = T_2 + V_2$$

where   $T$ = kinetic energy
     $V$ = total potential energy = $V_g + V_e$
     $V_g$ = gravitational potential energy
     $V_e$ = elastic potential energy

Section 20-5

**Power**

**TRANSLATION:**

$$\text{Power} = \frac{dU}{dt} = \mathbf{F} \cdot \mathbf{v} = \dot{T}$$

**ROTATION**

$$\text{Power} = \frac{dU}{dt} = M\omega \qquad \text{(for a constant moment } \mathbf{M} \text{ parallel to } \boldsymbol{\omega})$$

Section 20-6

**Impulse and momentum of a rigid body**

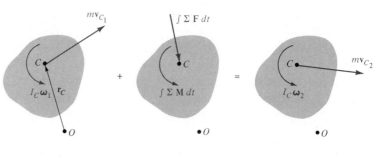

(a) Initial linear and      (b) Impulses of          (c) Final linear and
angular momenta        external forces          angular momenta
                          and moments

**LINEAR IMPULSE AND MOMENTUM:**

$$\int_{t_1}^{t_2} \sum \mathbf{F} \, dt = m(\mathbf{v}_{C_2} - \mathbf{v}_{C_1})$$

where $t$ is time, $C$ is the mass center, and $\sum \mathbf{F}$ includes *all* external forces.

**ANGULAR IMPULSE AND MOMENTUM:**

$$\int_{t_1}^{t_2} \sum \mathbf{M}_C \, dt = \mathbf{H}_{C_2} - \mathbf{H}_{C_1}$$

For plane motion, if $\sum \mathbf{M}$ is parallel to $\omega$,

$$\int_{t_1}^{t_2} \sum M_C \, dt = I_C(\omega_2 - \omega_1)$$

$$\int_{t_1}^{t_2} \sum M_O \, dt = I_O(\omega_2 - \omega_1) \qquad \text{for rotation about a fixed point } O$$

## Section 20-7

### Impulse and momentum of a system of rigid bodies

    (a) Initial momenta          (b) External impulses          (c) Final momenta

**PROCEDURES:**

1. Apply the principle of impulse and momentum to each rigid member separately. The mutual forces acting between members must be included in the formulation of the solution; *or*
2. Apply the principle of impulse and momentum to the entire system of bodies, ignoring the mutual forces between members.

## Section 20-8

**CONSERVATION OF LINEAR MOMENTUM** (time interval $t_1$ to $t_2$):

$$\sum \mathbf{F} = 0 \Rightarrow \Delta \mathbf{G} = 0$$

$$\mathbf{v}_{C_1} = \mathbf{v}_{C_2}$$

**CONSERVATION OF ANGULAR MOMENTUM:**

$$\sum \mathbf{M} = 0 \Rightarrow \Delta \mathbf{H}_C = 0$$

$$I_C \boldsymbol{\omega}_1 = I_C \boldsymbol{\omega}_2$$

For a system of rigid bodies, use the same *fixed reference point O* for all parts of the system:

$$\Delta \mathbf{H}_O = 0$$

**20–71** The pin-connected frame consists of two uniform members *AB* and *OD* of mass 10 kg each and member *BD* of mass 20 kg. A spring with an unstretched length of 4 m is fastened at pins *A* and *D*. A constant couple of 50 N·m magnitude is applied at pin *O* and the frame moves in a vertical plane. If the system is initially at rest in the position shown, determine the angular velocity of member *OD* after it has rotated clockwise through an angle of 90°.

**Fig. P20-71**

**20–72** The composite body *OABD* is free to rotate about point *O* starting from rest in the position shown. Cylinder *D* weighs 161 lb, the uniform rod *OA* weighs 32.2 lb, and the uniform rod *AB* weighs 64.4 lb. The spring's unstretched length is 4 ft. Determine the angular velocity of the system after it has rotated 90° clockwise from the position shown.

**Fig. P20-72**

**20–73** A designer of new roller skates and skateboards wishes to consider wheels of different geometry. The basic shapes of interest are the cylinder, hoop, and sphere. A useful thought experiment concerning the dynamic behavior of the wheels is to imagine that they have the same mass *m*, the same radius *r*, and individually start from rest on an incline of angle $\theta$ with the horizontal. Determine the linear speed of each wheel after rolling the same distance *l* without slipping.

**20–74** A 70-kg person starts on a level road ($\theta = 0$) on a 12-kg bicycle. Each wheel is essentially a thin hoop of 1.5-kg mass and 0.74-m diameter. What is the linear acceleration of the person if 0.4 hp is transmitted to the rear wheel which rolls without slipping? The linear speed at the given instant is 5 m/s.

**Fig. P20-74**

**20–75** Solve Prob. 20–74 with $\theta = 10°$ and the person heading uphill.

**20–76** Crate *A* of mass 250 kg is being dragged up an inclined plane by a rotating drum mechanism *B*. The drum of mass 50 kg may be modeled as a homogeneous cylinder of 2 m radius. The coefficient of friction between the crate and the inclined plane is $\mu = 0.3$. The drum has a torque applied to it given by $M = 200(6t^2 - 5)$ N·m, where *t* is in seconds. At time $t = 0$ the crate is sliding up the plane at a speed of 5 m/s. Determine the speed of the crate at $t = 4$ s.

**Fig. P20-76**

**20-77** A 2000-kg spacecraft with radius of gyration of 0.3 m is rotating about its longitudinal axis at a constant speed of 0.3 rad/s. A 1-kg rotor of 0.04-m radius of gyration starts from rest (with respect to the spacecraft) and attains a speed of $\omega_1 = 1257$ rad/s in a time of 40 s. The vectors $\omega_y$ and $\omega_1$ are parallel and in the same direction. Calculate the change in $\omega_y$ caused by $\omega_1$ of the motor.

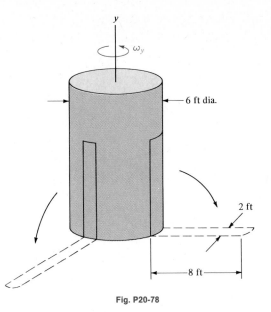

Fig. P20-78

**20-79** A truck skidding sideways and hitting a curb is modeled as a uniform square block of mass $m$ and sides $a$. Its mass center has a speed $v$ just before the impact which causes no rebound. Determine the angular speed of the block immediately after the impact. Assume that the curb applies a force at the bottom of the wheel.

Fig. P20-77

Fig. P20-79

**20-78** The total weight of a cylindrical spacecraft is 7,000 lb. Its radius of gyration is 2 ft when the four symmetrically arranged solar panels are parallel to the $y$ axis; $\omega_y = 0.8$ rad/s in this configuration. Determine the final angular speed after the 20-lb uniform panels are dropped to their positions normal to the $y$ axis.

# 21 Kinetics of Rigid Bodies in Three Dimensions

The forces and moments of forces acting on a rigid body may cause complex three-dimensional motion of that body. For a demonstration of this, experiment with toy tops or gyroscopes. Note that when an external force is applied to a rotating body, the resulting displacement may be in a different direction from that of the force. The reason for this phenomenon is that *the change in direction of the angular velocity vector involves an angular acceleration*. This concept was introduced in Sec. 18-7, dealing with the kinematics of rigid bodies, and is used here for analyzing the motion of bodies when the instantaneous axes of rotation change.

Many topics of spatial kinetics of rigid bodies are rather difficult for the average student and are often deferred to courses in advanced dynamics. However, the subject is important in practice, and ambitious students will find this chapter useful in further work in dynamics. The individual topics are described in perspective as follows.

**SECTION 21-1** presents qualitative considerations and basic mathematical methods for the analysis of angular momentum in three dimensions. These methods are essential for studying the other topics in this chapter.

**SECTION 21-2** extends the methods of impulse and momentum of systems of particles (Sec. 17-10) to rigid bodies. This application is straightforward after a thorough study of Sec. 21-1.

**SECTION 21-3** presents the derivation of the expression for the kinetic energy of a rigid body on the basis of Sec. 17-4. This useful expression is also readily applied after studying Sec. 21-1.

**SECTIONS 21-4 AND 21-5** present the equations of motion of a rigid body in three dimensions. The derivations are based on the results obtained in Secs. 17-7, 17-8, 18-8, and 21-1. These topics are intended only for the most ambitious students.

**SECTION 21-6** describes the scalar governing equations for general three-dimensional motion of a rigid body. Relatively simple equations for special cases are also discussed. These are useful in machine design, but they have particular limitations.

**SECTIONS 21-7 AND 21-8** apply the equations of three-dimensional motion to analyze gyroscopic motion. These important but rather difficult topics are intended only for the most ambitious and advanced students who wish to take a course in advanced dynamics.

## ANGULAR MOMENTUM IN THREE DIMENSIONS

### Qualitative Considerations

A little experimentation with a rapidly spinning top reveals the following remarkable facts. If the axis of the top is vertical, any arbitrary vertical displacement of the bearing support ($O$ in Fig. 21-1) has no effect on the motion of the top. Thus, the angular momentum vector $\mathbf{H}$ appears to be constant. A more curious phenomenon is observed when the spinning top is placed on its bearing so that the axis of rotation is not vertical, as in Fig. 21-2a. In this case the shaft of the top appears to have a motion that would sketch out a cone which is called the *space cone*. The rotation of the angular velocity vector $\boldsymbol{\omega}$ about the $y$ axis is called *precession*. The angular velocity of the precession of $\boldsymbol{\omega}$ (and of the vector $\mathbf{H}$) is denoted by the vector $\boldsymbol{\Omega}$ in Fig. 21-2a. Since the direction of the angular momentum vector $\mathbf{H}$ changes in the case of precession, $\mathbf{H}$ is not constant even if the magnitude $\omega$ for rotation about the top's own axis is a constant.

**FIGURE 21-1**

Top rotating about its axis of symmetry

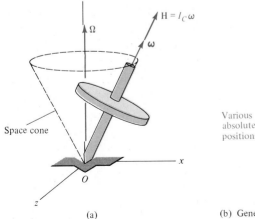

(a)

**FIGURE 21-2**

Motion of an inclined, spinning top

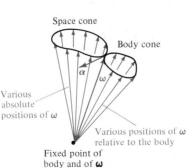

(b) General relation of body and space cones

The angular velocity vector $\omega$ also has a motion relative to the body. This motion would sketch out a cone which is called the *body cone*. For a rigid body with a fixed point, the instantaneous axis of rotation always passes through the fixed point of the body. This is shown schematically in Fig. 21-2b, where arbitrary space and body cones are also shown. The locus of the absolute positions of the instantaneous axis of rotation is the *space cone*. The locus of the positions of the instantaneous axis relative to the body is the body cone. These cones may be circular or irregular cones, but they are tangent along $\omega$ for the instant considered. The body cone appears to roll on the space cone analogously to the body and space centrodes introduced in Sec. 18-5.

It is useful to remember that $\alpha = d\omega/dt$. This means that, for motion about a fixed point, the angular acceleration vector $\alpha$ is tangent to the path of the tip of the angular velocity vector $\omega$ in the space cone as in Fig. 21-2b. $\alpha$ is normal to $\omega$ if the magnitude $\omega$ is constant.

## Analysis of Angular Momentum

For the analysis of rotational motion in three dimensions, consider a rigid body with a reference frame $xyz$ of fixed orientation attached at its mass center $C$ (Fig. 21-3). The motion of the body is observed from point $C$, which may be moving with respect to the fixed frame $XYZ$ (for details, see Sec. 17-8). The angular momentum of particle $P_i$ of mass $m_i$, position vector $\mathbf{r}'_i$, and velocity $\mathbf{v}'_i$ is $\mathbf{H}_i = \mathbf{r}'_i \times m_i\mathbf{v}'_i$ (all with respect to the $xyz$ axes which can only translate). Since the $xyz$ axes are attached at the mass center $C$, the only possible motion of a point $P_i$ relative to $C$ is due to rotation. Hence, $\mathbf{v}_i = \omega \times \mathbf{r}'_i$ and $\mathbf{H}_i$ can be rewritten as $\mathbf{H}_i = \mathbf{r}'_i \times m_i(\omega \times \mathbf{r}'_i)$. The total angular momentum of the body about $C$ is

$$\mathbf{H}_C = \sum_{i=1}^{n} \left[\mathbf{r}'_i \times (\omega \times \mathbf{r}'_i)m_i\right] \qquad \boxed{21\text{-}1}$$

Equation 21-1 can be written in determinant form,

$$\mathbf{H}_C = \sum_{i=1}^{n} \begin{vmatrix} \mathbf{i} & \mathbf{j} & \mathbf{k} \\ r'_{i_x} & r'_{i_y} & r'_{i_z} \\ (\omega \times \mathbf{r}'_i)_x & (\omega \times \mathbf{r}'_i)_y & (\omega \times \mathbf{r}'_i)_z \end{vmatrix} m_i \qquad \boxed{21\text{-}2}$$

which can be expanded using $\mathbf{r}'_i = x_i\mathbf{i} + y_i\mathbf{j} + z_i\mathbf{k}$ and $\omega = \omega_x\mathbf{i} + \omega_y\mathbf{j} + \omega_z\mathbf{k}$. First, $\omega \times \mathbf{r}'_i$ is expressed as $\omega \times \mathbf{r}'_i = (\omega_y z_i - \omega_z y_i)\mathbf{i} + (\omega_z x_i - \omega_x z_i)\mathbf{j} + (\omega_x y_i - \omega_y x_i)\mathbf{k}$. Then the $x$ scalar component of $H_C$ is written in detail,

$$\begin{aligned} H_x &= \sum_{i=1}^{n} \left[y_i(\omega_x y_i - \omega_y x_i) - z_i(\omega_z x_i - \omega_x z_i)\right]m_i \\ &= \omega_x \sum_{i=1}^{n} (y_i^2 + z_i^2)m_i - \omega_y \sum_{i=1}^{n} x_i y_i m_i - \omega_z \sum_{i=1}^{n} x_i z_i m_i \qquad \boxed{21\text{-}3} \end{aligned}$$

Similar expressions can be obtained for $H_y$ and $H_z$. Integrating over all infinitesimal particles of the body, the summation terms of Eq. 21-3 become

$$\int (y^2 + z^2)\, dm = I_x \qquad \int xy\, dm = I_{xy} \qquad \int xz\, dm = I_{xz}$$

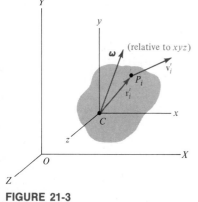

**FIGURE 21-3**

Schematic for the analysis of angular momentum

where $I_x$ is the *centroidal mass moment of inertia* of the body about the $x$ axis, and $I_{xy}$ and $I_{xz}$ are *centroidal mass products of inertia* with respect to the given axes.

The same procedure can be followed in completely evaluating Eq. 21-2. The resulting concise expressions for the three scalar components of the total angular momentum are

$$\boxed{\begin{aligned} H_x &= I_x\omega_x - I_{xy}\omega_y - I_{xz}\omega_z \\ H_y &= -I_{yx}\omega_x + I_y\omega_y - I_{yz}\omega_z \\ H_z &= -I_{zx}\omega_x - I_{zy}\omega_y + I_z\omega_z \end{aligned}} \qquad \boxed{\text{21-4}}$$

where all double subscripts denote products of inertia. The arrangement of the terms in Eq. 21-4 facilitates checking them for correctness by permutation of the subscripts. Note that Eq. 21-4 is valid for a particular position of the body; as the body rotates relative to $xyz$, the moments of inertia and products of inertia may in general change.

The mass moments of inertia and products of inertia in Eq. 21-4 are sometimes arrayed by themselves as

$$\begin{pmatrix} I_x & -I_{xy} & -I_{xz} \\ -I_{yx} & I_y & -I_{yz} \\ -I_{zx} & -I_{zy} & I_z \end{pmatrix} \qquad \boxed{\text{21-5}}$$

which is called the *inertia matrix* or *inertia tensor* $\mathbf{I}$. By definition, $I_{xy} = I_{yx}$, and so on. This matrix depends on the $xyz$ axes used in the analysis. However, although the elements of the inertia matrix change with choice of axes, the results of the matrix equation $\mathbf{H}_C = \mathbf{I}\boldsymbol{\omega}$ do not. Consequently, the centroidal angular momentum $\mathbf{H}_C$ for a given angular velocity $\boldsymbol{\omega}$ of a rigid body, is independent of the choice of axes. This fact makes it reasonable to use axes that simplify the inertia matrix. The ideal set of centroidal axes is the *principal axes of inertia* $x'y'z'$ for which all products of inertia of the given body are zero (see Ch. 9 in the statics volume). In that case the array in Eq. 21-5 becomes (for that instant only, in most cases)

$$\begin{pmatrix} I_{x'} & 0 & 0 \\ 0 & I_{y'} & 0 \\ 0 & 0 & I_{z'} \end{pmatrix} \qquad \boxed{\text{21-6}}$$

where $I_{x'}$, $I_{y'}$, and $I_{z'}$ are the *principal centroidal moments of inertia*. Two of these three values are the maximum and minimum mass moments of inertia of the given body. With these simplifications Eq. 21-4 becomes

$$H_{x'} = I_{x'}\omega_{x'} \qquad H_{y'} = I_{y'}\omega_{y'} \qquad H_{z'} = I_{z'}\omega_{z'} \qquad \boxed{\text{21-7}}$$

Note that in general the vectors $\boldsymbol{\omega}$ and $\mathbf{H}_C$ do not have the same directions. There are two special cases when these vectors are in the same direction. One is when $\omega_{x'}$, $\omega_{y'}$, and $\omega_{z'}$ are in the same ratios as $I_{x'}$, $I_{y'}$, and $I_{z'}$. The other special case is when $\boldsymbol{\omega}$ is collinear with a principal axis of inertia, which means that two of the angular velocity components in Eq. 21-7 are zero, as in Chs. 19 and 20.

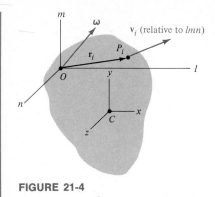

**FIGURE 21-4**

Motion of a rigid body about a fixed point

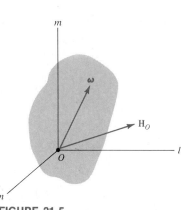

**FIGURE 21-5**

Generally, the vectors $\omega$ and $\mathbf{H}_O$ are in different directions

## Angular Momentum about an Arbitrary Fixed Point

The concepts of analysis described above are also applicable when the rigid body is rotating about an arbitrary fixed point $O$ as in Fig. 21-4. Here the frame $lmn$ is attached at the fixed point which is the most convenient for reference. The total angular momentum of the body about point $O$ is obtained similarly to Eq. 21-1,

$$\mathbf{H}_O = \sum_{i=1}^{n} \left[ \mathbf{r}_i \times (\boldsymbol{\omega} \times \mathbf{r}_i) m_i \right] \qquad \boxed{21\text{-}8}$$

where $r_i$ is the position vector of particle $P_i$ in the fixed $lmn$ frame. The $lmn$ components of the total angular momentum $\mathbf{H}_O$ are obtained using the same procedure as for the centroidal reference frame $xyz$,

$$\boxed{\begin{aligned} H_i &= I_l \omega_l - I_{lm} \omega_m - I_{ln} \omega_n \\ H_m &= -I_{ml} \omega_l + I_m \omega_m - I_{mn} \omega_n \\ H_n &= -I_{nl} \omega_l - I_{nm} \omega_m + I_n \omega_n \end{aligned}} \qquad \boxed{21\text{-}9}$$

The mass moments of inertia $I_l$, $I_m$, $I_n$, and the mass products of inertia $I_{lm}$, $I_{mn}$, $I_{ln}$ are evaluated with respect to the $lmn$ axes according to the definition of these quantities. Note that Eq. 21-9 is identical to Eq. 21-4 when the $lmn$ axes are attached at the center of mass and the body rotates about that point. In the general case of rotation about an arbitrary fixed point $O$, the vectors $\boldsymbol{\omega}$ and $\mathbf{H}_O$ are again possibly in different directions, as illustrated in Fig. 21-5.

## 21-2    IMPULSE AND MOMENTUM OF A RIGID BODY

The results obtained in Secs. 17-7 and 17-8 concerning the momenta of a system of particles are valid for a rigid body. The linear momentum vector of the body is given by

$$\mathbf{G} = m\mathbf{v}_C \qquad \boxed{21\text{-}10}$$

where $m$ is the total mass of the body and $\mathbf{v}_C$ is the velocity of the center of mass. The components of the angular momentum vector $H_C$ are determined from Eq. 21-4 or 21-7. In a general case, the total momentum of the body consists of linear and angular momenta, stated as

$$\text{system momenta} = \begin{cases} \text{linear momentum of mass center } (\mathbf{G}) \\ \text{angular momentum about mass center } (\mathbf{H}_C) \end{cases} \qquad \boxed{21\text{-}11}$$

One must remember that $G$ and $H$ have different units. The vectors $\mathbf{G}$ and $\mathbf{H}_C$ are analogous to a force and a couple, respectively, and this analogy is advantageous in solving problems where the principle of impulse and momentum can be used. This principle is based on the concept that changes in momentum are easily determined when they are caused by external impulses. In all such cases the final total momentum of a system equals the initial total momentum plus the impulses during the interval considered. From time 1 to time 2,

$$\boxed{\begin{array}{l} \mathbf{G}_2 = \mathbf{G}_1 + \text{(external linear impulses)}\big|_1^2 \\ \mathbf{H}_{C_2} = \mathbf{H}_{C_1} + \text{(external angular impulses)}\big|_1^2 \end{array}}$$

21-12a

21-12b

**FIGURE 21-6**

Example for the analysis of impulse and momentum

For the simple case in Fig. 21-6 where the body is initially at rest and $\mathbf{F}$ acts in the $y$ direction at point $A$ of coordinates $(-a, -b, c)$, the scalar components of Eq. 21-12 can be written as follows:

**LINEAR IMPULSE AND MOMENTA**

$$mv_{2_x} = mv_{1_x} + F_x \Delta t \quad mv_{2_x} = m(0) + (0)\Delta t = 0$$

$$mv_{2_y} = mv_{1_y} + F_y \Delta t \quad mv_{2_y} = m(0) + F \Delta t = F \Delta t$$

$$mv_{2_z} = mv_{1_z} + F_z \Delta t \quad mv_{2_z} = m(0) + (0)\Delta t = 0$$

**ANGULAR IMPULSE AND MOMENTA:**

$$H_{x_2} = H_{x_1} - F(c)\Delta t$$

$$\begin{aligned} H_{x_2} &= 0 - F(c)\Delta t \\ &= -F(c)\Delta t \end{aligned}$$  (the minus sign represents a moment of $\mathbf{F}$ in the $-\mathbf{i}$ direction since $c$ is positive; check using the right-hand rule)

$$H_{y_2} = H_{y_1} + F(0)\Delta t = 0$$  (since $\mathbf{F}$ has no moment about the $y$ axis, and $H_{y_1} = 0$)

$$H_{z_2} = H_{z_1} - F(a)\Delta t = -F(a)\Delta t$$

The angular velocity components can be determined from the angular momentum components using Eq. 21-4 and solving the equations above simultaneously. When a rigid body is rotating about a noncentroidal fixed point $O$, the impulse of the reaction must be included in the analysis if the center of mass is used for reference. It is generally more convenient to use the fixed point $O$ for reference which requires taking moments of the momenta and of the impulses about $O$. The reaction at $O$ has no moment about $O$ and thus it is ignored. In this case Eq. 21-9 should be used to relate the angular velocities and momenta.

## KINETIC ENERGY OF A RIGID BODY IN THREE-DIMENSIONAL MOTION

21-3

The kinetic energy of a rigid body in general motion is derived using the method developed in Sec. 17-4. The total kinetic energy $T$ of a system of particles is given according to Eq. 17-13,

$$T = \underbrace{\frac{1}{2}mv_C^2}_{\substack{\text{motion of total mass} \\ \text{imagined to be concen-} \\ \text{trated at mass center}}} + \underbrace{\frac{1}{2}\sum_{i=1}^{n} m_i v_i'^2}_{\substack{\text{motion of all particles} \\ \text{relative to mass center}}} \qquad 21\text{-}13$$

where $v_C$ is the speed of the mass center and $v_i'$ is the speed of particle $P_i$ in the $xyz$ frame, which is attached at the center of mass and has a fixed orientation (Fig. 21-7). The translational part $(\frac{1}{2}mv_C^2)$ of the kinetic energy is in the simplest possible form, but the second part involving a summation over all particles can be further simplified for rigid bodies. This latter part of the kinetic energy can result only from the particles' rotation about the center of mass in the case of a rigid body. Therefore, $\mathbf{v}_i'$ may be replaced by $\boldsymbol{\omega} \times \mathbf{r}_i$, and the last term in Eq. 21-13 can be written as

$$T_{\text{rot}} = \frac{1}{2} \sum_{i=1}^{n} m_i v_i'^2 = \frac{1}{2} \sum_{i=1}^{n} m_i (\boldsymbol{\omega} \times \mathbf{r}_i')^2 \qquad \boxed{21\text{-}14}$$

where $\boldsymbol{\omega} = \omega_x \mathbf{i} + \omega_y \mathbf{j} + \omega_z \mathbf{k}$ and $\mathbf{r}_i' = x_i \mathbf{i} + y_i \mathbf{j} + z_i \mathbf{k}$. Working with the rectangular components of the vector product and integrating over the whole body, Eq. 21-14 is restated as

$$T_{\text{rot}} = \frac{1}{2} \int [(\omega_y z - \omega_z y)^2 + (\omega_z x - \omega_x z)^2 + (\omega_x y - \omega_y x)^2]\, dm$$

After squaring and rearranging,

$$2T_{\text{rot}} = \omega_x^2 \int (y^2 + z^2)\, dm + \omega_y^2 \int (x^2 + z^2)\, dm$$
$$+ \omega_z^2 \int (x^2 + y^2)\, dm - 2\omega_x \omega_y \int xy\, dm$$
$$- 2\omega_y \omega_z \int yz\, dm - 2\omega_z \omega_x \int zx\, dm$$

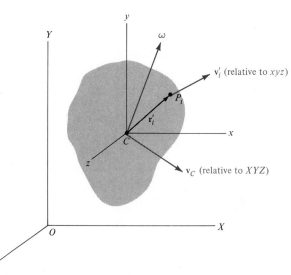

**FIGURE 21-7**
Schematic for the analysis of kinetic energy

The integrals are recognized as the centroidal mass moments of inertia and products of inertia used in deriving Eq. 21-4. Thus, the total kinetic energy of a rigid body can be expressed as

$$T = \frac{1}{2}mv^2 + \frac{1}{2}(I_x\omega_x^2 + I_y\omega_y^2 + I_z\omega_z^2) - I_{xy}\omega_x\omega_y \qquad \boxed{21\text{-}15}$$

$$- I_{yz}\omega_y\omega_z - I_{zx}\omega_z\omega_x$$

When the chosen $xyz$ frame coincides with the principal axes of inertia $x'y'z'$, Eq. 21-15 becomes

$$T = \frac{1}{2}mv_C^2 + \frac{1}{2}(I_{x'}\omega_{x'}^2 + I_{y'}\omega_{y'}^2 + I_{z'}\omega_{z'}^2) \qquad \boxed{21\text{-}16}$$

where $\omega$ is the same for any reference frame but its components $\omega_{x'}$, and so on, depend on the appropriate axes. An equivalent, concise form of Eqs. 21-15 and 21-16 is

$$\boxed{T = \frac{1}{2}mv_C^2 + \frac{1}{2}\omega \cdot \mathbf{H}_C} \qquad \boxed{21\text{-}17}$$

where $\mathbf{H}_C$ is determined from Eq. 21-4 or 21-7, as necessary.

Kinetic Energy of a Rigid Body
Rotating about a Fixed Point

A special case of three-dimensional motion is when the body has a fixed point. This means that all particles of the body are in rotational motion about that point. Accordingly, the total kinetic energy of the body may be determined by evaluating only the rotational kinetic energy using the fixed point $O$ for reference. Assume that a set of axes $lmn$ is attached at the fixed point $O$. In this case Eq. 21-15 becomes

$$T = \frac{1}{2}(I_l\omega_l^2 + I_m\omega_m^2 + I_n\omega_n^2) - I_{lm}\omega_l\omega_m \qquad \boxed{21\text{-}18}$$

$$- I_{mn}\omega_m\omega_n - I_{nl}\omega_n\omega_l$$

With principal axes $l'm'n'$ at the fixed point $O$, Eq. 21-18 is further simplified to give

$$T = \frac{1}{2}(I_{l'}\omega_{l'}^2 + I_{m'}\omega_{m'}^2 + I_{n'}\omega_{n'}^2) \qquad \boxed{21\text{-}19}$$

The most concise form of Eqs. 21-18 and 21-19 is

$$\boxed{T = \frac{1}{2}\omega \cdot \mathbf{H}_O} \qquad \boxed{21\text{-}20}$$

where care must be taken to evaluate $\mathbf{H}_O$ with respect to the fixed point of the rotating body.

# EXAMPLE 21-1

The 15-kg wheel $W$ in Fig. 21-8a has a 0.6-m diameter. Assume that the wheel is a homogeneous disk of very small thickness mounted on its axle at point $A$ of coordinate $0.5\mathbf{k}$ m (Fig. 21-8b). The axle is freely pivoted in a ball-and-socket joint at point $O$. The wheel has angular velocities $\boldsymbol{\omega}_z = 15\mathbf{k}$ rad/s and $\omega_x = 2\mathbf{i}$ rad/s. Determine **(a)** the linear and angular momenta of the wheel about point $O$, and **(b)** the kinetic energy of the wheel for the position shown.

## SOLUTION

**(a)** The linear momentum of the wheel is calculated from Eq. 21-10 involving only the velocity of the center of mass,

$$\mathbf{v}_A = \boldsymbol{\omega} \times \mathbf{r}_{OA} = (15\mathbf{k} + 2\mathbf{i})\ \text{rad/s} \times (0.5\mathbf{k}\ \text{m})$$
$$= -1\mathbf{j}\ \text{m/s}.$$

$$\mathbf{G} = m\mathbf{v}_A = (15\ \text{kg})(-1\mathbf{j}\ \text{m/s}) = -15\mathbf{j}\ \text{kg·m/s}$$

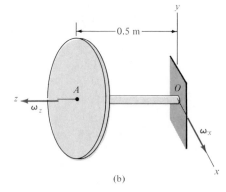

(b)

**FIGURE 21-8**

(a) Multiple-exposure photograph of a wheel in a complex test (courtesy MTS Systems Corporation, Minneapolis, Minnesota); (b) model of the wheel as a homogenous disk

854    KINETICS OF RIGID BODIES IN THREE DIMENSIONS

The angular momentum of the wheel at this instant is calculated from Eq. 21-7,

$$\mathbf{H}_O = I_x \omega_x + I_y \omega_y + I_z \omega_z$$

Using $\omega_y = 0$, $I_z = \frac{1}{2} mr^2$, and $I_x = (mr^2/4) + m(OA)^2$ gives

$$\mathbf{H}_O = 15\,\text{kg} \left[ \frac{(0.3\,\text{m})^2}{4} + (0.5\,\text{m})^2 \right] (2\,\text{rad/s})\,\mathbf{i} + 0 + \frac{1}{2}(15\,\text{kg})(0.3\,\text{m})^2(15\,\text{rad/s})\,\mathbf{k}$$

$$= (8.18\mathbf{i} + 10.12\mathbf{k})\,\text{kg·m}^2/\text{s}$$

**(b)** Using the principal body axes $(x', y', z')$ attached at $A$, which are parallel to the reference axes $(x, y, z)$, Eq. 21-16 can be used to determine the kinetic energy,

$$T = \frac{1}{2} m v_0^2 + \frac{1}{2}(I_{x'}\omega_{x'}^2 + I_{y'}\omega_{y'}^2 + I_{z'}\omega_{z'}^2)$$

$$= \frac{1}{2}(15\,\text{kg})(-1\,\text{m/s})^2$$

$$+ \frac{1}{2}\left[ \frac{1}{4}(15\,\text{kg})(0.3\,\text{m})^2(2\,\text{rad/s})^2 + 0 + \frac{1}{2}(15\,\text{kg})(0.3\,\text{m})^2(15\,\text{rad/s})^2 \right]$$

$$= 7.5\,\text{kg·m}^2/\text{s}^2 + \frac{1}{2}(1.35 + 151.9)\,\text{kg·m}^2/\text{s}^2$$

$$= 84.1\,\text{kg·m}^2/\text{s}^2 = 84.1\,\text{N·m}$$

Alternatively, the $xyz$ axes can be used as the reference frame, using Eq. 21-20,

$$T = \frac{1}{2}\boldsymbol{\omega} \cdot \mathbf{H}_O = \frac{1}{2}(2\mathbf{i} + 15\mathbf{k})\,\text{rad/s} \cdot (8.18\mathbf{i} + 10.12\mathbf{k})\,\text{kg·m}^2/\text{s}$$

$$= \frac{1}{2}(16.36 + 151.8)\,\text{kg·m}^2/\text{s}^2 = 84.1\,\text{N·m}$$

## EXAMPLE 21-2

In Ex. 21-1 a force $\mathbf{F} = (800\mathbf{i} + 500\mathbf{k})$ N is applied to the wheel $W$ at point $A$ for a time of 0.03 s. $\mathbf{F}$ is acting in the $xz$ plane. Determine the angular momentum of the wheel about point $O$ just after the action of the force $\mathbf{F}$.

### SOLUTION

Since the rotation is about a fixed point not at the center of mass, the new angular momentum about point $O$ can be calculated using Eq. 2-12. For this, the initial value of $\mathbf{H}_O$ calculated in Ex. 21-1 is used, and it is assumed that the change in orientation of the wheel is negligible during the short period considered ($I_{x'}$, $I_{y'}$, $I_{z'}$ remain constant).

$$\mathbf{H}_{O_2} = \mathbf{H}_{O_1} + \mathbf{M}_O\,\Delta t$$
$$= (8.18\mathbf{i} + 10.12\mathbf{k})\,\text{kg·m}^2/\text{s} + (0.5\mathbf{k}\,\text{m}) \times (800\mathbf{i}\,\text{N} + 500\mathbf{k}\,\text{N})(0.03\,\text{s})$$
$$= (8.18\mathbf{i} + 10.12\mathbf{k})\,\text{kg·m}^2/\text{s} + 12\mathbf{j}\,\text{N·m·s}$$
$$= (8.18\mathbf{i} + 12\mathbf{j} + 10.12\mathbf{k})\,\text{kg·m}^2/\text{s}$$

EXAMPLE 21-2     855

EXAMPLE 21-3

A crankshaft is modeled as a thin rod bent into the plane figure shown in Fig. 21-9a. It weighs 0.3 lb per linear inch and rotates freely in the bearings at $A$ and $B$ with a constant $\omega_x = 105$ rad/s. Determine the angular momenta of the crankshaft about points $A$ and $B$ when the crankshaft is in the $xy$ plane.

SOLUTION

First the relevant moments and products of inertia are determined using the parallel-axis theorem for a composite body. The crankshaft is analyzed in sections as shown in Fig. 21-9b. Using the subscript $C$ for centroidal quantities,

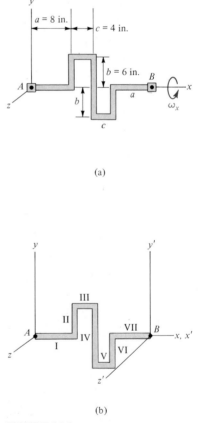

(a)

(b)

**FIGURE 21-9**

Model of a crankshaft

$$I_x = \sum (I_{C_x} + r_C^2 m)_i$$

$$= \left\{ (0)_I + \left[ \frac{1}{12}(6\,\text{in.})^2 + (3\,\text{in.})^2 \right](1.8\,\text{lb})_{II} + [0 + (6\,\text{in.})^2(1.2\,\text{lb})]_{III} \right.$$

$$+ \left[ \frac{1}{12}(3.6\,\text{lb})(12\,\text{in.})^2 + 0 \right]_{IV} + [0 + (-6\,\text{in.})^2(1.2\,\text{lb})]_V$$

$$\left. + \left[ \frac{1}{12}(6\,\text{in.})^2 + (-3\,\text{in.})^2 \right](1.8\,\text{lb})_{VI} + (0)_{VII} \right\} \frac{1}{32.2\,\text{ft/s}^2}$$

$$I_x = 5.36\,\text{slug}\cdot\text{in}^2$$

$$(I_{xy})_A = \sum (I_{Cxy} + xym)_i$$

$$= \{ (0)_I + [0 + (8\,\text{in.})(3\,\text{in.})(1.8\,\text{lb})]_{II} + [0 + (10\,\text{in.})(6\,\text{in.})(1.2\,\text{lb})]_{III}$$

$$+ [0 + (12\,\text{in.})(0\,\text{in.})(3.6\,\text{lb})]_{IV} + [0 + (14\,\text{in.})(-6\,\text{in.})(1.2\,\text{lb})]_V$$

$$+ [0 + (16\,\text{in.})(-3\,\text{in.})(1.8\,\text{lb})]_{VI} + (0)_{VII} \} \frac{1}{32.2\,\text{ft/s}^2}$$

$$(I_{xy})_A = -2.24\,\text{slug}\cdot\text{in}^2$$

Using Eq. 21-9 for the angular momentum about $A$ with $\omega_x = 105\,\text{rad/s}$, and $\omega_y = \omega_z = 0$ gives

$$H_x = I_x\omega_x - I_{xy}\omega_y - I_{xz}\omega_z$$

$$= (5.36\,\text{slug}\cdot\text{in}^2)(105\,\text{rad/s}) - 0 - 0$$

$$= 562\,\frac{\text{slug}\cdot\text{in}^2}{\text{s}}\left( \frac{1\,\text{ft}^2}{144\,\text{in}^2} \right)$$

$$= 3.90\,\text{lb}\cdot\text{ft}\cdot\text{s}$$

$$H_y = -I_{xy}\omega_x + I_y\omega_y - I_{yz}\omega_z$$

$$= -(-2.24\,\text{slug}\cdot\text{in}^2)(105\,\text{rad/s}) + 0 - 0$$

$$= 235\,\frac{\text{slug}\cdot\text{in}^2}{\text{s}} = 1.63\,\text{lb}\cdot\text{ft}\cdot\text{s}$$

$$H_z = 0 \qquad \text{(since } I_{xz} \text{ is zero by definition)}$$

Then

$$\mathbf{H}_A = (3.90\mathbf{i} + 1.63\mathbf{j})\text{lb}\cdot\text{ft}\cdot\text{s}$$

To calculate the angular momentum about $B$, the moments and products of inertia must be determined with respect to a reference system $x'y'z'$ with its origin at $B$. From symmetry, however,

$$I_{x'} = I_x = 5.36\,\text{slug}\cdot\text{in}^2$$

$$I_{x'y'} = I_{xy} = -2.24\,\text{slug}\cdot\text{in}^2$$

Thus,

$$H_{x'} = 3.90\,\text{lb}\cdot\text{ft}\cdot\text{s}$$

$$H_{y'} = 1.63\,\text{lb}\cdot\text{ft}\cdot\text{s}$$

$$\mathbf{H}_B = (3.90\mathbf{i} + 1.63\mathbf{j})\text{lb}\cdot\text{ft}\cdot\text{s}$$

EXAMPLE 21-3    857

**21-1** A slender rod $l$ of 2 lb weight and 2 ft length is firmly attached to a vertical shaft that rotates in bearings $A$ and $B$ at $\omega_y = 20$ rad/s. Determine the angular momentum of the rod about point $O$ for $\theta = 30°$ when the rod is in the $xy$ plane.

**Fig. P21-1**

**21-2** Determine the kinetic energy of the rod $l$ in Prob. 21-1 for $\theta = 0$ and $45°$.

**21-3** A force $\mathbf{F} = 30\mathbf{k}$ lb is acting for 0.01 s at the free end of the rod $l$ in Prob. 21-1 when the rod is at rest in the $xy$ plane. Determine the angular momentum of the rod about point $O$ immediately after the action of the force, assuming now that $O$ is a ball-and-socket joint.

**21-4** A slender rod of 1 kg mass and $2l = 1$ m length is firmly attached to a horizontal shaft that rotates in bearings $A$ and $B$ at $\omega_x = 10$ rad/s. Calculate the angular momentum of the rod about point $O$ for $\theta = 30°$ for the position shown. $\theta$ is measured in the $xy$ plane.

**Fig. P21-4**

**21-5** A force $\mathbf{F} = -2\mathbf{k}$ kN is acting for 0.002 s at the upper free end of the rod in Prob. 21-4 when the rod is at rest in the $xy$ plane. Calculate the final angular momentum of the rod about point $O$, assuming now that $O$ is a ball-and-socket joint.

**21-6** Assume that the crankshaft in Ex. 21-3 is changed by adding a 0.1-lb weight at point $P$. Determine the angular momenta of the system about points $A$ and $B$.

**21-7** A thin peg of mass $m$ and length $l$ is mounted at a distance $r$ from the center $O$ of a uniform disk. The peg is parallel to the shaft of the disk, which is rotating at angular speed $\omega_y$. Determine the angular momentum of the peg about point $O$ when it is in the $xy$ plane.

**Fig. P21-7**

**21-8** A uniform thin plate of mass $m$ is rotating about the fixed $x$ axis at angular speed $\omega_x$. Calculate the angular momentum of the plate about point $O$ and the kinetic energy of the plate for the position shown.

**Fig. P21-8**

**21-9** A uniform thin plate of mass $m$ is suspended from two inextensible wires at $A$ and $B$. An impulse $F\mathbf{k}\,\Delta t$ is applied at corner $D$. Determine the velocities of points $C$, $D$, and $E$ immediately after the impact.

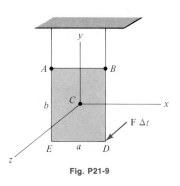

Fig. P21-9

**21-10** Calculate the tensions in wires $A$ and $B$ in Prob. 21-9 immediately after the impact. Note that all forces acting on the plate must be considered during the same time interval $\Delta t$.

**21-11** Calculate the kinetic energy of the plate in Prob. 21-9 after the impact.

**21-12** A thin uniform plate of mass $m$ is rotating about the $y$ axis at angular speed $\omega_y$. Determine the triangular plate's angular momentum about its mass center and about point $O$ when the plate is in the $xy$ plane.

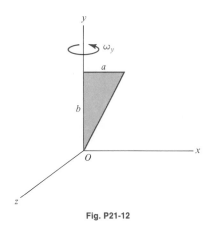

Fig. P21-12

**21-13** Calculate the kinetic energy of the plate in Prob. 21-12.

**21-14** A thin uniform plate of mass $m = m_1 + m_2$ is rotating about the $y$ axis at angular speed $\omega_y$. Determine the angular momentum of the plate about point $O$, for the position shown.

Fig. P21-14

**21-15** Calculate the kinetic energy of the bent plate in Prob. 21-14.

**21-16** The photograph shows a working model of a retractable-bladed Savonius rotor for a modern windmill. Assume that one of the two blades breaks off, leaving a half-cylindrical shell rotating about the $y$ axis at angular speed $\omega_y$ as sketched in Fig. 21-16b. The thin shell has a mass $m$, height $h$, and diameter $d$. Determine its angular momentum about point $O$ for the position shown.

Fig. P21-16

**21–17** Calculate the kinetic energy of the single blade in Prob. 21–16.

**21–18** A thin uniform disk of mass $m$ is rolling without slipping on the $xz$ plane at angular speed $\omega_y$ about the $y$ axis. Its shaft of length $l$ is always parallel to the $xz$ plane. Determine the angular momentum of the disk about point $A$ when the shaft is parallel to the $z$ axis.

**Fig. P21-20**

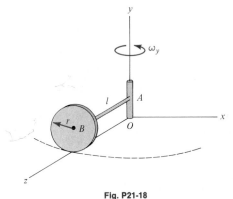

**Fig. P21-18**

**21–21** Calculate the final kinetic energy of the cylinder in Prob. 21–20.

**21–22** A rotor of a jet engine is modeled as a uniform cylinder of mass $m$, length $l$, and radius $r$. It has angular speeds $\omega_x$, $\omega_y$, and $\omega_z$ about the respective axes in the position shown. Determine the angular momentum about point $A$.

**21–19** Calculate the kinetic energy of the disk in Prob. 21–18.

**21–20** A 2000-kg spacecraft is modeled as a homogeneous cylinder of length $l = 6\,\text{m}$ and radius $r = 1\,\text{m}$. It is spinning about the $z$ axis at an angular speed of 100 rpm when a force $\mathbf{F} = 200\mathbf{i}$ N is applied to it for 0.1 s by a control jet at point $A$ in the $yz$ plane. Determine the angular momentum of the cylinder about point $O$ immediately after the impulse.

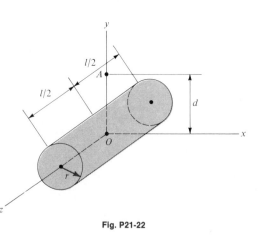

**Fig. P21-22**

The fundamental equations of motion given by Eqs. 17-20 and 17-29 are valid for any motion of a rigid body, so they are restated here:

$$\sum \mathbf{F} = \dot{\mathbf{G}} = m\mathbf{a}_C \qquad \boxed{21\text{-}21}$$

$$\sum \mathbf{M}_C = \dot{\mathbf{H}}_C \qquad \boxed{21\text{-}22}$$

where $\mathbf{a}_C$ is the acceleration of the center of mass and $\mathbf{H}_C$ is the angular momentum of the body about its center of mass. Equation 21-22 may also be used with any fixed point $O$ for a reference if both $\sum \mathbf{M}$ and $\mathbf{H}$ are evaluated about that point.

The greatest difficulty in using these apparently simple equations is in obtaining derivatives of the angular momentum $\mathbf{H}$. Recall that $\mathbf{H}$ was defined for a particular position of the body with respect to axes of fixed orientation. At a different time and position, the body's mass moments and products of inertia may be different with respect to the same fixed frame. Thus, differentiation of $\mathbf{H}$ is possible only if it is given as a function of time. This problem is solved using the following approach.

1. Attach axes $XYZ$ at the mass center $C$ of the body. These axes have fixed orientations in space regardless of the body's motion. The origin of $XYZ$ remains at $C$ (Fig. 21-10) during the motion.

2. Attach axes $xyz$ at the mass center of the body and in fixed orientations with respect to the body. It is often simplest to align the $xyz$ axes with the principal inertia axes of the body.

3. Establish $\boldsymbol{\omega}$, the angular velocity vector of the body and of the $xyz$ axes (since they are fixed in the body) with respect to the $XYZ$ frame. This $\boldsymbol{\omega}$ has components $\omega_x$, $\omega_y$, and $\omega_z$ along the $x$, $y$, and $z$ axes, respectively.

4. The angular momentum vector $\mathbf{H}_C$ is defined here with respect to the fixed $XYZ$ frame (similar to the derivation in Sec. 21-1, where $xyz$ is fixed) and is determined using Eq. 21-4. The vector $\mathbf{H}_C$ can be resolved into components along the rotating $xyz$ axes in Fig. 21-10 using $\mathbf{i}$, $\mathbf{j}$, and $\mathbf{k}$ unit vectors in the $xyz$ frame.

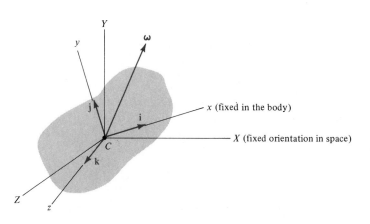

**FIGURE 21-10**

Schematic for deriving the equations of motion

**5.** The time derivative of the vector $\mathbf{H}_C$ in the rotating $xyz$ frame using the corresponding unit vectors is

$$(\dot{\mathbf{H}}_C)_{xyz} = \dot{H}_x\mathbf{i} + \dot{H}_y\mathbf{j} + \dot{H}_z\mathbf{k} \qquad \boxed{21\text{-}23}$$

**6.** The time derivatives of the angular momentum vector $\mathbf{H}_C$ in the fixed $XYZ$ and rotating $xyz$ frames are related to one another using Eq. 18-30,

$$(\dot{\mathbf{H}}_C)_{XYZ} = (\dot{\mathbf{H}}_C)_{xyz} + \boldsymbol{\omega} \times \mathbf{H}_C \qquad \boxed{21\text{-}24a}$$

In more detail, using

$$\dot{\mathbf{H}}_C = \frac{d}{dt}(H_x\mathbf{i} + H_y\mathbf{j} + H_z\mathbf{k})$$

$$= \dot{H}_x\mathbf{i} + \dot{H}_y\mathbf{j} + \dot{H}_z\mathbf{k} + H_x\dot{\mathbf{i}} + H_y\dot{\mathbf{j}} + H_z\dot{\mathbf{k}}$$

Equation 21-24a gives

$$\underbrace{(\dot{\mathbf{H}}_C)_{XYZ}}_{\substack{\text{rate of change} \\ \text{in magnitude} \\ \text{and direction} \\ \text{of } \mathbf{H}_C}} = \underbrace{(\dot{H}_x\mathbf{i} + \dot{H}_y\mathbf{j} + \dot{H}_z\mathbf{k})}_{\substack{\text{rate of change in} \\ \text{magnitude of } \mathbf{H}_C}} + \underbrace{\boldsymbol{\omega} \times (H_x\mathbf{i} + H_y\mathbf{j} + H_z\mathbf{k})}_{\substack{\text{rate of change in} \\ \text{direction of } \mathbf{H}_C}} \qquad \boxed{21\text{-}24b}$$

where $\boldsymbol{\omega} = \omega_x\mathbf{i} + \omega_y\mathbf{j} + \omega_z\mathbf{k}$ is the angular velocity of the body and of the $xyz$ frame fixed in the body. Note that $\mathbf{i}$, $\mathbf{j}$, and $\mathbf{k}$ are unit vectors along the instantaneous $xyz$ axes.

Equation 21-22 can be written using Eq. 21-24 as

$$\boxed{\sum \mathbf{M}_C = (\dot{\mathbf{H}}_C)_{xyz} + \boldsymbol{\omega} \times \mathbf{H}_C} \qquad \boxed{21\text{-}25}$$

In expanded form,

$$\sum \mathbf{M}_C = (\dot{H}_x + \omega_y H_z - \omega_z H_y)\mathbf{i} + (\dot{H}_y + \omega_z H_x - \omega_x H_z)\mathbf{j} \qquad \boxed{21\text{-}26}$$
$$+ (\dot{H}_z + \omega_x H_y - \omega_y H_x)\mathbf{k}$$

In certain problems it may be more advantageous to express these terms using components along a reference frame other than the $xyz$ or $XYZ$ frame. Assume that the new reference frame has an angular velocity $\boldsymbol{\Omega}$, and $\mathbf{M}_C$, $\mathbf{H}_C$, and $\dot{\mathbf{H}}_C$ are expressed using $\mathbf{i}$, $\mathbf{j}$, and $\mathbf{k}$ components along the new reference axes. The terms in the unit vector derivatives $\dot{\mathbf{i}}$, $\dot{\mathbf{j}}$, and $\dot{\mathbf{k}}$ above can be expressed as $\boldsymbol{\Omega} \times \mathbf{H}_C$, and Eq. 21-25 becomes

$$\boxed{\sum \mathbf{M}_C = (\dot{\mathbf{H}}_C)_{xyz} + \boldsymbol{\Omega} \times \mathbf{H}_C} \qquad \boxed{21\text{-}27}$$

where $\boldsymbol{\Omega}$ is the angular velocity of the rotating $xyz$ frame relative to the $XYZ$ frame. $(\dot{\mathbf{H}}_C)_{xyz}$ is the time rate of change of $\mathbf{H}_C$ in the rotating $xyz$ frame, and is evaluated using Eqs. 21-4 and 21-23. Note that both Eqs. 21-25 and 21-27 may be used in reference to an arbitrary fixed point $O$ besides the mass center $C$, if done consistently.

## 21-5   EULER'S EQUATIONS OF MOTION

The $xyz$ axes fixed in the body and rotating with it may be chosen to coincide with the principal axes of inertia $x'y'z'$ with respect to the mass center $C$ or a fixed point $O$. In that case the inertia matrix contains only

$I_{x'} = I_x$, $I_{y'} = I_y$, and $I_{z'} = I_z$ according to Eq. 21-6. Using Eq. 21-7, Eq. 21-26 can be written as three scalar equations,

$$\sum M_x = I_x \dot{\omega}_x - (I_y - I_z)\omega_y\omega_z$$
$$\sum M_y = I_y \dot{\omega}_y - (I_z - I_x)\omega_z\omega_x$$
$$\sum M_z = I_z \dot{\omega}_z - (I_x - I_y)\omega_x\omega_y$$

$\boxed{21\text{-}28}$

These expressions are known as *Euler's equations* after Leonhard Euler (1707–1783), a Swiss mathematician. Note that all moments, moments of inertia, and angular velocities in Eq. 21-28 must be evaluated with respect to the appropriate principal axes of the given body.

## SOLUTION OF BASIC PROBLEMS IN THREE-DIMENSIONAL MOTION

**21-6**

The maximum number of independent scalar equations of motion of a rigid body is six. They are presented here as a group of governing equations for convenience.

$$\sum F_x = ma_{C_x} \quad \sum F_y = ma_{C_y} \quad \sum F_z = ma_{C_z}$$

$\boxed{21\text{-}29}$

$$\sum M_x = \dot{H}_x + \omega_y H_z - \omega_z H_y$$
$$\sum M_y = \dot{H}_y + \omega_z H_x - \omega_x H_z$$
$$\sum M_z = \dot{H}_z + \omega_x H_y - \omega_y H_x$$

$\boxed{21\text{-}30}$

where $a_C$ refers to the linear acceleration of the center of mass, and the angular momenta are determined according to Eq. 21-4. These equations are valid *in general*. The following categories of problems are stated with a few comments to facilitate the analysis in common situations.

**Unconstrained motion.** The six governing equations, Eqs. 21-29 and 21-30, should be used with $xyz$ axes attached at the center of mass of the body.

**Motion of a body about a fixed point.** The governing equations are valid for a body rotating about a noncentroidal fixed point $O$. The reference axes $xyz$ must go through that fixed point to allow working with a set of moment equations (Eq. 21-30) that do not involve the unknown reaction at $O$.

**Motion of a body about a fixed axis.** This is the generalized form of plane motion of an arbitrary rigid body. Such a motion is very common since it includes the plane motion of unbalanced wheels and shafts. For the purpose of simplifying the governing equations, consider a body that is constrained to rotate about the fixed $Z$ axis as in Fig. 21-11. Also, the body cannot slide along $Z$. The $xyz$ axes are fixed in the body (with $z = Z$, but with point $O$ not necessarily at the mass center) and are rotating with the body at an angular velocity $\boldsymbol{\omega} = \omega \mathbf{k}$. Thus, $a_{C_z} = 0$, and $\omega_x = \omega_y = 0$ for this situation. The $xyz$ scalar components of the angular momentum are obtained from Eq. 21-4,

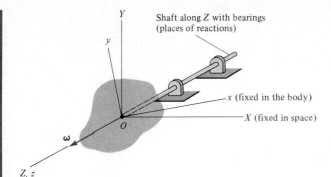

Shaft along $Z$ with bearings
(places of reactions)

$Y$

$y$

$x$ (fixed in the body)

$X$ (fixed in space)

$O$

$Z, z$

$\omega$

**FIGURE 21-11**

Rotation about a fixed axis

$$H_x = -I_{xz}\omega_z \quad H_y = -I_{yz}\omega_z \quad H_z = I_z\omega_z$$

Consequently, the governing equations of Eq. 21-29 and 21-30 are simplified to

$$\boxed{\begin{aligned} \sum F_x = ma_{C_x} \quad \sum F_y = ma_{C_y} \quad \sum F_z = 0 \\ \sum M_x = -I_{xz}\dot{\omega}_z + I_{yz}\omega_z^2 \\ \sum M_y = -I_{yz}\dot{\omega}_z - I_{xz}\omega_z^2 \\ \sum M_z = I_z\dot{\omega}_z \end{aligned}}$$

$\boxed{21\text{-}31}$

where $\dot{\omega}_z = \alpha_z$, the angular acceleration about the fixed axis. It is emphasized that Eqs. 21-31 are valid for the special case of motion about a fixed axis only.

In practice it is frequently possible to determine $\dot{\omega}_z$ directly from the last moment equation, and $\omega_z$ by integration of $\dot{\omega}_z$. The bearing reactions on the fixed axis of rotation can then be calculated.

**1. Dynamically balanced bodies.** The last set of governing equations is further simplified *if the mass distribution of the body is symmetrical with respect to the xy plane.* In that case $I_{xz} = I_{yz} = 0$, and the moment equations become

$$\sum M_x = 0 \quad \sum M_y = 0 \quad \sum M_z = I_z\dot{\omega}_z \qquad \boxed{21\text{-}32}$$

This means that, for $\omega_z = $ constant, no moments are involved with the rotation of the body in any direction. The body is said to be *dynamically balanced*. (Static balance requires the absence of moments when $\omega_z = 0$.)

**2. Dynamically unbalanced bodies.** A common problem concerning rotating rigid bodies is when $I_{xz} \neq 0$ and $I_{yz} \neq 0$, even with $\omega_z = $ constant according to the illustrative situation in Fig. 21-11. In that case,

$$\sum M_x = I_{yz}\omega_z^2 \quad \sum M_y = -I_{xz}\omega_z^2 \quad \sum M_z = 0 \qquad \boxed{21\text{-}33}$$

The mass products of inertia are constant in the rotating $xyz$ reference frame, but they are constantly changing in the fixed $XYZ$ frame as the body rotates. This creates vibratory reactions in the stationary supporting structure (transmitted through bearings) of the rotating member. It is worth remembering that the peak magnitudes of these reactions are directly proportional to $\omega^2$ according to Eq. 21-33. Dynamic unbalance of machine parts is normally undesirable (requiring corrective measures to rearrange the mass distribution), but sometimes it is a design feature of machines that must produce vibrations.

EXAMPLE 21-4

The turbine disk of a jet engine is modeled as a uniform thin disk of 5 kg mass and 0.2 m radius. It is rotating at constant angular velocity $\omega_z = 3000\mathbf{k}$ rad/s, and $\omega_x = -0.5\mathbf{i}$ rad/s when it is in the $XY$ plane (Fig. 21-12). Determine the moment required to maintain the axle of the disk in the $YZ$ plane.

## SOLUTION

Using $xyz$ axes coincident with $XYZ$, the moments of inertia are calculated,

$$I_x = I_y = \frac{1}{4}\,mr^2 = \frac{1}{4}(5\text{ kg})(0.2\text{ m})^2 = 0.05\text{ kg·m}^2$$

$$I_z = \frac{1}{2}\,mr^2 = \frac{1}{2}(5\text{ kg})(0.2\text{ m})^2 = 0.1\text{ kg·m}^2$$

The angular acceleration is calculated using Eq. 18-25,

$$\boldsymbol{\alpha} = \boldsymbol{\Omega} \times \boldsymbol{\omega} = \boldsymbol{\omega}_x \times \boldsymbol{\omega}_z$$
$$= (-0.5\mathbf{i}\text{ rad/s}) \times (3000\mathbf{k}\text{ rad/s})$$
$$= 1500\mathbf{j}\text{ rad/s}$$

Since $xyz$ are principal axes of inertia, Euler's equations of motion (Eq. 21-28) can be used. For this problem $\dot{\omega}_x = \dot{\omega}_z = \omega_y = 0$ and $\dot{\omega}_y = 1500$ rad/s; therefore,

$$\sum M_x = I_x\dot{\omega}_x - (I_y - I_z)\omega_y\omega_z = 0 - 0 = 0 \tag{a}$$

$$\sum M_y = I_y\dot{\omega}_y - (I_z - I_x)\omega_z\omega_x \tag{b}$$
$$= (0.05\text{ kg·m}^2)(1500\text{ rad/s}) - (0.1 - 0.05)\text{ kg·m}^2\,(3000\text{ rad/s})(-0.5\text{ rad/s})$$
$$= 75\text{ kg·m}^2/\text{s}^2 + 75\text{ kg·m}^2/\text{s}^2$$
$$= 150\text{ N·m}$$

$$\sum M_z = I_z\dot{\omega}_z - (I_x - I_y)\omega_x\omega_y = 0 - 0 = 0 \tag{c}$$

The required moment is

$$\mathbf{M} = 150\mathbf{j}\text{ N·m}$$

Note that even though rotations are occurring about the $x$ and $z$ axes, it is a moment about the $Y$ axis that is necessary for the desired motion!

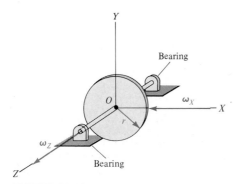

**FIGURE 21-12**

Model of a turbine disk; the bearings are fixed in the engine, not in the $XYZ$ frame

EXAMPLE 21-4    865

EXAMPLE 21-5

Determine the bearing reactions at $A$ and $B$ for the crankshaft in Ex. 21-3.

### SOLUTION

This problem involves motion of a body about a fixed axis (here the $x$ axis). From Ex. 21-3,

$$H_x = 3.90 \, \text{lb} \cdot \text{ft} \cdot \text{s}$$

$$H_y = 1.63 \, \text{lb} \cdot \text{ft} \cdot \text{s}$$

$$H_z = 0$$

Applying Eqs. 21-25 and 21-26 gives

$$\sum \mathbf{M}_A = \dot{\mathbf{H}}_A + \boldsymbol{\omega} \times \mathbf{H}_A \qquad (\text{point } A \text{ may be used since it is fixed})$$

$$= (\dot{H}_x + \omega_y H_z - \omega_z H_y)_A \mathbf{i} + (\dot{H}_y + \omega_z H_x - \omega_x H_z)_A \mathbf{j}$$
$$+ (\dot{H}_z + \omega_x H_y - \omega_y H_x)_A \mathbf{k}$$
$$= (0 + 0 - 0)\mathbf{i} + (0 + 0 - 0)\mathbf{j} + [0 + (105 \, \text{rad/s})(1.63 \, \text{lb} \cdot \text{ft} \cdot \text{s}) - 0]\mathbf{k}$$
$$= 171\mathbf{k} \, \text{lb} \cdot \text{ft}$$

From Fig. 21-13, the only reaction component that can cause a moment about the $z$ axis is the force $B_y$.

$$\sum M_z = B_y(2 \, \text{ft}) - W(1 \, \text{ft}) = 171 \, \text{lb} \cdot \text{ft}$$

$$B_y = \frac{1}{2 \, \text{ft}} \left[ 171 \, \text{lb} \cdot \text{ft} + (14.4 \, \text{lb})(1 \, \text{ft}) \right]$$

$$= 92.7 \, \text{lb} \qquad (\text{the direction assumed in Fig. 20-13 is correct})$$

$$\sum F_y = ma_y = 0$$

$$A_y + B_y - W = 0$$

$$A_y = 14.4 \, \text{lb} - 92.7 \, \text{lb}$$
$$= -78.3 \, \text{lb} \qquad (A_y \text{ should be downward in Fig. 21-13})$$

**FIGURE 21-13**

Model of a crankshaft

**21–23** A well-drilling rod is modeled as a slender rod of mass $m$ and length $l$ moving without resistance. The rod rotates at angular speed $\omega_y$ about the $y$ axis and can freely pivot about the horizontal $z$ axis. Determine the value of $\omega$ at which the rod would start to swing away from a nearly vertical position (from $\theta \approx 0$ to $\theta > 0$).

**Fig. P21-23**

**21–24** Add a concentrated mass $M$ to the tip of the rod in Prob. 21–23. Write an expression for the angle $\theta$ of the rod. Does $M$ allow an increased operating speed $\omega$?

**21–25** A proposed amusement ride is modeled as a thin, uniform disk of mass $m$ and radius $r$, firmly attached to a slender rod of length $l$ and negligible mass. The rod rotates at angular speed $\omega_y$ about the $y$ axis and can freely pivot about the horizontal $z$ axis. Write an expression for the angle $\theta$ of the rod.

**Fig. P21-25**

**21–26** Solve Prob. 21–25 including the mass $M$ of the rod.

**21–27** Determine the rate of change $\dot{\mathbf{H}}_O$ of the angular momentum of the rod in Prob. 21–4.

**21–28** Determine the reactions at bearings $A$ and $B$ in Prob. 21–4 for the position shown. $AB = 0.3\,\text{m}$, $BO = 0.2\,\text{m}$ and $\dot{\omega}_x = 0$.

**21–29** Determine the rate of change $\dot{\mathbf{H}}_O$ of the angular momentum of the peg in Prob. 21–7 for $\dot{\omega}_y = 0$.

**21–30** The thin plate of mass $m$ is attached to a shaft with bearings $A$ and $B$ at $x = 0$ and $2a$, respectively. It rotates about the fixed $x$ axis at a constant angular speed $\omega_x$. Determine the reactions at $A$ and $B$ for the position shown.

**Fig. P21-30**

**21–31** The thin plate of mass $m$ is attached to a shaft with bearings at $A$ and $B$. It rotates about the fixed $y$ axis at a constant angular speed $\omega_y$. Determine the reactions at $A$ and $B$ for the position shown.

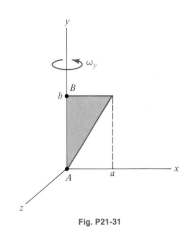

**Fig. P21-31**

**21–32** The thin bent plate of mass $m_1 + m_2$ is rotating about the fixed $y$ axis at angular speed $\omega_y$. Determine the rate of change $\dot{\mathbf{H}}_A$ of the angular momentum for $\omega_y = $ constant and also for $\omega_y$ changing at the rate of $\dot{\omega}_y$ for the position shown.

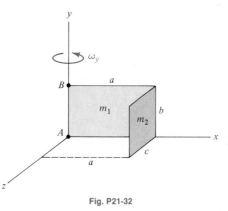

**Fig. P21-32**

**21–33** The bent plate in Prob. 21–32 is connected to bearings at $A$ and $B$. Determine the reactions at $A$ and $B$ for the position shown and $\omega_y = $ constant.

**21–34** Consider the semicylindrical blades of a Savonius rotor (Prob. 21–16) of total mass $m$ which are rigidly connected to a shaft at the $y$ axis. Corners $A$, $B$, and $C$ (and identical lower ones) are in the $yz$ plane. Determine the reactions at bearings $O$ and $A$ for this position and $\dot{\omega}_y = 0$.

**Fig. P21-34**

**21–35** The 50-kg wheel is modeled as a uniform disk of 0.6 m radius and 0.4 m thickness. Its center is located at $x = -0.3$ m, $y = z = 0$, and pivoted about $O$ ($x = y = z = 0$). The wheel is parallel to the $yz$ plane when its angular velocity is a constant $\boldsymbol{\omega} = (10\mathbf{i} - 3\mathbf{k})$ rad/s. Determine the required couple $\mathbf{M}_O$ about point $O$ to keep the wheel's axle in the $xy$ plane.

**Fig. P21-35**
Complex test of a tire (Courtesy MTS Systems Corporation, Minneapolis, Minnesota)

**21–36** A 5-lb, thin, uniform disk of 8 in. radius is rotating about a 15-in. arm which is pivoted at the fixed point $A$. Arm $AB$ of negligible weight is always parallel to the $xz$ plane and rotates about the $y$ axis at $\omega_y = 20$ rad/s as the disk rolls on the $xz$ plane. Calculate the rate of change $\dot{\mathbf{H}}_A$ of the angular momentum of the disk when arm $AB$ is parallel to the $z$ axis.

**Fig. P21-36**

**21-37**  Determine the normal force exerted by the disk on the floor in Prob. 21–36.

**21-38**  A thin plate of mass $m$ has bearings at points $O$ and $A$. A hole of radius $r$ is drilled at $x = 3a/4$ and $y = b/2$. Determine the reactions at $O$ and $A$ in the position shown for a constant angular speed $\omega_x$ about the $x$ axis.

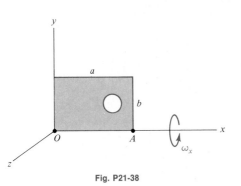

Fig. P21-38

**21-39**  A thin rod of mass $m$ per unit length is bent into the shape of a semicircle and firmly attached to the shaft which is in bearings $A$ and $B$. Determine the rate of change $\dot{\mathbf{H}}_O$ of the angular momentum when the angular speed about the $x$ axis is a constant $\omega_x$ for the position shown.

Fig. P21-39

**21-40**  Determine the bending moment in the rod at point $O$ in Prob. 21–39. Also determine the reactions at $A$ and $B$ when the member is in the $xy$ plane.

**21-41**  A two-bladed propeller of an airplane has a total mass of 12 kg and a radius of gyration of 0.8 m. The propeller is rotating at a speed of 1400 rpm when the airplane is flying at a 300-km/h speed in a circular path of 200 m radius. Determine the magnitude of the moment acting on the propeller shaft because of these motions.

**21-42**  The two main rotor blades of a helicopter have a total weight of 200 lb and a radius of gyration of 10 ft. The rotor axis is rotating at 300 rpm and is tilted forward at an angle $\theta = 10°$ when the helicopter is traveling in the $x$ direction. Calculate the moment acting on the rotor shaft when the helicopter begins to rotate about the $y$ axis at the rate of 2 rad/s.

Fig. P21-42

# GYROSCOPIC MOTION

The basic concepts of analyzing the three-dimensional motion of a rigid body were presented in the preceding sections. These concepts are extended here for an introductory analysis of gyroscopes which include navigational devices in aircraft and stabilizing machinery in ships.

## Precession

It is worthwhile to review qualitatively the appreciable differences that are possible between two-dimensional and three-dimensional motions of a rigid body. Consider first a uniform disk on a rigid axle at rest. Assume that a couple **M** about the $x$ axis is applied to the shaft in Fig. 21-14. With the disk at rest initially, the shaft rotates about the $x$ axis only. It can be said that the system changes its angular momentum (in this case, from $\mathbf{H}_O = 0$) in the direction of the applied couple **M**. This is analogous to the change in linear momentum of a particle in the direction of the force acting on it.

Next, assume that the disk has an initial angular velocity $\boldsymbol{\omega}_z$ about the $z$ axis as in Fig. 21-15a. This velocity is called the spin velocity, and it results in the initial angular momentum $\mathbf{H}_O$ about point $O$. A couple **M** about the $x$ axis causes this angular momentum vector to rotate toward the $x$ axis as shown in Fig. 21-15b. The resulting angular velocity $\boldsymbol{\omega}_y$ of **H** is called the precession velocity. The precession should be viewed as a change in angular momentum in the direction of the applied couple vector **M** (again analogous to the linear motion of a particle). The applied moment and the rate of precession are related according to Eq. 21-26. Precession is always slow relative to the spin rate. This particular motion may be readily experienced by gradually tilting the axis of a high-speed rotary tool while *very carefully* holding it in the hand.

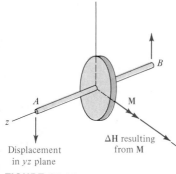

**FIGURE 21-14**

Analysis of a nonspinning disk

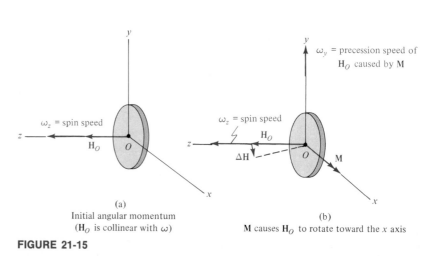

(a)
Initial angular momentum
($\mathbf{H}_O$ is collinear with $\omega$)

(b)
**M** causes $\mathbf{H}_O$ to rotate toward the $x$ axis

**FIGURE 21-15**

Analysis of a spinning disk

## Euler Angles

The precession axis of a rotating body may itself change orientation. Analysis of the motion including this additional complication is made using the model of a gyroscope in Fig. 21-16. In part (a) of the figure the shaft of a disk is pivoted at *aa* to spin freely in the *inner gimbal* ring *A*, which is pivoted at *bb* to rotate freely with respect to the *outer gimbal*

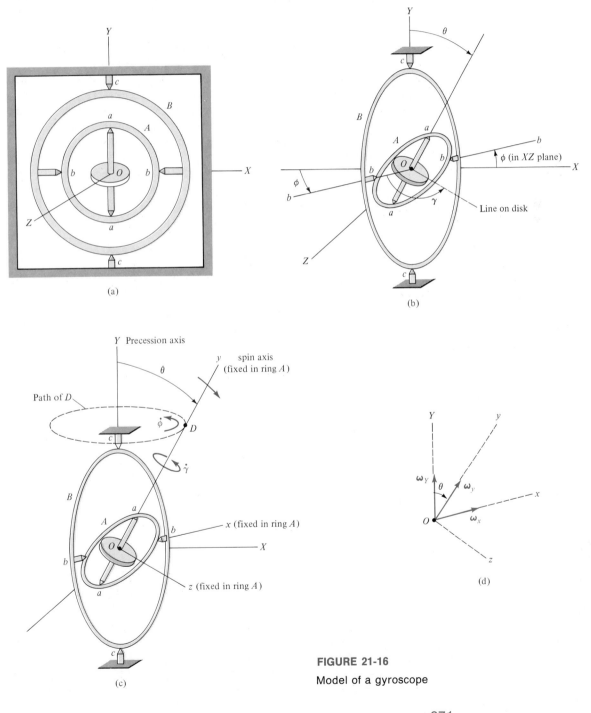

(a)

(b)

(c)

**FIGURE 21-16**

Model of a gyroscope

(d)

ring $B$. The latter is pivoted at $cc$ to rotate freely about the $Y$ axis of a fixed reference frame $XYZ$. A general position of these parts is shown in Fig. 21-16b. Angle $\phi$ specifies the position of line $bb$ from the fixed $X$ axis (line $bb$ always moves in the $XZ$ plane). Angle $\theta$ represents the position of line $aa$ of the inner gimbal from the fixed $Y$ axis. Angle $\gamma$ gives the position of a radial line on the disk measured from line $bb$ ($\gamma$ is measured in the $XZ$ plane when $\theta = 0$). The angles $\phi$, $\theta$, and $\gamma$ are known as the *Euler angles*. These angles completely define the position of the gyroscope at a given instant. Their time derivatives are defined as follows: $\dot{\phi}$ = rate of *precession*, $\dot{\theta}$ = rate of *nutation*, and $\dot{\gamma}$ = rate of *spin*.

For further analysis of the angular velocities, consider a rotating reference frame $xyz$ which is attached to the inner gimbal $A$ as in Fig. 21-16c. Thus, at the instant indicated, the disk spins about the $y$ axis at angular speed $\dot{\gamma}$, while the $y$ axis itself has two motions. The $y$ axis is rotating away from the fixed $Y$ axis at angular speed $\dot{\theta}$, and it is precessing about the $Y$ axis at angular speed $\dot{\phi}$ (the dashed line shows the circular path of point $D$ on the $y$ axis caused by the precession).

The absolute angular velocity $\boldsymbol{\omega}$ of the gyroscope is the vector sum of its three angular velocities. These are shown schematically in Fig. 21-16d. Considering the fixed unit vectors $\mathbf{I}$, $\mathbf{J}$, and $\mathbf{K}$, and the $\mathbf{i}$, $\mathbf{j}$, and $\mathbf{k}$ unit vectors of the rotating $xyz$ frame,

$$\boldsymbol{\omega} = \boldsymbol{\omega}_Y + \boldsymbol{\omega}_x + \boldsymbol{\omega}_y = \dot{\phi}\mathbf{J} + \dot{\theta}\mathbf{i} + \dot{\gamma}\mathbf{j} \qquad \boxed{21\text{-}34}$$

This equation should be transformed to use entirely either the fixed of the rotating frame. The latter is more convenient (both here and eventually when dealing with moments), and this is accomplished realizing that the $Y$, $y$, and $x$ axes are in the same plane, so

$$\mathbf{J} = \cos\theta\,\mathbf{j} - \sin\theta\,\mathbf{k} \qquad \boxed{21\text{-}35}$$

With this expression, Eq. 21-34 becomes

$$\boldsymbol{\omega} = \dot{\theta}\mathbf{i} + (\dot{\gamma} + \dot{\phi}\cos\theta)\mathbf{j} - \dot{\phi}\sin\theta\,\mathbf{k} \qquad \boxed{21\text{-}36}$$

where all components are defined in the rotating frame.

## Equations of Motion

The effects of external moments on the motion of a gyroscope are determined using Eq. 21-27. Here the center of mass $C$ is at the origin $O$ of the rotating frame. The concise form of the equation of motion is

$$\boxed{\sum \mathbf{M}_O = (\dot{\mathbf{H}}_O)_{xyz} + \boldsymbol{\Omega} \times \mathbf{H}_O} \qquad \boxed{21\text{-}37}$$

where $\boldsymbol{\Omega}$ is the angular velocity of the $xyz$ frame. For the configuration in Fig. 21-16c and d,

$$\boldsymbol{\Omega} = \boldsymbol{\omega}_Y + \boldsymbol{\omega}_x = \dot{\theta}\mathbf{i} + \dot{\phi}(\cos\theta)\mathbf{j} - \dot{\phi}(\sin\theta)\mathbf{k} \qquad \boxed{21\text{-}38}$$

The angular momentum vector $\mathbf{H}_O$ and its derivative are evaluated (neglecting the mass of the two gimbals) using Eqs. 21-4 and 21-23. A considerable simplification can be made here since the $xyz$ axes are principal axes of inertia of the gyroscope, which is always symmetrical with respect to its spin axis. Thus, only two moments of inertia have to be used, $I_1$ about the spin axis and $I_2$ about any centroidal line transverse to the spin axis. With these moments of inertia, the angular momentum is expressed using Eqs. 21-4 and 21-36 as

$$\mathbf{H}_O = I_2\dot{\theta}\mathbf{i} + I_1(\dot{\gamma} + \dot{\phi}\cos\theta)\mathbf{j} - I_2\dot{\phi}\sin\theta\,\mathbf{k} \qquad \boxed{21\text{-}39}$$

After performing the mathematical manipulations indicated in Eq. 21-37 (noting that $\mathbf{i}$, $\mathbf{j}$, and $\mathbf{k}$ are fixed in the $xyz$ frame), the $xyz$ components of the moment are written as

$$\boxed{\begin{aligned}
\sum M_x &= I_1\dot{\phi}(\sin\theta)(\dot{\gamma} + \dot{\phi}\cos\theta) \qquad \boxed{21\text{-}40} \\
&\quad + I_2(\ddot{\theta} - \dot{\phi}^2\sin\theta\cos\theta) \\
\sum M_y &= I_1(\ddot{\gamma} + \ddot{\phi}\cos\theta - \dot{\phi}\dot{\theta}\sin\theta) \\
\sum M_z &= I_1\dot{\theta}(\dot{\gamma} + \dot{\phi}\cos\theta) \\
&\quad - I_2(\ddot{\phi}\sin\theta + 2\dot{\phi}\dot{\theta}\cos\theta)
\end{aligned}}$$

The details of the derivation are left as an exercise for the student.

Equations 21-40 are the differential equations of motion of a gyroscope. Note in the above analysis that the $xyz$ frame is attached at the mass center of the gyroscope, where it is effectively supported by gimbals of negligible mass. These equations are also valid for a noncentroidal $xyz$ frame which is attached at any fixed point $O$ located on the spin axis of the gyroscope. The reason is that the Euler angles can be identically defined for centroidal and noncentroidal rotating frames, and the symmetry of the body allows using only two moments of inertia in either case. The position of point $O$ with respect to the mass center does not affect the value of $I_1$ in Eq. 21-40, but it affects $I_2$ (the parallel-axis theorem must be used).

## SPECIAL CASES OF GYROSCOPIC MOTION

The solution of the nonlinear equations of motion (Eq. 21-40) are very complex in general. In a few particular cases they are relatively simple and can be considered at the introductory level to gain further insight into three-dimensional motion.

### Steady Precession

Assume that a gyroscope precesses under the following conditions:

$$\text{angular speed of spin } \dot{\gamma} = \text{constant}, \ddot{\gamma} = 0$$
$$\text{angular speed of precession } \dot{\phi} = \text{constant}, \ddot{\phi} = 0$$
$$\text{angle } \theta \text{ to spin axis} = \text{constant}, \dot{\theta} = \ddot{\theta} = 0$$

In this case, Eqs. 20-40 reduce to

$$\sum M_x = I_1\dot{\phi}\sin\theta(\dot{\gamma} + \dot{\phi}\cos\theta) - I_2\dot{\phi}^2\sin\theta\cos\theta \qquad \boxed{21\text{-}41}$$
$$\sum M_y = 0$$
$$\sum M_z = 0$$

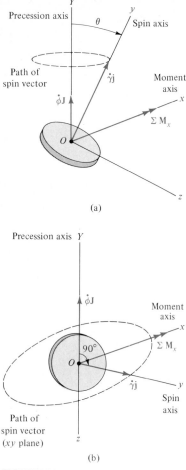

Precession axis *Y*

Path of spin vector

Moment axis *x*

$\dot\phi J$

$\dot\gamma j$

$\Sigma M_x$

$\theta$

Spin axis *y*

$O$

*z*

(a)

Precession axis *Y*

$\dot\phi J$

Moment axis *x*

$\Sigma M_x$

90°

$O$

$\dot\gamma j$

Spin axis *y*

Path of spin vector (*xy* plane)

*z*

(b)

**FIGURE 21-17**

Special case of a gyroscope in steady precession

The same result can be obtained directly from Eq. 21-37 after realizing that $H_O$ is constant in steady precession when observed from the rotating *xyz* frame, so its time derivative is zero. Equation 21-37 reduces to

$$\sum \mathbf{M}_O = \mathbf{\Omega} \times \mathbf{H}_O \qquad \boxed{21\text{-}42}$$

which is equivalent to Eq. 21-41 after the vector product is evaluated.

An important observation concerning the direction of the external moment acting on a steadily precessing gyroscope can be made from Fig. 21-16c and Eq. 21-41. It is noted that *the moment vector should be perpendicular to the spin axis and the precession axis to maintain steady precession of a gyroscope.* This concept is illustrated in Fig. 21-17a for the general case of steady precession. A spinning top with its weight causing the moment is in this category.

### Precession and Spin Axes Are at a Right Angle

In many engineering problems involving steady precession of a gyroscope the precession axis is normal to the spin axis. This allows a further simplification of Eq. 21-41 since $\theta = 90°$. The resulting single equation is

$$\sum M_x = I_1 \dot\phi \dot\gamma \qquad \boxed{21\text{-}43}$$

which is illustrated in Fig. 21-17b. The general meaning of this specific situation can be summarized as follows:

1. The gyroscope precesses about an axis perpendicular to both the spin axis and the moment axis.
2. The rate of precession $\dot\phi$ is inversely proportional to the rate of spin $\dot\psi$ for a given moment acting on the gyroscope.
3. Large moments are required if the orientation of a rapidly spinning gyroscope is to be changed.

### Steady Precession of Axisymmetric Bodies in the Absence of Moments

Projectiles and spacecraft are commonly given spin velocities to achieve stabilization. Such motions may be analyzed using the methods developed so far.

Assume that the body is an axisymmetric rigid body spinning about its longitudinal axis, and that no external moments are acting on it. In this case $\sum \mathbf{M}_C = \dot{\mathbf{H}}_C = 0$ from Eq. 21-22. Thus, the angular momentum $\mathbf{H}_C$ is constant in magnitude and direction. It is convenient to set up the fixed frame $XYZ$ with respect to this angular momentum vector. In particular, the axis of precession $Y$ is chosen to coincide with the vector $\mathbf{H}_C$ as in Fig. 21-18a. Axis *y* is the spin axis of the body as before, while *x* is chosen to be in the $Yy$ plane.

The total angular velocity $\omega$ is the vector sum of the spin velocity $\dot\gamma \mathbf{j}$ and the precession velocity $\dot\phi \mathbf{J}$. Since the angle $\theta$ between the $Y$ and *y*

**FIGURE 21-18**

Special case of a steadily precessing
axisymmetric body

(a)

(b)

axes is generally not a right angle (but $\theta = $ constant as shown below*),
components of $\omega$ are different from the precession and spin velocities,
respectively. These distinctions are shown in Fig. 21-18b. The $xyz$ scalar
components of $\mathbf{H}_C$ and $\omega$ depend on the angles $\theta$ and $\alpha$ according to
Fig. 21-18,

$$H_x = -H_C \sin\theta \quad H_y = H_C \cos\theta \quad H_z = 0 \qquad \boxed{21\text{-}44}$$

$$\omega_x = \omega \sin\alpha \qquad \omega_y = \omega \cos\alpha \qquad \omega_z = 0 \qquad \boxed{21\text{-}45}$$

where the magnitudes $H_C$ and $\omega$ are constant. Note that there are no components in the $z$ direction, which is perpendicular to the plane of $\mathbf{H}_C$ and
$\omega$. The scalar components of angular momentum and angular velocity are
simply related to one another since the $xyz$ axes are principal axes of
inertia,

$$H_x = I_2\omega_x \quad H_y = I_1\omega_y \quad H_z = 0 \qquad \boxed{21\text{-}46}$$

where $I_1$ and $I_2$ are the moments of inertia about the spin axis and a
transverse axis through $C$, respectively. From Eqs. 21-44 and 21-46,

$$\omega_x = -\frac{H_C \sin\theta}{I_2} \quad \omega_y = \frac{H_C \cos\theta}{I_1} \quad \omega_z = 0 \qquad \boxed{21\text{-}47}$$

Since $H_C$, $I_1$, $I_2$, and $\theta$ are all constant, $\omega_x$ and $\omega_y$ are constant. These
are the mathematical expressions of *steady precession* of the body about the
$Y$ axis, with a constant angle $\theta$ to the spin axis $y$ (since $\omega_z = 0$).

The angle $\alpha$ between the spin axis and the vector $\omega$ can be determined
from Fig. 21-18b and Eq. 21-47, from which

$$\boxed{\tan\alpha = -\frac{\omega_x}{\omega_y} = \frac{I_1}{I_2}\tan\theta} \qquad \boxed{21\text{-}48}$$

Thus, $\alpha$ is a constant angle in this motion.

---

\* First assume that in the fixed $XYZ$ frame $\omega = \omega \sin\alpha\,\mathbf{i} + \omega\cos\alpha\,\mathbf{j} + \dot\theta\mathbf{k}$. However,

$$\mathbf{H}_C = H_C \sin\theta\,\mathbf{i} + H_C \cos\theta\,\mathbf{j} + 0\mathbf{k}$$

since $\mathbf{H}_C$ is in the $xy$ plane. Hence, $\dot\theta = 0$ and $\theta = $ constant.

The rate of precession is obtained from Eq. 21-41. With $\sum \mathbf{M} = 0$,

$$\dot{\phi} = \frac{I_1\dot{\gamma}}{(I_2 - I_1)\cos\theta} \qquad \boxed{21\text{-}49}$$

The occurrence of precession depends on the initial rotational motion of the body. The direction of precession depends on the relative magnitudes of $I_1$ and $I_2$ according to Eq. 21-49. Four particular cases are worth noting as outlined in the following:

1. The initial spin is entirely about the axis of symmetry as in Fig. 21-19a. There is no precession since $\omega_x$, $H_x$, and $\sum \mathbf{M}$ are all zero.
2. The initial spin is entirely about a transverse axis (with respect to the axis of symmetry) as in Fig. 21-19b. There is no precession since $\omega_y$, $H_y$, and $\sum \mathbf{M}$ are all zero.
3. There is initial spin and precession as illustrated in Figs. 21-20a and 21-21a for two distinct possibilities.
   a. $I_2 > I_1$; the body is elongated along the spin axis $y$. $\theta > \alpha$ at all times according to Eq. 21-48. The body cone (Sec. 21-1) appears to roll on the outside of the space cone as shown in Fig. 21-20b. This is called *direct precession* because the spin and the precession have the same sense when viewed from the positive $Y$ or $y$ axes.
   b. $I_1 > I_2$; the body is short along the spin axis $y$. $\theta < \alpha$ at all times according to Eq. 21-48. This is possible if the spin and precession vectors have components in opposite directions as in Fig. 21-21a. The space cone is interior to the body cone, and the spin and the precession have opposite senses when viewed from the positive $Y$ or $y$ axes as in Fig. 21-21b. This precession is said to be *retrograde*.

(a) No precession

(b) No precession

**FIGURE 21-19**

Special cases involving no precession

(a)

(b) $I_2 > I_1$, direct precession

**FIGURE 21-20**

Schematic of direct precession

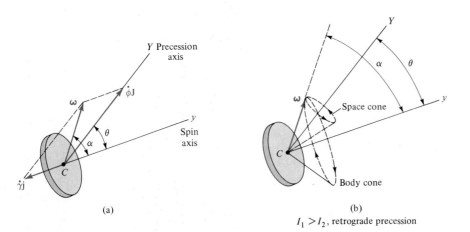

(a)

(b)

$I_1 > I_2$, retrograde precession

**FIGURE 21-21**

Schematic of retrograde precession

# EXAMPLE 21-6

A 2-kg disk of 0.3 m diameter is attached to a light axle of length $l = 0.05$ m. It is released with $l$ in the $XY$ plane and $\beta = 60°$ (Fig. 21-22a), and spinning about $l$ at the rate of $\dot{\gamma} = 83.8$ rad/s. Determine the rate of precession $\dot{\phi}$ of the disk.

## SOLUTION

Ignoring the weight of the axle, the only force acting on the disk is its weight, which causes a moment about the $z$ axis. By Eq. 21-42 and using the $xyz$ frame as the reference system (Fig. 21-22b),

$$\sum \mathbf{M}_O = \mathbf{\Omega} \times \mathbf{H}_O$$

where

$$\sum \mathbf{M}_O = Wl(\cos \beta)\mathbf{k}$$

$$\mathbf{\Omega} = \Omega_x \mathbf{i} + \Omega_y \mathbf{j} + \Omega_z \mathbf{k}$$
$$= (\dot{\phi} \cos \beta)\mathbf{i} + (\dot{\phi} \sin \beta)\mathbf{j} \qquad \text{(angular velocity of } xyz \text{ system)}$$

$$\mathbf{H}_O = I_x \omega_x \mathbf{i} + I_y \omega_y \mathbf{j} + I_z \omega_z \mathbf{k} \qquad (\omega = \text{angular velocity of the body})$$
$$= I_x \dot{\phi}(\cos \beta)\mathbf{i} + I_y(\dot{\gamma} + \dot{\phi} \sin \beta)\mathbf{j}$$

Evaluating the cross product yields

$$Wl(\cos \beta)\mathbf{k} = (I_y \Omega_x \omega_y - I_x \Omega_y \omega_x)\mathbf{k}$$
$$= I_y \dot{\phi}(\cos \beta)(\dot{\gamma} + \dot{\phi} \sin \beta)\mathbf{k} - I_x(\dot{\phi} \sin \beta)(\dot{\phi} \cos \beta)\mathbf{k}$$

Dividing through by $\cos \beta$, and realizing that moments of inertia must be calculated with respect to point $O$ using the parallel-axis theorem, gives

$$mg\,(0.05\,\text{m}) = \frac{1}{2} m\,(0.15\,\text{m})^2 \left( 83.8\,\frac{\text{rad}}{\text{s}}\,\dot{\phi} + 0.866\dot{\phi}^2 \right)$$
$$- \left[ \frac{1}{4} m\,(0.15\,\text{m})^2 + m\,(0.05\,\text{m})^2 \right](0.866)\dot{\phi}^2$$

Dividing through by the mass $m$, a quadratic in $\dot{\phi}$ is obtained. Using the quadratic formula, the solutions are

$$\dot{\phi} = 0.5195\,\text{rad/s} \qquad \text{and} \qquad -348.9\,\text{rad/s}$$

Of these, the positive root satisfies the condition for the change of angular momentum to be in the direction of the external moment.

(a)

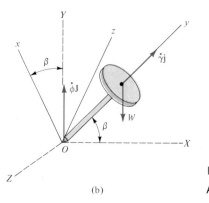

(b)

**FIGURE 21-22**

Analysis of a spinning disk

EXAMPLE 21-7

A 2000-lb spacecraft is modeled as a uniform cylinder of 6 ft diameter and 10 ft length (Fig. 21-23a). It is spinning at the rate of $\dot{\gamma} = 1$ rad/s. Small jets apply a couple $\mathbf{M}_x = 30\mathbf{i}$ lb·ft for a time of 2 s to the cylinder. Determine the resulting precession and sketch the body and space cones for the motion of the spacecraft after the burn.

## SOLUTION

Since the reference axes are principal inertial axes and initially $\omega_x = \omega_z = 0$, the initial angular momentum is simply

$$\mathbf{H}_1 = I_y \omega_y = \left(\frac{1}{2} mr^2\right)\omega_y$$

$$= \frac{1}{2}\left(\frac{2000\,\text{lb}}{32.2\,\text{ft/s}^2}\right)(3\,\text{ft})^2\,(1\mathbf{j}\,\text{rad/s})$$

$$= 279.5\mathbf{j}\,\text{ft·lb·s}$$

From Eq. 21-12 for angular impulse and momentum,

$$\mathbf{H}_2 = \mathbf{H}_1 + \sum \mathbf{M}_C\,\Delta t$$
$$= 279.5\mathbf{j}\,\text{ft·lb·s} + (30\mathbf{i}\,\text{ft·lb})(2\,\text{s})$$

where $\mathbf{i}$ and $\mathbf{j}$ are unit vectors in the rotating $xyz$ reference frame ($\mathbf{M}_x$ is always about the same axis, and moments of inertia remain the same and products of inertia are still zero). Setting components equal, from Eq. 21-4, yields

$$\omega_{y_2} = \omega_{y_1} = 1\,\text{rad/s}$$

$$I_x \omega_{x_2} = \frac{1}{12} m(3a^2 + L^2)\omega_{x_2} = 60\,\text{ft·lb·s}$$

$$\omega_{x_2} = \frac{60\,\text{ft·lb·s}}{\dfrac{1}{12}\left(\dfrac{2000\,\text{lb}}{32.2\,\text{ft/s}^2}\right)[3(3\,\text{ft})^2 + (10\,\text{ft})^2]}$$

$$= \frac{60\,\text{ft·lb·s}}{657.4\,\text{ft·lb·s}^2}$$

$$= 0.0913\,\text{rad/s}$$

(a)

From Eq. 21-48,

$$\tan\alpha = -\frac{\omega_x}{w_y} = -\frac{0.0913}{1}$$

$$\alpha = -5.2°$$

$$\tan\theta = \frac{I_2}{I_1}\tan\alpha = \frac{657.4}{279.5}(-0.0913)$$

$$\theta = -12.1°$$

The rate of precession is obtained from Eq. 21-49,

$$\dot{\phi} = \frac{I_1\dot{\gamma}}{(I_2 - I_1)\cos\theta} = \frac{279.5\,(1\,\text{rad/s})}{(657.4 - 279.5)\cos(-12.1°)}$$

$$= 0.756\,\text{rad/s}$$

The body and space cones are sketched in Fig. 21-23b.

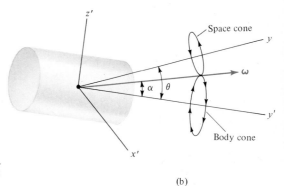

(b)

**FIGURE 21-23**

Model of a spinning spacecraft

EXAMPLE 21-7    879

**21–43** Plot the rate of precession of the disk in Ex. 21-6 for $\beta = 0$, 30°, and 60° and for $\dot{\gamma} = 62.8$ rad/s, 83.8 rad/s, and 104.7 rad/s.

**21–44** The disk in Ex. 21-6 is precessing at $\dot{\phi} = -0.1\mathbf{J}$ rad/s. Determine the rate of spin $\dot{\gamma}$.

**21–45** A 3-lb disk of 8 in. diameter is attached to a light rod which freely rotates in the ball-and-socket joint at $O$. Determine the rate of precession if the rate of spin $\dot{\gamma}$ about $OA$ is 104.7 rad/s and $\beta = 45°$.

Fig. P21-45

**21–46** Solve Prob. 21–45 assuming that the member $OA$ is a uniform 2-lb rod of 0.6 in. diameter which rotates with the disk.

**21–47** A thin-walled tube of mass $m$, length $l$, and diameter $d$ is spinning at angular speed $\dot{\gamma}$ about a smooth weightless rod $OA$ which is hinged at $O$. Write an expression for the rate of precession $\dot{\phi}$ of the rod if it is released in the horizontal position.

Fig. P21-47

**21–48** The tube in Prob. 21–47 has $m = 0.5$ kg, $l = 0.3$ m, $d = 0.04$ m, and $a = 0.1$ m. Determine $\dot{\gamma}$ if the tube is precessing at $\dot{\phi} = 0.2\mathbf{J}$ rad/s.

**21–49** The disk of 0.1 m diameter and 1 kg mass supported on a light rod is spinning in the upright position at $\dot{\gamma} = 1200\mathbf{J}$ rad/s. A force $\mathbf{F} = 20\mathbf{K}$ N is applied to the disk for a time of 0.1 s. Determine the resultant rate of precession if the spin is unchanged.

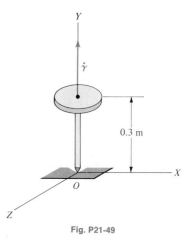

Fig. P21-49

**21–50** The takeoff speed of an aircraft is 60 m/s. The wheels freely spin at that time. Each 20-kg wheel has a 0.4 m radius and a 0.2 m radius of gyration about its axle. The main strut is pivoted at $O$, 1.1 m above the wheel axle. Determine the torsional moment on the main strut caused by gyroscopic action when the landing gear is retracted (soon after takeoff) at the rate of $\omega = 1$ rad/s.

Fig. P21-50

**21–51** Two identical 5-lb disks of 1 ft diameter are spinning at the same rate of $\dot{\gamma} = 104.7$ rad/s about a light shaft which is supported at a ball-and-socket joint at $O$. Determine the precession of the shaft if it is released in the horizontal position.

Fig. P21-51

2 ft     1 ft

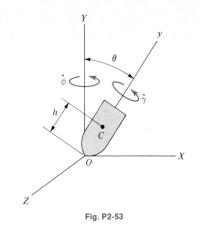

Fig. P2-53

**21–52** A wheel in a test is modeled as a thin hoop of 10 lb weight and 1 ft radius. The light guiding arm $AB$ has a length $l = 6$ ft and is pivoted at $A$. The wheel rolls on the horizontal plane at the rate of $\omega_Y = 5.24$ rad/s. Calculate the normal force between the wheel and the surface.

**21–54** The 50-kg rotor of a jet engine has a 0.17 m radius of gyration about its axle and rotates counterclockwise about the $x$ axis at 35 000 rpm. A maneuver causes a rotation of $\mathbf{\Omega} = 1\mathbf{k}$ rad/s. Determine the moment reaction caused by these motions at point $A$ of the supporting pylon.

Fig. P21-52

0.8 m

1 m

Fig. P21-54

**21–53** A top of weight $W$ is supported at point $O$ and is very rapidly spinning about its axis of symmetry $y$. It has a moment of inertia $I$ about its longitudinal axis. Show that the rate of a slow, steady precession is

$$\dot{\phi} \simeq \frac{Wh}{I\dot{\gamma}}$$

**21–55** An electric car has a transverse-mounted rotor of 12 lb weight and 3 in. radius of gyration about its axle. The rotor is turning at the rate of 1800 rpm (constant) when the car enters a 500-ft-radius curve on a level road at a speed of 80 ft/s (constant). Determine the vertical reactions at bearings $A$ and $B$.

Top view of car

$\omega$   2 ft

Fig. P21-55

$R = 500$ ft

**21–56** The 20-lb turbine rotor of a hydroplane boat has a 5-in. radius of gyration about its axle and is rotating counterclockwise about the $x$ axis at 20,000 rpm (constant). The boat is turning left on a circular path of 200 ft radius while traveling at 100 ft/s (constant). Determine the vertical reactions at bearings $A$ and $B$.

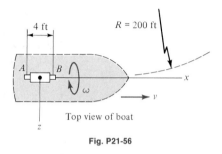

Top view of boat

**Fig. P21-56**

**21–57** The wheels of a motorcycle are modeled as thin hoops of 6 kg mass and 0.7 m diameter. The motorcycle is traveling at 100 km/h when the rider applies a couple $\mathbf{M}_y = 10\mathbf{j}$ N·m to the front wheel. Determine the resulting precession assuming that the front wheel is a free body.

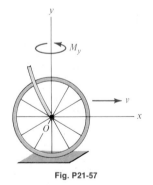

**Fig. P21-57**

**21–58** A uniform, 2-kg cylinder is spinning at the rate of 300 rpm about its axis of symmetry $y$, which itself has a slight precession. Determine if the precession is direct or retrograde for $l = 2d = 0.2$ m.

**Fig. P21-58**

**21–59** Determine the ranges of $l/d$ in Prob. 21–58 for direct and retrograde precessions.

**21–60** An amusement ride is modeled as a thin ring of 12,000 lb weight and 20 ft radius. It is spinning at the rate of $\dot{\gamma} = 1.5$ rad/s about $A$ on the rigid arm $OA$, which is pivoted at $O$. The arm and the ring are tilted upward about the $Z$ axis through $O$ at the rate of $\omega = 0.4$ rad/s. Determine the couple $\mathbf{M}_O$ caused by gyroscopic action when $\theta = 30°$.

**Fig. P21-60**

**21–61** A thin disk of mass $m$ and radius $r$ is spinning about its axis of symmetry $y$ at the rate $\dot{\gamma}$. The $y$ axis has a slight precession at the rate of $\dot{\phi} = 1.047$ rad/s. Calculate $\dot{\gamma}$.

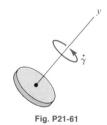

**Fig. P21-61**

**Angular momentum in three dimensions**

**PRECESSION:**  Rotation of angular velocity vector $\omega$ ($\Omega$ is angular velocity of $\omega$ and $\mathbf{H}$).

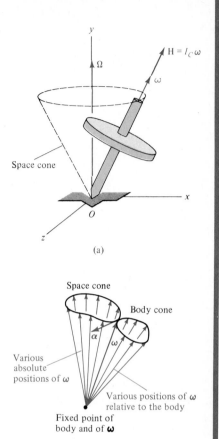

(a)

The $xyz$ frame is attached at the mass center $C$ and has fixed orientation. The body rotates relative to $xyz$.

**RECTANGULAR COMPONENTS OF TOTAL ANGULAR MOMENTUM $\mathbf{H}_C$:**

$$H_x = I_x\omega_x - I_{xy}\omega_y - I_{xz}\omega_z$$

$$H_y = -I_{xy}\omega_x + I_y\omega_y - I_{yz}\omega_z$$

$$H_z = -I_{zx}\omega_x - I_{zy}\omega_y + I_z\omega_z$$

These are valid for a particular position of the body.

For an arbitrary point $O$ of a fixed $lmn$ frame, $\mathbf{H}_O$ is determined in a similar procedure ($l$, $m$, and $n$ replace $x$, $y$, and $z$, respectively), but all $I$ must be evaluated with respect to the $lmn$ frame

(b) General relation of body and space cones

**Impulse and momentum of a rigid body in three-dimensional motion**

$$\text{System momenta} = \begin{cases} \text{linear momentum of mass} \\ \text{center } (\mathbf{G}) \\ \text{angular momentum about} \\ \text{mass center } (\mathbf{H}_C) \end{cases}$$

**APPLICATION OF PRINCIPLE OF IMPULSE AND MOMENTUM FROM TIME $t_1$ TO TIME $t_2$:**

$$\mathbf{G}_2 = \mathbf{G}_1 + (\text{external linear impulses})\big|_1^2$$

$$\mathbf{H}_{C_2} = \mathbf{H}_{C_1} + (\text{external angular impulses})\big|_1^2$$

**Kinetic energy of a rigid body in three-dimensional motion**

$$T = \underbrace{\frac{1}{2}mv_C^2}_{\substack{\text{translation} \\ \text{of mass center}}} + \underbrace{\frac{1}{2}\boldsymbol{\omega}\cdot\mathbf{H}_C}_{\substack{\text{rotation about} \\ \text{mass center}}}$$

For a rigid body that has a fixed point $O$,

$$T = \frac{1}{2}\boldsymbol{\omega}\cdot\mathbf{H}_O$$

**Equations of motion in three dimensions**

$$\sum\mathbf{F} = m\mathbf{a}_C$$

$$\sum\mathbf{M}_C = \dot{\mathbf{H}}_C = (\dot{\mathbf{H}}_C)_{xyz} + \boldsymbol{\Omega}\times\mathbf{H}_C$$

where  $\mathbf{a}_C$ = acceleration of mass center
$\mathbf{H}_C$ = angular momentum of the body about its mass center
$xyz$ = frame fixed in the body with origin at the mass center
$\boldsymbol{\Omega}$ = angular velocity of the $xyz$ frame with respect to a fixed $XYZ$ frame

An arbitrary fixed point $O$ may be used for reference (rather than point $C$) if used consistently.

**Euler's equations of motion.**  If the $xyz$ axes (fixed in the body) coincide with the principal axes of inertia of the body,

$$\sum M_x = I_x\dot{\omega}_x - (I_y - I_z)\omega_y\omega_z$$

$$\sum M_y = I_y\dot{\omega}_y - (I_z - I_x)\omega_z\omega_x$$

$$\sum M_z = I_z\dot{\omega}_z - (I_x - I_y)\omega_x\omega_y$$

where all quantities must be evaluated with respect to the appropriate principal axes.

**Solution of basic problems in three-dimensional motion.** Apply six independent scalar equations to analyze the motion of an unconstrained rigid body:

$$\sum F_x = ma_{C_x} \quad \sum F_y = ma_{C_y} \quad \sum F_z = ma_{C_z}$$

$$\sum M_x = \dot{H}_x + \omega_y H_z - \omega_z H_y$$

$$\sum M_y = \dot{H}_y + \omega_z H_x - \omega_x H_z$$

$$\sum M_z = \dot{H}_z + \omega_x H_y - \omega_y H_x$$

The equations of motion are simplified if the body has a fixed point or a fixed axis.

Section 21-7

**Gyroscopic motion**

**EQUATION OF MOTION:**

$$\sum \mathbf{M}_O = (\dot{\mathbf{H}}_O)_{xyz} + \mathbf{\Omega} \times \mathbf{H}_O$$

where $\mathbf{\Omega}$ is the angular velocity of the $xyz$ frame, which is fixed in the inner gimbal of the gyroscope.

Section 21-8

**Special cases of gyroscopic motion**

**STEADY PRECESSION:**

$$\sum \mathbf{M}_O = \mathbf{\Omega} \times \mathbf{H}_O$$

**PRECESSION AND SPIN AXES ARE AT A RIGHT ANGLE IN STEADY PRECESSION:**

$$\sum M_x = I_1 \dot{\phi} \dot{\gamma}$$

where   $x$ = axis fixed in inner gimbal, perpendicular to the spin axis
     $I_1$ = moment of inertia about the spin axis
     $\dot{\phi}$ = rate of precession
     $\dot{\gamma}$ = rate of spin

**21–62** A model aircraft is suspended at its center of mass $C$ by taut vertical wires. The fuselage is assumed to be a uniform cylinder of mass $M$, radius $r$, and length $l$, with the nose of the craft at $x = 0.5l$. The horizontal wings are assumed to be uniform slender rods of mass $m$ and length $0.6l$, attached to the fuselage at $x = 0.1l$ and $y = r$, and swept back at $45°$. The model is nosing over at the rate of $\omega$ when it is in the horizontal position. Determine its angular momentum about point $C$.

**Fig. P21-65**

**Fig. P21-62**
Model aircraft in a wind tunnel (Courtesy The Boeing Company, Seattle, Washington)

**21–66** A tire tester has two identical 30-lb wheels in a symmetric arrangement. Axles $AB$ and $AC$ have negligible weight and rotate at a constant angular speed of 100 rpm about the $y$ axis. The wheels are modeled as uniform disks of 1-ft radius rolling on the horizontal $xz$ plane. Determine the vertical reactions on each wheel if joint $A$ allows rotations of the axles in vertical planes.

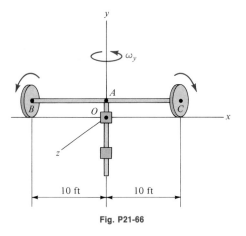

**Fig. P21-66**

**21–63** The model in Prob. 21–62 is at rest when it receives an impulse $\mathbf{F}\,\Delta t$ in the positive $x$ direction. $\mathbf{F}$ acts at $x = -0.2l$, $y = 0$, and $z = 0.2l$. Determine the resulting angular velocity of the model.

**21–64** Calculate the final kinetic energy of the model in Prob. 21–63.

**21–65** The flywheel of an experimental car is modeled as a uniform disk of 15-kg mass and 0.3 m radius. It is rotating at $\omega_y = 1257\,\text{rad/s}$ when braking of the car tilts $\omega_y$ toward the $x$ axis at the rate of 1 rad/s. Determine the couple acting on the disk's shaft caused by the braking.

**21–67** A space station is modeled as a thin ring of mass $m$ and average radius $r = 200\,\text{ft}$. It is rotating about its axis of symmetry $y$ at 2 rpm. A meteorite of mass $m_1 = m/10^5$ is traveling in the $xy$ plane, in the negative $y$ direction, at $v = 20{,}000\,\text{ft/s}$ relative to the station. The meteorite is embedded in the ring at $A$. Determine the angular velocity and angular momentum vectors of the space station immediately after the impact.

Fig. P21-67

relative to the spacecraft. The meteorite is embedded in the cylinder at point $A$. Determine the angular velocity and angular momentum vectors of the spacecraft immediately after the impact.

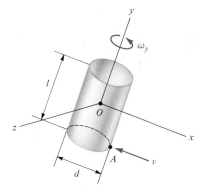

Fig. P21-68

**21–68** A spacecraft is modeled as a uniform cylinder of 2000-kg mass, 3-m length, and 2-m diameter. It is rotating about its axis of symmetry $y$ at 30 rpm. A 0.1-kg meteorite is traveling in the $xy$ plane, in the negative $x$ direction, at $v = 3000$ m/s

# 22 Vibrations

## OVERVIEW

A rigid body or a system of bodies is in vibration when its motion involves repeated oscillations about an equilibrium position. The oscillation may be repeated uniformly, or change with time. Vibrations in machines and structures are quite common and undesirable in most cases. Their undesirable effects may be classified with respect to human characteristics and damage to engineering structures. An extreme example is a slender skyscraper whose wind-induced oscillations are entirely safe for the structure yet unpleasant to the occupants on the upper floors. At the other extreme, certain vibrations in an airplane may be unnoticeable to the passengers yet cause damage (fatigue) with catastrophic consequences. Naturally, the simultaneous occurrence of unpleasant and structurally damaging vibration, such as those in cars and trucks, is the most common. Vibration is useful in certain testing machines, compactors, and pile drivers.

## STUDY GOALS

**SECTIONS 22-1 AND 22-2** present the basic characteristics and definitions of vibrating systems. The governing differential equation of motion of a simple system is described in detail. The student of vibrations should acquire skill in setting up and solving a variety of governing differential equations of motion, so it is essential to learn the basic method of analysis.

**SECTION 22-3** extends the concepts of vibrating particles to vibrating rigid bodies. It should be noted that the governing equation for a rigid body

888

depends not only on the mass of that body and the elastic members of the system, but also on the configuration of the particular system.

**SECTION 22-4** describes the analysis of undamped vibrating systems using the principle of conservation of energy. This method is advantageous when the maximum displacement and velocity of a body are involved in the analysis.

**SECTION 22-5** presents forced vibrations without damping, a useful approach in machine design and structural engineering work. The condition of resonance is a particularly important topic of study. Readers of this section may note that analyzing various vibration problems requires solving differential equations with different forcing functions.

**SECTIONS 22-6 THROUGH 22-8** present the analyses of vibrating systems with viscous damping, which frequently must be considered in engineering practice. The student should note the similarities and differences between damped and undamped vibrating systems of otherwise identical components.

**SECTION 22-9** describes electrical analogues of mechanical vibrations. This topic is not essential for the mechanics analysis of vibrations, but it is relevant to design practice, and it shows the value of a broad engineering education.

## SIMPLE HARMONIC MOTION 22-1

The basic characteristics of all vibrating systems are introduced using a simple system that consists of a mass $m$ and a spring of stiffness $k$. Figure 22-1a shows this system in equilibrium where the spring exerts no force on the block. Assume that the block is displaced to position $x$ on the frictionless surface by a *disturbing force* **P** (also called *shaking force* or *exciting force*). This force is opposed by the *restoring force* **F**, as shown

(a) Equilibrium position (force of spring is zero here)

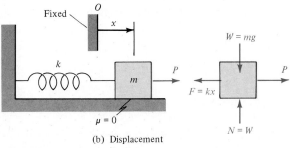

(b) Displacement

**FIGURE 22-1**

Model of a vibrating system

in the free-body diagram in Fig. 22-1b. The restoring force is always directed toward the equilibrium position. When the block is released in a position of displacement, the magnitude of the disturbing force **P** becomes zero and the restoring force **F** accelerates the block toward its equilibrium position. The equation of motion is written according to Newton's second law, with the positive direction taken to the right,

$$\sum F_x = ma_x$$
$$-kx = m\ddot{x}$$

22-1

This equation indicates that the block in Fig. 22-1b accelerates to the left when $P = 0$, as expected.

As the block is pulled to the left by the stretched spring, it has kinetic energy upon reaching the equilibrium positions $O$, overshoots this position, and compresses the spring. Now the restoring force is to the right, and the block accelerates to the right. Such changes in the direction of the restoring force cause the repeated oscillations of the block.

For a detailed analysis of the motion, Eq. 22-1 is written as

$$\ddot{x} + \omega^2 x = 0$$

22-2

where

$$\omega = \sqrt{k/m}$$

22-3

$\omega$ is called the *natural circular frequency* of the vibration. A motion for which Eq. 22-2 is valid is called *simple harmonic motion*. The same equation may also be obtained by considering a mass suspended by a spring and oscillating in the vertical direction. In that case the spring has an initial deformation in the equilibrium position.

Equation 22-2 is a linear differential equation of the second order with constant coefficients. The general solution of this equation is

$$x = C_1 \sin \omega t + C_2 \cos \omega t$$

22-4

where $t$ is time and $C_1$ and $C_2$ are arbitrary constants. The speed $v$ and acceleration $a$ of the vibrating mass are obtained by differentiating Eq. 22-4,

$$v = \dot{x} = C_1 \omega \cos \omega t - C_2 \omega \sin \omega t$$

22-5

$$a = \ddot{x} = -C_1 \omega^2 \sin \omega t - C_2 \omega^2 \cos \omega t$$

22-6

Substitution of Eqs. 22-4 and 22-6 into Eq. 22-2 is used to prove that Eq. 22-4 is the solution of the equation of motion. The constants $C_1$ and $C_2$ can be determined from the *initial conditions* of the motion.

For convenience, Eq. 22-4 is expressed as a single sine function by introducing new constants $A$ and $\phi$ such that $C_1 = A \cos \phi$ and $C_2 = A \sin \phi$. With these, Eq. 22-4 becomes

$$x = A \cos \phi \sin \omega t + A \sin \phi \cos \omega t$$

This expression is simplified by using the trigonometric identity $\sin (\alpha + \beta) = \sin \alpha \cos \beta + \cos \alpha \sin \beta$, resulting in

$$x = A \sin(\omega t + \phi)$$

22-7

**FIGURE 22-2**

Graph of simple harmonic displacements

The plot of this equation as $x$ vs. $\omega t$ in Fig. 22-2 is the typical graphic representation of vibratory motion. The maximum displacement of the vibrating mass from its equilibrium position ($x = 0$) is $A$, the amplitude of the vibration. The constant $\phi$ is called the *phase angle*, which allows defining the initial displacement $A \sin \phi$ of the mass at time $t = 0$. The time $\tau$ required to complete a *cycle* of the motion is the *period* of the motion. A complete cycle occurs when $\omega t = 2\pi$, so

$$\text{period} = \tau = \frac{2\pi}{\omega} = 2\pi \sqrt{\frac{m}{k}} \quad \text{(seconds per cycle)} \qquad \boxed{22\text{-}8}$$

after substituting for $\omega$ from Eq. 22-3. An important characteristic of a vibrating system is the rate at which complete cycles of the motion are repeated. This is defined by the *frequency* of the motion which represents the number of cycles that occur in a unit of time. The frequency $f$ is the the reciprocal of the period,

$$\text{frequency} = f = \frac{1}{\tau} = \frac{\omega}{2\pi} = \frac{1}{2\pi} \sqrt{\frac{k}{m}} \qquad \boxed{22\text{-}9}$$

The frequency $f$ is expressed in cycles per second (cps), with units of $1/\text{s}$. Frequency is denoted by *hertz* (Hz) in the SI system of units. The natural circular frequency $\omega$ has units of rad/s.

## Graphic Recordings of Simple Harmonic Motion

It is illustrative to consider two experiments for obtaining graphic vibration patterns. First, assume that a mass $m$ is suspended by a spring of stiffness $k$ (Fig. 22-3). A strip of paper can be moved near the mass to continually record the position of the mass. Assume that the mass is released at position $-x_0$ with the spring stretched when time $t = 0$. The paper moves at a constant speed $v$, and the oscillations of the mass are

**FIGURE 22-3**

Schematic of a vibration experiment with recording of displacements

recorded as in Fig. 22-3. The amplitude of the vibration is $x_0$, and time is measured to the right on the paper (as by an observer moving with the paper). It should be noted that the recorded amplitude is not dependent on the chart speed $v$, but the spacing of the oscillations is dependent on $v$. Thus, $v$ must be known for measuring the frequency from the paper. The recorded pattern for simple harmonic motion is always a sine wave, even though this may not be obviously at very low or very high chart speeds.

For another illustration, assume that a particle $P$ is moving at a constant speed on a fixed circular path as in Fig. 22-4. The vertical position of $P$ is recorded by its shadow on a horizontally moving chart. The recorded trace is a sine wave of amplitude $A$ (radius of the circle) when $\omega$ and $v$ are constant. This shows (in comparison with Fig. 22-3) that uniaxial vibrations may be associated with rotating members. This model also provides an interpretation for the meaning of the natural circular frequency $\omega$ since a complete cycle on the chart corresponds to the radius $A$ sweeping through an angle of $2\pi$ radians.

**FIGURE 22-4**

Illustration of simple harmonic motion

The basic concepts of analyzing a vibrating system that were developed in Sec. 22-1 are for a particular class of problems. The following list gives the characteristics of these problems and some practical considerations.

**1.** *One degree of freedom.* The motion of a mass *m* is defined by a single coordinate as a function of time (such as $x$ in Eqs. 22-2 and 21-7) Such a system is said to have *one degree of freedom.*

**2.** *Free vibration.* Equation 22-2 is valid when a disturbing force is applied only once to give the mass an initial displacement (as in plucking a taut string). The mass is in *free vibration* when only two kinds of forces are acting on it: (a) the elastic restoring force within the system, and (b) gravitational or other constant forces (such as an electric field) that cause no displacement from the equilibrium position of the system.

**3.** *Undamped vibration.* In the absence of dissipative forces acting on a vibrating mass, the amplitude of the vibration is constant, and the motion is said to be *undamped.* This is an idealization because all engineering systems have some dissipative forces (at least internal friction in the spring material). Nevertheless, Eq. 22-2 is useful to analyze many systems that oscillate in air.

**4.** *Natural frequency.* Each mass–spring system vibrates at a characteristic frequency in free vibration. This is known as the *natural frequency* of the system. It is readily calculated using Eq. 22-9,

$$f = \frac{1}{2\pi} \sqrt{\frac{k}{m}}$$

**5.** *Lumped parameters.* Strictly speaking, Eq. 22-2 is valid only for a particle of mass *m* and a spring of no mass and spring constant *k*. In practice this equation is used for other one-degree-of-freedom systems. For example, the mass of a translating rigid body is assumed to be concentrated in a particle, and the mass of the spring (whose mass center also moves) is often ignored resulting in a small error. Occasionally, several springs are connected to a single vibrating body. It is generally possible to represent these by a single equivalent spring constant.

## FREE VIBRATIONS OF RIGID BODIES

22-3

Mass-spring systems involving a rigid body vibrating in rotational motion about a fixed axis or a wheel vibrating in rolling motion along a flat surface are further examples of one-degree-of-freedom systems. In these cases, however, the particles of the body are in nonuniform motion, and it is necessary to derive an equation of motion which is appropriate for the particular case. For example, consider a concentrated mass $m_1$ on a uniform bar of mass $m_2$ which is hinged at point *A* and supported by a spring at point *B* as shown in Fig. 22-5a. The system is oscillating about the horizontal axis at *A*. The free-body diagram of the system in static

(a)

(c)

$A_y$

$A_x$

$B$   $C$

$A$

$m_2g$     $m_1g$

(b)

**FIGURE 22-5**

Model of a vibrating rigid body

equilibrium is given in Fig. 22-5b assuming that the bar is horizontal in equilibrium. $\delta$ is the static deflection of the spring. The force $k\delta$ is obtained after summing moments about $A$ and dividing by $l$,

$$k\delta = \frac{3}{2}m_1g + m_2g \qquad \boxed{22\text{-}10}$$

The free-body diagram for an arbitrary position of the bar during its oscillations is shown in Fig. 22-5c. If $\theta$ is always small, $\sin\theta \simeq \theta$ (in radians), and the force of the spring may be written as $k(\delta + l\theta)$.

The equation of motion is

$$\sum M_A = I_A\alpha = I_A\ddot{\theta} \qquad \boxed{22\text{-}11}$$

Assume that clockwise is positive in Fig. 22-5c. After substituting for the moments $M_A$ and the moments of inertia $I_A$,

$$\underbrace{-k(\delta + l\theta)l}_{\text{force arm}} + \underbrace{m_1g\left(\frac{3}{2}l\right)}_{\text{force arm}} + \underbrace{m_2g}_{\text{force arm}}\ (l)$$

$$= \left[\underbrace{m_2\frac{4l^2}{3}}_{(I_{\text{bar}})_A} + \underbrace{m_1\left(\frac{3}{2}l\right)^2}_{(I_{m_1})_A}\right]\theta$$

But from Eq. 22-10,

$$\left(\frac{3}{2}m_1g + m_2g\right)l - (k\delta)l = 0$$

Hence,

$$\left(\frac{9}{4}m_1 + \frac{4}{3}m_2\right)\ddot{\theta} + k\theta = 0 \qquad \boxed{22\text{-}12}$$

which is similar in form to Eq. 22-1. The term in parentheses in Eq. 22-12 is called the *effective mass* of the system. In general there may also be an *effective spring constant* which is numerically different from $k$. The details of these effective coefficients depend on the configuration of the system. The natural frequency of the system is determined by solving Eq. 22-9 using the effective mass and spring constant. Mass-spring systems having more than one degree of freedom are beyond the scope of this text and are covered in vibrations courses.

EXAMPLE 22-1

A demonstration model for power-line galloping (wind-induced vibrations of aerial electric wires) consists of a uniform bar of mass $m = 1\,\text{kg}$ and four springs of stiffness $k = 100\,\text{N/m}$ arranged as in Fig. 22-6a. Write the equation of motion for free vertical vibration of this system and determine its natural frequency.

## SOLUTION

The static equilibrium position is shown in Fig. 22-6b. With a static deformation $\delta$ of each spring,

$$\sum F_y = -4k\delta + mg = 0$$

It is seen here that the four springs have a resultant stiffness equivalent to a single spring of (effective) stiffness $K = 4k$. For a positive displacement $y$ (down) as shown in Fig. 22-6c,

$$\sum F_y = ma_y$$
$$-4k(y + \delta) + mg = m\ddot{y}$$

From the equilibrium equation, $-4k\delta + mg = 0$; hence, the equation of motion becomes

$$m\ddot{y} + 4ky = 0$$

or

$$\ddot{y} + \frac{4k}{m}\,y = 0$$

From Eq. 22-2,

$$\omega = \sqrt{\frac{4k}{m}} = \sqrt{\frac{4(100\,\text{N/m})}{1\,\text{kg}}} = \sqrt{400\left(\frac{\text{kg}\cdot\text{m}}{\text{s}^2}\right)\left(\frac{1}{\text{kg}\cdot\text{m}}\right)}$$

$$= 20\,\text{rad/s}$$

$$f = \frac{1}{2\pi}\,\frac{\text{cycle}}{\text{rad}}\,\omega = 3.18\,\text{cycles/s}$$

Note that the equilibrium deformation force $k\delta$ and weight $mg$ are again not involved in the final equation of motion. For simple harmonic motion and small deformations this is always true. Therefore, such forces can be (carefully) ignored when writing the equation of motion.

(a)

(b) Equilibrium position

(c)

**FIGURE 22-6**

Model for power-line galloping

EXAMPLE 22-1    895

## EXAMPLE 22-2

A simple pendulum consists of a concentrated mass $m$ suspended on a light string of length $l$ (Fig. 22-7a). Write the equation of motion of the pendulum for small amplitudes of displacement $\theta$. Determine how the period of oscillations depends on $m$ and $l$.

### SOLUTION

The appropriate free-body diagram is sketched in Fig. 22-7b. Here the weight is not involved in any initial deflections and therefore must be included in the equation of motion (see Ex. 21-1).

$$\sum F_t = ma_t$$

$$-W \sin \theta = m(l\ddot{\theta})$$

$$ml\ddot{\theta} + mg \sin \theta = 0$$

But, $\sin \theta \simeq \theta$ (in radians) is often used up to approximately 30° or 0.5 rad. Therefore, the equation of motion becomes

$$\ddot{\theta} + \left(\frac{g}{l}\right)\theta = 0$$

$$\omega = \sqrt{\frac{g}{l}}$$

$$\tau = \frac{2\pi}{\omega} = 2\pi\sqrt{\frac{l}{g}}$$

The period depends only on the length $l$ (and $g$) and not on the mass. The period increases with increasing $l$, which agrees with common experience.

It might be noted that the equation of motion could have also been obtained by summing moments about the support $A$,

$$\sum M_A = I_A \alpha$$

$$-mgl \sin \theta = (ml^2)\ddot{\theta}$$

$$\ddot{\theta} + \left(\frac{g}{l}\right)\theta = 0$$

which agrees with the previous result.

(a)

(b)

**FIGURE 22-7**

Analysis of an oscillating pendulum

EXAMPLE 22-3

The torsion pendulum of a satellite-tracking radar station is modeled as a 0.1-lb uniform disk of 2 in. diameter suspended on a thin rod of torsional stiffness $k$ (Fig. 22-8a). Determine $k$ in units of lb·in/rad if the natural frequency of the system should be 5 cycles per second.

## SOLUTION

First the equation of motion is written in general terms. A torsion spring resists twists with a torque that depends on the angle turned. Hence, $M_{sp} = k\theta$, and from the free-body diagram in Fig. 22-8b,

$$\sum M_A = I_A \alpha$$

$$-k\theta = I_A \ddot{\theta}$$

$$\ddot{\theta} + \left(\frac{k}{I_A}\right)\theta = 0$$

$$\omega = \sqrt{\frac{k}{I_A}}$$

$$f = \frac{1}{2\pi}\sqrt{\frac{k}{I_A}}$$

$$k = (2\pi f)^2 I_A = [2\pi(5 \text{ cps})]^2 I_A = \left(10\pi \frac{1}{s}\right)^2 I_A$$

$$= \left(100\pi^2 \frac{1}{s^2}\right)\left[\frac{1}{2}\left(\frac{0.1 \text{ lb}}{32.2 \text{ ft/s}^2}\right)\left(\frac{1 \text{ ft}}{12 \text{ in.}}\right)(1 \text{ in}^2)\right]$$

$$= 0.128 \text{ lb·in/rad}$$

## JUDGEMENT OF THE RESULT

The expression for the frequency appears reasonable since $f$ increases with $k$ and decreases with $I$. A stiffer spring will apply a larger torque for a given rotation hence it will cause a large angular acceleration and therefore more cycles per time period. An increase in the inertia $I$ can be considered as an increase in resistance to rotation. In other words, the system will be more "sluggish," resulting in less cycles per time period for an increase in $I$.

(a)

(b) Top view of disk

**FIGURE 22-8**

Analysis of an oscillating torsion pendulum

EXAMPLE 22-3    897

**22-1** A 1-kg mass is suspended from a spring of stiffness $k = 2\,\text{kN/m}$. Determine the period of oscillations of the mass.

**Fig. P22-1**

**22-2** Determine the natural frequencies of the system in Prob. 22–1 for the mass moving without friction in the (a) vertical direction, and (b) horizontal direction.

**22-3** A 2-lb block is supported by a spring of stiffness $k$. Determine the required stiffness to obtain the frequency of $5\,\text{s}^{-1}$.

**Fig. P22-3**

**22-4** Calculate the maximum velocity and the maximum acceleration of the block in Prob. 22–3 for an amplitude of 0.1 ft.

**22-5** A tuned mass damper has a mass of 300 000 kg and should have a 7-s period in free vibration. Calculate the required spring constant of the system (for a description of these, see Sec. 1-7 in the statics volume).

**Fig. P22-5**
Tuned mass damper in motion (Courtesy MTS Systems Corporation, Minneapolis, Minnesota

**22-6** A uniform, 10-kg bar is suspended by two springs whose spring constants are $k = 5\,\text{kN/m}$. Calculate the natural frequency of the bar in uniform vertical motion.

**Fig. P22-6**

**22-7** It is found that the average fruit freely vibrating in its container moves with an amplitude of 0.5 in. and a period of 0.3 s. Assuming that the motion is simple harmonic, calculate the maximum vertical acceleration of a fruit.

**22-8** A 5-lb block slides without friction on a horizontal surface. The two springs of spring constant $k = 500\,\text{lb/in.}$ are attached to the block but apply no force to it in the equilibrium position. Determine the natural frequency of the block.

**Fig. P22-8**

**22-9** Derive the equation for the natural frequency of the mass supported by two springs in series.

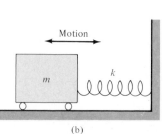

**Fig. P22-9**

Motion

(b)

**22–10** The pendulum has a length $l = 1$ m. Its angular speed is $\dot{\theta} = 0$ at $\theta_0 = 0.1$ rad. Calculate the period of the pendulum and write the equation for $\theta$ as a function of time if time is zero at $\theta_0$.

**Fig. P22-10**

**22–11** An elastic bar of negligible mass with a concentrated mass $m$ on its free end is firmly gripped at $A$. The static deflection of the free end is $\Delta$. Derive the equation for the natural frequency of the system.

**Fig. P22-11**

**22–12** An elastic bar of negligible mass with a concentrated mass $m$ on it is hinged at the ends. The static deflection of the bar is $\Delta$. Derive the equation for the natural frequency of the system.

**Fig. P22-12**

**22–13** A uniform, slender, rigid bar of 3 kg mass and 1 m length is hinged at $O$. Calculate the period of small oscillations of the bar.

**Fig. P22-13**

**22–14** Determine the maximum angular speed $\dot{\theta}$ and angular acceleration $\ddot{\theta}$ of the bar in Prob. 22–13 if the angular amplitude is 0.1 rad.

**22–15** A uniform, slender, rigid bar of weight $W_1 = 6$ lb and length $l = 4$ ft is hinged at $O$. A small block of weight $W_2 = 10$ lb is attached to the bar at $d = 3$ ft. Determine the required angular amplitude of this pendulum if the maximum speed of $W_2$ should be 1 ft/s.

**Fig. P22-15**

**22–16** A uniform, slender, rigid bar of 10 lb weight and 3 ft length is hinged at $O$. A torsion spring of stiffness $k = 500$ lb·ft/rad is attached between the bar and its support. The spring applies no moment to the bar when $\theta = 0$. Calculate the natural frequency of the system.

**Fig. P22-16**

**22–17** A uniform, slender, rigid bar of 4 kg mass and 1 m length is hinged at $O$ and supported by a spring of spring constant $k = 5$ kN/m at $d = 0.75$ m. Determine the natural frequency of the system assuming that the bar is attached to the spring.

**Fig. P22-17**

**22–18** A uniform, slender, rigid bar of 10 kg mass and length $2l = 1.6$ m is hinged at $O$ and is supported in the horizontal position by a spring of spring constant $k = 4$ kN/m. Determine the natural frequency of the system.

**Fig. P22-18**

**22–19** A uniform, thin plate of mass $m$ is hinged at $O$. Determine the period of small oscillations of the plate.

**Fig. P22-19**

**22–20** The modern building is modeled as a uniform, rigid bar of mass $m$, cross section $a \times a$, and height $h$. It is assumed to be hinged at $O$ with a torsion spring of stiffness $k$ restraining the building motions. Write the equation for the natural frequency of the system.

**Fig. P22-20**

**22–21** A slender rod of 1 kg mass and 0.4 m length is suspended at its center by a wire. Determine the torsional spring constant $k$ of the wire if the period of torsional vibrations is 2 s.

**Fig. P22-21**

**22–22** A wheel is suspended by a wire of torsional stiffness $k = 3$ lb·in/rad. The period of torsional vibration is 12 s. Calculate the moment of inertia of the wheel.

**Fig. P22-22**

**22–23** A 5-lb block is suspended from a wire that passes over a uniform disk of 6 lb weight. The disk rotates freely about its axle, and the spring constant is $k = 100$ lb/in. Determine the natural frequency of the system if the wire does not slip on the disk.

**Fig. P22-23**

**22–24** An arbitrary body of mass $m$ and mass center at $C$ is suspended by the hinge at $O$. The radius of gyration about $C$ is $r_g$. Determine the period of small oscillations.

**Fig. P22-24**

Systems that vibrate freely without damping are conservative systems that can be analyzed using the principle of conservation of energy. Thus, the total mechanical energy of the system is constant, total energy = kinetic energy + potential energy = constant,

$$T + V = \text{constant}$$

The principle can be used to determine the equation of vibratory motion and the natural frequency of a system as illustrated in the following. Consider a mass–spring system where the mass can move without friction on a horizontal surface (Fig. 22-9). The total energy of this system with the mass $m$ at position $x$ (measured from the equilibrium position) and moving at speed $\dot{x}$ is

$$T + V = \frac{1}{2} m\dot{x}^2 + \frac{1}{2} kx^2 = \text{constant} \qquad \boxed{22\text{-}13}$$

Differentiating this equation with respect to time gives

$$m\dot{x}\ddot{x} + kx\dot{x} = 0$$

This equation holds when $\dot{x} = 0$ but it is true for all time $t$ when

$$\boxed{m\ddot{x} + kx = 0} \qquad \boxed{22\text{-}14}$$

This result agrees with Eq. 22-1. The natural circular frequency is $\omega = \sqrt{k/m}$, as in Eqs. 22-3 and 22-9.

The energy method of vibration analysis is particularly advantageous when those positions of the mass are considered where the total energy consists entirely of potential energy or kinetic energy. These posi-

**FIGURE 22-9**

Model of a vibrating system

tions are shown schematically in Fig. 22-10 and further explained as follows.

1. At every peak displacement $A$ the kinetic energy $T$ is zero as $\dot{x}$ changes direction. At the same instants, the potential energy $V$ is maximum and represents the total energy of the system.
2. At every instant $B$ when the mass passes through its equilibrium position $x = 0$, the potential energy $V$ is zero. At the same instants, the kinetic energy $T$ is maximum since the speed $\dot{x}$ is maximum. This value of $T$ represents the total energy of the system.

The analysis of the motion is facilitated by using the simple expressions that can be written for the maximum displacement $x_{\text{max}}$ and speed $\dot{x}_{\text{max}}$. From Eq. 22-7,

$$x = A \sin (\omega t + \phi)$$ 22-15

$$x_{\text{max}} = A$$

$$\dot{x} = A\omega \cos (\omega t + \phi)$$ 22-16

$$\dot{x}_{\text{max}} = A\omega$$

Using these results gives

$$V_{\text{max}} = \frac{1}{2} kA^2 \qquad \text{(the total energy of the system)}$$ 22-17

$$T_{\text{max}} = \frac{1}{2} mA^2\omega^2 \qquad \text{(also the total energy of the system)}$$ 22-18

Since $V_{\text{max}} = T_{\text{max}}$, the natural circular frequency is

$$\omega = \sqrt{\frac{k}{m}}$$

which agrees with Eq. 22-3.

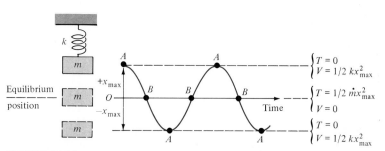

**FIGURE 22-10**

Kinetic and potential energies at equilibrium and peak displacements

EXAMPLE 22-4

A rocker arm is modeled as a uniform slender rod of mass $m$ hinged at point $O$ as in Fig. 22-11a. The spring touches the rod but applies no force to it when the rod is vertical. Derive an expression for the circular frequency $\omega$ of the system for small displacements.

## SOLUTION

Two positions of the bar are of interest. Assume that position 1 in Fig. 22-11b represents a maximum displacement where the bar has no velocity. In position 2 the velocity of the bar is maximum but the system's potential energy is chosen to be zero. Using $x_{max} = 2l\theta_{max}$,

$$T_1 = 0$$

$$V_1 = \frac{1}{2}k(2l)^2\theta_{max}^2 + \frac{1}{2}Wl\theta_{max}^2$$

where the last term is obtained using

$$1 - \cos\theta_{max} = 2\sin^2\frac{\theta_{max}}{2} \simeq 2\left(\frac{\theta_{max}}{2}\right)^2 = \frac{\theta_{max}^2}{2}$$

Also,

$$V_2 = 0$$

$$T_2 = \frac{1}{2}I_O\dot\theta_{max}^2 \qquad \text{(note: } \omega \neq \dot\theta \text{ here)}$$

since the rigid bar is in pure rotation about point $O$. Using $V_{max} = T_{max}$ yields

$$\frac{1}{2}k(2l)^2\theta_{max}^2 + \frac{1}{2}Wl\theta_{max}^2 = \frac{1}{2}I_O\dot\theta_{max}^2$$

From Eqs. 22-15 and 22-16,

$$x_{max} = A = 2l\theta_{max} \qquad \theta_{max} = \frac{A}{2l}$$

$$\dot x_{max} = A\omega = 2l\dot\theta_{max} \qquad \dot\theta_{max} = \frac{A\omega}{2l}$$

With $I_O = \frac{1}{2}m(2l)^2$,

$$\omega = \sqrt{\frac{2k}{m} + \frac{g}{2l}}$$

## JUDGMENT OF THE RESULT

The quantities $k$, $m$, $g$, and $l$ are all in reasonable places in the equation according to previous results for oscillating systems. For comparison, see Exs. 22-1 and 22-2.

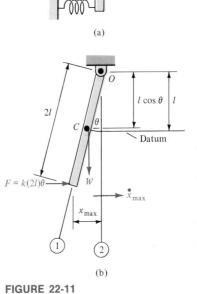

**FIGURE 22-11**

Model of a rocker-arm system

EXAMPLE 22-4    903

# PROBLEMS

In Probs. 22-25 to 22-36, solve the indicated problem using energy methods.

**22-25**  Prob. 22-1.

**22-26**  Prob. 22-2.

**22-27**  Prob. 22-6.

**22-28**  Prob. 22-8.

**22-29**  Prob. 22-13.

**22-30**  Prob. 22-15.

**22-31**  Prob. 22-16.

**22-32**  Prob. 22-17.

**22-33**  Prob. 22-18.

**22-34**  Prob. 22-19.

**22-35**  Prob. 22-21.

**22-36**  Prob. 22-23.

**22-37**  A uniform disk of mass $m_1 = 5$ kg and radius $r = 0.2$ m is supported at its axle $O$. A torsion spring of stiffness $k = 800$ N·m/rad causes angular oscillations of the disk. Two concentrated masses $m_2 = 1$ kg are placed diametrically on the disk. Calculate the period of vibrations of the system.

**Fig. P22-37**

**22-38**  The axle of a 10-lb uniform disk is connected to a spring of stiffness $k = 50$ lb/in.; $r = 8$ in. Determine the period of vibrations of the system for $\theta = 0$. The disk rolls without slipping.

**Fig. P22-38**

**22-39**  Determine the natural frequency of the system in Prob. 22-38 for $\theta = 30°$.

**22-40**  A thin, circular hoop of mass $m$ and radius $r$ is suspended at $O$. Determine the natural frequency of small oscillations in the $xy$ plane.

**Fig. P22-40**

**22-41**  Solve Prob. 22-40 for oscillations of the hoop's center $C$ in the $yz$ plane.

**22-42**  Fluid of mass $m$ and density $\rho$ occupies a segment of length $l$ of the U-tube. The column of fluid is displaced to $x_0$ from the equilibrium position and starts to oscillate with negligible friction. Determine the period of oscillations.

**Fig. P22-42**

In many important vibration problems encountered in engineering work, the exciting force is applied periodically during the motion. These problems are called *force vibration problems*. The most common periodic force is a *harmonic function of time* such as

$$P = P_0 \sin \Omega t \qquad \boxed{22\text{-}19}$$

where $P_0$ is a constant, $\Omega$ is the *forcing frequency*, and $t$ is time. The motion is analyzed using Fig. 22-12, where the magnitude of the force **P** is generally not equal to the magnitude of the restoring force **F** of the spring. Summing forces in the horizontal direction with positive $x$ to the right, the equation of motion in terms of the forcing function $P$ is

$$\boxed{m\ddot{x} + kx = P_0 \sin \Omega t} \qquad \boxed{22\text{-}20}$$

Considering different problems, a forced vibration may be described in terms of the displacement $d$ of a foundation or primary mass $M$ to which the vibrating system is attached. This type of problem is illustrated in Fig. 22-13, where $d$ is measured with respect to a fixed frame $A$. Here the magnitude of the restoring force **F** depends on the difference of the displacements $x$ and $d$. The equation of motion is written according to Newton's second law with the positive direction to the right. From $F = m\ddot{x}$,

$$m\ddot{x} + k(x - d_0 \sin \Omega t) = 0 \qquad \boxed{22\text{-}21}$$
$$m\ddot{x} + kx = kd_0 \sin \Omega t$$

where $d_0$ is the amplitude of vibration of the primary mass $M$, $\Omega$ is the frequency of motion of $M$, and $t$ is time. Equations 22-20 and 22-21 are of the same form, and they have the same solution if $P_0 = kd_0$.

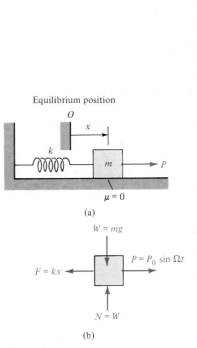

(a)

(b)

**FIGURE 22-12**

Model for analyzing forced vibrations

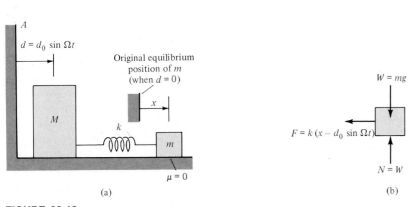

(a)

(b)

**FIGURE 22-13**

Forced vibration in terms of displacement of a primary mass

The *general solution* of Eq. 22-20 or 22-21 (nonhomogeneous, second-order differential equations) consists of two parts, $x = x_c + x_p$, where $x_c$ is the *complementary solution* and $x_p$ is the *particular solution*. The complementary solution is obtained by setting the right side of Eq. 22-20 or 22-21 equal to zero. The solution of this equation is Eq. 22-24 for a freely vibrating system,

$$x = C_1 \sin \omega t + C_2 \cos \omega t \qquad \boxed{22\text{-}22}$$

where $\omega = \sqrt{k/m}$ and $C_1$ and $C_2$ are arbitrary constants.

For the particular solution, it is reasonable to assume a harmonic function since the motion is harmonic. Let

$$x_p = A \sin \Omega t \qquad \boxed{22\text{-}23}$$

where $A$ is the amplitude, a constant. The value of this amplitude is determined by substituting Eq. 22-23 into Eq. 22-20,

$$-m\Omega^2 A \sin \Omega t + kA \sin \Omega t = P_0 \sin \Omega t$$

From this equation with $\omega = \sqrt{k/m}$,

$$\boxed{A = \frac{P_0}{k - m\Omega^2} = \frac{P_0/k}{1 - (\Omega/\omega)^2}} \qquad \boxed{22\text{-}24}$$

where $\omega$ is the natural circular frequency of the system of $m$ and $k$, and $\Omega$ is the forcing frequency.

The *general solution* of the forced vibration without damping is $x = x_c + x_p$,

$$\boxed{x = \underbrace{C_1 \sin \omega t + C_2 \cos \omega t}_{\substack{\text{free vibration} \\ \text{(transient)}}} + \underbrace{A \sin \Omega t}_{\substack{\text{forced vibration} \\ \text{(steady-state)}}}} \qquad \boxed{22\text{-}25}$$

The two terms of free vibration are dependent only on properties $m$ and $k$ of the system and on the initial conditions that affect the constant $C_1$ and $C_2$. These define the initial natural vibration of the system (e.g., after the forcing function suddenly begins or stops). This is also called *transient vibration* because in a real system it is damped out by friction.*

The single term representing forced vibration in Eq. 22-25 depends on the amplitude of the applied force or displacement, on the forcing frequency $\Omega$, and on the circular frequency $\omega$ of the system (because $A$ depends on $\Omega/\omega$ according to Eq. 22-24). This is also called *steady-state* vibration since it is the motion of the system after a transient vibration is dissipated.

---

* Some readers may notice the contradiction that the equation of free vibration was obtained assuming negligible damping, while here the transient vibration is said to be damped out by friction. This is a difficult subject to discuss, but here it may suffice to say that even small friction has a cumulative effect in reducing the amplitudes of the free vibration. Thus, the latter eventually becomes negligible *in comparison* with the forced component of the vibration.

## Resonance

A particularly important aspect of force vibrations is that the amplitude of the oscillations depends on the frequency ratio $\Omega/\omega$. According to Eq. 22-24, the amplitude becomes infinite when $\Omega = \omega$. This condition is called *resonance*. Of course, the amplitude does not become infinite in practice because of damping or physical constraints, but the condition is often a dangerous one (it can cause fractures). Dangerously large amplitudes may also occur at other frequency ratios near the resonant (natural) frequency. The various conditions can be readily compared graphically, as shown in the following.

Consider the ratio of the amplitude $A$ of the steady-state vibration to the maximum static deflection $P_0/k$ that the exciting force **P** would cause. This ratio is called the *magnification factor* and can be expressed in three ways using Eqs. 22-20, 22-21, and 22-24:

$$\text{magnification factor} = \frac{A}{P_0/k} = \frac{A}{d_0} = \frac{1}{1 - (\Omega/\omega)^2} \qquad \boxed{22\text{-}26}$$

This quantity is plotted as a function of the frequency ratio $\Omega/\omega$ in Fig. 22-14. Several items are of particular interest in this diagram:

**1. Static loading.** $\Omega = 0$; $A/(P_0/k) = 1$. The loading is also essentially static when $\Omega \ll \omega$, for then $A/(P_0/k) \simeq 1$.

**2. Resonance.** $\Omega = \omega$; $A/(P_0/k) = \infty$

**3. High-frequency excitation.** $\Omega \gg \omega$; $A/(P_0/k) \simeq 0$; the mass remains essentially stationary because of its inertia (the system cannot respond quickly enough).

**4. Phase relationships.** The sign of the magnification factor indicates whether the direction of motion of the vibrating mass is the same as that of the exciting force or displacement. The vibration is *in phase* for $\Omega < \omega$, and it is *180° out of phase* for $\Omega > \omega$.

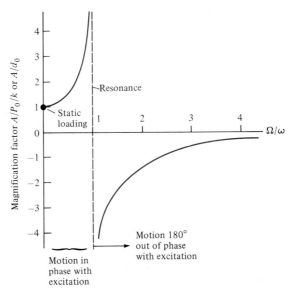

**FIGURE 22-14**

Magnification factor in forced vibrations

EXAMPLE 22-5

A 1500-kg truck cab is assumed to be a uniform block symmetrically supported by four springs, each with a stiffness of 120 kN/m, on top of four hydraulic actuators $A$ (Fig. 22-15). Determine the resonance frequency of the cab in units of Hz, and the amplitude of the vibration if the displacement input of each actuator is $d = 0.05 \sin 6t$ m.

## SOLUTION

For this motion, with $x$ measured from the static equilibrium position,

$$\sum F_x = ma_x$$

$$-4k(x - d) = m\ddot{x}$$

$$\ddot{x} + 4kx = 4k(0.05 \sin 6t) \text{ m}$$

This is now in the form of Eq. 22-21, with an effective stiffness, $k_{\text{eff}} = 4k$.

$$\omega = \sqrt{\frac{k_{\text{eff}}}{m}} = \sqrt{\frac{4k}{m}}$$

$$f = \frac{\omega}{2\pi} = \frac{1}{2\pi} \sqrt{\frac{4(120 \times 10^3 \text{ N/m})}{1500 \text{ kg}}} = 2.85 \text{ s}^{-1}$$

From Eq. 22-24,

$$\frac{P_0}{k_{\text{eff}}} = (4k)\frac{d_0}{4k} = d_0 = 0.05 \text{ m}$$

$$A = \frac{d_0}{1 - (\Omega/\omega)^2} = \frac{0.05 \text{ m}}{1 - (6/17.9)^2}$$

$$= 0.0563 \text{ m}$$

**FIGURE 22-15**

Road-simulation test of a truck cab (Courtesy MTS Systems Corporation, Minneapolis, Minnesota)

**22–43** A periodic force of magnitude $P = 20 \sin 30t$ lb is acting on a 10-lb block which is supported by a spring of stiffness $k = 150$ lb/in. Calculate the amplitude of vibration of the block.

**Fig. P22-43**

Motion

$k$

$W$

$P = P_0 \sin \Omega t$

**22–44** In Prob. 22–43, determine the value of $\Omega$ for which the amplitude of the vibration is 0.2 in.

**22–45** A 200-kg shake table $T$ moves horizontally with negligible friction on its tracks. The actuator rod $A$ moves under displacement control defined by $d = 0.15 \sin 12t$ m. Plot the amplitude of vibration of the table if it is connected to rod $A$ through a spring of stiffness $k = 10$ kN/m, 100 kN/m, and 1 MN/m.

**Fig. P22-45**

**22–46** An accelerometer consists of a 0.01-lb block on a spring of stiffness $k$. Determine $k$ if the amplitude $A$ of the block's vibration should be 0.1 in. when the base $O$ has vertical cyclic displacements of $d = 0.005 \sin 500t$ in.

Motion

$W$

$k$

$O$

**Fig. P22-46**

**22–47** Determine the amplitude of the block's vibration in Prob. 22–46 for the forcing frequency reduced by 50%, using the value of $k$ calculated in Prob. 22–46.

**22–48** Determine the range of values of the forcing frequency $\Omega$ for which the amplitude of the vibration is less than 1% of the static deflection caused by $P_0$, a constant force.

$k$

$m$

$P = P_0 \sin \Omega t$

**Fig. P22-48**

**22–49** The machine on the shake table has a component that is modeled as a 0.1-kg mass on the end of a light cantilever beam. In a test the machine is vibrated vertically with displacements of $d = 0.03 \sin 40t$ m. The resulting amplitude $A$ of the small mass is 2 mm. Calculate the spring constant $k$ of the beam.

**Fig. P22-51**
NASA-Langley three-degree-of-freedom (vertical, horizontal, roll) human factors research motion simulator with 7 ft × 7 ft welded aluminum table (Courtesy MTS Systems Corporation, Minneapolis, Minnesota)

(a)

(b)

**Fig. P22-49**
Vibration test of a gardening implement (Photograph courtesy MTS Systems Corporation, Minneapolis, Minnesota

**22–52** An interesting problem of tuned-mass dampers of skyscrapers (see Sec. 1-7 in the statics volume) is how to test them mechanically. The best method is to wait for a windless day and force the heavy block to oscillate with respect to the building. This eventually causes substantial vibrations of the building if the system is properly designed and constructed. Consider a 600,000-lb block with a spring of stiffness $k = 1500$ lb/in. Determine the magnitude of the required force **P** of the hydraulic actuator to move the block with an amplitude of 50 in. and a period of 7 s in a perfectly rigid building.

**22–50** The shake table in Fig. P22-49 consists of a 50-kg platform with the test piece and a 1200-kg base on four air springs. It is desired to apply a vertical force of magnitude $P = \sin 15t$ kN to the platform while limiting the amplitude of vibration of the base to 1 mm. Determine the required spring constant of the air springs. There is no spring between the upper platform and the heavy base.

**22–51** A 70-kg astronaut is belted to a chair of stiffness $k = 10$ kN/m. The vertical acceleration of the astronaut should not exceed 3 g. Determine the allowable amplitude $d_0$ of the massive platform at a frequency of 10 Hz.

**Fig. P22-52**
Model of a tuned-mass damper system (Courtesy MTS Systems Corporation, Minneapolis, Minnesota)

## DAMPING OF VIBRATIONS

The ever-present damping of vibrating systems is one of the most easily demonstrated phenomena in dynamics when used in free vibrations. The decaying amplitudes of plucked strings and small cantilever beams show that the initial mechanical energy of the system (potential and kinetic energy) is gradually dissipated. The mechanical energy may be directly converted into heat by *dry friction* (also called *Coulomb friction*) between rigid bodies, or by *internal friction* during deformation of a solid material. The mechanical energy of a vibrating system may also be dissipated by *fluid friction* in a gas or a liquid which temporarily acquires some kinetic energy. For all practical purposes, the mechanical energy in vibrations is ultimately converted into heat.

The damping caused by fluid friction is called *viscous damping*. This is quite common in engineering work and is relatively easy to analyze using the idealization that *viscous damping is directly proportional to the speed* of the body in the fluid at low speeds. The presence of this damping is modeled by a *dashpot*, which consists of a piston $A$ moving in a cylinder $B$ as illustrated in Fig. 22-16a. The relative motion of the piston with respect to the cylinder is caused by the external force $\mathbf{F}$ and resisted by the fluid friction between the moving parts. The elemental friction force $f$ is uniformly distributed on the circumference of the piston as implied in Fig. 22-16b. As stated previously, the resultant friction force is proportional to velocity and is denoted by $c\dot{x}$ in Fig. 22-16c, where $\dot{x}$ is the relative velocity of the piston in the cylinder, and the constant $c$ is the *coefficient of viscous damping*. The value of $c$ depends on the fluid and its temperature (inverse dependence on temperature) and on the construction of the dashpot. The units of the coefficient $c$ are N·s/m or lb·s/ft. Dashpots are real devices such as the shock absorbers of automobiles, but sometimes they are also used in the analytical modeling of systems that do not have distinct damping devices. The mass of the dashpot is generally neglected.

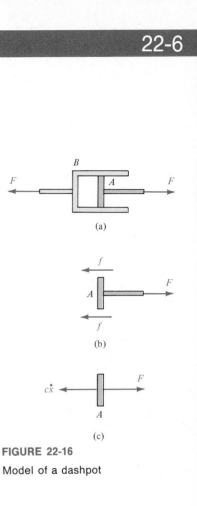

**FIGURE 22-16**

Model of a dashpot

## DAMPED FREE VIBRATIONS

The analysis of a damped, freely vibrating system is illustrated in Fig. 22-17. Assume that the mass $m$ is moving to the right so that the friction force $c\dot{x}$ is to the left. With $x$, $\dot{x}$, and $\ddot{x}$ taken as positive to the right, the equation of motion of the mass from the free-body diagram in Fig. 22-17b is

$$-c\dot{x} - kx = m\ddot{x}$$

according to Newton's second law. This is rewritten as

$$\boxed{m\ddot{x} + c\dot{x} + kx = 0} \qquad \text{22-27}$$

which is a linear, second-order, homogeneous, differential equation. It has solutions of the form

$$x = e^{\lambda t} \qquad \text{22-28}$$

where $e$ is the base of the natural logarithm, $\lambda$ is a constant, and $t$ is time. Substituting this solution in Eq. 22-27 and dividing the resulting equation by $e^{\lambda t}$,

$$m\lambda^2 + c\lambda + k = 0 \qquad \text{22-29}$$

**FIGURE 22-17**

Model of a damped vibrating system

which has two roots according to the quadratic formula,

$$\lambda = \frac{-c}{2m} \pm \sqrt{\left(\frac{c}{2m}\right)^2 - \frac{k}{m}}$$

22-30

Thus, the general solution of Eq. 22-27 is the sum of two exponentials with $\lambda_1$ and $\lambda_2$ defined by Eqs. 22-28 and 22-30, as discussed shortly.

An important parameter for describing free vibrations is obtained when the term under the radical in Eq. 22-30 is zero. The value of $c$ that satisfies this condition is called the *critical damping coefficient* $c_c$. It is determined from Eq. 22-30 that

$$\left(\frac{c_c}{2m}\right)^2 = \frac{k}{m} \qquad \text{or} \qquad c_c = 2m\sqrt{\frac{k}{m}} = 2m\omega$$

22-31

where $\omega$ is the natural circular frequency of the system without damping. There are three special cases of damping that can be distinguished with respect to the critical damping coefficient.

**1. Overdamped system.** $c > c_c$. The roots $\lambda_1$ and $\lambda_2$ of Eq. 22-29 are real and distinct. The general solution of Eq. 22-27 is

$$x = Ae^{\lambda_1 t} + Be^{\lambda_2 t}$$

22-32

where $A$ and $B$ are constants. The exponents $\lambda_1$ and $\lambda_2$ are both negative, so $x$ decreases from its initial value $x_0 = A + B$ toward zero as time $t$ increases. This motion is *nonvibratory* (or *aperiodic*), as illustrated in Fig. 22-18.

**2. Critically damped system.** $c = c_c$. In this case $\lambda_1 = \lambda_2 = -c_c/2m = -\omega$. The general solution of Eq. 22-27 is

$$x = (A + Bt)e^{-\omega t}$$

22-33

This motion is also nonvibratory but it is of special interest because $x$ decreases at the fastest possible rate without oscillation of the mass.

**3. Underdamped system.** $c < c_c$. The roots $\lambda_1$ and $\lambda_2$ of Eq. 22-29 are complex numbers. It can be shown that the general solution of Eq. 22-27 is

$$x = Ae^{-(c/2m)t} \sin(\omega_d t + \phi)$$

22-34

where $A$ and $\phi$ are constants depending on the initial conditions of the problem (similar to those in Eq. 22-7). The constant $\omega_d$ is defined as the *damped natural frequency* of the system and is expressed as

$$\omega_d = \omega\sqrt{1 - \left(\frac{c}{c_c}\right)^2}$$

22-35

where $\omega = \sqrt{k/m}$, the natural frequency for undamped vibration. The ratio $c/c_c$ is called the *damping factor* $\zeta$.

Equation 22-34 defines harmonic oscillations of diminishing amplitude as shown in Fig. 22-19. The amplitude is $Ae^{-(c/2m)t}$.

The *period of the damped vibration* $\tau_d$ is defined as the time interval between successive humps where they touch one of the exponential envelopes of the oscillations. It is interesting to note that the period is constant even though the amplitude decreases,

$$\tau_d = \frac{2m}{\omega_d}$$

22-36

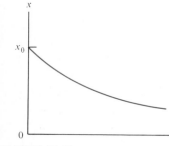

**FIGURE 22-18**

Nonvibratory motion of an overdamped system

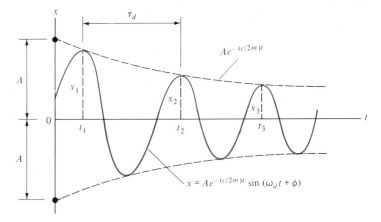

FIGURE 22-19

Oscillations of an underdamped system

The period of damped vibration is always larger than the period of the same system without damping, according to Eqs. 22-35 and 22-36.

## DAMPED FORCED VIBRATIONS

Assume that a periodic force of magnitude $P = P_0 \sin \Omega t$ is acting horizontally on mass $m$ in Fig. 22-20. The appropriate equation of motion in this case becomes

$$m\ddot{x} + c\dot{x} + kx = P_0 \sin \Omega t \qquad \boxed{22\text{-}37}$$

A similar equation is applicable when a support of the system moves periodically with displacement $d = d_0 \sin \Omega t$ as in Sec. 22-5. That requires replacing $P_0$ in Eq. 22-37 by $kd_0$.

Equation 22-37 is a nonhomogeneous differential equation, so its general solution consists of a complementary solution $x_c$, and a particular solution $x_p$. The complementary solution is the same as for Eq. 22-27, and it is given by Eq. 22-32, 22-33, or 22-34, depending on the damping of the system. These are *transient vibrations* and they rapidly diminish in most cases. The particular solution that applies to the *steady-state vibration* of the system should be a harmonic function of time, such as

$$x_p = A \sin (\Omega t - \phi) \qquad \boxed{22\text{-}38}$$

where $A$ and $\phi$ are constants. Substituting $x_p$ into Eq. 22-37 gives

$$-m\Omega^2 A \sin (\Omega t - \phi) + c\Omega A \cos (\Omega t - \phi)$$
$$+ kA \sin (\Omega t - \phi) = P_0 \sin \Omega t$$

The constant coefficients $A$ and $\phi$ in this equation can be determined by first setting $\Omega t - \phi = 0$ and then setting $\Omega t - \phi = \pi/2$, giving

$$c\Omega A = P_0 \sin \phi \qquad \boxed{22\text{-}39}$$

$$(k - m\Omega^2)A = P_0 \cos \phi \qquad \boxed{22\text{-}40}$$

The *phase angle* $\phi$ (phase difference between the applied force and the resulting vibration) is determined from the ratio of Eqs. 22-39 and 22-40,

$$\tan \phi = \frac{c\Omega}{k - m\Omega^2} \qquad \boxed{22\text{-}41}$$

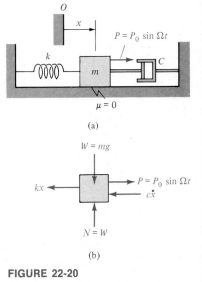

FIGURE 22-20

Model for forced vibrations of a damped system

The sine and cosine functions in Eqs. 22-39 and 22-40 can be eliminated by summing the squares of both equations,

$$A^2[(c\Omega)^2 + (k - m\Omega^2)^2] = P_0^2$$

22-42

The amplitude $A$ is expressed as

$$A = \frac{P_0}{\sqrt{(c\Omega)^2 + (k - m\Omega^2)^2}}$$

22-43

A magnification factor for the amplitude of the vibration can be defined as in the case of undamped forced vibration (Eq. 22-26). Using $c_c = 2m\omega$ and $\omega = \sqrt{k/m}$ gives

$$\begin{aligned}\text{magnification factor} &= \frac{A}{P_0/k} = \frac{A}{d_0} \\ &= \frac{1}{\sqrt{[2(c/c_c)(\Omega/\omega)]^2 + [1 - (\Omega/\omega)^2]^2}}\end{aligned}$$

22-44

This quantity is plotted as a function of the frequency ratio $\Omega/\omega$ for different damping factors in Fig. 22-21. The following items are of interest in this diagram:

**1. Static loading.** $\Omega = 0$; $A/(P_0/k) = 1$, independent of the damping (agrees with Eq. 22-24).

**2. Resonance.** $\Omega = \omega$; the amplitude is magnified substantially when the coefficient of viscous damping $c$ is low.

**3. High-frequency excitation.** $\Omega \gg \omega$; $A/(P_0/k) \simeq 0$; the mass remains essentially stationary because of its inertia, regardless of any damping of its motion.

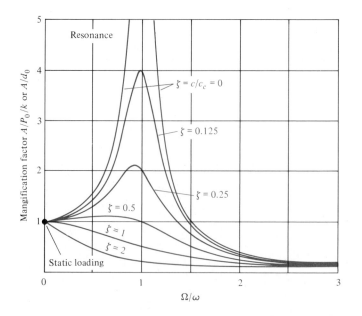

**FIGURE 22-21**

Magnification factor in damped forced vibrations

**4. Large coefficient of viscous damping.** The amplitude of the vibration is reduced at all values of $\Omega/\omega$ as the coefficient of viscous damping $c$ is increased in a particular system.

The phase angle $\phi$ can also be determined using the same parameters as in Eq. 22-44,

$$\tan\phi = \frac{2(c/c_c)(\Omega/\omega)}{1 - (\Omega/\omega)^2} \qquad \boxed{22\text{-}45}$$

## ELECTRICAL ANALOGUES OF MECHANICAL VIBRATIONS

**22-9**

Electrical circuits with transient or steadily oscillating currents are common. Interestingly, for each mechanical system there is an electrical circuit analogy. The fact that the oscillations of the two analogous systems can be analyzed using the same differential equation is of practical value because an electrical circuit is generally more easily constructed and modified than the corresponding mechanical system. Thus, a mechanical system can be designed, evaluated, and refined using an appropriate electrical circuit.

The analogy between electrical and mechanical systems is illustrated using the simple series circuit in Fig. 22-22. The source of alternating voltage $E = E_0 \sin\Omega t$ generates a current $i = dq/dt$ in the circuit, where $q$ denotes electric charge. The drop in electric potential (voltage) around the circuit is $L(di/dt)$ across the inductor, $Ri$ across the resistor, and $q/C$ across the capacitor. The sum of the voltage drops around the circuit equals the applied voltage according to Kirchhoff's second law,

$$L\ddot{q} + R\dot{q} + \frac{1}{C}q = E_0 \sin\Omega t \qquad \boxed{22\text{-}46}$$

which is identical in form to Eq. 22-37 since $L$, $R$, and $C$ are constants. The mechanical–electrical analogues can be established by comparing these two equations. The results are given in Table 22-1. These analogues allow the student to perform at least simple experiments to simulate the behavior of mechanical systems using electrical circuits.

**FIGURE 22-22**

Series electric circuit

**TABLE 22-1**

Mechanical–Electrical Analogues

| MECHANICAL SYSTEM | | ELECTRICAL SYSTEM | |
| --- | --- | --- | --- |
| Mass | $m$ | Inductance | $L$ |
| Coefficient of viscous damping | $c$ | Resistance | $R$ |
| Spring constant | $k$ | Reciprocal of capacitance | $1/C$ |
| Displacement | $x$ | Charge | $q$ |
| Velocity | $\dot{x}$ | Current | $i$ |
| Applied force | $P$ | Applied voltage | $E$ |

# EXAMPLE 22-6

An automobile tested in a laboratory is modeled as a 1500-kg mass on a spring of stiffness $k = 200\,\text{kN/m}$. The system has a damping factor of $c/c_c = 0.4$. Assume that the spring and dashpot are attached to the base whose vertical displacements are defined by $d = 0.04 \sin 6t$ m (Fig. 22-23a). Write the equation of motion of $m$ for steady-state vibration. Determine the magnification factor of the amplitude of vibration, the amplitude $A$, and the phase angle $\phi$.

## SOLUTION

Summing forces on the automobile, and noting that the weight and static spring deflection force cancel, gives

$$\sum F_y = ma_y$$

$$-k(y - d) - c\dot{y} = m\ddot{y}$$

$$m\ddot{y} + c\dot{y} + k(y - d_0 \sin \Omega t) = 0$$

$$m\ddot{y} + c\dot{y} + ky = kd_0 \sin \Omega t$$

This equation is of the form of Eq. 22-37. The natural circular frequency is

$$\omega = \sqrt{\frac{k}{m}} = \sqrt{\frac{200 \times 10^3\,\text{N/m}}{1500\,\text{kg}}}$$

$$= 11.5\,\text{s}^{-1}$$

From Eq. 22-44,

$$\text{magnification factor} = \frac{A}{d_0} = \frac{1}{\sqrt{(2c\Omega/c_c\omega)^2 + [1 - (\Omega/\omega)^2]^2}}$$

$$= \frac{1}{\sqrt{[(2)(0.4)(6/11.5)]^2 + [1 - (6/11.5)^2]^2}}$$

$$= 1.19$$

$$A = (1.19)d_0 = (1.19)(0.04\,\text{m}) = 0.0476\,\text{m}$$

$$c = 0.4c_c = 0.4(2m\omega) = 13{,}800\,\frac{\text{N·s}}{\text{m}}$$

Solving for the phase angle using Eq. 22-41,

$$\phi = \tan^{-1}\left(\frac{c\Omega}{k - m\Omega^2}\right)$$

$$= \tan^{-1}\left[\frac{13{,}800\,\dfrac{\text{N·s}}{\text{m}}(6\,\text{s}^{-1})}{200 \times 10^3\,\text{N/m} - 1500\,\text{kg}\,(6\,\text{s}^{-1})^2}\right]$$

$$= \tan^{-1}(0.567)$$

$$= 29.6°$$

$d = d_0 \sin \Omega t$

(a)

$+y$

$c\dot{y}$

$k(y - d_0 \sin \Omega t)$

(b)

**FIGURE 22-23**

Model of a damped system in forced vibrations

**22–53** A 2-kg mass is in series with a spring and a dashpot. The spring stiffness is $k = 500 \, \text{N/m}$ and the coefficient of viscous damping is $c = 100 \, \text{N} \cdot \text{s/m}$. The mass is released from rest at an initial displacement $x_0 = 0.05 \, \text{m}$. Determine the general solution of the vibrations.

**22–54** A 5-lb block is in series with a spring and a dashpot. The spring has stiffness $k = 100 \, \text{lb/in.}$ and the coefficient of viscous damping is $c = 20 \, \text{lb} \cdot \text{s/ft}$. The block has initial speed $\dot{x}_0 = 3 \, \text{ft/s}$ at the initial displacement $x_0 = 0$. Determine the general solution of the vibrations.

**22–55** Determine the period of damped vibrations in Prob. 22–54 for $k = 100 \, \text{lb/in.}$ and $500 \, \text{lb/in.}$

**22–56** A 100-kg instrument is resting on four springs whose static deflection is $0.02 \, \text{m}$. The mass is deflected an additional $0.03 \, \text{m}$ by an external force and released. The damping factor is $c/c_c = 0.1$. Determine the general solution of the vibrations.

**22–57** Calculate the time $t_1$ in which the amplitude of vibrations in Prob. 22–56 diminishes to 1% of its initial value

**22–58** A 2000-lb vehicle is designed with an energy-absorbing, self-recovering bumper. The light bumper is positioned by two springs of stiffness $k = 600 \, \text{lb/in.}$ Two shock absorbers of viscous damping $c$ also resist the deflection of the springs in an impact. Determine the coefficient $c$ that allows the most rapid compression of the springs without tending to cause oscillations. Assume that the bumper becomes wedged in a massive body during impact.

Top view of car

**Fig. P22-58**

**22–59** Consider an overdamped system of a mass $m$, spring stiffness $k$, and damping coefficient $c$ ($c > c_c$). Show that $m$ never passes its equilibrium position $x = 0$ after being released from rest at an arbitrary deflection $x_0$.

**22–60** Show that the result of Prob. 22–59 is also valid if $m$ has an arbitrary initial speed $\dot{x}_0$ at its equilibrium position $x = 0$.

**22–61** Consider an underdamped system of mass $m$, spring stiffness, $k$, and damping coefficient $c$ ($c < c_c$). Figure 22-19 applies in this case. Assume that $x_1, x_2, \ldots$ measured to the exponential envelope represent maximum displacements. Show that $x_1/x_2 = x_2/x_3 = \text{constant}$, and that for the $n$th cycle,

$$D = \ln \frac{x_n}{x_{n+1}} = \frac{2\pi(c/c_c)}{\sqrt{1 - (c/c_c)^2}}$$

The quantity $D$ is known as the *logarithmic decrement*.

**22–62** Prove the alternative expression for the logarithm decrement in Prob. 22–61,

$$D = \frac{1}{N} \ln \frac{x_a}{x_b}$$

where $N$ is the number of cycles between cycle $a$ and cycle $b$.

**22–63** Use the result of Prob. 22–62 to determine the damping factor $c/c_c$ of a system consisting of a 10-kg mass and a spring of stiffness $k = 5 \, \text{kN/m}$. The amplitude of the vibration decreases from 20 mm to 10 mm in 100 cycles.

**22–64** Solve Prob. 22–43 assuming that a dashpot of damping coefficient $c = 3 \, \text{lb} \cdot \text{s/ft}$ is connected parallel with the spring.

**22–65**  A 20-kg mass is shaken by a force of magnitude $P = 30 \sin 40t$ N. $k = 500$ N/m, and $c = 10$ N·s/m. Determine the amplitude of the steady-state vibrations.

Fig. P22-65

**22–66**  Solve Prob. 22–51 assuming that the chair has viscous damping of $c = 100$ N·s/m.

**22–67**  A simple fatigue testing machine is modeled as a mass $M = 20$ kg which is constrained to move vertically. A mass $m = 1$ kg is rotating at 1800 rpm on a 0.2-m-diameter path centered at point $O$ on block $M$. Calculate the amplitude of vibration of $M$ if $k = 20$ kN/m and $c = 300$ N·s/m.

Fig. P22-67

In Problems 2–68 to 2–72 determine the analogous electrical circuit and write the differential equation for it.

**22–68**

$P = P_0 \sin \Omega t$  Fig. P22-68

**22–69**

Fig. P22-69

**22–70**

Fig. P22-70

**22–71**

Fig. P22-71

**22–72**

Fig. P22-72

For a system of a particle (mass $m$) and spring (spring constant $k$) unless otherwise noted.

Section 22-1

**Simple harmonic motion**

**NATURAL CIRCULAR FREQUENCY:** $\omega = \sqrt{\dfrac{k}{m}}$

**PERIOD:** $\tau = \dfrac{2\pi}{\omega} = 2\pi\sqrt{\dfrac{m}{k}}$

**FREQUENCY:** $f = \dfrac{1}{\tau} = \dfrac{\omega}{2\pi} = \dfrac{1}{2\pi}\sqrt{\dfrac{k}{m}}$

Section 22-2

**Free vibrations of particles without damping.** Free vibration occurs when only two kinds of forces act on a mass: (a) elastic restoring force, and (b) constant force.

**NATURAL FREQUENCY:** $f = \dfrac{1}{2\pi}\sqrt{\dfrac{k}{m}}$

Section 22-3

**Free vibrations of rigid bodies.** The differential equation of motion for a rigid body depends on the mass of that body, the elastic members of the system, and on the configuration of the particular system.

Section 22-4

**Energy methods in vibration analysis.** For an undamped system,

$$T + V = \text{constant}$$

**DISPLACEMENT:** $x = A\sin(\omega t + \phi)$

$$x_{\max} = A$$

$$\dot{x}_{\max} = A\omega$$

$$V_{\max} = \frac{1}{2}kA^2 \qquad T_{\max} = \frac{1}{2}mA^2\omega^2 \qquad \omega = \sqrt{k/m}$$

Section 22-5

**Forced vibrations without damping**

**PERIODIC FORCE:** $P = P_0\sin\Omega t$

**PERIODIC DISPLACEMENT OF PRIMARY MASS:** $d = d_0\sin\Omega t$

where $\Omega$ is the forcing frequency.

**AMPLITUDE:** $A = \dfrac{P_0}{k - m\Omega^2} = \dfrac{P_0/k}{1 - (\Omega/\omega)^2}$

where $\omega = \sqrt{k/m}$.

**MAGNIFICATION FACTOR:** $\dfrac{A}{P_0/k} = \dfrac{A}{d_0} = \dfrac{1}{1 - (\Omega/\omega)^2}$

**RESONANCE:** $\Omega = \omega$

Section 22-6

**Damping of vibrations**

$$\text{Coefficient of viscous damping} = c$$

$$\text{Friction force in viscous damping} = c\dot{x}$$

Section 22-7

**Damped free vibrations**

$$\text{Critical damping coefficient} = c_c$$

**DAMPING FACTOR:** $\zeta = \dfrac{c}{c_c}$

**OVERDAMPED SYSTEM:** $c > c_c$    (nonvibratory motion)

**CRITICALLY DAMPED SYSTEM:** $c = c_c$    (nonvibratory motion)

**UNDERDAMPED SYSTEM:** $c < c_c$    (vibratory motion)

**DAMPED NATURAL FREQUENCY:** $\omega_d = \omega \sqrt{1 - \left(\dfrac{c}{c_c}\right)^2}$

Section 22-8

**Damped forced vibrations**

**AMPLITUDE:** $A = \dfrac{P_0}{\sqrt{(c\Omega)^2 + (k - m\Omega^2)^2}}$

**MAGNIFICATION FACTOR:** $\dfrac{A}{P_0/k} = \dfrac{A}{d_0} = \dfrac{1}{\sqrt{[2(c/c_c)(\Omega/\omega)]^2 + [1 - (\Omega/\omega)^2]^2}}$

**RESONANCE:** $\Omega = \omega$

For phase angle $\phi$,   $\tan \phi = \dfrac{c\Omega}{k - m\Omega^2} = \dfrac{2(c/c_c)(\Omega/\omega)}{1 - (\Omega/\omega)^2}$

Section 22-9

**Electrical analogues of mechanical vibrations.** Compare Eq. 22-37 (mechanical system) and Eq. 22-46 (electrical system). Use Table 22-1 to establish an analogous system.

**22–73** A 1000-kg rocket is suspended by a hinge at $O$. Determine the centroidal moment of inertia of the rocket if its period of small oscillations is 8 s.

Fig. P22-73

**22–74** A 4000-lb helicopter is suspended by a hinge at $O$. Determine the centroidal moment of inertia (transverse axis) of the helicopter if its period of small oscillations is 12 s.

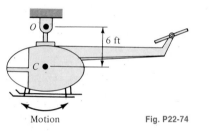

Motion          Fig. P22-74

**22–75** An amusement ride is modeled as a small block of mass $m$ sliding inside a smooth surface of radius $R$. Determine the period of small oscillations. Use energy methods.

Fig. P22-75

**22–76** An amusement ride is modeled as a uniform disk of mass $m$ and radius $r$ rolling without slipping inside a curved track of radius $R$. Determine the period of small oscillations. Use energy methods.

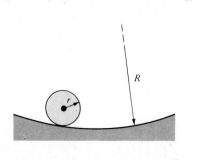

Fig. P22-76

**22–77** A vibrating pile driver is modeled as a 200-lb block on two springs of stiffness $k = 400\,\text{lb/in}$. The desired peak displacement of the base is $d_0 = 0.05\,\text{in}$. when the device is operated at 10 cycles per second. Determine the amplitude of vibrations of the block.

Fig. P22-77

**22–78** Determine the required forcing amplitude $P_0$ in Prob. 22–77.

**22–79** The pile driver in Prob. 22–77 has a coefficient of viscous damping $c = 30\,\text{lb·s/ft}$. Determine the amplitude of vibrations of the block.

Fig. P22-79

**22–80** Determine the analogous electrical circuit of the system and write the differential equation for it.

Fig. P22-80

# A Vector Derivatives

The student of dynamics will find the following vector analysis of general interest. Consider a vector **A** defined at time $t$ as

$$\mathbf{A} = A\mathbf{n}_A$$

where $A$ is the magnitude of the vector and $\mathbf{n}_A$ is a unit vector in the direction of **A**. Assume that the vector **A** changes to $\mathbf{A} + \Delta\mathbf{A}$ in the time interval $\Delta t$, as shown in Fig. A-1. The unit vector $\mathbf{n}_n$ is normal to $\mathbf{n}_A$ and lies in the plane of **A** and $\mathbf{A} + \Delta\mathbf{A}$. $\Delta\theta$ is the angular change of vector **A** in this plane. At the limit, the rotation $\Delta\theta$ is represented by a vector $d\boldsymbol{\theta}$ normal to $\mathbf{n}_A$ and $\mathbf{n}_n$ according to the right-hand rule. Note that the infinitesimal rotation $d\boldsymbol{\theta}$ is a vector whereas the finite rotation $\Delta\theta$ is not a vector. This is due to the fact that infinitesimal angular displacements are commutative and hence add according to the parallelogram law, but that finite angular displacements do not. This difference was explained in Ch. 10 of the accompanying *Statics* volume. It is repeated here for convenience with the help of Fig. A-2. First consider $\Delta\theta_x = -90°$ and $\Delta\theta_z = +90°$ rotations about the $x$ and $z$ axes, respectively. These are shown according to the right-hand rule in Fig. A-2a. Point $P$ rotates to different positions $P_1$ and $P_2$ depending on the order of the $\Delta\theta_x$ and $\Delta\theta_z$ rotations. $P_1$ is obtained when $\Delta\theta_x$ is followed by $\Delta\theta_z$ (the latter does not move the point to a new position). $P_2$ is the final location when $\Delta\theta_z$ is followed by $\Delta\theta_x$ (the latter does not move the point to a new position).

The infinitesimal rotations are shown in Fig. A-2b. Clearly, $P$ moves

**FIGURE A-1**

922

to the same point $P_2$ whether the sequence of rotations is $d\theta_x + d\theta_z$ or $d\theta_z + d\theta_x$. Thus, these rotations are commutative. Consequently, an infinitesimal rotation $d\theta$ is a vector, but a finite rotation, $\Delta\theta$ is not a vector. The infinitesimal angular displacements $d\theta$ add according to the parallelogram law.

The time derivative of vector $\mathbf{A}$ is expressed using the chain rule as

$$\dot{\mathbf{A}} = \frac{d\mathbf{A}}{dt} = \frac{d}{dt}(A\mathbf{n}_A) = \dot{A}\mathbf{n}_A + A\dot{\mathbf{n}}_A \qquad \boxed{\text{A-1}}$$

This can be expressed in a more convenient form after noting the following:

$$\dot{\mathbf{n}}_A = \frac{d\mathbf{n}_A}{dt} = \lim_{\Delta t \to 0}\left[\frac{(\mathbf{n}_A + \Delta\mathbf{n}_A) - \mathbf{n}_A}{\Delta t}\right] = \lim_{\Delta t \to 0}\left[\frac{\Delta\mathbf{n}_A}{\Delta t}\right] \qquad \boxed{\text{A-2}}$$

$$|\mathbf{n}_A| = |\mathbf{n}_A + \Delta\mathbf{n}_A| = 1 \qquad \text{by definition of the} \qquad \boxed{\text{A-3}}$$
$$\qquad\qquad\qquad\qquad\qquad\qquad \text{unit vector}$$

$$|\Delta\mathbf{n}_A| \simeq |\mathbf{n}_A|\,\Delta\theta = \Delta\theta \qquad \boxed{\text{A-4}}$$

Since $\Delta\mathbf{n}_A$ is perpendicular to $\mathbf{n}_A$ at the limit,

$$\dot{\mathbf{n}}_A = \lim_{\Delta t \to 0}\left[\frac{\Delta\theta}{\Delta t}\right]\mathbf{n}_n = \dot{\theta}\mathbf{n}_n \qquad \boxed{\text{A-5}}$$

Define the angular velocity vector of $\mathbf{A}$ as

$$\boldsymbol{\omega}_A = \frac{d\boldsymbol{\theta}}{dt}$$

which is in the same direction as $d\theta$ as shown in Fig. A-3. Then Eq. A-5 becomes

$$\dot{\mathbf{n}}_A = \dot{\theta}\mathbf{n}_n = \boldsymbol{\omega}_A \times \mathbf{n}_A \qquad \boxed{\text{A-6}}$$

With these considerations, Eq. A-1 is written as

$$\dot{\mathbf{A}} = \underbrace{\dot{A}\mathbf{n}_A}_{\text{magnitude change}} + \underbrace{\boldsymbol{\omega}_A \times \mathbf{A}}_{\text{direction change}} \qquad \boxed{\text{A-7}}$$

where all quantities are defined in the same arbitrary reference frame. Equation A-7 is one of the most important and most generally useful equations in dynamics.

## Special Cases

1. Vector $\mathbf{A}$ has constant direction. $\boldsymbol{\omega}_A = 0$, and

$$\dot{\mathbf{A}} = \dot{A}\mathbf{n}_A \qquad \boxed{\text{A-8}}$$

2. Vector $\mathbf{A}$ has constant magnitude. $\dot{A} = 0$, and

$$\dot{\mathbf{A}} = \boldsymbol{\omega}_A \times \mathbf{A} \qquad \boxed{\text{A-9}}$$

   Here the vector $\dot{\mathbf{A}}$ is perpendicular to the vector $\mathbf{A}$.

3. Arbitrary unit vector $\mathbf{n}$. Since the magnitude of $\mathbf{n}$ is always constant,

$$\dot{\mathbf{n}} = \boldsymbol{\omega}_n \times \mathbf{n} \qquad \boxed{\text{A-10}}$$

where $\boldsymbol{\omega}_n$ is the angular velocity of $\mathbf{n}$.

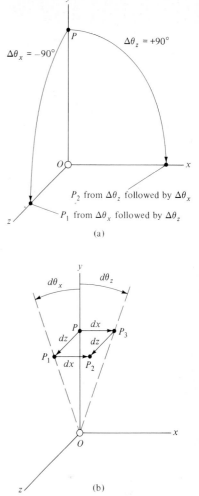

(a)

(b)

**FIGURE A-2**

Distinction of infinitesimal and finite rotations

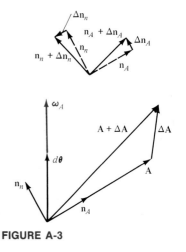

**FIGURE A-3**

# B Comments on the Use of Units

The two correct systems of units to be used in kinetics are described in Sec. 13-2. In addition, the reader should note that several other systems of units have been widely used until recently, and these may be encountered from time to time. In that case it is necessary to seek clarification of the precise meaning of each unit and use appropriate conversion factors. An even greater problem arises in commercial practice, as discussed in the following.

The units of mass and weight are often not distinguished in commercial practice. This is true throughout the world, regardless of the officially adopted technical system of units. For example, the kilogram is widely used as a measure of weight in many countries. In the study of dynamics it is necessary to understand the difference between mass and weight and use the appropriate units and conversion factors if necessary. An overview of the relation between mass and weight in the two systems is shown in Fig. B-1. Parts (a) and (b) of this figure show the *correct* units and terminologies to be used in engineering work. Parts (c) and (d) show the technically *incorrect* systems that have widespread usage in everyday life. Kilogram as force is sometimes denoted by kgf, and pound as mass is sometimes denoted by lbm. Unfortunately, even these clarifying notations are often omitted.

**FIGURE B-1**

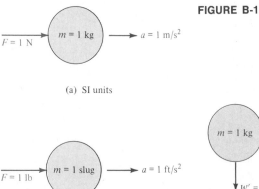

(a) SI units

(b) U.S. customary units

(c) kg as a commercial unit of weight, not to be used in technical work

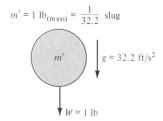

(d) lb as a commercial unit of mass, not to be used in technical work

# Inertia Force and Dynamic Equilibrium C

The French philosopher D'Alembert (1717–1783) suggested an intriguing concept in dynamics by making an apparently trivial restatement of Newton's second law. He wrote from $\mathbf{F} = m\mathbf{a}$,

$$\mathbf{F} - m\mathbf{a} = 0 \qquad \boxed{\text{C-1}}$$

The implication of this is that if $-m\mathbf{a}$ is considered a force and added to all of the other forces acting on the particle, the analysis is reduced to one of statics. This idea is illustrated using the force triangle in Fig. C-1. The triangle is closed when the fictitious force $-m\mathbf{a}$ is taken into account, creating the dynamic equivalent of static equilibrium, called *dynamic equilibrium*.

  The quantity $-m\mathbf{a}$ is called the *inertia force*. Its direction is always opposite to the direction of the particle's acceleration. The philosophical merits of this concept are controversial since cause and effect are debatable (the inertia force is the reaction to the force which is causing the acceleration of the body). The concept of the inertia force is further illustrated in the following two examples.

  **1. Inertia force in linear motion.** Consider a vehicle traveling on a straight path and accelerating. The passengers feel that they are forced in a direction opposite to the direction of the acceleration of the vehicle.

  **2. Centrifugal force.** Consider a vehicle traveling at a constant speed on a circular path. The force that keeps each passenger on the curved path is the *centripetal force*, which is directed toward the center of the path. The passengers feel that they are forced away from the center of the path. This fictitious force is called *centrifugal force*.

  Again, the inertia force felt by the people is in a direction opposite to the acceleration of the vehicle, which is entirely toward the center of the path if the angular velocity is constant (Fig. C-2). In spite of such human sensations, the fictitious inertia forces and dynamic equilibrium should not be used in elementary dynamics for several reasons:

  1. The application of dynamic equilibrium is confusing to most students because it is fictitious.

  2. Imagining dynamic equilibrium is not necessary to solve any problems.

(a) $m$ accelerates
($\mathbf{F}_{res} = \mathbf{F}_1 + \mathbf{F}_2$)

(b) $m$ accelerates

(c) Dynamic equilibrium

**FIGURE C-1**

3. The method of dynamic equilibrium becomes increasingly cumbersome (even mind-boggling) as the complexity of the problems increases.

4. Most dynamicists prefer to use the basic method of drawing free-body diagrams with real forces and applying the equations of motion according to these drawings. This is the simplest method in solving the majority of problems.

$F_n$ (centripetal force)

$F_n = \dfrac{mv^2}{\rho}$

$-F_n$ (centrifugal force) (fictitious force)

Path

**FIGURE C-2**

The method of dynamic equilibrium is illustrated in solving Ex. 19-1 as follows.

## EXAMPLE C-1

The combined mass of the motorcycle and its rider in Fig. C-3a is 250 kg. The mass center is at $C$. The wheels are locked when the brakes are applied at a high speed. Determine the linear acceleration and the normal and tangential forces acting on each wheel assuming that the coefficient of kinetic friction is 0.5, and that the angular acceleration is zero.

### SOLUTION

The free-body diagram is drawn in Fig. C-3b, showing both the actual forces and the inertial force $-m\mathbf{a}_{C_x}$ applied to the center of mass. The constraints of motion are that $a_{C_y} = 0$ and $\alpha = 0$. Applying the static equilibrium equations directly to Fig. C-3b,

$$\sum F_x = \mathscr{F}_A + \mathscr{F}_B - ma_{C_x} = 0 \tag{a}$$

$$\sum F_y = N_A + N_B - W = 0 \tag{b}$$

$$\sum M_C = \mathscr{F}_A\,(0.7\,\text{m}) - N_A\,(0.8\,\text{m}) + \mathscr{F}_B\,(0.7\,\text{m}) + N_B\,(0.5\,\text{m}) = 0 \tag{c}$$

Since the wheels are locked and sliding, the magnitudes of the friction forces are $\mathscr{F}_A = \mu_k N_A$ and $\mathscr{F}_B = \mu_k N_B$. Hence, from Eqs. (b) and (c),

$$N_A + N_B = W = (250\,\text{kg})(9.81\,\text{m/s}^2) = 2453\,\text{N}$$

$$\mu_k N_A\,(0.7\,\text{m}) - N_A\,(0.8\,\text{m}) + \mu_k N_B\,(0.7\,\text{m}) + N_B\,(0.5\,\text{m}) = 0$$

$$N_A\,(0.35 - 0.8)\,\text{m} + N_B\,(0.35 + 0.5)\,\text{m} = 0$$

Solving these equations simultaneously yields

$$N_A = 1604\,\text{N} \qquad\qquad N_B = 849\,\text{N}$$

$$\mathscr{F}_A = \mu_k N_A = 802\,\text{N} \qquad \mathscr{F}_B = \mu_k N_B = 425\,\text{N}$$

Hence, from Eq. (a),

$$802\,\text{N} + 425\,\text{N} - (250\,\text{kg})a_{C_x} = 0$$

$$a_{C_x} = \frac{1}{250\,\text{kg}}\,(802 + 425)\,\text{N} = 4.91\,\text{m/s}^2 \qquad \text{as in Ex. 19-1}$$

It should be noted that for the moment equation moments could have been summed about one of the contact points instead of the mass center $C$. For example,

$$\sum M_A = N_B(1.3\,\text{m}) - W(0.8\,\text{m}) + (ma_{C_x})(0.7\,\text{m}) = 0$$

$0.7$ m

$0.4$ m

$0.5$ m

$1.3$ m

(a)

$-ma_{C_x}$

$\mathscr{F}_A$   $W$   $\mathscr{F}_B$

$N_A$   $N_B$

(b)

**FIGURE C-3**

926   INERTIA FORCE AND DYNAMIC EQUILIBRIUM

# Coriolis Effects D

For a qualitative understanding of the *Coriolis acceleration*, consider Fig. D-1. Assume that persons $A$ and $B$ are standing on a disk which rotates at a constant angular speed $\Omega$ in a horizontal plane. Person $A$ throws a ball perfectly aimed in the direction of person $B$, as shown in the top view of the disk in Fig. D-1b. In this situation the ball would miss person $B$ for two reasons.

First, the ball has not only the relative velocity $\mathbf{v}_{rel}$ from $A$ to $B$ at the instant of release, but also a rotational velocity $\mathbf{v}_{rot}$ because $A$ is rotating with the disk. Thus, the actual direction of the ball is determined by the vector sum $\mathbf{v}_{abs} = \mathbf{v}_{rel} + \mathbf{v}_{rot}$. The resulting error in the ball's path to the position $B_1$ of person $B$ at the instant of release is the *Coriolis drift* 1. An additional error, Coriolis drift 2, occurs since $B$ moves into a new position $B'$ while the ball flies (Fig. D-1b).

The Coriolis drift can be eliminated if the ball moves in a groove or along a rail from $A$ to $B$. Such constraints accelerate the ball to correct its path, and it will reach person $B$.

**FIGURE D-1**

Notes: ① = Coriolis drift 1 (ball would miss fixed point $B_1$ because of $\mathbf{v}_{rot}$ of $A$ at instant ball is released)

② = Coriolis drift 2 (ball would further lag behind $B$ because $B$ moves into new position $B'$ while ball flies)

In general, ① ≠ ②

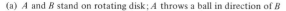

• $B_1$ (fixed, temporarily under $B$ when ball is released)

(a) $A$ and $B$ stand on rotating disk; $A$ throws a ball in direction of $B$

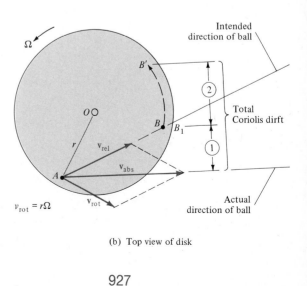

(b) Top view of disk

927

Note that the two components of the Coriolis drift discussed above are meant to clarify the physical situation, while the factor 2 in $\mathbf{a}_{Cor} = 2\boldsymbol{\Omega} \times \mathbf{v}_{rel}$ (such as in Eq. 18-36) is obtained from the mathematical derivation using a rotating reference frame. However, it should be clear that a larger $\Omega$ would increase the drift and would necessitate a larger acceleration to correct the path. Also, a larger $v_{rel}$ means a larger magnitude of the velocity vector $\mathbf{v}_{abs}$; hence, a larger acceleration is necessary to change the direction toward $B'$, the actual target (Fig. D-1b).

## EXAMPLE D-1

Assume that person $A$ in Fig. D-1 is standing at the center of the disk while throwing the ball. Discuss the Coriolis drift in this situation.

### SOLUTION

Consider two reference systems: the fixed $XY$ frame, and the $xy$ axes which are painted on the rotating disk. Let $\mathbf{v}_{rel} = v_{rel}\mathbf{i}$ when the ball is released by $A$, as shown in Fig. D-2a. In this case $\mathbf{v}_{rel} = \mathbf{v}_{abs}$, and the path of the ball with respect to the rotating $xy$ frame is the curved line in Fig. D-2b. Note that in this special case the Coriolis drift 1 is zero since $r = 0$ in $v_{rot} = r\Omega$ (Fig. D-1b). Thus, the entire drift consists of Coriolis drift 2: the target $B$ moves to $B'$ as the ball travels to the fixed point $B_1$ (Fig. D-2b).

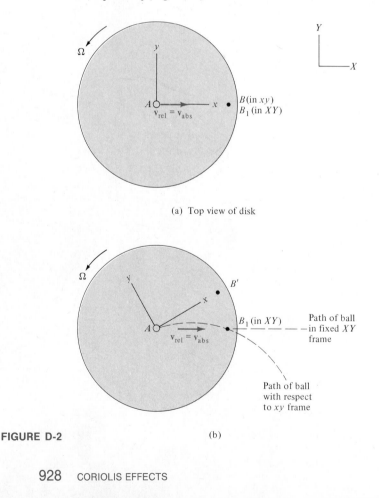

(a) Top view of disk

(b)

**FIGURE D-2**

928   CORIOLIS EFFECTS

# Answers to Even-Numbered Problems

## Chapter 2

**2-2**   $R = 1130\,\text{lb} \,\angle\, 8°$
**2-4**   $F_2 = 490\,\text{lb}; R = 750\,\text{lb}$
**2-6**   (a) $R = 31.8\,\text{kN} \,\angle\, 88.8°$
         (b) $\theta = 68°$
**2-8**   $R = 1900\,\text{N} \,\angle\, 77°$
**2-10**  $R = 12.1\,\text{kN} \,\angle\, 17.6°$
**2-12**  $R = 2.24 \times 10^5\,\text{lb} \,\angle\, 49.8°$
**2-14**  $R = 9.35\,\text{MN} \,\angle\, 76.8°$
**2-16**  $F_H = 19{,}319\,\text{lb}; F_V = 5176\,\text{lb}$
**2-18**  $F_{\text{plane}} = 34.1\,\text{N}; F_{\text{perp}} = 193.2\,\text{N}$
**2-20**  $\mathbf{F}_{AB} = 773\,\text{lb} \,\angle\, 75°; \mathbf{F}_{AC} = 773\,\text{lb} \,\angle\, 75°$
**2-22**  $\mathbf{R} = (150\mathbf{i} - 1126\mathbf{j})\,\text{lb}$
**2-24**  $F_2 = 490\,\text{lb}$
**2-26**  (a) $\mathbf{R} = 31.8\,\text{kN} \,\angle\, 88.8°$
         (b) $\theta = 68°$
**2-28**  $\mathbf{R} = (1812\mathbf{i} - 439\mathbf{j})\,\text{N}$
**2-30**  $R = 12.15\,\text{kN}; \alpha = -72.4°$
**2-32**  $\mathbf{R} = (-0.40\mathbf{i} + 2.22\mathbf{j}) \times 10^5\,\text{lb}$
**2-34**  $\mathbf{R} = (2.7\mathbf{i} - 8.9\mathbf{j})\,\text{MN}$
**2-36**  $\mathbf{R} = (1036\mathbf{i} + 501\mathbf{j})\,\text{lb}$
**2-38**  $\mathbf{R} = (2.71\mathbf{i} + 2.71\mathbf{k})\,\text{kN}$
**2-40**  $\mathbf{T} = (4.98\mathbf{i} + 2.99\mathbf{j} + 1.49\mathbf{k})\,\text{kN}$
**2-42**  $\mathbf{T} = (-79.3\mathbf{i} - 198.1\mathbf{j} + 277.4\mathbf{k})\,\text{lb}$
**2-44**  $n_x = -0.254; n_y = 0.906; n_z = 0.338$
**2-46**  $\theta_x = 114.9°; \theta_z = 70.3°$
**2-48**  $\mathbf{P} = (5.999\mathbf{i} + 0.075\mathbf{j})\,\text{kN}$
**2-50**  For wire $AB$: $\theta_x = 138°; \theta_y = 56.1°; \theta_z = 68.2°$
         For wire $AC$: $\theta_x = 138°; \theta_y = 56.1°; \theta_z = 111.8°$
**2-52**  $\mathbf{R}_H = 1.072\mathbf{k}\,\text{kN}; \mathbf{R}_V = -3.58\mathbf{j}\,\text{kN}$
**2-54**  $\mathbf{R}_{xy} = -0.3F\mathbf{i} + 0.45F\mathbf{k}$
**2-56**  $\mathbf{R} = (-2\mathbf{i} - 9.5\mathbf{j}) \times 10^4\,\text{lb}$
**2-60**  $T'_{CA} = 1.98\,\text{kN}$
**2-62**  $F' = 200\,\text{lb}$
**2-64**  $T'_{AB} = 1.244\,\text{kN}$

**2-66**  $T'_{BA} = 394\,\text{N}; T'_{CA} = 353\,\text{N}$
**2-68**  $\cos^{-1}[-b/(a^2 + b^2 + c^2)^{1/2}]$
**2-70**  $P' = Pa/(a^2 + b^2 + c^2)^{1/2}$
**2-72**  $T = 1\,\text{kN}$
**2-74**  $T = 75\,\text{lb}$
**2-76**  $T = 1667\,\text{N}$
**2-78**  $F_C = 2.67\,\text{kN} \uparrow; F_D = 5.33\,\text{kN} \uparrow$
**2-80**  $F_C = 800\,\text{N} \uparrow; F_D = 800\,\text{N} \downarrow; F_{BE} = 1600\,\text{N} \uparrow$
**2-82**  $T_{AC} = T_{BC} = 718\,\text{N}$
**2-84**  $T_{AC} = 228\,\text{lb}; T_{BC} = 233\,\text{lb}$
**2-86**  $C_x = -222\,\text{N}; C_y = 897\,\text{N}$
**2-88**  $F = 17.05\,\text{N}$
**2-90**  $\mathbf{O} = (-3790\mathbf{i} - 1669\mathbf{j})\,\text{lb}$
**2-92**  $P = W\ell/3h$
**2-94**  $T = 75\,\text{lb}$
**2-96**  $T = 50.7\,\text{N}$
**2-98**  $\mathbf{F}_{\text{pole}} = (-56.1\mathbf{i} + 859.2\mathbf{j} - 111.0\mathbf{k})\,\text{lb}$
**2-100** $F = 714\,\text{lb}$
**2-102** $T_{AB} = 1620\,\text{N}; T_{AC} = 1620\,\text{N}; m = 330\,\text{kg}$
**2-104** (a) $F_H = 1970\,\text{lb}; F_V = 347\,\text{lb}$
         (b) $F_n = 1532\,\text{lb}; F_t = 1285\,\text{lb}$
**2-106** $\theta = 27.7°; F' = 782\,\text{lb}$
**2-108** $\theta_{AB} = 55.2°; \mathbf{R} = (-103\mathbf{j} - 3\mathbf{k})\,\text{kN}$
**2-110** $\mathbf{P} = 9\,\text{lb} \,\angle\, 70°$

## Chapter 3

**3-2**   $M_O = 1299\,\text{lb}\cdot\text{ft} \,\circlearrowleft$
**3-4**   $M_A = 8.19\,\text{kN}\cdot\text{m} \,\circlearrowright; M_B = 9.96\,\text{kN}\cdot\text{m} \,\circlearrowright$
**3-6**   $M_A = 11.28\,\text{kN}\cdot\text{m} \,\circlearrowleft; M_C = 13.65\,\text{kN}\cdot\text{m} \,\circlearrowright$
**3-8**   $M_A = 8660\,\text{lb}\cdot\text{ft} \,\circlearrowright; M_B = 10{,}617\,\text{lb}\cdot\text{ft} \,\circlearrowright$
**3-10**  $(M_A)_1 = 1.73\,\text{kN}\cdot\text{m} \,\circlearrowleft;$
         $(M_A)_2 = 2.95\,\text{kN}\cdot\text{m} \,\circlearrowright; (M_A)_R = 4.69\,\text{kN}\cdot\text{m} \,\circlearrowleft$
**3-12**  $(M_O)_F = 1500\,\text{N}\cdot\text{m} \,\circlearrowleft;$
         $(M_O)_W = 196.2\,\text{N}\cdot\text{m} \,\circlearrowright; (M_O)_R = 1305\,\text{N}\cdot\text{m} \,\circlearrowleft$

| | |
|---|---|
| 3–14 | $M_O = 90.6\,\text{lb·in}; M_A = 66.0\,\text{lb·in};$ $M_B = 16.0\,\text{lb·in}$ |
| 3–16 | $M_A = -0.34(W_1 AB + W_2 AC) + 0.98F(AD)$ |
| 3–20 | $\mathbf{M}_O = 250\mathbf{k}\,\text{N·m}$ |
| 3–22 | $\mathbf{M}_A = -14.5\mathbf{k}\,\text{lb·in}$ |
| 3–24 | $\mathbf{M}_O = 150\mathbf{k}\,\text{lb·ft}$ |
| 3–26 | $\mathbf{M}_A = -8.19\mathbf{k}\,\text{kN·m}; \mathbf{M}_B = -9.96\mathbf{k}\,\text{kN·m}$ |
| 3–28 | $\mathbf{M}_A = -2780\mathbf{k}\,\text{kN·m}; \mathbf{M}_B = -1580\mathbf{k}\,\text{N·m}$ |
| 3–30 | $\mathbf{M}_O = (-600\mathbf{i} - 280\mathbf{j} + 350\mathbf{k})\,\text{lb·ft}$ |
| 3–32 | $\mathbf{M}_O = (-75\mathbf{i} - 370\mathbf{j} - 1390\mathbf{k})\,\text{lb·ft}$ |
| 3–34 | $\mathbf{M}_O = (1200\mathbf{i} + 300\mathbf{j} + 1400\mathbf{k})\,\text{lb·ft}$ |
| 3–36 | $\mathbf{M}_C = (80{,}000\mathbf{i} + 360{,}000\mathbf{j})\,\text{lb·ft}$ |
| 3–38 | $\mathbf{M}_A = (-7\mathbf{i} + 52\mathbf{j} - 8\mathbf{k})\,\text{N·m};$ $\mathbf{M}_O = (-91\mathbf{i} + 65\mathbf{j} - 104\mathbf{k})\,\text{N·m}$ |
| 3–40 | $M_x = -3600\,\text{lb·ft}; M_y = 1000\,\text{lb·ft};$ $M_z = -700\,\text{lb·ft}$ |
| 3–42 | $M_x = -4200\,\text{lb·ft}; M_y = 2800\,\text{lb·ft};$ $M_z = 7290\,\text{lb·ft}$ |
| 3–44 | $\mathbf{V} = -16ab^2\mathbf{i} - 12a^2b\mathbf{j} + (60a^2c - 136b^2c)\,\mathbf{k}$ |
| 3–46 | $-14$ |
| 3–48 | $M_x = -720\,\text{lb·ft}; M_y = -290\,\text{lb·ft};$ $M_z = 540\,\text{lb·ft}$ |
| 3–50 | $M_{CD} = 599\,\text{lb·ft}$ |
| 3–52 | $M_x = 2.94\,\text{N·m}; M_y = 0; M_z = -7.35\,\text{N·m}$ |
| 3–54 | $M_\ell = -277\,\text{N·m}$ |
| 3–56 | $M_\ell = 36.8\,\text{N·m}$ |
| 3–60 | $\mathbf{M}_{net} = -(P + F/2)c\,\mathbf{i}$ |
| 3–62 | $\mathbf{M}_{net} = Fc\mathbf{j} - Pa\mathbf{k}$ |
| 3–64 | $\mathbf{M}_O = -5M\mathbf{i}$ |
| 3–66 | $\mathbf{M}_O = (-250\mathbf{i} - 50\mathbf{k})\,\text{N·m}; F_{min} = 510\,\text{N}$ |
| 3–68 | $F = 43.3\,\text{kN}$ |
| 3–70 | $d = 1.88\,\text{ft}$ |
| 3–72 | $M = 664\,\text{lb·ft}$ |
| 3–74 | $\mathbf{F}_O = (-3\mathbf{i} + 0.5\mathbf{j} - \mathbf{k})\,\text{kN};$ $\mathbf{M}_O = (-0.65\mathbf{i} + 0.85\mathbf{j} + 2.38\mathbf{k})\,\text{kN·m};$ |
| 3–76 | $\mathbf{F}_A = (730\mathbf{i} + 192.8\mathbf{j})\,\text{N}; \mathbf{M}_A = -451\mathbf{k}\,\text{N·m}$ |
| 3–78 | $\mathbf{R} = (F - P)\,\mathbf{i}; \mathbf{M}_R = [Pa - F(a + b)]\,\mathbf{k}$ |
| 3–80 | $\mathbf{R} = -(F + P)\,\mathbf{j}; \mathbf{M}_R = -(Qb + Pa + 3Fa)\,\mathbf{k}$ |
| 3–82 | $\mathbf{R} = (15.83\mathbf{i} - 0.523\mathbf{j})\,\text{kN}; \mathbf{M}_R = -8.07\mathbf{k}\,\text{kN·m}$ |
| 3–84 | (a) $\mathbf{R} = (98.5\mathbf{i} - 47.4\mathbf{j})\,\text{lb}; \mathbf{M}_R = -406\mathbf{k}\,\text{lb·ft}$ (b) $d = 3.72\,\text{ft}$ |
| 3–86 | $\mathbf{R} = 8500\mathbf{j}\,\text{lb}; \mathbf{M}_R = 95{,}000\mathbf{k}\,\text{lb·ft}; d = 11.2\,\text{ft}$ |
| 3–88 | (a) $\mathbf{R} = (350\mathbf{i} - 1056\mathbf{j})\,\text{lb};$ $\mathbf{M}_R = -6{,}650\,\text{lb·ft}$ (b) $\mathbf{F}_O = (1458\mathbf{i} - 1056\mathbf{j})\,\text{lb}; \mathbf{F}_B = -1108\mathbf{i}\,\text{lb}$ |
| 3–90 | $\mathbf{R} = (50\mathbf{i} + 750\mathbf{j} + 40\mathbf{k})\,\text{lb};$ $\mathbf{M}_R = (-68\mathbf{i} + 100\mathbf{j} - 270\mathbf{k})\,\text{lb·ft}$ |
| 3–92 | $M_x = 59\,\text{kN·m}; M_y = 20\,\text{kN·m};$ $M_z = -100\,\text{kN·m}$ |
| 3–94 | $\mathbf{M}_A = -19.62\mathbf{k}\,\text{N·m}; \mathbf{M}_B = 19\mathbf{k}\,\text{N·m}$ |
| 3–96 | $\mathbf{M}_O = 85{,}980\,\text{lb·ft}$ |
| 3–98 | $M_O = -Fr\ell(\sin\theta)/(\ell^2 + r^2 - 2\ell r\cos\theta)^{1/2}$ |
| 3–100 | $M_{AC} = 6.85\,\text{N·m}$ |
| 3–102 | $M_x = 36\,\text{kN·m}; M_y = 11\,\text{kN·m};$ $M_z = 92\,\text{kN·m}$ |

| | |
|---|---|
| 3–104 | (a) $\mathbf{R} = (-268\mathbf{i} - 517\mathbf{j})\,\text{N}; \mathbf{M}_R = 1155\mathbf{k}\,\text{N·m}$ (b) $d = 4.31\,\text{m}$ |

## Chapter 4

| | |
|---|---|
| 4–2 | $F_{BC} = 141.4\,\text{lb}; F_{AD} = 141.4\,\text{lb};$ partially constrained |
| 4–4 | $F_{CD} = 0; F_{AF} = 150\,\text{N}; F_{BE} = 150\,\text{N};$ completely constrained |
| 4–6 | $A_y = 0;$ partially constrained |
| 4–8 | $F_{CG} = 2\,\text{kN}; F_{BF} = -0.3\,\text{kN}; F_{AE} = 0.3\,\text{kN};$ completely constrained |
| 4–10 | $T_2 = T_1; C_x = T_1(\sin\theta - \sin\phi);$ $C_y = T_1(\cos\theta + \cos\phi)$ |
| 4–12 | $(T_1)_{max} = (T_2)_{max} = 790\,\text{lb}$ |
| 4–14 | $B_y = 2.38\,\text{kN}; A_y = 3.62\,\text{kN}$ |
| 4–16 | $A_x = 500\,\text{lb} \leftarrow; A_y = 1066\,\text{lb} \uparrow;$ $M_A = 7730\,\text{lb·ft} \circlearrowright$ |
| 4–18 | $A_y = 0; A_x = 8.24\,\text{kN}$ |
| 4–20 | (a) four unknowns, three independent equations (b) $T_B = 3.67\,\text{kN}; A_x = -6.90\,\text{kN};$ $A_y = -1.10\,\text{kN}$ |
| 4–22 | $\mathbf{F} = (492\mathbf{i} - 179\mathbf{j})\,\text{lb}; O_x = 8.5\,\text{lb}; O_y = 529\,\text{lb}$ |
| 4–24 | $A_x = -200\,\text{lb}; A_y = 100\,\text{lb}; E_x = 200\,\text{lb}; E_y = 0$ |
| 4–26 | $A_y = 125\,\text{N}; B_y = -125\,\text{N}$ |
| 4–28 | (a) $\mathbf{0} = (-0.5\mathbf{i} - 1.33\mathbf{j})\,\text{kN}; A_y = 3.33\,\text{kN}$ (b) $\mathbf{P}_{max} = 2\mathbf{j}\,\text{kN}$ |
| 4–30 | $B_y = 350\,\text{N}; A_x = 174\,\text{N}; A_y = 635\,\text{N}$ |
| 4–32 | $A_x = 0$ and $B_y = W - F; A_x = F\tan\theta,$ $B_y = -F, B_x = -F\tan\theta$ |
| 4–34 | $A_y = 492\,\text{N}; A_x = 2.55\,\text{kN}$ |
| 4–36 | $T = 86.9\,\text{kN}; A_x = 43.5\,\text{kN}; A_y = 105.2\,\text{kN}$ |
| 4–38 | $A_x = 53.6\,\text{N} \rightarrow; B_x = 53.6\,\text{N} \leftarrow; B_y = 294\,\text{N} \uparrow$ |
| 4–40 | $T = 5.05\,\text{kN}; A_y = 3.01\,\text{kN} \nwarrow; B_y = 3.01\,\text{kN} \nwarrow$ |
| 4–42 | $F = 1042\,\text{lb}; A_y = 3370\,\text{lb} \nwarrow; B_y = 2540\,\text{lb} \nwarrow$ |
| 4–44 | $\mathbf{F}_A = -2\mathbf{k}\,\text{kN}; \mathbf{M}_A = (0.70\mathbf{i} + 1.0\mathbf{j})\,\text{kN·m}$ |
| 4–46 | $\mathbf{F}_A = (130\mathbf{j} + 200\mathbf{k})\,\text{lb};$ $\mathbf{M}_A = (-1000\mathbf{j} + 650\mathbf{k})\,\text{lb·ft};$ $\mathbf{F}_O = (430\mathbf{j} + 200\mathbf{k})\,\text{lb};$ $\mathbf{M}_O = (4400\mathbf{i} - 1000\mathbf{j} + 650\mathbf{k})\,\text{lb·ft}$ |
| 4–48 | $\mathbf{F}_O = -30\mathbf{j}\,\text{lb}; \mathbf{M}_O = (-70\mathbf{j} - 150\mathbf{k})\,\text{lb·ft}$ |
| 4–50 | $\mathbf{F}_O = (10\mathbf{j} + 20\mathbf{k})\,\text{N}; \mathbf{M}_O = (6\mathbf{i} - 14\mathbf{j} + 2\mathbf{k})\,\text{N·m}$ |
| 4–52 | $A_x = B_x = 0; A_y = 75\,\text{lb}; B_y = -8.3\,\text{lb}$ |
| 4–54 | $B_y = 50\,\text{lb}; A_y = 50\,\text{lb}; B_x = -58.3\,\text{lb}; A_x = 25\,\text{lb}$ |
| 4–56 | $A_y = 1.414\,\text{kN}; B_y = -943\,\text{kN}; A_z = -409\,\text{N};$ $B_z = 613\,\text{N}$ |
| 4–58 | $\mathbf{F}_O = (184\mathbf{j} + 1709\mathbf{k})\,\text{N}; T_{AB} = 1257\,\text{N};$ $T_{AC} = 1081\,\text{N}$ |
| 4–60 | $T_{DB} = T_{DA} = 12{,}370\,\text{lb}; O_x = -1250\,\text{lb};$ $O_y = 17{,}490\,\text{lb}; O_d = 0$ |
| 4–62 | $O_x = 3710\,\text{lb}; O_y = 150\,\text{lb}; O_z = 0$ |
| 4–64 | For $x_A = 0.3\,\text{m}$ and $x_B = 1.1\,\text{m}: A_y = 155.9\,\text{N},$ $B_y = 106.8\,\text{N}, A_z = 31.2\,\text{N}, B_z = -8.5\,\text{N}; x_A = 0$ and $x_B = 1.4\,\text{m}$ |

**4–66**  $B_x = 0$; $B_y = W \uparrow$; $M_B = Wd \cos \phi$; $A_x = 0$;
$A_y = W \uparrow$; $M_A = W(\ell \sin \theta - d \cos \phi)$

**4–68**  On arm $BD$: $\mathbf{F}_D = (800\mathbf{i} - 900\mathbf{j} + 400\mathbf{k})$ N;
$\mathbf{M}_D = (-240\mathbf{i} + 480\mathbf{k})$ N·m;
$\mathbf{F}_B = (-800\mathbf{i} + 2500\mathbf{j} - 400\mathbf{k})$ N;
$\mathbf{M}_B = (1440\mathbf{i} + 280\mathbf{j} - 1610\mathbf{k})$ N·m;
On arm $OB$: $\mathbf{F}_B = (800\mathbf{i} - 2500\mathbf{j} + 400\mathbf{k})$ N;
$\mathbf{M}_B = (-1440\mathbf{i} - 280\mathbf{j} + 1610\mathbf{k})$ N·m;
$\mathbf{F}_O = (-800\mathbf{i} + 3700\mathbf{j} - 400\mathbf{k})$ N;
$\mathbf{M}_O = (240\mathbf{i} + 680\mathbf{j} + 3890\mathbf{k})$ N·m

**4–70**  $M_A = 1000$ lb·in

**4–72**  $D_y = 5.93$ kN $\uparrow$; $E_y = 11.12$ kN $\uparrow$; $F_y = 4.67$ kN $\uparrow$;
$H_y = 2.33$ kN $\downarrow$

**4–74**  $\mathbf{P} = -4.83\mathbf{j}$ kN

**4–76**  $O_x = -400$ lb; $O_y = 154.8$ lb; $M_O = 176.6$ lb·ft

**4–78**  $F_B = M/\ell \sin 2\theta$; $F_C = M/\ell \sin 2\theta$

**4–80**  $A_x = 2.17$ kN $\leftarrow$; $F_{BD} = 3.34$ kN $\searrow$ 45°;
$A_y = 2.59$ kN $\uparrow$

**4–82**  $A_x = 84.4$ lb; $A_y = 200$ lb; $A_z = 14.9$ lb;
$D_x = -84.4$ lb; $D_z = -14.9$ lb

**4–84**  On member $DE$: $\mathbf{D}_y = 98.1\mathbf{j}$ N, $\mathbf{E}_y = -98.1\mathbf{j}$ N;
$\mathbf{M}_D = 54\mathbf{k}$ N·m; On member $BC$: $\mathbf{B}_y = 98.1\mathbf{j}$ N,
$\mathbf{C}_y = -98.1\mathbf{j}$ N, $\mathbf{M}_B = (10\mathbf{i} + 120\mathbf{k})$ N·m,
$\mathbf{M}_C = (-10\mathbf{i} - 120\mathbf{k})$ N·m; On member $OA$:
$\mathbf{O}_y = 98.1\mathbf{j}$ N, $\mathbf{M}_O = (-88.3\mathbf{i} + 107.9\mathbf{k})$ N·m,
$\mathbf{A}_y = -98.1\mathbf{j}$ N, $\mathbf{M}_A = (39.3\mathbf{i} - 107.9\mathbf{k})$ N·m

## Chapter 5

**5–2**  At $y = 0$: $V_O = -200$ lb, $M_O = 600$ lb·ft;
At $y = 1$ ft: $V = -200$ lb, $M = 400$ lb·ft;
At $y = 2$ ft: $V = -200$ lb, $M = 200$ lb·ft

**5–4**  At $x = 0$: $V_O = 4$ kN, $M_O = -6$ kN·m;
At $x = 0.5$ m: $V = -4$ kN, $M = -4$ kN·m;
At $x = 1.5$ m: $V = -2$ kN: $M = -1$ kN·m

**5–6**  For all $x_C$: $V_O = -2$ kN; For $x_C = 1$ m:
$M_O = 5$ kN·m; For $x_C = 2$ m: $M_O = 4$ kN·m;
For $x_C = 4$ m: $M_O = 2$ kN·m

**5–8**  For all $x_C$: $V_O = 500$ lb; For $x_C = 3$ ft:
$M_O = -1500$ lb·ft; For $x_C = 6$ ft:
$M_O = -3000$ lb·ft; For $x_C = 9$ ft:
$M_O = -4500$ lb·ft

**5–10**  $A_x = 40$ lb; $B_x = 60$ lb; At $y = 8$ in.: $V_D = 60$ lb,
$M_D = 120$ lb·in; At $y = 2$ in.: $V_E = -40$ lb,
$M_E = 80$ lb·in

**5–12**  At $x_C = 1$ ft: $V_D = -50$ lb, $M_D = 250$ lb·ft; At
$x_C = 2.5$ ft: $V_D = -125$ lb, $M_D = 625$ lb·ft; At
$x_C = 4$ ft: $V_D = -200$ lb, $M_D = 1000$ lb·ft

**5–14**  $M_C = 3.1$ kN·m; $M_D = 4.75$ kN·m;
$V_E = -1.9$ kN; $M_E = 1.9$ kN·m

**5–16**  $V_C = 75$ lb; $M_C = -1425$ lb·ft; $V_D = 75$ lb;
$M_D = -1275$ lb·ft

**5–18**  $V_A = 5$ kN; $M_A = -8$ kN·m; $V_D = 5$ kN;
$M_D = -3$ kN·m

**5–20**  $V_C = 233$ lb; $M_C = 16{,}630$ lb·ft;
$M_D = 16{,}870$ lb·ft; $M_E = 16{,}400$ lb·ft

**5–22**  For $x_C = 50$ ft: $N_E = 23{,}330$ lb, $V_E = 3750$ lb,
$M_E = 18{,}750$ lb·ft; For $x_C = 80$ ft: $N_E = 37{,}730$ lb,
$V_E = 0$, $M_E = 0$; For $x_C = 110$ ft: $N_E = 51{,}320$ lb,
$V_E = -3750$ lb, $M_E = -18{,}750$ lb·ft

**5–24**  At $d = 2$ m: $V_E = 333$ N, $V_D = -333$ N,
$M_E = 1$ kN·m, $M_D = 1$ kN·m; At $d = 4$ m:
$V_E = 667$ N, $V_D = -667$ N, $M_E = 2$ kN·m,
$M_D = 2$ kN·m; At $d = 6$ m: $V_E = 1000$ N,
$V_D = -1000$ N, $M_E = 3$ kN·m, $M_D = 3$ kN·m

**5–26**  $M_{B,B'} = -Wd$; $M_{C,C'} = -2Wd$; $M_{D,D'} = -3Wd$

**5–28**  $V_F = 7.43$ kN; $M_F = -62$ kN·m

**5–30**  At $x = a/2$: $V_1 = 0.9W$, $M_1 = -(0.45a - 0.1b)W$;
At $x = a + b/2$: $V_2 = -0.1W$, $M_2 = 0.05bW$

**5–32**  $V = -200$ lb; $M = (600 - 200y)$ lb·ft

**5–34**  For $0 < x < 1$ m: $V = -4$ kN,
$M = (4x - 6)$ kN·m; For $1$ m $< x < 2$ m:
$V = -2$ kN, $M = (2x - 4)$ kN·m

**5–36**  For $0 < x < 1.9$ m: $V = 4$ kN,
$M = (4x - 7.9)$ kN·m; For $1.9$ m $< x < 2.0$ m:
$V = 3$ kN, $M = (3x - 6.0)$ kN·m

**5–38**  For $0 < x < 4$ ft: $V = 500$ lb,
$M = (500x - 2500)$ lb·ft; For $4$ ft $< x < 6$ ft:
$V = 250$ lb, $M = (250x - 1500)$ lb·ft;
For $x > 6$ ft: $V = 0$, $M = 0$

**5–40**  For $0 < y < 6$ in.: $V = -40$ lb, $M = 40y$ lb·in;
For $6$ in. $< y < 10$ in.: $V = 60$ lb,
$M = (-60y + 600)$ lb·in

**5–42**  For $0 < x < 2a$: $V = -(7/6)P$, $M = -(7/6)Px$;
For $2a < x < 3a$; $V = -(1/6)P$,
$M = -(1/6)Px - 2Pa$; For $3a < x < 6a$:
$V = (5/6)P$, $M = (5/6)Px - 5Pa$

**5–44**  For $0 < x < 1$ m: $V = 3.1$ kN, $M = 3.1x$ kN·m;
For $1$ m $< x < 2.5$ m: $V = 1.1$ kN,
$M = (1.1x + 2)$ kN·m; For $2.5$ m $< x < 5$ m:
$V = -1.9$ kN, $M = (-1.9x + 9.5)$ kN·m

**5–46**  For $0 < x < 3$ ft: $V = -500$ lb, $M = -500x$ lb·ft;
For $3$ ft $< x < 7$ ft: $V = 75$ lb,
$M = (75x - 1725)$ lb·ft; For $7$ ft $< x < 9$ ft:
$V = 600$ lb, $M = (600x - 5400)$ lb·ft

**5–48**  For $0 < x < 2$ m: $V = 5$ kN, $M = (5x - 8)$ kN·m;
For $2$ m $< x < 3$ m: $V = 0$, $M = 2$ kN·m

**5–50**  For $0 < x < 6$ ft: $V = 2733$ lb, $M = 2733x$ lb·ft;
For $6$ ft $< x < 8$ ft: $V = 233$ lb,
$M = (233x + 15{,}000)$ lb·ft; For $8$ ft $< x < 15$ ft:
$V = -2267$ lb, $M = (-2267x + 35{,}000)$ lb·ft

**5–52**  For $0 < x < a$: $V = -P_1$, $M = -P_1x$; For
$a < x < a + b$: $V = -(P_1 + P_2)$,
$M = -(P_1 + P_2)x + P_2a$

**5–54**  For $0 < x < 1$ ft: $V = 500$ lb,
$M = (500x - 1500)$ lb·ft; For $1$ ft $< x < 2$ ft:
$V = 400$ lb, $M = (400x - 1400)$ lb·ft;
For $2$ ft $< x < 3$ ft: $V = 300$ lb,
$M = (300x - 1200)$ lb·ft; For $3$ ft $< x < 4$ ft:

$V = 200\,\text{lb}$, $M = (200x - 900)\,\text{lb}\cdot\text{ft}$;
For $4\,\text{ft} < x < 5\,\text{ft}$: $V = 100\,\text{lb}$,
$M = (100x - 500)\,\text{lb}\cdot\text{ft}$

**5–56**   At point $A$: $V = -W$, $M = 0$; At point $B$:
$V = -0.943W$, $M = -Wd$: At point $C$:
$V = -0.745W$, $M = -2Wd$; At point $D$: $V = 0$,
$M = -3Wd$

**5–58**   On member $AB$: $M = (3.54x - 46.2)\,\text{kN}\cdot\text{m}$;
On member $BC$: $M = (5y - 25)\,\text{kN}\cdot\text{m}$

**5–60**   For $0 < x < a$: $V = 0.9W$,
$M = 0.9Wx - 0.9Wa + 0.1Wb$:
For $a < x < a + b$: $V = -0.1W$,
$M = -0.1Wx + 0.1W(a + b)$

**5–62**   For $0 < x < a$: $V_1 = F$, $M_1 = Fx - Fa - M$;
For $a < x < 2a$: $V_2 = 0$, $M_2 = 0$

**5–64**   For $0 < x < 5\,\text{m}$: $V = -1902\,\text{kN}$,
$M = (-1902x + 17\,210)\,\text{kN}\cdot\text{m}$;
For $5\,\text{m} < x < 9\,\text{m}$: $V = -1951\,\text{kN}$,
$M = (-1951x + 17\,460)\,\text{kN}\cdot\text{m}$;
For $9\,\text{m} < x < 11\,\text{m}$: $V = 48.9\,\text{kN}$,
$M = (48.9x - 540)\,\text{kN}\cdot\text{m}$; For $x > 11\,\text{m}$:
$V = 0$, $M = 0$

**5–66**   For $0 < x < 15\,\text{m}$: $V = -1000\,\text{kN}$,
$M = -1000x\,\text{kN}\cdot\text{m}$; For $15\,\text{m} < x < 30\,\text{m}$:
$V = 0$, $M = -15\,000\,\text{kN}\cdot\text{m}$

**5–68**   $\mathbf{V}_G = 10\mathbf{i}\,\text{N}$; $\mathbf{M}_G = 0.80\mathbf{j}\,\text{N}\cdot\text{m}$; $\mathbf{V}_H = 10\mathbf{i}\,\text{N}$;
$\mathbf{M}_H = -4.5\mathbf{k}\,\text{N}\cdot\text{m}$

**5–70**   $\mathbf{V}_J = 10\mathbf{i}\,\text{N}$; $\mathbf{M}_J = (-0.32\mathbf{i} - 2.5\mathbf{k})\,\text{N}\cdot\text{m}$

**5–72**   $M_B = 1600\,\text{lb}\cdot\text{in}$; $M_C = 1600\,\text{lb}\cdot\text{in}$;
$M_D = 1600\,\text{lb}\cdot\text{in}$; $M_E = 533\,\text{lb}\cdot\text{in}$;
$M_F = -800\,\text{lb}\cdot\text{in}$

**5–74**   For $0 < x < \ell$: $V_1 = 0$, $M_1 = -M$;
For $\ell < x < 3\ell/2$: $V_2 = 0$, $M_2 = M$;
For $3\ell/2 < x < 2\ell$: $V_3 = F$,
$M_3 = -M + F(x - 3\ell/2)$

## Chapter 6

**6–2**   $F_{AB} = 1055\,\text{lb T}$; $F_{AC} = 986\,\text{lb T}$;
$F_{BC} = 1322\,\text{lb C}$

**6–4**   $F_{AB} = 1099\,\text{lb C}$; $F_{AC} = F_{BC} = 0$;
$F_{AD} = F_{BD} = 1170\,\text{lb T}$

**6–6**   $F_{AC} = 10.20\,\text{kN C}$

**6–8**   $F_{AB} = 2144\,\text{lb T}$; $F_{AD} = 670\,\text{lb T}$; $F_{BC} = 470\,\text{lb T}$;
$F_{BD} = 2247\,\text{lb C}$; $F_{CD} = 171\,\text{lb T}$

**6–10**   $F_{AB} = 10\,\text{kN T}$; $F_{BE} = 5\,\text{kN T}$; $F_{EF} = 5\,\text{kN C}$

**6–12**   $F_{AB} = 12.86\,\text{kN T}$; $F_{BD} = 1\,\text{kN T}$;
$F_{BE} = 8.72\,\text{kN C}$

**6–14**   $F_{AF} = 2121\,\text{lb T}$; $F_{BF} = 1500\,\text{lb C}$;
$F_{EF} = 3000\,\text{lb C}$; $F_{FG} = 4500\,\text{lb C}$

**6–16**   $F_{AB} = 3.46\,\text{kN T}$; $F_{AE} = 1.73\,\text{kN T}$;
$F_{BC} = 3.46\,\text{kN C}$; $F_{BE} = 3.46\,\text{kN T}$;
$F_{CD} = 3.46\,\text{kN C}$; $F_{CE} = 3.46\,\text{kN T}$;
$F_{DE} = 1.73\,\text{kN T}$

**6–18**   $F_{AL} = F_{KL} = 10,020\,\text{lb T}$; $F_{BL} = 5000\,\text{lb T}$;
$F_{BK} = 9610\,\text{lb T}$; $F_{CK} = 2500\,\text{lb C}$;

**6–20**   $F_{JK} = F_{IJ} = 16,030\,\text{lb T}$; $F_{CJ} = F_{EJ} = 3205\,\text{lb T}$;
$F_{DJ} = 0$
$F_{BC} = 12,016\,\text{lb C}$; $F_{LK} = 9996\,\text{lb T}$;
$F_{KJ} = 8494\,\text{lb T}$

**6–22**   $BP$, $CP$, $DN$, $CN$, $EM$, $HJ$, $GJ$, $FL$, $GL$

**6–26**   $F_{AB} = 1054\,\text{lb}$; $F_{BC} = 1323\,\text{lb}$

**6–28**   $F_{AB} = 1099\,\text{lb}$

**6–30**   $F_{AC} = 10\,\text{kN}$

**6–32**   $F_{AB} = 2144\,\text{lb}$

**6–34**   $F_{BC} = 5\,\text{kN}$

**6–36**   $F_{AB} = 12.86\,\text{kN}$; $F_{AE} = 5\,\text{kN}$; $F_{BC} = 6.97\,\text{kN}$;
$F_{BD} = 1\,\text{kN}$; $F_{BE} = 8.71\,\text{kN}$; $F_{CD} = 5.71\,\text{kN}$;
$F_{DE} = 5.71\,\text{kN}$

**6–38**   $F_{AB} = 0$; $F_{AC} = 10.2\,\text{kN C}$; $F_{CD} = 10.0\,\text{kN T}$

**6–40**   $F_{AB} = 2144\,\text{lb T}$; $F_{BD} = 2247\,\text{lb C}$;
$F_{CD} = 171\,\text{lb T}$

**6–42**   $F_{AB} = 12\,\text{kN T}$; $F_{BF} = 7.21\,\text{kN C}$; $F_{EF} = 6\,\text{kN C}$

**6–44**   $F_{AB} = 12.86\,\text{kN T}$; $F_{BE} = 8.72\,\text{kN C}$;
$F_{DE} = 5.72\,\text{kN C}$

**6–46**   $F_{BC} = 1500\,\text{lb T}$; $F_{BE} = 2121\,\text{lb T}$;
$F_{EF} = 3000\,\text{lb C}$

**6–48**   $F_{BC} = 3.46\,\text{kN C}$; $F_{CE} = 3.46\,\text{kN T}$;
$F_{DE} = 1.73\,\text{kN T}$

**6–50**   From 6–49: $F_{CD} = 12,014\,\text{lb C}$, $F_{DK} = 3354\,\text{lb T}$,
$F_{JK} = 8493\,\text{lb T}$; After changes: $F_{CD} = 10,212\,\text{lb C}$,
$F_{CJ} = 2497\,\text{lb C}$; $F_{JK} = 9995\,\text{lb T}$

**6–52**   $F_{DE} = 46.56\,\text{kN C}$; $F_{EJ} = 14.19\,\text{kN C}$;
$F_{IJ} = 51.66\,\text{kN T}$

**6–54**   $F_{BC} = 341\,\text{kN C}$; $F_{CH} = 150.5\,\text{kN T}$;
$F_{GH} = 195.8\,\text{kN T}$

**6–56**   $F_{CK} = 0$; $F_{DK} = 93.7\,\text{kips C}$; $F_{EK} = 20\,\text{kips C}$;
$F_{JK} = 72\,\text{kips C}$; $F_{KL} = 80\,\text{kips C}$

**6–58**   $F_{CR} = 13\,\text{kips T}$; $F_{DR} = 10\,\text{kips T}$;
$F_{ER} = 16\,\text{kips C}$; $F_{QR} = 25.5\,\text{kips T}$; $F_{RS} = 0$

**6–60**   $A_x = 16.25\,\text{kN} \leftarrow$; $A_y = 0$; $C_x = 16.25\,\text{kN} \leftarrow$;
$C_y = 10.0\,\text{kN} \uparrow$; $D_y = 10.0\,\text{kN} \downarrow$;
$E_x = 16.25\,\text{kN} \rightarrow$; $E_y = 5.0\,\text{kN} \uparrow$

**6–62**   $A_x = 6\,\text{kN} \leftarrow$; $A_y = 12\,\text{kN} \downarrow$; $B_x = 12\,\text{kN} \rightarrow$;
$C_x = 12\,\text{kN} \leftarrow$; $C_y = 12\,\text{kN} \uparrow$

**6–64**   $B_x = 7.5\,\text{kN} \rightarrow$; $B_y = 6.0\,\text{kN} \uparrow$; $C_x = 10.5\,\text{kN} \leftarrow$;
$C_y = 3.0\,\text{kN} \downarrow$

**6–66**   $F_{BG} = F_{DF} = 0$; $F_{AB} = 17.32\,\text{kips}$ ⟋ $60°$;
$F_{CG} = 17.32\,\text{kips}$ ⟋ $60°$;
$F_{CF} = 17.32\,\text{kips}$ ⇄ $60°$;
$F_{DE} = 17.32\,\text{kips}$ ⟍ $60°$

**6–68**   $A_x = 16.0\,\text{lb} \rightarrow$; $A_y = 7.5\,\text{lb} \downarrow$; $C_x = 16.0\,\text{lb} \leftarrow$;
$C_y = 0$; $D_y = 7.5\,\text{lb} \uparrow$

**6–70**   $F_x = 10.75\,\text{lb} \rightarrow$; $F_y = 2\,\text{lb} \uparrow$; $F_{BC} = 6.75\,\text{lb T}$;
$D_x = 6.78\,\text{lb} \rightarrow$; $D_y = 4.78\,\text{lb} \uparrow$

**6–72**   $F_{DG} = 798\,\text{lb}$; $F_{EF} = 1066\,\text{lb}$

**6–74**   $F = 29.9\,\text{kN}$

**6–76**   $F = 8.25\,\text{kN}$

**6–78**   $B_x = 333\,\text{lb} \rightarrow$; $B_y = 250\,\text{lb} \uparrow$; $C_x = 666\,\text{lb} \leftarrow$;
$C_y = 0$

**6–80**   $A_x = 205\,\text{kN} \leftarrow$; $A_y = 0$; $F_x = 26\,\text{kN} \leftarrow$;
$F_y = 52\,\text{kN} \uparrow$; $I_x = 48\,\text{kN} \leftarrow$; $I_y = 14.23\,\text{kN} \downarrow$;

$F_{BC} = 210 \text{ kN} \swarrow 12.65°;$
$F_{EG} = 115 \text{ kN} \searrow 50°$

6–82    $G_x = 17.58 \text{ kN} \rightarrow; G_y = 15.57 \text{ kN} \downarrow;$
$K_x = 33 \text{ kN} \leftarrow; K_y = 67 \text{ kN} \downarrow;$
$F_{FH} = 18.17 \nearrow 14°; F_{IJ} = 93 \text{ kN} \nwarrow 69.3°$

6–84    $y_B = 2.4 \text{ ft}; T_{AB} = 1690 \text{ lb}; T_{BC} = 1671 \text{ lb};$
$T_{CD} = 1703 \text{ lb}$

6–86    $y_C = 3.83 \text{ ft}; y_D = 2.26 \text{ ft}$

6–88    $T_{max} = 2 \text{ kN}$

6–90    $y_C = 7.50 \text{ m}; y_D = 8.38 \text{ m}$

6–92    $F_{CG} = 100 \text{ kN T}; F_{FG} = F_{GH} = 194 \text{ kN T};$
$F_{AH} = 290 \text{ kN T}; F_{BH} = 15 \text{ kN C};$
$F_{CH} = 150 \text{ kN T}$

6–94    $F_{SB} = F_{SC} = F_{SR} = F_{ST} = 0; F_{RC} = 13 \text{ kips T};$
$F_{RD} = 10 \text{ kips T}; F_{RE} = 16 \text{ kips C};$
$F_{RQ} = 25.5 \text{ kips T}; F_{QE} = 10 \text{ kips T};$
$F_{QF} = 16 \text{ kips C}; F_{QP} = 38 \text{ kips T}$

6–96    (a) $AT, ST, BT, SB, CS, SR, KL, LM, JL, JM,$
$IM, MN$
(b) $EF, DE, AJ$

6–98    $A_x = 0; A_y = 80.0 \text{ kips};$
$F_{BC} = F_{CD} = F_{AJ} = F_{AH} = 0; F_{AB} = 80.0 \text{ kips T};$
$F_{BD} = 112.5 \text{ kips T}; F_{BH} = 62.4 \text{ kips C};$
$F_{DH} = 42.1 \text{ kips C}; F_{GH} = 99.8 \text{ kips C};$
$F_{HJ} = 120.0 \text{ kips C}$

6–100   $P/2 = 25 \text{ lb} \downarrow; W = 200 \text{ lb} \downarrow; D_x = 60 \text{ lb} \rightarrow;$
$D_y = 50 \text{ lb} \uparrow; F_x = 60 \text{ lb} \leftarrow; G_y = 175 \text{ lb} \uparrow$

6–102   For $y_C = 2.5a$: $T_{AB} = 2.50F, T_{CD} = 1.34F;$
For $y_C = 2.75a$: $T_{AB} = 2.37F, T_{CD} = 1.15F;$
For $y_C = 3a$: $T_{AB} = 2.24F, T_{CD} = F$

## Chapter 7

7–2    $\bar{x} = 0.5a^2/(a + b); \bar{y} = b(a + 0.5b)/(a + b)$

7–4    $\bar{x} = (ab + 0.5a^2)/(a + 3b); \bar{y} = 2.5b^2/(a + 3b)$

7–6    $\bar{x} = (0.5a^2 + 0.433b^2)/(a + b); \bar{y} = 0.25b^2/(a + b)$

7–8    $\bar{x} = \bar{y} = 0.636r$

7–10   $\bar{x} = 0.389b; \bar{y} = 0.222h$

7–12   $\bar{x} = \dfrac{(b - c)d^2 + ca^2}{2[(b - c)d + ca]}, \bar{y} = \dfrac{b^2d + c^2(a - d)}{2[(b - c)d + ca]}$

7–14   $\bar{x} = 6.83 \text{ in.}; \bar{y} = 6.55 \text{ in.}$

7–16   $\bar{x} = 0.492a; \bar{y} = 0.535b$

7–18   $\bar{x} = 0, \bar{y} = 0.621r_O; \bar{x} = 0, \bar{y} = 0.637r_O$

7–20   $\bar{x} = 0.6a; \bar{y} = 0.375b$

7–22   $\bar{x} = 0.75a; \bar{y} = 0.3b$

7–24   $\bar{x} = -0.38 \text{ mm}; \bar{y} = -0.656 \text{ mm}$

7–26   $\bar{x} = \dfrac{a^2 + 1.571ab + 0.667b^2}{2a + 1.571b}; \bar{y} = b$

7–28   $\bar{y} = 0.161a, \bar{x} = \bar{z} = 0$

7–30   $\bar{y} = 0.769h, \bar{x} = \bar{z} = 0$

7–32   $\bar{y} = 5.96 \text{ mm}, \bar{x} = \bar{z} = 0$

7–34   $V = 2\pi^2 r^2(R + r)$

7–36   $A = 2\pi r(0.667\pi R + r)$

7–38   $V = 0.5\pi^2 r^2(R + 0.424r)$

7–40   $A = \pi r(r + \sqrt{h^2 + r^2}), V = 0.333\pi r^2 h$

7–42   $V = 3560 \text{ mm}^3$

7–44   $A = 13.32 \text{ m}^2$

7–46   $A = 2\pi r(4h + 1.5\sqrt{h^2 + r^2} + 1.5r)$

7–48   $A = 3934 \text{ m}^2$

7–50   $A_y = 100 \text{ lb} \uparrow; B_y = 2300 \text{ lb} \uparrow$

7–52   $A_y = 1.8 \text{ kN} \uparrow; M_A = 4.4 \text{ kN·m} \circlearrowright$

7–54   $B_y = 925 \text{ lb} \uparrow; C_y = 625 \text{ lb} \uparrow$

7–56   From $C$ to $B$: $V(x) = -wx, M(x) = -wx^2/2;$
From $B$ to $A$: $V(x) = -2wx + w\ell,$
$M(x) = -wx^2 + w\ell x - w\ell^2/2$

7–58   For $0 < x < 4 \text{ ft}$: $V(x) = 200 \text{ lb},$
$M(x) = 200x - 1400 \text{ (lb·ft)};$
For $4 \text{ ft} < x < 10 \text{ ft}$: $V(x) = 333.3 - 33.3x \text{ (lb)};$
$M(x) = -16.7x^2 + 333.4x - 1666.4 \text{ (lb·ft)}$

7–60   For $0 < x < 2 \text{ m}$: $M(x) = -3000x - wx^2/2 \text{ (N·m)};$
For $2 \text{ m} < x < 4 \text{ m}$:
$M(x) = -14\,354 + 4177x - wx^2/2 \text{ (N·m)}$

7–62   For $0 < x < L/3$: $V(x) = -w_B x^2/L - w_B x,$
$M(x) = w_B x^3/3L + w_B x^2/2;$
For $L/3 < x < 2L/3$:
$V(x) = -w_B x^2/L - w_B x + w_B L/8,$
$M(x) = w_B x^3/3L + w_B x^2/2 - w_B Lx/8 + w_B L^2/24$

7–64   $V(x) = w(a + b - c - x) - w[(b - c)^2 - a^2]/2b;$
$M(x) = \dfrac{w[(a + b - c) - (b - c)^2 - a^2](x - a)}{2b}$
$\qquad - \dfrac{wx^2}{2}$

7–66   $V(x) = 2wL + wx; M(x) = -2wLx - wx^2/2$

7–68   For $0 < x < 0.4L$: $V(x) = 1.967w_C L - 0.5w_C x^2/L,$
$M(x) = 1.967w_C Lx - 0.1667w_C x^3/L$
For $0.4L < x < L$:
$V(x) = -1.033w_C L - 0.5w_C x^2/L,$
$M(x) = 1.2w_C L^2 - 1.033w_C Lx - 0.1667w_C x^3/L$

7–70   $V(x) = -wx; M(x) = 3wL^2 - wx^2/2$

7–72   $V(x) = \dfrac{-w}{(a + b)}\left[\dfrac{3a^2}{2} - \dfrac{b^2}{2} - ab\right] - wx;$
$M(x) = 2wa^2 + \dfrac{-w}{(a + b)}\left[\dfrac{3a^2}{2} - \dfrac{b^2}{2} - ab\right]x - \dfrac{wx^2}{2}$

7–74   $T_{max} = 67.3 \text{ kN}$

7–76   $T_A = 2.39 \times 10^5 \text{ lb}$

7–78   $T_B = 342 \text{ kN}$

7–80   $T_{min} = 550 \text{ lb}; T_{max} = 586 \text{ lb}; y = 4.56 \text{ ft}$

7–82   $T_{max} = 357 \text{ lb}$

7–84   $T_{max} = 636 \text{ N}$

7–86   $P_x = 58.9 \text{ kN}; P_y = 21.4 \text{ kN}; \bar{y} = 1.33 \text{ m};$
$\bar{x} = 0.485 \text{ m}$

7–88   $P = 1174 \text{ lb}$ at $0.934 \text{ ft}$ from $C$

7–90   $P = 15.42 \text{ kN}$

7–92   $P = 10,560 \text{ lb}; d = 3.6 \text{ ft}$ below top of $AB$

7–94   $F_{BC} = 5354 \text{ lb}; A_x = 8152 \text{ lb} \leftarrow; A_y = 1397 \text{ lb} \downarrow$

| | | | |
|---|---|---|---|
| 7–96 | $\mathbf{R} = 10.32\,\text{kN} \searrow 13.17°$ at 0.591 m above $A$ | 8–44 | $\mu_{AB} = 0.667$ |
| 7–98 | $W = 1190\,\text{lb}$ | 8–46 | $M = 1.137\,\text{kN·m}$ |
| 7–100 | $d = 3.43\,\text{in.}$ | 8–48 | $M_{\text{tightening}} = 188.9\,\text{lb·ft}$; $M_{\text{loosening}} = 144.7\,\text{lb·ft}$ |
| 7–102 | $h = 0.43R$ | 8–50 | $F_{\text{clamp on pulley}} = 18.62\,\text{kips}$ |
| 7–104 | $F = 945\,\text{kN}$ | 8–52 | $p = 1.885\,\text{m}$ |
| 7–106 | $P = 134.7\,\text{lb}$ | 8–54 | $F_B = 125\,\text{lb}$ |

7–96    $\mathbf{R} = 10.32\,\text{kN} \searrow 13.17°$ at 0.591 m above $A$

7–98    $W = 1190\,\text{lb}$

7–100   $d = 3.43\,\text{in.}$

7–102   $h = 0.43R$

7–104   $F = 945\,\text{kN}$

7–106   $P = 134.7\,\text{lb}$

7–108   $\tau_\theta = 0;\ \sigma_\theta = 5000\,\text{psi}$

7–110   $\tau_\theta = 0;\ \sigma_\theta = -20\,\text{MPa}$

7–112   $\tau_\theta = -190.2\,\text{psi};\ \sigma_\theta = -851\,\text{psi}$

7–114   $\tau_\theta = 58.3\,\text{MPa};\ \sigma_\theta = -49.0\,\text{MPa}$

7–116   $\tau_\theta = -7577\,\text{psi};\ \sigma_\theta = -13,196\,\text{psi}$

7–118   $\bar{x} = 87.8\,\text{mm};\ \bar{y} = 19.97\,\text{mm}$

7–120   $A = 2\pi[6(0.5r + R)\sqrt{\ell^2 + (R - r)^2} + r^2]$

7–122   $C_x = 1.213\,\text{N} \rightarrow;\ C_y = 2.72\,\text{N} \uparrow$

7–124   $V(x) = -9x + 3x^2 - 0.333x^3 + 9$ (kN); $M(x) = -6.75 + 9x - 4.5x^2 + x^3 - 0.0833x^4$ (kN·m)

7–126   $T_{\max} = 23.2\,\text{kips}$

7–128   $d = \dfrac{L(R^2 + 2Rr + 3r^2)}{4(R^2 + Rr + r^2)}$

7–130   $\tau_\theta = -295\,\text{psi},\ \sigma_\theta = -207\,\text{psi}$

7–128   $d = \dfrac{L(R^2 + 2Rr + 3r^2)}{4(R^2 + Rr + r^2)}$

7–130   $\tau_\theta = -295\,\text{psi},\ \sigma_\theta = -207\,\text{psi}$

## Chapter 8

8–2    Block slides

8–4    $P = 364\,\text{N}$

8–6    $W = 6.19P$

8–8    $F = 237\,\text{lb}$

8–10   $P = 10\,\text{lb};$ slipping will occur at floor

8–12   Ladder remains at rest

8–14   $F > 100.9\,\text{N}$ will cause motion

8–16   Beam remains in equilibrium; Total friction force $= 20.5\,\text{lb}$

8–18   (a) $F = 100\,\text{lb};$   (b) $F = 89.6\,\text{lb}$

8–20   Stays in equilibrium; $F = 13.18\,\text{lb};$ $\mathbf{N}$ at 1.58 ft up from $A$ along incline

8–22   $\mu_s < 0.558$

8–24   $M_C = 166.8\,\text{N·m}$

8–26   $M_C = 1176\,\text{lb·in}$

8–28   $F = 78.6\,\text{N}$

8–30   $M = 0.1158Wr$

8–32   $P = 799\,\text{lb}$

8–34   $F = 200\,\text{lb}$ not large enough

8–36   (a) $P = 665\,\text{lb};$   (b) $P = 741\,\text{lb}$

8–38   $F = 627\,\text{N}$

8–40   $F = 718\,\text{N}$

8–42   $P = \dfrac{2\mu W \tan\dfrac{\theta}{2}}{1 - \mu \tan\dfrac{\theta}{2}}$

8–44   $\mu_{AB} = 0.667$

8–46   $M = 1.137\,\text{kN·m}$

8–48   $M_{\text{tightening}} = 188.9\,\text{lb·ft};\ M_{\text{loosening}} = 144.7\,\text{lb·ft}$

8–50   $F_{\text{clamp on pulley}} = 18.62\,\text{kips}$

8–52   $p = 1.885\,\text{m}$

8–54   $F_B = 125\,\text{lb}$

8–56   For $n = 1,\ P = 58.2\,\text{kN};$ For $n = 2,\ P = 51.4\,\text{kN};$ For $n = 3,\ P = 46.0\,\text{kN}$

8–58   $D = 8\,\text{in.}$

8–60   $M = 38.8\,\text{lb·ft}$

8–62   $M_{\text{total}} = 74\,\text{lb·ft}$

8–64   $M = 35\,\text{lb·ft}$

8–66   $M = 2\,\text{N·m}$

8–68   $M = 2.25\,\text{N·m}$

8–70   $M = 37.5\,\text{lb·in}$

8–72   $P = 1.840\,\text{kN} \downarrow$

8–74   $T_A = 198\,\text{lb};\ T_B = 203\,\text{lb};\ T_C = 197\,\text{lb}$

8–76   $M_{\text{total}} = 40.6\,\text{N·m}$

8–78   $F = 40\,\text{lb}$ per wheel

8–80   $\theta = 0.286°$

8–82   $T_1 = 155.9\,\text{N}$

8–84   $T_1 = 1.151\,\text{kN}$

8–86   $T = 513\,\text{lb}$

8–88   $T_2 = 0.204\,\text{lb}$

8–90   equation is independent of radius of curvature of contour

8–92   $T_2 = 14.65T_1$

8–94   For clockwise rotation: $T_A = 190.8\,\text{N};$ $T_B = 670\,\text{N};$ For counterclockwise rotation: $T_A = 1186\,\text{N},\ T_B = 338\,\text{N}$

8–96   $T_{\max} = 8.67\,\text{kN}$

8–98   $F = 0.313v^2$

8–100   $P = 289\,\text{lb}$

8–102   $m_B = 891\,\text{kg}$

8–104   $M = 23.4\,\text{lb·ft}$

8–106   $M = 200\,\text{lb·ft}$

8–108   $M = \dfrac{2\mu_K PR}{3 \sin\theta}$

8–110   $T_{\text{top}} = 81.2\,\text{lb};\ T_{\text{bottom}} = 660\,\text{lb}$

## Chapter 9

9–2    $I_{y'} = \frac{1}{12}ML^2$

9–4    $I_x = \frac{1}{2}MR^2;\ I_y = MR^2$

9–6    $I_x = at\rho/3\,(b^3 - 3b^2d + 3bd^2);$ $I_y = bt\rho/3\,(a^3 + 3a^2c + 3ca^2)$

9–8    $J_C = \frac{1}{12}ab(a^2 + b^2)$

9–10   $I_y = ba^3/12$

9–12   $I_x = \rho tab^3/12$

9–14   $I_x = 104.2;\ (r_g)_x = 2.04$

9–16   $I_x = \pi r^4/8,\ I_y = \frac{5}{8}\pi r^4$

9–18   $I_y = \frac{5}{8}\rho tr^4$

9–20   $I_x = I_y = \pi r^4/4$

9–22   $J_O = 54.5 \times 10^6\,\text{mm}^4$

9–24 $I_z = ML^2/3$

9–26 $I_x = a(0.333b^3 - b^2d + bd^2)$;
$I_y = b(0.333a^3 + a^2c + ac^2)$

9–28 $I_\ell = 5.93Mc^2$

9–30 $J_O = \frac{1}{12}ab(a^2 + b^2)$

9–32 $I_{x1} = 2MR^2$

9–34 $I_{x'} = 0.11r^4$

9–36 $I_{y1} = 4.25Mr^2$

9–38 $I_x = 12.63r^4$; $I_y = 43.1r^4$

9–40 $I_x = 358 \text{ in}^4$

9–42 $I_y = 1.039ma^2$; $(r_g)_y = 1.093a$

9–44 $I_x = \rho t\left[\dfrac{he^3}{6} + 2he\left(d + \dfrac{e}{2}\right)^2 + \dfrac{bd^3}{3}\right]$

9–46 $I_y = 723$

9–48 $I_{xy} = 0.25a^2b^2\rho t$;
$I_{uv} = ab\rho t[0.144(b^2 - a^2) + 0.125ab]$

9–50 $I_{xy} = -196.9a^4$; $I_{uv} = -265a^4$

9–52 $I_{max} = 645a^4$; $I_{min} = 115.1a^4$; $I_u = 371a^4$;
$I_v = 390a^4$

9–54 $I_{uv} = 0.036\rho t\{cb(b^2 - c^2) +$
$a(c - 2d)[(a - 2d)^2 + a^2]\}$

9–56 $I_{uv} = 2710\rho ta^4$

9–58 $I_{xy} = 11.56 \times 10^6 \text{ mm}^4$

9–60 $I_{xy} = -26.8\rho t \text{ slug·in}^2$

9–62 $I_{max} = 5.93 \times 10^6 \text{ mm}^4$, $I_{min} = 3.23 \times 10^6 \text{ mm}^4$,
$\theta_P = 13.3°$ or $-76.7°$

9–64 $I_x = 0.00485 \text{ slug·in}^2$; $I_y = 0.00594 \text{ slug·in}^2$

9–66 $I_y = \frac{1}{5}\rho tba^3$

9–68 $I_x = 0.0476ab^3$; $(r_g)_x = 0.378b$

9–70 $I_x = 47.2 \times 10^6 \text{ mm}^4$; $I_y = 10.95 \times 10^6 \text{ mm}^4$

9–72 $I_{xy} = 313$

## Chapter 10

10–2 $U_{AB} = -2067 \text{ ft·lb}$

10–4 $U_{AB} = 491 \text{ N·m}$

10–6 $U_{ABA} = 13.08 \text{ kN·m}$

10–8 $\dfrac{U_{AB}}{U_{BA}} = \dfrac{(\mu \cos 20° - \sin 20° + \mu)}{(\mu \cos 20° + \sin 20° + \mu)}$

10–10 $U_{AB} = 4510 \text{ ft·lb}$

10–12 $U_{AB} = 11.06 \text{ kN·m}$

10–14 $U = W\ell/2$

10–16 $U = 246 \text{ N·m}$

10–18 $U = 1675 \text{ ft·lb}$

10–20 $U = 1.152 \text{ kN·m}$

10–22 $U = 236 \text{ ft·lb}$

10–24 $M_O = 13Fd$

10–26 $P = 0.1986F$

10–28 $F = 0.5W$

10–30 $P = 0.75F \cot \theta$

10–32 $\eta = 5.73$

10–34 $\eta = \dfrac{\sin 20°}{\mu_k \cos 20° - \sin 20° + \mu_k}$

10–36 $U_{\text{friction}} = 19.5 \text{ kN·m}$

10–38 $\Delta V = 7.21 \text{ N·m}$

10–40 $\Delta V = 630 \text{ ft·lb}$

10–42 $\Delta V = 2.28 \text{ N·m}$

10–44 $\theta = 33.7°$, $\phi = 63.4°$

10–46 $x = 0.261 \text{ m}$, stable

10–48 Stable

10–50 $k = 288 \text{ kN/m}$, stable

10–52 $\theta = 0.463°$

10–54 Stable

10–56 $F = 17,320 \text{ lb}$

10–58 $\eta = 0$

## Chapter 11

11–2 $z = -1.089 \text{ m}$, $a = 4.9 \text{ m/s}^2$

11–4 $v(t_1) = -4.16 \text{ in/s}$, $v(t_2) = -6.54 \text{ in/s}$,
$a(t_1) = -18.19 \text{ in/s}^2$, $a(t_2) = 15.14 \text{ in/s}^2$

11–6 (a) $v = -4 \text{ m/s}$, $a = 4 \text{ m/s}^2$,
(b) maximum speed $= 5.33 \text{ m/s}$ (in negative direction)

11–8 $y = 10.37 \text{ m}$, $a = 1.632 \text{ ft/s}^2$

11–10 $x = 67.2 \text{ in.}$, $a = 13 \text{ in/s}^2$

11–12 $a = 12,800 \text{ ft/s}^2$

11–14 $x(t) = (3t^2 - 18t + 4) \text{ m}$, $v(t) = (6t - 18) \text{ m/s}$

11–16 $v_2 = 25 \text{ in/s}$

11–18 (a) $v = 6.11 \text{ m/s}$, (b) $x = 5.48 \text{ m}$

11–20 $x = 500 \text{ m}$

11–22 (a) $t = 0.358 \text{ s}$,
(b) No real time exists for particle to come to rest

11–24 $v = \dfrac{1}{k}\sqrt{1 - \exp(-2k^2gy)}$

11–26 $t_f = 102 \text{ s}$, $v_f = 500 \text{ m/s}$

11–28 (a) $y(t) = (30t - 4.905t^2) \text{ m}$,
$v(t) = (30 - 9.81t) \text{ m/s}$
(b) $y_{max} = 45.9 \text{ m}$, (c) $v = -30 \text{ m/s}$,
(d) $v = -33.1 \text{ m/s}$

11–30 $v_{A/B} = -18 \text{ ft/s}$, $a_{A/B} = -14 \text{ ft/s}^2$

11–32 $x_{B/A} = -161.7 \text{ m}$, $a_{B/A} = -38 \text{ m/s}^2$

11–34 $x_{B/A} = 3.5 \text{ ft}$, $a_{B/A} = 3 \text{ ft/s}^2$

11–36 $x_{B/A} = 450 \text{ m}$

11–38 (a) $v_P = 120 \text{ mi/h}$, $x_P = 0.367 \text{ mi}$,
(b) $x_{P/A} = -480 \text{ ft}$, $v_{P/A} = -8 \text{ ft/s}$, $a_{P/A} = 8 \text{ ft/s}^2$

11–40 $t = 18.25 \text{ s}$

11–42 $x_{B/A} = 0.527 \text{ ft}$, $v_{B/A} = 0.145 \text{ ft/s}$

11–44 $y_{A/B} = 23 \text{ ft}$, $v_{A/B} = 24 \text{ ft/s}$

11–46 $y = 116.6 \text{ ft}$, $v = 54 \text{ ft/s}$, $a = 12.5 \text{ ft/s}^2$

11–48 (a) $v = 40.1 \text{ ft/s}$ (b) $v = 40.1 \text{ ft/s}$

11–50 $t = 0.36 \text{ s}$

# Chapter 12

**12-2**  $\mathbf{v} = (150\mathbf{i} + 40\mathbf{j})$ ft/s, $\mathbf{a} = (60\mathbf{i} + 8\mathbf{j})$ ft/s$^2$

**12-4**  $\mathbf{r} = (6\mathbf{i} + 17\mathbf{j})$ m, $\mathbf{a} = (2\mathbf{i} + 12\mathbf{j})$ m/s$^2$

**12-6**  $\mathbf{v} = (83.3\mathbf{i} - 50\mathbf{j})$ ft/s

**12-8**  $h = 38.8$ ft, $d = 269.3$ ft

**12-10**  $v_0 = 102.2$ m/s

**12-12**  $v_0 = 10.51$ m/s

**12-14**  $v_0 = 46.95$ ft/s

**12-16**  $h = 16.1$ m

**12-18**  $\alpha_1 = 89.9$, $t_1 = 217$ s,
$\alpha_2 = 10.2$, $t_2 = 0.384$ s

**12-20**  (a) $\mathbf{v}_f = a\mathbf{i} - \sqrt{b^2 - 2gh}\,\mathbf{j}$,
(b) $\mathbf{v}_f' = -a\mathbf{i} - b\mathbf{j}$, yes

**12-22**  $\mathbf{a} = (-85.6\mathbf{i} - 51.7\mathbf{j})$ ft/s$^2$

**12-24**  $v_{\max} = 26.5$ m/s

**12-26**  $a_t = -23.5$ m/s$^2$

**12-28**  $a_B = 3.28$ m/s$^2$

**12-30**  $\rho = 509$ m

**12-32**  $a_0 = 189\mathbf{i}$ m/s$^2$, $\mathbf{a}_1 = (-6.38\mathbf{i} + 192.7\mathbf{j})$ m/s$^2$

**12-34**  $\mathbf{a}_0 = 0$, $\rho_0 = \infty$, $\mathbf{a}_5 = (30\mathbf{i} + 150\mathbf{j})$ ft/s$^2$,
$\rho = 1.842 \times 10^6$ ft

**12-36**  $\mathbf{a} = (-12.5\mathbf{n}_r + 3\mathbf{n}_\theta)$ m/s$^2$

**12-38**  $v = 827.1$ ft/s, $a = 59.9$ ft/s$^2$

**12-40**  $v_{\text{hor}} = 46.9$ m/s, $a_{\text{hor}} = -1.051$ m/s$^2$

**12-42**  $\dot{r} = 90.1$ m/s, $\dot{\theta} = -0.06$ rad/s,
$\ddot{r} = 11.18$ m/s$^2$, $\ddot{\theta} = 0.003$ rad/s$^2$

**12-44**  $\mathbf{v} = (72\,700\mathbf{i} - 42\,000\mathbf{j})$ m/s,
$\mathbf{a} = (-133\,500\mathbf{i} - 213\,700\mathbf{j})$ m/s$^2$

**12-46**  $\mathbf{v} = (11.6\mathbf{i} + 23.1\mathbf{j})$ ft/s, $\mathbf{a} = (339\mathbf{i} - 87.6\mathbf{j})$ ft/s$^2$

**12-48**  $\dot{\theta} = 0.15$ rad/s, $\ddot{\theta} = -0.139$ rad/s$^2$

**12-50**  $\dot{\theta} = -12$ rad/s, $\ddot{\theta} = 166.4$ rad/s$^2$

**12-52**  $\mathbf{a} = (-20\mathbf{n}_r + 7.29\mathbf{n}_\theta)$ ft/s$^2$

**12-54**  $\mathbf{v} = (150\mathbf{n}_\theta + 5\mathbf{k})$ ft/s,
$\mathbf{a} = (-1500\mathbf{n}_r + 30\mathbf{n}_\theta + 10\mathbf{k})$ ft/s$^2$

**12-56**  $\mathbf{a} = (-1000\mathbf{n}_r - 10\mathbf{n}_\theta + 0.048\mathbf{k})$ in/s$^2$

**12-58**  $\mathbf{a} = -12.95\mathbf{n}_r$ m/s$^2$

**12-60**  $\mathbf{a} = (-86.1\mathbf{n}_r + 4.35\mathbf{n}_\theta + 2.47\mathbf{k})$ ft/s$^2$

**12-62**  $\mathbf{a} = (-32\mathbf{n}_r - 2\mathbf{k})$ m/s$^2$

**12-64**  $\mathbf{a} = (-0.771\mathbf{n}_r - 0.281\mathbf{n}_\phi + 19.64\mathbf{n}_\theta)$ m/s$^2$

**12-66**  $\mathbf{a} = (-0.05\mathbf{n}_r - 0.0433\mathbf{n}_\phi + 0.0866\mathbf{n}_\theta)$ ft/s$^2$

**12-68**  $\mathbf{a}_{P2} = (-281.3\mathbf{n}_r - 166.9\mathbf{n}_\phi - 2.6\mathbf{n}_\theta)$ ft/s$^2$

**12-70**  $\mathbf{v}_{A/C} = \mathbf{v}_{A/B} - \mathbf{v}_{C/B}$

**12-72**  $\mathbf{v}_{B/A} = 17\mathbf{j}$ m/s, $\mathbf{a}_{B/A} = 10\mathbf{j}$ m/s$^2$

**12-74**  $\mathbf{v}_A = 130\mathbf{j}$ km/h

**12-76**  $\mathbf{v}_A = (-8.08\mathbf{i} - 9.02\mathbf{j})$ m/s

**12-78**  $\mathbf{v}_A = (37.4\mathbf{i} + 1.84\mathbf{j})$ m/s,
$\mathbf{a}_A = (-3.85\mathbf{i} - 14.82\mathbf{j})$ m/s$^2$

**12-80**  $a_{A/P} = 25.1$ ft/s$^2$

**12-82**  $\alpha = 0.069$ rad/s$^2$

**12-84**  $\mathbf{v}_C = (-0.468\mathbf{i} + 0.152\mathbf{j})$ m/s,
$\mathbf{a}_C = (-0.378\mathbf{i} - 0.1215\mathbf{j})$ m/s$^2$

**12-86**  (a) $v_0 = 53.1$ ft/s, (b) $h = 11.1$ ft

**12-88**  (a) $\rho = 12.4$ ft, (b) $\rho = 24.2$ ft

**12-90**  (a) $v = (100\dot{\theta} \sec^2 \theta)$ m/s,
$a = (100\ddot{\theta} \sec^2 \theta + 200\dot{\theta}^2 \tan \theta \sec^2 \theta)$ m/s$^2$,
(b) $v = 133$ m/s, $a = 221$ m/s$^2$

**12-92**  $\mathbf{v} = (6.93\mathbf{n}_r + 26\mathbf{n}_\theta + 4\mathbf{k})$ m/s,
$\mathbf{a} = (-258\mathbf{n}_r + 139\mathbf{n}_\theta + \mathbf{k})$ m/s$^2$

**12-94**  $\mathbf{a} = (-8.32\mathbf{n}_r - 13.86\mathbf{n}_\phi - 5.54\mathbf{n}_\theta)$ m/s$^2$

**12-96**  $\mathbf{a}_{B/A} = (-15\mathbf{i} + 18\mathbf{k})$ ft/s$^2$

# Chapter 13

**13-2**  $F_r = m(\ddot{r} - r\dot{\phi}^2 - r\dot{\theta}^2 \sin^2 \phi)$,
$F_\phi = m(r\ddot{\phi} + 2\dot{r}\dot{\phi} - r\dot{\theta}^2 \sin \phi \cos \phi)$,
$F_\theta = m(r\ddot{\theta} \sin \phi + 2\dot{r}\dot{\theta} \sin_\phi + 2r\dot{\phi}\dot{\theta} \cos \phi)$

**13-4**  $m = 0.466$ slug, $W_{\text{moon}} = 2.47$ lb

**13-6**  $R_x = 210$ N, $R_y = 2437$ N

**13-8**  $R_y = 28\,223$ N $\uparrow$

**13-10**  $\mu_k = 0.16$

**13-12**  $v_0 = 54.7$ mph

**13-14**  $\mathbf{a} = -35.6\mathbf{i}$ m/s$^2$

**13-16**  (a) $x - x_0 = 582$ ft, (b) $x - x_0 = 521$ ft

**13-18**  $s = 76.4$ m

**13-20**  $a_{\text{hor}} = 11.28$ ft/s$^2$

**13-22**  $x = 92.4$ m

**13-24**  $s = 2981$ ft, $F_1 = 398$ lb C, $F_2 = 202$ lb C

**13-26**  $T_A = T_C = T_D = 1104$ N, $T_B = 2208$ N

**13-28**  $P = 1964$ N

**13-30**  $F_x = 654$ lb

**13-32**  $F_{\text{hor}} = 10.3$ kN at $\diagup$ 29° to normal

**13-34**  $\theta \simeq 25°$

**13-36**  $v_{\max} = 24.7$ m/s

**13-38**  no sliding

**13-40**  $\alpha = 49.7°$

**13-42**  $(\omega_0)_{\min} = 2.65$ rad/s

**13-44**  $N = 479$ lb $\uparrow$

**13-46**  $R = 13.67$ kN $\diagup$ 85.9°

**13-48**  $T_{\max} = 444$ N, $T_{\min} = 430$ N

**13-50**  $v_A = 28.4$ ft/s

**13-52**  $F_{\text{hor}} = 6460$ lb $\diagdown$ 43.2°

**13-54**  $R_x = -18.52$ N, $R_y = 17.67$ N

**13-56**  $v = 10.2$ m/s, $a = 10.93$ m/s$^2$

**13-58**  $\omega = 14$ rad/s, $T = 3.92$ N

# Chapter 14

**14-2**  $h = 24.8$ ft

**14-4**  $\mathbf{v}_f = (20\mathbf{i} - 31.7\mathbf{j})$ m/s

**14-6**  $v_B = 9.12$ m/s

**14-8**  $F = 7.6$ lb

**14-10**  $A$ at 33.8 ft up slope from $B$

**14-12**  $v_B = 60.8$ m/s

**14-14**  $s = 125.4$ ft

**14-16**  $d_{\text{vert}} = 206$ ft, $d_{\text{hor}} = 180.4$ ft

**14-18**  $v = 2.63$ m/s

**14–20**  $v = 17.47\,\text{ft/s}$, $P = 6.35\,\text{hp}$

**14–22**  $v = 0.612\,\text{m/s}$

**14–24**  $P = 9.81\,\text{kW}$

**14–26**  (a) $v_5 = 4.49\,\text{ft/s}$,  (b) $y = 20\,\text{in.}$

**14–28**  $v = 1\,\text{m/s}$

**14–30**  $l = 0.745\,\text{m}$

**14–32**  $v_f^2 = v_0^2 - (2\mu_k g)^2 (m/k)\{\sqrt{1 + (v_0/\mu_k g)^2 (k/m)} - 1\}$

**14–34**  $v = 13.62\,\text{ft/s}$

**14–36**  $\mu_s = 0.1763$, $\mu_k = 0.1325$

**14–38**  $V = -(\frac{1}{2}x^2 + y^2) + C$

**14–40**  $V = \frac{2}{3}x^3 - \frac{5}{4}y^4 + C$

**14–42**  **F** not conservative

**14–44**  **F** not conservative

**14–46**  $v_0 = 22.7\,\text{ft/s}$

**14–48**  $v_B = 6.02\,\text{m/s}$

**14–50**  $v_A = 21.2\,\text{ft/s}$

**14–52**  $v_A = 9.9\,\text{m/s}$

**14–54**  $0.33\%$

**14–56**  $k = 22.5\,\text{lb/in}$

**14–58**  (a) $\delta = 123\,\text{mm}$,  (b) $\delta = 172.6\,\text{mm}$

**14–60**  $m = 32.8\,\text{kg}$

**14–62**  $(v_A)_2 = 4.6\,\text{m/s}$, $B_y = C_y = 1\,\text{kN}$

**14–64**  $F_{\text{max}} = 96.96\,\text{lb}$

**14–66**  $h = 14.52\,\text{in.}$

**14–68**  $P_{\text{max}} = 2.39\,\text{W}$

**14–70**  $x_T = 290\,\text{m}$

**14–72**  (a) $h = 31.6\,\text{ft}$,  (b) $v_{\text{max}} = 30\,\text{mi/h}$

**14–74**  (a) $v = 19.66\,\text{ft/s}$,  (b) $x_{\text{max}} = 1\,\text{ft}$

**14–76**  (a) $v_C = 25.3\,\text{ft/s}$,  (b) $d = 16.9\,\text{in.}$

## Chapter 15

**15–2**  $\mathbf{H}_O = -2.51\mathbf{j}\,\text{kg}\cdot\text{m}^2/\text{s}$

**15–4**  $\mathbf{M} = -52\mathbf{j}\,\text{lb}\cdot\text{ft}$

**15–6**  $\mathbf{H}_O = -6.43\mathbf{k}\,\text{kg}\cdot\text{m}^2/\text{s}$, $\dot{\mathbf{H}}_O = -43.13\mathbf{k}\,\text{N}\cdot\text{m}$

**15–8**  (a) $(M_O)_1 = 28\,\text{N}\cdot\text{m}$,

(b) $(M_O)_2 = 42\,\text{N}\cdot\text{m}$

**15–10**  $a = 25.5\,\text{ft/s}^2 \downarrow$

**15–12**  $\ddot{\theta} = -0.491\,\text{rad/s}^2$

**15–14**  $T = 17.2\,\text{N}$

**15–16**  $M_O = 37.1\,\text{lb}\cdot\text{ft}$

**15–18**  $v = 1\,\text{m/s}$

**15–20**  $\dot{\theta}_1 = 1.333\dot{\theta}_0$

**15–22**  $t = 27.5\,\text{days}$

**15–24**  $r = 42\,230\,\text{km}$, $v = 1.106 \times 10^4\,\text{km/h}$

**15–26**  $a = 10\,317\,\text{km}$

**15–28**  $h_2 = 18\,155\,\text{km}$

**15–30**  $v_A = 22{,}295\,\text{mph}$

**15–32**  $v_{\text{esc}} = 32\,121\,\text{km/h}$

**15–34**  $v_{\text{esc}} = 9448\,\text{mph}$

**15–36**  $\mathbf{v}_2 = 211\mathbf{j}\,\text{ft/s}$

**15–38**  $\mathbf{v}_2 = 390\mathbf{i}\,\text{m/s}$

**15–40**  $\mathbf{v}_2 = (20\mathbf{i} + 2624\mathbf{j})\,\text{ft/s}$

**15–42**  $F_{\text{avg}} = 400\,\text{kN}$

**15–44**  $T = 36.3\,\text{kN}$

**15–46**  $\Delta t = 3.98\,\text{s}$

**15–48**  $F = 3270\,\text{N C}$

**15–50**  $\text{Imp}_{\text{to}} - \text{Imp}_{\text{land}} = 47\,\text{lb}\cdot\text{s}$, approximately same in magnitude

**15–52**  $F = 496\,\text{lb}$

**15–54**  $v_G = 8.75\,\text{m/s}$

**15–56**  $v_F = 0.48\,\text{km/h} \leftarrow$

**15–58**  $v_f = 1\,\text{m/s}$

**15–60**  approximately 8 bumps

**15–62**  $v_f = 2.57\,\text{ft/s}$, $\Delta T = -2286\,\text{ft}\cdot\text{lb}$

**15–64**  $\mathbf{v}_f = 1\mathbf{i}\,\text{km/h}$

**15–66**  $(v_B)_f = 60\,\text{km/h} \leftarrow$, $(v_A)_f = 80\,\text{km/h} \leftarrow$

**15–68**  $e = \sqrt{h_f/h_i}$

**15–70**  $W_R = 400\,\text{lb}$

**15–72**  $v_f = -ev_0$, $\phi = \theta$

**15–74**  (a) $\mathbf{v}_B = (-200\mathbf{i} + 1.2\mathbf{j})\,\text{km/h}$,

(b) $\mathbf{v}_B = (-200\mathbf{i} + 1.199\mathbf{j})\,\text{km/h}$

**15–76**  $\mathbf{M}_O = (-6911\mathbf{n}_\phi - 18\,530\mathbf{n}_\theta)\,\text{N}\cdot\text{m}$

**15–78**  (a) $v_{\text{min}} = 3120\,\text{mi/h}$,

(b) $\dfrac{dA}{dt} = 4.74 \times 10^7\,\text{mi}^2/\text{h}$,  (c) $r_p = 7240\,\text{mi}$,

(d) $a = 17{,}240\,\text{mi}$, $b = 11{,}200\,\text{mi}$

**15–80**  $\mathbf{v}'_A = -0.0526v_A\mathbf{i} - 0.394v_A\mathbf{j}$,

$\mathbf{v}'_B = 0.702v_A\mathbf{i} + 0.262v_A\mathbf{j}$

**15–82**  $(\mathbf{v}_A)_f = (1\mathbf{i} + 0.8\mathbf{j})\,\text{ft/s}$, $(\mathbf{v}_B)_f = 1\mathbf{i}\,\text{ft/s}$

## Chapter 16

**16–2**  $\delta_{\text{max}} = 1.05\,\text{ft}$

**16–4**  $N = 596\,\text{lb} \leftarrow$

**16–6**  $v_f = 8.37\,\text{m/s}$

**16–8**  $d = 1.24\,\text{m}$

**16–10**  allowed space $= 1.532\,\text{ft}$

**16–12**  (a) $h = 4.9\,\text{mm}$,  (b) $F = 7.85\,\text{kN}$

**16–14**  $D = 13.94\,\text{m}$

**16–16**  (a) $D = 53.5\,\text{ft}$,  (b) $t_s = 2.58\,\text{s}$

**16–18**  $x = 1.43 \times 10^{-5}\,\text{m}$

**16–20**  Impulse $= 4796\,\text{lb}\cdot\text{s}$

## Chapter 17

**17–2**  $\mathbf{r}_C = (3.13\mathbf{i} + 0.89\mathbf{j})\,\text{ft}$, $\mathbf{a}_C = -3\mathbf{j}\,\text{ft/s}^2$

**17–4**  $\mathbf{F} = (1.36\mathbf{i} - 1.92\mathbf{j} + 1.32\mathbf{k})\,\text{N}$

**17–6**  $r_F = 75\,\text{ft}$

**17–8**  $v_1 = 552\,\text{m/s}$, $v_2 = 158.2\,\text{m/s}$

**17–10**  $m_2 = 1436\,\text{kg}$, $m_3 = 1309\,\text{kg}$, $m_T = 4345\,\text{kg}$

**17–12**  $v_A = 53\,\text{ft/s}$

**17–14**  $v_B = 13.12\,\text{m/s}$

**17–16**  $h_B = 89.4\,\text{ft}$

**17–18**  $U_{12} = -41{,}120\,\text{ft}\cdot\text{lb}$

**17–20**  $T = 955\,\text{ft}\cdot\text{lb}$

17–22   $T_{1,2} = T_{3,4} = 566\,\text{N}\cdot\text{m}$

17–24   $\mathbf{H}_O = -11.8\mathbf{k}\,\text{kg}\cdot\text{m}^2/\text{s}$, $\mathbf{H}_C = -3.46\mathbf{k}\,\text{kg}\cdot\text{m}^2/\text{s}$

17–26   (a) $\mathbf{G} = (1.24\mathbf{i} + 1.55\mathbf{j})\,\text{lb}\cdot\text{s}$, $\dot{\mathbf{G}} = -7\mathbf{j}\,\text{lb}$,

       (b) $\mathbf{H}_O = -2.18\mathbf{k}\,\text{ft}\cdot\text{lb}\cdot\text{s}$, $\dot{\mathbf{H}}_O = -1295\mathbf{k}\,\text{lb}\cdot\text{ft}$,

       $\mathbf{H}_C = 5.78\mathbf{k}\,\text{ft}\cdot\text{lb}\cdot\text{s}$, $\dot{\mathbf{H}}_C = -1290\mathbf{k}\,\text{lb}\cdot\text{ft}$

17–28   $\mathbf{H}_C = (-2033\mathbf{i} + 1126\mathbf{j} - 46\mathbf{k})\,\text{kg}\cdot\text{m}^2/\text{s}$,

       $\mathbf{H}_O = (-970\mathbf{i} + 2590\mathbf{j} - 615\mathbf{k})\,\text{kg}\cdot\text{m}^2/\text{s}$,

17–30   $\mathbf{H}_O = 157.1\mathbf{k}\,\text{kg}\cdot\text{m}^2/\text{s}$; after fracture,

       $\mathbf{H}_O = 153.9\mathbf{k}\,\text{kg}\cdot\text{m}^2/\text{s}$

17–32   $(\mathbf{G}_P)_f = (-35,528\mathbf{i} + 88.2\mathbf{j})\,\text{lb}\cdot\text{s}$,

       difference $= 16.8\mathbf{j}\,\text{lb}\cdot\text{s}$

17–34   $\mathbf{G}_O = m_T\mathbf{v}_A + \mathbf{G}_A$

17–36   $\mathbf{M}_C = 10,284\mathbf{j}\,\text{lb}\cdot\text{ft}$

17–38   $\dot{\mathbf{H}}_O = 21,435\mathbf{k}\,\text{lb}\cdot\text{ft}$

17–40   $\mathbf{M}_A = 3220\mathbf{k}\,\text{N}\cdot\text{m}$, $\mathbf{M}_B = 3220\mathbf{k}\,\text{N}\cdot\text{m}$,

       $\mathbf{M}_O = 79\,240\mathbf{k}\,\text{N}\cdot\text{m}$

17–42   $\mathbf{H}_O = -148.9mvR\,\mathbf{k}$

17–44   $\mathbf{H}_O = (2795\mathbf{j} + 7453\mathbf{k})\,\text{ft}\cdot\text{lb}\cdot\text{s}$

17–46   Thrust $= 6211\,\text{lb} \leftarrow$

17–48   $P = 21.9\,\text{N} \rightarrow$

17–50   (a) $F_T = 47.6\,\text{lb} \leftarrow$,

       (b) Thrust increases by a factor of four if wind

           speed doubles

17–52   Force on engine $= (-5740\mathbf{i} - 6\mathbf{j})\,\text{N}$

17–54   (a) $\mathbf{F} = [1000(1 - \cos\theta)\,\mathbf{i} + 1000\sin\theta\mathbf{j}]\,\text{N}$,

       (b) $F_{45} = 765\,N$, $F_{90} = 1414\,\text{N}$,

           $F_{135} = 1848\,\text{N}$

17–56   $F_{\text{hor}} = 38.8\,\text{N} \rightarrow$

17–58   Thrust of engine $= 47.2\,\text{kN} \leftarrow$

17–60   Force on tail section $= 7379\,\text{lb} \leftarrow$,

       $d' = 3.33\,\text{ft}$ below $\mathbf{v}_1$

17–62   $v = 67.6\,\text{ft/s}$

17–64   Braking force $= 5008\,\text{lb} \rightarrow$

17–66   $F = 55.6\,\text{N}$, independent of time

17–68   $F = 0.00373\,\text{lb}$

17–70   (a) $F = 0$,   (b) $F = 37.3\,\text{lb} \leftarrow$

17–72   $\dot{m} = 667\,\text{kg/s}$

17–74   $v_f = 5818\,\text{ft/s}$

17–76   15 m horizontally

17–78   $v_A = 2\,\text{m/s}$, $v_B = 3\,\text{m/s}$, $v_C = 1.73\,\text{m/s}$

17–80   $T = 2694\,\text{N}\cdot\text{m}$

17–82   $\omega_A = 1.036\,\text{rad/s}$, $\dot{\theta}_{OA} = 0.0314\,\text{rad/s}$

17–84   $T = 6985\,\text{N} \rightarrow$

17–86   $\dot{m} = 2.996 \times 10^{-9}\,\text{kg/s}$


## Chapter 18

18–2   $\mathbf{a}_A = (-32\mathbf{i} - 1\mathbf{j})\,\text{m/s}^2$, $\mathbf{a}_B = (-64\mathbf{i} - 2\mathbf{j})\,\text{m/s}^2$

18–4   $\mathbf{a}_A = (-3\mathbf{i} - 225\mathbf{j})\,\text{ft/s}^2$,

      $\mathbf{a}_B = (82.4\mathbf{i} - 147.3\mathbf{j})\,\text{ft/s}^2$

18–6   $\mathbf{a}_A = (1.5\mathbf{i} + 7\mathbf{j})\,\text{m/s}^2$

18–8   $\mathbf{a}_A = \mathbf{a}_B = (3.4\mathbf{i} - 11.9\mathbf{j})\,\text{ft/s}$

18–10   $\mathbf{v}_A = (25.9\mathbf{i} - 33.1\mathbf{j})\,\text{m/s}$,

       $\mathbf{a}_A = (4644\mathbf{i} + 3606\mathbf{j})\,\text{m/s}^2$

18–12   $\alpha = -31.4\,\text{rad/s}^2$

18–14   $t = 69.8\,\text{min} = 1.163\,\text{h}$, $\Delta\theta = 27,900\,\text{rev}$

18–16   $\omega_{\text{roll}} = 100\,\text{rad/s}$

18–18   $\alpha_2 = 0.75\,\text{rad/s}^2$, $\Delta\theta_1 = 5.85\,\text{rev}$,

       $\Delta\theta_2 = 2.92\,\text{rev}$

18–20   $\alpha_A = 2.56\,\text{rad/s}^2$

18–22   $(\omega_B)_1 = 20.6\,\text{rad/s}$, $(\omega_B)_2 = 25.7\,\text{rad/s}$,

       $(\omega_B)_3 = 34.4\,\text{rad/s}$,

       $(\Delta\theta_B)_1 = 2.6\,\text{rev}$, $(\Delta\theta_B)_2 = 3.25\,\text{rev}$,

       $(\Delta\theta_B)_3 = 4.33\,\text{rev}$

18–24   $\omega_C = 40\,\text{rad/s}$, $\alpha_C = 25\,\text{rad/s}^2$, $\Delta\theta = 5.09\,\text{rev}$

18–26   $\mathbf{v}_A = (27.8\mathbf{i} + 27.8\mathbf{j})\,\text{m/s}$,

       $\mathbf{v}_B = (47.5\mathbf{i} - 19.6\mathbf{j})\,\text{m/s}$

18–28   $v_{\text{cable}} = 5\,\text{ft/s}$

18–30   $v_{\text{cable}} = 0.6\,\text{m/s}$

18–32   $v_{\text{max}} = 2660\,\text{ft/s}$, $v_{\text{min}} = 2460\,\text{ft/s}$

18–34   $\omega_{\text{sh}} = 400\,\text{rad/s}$

18–36   $v_{\text{max}} = 51.021\,\text{m/s}$, $v_{\text{min}} = 48.979\,\text{m/s}$

18–38   $v_{C/E} = 3.36\,\text{ft/s}$

18–40   $\mathbf{v} = -0.508\mathbf{j}\,\text{m/s}$

18–42   $v_C = 0.735\,\text{m/s}$, $\dot{\phi}_{BC} = 0.522\,\text{rad/s}$,

18–44   $v_B = 18.17\,\text{ft/s} \;\searrow\; 16.7°$

18–46   $v_C = 8.37\,\text{m/s} \downarrow \;\circlearrowright$

18–48   $\omega_{CD} = 6.7\,\text{rad/s} \;\circlearrowright$

18–50   $\omega_B = 78.6\,\text{rad/s} \;\circlearrowright$

18–52   $r_I = 7.07\,\text{m}$ below center of wheel,

       $v_{\text{max}} = 23.1\,\text{m/s}$

18–54   $r_I = 0.178\,\text{in.}$ up from $A$,

       $\mathbf{v}_1 = (0.494\mathbf{i} - 4.495\mathbf{j})\,\text{ft/s}$, $\mathbf{v}_2 = 5\mathbf{i}\,\text{ft/s}$,

18–56   $\mathbf{a}_C = (4980\mathbf{i} - 20\mathbf{j})\,\text{ft/s}^2$,

       $\mathbf{a}_D = (-5020\mathbf{i} + 20\mathbf{j})\,\text{ft/s}^2$

18–58   $a_B = 30.5\,\text{m/s}^2$ to the left

18–60   $a = 57,116\,\text{ft/s}^2 \rightarrow$ at the shaft,

       $a = 57,116\,\text{ft/s}^2 \leftarrow$ at the outer case

18–62   $v_D = 1.92\,\text{m/s} \rightarrow$, $a_D = 10.31\,\text{m/s}^2 \;\nearrow\; 83.3°$

18–64   $\mathbf{a}_C = (-0.427\mathbf{i} - 0.911\mathbf{j})\,\text{m/s}^2$

18–66   $a_C = 15\,\text{ft/s}^2 \downarrow$

18–68   $\mathbf{a}_B = (169.1\mathbf{i} - 1669\mathbf{j})\,\text{ft/s}^2$

18–70   $a_D = 1291\,\text{ft/s}^2 \;\nearrow\; 81°$

18–72   $a_C = 412\,\text{ft/s}^2 \rightarrow$

18–74   $\mathbf{a}_D = (-1000\mathbf{j} + 24\mathbf{k})\,\text{m/s}^2$

18–76   $\alpha = (-0.893\mathbf{i} - 0.214\mathbf{j} + 0.375\mathbf{k})\,\text{rad/s}^2$

18–78   $\mathbf{a}_B = (6.93\mathbf{i} - 3.2\mathbf{j} - 3.2\mathbf{k})\,\text{m/s}^2$

18–80   $\alpha = (23.6\mathbf{j} + 0.02\mathbf{k})\,\text{rad/s}^2$

18–82   $\alpha = (-6\mathbf{j} + 12\mathbf{k})\,\text{rad/s}^2$

18–84   $\alpha = (-42\mathbf{j} - 168\mathbf{k})\,\text{rad/s}^2$

18–86   $\alpha = (48\mathbf{i} + 36\mathbf{j})\,\text{rad/s}^2$,

       $\mathbf{a}_P = (0.168\mathbf{i} + 140,000\mathbf{j} - 123.2\mathbf{k})\,\text{ft/s}^2$

18–88   $\mathbf{a}_A = -187.3\mathbf{i}\,\text{km/s}^2$

18–90   $\alpha = (0.005\mathbf{i} - 0.01\mathbf{k})\,\text{rad/s}^2$

18–92   $\mathbf{v}_C = (-0.703\mathbf{i} + 0.228\mathbf{j})\,\text{m/s}$,

       $\mathbf{a}_C = (-0.0534\mathbf{i} - 0.1849\mathbf{j})\,\text{m/s}^2$

18–94   $\mathbf{a}_A = (4.22\mathbf{i} - 8.41\mathbf{j})\,\text{ft/s}^2$, $\mathbf{a}_B = (4.63\mathbf{i} - 8\mathbf{j})\,\text{ft/s}^2$

18–96   $(\omega_A)_{\text{max}} = 1.69 \;\circlearrowright$ or $2.74 \;\circlearrowright$ rad/s

18–98   $\mathbf{a}_4 = (-20.14\mathbf{i} + 17.96\mathbf{j})\,\text{ft/s}^2$

18–100   $r = 2.4\,\text{ft/s}$

18–102   $x = 2.71\,\text{m/s}$

**18–104**   $\mathbf{a}_A = (-224\mathbf{i} - 1615\mathbf{j})$ m/s$^2$,
$\mathbf{a}_B = (-292\mathbf{i} - 1615\mathbf{j})$ m/s$^2$

**18–106**   $\dot{\phi} = 1.744$ rad/s, $\ddot{\phi} = 76.9$ rad/s$^2$

**18–108**   (a) $\alpha_{max} = 20$ rad/s$^2$ ⟳,   (b) $T = 783$ lb

**18–110**   $\mathbf{v}_D = (0.5\mathbf{i} - 1.5\mathbf{k})$ ft/s, $\mathbf{v}_C = (-2\mathbf{j} - 1.5\mathbf{k})$ ft/s,
$\mathbf{a}_D = (3.25\mathbf{i} + 10\mathbf{j} - 2.25\mathbf{k})$ ft/s$^2$,
$\mathbf{a}_C = (6.25\mathbf{i} + 5\mathbf{j} - 15\mathbf{k})$ ft/s$^2$

**18–112**   $\mathbf{v}_{A_1} = (4.34\mathbf{i} + 5.66\mathbf{j})$ m/s,
$\mathbf{r}_I = (1.2\mathbf{i} + 2.08\mathbf{j})$ m, $\omega_{AA_1} = -5\mathbf{k}$ rad/s

**18–114**   $\omega_{OBC} = -1.5\mathbf{k}$ rad/s, $\omega_{AD} = 1\mathbf{k}$ rad/s,
$\boldsymbol{\alpha}_{OBC} = -4.44\mathbf{k}$ rad/s$^2$, $\boldsymbol{\alpha}_{AD} = 1.625\mathbf{k}$ rad/s$^2$

**18–116**   $a_A = 24.2$ in/s$^2 \rightarrow$

**18–118**   $\mathbf{a}_P = (-3.73\mathbf{i} - 18\,720\mathbf{j})$ m/s$^2$

**18–120**   $\dot{r} = 685{,}900$ ft/s

## Chapter 19

**19–2**   $N_A = 82$ N

**19–4**   $N_A = 0$, $N_B = 492$ lb, $F_A = 0$, $F_B = 443$ lb

**19–6**   $\mu_k = 2.67$

**19–8**   $a = 3.82$ m/s$^2$

**19–10**   $T = 3964$ lb

**19–12**   $B_y = (m/d)(2dg + 10ha_x)\uparrow$,
$A_y = (m/d)(2dg - 10ha_x)\uparrow$, $Ax = 4ma_x \rightarrow$

**19–14**   $x = 1.419$ m

**19–16**   $0.22 < h < 2.18$ m

**19–18**   Front wheel drive: $a_{max} = 8.39$ ft/s$^2$,
Rear wheel drive: $a_{max} = 7.58$ ft/s$^2$

**19–20**   $N_A = 3488$ N, $N_B = 5793$ N

**19–22**   $\theta = 21.7°$

**19–24**   $v = 15.66$ m/s, $\theta = 63.4°$

**19–26**   $a_P = 21.6$ ft/s$^2$ ⦨ 70°, $a_C = 7.39$ ft/s$^2 \downarrow$

**19–28**   $M = 26.97$ lb·ft ⟳

**19–30**   $F = 20$ N

**19–32**   $M = 0.016$ N·m ⟲

**19–34**   $T = 2889$ N·m ⟳

**19–36**   $\alpha = 5.72$ rad/s$^2$ ⟳, $T = 322$ lb

**19–38**   $a_t = 25$ m/s$^2$

**19–40**   $\alpha = 64.2$ rad/s$^2$ ⟲

**19–42**   $\alpha = 1.94$ rad/s$^2$ ⟲

**19–44**   $a_C = 8.57$ ft/s$^2 \rightarrow$, $B_x = 522$ lb $\rightarrow$, $B_y = 720$ lb $\uparrow$

**19–46**   $\boldsymbol{\alpha} = 21.2\mathbf{k}$ rad/s$^2$, $O_x = 12.72$ N, $O_y = 17.16$ N

**19–48**   $O_x = 49.7$ lb, $O_y = 105.3$ lb

**19–50**   $F = 4496$ lb, $O_x = 2857$ lb, $O_y = 2293$ lb

**19–52**   $\boldsymbol{\alpha} = 12.12\mathbf{k}$ rad/s$^2$, $A_x = -521$ N, $A_y = -551$ N

**19–54**   $\alpha = 1.38$ rad/s$^2$ ⟳, $a_{hor} = 1.104$ m/s$^2 \rightarrow$,
$a_{vert} = 2.55$ m/s$^2 \downarrow$

**19–56**   $\alpha = 0.525g/R$ ⟳, $O_x = 0$, $O_y = 0.09\rho t g \pi R^2$

**19–58**   $F_{axial} = 1342$ lb, $F_{transverse} = 58.5$ lb

**19–60**   $a = 1.68$ m/s$^2$ ⦨

**19–62**   $a = 9.91$ ft/s$^2$ ⦨

**19–64**   $\theta = 68.7°$

**19–66**   $a = 6.54$ m/s$^2 \downarrow$, $T = 65.4\uparrow$

**19–68**   $(a_C)_x = 10.73$ ft/s$^2$, $\mu = 0.19$

**19–70**   $a = 20.1$ m/s$^2 \rightarrow$

**19–72**   $P = 106.1$ lb

**19–74**   $\alpha = 0.0013$ rad/s$^2$ ⟳

**19–76**   $\alpha = 1.61$ rad/s$^2$ ⟳, $T = 198.8$ lb

**19–78**   $B_x = -66$ N, $B_y = 2925$ N, $\mathbf{M}_B = 1540\mathbf{k}$ N·m

**19–80**   $N_A = 20$ lb $\rightarrow$, $N_B = 49$ lb $\uparrow$

**19–82**   $A_y = 154.8$ N $\downarrow$, $B_x = 53.4$ N $\rightarrow$,
$a_B = 81.6$ m/s$^2 \downarrow$

**19–84**   $B_x = 689.8$ lb, $B_y = -536$ lb

**19–86**   $\mathbf{a}_O = 1.32\mathbf{i}$ m/s$^2$

**19–88**   $F = 161.4$ lb

**19–90**   $\alpha = 66.4$ rad/s$^2$ ⟳

**19–92**   $\boldsymbol{\alpha} = 0.05\mathbf{j}$ rad/s$^2$, $\mathbf{a}_z = -2.5 \times 10^{-3}\,\mathbf{k}$ m/s$^2$

**19–94**   $a_A = 7.49$ m/s$^2$ ⦨ 2.65°

**19–96**   (a) $a_{max} = 10.71$ ft/s$^2$,   (b) $\mu_{min} = 0.333$

**19–98**   (a) $a = 1.635$ m/s$^2$, $T = 654$ N,
(b) $F_D = 2289$ N, $F_E = 1635$ N

**19–100**   $\mathbf{a}_{swing} = (7.62\mathbf{i} - 4.4\mathbf{j})$ m/s$^2$,
$\mathbf{a}_{child} = (1.62\mathbf{i} - 4.4\mathbf{j})$ m/s$^2$

**19–102**   $E_x = 27.8$ N $\rightarrow$, $E_y = 64.4$ N $\downarrow$

**19–104**   $t = 0.34$ s

**19–106**   $\alpha_{sphere} = 14.29$ rad/s$^2$ ⟳, $\alpha_{cyl} = 16.67$ rad/s$^2$ ⟳

**19–108**   (a) $M = 53.7$ lb·ft ⟳,   (b) $a_O = 30.8$ ft/s $\rightarrow$

## Chapter 20

**20–2**   $\omega = 5.08$ rad/s

**20–4**   $\omega = 5.42$ rad/s

**20–6**   $v_A = 11.04$ m/s

**20–8**   $v_P = 7.6$ ft/s

**20–10**   $M = 524$ N·m

**20–12**   $\omega = 4.42$ rad/s, $O_x = -96.9$ lb, $O_y = 214$ lb

**20–14**   $v_1 = 5.85$ m/s

**20–16**   $v_C = 72.9$ ft/s

**20–18**   $\theta = 173°$

**20–20**   $v_C = 11.57$ ft/s

**20–22**   $v_C = 17$ m/s

**20–24**   (a) $v_C = 35$ ft/s,   (b) $d = 76.1$ ft

**20–26**   $\theta = 23.8°$

**20–28**   $P = 19.4$ N·m/s

**20–30**   $F = 10.92$ N

**20–32**   $v_1 = 8.63$ ft/s

**20–34**   $U = 52.8$ kN·m

**20–36**   $P_{max} = 353$ kN·m/s

**20–38**   $\omega = 488$ rad/s

**20–40**   $M = 0.0015$ lb·ft

**20–42**   $M_{avg} = 0.74$ N·m

**20–44**   $v = 107.3$ ft/s ⦨, $\mu = 0.1926$

**20–46**   $\omega = 0.368$ rad/s

**20–48**   $\omega = 2.25$ rad/s

**20–50**   $\omega = 0.288$ rad/s

**20–52**   $v_B = 1.361$ ft/s $\downarrow$

**20–54**   $v = 0.686$ m/s

**20–56**   $v = 1.345$ ft/s

**20–58**   $v = 5.03$ m/s $\downarrow$

**20–60**  $M = 307\,\text{lb·ft}\;\circlearrowright$

**20–62**  $T = 601\,\text{N}$

**20–64**  (a) $v_P = (Mg\sin\theta + 1.5\,mg\sin\theta)\Delta t/$
$(M + 1.125m)$, (b) $v_P = (Mg\sin\theta\,\Delta t)/2m$

**20–66**  $M = 372\,\text{lb·ft}$, $T_U = 49{,}814\,\text{lb}$, $T_L = 50{,}186\,\text{lb}$

**20–68**  $\omega = 7.31\,\text{rad/s}$, $O_x = 5.3\,\text{lb}\leftarrow$, $O_y = 10\,617\,\text{N}\downarrow$

**20–70**  $\Delta v = -7\,\text{ft/s}$, $\omega = 0.55\,\text{rad/s}\;\circlearrowright$

**20–72**  $\omega = 3.11\,\text{rad/s}$

**20–74**  $a = 0.714\,\text{m/s}^2\leftarrow$

**20–76**  $v = 17.17\,\text{m/s}$

**20–78**  $\omega = 0.705\,\text{rad/s}$

## Chapter 21

**21–2**  $T_0 = 16.56\,\text{ft·lb}$, $T_{45} = 8.28\,\text{ft·lb}$

**21–4**  $\mathbf{H}_O = (-0.361\mathbf{i} + 0.624\mathbf{j})\;\text{kg·m}^2/\text{s}$

**21–6**  $\mathbf{H}_A = (3.99\mathbf{i} + 1.792\mathbf{j})\,\text{lb·ft·s}$,
$\mathbf{H}_B = (3.99\mathbf{i} + 1.472\mathbf{j})\,\text{lb·ft·s}$

**21–8**  $T = \frac{1}{6}mb^2\omega_x^2$

**21–10**  $T_A = T_B = mg/2$

**21–12**  $\mathbf{H}_O = -\frac{1}{4}mab\omega_y\mathbf{i} + \frac{1}{6}ma^2\omega_y\mathbf{j}$,
$\mathbf{H}_C = -\frac{5}{36}mab\omega_y\mathbf{i} - \frac{1}{18}ma^2\omega_y\mathbf{j}$

**21–14**  $\mathbf{H}_O = -\frac{1}{4}m_1ab\omega_y\mathbf{i}$
$+ [\frac{1}{3}m_1a^2 + m_2(a^2 + \frac{1}{3}c^2)]\omega_y\mathbf{j} - \frac{1}{4}m_2bc\omega_y\mathbf{k}$

**21–16**  $\mathbf{H}_O = (md\omega_y/2)\left(\dfrac{h}{\pi}\mathbf{i} + d\mathbf{j} - \dfrac{h}{2}\mathbf{k}\right)$

**21–18**  $\mathbf{H}_A = m\left(\dfrac{r^2}{4} + l^2\right)\omega_y\mathbf{j} - \dfrac{1}{2}mrl\omega_y\mathbf{k}$

**21–20**  $\mathbf{H}_{O_2} = (60\mathbf{j} + 10\,450\mathbf{k})\;\text{N·m·s}$

**21–22**  $\mathbf{H}_A = \frac{1}{12}m(3r^2 + l^2)\omega_x\mathbf{i}$
$+ \frac{1}{12}m(3r^2 + l^2)\omega_y\mathbf{j} + \frac{1}{2}mr^2\omega_z\mathbf{k}$

**21–24**  $\theta = \cos^{-1}\left(\dfrac{3g(m + 2M)}{2I_y(m + 3M)\omega^2}\right)$, $M$ does not allow
an increase in operating speed

**21–26**  $\theta = \cos^{-1}\left(\dfrac{6(2mg\ell + Mg\ell)}{(12m\ell^2 + 4M\ell^2 - 3mr^2)\omega^2}\right)$

**21–28**  $A_z = B_z = 0$, $\mathbf{A}_y = -42.3\mathbf{j}\,\text{N}$, $\mathbf{B}_y = 74.5\mathbf{j}\,\text{N}$

**21–30**  $A_z = B_z = 0$, $A_y = \frac{3}{4}mg - \frac{1}{8}mb\omega_x^2$,
$B_y = \dfrac{m}{8}(2g - b\omega_x^2)$

**21–32**  $\dot{\mathbf{H}}_A = -\dfrac{1}{4}m_2bc\omega_y^2\mathbf{i} + \dfrac{1}{4}m_1ab\omega_y^2\mathbf{k}$,

$\dot{\mathbf{H}}_A = -\dfrac{b}{4}(m_1a\dot\omega_y + m_2c\omega_y^2)\,\mathbf{i}$

$+ \left[\dfrac{1}{3}m_1a^2 + m_2\left(a^2 + \dfrac{1}{3}c^2\right)\right]\dot\omega_y\mathbf{j}$

$+ \dfrac{b}{4}(m_1a\omega_y^2 - m_2c\dot\omega_y)\,\mathbf{k}$

**21–34**  $\dot{\mathbf{H}}_O = 0$

**21–36**  $\dot{\mathbf{H}}_A = -28.3\mathbf{i}\,\text{lb·ft}$

**21–38**  $\mathbf{A}_z = \mathbf{O}_z = 0$,

$\mathbf{A}_y = \left\{mg\left[\dfrac{1}{2} - \dfrac{3\pi r^2}{4ab}\right] + \dfrac{3}{8}\dfrac{\pi r^2 m}{a}\omega_x^2 - \dfrac{1}{4}mb\omega_x^2\right\}\mathbf{j}$

$\mathbf{O}_y = \left\{mg\left(\dfrac{1}{2} - \dfrac{\pi r^2}{4ab}\right) + m\omega_x^2\left(\dfrac{\pi r^2}{8a} - \dfrac{b}{4}\right)\right\}\mathbf{j}$

**21–40**  $\mathbf{M}_O = mr^2(\pi g - 2r\omega_x^2)\mathbf{k}$,

$\mathbf{A}_y = \left[2mr^2\omega_x^2\left(\dfrac{r}{d} + 1\right) - \pi rmg\left(\dfrac{r}{d} + 1\right)\right]\mathbf{j}$,

$\mathbf{B}_y = -\left[2mr\omega_x^2\left(\dfrac{r}{d} + 2\right) - \pi rmg\left(\dfrac{r}{d} + 2\right)\right]\mathbf{j}$

**21–42**  $\mathbf{M}_O = 7204\mathbf{k}\,\text{lb ft on shaft}$

**21–44**  $\dot\gamma = -436\,\text{rad/s}$

**21–46**  $\dot\phi = 0.7009$ or $-47.17\,\text{rad/s}$

**21–48**  $\dot\gamma = 30{,}656\mathbf{I}\,\text{rad/s}$

**21–50**  $M_y = 240\,\text{N·m}$

**21–52**  $N = 18.53\,\text{lb}$

**21–54**  $\mathbf{M}_A = (5296\mathbf{j} - 491\mathbf{k})\;\text{N·m}$

**21–56**  $\mathbf{A}_y = -18.25\mathbf{j}\,\text{lb}$, $\mathbf{B}_y = 38.25\mathbf{j}\,\text{lb}$

**21–58**  direct precession

**21–60**  $\mathbf{M}_O = (77{,}500\mathbf{i} + 44{,}700\mathbf{j} - 207{,}800\mathbf{k})\;\text{lb·ft}$

**21–62**  $\mathbf{H}_C = [(0.0833M + 0.0551m)l^2$
$+ (0.25M + 2m)r^2]\omega\mathbf{k}$

**21–64**  $T = \dfrac{(F\,\Delta t)^2 l^2}{8.75ml^2 + 12.5Mr^2 + 4.17Ml^2}$

**21–66**  $N = 30\,\text{lb}$

**21–68**  $(\mathbf{H}_O)_2 = (91{,}400\mathbf{j} + 40m\mathbf{k})\;\text{lb·ft/s}$,
$\omega = (0.209\mathbf{j} + 0.002\mathbf{k})\;\text{rad/s}$

## Chapter 22

**22–2**  (a) $\omega = 44.7\,\text{rad/s}$,  (b) $\omega = 44.7\,\text{rad/s}$

**22–4**  $\dot y_{max} = 3.14\,\text{ft/s}$, $\ddot y_{max} = 98.7\,\text{ft/s}^2$

**22–6**  $\omega = 31.6\,\text{rad/s}$

**22–8**  $\omega = 278\,\text{rad/s}$

**22–10**  $\theta = 0.1\sin(3.13t + \pi/2)$

**22–12**  $\omega = \sqrt{g/\Delta}$

**22–14**  $\dot\theta_{max} = 0.38\,\text{rad/s}$, $\ddot\theta_{max} = 1.47\,\text{rad/s}^2$

**22–16**  $\omega = 23.5\,\text{rad/s}$

**22–18**  $\omega = 8.66\,\text{rad/s}$

**22–20**  $\omega = 2.45\sqrt{\dfrac{2k - mgh}{m(4h^2 + a^2)}}$

**22–22**  $I_O = 10.94\,\text{in·lb·s}^2$

**22–24**  $\tau = 2\pi\sqrt{(r_g^2 + d^2)/dg}$

**22–26**  $k = 61.3\,\text{lb/ft}$

**22–28** $\omega = 278\,\text{rad/s}$

**22–30** $\theta_{\max} = 0.1\,\text{rad}$

**22–32** $\omega = 45.9\,\text{rad/s}$

**22–34** $\tau = 2\pi\sqrt{\dfrac{4b^2 + a^2}{6bg}}$

**22–36** $\omega = 69.5\,\text{rad/s}$

**22–38** $\tau = 0.175\,\text{s}$

**22–40** $\omega = \sqrt{g/2r}$

**22–42** $\tau = 2\pi\sqrt{1/2g}$

**22–44** $\Omega = 43.95\,\text{rad/s}$

**22–46** $k = 81.7\,\text{lb/ft}$

**22–48** $0 < \Omega < 10.05\sqrt{k/m}$

**22–50** $k_i = 1281\,\text{kN/m}$ (for each of four springs)

**22–52** $P_O = 12{,}391\,\text{lb}$

**22–54** $x = 0.05e^{-64.4t}\sin(59.8t)$

**22–56** $x = 0.03e^{-0.221t}\left[\sin(22.04t + 1.56)\right]$

**22–58** $c = 946\,\text{lb}\cdot\text{s/ft}$

**22–60** not possible

**22–64** $A = 1.034 \times 10^{-4}\,\text{in.}$

**22–66** $d_o = 0.129\,\text{m}$

**22–68** $L\ddot{q} + q/c = F_O \sin\Omega t$

**22–70** $L\ddot{q} + R\dot{q} + q/c = E_O \sin\Omega t$

**22–72** $\ddot{q} + \dfrac{R}{L}\dot{q} + \left(\dfrac{C_1 + C_2}{C_1 C_2 L}\right)q = 0$

**22–74** $I_C = 83{,}069\,\text{lb}\cdot\text{ft}\cdot\text{s}^2$

**22–76** $\tau = 2\pi\sqrt{3(R - r)/2g}$

**22–78** $P_O = 20\,\text{lb}$

**22–80** $L\ddot{q} + R\dot{q} + \dfrac{2q}{C} = \dfrac{d_o}{C}\sin\Omega t$

## Appendix E

**E–2** $I_y = \frac{1}{12}ML^2$

**E–4** $I_x = \frac{1}{2}MR^2,\ I_y = MR^2$

**E–6** $I_x = at\rho/3(b^3 - 3b^2d + 3bd^2),$
$I_y = bt\rho/3(a^3 + 3a^2c + 3ca^2)$

**E–8** $J_C = \frac{1}{12}ab(a^2 + b^2)$

**E–10** $I_y = ba^3/12$

**E–12** $I_x = \rho tab^3/12$

**E–14** $I_x = 104.2,\ (r_g)_x = 2.04$

**E–16** $I_x = \pi r^4/8,\ I_y = 5/8\pi r^4$

**E–18** $I_y = 5/8\pi\rho tr^4$

**E–20** $I_x = I_y = \pi r^4/4$

**E–22** $J_O = 54.5 \times 10^6\,\text{mm}^4$

**E–24** $I_z = ML^2/3$

**E–26** $I_x = a(0.333b^3 - b^2d + bd^2),$
$I_y = b(0.333a^3 + a^2c + ac^2)$

**E–28** $I_l = 5.93Mc^2$

**E–30** $J_O = 1/12ab(a^2 + b^2)$

**E–32** $I_{x_1} = 2MR^2$

**E–34** $I_{x'} = 0.11r^4$

**E–36** $I_{y_1} = 4.25Mr^2$

**E–38** $I_x = 12.63r^4,\ I_y = 43.1r^4$

**E–40** $I_x = 358\,\text{in}^4$

**E–42** $I_y = 1.039ma^2,\ (r_g)_y = 1.093a$

**E–44** $I_x = \rho t\left[\dfrac{he^3}{6} + 2he\left(d + \dfrac{e}{2}\right)^2 + \dfrac{bd^3}{3}\right]$

**E–46** $I_y = 723$

**E–48** $I_{xy} = 0.25a^2b^2\rho t,$
$I_{uv} = ab\rho t[0.144(b^2 - a^2) + 0.125ab]$

**E–50** $I_{xy} = -196.9a^4,\ I_{uv} = -265a^4$

**E–52** $I_{\max} = 645a^4,\ I_{\min} = 115.1a^4,$
$I_u = 371a^4,\ I_v = 390a^4$

**E–54** $I_{uv} = 0.036\rho t\{cb(b^2 - c^2)$
$+ a(c - 2d)[(a - 2d)^2 + a^2]\}$

**E–56** $I_{uv} = 2710\rho ta^4$

**E–58** $I_{xy} = 11.56 \times 10^6\,\text{mm}^4$

**E–60** $I_{xy} = -26.8\rho t\,\text{slug}\cdot\text{in}^2$

**E–62** $I_{\max} = 5.93 \times 10^6\,\text{mm}^4,\ I_{\min} = 3.23 \times 10^6\,\text{mm}^4,$
$\theta_P = 13.3°$ or $-76.7°$

**E–64** $I_x = 0.00485\,\text{slug}\cdot\text{in}^2,\ I_y = 0.00594\,\text{slug}\cdot\text{in}^2$

**E–66** $I_y = 1/5\rho tba^3$

**E–68** $I_x = 0.0476ab^3,\ r_g = 0.378b$

**E–70** $I_x = 47.2 \times 10^6\,\text{mm}^4,\ I_y = 10.95 \times 10^6\,\text{mm}^4$

**E–72** $I_{xy} = 313$

# List of Symbols

| QUANTITY REPRESENTED | SYMBOL |
|---|---|
| Acceleration | $\mathbf{a}$ |
| Distance; magnitude of acceleration; semi-major axis of ellipse | $a$ |
| Acceleration of mass center | $\mathbf{a}_C$ |
| Acceleration of $B$ relative to frame at $A$ in translation | $\mathbf{a}_{B/A}$ |
| Coriolis acceleration | $\mathbf{a}_{\text{Cor}}$ |
| Vectors; reactions at supports or connections | $\mathbf{A}, \mathbf{B}, \mathbf{C}, \ldots$ |
| Points in space or on rigid body | $A, B, C, \ldots$ |
| Planar area; amplitude of vibration | $A$ |
| Width; distance; semi-minor axis of ellipse | $b$ |
| Distance; coefficient of viscous damping | $c$ |
| Critical damping coefficient | $c_c$ |
| Constant; center of mass; capacitance | $C$ |
| Distance; moment arm; diameter; periodic displacement | $d$ |
| Constant | $d_O$ |
| Drag force | $D$ |
| Natural logarithm base; coefficient of restitution | $e$ |
| Total mechanical energy; voltage | $E$ |
| Frequency | $f$ |
| Force (vector form) | $\mathbf{F}$ |
| Magnitude of force | $F$ |
| Friction force | $\mathscr{F}$ |
| Gravitational acceleration | $g$ |
| Linear momentum | $\mathbf{G}$ |
| Constant of gravitation; geometric center; magnitude of linear momentum | $G$ |
| Height | $h$ |
| Angular momentum about mass center | $\mathbf{H}_C$ |
| Angular momentum about point $O$ | $\mathbf{H}_O$ |

| SYMBOL | QUANTITY REPRESENTED |
|---|---|
| $\dot{\mathbf{H}}_O$ | Time rate of change of angular momentum about point $O$ with respect to fixed $XYZ$ frame |
| $(\dot{\mathbf{H}}_O)_{xyz}$ | Time rate of change of angular momentum about point $O$ with respect to rotating $xyz$ frame |
| $\mathbf{i, j, k}$ | Unit vectors along coordinate axes $x, y, z$, respectively |
| $i$ | Current |
| $I$ | Instantaneous center |
| $I_C$ | Centroidal mass moment of inertia |
| $I_x, I_y, I_z$ | Mass moment of inertia about $x, y, z$, axes, respectively |
| $I_{x'}, I_{y'}, I_{z'}$ | Principal centroidal moments of inertia |
| $I_{xy}, I_{yz}, I_{xz}$ | Mass products of inertia |
| $I_1$ | Principal moment of inertia about spin axis |
| $I_2$ | Mass moment of inertia about centroidal axis transverse to spin axis |
| $J$ | Polar moment of inertia |
| $k$ | Spring constant |
| $K$ | General cross section of beam; constant |
| $l$ | Length |
| $l, m, n$ | Direction cosines |
| $L$ | Span; length; lead of screw; inductance |
| $m$ | Mass |
| $\mathbf{M}$ | Moment or couple (vector form) |
| $M$ | Magnitude of moment or couple; mass |
| $\mathbf{M}_{AB}$ | Moment about axis $AB$ |
| $\mathbf{M}_O$ | Moment about point $O$ |
| $\mathbf{M}_R$ | Resultant moment |
| $\mathbf{n}$ | Unit vector |
| $\mathbf{n}_n, \mathbf{n}_t$ | Unit vectors in normal, tangential directions |
| $\mathbf{n}_r, \mathbf{n}_\theta$ | Unit vectors in radial, transverse directions |
| $n$ | Multiplicity of threads in screw |
| $N$ | Force normal to a surface |
| $O$ | Origin of coordinate system |
| $p$ | Pressure; pitch of screw; distance from fixed axis to center of percussion |
| $\mathbf{P}$ | Force |
| $P$ | Point in space; power; center of percussion |
| $P_O$ | Constant |
| $q$ | Electric charge |
| $\mathbf{Q}$ | Vector |
| $Q$ | Quantity of matter in fluid flow |
| $(\dot{\mathbf{Q}})_{XYZ}$ | Time rate of change of vector $\mathbf{Q}$ with respect to fixed $XYZ$ frame |
| $(\dot{\mathbf{Q}})_{xyz}$ | Time rate of change of vector $\mathbf{Q}$ with respect to rotating $xyz$ frame |
| $\mathbf{r}$ | Position vector |
| $\mathbf{r}_C$ | Position vector of mass center |
| $r$ | Radius; radial coordinate |
| $r_C$ | Nearest distance between fixed axis and mass center |
| $r_g$ | Radius of gyration about mass center |

| QUANTITY REPRESENTED | SYMBOL |
|---|---|
| Radius of gyration about point $O$ | $r_O$ |
| Resultant force; reaction | $\mathbf{R}$ |
| Radius of earth; resistance | $R$ |
| Length of curve; displacement | $s$ |
| Time; thickness | $t$ |
| Kinetic energy; tension; thrust | $T$ |
| Velocity | $\mathbf{u}$ |
| Work | $U$ |
| Work of non-conservative forces | $U'$ |
| Velocity | $\mathbf{v}$ |
| Velocity of $B$ with respect to frame at $A$ in translation | $\mathbf{v}_{B/A}$ |
| Velocity of mass center | $\mathbf{v}_C$ |
| Potential energy; shear; volume | $V$ |
| Elastic potential energy | $V_e$ |
| Gravitational potential energy | $V_g$ |
| Weight per unit length | $w$ |
| Weight | $\mathbf{W}, W$ |
| Rectangular coordinates; distances | $x, y, z$ |
| Principal axes of inertia | $x', y', z'$ |
| Rectangular coordinates of fixed frame | $X, Y, Z$ |
| Angular acceleration | $\boldsymbol{\alpha}, \alpha$ |
| Angle | $\alpha$ (alpha) |
| Angle | $\beta$ (beta) |
| Specific weight; angle; Eulerian angle | $\gamma$ (gamma) |
| Rate of spin | $\dot{\gamma}$ |
| Change in length; static deflection of spring; angle | $\delta$ (delta) |
| Damping factor | $\zeta$ (zeta) |
| Efficiency | $\eta$ (eta) |
| Angular coordinate; angle; Eulerian angle | $\theta$ (theta) |
| Angle of repose; angle of static friction | $\theta_C$ |
| Angular speed; rate of nutation | $\dot{\theta}$ |
| Constant | $\lambda$ (lambda) |
| Coefficient of friction | $\mu$ (mu) |
| Coefficient of kinetic friction | $\mu_k$ |
| Coefficient of static friction | $\mu_s$ |
| Density; radius of curvature | $\rho$ (rho) |
| Angle; Eulerian angle; phase angle | $\phi$ (phi) |
| Angular speed; rate of precession | $\dot{\phi}$ |
| Normal stress | $\sigma$ (sigma) |
| Period; shear stress | $\tau$ (tau) |
| Period of damped vibration | $\tau_d$ |
| Angular velocity of rigid body | $\boldsymbol{\omega}$ |
| Angular speed; natural circular frequency | $\omega$ (omega) |
| Damped natural frequency | $\omega_d$ |
| Angular velocity of rotating coordinate system | $\boldsymbol{\Omega}$ |
| Angular speed; forcing frequency | $\Omega$ (omega) |

# Index

**A**bsolute acceleration, 731–32, 749–51
Absolute motion of a particle, 515, 524
Absolute system of units, 10, 530
Absolute velocity, 719–21, 749–51
Acceleration:
absolute, 731–32, 749–51
analysis:
general motion, 741–42
general plane motion, 731–33
rotating reference frame, 749–51
scalar, 452–58
vector, 486
angular, *see* Angular acceleration
average, 453, 479, 486
basic expressions, 457, 480
components:
cylindrical, 501
radial and transverse, 500
rotational, 731
spherical, 503
tangential, and normal, 488
translational, 731
constant, 457–58, 481
Coriolis, 590, 750, 927, example, 928
gravitational, 6
instantaneous, 453, 479, 486
Newton's second law, 529, 531, 533, 548, 638
relative, 470, 481, 750
variable, 458, 481
Acceleration-time curve, 455
Action and reaction, 6, 167, 237, 261, 263, 600
Actuator:
hydraulic, 15, 70, 71, 157, 421–22
Addition:
of forces:
concurrent, 18, 36
vector, 18–19, 27
Aerostatics, 314
Amplitude, 13
of vibration, 891, 906, 920
Analogues: electrical, 915, 920
Analysis:
assumptions, 12
modeling, 12
progressive, 13–16
Angle:
phase, 891, 913, 915, 919
of repose, 340, 375

Angular acceleration:
radial line, 500, 522, 711
rigid body, 710, 731
rotating frame, 749
vector, 500, 710, 741, 742, 749, 761
Angular coordinates, 709, 711. *See also* Polar coordinates
Angular impulse, 677, 701, 825–29, 843, 851, 883
Angular momentum:
particles, 592–93, 632
conservation, 594, 632
general plane motion, 592–93, 632
general space motion, 593, 632
time rate of change, 593–94, 632
rigid bodies, 825–27, 843
analysis, 848–49
about an arbitrary fixed point, 850
conservation, 829, 843
time rate of change, 826, 861–62
system of particles, 637, 700
about mass center, 673–75, 700
useful expressions, 676, 700
Angular speed, 499, 522, 711, 758–59
Angular velocity:
gyroscope, 872
rigid body, 709, 720, 722, 731, 874–75
rotating frame, 749, 750, 861–62, 872
spherical coordinates, 502
unit vectors time derivative, 503, 592
vector, 502, 709, 741, 742, 846–50
Archimedes:
principle of buoyancy, 316, 333
Areas:
centroids, 271–72, 329, (*table*), inside back cover
moment of inertia, 385, 393, 402–5, 414
polar, 386, 414
principal axes, 402–5
products of inertia, 400, 415
Associative property:
scalar product, 44
Axial tension, 175, 254
Axis of precession, 871–72, 874, 885
Axis of rotation, 706, 709–10, 758
Axis of symmetry, 404
Axisymmetric body: steady precession, 874–77
Axle friction, 361, 378

**B**all-and-socket joints, 128
Ball bearings, 128
Beams:
cantilever, 130–31, 172–74
concentrated forces, 171–206
concentrated load, 289
design, 187
distributed loads, 289, 292–93, 330–31
examples and problems, 290–91, 294, 304
simply supported, 129, 172
reactions (*table*), 172, 202
Belts, 369–71
Bending moment, 173, 292–93
diagrams, 187–88, 204
examples and problems, 189–201, 294–304
review problems, 205–6
sign conventions, 176, 203
Biaxial test, 162–63
Body centrode, 723
Body cone, 847–48
Bow's notation, 221, 262
Buoyancy theory, 316, 333

**C**ables:
analysis, 254, 305
catenary, 308–9, 332
concentrated loads, 254–55, 263
examples and problems, 256–60
differential equation, 305, 332
distributed loads, 305–9
examples and problems, 310–12
parabolic, 306–7, 332
Catenary, 308
Center of buoyancy, 316
Center of curvature, 488, 532
Center of gravity, 71, 267–69, 272–73, 278, 329–30, 664, 811
composite body, 272–73, 278, 330
location, 268–69, 278
three-dimensional body, 268–69
Center of mass, 267, 358–59, *See also* Mass center
Center of percussion, 781–82
Center of pressure, 315, 332
Center of rotation:
instantaneous, 722–23, 759
Central force, 594
Central impact, 620–25, 634

Centrifugal force, 533, 925–26
Centripetal force, 533, 925–26
Centrode, 723
Centroidal axis, 781
Centroids:
  areas, 271–72, 329 (*table*), inside back cover
  composite objects, 272–73, 329–30
  coordinates of lines and areas, 270–71,
    329–30
    (*table*), inside back cover
  examples and problems, 274–81, 284
  geometrical property, 269
  lines, 270, 329, (*table*), inside back cover
  volumes, 271, 329–30 (*table*), inside back
    cover
Circle of friction, 361
Circular frequency, 890–99, 902, 906, 912
Circular motion, 500, 522, 532
Closed path, 574, 587
Coefficient:
  critical damping, 912, 920
  friction, 338–40, 375, 377, 378, 379
  restitution, 623, 624
  viscous damping, 911, 915, 920
Collision, *see* Impact
Commutative property:
  scalar product, 44
  vector product, 77
Components:
  *see* Normal and tangential components, Radial
    and transverse components, Rectangular
    components, Scalar components, or
    Vector components
Compression, 132, 134, 187, 211, 212
Conservation:
  angular momentum, *see* Angular momentum
  energy:
    particles, 575–76, 587, 639
    rigid bodies, 811–12, 842
    system of particles, 667, 700
  linear momentum, *see* Linear momentum
    particles, 591, 594, 612–13, 632, 633, 634
    rigid bodies, 828–29, 843
    system of particles, 676, 701
Conservative forces, 437, 556, 573–74, 811
Conservative systems, 556, 573–76, 587, 811,
    901
  kinetic energy, 811
  potential energy, 811
Constant acceleration, 457–58, 481
Constant of gravitation, 600, 633
Constrained motion:
  rigid bodies, 770
    general plane, 791–92
Constraint:
  complete, 133, 134
  improper, 134
  partial, 133, 134
  redundant, 132–34
Control volume, 685
Coordinate systems:
  translation, 514–15, 528
Coordinates:
  angular, 709, 711
  cylindrical, 501, 523
  polar, *see* Polar coordinates
  position, 452
  rectangular, cartesian, 484, 520
  spherical, 501–3, 523
Coplanar forces, 109
Coplanar vectors, 19
Coriolis effects, 927–28
  acceleration, 540, 750, 927
  drift, 927–28
  example, 928
Coulomb friction, 911
Couple, 96–97, 119

components and addition, 99–100, 119
  equivalent, 98–99
Couple vector, 97
Critical damping coefficient, 912
Critical locations, 187
Cross-head, 157
Cross product, *see* Vector product
Curvature: center, 488, 532
  path, 487
  radius, 487, 493, 521, 606
Curvilinear motion:
  circular, 500, 522, 532
  cylindrical coordinates, 501, 523
  polar coordinates, 498–500, 522
  rectangular coordinates, 484–85, 520
  spherical coordinates, 501–3, 523–24
  tangential and normal components, 486–88,
    521
    plane, 487–88
    space, 488
  velocity and acceleration: physical aspects,
    485–86, 520–21
Curvilinear translation, 706
Customary units, *see* U.S. customary units
Cylindrical coordinates, 501, 523

D'Alembert's definition of dynamic equilibrium,
    925–26
Damped vibrations, *see* Vibrations
Damper mass, 14
Damping: 911, 920
  critical, 912, 920
  overdamped, 912, 920
  underdamped, 912, 920
  viscous, 911
Damping factor, 912, 920
Dashpot, 911
Deceleration, 453
DeCoriolis, 750
Decrement: logarithmic, 917
Deformable bodies, 5
  collision, 420–25
  *See also* Strength of materials
Deformation:
  elastic, 620–21
  plastic, 620–21
Degrees of freedom, 134, 471, 481, 893
Density, 317
Dependent relative motion, 471
Derivatives: time:
  rotating frame, 748, 761
  vector function, 487
  vector products, 593–94, 731
  vectors, 487, 499, 922–23
Determinant form:
  angular momentum, 593
  moment of a force:
    about a line, 91
    about a point, 84, 92
  triple mixed product, 89
  vector product, 78
Differential equations:
  particle motion, 754
  vibrations, 890, 905, 911, 913
Direct central impact, 620–23, 634
Direct precession, 876, 877
Direction cosines, 34–35
Displacement:
  angular, 124
  cycle, 574
  function:
    variable acceleration, 458, 481
  linear, 124
  net, 455–56
  virtual, 429
Distributed forces:

basic concepts, 226
review problems, 334–35
Distributed loads:
  beams, 289, 292–93, 330–31
    examples and problems, 290–91, 294–304
  cables, 305–9
    examples and problems, 310–12
Distributive property:
  scalar product, 44
  vector product, 77, 80
Disturbing force, 889
Dot product, 44, 65
Drag, 336, 371
Drift, 927–28
Dry friction, 911
Dynamic balance, 864
Dynamic equilibrium:
  D'Alembert's definition, 925–26
  example, 926
Dynamics: definition, 451
Dynamic unbalance, 864

Eccentric impact, 620
Efficiency: mechanical, 430, 558–59, 586
Elastic deformation, 620–21
Elastic impact, 628
Elastic members:
  potential energy, 436
Electrical analogue, 915, 920
Elliptic orbit, 601, 603, 604
Energy:
  conservation:
    particles, 575–76, 587, 639
  kinetic, *see* Kinetic energy
  potential, *see* Potential energy
  principle of work and, *see* Principle
  total mechanical, 576
Energy methods:
  applications, 559–60, 586–87, 638
  particles, 551–89
  rigid bodies, 805
  vibration analysis, 901–2, 919
English system, 8, 10 (*See also* U.S. customary
    units)
Equations of motion:
  Euler's, 862–63, 884
  gyroscopic, 872–73, 885
  particle, 531
    Newton's second law, 548, 639
    normal and tangential components, 532, 549
    radial and transverse components, 532, 549
    rectangular components, 532, 549
  rigid bodies:
    plane: independent scalar, 769, 790, 800,
      801
    center of percussion, 782
    rolling/sliding wheels, 792, 801
    rotation about fixed axis, 781, 800
    rotation about mass center, 769, 781, 800
    translation, 769, 800
    rotational, 767, 768, 800
    three-dimensional, 861, 884
      independent scalar, 863, 885
    translational, 766, 800
  system of particles, 658–59
Equilibrium,
  definition, 50, 124
  dynamic, *see* Dynamic equilibrium
  particle, 50–52, 429–30
    problems, 58—62
  potential energy, 437
  rigid bodies, 123–70, 429–30
  stability, 438–39, 447
  two-force member, 132
  virtual work, 429
Equilibrium equations, 429

maximum for one body, 166
particles, 50, 66
rigid bodies, 124–25, 166
Equivalent forces, 72, 118
Escape velocity, 604; example, 606
Euler angles, 871–72
Euler, Leonhard, 863
Euler's equations of motion, 862–63, 884
Exciting force, 889
External forces, 71–72, 118
rigid bodies: impulse, 825
system of particles, 657–58, 667, 672
moments, 675
trusses, 221
External loads, 188
External moments:
rigid bodies: impulse, 826

**First moment, 270, 271**
Fixed axis rotation, 706–7, 709–10, 758, 780–82, 800, 809–10, 841
Fixed members, 126–28
Fixed reference point: angular momentum, 673–77, 700, 826–28, 843
Fluid flow, 685–90
examples and problems, 91–98
incompressible, 685, 702
momentum change, 686–87
pressures, 687
steady flow, 685–89, 701–2
applications, 688–89
variable, 685, 689–90, 702
special cases, 690
velocity change, 686
Fluid friction, 911
Fluid mechanics, 5, 313–22, 332–33, 371, 379
Fluid statics, 313–16, 332
examples and problems, 317–22
Foot, 10
Force, 5
components:
scalar, 27–28, 34–36, 64
vector, 19–20, 27–28
conversion, 11
direction cosines, 34
drag, 371
measurement system, 8–12
rectangular components, 20, 34–35
resultants, 18–19, 34–36
transformation, 100–101
Force-couple system, 100–101
in equilibrium, 124
examples and problems, 103, 106–7, 111–17
transformations, 120
Forced vibrations, 905
damped, 913–15, 920
without damping, 905–7, 919–20
Force polygons, 222, 262, 351
Forces:
axial, 132
on beams, 175
central, 594, 601
centrifugal, 533, 925–26
centripetal, 533, 925–26
common:
in free-body diagrams (table), 125–28
concurrent, 109, 120, 136
conservative, 437, 556, 573–74
coplanar, 109, 120
equivalent, 72, 118
external, see External forces
in free-body diagrams, 51, 125
gravitational, 600, 633
impulsive, 609
inertia, 925
internal, 71, 173–75, 657–58, 667

nonconservative, 575, 667
normal, 55, 336
parallel, 110, 120, 136
on particles, 17–49
review problems, 67
without work, 553, 560
on rigid bodies, 69–120
review problems, 121–22
without work, 863
simplification of many, 108
on a system of particles, 657–58, 667
unknown:
determination, 51, 129–31, 148
Force systems:
common simple, 109–10, 120
complex, 108, 120
examples and problems, 111–17
equivalent, 108, 120
parallel, 110
three-dimensional, 34–36, 108
examples and problems, 38–43
two-dimensional, 18–19
examples and problems, 21–26
Force triangles, 222, 262, 351
Force vector, 7
method for determining, 36–37, 65
Forcing frequency, 905
Frame of reference, see Reference frame
Frames, 36–37, 65
examples and problems, 240–53
Free-body diagrams, 51–52, 71–72, 125–29, 166
particles, 533, 549
frames and machines, 237–38
interrelated, 156–58, 167–68, 175
rigid bodies:
plane motion, 769
system of, 770, 800, 810
schematic drawings, 51, 71–72, 126, 128, 156–57, 172
trusses, 210–12, 226
Freedom: degrees of, 134, 471, 481, 893
Free fall, 490
Free vibration:
damped, 911–15, 920
without damping, 893–94, 919
Frequency: 891, 919
circular, 890–91
forcing, 905, 919
natural, 893, 919
natural circular, 890, 892, 902, 906, 912, 919
Friction:
angles, 39–40
axle, 361, 378
belt, 369–71, 379
examples, 372
coefficient, 338–40, 375, 377, 378, 379
Coulomb, 911
disk, 360, 378
dry, 337–39, 375
classification (table), 341, 376
examples and problems, 342–50
fluid, 336, 371, 379
forces, 575, 576
internal, 336, 362
kinetic, 339
review problems, 380–81
static, 338
coefficients (table), inside back cover
surface, 362
Friction force, 336–37, 375, 430
Friction problems:
procedures for solving, 341, 376

**Gage pressure, 314**
Galilei, Galileo, 451, 529
Gases, 313

General motion: rigid body, 707
kinematics:
analysis, 741–42
examples and problems, 743–47
kinetics:
analysis, 861–64
examples and problems, 865–69
Geneva mechanism, 757
Gravitation, 6
constant, 600, 633
force, 600, 633
Newton's law, 6, 600, 633
Gravitational acceleration, 6, 458, 490, 600–601, 633
Gravitational potential energy, 554, 556, 576, 585, 664, 667, 699, 811, 845
Gravitational system of units, 10, 531
Gravitational work, 446
Gravity, 435
Guldinus: theorems, 282, 330
Gun: ballistics test, 614–15
Gyration: radius, 385, 386, 414, 415
Gyroscopes, 870–77
Gyroscopic motion, 870–77, 885
differential equations of motion, 873
equation of motion, 872, 885
examples and problems, 878–882
precession, 847, 870–77, 885
special cases, 873–77
precession and spin axes at right angle, 874, 885
steady precession, 873–77, 885
axisymmetric body: absence of moments, 874–77

**Hancock Tower, Boston, 14**
Harmonic function of time, 905
Harmonic motion, see Simple harmonic motion
Hertz (unit), 891
High-frequency excitation, 907, 914
Hodograph, 486
Hooke, Robert, 554
theory of springs, 436
Horsepower, 558, 812
Hydraulic ram, see Actuator, hydraulic
Hydrostatic forces, 314–16, 332
Hyperbolic orbit, 603

**Impact, 620–25**
central, 620–25
direct, 620–23, 634
oblique, 224–25, 634
collinear, 620
eccentric, 620
elastic, 623
examples and problems, 626–31
line, 620
plastic, 623
real materials, 624
Impulse:
approximation, 611
and momentum:
particles, 609–11, 633
system of particles, 677, 701
rigid body, 825–28, 842–43
angular, 825–26, 843
external forces, 825
external moment, 826
linear, 825, 842
three-dimensional, 850–51, 883
system of rigid bodies, 827–28, 843
Impulsive force, 609
Impulsive motion, 609–10

Inertia, 6, 382–83, 529
  centroidal mass moment, 8, 849
  centroidal mass products, 849
  moments, *see* Moments of inertia
  principal axes, 849
  principal centroidal moments, 849
  products, *see* Products of inertia
Inertia force, 925
  centrifugal, 925–26
  linear motion, 925–26
Inertial reference frame, 530
Inertia matrix, 849
Inertia tensor, 849
Initial conditions, 456, 769, 890
Instantaneous center of rotation, 722–23, 759
Instantaneous velocity, 453, 479, 485
Internal forces, 71–72, 118, 173–75, 202–3,
    657–58, 667
  examples and problems, 177–86
  sign conventions, 176, 203
  trusses, 221
Internal friction, 911
Internal reactions, 173, 292–93
Internal shear, 173–75, 292–93
International system of units, *see* SI system of
    units

**J**oints:
  formal analysis procedure, 221–24
  special cases, 213
Journal bearings, 213

**K**epler, Johannes, 601
Kepler's laws: 601–3, 633
  first, 601–3, 633
  second, 601–3, 633
Kilogram, 530
Kinematics, 451
  particles:
    scalar analysis, 452–58
      examples and problems, 459–69
      relative motion, 470–78
      review problems, 482
    vector analysis, 483–527
      examples and problems, 489–97, 504–13
      relative motion, 514–19
      review problems, 525–26
    *see also* Curvilinear motion; Particles
  rigid bodies, 705–7
    general motion, 707, 741–42, 760–61
      examples and problems, 743–47
      general plane motion:
        acceleration, 731–33, 760
          examples and problems, 734–40
        velocity, 719–23, 759
          examples and problems, 724–30
      rotating reference frames, 748–51, 761
        examples and problems, 752–57
      review problems, 762–64
      rotation, 711, 758–59
      rotation about a fixed axis, 706–7, 709–10,
        758
      translation, 705–6, 708, 758
Kinetic energy:
  definition, 557, 586, 639
  particle, 557–58
    conservation of mechanical energy, 576,
      587, 639
    work and energy: principle, 557, 586, 639
  rigid body, 808–10, 841
    rotation, 809–10, 841
    three-dimensional motion, 884
      derivation, 851–53
      rotation about a fixed point, 853, 884
    translation, 808–9, 841

system of particles, 664–66, 699
Kinetics, 451, 528
  particles:
    momentum methods, 590–636, 638
      examples and problems, 595–99, 606–8,
        614–19, 626–31
      review problems, 635–36
    Newton's second law, 522–30, 548, 638
      examples and problems, 534–47
      equations of motion, 531–33, 549
      problem-solving, 533, 549
      review problems, 550
    overview of methods, 637–55
      equations (*table*), 639
      examples, 640–51
      problems, 625–55
    work and energy methods, 551–59, 638
      applications, 551, 559–60
      examples, 561–67, 577–81
      problems, 568–72, 582–84
      review problems, 588–89
  rigid bodies:
    plane motion, 765–804
      examples and problems, 771–79, 783–89,
        793–99
      review problems, 802–4
    three-dimensional, 846–87
      examples and problems, 854–60, 865–69,
        878–82
      review problems, 886–87
  systems of particles, *see* Systems

**L**aws:
  Kepler's, 601–3, 633
  Newton's, *see* Newton's laws
  Pascal's law, *see* Pascal's law
Length: conversion, 11
Lever arm, 72
Line of action, 7, 37
Linear impulse, 609, 633, 677, 701, 825, 842,
    851
Linear momentum:
  particles, 591, 632
    conservation, 591, 632
    moment, 592
    time rate of change, 591, 632
  rigid bodies, 825, 842, 850–51
    conservation, 828–29, 843
    time rate of change, 825
  system of particles, 672, 701
  system of rigid bodies, 827–28, 843
  *See also* Translation
Lines:
  centroids, 270, 329 (*table*), inside back cover
Liquids, 313
Load cell, 157
Load frame, 157–58
Logarithmic decrement, 984

**M**achines, 237, 239, 263
  ideal, 558
  real, 558–59
Magnification factor, 907, 914, 920
Mass, 5, 10, 530, 531
  conversion, 11
  effective, 894
  moment of inertia, 384–85, 405–6, 414, 416
  products of inertia, 401, 415
  steady flow, 685–89, 701–2
  units in commercial practice, 924
  variable, fluid systems, 685, 689–90, 702
    special cases, 690
Mass center:
  system of particles, 658–59
    angular momentum, 673–75, 700

gravitational potential energy, 664, 699
  motion, 658–59, 699
  reference frame, translating: kinetic energy,
    665–66, 699
Maxwell's diagram, 222–24
Measurement systems, 8–12
  *See also* SI system of units; U.S. customary
    units
  absolute, 10
  gravitational, 10
Mechanical efficiency, 430, 446, 558–59, 586
Mechanical energy, 576
  conservation of total, 576, 587
Mechanical systems, 576, 915, 920
Mechanics, 4–5 (*See also* Fluid mechanics;
    Newtonian mechanics)
  Newtonian, 5
  principles, 69
Mechanics of materials, 3, 5, 171, 323
Meter, 530
Method of joints, 210–13, 261
  examples and problems, 214–20
Method of sections, 226–27, 262–63
  examples and problems, 214–20
Mixed triple product, 89–90, 119
MSK system, *see* English system
Modeling, 12–13
  assumptions, 13
  beams, 171
  cables, 254, 306
  forces on bodies, 17
  rigid bodies, 69–70
Mohr's circle, 403–5
Moment arm, 72
Moment of a couple, 96–97, 119
Moment of a force:
  conversion, 12
  definition, 72, 79
  determination:
    examples and problems, 73–76, 81–88
    about a line, 89–91, 119
    examples and problems, 92–95
    rectangular components, 80, 119
    scalar definition, 72, 118
    vector definition, 79, 118
Moment of linear momentum, 592, *see also*
    Angular momentum
Moments: (*See also* Bending moment)
  bending, 173
  internal, 173–75, 292–93
  sign conventions, 176
  twisting, 173
  unknown, 12
    calculation, 129–31, 148
Moments of forces:
  system of particles, 675, 700
Moments of inertia:
  area, 385, 402–5, 414
  complex areas and bodies, 393, 415
  determination, by integration, 384–86
  mass, 384–85, 405–6, 414, 416
  parallel-axis theorem, 392–93, 414–15
  polar, 386, 414
  principal axes, 402–5, 416
  review problems, 417
  (*table*), inside back cover
Momentum:
  angular, *see* Angular momentum
  conservation, *see* Angular momentum; Linear
    momentum
  linear, *see* Linear momentum
  particles:
    examples and problems, 595–99
    review problems, 635–36
  rigid bodies:
    plane motion, 825–29, 842–43
      examples and problems, 830–40

three-dimensional, 847–50, 883
system of particles, 672–77
  examples and problems, 678–84
  *See also* Systems
system of rigid bodies, 827–28, 843
Motion:
absolute, 515
curvilinear, 484–86
  *See also* Curvilinear motion
equations, *see* Equations of motion
gyroscopic, *see* Gyroscopic motion
impulsive, 609–10
Newton's second law, 5, 529–30
particle, 452–54, 484–88, 498–503
planetary, 601–5
rectilinear, 452–56
relative, 470–71, 481, 514–15, 524
  *See also* Relative motion
rigid bodies:
  fixed point, 707, 742, 853, 863
  general, 707, 741–42
  general plane, 707, 719–23, 731–33
  *See also* Rigid bodies
simple harmonic, *see* Simple harmonic motion
Multiforce members, 237

**N**atural frequency, 890–91
  damped, 912, 920
Newton (*unit*), 530
Newton, Isaac, 451, 528
Newtonian mechanics, 5–6
Newtonian reference frame, 530, 548
Newton's laws, 5–6
  gravitation, 600, 633
  second, 529–30, 548, 551, 557, 560, 587,
    590, 591, 609, 637–39, 765, 766–67,
    890, 911
  system of particles, 657–58, 699
  third, 156, 167, 175, 212, 237, 261, 263
Normal and tangential components, 488
Normal force, 55, 336
Normal stress, 323
Nutation, 872

**O**blique central impact, 624–25, 634
Orbital geometries, 605
Orbits, 601, 602, 603, 604
Oscillation, *see* Vibrations
Osculating plane, 488
Output work, 430

**P**appus, theorems, 282, 330
Parabolis orbit, 603
Parallel-axis theorem, 392, 414–15
  examples and problems, 394–99
  products of inertia, 400–401, 416
Parallelogram law, 18, 63
Partial derivatives, 574, 587
Particles, 5–6
  definition, 17
  equilibrium, 50–52, 66
  forces on:
    review problems, 67–68
  kinematics:
    scalar analysis, 451–82
      examples and problems, 459–69
      relative motion, 470–71
        examples and problems, 472–78
      review problems, 482
    vector analysis, 483–527
      cylindrical coordinates, 501, 523
      polar coordinates, 498–500, 522
      rectangular coordinates, 484–85, 520
      relative motion, 514–15

examples and problems, 516–19
review problems, 525–27
spherical coordinates, 501–3, 523–24
tangential and normal components,
  486–88, 521
kinetics:
  momentum methods, 590–636, 637–39
  Newton's second law, 528–30, 637–39
  overview of methods, 637–55
    equations (*table*), 639
    examples, 640–51
    problems, 652–55
  work and energy methods, 551–89, 637–39
plane motion, *see* Plane motion
speed, 453, 485
systems, *see* Systems
vibrations, *see* Vibrations
Pascal's law, 313, 332
Path of particle, 484–88, 521
Pendulum: conical, 651; simple, 896; torsion,
  897
Percussion, center of, 781–82
Period:
  of vibration, 891, 919
Perpetuum mobiles, 435
Phase angle, 891, 913, 915, 919
Phase relationships, 907
Pin connections, 126–27
Plane motion:
  particle: kinematics:
    circular, 500
    curvilinear, 484, 486–88, 521
    polar coordinates, 498–500, 522
    rotating reference frame, 751
  rigid bodies:
    energy and momentum methods, 805–45
    kinematics, 705–40
      categories (*table*), 706
    kinetics, 765–804
      review problems, 802–4
      rotation about a fixed axis, 780–89
Planetary motion:
  Kepler's laws, 601–3, 633
Plane trusses, 208
Plastic deformation, 620–21
Plastic impact, 623
Point masses, 268
Polar coordinates, 498–500, 522
  acceleration, 500, 522
  angular acceleration, 500, 522
  angular speed, 499, 522
  circular motion, 500, 522
    conversion to tangential and normal
      coordinates, 500, 522
  unit vectors, 499, 522
  velocity, 499, 522
Polygon rule, 19, 63
Position coordinates, 452
Position vector, 79, 484–85, 499, 520
  relative, 515, 524
Potential energy, 435–37, 446–47
  elastic, 554–56, 586, 667, 811, 842
  examples and problems, 440–44
  gravitational, 554, 556, 585, 664, 667, 699,
    811, 842
  mechanical system, 576, 811
  notation, 556
  total, 556, 811, 842
  work, 554–55, 585–86
    transformation, 576
Potential functions, 573–75, 587
Pound, 10
Power:
  particles, 558, 586
  rigid bodies, 812, 842
Precession, 847, 870–77, 885
  axis, 871–72, 874, 885

direct, 876–77
retrograde, 876–77
steady, 873–77, 885
Pressure:
  center of, 315
  gage, 314
  hydrostatic, 314–15
Principal axes of inertia, 849
Principal moments of inertia, 402–6, 416
  area, 403, 416
  mass, 405, 416
  (*table*), inside back cover
Principle:
  conservation of energy, 576, 587, 667, 700,
    811, 842
  impulse and momentum:
    deformable bodies, 620–25
      examples and problems, 626–31
    particles, 609, 633
    rigid bodies, 625–28, 842–43, 850–51, 883
    system of particles, 677, 701
  transmissibility, 7, 72
  work and energy, 558, 586, 667, 700, 810,
    841
    particle, 557–58, 586
    rigid bodies, 810, 841
    system of particles, 667, 700
Products of inertia, 400–401, 415–16
  area, 400, 415
  mass, 401, 415
  parallel-axis theorem, 400–401, 416
    areas, 952
    masses, 953
Projectiles: equations, 491
  example, 490
  problems, 494–96

**R**adial and transverse components:
  acceleration, 500, 522
  equations of motion, 533, 549
  unit vectors, 449, 522
  velocity, 499, 522
Radius:
  curvature, 487, 493, 521, 606
  gyration, 385, 386, 414, 415
Rate of change, *see* Time rate of change
Reactions, 51, 71, 125, 289
Real systems:
  mechanical efficiency, 430, 446
Rectangular components:
  angular momentum, 593, 849
  couples, 100, 119
  equations of motion, 532, 549, 769, 800, 863,
    883, 885
    fixed axis rotation, 864
    Euler's equations, 863, 884
  force, 20, 34–35
  moment, 80, 119
  vector product, 78
Rectangular coordinates, 452
  curvilinear motion, 484–85, 520
Rectilinear motion: particle, 452–69
  graphical analysis, 454–55
Rectilinear translation, 705–6
Reference frame, 5
  fixed, 530, 673–75, 700
  inertial, 530
  Newtonian, 530, 548
  translating, 514–15, 524, 665–66, 673–75,
    699, 700
  rotating, 748–51, 761
Relative acceleration,
  two particles, 470, 481, 515, 524
  two points on rigid body, 719–21
Relative motion:
  dependent:

Relative motion: (*cont.*)
  degrees of freedom, 471, 481
  translation: coordinate axes, 514–15, 524
  two particles, 470–71, 481, 514–15, 524
  examples and problems, 472–78, 516–19
Relative position, 470, 481, 515, 524
Relative velocity:
  two particles, 470, 481, 515, 524
  two points on rigid body, 731–32
Resolving of a force:
  into vector components, 19–20
Resonance, 907, 914, 920
Resonant frequency, 907
Restitution, 621
  coefficient, 623, 634
    schematic variation, 623
Restoring force, 889
Resultant couple, 109
Retrograde precession, 876–77
Right-hand rule, 77
Righting couple, 316
Rigid bodies, 5
  constrained, 770
  equilibrium:
    two dimensions, 129–31
      examples and problems, 142–47
    three dimensions, 148
      examples and problems, 152–55
    review problems, 169–70
  forces on, 69–120
    review problems, 121–22
  kinematics, 705–64
    *See also* Kinematics
  kinetics, *see* Kinetics
  plane motion, 765
    categories, (*Fig.*), 706
    energy methods, 805–12, 841–42
      examples and problems, 813–24
    momentum methods, 825–29, 842–43
      examples and problems, 830–40
    *see also* Plane motion
  three-dimensional motion, 846
    derivation, 847–53, 861–64, 883–85
    about a fixed axis, 863
    about a fixed point, 863
    unconstrained, 863
  unconstrained, 769
Rigid frames, 239
Road simulation, 15–16, 131
Rolling contacts, 127
Rolling motion, *see also* Wheels,
  kinematics examples, 725, 735, 736
  kinetics examples, 793, 794
Rolling resistance, 362, 379
Rotating reference frame, 748–51, 761
  velocities and accelerations:
    analysis, 749–51, 761
Rotation: rigid body, 711, 758–59, 767, 800
  center of percussion, 781–82
  about a centroidal axis, 769, 781
  about a fixed axis, 706–7, 709–10, 758,
    780–82, 800, 863
  instantaneous center, 722–23, 759
  kinetic energy, 809–10, 841
  system of particles, 700
  three-dimensional, fixed point, 742, 863
  time function, 711, 758–59
  and translation, simultaneous, 706, 707,
    719–21, 731–32
  *See also* Angular momentum
Rotational components:
  acceleration, 731
Rotational motion:
  derivation of equation, 767–69, 800
Rotations, 124, 423

**S**atellites, 606, example, 607–8

Scalar components, 27–28, 34, 64
  sign conventions, 28
Scalar definition of a moment, 72, 118
Scalar equations:
  equilibrium, 50, 66, 124–25, 166, 227
Scalar product, 44, 65
Scalars, 7
Screws, 352–53
  examples and problems, 356, 358–59
Second, 530, 531
Shaking force, 889
Shear force, 173, 292–93
  diagrams, 187–88, 204
    examples and problems, 189–201, 294–304
  review problems, 205–6
  sign conventions, 176, 203
Shear stress, 323
SI system of units, 8–10, 530–31, 548
  (*table*), inside front cover
  *See also* Measurement; Units
Sign conventions, 28, 64, 176, 203
Simple harmonic motion, 889–92, 919
  definition, 890
  graphic recordings, 891, 892
Simple truss, 209, 210, 261
Sleeve bearings, 128
Sliding bearings, 361
Sliding vectors, 7
Slug, 10, 531
Solution:
  complementary, 906
  general, 906
  particular, 906
Space, 5
Space centrode, 723
Space cone, 847–48
Space motion:
  curvilinear, 488
  cylindrical coordinates, 501, 523
  particle, 488, 593, 632
  rigid body, 707, 741–42, 760–61
  spherical coordinates, 501–3, 523–24
Space trusses, 208, 227, 263
  example 232–33
Speed:
  angular, 499, 522, 711, 758–59
  linear, 453, 485
Spherical coordinates, 501–3, 523–24
Spin, 872
Spring constant, 436, 554; effective, 894
Spring stiffness, 554
Springs:
  potential energy, 436
  theory, 554–55
  torsion, 436
Stability: equilibrium, 438
Static loading, 907–14
Statically determinate frames, 239
Statically determinate members, 130
Statically determinate space truss, 227
Statically indeterminate frames, 239
Statically indeterminate members, 130
Statically indeterminate truss, 209
Statics, 3
  fluid, 313–16, 332–33
  importance, 207
Steady precession, 873–76, 885
  axisymmetric bodies:
    absence of moments, 874–77
Steady-state vibration, 906, 913
Strength of materials, 5
  application, 176
  importance, 207
  *See also* Deformable bodies
Stress, 323–25
  definitions, 323, 333
  normal, 323

shear, 323
Stress analysis:
  elements, 323–25
  examples and problems, 326–28
  wedge method, 325, 333
Structural mechanics:
  relative motion, 470
Structures:
  basic concepts, 207
    review problems, 264–65
Supports,
  common for beams, 172, 202
  redundant, 130
  two-dimensional (*table*), 126–27
  three-dimensional (*table*), 128
Surface:
  frictionless, 126–27
  low friction, 127
  rough, 126–28, 340
Surfaces:
  hydrostatic forces, 314–16
Surfaces of revolution, 282
Symbols (*list*), L-1–L-3
Symmetry: axis, 271, 400; plane, 271
Systems:
  conservative, *see* Conservative systems
  coordinate: translation, 514–15, 524
  fluid, 685
  particles:
    angular momentum, 673–76, 700
    conservation of energy, 667, 700
    conservation of momentum, 676, 701
    continuous, 656
    discrete, 656
    equations of motion, 658–59, 699
    gravitational potential energy, 664, 699
    impulse and momentum, 677, 701
    kinetic energy, 664–66, 699
    linear momentum, 672, 700
    moments of forces, 672, 673, 675, 700
    Newton's second law, 657–58, 699
    work and energy, 667, 700
  rigid bodies, 770, 800, 810
    impulse and momentum, 827–28, 843
      problem solutions, 827
      fixed reference point, 828
  vibrating, 889
    *See also* Vibrating systems

**T**angential components, *see* Normal and
  tangential components
Tension, 23, 132, 134, 187, 211, 222, 261, 263
  axial, 175, 254
  cables, 254, 305–9
Tensor: inertia, 849
Theory of springs, 554–55
Three-dimensional body:
  center of gravity, 268–69
Three-dimensional line:
  centroid, 270
Three-dimensional stress, 324
Three-force member, 136
Thrust bearings, 128
Time, 5, 530–31
  derivative of a vector, 489, 499, 922–23
    rotating frame, 748, 761
  function, 454–56, 479–80
    acceleration, 456, 458
    harmonic function, 905
    position of a particle, 454–56
    rotational motion, 711, 758–59
    velocity, 956
Time rate, 558
Time rate of change:
  angular momentum:
    particles, 593–94, 632

rigid bodies, 826
linear momentum:
    particles, 591, 632
    rigid bodies, 825
Top, 846, 847
Torque, *see* Moment of a force
Torsional vibrations, 897
Total distance traveled, 455–56
Trajectory, *see* Orbits; Path; Projectiles
Transient vibration, 906, 913
Translation, 124, 705–6, 708, 758
    coordinates, axes, 514–15, 524
    curvilinear, 706, 708
    rectilinear, 706, 708
    rigid bodies: kinetics, 769
    and rotation, simultaneous, 706, 707, 719–21,
        731–32
    reference frames, 665–66, 673–75, 699, 700
    *See also* Linear momentum
Translation components:
    acceleration, 731
Translational motion, 705–6, 766–67, 800
Transmissibility: principle, 7
Transverse components, *see* Radial and
        Transverse components
Triangle rule, 18, 63
Triple mixed products, 89–90, 119
Trusses, 208–9, 261
    analysis, 261–63
    formal procedure, 221–22, 262
        examples and problems, 223–25
    equilibrium, 210, 226
    free-body diagram, 210, 226
    graphical analysis, 222, 262
    method of joints, 210–13, 261
    method of sections, 226–27, 262,
    rigid, 209, 261
    unknown forces, 210, 226
Tuned mass damper system, 13–15
    design, 14
    forces, 108
    photograph, 99
Two-force members, 126, 131–34, 167
    example, 140

**U**ndamped vibration, 893, 905–7, 919–20
Unit vectors, 548–50
    derivatives, 923
Units, 8–12 (*See also* Specific systems of
        measurement)
    conversions, 11–12
    displaying, 12
Unit coordinate vectors, 27–28, 34, 64
    examples and problems, 29–33
    parallel coordinate systems, 28
    scalar products, 44
    sign conventions, 28
Unit vectors, 27, 34–37, 64
    normal and tangential, 487
    radial and transverse, 499
    rectangular cartesian, 484
U.S. customary units, 531, 548, 558
    conversion table, inside front cover
    *See also* Measurement; SI system; Units

**V**-belts, 370–71, 379

Varignon's theorem, 80
Vector components, 19–20, 36, 63
Vector definition of moment, 79, 118
Vector equation, 18
Vector function, 487
Vector notation, 7, 18, 156, 237
Vector products, 77–79, 118
    calculated from determinant, 78–79
    differentiation, 593–94, 731
Vectors, 7 (*See also* Unit vectors)
    addition, 18–19, 27, 63
    angles between, 44–45, 66
        example, 46
    classification, 7–8
    concurrent, 18
    coplanar, 19
    fixed, 8
    free, 7, 97
    magnitude, 7, 26, 34, 64
    notation scheme, 27, 37 (*table*), 28
    projection, 45, 66
        example and problems, 47–49
    representation, 7
    sliding, 7, 72
    subtraction, 19, 27, 63
    time derivatives, 487, 499, 922–23
    rotating frames, 748, 761
Vector summation, 28, 64
Vector triple products, 89, 119
Velocity:
    absolute, 719–20, 749, 750
    analysis:
        rotating reference frames, 749–51, 761
    angular, *see* Angular velocity
    average, 452, 479, 485
    escape, 604
    function: variable acceleration, 458, 481
    instantaneous, 453, 479, 485
    particle, 452–54, 485–86
    radial: scalar component, 498
    relative, 470, 481, 515, 524, 719–21
    resistance to changes in, 529
    rigid bodies:
        instantaneous center, 722–23, 759
        general motion, 741, 760
        general plane motion, 719–23, 759
        rotation about a fixed axis, 709, 758
        translation, 708, 758
    rotational, 500
    scalar analysis, 452–58
    transverse: scalar component, 498
    units, 453
Velocity-time curve, 455
Vibrating systems:
    examples and problems, 895–900
    model, 889
Vibrations, 14, 888 (*See also* Tuned mass
        damper system)
    energy methods in analysis, 901–2, 919
        examples and problems, 903–4
    frequency, 891, 919
    graphic patterns, 891–92
    mechanical, electrical analogues, 915, 920
    natural circular frequency, 890, 919
    natural frequency, 893, 919
    particles:
        damped, 911, 920
            example and problems, 916–18
            forced, 913–15, 920

        free, 911–13, 920
        period, 912
        without damping:
            forced, 905–7, 919–20
            free, 893, 919
    period, 891, 919
    review problems, 921
    rigid bodies: free, 893–94, 919
    simple harmonic motion, 889–92, 919
    *See also* Simple harmonic motion
    steady state, 906, 913
    torsional, 897
    transient, 906, 913
Virtual displacement, 429
Virtual work, 429–30, 445
    examples and problems, 431–34
Viscous damping, 911, 920
Volumes:
    body of revolution, 282, 330
    centroids, 271–72, 329 (*table*), inside back
        cover

**W**att, 558, 812
Wave board, 69–72, 156–57, 314–15
Wedge method, 325, 333
Wedges, 351–53
    examples and problems, 354–59
    free-body diagram, 351
Weight, 6, 10, 267, 530–31
    in free-body diagram, 71
    units in commercial practice, 924
Wheels:
    plane motion:
        kinematics, 706, 707, 721, 723, 733
        kinetics:
            pure rolling, 791, 801
            rolling and sliding, 792, 801
                analysis, 792
                examples and problems, 793–96, 798
        space motion, 742
        unbalanced, 792, 801
        vibrations, 893
Work, 422–23, 552
    of a constant force, 808
    of a constant moment, 808
    of a couple, 422–23, 445
        examples, 425
        on a rigid body, 807, 841
    definition, 419
    and energy, principle, *see* Principle
    force, 419–22, 445, 552–53, 573, 585
        examples, 424
        concurrent, 553
        on a rigid body, 806–7, 841
    force and a couple:
        problems, 426–28
    gravitational, 446
    output, 430
    review problems, 448
    scalar quantity, 553
    total: special aspects, 553
    transformation:
        potential energy, 576
    virtual, *see* Virtual work

**Z**ero-force member, 213

## TABLE 3

## Centroids of Lines, Areas, and Volumes

| | Length of line $L$ Area $A$ | $\bar{x}$ | $\bar{y}$ | | Volume $V$ | $\bar{x}$ | $\bar{y}$ |
|---|---|---|---|---|---|---|---|
| **Circular arc** | $L = 2\alpha r$ | $\dfrac{r \sin \alpha}{\alpha}$ | $0$ | **Hemisphere** | $V = \dfrac{2}{3}\pi r^3$ | $0$ | $\dfrac{3r}{8}$ |
| **Semicircular arc** | $L = \pi r$ | $0$ | $\dfrac{2r}{\pi}$ | **Cone** | $V = \dfrac{\pi r^2 h}{3}$ | $0$ | $\dfrac{h}{4}$ |
| **Triangular area** | $A = \dfrac{bh}{2}$ | | $\dfrac{h}{3}$ | **Pyramid** | $V = \dfrac{abh}{3}$ | $0$ | $\dfrac{h}{4}$ |
| **Semicircular area** | $A = \dfrac{\pi r^2}{2}$ | $0$ | $\dfrac{4r}{3\pi}$ | | | | |